Computing and Systems Technology Division

Presentations at the 2008 AIChE Annual Meeting

Philadelphia, Pennsylvania
16 - 21 November 2008

Volume 1 of 2

ISBN: 978-1-61567-225-7

Printed from e-media with permission by:

Curran Associates, Inc.
57 Morehouse Lane
Red Hook, NY 12571

CURRAN ASSOCIATES INC. proceedings.com

Some format issues inherent in the e-media version may also appear in this print version.

Printed by Curran Associates, Inc. (2009)

AIChE
3 Park Avenue
New York, NY 10016-5991

Phone: (203) 702-7660
Fax: (203) 775-5177

www.aiche.org

Additional copies of this publication are available from:

Curran Associates, Inc.
57 Morehouse Lane
Red Hook, NY 12571 USA
Phone: 845-758-0400
Fax: 845-758-2634
Email: curran@proceedings.com
Web: www.proceedings.com

TABLE OF CONTENTS

Volume 1

Volume 2

19a Opening Remarks

Francis J. Doyle III, Department of Chemical Engineering, University of California, Santa Barbara, Santa Barbara, CA 93106-5080

Introductory remarks will be given for the CACHE Special Session.

Computation in Undergraduate Chemical Engineering Education

Bruce A. Finlayson
Professor Emeritus
Department of Chemical Engineering
University of Washington

ASEE-CACHE Award for Excellence in Computing in Chemical Engineering Education, 2008

AIChE meeting, paper 19b, Nov. 17, 2008

Status in 1967 when I started my career

- Perry's Chemical Engineers' Handbook, 4th Ed. (1963) (nothing in earlier editions):
  - To solve ODEs: Euler, Adams, simple Runge-Kutta methods
  - To solve PDEs: diffusion/conduction steady problems in 2D (finite difference) or unsteady problems in 1D
  - None of this was reflected in Sections on Fluid Flow or Heat Transmission
- Luther, Carnahan and Wilkes, Applied Numerical Methods (1963)
  - Detailed treatment of numerical analysis, but only explicit techniques with specified time steps

Numerical Analysis is now used to solve problems ranging from the orientation of nanoparticles to predicting global climate change.

It wasn't always that way.

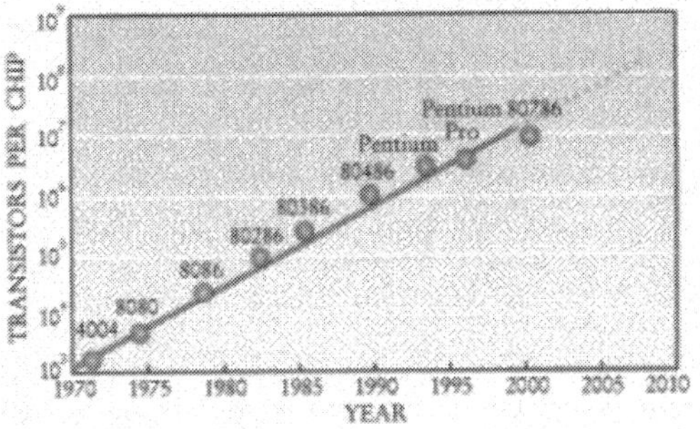

Physics Today, Jan. 2000, p. 40

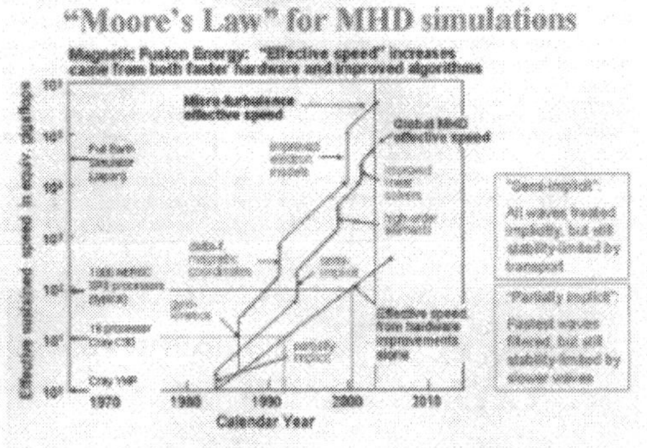

Algorithms help, too! MHD Simulations, faster hardware and improved algorithms, SIAM Newsletter

Process Simulation: 1970s

- 1971-*Chemical Plant Simulation*, C. M. Crowe, A. E. Hamielec, T. W. Hoffman, A. I. Johnson, D. R. Woods - based on PACER, written at Purdue
- 1973 - *Computation for Process Engineers*, G. L. Wells, and P. M. Robson - wrote own code (in back)
- 1979 - *Process Flowsheeting*, A. W. Westerberg, H. P. Hutchison, R. L. Motard, P. Winter - discussed flowsheeting on computer, degrees of freedom, sequential solution, tearing
- ~1986 - ASCEND - A. W. Westerberg - simultaneous solution methods

1970s - Mass and Energy Balance Course

- As a young professor, eager to show students what marvelous things they could do on the computer, I taught them how to prepare a deck of cards, carry them to the computer center and solve the following energy balance:

$$\Delta H = \int_{T_1}^{T_2} C_p(T)dT, \quad C_p = \sum_{i=1}^{n} C_{pi}$$

$$C_{pi} = \alpha_i + \beta_i T + \chi_i T^2 + \delta_i T^3$$

CHESS Program
Rudy Motard, University of Houston

- Process simulator using card input, included short-cut distillation, thermodynamics using solubility parameters, but quite powerful within its area of application.
- 1973 - I volunteered to teach a one-credit course as an overload; the students would use the program to simulate a process.
- 1977 - expanded into a 3-credit course, now an elective and not an overload.
- 1982 - moved FLOWTRAN into the regular design course when older professors retired.

FLOWTRAN, Monsanto, in 1980s

- Monsanto agreed to let Universities use their process simulator, FLOWTRAN. This was the program being used in industry, and this gave a great impetus to the use of computers in chemical engineering education.
- FLOWTRAN formed the basis of AspenPlus.
- 1985 - I prepared the load module for FLOWTRAN when it was to run on CYBER computers; installed at 13 Universities

Textbooks for Process Design (sample)

- 1968 - Peters and Timmerhaus, 2nd Ed., Gives costs on graphs and includes a table of discount factors to calculate DCFROR
- 1983 - Valle-Riestra, still no computer use, but includes difference formulas so that students can derive discount factors
- 1985 - Coulson and Richardson, includes a FORTRAN program MASSBAL printed in the back of the book
- 1988 - Douglas, focused on design concepts, but includes FLOWTRAN input forms, line by line
- 1998, 2003 - Turton, Bailie, Whiting, Shaeiwitz, comes with a CD containing CAPCOST to allow efficient costing of a process, with automatic cash flow analyses; book oriented to process design using computer programs.
- 2003 - Product and Process Design Principles: Synthesis, Analysis, and Evaluation, Warren D. Seider, J. D. Seader, and Daniel R. Lewin

Things I've learned from process simulation

Early 1970s, 1-credit lab course using CHESS to model chemical processes. One student, GPA 2.1, was extremely creative when he had a computer to do the calculations for him. His grade problem had been that he lacked the skills to do the calculations fast; he certainly was creative. He learned by trying (inductive learning).

In the late 1970s I took elements from CHESS and made an interactive program, DISTILL, which ran on computer terminals (now referred to as dumb terminals, with question/answer input). The user would specify the distillation problem and get the output and cost information. An inner loop allowed them to change the pressure, fraction of light and heavy key, and ratio of reflux to minimum reflux, and get another design, with all capital and operating costs displayed. Students would run case after case and learn about the economic effect of their decisions. That was when I learned that students can learn by repetition. A McCabe-Thiele diagram is still useful, but nothing beats rapid feedback about the cost of your decisions. Again, inductive learning. Copies made available for Mac and PC (they still work!)

While this activity was going on in process simulation, numerical analysis was advancing, too. Stiffly-stable methods for ODEs

- Runge-Kutta methods existed with error control and automatic step-size adjustment.
- Most engineers used Crank-Nicolson methods, but had to guess a stable step size.
- Gear, 1971; Hindmarsh, 1975, GEARB, later LSODE
  - –When different time constants are important - you want to resolve something occurring on a fast time scale but need to do so over a long time - explicit (RK) methods take a long time.
  - –Implicit methods can be 1000 times faster.
  - –Gear's method allowed for automatic step size adjusment, automatic change of order if that was useful, and basically automatic solution of ordinary differential equations (IVP)

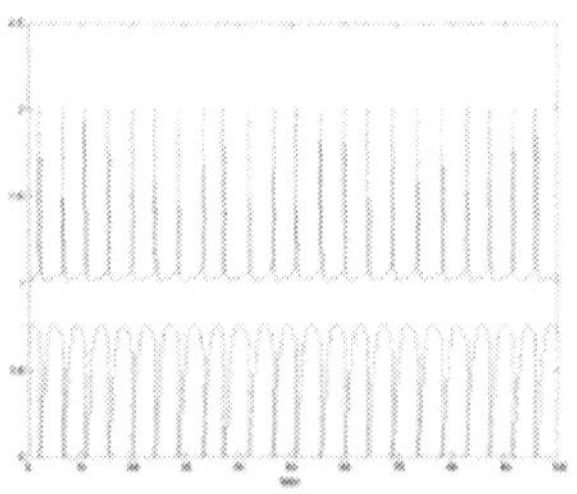

Stirred tank reactor example with a limit cycle

But, the methods are useful for partial differential equations, too!

1980 - orthogonal collocation, finite difference, finite element, with programs (still available at www.ravennapark.com)

Orthogonal Collocation - a good idea

Lanczo, 1938 - collocation method with orthogonal polynomials

Villadsen and Stewart, 1967 - solved in terms of value at collocation nodes rather than coefficients - the programming is much simpler

$$c(y,t) = \sum_{i=1}^{N+2} a_i(t) P_{i-1}(y)$$

$$\frac{\partial c}{\partial t} = D\frac{\partial^2 c}{\partial y^2} - \phi^2 R(c) \quad \text{becomes} \quad \frac{dc_j}{dt} = \frac{D}{h^2}\sum_{i=1}^{N+2} B_{ji} c_i - \phi^2 R(c_j)$$

Stiff methods essential for partial differential equations

Depends upon the eigenvalues of the matrix of the Jacobian.

$$\frac{dT_i}{dt} = \sum_{j=1}^{N+2} B_{ij} T_j$$

One eigenvalue is due to the problem (diffusion) and the other is due to the method.

$$As \;\; N \to \infty \;\; or \;\; \Delta x \to 0, \;\; \lambda_{l\arg est} \;\; gets \;\; big.$$

The more accurate your model, the stiffer the problem.

Application to catalytic converter

Involves unsteady heat and mass transport with a complicated rate expression, perhaps eased by occurring in a thin layer of catalyst. The problem may be only one-dimensional, but it must be solved thousands of times in a simulation, even if in steady state. The solid heat capacity makes the time scales very different. Orthogonal collocation models were "4 to 40 times faster (Chem. Eng. J. **1**, 327 (1970).

$$\varepsilon \frac{\partial c}{\partial t} = \frac{D_e}{r^2} \frac{\partial}{\partial r}(r^2 \frac{\partial c}{\partial r}) - kR(c,T)$$

$$\left(\varepsilon \rho C_{pf} + (1-\varepsilon)\rho_s C_{ps}\right)\frac{\partial T}{\partial t} = \frac{k_e}{r^2}\frac{\partial}{\partial r}(r^2 \frac{\partial T}{\partial r}) + (-\Delta H_{rxn})kR(c,T)$$

Catalytic Converter

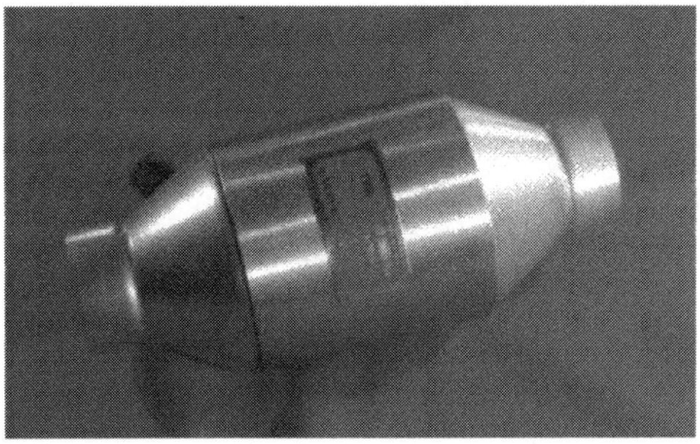

Phenomena included:

Chemical reaction

Flow

Axial conduction of heat

Diffusion

Geometry

What is the importance of the shape of channel?

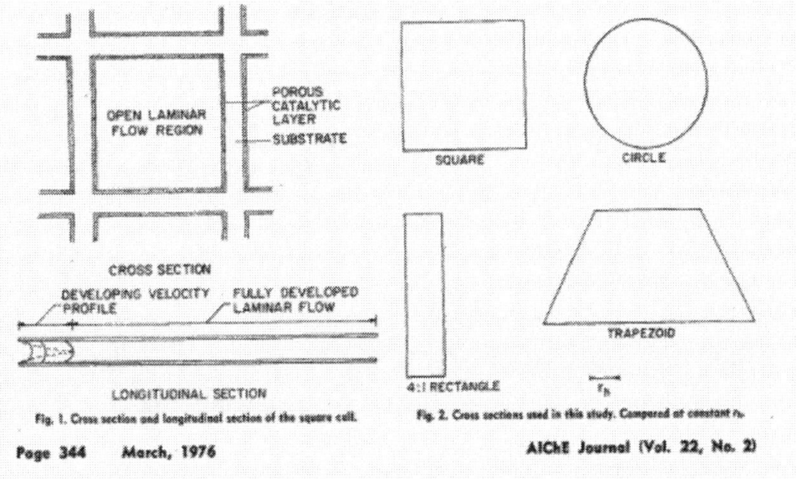

Model I-A is lumped
Model II-A is distributed, using orthogonal collocation on finite elements

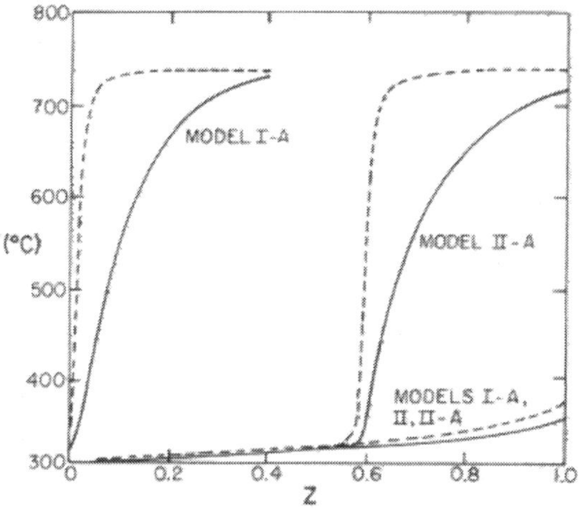

Fig. 8. Steady state model predictions for square geometry. Inlet T = 315°C, y_{CO} = 4%, $\hat{G}$ = 0.0151 m³/s, D^f_{CO}/D^s_{CO} = 70, ξ = 0.0254 mm. - - - - - Solid temperature, ——— Fluid average temperature. P2 kinetics, model I-A: $Nu = Sh = 4.0$.

But, how is this impacting undergraduate education?

Reprinted from CHEMICAL ENGINEERING PROGRESS, February 1984

The Impact of Computers On Undergraduate Education

Personal computers enable the curriculum to capitalize on the greater motivation that comes with open-ended, design-type problems in each course, rather than constraining them to a "design" course.

Bruce A. Finlayson, Univ. of Washington, Seattle, Wash. 98195

- Article went on to discuss
  - Databases
  - Graphical capabilities
  - Spreadsheets (might be used for engineering calculation)
  - Word processing (enhances written communication)
  - Data acquisition
  - A new scheme Xerox developed: Windows

Changes brought about by the microcomputer

- Mechanize many tedious calculations
- Availability of software (more machines means bigger markets)
- User-friendly
- My 1984 article explained bit, byte, floppy diskette, networking
- For many years I edited a column in CACHE News in which computer programs provided by professors were reviewed for the chemical engineering education community.

Learn by induction, CACHE News, Fall, 1987

- Graduate students were given a program CDEQN (available through CACHE at the time) to solve the diffusion equation. We had worked out the stability limitation on the time step size for the finite difference method, and I asked them to deduce the limit for the finite element method. The program used a GUI, solved and plotted the solution; by choosing several dx and dt they found it was 1/3 that of the finite difference method.

$$\frac{\partial c}{\partial t} = D\frac{\partial^2 c}{\partial x^2} \qquad D\frac{\Delta t}{\Delta x^2} \le \frac{1}{6}$$

- Then I asked them to do the same thing for the problem with convection. Many of them determined there was something like a Courant number limit.

$$\frac{\partial c}{\partial t} + u\frac{\partial c}{\partial x} = D\frac{\partial^2 c}{\partial x^2} \qquad u\frac{\Delta t}{\Delta x} \le 1$$

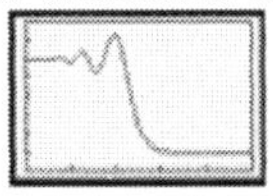

- Finally, one student persisted and said - there are two stability limits: one says it is stable the other unstable. What he had found was the difference between a von Neumann analysis (does the disturbance continue growing?) and the eigenvalue analysis (does the solution at a point eventually reach a steady state?). There are articles in the engineering literature discussing this difference, derived using lots of trigonometry and linear algebra, but the student deduced it by trial and error.

Chemical Reactor Design Tool (CRDT) - 1992

- Models CSTR, Plug-flow reactors, reactors with axial dispersion, reactors with radial dispersion.
- User supplies only a FORTRAN reaction rate expression.
- Allows up to 20 components plus temperature.
- Won the Martin Award, 1994, ASEE, ChE

$Cl_2 + C_3H_6 \rightarrow CH_2{=}CHCH_2Cl + HCl$
allyl chloride

$Cl_2 + CH_2{=}CHCH_2Cl \rightarrow CHCl{=}CHCH_2Cl + HCl$
allyl chloride 1,3-dichloropropene

$Cl_2 + C_3H_6 \rightarrow CH_2ClCHClCH_3$
1,2-dichloropropane

$$r_1 = A_1 \exp\left(-\frac{15{,}840}{RT}\right) C_{Cl_2} C_{C_3H_6}$$

$$r_2 = A_2 \exp\left(-\frac{23{,}760}{RT}\right) C_{Cl_2} C_{allyl\ chloride}$$

$$r_3 = A_3 \exp\left(-\frac{7920}{RT}\right) C_{Cl_2} C_{C_3H_6}$$

B. M. Rosendall and B. A. Finlayson, "Reactor/Transport Models for Design: How to Teach Students & Practitioners to Use the Computer Wisely," FOCAPD, 1999, AIChE Symposium Series No. 323, 96 176-191 (2000).

$$\frac{dF_i}{dV} = Da_I RA_i,\ F_i = F_{i0} \text{ at } V = 0$$

$$\sum_{j=1}^{NC} F_j C_{pj} \frac{dT}{dV} = Da_{III} RT - St\,(T - T_c)$$

$$T = T_{in} \text{ at } V = 0$$

$$St = \frac{UA}{F_s C_{ps}}, \text{ where } F_s \text{ and } C_{ps} \text{ are standard}$$

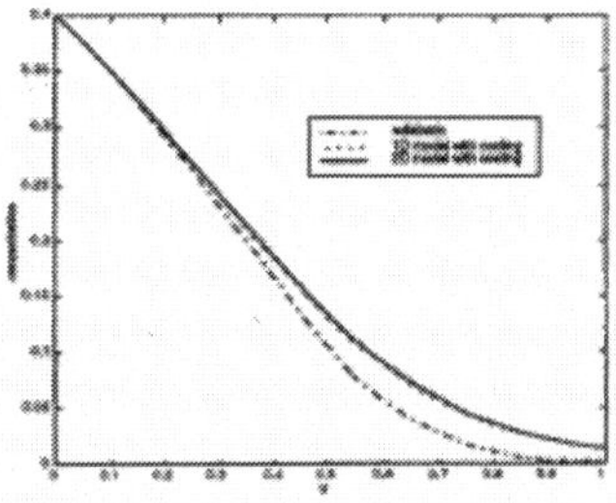

Figure 13. Average Concentration of Chlorine

$$\frac{\partial F_i}{\partial V} = \alpha_i \nabla^2 C_i + Da_I RA_i$$

$$\sum_{j=1}^{NC} F_j C_{pj} \frac{\partial T}{\partial V} = \alpha_T \nabla^2 T + Da_{III} RT$$

$$-\frac{\partial T}{\partial r} = Bi_w\,(T - T_c) \text{ at } r = 1$$

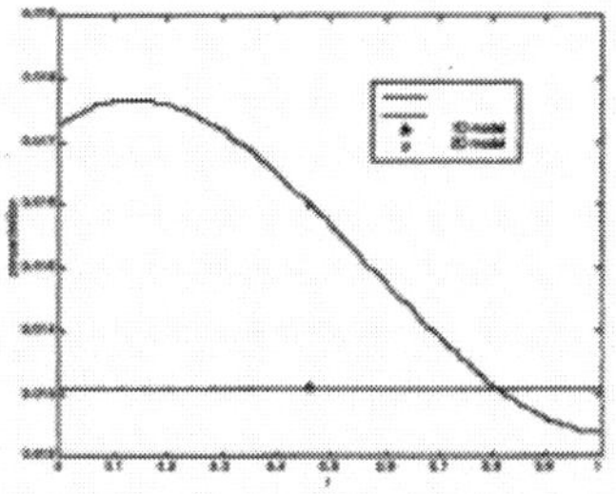

Figure 15. Chlorine Concentration at Exit

Finite Element Method

Began in Civil Engineering for structural problems. The finite elements were beams and rods. It solved the same kind of problems done in Physics 101, except in more complicated structures. Then it was expanded to differential equations.

The dependent variable was expanded in known functions.

$$T(x) = \sum_{i=1}^{N+2} a_i P_{i-1}(x)$$

Key ideas in Finite Element Method

- Cover domain with small triangles or rectangles, or their 3D equivalents.
- Approximate the solution on that triangle using low order polynomials.
- Use Galerkin method to find solution at nodal points.
- Can use higher order polynomials.
- Requires lots of memory, fast computers.

Hole Pressure Problem, Nancy Jackson

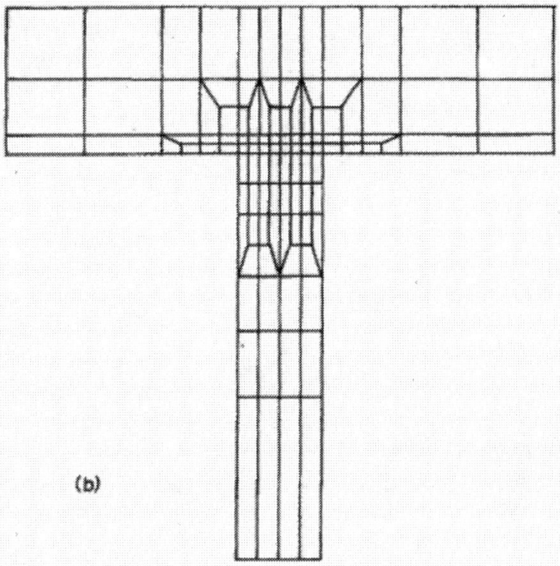

Fig. 3. Finite element mesh: (a) 29 elements; (b) 110 elements.

Extrudate Swell, Tom Patten, 1977

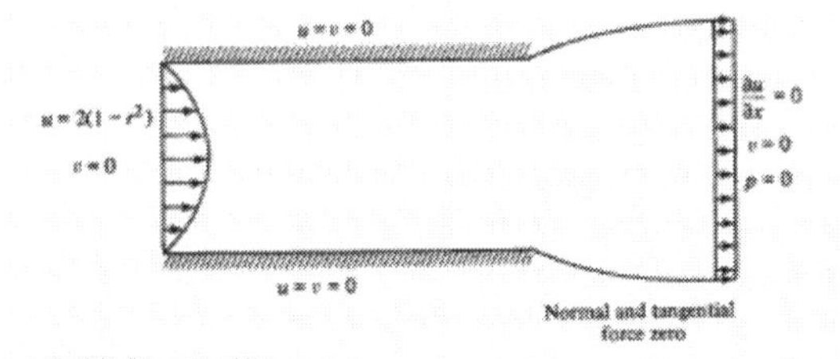

Fig. 6-16 Die swell problem.

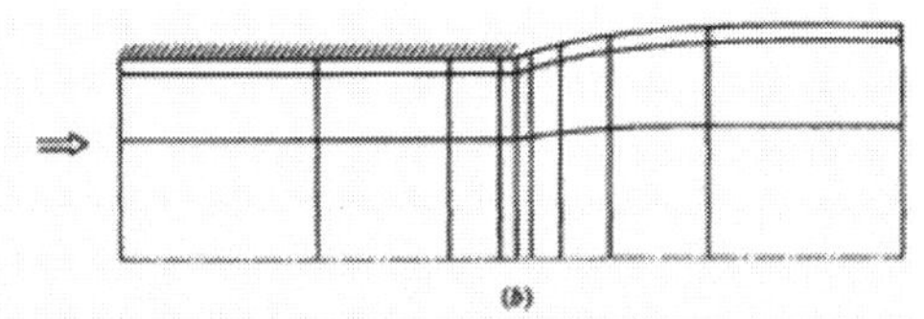

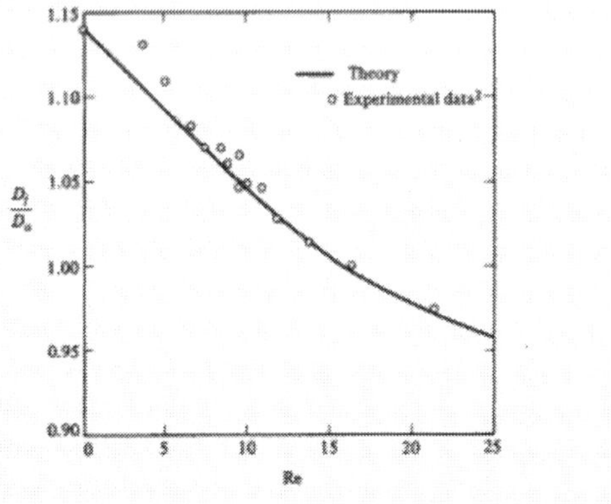

Figure 6-18 Die swell dependence on Reynolds number.

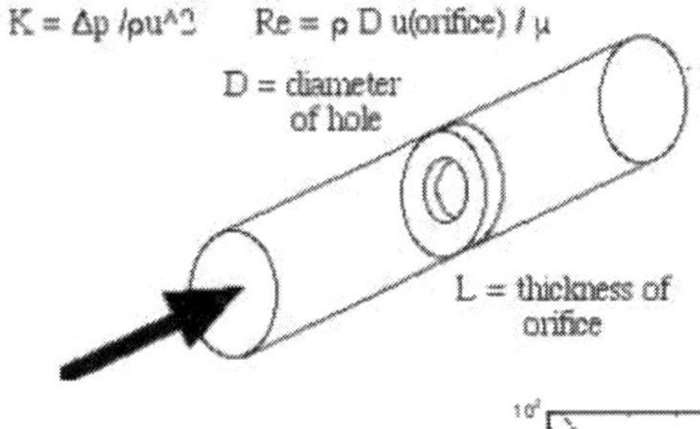

Later, undergraduates could solve harder problems using Comsol Multiphysics (FEMLAB).

Kusmanto, Jacobsen, and Finlayson, Phys. Fluids **16** 4129 (2004)

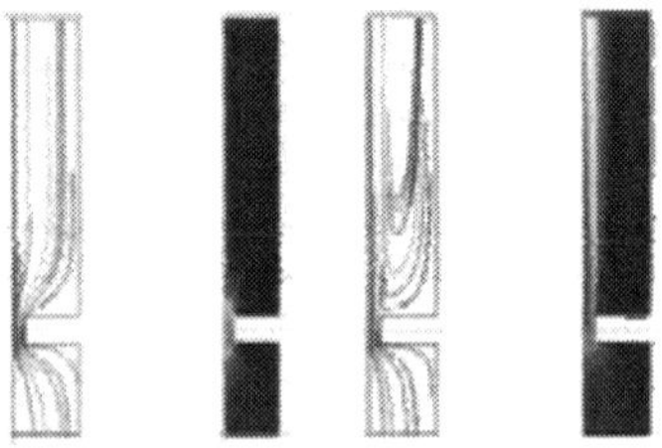

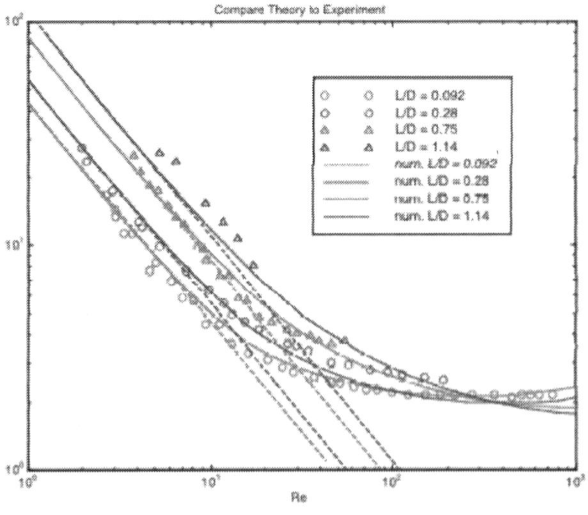

Convective Instability, Michael Harrison

Heat transfer between flat plates, heated from below

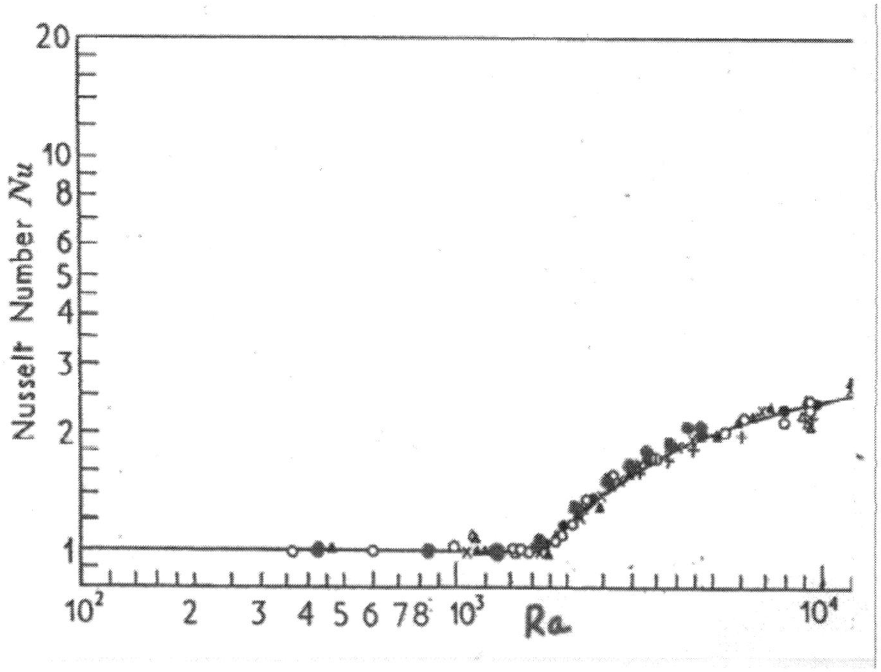

Thermal diffusion, Pawel Drapala

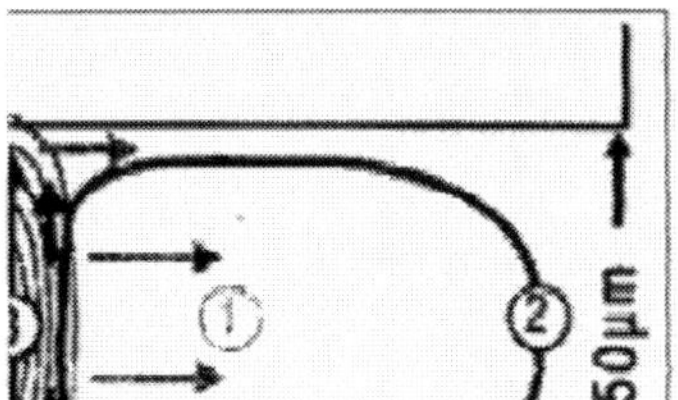

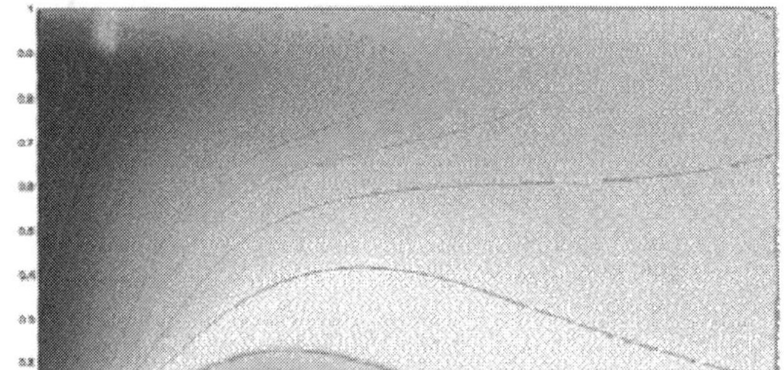

Figure 9. FEMLAB's concentration profile at a final transient value of 5 seconds.

Patterned after experiments by Braun and Libchaber, *Phy. Rev. Letters* **89** 188103 (2002).

Mixing in a Serpentine Microfluidic Mixer

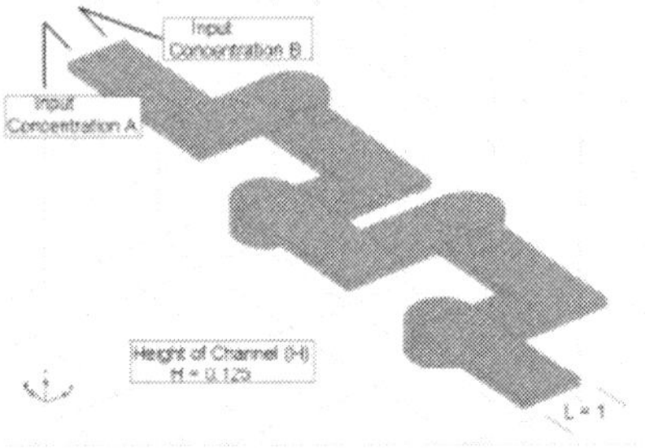

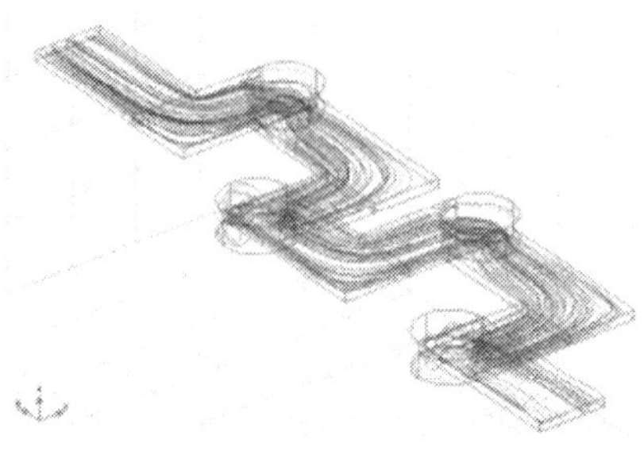

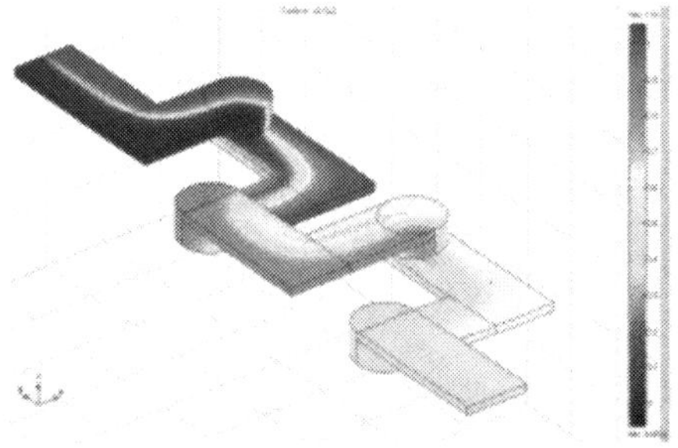

Published in Neils, Tyree, Finlayson, Folch, *Lab-on-a-Chip* **4** 342 (2004)

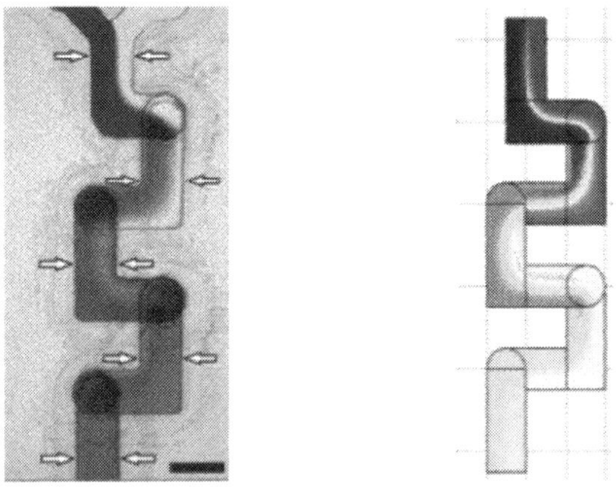

Figure 8. Comparison with experiment (6)

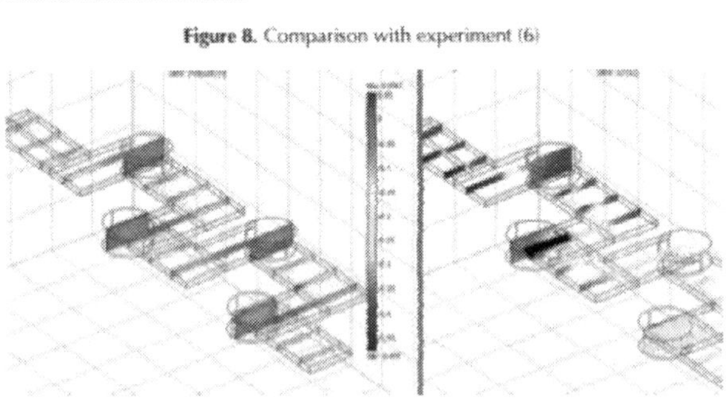

Figure 9. Velocity profiles and concentration profiles inside serpentine mixer

For Re = 1 or so, the flow problem is easy. But, the Peclet number can be large (2000). Then the mesh for the concentration problem has to be refined significantly. Comsol allows solution of the flow problem and the convective diffusion problem on different meshes, *thus* speeding up the solution time.

Mixing in Microfluidic Devices
(11 undergraduate projects)

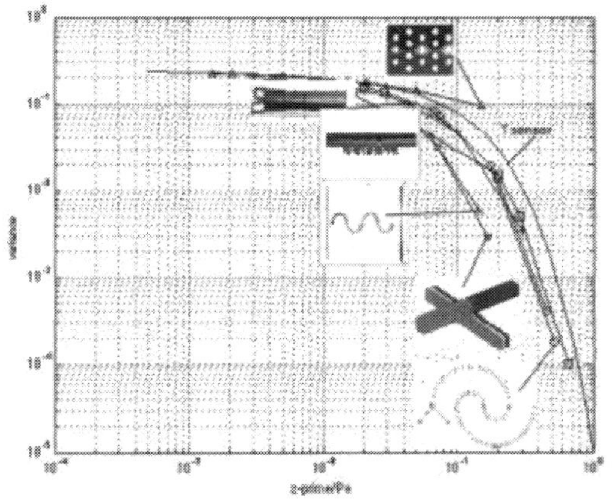

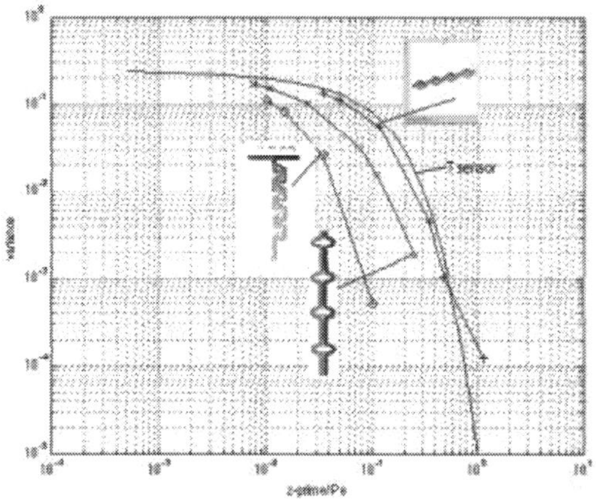

Problem Centered Course n Numerical Methods

- Created model of catalytic converter in summer of 1995.
- While doing so, realized that every type of equation that was covered in my numerical analysis class was in the model somewhere.
- Reorganized the course to be a problem-centered course - everything had to do with catalytic converters.

- Write short theme in first week.
- Solve succession of numerical problems.
- Work in a team to design a catalytic converter for the 'two minute problem'.
- Prepare a web lesson about their work, one suitable for high-school seniors, one for undergraduate chemical engineering students.
- Won Award from DOE in 1996 for Undergraduate Computational Science and Engineering

In successive years:

- Catalytic converter
- Plastic garbage bags from natural gas
- Membrane Separation
- Microreactors
- Fuel cells
- This was the course in which students taught me that the computer can differentiate!

But, it also led me to a new paradigm for teaching, and my next book.

Programs

- Microsoft Excel ®
- MATLAB®
- Aspen Plus ®
- FEMLAB ®

➢Philosophy - students can be good chemical engineers without understanding the details of the numerical analysis.

➢By using modern programs with good GUIs, the most important thing is to check your results.

➢Instead of teaching a small fraction of the class numerical methods, I now teach all the class to use the computer wisely.

Introduction to Chemical Engineering Computing, process applications

- Cubic equations of state
- Vapor-liquid equilibria
- Chemical reaction equilibria
- Mass balances with recycle, vapor-liquid equilibria, and chemical reaction equilibria
- Process simulation, including proper choice of thermodynamic model and use of Aspen Plus

Computing Class - one objective: how do you check your work?

http://courses.washington.edu/checomp

- First assignment using Aspen asked them just to design a once-through process (given) for making ethyl chloride. They were asked to calculate the cost of the utilities (only). Part of the discussion the next week was about the cost of raw materials, how the form of them affected the process, how the cost of utilities could be reduced, how the price dropped precipitously when lead was phased out of gasoline.
- Second week they added a recycle stream and redid the problem, leading to different results because the separations were different when there was a recycle stream and other options could be used.

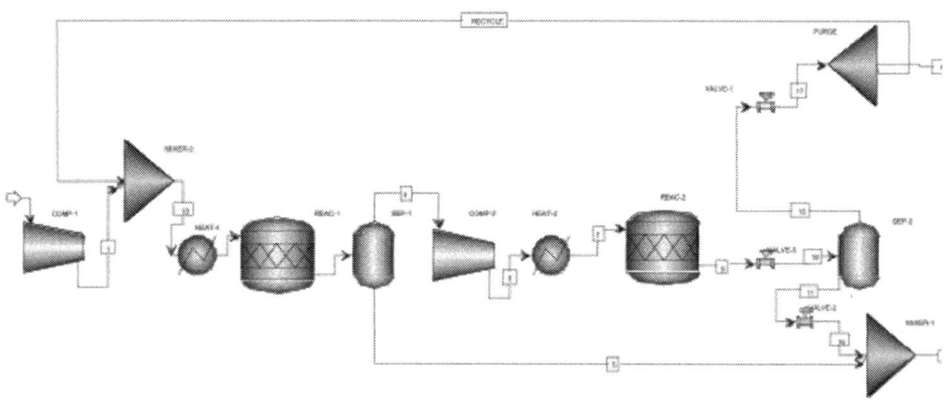

Class was asked: when you are the supervisor and one of your employees brings a simulation to you, what questions will you ask?

- Do the results seem reasonable?
- What assumptions did you make?
- What equations or models did you use and why?
- How do the answers compare to other simulations?
- Have you done an error analysis? Shows?
- Where or how did you get your data?
- What is the biggest source of uncertainty?
- Compare trends to literature values.

Introduction to Chemical Engineering Computing, transport applications

➢Chemical reactor models with radial dispersion, axial dispersion

➢Catalytic reaction and diffusion

➢One-dimensional transport problems in fluid mechanics, heat and mass transfer

- Newtonian and non-Newtonian
- Pipe flow, steady and start-up
- adsorbtion

➢Two- and three-dimensional transport problems in fluid mechanics, heat and mass transfer - focused on microfluidics and laminar flow

- Entry flow
- Laminar and turbulent
- Microfludics, high Peclet number
- Temperature effects (viscous dissipation)
- Proper boundary conditions

Checklist for Transport Problems

- 1. Say what problem you are solving;
- 2. Give the shape and dimensions, number of elements, degrees of freedom;
- 3. Give the parameters in the equation and identify the boundary conditions;
- 4. Tell how you solved it;
- 5. Give checks to your answer (previous similar results, etc.);
- 6. Give your results, including pertinent plots and integrals;
- Your report should have an Appendix with sample calculations.

- Some of these steps verify what choices they made in the program to verify that they are solving the right problem.
- Some of them verify the accuracy of the solution.
- If they use complicated expressions (like kinetic expressions) they must do a hand calculation to verify that they typed it correctly.

Steps in Solution

from *Introduction to Chemical Engineering Computing*, Bruce A. Finlayson, Wiley (2006)

- Open Comsol Multiphysics
- Draw domain
- Physics/Subdomain Settings
- Physics/Boundary Settings
- Mesh (Need to solve one problem on at least three meshes, each more refined than the last, to give information about the accuracy.)
- Solve (Can solve multiple equations together or sequentially; can use parametric solver to enhance convergence of difficult non-linear problems.)
- Post-processing (Plot solution, gradients, calculate averages, calculate or plot any expressions you've defined.

Determine Pressure Drop Coefficients for Slow Flow (to mimic those available for turbulent flow)

$$\frac{\Delta p\, x_s}{\eta v_s} = K$$

Table III. Coefficient K_L for contractions and expansions for Re negligbily small

	Picture	K_L
2:1 pipe/planar		7.3/3.1
3:1 pipe/planar		8.6/4.1
4:1 pipe/planar		9.0/4.5
45 degrees tapered, planar, 3:1		4.9
28.07 degrees tapered, planar, 3:1		10.8
3:1 square (quarter of the geometry)		8.1

v_s = average velocity, x_s = thickness or diameter, both in the small section

Table in Ch. 8, "Micro-component flow characterization," Koch, Vanden Bussche, Chrisman (ed), Wiley (2007). The chapter has 11 authors, 10 UW undergraduates plus Finlayson.

Three-dimensional hole pressure
(work done by junior Stephanie Yuen)

Comparing 2D and 3D calculations. Hole pressure is used in rheology to measure the first normal stress difference.

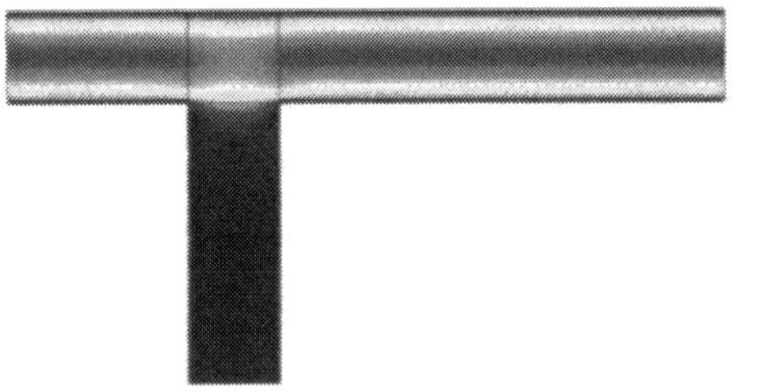

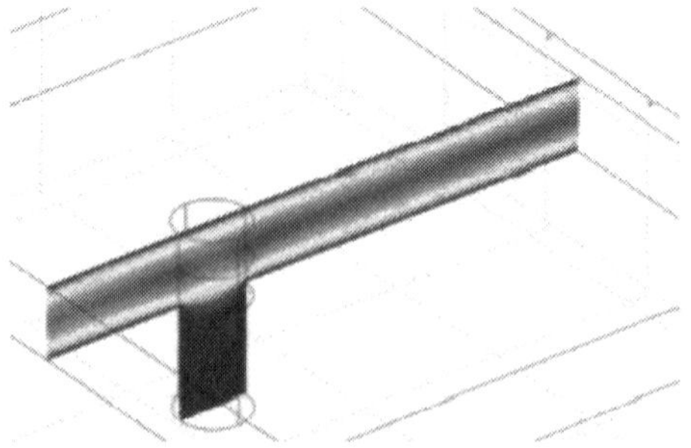

Mixing in a Three-dimensional T
(work done by junior Daniel Kress)

$$c_{\text{mixing cup}} = \frac{\int_A c\,\mathbf{u} \bullet d\mathbf{A}}{\int_A \mathbf{u} \bullet d\mathbf{A}}, \quad variance = \frac{\int_A (c - c_{\text{mixing cup}})^2\,\mathbf{u} \bullet d\mathbf{A}}{\int_A \mathbf{u} \bullet d\mathbf{A}}$$

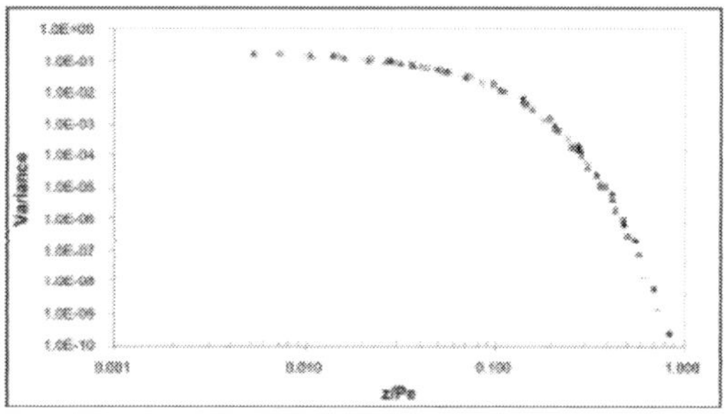

Variance as a function of length in the outlet leg

The work showed that the 3D case followed the same curve as the 2D case (T-sensor).

H-sensor - used to separate chemicals by diffusion (solutions by Krassen Ratchev)

$D = 10^{-9}\ m^2/s$

$D = 10^{-11}\ m^2/s$

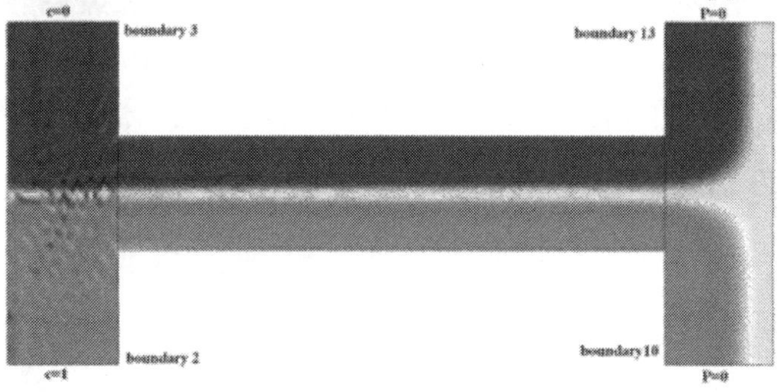

Another approach: FLOWLAB, Jennifer Curtiss, University of Florida

- A special, academic version of FLUENT has been made by the company. It can only solve specific problems, but students can use it to explore the details of fluid flow.
- Problems:
  - Turbulent entry flow into a pipe
  - Boundary layer flow past a flat plate
  - Heat transfer to a fluid in a pipe
  - Drag with flow past a cylinder
  - Flow through an orifice
  - Flow in a sudden expansion
- Limited geometries
- Usually turbulent

New CACHE-CFD Taskforce is developing model problems to use in chemical engineering curriculum.

Conclusions

- Computer usage in chemical engineering education has advanced from non-existent to the solution of very complicated problems.
- The emphasis now is more on how to solve chemical engineering problems (and verify that) than on writing the computer code.
- The computer programs are powerful enough that they permit inductive learning.
- Comsol Multiphysics is so powerful it can be mis-used.
- Introduction in a step-by-step way helps undergraduates learn to solve the problem and show they have solved the problem correctly.
- The 2D and 3D nature of the problems provides motivation beyond the simplified problems solved in textbooks.

19c Smart Process Manufacturing – Results from An NSF-Sponsored Workshop

Thomas F. Edgar, Chemical Engineering, Dept. of Chem. Eng.,The University of Texas at Austin, 1 University Station Stop C0400, Austin, TX 78712 and **Jim Davis**, Chemical Engineering, University of California - Los Angeles, Los Angeles, CA 90095.

There are a number of forces driving the U.S. chemical industry in the 21st century, including shareholder return, globalization, efficient use of capital, faster product development, minimizing environmental impact, improved return on investment, improved and more efficient use of research, and efficient use of people. To achieve these goals, the chemical industry is investigating the expanded use and application of new computational technologies employed in areas such as modeling, computational chemistry, design, control, instrumentation, and operations. The key technology driver over the past 20 years has been the continuing advances in digital computing. A new paradigm that encompasses the process systems engineering (PSE) topics mentioned above is smart process manufacturing, which will require the development of research environments to support such activities, also known as cyberinfrastructure. A research agenda must be developed to provide a coherent and orchestrated effort to achieve the vision of smart process manufacturing. To that end an NSF-sponsored workshop was held on April 21-22, 2008 under a grant to the CACHE Corporation. Details on the research agenda and how the PSE community can participate will be covered in this presentation.

19d Integration of Computational Fluid Dynamics (CFD) into the Chemical Engineering Undergraduate Curriculum

Richard D. LaRoche, DEM Solutions, 24 Hanover Street, Suite 10, Lebanon, NH 03766

As computational fluid dynamics (CFD) has gained acceptance as an industrial tool for the chemical process industries, there has been an increasing demand from industry for B.S. chemical engineers with some formal training in using CFD software. Several prominent chemical engineering faculty have led successful projects in developing and disseminating CFD teaching modules for undergraduate chemical engineering education. Unfortunately, the use of CFD teaching modules is only present in a few U.S. chemical engineering departments. The CACHE CFD Task Force is a coalition of academic and industrial CFD researchers and practitioners whose aim is to assist in building the infrastructure necessary to insure that these CFD teaching modules become standard practice in the U.S. chemical engineering undergraduate curriculum.

19e The Cache Systems Biology Task Force

Michael H. Henson, Department of Chemical Engineering, University of Massachusetts Amherst, 686 N. Pleasant Street, Amherst, MA 01003

The Systems Biology Task Force was established by CACHE to develop educational materials for and to promote educational activities in systems biology with particular emphasis on engineering approaches and tools. This talk will provide an update on task force activities, which include the development of teaching modules for the core chemical engineering curriculum and the continuing development of the Foundations of Systems Biology in Engineering (FOSBE) conference series.

19f Molecular Modeling Task Force Update

David A. Kofke, Chemical and Biological Engineering, University of Buffalo, 303 Furnas Hall, SUNY at Buffalo, Buffalo, NY 14260

We review recent activities of the CAChE Molecular Modeling Task Force. One such activity is the development of interactive molecular simulation modules for instructional purposes, and we review the project and discuss some of the recently completed modules as well as others still under development. We also describe an effort to develop a molecular simulation Application Programming Interface that can be used to facilitate interoperability of different molecular simulation code bases.

CACHE Modules on Energy In the Curriculum: Fuel Cells

Jason M. Keith, Michigan Technological University, Department of Chemical Engineering, 1400 Townsend Drive, Houghton, MI 49931-1295
Michael D. Gross, Department of Chemical Engineering, Bucknell University, Lewisburg, PA 17837 Donald J. Chmielewski, Chemical and Biological Engineering, Illinois Institute of Technology, Suite 127, Perlstein Hall, 10 West 33rd Street, Chicago, IL 60616
H. Scott Fogler, The University of Michigan, 2300 Hayward St., Ann Arbor, MI 48109-2136

In this work we describe the development of modules that can be used to introduce fuel cells into existing chemical engineering courses. Course specific modules have been developed that apply fundamental chemical engineering principles to the analysis of fuel cell systems. The modules are currently located at the following website: http://www.chem.mtu.edu/~jmkeith/fuel_cell_curriculum, and after beta testing, will be available through the CACHE website http://www.cache.org. Each module contains a problem motivation, reference to material from textbooks widely used in the chemical engineering curriculum, an example problem statement, example problem solution, home problem statement, and home problem solution. To date modules have been developed for the following chemical engineering core courses: mass and energy balances, thermodynamics, fluid mechanics, heat and mass transfer, and kinetics and reaction engineering. In addition to presenting the modules, we will present some preliminary assessment on these educational modules.

19h Current Attributes and Future Prospects for Polymath Software

Michael B. Cutlip, University of Connecticut, 191 Auditorium Rd., Rm. 204, Storrs, CT 06269-3222 and Mordechai Shacham, Ben Gurion University of the Negev, Chemical Engineering Department, Beer-Sheva, 84105, Israel.

POLYMATH Software has been a product that the CACHE Corporation has supported and made available through inexpensive departmental site licenses since 1990. This year the number of academic sites licenses has reached well over 100 world wide for the first time.

The authors of POLYMATH will briefly outline some of the current novel aspects of this software and will discuss possible enhancements for the next version.

http://www.polymath-software.com/

19i Education Modules in Process Design and Operability

Thomas E. Marlin, Department of Chemical ENgineering, McMaster University, 1280 Main Street West, Hamilton, ON L8S 4L7, Canada

Keywords: Engineering education, process design, robust design, operability

Undergraduate students in Chemical Engineering are introduced to the concept of operability in their design course. Often, the term is used in a narrow sense that concentrates on the dynamic responses of variables. In this paper, the term is given a broader meaning consistent with engineering practice. In addition, examples are provided to demonstrate the value of teaching operability in (1) reinforcing engineering sciences and (2) extending design from a "point" to successful operation over a region.

Current textbooks provide limited coverage of selected operability topics. For example, Turton et al (1998) provide a brief coverage of process control and trouble shooting. In contrast, Biegler et al. (2000) provide detailed coverage of the operating window. Also, Seider et al (2004) address dynamic simulation and plant-wide loop pairing. Finally, older books such as Peters and Timmerhaus (1991) do not address operability.

This paper proposes a broader definition of the design problem to explicitly include operability, which consists of the following topics.

* Operability Window, including methods for increasing the window “frame” and determining that operation is possible at all points within the window

* Reliability, including reliability calculations for simple process structures and methods for improving the reliability of a plant through redundant equipment, intermediate inventory, etc.

* Flexibility, including tests to ensure that flexibility is located correctly to achieve controllability, with examples of methods for expanding operability

* Safety, including the six levels of safety protection (AIChE) and HAZOP studies with details on the first four levels (BPCT, alarms, safety instrumented systems, and pressure relief).

* Efficiency, including designs to improve operating profit

* Operation during transitions, including batch, startup, grade changes, etc.

* Dynamic performance, including designs to give good feedback performance

* Monitoring and Diagnosis, including sensors for monitoring and a systematic plant trouble shooting method with many exercises

The integrated coverage of the topics allows the students to recognize that flexibility, provided for reliability or high capacity, provides opportunities for affecting efficiencies and dynamic performance within the operating window.

Some excellent resources are available for some of these topics, but they are neither identified in a common source nor integrated by highlighting common approaches. The proposal in this paper is to establish an engineering module for instructors (and students) to share WEB-based resources for teaching and engineering practice.

67a Model-Based Innovations In Process Development and Design

Constantinos C. Pantelides, Process Systems Enterprise Ltd, Bridge Studios, 107a Hammersmith Bridge Road, London, W6 9DA, United Kingdom

The process industries are going through a period of great opportunities and challenges. Unprecedented economic growth in Asia and elsewhere is leading to enhanced demand for their products, and this is combined with increasing customer expectations on product performance. All this has to be achieved against decreasing availability of primary raw materials and global concerns relating to energy consumption, safety and environmental impact.

Innovation is key in addressing the above challenges in a manner which establishes and maintains competitive advantage. Unsurprisingly, the industry is investigating a host of new ideas in diverse areas such as the design of new types of processing equipment for reaction and separation, the exploration of new chemistry for the production of old and new chemicals, the utilisation of new catalysts, the reconsideration of old chemistry which had been tried and abandoned in the past, and radical changes in operating procedures in processes which, because of either regulation or plain conservatism, had been left untouched for decades. Moreover, there have been major changes in the scale of the processes being designed, both upwards (e.g. towards "world scale" plants for basic chemicals such as ethylene and methanol) and downwards (as is the case for new energy-related technologies and devices such as fuel cells).

By its very nature, innovation involves making decisions in the absence of complete information, and this inevitably leads to risks – both technological and commercial. In this paper, we argue that a major role of process modelling is to manage technological risk in innovation. We provide an overview of some of the key developments in process modelling in the recent past and how they relate to model-based innovation. We focus on four aspects: the increasing detail in modelling complex processes which are distributed with respect to several spatial and other dimensions; the increasingly tighter links between modelling and experimentation; model-based techniques for process scale-up; and the use of modelling for the quantification of technological risk.

67b Advances in Process Synthesis - Energy & Environment

Robin Smith, Centre for Process Integration, The University of Manchester, PO Box 88, M60 1QD, Manchester, United Kingdom

The origins of much of the research into process synthesis in the late 1970's and early 1980's were rooted in energy conservation, and in particular the design of heat exchanger networks. Today, the activity in process synthesis covers a much wider area than energy conservation. However, significantly increasing energy prices and new pressures to deal with environmental problems, such as the emission of greenhouse gases, has led to a resurgence of interest in the application of process syntheses for energy and environment.

During the 1980's, tools for process synthesis based on thermodynamic and graphical methods were developed. This process synthesis technology for energy efficiency has seen widespread application throughout the world. There are now many thousands of successful applications in a wide range of industries from petroleum, petrochemical, chemical, fine chemical, through to steel manufacture, food and drink, pulp and paper. In parallel with the thermodynamic and graphical approaches that were developed, alternative approaches based on the optimisation of superstructures were also developed. In this approach a superstructure is first created that has embedded within it all feasible operations and all feasible interconnections that are candidates for an optimal design. The problem is formulated mathematically and the superstructure optimised in a structural and parameter optimisation. The optimisation then removes redundant features, reducing the complexity of the design and setting operating conditions. Thus, the decision making aspect of design are carried out by automated procedures in software. Commercial software is now available that follow this approach.

Process synthesis techniques for heat exchanger network retrofit have also improved significantly. This problem was difficult to address using the thermodynamic methods developed in the 1980's, since when they were applied to retrofit, they effectively considered the problem as a pseudo new design. The more recent methods of heat exchanger network retrofit start from the existing network and gradually evolve towards a new design in steps that allow the process designer to maintain control over the complexity of the project. Advances in heat transfer enhancement in heat exchanger technology also allow simpler retrofit designs to be developed by minimising the number of structural modifications necessary to the retrofit of heat exchanger network.

To complement these network design tools, new methods for utility system design and optimisation have been developed. These allow a more complete understanding of the interactions between the processes and the utility system on a site, but also allow a better understanding of the interactions between different processes through the utility system.

Even though these tools are well established, with commercial software available, the understanding of such tools is still patchy around the process engineering community as a whole. Whilst simulation has moved out of the realm of specialists in the last 20 years, process synthesis is still largely carried out by specialists.

The reduction of flue gas emissions as a result of combustion of fuel to produce heat and power can be brought about by:

- Changing fuels
- Changes to utility systems
- Improved heat recovery
- Process changes
- Waste heat export
- CO2 capture and sequestration
- Carbon trading

Process synthesis techniques have significant role to play in developing an effective strategy for minimisation of emissions that might call for an integrated combination of these actions to achieve environmental targets at minimum cost. Finally, process synthesis has also had a significant role to play through the minimisation of water use. Techniques for minimisation of water use have been developed, but are as yet not widely practised, due to the low cost of water, despite water shortages being problematic in certain parts of the world.

67c Process Synthesis of Separation Systems

Michael F. Doherty, University of California Santa Barbara, Department of Chemical Engineering, Santa Barbara, CA 93106 and Michael F. Malone, Chemical Engineering, University of Massachusetts, Amherst, MA 01002.

Process synthesis is a body of knowledge as essential to chemical engineering as fluid mechanics, heat & mass transfer, reaction engineering, and the other foundation subjects of our profession. Process synthesis of separation systems has matured to the point that if sufficient physical and chemical property data are available, vast numbers of process alternatives can be systematically evaluated to find the best process configuration for the task. We will discuss the present state of the art and how we arrived there during the last century. The new century, however, calls for something different that builds from our current knowledge but focuses on the very early stages of process development where business ventures have not yet been selected from among the long list of possible investment opportunities. At this early stage there is limited knowledge of all aspects of the business case. To make good early decisions, synthesis methods are needed based on preliminary experimental data, property estimation, qualitative and semi-quantitative data, heuristics, and numeric calculations, to rapidly generate manufacturing schematics and to estimate the chances for both technical and business success. The desired methodology will not only suggest a good processing route, but should also provide guidance to project managers for planning subsequent experiments to be performed by the chemists and engineers leading the early development work.

67d In the Spirit of Carnot: Theoretical Limits to Steady-State Productivity

Yangzhong Tang, Systems Biology, Harvard Medical School, Alpert 536, 200 Longwood Ave., Boston, MA 02115, Phillipp Ellison, ConocoPhillips Company, 1000 South Pine, Ponca City, OK 74602-1267, and **Martin Feinberg**, Ohio State University, 140 W. 19th Avenue, Columbus, OH 43210.

Although this talk will not be about thermodynamics, it is useful to recall the spirit of the classical Carnot analysis of heat engines. The Carnot analysis tells us that, for specified constraints (upper and lower operating temperatures), there is a limit to how much of a supplied entity (heat) can be converted into a desired product (work). Although highly idealized Carnot engines cannot be realized in practice, they nevertheless play an essential role in the analysis, not least in establishing that the derived bound on conversion of heat to work is sharp: To the extent that Carnot engines might be approximated in the limit by real engines, the theoretical Carnot bound on conversion can be realized arbitrarily closely. This is of enormous importance, for the Carnot analysis offers a theoretical benchmark against which all engines, subject to the same operating constraints, can be measured.

Process designs (like engine designs) are finally judged on practical grounds, but the Carnot story is an instructive one, suggesting a spirit in which reactor-separator synthesis might be approached. Can we know when a candidate steady-state design is, in some theoretical sense, efficient in its production rate relative to all other designs consistent with the same commitment of resources? In particular, is there a theoretical limit to what might have been produced from the same feed in any steady-state design that utilizes the same reactor size as the candidate design, that respects the same pressure-temperature bounds within reactor units, and that respects the same constraints on the production of unwanted side products? More generally, for specified temperature-pressure bounds in reactor units is there a minimum reactor size, independent of design, that is required to achieve a target productivity, and, if so, how might it be calculated? And are there certain universal Carnot-like configurations that invariably achieve maximum productivity subject to process constraints?

Some thoughts on these questions will be presented.

67e One History of Industrial Process Synthesis

Jeffrey J. Siirola, Eastman Research Division, Eastman Chemical Company, PO Box 1972, 200 South Wilcox Drive, Kingsport, TN 37662-5150

This presentation will trace the development and use of process synthesis techniques in one company over more than three decades. Adventures in task-oriented systematic generation approaches, heat integration developments, energy-efficient distillation sequencing, mass separating agent synthesis, process intensification, nonideal separation synthesis algorithms, and process network superstructure approaches will be discussed, illustrated by actual industrial examples.

110a In Honor of the 60th Birthday of George Stephanopolus

Gregory N. Stephanopoulos, Department of Chemical Engineering, Massachusetts Institute of Technology, 77 Massachusetts Ave., 56-439, Cambridge, MA 02139

Text Not Available.

110b Process Dynamics and Control

Manfred Morari, Control Laboratory, Swiss Federal Institute of Technology Zurich, ETHZ, ETH-Centrum, Zurich, 8092, Switzerland

Text Not Available.

110c Process Synthesis and Optimization

Ignacio E. Grossmann, Dept of Chemical Engineering, Carnegie Mellon University, 5000 Forbes Ave, Pittsburgh, PA 15213

Text Not Available.

110d Biomolecular Separations and Downstream Processing

Georges Belfort, Howard P. Isermann Department of Chemical and Biological Engineering, Rensselaer Polytechnic Institute, 110 8th Street, Troy, NY 12180-3590

Text Not Available.

110e Thermodynamics and Statistical Mechanics

Pablo G. Debenedetti, Department of Chemical Engineering, Princeton University, Engineering Quadrangle, Princeton, NJ 08544

Text Not Available.

110f High Throughput Heterogeneous Catalyst Research

Henry Weinberg, Symyx Technologies, 3100 Central Expressway, Santa Clara, CA 95051

With the discovery of abundant and low cost crude oil in the early 1900's came the need to create efficient conversion processes to produce low cost fuels and basic chemicals. Enormous investment over the last century has led to the development of a set of highly efficient catalytic processes which define the modern oil refinery and which produce most of the raw materials and fuels used in modern society. Process evolution and development has led to a refining infrastructure that is both dominated and enabled by modern heterogeneous catalyst technologies. Refineries and chemical manufacturers are currently under intense pressure to improve efficiency, adapt to increasingly disadvantaged feedstocks including biomass, lower their environmental footprint, and continue to deliver their products at low cost. This pressure creates a demand for new and more robust catalyst systems and processes that can accommodate them.

Traditional methods of catalyst synthesis and testing are slow and inefficient, particularly in heterogeneous systems where the structure of the active sites is typically complex and the reaction mechanism is at best ill-defined. While theoretical modeling and a growing understanding of fundamental surface science help guide the chemist in designing and synthesizing targets, even in the most well understood areas of catalysis, the parameter space that one needs to explore experimentally is vast. The result is that the chemist using traditional methods must navigate a complex and unpredictable diversity space with a limited data set to make discoveries or to optimize known systems.

We describe here a mature set of synthesis and screening technologies that together form a workflow that breaks this traditional paradigm and allows for rapid and efficient heterogeneous catalyst discovery and optimization. We exemplify the power of these new technologies by describing their use in the development and commercialization of a novel catalyst for the hydrodesulfurization of gasoline distillates having 50% more selectivity and 30% more activity for sulfur removal than the state-of-the-art commercial reference.

177a Process Operations for Batch and Continuous Processes

G. V. Reklaitis, Chemical Engineering, Purdue University, Forney Hall of Chemical Engineering, Room 1019, 480 Stadium Mall Drive, West Lafayette, IN 47907-2100

Text Not Available.

177b Opportunities for Process Development in the Conversion of Biomass to Fuels and Chemicals

James A. Dumesic, Department of Chemical Engineering, University of Wisconsin - Madison, 1415 Engineering Dr., Madison, WI 53706

Text Not Available.

177c Systems Biology: An Era of Reconstruction and Interrogation

Bernhard O. Palsson, University of California, San Diego, 9500 Gilman Dr. Dept. 0412, La Jolla, CA 92093-0412

The emerging systems biology paradigm (components -> networks -> computational models -> physiological states) involves two key steps; first, the reconstruction of a network from massive amounts of disparate data, and, second, an assessment of its functional states. We will detail both steps from a bottom-up perspective. The reconstruction process fundamentally the assembly of biochemical, genetic and genomic (BiGG) data in a mathematically structured format: basically the formation of a BiGG knowledge base. The in silico analysis methods that have been developed to compute the functional states of the reconstructed network are fundamentally query tools to interrogate the contents of the BiGG knowledge base.

177d Creating Technologies for New Businesses: Insights on Innovation

Gary Calabrese, Science & Technology, Corning Incorporated, Corning, NY 14831

Text Not Available.

196a Graphics Processing Units for High-Performance Computing in Bioinformatics

Panagiotis Vouzis[1], Joe Elble[2], and Nick Sahinidis[1]. (1) Chemical Engineering, Carnegie Mellon University, 5000 Forbes Avenue, Pittsburgh, PA 15213, (2) Industrial and Systems Enterprise Engineering, University of Illinois, Urbana, IL 61801

Bioinformatics emerged under the need to analyze proteomics and DNA data in the early 1980's. Although computers could deal efficiently with the amount of sequence data that was available until the 1990's, the amount of bioinformatics data currently available overwhelms existing computing power and algorithms for tasks such as protein alignment, motif finding, and structure calculations. Supercomputers can and have been used to reduce computing times in bioinformatics but their high cost makes them inaccessible to the average researcher.

The main goal of this work is to demonstrate that bioinformatics calculations stand to benefits tremendously from parallel algorithms executed on a GPU-enabled desktop workstation with over one TFLOPS of peak-performance and a cost under $3000. In particular, we will show how the scientific-computing-based C870 Tesla GPU [1] can be used on bioinformatics optimization algorithms, such as the Smith-Waterman dynamic programming algorithm for sequence alignment [2, 3], that lend themselves to parallelization. Although GPUs have been available for a couple of decades, they have been utilized extensively for general-purpose computing only in the last few years. This trend was boosted by the introduction of CUDA [4, 5] that facilitates the use of GPUs for parallel scientific calculations. We will demonstrate the software development process using CUDA, and the emerging challenges of parallelizing an algorithm to run efficiently on a GPU. Comparative studies and experimental performance results will demonstrate how users of bioinformatics software benefit from these emerging computing environments.

[1] www.nvidia.com

[2] M. Farrar, "Striped Smith-Waterman Speeds Database Searches Six Times over Other SIMD Implementations," Bioinformatics, pp. 156–161, Jan. 2007.

[3] S. A Manavski and G. Valle, "CUDA Compatible GPU Cards as Efficient Hardware Accelerators for Smith-Waterman Sequence Alignment," BMC Bioinformatics, March 2008.

[4] J. Nickolls, I. Buck, M. Garland, K. Skardon, "Scalable Parallel Programming with CUDA," ACM Queue Magazine, pp. 41–53, March/April 2008.

[5] D. Blythe, "Rise of the Graphics Processor," IEEE Proceedings, Vol. 96, No. 5, pp. 761–778, May 2008.

196b Fully-Implicit Finite Element Formulations for Stability and Bifurcation Analysis of Large-Scale Resistive Magneto-Hydrodynamic Systems

Roger P. Pawlowski and John N. Shadid. Computational Sciences, Sandia National Laboratories, PO Box 5800-0316, Albuquerque, NM 87114

Ionized fluids with strong electromagnetic effects occur frequently in nature and are critical for many important technological applications. Examples include stellar interiors, gaseous nebula, the earth's magnetoshpere, and Tokamak and Z-pinch physics. These systems are described by a set of partial differential equations that conserve momentum, mass, charge, and energy along with magnetic flux for the electric and magnetic fields (Maxwell's equations). The resulting magnetohydrodynamics (MHD) equations are strongly coupled, highly nonlinear, and span a large range of time and length scales, making the scalable, robust, and accurate solution of such systems extremely challenging.

In this presentation, we will discuss the development of multiple MHD formulations based on unstructured stabilized finite element methods. The formulations are designed for weak enforcement of the solenoidal constraint (divergence–free magnetic field). The formulations include a 2D vector potential formulation, a 3D projection formulation and a 3D Lagrange multiplier formulation using variational multi-scale stabilization. The resulting set of nonlinear equations are solved using a fully-coupled Newton-Krylov solver with nonlinear globalization techniques. Linear systems are solved using a multi-level preconditioned GMRES iterative technique. We will present numerical performance, accuracy, and scalability studies of the formulations. Additionally we will present an application of our solvers to perform a stability and bifurcation analysis of the hydromagnetic Rayleigh-Bernard problem. The bifurcation analysis uses the minimally augmented bifurcation formulation that has been extended to large-scale systems.

*This work was partially funded by the DOE Office of Science AMR Program, and was carried out at Sandia National Laboratories operated for the U.S. Department of Energy under contract no. DE-ACO4-94AL85000

Model Reduction-Based Constrained Optimization for Large-Scale Steady State Systems Using Black-Box Simulators

*Ioannis Bonis and Constantinos Theodoropoulos**
School of Chemical Engineering and Analytical Science, The University of Manchester, Manchester M60 QD, United Kingdom

Abstract

A novel model reduction-based deterministic optimization algorithm for large scale systems modeled by black-box simulators is presented. The proposed algorithm relies solely on the computation of low-dimensional Jacobian and reduced Hessian matrices, which correspond to the dominant modes of the system at hand. A basis for the dominant subspace of the system is computed using subspace iterations and is exploited for the calculation of the reduced Jacobian matrices through a small number of numerical directional perturbations. The reduced Hessian matrices are calculated from a 2-step projection scheme, firstly onto the dominant subspace of the system and secondly onto the subspace of the decision variables. To illustrate the efficiency of the proposed algorithm, we have applied it to the cases of the optimization of a tubular reactor, as well as a Counterflow Jet Reactor which is modeled using MPSalsa.

Introduction

The design of all industrial systems involves the concept of optimization. Many, if not most, of those systems are complex and are typically described accurately by a set of partial differential equations (PDEs). The latter are discretized over a mesh for the numerical simulation of the system at hand, which leads to a large-scale system. Steady state simulators employ iterative methods for the solution of those systems. Optimization of those processes can be based on deterministic [1], or stochastic/meta-heuristic [2] methods. Stochastic methods perform a large number of function evaluations, invoking the system simulator. Thus they are more appropriate for moderate-sized problems [2]. On the other hand, the application of deterministic optimization methods to large-scale systems with constraints is often problematic or even unrealistic, having increased requirements regarding computing power and memory size. The task is deemed even more tedious in the case of black-box solvers (commercial simulators or legacy codes), since the system equations are not available to the user.

In recent years we have developed a model reduction-based framework for gradient-based steady-state [3] and dynamic [4] optimization that employs input/output dynamic simulators. Here we extend this work by presenting a novel framework for steady-state optimization which uses black-box steady-state simulators based on solvers employing iterative linear algebra methods. We demonstrate the behavior of the algorithm by applying it for the optimization of a tubular reactor [5] and its performance and efficiency in handling large-scale input/output simulators through another illustrative example: the optimization of a Counter-Flow Jet Reactor [6] which is simulated using the state-of-the-art massively parallel finite element code MPSALSA developed at SANDIA National Laboratories [7].

---

* Corresponding author, e-mail: k.theodoropoulos@manchester.ac.uk

The proposed algorithm

The optimization problem considered here is:

$$\min f(\mathbf{x}) \quad \text{s.t. } \mathbf{G}(\mathbf{x}) = 0\,,\ \mathbf{x}^L \le \mathbf{x} \le \mathbf{x}^U \tag{1}$$

where $\Re^{N+dof} \to \Re$ is the objective function and $\mathbf{x} \in \Re^{N+dof}$ is the joint vector of the dependent $\mathbf{u} \in \Re^N$ and independent variables $\mathbf{z} \in \Re^{dof}$:

$$\mathbf{x} = \begin{bmatrix} \mathbf{u} \\ \mathbf{z} \end{bmatrix} \tag{2}$$

Feasible states are calculated by solving the equation:

$$\mathbf{G}(\mathbf{x}) = 0 \tag{3}$$

using the input/output, black-box simulator.

If the dimension of $\mathbf{x}$ is small to medium, the optimization problem can be efficiently handled by the Sequential Quadratic Programming method (SQP). The SQP is equivalent to solving the Kuhn-Karush-Tucker optimality conditions using the Newton-Raphson method. In every step of this method, a QP subproblem is solved:

$$\min_{\mathbf{d}} \frac{1}{2}\mathbf{d}^T B\mathbf{d} + \nabla f(\mathbf{x})^T \mathbf{d}, \quad \text{s.t. } \nabla\mathbf{G}(\mathbf{x})^T \mathbf{d} = -\mathbf{G}(\mathbf{x}), \quad (\mathbf{x}^L - \mathbf{x}) \le \mathbf{d} \le (\mathbf{x}^U - \mathbf{x}) \tag{4}$$

Where $B \in \Re^{n+dof \times n+dof}$ is the Hessian of the system and $\mathbf{d}$ is the search direction $\left(\mathbf{d} \in \Re^{n+dof}\right)$.

For large scale systems that include a small number of degrees of freedom, the reduced Hessian methods are more appropriate [8]. They consider decomposing the space of the search direction in two subspaces, one being the tangent space of the constraints $\nabla\mathbf{G}(\mathbf{x})$ and the other being its complement. Let $\mathbf{Z}$ and $\mathbf{Y}$ non-orthonormal bases for the two subspaces correspondingly:

$$Z = \begin{bmatrix} -\left(\nabla_u \mathbf{G}^T\right)^{-1} \nabla_z \mathbf{G}^T \\ I \end{bmatrix}, \quad Y = \begin{bmatrix} I \\ 0 \end{bmatrix} \tag{5}$$

This way, the search direction can be decomposed as follows:

$$\mathbf{d} = Y\mathbf{p}_Y + Z\mathbf{p}_Z \tag{6}$$

Following this formulation, the QP subproblem defined in Eq. 4 can be transformed as follows:

$$\min_{\mathbf{p}_z} \left(Z^T \nabla f + Z^T BY\mathbf{p}_z\right)^T \mathbf{p}_z + \frac{1}{2}\mathbf{p}_z^{\ T}\left(Z^T BZ\right)\mathbf{p}_z \quad \text{s.t.} \quad (\mathbf{x}^L - \mathbf{x}) \le Y\mathbf{p}_Y + Z\mathbf{p}_Z \le (\mathbf{x}^U - \mathbf{x}) \tag{7}$$

The matrix $Z^T BZ = B_R$, $B_R \in \Re^{dof \times dof}$ s the reduced Hessian and can be calculated numerically, generally through a BFGS update procedure. The Lagrange multipliers needed for this calculation can be calculated from:

$$Y^T BY\mathbf{p}_Y + Y^T BY\mathbf{p}_Z + Y^T \nabla\mathbf{G}\lambda = -Y^T \nabla f \tag{8}$$

Note here that the term $\mathbf{p}_y$ is zero if feasible points (steady-states) are calculated at each iteration [8].

The calculation and inversion of the Jacobian for the computation of the basis Z as defined by Eq. 5 is expensive for large problems. Especially for the case of black-box simulators, the Jacobian may not be available to the user, or even not calculated if the solver employs linear algebra procedures. To tackle this problem, we consider using a reduced Jacobian rather than the full one.

Let **P** be the invariant subspace of the Jacobian belonging to the m right-most eigenvalues. Those are the dominant eigenmodes of the system at hand which govern its behavior. In many engineering problems m << n. This subspace can be efficiently identified through its basis $\hat{Z} \in \Re^{n\times m}$, using subspace iterations. Here we used the subroutine EB22, part of the HSL library [9]. The Jacobian itself is not explicitly provided. Only products of the Jacobian with given vectors are needed, which are calculated using numerical directional perturbations on the constrains. The reduced Jacobian, $\mathbf{H} \in \Re^{m\times m}$, is a projection of the original one onto **P**:

$$\mathbf{H} = \hat{Z}^T \mathbf{G}_u \hat{Z} \tag{9}$$

The dominant subspace of the system can be easily extended to include the independent variables. The basis for the new subspace is $\hat{Z}_{ext} \in \Re^{(N+dof)\times(m+dof)}$:

$$\hat{Z}_{ext} = \begin{bmatrix} \hat{Z} & 0 \\ 0 & I \end{bmatrix} \tag{10}$$

An approximation of a basis for the subspace of the independent variables, Z_r, can be constructed using the reduced Jacobian, **H**:

$$Z_r = \begin{bmatrix} -\mathbf{H}^{-1}\hat{Z}^T \nabla_z \mathbf{G}^T \\ I \end{bmatrix} \tag{11}$$

Hence, a basis for the subspace that stems from the 2-step projection scheme is:

$$Z^* = \hat{Z}_{ext} Z_r = \begin{bmatrix} \hat{Z} & 0 \\ 0 & I \end{bmatrix} \begin{bmatrix} -\mathbf{H}^{-1}\hat{Z}^T \nabla_z \mathbf{G}^T \\ I \end{bmatrix}^* = \begin{bmatrix} -\hat{Z}\mathbf{H}^{-1}\hat{Z}^T \nabla_z \mathbf{G}^T \\ I \end{bmatrix} \tag{12}$$

Using this base, the reduced Hessian of the Lagrangian, $B \in \Re^{dof\times dof}$, can be calculated with a few numerical directional perturbations:

$$\hat{\mathbf{B}}_R = Z^{*T} \mathbf{B} Z^* = Z_r^T \left(\hat{Z}_{ext}^T B \hat{Z}_{ext} \right) Z_r \tag{13}$$

The Lagrange multipliers, ϕ, needed for this calculation are reduced and are projections of the original ones onto the dominant subspace: $\phi = Z^T \lambda$. Those are calculated from Eq. 14:

$$\mathbf{H}\phi = -\hat{Z}^T Y^T \nabla f \tag{14}$$

It is straightforward to show that the QP subproblem of Eq. 7 becomes:

$$\min_{\mathbf{p}_z} \left(Z^T \nabla f_z \right)^T \mathbf{p}_z + 1/2 \mathbf{p}_z^T B_R \mathbf{p}_z, \text{ s.t. } (\mathbf{x}^L - \mathbf{x}) \le Z\mathbf{p}_z \le (\mathbf{x}^U - \mathbf{x}) \tag{15}$$

The proposed algorithm is depicted in Figure 1.

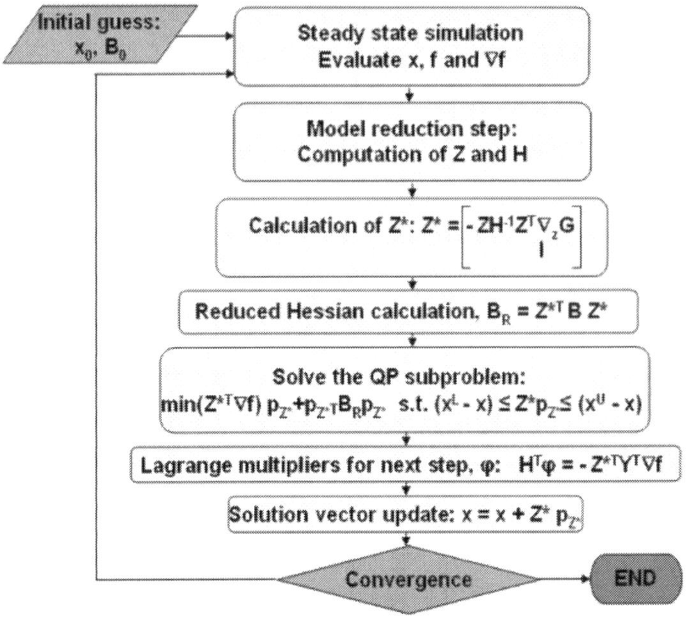

Figure 1: The proposed algorithm.

Case Study I: Optimization of a tubular reactor with 3 independent variable

To illustrate the behaviour of the proposed algorithm, we have chosen the case of the optimization of a tubular reactor with axial dispersion, where an elementary first order irreversible exothermic reaction takes place (A → B), as it is known to exhibit a rich parametric behaviour.

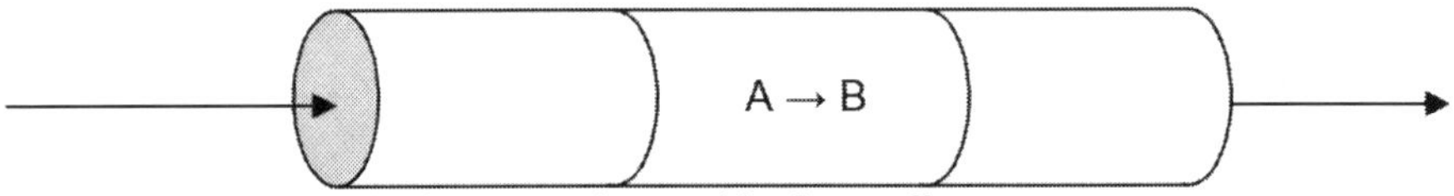

Figure 2: Schematic of a the tubular reactor, with 3 cooling zones

The model of the tubular reactor consists of two Partial Differential Equations, which for steady state become [5]:

$$\frac{1}{Pe_1}\frac{\partial^2 \mathbf{x}_1}{\partial y^2} - \frac{\partial \mathbf{x}_1}{\partial y} + Da(1-\mathbf{x}_1)\exp\left(\frac{\mathbf{x}_2}{1+\mathbf{x}_2/\gamma}\right) = 0 \tag{16}$$

$$\frac{1}{LePe_2}\frac{\partial^2 \mathbf{x}_2}{\partial y^2} - \frac{1}{Le}\frac{\partial \mathbf{x}_2}{\partial y} - \frac{\beta}{Le}\mathbf{x}_2 + \frac{C}{Le}Da(1-\mathbf{x}_1)\exp\left(\frac{\mathbf{x}_2}{1+\mathbf{x}_2/\gamma}\right) + \frac{\beta x_{2w}}{Le} = 0 \tag{17}$$

Here $\mathbf{x}_1$ and $\mathbf{x}_2$ are the dimensionless concentrations and temperatures and x_{2_w} is the dimensionless wall temperature and y the dimensionless longitudinal coordinate. We consider three cooling zones on the jacket of the reactor, each one of which has a temperature that can be controlled independently. In this case, the dimensionless temperature at the jacket of the reactor is given by Eq. (18)

$$x_{2_w}(y) = \sum_{j=1}^{3}[H(y-y_{j-1}) - H(y-y_j)]x_{2_w j}\text{, } y_1 = ⅓\text{, } y_2 = ⅔\text{, } y_3 = 1 \tag{18}$$

All the parameters are described in [5]. The boundary conditions chosen for this case are:

$$\frac{\partial \mathbf{x}_1}{\partial y} - Pe_1\mathbf{x}_1 = 0, \quad \frac{\partial \mathbf{x}_2}{\partial y} - Pe_2\mathbf{x}_2 = 0 \quad \text{at} \quad y = 0, \quad \frac{\partial \mathbf{x}_1}{\partial y} = 0, \quad \frac{\partial \mathbf{x}_2}{\partial y} = 0 \quad \text{at} \quad y = 1 \qquad (19)$$

The values chosen for the rest of the parameters are Le = 1.0, $Pe_1 = Pe_2$ = 5.0, γ = 20.0, β = 1.50, C = 12.0. The Finite Difference (FD) method was used for the discretization of the equations at hand and a number of nodes of 250 was chosen, leading to 500 algebraic equations.

The objective is to maximize the dimensionless concentration, x_1, at the output of the reactor with respect to the values of the dimensionless temperatures of the 3 cooling zones and subject to Eqs. (16), (17) and (19) being satisfied.

The termination criterion for the algorithm was $\|Z^* p_z\| \leq 10^{-6}$ and the size of the dominant subspace chosen was m = 10. The lower bounds for all variables were 0, the upper bounds were 1 for x_1, 8 for x_2 and 4 for x_{2_w}. The simulator of the tubular reactor was treated as a black-box simulator and all Jacobians and Hessians were calculated by employing numerical directional perturbations. Convergence was achieved in 18 iterations and the optimal point found is $x_{2w,1}$ = 2.4826, $x_{2w,2}$ = 0.52539, $x_{2w,3}$ = 4.0000. For those values of the independent variable, a yield of 99.868% is achieved. In Fig. 3 the concentration and temperature profiles at the optimum for the reactor and for the cooling zones are presented.

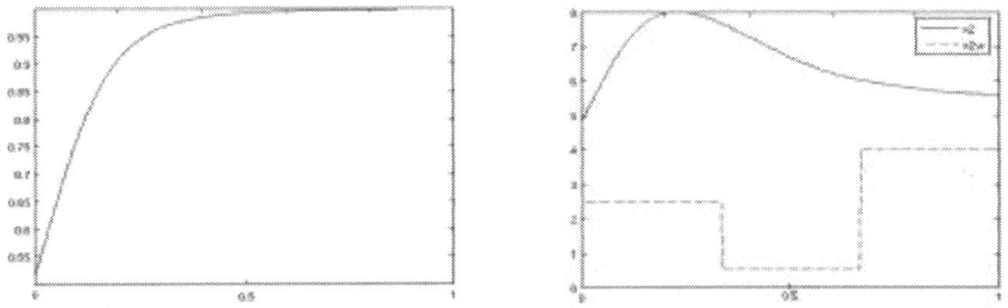

Figure 3: Solution profiles at the optimum point for dimensionless concentration (a) and dimensionless temperature (b) both for the reactor and for the 3 cooling zones.

Case Study II: Optimization of a Counter Flow Jet Reactor simulated using MPSalsa

To illustrate the efficiency of the proposed algorithm in handling large scale optimization problems using input/output simulators, we have applied it for the optimization of a Counter Flow Jet Reactor used for the decomposition studies of tertiary butyl arsine (TBA). The model of the reactor was setup using MPSalsa [7], which is state-of-the-art massively parallel CFD software, developed at SANDIA national laboratories. It implements the Finite Element Method, employing unstructured meshes and inexact Newton methods with iterative linear solvers. In this formulation, MPSalsa was used by our in-house developed optimization code as a black box.

A schematic of the reactor, as well as the formulation of the model and the pathways for the decomposition of TBA is presented in Fig. 4. The model considered is a 2D symmetric one. The momentum equation is:

$$\frac{\partial U}{\partial t} = \rho(U \cdot \nabla)U - \nabla \cdot \tau - g\rho e_g \tag{20}$$

where U is the velocity vector, ρ is the mixture density, which is a function of temperature and composition, g is the gravity vector and τ is the stress tensor, which for a Newtonian fluid can be calculated from Eq. (21):

$$\tau = -pI - \frac{2}{3}\mu(\nabla \cdot U)I + \mu(\nabla U + \nabla U^T) \tag{21}$$

where p is the operating pressure, μ is the mixture viscosity and I is the unitary tensor. The total mass equation is

$$\frac{\partial \rho}{\partial t} = \nabla \cdot (\rho U) \tag{22}$$

The energy balance is:

$$\frac{\partial T}{\partial t} = \rho C_p(U \cdot \nabla)T - \nabla \cdot (\kappa \nabla T) \tag{23}$$

where T is the temperature, C_p is the mixture heat capacity and κ is the thermal conductivity. The mass conservation equation for species i, i = 1, ..., N_c– 1, N_c being the number of components of the mixture.

The model of the reactor consists of 19040 dependent variables (velocities, temperatures, pressures and concentrations for each point of the mesh) and 1 degree of freedom: the velocity of the upper stream.

The objective in this case is to maximize the production of AsH, thus maximizing the decomposition of TBA, while minimizing the production of the highly toxic AsH_3.

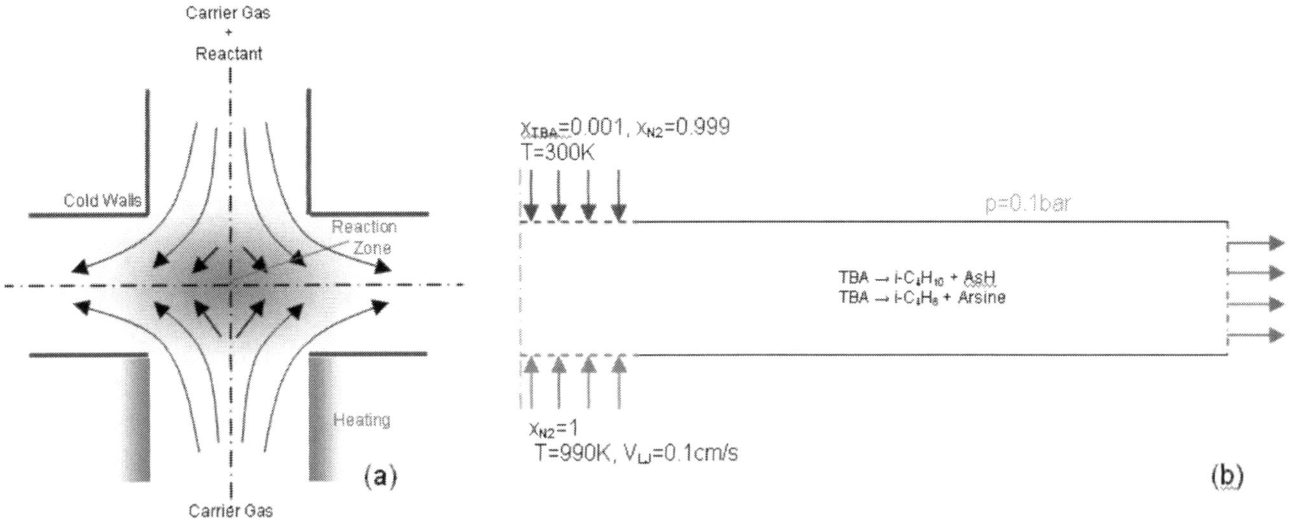

Figure 4: Schematic of the Counter Flow Jet Reactor (a) and formulation of the model (b).

The size of the dominant subspace chosen was 12. Convergence was achieved in 4 iterations and the optimal velocity found was 1.348cm/s, which results to a yield of 78.85%. Fig. 5 shows the temperature, velocity, pressure and concentration profiles at the optimum.

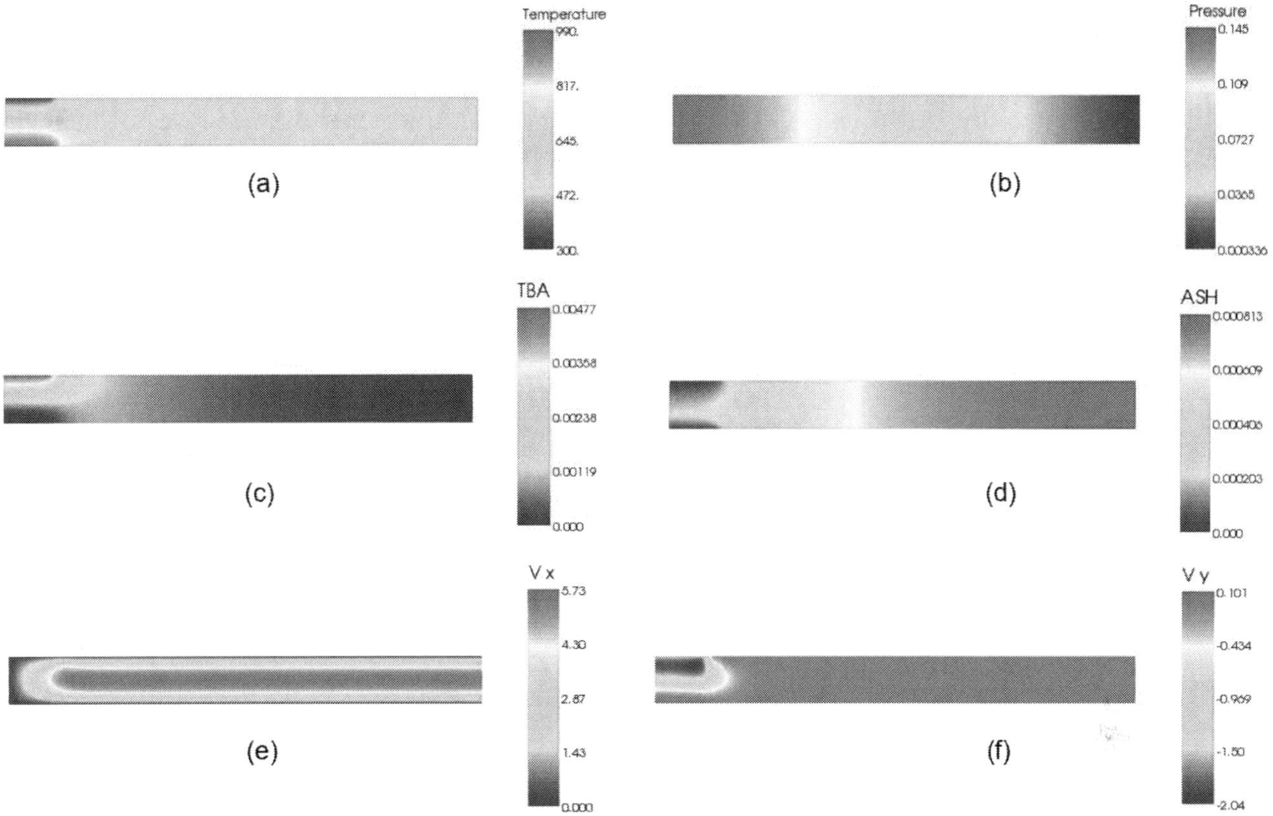

Figure 5: Schematic Temperature (a), pressure (b), TBA mass fraction (c), AsH mass fraction (d), and velocity (e and f) of the Counter reactor at the optimum point.

Acknowledgements
The financial support from the EU program CONNECT (COOP-CT-2006-031638) is gratefully acknowledged.

References

[1] T.F. Edgar, D.M. Himmelblau, 1988, Optimization of Chemical Processes, McGraw-Hill, NY
[2] C. Blum and A Roli, 2003, Metaheuristics in Combinatorial Optimization: Overview and Conceptual Comparison, ACM Computing Surveys, 35: 268-308.
[3] E. Luna-Ortiz and C. Theodoropoulos, 2005, An input/output model reduction-based optimization scheme for large-scale systems, Multiscale Modeling & Simulation 4: 691-708.
[4] C. Theodoropoulos and E. Luna-Ortiz, 2006, "A Reduced Input/Output Dynamic Optimisation Method for Macroscopic and Microscopic Systems", in Model Reduction and Coarse-Graining Approaches for Multiscale Phenomena, A. Gorman, N. Kazantzis, I.G. Kevrekidis, H.C. Ottinger and C. theodoropoulos (eds) pp. 535-560.
[5] K. F. Jensen and W. H. Ray, 1982, The Bifurcation Behavior of Tubular Reactors, Chemical Engineering Science 37: 199-222.
[6] Safvi, S.A. and T.J. Mountziaris, "A new reactor for purely homogeneous kinetic studies of endothermic reactions", AIChE Journal, 1994. 40: 1535-1548.

[7] Shadid J, Hutchinson S, Hennigan G, Moffat H, Devine K, Salinger AG, "Efficient parallel computation of unstructured finite element reacting flow solutions", Parallel Computing 1997. 23: 1307-1325.
[8] L.T. Biegler, J. Nocedal and C. Schmidt, 1995, A Reduced Hessian Method for Large-Scale Constrained Optimization, Siam Journal on Optimization 5: 314-347.
[9] HSL, A Collection of Fortran codes for large-scale scientific computation. See http://www.hsl.rl.ac.uk, (2007).

196d Canonical Primal-Dual (CPD) Formulation of Process Models

Cliff Amundsen and **Ross Swaney**. Dept. of Chemical & Biological Engineering, University of Wisconsin, 1415 Engineering Drive, Madison, WI 53706

This paper presents a canonical form for chemical process systems models for steady state. The approach is motivated by the fact that current mainstream technologies for process modeling at the level of process flowsheet simulation and similar "rigorous" fidelity models have seen little fundamental change in more than 20 years. The sequential modular architecture is still relied upon. The successor "equation oriented" architecture has afforded new capabilities, but has not realized its full promise. The skilled modeling specialist can accomplish results if the required time can be invested, but the capabilities available to the mainstream user/modeler have improved little. The drawbacks of the standard methods are well known and remain.

The CPD modeling approach is intended to provide a means to mitigate or avoid the technical issues that to date make modeling software development a demanding task, and model use problematic. The core issues to be resolved are (I) Numerical convergence of the inherently nonlinear models; (II) Adequacy/ consistency/specification of complex formulations and positive diagnosis of misformulation; and (III) Flexibility to embody all relevant physical phenomena, extending to applications beyond the conventional.

The formulation employs a single, uniform basic submodel structure that instantiates to form all parts of the entire large scale process model. The complete system is a collected composite of these coupled subsystems, possibly in nested hierarchy. Such a uniform representation is possible because the basic submodel is derived directly from the fundamental physics of the phenomenology.

Fundamental conservations are identified according to the underlying symmetries of the physical behavior. In addition to strain, momentum, and energy, relevant species are identified along with allowable interconversions. Nonconserved species such as photons are also included as needed. Then all physical phenomena are expressed as variational problems subject to the conservation constraints.

Physical modes are partitioned into microscopic and macroscopic. The microscopic modes are treated with a local equilibrium (LE) approximation. This partition of the subsystem evolves the thermodynamic behavior and the macroscopic transport diffusivities. The macroscopic modes are governed by a convection/diffusion/conduction model. Significantly, all these macroscopic phenomena are captured by a single variational problem.

Formulation in terms of variational principles is a very important feature. It allows the convexity characteristics of the governing physical principles to be recognized and exploited. When representing process models simply as algebraic equations it is difficult or impossible to do this - perhaps this partly explains the numerical difficulties that frequent the modeling-via-equations paradigm.

The main characteristics of the CPD formulation are summarized:

1. The physical phenomena are formulated in a single basic subsystem that applies to all points in space.

2. The phenomena are represented by a pair of coupled macroscopic-microscopic variational principles with convexity properties that are established a priori.

3. The macroscopic variational principle represents the behavior in terms of a set of primal fields (generalized densities) and dual fields (potentials).

4. The macroscopic behavior is integrated over the geometry of macroscopic volumes to obtain low-dimensional characterizations of equipment regions. Simplified basis functions are used to approximate the primal and dual fields. These could be empirically determined, for instance from pressure drop, heat, and mass transfer correlations for turbulent flow. Alternatively, methods ranging from simple approximate calculations to detailed CFD might be employed. The LE material properties are averaged over space by associated simple quadratures.

5. All equipment subsystems inherit the convexity characteristics of the intrinsic physical phenomena. They are interconnected at physical interfaces by continuity of both the primal flows and dual field quantities. Geometric congruence of the interconnections is unambiguous.

6. Model feasibility can be determined by analyzing a global linear system. Misformulation diagnosis is obtained using standard linear algebraic analysis.

7. The convexity properties of the variational principles rigorously guarantee that the combined system can be solved via homotopy.

This paper will explain the basic subsystem model that provides the core of the canonical primal-dual formulation. It will also briefly outline the significant properties mentioned above and their consequences for modeling large scale nonlinear process systems.

196e Hybrid Niche Algorithm for Problem Inversion of Distributed Systems with Solution Multiplicity

Jeonghwa Moon[1], Libin Zhang, and Andreas A. Linninger[2]. (1) Chemical Engineering, University of Illinois at Chicago, 851 S. Morgan St., Room 218, Chicago, IL 60607, (2) Bioengineering, University of Illinois at Chicago, 851 S. Morgan St., Room 218, Chicago, IL 60607

Many important chemical reaction systems have strongly coupling in heat and mass transfer. These systems are described with partial differential equations (PDEs) and method of the transport. We have introduced a kinetic inversion problem (TKIP) method to estimate the parameters for these distributed models. However, the inversion problems are characterized multiple solutions and many conventional numerical methods fail to find all solutions. In order to determine which solution constitutes the actual physical response or which parameter set best explains the actual measurements, we propose to identify first all locally optimum solutions. In the second stage, we revisit all local minima to select most physically meaningful outcome. Therefore it is necessary to deploy inversion algorithms which we can reliably identify all local minima for distributed inversion problems. For this reason, we developed 'a hybrid niche algorithm', based on the sequential niche technique (Beasley et al 93) in combination with deterministic problem inversion. In this method, we combined a genetic search with a deterministic method to accelerate locating multiple solutions to the parameter inversion problem. The method uses a self-adjusting niche radius to overcome the problem of clustering in the search space. Thus every niche area has a different size and updated. In this presentation, we will show the efficiency and robustness of our algorithms with several examples.

196f Taylor Based Pseudo-Analytical Solution of the Power Law Model for Non-Newtonian Fluid In FREE Convective Cells

Mario Oyanader and Pedro E. Arce. Chemical Engineering, Tennessee Technological University, Department of Chemical Engineering, P.O. Box 5013, Cookeville, TN 38505

Not a few numerical methods have been developed to empower modeler in solving complex mathematical expressions and in conducting analysis. Nevertheless, even more complex system conditions, fluid behavior and mathematical models may outperform the goodness of any given methodology. This is the case of the so called Non-Newtonian fluids under the influence of buoyancy driven forces. Motivated by the need of understanding the hydrodynamic developed by non- Newtonian fluids in non-isothermal conditions, this contribution focuses on the application of the Taylor's expansion series as a strategic solution of the "Power Law hydrodynamic model" under free convective forces for different types of non-Newtonian fluids. The implication of this contribution can be linked to the search of new chemical products, the exploration of their physicochemical properties, and the design of the most basic unit operation. Furthermore, the study concentrates on the comparison of the numerical solution and the Taylor's solution to validate the existence of multiple solutions as predicted by this last method. Several cases of Joule heating generation are examined and its influence on the velocity profile. Especial attention is given to fact that traditional numerical methods do not facilitate the reconciliation of the material balance under batch conditions. Illustrative results involving different geometries are presented to demonstrate the validation of the proposed approach when compared with typical numerical methods.

A new efficient eigenvalue bounding method for convexity detection with applications in global optimization and control

M. Mönnigmann
Department of Mechanical Engineering
Technische Universität Braunschweig, Braunschweig, Germany

September 17, 2008

We introduce a new method for the calculation of bounds on the eigenvalues of Hessian matrices of twice continuously differentiable functions. Eigenvalue bounds of Hessian matrices arise in a number of notoriously difficult tasks in computational chemical engineering. For example, Hessian matrix eigenvalue bounds are used in global nonlinear optimization, global convexity/concavity analysis in convex optimization, and global positive/negative invariance analysis in nonlinear control.

We stress that the improvements in computational complexity to be stated below are only possible, because the desired Hessian matrix eigenvalue bounds are calculated without ever calculating the Hessian matrix itself. To the author's knowledge the proposed method is the first Hessian-matrix-free approach to bounding Hessian matrix eigenvalues.

We start with a more precise problem statement in the next section, summarize the methodological advances in a subsequent section, and turn to applications in the last section.

Problem statement Two variants of the proposed method must be carefully distinguished from one another. Both variants apply to twice continuously differentiable functions $f : U \subseteq \mathbb{R}^n \to \mathbb{R}$. Let $x_0 \in U$ be an arbitrary point and let $S = [\underline{x}_1, \bar{x}_1] \times \cdots \times [\underline{x}_n, \bar{x}_n]$ be an arbitrary closed hyperrectangle. With the first variant we calculate bounds on the eigenvalues of a Hessian matrix that has been evaluated ***at a point***. More precisely, we determine $\underline{\lambda} \in \mathbb{R}$ and $\bar{\lambda} \in \mathbb{R}$ such that $\lambda \in [\underline{\lambda}, \bar{\lambda}]$ for all eigenvalues λ of $\nabla^2 f(x)$ evaluated at an arbitrary but fixed $x_0 \in U$. The second variant provides bounds that apply to all eigenvalues of the Hessian $\nabla^2 f(x)$ evaluated ***on a hyperrectangle***. More precisely, we calculate bounds $\underline{\lambda} \in \mathbb{R}$ and $\bar{\lambda} \in \mathbb{R}$ such that $\lambda \in [\underline{\lambda}, \bar{\lambda}]$ for all eigenvalues λ of $\nabla^2 f(x)$ for all $x \in S$. For brevity we refer to these two cases as the real and the interval variant of the method, respectively. These names are chosen because the first variant applies to real matrices $\nabla^2 f(x) \in \mathbb{R}^{n \times n}$ while Hessian matrices on hyperrectangles are often approximated by interval Hessian matrices.

Advances and relation to existing methods Both the real and the interval variant of the new method are compared to established methods, specifically the method due to Gershgorin [3] and the method due to Hertz [4] and Rohn [6], respectively. Hertz and Rohn's method is known to provide *tight*, i.e. the best possible, bounds on the eigenvalues of interval matrices. We note that Hertz and Rohn's method is exponentially complex, however. This complexity is not surprising, since the problem of calculating tight eigenvalue bounds for interval matrices is known to be NP-hard [1].

In a formal proof the real variant of the method can be shown to be one order of magnitude computationally less complex than the method due to Gershgorin. More precisely, the proposed method provides bounds at a complexity

$$\mathcal{O}(n)\, N(f_i) \tag{1}$$

where $N(f_i)$ denotes the number of operations necessary to evaluate the function f_i at a point x_0, where $i \in \{1, \ldots n\}$. In contrast, the calculation of the Hessian matrix and its eigenvalue bounds with Gershgorin's approach can be shown to require $\mathcal{O}(n^2)\, N(f_i)$ operations. Despite its lower complexity, nontrivial examples exist for which the proposed method results in bounds that are tighter than Gershgorin's bounds.

The results for the interval variant of the method are somewhat surprising and have to be summarized more carefully. First we note that the interval variant of the method can be shown to belong to the same polynomial complexity class (1) as its real counterpart. Furthermore, nontrivial examples of functions exist for which the new method provides bounds on the eigenvalues of the Hessian matrix on hyperrectangles that are *tighter* than the bounds obtained with Hertz and Rohn's method for the interval Hessian matrix. At first sight, this claim seems to be a contradiction to both the NP-hardness and the tightness of Hertz and Rohn's method. Closer inspection reveals that the restrictions that apply to methods for interval Hessians do *not* apply here, because neither real nor interval Hessian matrices are ever calculated in the proposed method.

Due to its polynomial complexity the interval variant of the new approach can be expected to be of use in cases where the exponential complexity of Hertz and Rohn's method is prohibitive.

Applications Bounds on eigenvalues of Hessian matrices can be used in a variety of applications. Eigenvalues of real Hessian matrices can be used to check sufficient optimality conditions in nonlinear programming, for example. Bounds on eigenvalues of Hessians on hyperrectangles can be used in deterministic approaches to global nonlinear optimization to create convex underestimators. A famous example for such an approach is the αBB method developed by Floudas and coworkers [2].

In the present contribution we use an application from stability and control theory to illustrate the use of the new method for the calculation of eigenvalue bounds. In stability and control theory, positive definiteness, or equivalently convexity, of quadratic forms appears in various criteria that are based on Lyapunov functions (see e.g. [5]).

Here we focus on the use of Hessian matrices on hyperrectangles for the search of positive invariant sets of dynamical systems. Essentially, a set in the domain of a nonlinear dynamical system is called positive invariant if the system never leaves it once it entered it. We state two simple criteria for positive invariance, which are based on the first and second order Taylor approximation of the nonlinear dynamical system. We show that examples exist where positive invariance can be established with the second order but not the first order criterion. In order to efficiently evaluate the second order criterion, bounds on eigenvalues of Hessian matrices are calculated with the new method proposed here.

References

[1] V. D. Blondel and J. N. Tsitsiklis. A survey of computational complexity results in systems and control. *Automatica*, 36:1249–1274, 2000.

[2] C. A. Floudas. *Deterministic Global Optimization: Theory, Methods, and Applications*. Kluwer, 1999.

[3] S. Gerschgorin. Über die Abgrenzung der Eigenwerte einer Matrix. *Izv. Akad. Nauk SSSR, Ser. fizmat.*, 6:749–754, 1931.

[4] D. Hertz. The extreme eigenvalues and stability of real symmetric interval matrices. *IEEE Transactions on Automatic Control*, 37(4):532–535, 1992.

[5] D. Hinrichsen and A. Pritchard. *Mathematical Systems Theory I*. Springer, 2005.

[6] J. Rohn. Positive definiteness and stability of interval matrices. *SIAM Journal on Matrix Analysis and Applications*, 15(1):175–184, 1994.

198a Computational Comparison of Piecewise Linearization Schemes in Gas Lifting and Pooling Optimization Problems

Ruth Misener, Chrysanthos E. Gounaris, and Christodoulos A. Floudas. Chemical Engineering, Princeton University, Princeton, NJ 08544

Continuous gas lifting is the process of injecting compressed natural gas into the production tubing of an oil well, in order to increase the pressure of the reservoir and enhance its production [1]. The allocation of the limited amount of available gas among a set of neighboring oil wells constitutes an interesting optimization problem. Due to the piecewise linear nature of the gas injection vs. oil production curves, the problem can be formulated as an MILP, and solved to global optimality [2]. In this study, we exploit four different mathematically equivalent representations of the problem, utilizing different applicable formulations presented in the literature, such as the linear segmentation [3] and convex hull [4,5] formulations. All variants succeed in solving the case studies to global optimality, including large cases involving 200 wells. It is shown that the novel formulation of [6] outperforms the competition in terms of both number of iterations and CPU time.

Pooling problems correspond to maximizing the profit of operating a network of blending pools, while satisfying various flow rate and material composition constraints. These problems appear in a wide variety of fields, including oil refining and water treatment, and their efficient solution has been an active field of research for many years. The pooling problem is formulated as a bilinear NLP, and its relaxations are typically constructed with use of bilinear convex envelopes [7]. We present here piecewise linearization schemes that can be employed to refine the tightness of these standard relaxations. In particular, we compare the computational performance of various big-M [8], convex hull [9], and incremental cost [10] formulations on a variety of benchmark problems. Despite the fact that all formulations eventually result into the same root node lower bound, our study demonstrates that some of them are consistent in exhibiting superior computational performance, and should therefore be preferred in the construction of pooling problem relaxations.

[1] T. L. Mason, A. Bagirov, J. Klunder, and J. van Berkel. Development and application of the discrete gradient method for non-smooth non-convex production optimization. In International Oil Conference and Exhibition, Veracruz, Mexico, 2007. Society of Petroleum Engineers.

[2] V. D. Kosmidis, J. D. Perkins, and E. N. Pistikopoulos. Optimization of well oil rate allocations in petroleum fields. Ind. Eng. Chem. Res., 43(14):3513–3527, 2004.

[3] C. A. Floudas. Nonlinear and Mixed-Integer Optimization: Fundamentals and Applications. Oxford University Press, New York, NY, 1995.

[4] H. D. Sherali. On mixed-integer zero-one representations for seperable lower-semicontinuous piecewise-linear functions. Operations Research Letters, 28(4):155–160, 2001

[5] E. Camponogara and P. Nakashima. Optimizing gas-lift production of oil wells: piecewise linear formulation and computational analysis. IIE Transactions, 38(2):173–182, 2006.

[6] A. B. Keha, I. R. de Farias Jr., and G. L. Nemhauser. Models for representing piecewise linear cost functions. Operations Research Letters, 32(1):44–48, 2004.

[7] F.A. Al-Khayyal and J.E. Falk. Jointly constrained biconvex programming. Mathematics of Operations Research, 8(2):273–286, 1983.

[8] C.A. Meyer and C.A. Floudas. Global optimization of a combinatorially complex generalized pooling problem. AIChE Journal, 52(3):1027–1037, 2006.

[9] M.L. Bergamini, P. Agguirre, I.E. Grossmann. Logic-based outer approximation for globally optimal synthesis of process networks. Computers & Chemical Engineering, 29(9):1914–1933, 2005.

[10] D.S. Wicaksono and I.A. Karimi. Piecewise MILP under- and overestimators for global optimization of bilinear programs. AIChE Journal, 54(4):991–1008, 2008.

198b Global Solution of Bilevel Mixed-Integer Nonlinear Programs

Alexander Mitsos, Aices, RWTH Aachen, Pauwelstrasse 12, Aachen, Germany

A deterministic algorithm for the global solution of bilevel mixed-integer nonlinear programs is presented. The algorithm is rigorous and terminates finitely to a point that satisfies ε-optimality in the upper- and lower-level programs. This is the first deterministic algorithm to do so.

Bilevel programs are programs where one optimization problem (upper-level program) is constrained by the (global) solutions of another optimization problem (lower-level program). Bilevel programs represent hierarchical decision making where two decision makers (optimizers) control a subset of the decision variables. Bilevel programs are used in many applications of (bio-)chemical engineering, e.g., [1,2,3,4]. A major challenge in bilevel programs is nonconvexity of the lower-lever program, in which case traditional methods based on the KKT necessary conditions typically obtain infeasible points. The presence of integer variables gives significant more flexibility in modeling, but further complicates the solution, especially when integer variables are present in the lower-level program. Moore and Bard [5,6] discuss difficulties of the mixed-integer case, in particular that removing the integrality constraints does not necessarily give a relaxation of the bilevel program. Reformulating integer variables using MPEC constraints would not be efficient.

The algorithm presented is based on a recent proposal for continuous bilevel programs [7], however the generalization to mixed-integer bilevel programs is far from trivial. The assumptions necessary for convergence are compact host sets, continuous functions and a suitable regularity condition of the lower-level program. The algorithm proceeds similar to the algorithm by Blankenship and Falk [8] for semi-infinite programs, by adding cuts to the lower bounding problem until the upper bound is guaranteed to generate an ε-optimal point. The upper bounding problem is based on probing the solution obtained by the lower bounding procedure. The cuts used are generated by parametric upper bounds to the lower-level program. In the special case that the lower-level program satisfies a constraint qualification for fixed integer values, the KKT necessary conditions can be used to tighten the lower bound and reduce the number of iterations. Branching can be performed, but is not necessary for convergence.

The presentation gives a motivation and statement for the algorithm along with a proof outline for a finite termination theorem. Then illustrative examples are discussed as well as numerical results from literature examples and new test problems. As is demonstrated in the numerical examples, the proposed algorithm can solve literature problems for which previous contributions have terminated with suboptimal or infeasible points.

[1] P. A. Clark and A. W. Westerberg. Bilevel programming for steady-state chemical process design. 1. Fundamentals and algorithms. Computers & Chemical Engineering, 14(1):87-97, 1990.

[2] C. A. Floudas, Z. H. Gumus, and M. G. Ierapetritou. Global optimization in design under

uncertainty: Feasibility test and flexibility index problems. Industrial & Engineering Chemistry Research, 40(20):4267-4282, 2001.

[3] I. E. Grossmann and K. P. Halemane. Decomposition strategy for designing flexible chemical-plants. AIChE Journal, 28(4):686-694, 1982.

[4] A. P. Burgard and C. D. Maranas. Optimization-based framework for inferring and testing hypothesized metabolic objective functions. Biotechnology and Bioengineering, 82(6):670-677, 2003.

[5] J. T. Moore and J. F. Bard. The mixed integer linear bilevel programming problem. Operations Research, 38(5):911--921, 1990.

[6] J. F. Bard. Practical Bilevel Optimization: Algorithms and Applications. Nonconvex Optimization and Its Applications. Kluwer Academic Publishers, Dordrecht, 1998.

[7] A. Mitsos, P. Lemodinis and P. I. Barton. Global solution of bilevel programs with a nonconvex inner program. In press: Journal of Global Optimization, October 12, 2007.

[8] J. W. Blankenship and J. E. Falk. Infinitely constrained optimization problems. Journal of Optimization Theory and Applications, 19(2):261--281, 1976.

198c Global Optimization of Mixed-Integer Bi-Level Programming Problems Via Multi-Parametric Programming

Luis F. Dominguez and Efstratios N. Pistikopoulos. Centre for Process Systems Engineering, Imperial College London, Roderic Hill Building, South Kensington Campus, London, SW7 2AZ, United Kingdom

Bi-level (BLPP) and multi-level programming problems (MLPP) arise very often in areas of engineering, control and economics (Vicente, 2001). A key feature of such problems from a mathematical viewpoint is that even for the simplest linear case, a global optimization approach is typically necessary (Floudas, 2001; Visweswaran, 2001). On the other hand, a multi-parametric programming approach applied to certain classes of bi-level programs overcomes the need for global optimization (Faísca et al. 2007).

In this work, we present a multi-parametric programming based approach for mixed-integer bi-level programming problems. We show that, in this case, there is indeed a need for a global optimization approach due to the presence of integer/0-1 variables in the inner/follower's problem. We first employ, a reformulation-linearization technique (Sherali and Adams, 1994) to replace the inner constraint set by its convex hull and then apply a multiparametric programming framework to reach the global optimal solution. Numerical examples and an engineering control application will be provided to illustrate our approach.

References:

Vicente, L., Bi-level programming: Introduction, History and Overview, Encyclopedia of Optimization, Kluwer Academic, Foudas, C. A. & Pardalos, P. M. (ed.), 2001, I, 178-180.

Faísca, N. P.; Dua, V.; Rustem, B.; Saraiva, P. M. & Pistikopoulos, E. N., Parametric global optimisation for bi-level programming, J. Glob. Opt., 2007, 38, 609-623.

Floudas, C. A., Deterministic global optimization: theory, methods and applications, The GOP approach in bi-level linear and quadratic Problems, Kluwer Academic Publishers, 2000, 173-191.

Sherali, H. D. & Adams, W. P., A hierarchy of relaxations and convex hull characterizations for mixed-integer zero-one programming problems, Discrete Applied Mathematics, 1994, 52, 83-106.

Visweswaran, V., Bi-level programming: global optimization, Encyclopedia of Optimization, Kluwer Academic, Floudas, C. A. & Pardalos, P. M. (ed.), 2001, I, 163-167.

Web Page: www.imperial.ac.uk/people/luis.dominguez06

198d Relaxations of Nonconvex, Quadratically-Constrained Quadratic Programs

Xiaowei Bao, Chemical and Biomolecular Engineering, University of Illinois at Urbana-Champaign, 2725 NW OVERLOOK DR APT 1928, Hillsboro, OR 97124 and Nick Sahinidis, Chemical Engineering, Carnegie Mellon University, Doherty Hall 4210C, 5000 Forbes Avenue, Pittsburgh, PA 15213.

Quadratic programming problems arise frequently in various industrial application settings, including facility location, multiperiod refinery, and circle packing problems. The general nonconvex quadratically constrained quadratic programming (QCQP) problem is NP hard and presents a significant challenge. We present a tight linear relaxation scheme for QCQP that can be used in the context of a branch-and-bound global optimization algorithm for this problem.

Most of the current approaches rely on convex relaxations of the QCQP problem which are either nonlinear [7, 2] or relaxing each nonconvex term separately [1, 5]. The main idea of our approach is to relax the entire quadratic constraints by linear underestimations. We propose a relaxation that is tighter than standard linear relaxation based on the convex envelope of multilinear functions [4, 3, 6] and develop an efficient mechanism to tighten the standard relaxation via a class of multilinear cutting planes. Computational experience shows the favorable features of the proposed relaxation.

[1] F. A. Al-Khayyal, C. Larsen, and T. V. Voorhis. A relaxation method for nonconvex quadratically constrained quadratic programs. Journal of Global Optimization, 6:215–230, 1995.

[2] S. Burer and D. Vandenbussche. A finite branch-and-bound algorithm for nonconvex quadratic programming via semidefinite relaxations. Mathematical Programming, 2006. Published on-line, 12 December 2006, DOI 10.1007/s10107-006-0080-6.

[3] A. D. Rikun. A convex envelope formula for multilinear functions. Journal of Global Optimization, 10:425–437, 1997.

[4] H. D. Sherali. Convex envelopes of multilinear functions over a unit hypercube and over special discrete sets. Acta Mathematica Vietnamica, 22:245–270, 1997.

[5] H. D. Sherali and C. H. Tuncbilek. A reformulation-convexification approach for solving nonconvex quadratic programming problems. Journal of Global Optimization, 7:1–31, 1995.

[6] M. Tawarmalani and N. V. Sahinidis. Convex extensions and convex envelopes of l.s.c. functions. Mathematical Programming, 93:247–263, 2002.

[7] N. V. Thoai. Duality bound method for the general quadratic programming problem with quadratic constraints. Journal of Optimization Theory and Applications, 107:331–354, 2000.

198e Polyhedral Results for a Discrete-Time Mip Formulation for Production Planning

Christos T. Maravelias, University of Wisconsin - Madison, 1415 Engineering Drive, Madison, WI 53706

The scheduling and production planning of multi-product facilities has received considerable attention in the process systems engineering literature. The focus of most of the proposed approaches has been on the development of mixed-integer programming (MIP) formulations that are shown to solve well a class of problems. The quality of such formulations is typically assessed from the computational requirements for the solution of a set of example problems. In general, it can be argued that the development of MIP models for the scheduling of chemical processes has been guided by empirical (computational) rather than theoretical results. This is in sharp contrast with efforts on production planning in the operations research and mathematical programming communities, where the development of reformulations and the derivation of theoretical results concerning the structure and tightness of MIP formulations is the method of choice. The advantage of this more rigorous approach is that these results and reformulations point the way to effective decomposition methods. In this work, we follow the latter approach. We study the polyhedral properties of a discrete-time MIP formulation for the production planning of multi-stage multi-product continuous processes.

First, we review concepts and known results on polyhedral theory and mixed-integer programming. We introduce the basic concepts about polyhedra and polytopes (Cook et al., 1997), discuss how total unimodularity ensures the integrality of polyhedra (Nemhauser and Wolsey, 1988; Hoffman and Kruskal, 1956), and present properties of totally unimodular matrices. We then discuss total k-modularity and k-regularity, the two generalizations of total unimodularity to general matrices [Kotnyek, 2002; Appa and Kotnyek, 2004], and review existing results and prove one new result for the characterization of k-regular matrices. We close this part with a short discussion of decomposition methods and reformulations (Pochet and Wolsey, 2006).

Second, we develop new results for the discrete-time production planning MIP formulation (Maravelias, 2006). The original problem is decomposed into two subproblems, one that involves only assignment variables and constraints, and one that involves assignment and inventory variables and material balance constraints. We show that the constraint matrix of the first subproblem is an "interval" matrix (Fulkerson and Gross, 1965), and thus the feasible region of the LP-relaxed problem is integral. We next derive results for the second subproblem:

1) We show that the constraint matrix is k-regular for integral data.

2) We determine the length of the planning period for which the k-regularity of the matrix implies the integrality of the LP-relaxed feasible region.

3) We show that the LP-relaxed feasible region of the second subproblem is bounded (i.e. it is a polytope).

4) We generalize our results for rational data.

The aforementioned polyhedral results enable us to develop the tightest possible formulation for the second subproblem, thus leading to the tightest MIP formulation for the original problem.

Third, we show how the proposed results can be used to address large scale production planning problems. We study a problem with three stages, five processing units, 11 chemicals and 13 tasks over a planning period of eight weeks. We study the effect of the length of the planning period and show how our results enable us to select the planning period that yield a MIP formulation with zero integrality gap that is solved in few seconds. We also present results for larger instances, where planning horizons of 6-12 weeks are divided into planning periods of 3-6 hours, resulting in MIP formulations with thousands of binary variables and constraints. We show how our results can be used to address these problems to optimality in reasonable computational time.

References:

Appa, G.; Kotnyek, B. (2004). Rational and integral k-regular matrices. Discrete Mathematics, 275, 1-15.

Cook, W.J.; Cunningham, W.H.; Pulleyblank, W.R.; Schrijver, A. Combinatorial Optimization. Wiley-Interscience, 1997.

Hoffman, A. J.; Kruskal, J. B. Integral Boundary Points of Convex Polyhedra. In Linear Inequalities and Related Systems (Editors: Kuhn, H. W.; Tucker, A. W.), Princeton University Press, Princeton, 1956. (223-246).

Kotnyek, B. (2002). A generalization of totally unimodular and network matrices. PhD Thesis, London School of Economics and Political Science, London, UK.

Maravelias, C. T. (2005). Mixed Time Representation for State-Task Network Models. Ind. Eng. Chem. Res., 44 (24), 9129-9145.

Nemhauser, G.L.; Wolsey, L.A. Integer and Combinatorial Optimization, John Willey and Sons, Inc., New York, 1989.

Pochet, Y.; Wolsey, L.A. Production Planning by Mixed Integer Programming. Springer, 2006.

198f A Kriging Optimization Algorithm Incorporating Efficient Sampling for Problems Containing Black-Box Models

Eddie Davis and **Marianthi Ierapetritou**. Chemical & Biochemical Engineering, Rutgers University, 98 Brett Road, Piscataway, NJ 08854

When accurate closed-form descriptions of chemical processes are unavailable, gradient-based optimization methods can fail due to the lack of reliable derivative information for these black-box (BB) systems. To overcome this problem, direct search techniques can be used, but convergence to an optimum can be slow. Convergence can be accelerated using surrogate models, but it is possible for misleading search directions to be identified and/or for premature termination to occur, as can happen when the input-output data are noisy. Furthermore, the value of the information obtained must not be outweighed by model building costs for the nonlinear program (NLP). This last consideration is very important if integer variables are also present, as the solution of many relaxed NLP subproblems may be required. In addition, a global optimum must be obtained as the solution to the corresponding NLP in order to guarantee a lower or upper bound; early termination at a local optimum can lead to longer search and/or termination of the original problem at a suboptimal solution. Local methods ensure a global optimal solution only under convexity conditions, and deterministic surrogate models may be inaccurate when noisy data are present. To overcome these problems, an iteratively refined stochastic data-driven global model generated using kriging can instead be used to: 1) describe noisy system behavior [1,2], and 2) identify the set of local optima [5-7]. Each local optimum is then refined by sequentially applying response surface methodology and direct search in order to optimize the continuous and integer variables, respectively, appearing inside the black-box system [3-5,7]. When a problem formulation contains integer variables appearing outside a black-box system, as in the case of process synthesis problems, Branch-and-Bound is applied in order to efficiently find the optimal solution for this second class of integer variables [6]. Given a relaxed NLP at any node of a Branch-and-Bound tree, a kriging predictor is constructed in terms of the input variables with respect to the corresponding objective function. The NLP solution is identified after confirming the best model optima using sampling, and improved models can be generated using additional sampling information. However, accurate model generation, especially in the case of high-dimension problems, may necessitate a high sampling cost. Sampling is required for the initial model-building, and the iterative model refinement. In our previous work [4-7], random sampling has been employed for initial model construction; however inaccurate models may be generated from a poorly chosen sampling set. Heuristic sampling is employed for iterative model refinement whereby a new sampling set is comprised of test points proportionally distributed among regions of high model variance, prediction optima, and where model predictions significantly change between iterations. However, a method for determining the optimal number of additional sampling experiments to perform at each iteration, i.e. the optimal sampling set size, has not yet been developed. In our previous work [4-7], the sampling size has also remained an arbitrarily chosen user-determined parameter. If the number of additional sampling points is low, model improvement may only be incremental. Conversely, extensive sampling may be similarly undesirable, not only because the cost is high, but also because model improvement may be similarly small if the information value obtained from a poorly chosen sampling set fails to yield new global system information. The limitations of these two techniques thereby motivate the subject of this work, which is the development and implementation of a new sampling strategy employed for both the initial and iterative stages of model-building. The strategy is comprised of five steps: 1) subdivision of the feasible region into nonoverlapping convex hulls, 2) construction of local kriging models based on sampling at convex hull vertices, 3) sampling at optima residing inside each convex hull subregion, 4) subdivision of each of convex hull subregions into smaller convex hulls whose vertices now include the new sampling points, and 5) iteration of steps 2) – 4), until no better optima are found. In step 1, the initial vertices of the feasible region are defined by low/high operating ranges for the process variables. Sampling at the feasible region vertices ensures that the subsequent local models are built using data covering the entire feasible region. This sampling method can be contrasted with random sampling, in which a sampling set may fail to contain vectors located at low/high operating ranges. Extrapolation of the kriging predictor to the feasible region extremes may result in the generation of correspondingly inaccurate models. Nonoverlapping convex hulls are employed in order to minimize the number of localized kriging predictors needed. In order for optima to be identified in Step 3 using analytical kriging predictors, the solution of a set of complex nonlinear gradient equations is needed. In step 2, local models are generated by making predictions at a test point set within each subregion. Each prediction is generated as a sum of weighted sampling data, where a unique weight is generated for each sampled function value at a given convex hull vertex. The weights are easily obtained for an arbitrary test point based on the solution of a small linear program. In step 3, optima are identified either by determining the best predictions from a corresponding set obtained over a grid of test points, or applying gradient methods to an analytical form of the kriging model. In step 4, the purpose of convex hull splitting is to enable the generation of more accurate predictors using increasingly localized sampling information. In step 5, the repetition of steps 2 – 4 enables more accurate optima to be identified due to the higher number of more localized models being employed. The number of NLP problems solved at each iteration, to find model optima, is higher than the corresponding number at the previous iteration. However, the additional computational expense is balanced against the possible discovery of improved optima. The procedure terminates when further improvement in sampled optima cannot be accomplished, and at this point response surface methodology, direct search, and branch-and-bound are used to further refine the best kriging solutions. Numerical and industrial case study examples are both presented to provide proof of concept.

References:

[1] Goovaaerts, P., 1997, Geostatistics for Natural Resources Evolution.

[2] Isaaks, E., Srivistava, R., 1989, An Introduction to Applied Geostatistics.

[3] Myers, R. and D. Montgomery, 2002, Response Surface Methodology.

[4] Davis, E. and M. Ierapetritou (2007) Adaptive Optimization of noisy black-box functions inherent in microscopic models, Comp. Chem. Eng. 31, 466-476.

[5] Davis, E. and M. Ierapetritou (2007) A Kriging Method for the Solution of Nonlinear Programs with Black-Box Functions, AIChE J., 53, 2001-2012.

[6] Davis, E. and M. Ierapetritou (2007) A Kriging Based Method for the Solution of Mixed-Integer Nonlinear Programs Containing Black-Box Functions, JOGO, DOI 10.1007/s10898-007-9217-2.

[7] Davis, E. and M. Ierapetritou (2008) A Kriging-Based Approach to MINLP containing Black-Box Models and Noise, Ind. Eng. Chem. Res., Accepted for Publication.

215a Distributed Optimization and Control of Offshore Oil Production: The Intelligent Platform

Michael R. Wartmann, Carnegie Mellon University, 5000 Forbes Avenue, Pittsburgh, PA 15213 and B. Erik Ydstie, Department of Chemical Engineering, Carnegie Mellon University, 5000 Forbes Ave, Pittsburgh, PA 15213.

Information technologies and innovative hardware solutions provide novel opportunities for systematic reservoir management. Downhole sensors and remotely operated wells in combination with advanced software for data processing combined with oil reservoir models allow for monitoring and control of reservoir fluid dynamics. Application of advanced optimization and control techniques give improved production [1].

However, reservoir management includes complex decision making on different hierarchical levels. Time scales range from real-time production optimization on an hourly or daily basis to long-term decisions on drainage strategies where the life-time of an asset comes into play. The complexity of decision-making for a large production system exceeds by far the magnitude in which an all-encompassing global optimization problem can be formulated and solved. Although model-based optimization algorithms are utilized to some degree, the state of the art in operations of oil and gas fields still relies on heuristics. Valuable resources are lost because of poor coordination [2].

We describe a novel approach to distributed optimization and control of offshore oil production systems. The model incorporates a complex pipeline network. Oil and gas production systems are represented as a network of connected hierarchical structures of sub sea wells, manifolds and clusters. We consider multiphase flow of water, gas, and oil in the pipelines, and account for discrete switching and typical inflow characteristics of the sub sea wells. Network methods based on variational calculus provide a modeling framework for decentralized optimization and control. Conservation laws and the second law of thermodynamics combined with the passivity theory of nonlinear control lead to conditions for stability and optimality. We describe interconnections in networks through matrix representations that capture a network's topology. Control strategies are derived from the model, and stability and convergence to the optimal solution follows from the passivity conditions. The proposed distributed controller network can be seen as a special case of a Multi Agent System [3].

[1] Jansen, J.D., (2007), Model-based control of subsurface flow, DYCOPS 2007, Cancun, Mexico.

[2] Kosmidis, V. D., Perkins, J. D., Pistikopoulos, E. N. (2005), A mixed integer optimization formulation for the well scheduling problem on petroleum fields, Comput. Chem. Eng., 29 1523

[3] Rawlings, J.B., Stewart, B.T. (2007), Coordinating Multiple Optimization-based Controllers: New Opportunities and Challenges, DYCOPS 2007, Cancun, Mexico.

215b Cogeneration Power Plant Operation: A Novel Controller Assessment Method for Improving Disturbance Rejection Performance

Rachelle Howard and Douglas Cooper. Chemical Engineering Program, CMBE Department, University of Connecticut, 191 Auditorium Rd., U-3222, Storrs, CT 06269-3222

Cogeneration power plants offer diverse operating challenges because of the variety of units involved, including processes for electricity generation, steam production, and steam and chilled water distribution for both heating and cooling local buildings and heating and cooling unit operations. The plants must meet demands for electrical energy and heating and cooling, while also adhering to constraints imposed by operating costs, environmental emissions and plant safety.

The cogeneration power plant on the Storrs campus of the University of Connecticut (UConn Cogen) has a capacity of 25 MW of electricity and 200,000 lbs/hr of steam. The electric demand of the campus is met with water-cooled electric generators on three 7.5 MW gas turbines and one 5 MW steam turbine. Three heat recovery steam generators (HRSGs) located after the gas turbine generators each contain 600 psi steam systems and 125 psi steam systems. The 600 psi steam is used to run the steam turbine generator, while the 125 psi steam is used for campus heat in the winter and running chillers for campus cooling in the summer.

For additional steam production, gas fired duct burners can be lit in each HRSG and five auxiliary boilers are used as needed. NOx stack emissions are controlled with a vaporized liquid ammonia injection system. To provide cooling, the plant contains three steam driven chillers, two electric chillers, and two gas turbine chillers. Cooling towers on the plant roof provide cooled water for both the generators and the chillers.

The cogeneration facility is capable of producing more electrical energy than the campus uses, but state regulations will not allow excess electricity to be exported onto the power grid. So the challenge of keeping up with changing electrical, heating and cooling demands due to the New England climate and academic cycle is heightened because the plant must operate as efficiently as possible without producing excess electricity.

The UConn Cogen plant contains approximately 150 control loops to regulate process operation. These must be properly designed and tuned to maintain production within the changing operational constraints mentioned. To determine where and how to improve plant operation, controller performance assessment (CPA) tools are being applied. Plant data is continuously being recorded to an OSI Process Information data historian. Using the process variable (PV) and the controller output (CO) signals available during regular operation, CPA benchmarks have been developed to indicate which control loops require attention.

Many benchmarks compare current performance to desired performance, but do not indicate the nature of the controller misbehavior when it has been determined that performance is not as desired [1]. Published methods with diagnostic capabilities for classifying the full range of controller behavior, from sluggish to aggressive, require that the data be tracked for isolated disturbance events [2, 3].

In assessing the performance of the UConn Cogen power plant, a new CPA method was applied that is capable of diagnosing the full range of tuning-related controller behaviors without isolating disturbances. The new method is a novel automated application of the autocorrelation function, ACF, which uses regulatory data and does not require *a priori* information about the process. The relationship of the ACF to the closed loop transfer function has been established by Box and Jenkins [4] and Harris [5]. Several researchers [1, 6, 7] have noted that a plot of the ACF can be used as a first estimate of controller performance and ACF plots are included as a tool in most CPA software packages.

The work proposed for presentation uses the ACF as an automated process monitoring tool. The characteristic shapes of the ACF are extracted with a "pattern recognition" model and a new performance metric, the Relative Damping Index (*RDI*), is computed from the model parameters. The *RDI* provides an easy-to-interpret measure of the nature of the controller behavior from sluggish to aggressive. The new ACF analysis also provides a means for determining how to correct proportional and integral (PI) controller tuning when it is indicated that the controller performance is unacceptable.

The pattern recognition model is a general second order underdamped time domain form. The model provides a period of oscillation, τ_n, and a damping factor, ξ, for classifying the predicted disturbance rejection response of the controller. A method based on an initial estimate of τ_n and confidence intervals is presented as a means for selecting the length of the ACF to use in the least-squares model fit.

A user-defined range of acceptable controller performance based on ξ is combined with the computed ξ to calculate the *RDI*. $RDI = (\xi_{act} - \xi_{agg})*(\xi_{slug} - \xi_{act})^{-1}$, where ξ_{act} is the damping factor computed from the ACF, ξ_{agg} is the limit of acceptable aggressive behavior, and ξ_{slug} is the limit of acceptable sluggish behavior. The *RDI* is positive when ξ_{act} is within the user-defined range. When the controller response is too aggressive, the *RDI* is negative and less than one, and when the response is too sluggish, the *RDI* is negative and greater than one. In this work, methods are included which provide a means of making an informed selection of the acceptable performance range.

The final contribution of the new CPA method is a guide for correcting controller tuning when it has been determined that the performance is not acceptable. A PI tuning map method is provided as a means to choose new controller gain and integral time parameters. The map provides a visual guide with reference points established by ξ and τ_n.

Control loops throughout UConn Cogen were analyzed in this study. By identifying underperforming loops (both too sluggish and overly aggressive), the controller tuning could be corrected and the performance improved. The improvements are verified not only with the ACF approach, but also by isolating disturbances to show the change in disturbance rejection, showing reduced variance in the process variable and controller output, and by testimonial from operators working with the process daily.

[1] Jelali, M. (2006). "An overview of control performance assessment technology and industrial applications." Control Engineering Practice 14(5):441-466.

[2] Salsbury, T. I. (2005). "A practical method for assessing the performance of control loops subject to random load changes." Journal of Process Control 15(4): 393-405.

[3] Visioli, A. (2006). "Method for proportional-integral controller tuning assessment." Industrial and Engineering Chemistry Research 45(8): 2741-2747.

[4] Box, G. and G. Jenkins (1976). Time Series Analysis: Forecasting and Control. San Francisco, Holden-Day.

[5] Harris, T. J. (1989). "Assessment of control loop performance." Canadian Journal of Chemical Engineering 67(5): 856-861.

[6] Shah, S. L., R. Patwardhan, et al. (2001). "Multivariate controller performance analysis: Methods, applications and challenges." CPC-6 98: 187-219.

[7] Hugo, A. (2001). "Process controller performance assessment." Hydrocarbon Processing 80(4): 85-90

215c Optimal Scheduling with Energy Constraints

Pedro Castro, Department of Process Modeling and Simulation, INETI, Lisbon, Portugal, Iiro Harjunkoski, Corporate Research, ABB, Ladenburg, Germany, and Ignacio E. Grossmann, Center for Advanced Process Decision-making, Dept of Chemical Engineering, Carnegie Mellon University, 5000 Forbes Avenue, Pittsburgh, PA 15213.

Although significant progress has occurred in the last few years, there are still some challenges that need to be addressed by scheduling formulations. One of them is related to the optimal policy of energy consumption in process operations. On the one hand, the price of electricity changes from week days to weekends and even within an operating day, and it is desired to execute tasks during the low-cost hours. On the other hand, in the planning stage, there is sometimes the need to make a commitment to certain energy levels. These come at a certain cost, which is not lowered if the actual operation consumes less energy. If, on the contrary, the levels are exceeded, the plant incurs on a large penalty that should be minimized. Efficient scheduling tools that can cope with such constraints can have an important impact on the energy bill.

Modeling energy costs/profiles that change during the day is relatively straightforward with a general RTN/STN discrete-time formulation. There are however some well-known issues related to discrete-time models. Large mathematical problems result, due to high number of time points that normally needs to be used to represent the problem data accurately. While this may not be that much of an issue for batch plants, for continuous plants, even a very fine time grid may be inappropriate. For these type of plants, using a continuous-time formulation is generally a better approach. One of the problems with continuous-time is that the number of event points of the single or multiple time grids that are used, has a profound effect on computational effort, and going beyond 10 may render the mathematical problem intractable. Thus, postulating a large number is simply not an option.

In order to model variable energy costs/profiles with a continuous-time formulation, we need to place a subset of the time points at certain real times. Although the effect of using fixed instead of variable time points has received little or no attention in the literature, a lower impact in computational effort is expected so hopefully, the number of event points can be increased. However, we do need to place the other event points so fixing some of them is actually leading to another problem. How many event points should be placed between two consecutive fixed points?

In this paper, we address a real scheduling problem from a continuous multiproduct plant. It is basically single stage with a few processing units in parallel sending material to some storage units. Although storage is shared, only one product can be stored at a particular unit at a given time. In order to model variable energy costs and profiles of energy availability, an algorithm is developed that relies on a fairly accurate discrete-time formulation to predict the number and location of the variable event points, and on a continuous time formulation to find an exact solution. A comparison to the stand alone approaches is also provided.

215d Integrated Hedging and Operations for Complex Energy Utilities

Jeffrey Kantor, Chemical and Biomolecular Engineering, University of Notre Dame, Notre Dame, IN 46556, Anees Attarwala, Chemical and Biomolecular Engineering, Houlihan Lokey Howard & Zukin, 245 Park Avenue, New York, NY 10167-0001, and Patrick Mousaw, Department of Chemical & Biomolecular Engineering, University of Notre Dame, Notre Dame, IN 46556.

The economic operation of complex energy utilities involves a tradeoff among operational constraints, energy costs, and the financial risks due to price and demand uncertainty. Examples of complex energy utilities are campus and industrial scale facilities co-generating power, heat, chilling, and possibly other commodities. Such utilities often have some degree of flexibility in energy sources and operating conditions, and normally operators hedge at least a portion of their financial risk through trading in the energy markets. The market hedging operations, however, are generally executed separately from actual process operations, and without the benefit of more detailed process models. Here we wish to explore methods and the benefits that accrue from integrating operational and financial hedging for complex energy utilities.

In this work we propose a novel framework for integrating financial and process operations for complex energy utilities using thermodynamically consistent models. This work demonstrates four specific ideas:

1. We propose a modeling framework for energy conversion networks applying principles from finite time thermodynamics. By incorporating finite resistances to heat transport, and non-isentropic energy conversions, these models have been shown accurately bound the behavior of complex utilities. In this work, we generalize the existing literature by extending finite time thermodynamics to "energy conversion networks." We show energy conversion networks can be modeled with a system of bilinear matrix inequalities in which entropy fluxes associated with energy conversion devices constitute a system of decision variables. We have developed a library of model elements, and demonstrate application to a realistically scaled model of a campus energy utility.

2. We show the problem of optimal economic operation over a single period is a non-convex, bilinear optimization problem. Global optimal solution of these problems is found using convex relaxations with a branch-and-bound strategy exploiting the thermodynamic origin of these problems. Of independent interest, there are several interesting special cases leading to bilinear eigenvalue problems, and even conventional generalized algebraic eigenvalue problems.

3. In order to provide valuation of energy utilities subject to price uncertainty, we extend the model to multiple periods. Using standard stochastic models for the prices of commodity goods, we discretize on trinomial lattices to capture the behavior and correlations among prices of multiple fuels. For utilities without storage, we show valuation is also a bilinear optimization problem.

4. Finally, we consider the optimal hedging of energy utilities. We define an optimal hedge as the

purchase of a cash flow that minimizes the expected value of a convex utility function in the presence of price and demand uncertainty. From second-law considerations, we show that such a cash-flow is not a linear function of underlying commodity pricing.

Web Page: www.nd.edu/~jeff/Publications/Presentations/ECN_Presentation.pdf

215e Life Cycle Assessment Coupled with Multi-Objective Optimization for Environmentally Conscious Planning of Energy Production Systems

Gonzalo Guillén-Gosálbez, Chemical Engineering, University Rovira i Virgili, Av. Països Catalans, 26, Tarragona, 43002, Spain and **Metin Turkay**, Industrial Engineering, Koc University, RUMELIFENERI YOLU, SARIYER, Istanbul, 34450, Turkey.

During the last decade, there has been a growing awareness of the importance of incorporating environmental concerns along with traditional economic criteria in the optimization of industrial processes. This trend has been motivated by several issues, a major one being the tighter government's regulations. Additionally, the need to improve the perception of the firm by customers for being more environmentally conscious, which may eventually lead to higher product sales, has also contributed to this business trend.

Particularly, nowadays a high percent of the total human-originated environmental impact is energy related [1]. For this reason, there has been extensive research on the area of design and planning of efficient energy systems considering both financial and environmental objectives [2,3]. A typical energy production system consists of (1) storage tanks to store raw materials, (2) boilers that convert fuel into steam at high pressure, (3) turbines that expand higher pressure steam into lower pressure steam in order to generate electricity and (4) mixing equipments for mixing compatible materials originated from different sources in the system. The goal of the environmentally conscious modeling tools that address the optimization of these systems is to reduce their environmental impact without having to sacrifice the economic performance of the process.

Unfortunately, despite the effort made so far, the majority of the strategies devised so far focused on the manufacturing stage, and for this reason their scope is rather limited. Because of this, they can sometimes lead to solutions that decrease the impact locally at the expense of increasing certain negative effects in other stages of the life cycle of the process, in such a manner that the overall environmental damage is increased. This drawback can be overcome by expanding the boundaries of the study beyond the production stage in order to include a wider range of production and logistic activities. Thus, in the past decade it has become clear that the environmental issues must be considered throughout the entire production chain [4].

With the aim to expand the scope of the modeling strategies for energy systems, this work presents a novel approach that relies on the combined use of multi-objective optimization and life cycle assessment (LCA) [5]. In our framework, the design and planning of energy systems is formulated as a multi-objective mixed integer lineal problem (mo-MILP) that simultaneously accounts for the minimization of cost and environmental impact. The latter objective is explicitly captured within the model by making use of the Eco-indicator 99 [6], which incorporates the recent advances made in LCA. The solution of the proposed formulation, which can be obtained via standard techniques for multi-objective optimization, provides as output the set of Pareto points that represent the optimal trade-off between the criteria considered in the analysis.

The capabilities of the proposed modeling framework and solution strategy are illustrated through a case study based on a real industrial scenario. In this context, our model is used to investigate a number of alternatives to produce and deliver electricity. The environmental data for the case study was retrieved from the Eco-invent database [7], which contains information regarding the main emissions associated with a wide range of industrial processes used in Europe. In this case study, it is clearly shown how the solely minimization of the environmental impact measured within the boundaries of the system leads to solutions where the environmental problem is just transferred to other stages of the energy supply chain (i.e., raw materials extraction, generation of utilities, etc.). The obtained results demonstrate the benefits of using a holistic environmental assessment method that covers the entire production chain (i.e., LCA), starting from the extraction of raw materials and ending with the delivery of energy to the final customers. The proposed method is intended to guide the decision-makers towards the adoption of more sustainable production patterns for energy generation, thus leading to a reduction of the overall impact caused to the environment.

References:

[1] EIA, Annual Energy Review 2005, Energy Information Administration, Report No. DOE/EIA-0384, 2005.

[2] Türkay, M., Oruç, C., Fujita, K., Asakura, T. 2004. *Computers and Chemical Engineering*, 28, 985-992.

[3] Solyu, A., Oruç, C., Türkay, M., Fujita, K., Asakura, T. 2006. *European Journal of Operations Research*, 174(1), 387-403.

[4] Srivastava, S. K. 2007. *International journal of management reviews*, 9(1), 53-80.

[5] Guinée, J. B.; Gorrée, M.; Heijungs, R.; Huppes, G.; Kleijn, R.; de Koning, A.; van Oers, L.; Sleeswijk, A. W.; S. Suh, S.; de Haes, H. A. U.; de Bruijn, H.; van Duin, R.;Huijbregts, M. A. J. Handbook on Life Cycle Assessment. Operational Guide to the ISO Standards; Kluwer Academic Publishers: Dordrecht, 2002.

[6] PRé-Consultants, "The Eco-indicator 99, A damage oriented method for life cycle impact assessment. Methodology Report and Manual for Designers", Technical Report, PRé Consultants, Amersfoort, The Netherlands, 2000.

[7] PRé-Consultants, SimaPro 6 LCA software; The Netherlands (www.pre.nl/simapro/default.htm), 1998.

215f Simplification of the Inviscid Navier-Stokes Equations for Dynamic Simulation of Nonisothermal Gas Flow Networks

Patricio D. Ramirez-Munoz[1], Mujid S. Kazimi[2], and Paul I. Barton[1]. (1) Chemical Engineering, Massachusetts Institute of Technology, 77 Massachusetts Ave., RM 66-363, Cambridge, MA 02139, (2) Center for Advanced Nuclear Energy Systems, Department of Nuclear Science and Engineering, Massachusetts Institute of Technology, 77 Massachusetts Ave., RM 24-219, Cambridge, MA 02139

Accurate gas dynamics representation is required to simulate different processes in chemical engineering, some examples are the formaldehyde production process [5] and the nuclear hydrogen production process [6]. A very accurate representation of gas dynamics in a pipe can be achieved by using the inviscid Navier-Stokes equations with a friction term [2]. However, this system of equations produces computationally expensive simulations not suitable for the simulation of chemical process operations involving nonisothermal gas flow networks. The problem lies in the fact that information is propagated through the gas at widely different time scales. The time scales depend on the characteristics of the hyperbolic system. Two of these characteristics are related to the speed of sound and the speed of the gas, and one characteristic is related only to the speed of the gas. The characteristics related to the speed of sound are very fast (transients take less than 0.1 s) and an accurate dynamic simulation of the system requires the use of very small time steps with an explicit integrator and high-resolution methods [2].

One way to avoid this complicated and computationally expensive implementation is to simplify the equations taking into account only the relevant time scales for the simulated phenomena [3]. The fast time scale phenomena are often not of importance when studying the dynamics of chemical process operations, because discrete disruptions in the flow of the gas are not expected. As changes in pressure propagate with the speed of sound, we can make a quasi-equilibrium approximation (singular perturbation) for the equations related to the fast dynamics. These equations are the equations for mass conservation and momentum conservation. This simplified system with a dynamic energy balance can be solved by using a cheaper and faster implicit integrator like JACOBIAN® [4]. Moreover, the equations are much easier to integrate with a plant-wide dynamic simulator.

This paper presents the implementation of this framework to represent the transients in a heat transfer loop using helium. A heat transfer loop is used in a nuclear hydrogen production facility to transfer the heat generated in the nuclear reactor to the hydrogen production plant. The nuclear reactor and the hydrogen plant are at least 90m apart [1] and an accurate representation of the system is required to understand the behavior of the system. The important transients have a time scale larger than 1s and no sudden changes are expected to occur. Therefore, the simplified system of equations can be used to simulate the helium behavior in the system. Additional models for compressors and heat exchangers were considered, and the whole system of models was implemented in JACOBIAN®[4]. This framework allows the accurate representation of gas dynamics in complex chemical processes without dramatically increasing the computational cost.

1. C. B. Davis, R. B. Barner, S. R. Sherman, D. F. Wilson. Thermal-Hydraulic Analyses of Heat Transfer Fluid Requirements and Characteristics for Coupling a Hydrogen Product Plant to a High-Temperature Nuclear Reactor. Technical Report INL/EXT-05-00453, Idaho National Laboratory, 2005.

2. R. J. Leveque. Finite Volume Methods for Hyperbolic Problems. Cambridge University Press, Cambridge, U. K., 2002.

3. J. E. Meyer. Hydrodynamic Models for the Treatment of Reactor Thermal Transients. Nuclear Science and Engineering, 10:269-277, 1961.

4. Numerica Technology LLC, 2005 Numerica Technology LLC (2005). JACOBIAN dynamic modeling and optimization software. http://www.numericatech.com/.

5. D. Sedes. Modelling, Simulation and Process Safety Analysis. A case study: The formaldehyde process. Technical Report, Massachusetts Institute of Technology, 1994.

6. R. B. Vilim. Dynamic Modeling Efforts for System Interface Studies for Nuclear Hydrogen Production. Argonne National Laboratory, ANL-07/16, 2007.

239a Classification of Chemical Processes. A Graphical Approach Using Heat Engine Concept

David Glasser and Diane Hildebrandt. Centre of Materials and Process Synthesis (COMPS), University of the Witwatersrand, Private Bag 3, WITS 2050, Johannesburg, South Africa

In order to reduce carbon dioxide emissions for chemical processes one would like to make them as reversible as possible. Patel ET. Al have shown that in some cases one can analyse processes in terms of their work and heat requirements. In particular for processes there exist a temperature that is called 'the Carnot Temperature' at which one can satisfy the work requirement for the process using the heat that needs to be added or removed via a heat engine.

The analogy of heat engine and Carnot temperature is applied to chemical processes using a graphical approach. The approach looks at chemical processes holistically, where only the inlet and outlet streams are considered, and represents them in the space. Chemical processes are classified in different thermodynamics regions as defined in the space and their feasibility in terms of heat and work requirement is discussed. This approach allows determining whether heat at an appropriate temperature is sufficient to meet the work requirement of a chemical process or other means should be considered. The approach is used to investigate and discuss the possibility of combining chemical processes classified in different thermodynamic regions in the space, with the purpose to make infeasible processes possible or, minimize or even nullify the work requirement of the combined process.

239b Design Heuristics for Semicontinuous Separations with Reaction

Thomas A. Adams II and Warren D. Seider. Chemical and Biomolecular Engineering, University of Pennsylvania, 352 Towne Building, Philadelphia, PA 19104

Semicontinuous separation processes combine a single column with one or more middle vessels to perform multi-species separations which normally require multiple columns using traditional continuous methods. The semicontinuous process can be further intensified by incorporating reaction, either in the column or in one or more middle vessels. With this novel process configuration, significant capital cost and/or energy savings may be achieved over traditional batch or continuous designs, particularly at intermediate production rates. Examples of such processes include ternary semicontinuous distillation [1], semicontinuous distillation with auxiliary reaction [2], semicontinuous distillation with reaction in a middle vessel [3], semicontinuous reactive extraction coupled with semicontinuous reactive distillation [4], and others.

Based upon these processes, a set of generalized design heuristics are presented in this work. These include:

a. The classification of semicontinuous processes into three design schemes, and conditions preferable for each.

b. Heuristics for integrating common chemical reaction systems into semicontinuous processes (that is, in the column, in a middle vessel, etc.)

c. Heuristics for handling controllability issues, such as configurations for dynamic set-point tracking and the prevention of flooding and weeping in distillation columns.

d. Process configurations for the transitional modes, when the columns shift from one operation to another.

The heuristics presented in this work will help engineers to determine the configuration, operational recipe, and control structure of semicontinuous processes in the conceptual design stages. Properties such as the order of the relative volatilities of the chemical species, the structure of the chemical reactions, the presence of azeotropes, etc., are the basis for these heuristics.

References

1. Phimister, J. R. and W. D. Seider, "Semicontinuous, middle-vessel distillation of ternary mixtures." AIChE J., 46, 8 (2000): pp. 1508-1520.

2. Adams II, T. A., and W. D. Seider, "Ethyl Lactate Production Using Semicontinuous Distillation with Reaction in an Auxiliary Vessel and Pervaporation," 2007 AIChE Spring Nat'l. Meet., Houston, TX, Apr 23-27 2007.

3. Adams II, T. A., and W. D. Seider, "Semicontinuous Reactive Distillation for Specialty Chemical Production: Economic Comparison with Batch and Continuous Processing," 2005 AIChE Fall Nat'l. Meet., Cincinnati, OH, Oct 30-Nov 2 2005; p 6251-6257. Plenary session.

4. Adams II, T. A., and W. D. Seider, "Semicontinuous Process Intensification: Combining Reactive Extraction, Reactive Distillation, And Binary Distillation For Production Of 1,3-Propanediol", 2007 AIChE Annual Meet., Salt Lake City, UT, Nov 4-9, 2007.

Translation of variables for different applications in process synthesis

Marcel Ropotar[a,b,+], *Zdravko Kravanja*[a,*]
[a]*University of Maribor, Faculty of Chemistry and Chemical Engineering, P.O. Box 219, 2000 Maribor/Slovenija,*
[b]*Tanin Sevnica kemična industrija d.d., Hermanova cesta 1, Sevnica, SI - 8290 Slovenija*
[+]marcel.ropotar@uni-mb.si
[*]zdravko.kravanja@uni-mb.si

Abstract

The aim of this contribution is to present a special translation of variables, which can be applied to logic-based and mixed-integer programming problems in process synthesis. Our research focused on the development of an alternative convex-hull representation for a logic-based outer-approximation algorithm (OA), implemented in an automated process synthesizer MIPSYN. Three examples were solved in order to compare models with and without translation of variables.

1. Introduction

Recent developments in logic-based optimization (e.g. Grossmann and Biegler, 2004) are regarded as some of the most important achievements for effectively modeling and solving discrete-continuous synthesis problems. One of the possible representations of discrete-continuous problems is the Generalized Disjunctive Programming (GDP), which was developed by Raman and Grossmann (1994) as an extension of the disjunctive programming paradigm developed by Balas (1974). GDP problems could be solved either by transforming it into mixed-integer programs or by the development of specific solution methods, e.g. the branch and bound algorithm with convex relaxation by Lee and Grossmann, 2000. Sawaya (2006) showed that the tightness of convex hull representation of disjunctions could be significantly improved by moving global constraints into representation of disjunctions.

In this contribution we present another idea – a variable translation in order to have a narrower space of variables, which may then further increases efficiency when solving discrete-continuous problems.

2. Variable translation

In synthesis problems, continuous variables v^{s} are usually defined within zero lower and non-zero upper bounds and constrained by

$$v^{\mathrm{LO}} y \leq v^{s} \leq v^{\mathrm{LO}} y \tag{1}$$

in order to force non-zero bounds when alternatives are selected ($y = 1$). The main idea is to substitute a zero-lower-bounded variable v^{s} of alternatives ($0 \leq v^{s} \leq v^{s,\mathrm{UP}}$) by a non-zero-lower-bounded variable ($v^{\mathrm{LO}} \leq v \leq v^{\mathrm{UP}}$), through the use of the following translation equation:

$$v^{s} = v - v^{f}(1 - y) \tag{2}$$

where v^f is an arbitrarily-forced scalar from the interval (v^{LO},v^{UP}) and y is a corresponding binary variable. When an alternative is selected, an integer term $v^f(1 - y)$ becomes zero and v^s becomes equal to v, and when it is rejected, a value v^f is subtracted from the variable v. When eq. (2) is applied to eq. (1) we obtain

$$v^f + (v^{LO} - v^f)y \le v \le v^f + (v^{UP} - v^f)y \quad (3)$$

Note that when $y = 1$, it follows that v is constrained within its non-zero bounds, and when $y = 0$, v becomes v^f. In this way the original space of variable is preserved irrespective of discrete decisions. Intuitively, one could expect that retaining within the narrower original space of variables would increase the efficiency of the GDP. Two additional types were obtained besides a mixed integer type of variable translation (eq. (2)). The following relaxed translation formula is obtained if a binary variable y is relaxed to a continuous variable λ defined between 0 and 1.

$$v^s = v - v^f(1 - \lambda) \quad (4)$$

A logic-based form of the variable translation is:

$$[Y: v^s = v] \vee [\neg Y: v^s = v - v^f] \quad (5)$$

for a Boolean variable $Y = true$ it follows that $v^s = v$ and for $Y = false$ $v^s = v - v^f$.

2.1. Alternative logic-based OA algorithm and MILP transformation

Selection of a different variable translation type depends on a type of model (MIP – eq. (2), relaxed MIP – eq. (4), logic-based – eq. (5)).

A convex hull representation is the tightest relaxation of disjunctions in Generalized Disjunctive Programming − GDP problem. It is generated from taking the linear combination of all points in feasible regions of disjunctions. By applying the convex hull relaxation to the GDP problem the (CHRP) problem given bellow is obtained. Since the problem is relaxed, the relaxed form of the variable translation (eq. (4)) is applied in order to translate its zero-lower-bounded variables into nonzero-lower-bounded variables. The following alternative (A-CHRP) problem is obtained:

(CHRP):

$$\min Z = \sum_k \sum_i \left(\gamma_{ik}\lambda_{ik} + f_{ik}^{\mathrm{a}}(v_{ik})\right) + f^{\mathrm{g}}(\boldsymbol{x})$$

$$\text{s.t.} \quad h(\boldsymbol{x}) \le 0$$

$$A^{\mathrm{g}}(\boldsymbol{x}) \le b^{\mathrm{g}}$$

$$\boldsymbol{x} = \sum_{i \in D_k} v_{ik}, \quad k \in SD$$

$$\sum_{i \in D_k} \lambda_{ik} = 1, \quad 0 \le \lambda_{ik} \le 1$$

$$E\boldsymbol{\lambda} \le e$$

$$\lambda_{ik} VLO_{ik} \le v_{ik}, \quad i \in D_k, k \in SD$$

$$v_{ik} \le \lambda_{ik} v_{ik}^{\mathrm{UP}}, \quad i \in D_k, k \in SD$$

$$\lambda_{ik} g_{ik}\left(\frac{v_{ik}}{\lambda_{ik}}\right) \le 0, \quad i \in D_k, k \in SD$$

$$\boldsymbol{x}^{\mathrm{LO}} \le \boldsymbol{x} \le \boldsymbol{x}^{\mathrm{UP}}$$

$$0 \le v_{ik} \le v_{ik}^{\mathrm{UP}}, \quad i \in D_k, k \in SD$$

$$\xrightarrow{v^{\mathrm{s}} = v - v^{\mathrm{f}}(1-\lambda)}$$

(A-CHRP):

$$\min Z = \sum_k \sum_i \left(\gamma_{ik}\lambda_{ik} + f_{ik}^{\mathrm{a}}\left(v_{ik} - v_{ik}^{\mathrm{f}}(1-\lambda_{ik})\right)\right) + f^{\mathrm{g}}(\boldsymbol{x})$$

$$\text{s.t.} \quad h(\boldsymbol{x}) \le 0$$

$$A^{\mathrm{g}}(\boldsymbol{x}) \le b^{\mathrm{g}}$$

$$\boldsymbol{x} = \sum_{i \in D_k} v_{ik} - v_{ik}^{\mathrm{f}}(1-\lambda_{ik}), k \in SD$$

$$\sum_{i \in D_k} \lambda_{ik} = 1, \quad 0 \le \lambda_{ik} \le 1$$

$$E\boldsymbol{\lambda} \le e$$

$$v_{ik}^{\mathrm{LO}}\lambda_{ik} + v_{ik}^{\mathrm{f}}(1-\lambda_{ik}) \le v_{ik}, \quad i \in D_k, k \in SD$$

$$v_{ik} \le v_{ik}^{\mathrm{UP}}\lambda_{ik} + v_{ik}^{\mathrm{f}}(1-\lambda_{ik}), \quad i \in D_k, k \in SD$$

$$\lambda_{ik} g_{ik}\left(\frac{v_{ik} - v_{ik}^{\mathrm{f}}(1-\lambda_{ik})}{\lambda_{ik}}\right) \le 0, i \in D_k, k \in SD$$

$$\boldsymbol{x}^{\mathrm{LO}} \le \boldsymbol{x} \le \boldsymbol{x}^{\mathrm{UP}}$$

$$v_{ik}^{\mathrm{LO}} \le v_{ik} \le v_{ik}^{\mathrm{UP}}, \quad i \in D_k, k \in SD$$

where v_{ik} are disaggregated variables and VLO_{ik} nonzero scalars in the problem (CHRP) forcing the nonzero lower bounds when an alternative is selected. Note that in the alternative formulation (A-CHRP), nonzero lower bounds can now be applied directly to the disaggregated variables v_{ik}. In both convex hull representations, global variables can in principle be defined between nonzero lower and upper bounds. However, in the conventional (CHRP) problem, disaggregated variables should always have zero lower bounds so that they can obtain zero values in the bounding constraints when λ_{ik} becomes zero. In the case of the alternative problem (A-CHRP), when λ_{ik} takes zero value, the corresponding variables v_{ik} in the bounding constraints are set to v_{ik}^{f}. At the same time the terms v_{ik} *and* $v_{ik}^{\mathrm{f}}(1-\lambda_{ik})$ in the balance equation and the objective function precisely cancel each other out, which is equivalent to obtaining zero values for v_{ik} in the original problem (CHRP). Note that if the continuous variables λ_{ik} are replaced by integer variables y_{ik}, MINLP reformulation is obtained.

Lee and Grossmann (2000) showed that applying the outer-approximation method to the MINLP reformulation of the convex hull relaxation regarding problem (GDP) reduces to the logic-based OA method by Turkay and Grossmann (1996). Logic-based OA problems are usually solved through MILP transformation where Boolean variables *Y*s are replaced by binary variables *y*s, logical relations are formulated as integer constrains and disjunctives are represented either by big-M or convex hull representation. When a convex hull representation is considered, the following MILP master problem (CCH-MILP) is obtained and when it is reformulated by the translation of variables, using eq. (2), the following alternative MILP master problem (ACH-MILP) is obtained:

(CCH-MILP):

$$\min\ Z = \sum_i \sum_k \left(c_{ik} y_{ik} + \alpha_{ik}^{a}\right) + \alpha^{g}$$

s.t.

$$\left.\begin{array}{l} \alpha^{g} \geq f(x^{l}) + \nabla_x f(x^{l})^{T}(x - x^{l}) \\ h(x^{l}) + \nabla_x h(x^{l})^{T}(x - x^{l}) \leq 0 \end{array}\right\}, \quad l = 1,\ldots, L$$

$$A^{g}(x) \leq b^{g}$$

$$E^{g}(y) \leq e^{g}$$

$$\boldsymbol{xLOy}_{ik} \leq \boldsymbol{x}$$

$$\boldsymbol{x} \leq \boldsymbol{x}^{UP} y_{ik}$$

$$A^{ik}(\boldsymbol{x}) \leq b_{ik} y_{ik}$$

$$\nabla_x f_{ik}^{a}(\boldsymbol{x}^{l})^{T} \boldsymbol{x} - \alpha_{ik}^{a} \leq \left[\nabla_x f_{ik}^{a}(\boldsymbol{x}^{l})^{T} \boldsymbol{x}^{l} - f_{ik}^{a}(\boldsymbol{x}^{l})\right] y_{ik}$$

$$\nabla_x h_{ik}(\boldsymbol{x}^{l})^{T} \boldsymbol{x} \leq \left[\nabla_x h_{ik}(\boldsymbol{x}^{l})^{T} \boldsymbol{x}^{l} - h_{ik}(\boldsymbol{x}^{l})\right] y_{ik}$$

$$x \in R^{n},\ y_{ik} \in \{0,1\}^{m}$$

$$0 \leq \alpha^{g}, \alpha_{ik}^{a} \qquad i \in D_k, k \in SD$$

$$0 \leq \boldsymbol{x} \leq \boldsymbol{x}^{UP}$$

$$\xrightarrow{x^{s} = x - x^{f}(1 - y)}$$

(ACH-MILP):

$$\min\ Z = \sum_i \sum_k \left(c_{ik} y_{ik} + \alpha_{ik}^{a}\right) + \alpha^{g}$$

s.t.

$$\left.\begin{array}{l} \alpha^{g} \geq f(x^{l}) + \nabla_x f(x^{l})^{T}(x - x^{l}) \\ h(x^{l}) + \nabla_x h(x^{l})^{T}(x - x^{l}) \leq 0 \end{array}\right\}, \quad l = 1,\ldots, L$$

$$A^{g}(x) \leq b^{g}$$

$$E^{g}(y) \leq e^{g}$$

$$\boldsymbol{x}^{f} + (\boldsymbol{x}^{LO} - \boldsymbol{x}^{f}) y_{ik} \leq \boldsymbol{x} \qquad (6)$$

$$\boldsymbol{x} \leq \boldsymbol{x}^{f} + (\boldsymbol{x}^{UP} - \boldsymbol{x}^{f}) y_{ik} \qquad (7)$$

$$A^{ik}\left(\boldsymbol{x} - \boldsymbol{x}^{f}(1 - y_{ik})\right) \leq b_{ik} y_{ik}$$

$$\nabla_x f_{ik}^{a}(\boldsymbol{x}^{l})^{T} \boldsymbol{x} - \alpha_{ik}^{a} \leq \nabla_x f_{ik}^{a}(\boldsymbol{x}^{l})^{T} \boldsymbol{x}^{f} + \left[\nabla_x f_{ik}^{a}(\boldsymbol{x}^{l})^{T}(\boldsymbol{x}^{l} - \boldsymbol{x}^{f}) - f_{ik}^{a}(\boldsymbol{x}^{l})\right] y_{ik}$$

$$\nabla_x h_{ik}(\boldsymbol{x}^{l})^{T} \boldsymbol{x} \leq \nabla_x h_{ik}(\boldsymbol{x}^{l})^{T} \boldsymbol{x}^{f} + \left[\nabla_x h_{ik}(\boldsymbol{x}^{l})^{T}(\boldsymbol{x}^{l} - \boldsymbol{x}^{f}) - h_{ik}(\boldsymbol{x}^{l})\right] y_{ik} \qquad (8)$$

$$x \in R^{n},\ y_{ik} \in \{0,1\}^{m}$$

$$0 \leq \alpha^{g}, \alpha_{ik}^{a} \qquad i \in D_k, k \in SD$$

$$\boldsymbol{x}^{LO} \leq \boldsymbol{x} \leq \boldsymbol{x}^{UP}$$

where L, SD and D_k are sets of NLP solutions, disjunctives and terms in disjunctives, respectively. The key feature of the alternative outer approximations (ineq. (8)) is that they preserve feasibility when alternatives are not selected when $\boldsymbol{x}$ is set to $\boldsymbol{x}^{f}$. This enables the use of variables with non-zero lower bounds. Note that if $\boldsymbol{x}^{f}$ is set to $\boldsymbol{x}^{LO}$, ineq. (6) becomes redundant, if it is set to $\boldsymbol{x}^{UP}$, ineq. (7) becomes redundant. If the lower bounds are zero, the problem (ACH-MILP) reduces to the problem (CCH-MILP). The MILP model given above is implemented in the MINLP process synthesizer MIPSYN.

2.2. Implementation in a process synthesizer MIPSYN

Until recently only big-M models were used in MIPSYN, the successor of PROSYN-MINLP (Kravanja and Grossmann, 1994), to solve MINLP synthesis problems. Now, the conventional convex hull and the alternative convex hull formulations, using translation of variables, are implemented in MIPSYN, too. The models are formulated in the most generalized form using various capabilities of the high-level language of GAMS. Data-and-topology independent models were developed in this way.

3. Examples

Three synthesis problems of different sizes and complexities have been solved in order to test and compare the efficiencies of models with and without translation of variables.

Example 1: The first example is a network synthesis problem with a simple model but very large-scale combinatorics with 400 binary variables. This numerical problem is an extension of the small flowsheet problem by Kocis and Grossmann (1989). Additional pairs of reactors were added to the superstructure. The objective is to minimize total cost at the fixed demand of the final outflow; $\boldsymbol{x}^{\mathrm{f}}$ was set to $\boldsymbol{x}^{\mathrm{LO}}$.

$$\min z = c_{\mathrm{f}\,i}^{1} y_i^1 + c_{\mathrm{v}}^{1} V_i^1 \; + \; c_{\mathrm{f}\,i}^{2} y_i^2 + c_{\mathrm{v}\,i}^{2} V_i^2 \; + \; c_{\mathrm{o}\,i}^{1} x_i^1 + c_{\mathrm{o}\,i}^{2} x_i^2 + 5x_1 \quad i \in 200 \tag{9}$$

The solution statistics until the third major MINLP iteration are reported in Table 1. As can be seen in Table 1, it was impossible with big-M formulation to solve the problem within a reasonable time, whilst both convex hull representations enable the solving of this high-combinatorial problem very quickly. Note that with the same integrality gap and smaller number of constraints, the alternative formulation (ACH) could solve the problem in only a quarter of the CPU time needed to solve the problem using the conventional convex hull formulation (CCH).

Table 1. MILP solution statistics of the reactor network synthesis problem.

	Best NLP	**Int. gap, %**	**No. of eq./ No.of var.**	**No. of iterations**	**No. of nodes**	**CPU for 3 it., sec.**	**Nodes/s for 3 it.**
Big-M	n/a	n/a	3802/1801	n/a	n/a	n/a	n/a
CCH	183.87	0.868	3402/1801	23214	319	19.2	16.6
ACH	183.87	0.868	2202/1801	4696	293	5.5	53.6

CPLEX/GAMS version 21.7, processor PENTIUM 4 2.81 GHz, 512 MB of RAM.

Example 2: The second example is the synthesis of a heat exchanger network (HEN) comprising different types of exchangers. Each match in a stage-wise superstructure is comprised of a double pipe, a plate and frame, a shell and tube exchanger, and a by-pass (Figure 1). Consideration of different types of exchanger enables the simultaneous selection of exchanger types; however, it significantly increases the number of binary variables. The model thus exhibits moderate complexity and high combinatorics (249 binary variables).

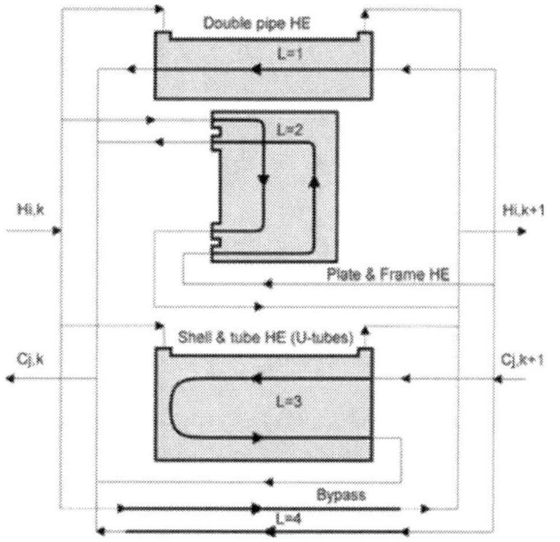

Figure 1. Match superstructure.

Table 2 shows the statistics when x^{f} was set to x^{LO}. With respect to the integrality gap, number of iterations, CPU time and number of nodes, both convex hull representations significantly outperform the big-M one, whilst the efficiency of the alternative convex hull formulation is approximately twice that of the conventional formulation.

Table 2. MILP solution statistics for the HEN synthesis problem.

	Best NLP	**Int. gap, %**	**No. of eq./ No.of var.**	**No. of iterations**	**No. of nodes**	**CPU for 15 it., sec.**	**Nodes/s for 15 it.**
Big-M	884.07	1.548	9214/5595	2248581	70257	500.9	140.3
CCH	818.69	0.607	6894/5595	612529	35150	163.3	215.2
ACH	818.69	0.607	4574/5595	321062	19411	83.8	231.6

CPLEX/GAMS version 21.7, processor PENTIUM 4 2.81 GHz, 512 MB of RAM.

Example 3: The last, allyl chloride example, is the synthesis of a reactor/separator network within an overall heat integrated process scheme, with a complex model and moderate-size combinatorics (184 binary variables). The reactor/separator superstructure (Figure 2) comprises a sequence of PFR/CSTRs with side streams and intermediate separators at different locations. Each PFR consists of a train of several alternative elements. The corresponding DAE system is modeled by the orthogonal collocation on finite elements. The overall model is highly nonlinear and nonconvex. Therefore, many numerical and other issues are present which makes any comparison between formulations harder, e.g. due to the effects of nonconvexities it is impossible to compare different formulations based on an integrality gap.

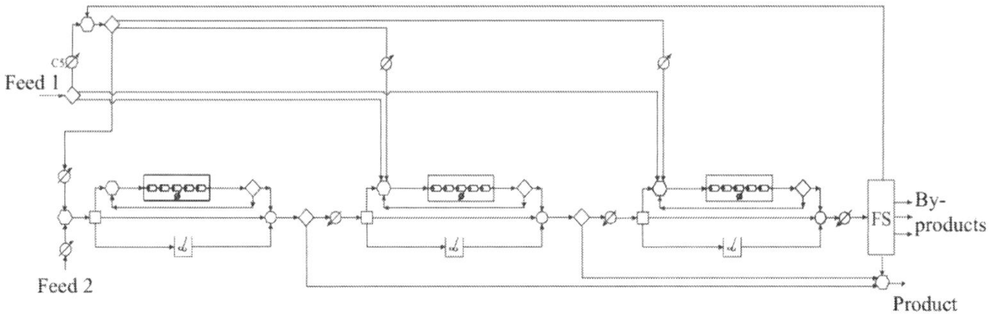

Figure 2. Reactor/separator superstructure of the allyl chloride problem.

Table 3 shows solution statistics until the 17th major MINLP iteration. As can be seen from Table 3, the efficiency of the ACH formulation is almost twice as good as that of CCH and four times better than that of Big-M. It should be noted that selection of the optimal final element in PFR is formulated by big-M constraints, so that the overall process ACH and CCH formulations are, in fact, combined ACH/Big-M and CCH/Big-M formulations.

Table 3. MILP solution statistics of the allyl chloride problem

	Best NLP	**Int. gap, %**	**No. of eq./ No.of var.**	**No. of iterations**	**No. of nodes**	**CPU for 17 it., sec.**	**Nodes/s for 17 it.**
Big-M	81.924	0.190	2307/2193	4525051	19830	2166.5	9.2
CCH	81.836	100.00	1743/1011	1503644	8053	960.3	8.4
ACH	81.769	0.343	1230/1313	854049	7700	529.2	14.6

CPLEX/GAMS version 21.7, processor PENTIUM 4 2.81 GHz, 512 MB of RAM.

4. Conclusions

By the use of translation of variables on the conventional convex hull representation the alternative convex hull representation was obtained. Initial experiences indicate that the alternative convex hull representation is generally more efficient when solving high-combinatorial problems than the conventional one and has the smallest model sizes. In spite of the above mentioned efficiency, ACH formulations exhibit stronger sensitivity to the effects of nonconvexities, and the model representations are more complicated. Hence, when solving process synthesis examples the use of translation of variables on models is especially worthwhile when models are generated automatically as in the case with the equation-oriented modular synthesizer MIPSYN, since the likelihood of achieving the best efficiency of the MINLP search is increased.

References

Balas, E. (1974). *Disjunctive programming: Properties of the convex hull of feasible points*, technical report MSRR 348, Carnegie Mellon University, 1974.

Grossmann, I. E. and L. T. Biegler (2004). *Part II. Future perspective on optimization*. Comput Chem Eng, 28(8): 1193-1218.

Kocis, G. R. and I. E. Grossmann (1989). *A Modeling and Decomposition Strategy for the Minlp Optimization of Process Flowsheets*. Comput. Chem. Eng., 13(7): 797-819.

Kravanja, Z. and I. E. Grossmann (1994). *New Developments and Capabilities in Prosyn - an Automated Topology and Parameter Process Synthesizer*. Comput Chem Eng, 18(11-12): 1097-1114.

Lee, S. and I. E. Grossmann (2000). *New algorithms for nonlinear generalized disjunctive programming*. Computers and Chemical Engineering, 24: 2125-2141.

Raman, R. and I. E. Grossmann (1994). *Modeling and Computational Techniques for Logic-Based Integer programming*. Comput Chem Eng, 18(7): 563-578.

Sawaya, N. (2006). *Reformulations, relaxations and cutting planes for generalized disjunctive programming*, Thesis (Ph.D), Carnegie Mellon University, Carnegie Mellon University, 2006.

Turkay, M. and I. E. Grossmann (1996). *Logic-based MINLP algorithms for the optimal synthesis of process networks*. Comput Chem Eng, 20(8): 959-978.

239d A Bilevel Framework for Process Design & Operation

Alexander Mitsos, Aices, RWTH Aachen, Pauwelstrasse 12, Aachen, Germany, **Panayiotis Lemonidis**, Department of Chemical Engineering, Massachusetts Institute of Technology, 66-363, 77 Massachusetts Avenue, Cambridge, MA 02139, George M. Bollas, Department of Chemical Engineering - Process Systems Engineering Laboratory, Massachusetts Institute of Technology, 77 Massachusetts Ave., RM 66-363, Cambridge, MA 02139, Benoît Chachuat, Automatic Control Laboratory, Ecole Polytechnique Federale de Lausanne (EPFL), Station 9, Lausanne, CH-1015, Switzerland, and Paul I. Barton, Chemical Engineering, Massachusetts Institute of Technology, 77 Massachusetts Ave., RM 66-464, Cambridge, MA 02139.

A modeling framework based on bilevel programs involving nonconvex functions and mixed-integer variables is presented. Bilevel programs are programs where one optimization problem (upper-level program) is constrained by the (global) solutions of another optimization problem (lower-level program). Bilevel programs represent hierarchical decision making where each of two decision makers (optimizers) controls a subset of the decision variables. After discussing existing applications from the literature, new formulations from process design and operations are proposed, including parameter estimation in thermodynamics, scenario-based design and decision making in monopoly markets. In many literature applications and all the proposed problems, the lower-level program is nonconvex and therefore can only be solved rigorously based on recent algorithmic advances [1].

Bilevel programs have been used extensively in operations research for several decades. In this discipline typically the optimizers represent actual decision makers, such as policy makers, military leaders or industrial managers. For instance, Bracken and McGill [2] consider the allocation of military resources; the optimizers model the decision makers of two sides in a war. Fortuny-Amat and McCarl [3] discuss the case of irrigation during limited water supply; here the upper-level program corresponds to the government entity which allocates water to the various farmers; the lower-level programs represent the farmers which given an allocation try to optimize their production. The first application proposed in this talk is decision making by pharmaceutical companies which have monopoly or oligopoly. It is shown that this problem can be cast as a mixed-integer bilevel program with nonconvex objective functions.

Often in operations research linear models are sufficient. However, this is not the case in process design. Clark and Westerberg [4] discuss multi-level and multi-objective optimization problems and show that design under thermodynamic equilibrium can be cast as a bilevel optimization problem. The upper-level optimizer is the designer, while the lower-level optimizer is nature which (at constant temperature and pressure) minimizes Gibbs free energy. Thus, the lower-level program is nonconvex and multi-modal, except for special cases such as ideal solutions [5]. Clark and Westerberg [4] recognize the difficulties of nonconvexity and introduce a variation of the original formulation where a solution is defined as a local solution; this change of what a solution means may be acceptable in some cases, but in many it is not. Another application of bilevel programs in process design is feasibility and flexibility analysis [6,7];

here the upper-level program is the designer, while the lower-level program is the maximization of constraint violation; here a global solution of the lower-level program is necessary, for otherwise an infeasible point may be obtained. Maranas and co-workers [8,9] consider optimal gene knockout; the upper-level program is the designer which tries to obtain maximal yield of a chemical from a microorganism, while the lower-level optimizer is the microorganism which will maximize its growth; for metabolic networks linear formulations are used, so nonconvexity is not an issue.

An extension of feasibility-flexibility analysis is scenario-integrated dynamic optimization. In this talk formulations by Abel and Marquardt [10] are analyzed and extended, and solution methods are proposed. The goal of the formulations in this section is to account for uncertainty in the operation of dynamic systems. During process operation events outside the control of the operator can occur, such as drastic price changes, change of weather or failure of a process component. These events change the dynamics of the system. If these events are not taken into account during the plant design phase the operation may become infeasible and result in catastrophic events. The upper-level objective is to optimize the nominal operation, i.e., the operation without the occurrence of the external events. The lower-level problem is that the scenario operation, i.e., the operation once the event has occurred, is feasible. The global solution of the lower-level programs is mandated similar to feasibility and flexibility problems.

A third application of bilevel programs is based on parameter estimation in thermodynamic equilibrium. In process design and operation typically the phase equilibria are calculated based on activity coefficient models such as the NRTL model. These models contain adjustable parameters which cannot be measured directly, but rather must be estimated based on measured phase-splits. In this talk it is shown that this parameter estimation naturally leads to a bilevel optimization problem. The upper-level program aims in minimizing the error in predictions by adjusting the model parameters, e.g., through a least squares error objective function. The lower-level programs represent the thermodynamic stability requirement, i.e., the predictions must be the global minimum of the Gibbs free energy for the chosen parameter values; equivalent stability criteria [11,12] can be used instead. The lower-level programs are nonconvex and their global solution is necessary.

[1] A. Mitsos, P. Lemonidis, and P. I. Barton. Global solution of bilevel programs with a nonconvex inner program. In press: Journal of Global Optimization, October 12, 2007.

[2] J. Bracken and J. T. McGill. Mathematical programs with optimization problems in constraints. Operations Research, 21(1):37-44, 1973.

[3] J. Fortuny-Amat and B. McCarl. A representation and economic interpretation of a two-level programming problem. Journal of the Operational Research Society, 32(9):783-792, 1981.

[4] P. A. Clark and A. W. Westerberg. Optimization for design problems having more than one objective. Computers & Chemical Engineering, 7(4):259-278, 1983.

[5] W. Robert Smith and R. W. Missen. Chemical Reaction Equilibrium Analysis: Theory and Algorithms. John Wiley & Sons, New York, 1982.

[6] I. E. Grossmann and K. P. Halemane. Decomposition strategy for designing flexible chemical-plants. AIChE Journal, 28(4):686-694, 1982.

[7] C. A. Floudas, Z. H. Gümüs, and M. G. Ierapetritou. Global optimization in design under uncertainty: Feasibility test and flexibility index problems. Industrial & Engineering Chemistry Research, 40(20):4267-4282, 2001.

[8] A. P. Burgard and C. D. Maranas. Optimization-based framework for inferring and testing hypothesized metabolic objective functions. Biotechnology and Bioengineering, 82(6):670-677, 2003.

[9] A. P. Burgard, P. Pharkya, and C. D. Maranas. OptKnock: A bilevel programming framework for identifying gene knockout strategies for microbial strain optimization. Biotechnology and Bioengineering, 84(6):647-657, 03.

[10] O. Abel and W. Marquardt. Scenario-integrated modeling and optimization of dynamic systems. AIChE Journal, 46(4):803-823, 2000.

[11] L. E. Baker, A. C. Pierce, and K. D. Luks. Gibbs energy analysis of phase equilibria. Soc. Petrol. Engrs. J., 22:731-742, 1982.

[12] A. Mitsos and P. I. Barton. A dual extremum principle in thermodynamics. AIChE Journal, 53 (8):2131-2147, 2007.

239e A Novel Approach for Global Optimization of Stage-Wise Heat Exchanger Network Models

Miguel J. Bagajewicz, Chemical, Biological and Materials Engineering, University of Oklahoma, 100 E. Boyd St., T-335, Norman, OK 73019, **Milos Bogataj**, Faculty of Chemistry and Chemical Engineering, University of Maribor, Smetanova 17, Maribor, Slovenia, and Debora C. Faria, School of Chemical Engineering and Material Science, University of Oklahoma, 100 East Boyd Street, T-133, Norman, OK 73019-0628.

In the last two decades different approaches using mathematical programming were presented for synthesis of heat exchanger networks. Of these, one can classify them as transportation-transshipment oriented and superstructure oriented. Among the superstructure-based models, the most popular method is a stage-wise superstructure approach (Yee and Grossmann, 1990). A comprehensive overview of the advances in this field of research is given in Furman and Sahinidis (2002).

One of the major advantages of the approaches based on mathematical programming is the ability of simultaneous optimization of energy and investment costs. However, this often requires solving a non-convex mixed-integer non-linear problem (MINLP), thus, the solutions obtained may not be globally optimal.

This paper presents a new approach to obtain globally optimal solutions to the Synheat model using recently developed discretization methodology (Bagajewicz and Faria, 2007).

The methodology relies on linearizations of nonlinear equations by discretizing the variables causing nonlinearities over the intervals bounded by their lower and upper bounds. With proper selection of the discrete values, the original non-convex MINLP is converted to a lower-bounding MILP. A proven globally optimal solution is then obtained through the use of an algorithm, which serves to reduce the gap between the lower bounding MILP and original non-convex MINLP.

REFERENCES

Bagajewicz M. J., and Faria D. C., A Discretized Approach to Design Multicomponent Water/ Wastewater Network Systems, AIChE Annual Meeting, Salt Lake City, 2007.

Furman, K. C., Sahinidis, N. V. (2002). A critical review and annotated bibliography for heat exchanger network synthesis in the 20th century. Industrial and Engineering Chemistry Research, 41, 2335–2370.

Yee, T. F., Grossmann, I. E. (1990). Simultaneous optimization models for heat integration II. Computers and Chemical Engineering, 14, 1165–1184.

239f Minimum Capital and Minimum Total Annualized Cost for a Power Plant Using the Ideas Framework

Jorge Pena Lopez, Chemical Engineering Department, University of California at Los Angeles, 5531 Boelter Hall, Los Angeles, CA 90095 and Vasilios I. Manousiouthakis, Chemical Engineering Department, UCLA,, 5531 Boelter Hall, Los Angeles, CA 90095-1592.

The advent of swift increases in both energy costs and tighter environmental regulations give rise to a necessity for redesigning existing engineering processes such as power generation plants. These new designs aim to lower capital input and maximize energy output. These goals can be attained by solving a variety of optimization problems, such as minimum capital cost (MCC), minimum total annualized cost (MTAC), minimum utility cost, maximum power generation and so on. Conventional methods approach the optimization design as a combinatorial problem by creating a pre-established network (superstructure) that is then solved. Such superstructure-based methods cannot guarantee global optimality.

In this work we apply the infinite-dimensional state-space (IDEAS) framework with the MCC and MTAC objective functions to the design of power generating plants. The aforementioned objective functions are chosen so as to reflect economies of scale and are thus reverse convex functions of the corresponding characteristics of the corresponding units (heat exchanger's heat exchange area, compressor's power consumption, turbine's power generation, etc). An infinite number of linear equations is used to describe the feasible region. Global solution of this infinite-dimensional optimization problem is pursued through a truncation, underestimation procedure that is guaranteed to converge in the limit to the global optimum. Software developed at UCLA will then be used to automatically generate a globally optimal flowsheet of a power generating plant.

Keywords: IDEAS, Power plant, minimum capital cost, minimum total annualized cost, infinite-dimensional linear program

240a A Metric for Monitoring Mpc Optimization Strategy

William Canney, Advanced Process Control, Aspen Technology, Inc., 2500 CityWest Blvd, Suite 1500, Houston, TX 77042

A key element of Model Predictive Control (MPC) applications is the determination of steady-state operating targets based on consideration of the full set of manipulation variables, control variable constraints, and process economics. Since these targets are generally determined as the result of a multivariable calculations, most often formulated as a linear program steady-state target calculation, it is frequently difficult for the user of MPC to detect and understand suttle changes in operating strategy. While active constraint sets can change frequently, the predominate strategies of the MPC application rarely change, thus monitoring constraints is a difficult method for monitoring changes in MPC strategy. This paper defines a new metric for monitoring MPC controller strategy and discusses simulation results, and experiences using this metric on an operating processing unit.

Computation of Arrival Cost for Moving Horizon Estimation via Unscented Kalman Filtering

Cheryl C Qu, Juergen Hahn*

Artie McFerrin Department of Chemical Engineering, Texas A & M University, College Station, TX 77843-3122

Abstract

Moving horizon estimation alleviates the computational burden of solving a full information estimation problem by considering a finite horizon of the measurement data, however, it is non-trivial to determine the arrival cost. A commonly used approach for computing the arrival cost is to use a first order Taylor series approximation of the nonlinear model and then apply an extended Kalman filter. In this paper, an approach to compute the arrival cost for moving horizon estimation based on an unscented Kalman filter is proposed. The performance of such a moving horizon estimator is compared with the one based on an extended Kalman filter and illustrated in a case study.

Keywords: Moving Horizon Estimation; Arrival Cost; Unscented Kalman Filter

1 Introduction

For the last two decades, optimization-based moving horizon estimation (MHE) has found widespread use for nonlinear constrained problems [1], [2], [3]. Compared to full information estimation (FIE), MHE reduces the computational burden by considering a finite horizon of the available measurements, however, it is non-trivial to summarize the effect of the discarded data on the current states, which is the so called *arrival cost*. For linear unconstrained systems, the standard Kalman filter covariance update formula can be used to express the arrival cost explicitly. However, for nonlinear or constrained systems, a general analytical expression for the arrival cost is rarely available. Tenny and Rawlings [4] estimate the arrival cost by approximating the constrained, nonlinear system as an unconstrained linear time-varying system and applying linearization and an extended Kalman filter (EKF).

Unscented Kalman filters (UKF), as proposed by Julier and Uhlmann [5], avoid the linearization in the Kalman filter update formula by an unscented nonlinear transformation. By

*Corresponding author. Tel: +1 979 845 3568; fax: +1 979 845 6446. *E-mail address*: hahn@tamu.edu (J. Hahn).

carefully choosing a set of sigma points, which capture the true mean and covariance of a given distribution and then passing the means and covariances of estimated states through a nonlinear transformation, UKF is capable of estimating the posterior means and covariances accurately to an order higher than two. Therefore UKF can improve upon EKF performance for nonlinear systems as has been demonstrated for some applications [5]. These advantages of UKF over EKF form the motivation of this work where a MHE with computation of the arrival cost via UKF is proposed and its performance is illustrated in a case study.

The paper is organized as follows: In Section 2, a brief review of MHE and UKF is presented. The MHE with the arrival cost determined by UKF for nonlinear constrained estimation is then proposed in Section 3. Section 4 compares the performance of the MHE via EKF to that via UKF for nonlinear state estimation, and concluding remarks are given in Section 5.

2 Preliminaries

2.1 Moving Horizon Estimation

From a perspective of Bayesian theory, the constrained state estimation problem can be formulated as the solution of a full information estimation problem [6]. However, real-time implementations of FIE are often not feasible because the size of the problem grows unbounded with the number of points in time considered. One strategy to make the estimation problem tractable is to bound the problem size by employing a moving horizon approach.

The optimization problem defining MHE is of the following form

$$\min_{x_0,\{w_k\}_{k=0}^{T-1}} \phi_T(x_0,\{w_k\}) = \min_{z,\{w_k\}_{k=T-N}^{T-1}} \sum_{k=T-N}^{T-1} v_k' R^{-1} v_k + w_k' Q^{-1} w_k + \theta_{T-N}(z). \tag{1}$$

subject to

$$\begin{aligned} x_k &= f(x_{k-1}, u_{k-1}, w_{k-1}) \\ y_k &= h(x_k, u_k) + v_k \\ x_k &\in \mathbb{X},\ w_k \in \mathbb{W},\ v_k \in \mathbb{V} \end{aligned} \tag{2}$$

where the sets $\mathbb{X}$, $\mathbb{W}$ and $\mathbb{V}$ are constrained, N is the size of horizon length, $x_k := x(k; z, \{w_j\}_{j=T-N}^{k-1})$ denotes the solution of the system (2) at time k when the initial state is z, $\{w_j\}_{j=T-N}^{k-1}$ is the process noise sequence from time $T-N$ to $k-1$ and $v_k := y_k - h(x_k, u_k)$. $\theta_{T-N}(z)$ is referred to as the arrival cost, which summarizes the effect of the data $\{y_k\}_{k=0}^{T-N-1}$ on the state x_{T-N}

and makes it possible to transform the optimization problem into one of lower dimension. However, the best choice of the arrival cost remains an open issue for MHE.

For unconstrained, linear systems, the arrival cost can be expressed explicitly since the MHE optimization simplifies to the Kalman filter and its covariance update formula can be used [7]. Subject to the initial condition Π_0 and assuming the matrix Π_{T-N} is invertible, the arrival cost can then be expressed as

$$\theta_{T-N}(z) = (z - \hat{x}_{T-N})'\Pi_{T-N}^{-1}(z - \hat{x}_{T-N}) + \phi_{T-N}^*. \tag{3}$$

where $\hat{x}_{T-N}$ denotes the optimal estimate at time $T-N$ given all of the measurements y_k from time 0 to $T-N-1$, ϕ_{T-N}^* represents the optimal cost at time $T-N$ and Π_{T-N} is computed from the Kalman filter covariance update

$$\Pi_T = A\Pi_{T-1}A^T + GQG^T - A\Pi_{T-1}C^T(C\Pi_{T-1}C^T + HRH^T)^{-1}C\Pi_{T-1}A^T. \tag{4}$$

The solution to the problem described by equations (1) and (3) is the unique optimal pair $(z^*, \{\hat{w}_k^*\}_{k=T-N}^{T-1})$ and it can be integrated to yield the optimal state estimates $\{\hat{x}_k^*\}_{k=T-N+1}^{T}$, where $\hat{x}_k^* := x(k; z^*, \{\hat{w}_j^*\}_{j=T-N}^{k-1})$ denotes the optimal estimate of the system at time k when the initial state is z^* and the estimated process noise sequence is $\{\hat{w}_j^*\}_{j=T-N}^{k-1}$.

For constrained, linear systems, general analytical expressions for the arrival cost are not available. One reasonable strategy is to approximate the arrival cost by the one for the unconstrained problem. The approximation is exact when the inequality constraints are inactive. For nonlinear systems, Tenny and Rawlings estimate the arrival cost by approximating a constrained, nonlinear system as an unconstrained, linear time-varying system [4]. In their work the model functions $f(.)$ and $h(.)$ in Eq.(2) are supposed to be sufficiently smooth so that a first-order Taylor series approximation of the model can then be applied, i.e. $A_k := \frac{\partial f}{\partial x}|_{\hat{x}_{k|k-1}}$, $C_k := \frac{\partial h}{\partial x}|_{\hat{x}_{k|k-1}}$, $G_k := \frac{\partial f}{\partial w}|_{w_k}$ and $H_k := \frac{\partial h}{\partial v}|_{v_k}$ can be obtained. The arrival cost $\theta_{T-N}(z)$ in Eq.(3) can be computed by solving the matrix Riccati Eq.(4) subject to the initial condition Π_0. This is called the MHE problem with an arrival cost computed by EKF.

The horizon length N is a tuning parameter for MHE. As a general rule, the larger the horizon length, the more accurate the estimation results will be, however, this comes at the expense of an increase of the computational burden. A practical rule of thumb is that the length of the horizon should not be less than the number of the system states. Rao and Rawlings recommend to choose the horizon length as twice the order of the system [7].

2.2 Unscented Kalman Filter

An unscented Kalman filter is the application of the unscented transformation to recursive estimation. In the unscented transformation procedure, a set of weighed sigma points are deterministically chosen from the statistics of the transformation so that certain properties of these points (e.g., a given mean and covariance) match those of the prior distribution. These sigma points are propagated through a nonlinear mapping and then weighted means and covariances are computed. One way to find a set of sigma points that have the same first two moments and all higher odd-ordered central moments as the given distribution is given by the following:

1. Augment the system state vector to an $n^a = n+q+r$ dimensional vector $x^a = [x^T \; w^T \; v^T]^T$ to obtain its augmented means and covariances,

$$\begin{aligned} \hat{x}^a_{k-1|k-1} &= \begin{pmatrix} \hat{x}_{k-1|k-1} \\ \mathbf{0}^{\mathbf{q}\times 1} \\ \mathbf{0}^{\mathbf{r}\times 1} \end{pmatrix} \\ P^a_{k-1|k-1} &= \begin{pmatrix} P_{k-1|k-1} & \mathbf{0}^{\mathbf{n}\times\mathbf{q}} & \mathbf{0}^{\mathbf{n}\times\mathbf{r}} \\ \mathbf{0}^{\mathbf{q}\times\mathbf{n}} & Q_{k-1} & P^{wv}_{k-1} \\ \mathbf{0}^{\mathbf{r}\times\mathbf{n}} & P^{vw}_{k-1} & R_{k-1} \end{pmatrix} \end{aligned} \tag{5}$$

where n is the dimension of the original state vector, q and r are the dimensions of the system and measurement noise vectors and P^{vw}_{k-1} and P^{wv}_{k-1} are the correlations between the system and measurement noise. For ease of computation, P^{vw}_{k-1} and P^{wv}_{k-1} are usually set to zero.

2. Generate a set of $2n^a + 1$ symmetric sigma points

$$\chi^a_{k-1} = \hat{X}^a_{k-1|k-1} + \left(0 \quad \sqrt{(n^a+\kappa)P^a_{k-1|k-1}} \quad -\sqrt{(n^a+\kappa)P^a_{k-1|k-1}} \right) \tag{6}$$

where $\hat{X}^a_{k-1|k-1}$ is the expanded $n^a \times (2n^a + 1)$ matrix with $\hat{x}^a_{k-1|k-1}$ as each column. $\kappa \in \Re$ is a parameter which could be any positive or negative number with the exception of $\kappa = -n_a$. κ can be used to fine tune the higher order moments of the approximation. The more higher order moments are taken into account, the less the overall prediction error will be. For a Gaussian distribution $x(k)$, a useful heuristic is to select $n^a + \kappa = 3$ [8].

After a set of sigma points is selected, each of them is propagated through the nonlinear model functions $f(.)$ and $h(.)$. Weighted means and covariances are then computed from the transformed set of points. Finally the Kalman filter gain is calculated from the covariances and the predicted states are updated based on the available measurements. This procedure results in the following equations defining an unscented Kalman filter:

Prediction equations:

$$
\begin{aligned}
\chi_k^x &= f(\chi_{k-1}^x, \chi_{k-1}^w, u_{k-1}) \\
\hat{x}_{k|k-1} &= \sum_{i=1}^{2n^a+1} W_i \chi_{i,k}^x \\
\gamma_k &= h(\chi_k^x, \chi_{k-1}^v, u_k) \\
\hat{y}_k &= \sum_{i=1}^{2n^a+1} W_i \gamma_{i,k}
\end{aligned} \tag{7}
$$

Update equations:

$$P_{k|k-1} = \sum_{i=1}^{2n^a+1} W_i [\chi_{i,k}^x - \hat{x}_{k|k-1}][\chi_{i,k}^x - \hat{x}_{k|k-1}]^T \tag{8}$$

$$P_{y,k} = \sum_{i=1}^{2n^a+1} W_i [\gamma_{i,k} - \hat{y}_k][\gamma_{i,k} - \hat{y}_k]^T \tag{9}$$

$$P_{xy,k} = \sum_{i=1}^{2n^a+1} W_i [\chi_{i,k}^x - \hat{x}_{k|k-1}][\gamma_{i,k} - \hat{y}_k]^T \tag{10}$$

$$K_k = P_{xy,k} P_{y,k}^{-1} \tag{11}$$

$$P_{k|k} = P_{k|k-1} - K_k P_{y,k} K_k^T \tag{12}$$

$$\hat{x}_{k|k} = \hat{x}_{k|k-1} + K_k (y_k - \hat{y}_k) \tag{13}$$

where $\chi^a = [(\chi^x)_{1\times n}^T \ (\chi^w)_{1\times q}^T \ (\chi^v)_{1\times r}^T]^T$ and W_i are weights calculated by

$$W_i = \begin{cases} \frac{\kappa}{2(n_a+\kappa)}, & \text{if i=1;} \\ \frac{1}{2(n_a+\kappa)}, & \text{otherwise.} \end{cases} \tag{14}$$

3 Computing Arrival Cost of MHE by UKF

As mentioned in Section 2.1, algebraic expressions for the arrival cost, which account for data not included in the estimation window, are not available for the majority of systems. Therefore an approximation of the arrival cost is required to implement an estimator.

The main idea behind the presented MHE algorithm is to make use of the advantages that UKF offers over EKF for approximating the arrival cost. A set of weighted sigma points χ_k are selected for computing the arrival cost based on two considerations: Firstly the means and covariances of the set of sigma points need to match those of the prior distribution at time $k = T - N$ in the absence of active bounds. Secondly the distribution of the selected sigma points should be within the feasible region if bounds are active. If $s_{k,i} = \pm(\sqrt{P_{k|k}^a})_i, i = 1, ..., n$ are considered as the directions along which the sigma points are selected and $r_{k,i}$ as the step sizes along these directions, then the step sizes for the selected sigma points in Eq.(6) for a general UKF are all equal to $\sqrt{n_a+\kappa}$. In other words all the sigma points chosen are located symmetrically around the current estimate. This selection of sigma points can be applied to the MHE via UKF when the constraints are inactive. However, such selection is not adequate when constraints are active. To better approximate the covariance and then the arrival cost in the presence of active constraints, the set of sigma points is chosen in each direction with a step size of

$$r_{k,i} = \min(\sqrt{n_a+\kappa}, (x_{U,i} - \hat{x}^a_{k|k,i})/s_{k,i}, (x_{L,i} - \hat{x}^a_{k|k,i})/s_{k,i}). \tag{15}$$

where $x_{U,i}$ and $x_{L,i}$ are the upper and lower bounds in the direction $s_{k,i}$. The central point $\hat{x}^a_{k|k,0}$ is the same as in the general UKF formulation. The rest of selected sigma points may be asymmetrically around the central point due to the presence of active constraints. These sigma points are identical to those for conventional UKF for an unconstrained problem or if the constraints are not active. However, under the condition that the current estimate is close to the bounds, the selection process takes the constraints into account so that none of the selected sigma points violate the constraints on the state variables.

Considering that weights of all sigma points sum up to unity and that these are the same as those for UKF in Eq.(14) in the absence of active bounds, weights W_i for each sigma point can be calculated as follows:

$$W_i = \begin{cases} \frac{\kappa}{2(n_a+\kappa)}, & \text{if i=1;} \\ ar_i + b, & \text{otherwise.} \end{cases} \tag{16}$$

where

$$\begin{aligned} a &= \frac{2\kappa - 1}{2(n_a+\kappa)(S_r - (2n_a+1)(\sqrt{n_a+\kappa}))} \\ b &= \frac{1}{2(n_a+\kappa)} - \frac{2\kappa-1}{2\sqrt{n_a+\kappa}(S_r - (2n_a+1)(\sqrt{n_a+\kappa}))} \\ S_r &= \sum_{i=1}^{2n_a} r_i. \end{aligned} \tag{17}$$

The procedure to compute the weights with constraints is similar to the work presented by Vachhani et al. [9], where a complete mathematical derivation of the weight equations albeit for different purposes can be found.

After a set of sigma points is obtained, each of these sigma points is instantiated through the nonlinear model functions $f(.)$ and $h(.)$ to obtain the transformed sets χ^x_k and γ_k.

$$\begin{aligned} \chi^x_k &= f(\chi^x_{k-1}, \chi^w_{k-1}, u_{k-1}) \\ \gamma_k &= h(\chi^x_k, \chi^v_{k-1}, u_k) \end{aligned} \tag{18}$$

The weighted means of system states and measurements and weighted covariances of process and observation noise are then computed from the transformed sets. It should be noted that the weighted predicted estimates $\hat{x}_{k|k-1}$ may not satisfy the constraints.

$$\begin{aligned}
\hat{x}_{k|k-1} &= \sum_{i=1}^{2n^a+1} W_i \chi_{i,k}^x \\
\hat{y}_k &= \sum_{i=1}^{2n^a+1} W_i \gamma_{i,k} \\
P_{k|k-1} &= \sum_{i=1}^{2n^a+1} W_i [\chi_{i,k}^x - \hat{x}_{k|k-1}][\chi_{i,k}^x - \hat{x}_{k|k-1}]^T \\
P_{y,k} &= \sum_{i=1}^{2n^a+1} W_i [\gamma_{i,k} - \hat{y}_k][\gamma_{i,k} - \hat{y}_k]^T \\
P_{xy,k} &= \sum_{i=1}^{2n^a+1} W_i [\chi_{i,k}^x - \hat{x}_{k|k-1}][\gamma_{i,k} - \hat{y}_k]^T
\end{aligned} \tag{19}$$

Finally the matrix P_k is calculated from the filter gain and is used to compute the arrival cost in Eq.(20). The MHE problem described in the Eq.(1) and (3) is solved with the approximated arrival cost to obtain the updated estimates $\hat{x}_{k|k}$ as the solutions:

$$\begin{aligned}
K_k &= P_{xy,k} P_{y,k}^{-1} \\
P_{k|k} &= P_{k|k-1} - K_k P_{y,k} K_k^T \\
\theta_k(z) &= (z - \hat{x}_k)' P_k^{-1} (z - \hat{x}_k) + \phi_k^*
\end{aligned} \tag{20}$$

The proposed approach to approximate the arrival cost for MHE does not require linearization of the system and measurement functions. As in any MHE filter, the matrices Q and R can be chosen to take uncertainty in the model and measurement noise into account and the size of the estimation horizon serves as an additional tuning parameter.

4 Case Study

To illustrate the performance of MHE based on UKF (uMHE) compared against the one based on EKF (eMHE), both algorithms have been applied to a variety of models and a large number of scenarios such as different operating conditions, different tuning parameters Q and R, and different process and measurement noise. Due to space constraints only one representative case study is shown in this section. Monte Carlo simulations with 50 sample points have been conduct for each procedure so as not to bias results to one set of data. The performance is evaluated by the overall mean-squared error (MSE). MSE is first averaged over all simulations for each time point and then over time to to take the behavior over the entire time horizon into account.

The model of this case study is a nonisothermal continuous stirred tank reactor with coolant jacket dynamics, where the following exothermic irreversible reaction between sodium thiosulfate and hydrogen peroxide is taking place:

$$2Na_2S_2O_3 + 4H_2O_2 \rightarrow Na_2S_3O_6 + Na_2SO_4 + 4H_2O \tag{21}$$

The capital letters A and B are used to denote the chemical compounds $Na_2S_2O_3$ and H_2O_2 in the following. The reaction kinetic law is reported in the literature to be [10]:

$$-r_A = k_0 e^{-E/RT} C_A C_B$$

where k_0 is the pre-exponential factor, E is the activation energy, R is the gas constant, T is the temperature, and C_A and C_B are the concentrations of species A and B, respectively. A stoichiometric proportion of species A and B in the feed stream is assumed which results in $C_B(t) = 2C_A(t)$. A mole balance for species A and energy balances for the reactor and the cooling jacket result in the following nonlinear process model:

$$\begin{aligned}\frac{dC_A}{dt} &= \frac{F}{V}(C_{Ain} - C_A) - 2k(T)C_A^2 \\ \frac{dT}{dt} &= \frac{F}{V}(T_{in} - T) + 2\frac{(-\Delta H)_R}{\rho c_p}k(T)C_A^2 - \frac{UA}{V\rho c_p}(T - T_j) \\ \frac{dT_j}{dt} &= \frac{F_w}{V}(T_{jin} - T_j) + \frac{UA}{V_w \rho_w c_{pw}}(T - T_j)\end{aligned} \tag{22}$$

where F is the feed flow rate, V is the volume of the reactor, C_{Ain} is the inlet feed concentration, T_{in} is the inlet feed temperature, F_w is the feed flow rate of the cooling jacket, V_w is the volume of the cooling jacket, T_{jin} is the inlet coolant temperature, c_p is the heat capacity of the reacting mixture, c_{pw} is the heat capacity of the coolant, ρ is the density of the reaction mixture, ρ_w is the density of the coolant, U is the overall heat-transfer coefficient, and A is the area over which the heat is transferred. The process parameter values are given in the work by Rajaraman et al. [11].

The initial conditions and filter parameters are as follows:

$$\begin{aligned}\hat{x}_0 &= \begin{bmatrix} 0.018 & 382 & 371.3 \end{bmatrix}^T, \qquad \hat{P}_0 = diag\{10^{-7}, 2.5, 2.5\}, \\ Q_0 &= diag\{10^{-8}, 0.25, 0.25\}, \qquad R_0 = 0.25.\end{aligned} \tag{23}$$

The dimension of the augmented state vector is 7 and the set of sigma points is composed of 15 elements. The additional tuning parameter of the UKF, κ, is set to -4 for the case study according to the heuristic mentioned in Section 2.2. For fairness of comparison, κ is

Table 1: MSE Average over 50 Monte Carlo Simulations for Varying Measurement Noise Levels and Horizon Lengths

MSE	N=3		N=4		N=6		N=10	
	eMHE	uMHE	eMHE	uMHE	eMHE	uMHE	eMHE	uMHE
R = 25	8.45	5.64	5.56	4.94	5.14	4.76	2.66	2.52
R = 0.25	2.69	1.45	1.43	1.21	0.89	0.84	0.87	0.70
R = 0.01	1.01	0.86	0.80	0.78	0.59	0.56	0.48	0.45

not further adjusted to fine tune the higher order moments of the approximation. A non-negative constraint is enforced on the concentration C_A for both MHE formulations.

Based on the overall mean-squared error, the performance of each MHE is evaluated for horizon lengths N=3, 4, 6, and 10. Further simulations have also been carried out by varying measurement noise parameters for a fixed horizon length. Table 1 provides a summary of the results. It can be seen that uMHE performs better than eMHE for all the investigated horizon lengths and measurement noise levels. The MSEs for both uMHE and eMHE are decreasing with increasing lengths of the horizons, i.e, the performance of both uMHE and eMHE improves as more data are included in a horizon. If N is chosen to be large, the arrival cost could be accurately computed with either approach. Therefore the advantages of uMHE over eMHE decrease for large N.

5 Conclusion

This paper presented a MHE formulation where the arrival cost is computed by UKF. The unscented transformation and a set of selected sigma points are used to compute the covariances and then the arrival cost. The selection procedure for the sigma points is the same as the one used for unscented Kalman filtering if the constraints are inactive, however, a modification is used that satisfies the state variable constraints when the constraints are active. Linearization of the model is not required for the presented approach.

While only a representative case study was included in this paper, the presented method performed slightly better than the commonly used eMHE for all investigated cases. Therefore, the method can be a promising alternative for approximating the arrival cost for MHE.

Acknowledgement

The authors gratefully acknowledge financial support from the Mary Kay O′Connor Process Safety Center and appreciate valuable feedback given by Dr. Sam Mannan.

References

[1] M. Liebman, T. Edgar, and L. Lasdon. Efficient data reconciliation and estimation for dynamic processes using nonlinear programming techniques. *Comput. Chem. Eng.*, 16:963–986, 1992.

[2] E. Gatzke and F. J. Doyle. Use of multiple models and qualitative knowledge for on-line moving horizon disturbance estimation and fault diagnosis. *J. Proc. Cont.*, 12(2):339–352, 1994.

[3] J. B. Rawlings and B R. Bakshi. Particle filtering and moving horizon estimation. *Comput. Chem. Eng.*, 30:1529–1541, 2006.

[4] M.J. Tenny and J. B. Rawlings. Efficient moving horizon estimation and nonlinear model predictive control. *Proceedings of the American Control Conference*, pages 4475–4480, 2002.

[5] S. J. Julier and J. K. Uhlmann. Unscented filtering and nonlinear estimation. *Proceedings of The IEEE*, 92(3):401–422, 2004.

[6] C.V. Rao, J. B. Rawlings, and D. Q. Mayne. Constrained state estimation for nonlinear discrete-time systems: Stability and moving horizon approximations. *IEEE Transactions on Automatic Control*, 48(2):246–258, 2003.

[7] C. V. Rao and J. B. Rawlings. Constrained process monitoring: Moving horizon approach. *AIChE J*, 48(1):97–109, 2002.

[8] S. J. Julier and J. K. Uhlmann. A new extension of the kalman filter to nonlinear systems. *In Proc. AeroSense: 11th Int. Symp. Aerospace/Defense Sensing, Simul. and Controls*, pages 182–193, 1997.

[9] P. Vachhani, S. Narasimhan, and R Rengaswamy. Robust and reliable estimation via unscented recursive nonlinear dynamic data reconciliation. *J. Proc. Cont.*, 16:1075–1086, 2006.

[10] S. A. Vejtasa and R. A. Schmitz. An experimental study of steady-state multiplicity and stability in an adiabatic stirred reactor. *AIChE J*, 16(3):410–419, 1970.

[11] S. Rajaraman, J. Hahn, and M.S. Mannan. A Methodology for Fault Detection, Isolation and Identification for Nonlinear Processes with Parametric Uncertainties. *Ind. Eng. Chem. Res.*, 43(21):6774–6786, 2004.

240c Nonlinear Stochastic Modeling for State Estimation of An Industrial Polymerization Process

Fernando V. Lima[1], James B. Rawlings[1], and Tyler A. Soderstrom[2]. (1) Department of Chemical and Biological Engineering, University of Wisconsin-Madison, 1415 Engineering Drive, 2011, Madison, WI 53706, (2) Core Process Control, ExxonMobil Chemical, 4500 Bayway Dr., Baytown, TX 77522

State estimation from plant measurements plays an important role in advanced process control. For example, the performance of a closed-loop control system is directly affected by the quality of the current state estimates. The challenge of state estimation is to determine good estimates in the face of noisy and incomplete output measurements. Moving Horizon Estimator (MHE) is an optimization based state estimator that is robust to process disturbances and model errors. Moreover, MHE naturally handles nonlinear models and incorporates hard state constraints, such as non-negative concentrations and pressures, in its formulation [1-4]. For optimal performance, this state estimator requires knowledge about the noise statistics affecting the states and measurements. If these noises are modeled as zero-mean Gaussian sequences, then their covariances are required to specify their statistics. These statistics are usually unknown, but can be estimated from process operating data [5-7]. The Autocovariance Least-Squares (ALS) technique has proven to effectively estimate the covariances of system disturbances from data for linear [5,6] and nonlinear systems [7], using both simulated and real process data. This technique uses routine process operating data and thus does not require input-output testing to be applied to the system.

This presentation introduces a design method for all of the steps of nonlinear state estimation, from nonlinear stochastic modeling to estimator implementation. The effectiveness of the developed method is illustrated with an industrial polymerization process from ExxonMobil Chemical Company.

First, the modeling task is performed by showing that a nonlinear stochastic model can be well represented by a nonlinear deterministic model, obtained from first principles, with an added noise component that can be estimated from operating data containing the measurement noise. Then, the covariances of the process (Q) and measurement (R) noises are estimated using the time-varying ALS technique formulated for nonlinear models [7]. For this estimation, two cases using different data sets are considered: 1) simulated data generated using a published model of an industrial gas-phase ethylene copolymerization process [8-10] and an assumed Q and R; 2) real industrial data of the same process provided by ExxonMobil Chemical Company. For the second case, the stochastic structure of the disturbance model is also identified from data, by determining the minimum number of independent disturbances affecting the states, using the ALS-SDP technique [5,6]. These estimated covariances correspond to the Q and R weights used in the MHE formulation [1] and are used to specify the noise statistics of the state estimator. Next, MHE is implemented in both scenarios (real and simulated data) to show that better state estimates can be obtained after the specification of its noise statistics. Comparisons of the implemented estimator are made to other commonly used estimation techniques, such as Extended Kalman Filter (EKF) and a typical industrial estimator (heuristic KF). Finally, the ability of MHE to deal with infrequent asynchronous lab measurements is analyzed and it is demonstrated that the covariance estimation of the measurement noises using ALS scales this covariance based on the frequencies of the measurements.

References:

[1] Christopher V. Rao and James B. Rawlings. Constrained process monitoring: Moving-horizon approach. AIChE J., 48(1):97-109, 2002.

[2] Douglas G. Robertson, Jay H. Lee, and James B. Rawlings. A moving horizon-based approach for least-squares state estimation. AIChE J., 42(8):2209-2224, 1996.

[3] Douglas G. Robertson and Jay H. Lee. On the use of constraints in least squares estimation and control. Automatica, 38(7):1113-1123, 2002.

[4] Christopher V. Rao, James B. Rawlings, and David Q. Mayne. Constrained state estimation for nonlinear discrete-time systems: Stability and moving horizon approximations. IEEE Trans. Auto. Cont., 48(2):246-258, 2003.

[5] Murali R. Rajamani and James B. Rawlings. Estimation of the disturbance structure from data using semidefinite programming and optimal weighting. Technical Report 2007-02, TWMCC, Dept. of Chem. and Bio. Eng., University of Wisconsin-Madison (Available at http://jbrwww.che.wisc.edu/tech-reports.html), 2007.

[6] Murali R. Rajamani and James B. Rawlings. Estimation of the disturbance structure from data using semidefinite programming and optimal weighting. Accepted for publication in Automatica, 2007.

[7] Murali R. Rajamani. Data-based Techniques to Improve State Estimation in Model Predictive Control. PhD thesis, Dept. of Chem. and Bio. Eng., University of Wisconsin-Madison (Available at http://jbrwww.che.wisc.edu/theses.html), 2007.

[8] K. B. McAuley, J. F. MacGregor, and A. E. Hamielec. A kinetic model for industrial gas-phase ethylene copolymerization. AICHE J., 36(6):837-850, 1990.

[9] S. A. Dadebo, M. L. Bell, P. J. McLellan, and K. B. McAuley. Temperature control of industrial gas phase polyethylene reactors. J. Proc. Cont., 7(2):83-95, 1997.

[10] Adiwinata Gani, Prashant Mhaskar, and Panagiotis D. Christofides. Fault-tolerant control of a polyethylene reactor. J. Proc. Cont., 17(5):439-451, 2007.

Web Page: jbrwww.che.wisc.edu/

240d Design of Experiments for Identification of Multivariable Models Satisfying Integral Controllability: The Dynamic Model Case

Mark Darby, Department of Chemical and Biomolecular Engineering, University of Houston, 4800 Calhoun Ave., Houston, TX 77204-4004 and Michael Nikolaou, Chemical and Biomolecular Engineering, University of Houston, 4800 Calhoun Ave., Hoston, TX 77204-4004.

Model-based multivariable control requires good-quality process models which, in addition to standard specifications for model quality, must also satisfy certain control-relevant conditions to allow for robust closed-loop operation. In recent work, we developed a rigorous mathematical framework for practical design of experiments that produce models satisfying the integral controllability condition for nxn systems. In this work we extend our framework, which previously employed a static description of the identified plant, to the dynamic case for multivariable finite impulse response (FIR) models. The extended mathematical framework directly suggests a methodology for experiment design that incorporates integral controllability, and satisfies maximum variance constraints on each input and output variable. We perform simulations on an industrial 5x5 model to demonstrate the effectiveness of the approach and compare it to previous results in literature.

240e Cause and Effect Modeling from Plant Data

Derrick K. Rollins Sr.[1], Swee-Teng Chin[2], and Nidhi Bhandari[1]. (1) Chemical and Biological Engineering, Iowa State University, 2114 Sweeney Hall, Ames, IA 50011, (2) Dow Chemical Company, 2301 N. Brazosport Blvd., B-1225, Freeport, TX 77541

Plant data bases are filled with information on the relationships of state (output) and input variables. However, due to natural high multi-collinearities of the inputs and low signal to noise ratios for the outputs, modelers are challenged in developing cause and effect relationships using plant data. As a result, dynamic models are commonly developed from plant tests which incur costs and risks to the operation. Thus, the purpose of this work is to introduce a modeling approach that is capable of developing accurate cause and effect models from plant data. The proposed method is a special application of the Wiener block-oriented system and the unique and powerful attributes of this approach over existing techniques are demonstrated on a simulated continuous stirred tank reactor (CSTR) and real distillation column data.

Using the proposed method, the models are developed using high input correlation in training and then tested under conditions of low to zero input correlation in testing. Excellent performance of the models in testing validate the cause and effect ability of the modeling approach. The proposed method is compared with empirical methods like NARMAX which fail to predict well in testing, indicating their inability to model causation under high input correlation.

Web Page: www.public.iastate.edu/~drollins/Publications

240f Data-Driven State and Parameter Estimation Using Bayesian Belief Networks

Ameneh Sahraneshin[1], Nasir Mehranbod[1], Reza Eslamlouian[1], and Masoud Soroush[2]. (1) Department of Chemical Engineering, Shiraz University, Shiraz, Iran, (2) Department of Chemical and Biological Engineering, Drexel University, Philadelphia, PA 19104

Effective control and monitoring of a plant require reliable and frequent real-time information on state variables of the plant. In practice, frequent measurements of all state variables of a process are rarely available, and in such cases, the missing frequent information on the un-measurable state variables can be obtained using a state estimator [1, 2]. Reliable state estimation requires an accurate mathematical model of the plant under consideration. However, in most cases, an accurate mathematical model is rarely available. One way to address this problem is to use a purely data-driven method to conduct the estimation.

In this paper, we present a data-driven approach for state and parameter estimation. The approach is based on Bayesian Belief Network (BBN) theory. In this approach, nodes of a BBN model represent the parameters and state variables of the process under study. The numerical-value-span of interest for each node is divided into equally spaced states. There is no limitation on number of states that can be assigned for each node in BBN theory. However, equal number of states has been considered in this work to achieve as accurate estimation as possible [3]. The necessary conditional probability data set associated with the model that captures correlation between pair of nodes in the model is determined by quantitative batch BBN training procedure [3, 4]. Evidence propagation by junction tree method [5] in the BBN model is used to update probability distribution of nodes representing un-measurable/ unavailable process states and parameters. Weighted averaging of updated probability distribution is used to calculate numerical values for such states and parameters. To show the application and performance of the method, two plants, a simple stirred heating tank and the Tennessee Eastman Process [6], are considered. The training data for the first plant is generated by conducting Monte-Carlo simulation of the plant described by a first-principles model. For the second plant, the training data is available for the Tennessee Eastman process [6]. The simulation results show that the proposed method is capable of providing reliable estimates of plant parameters and state variables.

Keywords: data-driven state estimation; data-driven parameter estimation; Bayesian Belief Networks; Tennessee Eastman Process

References

1. Bastin, G., and D. Dochain, Online Estimation and Adaptive Control of Bioreactors, New York: Elsevier (1990)

2. Soroush, M., "State and Parameter Estimations and Their Applications in Process Control," Comput.

& Chem. Eng., 23, pp.229 (1998a)

3. Mehranbod N., M. Soroush, M. Piovoso and B. A. Ogunnaike, "Probabilistic Model for Sensor Fault Detection and Identification," AIChE Journal, 49(7), pp. 1787-1802 (2003).

4. Mehranbod N., M. Soroush and C. Panjapornpon, "A method of sensor fault detection and identification, " Journal of Process Control, 15(3), pp. 321-339 (2005).

5. Spiegelhalter, David J., A. Philip Dawid, Steffen L. Lauritzen and Robert G. Cowell, "Bayesian analysis in expert systems" Statistical Science, 8(3), pp. 219-283 (1993)

6. Downs J. J.and E. F. Vogel, "A plant-wide industrial process control problem," Computers and Chemical Engineering, 17(3), pp. 245-255 (1993).

Modular Design of Discrete-Time Nonlinear Observers for State and Disturbance Estimation

Costas Kravaris

Department of Chemical Engineering, University of Patras, 26500 Patras, Greece

Paper No. 240g, AIChE Annual Meeting, Philadelphia, PA, Nov. 2008

Abstract—**This work studies the problem of design of discrete-time nonlinear observers in the presence of exogenous disturbances. In particular, the present work proposes a systematic design method for discrete-time nonlinear state and disturbance observers in modular form, consisting of an observer for the disturbance-free part of the system, along with a disturbance observer and a state-estimate corrector. The modular observer is first defined and characterized in a general setting and then, a systematic design method is developed on the basis of exact linearization with eigenvalue assignment. Necessary and sufficient conditions for feasibility of the proposed nonlinear modular observer design are derived.**

I. Introduction

TECHNICAL limitations and/or high cost of sensors result in the non-availability of all state variables for direct on-line measurement, and this creates the need for on-line state estimation. Furthermore, the operation of a process or plant is subject to time-varying disturbances, associated with changes in key process parameters or improper operation of sensing instruments. For this reason, in addition to monitoring the state variables, there is a definite practical need for detection and estimation of disturbances.

The problem of combined state and disturbance estimation can be conceptually formulated as a state estimation problem for an extended system. In the case of linear systems, the well-known Luenberger observer offers a comprehensive solution. More specifically, in industrial applications of combined state and disturbance estimation, the Luenberger observer is designed and implemented in a modular configuration, consisting of an observer for the disturbance-free part of the system and, on top of it, a disturbance estimator and a state-estimate corrector ([3]).

The purpose of the present work is to develop a systematic discrete-time nonlinear observer design method for state and disturbance estimation, so that the resulting observer possesses the modular configuration that is sought for in practice. The discrete-time nonlinear observer design problem will be formulated within the general framework of exact observer linearization ([1],[2],[5],[8]–[11]), and in particular, following the invariant-manifold formulation, originally proposed in [5] and further developed in [10],[11].

The theoretical developments and results of the present paper leading to the discrete-time modular observer design are exactly parallel to the continuous-time modular observer of [7].

After a review of basic results on discrete-time observer linearization and their implication on the problem of simultaneous state and disturbance estimation, the modular observer will be defined and characterized in terms of appropriate invariance conditions. Necessary and sufficient conditions for exact linearization with eigenvalue assignment will be derived, leading to a step-by-step design procedure for the modular observer. It will be shown that the proposed discrete-time modular observer design is feasible whenever the corresponding simultaneous design is feasible. Moreover, it will be shown that, when this observer is applied to linear systems, the proposed modular design generates the standard modular Luenberger observer.

II. Problem Formulation

Consider a discrete-time nonlinear dynamic system

$$\begin{aligned} x(k+1) &= f(x(k), w(k)) \\ y(k) &= h(x(k), w(k)) \end{aligned} \qquad (1)$$

that represents the dynamics of a process, where x is the process state vector, y is the vector of measurements and w is the vector of unmeasurable disturbances. The dynamics of the disturbances is generated by the exosystem

$$w(k+1) = s(w(k)) \qquad (2)$$

The problem of *state and disturbance estimation* is the one of estimating both the process states x and the disturbances w given on-line measurements of y. If one considers the extended system consisting of (1) and (2), i.e.

$$\begin{aligned} x(k+1) &= f(x(k), w(k)) \\ w(k+1) &= s(w(k)) \\ y(k) &= h(x(k), w(k)) \end{aligned} \tag{3}$$

the problem of state and disturbance estimation becomes a standard state estimation problem for system (3), where the state $\begin{bmatrix} x \\ w \end{bmatrix}$ of (3) must be estimated from on-line measurements of y.

The effect of disturbances is often neglected by engineers, whenever their magnitude is viewed to be small relative to the other variables affecting the operation of the process. In this case, process dynamics is approximately represented as

$$\begin{aligned} x^{df}(k+1) &= f\left(x^{df}(k), 0\right) \\ y(k) &= h\left(x^{df}(k), 0\right) \end{aligned} \tag{4}$$

where x^{df} represents the "disturbance-free" process state, i.e. the process state when $w = 0$.

When the effect of disturbances can be neglected, (4) can be used as the basis for designing an observer. This is the most convenient approach from an engineering point of view, since it simplifies the observer equations and implementation. When disturbances are large enough so that they must be accounted for, engineers still want to monitor process state estimates obtained from (4) ("disturbance-free part of the system"), together with disturbance estimates and estimates of the process states that account for disturbances ("corrected state estimates"). In practice, a typical industrial implementation of a linear Luenberger observer for state and disturbance estimation has the structure of Figure 1 ([3]).

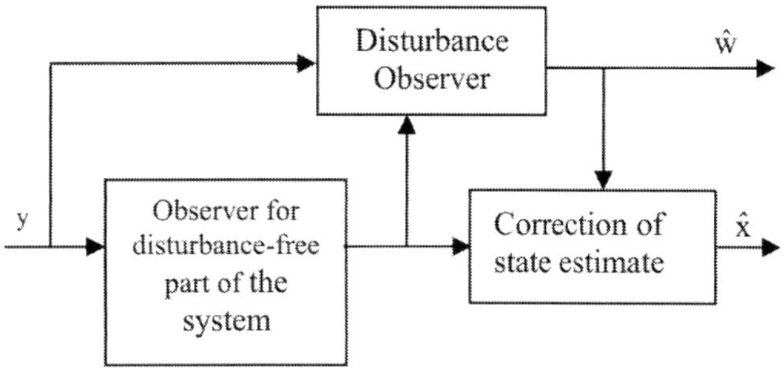

Fig. 1. Modular observer for state and disturbance estimation

The observer of Figure 1 has a modular structure: there is a basic observer for the disturbance-free part of the system and, on top of it, there is a disturbance observer. The disturbance estimate is used to correct the state estimate that was computed from the disturbance-free part of the system.

The modular configuration generates two sets of state estimates, one neglecting disturbances and another accounting for disturbances, and their comparison can be useful from a process monitoring point of view. Moreover, it gives the process engineer the flexibility of turning on or off the disturbance observer, or using alternative disturbance observers based on different assumptions on the nature of disturbances.

For continuous-time systems, the problem of simultaneous state and disturbance estimation in non-modular structure was recently studied in [6], in the context of exact observer linearization ([4]). A modular form of this continuous-time state and disturbance observer was developed in [7]. The present paper will develop a modular observer design method for discrete-time nonlinear systems, in a way that is parallel to the continuous-time modular observer in [7].

In the proposed discrete-time modular observer design method, the observer for the disturbance-free part of the system will be designed first and then the disturbance observer plus state-estimate corrector will be designed. The observer error linearization approach ([5]) will be utilized for both steps of the design, for eigenvalue assignment.

III. Background

Definition 1: *Given a dynamic system*

$$\begin{aligned} x(k+1) &= f\left(x(k)\right) \\ y(k) &= h\left(x(k)\right) \end{aligned} \tag{5}$$

where $f:\mathbb{R}^n \to \mathbb{R}^n$, $h:\mathbb{R}^n \to \mathbb{R}^p$, *the system*

$$\begin{aligned} \bar{x}(k+1) &= \Gamma(\bar{x}(k), y(k)) \\ \hat{x}(k) &= \Xi(\bar{x}(k)) \end{aligned} \tag{6}$$

where $\Gamma:\mathbb{R}^n \times \mathbb{R}^p \to \mathbb{R}^n$, $\Xi:\mathbb{R}^n \to \mathbb{R}^n$, *is called an* *<u>observer</u>* *for (5) if in the series connection*

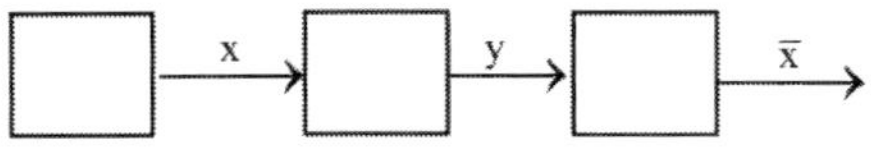

the overall dynamics

$$\begin{aligned} x(k+1) &= f(x(k)) \\ \bar{x}(k) &= \Gamma(\bar{x}(k), h(x(k))) \end{aligned} \tag{7}$$

has the property that $x = \Xi(\bar{x})$ *is an invariant manifold. In particular, when* Ξ *is the identity map, (6) is called an* *<u>identity observer</u>.*

In the above definition, the requirement that $x = \Xi(\bar{x})$ is an invariant manifold of (7), i.e. that $x(0) = \Xi(\bar{x}(0)) \Rightarrow x(k) = \Xi(\bar{x}(k)) \quad \forall k > 0$, translates to the following condition:

$$\Xi\left(\Gamma\left(\bar{x}, h\left(\Xi(\bar{x})\right)\right)\right) = f\left(\Xi(\bar{x})\right) \tag{8}$$

In the special case of identity observer, the above condition collapses to

$$\Gamma(\bar{x}, h(\bar{x})) = f(\bar{x}) \tag{9}$$

In the observer linearization design method, the observer is chosen so that it can be transformed to a linear system

$$z(k+1) = Az(k) + By(k) \tag{10}$$

with A, B being $n \times n$ and $n \times p$ matrices respectively and A having pre-assigned eigenvalues, under an appropriate coordinate transformation. Considering the case of identity observer for reasons of simplicity, and denoting by

$$z = \theta(\bar{x}) \tag{11}$$

a coordinate transformation that maps

$$\bar{x}(k+1) = \Gamma(\bar{x}(k), y(k)) \tag{12}$$

to (10), one immediately concludes that the transformation of (12) to (10) will be feasible if and only if

$$\theta(\Gamma(\bar{x}, y)) = A\theta(\bar{x}) + By \tag{13}$$

for every $\bar{x} \in \mathbb{R}^n$, $y \in \mathbb{R}^p$.

Applying condition (13) for $y = h(\bar{x})$ and using (9), it follows that $\theta(\bar{x})$ must satisfy

$$\theta(f(\bar{x})) = A\theta(\bar{x}) + Bh(\bar{x}) \tag{14}$$

Combining (13) with (14), it follows that

$$\theta(\Gamma(\bar{x}, y)) = \theta(f(\bar{x})) + B(y - h(\bar{x})) \tag{15}$$

from which

$$\Gamma(\bar{x}, y) = \theta^{-1}\left(\theta(f(\bar{x})) + B(y - h(\bar{x}))\right) \tag{16}$$

The following Proposition summarizes the result of the above derivation:

<u>Proposition 1</u>: *Given* $f:\mathbb{R}^n \to \mathbb{R}^n$, $h:\mathbb{R}^n \to \mathbb{R}^p$, *it is possible to find* $\Gamma:\mathbb{R}^n \times \mathbb{R}^p \to \mathbb{R}^n$ *that satisfies condition (9) and a coordinate transformation that maps (12) to (10) if and only if there exists an invertible function* $\theta:\mathbb{R}^n \to \mathbb{R}^n$ *that satisfies the linear functional equation (LFE)*

$$\theta(f(x)) = A\theta(x) + Bh(x) \tag{17}$$

Assuming that such a function θ *exists, the choice*

$$\Gamma(x, y) = \theta^{-1}\left(\theta(f(x)) + B(y - h(x))\right) \tag{18}$$

satisfies condition (9) and makes (12) transformable to (10) via the coordinate transformation $z = \theta(\overline{x})$.

The result of Proposition 1 reduces the problem of observer linearization to the one of solving the LFE (16). Alternative local solvability conditions for (17) are available in the literature ([5], [10]), depending on the smoothness assumptions on $f(x)$ and $h(x)$ and also, depending on the nature of the spectrum of $\frac{\partial f}{\partial x}(x)$ evaluated at the equilibrium point. The following Proposition provides local solvability conditions for the real-analytic Poincaré-domain case.

<u>Proposition 2</u> [5]: *Let* $f:\mathbb{R}^n \to \mathbb{R}^n$, $h:\mathbb{R}^n \to \mathbb{R}^p$ *be real analytic functions with* $f(0)=0$, $h(0)=0$ *and denote* $F = \frac{\partial f}{\partial x}(0)$, $H = \frac{\partial h}{\partial x}(0)$. *Also, denote by* $\sigma(F)$ *the set of eigenvalues of* F.

Suppose:

1. *There exists an invertible matrix* T *such that* $TF = AT + BH$
2. *All the eigenvalues of* A *are non-resonant with* $\sigma(F)$, *i.e. no eigenvalue* λ_j *of* A *is of the form* $\lambda_j = \prod_{i=1}^{n} \kappa_i^{m_i}$ *with* $\kappa_i \in \sigma(F)$ *and* m_i *nonnegative integers, not all zero.*
3. $\sigma(F)$ *is either a subset of the unit disc or disjoint with the unit disc.*

Then there exists a unique analytic solution of the LFE (17) locally around $x = 0$. *The solution has the property that* $\frac{\partial \theta}{\partial x}(0) = T$ *and so,* θ *is a local diffeomorphism.*

Note that assumptions 1 and 2 of the above Proposition imply that (H, F) is an observable pair. On the other hand, if (H, F) is an observable pair, it is always possible to find matrices A, B, T which satisfy the matrix equation of assumption 1, with T invertible and A having prescribed eigenvalues.

The results of Propositions 1 and 2 can now be applied to the problem of state and disturbance estimation, considering the extended system (3), where $x \in \mathbb{R}^m$ is the vector process states, $w \in \mathbb{R}^\ell$ is the vector of disturbances and $\begin{bmatrix} x \\ w \end{bmatrix} \in \mathbb{R}^{m+\ell}$ is the state of the entire system (3) that must be estimated. The corresponding observer LFE (17) for system (3) is

$$\theta(f(x,w),s(w)) = A\theta(x,w) + Bh(x,w) \tag{19}$$

and, assuming that it admits an invertible solution, the observer has the form

$$\begin{bmatrix} \hat{x}(k+1) \\ \hat{w}(k+1) \end{bmatrix} = \theta^{-1}\left(\theta(f(\hat{x}(k),\hat{w}(k)),s(\hat{w}(k))) + B\left(y(k) - h(\hat{x}(k),\hat{w}(k))\right)\right) \tag{20}$$

or equivalently

$$\theta\left(\hat{x}(k+1),\hat{w}(k+1)\right) = \theta\left(f(\hat{x}(k),\hat{w}(k)),s(\hat{w}(k))\right) + B\left(y(k) - h(\hat{x}(k),\hat{w}(k))\right) \tag{21}$$

Note that the computational effort in the observer design can be somewhat reduced if A is taken to be block-diagonal, with diagonal blocks of sizes $m \times m$ and $\ell \times \ell$. In particular, setting $A = \begin{bmatrix} A_0 & 0 \\ 0 & A' \end{bmatrix}$, $B = \begin{bmatrix} B_0 \\ B' \end{bmatrix}$, where A_0, A', B_0, B' are $m \times m$, $\ell \times \ell$, $m \times p$, $\ell \times p$ matrices respectively and partitioning $\theta(x,w)$ accordingly, $\theta(x,w) = \begin{bmatrix} \pi(x,w) \\ \eta(x,w) \end{bmatrix}$, with $\pi:\mathbb{R}^m \times \mathbb{R}^\ell \to \mathbb{R}^m$, $\eta:\mathbb{R}^m \times \mathbb{R}^\ell \to \mathbb{R}^\ell$, the observer LFE (19) breaks up into two uncoupled LFE's :

$$\pi(f(x,w),s(w)) = A_0\pi(x,w) + B_0h(x,w) \tag{22}$$

$$\eta(f(x,w),s(w)) = A'\eta(x,w) + B'h(x,w) \tag{23}$$

As an immediate consequence of Proposition 2, we have

<u>Proposition 3</u>: *Let* $f:\mathbb{R}^m \times \mathbb{R}^\ell \to \mathbb{R}^m$, $s:\mathbb{R}^\ell \to \mathbb{R}^\ell$ *and* $h:\mathbb{R}^m \times \mathbb{R}^\ell \to \mathbb{R}^p$ *be real analytic functions with*

$f(0,0)=0,\ s(0)=0,\ h(0,0)=0$ *and denote* $F=\frac{\partial f}{\partial x}(0,0),\ P=\frac{\partial f}{\partial w}(0,0),\ S=\frac{\partial s}{\partial w}(0),\ H=\frac{\partial h}{\partial x}(0,0),\ Q=\frac{\partial h}{\partial w}(0,0)$.

Also, denote by $\sigma(F)$ *and* $\sigma(S)$ *the sets of eigenvalues of* F *and* S *respectively.*

Suppose:

1. *There exists an invertible matrix* T *such that* $T\begin{bmatrix} F & P \\ S & 0 \end{bmatrix} = \begin{bmatrix} A_0 & 0 \\ 0 & A' \end{bmatrix} T + \begin{bmatrix} B \\ B' \end{bmatrix} [H \quad Q]$
2. *All the eigenvalues of* A_0 *and* A' *are non-resonant with* $\sigma(F) \cup \sigma(S)$.
3. $\sigma(F) \cup \sigma(S)$ *is either a subset of the unit disc or disjoint with the unit disc.*

Then the system of LFE's (22) and (23) admits a unique solution locally around $(x,w)=(0,0)$. *The solution has the property that* $\begin{bmatrix} \frac{\partial \pi}{\partial x}(0,0) & \frac{\partial \pi}{\partial w}(0,0) \\ \frac{\partial \eta}{\partial x}(0,0) & \frac{\partial \eta}{\partial w}(0,0) \end{bmatrix} = T$

IV. MODULAR OBSERVER

In the present section, a modular design of an observer for system (3) will be developed, as an alternative to the "simultaneous" design outlined in the previous section.

The proposed modular observer will consist of two distinct parts. An identity observer for the disturbance-free part of the system (system (4)), followed by a disturbance observer together with a corrector for the state estimate. The identity observer for the disturbance-free part of the system will have the form

$$\tilde{x}(k+1) = \Gamma_0(\tilde{x}(k), y(k)) \tag{24}$$

whereas the disturbance observer and corrector will have the form

$$\hat{w}(k+1) = \Gamma_w(\tilde{x}(k), \hat{w}(k), y(k)) \tag{25}$$

$$\hat{x}(k+1) = \Upsilon(\tilde{x}(k), \hat{w}(k)) \tag{26}$$

Figure 2 depicts the structure of the proposed modular observer:

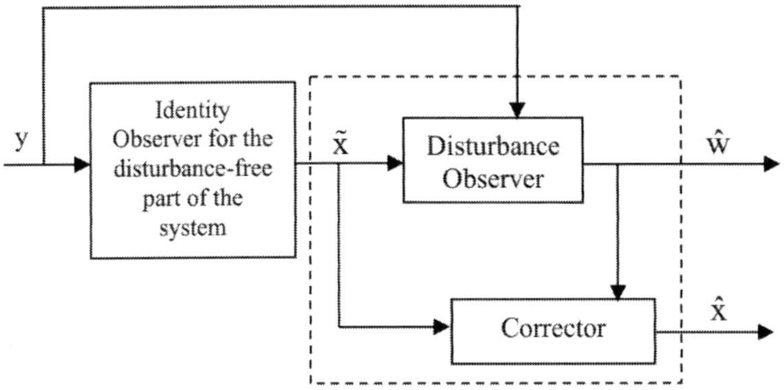

Figure 2: Structure of the modular observer (24) – (26).

The following Proposition is an immediate consequence of Definition 1, applied to the modular structure considered here.

<u>Proposition 4</u>: a) *System (24) will be an identity observer for (4) if*

$$\Gamma_0(\tilde{x}, h(\tilde{x},0)) = f(\tilde{x},0) \tag{27}$$

b) *The overall system (24) – (26) will be an observer for (3) if the following conditions are met*

$$\Upsilon(\Gamma_0(\tilde{x}, h(\Upsilon(\tilde{x},\hat{w}),\hat{w})), s(\hat{w})) = f(\Upsilon(\tilde{x},\hat{w}),\hat{w}) \tag{28}$$

and

$$\Gamma_w\left(\tilde{x},\hat{w},h\left(\Upsilon(\tilde{x},\hat{w}),\hat{w}\right)\right)=s(\hat{w}) \tag{29}$$

We can now proceed with the design of the modular observer via linearization and eigenvalue assignment. The rationale of the previous section will be followed, while adhering to the modular structure of (24)–(26).

Proposition 5: *It is possible to find* $\Gamma_0:\mathbb{R}^m\times\mathbb{R}^p\to\mathbb{R}^m$ *that satisfies (27) and a coordinate transformation that maps (24) to*

$$z_0(k+1)=A_0z_0(k)+B_0y(k) \tag{30}$$

where A_0 *and* B_0 *are* $m\times m$ *and* $m\times p$ *matrices respectively, if and only if there exists an invertible function* $\theta_0:\mathbb{R}^m\to\mathbb{R}^m$ *that satisfies the LFE*

$$\theta_0\left(f(x,0)\right)=A_0\theta_0(x)+B_0h(x,0) \tag{31}$$

Assuming that such a function θ_0 *exists, the choice*

$$\Gamma_0(x,y)=\theta_0^{-1}\left(\theta_0(f(x,0))+B_0(y-h(x,0))\right) \tag{32}$$

satisfies condition (27) and makes (24) transformable to (30) via the coordinate transformation $z_0=\theta_0(\tilde{x})$.

Proof: In order for $z_0=\theta_0(\tilde{x})$ to map (24) to (30), the condition

$$\theta_0\left(\Gamma_0(\tilde{x},y)\right)=A_0\theta_0(\tilde{x})+B_0y \tag{33}$$

is necessary and sufficient.

Applying (33) for $y=h(\tilde{x},0)$ and taking into account (27), it follows that $\theta_0(x)$ must satisfy (31).

Conversely, suppose that θ_0 satisfies (31). Then, the function $\Gamma_0(x,y)$ given by (32) automatically satisfies (27) and, moreover,

$$\theta_0\left(\Gamma_0(x,y)\right)=\theta_0(f(x,0))+B_0(y-h(x,0))$$

which, combined with (31), gives (33).

Proposition 6: *Assume that* Γ_0 *is given by (32), with* θ_0 *being an invertible function that satisfies (31).*

It is possible to find $\Gamma_w:\mathbb{R}^m\times\mathbb{R}^\ell\times\mathbb{R}^p\to\mathbb{R}^m$ *that satisfies (29) and a coordinate transformation that maps the dynamics of (24) and (25) to*

$$\begin{bmatrix}z_0(k+1)\\ z'(k+1)\end{bmatrix}=\begin{bmatrix}A_0 & 0\\ 0 & A'\end{bmatrix}\begin{bmatrix}z_0(k)\\ z'(k)\end{bmatrix}+\begin{bmatrix}B_0\\ B'\end{bmatrix}y(k) \tag{34}$$

if and only if there exists a function $\Phi:\mathbb{R}^m\times\mathbb{R}^\ell\to\mathbb{R}^\ell$ *such that the equation* $z'=\Phi(x,w)$ *is uniquely solvable with respect to* w *and*

$$\Phi\left(\Gamma_0\left(x,h\left(\Upsilon(x,w),w\right)\right),s(w)\right)=A'\Phi(x,w)+B'h(\Upsilon(x,w),w) \tag{35}$$

Assuming that such a function Φ exists, the implicit function $\omega=\Gamma_w(x,w,y)$ *defined as the solution of the algebraic equation*

$$\Phi\left(\Gamma_0(x,y),\omega\right)-\Phi\left(\Gamma_0\left(x,h\left(\Upsilon(x,w),w\right)\right),s(w)\right)=B'\left(y-h\left(\Upsilon(x,w),w\right)\right) \tag{36}$$

satisfies condition (29) and makes (24)–(25) transformable to (34) via the coordinate transformation $\begin{bmatrix}z_0\\ z'\end{bmatrix}=\begin{bmatrix}\theta_0(\tilde{x})\\ \Phi(\tilde{x},\hat{w})\end{bmatrix}$.

Proof: Since θ_0 invertible, the transformation $\begin{bmatrix}z_0\\ z'\end{bmatrix}=\begin{bmatrix}\theta_0(\tilde{x})\\ \Phi(\tilde{x},\hat{w})\end{bmatrix}$ will be invertible if and only if the equation $z'=\Phi(\tilde{x},\hat{w})$ is uniquely solvable with respect to $\hat{w}$.

Since Γ_0 is given by (32) with θ_0 satisfying (31), $z_0(k+1)=A_0z_0(k)+B_0y(k)$ holds, as a result of Proposition 5.

Therefore, the question is the transformation of (25) to $z'(k+1)=A'z'(k)+B'y(k)$. In order for this to happen, it is necessary and sufficient to have

$$\Phi\left(\Gamma_0(x,y),\Gamma_w(x,w,y)\right)=A'\Phi(x,w)+B'y \tag{37}$$

Applying (37) for $y = h(\Upsilon(x,w),w)$ and taking into account (29), it follows that Φ must satisfy LFE (35).

Conversely, suppose that Φ satisfies (35) and that the equation $z' = \Phi(x,w)$ is uniquely solvable with respect to w. Then, the function $\Gamma_w(x,w,y)$ defined as the solution of (36) automatically satisfies (29) and, moreover, $\Phi(\Gamma_0(x,y),\Gamma_w(x,w,y)) = \Phi(\Gamma_0(x,h(\Upsilon(x,w),w)),s(w)) + B'(y-h(\Upsilon(x,w),w))$ which, combined with (35), gives (37).

Combining the results of Propositions 4, 5 and 6, leads the following design procedure for the modular observer:

1. *Design of the identity observer for the disturbance-free part of the system:*
   $$\Gamma_0(x,y) = \theta_0^{-1}(\theta_0(f(x,0)) + B_0(y-h(x,0))) \quad (32)$$
   where $\theta_0(x)$ *satisfies the LFE*
   $$\theta_0(f(x,0)) = A_0\theta_0(x) + B_0 h(x,0) \quad (31)$$
2. *Design of the corrector:*
   $\Upsilon(x,w)$ *satisfies the functional equation (FE)*
   $$\Upsilon(\Gamma_0(x,h(\Upsilon(x,w),w)),s(w)) = f(\Upsilon(x,w),w) \quad (28)$$
3. *Design of the disturbance observer:*
   $\omega = \Gamma_w(x,w,y)$ *is the implicit function defined as the solution of the algebraic equation*
   $$\Phi(\Gamma_0(x,y),\omega) - \Phi(\Gamma_0(x,h(\Upsilon(x,w),w)),s(w)) = B'(y-h(\Upsilon(x,w),w)) \quad (36)$$
   where $\Phi(x,w)$ *satisfies the LFE*
   $$\Phi(\Gamma_0(x,h(\Upsilon(x,w),w)),s(w)) = A'\Phi(x,w) + B'h(\Upsilon(x,w),w) \quad (35)$$

The conclusion is that the modular design of the observer reduces to solvability of three FE's, (31) for θ_0, (28) for Υ and (35) for Φ.

Local solvability conditions for the LFE (31) immediately arise from Proposition 2. However, FE's (28) and (35) need further attention.

In the next section, it will be shown that if the LFE's (22) and (23) for simultaneous design are locally solvable, then FE's (28) and (35) for modular design are also solvable.

V. Modular Versus Simultaneous Design

The following proposition provides the solution of the modular observer design problem if the solution of the simultaneous design problem is available.

Proposition 7: *Suppose that the LFE's (22) and (23) are locally solvable in a neighborhood of* $(x=0, w=0)$ *and that the solution is differentiable, with* $\left[\frac{\partial \pi}{\partial x}(x,0)\right]$ *and* $\begin{bmatrix} \frac{\partial \pi}{\partial x}(x,w) & \frac{\partial \pi}{\partial w}(x,w) \\ \frac{\partial \eta}{\partial x}(x,w) & \frac{\partial \eta}{\partial w}(x,w) \end{bmatrix}$ *invertible matrices, in a neighborhood of* $(x=0, w=0)$. *Then*

i. *The function* $\theta_0(x) = \pi(x,0)$ *is an invertible function in a neighborhood of* $x=0$ *and satisfies the LFE (31).*
ii. *The algebraic equation* $\pi(\hat{x},w) = \pi(x,0)$ *is locally solvable with respect to* $\hat{x}$ *and its solution* $\hat{x} = \Upsilon(x,w)$ *satisfies the FE (28).*
iii. *The function* $\Phi(x,w) = \eta(\Upsilon(x,w),w)$ *satisfies the LFE (35). Also, the algebraic equation* $z' = \Phi(x,w)$ *is locally solvable with respect to* w.

Proof: i) Applying (22) for $w = 0$ leads to $\pi(f(x,0),0) = A_0\pi(x,0) + B_0 h(x,0)$, i.e. the function $\theta_0(x) = \pi(x,0)$ satisfies (31). Moreover, the function $\theta_0(x) = \pi(x,0)$ is locally invertible, since $\frac{\partial\pi}{\partial x}(x,0)$ is an invertible matrix in a neighborhood of the origin, as a consequence of the inverse function theorem.

ii) The equation $\pi(\hat{x}, w) = \pi(x,0)$ is locally solvable with respect to $\hat{x}$ since $\frac{\partial\pi}{\partial x}(0,0)$ is an invertible matrix, as a consequence of the implicit function theorem. The implicit function $\hat{x} = \Upsilon(x,w)$ will satisfy $\pi(\Upsilon(x,w),w) = \pi(x,0)$, i.e.

$$\pi(\Upsilon(x,w),w) = \theta_0(x) \tag{38}$$

Notice that the function $\Gamma_0(x,y)$ defined by (32) satisfies $\theta_0(\Gamma_0(x,y)) = \theta_0(f(x,0)) + B_0(y - h(x,0))$, which, substituting $y = h(\Upsilon(x,w),w)$, gives

$$\begin{aligned}
\theta_0(\Gamma_0(x, h(\Upsilon(x,w),w))) &= \theta_0(f(x,0)) + B_0(h(\Upsilon(x,w),w) - h(x,0)) \\
&= A_0\theta_0(x) + B_0 h(\Upsilon(x,w),w) && \text{[because of (31)]} \\
&= A_0\pi(\Upsilon(x,w),w) + B_0 h(\Upsilon(x,w),w) && \text{[because of (38)]} \\
&= \pi(f(\Upsilon(x,w),w), s(w)) && \text{[because of (22)]}
\end{aligned}$$

Hence,

$$\theta_0(\Gamma_0(x, h(\Upsilon(x,w),w))) = \pi(f(\Upsilon(x,w),w), s(w)) \tag{39}$$

On the other hand, equation (38), under the substitutions $x \to \Gamma_0(x, h(\Upsilon(x,w),w))$ and $w \to s(w)$, gives

$$\pi(\Upsilon(\Gamma_0(x, h(\Upsilon(x,w),w)), s(w)), s(w)) = \theta_0(\Gamma_0(x, h(\Upsilon(x,w),w)))$$

which, combined with (39), leads to

$$\pi(\Upsilon(\Gamma_0(x, h(\Upsilon(x,w),w)), s(w)), s(w)) = \pi(f(\Upsilon(x,w),w), s(w))$$

Taking into account the unique local solvability of the algebraic equation $\pi(x,w) = \zeta$ with respect to x, it follows that $\Upsilon(\Gamma_0(x, h(\Upsilon(x,w),w)), s(w)) = f(\Upsilon(x,w),w)$, which is exactly equation (28).

iii) The composite function $\Phi(x,w) = \eta(\Upsilon(x,w),w)$ satisfies

$$\begin{aligned}
\Phi(\Gamma_0(x, h(\Upsilon(x,w),w)), s(w)) &= \eta(\Upsilon(\Gamma_0(x, h(\Upsilon(x,w),w)), s(w)), s(w)) \\
&= \eta(f(\Upsilon(x,w),w), s(w)) && \text{[because of (28)]} \\
&= A'\eta(\Upsilon(x,w),w) + B'h(\Upsilon(x,w),w) && \text{[because of (23)]} \\
&= A'\Phi(x,w) + B'h(\Upsilon(x,w),w)
\end{aligned}$$

This proves (35). Moreover, since

$$\begin{aligned}
\frac{\partial\Phi}{\partial w}(x,w) &= \frac{\partial\eta}{\partial x}(\Upsilon(x,w),w)\frac{\partial\Upsilon}{\partial w}(x,w) + \frac{\partial\eta}{\partial w}(\Upsilon(x,w),w) \\
&= -\left[\frac{\partial\eta}{\partial x}(\Upsilon(x,w),w)\right]\left[\frac{\partial\pi}{\partial x}(\Upsilon(x,w),w)\right]^{-1}\left[\frac{\partial\pi}{\partial w}(\Upsilon(x,w),w)\right] + \left[\frac{\partial\eta}{\partial w}(\Upsilon(x,w),w)\right]
\end{aligned}$$

it follows that $\det\frac{\partial\Phi}{\partial w}(x,w) = \dfrac{\det\begin{bmatrix} \frac{\partial\pi}{\partial x}(\Upsilon(x,w),w) & \frac{\partial\pi}{\partial w}(\Upsilon(x,w),w) \\ \frac{\partial\eta}{\partial x}(\Upsilon(x,w),w) & \frac{\partial\eta}{\partial w}(\Upsilon(x,w),w) \end{bmatrix}}{\det\left[\frac{\partial\pi}{\partial x}(\Upsilon(x,w),w)\right]}$

Given the assumptions made, $\det\frac{\partial\Phi}{\partial w}(x,w) \neq 0$ in a neighborhood of $(x = 0, w = 0)$ and therefore the algebraic equation $z' = \Phi(x,w)$ is locally solvable with respect to w, as a result of the implicit function theorem.

VI. EXAMPLE

In the present section, the proposed observer design methodology will be illustrated through a simple example. Consider

$$
\begin{aligned}
x(k+1) &= \frac{x(k)-w(k)}{2+x(k)+\big(2x(k)+3\big)w(k)} \\
w(k+1) &= w(k) \qquad\qquad (40)\\
y(k) &= \frac{2x(k)}{1+x(k)}+\frac{w(k)}{1+w(k)}
\end{aligned}
$$

where x is the process state and w is a constant (step) disturbance, the objective being to design a nonlinear observer with linearizable dynamics with prescribed eigenvalues α_0 and α' ($|\alpha_0|<1$, $|\alpha'|<1$, $\alpha_0 \neq \alpha'$).

Applying the modular observer design procedure of Section IV, the following equations must be solved in a sequential manner:

$$\theta_0\left(\frac{x}{2+x}\right) = \alpha_0\theta_0(x) + b_0\frac{2x}{1+x} ,$$

$$\theta_0\left(\Gamma_0(x,y)\right) = \theta_0\left(\frac{x}{2+x}\right) + b_0\left(y-\frac{2x}{1+x}\right),$$

$$\Upsilon\left(\Gamma_0\left(x\,,\frac{2\Upsilon(x,w)}{1+\Upsilon(x,w)}+\frac{w}{1+w}\right),w\right) = \frac{\Upsilon(x,w)-w}{2+\Upsilon(x,w)+\big(2\Upsilon(x,w)+3\big)w} ,$$

$$\Phi\left(\Gamma_0\left(x\,,\frac{2\Upsilon(x,w)}{1+\Upsilon(x,w)}+\frac{w}{1+w}\right),w\right) = \alpha'\Phi(x,w)+b'\left(\frac{2\Upsilon(x,w)}{1+\Upsilon(x,w)}+\frac{w}{1+w}\right),$$

$$\Phi\big(\Gamma_0(x,y),\Gamma_w(x,w,y)\big) = \Phi\left(\Gamma_0\left(x\,,\frac{2\Upsilon(x,w)}{1+\Upsilon(x,w)}+\frac{w}{1+w}\right),w\right)+b'\left(y-\frac{2\Upsilon(x,w)}{1+\Upsilon(x,w)}-\frac{w}{1+w}\right).$$

It turns out that the results can be obtained in closed form:

$$\theta_0(x) = \frac{2b_0}{\frac{1}{2}-\alpha_0}\cdot\frac{x}{1+x},\quad \Gamma_0(x,y) = \frac{x+\left(\frac{1}{2}-\alpha_0\right)(1+x)\left(y-\frac{2x}{1+x}\right)}{2+x-\left(\frac{1}{2}-\alpha_0\right)(1+x)\left(y-\frac{2x}{1+x}\right)},\quad \Upsilon(x,w) = \frac{x-\frac{1}{2}\cdot\frac{\frac{3}{2}-\alpha_0}{1-\alpha_0}\cdot(1+x)\left(\frac{w}{1+w}\right)}{1+\frac{1}{2}\cdot\frac{\frac{3}{2}-\alpha_0}{1-\alpha_0}\cdot(1+x)\left(\frac{w}{1+w}\right)},$$

$$\Phi(x,w) = \frac{b'}{\frac{1}{2}-\alpha'}\left(\frac{2x}{1+x}+\frac{1}{2}\left(\frac{1}{1-\alpha'}-\frac{1}{1-\alpha_0}\right)\frac{w}{1+w}\right),\quad \Gamma_w(x,w,y) = \frac{\frac{w}{1+w}-2(1-\alpha')(1-\alpha_0)\left(y-\frac{2x}{1+x}+\frac{1}{2}\cdot\frac{1}{1-\alpha_0}\cdot\frac{w}{1+w}\right)}{\frac{1}{1+w}+2(1-\alpha')(1-\alpha_0)\left(y-\frac{2x}{1+x}+\frac{1}{2}\cdot\frac{1}{1-\alpha_0}\cdot\frac{w}{1+w}\right)}$$

The resulting modular observer is

$$\tilde{x}(k+1)=\frac{\tilde{x}(k)+\left(\frac{1}{2}-\alpha_0\right)(1+\tilde{x}(k))\left(y(k)-\frac{2\tilde{x}(k)}{1+\tilde{x}(k)}\right)}{2+\tilde{x}(k)-\left(\frac{1}{2}-\alpha_0\right)(1+\tilde{x}(k))\left(y(k)-\frac{2\tilde{x}(k)}{1+\tilde{x}(k)}\right)}$$

$$\hat{w}(k+1)=\frac{\frac{\hat{w}(k)}{1+\hat{w}(k)}-2(1-\alpha')(1-\alpha_0)\left(y(k)-\frac{2\tilde{x}(k)}{1+\tilde{x}(k)}+\frac{1}{2}\cdot\frac{1}{1-\alpha_0}\cdot\frac{\hat{w}(k)}{1+\hat{w}(k)}\right)}{\frac{1}{1+\hat{w}(k)}+2(1-\alpha')(1-\alpha_0)\left(y(k)-\frac{2\tilde{x}(k)}{1+\tilde{x}(k)}+\frac{1}{2}\cdot\frac{1}{1-\alpha_0}\cdot\frac{\hat{w}(k)}{1+\hat{w}(k)}\right)} \tag{41}$$

$$\hat{x}(k)=\frac{\tilde{x}(k)-\frac{1}{2}\cdot\frac{\frac{3}{2}-\alpha_0}{1-\alpha_0}\cdot(1+\tilde{x}(k))\left(\frac{\hat{w}(k)}{1+\hat{w}(k)}\right)}{1+\frac{1}{2}\cdot\frac{\frac{3}{2}-\alpha_0}{1-\alpha_0}\cdot(1+\tilde{x}(k))\left(\frac{\hat{w}(k)}{1+\hat{w}(k)}\right)}$$

The first state equation of the modular observer (41) is an identity observer based on the assumption of zero disturbance, the second equation is the disturbance observer, and the third equation is the corrector of the state estimates.
Alternatively, the modular observer can be designed by applying the result of Proposition 7. One must then solve the LFE's (22) and (23) which, for system (40), become

$$\pi\left(\frac{x-w}{2+x+(2x+3)w},\ w\right)=\alpha_0\pi(x,w)+b_0\left(\frac{2x}{1+x}+\frac{w}{1+w}\right) \tag{42}$$

$$\eta\left(\frac{x-w}{2+x+(2x+3)w},\ w\right)=\alpha'\eta(x,w)+b'\left(\frac{2x}{1+x}+\frac{w}{1+w}\right) \tag{43}$$

Again, the solutions of the LFE's can be obtained in closed form:

$$\begin{aligned}\pi(x,w)&=\frac{b_0}{\frac{1}{2}-\alpha_0}\left(\frac{2x}{1+x}+\frac{\frac{3}{2}-\alpha_0}{1-\alpha_0}\cdot\frac{w}{1+w}\right)\\ \eta(x,w)&=\frac{b'}{\frac{1}{2}-\alpha'}\left(\frac{2x}{1+x}+\frac{\frac{3}{2}-\alpha'}{1-\alpha'}\cdot\frac{w}{1+w}\right)\end{aligned} \tag{44}$$

from which, applying Proposition 7,

$$\theta_0(x)=\pi(x,0)=\frac{2b_0}{\frac{1}{2}-\alpha_0}\cdot\frac{x}{1+x},$$

$$\pi(\hat{x},w)=\pi(x,0)\ \Rightarrow\ \frac{b_0}{\frac{1}{2}-\alpha_0}\left(\frac{2\hat{x}}{1+\hat{x}}+\frac{\frac{3}{2}-\alpha_0}{1-\alpha_0}\cdot\frac{w}{1+w}\right)=\frac{2b_0}{\frac{1}{2}-\alpha_0}\cdot\frac{x}{1+x}\ \Rightarrow\ \hat{x}=\Upsilon(x,w)=\frac{x-\frac{1}{2}\cdot\frac{\frac{3}{2}-\alpha_0}{1-\alpha_0}\cdot(1+x)\left(\frac{w}{1+w}\right)}{1+\frac{1}{2}\cdot\frac{\frac{3}{2}-\alpha_0}{1-\alpha_0}\cdot(1+x)\left(\frac{w}{1+w}\right)},$$

$$\Phi(x,w)=\eta(\Upsilon(x,w),w)=\frac{b'}{\frac{1}{2}-\alpha'}\left(\frac{2\Upsilon(x,w)}{1+\Upsilon(x,w)}+\frac{\frac{3}{2}-\alpha'}{1-\alpha'}\cdot\frac{w}{1+w}\right)=\frac{b'}{\frac{1}{2}-\alpha'}\left(\frac{2x}{1+x}+\frac{1}{2}\left(\frac{1}{1-\alpha'}-\frac{1}{1-\alpha_0}\right)\frac{w}{1+w}\right)$$

i.e. the same results as before, leading to the same modular observer (41).

The system (40) and the observer (41) were simulated under the following initial conditions:

$x(0)=2, \qquad \tilde{x}(0)=1$

$w(0)=1, \qquad \hat{w}(0)=0$

These account for a step change in the disturbance w from 0 to 1, while the process state is initially away from equilibrium. The design parameters of the observer were selected as

$\alpha_0 = 0.4, \ \ \alpha' = 0.6$

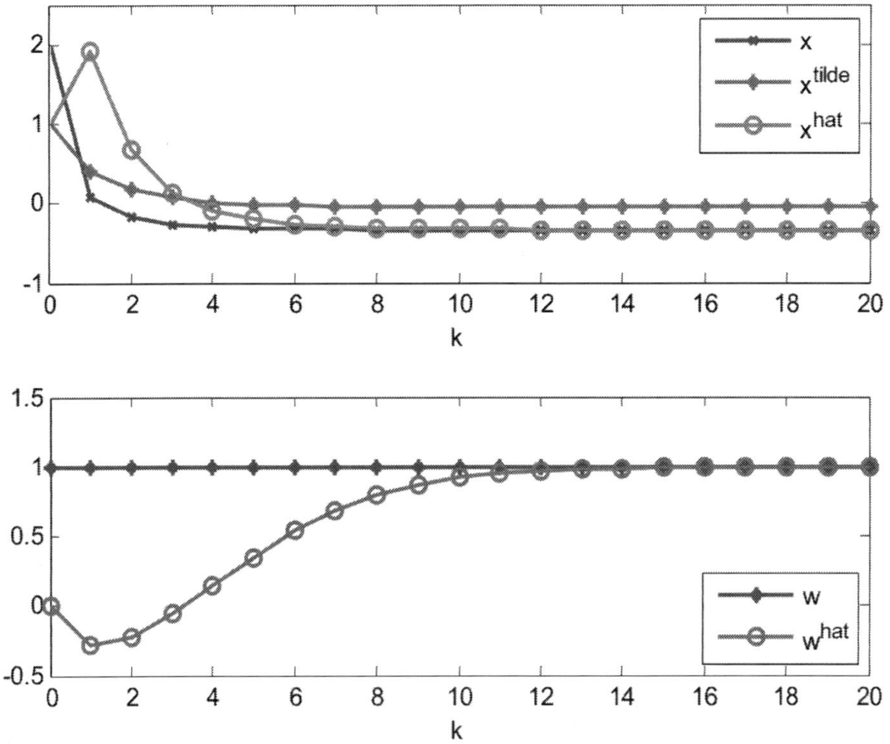

Figure 3: Representative responses for the example system (40) and the observer (41) with $\alpha_0 = 0.4$ and $\alpha' = 0.6$.

In the first plot in Figure 3, the response of the process state is compared with two alternative state estimates – one neglecting the disturbance ($\tilde{x}$), and one accounting for it ($\hat{x}$). Neglecting the disturbance gives rise to a permanent error (offset). The corrector part of the modular observer provides the necessary correction to completely eliminate the offset.

The second plot compares the applied step disturbance with the disturbance estimate. The disturbance observer part of the modular observer quickly detects the presence of the step disturbance and accurately estimates it without permanent error.

VII. Computational Issues

The example of the previous section is a rare case where the design equations can be solved in closed form. In general, the calculations have to be carried out following a numerical approximation scheme. The functional equations (31), (28) and (35) for the modular design or (22) and (23) for the simultaneous design, can be solved via a power series approach (postulating the solution in the form of power series with unknown coefficients, substituting it into the FE, equating coefficients of like terms and solving the resulting the algebraic equations for the unknown coefficients). Of course, the computations for the power series approach will have to be carried out symbolically and will have to terminate at a certain truncation order, large enough to provide an adequate approximation of the unknown solution. This approach has been successfully applied in [5] for the solution of the observer LFE in the absence of disturbances (equation (31)), which is of the same form as the observer LFE's for the simultaneous design in the presence of disturbances (equations (22) and (23)).

Questions of relative computational advantages of solving modular versus simultaneous design equations will arise in applications. At this point, there seems to be no clear general answer of whether direct solution of FE's (31), (28) and (35) is computationally advantageous over solution of LFE's (22) and (23) and subsequent application the of Proposition 7.

VIII. Modular Luenberger Observer for State and Disturbance Estimation in Linear Systems

In the linear case, where the process follows linear dynamics

$$\begin{aligned} x(k+1) &= Fx(k) + Pw(k) \\ y(k) &= Hx(k) + Qw(k) \end{aligned} \tag{45}$$

and the disturbances are generated by a linear exosystem

$$w(k+1) = Sw(k) \tag{46}$$

where F, P, H, Q, S are $m\times m$, $m\times \ell$, $p\times m$, $p\times \ell$, $\ell\times\ell$ matrices respectively, the modular form of the Luenberger observer is:

$$\begin{aligned} \tilde{x}(k+1) &= F\tilde{x}(k) + L_0\left(y(k) - H\tilde{x}(k)\right) \\ \hat{w}(k+1) &= S\hat{w}(k) + L'\left(y(k) - H\left(\tilde{x}(k) + V\hat{w}(k)\right) - Q\hat{w}(k)\right) \\ \hat{x}(k) &= \tilde{x}(k) + V\hat{w}(k) \end{aligned} \tag{47}$$

where V is the solution of the matrix equation

$$VS - (F - L_0 H)V = P - L_0 Q \tag{48}$$

and the gains L_0 and L' can be selected by eigenvalue assignment.

In particular, the choices

$$L_0 = T_0^{-1} B_0 \tag{49}$$

where T_0 is the solution of

$$T_0 F - A_0 T_0 = B_0 H \tag{50}$$

and

$$L' = R^{-1}(B' - T' L_0) \tag{51}$$

where T', R are the solutions of

$$\begin{aligned} T'F - A'T' &= B'H \\ RS - A'R &= (B' - T'L_0)(HV + Q) \end{aligned} \tag{52}$$

assign the observer eigenvalues to the ones of $\begin{bmatrix} A_0 & 0 \\ 0 & A' \end{bmatrix}$.

It is important to point out that the above standard modular observer design for linear systems emerges directly from the general design procedure developed in section IV for nonlinear systems. In fact, when the proposed design approach is applied to the system of (45) and (46), the result is

$$\theta_0(x) = T_0 x\,,\quad \Gamma_0(x,y) = Fx + L_0(y - Hx) \quad \text{where} \quad L_0 = T_0^{-1} B_0\,,$$

$$\Upsilon(x,w) = x + Vw$$

$$\Phi(x,w) = T'x + Rw\,,\quad \Gamma_w(x,w,y) = Sw + R^{-1}(B' - T'L_0)\left(y - H(x + Vw) - Qw\right)$$

where T_0, V, T', R are the solutions of (50), (48) and (52), i.e. exactly the same as the standard linear result.

IX. Conclusions

This work developed a systematic method for the design of discrete-time nonlinear observers for state and disturbance estimation, in the context of the observer linearization approach. The proposed design method leads to a nonlinear observer with modular structure. The results provide a direct generalization of the modular form of the Luenberger observer for linear systems.

Acknowledgements: Financial support from the General Secretariat of Research and Technology of Greece through grant 03ΕΔ605 is gratefully acknowledged. The author wishes to thank Georgios Savvoglides for his assistance in the calculations of Section VI.

REFERENCES

[1] C. Califano, S. Monaco and D. Normand-Cyrot, "On the observer design in discrete-time", *Systems Control Lett.*, Vol. 49, No.4, pp. 255-265, 2003.

[2] S. T. Chung and J. W. Grizzle, "Sampled-data observer error linearization", *Automatica*, Vol. 26, pp. 997-1007, 1990.

[3] B. Friedland, *Control System Design. An Introduction to State-Space Methods.* New York: McGraw-Hill, 1986.

[4] N. Kazantzis and C. Kravaris, "Nonlinear observer design using Lyapunov's auxiliary theorem," *Systems Control Lett.*, Vol. 34, No.5, pp. 241-247, 1998.

[5] N. Kazantzis and C. Kravaris, "Discrete-time nonlinear observer design using functional equations," *Systems Control Lett.*, Vol. 42, No.2, pp. 81-94, 2001.

[6] C. Kravaris, V. Sotiropoulos, C. Georgiou, N. Kazantzis, MQ. Xiao and A. J. Krener, "Nonlinear observer design for state and disturbance estimation," *Systems Control Lett.*, Vol. 56, No.11-12, pp. 730-735, 2007.

[7] C. Kravaris, "Modular design of nonlinear observers for state and disturbance estimation," in *Proc. 45th IEEE CDC*, San Diego, CA, pp. 4621-4626, 2006.

[8] W. Lee and K. Nam, "Observer design for autonomous discrete-time nonlinear systems," *Systems Control Lett.*, Vol. 17, pp. 49-58, 1991.

[9] W. Lin, and C. I. Byrnes, "Remarks on linearization of discrete-time autonomous systems and nonlinear observer design," *Systems Control Lett.*, Vol. 25, pp. 31-40, 1995.

[10] MQ. Xiao, N. Kazantzis, C. Kravaris and A. J. Krener, "Nonlinear discrete-time observer design with linearizable error dynamics," *IEEE Trans. Automat. Contr.*, Vol. 48, No.4, pp. 622-626, 2003.

[11] MQ. Xiao, "A direct method for the construction of nonlinear discrete-time observer with linearizable error dynamics," *IEEE Trans. Automat. Contr.*, Vol. 51, No.1, pp. 128-135, 2006.

Preventing Plant Problems Before They Occur

Douglas H. Lenz[1] (speaker), Daniel L. Chester[2]

[1] *FALCONEER Technologies LLC, 4455 Transit Road, Suite 3C, Williamsville, NY 14221*
[2]*Department of Computer & Information Sciences, University of Delaware, Newark, DE 19716*

Prepared for Presentation at the AIChE 2008 Annual Meeting
Information Technology: Smart Manufacturing Plants
Philadelphia, Pennsylvannia
November 17-21, 2008

ABSTRACT

The process industries are facing many challenges and are awash with data that is often wasted. The costs for raw materials, utilities and energy are rising rapidly with no end in sight. There is also global warming, shortage of technical skills, loss of knowledge from an aging workforce and other big challenges to face. To overcome these challenges, manufacturing sites must focus and invest in technologies to do more with the existing resources – data, people, and systems. Failure to do so may impair the organization's ability to survive. The need to get more value from their existing data is now a high priority.

The value of this data is well recognized because, as one BASF engineer says "the process (data) always tells on itself". FALCONEER Technologies smart plant solution takes this data and adds "wisdom" or intelligence to the raw information by using engineering models (dynamic mass and energy balances) to capture the plant's process knowledge and uniquely combine with the real-time process data. This helps operators, engineers, and managers to focus on the information that's most critical to preventing problems and optimizing operations and costs, making plants smarter, safer and more profitable.

A real case study from the power industry on both the benefits and barriers to this technology will be discussed. However, this case study is also relevant to many chemical, petrochemical, and other industries that own/operate their own steam plants or co-generation plants.

INTRODUCTION

FALCONEER Technologies has developed a patented smart plant technology called FALCONEER™ IV Process Performance Solution. This solution works with a wide variety of existing process control systems to increase the value of the information available to managers, supervisors, engineers and operators. This smart plant technology leverages the information used to design and operate the process with the real-time information from the process. FALCONEER™ IV takes the data from DCS and SCADA systems and adds wisdom or intelligence to the raw information. It further provides intelligence by using the plant's process knowledge combined with the raw process data. This intelligent system comprehensively accounts for ALL process information by concurrently auditing & alerting at the highest, process-wide level as well as the individual sensor level. The intelligent software manages all the information about how the process should run (design and/or optimum conditions) and currently is running (normal, problems, out of control or out of specification). It analyzes and presents this information continuously and in real-time to the engineer, supervisor, manager or operator, helping them to focus on the information that's most critical to optimal costs and operations. Tools that are normally used offline by these groups are incorporated into a comprehensive online auditing and advisory suite. These tools include alarming, charting, reporting, emails, real-time key performance indicators, and soft sensors. Identifying and preventing a single incident can often provide immediate payback for this software.

How do managing and analyzing this information and providing the right advice at the right time in a way that facilitates the decisions help the company and the engineer prevent plant problems before they occur or become significant? For example, companies have invested in systems (DCS, Data Historians, Networks, PCs) to generate, collect and save operational information while also using it to improve control over the their systems and plants. However, most of the time, this information is not used and simply wasted, thus losing the opportunity not only to increase the ROI on this capital investment, but to fully utilize the information to identify problems and abnormal conditions.

With a smart technology such as FALCONEER™ IV in place, most of this information management process is now automated, saving the engineer's time to best use his/her skills to provide the optimal energy utilization solution rather than manually sorting through data or sitting and watching the process to figure out what is happening. It is like having a virtual engineer or supervisor doing these tasks continuously to free up the real engineer or supervisor. The most acceptable smart technology automates the data COLLECTION and ANALYSIS using the existing knowledge of the process systems, not a black box approach. For example, FALCONEER™ IV uses the design equations, the mass and energy balance equations, key performance indicators. These equations and models capture the knowledge and expertise of the engineers, supervisors and operators. These are the equations and models that the engineer used to build the process and uses, when he/she has time, to evaluate the performance of the process or to identify how to optimize energy or other parameters and costs.

Finally, it is important that smart technology can also automate the process performance auditing and advising to MAINTAIN the improved and/or optimal conditions, for this example, energy utilization. In addition, when the process system is operating at desired or target conditions, the smart plant software will diligently audit and advise when deviations or problems or failures are occurring from these normal conditions. When detected, it will also identify the most likely causes so appropriate decisions and adjustments can be made to avoid problems and to maintain proper.

PATENTED AND PROVEN TECHNOLOGY

FALCONEER Technologies FALCONEER™ IV Real-Time Process Performance Solution (*FALCONEER™*) includes sensor and process condition validation, continuous statistical process control monitoring, and predictive fault analysis. This intelligent software thus provides a single source means for complete, multi-level monitoring and analyzing all manufacturing & environmental process data in real time to identify and prevent plant problems or to optimize current process performance. This enhanced capability for process control and information systems helps improve reliability, yield and quality, avoid failures and accidents, and reduce costs. The patented smart technology is designed with many automated configuration features to significantly reduce the cost, time, and resource commitment to implement and to maintain.

FALCONEER™ consists of several software modules with display screens. These include the State Identification (**State ID**) module, the **S**ensor **V**alidation and **P**roactive **F**ault **A**nalysis (**SV&PFA**) module, the **V**irtual **S**tatistical **P**rocess **C**ontrol (**VSPC**) module and a configuration interface. This application can operate as a standalone system or run as Web service with alarm screens and information screens as Web clients with the capability of OPC messaging to notify pagers, cell phones, etc. Below is a summary of the purpose and operation of the different modules that add intelligence to the control systems and instrumentation.

State ID: The software program first determines whether the process is operating within standard conditions or not. This is accomplished with the State Identification (State ID) Module. *FALCONEER* is idle if the process is not operating intentionally in normal or standard conditions. The program begins its analysis of sensor measurements once startup is complete. It continues this analysis until the process is shutdown and then is idle again until the next startup completes. The process performance solution thus runs continuously and adjusts its analysis to current process operations accordingly.

SV&PFA: The Sensor Validation and Proactive Fault Analysis (**SV&PFA**) software module monitors current process sensor measurements to determine if they are either valid or incorrect. It also determines if certain processing faults (leaks, pump failures, controller malfunctions, etc.) are occurring or not. It does this by evaluating engineering models that describe normal process operation. By definition, when the process system is operating normally (i.e., fault free), the engineering models describing normal process operation should all close. When they don't close, this is evidence that something is going wrong in the process system. By looking at the patterns of all this evidence it is possible to infer the underlying fault(s) in the process that could cause such behavior. These results are then given as alerts to the users. If not found to be faulty, the sensor measurements are considered validated (i.e., trustworthy).

It is very common that the SV&PFA module reports more than one plausible fault at the same time, i.e., many alerts occur simultaneously. This is referred to as the diagnostic resolution of the analysis. In those cases all alerts need to be checked out further to identify the actual process problem occurring. In other words, although each alert given is a plausible explanation of the current process situation, it still needs to be verified as actually occurring or not by the user. Thus, alerts are meant to identify all the possible operating problems that can explain the process behavior currently being encountered. These alerts are continuously updated as process conditions warrant.

The evidence determined by evaluating the engineering models is combined together using a Fuzzy Logic calculation. This calculation determines a certainty factor associated with each fault hypothesis. These certainty factors can range from 0 (at least some evidence does not support that fault hypothesis) to 1.0 (that fault hypothesis is a highly plausible explanation of the current process behavior). Appropriate alerts are given if these certainty factors exceed the chosen alert thresholds. These alerts or alarms indicate that process operations are no longer considered to be normal and the identified fault may be the reason why. These threshold values, which are configurable, are chosen to be highly certain that those associated alerts are real. They make the SV&PFA module results conservative by not alerting until process conditions are well outside their normal operating ranges. The SV&PFA module should thus give few false alarms.

VSPC: The **V**irtual **S**tatistical **P**rocess **C**ontrol (**VSPC**) software module augments the SV&PFA module, providing an independent but complementary analysis of sensor measurements. The VSPC module uses Exponentially Weighted Moving Averages (EWMA's) as well as more traditional SPC approaches to determine in real-time if individual process sensors and process conditions are in control, are going out of control, or are definitely out of control. It directly flags out of control sensors and process conditions in real-time rather than after-the-fact. These alerts can also occur at levels that may allow the process operators to take appropriate control actions to reduce or eliminate disruptions to process operations or product quality. This method is considered virtual because the analysis is done automatically without the need for the operators to collect and chart any process sensor readings.

Statistical Process Control (SPC) is a tool used to assess whether a process is currently under or out of control. Various techniques exist for doing this analysis depending upon the nature of the process being monitored. In continuous processes (as opposed to the manufacture of discrete, individual units), process data collected at a particular moment in time is not completely independent of its previous data. This phenomenon is referred to as auto-correlation between the data. SPC techniques for continuous single point samples that handle auto-correlation in the data include Cumulative-Sum (CUSUM) as well as EWMA. These techniques can better handle the effects of auto-correlation and allow small but statistically significant shifts in an observed variable's value to be readily detected. With this smart technology for continuous processes, the EWMA calculation is the preferred method because of its ability to not only monitor current process operations but to forecast where the process is headed. There are two classes of sensor variables that can be used by this module: sensors that are being directly controlled and those that are not.

Since this analysis is based solely on statistical calculations, false alarms are possible. Measurements may be identified as out of control and then rapidly determined to be in control once again. Such spurious alerts can be minimized by choosing update times for the analysis that are close to one time constant for the process being monitored. However, they cannot be totally eliminated. Therefore, VSPC alerts also need to be further confirmed by the users before any control actions based on them are taken.

Combined together in an integrated smart solution, the modules provide a very complete diagnostic tool for identifying likely causes of pending problems in a timely manner so that they can be prevented. Certain problems are more easily detected by one or the other module depending on its sensitivity to the sensor measurement. In the case where the SVPFA module may identify several possible root causes, most often the correct root cause is also concurrently identified by the VSPC module.

POWER UTILITY PLANT DISCUSSION AND RESULTS

The most persuasive way to demonstrate the power of this smart technology for identifying and preventing plant problems is with actual examples. For this paper, an example will be used from a power plant that generates steam to operate turbines for electricity generation. This case study is also relevant to many chemical, petrochemical, and other industries that own/operate their own steam plants or co-generation plants. *FALCONEER™* was configured and installed on the incinerator and boiler units at the Power Utility Customer's facility as shown in the figure below.

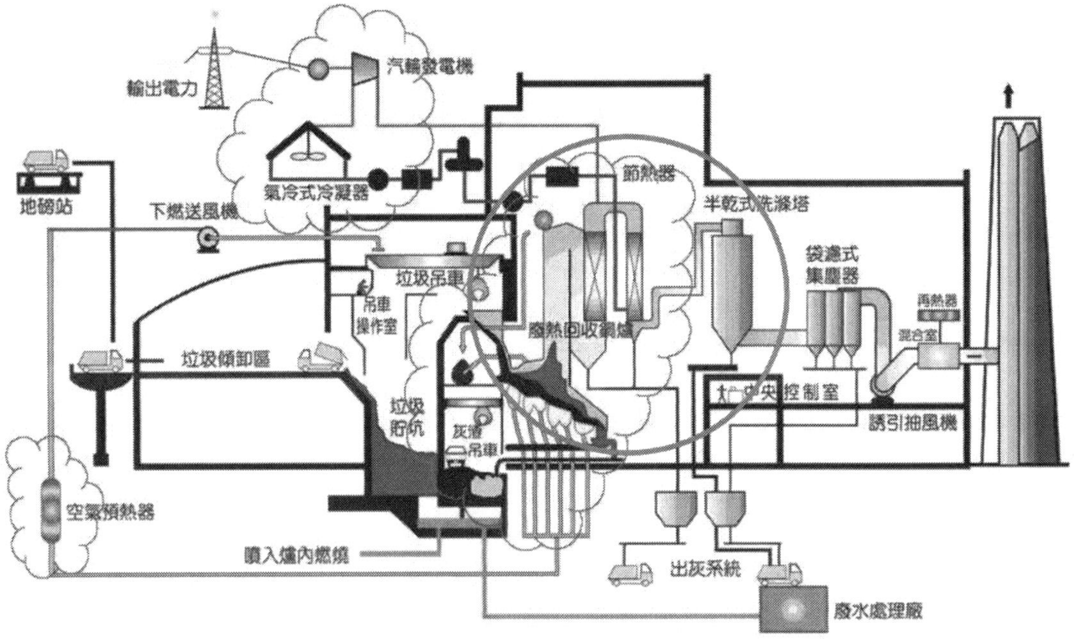

The Power Utility Customer's facility uses a Yokogawa CS3000 DCS system for control and data collection. *FALCONEER™* pulled the data from this system into its OPC Client/Server for analysis every 5 minutes.

For the pilot, *FALCONEER™* targeted 4 process areas including the incinerator, boiler, economizer and turbines to help the Power Utility Customer in a number of ways, including:

- Saving operational costs by improved real-time on-line monitoring of key dynamic performance indicators, such as energy & natural gas costs and raw material usages. Examples include
  - Tracking Low Heating Value (LHV) KPI deviations for burning control for efficiency and stability improvements due to fuel quality.
  - Tracking Urea and Lime usage KPIs, which may help optimize raw material use and reduce costs for unnecessary chemical addition.
  - Tracking Power Generation Efficiency KPI.
- Providing timely alerts of pending sensor, equipment or process failures or abnormal conditions to avoid unscheduled shutdowns
  - Tracking Boiler Leak Detection KPI to identify possible tube leaks requiring scheduled maintenance and avoiding unplanned failures and shutdown.
  - Reducing pipe corrosion inside the furnace.
  - Tracking Turbine Performance KPIs for maintaining good equipment performance and operation conditions.
- Validating sensor data & process information to help with control decisions and to meet environmental and safety reporting requirements.
- Tracking Incinerator Operation Stability and its associated Indices related to Incinerator Temperature, Steam Production, Boiler Output O2, Stack CO, SOx, NOx, HCl
- Avoiding environmental release or process safety events from sensor or process failures.
- Saving costs from avoiding unnecessary preventative maintenance instrumentation calibration.
- Improving steam & power generation by avoiding significant fluctuations from uncontrolled process variables.

To date, the Power Utility Customer operations have already begun to realize these benefits in several important areas – 1) cost savings from KPI advising & reporting, 2) instrument fault and abnormal condition identification and reporting. The result is more power generated from the same amount of steam produced at a lower cost. More power to sell brings in higher revenues. Higher revenues with reduced costs yield greater profits!! This case study will only summarize the fault and abnormal condition identification and reporting, which are key to preventing problems before they occur or create significant issues.

Based on the P&ID diagrams and the generated historical data, models of mass balances, energy balances and process specific equations were created. Examples of these models include

- Ideal Gas Model Overfire Air Outlet
- Overall Mass Balance around Boiler Drum
- Overall Mass Balance First Preheater
- Steam & Water Mass Balance on Boiler
- Low Pressure Steam Overall Energy Balance
- Economizer Overall Energy Balance

With these models, process and instrument fault diagnosis and analysis is achieved for the combustion system, the boiler system, feed water system, and the economizer system, with total of 59 critical tags covered with both the SVPFA and VSPC modules. When an abnormal situation occurs (sensor failure, process upset, equipment problems) involving the measured tag, *FALCONEER™* uses the model residuals and patent pending fuzzy logic estimation results to distinguish the possible abnormal tags. At the same time, it can determine the faulty tags' estimated value for operators to review in order to prevent operators from continuing to use the wrong measured values. Using wrong values can cause operators to make wrong operational decisions leading to poor and costly operation and/or high emissions. Early detection of the abnormal issues for tags allows operations and maintenance time to investigate or adjust the corresponding tags' reading or related equipment and control values. This timely response helps reduce the likelihood of unscheduled (and costly) system shutdowns or outages. In all cases, plant problems can be identified sooner and mitigated or prevented

FALCONEER™ can be configured to include unmeasured variables such as leaks in process lines and tanks. Standard practice is to configure inlet stream leaks downstream from the flow meters and outlet stream leaks upstream from the flow meters. Such a leak detection system was created for the Power Utility Customer. For the purpose of boiler leak detection, *FALCONEER™* system setup virtual leak indicators as unmeasured variables and used these unmeasured variables in the mass balance models in order to help to identify possible boiler pipe leak conditions.

After completing the model configurations, several simulation tests, both offline and online, were performed to validate the fault diagnosis functions. These test results indicated that the *FALCONEER™* system can successfully provide timely and correct proactive fault diagnosis while validating instrument performance.

FALCONEER™ PROCESS VALIDATION & ABNORMAL CONDITION DETECTION RESULTS

For real-time validation and abnormal condition or fault detection, *FALCONEER™* is equivalent to a very intelligent engineer or operator looking at all the information and evidence. If the process conditions exist such that the evidence is clear and the cause of the fault can be found quickly, then *FALCONEER™* will find it as would a person looking through the data if they had the time. If the process conditions exist such that the evidence is weak or there are other problems that might be confusing or slowing down the analysis, *FALCONEER™* will eventually find the problem whereas a person might not or would take a long time to find the problem if they were actually looking.

Subsequently, the opportunity to use *FALCONEER™*'s additional troubleshooting and optimization capabilities was also evaluated. Live testing demonstrated *FALCONEER™*'s ability to effectively identify and communicate pending problems and abnormal conditions before the current control system alarms would detect these situations. During the duration of the pilot, *FALCONEER™* was able to validate critical process instrumentation and identify several examples of both abnormal conditions or failures or changes to the process.

In each case, faults and problems were identified correctly by one or more combinations of *FALCONEER™* integrated methods and modules:

- By Virtual Statistical Process Control (VSPC) Analysis
- By Sensor Validation & Proactive Fault Analysis (SVPFA)
- By both SVPFA & VSPC alarms when other possible causes do not have VSPC alarm

- By SVPFA & Estimated Value/Data Reconciliation analysis when other possible causes do not have reasonable estimated values assuming they were actual fault.

In all cases, detection and alarming occurred and was reported in less than an hour from the start of the problem or abnormal condition. This response is quicker than operations would likely have detected the problem unless they were lucky enough to be looking at that particular information at the right time and recognized the event before it triggered a High High or Low Low control system alarm.

A) Fault and Abnormal Condition Testing & Reporting

First, one important capability that the Power Utility Customer wanted was to identify and advise on instrumentation and process problems, failures, abnormal conditions in advance of control system High High or Low Low alarms that might also be associated with interlocks or lead to costly unplanned outages. The critical instruments, the conditions tested and the test objectives were the following.

Test	Test Conditions and Objectives	Tag
1	Simulate a leak from a pipe from boiler that doesn't contain any measurement instrument and hence, in the DCS operation control system, the balance of feed water and generated steam is messed up. The purpose for doing this is to have *FALCONEER™* generate both leak alarms and KPI3 leak detection BEFORE DCS generates alarm associated with the FIC-06101A (there are no DCS leak detection alarms).	**FIC-06101A**
2	Generate abnormal low feed water flow from slowly declining (due to plug, instrument drift, poor control, etc.) flow so that *FALCONEER™* can generate alarm before the DCS low-low alarm to occurs due to the low feed water flow. In addition *FALCONEER™* should estimate the actual flow value for operations to use.	**FIC-06101A**
3	Generate primary air outlet temperature (TI-05114) abnormal condition or fault (especially low temperature condition) and the related primary air preheater flow (FIC-06106) condition or fault. The purpose for doing this is to have *FALCONEER™* generated alarms for TI-05114 and FIC-06106 separately before the DCS alarms.	**TI-05114** **FIC-06106**
4	Generate the economizer flue gas outlet temperature (TICR-07113) abnormal high condition or fault. The purpose for doing this is to have *FALCONEER™* to point out the TICR-07113 problem and calculate its actual temperature before the TICR-07113 temperature causes the DCS high-high alarm to occur.	**TICR-07113**

Two types of testing were conducted.

Simulation Test:

The test simulates the process operating at well controlled or ideal process operating condition (all model are well behaved and close). This test is done by manually changing the desired test tag value on the OPC server (using KEPware Quick Client) and keeping all other tags' values unchanged. The purpose of this test is to show the *FALCONEER™* system's ability to diagnosis the fault tag and generate the SV&PFA and VSPC alarms as well as give a reasonable estimation or reconciliation of the actual field value. The simulation test results indicated that at present established models can successfully and quickly provide correct SVPFA diagnosis and VSPC analysis for all related problems.

Online Test:
Using current normal operating process conditions, the tag's value is changed manually on the OPC bridge (by multiplying the current DCS value by a factor to simulate abnormal differences from current DCS's value). The other tags' values vary as the current operating process conditions normally would behave. This testing is used to test whether or not *FALCONEER™* system can detect the fault and generate SV&PFA and VSPC alarms as well as give a reasonable estimation value under more realistic operational conditions.

TEST 1 (FIC-06101A: Boiler Feed Flow)

When FIC-06101A fault condition is created, this condition simulates both a flow or instrument problem and/or a possible boiler leak problem (if there is a pipe leak, then the control system would maintain the level balance and hence cause FIC-06101A to increase), *FALCONEER™* system's diagnosis result and simulation result is consistent – it identifies the fault correctly.

However, at the same time *FALCONEER™* generates LeakF06106, LeakF06102B as well as Leak F06101 SVPFA alarms. These alarms indicate that there are possible pipe leaks and simultaneously *FALCONEER™* system provides reasonable estimates of the size of these pipe leak quantities (around 3000~4000 Kg/Hr).

Test 1 results indicated that *FALCONEER™* system is capable to detect the boiler leak and or instrument problems, thus helping to prevent ensuing plant problems.

TEST 2 (FIC-06101A: Boiler Feed Flow)

When FIC-06101A slow declining failure condition is created, *FALCONEER™* system can immediate detect VSPC alarms. As the factor is decreased to simulate this abnormal condition, the SVPFA module's certainty factor is slowly increasing as it should. Finally, *FALCONEER™* system can generate the SV&PFA alarm and maintain this alarm state until the testing is over.

Simultaneously, *FALCONEER™* system can precisely detect the fault and no other fault alarms are generated. These alarms are well before DCS high and low alarm conditions, thus helping to prevent ensuing plant problems.

TEST 3 (TI-05114: Primary Air Outlet Temperature & FIC-06106: Primary Air Preheater Flow)

When TI-05114 high fault condition is created, *FALCONEER™* system can precisely detect the fault and generate the SV&PFA as well as VSPC alarms. However, when the TI-05114 low fault condition is created, *FALCONEER™* system can detect this fault, but at the same time generates other possible fault alarms that can also be explained by the current process condition evidence in the same process (secondary air heater).

At the same time, however, TI-05114 is the only tag that generates a VSPC alarm and these combined alarms correctly identify the most likely cause of the process behavior. So the TI-05114 low alarm, even with the other process fault alarms, can indicate possible process problem and abnormal condition issues.

When FIC-06106 is 15% higher than the real value, *FALCONEER™* system can precisely detect the fault and generate SV&PFA as well as VSPC alarms. When FIC-06106 value is only 10% lower than the real value, *FALCONEER™* system can detect the fault but also generate other process fault alarms (FIC-06101A and other Leak tags). Nevertheless, these alarms are consistent to the boiler system's pipe leaking condition; therefore, it can be thought as some valuable or useful alert to guide efficient troubleshooting.

Test 3 results indicated that *FALCONEER™* system can detect faults and abnormal conditions before the DCS High and Low alarms, thus helping to prevent ensuing plant problems.

TEST 4 (TICR-07113: Economizer Flue Gas Outlet Temperature)

When TICR-07113 high fault condition is created, *FALCONEER™* system can precisely detect the fault and generate SV&PFA as well as VSPC alarms. When a low fault condition is created, only a VSPC alarm is generated, which is sufficient. However, when a lower fault condition is created, *FALCONEER™* system can detect the fault and generate SV&PFA as well as VSPC alarms, but two additional fault alarms (AIR-07113) and (TI-06101A) have been generated that can also be explained by the current process condition evidence.

At the same time, however, TICR-07113 is the only tag that generates a VSPC alarm and these combined alarms correctly identify the most likely cause of the process behavior. So the TICR-07113 low alarm, even with the other process fault alarms, can indicate possibility of process problem and abnormal condition issues in a timely manner for problem prevention.

Test 4 results indicated that *FALCONEER™* system can achieve TICR-07113 high fault diagnosis, but the low fault's diagnosis precision needs further improvement if this condition becomes important for early detection (it currently is not critical).

Conclusion:

For the 4 critical conditions tested with the *FALCONEER™* system through simulation test and online test, *FALCONEER™* system can detect the related tag or process problems and generate VSPC as well as SV&PFA alarms. Simultaneously, it can calculate the estimated or reconciled value that is consistent enough to the real value to be used for control or manual operation decisions. Testing results proved that *FALCONEER™* system related functions or tools can help the Power Utility Customer save unnecessary expenses, create even higher efficiency, and help to reduce the Power Utility Customer safety and operational risks at the same time – all while helping to prevent problems before they occur or become significant.

B) Actual Examples of Fault or Abnormal Condition Detection and Reporting:

Finally, two specific examples are provided of how *FALCONEER™* features, dashboards and charts help identify abnormal conditions and problems in a timely manner so operators and engineers can adjust or address the situation.

In the first example, there was a significant change in the variability of FI-05117 (Low Pressure heating steam flow of steam air preheater) and FI-05118 (High Pressure heating steam flow of steam air preheater). *FALCONEER™* intelligent advisory alerts correctly identified this problem. It seemed to be directly related to a significant change in the variability of the "Total Heat Input to the Incinerator" – a calculated tag not previously available to operations, which is a factor for the incinerator control. This situation is shown in the Figure below using a *FALCONEER™* multi-point trending chart.

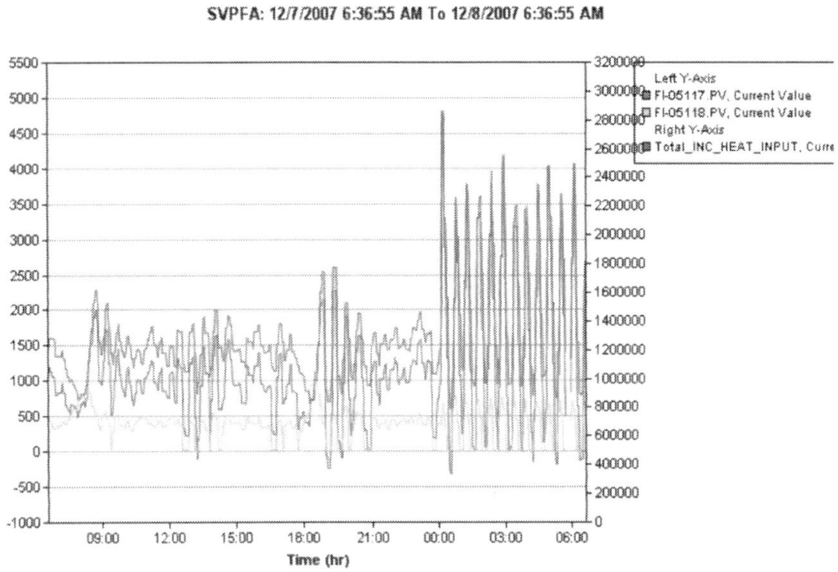

In the second example, one of three tags associated with a diagnostic model seemed to be significantly different in December and January than the original historical data set from July. The explanations for this condition were either a fault, failure, abnormal condition or approved changed from previous performance. Using a combination of dashboard alarms and alerts with charting and trending features, this situation was discovered by *FALCONEER™* for the following tags: FI-05151 - Air flow on overfire air fan inlet; FIC-05155 - Overfire air flow on BL side; FIC-05156 - Overfire air flow on BU side.

FALCONEER™ was telling operation that either FI-05151 was significantly or abnormally higher than expected or FIC-05155 and/or FIC-05156 were significantly or abnormally lower than expected or there was a leak in the system.

In this case, FI-05151 was significantly higher (10,000 to 11,000 NM3/hr) than its historic average (7,000 to 8,000 NM3/hr), since it should be the sum of the two controlled flows. In fact, the Power Utility Customer engineers therefore found this tag to be reading incorrectly and recalibrated it for proper operation.

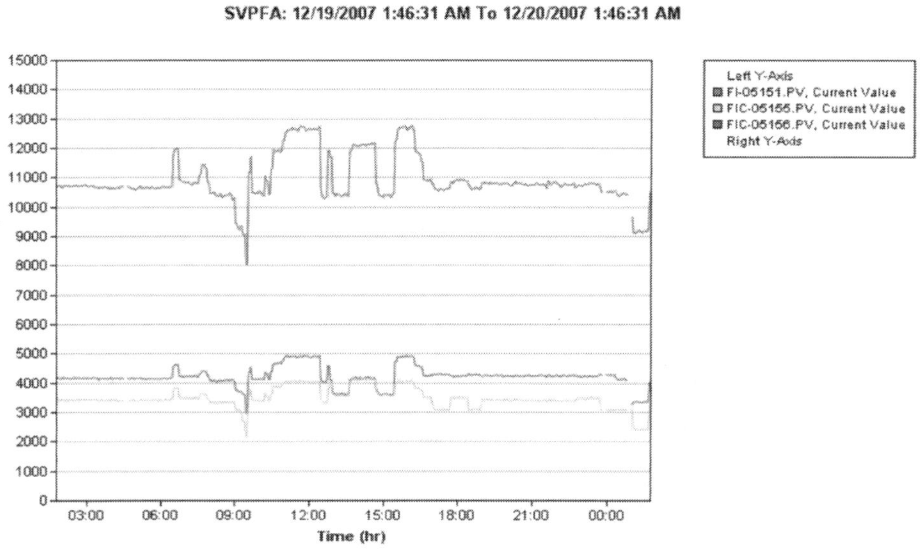

SUMMARY

The smart technology described comprises several modules providing continuous process performance auditing and optimization, sensor and process abnormal condition monitoring and detection, and sensor and process validation and reliability. All work together in a patented solution that can help to detect and prevent plant problems.

Simulated faults or problems were created for several critical instruments. In all cases, detection and alarming occurred and was reported in less than an hour from the start of the problem or abnormal condition. This response is quicker than operations would likely have detected the problem unless they were lucky enough to be looking at that particular information at the right time and recognized the event before it triggered a Hi Hi or Lo Lo control system alarm.

Several abnormal conditions have been detected and reported to operations. In some cases corrections or adjustments by maintenance or operations were made in a timely manner. Several instruments were found to require recalibration or adjustments to their control parameters compared to expected normal operational performance.

249b The Evolution of the Smart Plant In the Industrial Gas Business

Larry Megan, Praxair, 175 East Park Drive, PO Box 44, Tonawanda, NY 14150

Industrial gas plants have been in operation for over 100 years. The industrial gas business is extremely energy intensive - several billion dollars are spent annually to operate these plants. Today, more than ever, it's critical to operate these plants in a way that minimizes their electricity and natural gas consumption. Over the past 20 years, tremendous strides have been made towards the goal of the smart plant, including such things as automated start-up and shutdown, real time customer demand tracking, and the optimization of several plants within a network. However, there is still much to accomplish. This paper will review past achievements and future needs for the industrial gas industry and, more importantly, address the concept of the Smart Supply Chain - a tightly integrated network of energy suppliers, industrial gas suppliers and end customers.

249c Early Warning, Cloud Computing & Semantic Modeling in Smart Manufacturing

Son X. Huynh, Industrial Sector Solutions, IBM, 1475 Phoenixville Pike, West Chester, PA 19380 and **David L. Haake**, Global Business Services (GBS), IBM, Two Riverway, Suite 1500, Houston, TX 77056-1949.

Smart Manufacturing (SM) is regarded a vital step toward optimal and more cost efficient operations with respect to future offshore oil platforms, oil fields and chemical / refinery production. A reliable and functional Early Warning (EW) technology is a fundamental condition for a successful intelligent oilfield and smart-plant operations. Such event early warning system can give the operators several operational options and various degree of freedom to act in advance to identify false alerts and to prevent unplanned production deviations and shut downs.

Today, although large quantities of data are being collected at production facilities, it is difficult to obtain insight from this data in order to support actions to improve production. While manual approaches to visualizing, characterizing, and understanding data can in principle be effective, in practice a manual approach is far too labor intensive in the current data-rich environment. Automated methods are essential to help focus limited human resources and expertise and to improve decision making process.

In this paper, we will review how new emerging technologies in advanced data analytics, pattern recognition, semantic data modeling and knowledge management - can collect, capture, predict, prevent and correct plant problems even before they occur. It also examine the progress of Grid Computing to today's Cloud Computing environment of how they would impact IT Infrastructure in to manufacturing industries to deliver and support data-intensive applications.

At the conclusion, we will discuss on industry best practices to implement these the technologies with Business Process Modeling.

SMART PROCESS MANUFACTURING – A VISION OF THE FUTURE

James F. Davis, UCLA and Thomas F. Edgar University Texas-Austin

There are a number of forces driving the U.S. chemical industry in the 21st century, including shareholder return, globalization, efficient use of capital, faster product development, minimizing environmental impact, new ways to realize economic value, improved and more efficient use of research, workforce transitions and efficient use of people. As the chemical industry responds to these forces and tries to redefine and achieve these goals, it is investigating the expanded use and application of new computational technologies employed in areas such as modeling, computational chemistry, design, control, instrumentation, and operations. The key technology driver over the past 20 years has been the continuing advancements in digital computing. The 100-fold increase in computer speed each decade, the concomitant reductions in the cost of storing data, the significantly increased capability of high fidelity multiscale models, the advances in networking and connectivity, and the expansive potential of integration regardless of geographical location have tremendously increased the scope of computer applications in chemistry and chemical engineering. The new paradigm that encompasses the process systems engineering (PSE) topics mentioned above is *smart process manufacturing*, which will require the development of computational research environments to support such activities, also known as *cyberinfrastructure*.

In April 2008, a workshop comprised of industrial practitioners and academic researchers in process systems engineering was held at the NSF headquarters in Arlington, VA. The objective of the Smart Process Manufacturing workshop was to provide much-needed focused collaboration through facilitated assessment and creation of the base foundation for a roadmap. The workshop was sponsored by the National Science Foundation under a grant to the CACHE Corporation on behalf of the Smart Process Manufacturing Engineering Virtual Organization (SPM-EVO). The paper is drawn from the workshop documentation and reflects the collective views of the industrial and academic participants.

Smart process manufacturing (SPM) refers to a design and operational paradigm involving the integration of measurement and actuation, safety and environmental protection, regulatory control, high fidelity modeling, real-time optimization and monitoring, and planning and scheduling. SPM is the enterprise-wide application of advanced technologies, tools, and systems, coupled with knowledge-enabled personnel, to plan, design, build, operate, maintain, and manage process manufacturing facilities. This framework provides the basis for a strong predictive and preventive mode of operation with a substantially more rapid incident-response capability, ensuring safe and health-conscious operations. Driving toward zero incidents and zero emissions in the smart manufacturing paradigm recognizes that energy usage, energy production, safety, and environmental impact are inextricably linked.

What are the attributes of SPM? SPM systems have intelligent actions and possess a learning process to decide appropriate actions including the ability to actually implement that intelligent action. SPM systems are self-aware and proactive. They can adapt to new situations or perturbations (abnormal situations), and, by using feedback, smart systems can adapt or evolve for better performance. People are essential to SPM but play a more significant role than before. SPM systems integrate human expertise in new ways that allow for much more strategic, real-time decision making. In SPM pertinent information related to system performance is available, accessible, timely and appropriate and is understandable to the various parties or functions that need the information. Plant assets are integrated and self-aware (via sensors) of their state. Assets may be many things: plant, equipment, knowledge, experience, models, or data properties. Field devices have CPU and sensors, recognize their condition, and publish that information so they or other devices can take appropriate actions. A typical device receives, processes, translates, and publishes or shares information.

Specific process status variables can be related to larger business issues (profitability, zero incidents, zero emissions, etc.), i.e., beyond specific process and cross-industry impacts. Sustainable smart manufacturing includes reuse, with a life-cycle view of products and processes and a minimum environmental footprint (energy, water, emissions). Human resources (people) have to be knowledgeable, trained, empowered, connected (via cyber initiatives), and able to adapt/improve the system's performance.

The Figure below depicts a functional framework developed by PSE researchers from industry and academia for discussing SPM and how it is differentiated from current practices.

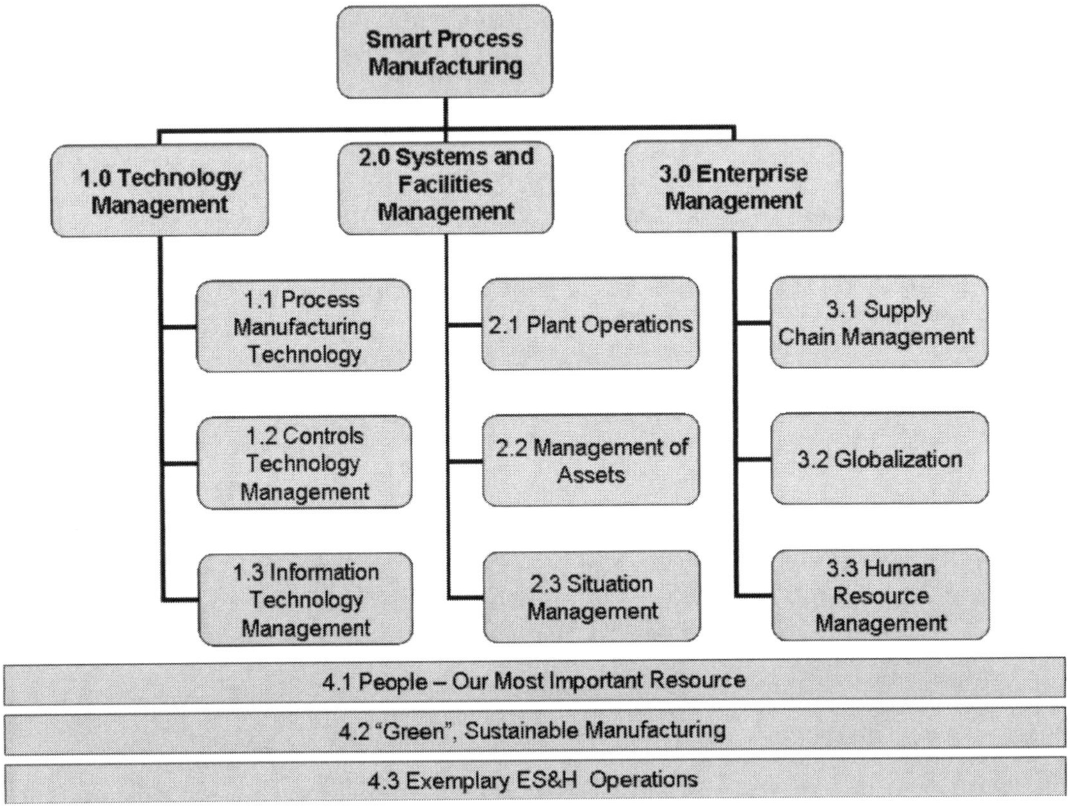

The Smart Process Manufacturing functional model provides a hierarchical, logical framework for analysis of technology research and development requirements.

Technology Management

Technology management addresses the determination of all the technological resources required to sustain, protect, and improve operations of the manufacturing enterprise, and includes the subareas of process manufacturing technology, control technology management, and information technology management. Typically, many types of technology are involved, and they act as enablers for both local and cross-enterprise/global issues. Key components are specialized processes and systems, tools, control systems, modeling capabilities, and the people whose knowledge and practical expertise make the processing successful.

Increased use of technology has led to mountains of data and associated data management problems. However, the relationships of data to underlying processes and material properties are not well understood; the data are not integrated between systems and so important data are not fully exploited. The challenge is to understand relevant technologies and their associated data well enough to target acquiring and processing only the needed data. There is, however, increasing use of models and efforts to grow their scope, interoperability, and applicability to production operations – not just design. Technology investment decisions tend to be made with too short-term a view, expecting a payoff too quickly. Operations decisions tend to be highly reactive instead of proactive.

Smart plants will be developed, designed, and operated using molecularly-informed engineering and will operate in a robust fashion. Models and all associated knowledge will be maintained and enriched as part of the plant's routine operation. Models will be developed to the level of detail needed, including multi-scale modeling to achieve the higher fidelity needed for some functions. Business goals will be directly translated into technology plans. This will enable flexibility of operations, robustness, agility, and the ability to change product streams quickly (especially in batch) or to make grade changes (for continuous processing).

Systems and Facility Management

Systems and facility management encompasses the oversight and assurance that the assets of the company and the enterprise are available to execute all needed functions within the defined operating envelope. Subareas include plant operations, asset management, and situation management.

One of the largest barriers to continuous improvement is the need to maintain current production and not change anything. There needs to be more of a balance between keeping the plant at constant operations vs. improving the profitability. The application of computing power to analyzing system performance has been limited to personal computers. Applying advanced technologies such as model predictive control and real-time optimization has been successful for many continuous petrochemical plants but is not applied as much to batch processes and more specialized products. There are concerns about the cost of supporting and maintaining such technologies. An awareness of the value of computing technology does not uniformly exist in process manufacturing and there needs to be more meaningful metrics for valuing different technologies.

Smart Process Manufacturing systems and facilities will be managed for optimum availability and performance. Continuous sensing and control, at the appropriate level based on the need, will assure safe, optimized performance with the realization of zero incidents and zero emissions. Models will accurately predict plant performance and will support automated and computer-assisted decision-making for assurance of operation within the envelope for safety, environmental responsibility, and cost optimization.

Enterprise Management

Enterprise management takes an integrated view of all enterprise activities, from process loops (perhaps in many facilities within a single company) up through strategic direction setting. It includes multiple plants working together, interaction between companies plus the global view, and integration and optimization of processing and business functions. Subareas include supply chain management, globalization, and human resource management. Business planning determines the right business/product mix and how products and technologies should evolve. Achieving smart manufacturing capabilities in the process manufacturing industry will require a much higher level of integration, flexibility, and adaptability in enterprise management than exists today. Although many business functions and many manufacturing functions have been integrated in large, complex software systems (ERP/ERM, PDM, etc.), there is still a lack of interoperability of business and technical functions across the enterprise. Furthermore, there is no direct connection between the manufacturing production goals and the larger business objectives of the enterprise. As the enterprise extends its relationship with supplier partners, there is a lack of consistent understanding of quality, safety and environmental standards across the supply chain, especially on a global basis. In addition, as companies become more closely integrated with their supply network and global partners, there is still a reluctance to share information freely and risk losing the company's intellectual property.

Another challenge to enterprise management is lost expertise as the industry's aging workforce has the potential of creating a loss of critical knowledge by attrition. Companies find it difficult to acquire needed skill sets and hard to attract skilled workers for some jobs, such as facility management. University curricula are not producing sufficient numbers of graduates with the needed practical and technical skills, so there is an ongoing dialogue on how U.S. universities should address this need.

In the future enterprise management tools, cost-effective technologies and cross-industry standard practices will enable U.S. manufacturing companies to successfully collaborate and compete in the global economy. Beneficial technologies and productivity improvements will spread across global industry via the development of industry-wide standards. Interoperability of systems will enable use of the complete set of plant and enterprise data in decision-making. ES&H concerns will be effectively addressed by collective cross-industry practices on a global basis, enabled by the better tools and technologies that comply with industry standards. Cybersecurity technology will automatically provide protection of intellectual property, and liability and trust issues will be resolved to the extent that cross-industry relationships are encouraged. The cost-effective, flexible and adaptive operation that results from these improved technologies will ensure U.S. competitiveness in the global market.

Green Sustainable Manufacturing

Green, Sustainable Manufacturing in an SPM facility make environmentally sound practices become automatic, which are part of the business drivers that guide all operations. The benefits of improved environmental performance on manufacturing operations can extend beyond improving the quality of the environment and the need to take a life cycle perspective.

Exemplary Environmental, Safety, and Health Operations

Exemplary ES&H Operations beyond ES&H regulatory compliance is a prominent business driver to "zero incident" operations, where the goal is to have no negative impact on personnel, surrounding communities, or the environment in general. Smart plants proactively prevent environmental, health, and safety problems while they at the same time seize opportunities to optimize operational and financial performance and monitor environmental conditions, and any aberrations are immediately noted and mitigated.

249e Smart Manufacturing Plants: Advances and Priorities

Jimmy L. Humphrey, J. L. Humphrey & Associates, 3605 Needles Dr., Austin, TX 78746, A. Frank Seibert, Separations Research Program, University of Texas at Austin, 10100 Burnet Road, Bldg 133/ CEER, Austin, TX 78758, J. Christopher Lewis, Technology R& D, ConocoPhillips, 166 PDC Bartlesville Technology Center, Highway 60 & 183, Bartlesville, OK 74004, and John P. Farone, Retired, 511 Fern Court, Cedar Park, TX 78613.

Smart Plants prevent problems before they occur. Advances and priorities for Smart Plants will be presented. Included will be a discussion of the importance of accurate real time material and heat balances around individual processes. A Smart Plant Case Study will be presented showing a quick pay-out for an energy saving opportunity.

249f Roundtable Discussion, "Overcoming Barriers to Smart Manufacturing Plants"

John Forgac, BP Naperville Campus, BP North American Products, MC 605-1E, 150 West Warrenville Road, Naperville, IL 60563-8460, Miguel J. Bagajewicz, Chemical, Biological and Materials Engineering, University of Oklahoma, 100 E. Boyd St., T-335, Norman, OK 73019, and Larry Megan, Praxair, 175 East Park Drive, PO Box 44, Tonawanda, NY 14150.

Smart manufacturing plants prevent problems before they occur. A key ingredient for Smart Plants is accurate real time material and heat balances around individual processes (e.g. reactors and distillation columns). Such balances are made possible by accurate real time plant measurements of process flow rates, compositions,temperatures, etc. This discussion will focus on overcoming two key barriers to Smart Plants: (1) How can accurate real time plant measurements be realized, even in older plants ? (2) How can the volumes of plant data that are created be quickly analyzed ?

264a A Barrier-Terrain Methodology for Global Optimization

Angelo Lucia[1], Rajeswar R. Gattupalli[1], Kedar Kulkarni[2], and Andreas A. Linninger[2]. (1) University of Rhode Island, Dept of Chem. Engineering, Kingston, RI 02881, (2) Bioengineering, University of Illinois at Chicago, 851 S. Morgan St., Room 218, Chicago, IL 60607

The global terrain methodology of Lucia and Yang (2003) is a method for finding multiple solutions to optimization problems that is based on the fact that stationary points (i.e., saddle points and minima) and singular points are generally connected along smooth valleys on the objective function surface. In this presentation, we show that there are cases where not all stationary and singular points to optimization problems necessarily lie in the same valley and that these valleys are not necessarily smoothly connected. Accordingly, logarithmic barrier functions are used to create smooth connections between distinct valleys so that the global terrain method is guaranteed to explore the entire feasible region. Once valleys are connected, different stationary and singular points in separate parts of the feasible region can be calculated, identified or characterized, and sequentially tracked as the barrier parameter is reduced. The proposed barrier-terrain methodology is used to successfully find all physically meaningful solutions to a number of small illustrative examples and a collocation model for a spherical catalyst pellet problem with 20 variables. The key contributions of this work are the discovery that barrier methods provide connections between valleys containing stationary points for intermediate barrier parameter values under mild conditions on the model equations and the robustness of the proposed barrier-terrain method. Many geometric illustrations are used to highlight key features of the barrier-terrain approach.

An Ordinal-Optimization-based Approach to Stochastic Mixed Integer Nonlinear Programming

Chengtao Wen, B. Erik Ydstie, Pittsburgh, PA, US

Introduction

The stochastic Mixed Integer Nonlinear Programming (MINLP) is an important and fundamental problem in the area of process system engineering. This has motivated the development of efficient algorithms for handling the large-scale, non-convex, highly combinatorial and strongly nonlinear problems.

The methods to solve the stochastic MINLP problems fall into two categories: numerical integration method and sample average method. The numerical integration method utilizes numerical integration techniques to calculate the multi-dimensional integrals, such as the Gaussian Quadratures/Cubatures method. The Monte Carlo simulation is one of the most widely used sample average methods. Acevedo and Pistikopoulos (1998) suggested the use of numerical integration methods for smaller dimension problems and sampling-based methods for larger problems [1].

The established algorithms to stochastic MINLP problems are successful in many applications. However, they are built on cardinal optimization, in which the optimums are obtained from the accurate calculation of cardinal values. The cardinal optimization is generally associated with inherent computational difficulties from the combinatorial natural of discrete variables and stochastic natural of uncertain parameters. It is easy to become intractable with the increase of the number of discrete variables and/or uncertain parameters.

The ordinal optimization (OO) is a novel optimization methodology [2]. It is designed to cope with hard problems such as problems with uncertainties, or problems with huge sample space that grows exponentially with respect to the problem size. The OO theory provides a way to obtain solutions of high quality with much less computation effort than the conventional cardinal optimization methods.

The OO theory is based on the following two tenets: 1) it is much easier to determine "order" than "value." To determine whether A is better or worse than B is a simpler task than to determine how much better is A than B (i.e., the value of A-B) especially when uncertainties exist. 2) Instead of asking the "best for sure," we seek the "good enough" solution with high probability. This goal softening technique makes the optimization problem much easier. The OO method has found successful applications in a variety of fields, e.g. optimal control, network topology design, planning and scheduling of manufacturing systems.

In this paper, an ordinal-optimization-based algorithm is developed to solve the stochastic MINLP problems. A novel approximation method is proposed, in which the expected performances are estimated by the intermediate NLP optimization solutions and Monte-Carlo simulations with a small sample size. The OO theory is then applied to form a subset of good enough decisions with high probability by ranking the crude performance estimations. The sub-optimal solution is obtained by searching the subset of selected good decisions and using the accurate performance values. A benchmark problem is illustrated to show the extra high computation efficiency and the quality of the obtained solutions.

Stochastic MINLP Problems

In this paper, we consider the following stochastic MINLP problem

$$\min_{x,y} E_\theta[f(x,y,\theta)] \tag{1}$$

$$s.t. \quad \begin{cases} h(x,y,\theta)=0 \\ g(x,y,\theta)\le 0 \end{cases} \tag{2}$$

where $x \in \Re^m$ is the continuous variables, $y \in Z^n$ represents the discrete variables, and $\theta \in \Re^k$ is the uncertain parameters. The h and g denote a set of equality and inequality constraints, respectively. The scalar function f represents the objective function. Typically, it is an expected value of an economic performance. For simplicity, we define y as the decision variable and x the operation variable. The objective of a stochastic MINLP problem is then described as finding the optimal decision and operation variables in presence of uncertain conditions.

Ordinal Optimization

The OO theory is based on two central ideas: order comparison and goal softening.

The order comparison is to evaluate the quality of the expected performances with their relative order rather than the exact cardinal values. According to [3], the relative order of expected performances converges exponentially fast. By comparison, the cardinal values converge at rate of $\frac{1}{\sqrt{N}}$. The order converges much faster than the cardinal values with respect to the number of samples.

The second basis is goal softening. The OO settles for the "good" solutions with high probability. This deviates from the conventional optimization algorithms, which ask for the best solution for sure. By softening our goal to the "good" solutions, e.g. top-g solutions in the search space, the number of good solutions in the set of estimated top-g decisions can be quite substantial even in presence of very large estimation errors in the performance values.

Suppose we simultaneously evaluate a large set of alternatives very approximately and order them according to the approximate evaluations. High probability can be anticipated that we can find the actual good alternatives by limiting ourselves to the top of the observed good choices.

Ordinal-Optimization-Based Algorithm to Stochastic MINLP

Approximation of Expected Performances

In order to relieve the computation burden, two methods are used to approximate the expected performance. The first method is a direct heritage from the classical OO theory. The short Monte Carlo simulation is used, whose sample size is at the magnitude of 10^2, i.e. $O(10^2)$. This is insufficient from the cardinal optimization point of view. But it is often enough for rank comparison.

The second method is to use the NLP solvers with a fixed number of iterations. This is a novel goal softening strategy. Instead of running a NLP solver until it converges to the optimum, we ask for an intermediate optimization results that are "good" enough with high probability. The intermediate results are defined as "good" if it contains enough information for rank comparison.

Theorem 1: Assume that the objective function $f(x \mid y_i)$ is Lipschitz continuous for a fixed y_i with $i \in \{1, \cdots, m\}$. Denote M_i^* as the number of iterations needed to calculate the optimum x^* of $f(x \mid y_i)$ with. Then there must exist m positive integers $M_i \le M_i^*$, such that

$$\min_{1 \le i \le m} \left\{ f\left(x^{M_i} \mid y_i\right) \right\} = \min_{1 \le i \le m} \left\{ f\left(x^* \mid y_i\right) \right\} \tag{3}$$

where x^{M_i} is the intermediate optimal results of $f(x \mid y_i)$ obtained from M_i iterations.

Proof: Please refer to [4] for detail proof.

Theorem 1 shows that the optimal decision can be obtained with less computation cost if the intermediate optimization results are fully utilized in term of the rank-based comparison. Using the order comparison and goal softening, the number of iterations of the NLP problems can be much decreased without affecting the final optimum significantly, because the order is very robust to approximation errors [3]. Therefore, the use of intermediate optimizations can lead to a significant reduction in computation burden when the time saving of a single NLP optimization is amplified by the sample size.

Selection of Good Enough Decisions

Given a representative set Π with M decisions, whose performances are estimated by using the approximation algorithm stated in previous section. Denote the true and crude expected performances as J_i and $\hat{J}_i$, respectively. We can get

$$J_i = \hat{J}_i + e_i \tag{4}$$

where e_i is the approximation error with $i = 1, \cdots, M$. There may be significant errors due to the rough approximations. The advantage of an OO-based method is its capability to separate the good from the bad even using very crude approximation. The performance "order" is relatively immune to large approximation errors. Therefore, even if the rough estimation is used to rank M selections' performances, some good enough decisions will be kept within the select set with high probability.

The major task in applying OO theory is to construct a selected subset S containing "good enough" decisions with high probability. The quantitative measure of the "good enough" is the alignment probability defined as

$$P(G, S, a) = P_A(|G \cap S| \ge a) \tag{5}$$

where G is the "good enough" set and a is called the alignment level. Intuitively the alignment probability is the probability of the event that there are at least a elements in the intersection set of G and S.

To select s good ones from M representatives, the crude approximated performances are calculated and ranked. The top s decisions are then selected as S and its size s is determined by a regressed nonlinear equation to satisfy certain confidence requirement. The value of s can be estimated by

$$s \approx Z(a,g) = e^{Z_0} a^{\rho} g^{\gamma} + \eta \qquad (6)$$

where $g = |G|$ is the size of G, and Z_0, ρ, γ, η are coefficients or parameters obtained by nonlinear regression.

The evaluation of M decisions using crude approximations is computationally efficient and the OO theory guarantees that good decisions will be among the selected set in high probability. More accurate but time-consuming evaluation is applied to evaluate the selected decisions. For each decision, the accurate performance is calculated using a long Monte-Carlo simulation and the real optimums of NLP problems. The best decision is then selected by evaluating those decisions in the selected subset based on the accurate performances. Since $s = |S|$ is much smaller than M, the OO method is extremely efficient in comparison with brute and force method by calculating M accurate performances.

Rank-Filter-Based Algorithm

The main steps of the rank-filter-based algorithm are summarized as follows:

1. Choose M decisions from the decision space randomly;
2. Run short simulations with a small sample size of N_1 for each decision;
3. Run the NLP solver with k iterations for each sample realization;
4. Calculate the approximated expectations of M decisions using the average of N_1 intermediate optimization solutions;
5. Rank the approximated expectations and choose the top-s ones, where s is calculated from (26);
6. Calculate the accurate performances of the selected s decisions using a big sample size of N_2 and optimal NLP solutions;
7. Get the sub-optimal solution by ranking the accurate performances of s decisions.

The parameters in this algorithm can be obtained from the corresponding deterministic MINLP problem, where the random parameters are substituted by their mathematical expectations. A typical set of parameters are shown in [4].

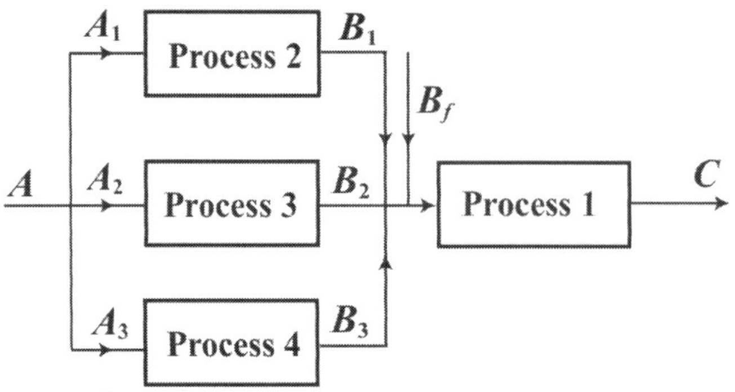

Fig. 1 Process superstructure

Numerical Examples

The example is a process flow-sheet synthesis problem, whose superstructure is shown in Fig. 1.

This problem is also studied by [5] and [6]. In this example, process 1 is fixed. This implies that 3 binary variables are needed to represent the existence of the process 2, 3 and 4. This example involves 7 random parameters. The normal distributions are assumed for 4 continuous random variables, i.e. the availability of raw material *A*, the demand of product *C*, the prices of material *A* and pure $B(B_f)$. Refer to [4] for the distribution data of 3 discrete random parameters, i.e. the kinetic constants of process 2, 3 and 4.

$$\max_{x,y} E_\theta\{550x_{10} - (\theta_3 x_1 + \theta_4 x_9) - [100 + 15(x_5 + x_9) + (80 + 5x_2) + (130y_3 + 15x_3) + (150y_4 + 5x_4)]\}$$

$$
\begin{aligned}
s.t.\quad & x_1 = x_2 + x_3 + x_4 \\
& x_5 = x_6 + x_7 + x_8 \\
& x_{10} = 0.9(x_5 + x_9) \\
& x_6 = 18\ln(1 + \frac{x_2}{\theta_5}) \\
& x_7 = 20\ln(1 + \frac{x_3}{\theta_6}) \\
& x_8 = 15\ln(1 + \frac{x_4}{\theta_7}) \\
& x_1 \le \theta_2, \quad 5 \le x_2 \le 25y_2 \\
& x_3 \le 20y_3, x_4 \le 20y_4 \\
& x_9 \le 15, \quad x_{10} \le \theta_1 \\
& x \in \Re^{10}, \quad y \in \{0,1\}^3 \\
& \theta_1 \in N(17.5, 2.5) \\
& \theta_2 \in N(27.5, 2.5) \\
& \theta_3 \in N(250, 10) \\
& \theta_4 \in N(300, 15)
\end{aligned}
$$

with $x = [A, A_2, A_3, A_4, B, B_2, B_3, B_4, B_f, C]^T$ and $y = [y_2, y_3, y_4]^T$.

The realizations of continuous random parameters are generated using a truncated range of $\mu \pm 3\delta$. To calculate the coarse expected performances, a sample size of 200 is used. The intermediate optimization results are calculated by running a Matlab NLP solver with 15 iterations. The top-2 designs are chosen to form the selected set *S*. The precise expected performances of the selected designs are calculated from 5000 optimums of NLP problems.

Table 1 lists the calculation results of the top-2 decisions. By comparing the accurate optimums, we can get the optimal decision vector $y^* = [1,1,1]^T$, with an expected performance of 4068.6. The same optimal decision is obtained by the decomposition-based algorithm in [6]. The overall calculation time is 282.7 CPUs. It is of high computation efficiency by considering the big sample size of 5000 for calculating the accurate expectations.

Table 1. Expected Performance of Selected Decisions

#	y	Approximated Optimal	Real Optimal
1	[1 1 1]	3848.3	4068.6
2	[1 1 0]	3873.2	4057.6

Conclusion

In this paper, we propose an ordinal-optimization-based algorithm to solve the stochastic MINLP problems. This algorithm is based on a novel approximation method and the OO-theory framework. It guarantees a sub-optimal solution of high quality with high probability. The proposed algorithm can ease several order of magnitude of computation expense without significant loss of solution precision compared to the conventional algorithms based on cardinal optimization. The ordinal-optimization-based algorithm is promising to find successful applications in a variety fields, such as the synthesis of large-scale chemical process, optimization of supply chain, and control of hybrid systems.

Reference

[1] Acevedo, J., & Pistikopoulos, E. N., "Stochastic optimization based algorithms for process synthesis under uncertainty." *Computers and Chemical Engineering*, vol. 22: 647-671, 1998.

[2] Ho, Y. C. "An explanation of ordinal optimization: Soft computing for hard problems." *Information Sciences*, 113(3-4), 169-192, 1999.

[3] Dai, L. Y., "Convergence properties of ordinal comparison in the simulation of discrete event dynamic systems." *Journal of Optimization Theory and Applications*, 91 (2): 363-388, 1996.

[4] Wen, Chengtao, & Ydstie, B. Erik, " An ordinal-optimization-based approach to stochastic MINLP programming." *AIChE Journal* (second submission).

[5] Duran, M. A., & Grossmann, I. E., "An outer-approximation algorithm for a class of mixed integer nonlinear programs." *Mathematical Programming*, 36, 307, 1986.

[6] Goyal, V. & Ierapetritou, M. G., "Stochastic MINLP optimization using simplicial approximation." *Computers and Chemical Engineering*, vol. 31: 1081-1087, 2007.

264c Global Optimization of Nonconvex Minlps Using Generalized Reduction Constraints Obtained from Alternative Model Formulations

Juan P. Ruiz, Dept of Chemical Engineering, Carnegie Mellon University, 5000 Forbes Ave, Pittsburgh, PA 15213 and Ignacio E. Grossmann, Center for Advanced Process Decision-making, Dept of Chemical Engineering, Carnegie Mellon University, 5000 Forbes Avenue, Pittsburgh, PA 15213.

This paper is concerned with the global optimization of Nonconvex Mixed Integer Nonlinear Programs. These problems arise in many diverse industrial applications, to mention some of them, in Process Network Synthesis, in Pooling Network Systems, in Plant Layout Design and in Power Distribution Systems. In order to improve the computational efficiency of the spatial branch and bound methods for solving these problems, this paper has as major objective to propose a methodology to generate formulations which lead to tighter convex relaxations. This is accomplished by systematically developing and combining a set of alternative model formulations using engineering knowledge and physical insight.

Most of the current approaches to obtain convex relaxations for nonconvex problems rely on the use of predefined convex envelopes or suitable linear outer approximations of the non convex terms that appear in the formulation, Tawarmalani & Sahinidis [2002]. This relaxation procedure leads to formulations where some of the physical laws are often violated (e.g. mass and energy balances are not met). Adding constraints that are redundant in the original formulation and which enforce such laws while keeping the program convex, can strengthen the relaxation.

Redundancy, however, is usually not considered a good practice in process modeling. However, we show that this often has a great impact in the tightness of the final relaxation without incurring in a significant increase of the problem size. The explanation behind this behavior can be obtained by realizing that even when alternative nonconvex representations may contain the same information, their relaxations do not. Sherali and Alameddine [1992] proposed a procedure to improve the representation of the problem when it is relaxed through the generation of redundant constraints in bilinear nonconvex programs. This is accomplished by using pairwise products of the linear inequalities originally present, previous to their linearization. Based on a similar idea Liberti [2004] and Liberti and Pantelides [2006] proposed the "method of reduction constraints" which improves the efficiency of the implementation. However, for the general case, the problem of finding useful redundant constraints systematically remains as a challenge. Revisiting the physical meaning behind the problem may still be required. Previous results in this respect were reported by Tawarmalani & Sahinidis [2002] in the p-q pooling problem and Quesada & Grossmann [1995] in Process Networks design. However, to the best of our knowledge, no systematic methodology based on this approach for more general nonconvex problems has been proposed. It is the aim of this work to formalize and extend this idea to any nonconvex MINLP problem whose physical meaning is perfectly defined.

The main idea of the approach proposed in this paper is to first generate all alternative model formulations based on engineering knowledge. These models are then combined to give rise to a unique formulation that has redundant constraints, which however, may be non redundant in the relaxation. We study the theoretical properties of these formulations, and for the particular case of Process Networks and Power Distribution Networks, a refined strategy is presented that exploits substructures to infer generalized reduction constraints (i.e. constraints that are redundant in the original formulation but not in its relaxation). We show with a number of computational results that in some cases, order of magnitude improvements in the efficiency of the spatial branch and bound method can be achieved when using the proposed formulation. We also compare the proposed approach with traditional modeling approaches that rely on the selection of a single monolithic nonconvex MINLP model.

Liberti L. (2004) Reduction Constraints for the Global Optimization of NLPs. International Transactions in Operation Research Vol 11, No 1, pp 33-41

Liberti L. & Pantelides (2006) An exact reformulation algorithm for large nonconvex NLPs involving bilinear terms. Journal of Global Optimization. Vol 36, Issue 2, pp. 161-189

Quesada & Grossmann (1995) Global Optimization of bilinear process networks with multicomponent flows. Computers Chem. Eng. Vol 19, No 12, pp 1219-1242

Sherali & Alameddine (1992) A new reformulation-linearization technique for bilinear programming problems. Journal of Global Optimization, 2:379-410

Tawarmalani. M & Sahinidis, N. (2002) Convexification and Global Optimization in Continuous and Mixed Integer Nonlinear Programming. Kluwer Academic Publishers

264d Mpec Strategies for Optimization of Chemical Process Dynamics

Brian T. Baumrucker and Lorenz T. Biegler. Department of Chemical Engineering, Carnegie Mellon University, 5000 Forbes Avenue, Pittsburgh, PA 15213

With the development and widespread use of large-scale nonlinear programming (NLP) tools for process optimization, there has been an associated application of NLP formulations with complementarity constraints in order to represent discrete decisions. Also known as Mathematical Programs with Equilibrium Constraints (MPECs), these formulations can be used to model certain classes of discrete events and can be more efficient than a mixed integer formulation. In this talk, we consider MPEC formulations and solution strategies for chemical engineering applications, particularly for a dynamic gas pipeline system. The results illustrate the effectiveness of MPEC strategies as well as some novel operating strategies for pipeline networks.

264e Cfd-Based Shape Optimization of Pressure-Driven Microchannels Via Adjoint Formulation

Osamu Tonomura, Manabu Kano, and Shinji Hasebe. Chemical Engineering, Kyoto University, Kyoto, 615-8510, Japan

With the advances in computational resources and algorithms, optimal shape design based on computational fluid dynamics (CFD) is an interesting filed for industrial applications. Although gradient based method is widely used to solve optimization problems, it requires the solution of a large linear system of equations to calculate the sensitivity of the cost function with respect to each design parameter. Thus, for n design parameters, n large linear programming problems have to be solved. Consequently, gradient based method requires enormous computation time when the number of design variables is large. Recently, the adjoint variable method attracts the attention as an efficient sensitivity analysis method, particularly for aeronautical design, since it allows to successfully obtain the shape gradient functions independently of the number of design variables.

In this research, an automatic shape optimization system based on the adjoint variable method is developed using C language on Windows platform. The flow equation is added to the cost function by introducing the set of Lagrange multipliers, because it must be fulfilled by the optimal solution. The principal steps of the optimization procedure are as follows: 1) assume an initial shape, 2) generate computational grids, 3) solve the flow equations, viz. the Navier-Stokes equation and the continuity equation, for deriving the flow velocity and the pressure, 4) solve the adjoint equation to obtain the set of Lagrange multipliers, 5) calculate the shape gradient functions, 6) obtain a new shape by moving each point on the boundary, and 7) go to step 2 unless the change in the cost function is smaller than a desired convergence parameter. Each design cycle requires the numerical solution of both the flow and the adjoint equations, whose computational time is roughly twice of that required to obtain the flow solution. In order to validate the effectiveness of the developed system, the optimal shape design problems of the pressure-driven microchannels are solved. The shape of the microchannels is an important design variable to achieve the desired function. As examples, pressure drop minimization problems of a 180° curved channel and a branched channel in incompressible flows under constant volume conditions are formulated. The atmospheric pressure is set at the outflow boundary, and a uniform velocity profile (Re = 10) is prescribed at the inflow boundary. 92 rounds of design cycles are required till convergence for 42 design variables, which are assigned on the curved channel surface. Each design cycle requires approximately 10 seconds. The pressure drop of the optimally designed curved channel is decreased by 27.6 % as compared with that of the initial shape. It is demonstrated that the adjoint variable method can be used to formulate computationally feasible procedures for the shape design of microchannels. Our future work will focus on the extension of the developed system to shape optimization problems of thermo-fluidic microdevices.

264f Coarse Regularization and Optimization of Microscopic Monte-Carlo Based Time-Steppers

Eduardo Luna-Ortiz[1], Santiago Martínez[2], and Susana Gómez[2]. (1) Centre for Process Systems Engineering, Department of Chemical Engineering, Imperial College London, South Kensington Campus, London, SW7 2AZ, United Kingdom, (2) Instituto de Investigaciones Matemáticas Aplicadas y Sistemas (IIMAS-UNAM), C.U., México D.F, Mexico

Nowadays, there is an increase number of industrial and biological processes whose coarse-grained (mean-field) model equations do not conceptually exist or simply can not capture the complex interactions in the process. Instead, a detailed microspic (often stochastic) description can be alternatively used (i.e. kinetic Monte Carlo [1]). In this scenario, while the optimization is performed on a set of macroscopic variables, the simulation operates on the microscopic level. In order to bridge the macroscopic optimization-microscopic simulation, the coarse time-stepper concept is exploited [2,3]. Due to the natural random noise in the stochastic coarse time-stepper, any optimization problem (optimal control or parameter estimation) will be mathematically ill-posed.

In this presentation, we use a regularization strategy aimed to control the noise propagation (coming from the microscopic simulation) that prevents convergence and in this way it overcomes the ill-posedness of the optimization problem at the macroscopic level. When the optimal regularization is used, convergence is achieved where no further precision in the control variables is possible. The performance of the strategy is evaluated using a stochastic reacting system as a case study and using either local (deterministic) and global (stochastic) optimization methods

[1] Gillespie, D. T. (1977) J. Phys. Chem., 81(25), 2340-2361

[2] Theodoropoulos, C. et al. (2000) PNAS USA, 97(18), 9840-9843

[3] Kevrekidis, I. G. et al. (2003) Comm. Math. Sci., 1(14), 715-762

264g On the Global Optimality of a Class of Multilevel QP Problems in Dynamic Operability Applications

Rhoda Baker and **Christopher L. E. Swartz**. Chemical Engineering, McMaster University, 1280 Main Street West, Hamilton, ON L8S 4L7, Canada

The steady-state economic optimum of process plants generally lies at the intersection of constraints (Lee and Weekman, 1976). However, in order for operation to remain feasible in the face of disturbances, the steady-state operating point needs to be backed-off from the constraints into the feasible region. The calculation of the optimal back-off amount has been the subject of several studies, typically involving the solution of a dynamic optimization problem of some form (Figueroa et al., 1996). This required amount of back-off can be related to the plant economics, and can also be used as a measure of dynamic operability (Narraway et. al, 1991; Figueroa et al., 1996).

Most studies considered either no control or linear controllers in back-off calculations. Recent studies by Soliman et al. (2008) and Baker and Swartz (2008) considered constrained model predictive control (MPC) within this framework. The resulting optimization problem is multilevel in nature, since the control input at each sampling interval over the closed-loop horizon requires the solution of a quadratic programming (QP) problem. In both studies, the inner level QPs were replaced by their equivalent Karush-Kuhn-Tucker (KKT) optimality conditions to give a single-level mathematical program with complementarity constraints (MPCC). In Soliman et al. (2008) the complementarity constraints are replaced by mixed-integer linear constraints, similar to the approach in Fortuny-Amat and McCarl (1981), resulting in a mixed-integer quadratic programming problem. Baker et al. (2008) instead solved the MPCC problem directly, using an interior-point algorithm specially tailored to handle complementarity constraints (Raghunathan and Biegler, 2003). Solution times were found to be significantly faster than the MIQP approach, and moreover, the solution was found to coincide with that of the MIQP when a solution to the MIQP was obtained (guaranteed to be the global optimum). Since the MPCC formulation is nonlinear and nonconvex, some analysis is required to determine whether the global optimum can be reliably obtained for these problems using this approach. This is the subject of the present study.

Small-scale numerical experiments were performed with a global optimizer to detect the presence of multiple local optima; only single optima were found. We considered then a bilevel LP problem, and recognized a condition under which only a single local optimum is obtained. Key to this is that the upper level objective function does not involve lower-level variables. By extending bilevel optimization theory to the multilevel case, we explore the optimality characteristics of the SISO back-off problem under constrained MPC. We argue that convergence to a stationary point of an appropriate reformulation of the MPCC as an NLP yields the optimal solution to the multilevel problem (Fletcher et al., 2006).

Case studies include the demonstration of the existence of a single optimum for a small example problem, as well as a counter-example that demonstrates that this result does not apply to all multilevel quadratic problems. Additional case studies demonstrating the difference in performance of solving the problem as a MPCC compared to solving it as a MIQP are also presented. Since the results apply to a general class of multilevel quadratic programs rather than the specific case of back-off problems with constrained MPC, additional chemical engineering applications that can be reformulated in this fashion are being explored.

References

Baker, R. and Swartz, C. L. E. (2008). Interior point solution of multilevel quadratic programming (QP) problems in constrained model predictive control (MPC) applications. Ind. Eng. Chem. Res., 47 (1), 81-91.

Figueroa, J.L., Bahri, P.A., Bandoni, J.A., and Romagnoli, J.A. (1996). Economic impact of disturbances and uncertain parameters in chemical processes – A dynamic back-off analysis. Comput. Chem. Eng., 20(4), 453-461.

Fletcher, R., Leyffer, S., and Ralph, D. (2006). Local Convergence of SQP methods for Mathematical Programs with Equilibrium Constraints. SIAM Journal Optimization, 17 (1), 259-286.

Fortuny-Amat, J. and McCarl, B. (1981). A representation and economic interpretation of a two-level programming problem. J. Operational Research Society, 32(9), 783-792.

Lee, W. and Weekman, W.V. (1976). Advanced control practice in the chemical process industry: A view from industry. AIChE J., 22(1), 27-38.

Narraway, L.T., Perkins, J.D., and Barton, G.W. (1991). Interactions between process design and process control: economic analysis of process dynamics. J. Proc. Control, 1, 243-250.

Raghunathan, A.U. and Biegler, L.T. (2003). Mathematical programs with equilibrium constraints (MPECs) in process engineering. Comput. Chem. Eng., 27 (10), 1381-1392.

Soliman, M., Swartz, C. L. E., and Baker, R. (2008). A mixed-integer formulation for back-off under constrained predictive control. In press, Comput. Chem. Eng. doi:10.1016/j.compchemeng.2008.01.004

266a Actuator Fault Detection and Isolation In Controlled Particulate Processes with Measurement Limitations

Arthi Giridhar, **Sathyendra Ghantasala**, and Nael H. El-Farra. Department of Chemical Engineering & Materials Science, University of California, Davis, One Shields Avenue, Davis, CA 95616

Particulate processes are found in a number of important processing industries including agricultural, chemical, food, minerals, and pharmaceuticals. The realization that the end-product quality in these processes is critically dependent on the ability to manipulate the particle size distribution (PSD) has motivated significant research work on the design of model-based feedback control systems that achieve PSDs with desired characteristics (e.g., see [1] for a recent survey of results in this area). Despite the significant and growing body of literature on control of particulate processes, the problems of designing and implementing fault detection and fault-tolerant control systems for particulate processes have received limited attention. These are important problems given the fact that the erosion of control authority caused by control system malfunctions can directly impact product quality and lead to substantial production losses if such faults are not properly diagnosed and handled.

One of the central issues in the design of model-based fault diagnosis schemes for particulate processes is the infinite-dimensional nature of particulate process models which are typically obtained through the application of population, material and energy balances and consist of systems of nonlinear partial integro-differential equations that describe the evolution of the PSD, coupled with systems of nonlinear ordinary differential equations that describe the evolution of the state variables of the continuous phase. The infinite-dimensional nature of population balance models precludes their direct usage for the synthesis of practically implementable controllers or fault diagnostic filters which need to be designed on the basis of appropriate low-order models to be suitable for practical implementation. This requires that the fault diagnosis filters be designed and implemented in a way that allows discriminating between approximation errors and faults.

To address this problem, we recently developed in [2] a methodology for the detection and handling of actuator faults in particulate processes on the basis of appropriate reduced-order models that capture the dominant process dynamics. The fault detection task was addressed by means of a filter that simulates the behavior of the fault-free, reduced-order model and uses the discrepancy from the behavior of the actual process as a residual signal. Failure compensation, on the other hand, was accomplished through a switching mechanism that reconfigures the control system based on the stability regions of the constituent control configurations in a way that preserves closed-loop stability in the event of fault detection. Using regular perturbation theory, appropriate fault detection thresholds and control reconfiguration criteria that account for model reduction and state estimation errors were derived for the implementation of the control architecture on the particulate process. These results were subsequently generalized in [3] to address the problems of fault isolation and model uncertainty.

The implementation of the above schemes requires the availability of accurate measurements of the process outputs which are needed to implement the feedback controllers and monitor the closed-loop system for FDI and supervisory control purposes. In practice, exact measurements are often unavailable due to the presence of measurement noise, the occurrence of sensor malfunctions that degrade the accuracy of the measurements, or the presence of inherent limitations on the capabilities of the sensing devices as in discrete sensors that provide only a limited (i.e., qualitative) information about the state of the system. These practical limitations on the number and accuracy of the available measurements can seriously erode the diagnostic and fault-tolerance capabilities of the fault-tolerant control architecture, if not explicitly accounted for in the design of each layer. Within the feedback control layer, for example, measurement errors can degrade the stability and performance properties of the nominal controllers and may render the closed-loop system unstable unless the controller is designed with a sufficient robustness margin. At the fault diagnosis level, the presence of measurement errors limits our ability to accurately monitor the actual evolution of the process to determine if and when a fault can be declared. Unless the FDI rules are re-designed to discriminate between those errors and faults, the FDI scheme may lead to false or missed alarms that trigger unnecessary control system reconfiguration. The lack of full or accurate state measurements also limits the size of the stability regions as well as the supervisor's knowledge of where the system trajectory is relative to those regions. This in turn complicates the actuator reconfiguration task in the event of faults.

Motivated by these considerations, we develop in this work a robust fault diagnosis and fault-tolerant control structure for particulate processes described by population balance models with control constraints, time-varying uncertain variables, actuator faults and a limited number of output measurements with limited accuracy. To facilitate the controller synthesis and fault diagnosis tasks, a finite-dimensional system that approximates the dominant dynamic modes of the process is initially derived using the method of weighted residuals and then decomposed into an interconnection of subsystems where each subsystem is excited directly by only one manipulated input. A robustly stabilizing bounded output feedback controller is then designed for each subsystem by combining a bounded Lyapunov-based robust state feedback controller with a state estimation scheme that relies on the available measurements of the principal moments of the PSD and the continuous phase variables to provide estimates of the dominant process modes. The controller synthesis procedure facilitates the derivation of (1) an explicit characterization of the fault-free behavior of each subsystem in terms of a time-varying bound on the dissipation rate of the corresponding Lyapunov function that accounts for the uncertainty and measurement errors, and (2) an explicit characterization of the robust stability region where constraint satisfaction and robustness with respect to uncertainty and measurement errors are guaranteed.

Using the fault-free Lyapunov dissipation bounds as thresholds for FDI, the detection and isolation of faults in a given actuator is accomplished by monitoring the evolution of the dominant modes within the corresponding stability region and declaring a fault when the threshold is exceeded. A key feature of this threshold is that it is dependent on the achievable degree of asymptotic uncertainty attenuation and thus can be made small by properly tuning the robust controller. Robustness of the FDI scheme to measurement errors is ensured by confining the FDI region to an appropriate subset of the stability region, and enlarging the FDI thresholds appropriately. It is shown that these safeguards can be tuned by appropriate selection of the sensor configuration. The robust FDI scheme is integrated with a controller reconfiguration strategy that orchestrates the transition from the faulty actuators to a well-functioning fall-back configuration following FDI. Appropriate FDI criteria are derived for the implementation of the fault-tolerant control architecture on the particulate process to ensure its robustness with respect to model reduction errors. Using regular perturbations theory, the criteria are expressed in terms of residual thresholds that capture the closeness of solutions between the fault-free reduced and full-order models. Finally, the proposed approach is applied to the problem of robust fault-tolerant control of the Crystal Size Distribution in a continuous crystallizer with a fines trap.

References:

[1] Christofides, P. D., N. H. El-Farra, M. H. Li and P. Mhaskar, ``Model-Based Control of Particulate Processes," Chemical Engineering Science, 63:1156-1172, 2008.

[2] El-Farra, N. H. and A. Giridhar, ``Detection and Management of Actuator Faults in Controlled Particulate Processes Using Population Balance Models," Chemical Engineering Science, 63:1185-1204, 2008.

[3] Giridhar, A. and N. H. El-Farra, ``A Unified Framework for Robust Fault Detection, Isolation and Compensation in Uncertain Particulate Processes," Proceedings of American Control Conference, to appear, Seattle, WA, 2008.

RELAXED LYAPUNOV CRITERIA FOR ROBUST GLOBAL STABILIZATION OF NONLINEAR SYSTEMS WITH APPLICATION TO CHEMOSTAT CONTROL

Iasson Karafyllis[*], Costas Kravaris[**] and Nicolas Kalogerakis[*]

[*]**Department of Environmental Engineering, Technical University of Crete, 73100, Chania, Greece**
[**]**Department of Chemical Engineering, University of Patras, 26500 Patras, Greece**

Paper No. 266b, AIChE Annual Meeting, Philadelphia, PA, Nov. 2008

Keywords: uniform robust global feedback stabilization, chemostat models.

Introduction

The motivation of the present work arises from control problems in continuous stirred microbial bioreactors, often called chemostats. In many applications, the design conditions for the chemostat represent a locally asymptotically stable steady state but the stability region could be too small to allow proper operation in the presence of disturbances. Thus, the need for control arises in the sense of enlargement of the stability region of the design steady state in the presence of disturbances ([8]). Under these circumstances, achieving global stability of the bioreactor in the presence of disturbances will be best possible outcome from the design of a control system.

The problem of designing of a feedback law that achieves robust global stabilization of a nonlinear system is closely related to the existence of a Robust Control Lyapunov Function. However, it may be a big challenge to derive a Control Lyapunov Function satisfying global properties for a given nonlinear system. In the present work, under appropriate hypotheses, we derive relaxed Lyapunov-like sufficient conditions for Uniform Robust Global Asymptotic Stability. The Lyapunov-like conditions will be "relaxed" in the sense that the Lyapunov differential inequality is not required to hold over the entire state space, but only over an appropriate absorbing set, having the property that every trajectory of the system enters the set in finite time.

The theoretical results will be applied to a chemostat stabilization problem, where the dynamics is adequately represented by a two-state model involving the microbial biomass and the limiting organic substrate, with manipulated input the dilution rate. The growth rate of the microorganisms will be assumed follow Haldane kinetics, whereas the death rate of the microorganisms as well as the substrate consumption for cell maintenance will be accounted for assuming first-order kinetics. Moreover, the biomass balance will involve a time-varying uncertainty, accounting for the adjustment of the biomass to changes in the substrate levels. Applying the theoretical results on relaxed Lyapunov crireria, a robust globally stabilizing state feedback control law will be derived. Simulation results will also be used to illustrate the robustness properties of the derived control law.

1. Motivation: Globally stabilizing control of a chemostat

Continuous stirred microbial bioreactors, often called chemostats, cover a wide range of applications. The dynamics of the chemostat is often adequately represented by a simple dynamic model involving two state variables, the microbial biomass X and the limiting organic substrate S (see [13]). For control purposes, the manipulated input is usually the dilution rate D. A commonly used (delay-free) mathematical model for microbial growth on a limiting substrate in a chemostat is of the form:

$$\begin{aligned}&\dot{X} = \mu(S)X - DX\\&\dot{S} = D(S_i - S) - \frac{1}{Y}\mu(S)X\\&X \in (0,+\infty),\, S \in (0,S_i),\, D \geq 0\end{aligned} \tag{1.1}$$

where S_i is the feed substrate concentration, $\mu(S)$ is the specific growth rate and $Y > 0$ is a biomass yield factor. Chemostat models with delays can be found in [4,14,15]. In most applications, Monod or Haldane or generalized Haldane models are used for $\mu(S)$.

The literature on control studies of chemostat models of the form (1.1) is extensive. In [3], feedback control of the chemostat by manipulating the dilution rate was studied for the promotion of coexistence. Other interesting control studies of the chemostat can be found in [1,6-10]. The stability and robustness of periodic solutions of the chemostat was studied in [11,12]. The problem of the stabilization of a given non-trivial steady state (X_s, S_s) of the chemostat model (1.1) was considered in [9], where it was shown that the simple feedback law $D = \mu(S)\frac{X}{X_s}$ or equivalently $D = \frac{\mu(S)X}{Y(S_i - S_s)}$ is a globally stabilizing feedback. See also the recent work [8] for the study of the robustness properties of the resulting closed-loop system under this feedback law in the presence of constant or time-varying errors in the inlet substrate concentration S_i.

In this work we consider the robust global feedback stabilization problem for the more general uncertain chemostat model

$$\dot{X} = \left(\mu(S) + \Delta(S,t)\right)X - DX - bX$$
$$\dot{S} = D(S_i - S) - \frac{1}{Y}\mu(S)X - mX \tag{1.2}$$
$$X \in (0,+\infty), S \in (0,S_i), D \geq 0$$

In the above,

- The term $-bX$ in the biomass balance represents the death rate of the cells in the chemostat. The parameter $b \geq 0$ is the cell mortality rate.
- The term $-mX$ in the substrate balance accounts for the rate of substrate consumption for cell maintenance ([2]) as well as the rate of release of substrate due to the death of the cells in the chemostat (which is proportional to bX). The parameter m is either positive or assumes a small negative value. The parameter m is related to the presence of variable apparent yield coefficient, which has been studied recently in [16].
- The term $\Delta(S,t)$ represents possible deviations of the specific growth rate of the biomass, primarily accounting for the adjustment of the biomass to changes in the substrate levels. It is assumed to be of the form $\Delta(S,t) = d_1(t)\left|S - S_s\right| - d_2(t)\max\left\{0, S_s - S\right\}$ with $d_1(t) \in [0,a]$, $d_2(t) \in [0,a]$, $a \geq 0$. Notice that at design steady state conditions $S_s \in (0, S_i)$, the uncertainty $\Delta(S,t)$ is assumed to vanish. See Figure 1 below for a sketch of the shape of the uncertainty range as a function of S_s.

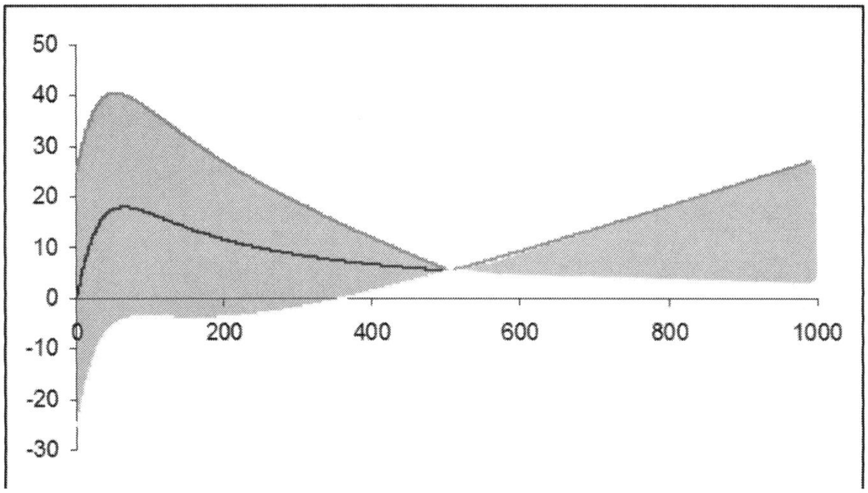

Figure 1: Indicative uncertainty range for the specific growth rate of the biomass, $\mu(S) + \Delta(S,t)$

(here $\mu(S) = \frac{75S}{100 + S + 0.025S^2}$, $a = 0.05$, $S_s = 506.72$)

Notice that model (1.2) becomes (1.1) if we set $a = b = m = 0$.

It is also important to notice that even in the case of zero uncertainty ($a = 0$), zero mortality rate b and for positive values for the constant m, the application of the feedback law $D = \mu(S)\frac{X}{X_s}$ does not necessarily lead to global stability. For example, for the Haldane model $\mu(S) = \frac{\mu_{max} S}{K_S + S + \frac{S^2}{K_I}}$, it is easy to verify that for arbitrarily small positive values for the constant m, the resulting closed-loop system under $D = \mu(S)\frac{X}{X_s}$ and $a = b = 0$ has two equilibrium points in the first quadrant $(S_s^{[1]}, X_s)$ and $(S_s^{[2]}, X_s)$, with $0 < S_s^{[1]} < S_s^{[2]}$. The equilibrium point $(S_s^{[2]}, X_s)$, is locally asymptotically stable with region of attraction the set $\{(S, X) : S > S_s^{[1]}, X > 0\}$. The stable manifold of the unstable equilibrium $(S_s^{[1]}, X_s)$ is the straight line $S = S_s^{[1]}$ and if the initial condition for the substrate is less than $S_s^{[1]}$ then the system is led to shut-down in finite time (i.e., there exists $T \geq 0$ such that $\lim_{t \to T^-} S(t) = 0$). Therefore, the feedback law needs to be modified in order to be able to guarantee global asymptotic stability for the desired equilibrium point.

This observation is illustrated by simulation in Figures 2 and 3. Figure 2 depicts the response of system (1.1) under the feedback law $D = \mu(S)\frac{X}{X_s}$ for the following parameter values, design steady state and initial conditions:

$$S_i = 10000\, mg/l,\ \frac{1}{Y} = 20\, mg/mg,\ \mu_{max} = 0.5\, d^{-1},\ K_S = 100\, mg/l, K_I = 4000\, mg/l$$
$$X_s = 474.66\, mg/l,\ S_s = 506.72\, mg/l,\ X(0) = 10\, mg/l,\ S(0) = 10\, mg/l$$

It is observed that the controller is very effective in bringing the system to its design steady state.

If, however, a term $-mX$ is present in the substrate balance, with $m = 0.1\, d^{-1} \cdot mg/mg$ and all the other parameters remain the same, the control law $D = \mu(S)\frac{X}{X_s}$ completely fails to bring the system to its design steady state; instead, it leads the system to shut down in finite time, as shown in Figure 3.

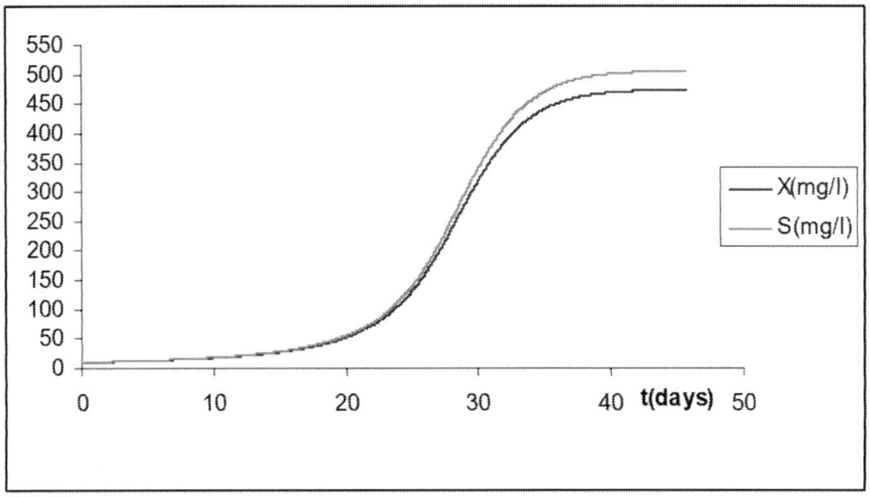

Figure 2: Evolution of states for system (1.2) with $a = b = m = 0$ under the control law $D = \mu(S)\frac{X}{X_s}$

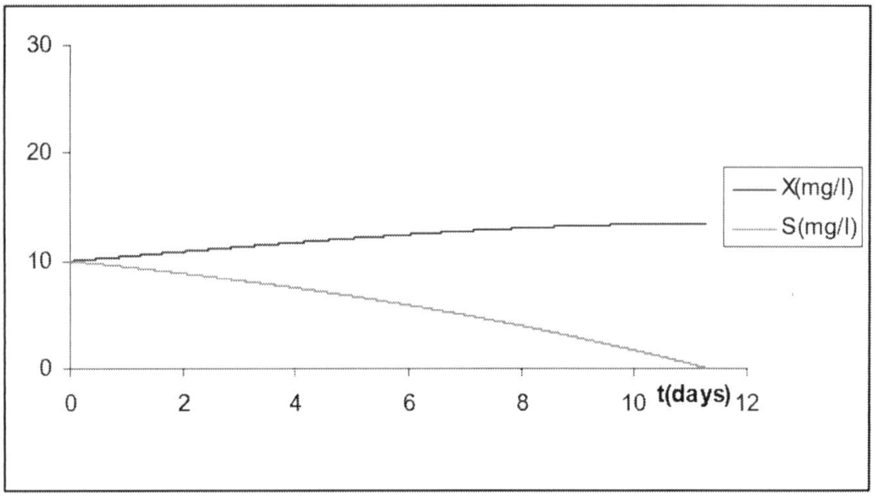

Figure 3: Evolution of states for system (1.2) with $a = b = 0$, $m = 0.1\ d^{-1} \cdot mg / mg$ under the control law $D = \mu(S)\dfrac{X}{X_s}$

2. Theory: Relaxed Lyapunov criteria for robust global stabilization of nonlinear systems

In this section, the main theoretical results of the present work are presented. We start with systems without manipulated input that are subject to disturbances d, and we review the notion of Uniform Robust Global Asymptotic Stability (URGAS). Consider a dynamical system of the form:

$$\begin{aligned} &\dot{x} = F(d,x) \\ &x \in \Re^n \,, d \in D \end{aligned} \tag{2.1}$$

We assume throughout this section that system (2.1) satisfies the following hypotheses:

(H1) $D \subset \Re^l$ is closed and bounded.

(H2) The mapping $D \times \Re^n \ni (d,x) \to F(d,x) \in \Re^n$ is continuous.

(H3) The vector field $F(d,x)$ is locally Lipschitz continuous with respect to $x \in \Re^n$.

Hypothesis (H3) guarantees that for every $(x_0, d) \in \Re^n \times M_D$, there exists a unique solution $x(t)$ of (2.1) with initial condition $x(0) = x_0$ corresponding to input $d \in M_D$. (Here by M_D we denote the class of all measurable and locally essentially bounded mappings $d : \Re^+ \to D$).

Definition 1 ([5]): *Let $x(t;x_0,d)$ denote the unique solution of (2.1) with initial condition $x(0) = x_0 \in \Re^n$ corresponding to input $d \in M_D$. Assume that hypotheses (H1-3) hold, with $F(d,0) = 0$ for all $d \in D$.*

We say that $0 \in \Re^n$ is Uniformly Robustly Globally Asymptotically Stable (URGAS) for (2.1) if the following properties hold:

- *for every $s > 0$, it holds that*
  $\sup\{|x(t;x_0,d)| \,; t \ge 0\,,\ |x_0| \le s \ , d \in M_D \} < +\infty$ *(Uniform Robust Lagrange Stability)*
- *for every $\varepsilon > 0$ there exists a $\delta := \delta(\varepsilon) > 0$ such that:*
  $\sup\{|x(t;x_0,d)| \,; t \ge 0\,,\ |x_0| \le \delta \ , d \in M_D \} \le \varepsilon$ *(Uniform Robust Lyapunov Stability)*
- *for every $\varepsilon > 0$ and $s \ge 0$, there exists a $\tau := \tau(\varepsilon, s) \ge 0$, such that:*
  $\sup\{|x(t;x_0,d)| \,; t \ge \tau\,,\ |x_0| \le s \ , d \in M_D \} \le \varepsilon$ *(Uniform Attractivity for bounded sets of initial states)*

Next we present relaxed Lyapunov-like sufficient conditions for URGAS. The Lyapunov-like conditions of the following theorem are "relaxed" in the sense that the Lyapunov differential inequality is not required to hold over the entire state space, but only for states that belong to an appropriate set of the state space, such that every trajectory of the system enters the set in finite time.

Theorem 1: *Consider system (2.1) under hypotheses (H1-3) with* $F(d,0)=0$ *for all* $d \in D$ *and suppose that there exists a set* $\Omega \subseteq \Re^n$ *with* $0 \in \Omega$, *functions* $V \in C^1(\Omega;\Re^+)$ *being positive definite and radially unbounded,* $T \in C^0(\Re^n;\Re^+)$, $G \in C^0(\Re^n;\Re^+)$, *which satisfy the following properties:*

(P1) *For every* $(d,x_0) \in M_D \times \Re^n$, *there exists* $\hat{t}(x_0,d) \in [0,T(x_0)]$ *such that the unique solution* $x(t;x_0,d)$ *of (2.1) satisfies* $x(t;x_0,d) \in \Omega$ *for all* $t \in [\hat{t}(x_0,d),t_{\max})$ *and* $|x(t;x_0,d)| \le G(x_0)$ *for all* $t \in [0,\hat{t}(x_0,d)]$, *where* $t_{\max} = t_{\max}(x_0,d)$ *is the maximal existence time of the solution,*

(P2) $\sup_{d \in D}\left(\nabla V(x) F(d,x)\right) < 0$ *for all* $x \in \Omega$, $x \ne 0$.

Then $0 \in \Re^n$ *is URGAS for (2.1).*

Remark 1: For disturbance-free systems, a set $\Omega \subseteq \Re^n$ that satisfies condition (P1) of Theorem 1 is called an <u>*absorbing set*</u>.

The following lemma provides sufficient conditions for the reachability condition (P1) of Theorem 1.

Lemma 1: *Consider system (2.1) under hypotheses (H1-3) and suppose that there exist locally Lipschitz functions* $h:\Re^n \to \Re$ *with* $h(0)<0$, $a:\Re^n \to \Re$ *being bounded from above with* $a(0)=0$, $W:\Re^n \to \Re^+$ *being radially unbounded, a continuous function* $\delta:\Re^+ \to (0,+\infty)$ *and constants* $K \ge 0$, $\varepsilon > 0$, *such that* $\{x \in \Re^n : 0 < h(x) < \varepsilon\} \ne \varnothing$ *and*

$$\sup_{d \in D} \nabla h(x) F(d,x) \le 0, \text{ for almost all } x \in \Re^n \text{ with } 0 < h(x) < b \tag{2.2a}$$

$$\sup_{d \in D}\left(\nabla h(x) - \nabla a(x)\right) F(d,x) \le -\delta(h(x)), \text{ for almost all } x \in \Re^n \text{ with } h(x) > 0 \tag{2.2b}$$

$$\sup_{d \in D} \nabla W(x) F(d,x) \le K W(x), \text{ for almost all } x \in \Re^n \text{ with } h(x) > 0 \tag{2.3}$$

Then for every $\hat{\varepsilon} \in (0,b)$ *there exist functions* $T \in C^0(\Re^n;\Re^+)$, $G \in C^0(\Re^n;\Re^+)$ *such that property (P1) of Theorem 1 holds with* $\Omega := \{x \in \Re^n : h(x) \le \hat{\varepsilon}\}$.

We next consider a nonlinear dynamic system with manipulated input u, of the form:

$$\begin{gathered} \dot{x} = f(d,x) + g(d,x)u \\ x \in \Re^n, d \in D, u \in U \end{gathered} \tag{2.4}$$

where $D \subset \Re^l$ is a closed and bounded set, $U \subseteq \Re^m$ a non-empty convex set with $0 \in U$, $f: D \times \Re^n \to \Re^n$, $g: D \times \Re^n \to \Re^{n \times m}$ are continuous mappings with $f(d,0)=0$ for all $d \in D$ and the vector field $f(d,x)+g(d,x)u$ is locally Lipschitz continuous with respect to $x \in \Re^n$. The problem is to construct a continuous state feedback law $u = k(x)$ with $k:\Re^n \to U$ and $g(d,0)k(0)=0$ for all $d \in D$, which achieves robust global stabilization of $0 \in \Re^n$ for (2.4), i.e., $0 \in \Re^n$ is uniformly robustly globally asymptotically stable for the closed-loop system $\dot{x} = f(d,x) + g(d,x)k(x)$.

In view of the foregoing analysis for systems of the form (2.1), we can develop "relaxed" Lyapunov-like conditions for the existence of a locally Lipschitz, globally stabilizing state feedback law $u = k(x)$ for system (2.4):
Instead of requesting the existence of a Robust Control Lyapunov Function (RCLF), i.e., the existence of a continuously differentiable, positive definite and radially unbounded function $V:\Re^n \to \Re^+$ with

$$\inf_{u \in U} \sup_{d \in D} \nabla V(x)\left(f(d,x) + g(d,x)u\right) < 0, \text{ for all } x \ne 0,\ x \in \Re^n$$

a "relaxed" condition

$$\inf_{u \in U} \sup_{d \in D} \nabla V(x)\left(f(d,x) + g(d,x)u\right) < 0, \text{ for all } x \ne 0,\ x \in \Omega$$

with $\Omega \subseteq \Re^n$, can be used to design a globally stabilizing state feedback. In particular, a continuous state feedback law $k:\Re^n \to U$ with $g(d,0)k(0)=0$ for all $d \in D$, which guarantees

$$\sup_{d\in D} \nabla V(x)\big(f(d,x)+g(d,x)k(x)\big)<0 \text{, for all } x \neq 0,\ x \in \Omega$$

with $\Omega \subseteq \Re^n$ such that every solution of (2.4) with $u=k(x)$ enters $\Omega \subseteq \Re^n$ in finite time, achieves robust global stabilization of $0 \in \Re^n$ for (2.4).

Using Theorem 1 as well as the sufficient conditions for the reachability condition (P1) from Lemma 1, we can show the following:

Theorem 2: *Consider system (2.4) and suppose that there exist continuously differentiable functions $h:\Re^n \to \Re$ with $h(0)<0$, $W:\Re^n \to \Re^+$ being radially unbounded, $V:\Re^n \to \Re^+$ being positive definite and radially unbounded, a continuous non-increasing function $\delta:\Re^+ \to (0,+\infty)$ and constants $K \geq 0$, $\varepsilon > 0$ such that $\{x \in \Re^n : h(x) \geq \varepsilon\} \neq \varnothing$ and the following properties hold:*

(R1) *For every $x \in \Re^n$ with $h(x) \geq 0$ there exists $u \in U$ with*

$$\sup_{d\in D} \nabla h(x)\big(f(d,x)+g(d,x)u\big) \leq -\delta(h(x)) \tag{2.5}$$

$$\sup_{d\in D} \nabla W(x)\big(f(d,x)+g(d,x)u\big) \leq KW(x) \tag{2.6}$$

(R2) *For every $x \neq 0$ with $h(x) \leq \varepsilon$ there exists $u \in U$ with*

$$\sup_{d\in D} \nabla V(x)\big(f(d,x)+g(d,x)u\big) < 0 \tag{2.7}$$

(R3) *For every $x \in \Re^n$ with $h(x) \in [0,\varepsilon]$ there exists $u \in U$ satisfying (2.5), (2.6) and (2.7).*

(R4) *There exists a neighbourhood $\mathcal{N}$ of $0 \in \Re^n$ and a locally Lipschitz mapping $\tilde{k}:\mathcal{N} \to U$ with $\tilde{k}(0)=0$ such that $\sup_{d\in D} \nabla V(x)\big(f(d,x)+g(d,x)\tilde{k}(x)\big)<0$ for all $x \in \mathcal{N}$, $x \neq 0$.*

Then there exists a locally Lipschitz mapping $k:\Re^n \to U$ with $k(0)=0$ such that $0 \in \Re^n$ is URGAS for the closed-loop system (2.4) with $u=k(x)$.

The theoretical results developed in this section can have the following types of applications:

i) in many cases, RCLFs are not available, while "relaxed" RCLFs can be found, leading to robust globally stabilizing controllers.while RCLFs are not available.

ii) even when a RCLF is known, the use of "relaxed" RCLFs can lead to alternative feedback designs that may have advantages over the ones from RCLF.

In the next section, the results will be applied to the problem of robust feedback stabilization of the chemostat.

3. Application to chemostat control

Consider again the general chemostat model (1.2) and assume that the specific growth rate function $\mu:\Re \to [0,\mu_{\max}]$ is a locally Lipschitz function with $\mu(S)=0$ for all $S \leq 0$ and $\mu(S)>0$ for all $S>0$. Additionally, we make the following assumptions:

(S1) *There exists an equilibrium point $(X_s,S_s) \in (0,+\infty)\times(0,S_i)$ with $\mu(S_s)=D_s+b$ and $\dfrac{D_s Y(S_i-S_s)}{D_s+b+Ym}=X_s$*

for a certain value of the dilution rate $D_s>0$.

Assumption is satisfied for Monod, Haldane and generalized Haldane kinetics, as long as the value of the dilution rate D_s is not too high (smaller than the washout dilution rate).

(S2) *There exists $S^+ \in (0,S_s)$ and $p>0$ such that $\frac{1}{Y}\mu(S)+m \geq p$ and $\mu(S)-b \geq 2p$ for all $S \in [S^+,S_i]$.*

Assumption is satisfied for Monod, Haldane and generalized Haldane kinetics, as long as $\min\left(\mu(S_s),\mu(S_i)\right) > \max\left(b,-Ym\right)$, in which case, any $S^+ > \max\left(b,-Ym\right)$ satisfies it.

The goal is the robust global stabilization of the non-trivial equilibrium point $(X_s, S_s) \in (0,+\infty)\times(0,S_i)$ with $\mu(S_s) = D_s + b$ and $\frac{D_s Y(S_i - S_s)}{D_s + b + Ym} = X_s$ involved in hypotheses (S1-2) for system (1.2).

As a first step, we apply the change of coordinates:

$$S = \frac{S_i \exp(x_1)}{c+\exp(x_1)} \;;\; \frac{X}{S_i - S} = G\exp(x_2) \tag{3.1}$$

and the input transformation:

$$D = D_s + u \tag{3.2}$$

where $c := \frac{S_i}{S_s} - 1$ and $G := \frac{D_s Y}{D_s + b + Ym}$.

The above coordinate change maps the strip $\{(X,S)\in\Re^2 : X>0,\ 0<S<S_i\}$ onto $\Re^2$.

Under the above transformations, the uncertain system (1.2) is transformed to the following:

$$\begin{aligned}
\dot{x}_1 &= \left(c\exp(-x_1)+1\right)\left(D_s + u - \left(\frac{1}{Y}\tilde{\mu}(x_1)+m\right)G\exp(x_2)\right)\\
\dot{x}_2 &= \left(\tilde{\mu}(x_1) + d_1\frac{cS_s}{c+\exp(x_1)}\left|\exp(x_1)-1\right| - d_2\frac{cS_s}{c+\exp(x_1)}\max\left(0,1-\exp(x_1)\right) - b\right) - \left(\frac{1}{Y}\tilde{\mu}(x_1)+m\right)G\exp(x_2) \qquad (3.3)\\
x &= (x_1,x_2)\in\Re^2, u\in U := [-D_s,+\infty), d=(d_1,d_2)\in[0,a]^2
\end{aligned}$$

where $\tilde{\mu}(x_1) := \mu\left(\frac{S_i\exp(x_1)}{c+\exp(x_1)}\right)$. Notice that $x = 0$ is an equilibrium point for the above system for $u \equiv 0$. Therefore, we seek for a locally Lipschitz feedback law $k:\Re^2\to\Re$ with $k(0)=0$ so that the origin is uniformly robustly globally asymptotically stable for the closed-loop system under $u = k(x)$ in the sense described in the previous section.

Insights for the solution of the feedback stabilization problem for (3.3) may be obtained by considering the transformation of the nominal system (1.1) (i.e. for $a = b = m = 0$):

$$\begin{aligned}
\dot{x}_1 &= \left(c\exp(-x_1)+1\right)\left(D_s + u - \tilde{\mu}(x_1)\exp(x_2)\right)\\
\dot{x}_2 &= \tilde{\mu}(x_1) - \tilde{\mu}(x_1)\exp(x_2) \qquad (3.4)\\
x &= (x_1,x_2)\in\Re^2, u\in U := [-D_s,+\infty)
\end{aligned}$$

For the control system (3.4), families of Control Lyapunov Functions (CLF) are known (see [8]). Let $\gamma:\Re\to\Re^+$, $\beta:\Re\to\Re^+$ be non-negative, continuously differentiable functions with $\gamma(0)=\beta(0)=0$ and such that

$$x\gamma'(x) > 0\,,\ x\beta'(x) > 0\,, \text{ for all } x \neq 0 \tag{3.5a}$$

$$\text{if } x\to\pm\infty \text{ then } \gamma(x)\to+\infty \text{ and } \beta(x)\to+\infty \tag{3.5b}$$

Then the family of functions:

$$V(x) = \gamma(x_1) + \beta(x_2) \tag{3.6}$$

are radially unbounded, positive definite and continuously differentiable, and constitute CLFs for system (3.4). The knowledge of the above family of CLFs allows us to obtain a family of stabilizing feedback laws for (3.4). The reader may verify that the following family of feedback laws:

$$k(x) := -D_s + \tilde{\mu}(x_1)\exp(x_2)\varphi(x_1) + q(x_1, x_2) \tag{3.7}$$

where $\varphi : \Re \to \Re^+$ is a locally Lipschitz, non-negative function with $\varphi(0) = 1$, $\varphi(x) < 1$ for $x > 0$ and $\varphi(x) > 1$ for $x < 0$, $q : \Re^2 \to \Re^+$ is a locally Lipschitz, non-negative function with $q(x_1, x_2) = 0$ for $x_1 \geq 0$, is a family of globally stabilizing feedback laws for (3.4).

Back-transforming (3.7) to the original variables, gives the following family of globally stabilizing control laws for the nominal system (1.1):

$$D = \frac{\mu(S)X}{Y(S_i - S)}\varphi(S) + q(S, X) \tag{3.8}$$

where

a) $\varphi : \Re^+ \to \Re$ is a locally Lipschitz function with $\varphi(S) \geq 0$ for all $0 < S < S_i$,

$\varphi(S) > 1$ for $0 < S < S_s$, $\varphi(S_s) = 1$, $\varphi(S) < 1$ for $S > S_s$.

b) $q : \Re^+ \times \Re^+ \to \Re^+$ is a locally Lipschitz, non-negative function with $q(S, X) \equiv 0$ for $S > S_s$,

Notice hat the special choices $\varphi(S) = \frac{S_i - S}{S_i - S_s}$ and $q(S, X) = 0$ lead to $D = \frac{\mu(S)X}{Y(S_i - S_s)}$, which is the feedback law in [9] (see also [8] and references therein).

The method of CLF can be extended to the uncertain system (1.2) or (3.4), leading to globally stabilizing controllers. However, the results are extremely complicated and are omitted here for brevity.

The method of "relaxed" Control Lyapunov Functions proposed in the previous section can now be applied to the problem in hand and give rise to simple feedback controllers. Details are omitted, but the main point is that the Lyapunov differential inequality is imposed over the set $\Omega := \{(x_1, x_2) \in \Re^2 : x_1 \geq x_1^*\}$, where $x_1^* = \ln\left(\frac{\frac{S_i}{S_s} - 1}{\frac{S_i}{S^+} - 1}\right)$, instead of the entire $\Re^2$.

Theorem 3: *Let $\psi : \Re \to \Re^+$ be a locally Lipschitz non-increasing function with $\psi(x) = 0$ for all $x \geq 0$ and $\psi(x) > 0$ for all $x < 0$ and let $L : \Re^2 \to (0, +\infty)$ be a locally Lipschitz function with $\inf\{L(x) : x \in \Re^2\} > 0$. Under hypotheses (S1-2), for every $a \geq 0$, $0 \in \Re^2$ is URGAS for the closed-loop system (3.3) with*

$$u = -D_s + \max(0, \frac{1}{Y}\tilde{\mu}(x_1) + m)G\exp(x_2 - x_1) + L(x_1, x_2)\psi(x_1) \tag{3.9}$$

Transforming the above result in terms of the original variables, we obtain:

Theorem 3′: *Let $\psi : \Re^+ \to \Re^+$ be a locally Lipschitz non-increasing function with $\psi(S) = 0$ for all $S \geq S_s$ and $\psi(S) > 0$ for all* for *$0 < S < S_s$ and let $L : \Re^+ \times \Re^+ \to (0, +\infty)$ be a locally Lipschitz function with $\inf\left\{L(X, S) : X \in \Re^+,\ 0 < S < S_i\right\} > 0$. For every $a \geq 0$, the origin is URGAS for the closed-loop system (1.2) under the feedback law*

$$D = \frac{S_s}{S_i - S_s} \cdot \frac{X}{S} \cdot \max\left(0, \frac{1}{Y}\mu(S) + m\right) + L(S, X) \cdot \psi(S) \tag{3.10}$$

Notice that the special choices $L(x_1, x_2) \equiv L > 0$ and $\psi(x_1) = \max\left(0, 1 - \exp(x_1)\right)$, or, in terms of the original variables, $L(S, X) \equiv L > 0$ and $\psi(S) = \frac{S_i}{S_i - S} \max\left(0, 1 - \frac{S}{S_s}\right)$, lead to the following more concrete form of feedback control law

$$D = \frac{S_s}{S_i - S_s} \cdot \frac{X}{S} \cdot \max\left(0, \frac{1}{Y}\mu(S) + m\right) + L \frac{S_i}{S_i - S} \max\left(0, 1 - \frac{S}{S_s}\right) \tag{3.11}$$

Finally, it is important to point out certain possible simplifications of the derived controller:

- for $m = 0$, control law (3.10) simplifies to $D = \frac{\mu(S)X}{Y(S_i - S_s)} \cdot \frac{S_s}{S} + L(S, X) \cdot \psi(S)$, which is of the form (3.8).
- for $m > 0$, control law (3.10) simplifies to $D = \frac{\left(\mu(S) + Ym\right)X}{Y(S_i - S_s)} \cdot \frac{S_s}{S} + L(S, X) \cdot \psi(S)$,

  or equivalently $D = \frac{\mu(S)X}{\breve{Y}(S) \cdot (S_i - S_s)} \cdot \frac{S_s}{S} + L(S, X) \cdot \psi(S)$, where

  $\breve{Y}(S)$ is an apparent variable yield factor defined by $\frac{1}{\breve{Y}(S)} = \frac{1}{Y} + \frac{m}{\mu(S)}$.

The performance of the derived controller is illustrated through simulations in Figures 4-6. Figures 4 and 5 depict the response of system (1.2) under the feedback law (3.11) and for the following parameter values and design steady state conditions:

$$S_i = 10000\ mg/l,\ \frac{1}{Y} = 20\ mg/mg,\ \ \mu_{\max} = 0.5\ d^{-1},\ K_S = 100\ mg/l, K_I = 4000\ mg/l$$

$$a = 0,\ b = 0,\ m = 0.1\ d^{-1} \cdot mg/mg,\ \ S_s = 506.72\ mg/l,\ X_s = 468.46\ mg/l$$

In Figure 4, the system's initial conditions were $X(0) = 10\ mg/l$, $S(0) = 10\ mg/l$, whereas in Figure 5, $X(0) = 1000\ mg/l$, $S(0) = 1000\ mg/l$. Note that the initial conditions were chosen to be very far from the design steady state conditions in order to test the capabilities of the proposed controller. It is observed that, despite the adverse conditions, the controller is very effective in bringing the system to its design steady state.

In Figure 6, parameter values, design steady state and initial conditions are as in Figure 5, except that an oscillating perturbation in the biomass growth rate is assumed to be present in system (1.2). Despite this persistent perturbation, the controller is very effective in bringing the system to its design steady state, even though the time response is not as fast.

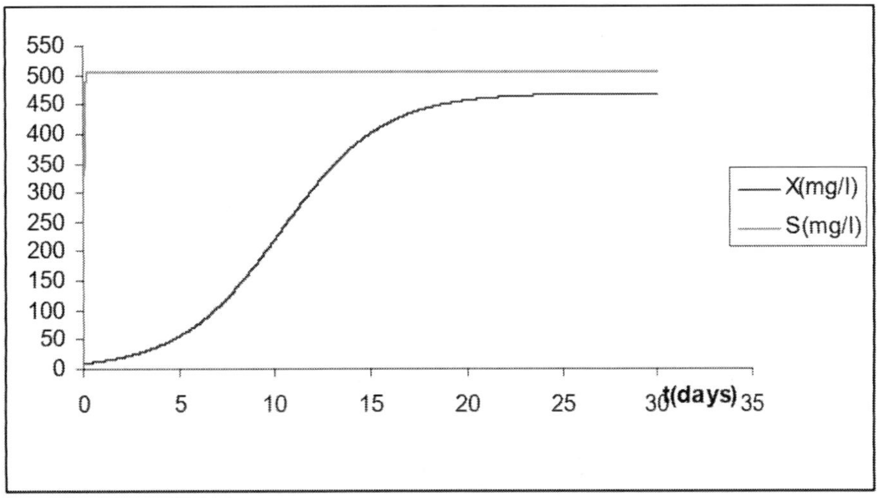

Figure 4: Evolution of states for system (1.2) with $a = b = 0$, $m = 0.1\ d^{-1} \cdot mg/mg$ under the control law (3.11) with $L = 1$ and initial conditions $X(0) = 10\ mg/l$, $S(0) = 10\ mg/l$

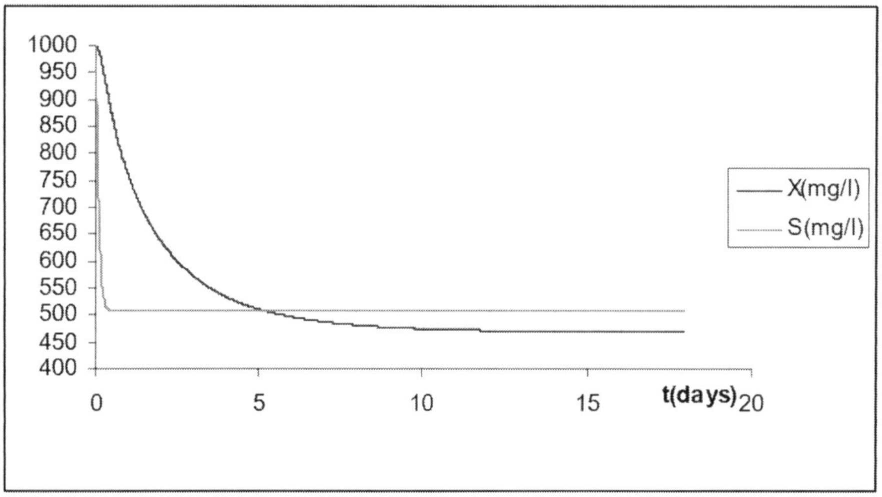

Figure 5: Evolution of states for system (1.2) with $a = b = 0$, $m = 0.1\ d^{-1} \cdot mg/mg$ under the control law (3.11) with $L = 1$ and initial conditions $X(0) = 1000\,mg/l$, $S(0) = 1000\,mg/l$

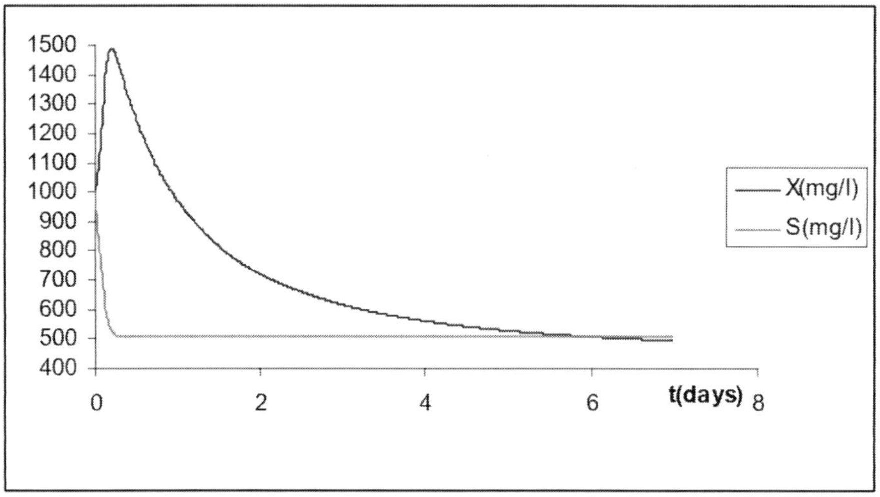

Figure 6: Evolution of states for system (1.2) with $a = 0.19$, $d_1(t) = |\sin(t)|$, $d_2(t) \equiv 0$, $b = 0$, $m = 0.1\ d^{-1} \cdot mg/mg$ under the control law (3.11) with $L = 1$ and initial conditions $X(0) = 1000\,mg/l$, $S(0) = 1000\,mg/l$

References

[1] Antonelli, R. and A. Astolfi, "Nonlinear Controller Design for Robust Stabilization of Continuous Biological Reactors", *Proceedings of the IEEE Conference on Control Applications*, Anchorage, AL, September 2000.

[2] Bailey, J. E. and D. F. Ollis, "Biochemical Engineering Fundamentals", 2nd Ed., McGraw-Hill, New York, 1986.

[3] De Leenheer, P. and H. L. Smith, "Feedback Control for Chemostat Models", *Journal of Mathematical Biology*, 46, 2003, 48-70.

[4] Freedman, H. I., J.W.H. So and P. Waltman, "Chemostat Competition with Time Delays", *IMACS Ann. Comput. and Appl. Math.*, 5, 1/4 (12th IMACS World Congr. Sci. Com., Paris, July 1988), 1989, 171-173.

[5] Freeman, R. A. and P. V. Kokotovic, "Robust Nonlinear Control Design- State Space and Lyapunov Techniques", Birkhauser, Boston, 1996.

[6] Gouze, J. L. and G. Robledo, "Robust Control for an Uncertain Chemostat Model", *International Journal of Robust and Nonlinear Control*, 16(3), 2006, 133-155.

[7] Harmard, J., A. Rapaport and F. Mazenc, "Output tracking of continuous bioreactors through recirculation and by-pass", *Automatica*, 42, 2006, 1025-1032.

[8] Karafyllis, I., C. Kravaris, L. Syrou and G. Lyberatos, "A Vector Lyapunov Function Characterization of Input-to-State Stability with Application to Robust Global Stabilization of the Chemostat", *European Journal of Control*, 14(2), 2008, 47-61.

[9] Mailleret, L. and O. Bernard, "A Simple Robust Controller to Stabilize an Anaerobic Digestion Process" in *Proceedings of the 8th Conference on Computer Applications in Biotechnology*, 2001, 213-218.

[10] Mazenc, F., M. Malisoff, J. Harmand, "Stabilization and Robustness Analysis for a Chemostat Model with Two Species and Monod Growth Rates via a Lyapunov Approach", *Proceedings of the 46th IEEE Conference on Decision and Control*, New Orleans, 2007.

[11] Mazenc, F., M. Malisoff and P. De Leenheer, "On the Stability of Periodic Solutions in the Perturbed Chemostat", Mathematical Biosciences and Engineering, 4(2), 2007, 319-338.

[12] Mazenc, F., M. Malisoff and J. Harmand, "Further Results on Stabilization of Periodic Trajectories for a Chemostat with Two Species", *IEEE Transactions on Automatic Control*, 53(1), 2008, 66-74.

[13] Smith, H. and P. Waltman, "The theory of the chemostat. Dynamics of Microbial Competition", Cambridge Studies in Mathematical Biology, 13, Cambridge University Press: Cambridge, 1995.

[14] Wang , L. and G. Wolkowicz, "A delayed chemostat model with general nonmonotone response functions and differential removal rates", *Journal of Mathematical Analysis and Applications*, 321, 2006, 452-468.

[15] Wolkowicz , G. and H. Xia, "Global Asymptotic Behavior of a chemostat model with discrete delays", *SIAM Journal on Applied Mathematics*, 57, 1997, 1019-1043.

[16] Zhu, L. and X. Huang, "Multiple limit cycles in a continuous culture vessel with variable yield", *Nonlinear Analysis*, 64, 2006, 887-894.

266c Lyapunov-Based Model Predictive Control of Nonlinear Systems Subject to Time-Varying Measurement Delays

Panagiotis D. Christofides[1], Jinfeng Liu[2], David Muñoz de la Peña[2], and James F. Davis[3]. (1) Department of Chemical and Biomolecular Engineering, Department of Electrical Engineering, University of California, Los Angeles, Los Angeles, CA 90095, (2) Department of Chemical and Biomolecular Engineering, University of California, Los Angeles, Los Angeles, CA 90095, (3) University of California - Los Angeles, Los Angeles, CA 90095

The problem of designing feedback control systems for nonlinear systems subject to time-varying measurement delays is a fundamental one and its solution can find significant application in a number of control engineering problems including, for example, design of networked control systems (NCS). NCS are control systems which operate over a communication network (wired or wireless) and can lead to significant improvements in the efficiency, flexibility, robustness and fault-tolerance of industrial control systems as well as to reduction of the installation, reconfiguration and maintenance costs. However, the design of NCS has to account for the dynamics introduced by the communication network which may include time-varying delays, data quantization or data losses. In addition to NCS, another source of time-varying delays in the feedback loop is measurement sensor delays, which are particularly important during the measurement of species concentrations and particle size distributions in process control applications.

Motivated by the above, this work deals with the design of predictive controllers for nonlinear system subject to time-varying measurement delay in the feedback loop. In particular, we modify the Lyapunov-based MPC schemes developed previously by our group to take into account time-varying measurement delays, both in the optimization problem formulation and in the controller implementation. In the LMPC scheme proposed in the present work, when measurement delays occur, the nominal model of the system is used together with the delayed measurement to estimate the current state, and the resulting estimate is used to evaluate the LMPC controller; at sampling times where no measurements are available due to the delay, instead of setting the control input to zero or to the last available value, the actuator implements the last optimal input trajectory evaluated by the controller (this requires that the controller must store in memory the last evaluated optimal control input trajectory). The proposed LMPC scheme inherits the stability and robustness properties in the presence of uncertainty and time-varying delay of the Lyapunov-based controller, while taking into account optimality considerations. Specifically, the proposed LMPC scheme allows for an explicit characterization of the stability region and guarantees that the closed-loop system in the presence of time-varying measurement delays is ultimately bounded in a region that contains the origin if the maximum delay is smaller than a constant that depends on the parameters of the system and the Lyapunov-based controller that is used to formulate the optimization problem. The application of the proposed Lyapunov-based model predictive control method is illustrated using a nonlinear chemical process example with asynchronous, delayed measurements and its stability and performance properties are demonstrated to be superior to the ones of two existing Lyapunov-based model predictive controllers.

1. Liu, J., D. Munoz de la Pena, P. D. Christofides and J. F. Davis, "Lyapunov-Based Model Predictive Control of Nonlinear Systems Subject to Time-Varying Measurement Delays," Int. J. Adapt. Contr. & Sign. Process., in press.

266d Agent-Based Control of RAFT Polymerization Systems In Cstr Networks

Derya Demirel, Derya Tetiker, Sinem Perk, Fouad Teymour, and Ali Cinar. Chemical and Biological Engineering, Illinois Institute of Technology, 10 W. 33rd Street, Chicago, IL 60616

Multi-agent systems provide a powerful platform for supervision and control of distributed and complex processes. This work focuses on controlling the product grade distribution in RAFT polymerization reactor networks for smooth transition between different product grades or maintaining the targeted product grade despite various disturbances using an agent based approach. The control algorithm is designed for control of the complete molecular weight distribution. The use of a distributed reactor network allows for flexibility in targeting almost any shape for the overall molecular weight distribution.

RAFT polymerization is one of the most popular living polymerization methods. Compared to the other living polymerization methods a wider range of monomers can be used for RAFT Polymerization with more flexible operating conditions. This system is studied in a network of interconnected continuous stirred tank reactors (CSTRs). Individual reactors produce polymer chains with different molecular weight distributions contributing to the overall molecular weight distribution of the network.

A number of centralized and decentralized agent-based control structures have been developed for controlling the distribution of autocatalytic species in CSTR networks [1, 2, 3, 4, 5]. These distributed structures consist of multiple layers of agents where in one layer, local controller agents work on the local objectives in parallel using local information in that region of the network together with heuristics. On the other hand, agents in higher layers coordinate local controller agents in order to achieve the global objective.

A similar agent-based structure is implemented to control the molecular weight distribution in RAFT polymerization in CSTR networks. Molecular weight distribution is one of the most important properties that characterize the application properties of polymeric materials which makes it a critical control issue. In this work two control problems will be studied. The first problem is to maintain the overall molecular weight distribution of the RAFT polymerization system in the CSTR network that is subject to a number of disturbances and the second one is to drive the system from one molecular weight distribution to another.

The overall distribution obtained for the network results from the blend of the individual reactor's distributions taking into account the material flow between the reactors. One can change the individual reactors' molecular weight distributions to control the overall network distribution. KR Method [6, 7, 8] is used for solving the corresponding population balance equations for obtaining the individual and overall molecular weight distributions. As a result, the distribution is represented by a number of pivot points that are used as the basis for controlling the system. A global agent will track the current overall molecular weight distribution and compare it with the desired one. In the case of a difference between the current and targeted distribution, local controller agents will calculate their individual contributions to the error at each pivot point and shift their reactor's molecular weight distribution to minimize the error, in accordance with their local objectives. The local controller agents have access to a number of manipulated variables such as the inlet monomer concentration, inlet initiator concentration, RAFT agent concentration and interconnection flow rates between reactors. Using local information available and heuristics, local controller agents decide on which manipulated variables to adjust to reach their local objectives.

The state equations of the individual CSTR's that are obtained by applying KR Method are solved by a DAE solver available in MATLAB. The agent based control framework is developed in REPAST Symphony which is an agent development platform based on Java. The connection between the control system and the process simulator is achieved by JMatLink library.

[1] E. Tatara, I. Birol, F. Teymour, and A. Cinar, "Agent-based Control of Autocatalytic Replicators in Networks of Reactors," Computers & Chemical Engineering, vol. 29, pp. 807–815, 2005.

[2] E. Tatara, C. Hood, F. Teymour and A. Cinar, "Adaptive Agent-based Control of Product Grade Transitions in Reactor", Prepr. IFAC Intl Symposium on Advanced Control of Chemical Processes (ADCHEM 03), Gramado, Brazil, April 2-5, 2006.

[3] E. Tatara, F. Teymour and A. Cinar, "Agent-based Control of Spatially Distributed Chemical Reactor Networks", AIChE Annual Meeting, Cincinnati, OH, November, 2005.

[4] M. D. Tetiker, A. Artel, E. Tatara, F. Teymour, M. North, C. Hood, A. Cinar, "Agent-based System for Reconfiguration of Distributed Chemical Reactor Network Operation", Proc. American Control Conf., June 14-16, 2006, Minneapolis, MN.

[5] M. D. Tetiker, A. Cinar, F. Teymour, M. North, "Decentralized Multi-Agent Control of Distributed Reactor Networks", AIChE Annual Meeting, San Francisco CA, November, 2007.

[6] S. Kumar, D. Ramkrishna, "On the Solution of Population Balance Equations by Discretization-I. A Fixed Pivot Technique", Chem. Eng. Sci. 1996, 51, 1311.

[7] S. Kumar, D. Ramkrishna, "On the Solution of Population Balance Equations by Discretization-I. A Fixed Pivot Technique", Chem. Eng. Sci. 1996, 51, 1333.

[8] S. Kumar, D. Ramkrishna, "On the Solution of Population Balance Equations by Discretization-I. A Fixed Pivot Technique", Chem. Eng. Sci. 1997, 52, 4659.

266e Dynamic Estimation and Control of Dissipative Pde Systems Using Extended Kalman Filter

Sivakumar Pitchaiah and Antonios Armaou. Department of Chemical Engineering, The Pennsylvania State University, University Park, PA 16802

Many processes relevant to the chemical process industry necessitate the consideration of transport phenomena (momentum, heat and mass transfer), often coupled with chemical reactions. Examples range from reactive distillation in petroleum processing to plasma enhanced chemical vapor deposition, etching and metallorganic vapor phase epitaxy (MOVPE) in semiconductor manufacturing. Mathematical descriptions of these transport-reaction processes can be derived from dynamic conservation equations and usually involve highly dissipative (typically parabolic) partial differential equation systems (PDEs).

The feedback control problem of these processes is nontrivial owing to the spatially distributed description of their dynamics. The state in the corresponding control problem is infinite dimensional when expressed in appropriate spaces. However the discrete nature of actuators, finite precision and limited memory capacity of the computers implementing the control action requires the approximation of the dynamics of the infinite dimensional PDE system by a finite dimensional dynamical system. The traditional approach to achieve this is based on discretization of the underlying PDEs using finite differences/elements to yield a set of ordinary differential equations (ODEs) which are subsequently used in formulating the feedback control problem for the processes. However, this approach in general leads to very high dimensionality of the controllers and especially cannot be applied to problems which have complicated geometries and/or have sharp variations of solutions, since use of finer discretization to accurately capture the dynamics is then required.

Some model reduction techniques capitalize on the fact that the eigenspectrum of the spatial differential operators in the parabolic PDE systems can be partitioned into a finite-dimensional slow subspace and an infinite-dimensional fast subspace. In other words, the

long-term dynamics of the dissipative PDEs are finite-dimensional and therefore a finite dimensional model would accurately capture the dynamics of the original infinite dimensional problem. One such approach is Galerkin's method, where the solution of the system is expanded using the eigenfunctions of the spatial differential operator. This yields a system of ordinary differential equations (ODEs) that accurately describe the dominant (slow) modes of the PDEs, and is subsequently used to design feedback controllers.

In this work, the problem of dynamic estimation and control of these PDEs when noisy measurements are available (from limited number of sensors) is investigated using Galerkin's method combined with an extended Kalman filter (EKF). We present an application of the above approach to control the temperature in a catalytic rod where an exothermic reaction is taking place. We assume that the noisy measurement data is available from three measurement sensors. The effect of location of these sensors on the performance of EKF is studied through Monte-Carlo simulations of the estimator. The role of process noise covariance and the measurement noise covariance towards the effectiveness of EKF is elucidated through numerical simulations. It is observed from numerical simulations that the 2-norm of the estimation error asymptotically converges to zero as more measurements from the process are made available to the estimator. A discrete time nonlinear feedback controller, based on the estimated states, is designed which successfully stabilizes the process to an open-loop unstable steady-state.

266f Nonlinear Quasi-Decentralized Control of Networked Process Systems

Yulei Sun and Nael H. El-Farra. Department of Chemical Engineering & Materials Science, University of California, Davis, One Shields Avenue, Davis, CA 95616

Quasi-decentralized control refers to a distributed control strategy in which most signals used for control are collected and processed locally within each unit (typically over dedicated networks), although some signals -- the total number of which is kept to a minimum -- still need to be transferred between the plant units and their local controllers over a shared (possibly wireless) communication network to account for the interactions between the different units and minimize the propagation of disturbances and process upsets. It is a compromise solution that overcomes the stability and performance limitations of decentralized control approaches while avoiding, at the same time, the complexity and lack of flexibility associated with traditional centralized control structures. A key problem that needs to be addressed in the design of quasi-decentralized control systems is the integration of communication issues and limitations in the formulation and solution of the plant-wide control problem. This is an important problem given the increased reliance in the process industries in recent years on distributed sensor and control systems that are accessed over shared communication networks rather than hardwired. The transition from dedicated, point-to-point connections to multi-purpose shared communication networks is driven in part by the reduced installation and maintenance time and costs as well as the flexibility and enhanced reconfigurability and fault-tolerance of networked control systems [1].

The design of a quasi-decentralized control strategy that enforces the desired closed-loop objectives with minimal cross communication between the component subsystems is an appealing goal since it reduces reliance on the communication medium and helps save on communication costs. This is an important consideration particularly when the communication medium is a potentially unreliable wireless sensor network where conserving network resources is key to prolonging the service life of the network. While the emerging paradigm of control over networks (see, for example, [2], [4] for surveys of results in this area) provides a natural framework to address the issues of control and communication integration, the majority of research studies on networked control systems have focused on single-unit processes using a centralized control architecture, which is not always the best choice for the structure of the controller in a plant-wide setting. By comparison, results on networked control of multi-unit plants with tightly interconnected units have been limited.

In an effort to address this problem, we recently developed in [3] a quasi-decentralized model-based networked control framework for multi-unit plants modeled by linear systems of differential equations. In this architecture, each unit in the plant has a local control system with its sensors and actuators connected to the local controller through a dedicated communication network. The local control systems in turn communicate with one another through a shared communication network. To satisfy the desired closed-loop stability and performance properties of the overall plant with minimal cross-communication between the plant units, we embed in the local control system of each unit a set of dynamic models that provide an approximation of the interactions between the given unit and its neighbors in the plant when communication is suspended and measurements are not transmitted through the plant-wide network. The state of each model is then updated using measurements from the corresponding unit when communication is re-established. The successful implementation of this architecture requires characterizing the maximum allowable update period (equivalently, the minimum transmission frequency) between the sensor suite of each unit and the local controllers of its neighbors, which is a measure of the extent of network resource utilization. Exploiting the linear structure of the plant and the controllers, both necessary and sufficient conditions for closed-loop stability were obtained.

When this architecture is implemented on a nonlinear plant, the update period predicted by linearization-based analysis can guarantee stability for sufficiently small initial conditions where the linearization is valid. Stabilization from initial conditions that are far from the desired equilibrium point requires increasing the frequency of measurement communication (i.e., reducing the update period) substantially which leads to additional network utilization. Since many chemical processes are characterized by strong nonlinear dynamics and need to operated over wide regions of the operating space for economic reasons, it is important to develop networked control approaches that account explicitly for the nonlinearities -- both in the control law and the communication logic designs -- and that lead to an explicit characterization of the minimum allowable cross-communication frequency that is independent of the initial condition.

Motivated by these considerations, we present in this work a nonlinear quasi-decentralized control structure for multi-unit nonlinear plants whose constituent units communicate over a shared communication network. The networked control structure accounts explicitly for the presence of nonlinearities and thus allows stabilization from large initial conditions without the unnecessary expenditure of additional network resources. To obtain an explicit characterization of the minimum stabilizing communication frequency, we consider a class of nonlinear plants where the local dynamics of each unit are nonlinear and its interactions with the other units depend linearly on the states of those units. This class arises commonly in the modeling of chemical plants in which the interconnections between different processing units are the result of material and energy stream exchanges. We synthesize for each unit a controller that consists of a nonlinear feedback component and a linear feed-forward component. The nonlinear component is designed using feedback linearization techniques and is responsible for stabilizing the unit in the absence of interconnections using the locally available measurements. The fact that these measurements are typically transmitted over a dedicated network (and can therefore be assumed to be continuously available to the local controller) allows the nonlinear controller to cancel the effects of the nonlinearities and enforce a linear closed-loop structure. In contrast to the local nonlinear controller, the linear ``feed-forward" component acts to compensate for the interconnections of the unit with its neighbors and receives measurements from those units over the shared communication network at discrete time instances only.

To reduce the frequency at which these measurements are transmitted over the shared network, we embed in each unit a set of nonlinear dynamic models that recreate the interactions of the unit with its neighbors when measurements are not available through the network. The state of each model is updated at discrete time instances when communication is re-established. By exploiting the specific structure of the plant, the properties of the hybrid linear/nonlinear controllers designed, and the rates at which the measurements are available at the local and plant-wide levels, we are finally able to formulate the overall networked closed-loop plant as a switched linear system. Both necessary and sufficient conditions for closed-loop stability are derived leading to an explicit characterization of maximum allowable update period in terms of model uncertainty and controller design parameters. Unlike linearization-based schemes, the resulting update period is independent of the initial condition and can be applied to the nonlinear plant globally without loss of stability. This feature allows expanding the operating range of the quasi-decentralized networked control structure without increasing communication requirements. The design and implementation of the proposed control structure are demonstrated through an application to a representative chemical plant.

References:

[1] Christofides, P. D., J. F. Davis, N. H. El-Farra, K. Harris, and J. N. Gibson, ``Smart plant operations: Vision, progress and challenges," AIChE J., 53: 2734-2741, 2007.

[2] Hokayem, P. F. and C. T. Abdallah, ``Inherent issues in networked control systems: A survey," Proceedings of the American Control Conference, pages 329--336, Boston, MA, 2004.

[3] Sun, Y. and N. H. El-Farra, ``Quasi-decentralized model-based control of networked process systems," Comp. & Chem. Eng., in press.

[4] Tipsuwan, Y. and M. Y. Chow, ``Control methodologies in networked control systems," Contr. Eng. Prac., 11:1099--1111, 2003.

266g Discrete Control of a Chaotic Reaction in a Stirred Tank Reactor

Ankur Mukherjee[1], Dominic P. Searson[2], and Mark Willis[2]. (1) School of Chemical Engineering and Advanced Materials, Newcastle University, Merz Court, Claremont Road, Newcastle upon Tyne, United Kingdom, (2) School of Chemical Engineering & Advanced Materials, Newcastle University, Merz Court, Claremont Road, Newcastle upon Tyne, United Kingdom

Temporal oscillations of species concentration in reactions are prevalent due to the nonlinear and often autocatalytic dependence of reaction kinetics on the species concentrations. Self heating in non-isothermal systems too is frequently responsible for inducing oscillations in the species temporal concentrations. Quantitative knowledge of oscillations in reaction systems is therefore imperative for the design, operation, control and optimisation of reactors. It is also crucial for prevention of thermal runaways and migration to bi-stable steady states. Chaos is a unique nonlinear dynamical phenomenon which induces unpredictable oscillations and extreme critical sensitivity to reaction conditions. The trajectories of state variables of chaotic systems diverge exponentially in the state space. Most trajectories of a chaotic system never return to the same point, although they eventually approach it to an arbitrarily small distance. Examples of chaotic reaction systems are the Belousov-Zhabotinsky reaction, combustion of H2, oxidation of CO on platinum, anaerobic fermentation of ethanol, polymerization of olefins in fluidized bed, conversion of heavy gas oil to olefins etc. Control of chaos in chemical reaction systems is crucial considering the necessity to mitigate the long term unpredictability and non-uniform system performance, e.g. irregular product concentration/selectivity at any given time. Chaos in chemical systems may be controlled by stabilising unstable periodic orbits using temporal perturbations to an accessible process parameter. We present an algorithm for control of chaos in reacting systems which implicitly takes into account limitations associated with infrequent sampling of product concentrations. This algorithm is shown to perform better than the Simple Proportional Feedback (SPF) algorithm when species concentration measurements are sparse and noise corrupted. This two dimensional algorithm (2D) SPF is further improved by proposing an inferential variant, which utilizes only a single easily observable secondary state variable for measurement and control. The state variable used for the algorithm is derived from the time series of the rate of heat evolution data from the reactor. The aim of this work is to demonstrate this novel design technique for the control of chaos in a mutating chemical reaction system. In this reaction model, the dynamic equations are unknown and only a single state variable is available for measurement and control.

Web Page: www.ncl.ac.uk/htt/

269a Probability Bounds Analysis In Modeling Nonlinear Ecosystem Dynamics

Joshua A. Enszer and Mark A. Stadtherr. Chemical and Biomolecular Engineering, University of Notre Dame, 182 Fitzpatrick Hall, Notre Dame, IN 46556

One major motivation for studying ecological systems with numerical models is the potential use of such models in studying environmental impacts from outside influences, such as the introduction of a pollutant, or other anthropogenic disturbances. This allows for experimentation under a wide variety of possible conditions without causing harm to real-world ecosystems. Food chains and webs in ecology may be modeled using systems of ordinary differential equations (ODEs), in which the state variables represent the population of different species within the ecosystem, and the equations capture the interactions between the different species. The effects of outside disturbances on ecosystem characteristics are generally not precisely known, but can often be bounded by a range of values. For example, introduction of a pollutant may cause the intrinsic death rates in an ecosystem to change, but the exact values of those parameters are uncertain. This means that the ecosystem model includes parameters that are not precisely known, but which can be bounded or described with some probability distribution. Thus, in the numerical simulation of ecosystem dynamics, it is desirable to be able to rigorously capture all possible results (population trajectories) over a range of parameter values. If a probability distribution is available for the uncertain parameters, then it is also desirable to be able to compute bounds on the resulting probability distribution for the population values at any particular time.

In this presentation, we demonstrate a method for the verified solution of nonlinear ODE models, thus computing rigorous bounds on the population of a species over a given time period, based on the ranges of uncertain values. The method is based on the general approach described by Lin and Stadtherr [1], which uses an interval Taylor series to represent dependence on time, and uses Taylor models to represent dependence on uncertain parameters and/or initial conditions. We also demonstrate an approach for the propagation of uncertain probability distributions in one or more model parameter and/ or initial condition through a population model. Assuming an uncertain probability distribution for each parameter and/or initial condition of interest, we use a method, based on Taylor models and probability boxes (p-boxes) and recently described by Enszer et al. [2], that propagates these distributions through the dynamic model. As a result, we obtain a p-box describing the probability distribution for each species population at any given time of interest. As opposed to the traditional Monte Carlo simulation approach, which may not accurately bound all possible results of nonlinear models under uncertainty, this Taylor model method for verified probability bounds analysis fully captures all possible system behaviors. The methods presented are tested on a series of small food chains or webs representable by nonlinear ODE systems, including the tritrophic Rosenzweig-MacArthur model. The results are compared to those obtained by Monte Carlo simulations.

[1] Lin, Y., Stadtherr, M.A. Validated Solutions of Initial Value Problems for Parametric ODEs. Applied Numerical Mathematics, 57: pp. 1145--1162, 2007.

[2] Enszer, J.A., Lin, Y., Ferson, S., Corliss, G.F., Stadtherr, M.A. Propagating Uncertainties in Modeling Nonlinear Dynamic Systems. In Proceedings of the 3rd International Workshop on Reliable Engineering Computing, Georgia Institute of Technology, Savannah, GA: pp. 89--105, 2008.

269b Simulation of the Silicon Ingot Slicing Process for Solar Cell Manufacturing

Raymond A. Adomaitis and Thomas Palathra. Chemical Engineering, University of Maryland, College Park, MD 20742

Multicrystalline (MC) silicon solar cells are manufactured from bread-loaf sized ingots of solar-grade silicon. These ingots are sliced by a multi-wire saw mechanism consisting of a single thin and extremely long stainless steel wire wound on constant-pitch wire grooves. The wire is wound over each groove to create a web consisting of 500-700 parallel wires. The wire is kept at a constant tension using feedback control and the wire speeds typically are 10-20 m/s. A high speed nozzle directs an aqueous slurry of oil and SiC particles to the top of the wire array and the crystal silicon ingot is pushed upwards against the wire array during the cut. The role of the cutting fluid is to transport the SiC abrasives and Si chips, reduce friction, and to provide cooling. In a typical wire saw system MC ingots are sliced with an area of 100x100 mm^2 and the latest wire saw systems can achieve thicknesses down to 300 microns. The goal of this modeling work is to understand the physical mechanisms that limit how thin the wafers can be cut and to determine the process operating conditions that minimize cutting time.

What makes this a challenging simulation problem is the wide range of timescales that characterize the overall cutting process. The slowest dynamics are associated with the evolution of the cut, which is described by a spatially dependent differential equation in time and in which the cutting rate is modeled much in the same manner as the Chemical Mechanical Planarization (CMP) process used in microelectronic device manufacturing. Cutting rate is a direct function of the distance between the wire and bottom of the cut, and so the wire deflection is modeled by approximating the wire as a static circular beam (the wire dynamics are orders of magnitude faster than cut evolution) which is subjected to a force in the axial direction (wire tension) as well as a distributed load in the transverse direction (the upward force of the ingot). We assume a linear stress-strain relationship, that the applied stress is below the elastic limit of the wire, and that wire deflection obeys Hooke's law. Solving for the equilibrium equations for the vertical forces and the sum of the moments at point ultimately leads to a fourth-order boundary-value problem.

The numerical solution for this fourth order BVP is found by employing a collocation spectral method to discretize the the di erential equation. This discretization procedure gives a set of algebraic equations which are solved Newton's method. Objected-oriented programming concepts were employed in MATLAB to reduce the complexity of these weighted residual and Newton numerical methods. Thus, a numerical solution for wire deflection can be found for the current applied force and other operating conditions, as well as the current ingot shape, and the solution then is used to advance the evolution of the cut. Representative solutions will be presented as well as the results of sensitivity analysis to show the e ects of process parameters on cutting rate.

Web Page: www.isr.umd.edu/Labs/CACSE/A-Team

269c Accurate Calculation of Fluid-Fluid Interfaces from X-Ray Tomography Images

Pradeep Bhattad, Chemical Engineering, Louisiana State University, 110 Chemical Engineering, South Stadium Road, Baton Rouge, LA 70803, Karsten E. Thompson, Gordon & Mary Cain Department of Chemical Engineering, Louisiana State University, 110 Chemical Engineering, South Stadium Rd, Baton Rouge, LA 70803, and Clinton S. Willson, Department of Civil & Environmental Engineering, Louisiana State University, Baton Rouge, LA 70803.

Fluid-fluid interfaces play an important role in mass transfer and multiphase flow in many chemical and environmental engineering processes. However, quantitative characterization of interfaces can be difficult, especially in complex geometries such as porous materials. We use X-ray computed tomography to image interfaces in a variety of different porous materials. Quantitative characterization of the interfaces is performed in a two-step process: Starting from a voxelized image of a fluid-fluid interface, we generate an initial surface triangulation using the signed distance function calculated from the segmented images. Since the real fluid interfaces are smooth in nature we use energy minimization to smooth the triangulated surface. The energy minimization consists of two parts, an elastic energy function that minimizes the surface area, and an image energy function to ensure that the triangulated interface respects the original voxelized image. The resulting smooth triangulated surface is a powerful tool for characterizing or modeling interfacial behavior. We report the accuracy of interfacial curvature and surface area calculations as a function of image resolution and initial surface mesh resolution for both artificially generated and real tomography images.

A HYBRID INTERVAL CONSTRAINT PROPAGATION AND MATHEMATICAL PROGRAMMING METHOD FOR ENERGY SYSTEMS SYNTHESIS

Mahmoud Bahy M. Noureldin and Majid M. Al-Gwiaz
mahmoudbahy.noureldin@aramco.com and majid.gwaiz@aramco.com

Process and Control Systems Department, Saudi Aramco
Dhahran, Saudi Arabia

November 2008

Abstract:

In the last three decades the simultaneous synthesis of process and utilities systems and process optimization and HEN synthesis has become very important research field in chemical engineering. These large systems are difficult to solve and a lot of simplifying assumptions are still needed to keep these problems mathematically tractable especially in industrial applications. In this paper a hybrid method in which we are integrating constraint propagation with mathematical programming on the stand-alone modeling level for solving large energy systems synthesis is introduced. Such integration is illustrated using new formulation for solving HEN synthesis targeting problem under all possible combinations of process modifications and a CP application numerical example in which CP is used as a preparatory step to NLP solution for the total annualized cost minimization of HEN synthesis problem.

Keywords: Interval Constraints Propagation; Constraint Programming, Mathematical Programming, Process Synthesis, Constraint Satisfaction

Introduction:

Worldwide industrial community is ravenous to producing more-with-less through cost effective but benign new products and processes. Chemical process industry is known as one of the most energy-intensive manufacturing sectors in the industrial community. Waste heat recovery applications, as old as the industry itself, are nowadays a must. Many techniques emanated in the early seventies for systematic heat integration has been in use since in the industrial community and are now gaining more momentum. Finding any small room for improvement in the newly designed processes that can bring some value towards energy saving and GHG emissions reduction will be more than welcomed. For non-constrained and non-thermodynamic constrained processes process design and operation alternatives are cumbersome and the number of alternatives that include different process structures, design parameters and operating conditions result in a combinatorial optimization problem.

Constraint Programming field, known to be a little younger than mathematical programming, is getting every day ground in reducing the computational effort needed to solve combinatorial optimization problems. A crucial component in Constraint Programming approach is the Constraint Propagation.

Constraint Propagation (CP) is a very efficient method for reducing the search space of combinatorial search and optimization problems and has become more important in the last two decades. The basic idea of constraint propagation methods is to detect and remove inconsistent variable(s) assignment(s) that can not participate in any feasible solution through the repeated analysis and evaluation of the variables, domains and constraints describing a specific problem instance.

Constraint logic Programming or sometimes called Propagation (CLP), as a technique used in Constraints Satisfaction (CS) applications, is a unique problem-solving paradigm that establishes a clear distinction between two pivotal aspects of any problem; a precise definition of the constraints that define the problem to be solved, and the algorithms and heuristics enabling to solve it. CLP is increasingly being used as a problem-solving tool for many engineering problems. It was originally developed for solving CS feasibility problems but has recently been used in optimization problems too.

Grossmann and co-workers have combined the power of both Constraint Programming and MILP on the algorithmic level to solve continuous time scheduling of multipurpose batch plants [1].They used MILP to optimize the high level decisions and Constraint Programming to obtain feasible detailed schedule. Hooker and co-workers in 2002, referenced by Grossmann et al., specified two ways to integrate Constraint Programming and mathematical programming. One by combining them on the modeling level and try to find the algorithm that can solve such hybrid model. The second by decomposing the original problem into two sub-problems; one is MILP and the other is Constraint Programming. Each model is solved separately and information obtained from one help solving the other.

Essentially, combinatorial optimization problems can be posed as multi-level optimization one. Sub-problems can be solved using constraints propagation for feasibility checks. The decision maker inputs and the solutions of the sub-problems insights can be fed to another MP formulation to find the details of the final solution.

The HEN synthesis problem under all possible combinations of process parameters modifications is a quite large problem. Upon decomposition, to targeting and synthesis sub-problems, it can be easily tackled using CS to find under all possible combinations of process changes the heating and cooling utilities minima and maxima targets.

In this paper a hybrid method in which we are integrating constraint propagation with mathematical programming for solving large energy systems synthesis is introduced. Such integration is illustrated using new formulation for solving HEN synthesis targeting problem under all possible combinations of process modifications sub-problem. A second example for integrating CP and MP; is using CP and NLP for total annualized cost minimization of HEN. It is a numerical example for solving minimum area HEN sub-problem to show the use of CP as a preparatory step to MP problem solution.

Interval constraint propagation at a Glance:

Interval mathematics

Because floating-point computational errors need to be estimated and controlled, interval analysis has grown considerably in the past 30 years. Interval arithmetic deals with processing intervals that bound real numbers. Consider a real variable, x, bounded by two other numbers, $xl \leq x \leq xu$. One can define an interval X such that $x \in X$ where X = [xl , xu]. Similarly, an interval Y can be defined to include a real variable y such that $y \in Y$. An interval arithmetic operation, *, (for example, addition, subtraction, multiplication, and division) is defined by:

$$X*Y = \{x*y: x \in X , y \in Y\} \qquad (1)$$

A particularly useful property of interval arithmetic operations is:

$$x*y \in X*Y \qquad (2)$$

which means that the sum, difference, product, and quotient of two real numbers belong to the same properties of the including intervals. Rules for interval operations include:

$$X + Y = [xl , xu] + [yl , yu] = [xl + yl , xu + yu] \qquad (3)$$

$$X - Y = [xl , xu] - [yl , yu] = [xl - yu , xu - yl] \qquad (4)$$

$$X\,Y = [xl , xu]\ [yl , yu] = [\min(xl\,yl, xu\,yu, xl\,yu , xu\,yl), \max(xl\,yl, xu\,yu, xl\,yu , xu\,yl)] \qquad (5)$$

$$X / Y = [xl , xu] / [yl , yu] = [xl , xu]\,[1/yu, 1/yl] \quad \text{if } 0 \notin \text{to } [yl , yu] \qquad (6)$$

Isotonicity is another useful property of interval operations. It states that for intervals X, Y, W, and Z

$$\text{if } X \subset W \text{ and } Y \subset Z \text{ then } X*Y \subset W*Z \qquad (7)$$

For each continuous function, f(x), where x is an n-dimensional vector and $x \in X$, one can use interval arithmetic to identify bounds on the range of the function. Consider a function f(x) whose range over interval X is defined as □ f(X), i.e. □ f(X) = {f(x): $x \in X$}.

An interval function F is called an inclusion function for f over interval X if

$$\square f(X) \subseteq F(X) \qquad (8)$$

Inclusion functions are extremely important in interval analysis as they provide bounds on ranges without exhaustive enumeration. There are two common methods for constructing inclusion functions: natural interval extensions and centered forms. A natural interval extension is an expression where each x in the various terms of f(x) is replaced with it including interval, X, and the mathematical operators are replaced with interval operations. Centered forms are inclusion functions that represent generalization of the algebraic centered forms for real variables. A particularly useful centered form is based on the natural interval extension of Taylor's expansion of the function. More details on interval analysis are given in [2-4].

Constraint logic propagation

Constraint logic propagation known as CLP is a problem-solving paradigm that establishes a clear distinction between two pivotal aspects of a problem: a precise definition of the constraints that define the problem to be solved, and the algorithms and heuristics enabling selection of the decisions to solve the problem. Because of these capabilities, constraint programming is increasingly being used as a problem-solving tool for many engineering problems [5,6]. Constraint programming was originally developed for solving feasibility problems but has recently been used in optimization problems [7]. Although constraint programming methods are efficient in solving feasibility/targeting problems, optimization depends entirely on building the correct constrained model. To show how constraint logic propagation with interval labels works, this paper gives numerical examples one is the "Waltz" algorithm, first introduced in the mid seventies [8].

Suppose the following relations model/define our problem:

$$x+y=z, \quad y \leq x \qquad (9)$$

and we can start with the following bounds:

$$x \in [1,10], \; y \in [3,8], \; z \in [2,7], \qquad (10)$$

This would be implemented in a data structure given by the following constraint network:

Constraints: (11)

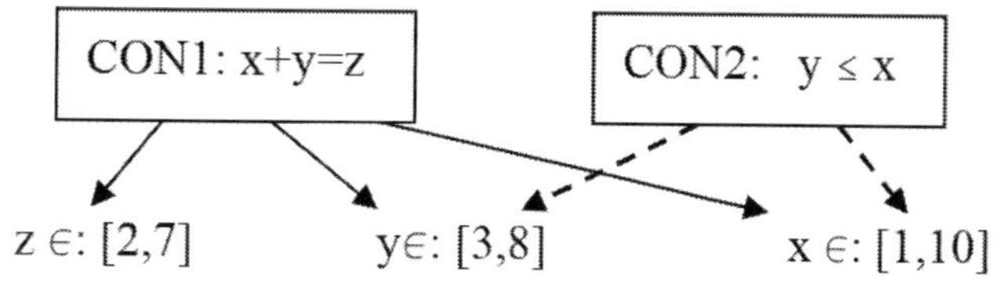

Nodes: (12)

The very famous old "Waltz" algorithm proceeds as follows:

The constraint queue begins with both constraints (CON1, CON2).

CON1 ($x + y = z$) is popped from the queue.

Since $x \geq 1$ and $y \geq 3$, CON1 gives $z \geq 4$; therefore reset the bounds of z to [4,7].

Since $z \leq 7$ and $y \geq 3$, CON1 gives $x \leq 4$; therefore reset the bounds of x to [1,4].

Since x and z have been changed, add CON2 to the queue.

CON2 ($y \leq x$) is popped from the queue.

Since $x \leq y$, CON2 gives $y \leq 4$; therefore reset the bounds of y to [3,4].

Since $y \geq 3$, CON2 gives $x \geq 3$; therefore reset the bounds of x to [3,4].

Since x and y have changed, add CON1 to the queue.

CON1 ($x + y = z$) is popped from the queue.

Since $x \geq 3$, $y \geq 3$, CON1 gives $z \geq 6$; therefore reset the bounds of z to [6,7].

Since only z has changed and z has no other constraints beside CON1, nothing is added to the queue.

Since the queue is empty, the algorithm quits.

This simple numerical example demonstrates how to use constraint logic propagation techniques in solving engineering problems that can be modeled using constraints and interval labels consisting of deterministic bounds [9].

Combined CP and MP method for energy systems synthesis:

A system that includes equality constraints such as mass, heat balances and performance equations, inequalities constraints such as the ones including operating windows and design parameters and logical expressions constraints of superstructures and/or operating philosophies can easily be modeled in CS environment as below:

X^2 + 6.0x=y-2^k;

kx+7.7y=2.4;

(k-1)^2<4;

(ln(y+2x+12)< (k+5)) or (y>k^2) → (x<0.0) and (y<1);

Where k is integer; x, y are real and → stands for implication

The solution of the problem above exhibits several solutions that all lay in these intervals of the three variables.
k = [0, 2];
x = [-6, -1e-10];
y = [0.311688, 1];

<u>The exact solutions are as follows:</u>
k = 0; x = -5.88301; y = 0.311688;
k = 0; x = -0.11701; y = 0.311688;
k = 1; x = -0.289055; y = 0.349227;
k = 2; x = -0.658479; y = 0.482721;

In real life industrial problems we can find many ways to limit the number of solutions or even push the problem further to having a unique solution via adding to the problem new constraints. Such constraints can be based upon the physico-chemical nature of the problem, using entropy balance for instance and through the insertion of decision maker's constraints and preferences such as defined ranges for objective function, operating window of some equipment, desired features in process structure and units design and so on.
In the above mentioned problem we can add one more constraint and the problem will be fully defined and has a unique solution [10].

X^2 + 6.0x=y-2^k;

kx+7.7y=2.4;

(k-1)^2<4;

(ln(y+2x+12)< (k+5)) or (y>k^2) → (x<0.0) and (y<1);

(x<2.5y) →(ky<=3) and (k>y+1);

This unique solution then is; k= 2, x= -0.658 and y= 0.4827

The essence of our method of integrating CS and MP in solving large optimization problems is to convert the large industrial combinatorial problem into multi-level optimization problem. We then solve using CP first at the lowest level of "followers" to define certain range of this level objective function and variables. We move from the lowest follower level up to a higher level and repeat this step of solving using CP till we reach to the upper/leader level. Again we use CP for the leader problem to tighten the range of both the variables and leader's objective function.

MP is then used to find the details of the whole problem solution after inserting bounded variables, objectives ranges of the followers.

The simultaneous HEN synthesis and process operations optimization problem can be represented as three levels optimization problems; minimize area cost subject to minimum number of units subject to minimum utility cost, or in two levels (BLPP) minimize capital cost subject to minimum energy cost.

Illustrating Examples:

In this section we introduce two examples, CP formulation for HEN synthesis energy targeting application and a numerical example in which we use CP solution of a HEN minimum capital cost synthesis as a preparatory step for the NLP application for finding the solutions details of the HEN synthesis.

The first example that shows how to formulate the CP model for solving energy targeting sub-problem under all possible combinations of process modifications, is detailed below.

Model Description and CP Formulation: (Example 1)

Hot streams temperatures shifted down one-by-one by the first set of desired minimum temperature differences, ΔT_{min}^{i}, between the hot and cold resource streams to form a set of possible discrete temperature values for a continuum of possible values for ΔT_{min}^{i}. ΔT_{min}^{i} represents the ΔT_{min} of hot stream (i) which refers to the minimum temperature approach between a specific hot stream and all other cold streams, can be taken here in this paper as equal to 1 F. The shifted supply and target (output) temperatures of resource hot streams, and the actual supply and the target cold streams temperatures obtained through this are then sorted in a descending order, with duplicates removed, with each successive temperature pair representing the boundaries of a temperature step and defining a new temperature step "S". Each supply temperature and target temperature input is in the form of intervals (e.g., a pair of range boundaries) and not single discrete numbers. Each supply temperature and target temperature input interval is then divided on an equal basis according to the desired temperature precision. The completion of this step results in "N" number of process temperature steps. While not discussed in details here creating the new heat carrier streams in the hot side or in the cold side can be done with the same concept used above of temperature range division. The heat carrier stream or pump around streams exist and no exist decisions can also be modeled with supply temperature defined by the user on a range and its target temperature and FCp too as variables defined in ranges. The target temperature range will be divided to equal temperature intervals again degree-by-degree or according to the required accuracy in temperature.

The total number of temperature steps is "N+1," where S varies from 0,1,2,...,N and the temperature step number "0" represents the external energy utility temperature step. In this step, known as the external energy step, energy output, initial values: $Q_{s=0}^{low\_output} = 0.0$ in "energy units" and $Q_{s=0}^{high\_output} = 0.0$ in "energy units."

Each temperature step, "S" greater than 0, where S =1,2,...,N, has energy surplus $Q_s^{surplus}$. Such energy surplus has two calculated values: $Q_s^{low\_surplus}$ and $Q_s^{high\_surplus}$. It also has energy output Q_s^{output} from one temperature step to another. Such energy output Q_s^{output} has also two calculated values: $Q_s^{low\_output}$ and $Q_s^{high\_output}$. These values for energy surplus $Q_s^{surplus}$ and energy output Q_s^{output} are calculated for S=1,2,...,N:

$$Q_s^{low\_surplus} = (\sum_{k=1}^{n_S} FCp_k^{low} - \sum_{j=1}^{m_S} FCp_j^{high})(Th_s - Tc_s)$$

$$Q_s^{high\_surplus} = (\sum_{k=1}^{n_S} FCp_k^{high} - \sum_{j=1}^{m_S} FCp_j^{low})(Th_s - Tc_s)$$

$$Q_s^{low\_output} = Q_{s-1}^{low\_output} + Q_s^{low\_Surplus}$$

$$Q_s^{high\_output} = Q_{s-1}^{high\_output} + Q_s^{high\_Surplus}$$

Where n_s and m_s are the number of the resource hot and cold streams respectively represented in the sth temperature step and Th_s and Tc_s are the higher and lower shifted temperatures for the hot streams, respectively, and actual temperature of the cold streams representing the temperature boundaries. Logic propositions such as "or, and,→" are used to enable modeling logics.

$[FCp_k]$ and $[FCp_j]$ are heat capacity flowrate intervals; defined as follows:

$FCp^{low}{}_k$: is the low value of the Heat Capacity Flowrate term resulted from the multiplication of the value of the flow F lower bound by the specific heat value Cp of the hot stream number k in flow-specific heat units.

$FCp^{high}{}_k$: is the high value of the Heat Capacity Flowrate term resulted from the multiplication of the value of the flow F upper bound by the specific heat value Cp of the hot stream number k in flow-specific heat units.

$FCp^{low}{}_j$: is the low value of the Heat Capacity Flowrate term resulted from the multiplication of the value of the flow F lower bound by the specific heat value Cp of the cold stream number j in flow-specific heat units.

$FCp^{high}{}_j$: is the high value of the Heat Capacity Flowrate term resulted from the multiplication of the value of the flow F upper bound by the specific heat value Cp of the cold stream number j in flow-specific heat units

$$\sum_{k \in Hs} [FCp_k] - \sum_{j \in Cs} [FCp_j] + [Qh_s] - [Qc_s] + [Q_s^{surplus}] + [Q_s^{output}] = 0.0; \quad S = 1,2,\ldots,N;$$

$$[Qh_s] \geq 0.0; [Qc_s] \geq 0.0; [Q_s^{output}] \geq 0.0; [Q_0^{surplus}] = 0.0; [Q_0^{output}] = 0.0;$$

Upon the calculation of utilities minima using the above formulation the selection of the heating and cooling utilities exact consumption can be decided by the decision maker from the resulted ranges. An interval collapsing algorithm using MP or other methods can now be used to select the operating conditions that render the desired level of energy consumption [11, 12].

The second example of integrating CLP and MP methods is a heat exchanger synthesis problem, to find the minimum area/cost network for desired energy consumption and defined stream matches. Table 1.0 Data for the CP Illustration Model

stream	Tin(K)	Tout(K)	FCp(kW/K)
H1	440	350	22
C1	349	430	20
C2	320	368	7.5

Exchanger minimum approach temperature is 10 K, the problem needs no heating and cooling utilities, the cost of heat exchanger is function in its area (cost = 1300*A^0.6 $/yr) and the network streams superstructure is shown below.(Fludas et al., 1996)

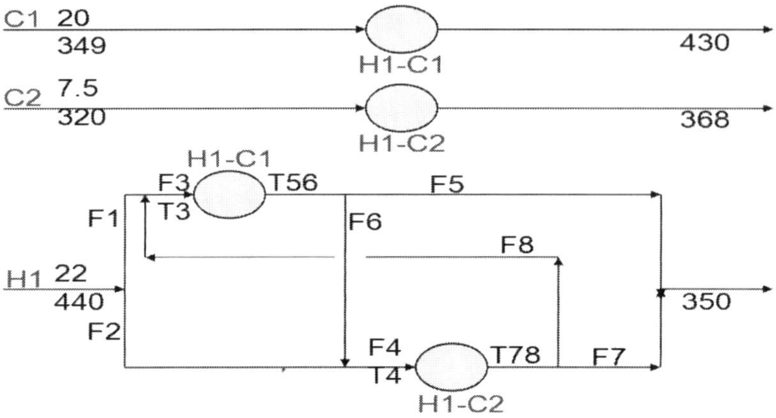

NLP can be used alone to solve this problem (Fludas et al., 1996). However combining the power of both CP and MP methods can make the solution even easier since the application of the CP first can confirm feasibility of desired objective function, reduce noticeably the search space since it rigorously define lower and upper bounds of each variable and/or parameters and enable the deletion of parts of the superstructure that are not feasible and/or undesired by the decision maker.

Note: Chen approximation for the log mean temperature difference is used.

CP Model Formulation:

```
1300*A1^0.6+ 1300*A2^0.6)-OF=0.0;
A1-(1620/DTA) = 0.0;
A2-(360/(0.5*DTB))=0.0;
ta-(T3- 430)=0.0;
tb-(T56-349)=0.0;
tab-(ta+tb)=0.0;
tc-(T4-368)=0.0;
td-(T78-320)=0.0;
tcd-(tc+td)=0.0;
((0.5*ta*tb*tab)^0.3334)-DTA=0.0;
((0.5*tc*td*tcd )^0.3334)-DTB=0.0;
F1+F2-22=0.0;
F3-F5-F6=0.0;
F4-F7-F8=0.0;
F3-F1-F8=0.0;
F4-F2-F6=0.0;
F1*440-F8*T78-F3*T3=0.0;
F2*440-F6*T56-F4*T4=0.0;
(F3*(T3-T56))-1620=0.0;
(F4*(T4-T78))-360=0.0;
T3-430>=0.1;
T56-349>=0.1;
```

T4-368>=0.1;
T78-320>=0.1;
T3<=440;
T56<=440;
T4<=440;
T78<=440;
F1; F2; F3; F4; F5; F6; F7; F8>=0.0;

CP Interval Model Solution:

A1 = [171.678, 3764.16]; A2 = [11.9785, 244.221]; F1 = [17.4659, 19.3347];
F2 = [2.66527, 4.53409]; F3 = [17.8218, 19.7287]; F4 = [3.0025, 5.10778];
F5 = [15.4664, 19.7287]; F6 = [0, 2.35543]; F7 = [0.739738, 5.10778]; F8 = [0, 2.26277];
T3 = [431.214, 440]; T4 = [390.581, 440]; T56 = [349.1, 357.886]; T78 = [320.1, 369.519];
OF = [34262.7, 216901];

The objective function *OF = [34262.7:216901]* $/yr lower and upper bounds can be tightened more and more using repeated bisectional method followed by feasibility checks. *OF* can then be inserted as a new constraint in the model. The decision maker for instance; can define desired range for his/her budget (*OF*), if he/she gets feasible solution it means his/her desired objective is achievable and he/she can continue then using NLP to find the details of the solution.

Now let us assume that the decision maker has a budget of *OF = [40000:45000]* $/yr.
Inserting new budget interval value in the CP model constraints list renders the following tightened bounds for system variables.
A1 = [180.476, 287.627]; A2 = [13.3803, 63.1493]; F1 = [17.9487, 19.2544];
F2 = [2.7456, 4.05127]; F3 = [18.1984, 19.5222]; F4 = [3.00929, 4.44035];
F5 = [16.5616, 19.5222]; F6 = [0, 1.63681]; F7 = [1.43581, 4.44035]; F8 = [0, 1.57349];
T3 = [433.964, 440]; T4 = [401.445, 440]; T56 = [350.981, 357.018]; T78 = [320.371, 358.925];

It is clear from the new bounds that the decision maker objective is achievable, HEN superstructure is reducible, F8 superstructure branch is close to zero and can be removed from the H1 stream superstructure, and the flow & temperature ranges have been reduced drastically, compared with normal monotonoicity checks. Such insights make the NLP formulation and solution easier especially for large industrial problems.

Conclusion:

Combining CP and mathematical programming in a stand-alone fashion on the modeling level creates more powerful environment for solving optimization problems. Logical constraints that are very common in process synthesis and operation problems are easily formulated in CP environment. Formulating and solving energy systems synthesis problems in CP environment results in insights and tight bounds on the problem variables and objective function before applying MP. CP as a preparation step to large optimization problem solution makes the step of obtaining the detailed solution of the problem using mathematical programming easier.

References:

[1] C.T. Maravelias, I.E. Grossmann, A hybrid MILP/CP decomposition approach for the continuous time scheduling of multipurpose batch plants, Computers and Chemical Engineering 28 (2004) 1921-1949.

[2] M. B. Noureldin and EL-Halwagi, M. M., Interval-Based Targeting for Pollution Prevention via Mass Integration, Computers & Chemical Engineering 23 (1999) 1527-1543

[3] H. Ratschek, J. Rokne, New computer methods for global optimization, Ellis Horwood/John Wiley and Sons, New York, USA, 1984.

[4] R.E. Moore, Reliability in computing: The role of interval methods in scientific computing, Academic Press, San Diego, USA, 1988.

[5] M.B. Noureldin, Systematic approach to chemical kinetics analysis using constraint logic propagation, Poster session: Kinetics, Catalysis and Reaction Engineering, AIChE annual meeting, Nevada, November 4-9, 2001.

[6] I. Harjunkoski, I.E. Grossmann, Decomposition techniques for multistage scheduling problems using mixed-integer and constraint programming methods, Computers and Chemical Engineering 26 (2002) 1533-1552.

[7] ILOG Solver 4.4 User's manual, ILOG Inc., 1999.

[8] D. Waltz, Understanding line drawings of scenes with shadows, The Psychology of Computer Vision, McGraw-Hill, N.Y., USA, 1975.

[9] E. Davis, Constraint propagation with interval labels, Artificial Intelligence 32 (1987) 99-118.

[10] UniCalc Solver for Mathematical Problems user guide version 3.41, 1999.

[11] M.B. Noureldin, Improved system and computer software for modeling energy consumption, New Zealand Patent No.527,244, 2003.

[12] M.B. Noureldin, **TEM_*icons*™** 1.2 User's manual report, Department of Materials and Process Engineering, University of Waikato, Hamilton, New Zealand, 2003

277a A Survey of Web 2.0 Applications in Chemical Engineering

Scott Butner, 902 Battelle Boulevard, Pacific Northwest National Laboratory, PO Box 999, MSIN K7-28, Richland, WA 99354

Technology pundit and publisher Tim O'Reilly first coined the term "Web 2.0" in 2004, to describe a convergence of technologies and business models that have had a significant impact on the way we view the World Wide Web. Social Networking sites such as FaceBook and LinkedIn, social bookmarking sites such as del.icio.us, and collaborative content tools such as Wikis and blogs, are all part of this transformation of the web from "top down" information models to "bottom-up" models, where information is built collaboratively by communities of interest. This paper will provide a current review of how Web 2.0 technologies are being used in chemical engineering and how they may change the nature of the profession. The author will also discuss the ongoing efforts by AIChE to integrate Web 2.0 concepts into its own web site.

277b Engineering Virtual Organization for Pharmaceutical Engineering and Science

Kamal Kuriyan[1], Ann Christine Catlin[2], and G. V. Reklaitis[1]. (1) Pharmaceutical Technology & Education Center, Purdue University, 203 S Martin Jischke Dr, W Lafayette, IN 47907, (2) Rosen Center for Advanced Computing, Purdue University, W Lafayette, IN 47907

The pharmaHUB Engineering Virtual Organization for Pharmaceutical Engineering and Science (EVO-PES) is an interdisciplinary resource for sharing and building the science and engineering base for pharmaceutical product development and manufacturing. pharmaHUB leverages the cyberinfrastructure technology developed by HUBzero to support the collaborative research and education activities of the interdisciplinary but widely scattered community of academic researchers in pharmaceutical engineering and science, and their governmental and industrial counterparts. pharmaHUB was created with a mission to facilitate sharing of information, knowledge and modeling tools. The site promotes the dissemination of educational materials and course modules for use in the education and training of engineers and pharmaceutical scientists in academia, industry and regulatory agencies. Most importantly, pharmaHUB offers powerful research simulation tools which can be accessed with a web browser. pharmaHUB also provides tools for management of structured knowledge including an ontology browser and editor which can be used to explore OWL ontology files.

Web Page: https://pharmahub.org/about

277c Application of Ontology in Pharmaceutical Product Development: The Pope Experience

Leaelaf M. Hailemariam[1], Venkat Venkatasubramanian[2], and **Arun Giridhar**[2]. (1) School of Chemical Engineering, Purdue University, 480 Stadium Mall Drive, West Lafayette, IN 47907, (2) Chemical Engineering, Purdue University, 480 Stadium Mall Drive, West Lafayette, IN 47907-2100

The information generated during pharmaceutical product development is voluminous and diverse [1]. Current applications that manage this information include applications based on relational databases [2] or document management systems [3]. In addition, mathematical models and heuristics about unit operations are frequently embedded in the source code [4], making it difficult to access and update. An efficient information management platform, necessary for the profitability of the enterprise, would require a common data model to consolidate and effectively model the disparate pieces of information from multiple sources in the domain.

Ontology is the explicit representation of concepts and their relationship [5]. Due to the potential of ontology to represent the concepts in a complex domain (pharmaceutical product development), the Purdue Ontology for Pharmaceutical Engineering (POPE) was developed. POPE has components to model phases, molecular structure, reactions, material properties, experiments, unit operations and equipment. Several applications that utilize POPE were developed; including an application for prediction of potential reactions a drug substance may undergo and an application to analyze experiments with respect to their settings and procedures. In this work, the Purdue Ontology for Pharmaceutical Engineering, as well as the applications that make use of it, are discussed.

References

1. Venkatasubramanian V., C. Zhao, G. Joglekar, A. Jain, L. Hailemariam, P. Suresh, V. Akkisetty, K. Morris and G.V. Reklaitis (2006) Ontological Informatics Infrastructure for chemical product design and process development. Computers and Chemical Engineering. CPC 7 Special Issue, 30(10-12) 1482-1496

2. Paszko C. and C. Pugsley (2000) Considerations in Selecting a Laboratory Information Management System (LIMS). American Laboratory. 9 38-42

3. Zall M. (2001) The Nascent Paperless Laboratory. Chemical Innovation. 31 2-9

4. Suresh P., C. Zhao, C. Joglekar and V. Venkatasubramanian. (2006) Informatics Based Approach for Mathematical Knowledge Modeling in Process Operations. AIChE Annual Meeting, November 12-17 San Francisco Hilton, San Francisco, CA

5. Gruber T. (1993) Toward principles for the design of ontologies used for knowledge sharing Internation al Journal of Human Computer Studies 43 907-928

277d My Chemical Engineering Web 2.0 Applications

S. A. D. T Subasinghe, Department of Chemical and Process Engineering, University of Moratuwa, Katubedda, Moratuwa, Sri Lanka

My Chemical Engineering Web 2.0 Applications In the beginning of 2007, when I came across blogger.com which belongs to Web 2.0 I thought it as a vital platform to share chemical engineering related knowledge with people around the world. I initiated a web blog on blogger.com with a title of "Chemical and Process Engineering" on the domain http://www.cpelanka.blogspot.com . This lengthy domain was annoying for many users, so I brought a domain name, www.chempro.org and host my blog on it. I published many posts on my web blog which is closely related to chemical engineering. It was quickly indexed with Google, Yahoo and many of world famous search engines and it became one of chemical engineering web 2.0 application. Some of my web blog visitors were kind enough to give feedback on my work. A Malaysian chemical engineer Mr. Zaki Yamani Zakariya was first one who came up with series of progressive comments. Prof. Ajith de Alwis, Ph.D.(Cambridge) who is my academic advisor and one of readers of my web blog gave me enough direction on topic selection and provided me with many resources. I maintain my web blog for more than one year and it has more than 3000 visitors each month reading my posts. I came across a free social network creator and initiated a social network for chemical engineers and placed that network on http://www.cpeng.snappville.com. It is fast growing chemical engineer's network and it is also dedicated to share information. But the approach of this social network is different. It is a platform to share user generated content with users. In my web blog I share content which is generated by my self. But in Chemical Engineers' network I am also one content make and there are many of others who contribute to generate chemical engineering related information. As an engineering student of developing country, Sri Lanka; I am doing this effort with lot of financial barriers. So these days I am try to monetize my blog and social network to make some money for my living cost. I have put Google Adsense on my blog and social network. However I have never never never never never give up my effort and I am continually working on my blog and social network to realize my vision "Sharing Chemical & Process Engineering knowledge with people around the world is my perspective."

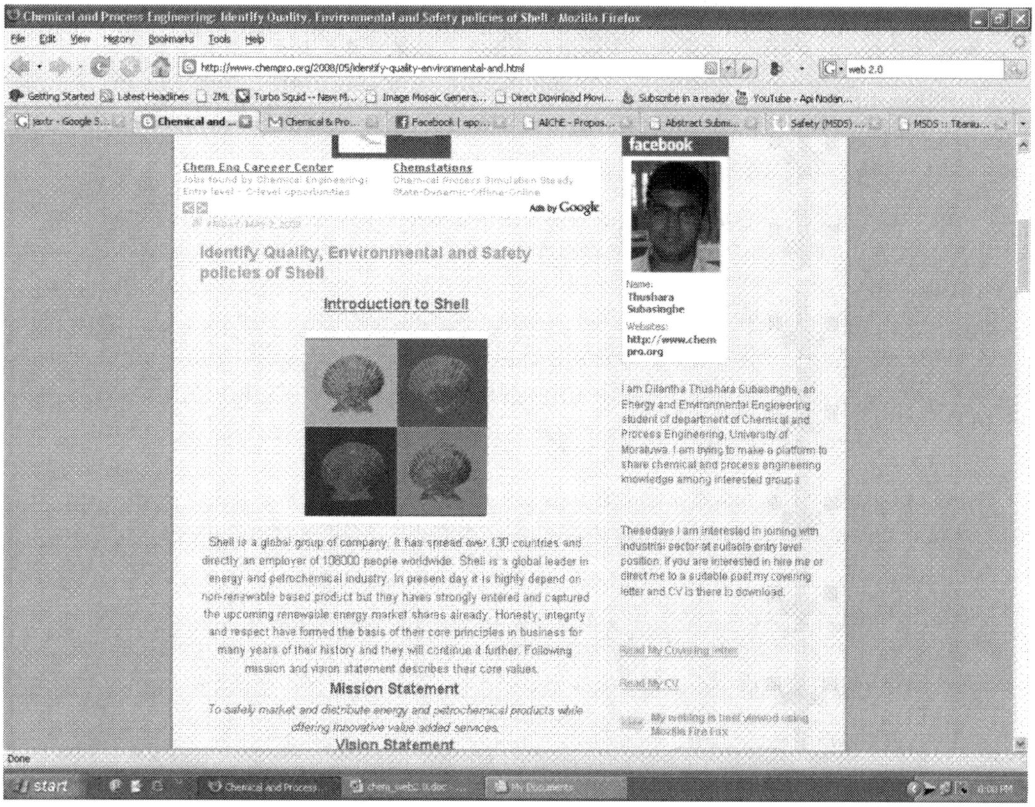

Web Page: www.chempro.org

277e Value Network Analysis: Tools and Paradigms for Building Healthy Ecosystems In Business

Jeff Lindsay, Innovation Edge, 1526 South Commercial St., #200, Neenah, WI 54956

While many engineers and managers are trained to consider business operations in terms of linear supply chains and value chains, the actual structures we deal with in the new knowledge economy are typically complex non-linear ecosystems. Value creation in these ecosystems involves transactions of both tangible items as well as intangibles, which typically are not considered in mapping business processes. Yet much of the value created in a business comes from the benefits created by intangible transactions - exchanges of knowledge, giving advice, factors that generate loyalty and build relationships, etc.

Value Network Analysis is an approach that quickly help a business understand the structure, behavior, and health of the ecosystems it participates in, and to understand where value is actually being created. It provides tools for analyzing tangible and intangible transactions. Developed by Verna Allee and others in the knowledge management community, Value Network tools are proving their value in improving the way engineers, managers, and others approach projects and business problems. We will discuss our experience in applying Value Network tools to several problems, and share insights that may be helpful to engineers and managers in their work.

Web Page: www.value-networks.com/

285a Fast Fault Detection and Identification Using Grey Box Models

Ivan Castillo, Chemical Engineering, The University of Texas in Austin, 1 University Station C0400, Austin, TX 78712, Thomas F. Edgar, University of Texas at Austin, 1 University Station C0400, Austin, TX 78712, and Ricardo Dunia, LabVIEW R&D, National Instruments, 11500 N. Mopac Expwy, Austin, TX 78759.

Fault detection and identification applications are mostly based on empirical models that quickly degenerate in time due to slow changes in what operators consider normal process conditions. The extensive use of such algorithms is always rare due to the variations that a traditional industrial facility undergoes in time. This work illustrates the use of a monitoring system that takes advantages of grey box models to distinguish the parameters that are expected to shift in time versus those that are expected to remain constant or invariant. A change in the invariant parameters indicates the existence of a potential fault. The identification of the fault is closed related to a parameter that has shifted in time. This detection and identification methodology is considerably faster than the traditional process in the sense that the direction in which a fault could bring the process operation is well-defined. A well-defined fault space allows the identification of such a fault in a more precise and decisive way than in cases where several samples are needed to corroborate the presence of abnormal operating conditions.

This technique is applied to an Air Heater Lab Experiment (AHLE) in which obstruction of the air flow, changes in the heat throughput and air leaks are considered faults that should be detected and identified, while variations in the ambient temperature or adjustments in the forced flow are expected to be part of normal operating changes.

285c Valve Stiction Detection and Quantification In Oscillatory Control Loops

Qinghua (Peter) He, Department of Chemical Engineering, Tuskegee University, Tuskegee, AL 36088 and Jin Wang, Department of Chemical Engineering, Auburn University, Auburn, AL 36849.

Valve stiction is one of the most common equipment problems that cause poor performance in control loops. Consequently, there is a strong need in the process industry for non-invasive methods that can not only detect but also quantify stiction [1]. In our previous work [2], a data-driven valve stiction model is presented and a stiction detection method is proposed based on curve-fitting approach. Once valve stiction is detected, it is desirable to quantify the severeness of the stiction so that the right action can be taken – leaving it alone, compensating by control, or requiring immediate valve replacement.

In this work, we propose a noninvasive method to quantify the severeness of valve stiction using the routine operating data obtained from the process, i.e., process variable (PV) and controller output (OP). The algorithm is proposed to estimate the stiction parameters, namely static friction f_S and dynamic friction f_D, without requiring the valve position (VP) signal. Identification is accomplished by using linear and nonlinear least-squares methods which are robust and easy to implement. Several simulation examples, including both self-regulating and integrating processes with different degrees of stiction, are used to demonstrate the effectiveness of the method. In the simulation examples, we show that the proposed method is applicable to different stiction models, including the one we developed previously [2] and the one developed by Choudhury [3] and Kano [4]. The proposed approach is also applied to several industrial data sets where stictions have been detected.

The proposed method is based on the following observation: when a stable limit cycle caused by valve stiction is detected, the output of a sticky valve resembles a rectangular wave as illustrated in Fig.1; while the case shown in Fig. 2 will not occur after the limit cycle stabilizes. In other words, the valve jumps only once before it changes direction. Based on this observation, we can see that $(f_S + f_D)$ or S (shown in Fig. 1 (a) and termed as the amount of apparent stiction [1]) is approximately the length of $(|AB|+|AC|)/2$, as $|AB| \leq (f_S + f_D)$ and $|AC| \geq (f_S + f_D)$; while $(f_S - f_D)$ or J (shown in Fig. 1. (a) and termed as slip jump [3]) is approximately the length of $|DE|$ which can be estimated based on measurements of PV. Because the location of O as shown in Fig. 1 (a) cannot be determined based on process data, the values of f_S and f_D cannot be estimated until the values of both $(f_S + f_D)$ and $(f_S - f_D)$ are estimated. By making use of both OP and PV measurements, valve stiction parameters can be estimated without requiring measurements of VP which is usually not available.

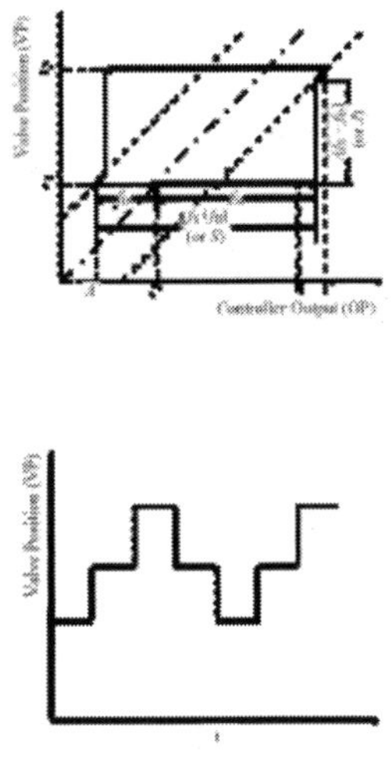

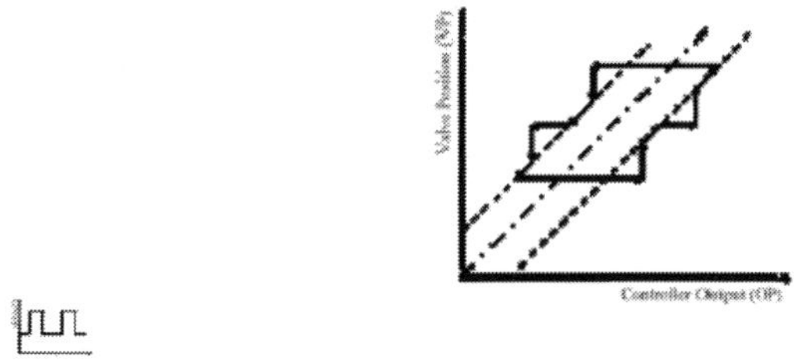

(a) (b)

Fig. 1 Valve stiction with single jump:

(a) VP-OP plot; (b) VP-t plot

(a) (b)

Fig. 2 Valve stiction with multiple jumps:

(a) VP-OP plot; (b) VP-t plot

· Estimation of $(f_S + f_D)$:

We use a self-regulating process to illustrate how to estimate $(f_S + f_D)$ based on the measurements of OP. Fig. 3 shows the time series plot of OP when a sticky valve is present in the control loop.

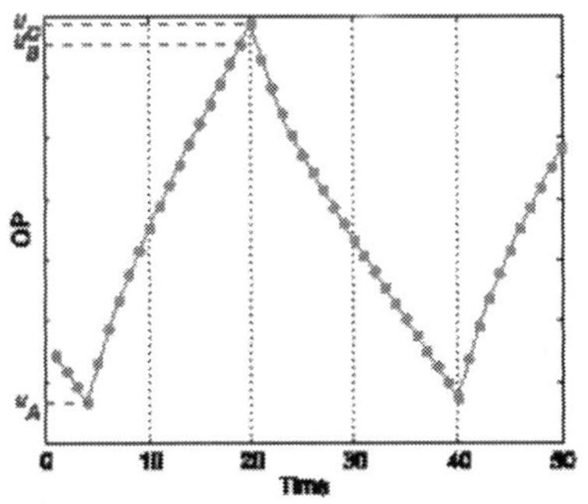

Fig. 3 OP measurements of a self-regulating process with a sticky valve

By comparing Fig. 3 with Fig. 1 (a), the points (A, B and C) in Fig.1 (a) can be identified and marked correspondingly in Fig.3 and the estimate of $(f_S + f_D)$ can be determined from the OP measurements by detecting peaks and valleys in the OP signal and we have

$$(f_S + f_D) \approx \frac{(|u_B - u_A| + |u_C - u_A|)}{2}$$

For an integrating process, the estimation procedure of $(f_S + f_D)$ is the same as that of a self-regulating process. The effects of noise and time delay on the estimate of $(f_S + f_D)$ and their handling are discussed in detail.

· Estimation of $(f_S - f_D)$:

For a self-regulating process, the process output (PV) can be approximated by a series of step responses. For a general FOPTD model, the step response is:

$$y(t) = K_p (f_S - f_D)\left(1 - e^{-\frac{t-\theta}{\tau}}\right)$$

where K_p is the process gain, $(f_S - f_D)$ is the magnitude of the rectangular pulse/step input, θ is the time delay, and τ is the time constant. For a given or estimated process gain K_p, a nonlinear least-squares method is proposed to estimate $(f_S - f_D)$ based on the PV response. In the case that the steady-state is approximately reached, $(f_S - f_D)$ can be estimated by the approximate steady-state value changes in PV:

$$(f_S - f_D) \approx \frac{|PV_+ - PV_-|}{K_p}$$

where PV_+ and PV_- are upper and lower bounds of PV.

For an integrating process, we show that

$$(f_S - f_D) \approx \frac{|slope(PV_\nearrow)| + |slope(PV_\searrow)|}{K_p}$$

where $PV_\nearrow$ and $PV_\searrow$ denote ascending portion and descending portion of the PV triangular wave caused by valve stiction and $(f_S - f_D)$ can be obtained via a simple least-squares method.

Notice that in the proposed method, both f_S and f_D (or S and J) can be estimated. In addition, the proposed method only requires limited process knowledge, i.e., estimated process gain and time delay, besides routine operating data OP and PV.

Key words: stiction quantification; control valve; identification and estimation; fault diagnosis

References:

1. Choudhury, M. S.; Shah, S.; Thornhill, N.; Shook, D. Automatic detection and quantification of stiction in control valves. *Control Engineering Practice* 2006, 14, 1395-1412.

2. He, Q.P., Wang, J., Pottmann, M. and Qin, S.J. A Curve fitting method for detecting valve stiction in oscillating control loops, *Industrial & Engineering Chemistry Research*, vol.46 (13), 4549-4560, 2007.

3. Choudhury, M. A. A. S.; Thornhill, N. F.; Shah, S. L. Modelling valve stiction. *Control Engineering Practice* 2005, 13, 641-658.

4. Kano, M.; Maruta, H.; Kugemoto, H.; Shimizu, K. Practical model and detection algorithm for valve stiction. In *Proceedings of the Seventh IFAC-DYCOPS Symposium*, Cambridge, MA, July 5-7, 2004.

Spurious Drop-out Detection In Subcutaneously Implanted Continuous Glucose Monitors

Quan Shen[*], S. Joe Qin[†]

Continuous glucose monitoring (CGM) is of importance for the management and control of diabetes mellitus, which is a prevalent chronic disease of no cure as yet. CGM can continuously inform diabetic people of the status of their glucose and warn of actual or impending hypo- or hyperglycemia. In the past decades, biosensors have been developed for CGM, among which the implanted wired enzyme biosensor [1] has advantages of consistency and accuracy [4, 3] over other subcutaneously implanted sensors.

It is found in the experiments of subcutaneously implanted sensors that there are some spurious drop-outs which do not reflect the true glucose level and will cause problems in subsequent glucose calibration and lag correction. These spurious drop-outs are usually very fast and abrupt. Our objectives are to (1) detect spurious drop-outs online for a single glucose sensor signal, since in practice only one subcutaneously implanted glucose monitor is carried by one person; (2) detect different sizes of spurious drop-outs; and (3) detect drop-outs as quickly as possible.

In this paper, we propose an online spurious drop-out detector based on the discrete wavelet transform (DWT) [2], a signal processing technique which can effectively extract the signal component at a specific frequency range and a specific time region with scaled

[*]Department of Chemical Engineering, The University of Texas at Austin, Austin, Texas 78712, USA. Email: quanshen@che.utexas.edu

[†]The Mork Family Department of Chemical Engineering and Materials Science, University of Southern California, Los Angeles, CA 90089, USA. Email: sqin@usc.edu

and shifted wavelets. Because of its multiscale feature, DWT is very useful in detecting singularities of a single sensor signal. In our study, the finest five level details ($D_1 \sim D_5$) are found to have the ability to capture the spurious drop-outs in the glucose signal. The online detection scheme is designed in a recursive way. At each new time point, a moving window is reconstructed. Then DWT is applied on the glucose data inside the window to extract the finest five level details and calculate D_c=sum(D_1:D_5). If D_c exceeds its confidence interval (CI), an abrupt drop-out is warned, which will be confirmed as a spurious drop-out or otherwise by a rule-based evaluator. The rule-based evaluator is a Boolean valued composite measure based on the boundaries, minimum, width of the drop-out, normal signal variations around the drop-out, etc. This evaluator is to exclude normal fast changes such as a real glucose rise after a meal.

The subcutaneously implanted wired enzyme glucose biosensor has been commercially available to diabetic people. In this study, we apply the proposed online spurious drop-out detection method to glucose data from non-diabetic pigs. The results demonstrate effectiveness of the proposed method.

References

[1] A. Heller, *Implanted electrochemical glucose sensors for the management of diabetes*, Annual Reviews of Biomedical Engineering **1** (1999), 153–175.

[2] R. M Rao and A. S. Bopardikar, *Wavelet transforms: Introduction to theory and applications*, Addison Wesley Longman Inc, 1998.

[3] R. L. Weinstein, S. L. Schwartz, R. L. Brazg, J. R. Bugler, T. A. Peyser, and G. V. McGarraugh, *Accuracy of the 5-day freestyle navigator continuous glucose monitoring system: comparison with frequent laboratory reference measurements.(emerging treatments and technologies)*, Diabetes Care **30** (2007), 1125–1130.

[4] D. M. Wilson, R. W. Reck, W. V. Tamborlane, M. J. Dontchev, C. Kollman, P. Chase, L. A. Fox, K. J. Ruedy, E. Tsalikian, and S. A. Weinzimer, *The accuracy of the freestyle navigator continuous glucose monitoring system in children with type 1 diabetes*, Diabetes Care **30** (2007), 59–64.

285e Actuator Placement for Fault Tolerant Control of Transport Reaction Processes

Paul. C. Russell[1], **Antonios Armaou**[1], and Michael A. Demetriou[2]. (1) Chemical Engineering, The Pennsylvania State University, University Park, PA 16802, (2) Mechanical Engineering, Worcester Polytechnic Institute, Worcester, MA 01609

When considering the control problem for transport reaction processes, an important issue is the actuator placement so that the system exhibits desired properties, such as enhanced controllability. A complexity in this endeavor lies in the spatial dependence of the concepts of controllability and observability which prompted a number of researchers to address the important topic of actuator and sensor placement. The current approach to actuator placement is to select the actuator locations based on open-loop considerations to ensure that the necessary controllability and reachability or power factor requirements are satisfied [1]. Actuator and sensor failures are a different issue that has often plagued chemical industries, as they may lead to deteriorating product quality or potentially dangerous process operation, such as runaway conditions. Motivated by the importance of the aforementioned failures, the issue of fault tolerant and fault accommodating controller design has been an active research topic in the chemical engineering community. However, while there has been extensive research from the control community on fault detection and diagnosis of finite dimensional systems using model-based robust and adaptive control techniques, the issue of fault tolerant control for distributed processes has been investigated only recently [2,3].

To address the issue of fault tolerance specifically for distributed processes, we consider a radically different problem than previously [1]. The actuator optimization metric is not defined in terms of enhanced controllability, improved performance or enhanced robustness with respect to disturbances or unmodeled dynamics, but with respect to fault tolerability. Taking advantage of the spatial variability that transport-reaction processes naturally benefit from, we explore an entirely new concept of identifying actuator locations by identifying groups of actuators that, while being physically apart, have the same authority on the process states. This new system is only applicable to distributed parameter systems wherein two or more different locations within the spatial domain of definition can provide the same level of controllability and at the same time the same feedback gain can be applicable to all such locations. The identification of such actuator groups has the potential to greatly facilitate the design of fault accommodating control structures, such that the profound simplicity of the controller design will significantly reduce the costs associated with process supervision and be accompanied by obvious performance advantages.

[1] A. Armaou and M. A. Demetriou, "Optimal Actuator/Sensor Placement for Linear Parabolic PDEs Using Spatial H2 Norm," Chem. Eng. Sci., 61, 7351-7367, 2006.

[2] M. A. Demetriou, "Utilization of LMI methods for fault tolerant control of a flexible cable with faulty actuators," Proceedings of the 2001 40th IEEE Conference on Decision and Control, pp. 1885-1890, vol. 2, Orlando, Florida, December 4-7, 2001.

[3] A. Armaou and M. A. Demetriou, "Robust Detection and Accommodation of Incipient Component and Actuator Faults in Nonlinear Distributed Processes," AIChE J., in press, 2008.

Harmonic Analysis – A Tool for Troubleshooting Plant-wide Disturbances.

Nahid Sanzida[1], *M. A. A. Shoukat Choudhury*[1], *Sirish L. Shah*[2]

[1] *Department of Chemical Engineering, Bangladesh University of Engineering and Technology, Dhaka – 1000.*

[2] *Department of Chemical and Materials Engineering, University of Alberta, Edmonton AB, Canada T6G 2G6*

Extended Abstract:

Large process plants, such as oil refineries, power plants and pulp mills, are complex integrated systems, containing thousands of measurements, hundreds or thousands of controllers and many recycle streams. This integration of energy and material flow, required for efficiency, means that the smallest upset in the process will quickly propagate throughout the plant. These fluctuations force the plant to operate farther from the economic optimum, and so cause decreased efficiency, lost production and often increased safety concerns. Because of the scale of operation of process plants, a small percentage decrease in productivity has large financial consequences.

It is a non-trivial task to pinpoint the exact cause of such fluctuations. In the most difficult case, the fluctuations are in the form of oscillations. The problem is that oscillations have no defined beginning and end, and so the cause cannot be isolated by standard techniques. Finding the cause of oscillations is a tedious, labour-intensive, often fruitless task. Once the cause or diagnosis is understood, removal of the oscillations is usually straightforward.

Figure 1 shows time trajectories of 20 refinery variables in the left panel, with the corresponding power spectra in the middle panel. An oscillation in the time trajectory shows up as a spike in the power spectrum. From the power spectra of 20 process variables it is clear that several measurements oscillate at a common frequency of 0.06 (17 samples per cycle). The questions that process engineers are routinely faced with is: 1) What is the source of these oscillations? 2) Can it be removed so that the process can be run in a productive and efficient manner?

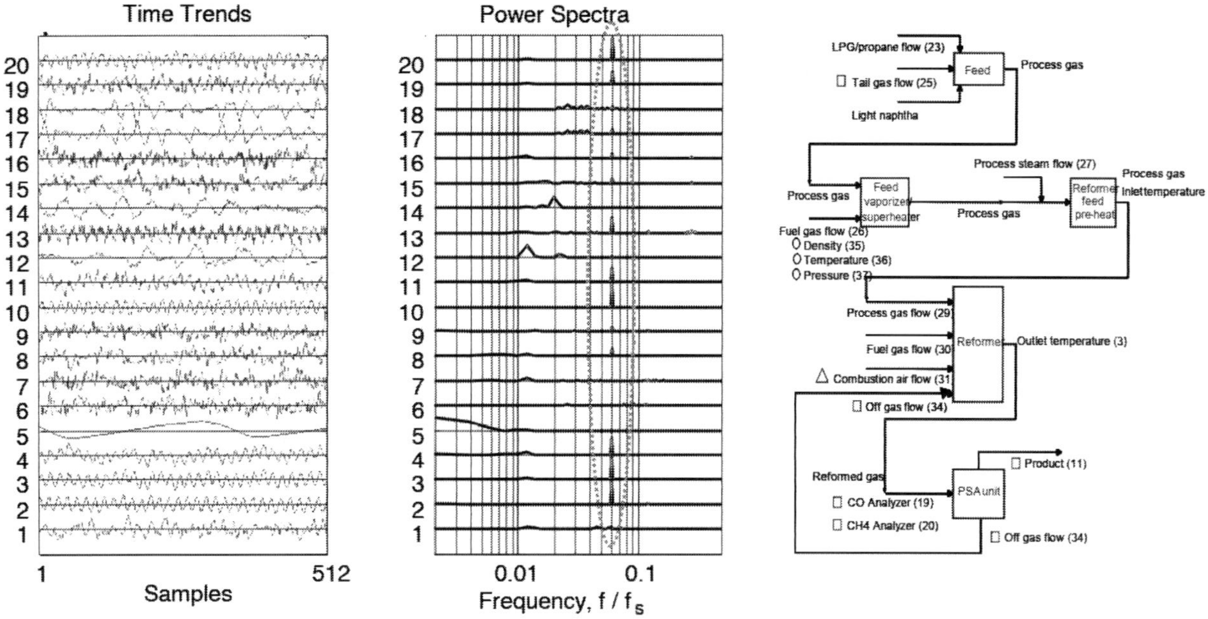

Figure 1: Illustration of plant wide oscillations in an industrial process. Temporal trends (left), Spectral trends (middle), and process schematic (right)

The objective of this presentation is to identify the root cause of such common oscillations in a plant. Once the root cause is known, the offending oscillations can be eliminated, and the process can then be run more efficiently.

Oscillations commonly originate in process plants for reasons such as sensor faults, valve faults, process faults, etc. Many of these faults manifest themselves as nonlinear elements in an otherwise locally linear control loop. Faults in the form of nonlinearity may produce oscillations with a fundamental frequency and corresponding higher order harmonics. Such faults in a control loop of a process unit can easily propagate to the adjacent units due to the material and energy flow paths or the interactions due to the multivariate nature of the process. It is well known that chemical processes are low-pass filters in nature. Therefore, when a fault propagates away from its origin or source, the higher order harmonics generally get filtered out. This presentation will present a method to estimate the amplitudes and phases of the fundamental oscillation and their harmonics and use this information to troubleshoot or pinpoint the root-cause of plantwide or unit-wide oscillations. The successful application of the method will be demonstrated on simulated and industrial data sets.

285g Unification of Contribution Analysis for Process Monitoring

Carlos F. Alcala and **S. Joe Qin**. Mork Family Department of Chemical Engineering and Materials Sciences, University of Southern California, 3710 McClintock Ave, RTH 420, Los Angeles, CA 90089

Abstract

Industrial processes have made successful use of statistical methods to perform detection and diagnosis of faults [5, 9]. These methods use statistical indices for fault detection, being the most popular the squared prediction error (SPE), Hotelling's T^2 statistic and a combination both indices [6].

Several methods have been developed for fault diagnosis. Among these methods, contribution plots are popular and are used to diagnose the cause of a fault by determining the contribution of each variable to the fault detection statistics calculated [4, 5, 8]. The assumption behind the contribution plot method is that faulty variables have high contributions to the fault detection index. However, contribution plots not always point to the right source of a fault, and it has been demonstrated that they do not guarantee correct diagnosis of simple faults, which are sensor faults with very large magnitudes [1].

Other proposed methods for fault diagnosis have been the reconstruction-based contribution (RBC) [1], the sensor validity index (SVI) [3], the identification index [2] and angle-information methods [7, 10]. The RBC, SVI and identification index use the reconstruction of a faulty index along a variable direction; in this work, they will be called reconstruction-based methods. The SVI is used to diagnose sensor faults; the identification index is a generalization of the SVI and can be used for diagnosis of sensor and process faults; the RBC method is also used for diagnosis of process and sensor faults. On the other hand, angle-information methods look at the angle between a faulty measurement and the direction of a given variable or fault; in this work, an angle-based contribution (ABC) will be defined and used for fault diagnosis.

Reconstruction-based and angle-based methods have been developed under different criteria; however, in this work, it will be shown that both kinds of methods are interrelated. Furthermore, since it has been proven that RBC guarantees correct diagnosis of simple faults, it will be shown that both reconstruction-based and angle-based methods guarantee correct diagnosis of simple faults. The diagnosability analysis will be performed in a unified way for the diagnosis methods, and a general fault detection index will be used instead of treating each index individually. Reconstruction-based and angle-based methods will be used to diagnose faults in an industrial polyester process and the results will be compared.

References

[1] C. Alcala and S. J. Qin. Reconstruction-based contribution for process monitoring. In *Proceedings of 17th IFAC World Congress*, Seoul, Korea, July 2008.

[2] R. Dunia and S. J. Qin. A unified geometric approach to process and sensor fault identification: the unidimensional fault case. *Comput. Chem. Eng.*, 22:927-943, 1998.

[3] R. Dunia, S. J. Qin, T. F. Edgar, and T. J. McAvoy. Identification of faulty sensors using principal component analysis. *AIChE J.*, 42:2797-2812, 1996.

[4] P. Miller, R.E. Swanson, and C.F. Heckler. Contribution plots: the missing link in multivariate quality control. In *Fall Conf. of the ASQC and ASA*, 1993. Milwaukee, WI.

[5] P. Nomikos and J.F. MacGregor. Multivariate SPC charts for monitoring batch processes. *Technometrics*, 37(1):41-59, 1995.

[6] S. J. Qin. Statistical process monitoring: Basics and beyond. *J. Chemometrics,* 17:480-502, 2003.

[7] A. Raich and A. Cinar. Statistical process monitoring and disturbance diagnosis in multivariate continuous processes. *AIChE J.*, 42:995-1009, 1996.

[8] J. A. Westerhuis, S.P. Gurden, and A.K. Smilde. Generalized contribution plots in multivariate statistical process monitoring. *Chemometrics Intell. Lab. Syst.*, 51:95-114, 2000.

[9] B.M. Wise and N.B. Gallagher. The process chemometrics approach to process monitoring and fault detection. *J. Proc. Cont.*, 6:329-348, 1996.

[10] S. Yoon and J.F. MacGregor. Fault diagnosis with multivariate statistical models, part I: using steady state fault signatures. *J. Proc. Cont.*, 11:387-400, 2001.

293a Solubility Measurement Setup for Rapid Determination of Solid-Liquid Equilibrium Behavior and Process Synthesis

Yik C. Chan[1], Ka M. Ng[1], and Ketan D. Samant[2]. (1) Department of Chemical Engineering, Hong Kong University of Science and Technology, Clear Water Bay, Kowloon, Hong Kong, China, (2) CWB Technology, Inc., 20311 Valley Blvd, Suite C, Walnut, CA 91789

Abstract

The synthesis of crystallization-based separation processes requires experimental solid-liquid equilibrium (SLE) behavior of the system under consideration, effective representation of such SLE data, selection of crystallization solvent, and determination of crystal quality under different experimental conditions. Often, because of the considerable time and labor required for a thorough investigation of a given system, some of these steps are omitted, resulting in a suboptimal process, or worse, completely missing a feasible, superior process alternative. In order to tackle this problem, a solubility measurement system was developed. With the use of multiple turbidity probes, online monitoring of the turbidity in multiple crystallization experiments was achieved. The data extracted from the system was fed to SLEEK – a solid-liquid equilibrium engineering software code, which allows for the visualization and interpretation of solid-liquid phase behavior. Several solid-liquid equilibrium phase diagrams of organic systems including adipic acid in water (binary), paracetamol in various organic solvents (binary), phenol/o-cresol/tert-butanol (ternary), bisphenol A/acetone/water (LL forming), and others were measured. These demonstrate that, taken together, this rapid solubility measurement setup and SLEEK, is highly effective in solubility data measurement, for identifying liquid-liquid region, solvent selection, and synthesizing crystallization-based separation processes.

SIMULATION OF AN INDUSTRIAL PLANT OF ANHYDROUS ETHANOL PRODUCTION FROM SUGAR CANE JUICE

Paper Number: 126848

J.P. Contreras, Chemical Engineer, Universidad de los Andes, Bogotá, Colombia
I.D. Gil, M.Sc. Chemical Engineer, Universidad Nacional de Colombia, Bogotá, Colombia

Introduction

Fuel ethanol has become an alternative of renewable energy source, because the less production to medium term of petroleum and contamination that the derivatives from the fossils fuel produce. The process of anhydrous alcohol production reunites to different stages from reaction and separation, which combined and with slight variations can get to produce different ethanol qualities. In this work a detailed study becomes of the process by means of simulation in Aspen Plus v2006. Established down the main operating conditions and different alternatives are evaluated to select the best one in some important stages. The simulated process involves from cane juice preparation for the fermentation to the process of ethanol dehydration, in this last process evaluate two techniques of dehydration, the extractive distillation with glycerol and adsorption by molecular sieves. The restriction of the simulation is the ethanol composition in the bottoms stream of the stripping, rectifying and recovery column, and the ethanol composition in distilled of the column the dehydrator. The results show that the kinetics reaction of glucose fermentation fits the power law of first order and that the process of dehydration of the mixture ethanol-water by extractive distillation with glycerol as separating agent is efficient from the energy consumptions and economic point of view.

1. Production process of Fuel Ethanol from Sugar Cane Juice.

The first step in obtaining alcohol fuel is the fermentation process where the sugar content in juices and honey of crops are transformed into alcohol using yeast. Subsequently fermented alcohol becomes a distillation columns where, through a process of evaporation is separated from the compound, obtaining alcohol purest, stillage and fusel oils. The final stage is dehydration, which withdrew water obtaining anhydrous alcohol. The process can be seen in the block flow diagram in Figure 1.

Fermentation

For the fermentation process used four reactors type stirred tank (CSTR) in series with cascade technology, with recirculation of yeast. It is fed continuously cane juice (14°Brix) to the first fermenter (Cardona, 2006). Likewise, feed crops of *Saccharomyces Cerevisiae* from the vats of reproduction and circulation to maintain cell population levels between 200-300 million cells per milliliter of solution in the fermenter. This fermentative process is under anaerobic conditions at a temperature of 32 to 35°C and a pH of 4.2 to 4.5 (Jacques, 1999).

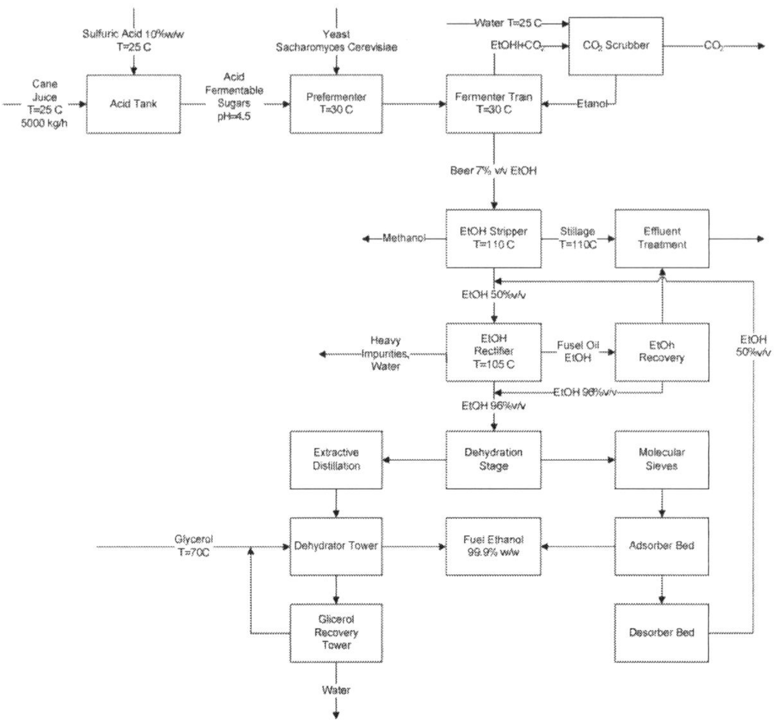

Figure 1 BFD Anhydrous Ethanol Production Process

Ethanol Recovery

The wine stripping, the fermentation product, which usually takes on average 7% v/v (Gil, 2006), is pumped into a preheated exchanger plates at a temperature of 80 ° C, then switched to a degassing column, which removes compounds with lower temperature of boiling that ethanol. For obtain ethanol 40-45% v / v (Gil, 2006), the mixture is sent to a Stripping column, which is powered by steam of 45 psig to separate alcohol (Gil, 2006), in the bottoms of the column leaves a byproduct called stillage which will be discussed in later sections. In the ethanol rectification stage, vapors with 40-50% ethanol concentration from the top of the stripping column are sent to a rectifier column (Cardona, 2006). The byproducts from column bottoms are called flemazas.

Ethanol Dehydration

The molecular sieves are materials that are characterized by their excellent ability to retain on its surface defined types of chemical species (Gomez, 2007). A key feature in operations involving the action of molecular sieves is that the amount of the substance to rowing-see through the sieve should be low. A synthetic zeolite 3Å type used in the vast majority of dehydrated ethanol, because their pores with a diameter of 3Å, while water molecules have a diameter of 2.8 Å and molecules of ethanol a diameter of 4.4 Å (Gil, 2006). Thus, water molecules are strongly attracted within the pores and the molecules of ethanol passes through a bed without experiencing any attraction (Gil, 2006).

Dehydration alcohol by extractive distillation consist in to add a solvent to the ethanol-water mixture (this mixture is impossible to separate by ordinary distillation) altering the relative volatility of the components and allows separation. But the solvent should be have low volatility, not even be a way evaporate in the column.

2. Process Simulation

The NRTL model fits best to equilibrium because the components involved in the process have characteristics of polarity and electrolytes, besides operating conditions in the process is less than 10 bar pressure (Carlson, 1996). The process assumes a pressure drop, which implements pump units, which discharged liquid flows at a pressure of 25 psi, this value was calculated using a heuristic and assumes a head of 20 ft (Seider, 2003).

Fermentation

The reaction is carried out in four reactors in series CSTR simulated in a module for calculating RCSTR modeling rigorously the reactor, manages kinetic reactions and equilibrium as well as the reactions involving solids. It can provide the kinetics of the reaction in the reactions models. To know the reaction kinetics of glucose fermentation, experimental data were taken at a temperature of 33°C and then uses a Integration Method (Fogler, 2001) to determine the rate constant k, which assumes Reaction order of one regarding glucose. Figure 2 shows the Flowsheet Diagram of simulation of fermentation

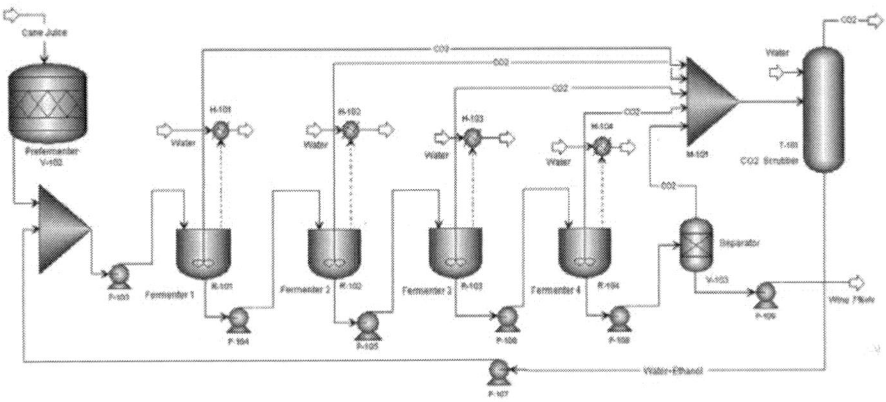

Figure 2 Flowsheet Diagram of simulation of fermentation

Alcohol Distillation

Figure 3(a) shows the Flowsheet diagram of the simulation of stripping stage. The stripping column (T-201) is simulated with a distillation column with steam saturated injected directly to a pressure of 45 psig; The Bottoms Stream, where leaves stillage, is used to preheat the feed stream to the stripping column, for this is used a module HeatX where specified temperature output desired in the cold stream in the exchanger (80°C) to separate the more volatile compounds. Figure 3(b) shows the Flowsheet diagram of simulation of rectification and recovery stage.

Much of the ethanol fed into the rectification column, leaves by the distillate with a mass fraction of 95%. In the column specifying a stream of fusel high side that corresponds to a mixture of alcohols rich in propanol and its extraction are performed five stages above the feed stage. On all the columns specifies a Murphree efficiency of 0.68 per stage (Kister, 1990) to adjust the results to reality and get a column size approximate of the industry Distilleries. The pressure drop in the stripping column is considered linear, (estimated by heuristic), with a value of 0,015 atm per stage (Seider, 2003).

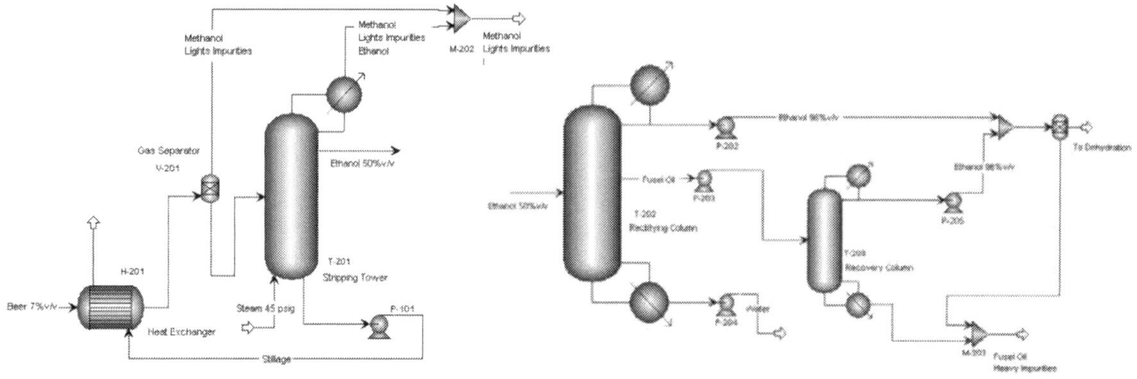

(a) (b)

Figure 3 Flowsheet Diagram of the Simulation of Ethanol Distillation

Ethanol Dehydration

The most used Techniques for alcohol dehydration industry are the extractive distillation and adsorption by molecular sieves. Figure 4(a) shows the Flowsheet diagram of dehydration by extractive distillation. The Feed stream to the dehydration column is a liquid saturated of azeotropic ethanol. When already know the flow rate of glycerol, realize a specification design such that the flow rate of solvent make up mixed with the flow rate of bottoms from the regenerate column are equal to the flow rate calculated above. It is done with the help of the tool Design Spec Flowsheeting option under the Options, there is created a new analysis, specifying the desired flow molar solvent input to the column and varies the flow rate of solvent make up within a consistent interval.

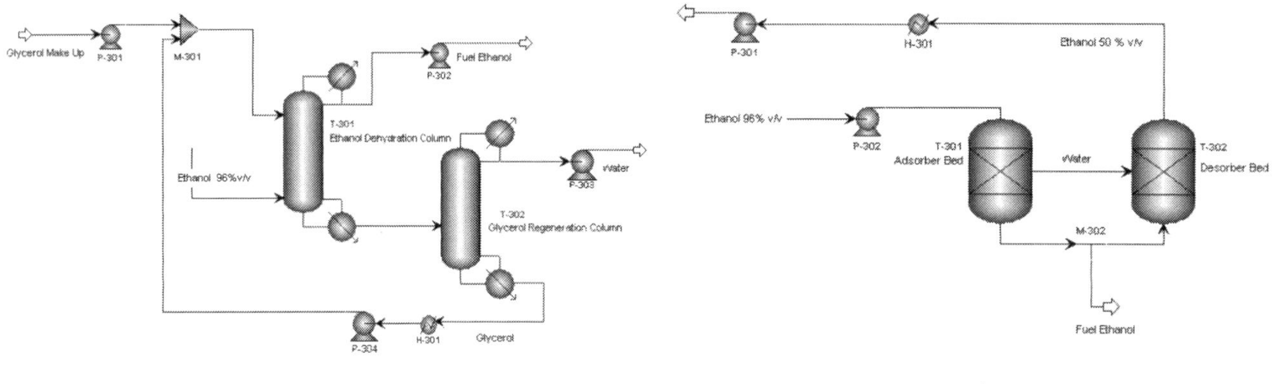

(a) (b)

Figure 4 Flowsheet Diagram of the Simulation of the Dehydration by Extractive Distillation.

The dehydration by adsorption with molecular sieves is a cyclical process of separation, i.e., while the first adsorbs the second regenerates. Consist of two packed columns with zeolites or carbon compounds adsorbents. The bed that is packed in adsorption operates at high pressure of 25 psig while the second screening is carried out the operation of regeneration to a reduced pressure of 26 inches of mercury (Gil, 2006), which is achieved by combining a condenser and a ring vacuum pump liquid. Figure 4(b) shows the Flowsheet diagram of the simulation process by adsorption molecular sieves.

3. Results

With the specifications given in each of the stages of the simulation process of producing ethanol are obtain the following results.

Fermentation

Table 1 presents the results of the simulation of the fermentation stage.

Table 1 Fermentation Stage Results

Fermenter	Residence Time (H)	Volume (m^3)	Heat Duty (kJ/s.)	Volumetric Water Flow (m^3/h)
1	3.94	953,38	-7536,24	270,481
2	6.19	3112,15	-8779,01	316,330
3	13.82	30459,16	-10870,79	390,382
4	48.24	49273,93	-10039,89	300,145
Total	72.19	118792	-37225.9	1279,338

Figure 5 shows the mass concentration profile in the fermenters simulated

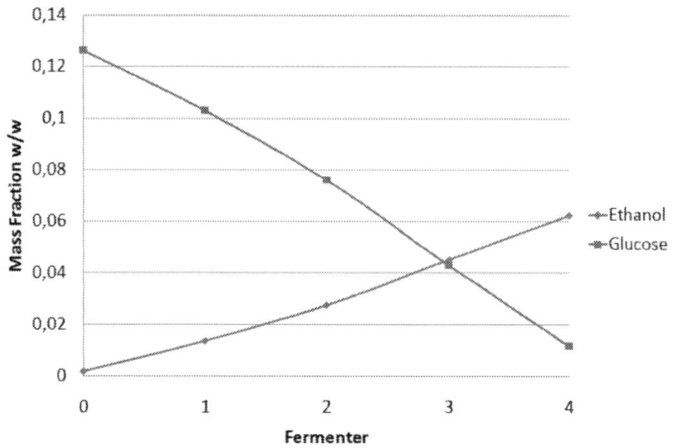

Figure 5 Ethanol Concentration in the fermentation.

It is observed as the stoichiometry conversion of glucose is not proportional to production; this is due to the generation of unwanted products of the reactions (2) to (8) as methanol and acetic acid, among others. It was further noted that the train fermentation, the concentration is increasing steadily, but the conversion of glucose is different for each reactor.

Ethanol Distillation

Figure 6 shows the profile results on the stripping column

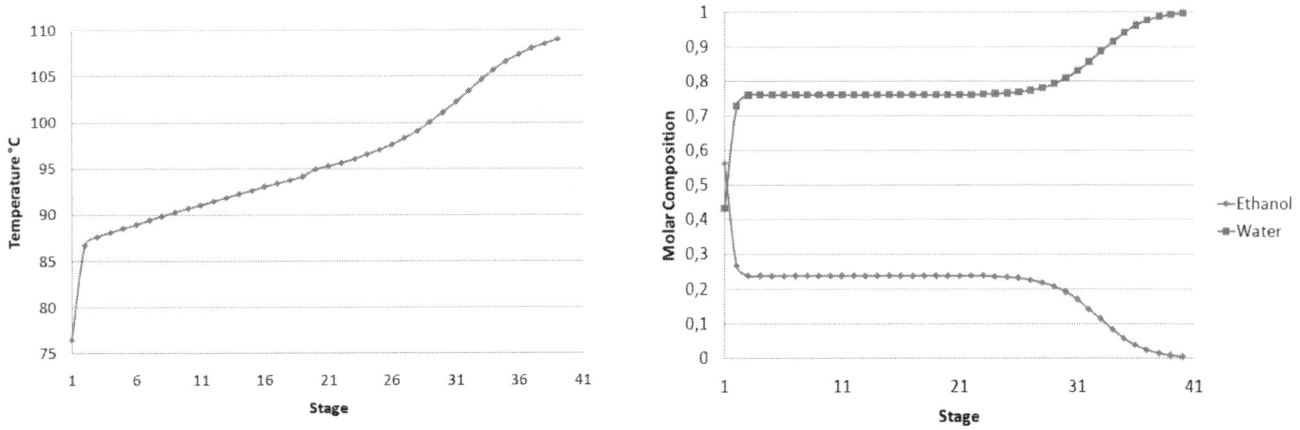

(a) (b)

Figure 6 (a) Temperature profile in Stripping Column. (b) Concentration profile in the Vapor (Stripping)

The temperature increases linearly as it increases the number of stage, given that water has a higher boiling point that ethanol and thus the temperature will be higher where the concentration of water and organic compounds is higher. The gradient of ethanol concentration is higher in the steam in the area of the bottom of the column to the stage where remains constant until Stage 2, but more volatile compounds are enriching in the vapor upper zone of the column. The ethanol fumes out through the flow side, while leaving the volatile components for the distillate, which increases the mole fraction in this stream.

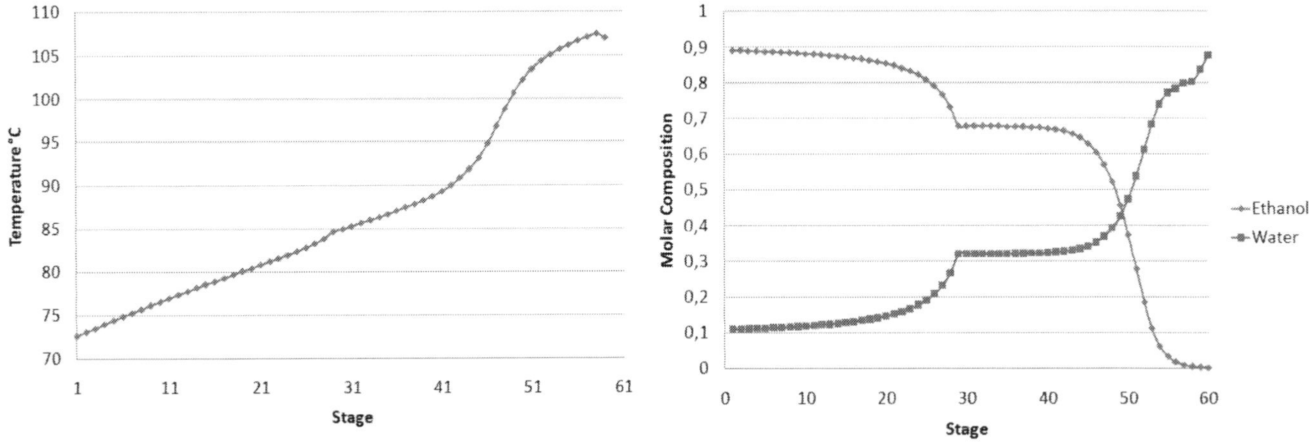

(a) (b)

Figure 7 (a) Temperature Profile in the Rectifier Column. (b) Concentration profile in the Vapor (Rectifier)

Figure 7 shows the profile results in rectifier column. The concentration of the less volatile components increase in the liquid, the temperature profile increases due to the concentration gradient components. The gradient of concentration of ethanol is higher in the zone below to the feed stage, due to temperature increase in stages near to the Reboiler. There is a change in the trend of the concentration of ethanol in the steam due to the feed stage. In the area above the stage feeding the driving force is much smaller because the ethanol is approaching its azeotropic point and is more difficult to separate the mixture.

Table 2 summarizes the results of the simulation of the ethanol distillation stage.

Table 2 Ethanol Distillation Results

Parameters	**Stripping Column (T-201)**	**Rectifier Column (T-202)**	**Recovery Column (T-203)**
Condenser Duty	-493,285kJ/s	-39682,97 kJ/s.	-3710,52kJ/s.
Reboiler Duty	---	30330,31kJ/s.	3699,14kJ/s.
Steam Flow (45 psi)	2.27 kg /kg_{Et}	1.55 kg/kg_{Et}	0.19 kg/kg_{Et}
Wt. % Ethanol Bottoms	0.05	0.03	0.03
Wt.% Ethanol Side Stream	48.1	0.78	---
Wt.% Ethanol Distillate	70.2	0.951	0.95

Ethanol Dehydration

Figure 8 shows the profile results on the column dehydration.

The ethanol concentration of through the stages of the column increases faster in the zone below to the feed stage, and then becomes constant because the few water that remains in the mixture. The temperature tends to decrease as the vapor flow is close to the column top, in the final stages there is a very high concentration end of heavy compounds (glycerol), representing a sharp rise in temperature . In the 50 stage show a temperature decrease because it corresponds to the feed stage. The temperature tends to decrease as the steam away from the stage of rehervidor, and in stage 2 reduces the temperature due to the condenser presence.

The glycerol of in the process, which complies the conditions for a good solvent because it is not evaporates and is removing water from the time they were fed, allows obtaining anhydrous ethanol, the less volatile compound (water) increases the concentration in the vapor in the section of depletion of the tower dehydration. In addition it appears that at the bottom of the column, the concentration of ethanol tends to zero, which suggests that the method is efficient.

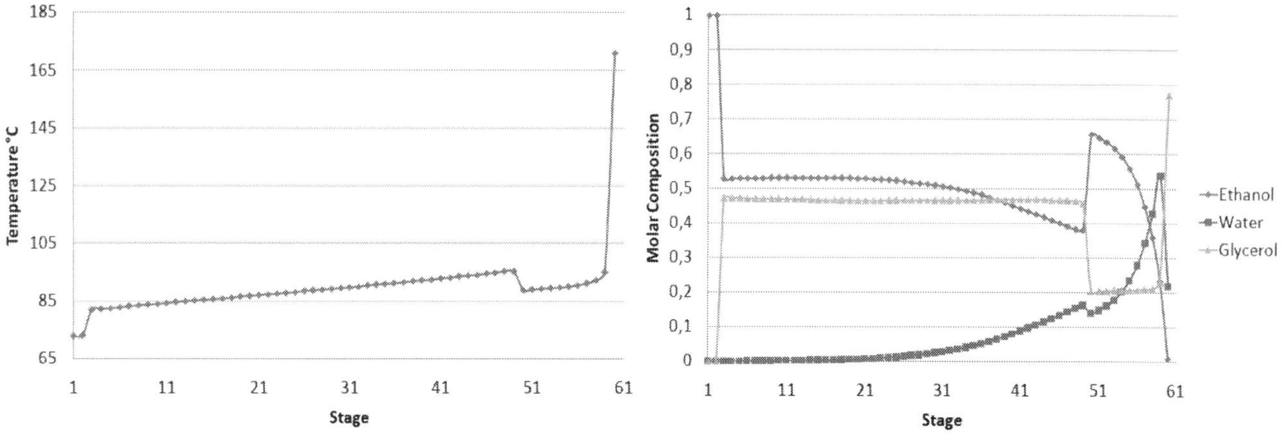

(a) (b)

Figure 8 (a) Temperature Profile in the Dehydration Column. (b) Molar concentration profile on the liquid in dehydration column

Table 3 summarizes the results of the simulation of ethanol dehydration by extractive distillation

Table 3 Dehydration Results by Extractive Distillation

Parameters	**Dehydration Column (T-301)**	**Regeneration Column (T-302)**
Distillate Molar Flow	227.4 Kmol/h	98.6 Kmol/h
Condenser Duty	-3605,76 kJ/s	-552,83 kJ/s
Volumetric Water Flow	63,49 m^3/h	8.79 m^3/h
Reboiler Duty	4299,42 kJ/s	661,55 kJ/s
Vapor Flow (145 psig)	0.44 Kg/Kg_{Et}	0.06 Kg/Kg_{Et}
Wt.% Ethanol Bottoms	0.098	0.06
Wt.% Ethanol Distillate	99.99	6.9

Table 4 summarizes the results of the simulation of ethanol dehydration by Adsorption with molecular sieves

Table 4 Dehydration Results by Molecular Sieves

Parameter	Adsorber Bed	Desorber Bed
Electric Duty	18828 kW/h	1368000 kW/h
Wt.% Ethanol Output	99.99	49

4. Results Analysis

The highest energy consumption stage is the fermentation and here is where it is most needed water process, and the stage with greater use of steam is the stripping, because of the large amount of ethanol that is separated from the fluid fermented. Noting the trend can be said that energy consumption is decreasing as it moves through the process, namely the stage of fermentation is the process that consumes more energy and services, while dehydration is the stage where energy and services consumption is lower. The energy consumption and raw materials, varies in each of the simulations of the dehydration methods, which allows comparing the best suited to the Colombian industry, with the lowest operating cost and capital.

Table 5 shows the results of energy consumption in each of the simulations

Table 5: Energy Consumption of each Simulation

Process Type	Energy Consumption (kJ/kg_{Et})	Steam Flow (kg/kg_{Et})	Volumetric Water Flow (m^3/h)
Extractive Distillation (Glycerol)	38027.1	3.95 (45 psig) 0.5 (145 psig)	2850.6
Molecular Sieves	35557.8	4.01 (45 psig)	2778.32

Comparing energy consumption is concluded that the dehydration technology of adsorption with molecular sieves and extractive distillation with glycerol reported the lowest values, which makes them competitive and relevant to their study and implementation. The glycerol, due to increased production of biodiesel, the purchase price is low, which significantly reduces operating costs of extractive distillation.

Table 12 presents the initial investment costs for all three technologies. The initial investment costs are found to be high for molecular sieves, almost four times regarding the azeotropic distillation and twice regarding the extractive distillation. These high costs are justified by the level of automation required to control the cycles of adsorption/desorption beds to make a continuous operation. In most cases, the initial investment in a plant with molecular sieves is 40% higher than a plant of equal capacity to use the distillation and quarrying (Gil, 2006).

Finally, the Extractive distillation constitutes a chance to study because it involves low energy consumption and industrial services moderate compared with other technologies. It is also important to note that investment costs are relatively high and not with the extractive distillation there is the option to make changes in the design of existing plants (revamping), with small capital investments to produce ethanol profitably in anhydrous.

Table 12 Investment Cost

Capacity	Extractive Distillation	Azeotropic Distillation	Molecular Sieves
300m^3/day	USD$650000	USD$370000	US$1170000
600m^3/day	USD$970000	USD$600000	US$1520000

Source: (Gil, 2007)

5. Conclusions

The simulation allowed the identification technique dehydration alcohol by extractive distillation with Glycerol role as the best option to implement the production method to the reality of the Colombian industry.

Allowed to hear three variables, which are the most influential throughout the proceedings, which are:

- The pH of juice fermentation.
- The temperature reactors.
- The steam conditions injected into the stripping column.

Since any change in any of these variables affect the production of alcohol. It is important to mention that the above variables are not the only ones able to affect the process, since the pressure, the relationship between reflux in the columns, the ratio of solvent-feeding among others, can positively or negatively affect the process.

The current trend in the design and implementation process is one of the prerequisites energy efficiency of the processing operations and separation. The processes of ethanol dehydration not escape this trend and, hence, energy consumption defendant in the production of one kilogram of anhydrous ethanol is one of the main parameters in the application of technology. Apart from energy consumption, another important factor in selecting the best alternative technological ethanol dehydration is the consumption of industrial services required, as well as investment costs incurred during initial deployment of technology.

However, when taking into account the consumption of industrial services, is the extractive distillation former glycerol excels at two other alternatives in terms of the required amounts of steam, water, electricity and agent of separation.

REFERENCES

1. Cardona, C.A., Sánchez, O.J. (2006). *Energy consumption analysis of integrated flowsheets for production of fuel ethanol from lignocellulosic biomass,* Energy No 31, pp. 2447-2459.

2. Cardona, C.A., Sánchez, O.J., Montoya, M.I, Quintero, J.A., *Simulación de los Procesos de Obtención de Etanol a partir de Caña de Azúcar y Maíz,* Scientia et Technica Año XI No 28 Octubre, pp. 187-192.

3. Carlson, E.C.,(1996). *Don't Gamble With Physical Properties for Simulations,* Chemical Engineering Process, Octubre de 1996, pp. 35-46.

4. Fogler, H.S., (2001). *Elementos de Ingeniería de las Reacciones Químicas.* Ed. Prentice-Hall, Capitulo 4, Capitulo 5.

5. Gil, I.D., (2006), *Diseño, montaje y puesta en marcha de una destilación extractiva para la producción de alcohol anhidro.* Tesis para obtener el título de Magister en Ingeniería Química, Universidad Nacional de Colombia.

6. Gil, I.D., Aguilar, J., Rodríguez, G., Caicedo, L.A., Uyazán, A.M. (2006), *Producción de Alcohol Carburante por Destilación Extractiva: Simulación del proceso con Glicerol.* Ingeniería e Investigación, abril, año/ Vol.26, numero 001, pp. 45-50.

7. Grisales, P.A., Ríos, L.A., Triana M. *Diseño de un proceso de producción de etanol anhidro a partir de jugo de caña.* Escuela de Ingeniería Química, Universidad del Valle

8. Honorato, Flavio., Rodrigues, Maria., Maugery Francisco., (1999). *Dynamic modelling, simulation and optimization of an extractive continuous alcoholic fermentation process.* Journal of Chemical Technology and Biotechnology No 74, pp. 176-182.

9. Jacques, K., Lyons, T.P., Kelsall, D.R., (1999). *The Alcohol Text Book.* Ed. Nottingham University Press, 1999, Chapter 5, Chapter 9, Chapter 16, Chapter 17, Chapter 18.

10. Kister,H., (1990). *Distillation Operation.* Ed. McGraw-hill, 1990, Chapter 2, chapter 4.

11. Seider, D.W., Seader, J.D., Lewin, L.W.(2003). *Product and Process Design Principles.* Ed. John Wiley & Sons, Chapter 3, Chapter 4.

A CLEAN PRODUCTION PROCESS OF CHROMIC OXIDE

Yi ZHANG, Zuo-Hu LI, Hong-Bin XU, Shi-Li ZHENG, Tao QI, Institute of Process Engineering, Chinese Academy of Sciences, Beijing, China

Introduction

Chromic oxide is an important basic chemical and finds many applications including metallurgical materials, green pigments, construction materials, refractory materials and catalysts. Presently, the industrial production of chromic oxide generally employs two processes (Ding, 2003): one is the reduction of sodium dichromate with ammonia sulfate; the other is the thermal decomposition of chromic anhydride. Both of the processes employ hexavalent chromium as the raw materials. However, the industrial production of hexavalent chromium compounds, including sodium dichromate and chromic anhydride, leads to serious environmental problem and low resource utilization.

Fig. 1 illustrated the traditional production process for hexavalent chromium compounds, as well as chromic oxide. It produces useful products of sodium dichromate and chromium anhydride, as well as several kinds of toxic residues, including chromium-containing residue, aluminum-enriched residue and chromium-containing Glauber's salt ($Na_2SO_4{\cdot}10H_2O$) and sodium bisulfate.

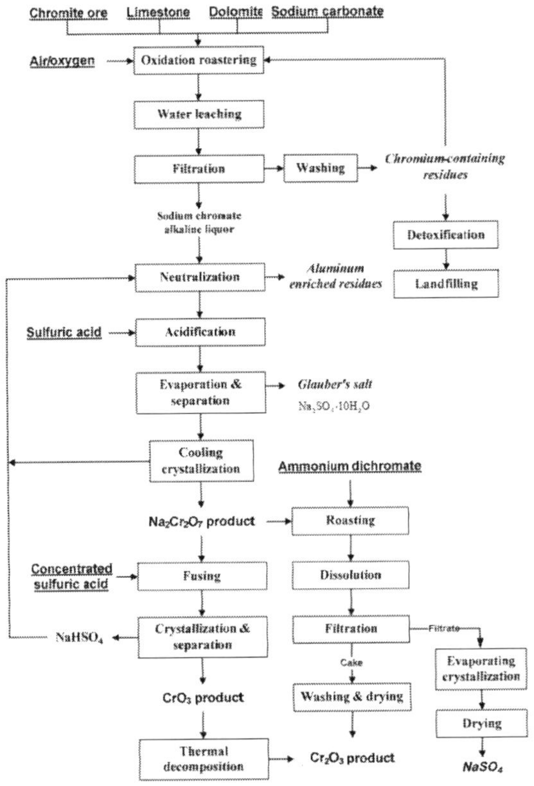

Fig. 1. Illustrative Flow Sheet of the Traditional Production Process for Chromium Compounds

In the traditional process, there are three main problems. The first is the environmental pollution. During production of one ton of chromium anhydride product, the chromate production plant discharges approximately 2.0 to 2.5 tons of toxic chromium-containing residues that are difficult to be detoxified and comprehensively used because of high content of hexavalent chromium. Also, the produced calcium chromate is highly toxic and carcinogenic. Furthermore, the discharge of large amount of chromium-containing gases and dusts creates serious pollution. The second problem is that the conversion efficiency of chromium is only 76%, which means that a considerable amount of chromium is discharged into the residue. The third problem with the process is the production of by-products that are not valuable. The chromium-containing Glauber's salt and sodium bisulfate produced are of little use and constitute a pollution source. Consequently, the total atom utilization efficiency of the traditional process is quite low.

In this paper, a clean production process, which was recently developed by the Institute of Process Engineering, Chinese Academy of Sciences, will be described in details (Zhang, 2005). With the design objective of eliminating pollution at the source, the clean process greatly improves the resource utilization efficiency and successfully achieves the zero emission of chromium-containing residues.

Overall Description of the Clean Process

Illustrative Flow Sheet

As shown in Fig. 2, the clean production process of chromic oxide mainly includes the liquid-phase oxidation in Sub-Molten Salt (SMS) medium, the multi-phase separation and evaporative crystallization, the hydrogen reduction of potassium chromate, and the activated sintering of the intermediate product (Li, 2008a).

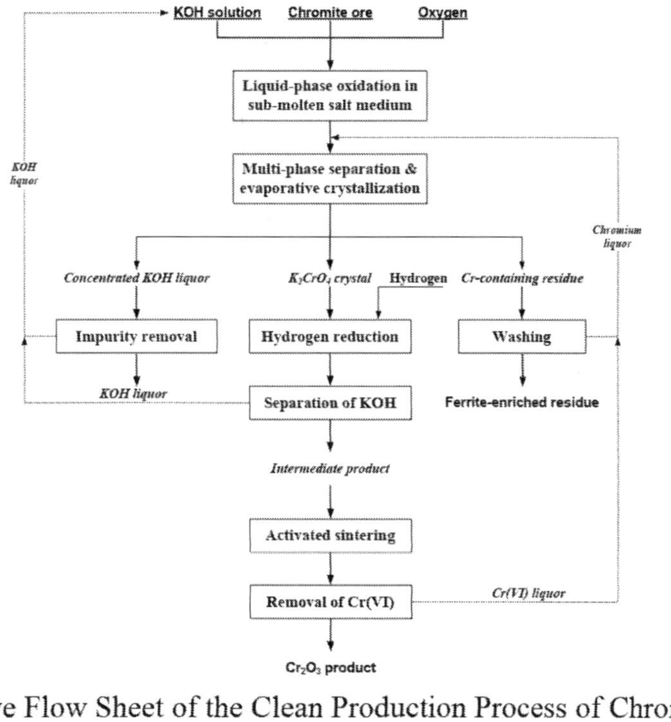

Fig. 2. Illustrative Flow Sheet of the Clean Production Process of Chromic Oxide

Main Reactions in the Clean Process

For the trivalent chromium in the chromite ore, it is firstly converted into hexavalent chromium as potassium chromate during the liquid-phase oxidation process.

$$Cr_2O_3 + 4KOH + 1.5O_2 \rightarrow 2K_2CrO_4 + 2H_2O \quad (1)$$

Then it is converted back into trivalent chromium as chromic oxide in the hydrogen reduction process.

$$2K_2CrO_4 + 3H_2 \rightarrow 4KOH + Cr_2O_3 + H_2O \qquad (2)$$

Since the final product, chromic oxide, bears no potassium, the KOH consumed in the liquid-phase oxidation process can be actually recovered and recycled inside the process. Therefore, no more raw materials than chromite ore, oxygen, and hydrogen are consumed in the clean process theoretically.

$$Cr_2O_3(\text{chromite ore}) + 3H_2 + 1.5O_2 \rightarrow Cr_2O_3(\text{final product}) + 3H_2O \qquad (3)$$

For the associate components in the chromite ore, including Al, Mg, Si, and Fe, they mainly enter the ferrite-enriched residue and are converted into the byproduct desulfurization agent.

Comparison of the Clean Process with the Traditional Process

Fig. 3 showed the comparison of the clean process with the traditional process for producing chromic oxide and four features of the clean process were also stated.

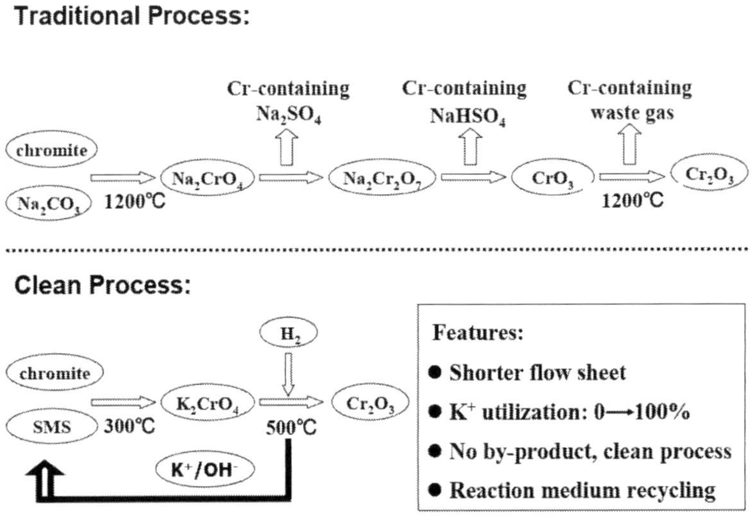

Fig. 3. Comparison of the Clean Process with the Traditional Process

From the raw material, chromite ore, to the final product, chromic oxide, only the intermediate product, potassium chromate is produced, the clean process achieves a shorter flow sheet than the traditional process. Since no potassium-bearing byproduct or waste is produced, the utilization of potassium can be as high as 100%. The reaction media, KOH in SMS, can be completely recovered in the hydrogen reduction unit. Moreover, since no chromium-containing byproduct or waste is produced, a clean process is thus achieved.

Lab Scale Test Results and Discussion

Typical Composition of the Chromite ore (Xu, 2005)

Table 1 listed the chemical analysis of the chromite ore (originating from Vietnam). Fig. 4 illustrated the X-ray diffraction and phase identification results.

Table 1. Chemical Analysis of the Chromite Ore Used in the Lab Scale Test

Component	Cr_2O_3	FeO	MgO	SiO_2	Al_2O_3	MnO_2
Content, %wt	41.40	22.31	8.88	12.73	11.94	0.08

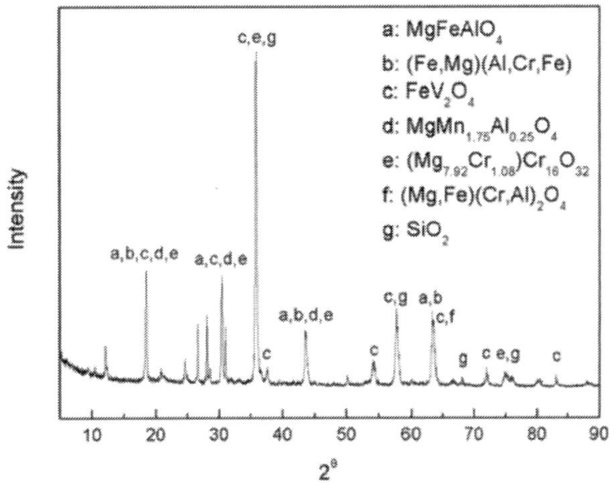

Fig. 4. The X-ray Diffraction and Phase Identification Results of the Chromite Ore Used in Lab Scale Test

Liquid-phase Oxidation (Xu, 2005)

In the liquid-oxidation process, the following reactions, besides Reaction (1), mainly occur:

$$2FeO \cdot Cr_2O_3 + 8KOH + 7/2O_2 \rightarrow Fe_2O_3 + 4H_2O + 4K_2CrO_4 \quad (4)$$

$$2MgO \cdot Cr_2O_3 + 8KOH + 3O_2 \rightarrow 2MgO + 4H_2O + 4K_2CrO_4 \quad (5)$$

$$Al_2O_3 + 2KOH \rightarrow 2KAlO_2 + H_2O \quad (6)$$

$$SiO_2 + 2KOH \rightarrow 2K_2SiO_3 + H_2O \quad (7)$$

The extraction sequence of the main components in the chromite ore, Cr, Al, and Si, was given in Fig. 5. And the X-ray diffraction and phase identification results of the chromite ore process residue was given in Fig. 6. It can be seen that the aluminum in the chromite ore is almost steadily leached into the liquor according to Reaction (6). However, the silicon in the chromite ore might firstly be leached into the liquor according to Reaction (7) and then react with aluminum and potassium hydroxide and produce potassium aluminosilicate ($KAlSiO_4$).

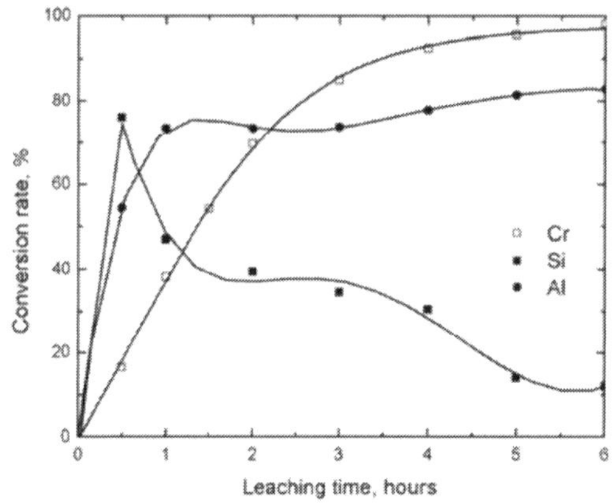

Fig. 5. The Extraction Sequence of Cr, Al, and Si during the Liquid-phase Oxidation Process

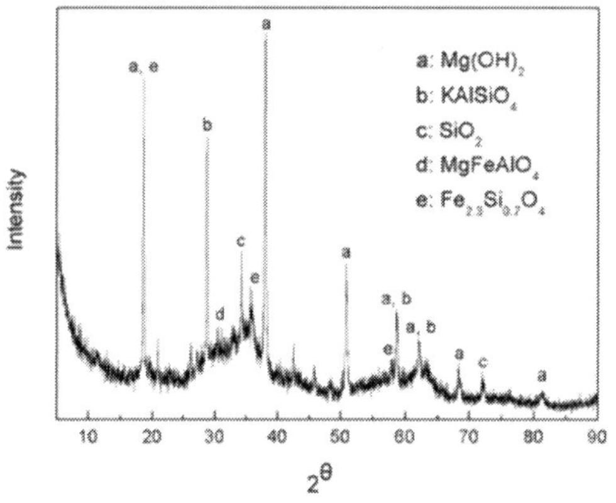

Fig. 6. The X-ray Diffraction and Phase Identification Results of the Chromite Ore Processing Residue

Fig. 7 illustrated the effect of reaction temperatures on the conversion of chromium for the combination of Reactions (1), (4), and (5). It can be found that, at a temperature higher than 320℃, the conversion of chromium can be achieved as high as 98%.

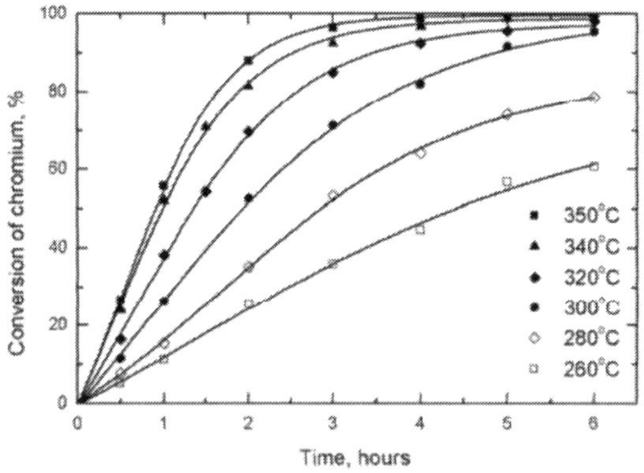

Fig. 7. Effect of Reaction Temperature on the Conversion of Chromium

Hydrogen Reduction (Bai, 2006b)

In the hydrogen reduction and activate sintering process, the following reactions actually occur. For the hydrogen reduction:

$$2K_2CrO_4 + 3H_2 \rightarrow 2KCrO_2 + 2KOH + H_2O \quad (8)$$

For the hydrolysis and separation of KOH process:

$$KCrO_2 + H_2O \rightarrow CrOOH + KOH \quad (9)$$

Therefore, the main composition of the intermediate product is CrOOH, as can be seen in Fig. 8. For the activated sintering process:

$$2CrOOH \rightarrow Cr_2O_3 + H_2O \quad (10)$$

$$2KCrO_2 + 1.5O_2 \rightarrow K_2Cr_2O_7 \quad (11)$$

Through Reaction (11), the deep removal of potassium from the finally product, chromic oxide, is made possible, as can be seen in Fig. 9. Fig. 10 showed the SEM pattern of the Cr_2O_3 final product (Li, 2008b).

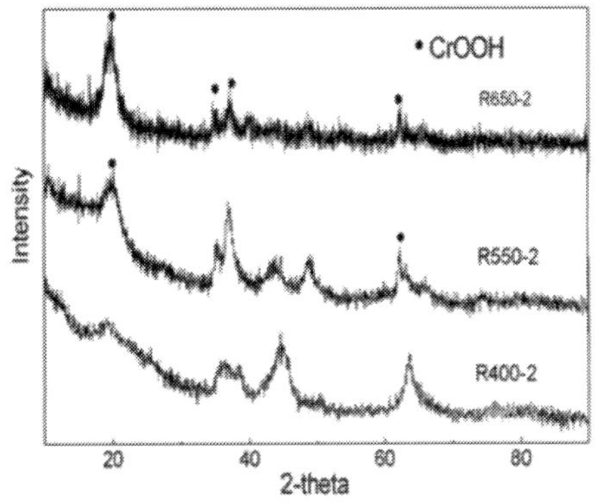

Fig. 8. XRD Pattern of the Intermediate at Different Reduction Temperatures

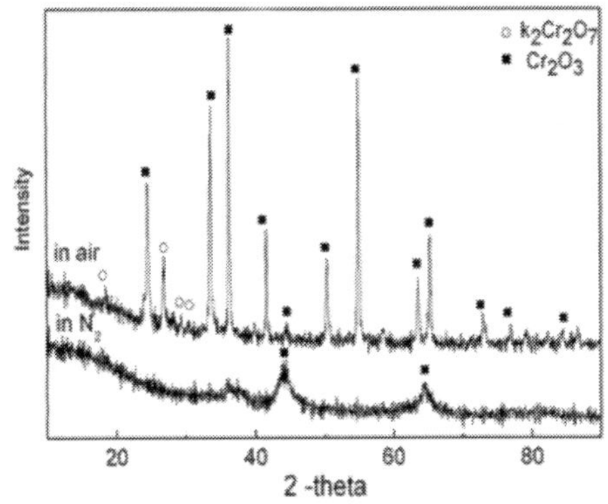

Fig. 9. XRD Pattern of the Activated Sintering Product in Different Atmospheres

Fig. 10. The SEM Pattern of the Cr_2O_3 Final Product

Table 2 listed the hydrogen reduction conversion under different reduction temperatures (Bai, 2006a). It can be seen that, at a temperature higher than 500℃, the conversion of hexavalent chromium can be achieved as high as 98% in a reaction time period of 90 min.

Table 2. Effect of Reducing Temperature on the Conversion of Hexavalent Chromium in 90 min

Temperature, ℃	350	400	450	500	600	700
Conversion, %	3.2	12.9	95.1	98.2	99.3	99.8

Industrial Demonstration

By use of the clean production described above, a demonstration plant with an annual production capability of 2,000 tons of chromic oxide has been built in China. The comparison of the industrial operation results of the clean process with those of the tradition process was given in Table 3 (Zhang, 2005). The clean process has exhibited promising results for the industrial production of chromic oxide.

Table 3. Comparison of the Industrial Results of the Clean Process with those of the Traditional Process

Item	Traditional process	Clean process
Chromium conversion, %	75	98
Reaction temperature, ℃	1200	300 ~ 350
Resource utilization, %	15	90
Chromium-containing residue amount, t/t	2.0	0.5 → 0 (final zero emission)
Energy consumption	--	Decreased by 20%
Production cost	--	Decreased by 17%

Conclusion

With the design objective of eliminating pollution at the source, the clean production process greatly decreases the reaction temperature, remarkably improves the resource utilization efficiency and successfully achieves the zero emission of chromium-containing residues. The industrial demonstration results have shown a promising prospect for the industrial production of chromic oxide.

Acknowledgements

The authors gratefully acknowledge the financial supports from the National Science and Technology Pillar Program of China (Grant No. 2006BAC02A05), the National Science Foundation of China (Grant No. 50234040), and the National Basic Research Program (973 Program) of China (Grant No. 2007CB613501).

References

1. Bai, Y. L. (2006a). "Fundamental Research on Preparation of Chromic Oxide by Hydrogen Reduction at Low Temperature in the Clean Production Process of Chromate", Ph. D. Thesis, Beijing: Institute of Process Engineering, Chinese Academy of Sciences.
2. Bai, Y. L. and *et al.* (2006b). "Reductive Conversion of Hexavalent Chromium in the Preparation of Ultra-fine Chromia Powder". *J. Phys. Chem. Solids*. 67 (12), pp. 2589-2595.
3. Ding, Y. and Ji, Z. (2003). Production and Utilization of Chromium Compounds. Beijing: Chemical Industry Press.
4. Li, P. and *et al.* (2008a). "A Green Process to Prepare Chromic Oxide Green Pigment", *Environ. Sci. Technol.*, Article in press, available online 08/29/2008.
5. Li, P. and *et al.* (2008b). "The Effects of Al and Ba on the Color Performance of Chromic Oxide Green Pigment", *Dyes and Pigments*, Article in press, available online 07/16/2008.
6. Xu, H. B. and *et al.* (2005). "Oxidative Leaching of a Vietnamese Chromite Ore in Highly Concentrated Potassium Hydroxide Aqueous Solution at 300°C and Atmospheric Pressure", *Minerals Engineering*, 18, pp. 527-535.
7. Zhang, Y. and *et al.* (2005). "Green Manufacturing Process of Chromium Compounds", *Environmental Progress*, 24 (1), pp. 44-50.

293d A Physical Absorption Process for the Capture of CO_2 from CO_2-Rich Natural Gas Streams Using Hydrocarbon Solvents, with Post-Capture CO_2 Re-Compression

Frances E. Pereira, Emmanuel Keskes, Claire S. Adjiman, Amparo Galindo, and George Jackson. Chemical Engineering, Imperial College London, South Kensington Campus, London, SW72AZ, United Kingdom

The increasing importance of natural gas as a source of energy poses difficult gas separation design challenges, as the high flow-rate streams recovered from gas fields are at high pressures (typically about 10MPa) and can contain a high proportion of CO_2 (up to 70%). Conventional separation techniques are usually restricted to low CO_2 content or low pressure feeds, and consequently there is a pressing need for an alternative process that is appropriate for such a scenario.

In addition, increasingly stringent regulation of CO_2 emissions has rendered release of CO_2 to the atmosphere more difficult. There is growing pressure for the captured CO_2 to be re-compressed before it may be re-injected and re-used on site, or transported for further use elsewhere. This step is energy intensive, and could potentially change the economic outlook of the separation process, and consequently the entire natural gas production operation.

A process for the separation of gaseous CO_2/CH_4 has been developed, capable of economically treating natural gas feeds at the high pressures and CO_2 concentrations mentioned. The process has been modelled for a system comprising methane, CO_2 and alkane solvent. The separation process model includes absorption and desorption units, as well as heat exchangers and a CO_2 re-compression stage. An advanced equation of state, SAFT-VR, has been employed in this study. This facilitates accurate representation of the complex thermodynamic and fluid-phase equilibrium behaviour of high pressure/temperature CO_2-hydrocarbon mixtures. The work has been carried out using the modelling-optimisation software gPROMS. The process is optimised by considering an economic objective over a 15-year lifetime. In addition to finding the best equipment sizes and operating conditions, the alkane chain length of the physical solvent is also treated as an optimisation variable, to achieve the most cost effective absorption for differing CO_2 content.

Several case studies are examined. In particular, the impact of including further impurities in the feed is assessed. The effect of recompression costs on the optimal design is also considered. Given the large variability in feedstock over the 15-year lifetime of the plant, process flexibility is quantified.

Vinyl Acetate from ethylene, acetic acid and oxygen Industrial Plant Simulation.

J.P. Contreras, J.C. Naranjo, S. Ramírez & D.M. Martínez
Chemical Engineer department, Los Andes University, Bogotá, Colombia.

I.D. Gil
Chemical Engineer department, National University of Colombia, Bogotá, Colombia.

Paper # 134650

ABSTRACT

The production of vinyl acetate monomer (VAM) represents one of the largest industries with a growth estimate of a near 2.7% per year starting in 2008. In recent times, thanks to different studies and the importance of said product, one process has been developed based on the reaction of ethylene, acetic acid, and oxygen using a Pd/Au catalyst. In this work, a detailed study was made based on a simulation of the process using the Aspen Plus v2006 program and establishing the correct operation conditions. The simulated process involves, from the preparation of the raw materials until the dehydration of the monomer. The restrictions set for the simulation corresponds to the VAM composition restriction in the extractive distillation column using glycerol.

INTRODUCTION

Vinyl acetate is a colorless, flammable liquid that also has a characteristic smell that can quickly become irritating. This monomer is used principally in the production of polyvinyl acetate (PVAc) and other vinyl acetate co-polymers. Polyvinyl acetate is a precursor of polyvinilyc alcohol and polyvinyl acetate resins (PVA). Vinyl acetate is also copolymerized as a minor raw material for vinyl chloride and ethylene to form commercial polymers and acrylic fibers.[1]

Vinyl acetate is completely soluble in organic liquids but not in water. At 20° C a saturated solution of the monomer in water can contain between 2-2.4% of vinyl acetate, while a saturated water solution in vinyl acetate contains 1% of water. At 50°C the solubility of the monomer in water increases in 0.1% in regards to 20°C, while the solubility of water in vinyl acetate doubles at 50°C

The most important reaction of vinyl acetate is the polymerization by free radicals which is fast and exothermic. Due to the fact that the vinyl acetate is an Esther, it presents the reactive properties of said group. Keeping in mind that the corresponding chain of alcohols is unsaturated, the compound presents some differences in regards to the Esther group. The speed of hydrolysis of the vinyl acetate is 1,000 times greater than its saturated analogous in a alkaline medium, and it was researched that its speed of hydrolysis is significantly lower at a p.H. of 4.4.

---

[1] Kirk-Othmer, (1978), Vol 23, pp 817.

VAM PRODUCTION PROCESS

The first available process for the synthesis of vinyl acetate was the acetoxilation of acetylene in gaseous form over a zinc acetate catalyst supported on carbon. This process consisted in the reaction of acetylene with acetic anhydride in a catalyzed medium and high temperature to form dietilene diacetate. This product passed through a cracking tower which in result gave as products acetic acid and vinyl acetate.[2] This type of reaction had a high production percentage (92-98%), but due to the increasing value of acetylene, new technologies and methodologies needed to be research in the 60's.

The process was modified until acetoxilation in a gaseous phase of ethylene over palladium and gold catalyst supported on silica gel was capable. The catalyst used potassium acetate to help the reaction in a temperature range of 423 – 463 K and a pressure range of 600 to 1,000 kPa. This is how ethylene acetoxilation with oxygen, acetic acid and Pd as a catalyst was developed. This process consists of two parts: a process in a homogeneous liquid phase that is used to produce 25% of the production of vinyl acetate, and another heterogeneous gaseous phase process used to produce the last 75% of the product.

The synthesis of vinyl acetate from ethylene, acetic acid and oxygen over a Palladium catalyst is a very important industrial process, but its selectivity is affected by the production of CO_2 due to the combustion of the ethylene. Various factors affect the synthesis of vinyl acetate, like for example, the dispersion of Pd, partial pressures from the reactives that are the catalyst additives (near 80% if it is Pd/SiO_2 and greater to 94% if it is $Pd\text{-}Au/SiO_2$) and contact time. The investigation done by Han, H. F., et al. (2004) shows that the ethylene is the principal cause for the production of CO_2 due to the fact that combustion kinetic does not change if the acetic acid is taken out of the medium.[3]

The process that involves ethylene, acetic acid, oxygen over a Pd catalyst medium proves to be the best option to develop. This is evident because for the last thirty years it has been applied in the industry. In the U.S.A. alone, the production of ethylene is around 11.533×10^3 tons per year, which more than 30% is used on the production of vinyl acetate.

Production Process using Ethylene, Acetic Acid and Oxygen.

The process is centered on the reaction between ethylene, oxygen and acetic acid in order to produce vinyl acetate and water. There is a second unwanted reaction in the process based on the ethylene oxidization to produce water and carbon dioxide. The gaseous effluents from the reactor are cooled trying to have a partial condensation of the stream. Next, the gas is separated from the liquids in a phase separator. The gas is washed with acetic acid from a recycle stream. This helps to recover the vinyl acetate in the absorber. Afterwards, there is a process to eliminate the CO_2 from the gas phase and recycle it back in to the reactor the Ethylene is necessary.

The liquids that come out from the separator and the acetic acid used in the washing process enter an azeotropic distillation tower. In this part of the process, the main objective is to separate the acetic acid (bottoms) form the water and vinyl acetate mixture (distillate). The distillate enters to a decantation process in which its organic product (mainly vinyl acetate) divides into a reflux stream that heads back to the distillation column

[2] Kirk-Othmer, Concise enciclopedia of Chemical Technology, John Wiley & Sons, 1985 (USA), pp. 1227.
[3] Han, Y.F., et al, Kinetics of ethylene combustion in the synthesis of vinyl acetate over Pd/SiO_2 catalyst, Journal of Catalyst, Vol. 224, 2004, pp. 60-68.

and another stream that enters a separation train process where the proper concentration level of vinyl acetate is achieved to comply with the wanted specifications. The aqueous stream from the decanter is discarded.

A portion of the bottoms stream from the distillation column (mainly acetic acid) is used in the washing process in the absorber, while the other portion of the bottom stream is mixed with pure acetic acid from the feed streams to later on enter to an evaporation process. The gaseous acetic acid obtained is mixed with the recycle gas, oxygen and ethylene to form the feed to the reactor.

For a detail analysis, the process has been divided in to three parts: reaction, distillation and dehydration. It is shown in figure 1.

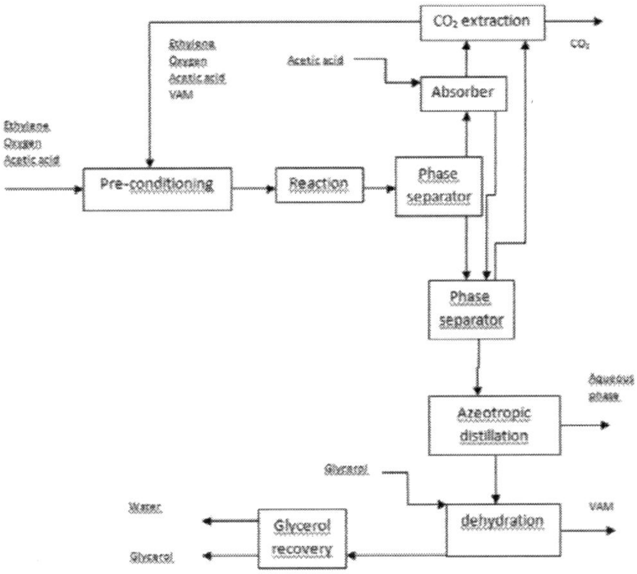

Figure 1. VAM production process flow diagram.

Reaction system

The reaction that produces the vinyl acetate takes place in a P.B.R. reactor where a gaseous mixture of ethylene, acetic acid, and air (with small presence of CO_2 and water from the recycle streams) are fed to the reactor. To obtain an annual production of 5,000 tons (amount determined for the Colombian market) of vinyl acetate, 2,197.3 mol per minute with a molar composition of 0.5 of ethylene, 0.07 of oxygen, 0.15 acetic acid and 0.263 of nitrogen have to be fed. An excess of ethylene over acetic acid (3:1) must be guaranteed. The concentration of oxygen must be kept below 8 % molar because at higher concentrations it breaches the explosive limit of the mixture.

The reactions that take place are:

$$C_2H_4 + C_2H_4O_2 + \frac{1}{2}O_2 \xrightarrow{Pd} C_4H_6O_2 + H_2O \qquad (1)$$

$$C_2H_4 + 3O_2 \xrightarrow{Pd} 2CO_2 + 2H_2O \qquad (2)$$

These reactions are highly exothermic and the standard heat of reaction for the first reaction is 42.1 Kcal per mol of vinyl acetate, while the heat of reaction for the second reaction is -316Kcal per mol of ethylene.

The heat of reaction is removed by generating vapor in the shell section of the tubes. This vapor is used in other part of the process. This is the same for the reaction catalyzed by palladium and gold supported by silica.

The temperature must be below 200°C because at this temperature the catalyst suffers irreversible mechanical damages. The catalyst is activated with a temperature over 140° C. This allows to take an inferior operation limit temperature of 150°C.

The recommended operation pressure is between 8 and 10 atm (Dimian y Bildea, 2008) because higher pressures favor the productivity, but affect negatively the selectivity due to the adsorption of the ethylene in to the active sites of the catalyst benefiting the unwanted combustion reaction.

The rate law for this reaction (1) is given by:

$$r_{VA} = k_1\, p_{Et}^{\alpha 1} p_{Ox}^{\beta 1} \qquad (3)$$

Where p_{Et} is the partial pressure of the ethylene and p_{Ox} is the partial pressure of the oxygen:

$$k_1 = A_1 \exp\left(E_1/(RT)\right) \qquad (4)$$

The values of the activation energy E_1 depend on the operation temperature. At temperatures between 120° C and 140° C (1st region) activation energies of 40 kJ/mol can be obtained keeping in mind a dependency of Arrhenius law. At temperatures between 140° C and 160° C (2nd region) the mean activation energy is 15 kJ/mol, there for reaction rates higher than those of the first region can be observed. This is the same for α_1, which values are 0.38, 0.35, and 0.35 for temperatures of 120° C, 140° C, and 160° C respectively. From this observation comes the preference for temperatures over 140° C in the process.

The values of β_1 reported are 0.20, 0.20, 0.21 for temperatures of 120° C, 140° C and 160° C respectively. A molar relation of ethylene/acid higher than 3 is used.

The reaction rates were reported like TOF (turnover frequency). This type of data indicates the number of reactions or catalytic cycles done by a unit of the activated center by unit of time (molVA/(molPd seg):

$$r\left(\frac{molVA}{KgCat}\right) = TOF\,\frac{w_{Pd} D}{PM_{Pd}} \qquad (5)$$

Were w_{Pd} is the fraction of palladium in the catalyst (0.01), D is the palladium dispersion coefficient (with which the fraction of atoms of active palladium can be calculated) and PM_{Pd} is the molecular weight of Pd in kg/mol. The value of D is reported with a value of 0.4 by Han et al. (2005), the value of A_1 calculated with the data reported by Han et al., (2005) is of 9.7×10^{-3} molVA/(kgCat seg) (For partial pressures expressed in kPa).

For reaction (2) the same authors report a similar rate reaction law (Han et al., 2004):

$$r_{CO_2} = k_2\, p_{Et}^{\alpha 2} p_{Ox}^{\beta 2} \qquad (6)$$

$$k_2 = A_2 \exp\left(E_2/(RT)\right) \qquad (7)$$

α_2 was determined with values of -0.27, -0.31 and -0.27 for temperatures of 140° C, 160° C and 180° C respectively. Besides this, an increase of 50% of the production of CO_2 was observed when the temperature increased from 140°C to 160°C. The values found for β_2 were 0.88, 0.82, 0.89 for temperatures of 140°C, 160°C and 180°C respectively. The activation energy E_2 obtained from data was 21 kJ/mol (higher than the activation energy of reaction (1) which is 15 kJ/mol). No dependency was observed by the partial pressure of the acetic acid in the rate law. This confirms that the production of CO_2 is due basically to the oxidation of the ethylene.

The value of A_2 calculated taking in to account the data reported by Han et al., (2004) resulted in a value of 5.13 x 10^{-4} molVA/(kgCat seg). (For partial pressures expressed in kPa).

VAM distillation and dehydration.

It is based on the outlet homogeneous gaseous reactor stream in which the high temperature and pressure guarantee its state, favoring the reaction conditions. This shows that through manipulating factors like temperature and pressure, it's possible to generate changes that allow the separation of the components, principally vinyl acetate. The effluent reactor stream is composed by 45,9% mol of ethylene, 6,7% of water, 5,7% of VAM, 5,4% of oxygen, 1% of Carbon Dioxide, 12,5% of Acetic Acid y 34,3% of Nitrogen.

The general description of this system consists of an effluent reactor stream that is cooled with cooling water with the objective of generating two phases: the gaseous phase which contains principally oxygen, ethylene, carbon dioxide, and ethane when the feed of ethylene is not pure. The second phase consistis of a liquid phase that contains principally vinyl acetate, water and acetic acid. The vapor phase passes to a compressor while the liquid phase passes to an azeotropic distillation column. The gas stream passes from the compressor to an adsorption column with the objective of extracting the remaining vinyl acetate, while the liquid stream exits the azeotropic distillation column in the form of distillates which later on are condensed to form two liquid phases in a decanter. The heavier of the two phases is recycled back in to the distillation column and the other passes on to the purification process to obtain the vinyl acetate, while the other is mainly water. The bottoms stream is fed back to the absorption tower. With these conditions, the column gets to generate three currents: the bottoms which are mainly acetic acid, the distillate which is composed by two immiscible phases that allow the separation of the VAM and the water. This is the most important process due to its difficulty to separate (water-VAM mixture). The liquid bottoms of the separator are mixed with the liquid stream of the separator so it can head later on to a distillation column.

On the other hand, the nearly free vinyl acetate gaseous stream goes to a carbon dioxide extraction unit in order to take out most of this component out of the stream. The stream that had the CO_2 extracted is divided in to a purge and a recirculation stream that mixes with the other portion of the stream that had no CO_2 extracted. This stream will be used to exchange heat with the outlet stream of the reactor and to serve as a base feed stream for the reactor, taking in to account that this streams is added through other feed streams that are needed for the reaction.

PROCESS SIMULATION

The development of this work began with a bibliographic compilation form which the optimum VAM production process and its conditions were established. Starting from said process, a simulation was done with the ASPEN PLUS v.2006 program from where a total process material and energy balance were established. Also, the final purity of the product and the possible modifications that allow a production of a greater quality and lower operation cost were determined.

Defined Properties

To determine the models that would be applied in the simulation, an analysis of two articles were used as strating points. These article were *"Don't gamble with physical properties for simulations"* and the article about thermodinamics of Galen J. Supples. Thanks to the information found in the text of Alexandre D. Dimian [1], it was deterimined that the model that best fitted the property's description was the SR-Polar model because it takes into account the interaccion between gaseous components.

This choice was supported by the fact of the liquid phase is mainly composed by vinyl acetate (apolar), water (polar), acetic acid (polar), and supposing that there are interaction parameters due to the fact that these are known compounds. Finally, the NRTL and UNIQUAC models and all their variations were taken into consideration in order to get presition with these types of components, but after researching the Alexandre D. Dimian book, the NRTL-HOC model was chosen.

Reaction

Figure 2 illustrates the preparation of the raw material simulation and reaction flow diagram.

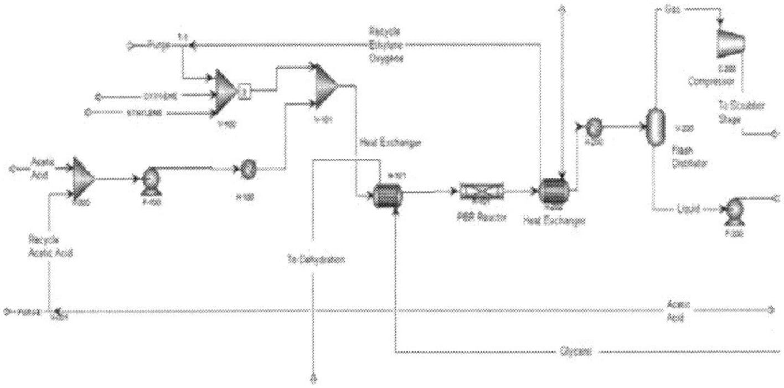

Figure 2 Reaction simulation flow diagram.

To design the reactor, it was established that it had to be done isothermally trying to prevent the sections that exceed 200°C which in turns lowers the catalyst activity. To control the temperature effect, the use of an inert gas was applied, in this case nitrogen. Besides this, the reactors are designed to achieve a low conversion of acetic acid by each pass (between 15 to 38% of acetic acid, and 8 to 10% of ethylene). It has to be taken in to account that activation energy of the unwanted reaction is greater than the activation energy of the wanted reaction, so at higher temperature the selectivity will be affected.

The reactor is a tube and shell reactor with a tube diameter of 3.7 cm, which have inside the catalyst. The calculations took in to account a refrigerant at a constant temperature of 5 degrees (° C) below the inlet temperature of the reactives and afterwards, through Aspen Plus v.2006, the simulation was made using saturated water in co-current through the shell at 418K and vapor was produced at 4 atm. The simulation complies with the specification of the reactor's design.

The global heat transfer coefficient calculated by Dimian y Bildea, (2008), was used (0.2KW/m^{2}K). Likewise, the physical properties of the catalyst are the ones reported in the US Patent No. 5808136: a bulk density of 0.5 kg/L, partial size of 5 mm and a porous volume of 0.8 L/kg. Thanks to this data, an empty fraction of 0.4, and a catalyst solid density of 0.833 kg/L were calculated. The operation inlet temperature for the reactor was 150°C.

Using all of the information, a 0.6 m/s rate was obtained which implies the use of 234 tubes and a pressure drop of 7.28% (and outlet pressure of 7.83 bar), achieving the needed conversion of acetic acid (35%). The previously mentioned parameters are suggested in this report. A reactor length of 6.01m was calculated taking in to account the data. The calculations were made always keeping in mind the volume in liters of the reactor. The volume per tube is 6.46 Lt, and the total reactor volume is 1,513 Lt.

VAM distillation

Figure 3 illustrates the distillation simulation flow diagram of the vinyl acetate monomer.

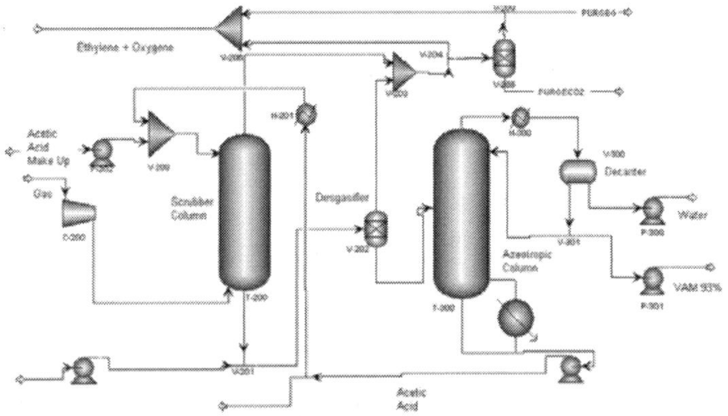

Figure 3. VAM distillation simulation flow diagram.

For the gas phase obtained from the first flash unit, a 20 stage VAM absorption tower was defined. It is in the range with what was found in the literature for two the scenarios mentioned earlier. Finally, the same analysis was conducted keeping in mind the stream temperature, although the final separation showed the same behavior as seen in the acid stream.

Finally in the carbon dioxide extraction stage, the most important factor that was taken in to account was the concentration of the absorbing agent because this stream is recycled to the reactor. It was necessary keeping a constant CO_2 concentration because it is fundamental, if not, possibly the selectivity and production could be affected. To do this, it is imperative to have a control over the concentration in this stage, so that the concentrate of the absorbing agent increases as the internal acidity increases. At an industrial level, the absorbing agents for carbon dioxide commonly used are amine solution (due to their basic characteristic) or Potassium carbonates (due to their affinity)

The liquid phase begins its separation using an azeotropic distillation column. For design purposes, the effect of the temperature over the quantity of VAM in the desired stream was researched and tested with a sensitivity analysis and was found that any effect was observed in temperature range of 0 to 1000 K. A similar situation occurs with the acid concentration in the distillate stream when the temperature is varied. On the other hand, based on the effect of the number of stages over the acid concentration in the distillate stream, it was found that a total of 23 stages is the optimum because it reaches a flow of 0.57 Kmol/h, lower than 0.85 that is generated with a larger number of stages. Finally, the effect of the recirculation flow over the vinyl acetate produced was studied and no difference was found when it was varied, so it was kept at 74.55 kmol/h for the VAM on the distillate stream.

The decanter is the unit in which one has to be more careful due the fact that is in this unit that the two phases are generated: organic and aqueous phases. Keeping in mind the temperature, and supposing that when the temperature rises the amount of vinyl acetate varies in the outlet stream, this had no effect. Therefore the decanter's inner temperature was varied with the objective of finding the optimum temperature that can generate the two phases. This temperature was found to be 150 f and a pressure of 1 atm.

It was established that the process conditions were going to be 1 atm and a temperature slightly lower than 150 f.

Dehydration

Figure 4 illustrates the VAM dehydration simulation flow diagram.

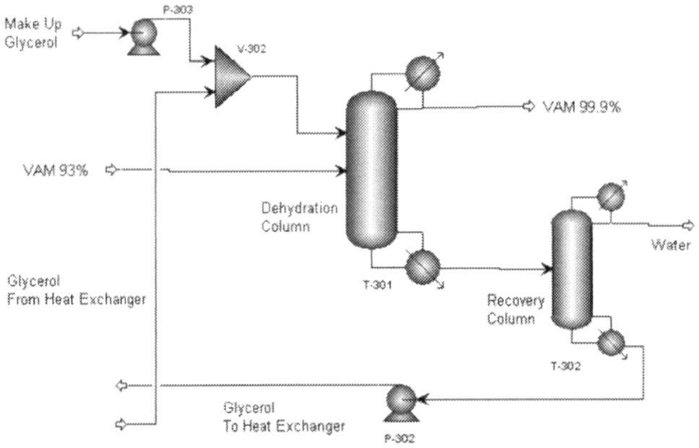

Figure 4: VAM dehydration simulation flow diagram.

The dehydration of the VAM via extractive distillation consists of adding a solvent (Glycerol) to the VAM-water mixture. This mixture is impossible to separate by normal distillation methods; it was necessary to alternate the relative volatility of the components allowing the separation. Therefore a dehydration column was design with 20 stages.

This dehydration column receives a stream of 93% pure VAM and a stream of glycerol composed of pure and recycled glycerol. This leads to a stream of 99% percent pure VAM. This stage is extremely important because when the humidity is retrieve from the product, it results in a better quality product and decreases the risk associated with the storage of the final product.

ENERGY INTEGRATION

Energy integration was used with the objective of using the energy provided by the streams with high temperature that need to be cooled or condensed so they can heat or vaporize the cold streams.

The analysis of the energy integration for the production process of vinyl acetate was done with Aspen Hx-Net program. The streams presented in the table below were analyzed. The condensers and reheaters from the distillation columns weren't taken in to account in this analysis because it was considered that their energetic requirements were going to be covered with the utilities intrinsic into the process.

Table 1. Stream to analyze

Inlet Streams	Outlet Streams	Destination	Effect
33	34	Mixer	Heating
35	32	Reactor	Heating

37	2	Vaporizer	Cooling
6	7	Absorber	Cooling
10	11	Separator	Cooling

Using the information of the streams and that of the Aspen Hx-Net program, a composite curve graph was made.

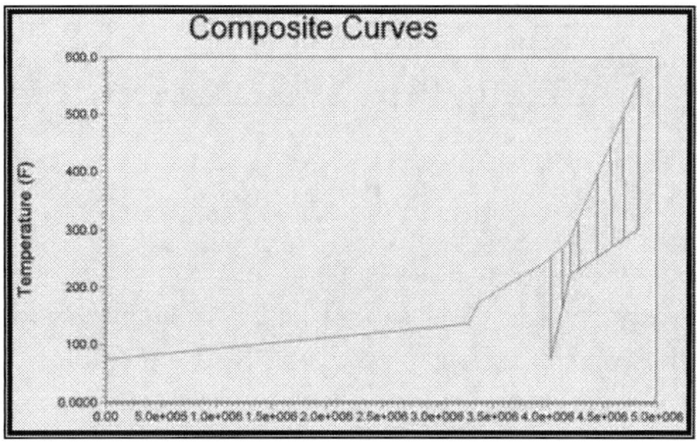

Figure 5 Composite curve

A preliminary analysis of the graph reveals that the process has an uncommon energy integration case; this case is known as threshold. This type of scenario consists of only a small portion of the energy is going to be covered by other streams and the rest of the energy requirements have to be met with outside utilities.

The Δt_{min} used in this analysis was of 10° C because it is the minimum standard for said analysis. In the composite curve graph it is illustrated the portion where stream interaction can take place.

Using the Aspen Hx-Net program it was possible to simulate the most optimum designs for the energy integration for the vinyl acetate process using cooling water as a refrigerant as a utility. After analyzing the options given by the program, a scenario was chosen due to its low consumption of energy. Its operation cost is lower than the rest. Taking these criteria in to account, it is clear that investment capital is higher but thanks to the low operation cost and low energy consumption represents a quick capital return.

Table 2. Process streams used in the analysis of the heat integration.

	Inlet Temp.	Outlet Temp.
Cooling water	20	25
High pressure vapor	250	249

Table 3. Heat integration result analysis

Design 1	Cost index ($)	Energy and Units
Heating (value/s)	5.96E-06	Heating (kj/h)
Cooling(value/s)	2.51E-04	Cooling (kj/h)
Operation (value/s)	2.57E-04	# units
Capital	2.243.005.036	# of shells
Total value/s	2.55E-03	Total area

Table 3 (continued). Heat integration result analysis

design 1	HEN		Cost index ($)
Heating(valor/s)	8.576.125.868	Heating value/h	2.14E-02
Cooling (valor/s)	4.255.681.722	Cooling value/h	9.04E-01
Operation(valor/s)	9	Operation value/h	9.26E-01
Capital	13		
Total value/s	4.088.872.109		
Total energy used (KW)	1.184.516.069	total	1.85E+00

CONCLUSIONS

With a thorough bibliographic revision it was possible to determine that the optimum process to produce VAM was the acetoxilation of ethylene using acetic acid and a palladium/gold catalyst. Said process was divided in to three: the reaction, the separation system and the VAM dehydration. In regards to the reaction it was evident that the temperature would determine the life of the catalyst and the selectivity of the process, even more so in this case which presents two parallel reactions.

Based on the separation system, it was evident the need to separate the liquid phase from the gaseous phase with the objective of purifying the final product and to recycle some raw material that did not react and are very expensive. To choose the most adequate properties was fundamental in order to observe and establish the correct separation characteristics of the components based on their immiscibility and azeotropic points through the simulation.

Finally the existence of an azeotropic point and immiscibility between water and vinyl acetate made it necessary to add and dehydration stage which consisted in a extraction column using glycerol as an extracting agent. Thanks to its polar properties, Glycerol allows the water to be extracted from the VAM

The analysis in regards with the total process production of VAM, it was necessary to determine the best way to use the energy in the process so it could be used to exchange with other streams and by doing this cutting back on energy expenses.

References

[1] Han, Y. F., et al., A kinetic study of vinyl acetate synthesis over Pd-based catalysts: kinetics of vinyl acetate synthesis over Pd-Au/SiO_2 and Pd/SiO_2 catalysts, Journal of catalysis, Vol. 232, pp. 467-475, 2005.

[2] McKetta, J., Encyclopedia of Chemical Processing and design, Vol. 62, Marcel Dekker, INC., USA, 1998, pp. 265-300.

[3] Han , Y. F. , Kumar , D. , Sivadinarayana C. , Goodman , D. W. , Kinetics of ethylene combustion in the synthesis of vinyl acetate , *J. Catal.* , *224* , 60 , 2004

[4] Dimian, A., Bildea, C. S., Chemical process design, Wiley-vch, (2008), pp. 287.

[5] Luyben , W. , Tyreus , B. , An industrial design/control study for vinyl acetate monomer plant , *Comput. Chem. Eng.* , *22* , 867 , 1998.

[6] Han, Y., et al., Particle size effects in Vinyl Acetate synthesis over Pd/SiO_2, Journal of catalysis, Vol. 230, pp. 353-358, 2005.

[7] Chen, M. S., et al., The nature of the active site for vinyl acetate synthesis over Pd-Au, Catalysis Today, Vol. 17. Pp. 37-45, 2006.

[8] Han, Y. F., et al., A kinetic study of vinyl acetate synthesis over Pd-based catalysts: kinetics of vinyl acetate synthesis over Pd-Au/SiO_2 and Pd/SiO_2 catalysts, Journal of catalysis, Vol. 232, pp. 467-475, 2005.

[9] Neurock, M., First principle analysis of the catalytic reaction pathways in the synthesis of Vinyl Acetate, Chemical Engineering science, Vol. 51, pp. 1691-1699, 1996.

[10] McKetta, J., Encyclopedia of Chemical Processing and design, Vol. 62, Marcel Dekker, INC., USA, 1998, pp. 265-300.

[11] Tacke, T., et al., Catalyst, method of its production and its use for the production of vinyl acetate monomer, U.S. Patent office No. 5,808,136, 15 of September of 1998.

[12] Ethylene oxide (EO) and Vinyl acetate monomer (VAM) production, VaporSep® Systems for Petrochemicals, Membrane Technology & Research, disponible en <www.mtrinc.com>, consultado el [25 de Enero de 2008].

[13] Jobson, S., et al., Integrated process for the production of vinyl acetate and/or acetic acid, U.S. Patent office No. 6,040,474, 21 of March of 2000.

[14] Schaeffer, W., et al., Production of Vinyl Acetate from Ethylene, U.S. Patent office No. 3,277,158, October 4 of 1966.

[15] Mazanec, T., et al., Process and apparatus for improved methods for making vinyl acetate monomer (VAM), International Publication Number WO2006/052615, May 18 of 2006.

[16] Copelin, H., et al., Cyclic method of preparing vinyl acetate, United States Patent Office, Patent No. 3.306.930, February 28 of 1967.

[17] Durston, P., et al., Production of Vinyl Acetate, United States Patent Office, Patent No. 3,592,840, July 13 of 1971.

[18] Kumar, D., et al., Synthesis of Vinyl Acetate on Pd-based catalysts, Catalysis Today Journal, Vol. 123, 2007, pp. 77-85.

[19] Macleod, N., et al., The effects of ageing a bimetallic catalyst under industrial conditions: a study of fresh and used Pd-Au-K/silica vinyl acetate synthesis catalysts, Applied Catalysis A, Vol. 261, pp. 37-46, 2004.

[20] Gao, F., et al., Probing reaction pathways on model catalyst surfaces: vinyl acetate synthesis and olefin metathesis, Journal of molecular catalysis A: Chemical (2007), doi: 10.1016/j.molcata.2007.08.026

[21] Kragten, D., et al., A Spectroscopic study of the homogeneous catalytic conversion of ethylene to vinyl acetate by Palladium Acetate, Inorg. Chem. J., Vol. 38, 1999, pp. 331-339.

[22] Tustin, G., et al., Synthesis of Vinyl Acetate monomer from synthesis gas, Catalysis Today J., Vol. 58, 2000, pp. 281-291.

[23] Spivey, J., et al., Direct utilization of Carbon dioxide in chemical synthesis: Vinyl acetate via methane carboxylation, Catalysis Communications J., Article in press, 2007, doi:10.1016/j.catcom.2007.08.023

[24] Neurock, M., et. Al., First principle analysis of the catalytic reaction pathways in the synthesis of Vinyl Acetate, Chemical Engineering Science J., Vol. 51, No. 10, pp. 1691, 1996.

[25] Han, Y.F., et al, Kinetics of ethylene combustion in the synthesis of vinyl acetate over Pd/SiO_2 catalyst, Journal of Catalyst, Vol. 224, 2004, pp. 60-68.

[26] Kirk-Othmer, Concise enciclopedia of Chemical Technology, John Wiley & Sons, 1985 (USA), pp. 1227.

[27] Kragten, D., et al., A density functional study of the acetoxylation of ethylene to Vinyl Acetate catalyzed by Palladium Acetate, J. Phys. Chem., Vol. 103, pp. 2756-2765, 1999.

[28] Acetato de vinilo, de http://www.telecable.es/personales/albatros1/quimica/industria/acevin.htm

[29] Renneke, R., et al., Development of high performance catalysts for the production of vinyl acetate monomer, Topics in catalysis, Vol. 38, No. 4, pp. 279-287, 2006.

[30] Austin, G., Shreve´s Chemical Process Industries, 5th Edition, McGraw Hill, 1984 , pp. 758.

293f Biodiesel – An Efficient Absorbent for the Recovery of Aromatic Hydrocarbons

Diethmar Richter, University of Technology Berlin, Straße des 17. Juni 135, Berlin, 10623, Germany, Holger Thielert, ThyssenKrupp Uhde / EnCoke GmbH, Friedrich-Uhde-Strasse 15, Dortmund, 44141, Germany, and Günter Wozny, Dynamik und Betrieb technischer Anlagen, University of Technology Berlin, Straße des 17. Juni 135, Berlin, 10623, Germany.

Introduction

This contribution is dealing with the applicability of biodiesel (RME) for the recovery of aromatic hydrocarbons as an alternative washing liquid to conventional absorbents, whereby mainly diesel oil and coal tar oil (CTO) are used. Based on studies on a typical large-scale, industrial process it will be shown that biodiesel is not only presenting an alternative washing liquid for the regarded separation task, moreover due to higher absorption rates it is also offering a possibility for the reduction of operational as well as investment costs. To exemplify the improvement when using biodiesel, the coke-oven gas purification process is regarded in this case study.

In conventional coke-oven plants aromatic hydrocarbons, i.e. benzene, toluene and xylene, are recovered from coke-oven gases by using complex polyaromatic mixtures as absorption liquids like coal tar oil. In order to simulate the regarded process of absorption as well as desorption more accurately entire studies are carried out concerning thermodynamical properties as well as phase equilibrium behaviour of coal tar oil which is more favourable than diesel oil in this process. Based on these studies a rate-based approach is implemented in FORTRAN which is validated using experimental data from a pilot plant. For an accurate comparison of biodiesel and coal tar oil, experimental as well as simulation studies are carried out in order to demonstrate the advantages of the alternative absorbent.

Experimental Studies

In previous publications it was shown that due to lower values of the volumetric related Henry coefficients, the solubility of aromatic hydrocarbons in biodiesel is about 30% higher than in coal tar oil. Therefore studies are carried out in a pilot plant for both washing liquids. For the experiments a column equipped with structured packing of 2.6m of height and a diameter of 0.1m is used for studying the absorption capacity under bench-scale conditions. Hereby the volumetric gas flow is adjusted to an F-factor of 1.5 $Pa^{1/2}$ and loaded with different inlet concentrations of benzene and toluene. During the experiments different liquid loads are studied, on the one side for the validation of the implemented models and on the other side in order to analyze the improvement in the absorption rate when using biodiesel. In Fig. 1 a comparison of the analyzed washing liquids is shown for identical liquid loads. It can be seen that in case of biodiesel the gas profile of benzene is shifted to lower concentrations compared to coal tar oil. In order to point out the improvement of absorption rate the liquid load of Biodiesel is reduced in the second case until the concentration profiles are matching each other. Fig. 2 is demonstrating the savings in case of using biodiesel which is leading to a reduction of 30% of liquid load whereby almost the same absorption rate is obtained.

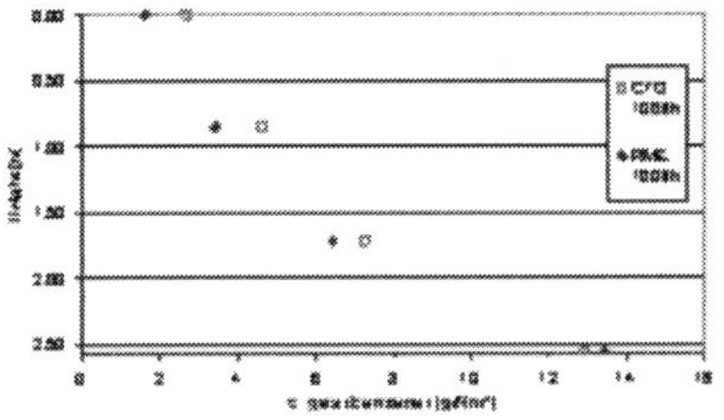

Fig.1: Gas concentration profiles of benzene for identical liquid loads

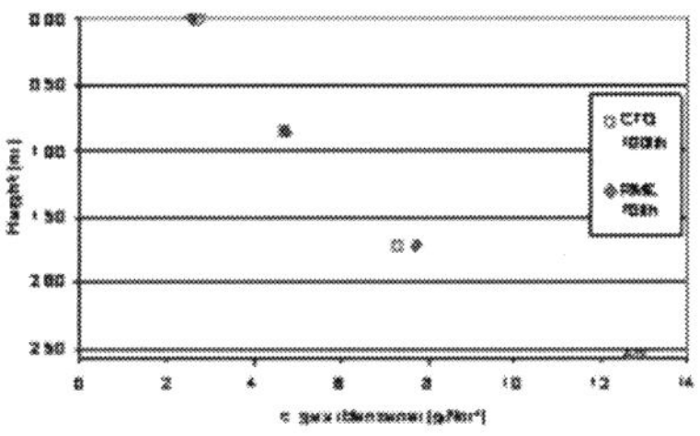

Fig.2: Gas concentration profiles of benzene for reduced liquid load

Simulation studies

In order to optimize operational and investment costs, the energy consuming desorption as well as heat exchanging processes are also regarded in the second part of this contribution. Caused by different thermodynamical properties like heat capacity and boiling range, an alternative process layout will be available. Thus the applicability of biodiesel will be studied regarding the whole process. Based on comparisons of cost functions, the advantages of the alternative absorption liquid are presented. Thereby savings in operational and investment costs are regarded.

INTEGRATING MPC AND RTO IN THE PROCESS INDUSTRY BY ECONOMIC DYNAMIC LEXICOGRAPHIC OPTIMIZATION; AN OPEN-LOOP EXPLORATION

Adrie E.M. Huesman, Okko H. Bosgra and Paul M.J. Van den Hof
Delft Center for Systems and Control, Delft University of Technology, the Netherlands

Introduction

The need for dynamic operation in the process industry is described by Backx et al. [1]. The economics are increasingly driven by the market and this requires flexible supply chains. In such an environment it is much harder to optimize economic performance while maintaining the required product quality and honoring operational constraints. Therefore it is expected that the current industrial state-of-the-art approach for process operation (see figure 1) will prove to be insufficient.

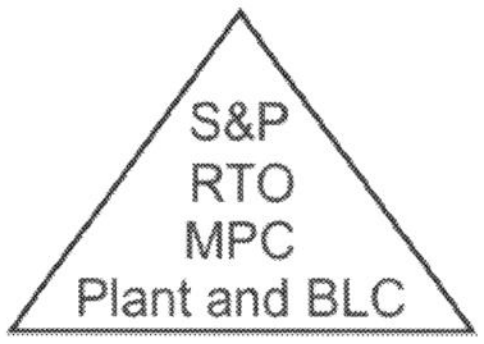

Figure 1 State-of-the-art approach.

The state-of-the-art approach consists of various layers. The lowest layer is the plant with the Base Layer Control (BLC). The latter typically consists of PID controllers and extensions like ratio, override and split range control. The next layer is Model Predictive Control (MPC), a form of model based control that can handle interaction and constraints. The third level is Real Time Optimization (RTO) that optimizes economic performance using steady state models. The final level is Scheduling and Planning (S&P). Planning typically generates economic forecasts and production goals while scheduling basically determines how to execute the chosen plan, the key issue being feasibility. More information on the state-of-the-art approach can be found in Luyben [7] or Marlin [8]. There are a few problems with the state-of-the-art approach which are all related to the fact that RTO uses steady state models. The first thing is that the approach is not general; RTO only supports continuous operation. The second problem is that RTO is slow; model updates can only occur at steady state and this can take a long time, see White [12]. And finally RTO is limited to only steady state improvements; transitions can not be optimized.

Therefore it has been argued by for example Backx et al. [1], Kadam et al. [6] and Rolandi and Romagnoli [10] that MPC and RTO should be integrated "to get the best of both worlds"; Economic Dynamic Optimization (EDO) or as it sometimes called Dynamic Real Time Optimization (DRTO). A considerable amount of work has been done in the area of dynamic process optimization; see for example Biegler [2], Diehl et al. [4] and Schlegel and Marquardt [11]. However this work concentrates on how to solve this type of optimization problems. What remains unanswered is the question what operation results from such an optimization.

This paper will address the last question in an open-loop setting. First we will formulate EDO in a general sense. Then this formulation is used to perform EDO on three process systems (experiments). It will be shown that for each experiment EDO leads to non-unique (multiple) solutions which are handled well by lexicographic optimization. The paper ends with conclusions and directions for future work.

Formulation

The open-loop formulation of EDO takes the shape of an optimization problem. So we must define an objective and a set of constraints. The starting point for the objective is profitability. Standard profitability measures like payback period, net present value or internal rate of return, see Brennan [3], cover the whole project period (initiation, design, construction, operation and dismantling) which means a few to 20 years. Because of these long periods the time value of money must be taken into account. However it will become clear in the next paragraph that the time horizon of EDO is less than four weeks. So there is no need to compensate for the time value of money and a straightforward profit *P* can be maximized:

$$P = \underbrace{\text{revenues}}_{\text{prices}\cdot\text{quantity products}} - \underbrace{\text{total costs}}_{\text{fixed+variable}} - \text{taxes}^{1} \tag{1}$$

By definition fixed and variable costs are respectively independent and dependent on product quantity. It should be noted that the fixed costs and taxes are not under the influence of operation. So in an optimization problem these terms are constants and maximization of equation (1) simplifies to the maximization of operational profit:

$$OP = \text{prices}\cdot\text{quantity products - variable costs} \tag{2}$$

With respect to the constraints two types are obvious; plant behavior and operational constraints. However, also so-called "scheduling constraints" are needed. In the simplest case this type boils down to the required product quantity and quality at a time t_f in the future. In the more elaborate case it can be a number of required quantities and qualities at times t_{f1}, t_{f2}...t_{fn}. The last point in the future t_{fn}, equals the scheduling horizon which is normally between one to four weeks. For this reason the time horizon of EDO is less than four weeks. The scheduling constraints provide the necessary link with scheduling and therefore possible interaction with other plants on the site (shared utilities, tanks etc.). To summarize; economic dynamic process optimization can be formulated as the following optimization problem:

$$\max\int_0^{t_f}(OP)dt \quad \text{s.t.}\begin{cases}\text{plant behavior}\\ \text{operational limitations}\\ \text{scheduling constraints}\end{cases} \tag{3}$$

If the required product quantity is not allowed to vary (so the corresponding constraint is an equality rather than an inequality) then problem (3) can be reduced to:

[1] According the viewpoint of the plant manager taxes have to be subtracted.

$$\min \int_0^{t_f} (OC)dt \quad \text{s.t.} \begin{cases} \text{plant behavior} \\ \text{operational limitations} \\ \text{scheduling constraints} \end{cases} \tag{4}$$

Here OC stands for operational costs. Problems (3) and (4) can also be expressed mathematically:

$$\min_u \int_0^{t_f} F_1 dt \quad \text{s.t.} \begin{cases} dx/dt = f(x,y,u,t),\ x(0) = x_0 \\ h(x,y,u,t) = 0 \\ g(x,y,u,t) \geq 0 \end{cases} \tag{5}$$

Here F_1 stands for *-OP* (note the minus sign) or *OC*, *x* for state variables, *y* for algebraic variables and *u* for inputs or Degrees Of Freedom (DOF). All variables are functions of time. Typically all relationships in problem (5) are linear except for those that are related to plant behavior. So this problem constitutes a Non-Linear Program (NLP).

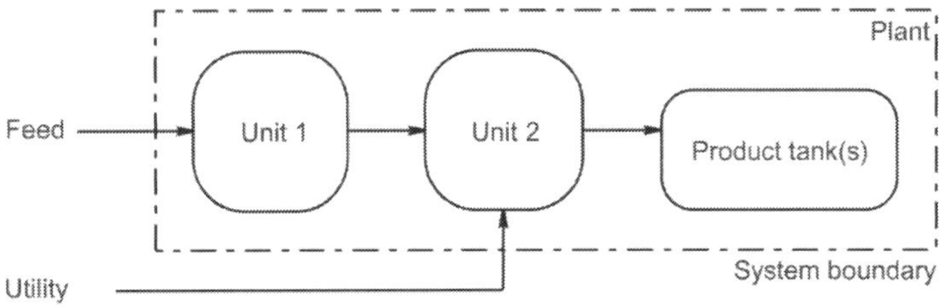

Figure 2 A plantwide system boundary.

The economic objective has consequences for the system boundary. Normally prices of flows between feed and product, so-called intermediate prices, are unknown. See for example figure 2, the price of the flow between unit 1 and 2 is normally not known. The system boundary should be chosen such that intermediate prices are not needed. This means that a plantwide system boundary is adopted. It should be noted that the boundary also includes product tank(s). This allows for dynamic improvement; the product flow to the tank may vary freely between operational limits and the same might even apply to the product quality.

Experiments

Single Tank

The first system is just a product tank (see figure 3).

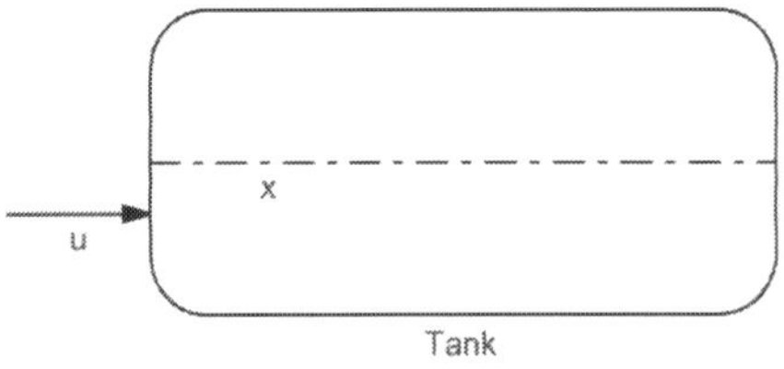

Figure 3 A product tank.

The tank should be filled at minimum cost to at least a certain minimum level given some input constraints:

$$\min_u \int_0^2 u(t)dt \quad \text{s.t.} \begin{cases} dx/dt = u(t),\ x(0)=0 \\ 0 \le u(t) \le 10 \\ x(2) \ge 5 \end{cases} \qquad (6)$$

Problem (6) was solved by means of the simultaneous approach. The transcription is based on implicit Euler. The implementation was done in the algebraic language GAMS, the used solver was CONOPT. The result is shown in figure 4a.

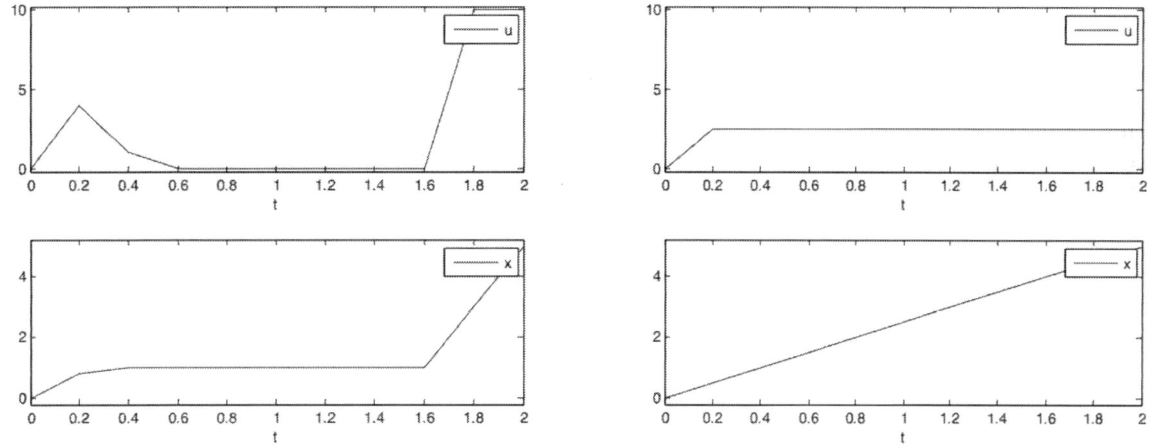

Figure 4a (left) and b (right) The result of the 1st and 2nd stage single tank problem.

Looking at the result the question rises if this is the only possible solution. A simple way to check this is by changing the initial guess for the input and/or state trajectories. The effect is that these trajectories change but that the objective value stays the same. This implies that the optimal trajectories are not unique; there exist multiple solutions.

Stirred Tank Reactor and Tank

This system includes a Stirred Tank Reactor (STR)[2] besides a tank (see figure 5).

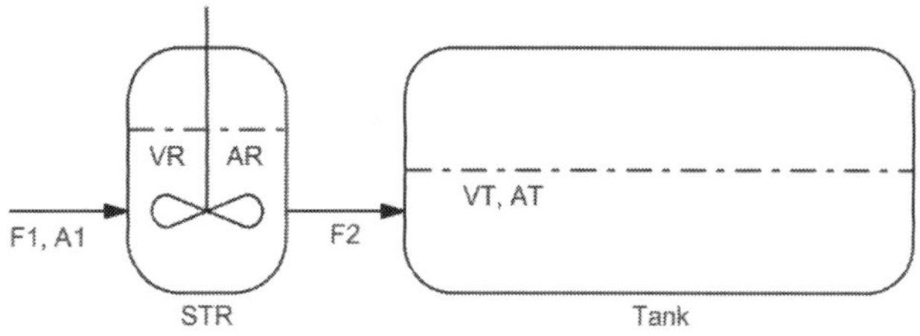

Figure 5 A STR and a product tank.

[2] A STR is a continuous stirred tank reactor of which the level is allowed to vary.

The discussion of this experiment will be quite brief since it was already described in detail in Huesman et al. [5]. In the STR the reaction A to B takes place. The product B is stored in the tank. The following assumptions are made:

- The reaction is first order in A.
- The density is constant.
- The STR and tank are both well-mixed.

This system is described by four nonlinear differential equations with four state variables (VR, VT, AR and AT) and two DOF (F1 and F2). The objective is to minimize the integrated value of (vF1 + VR) over 2 time units while producing a certain amount of product; VT(2) = 2.1 of a certain quality; AT(2) = 0.05. The objective reflects the operational costs; feed plus stirring (the stirring power is assumed to be proportional to the reactor volume). The parameter v represents the ratio cost of feed over cost of stirring. The optimization was solved in the same way at the single tank experiment. The result is shown in figure 6a.

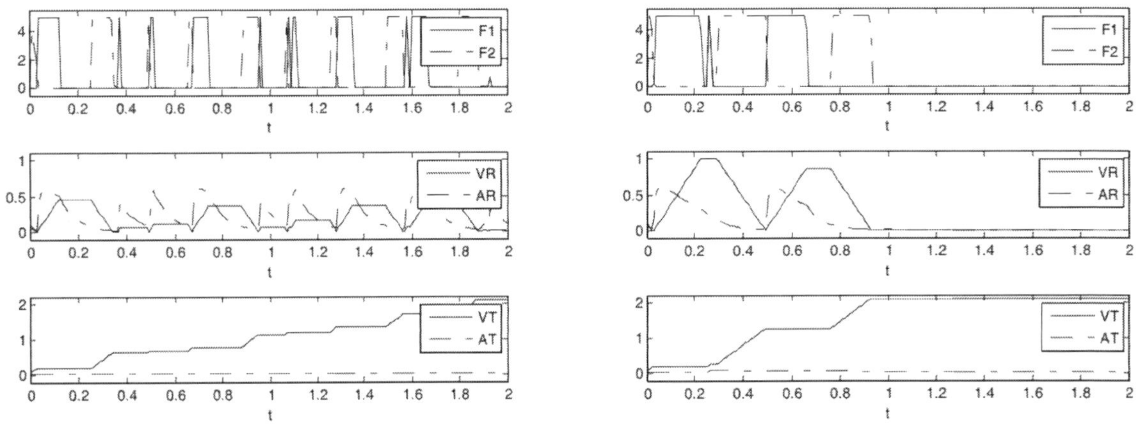

Figure 6a (left) and b (right) The result of the 1st and 2nd stage STR and tank problem.

The STR is filled and emptied several times; so EDO prefers batch operation. From a reactor engineering point of view this result was to be expected, batch operation leads to a high average reactant concentration which results in a low average reactor volume and therefore low cost. The timing of the batches looks arbitrary. And also in this case changing the initial guess for the input and/or state trajectories reveals the existence of multiple solutions.

Distillation Column and Tank

In the last system the STR is replaced by a Distillation Column (DC) with eight trays, see figure 7.

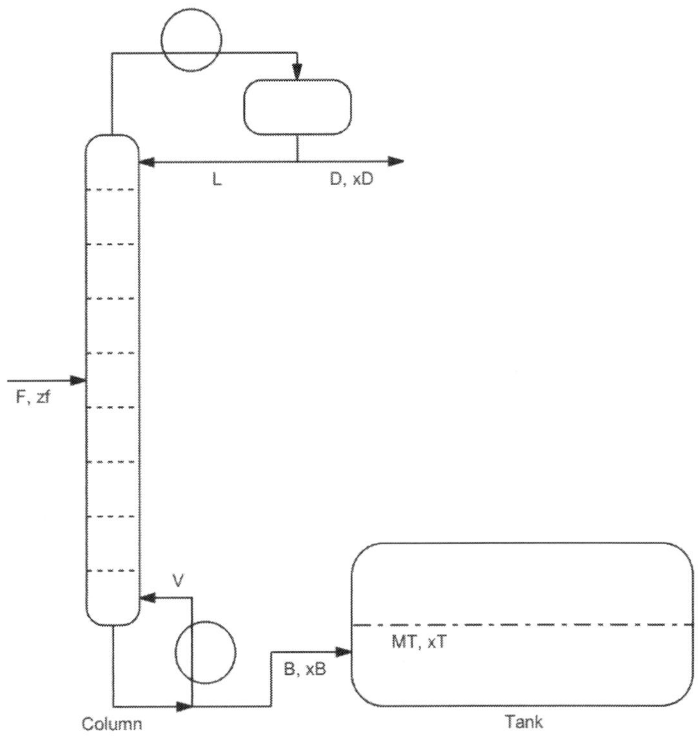

Figure 7 A DC and a product tank.

Also this experiment is described in detail in Huesman et al. [5]. The heavy component (product) is stored in a tank while the light component leaves the system. The following assumptions are made:

- The vapor phase can be neglected.
- Trays, top and bottom inventory are well-mixed.
- Outgoing flows on a tray are in equilibrium which is described by constant relative volatility.
- The liquid molar hold-up on each tray and in top and the bottom inventory is constant.
- The liquid and vapor molar flows are constant (constant molar overflow).
- The tank is well-mixed.

Trays, top and bottom inventory are all considered compartments. So in total there are ten compartments. The compartments are numbered from the bottom to the top, so xB = x1 and xD = x10. This system is described by 12 nonlinear differential equations with 12 state variables (x1 to x10, xT and MT) and three DOF (F, V and L). The objective is to minimize the integrated value of (wF + V) over 2 time units while producing a certain amount of product; MT(2) = 1.1 of a certain quality; xT(2) = 0.2. Again the objective reflects the operational costs; feed plus utility, the parameter w represents the ratio cost of feed over cost of steam. This dynamic optimization was solved in the same way as the single tank experiment. The result is shown in figure 8a.

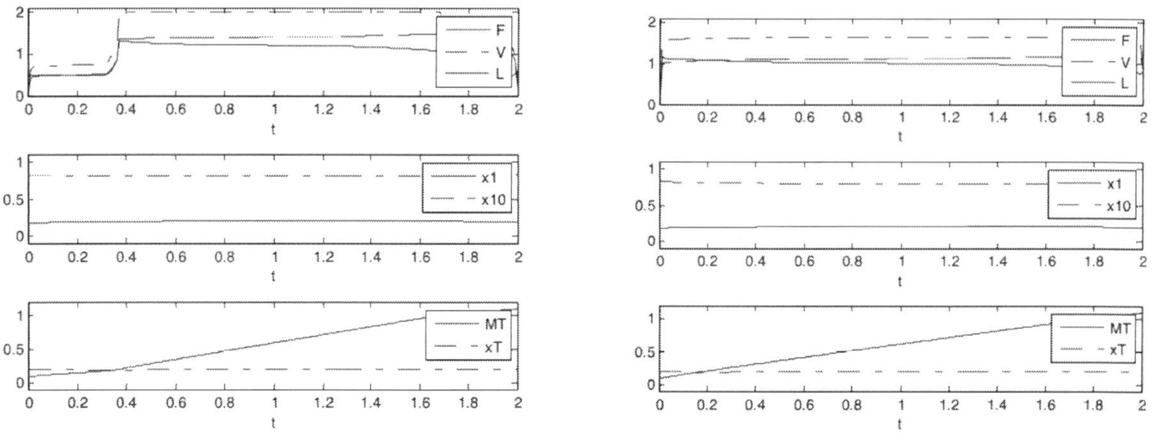

Figure 8a (left) and b (right) The result of the 1st and 2nd stage DC and tank problem.

It is clear that F rises rapidly. Closer inspection reveals that this also happens to V and L. As a matter a fact the ratios V over F and L over F remain constant over time. So EDO prefers constant ratios and from a distillation point of view this is not a complete surprise. Repeating the exercise of changing the initial guess for the input and/or state trajectories again reveals the existence of multiple solutions.

Lexicographic Optimization

Figure 9 shows the graph of a function with a unique and multiple solutions. For one decision variable the existence of multiple solutions seems artificial however Huesman et al. [5] show that for two and more decision variables cases are easy to construct.

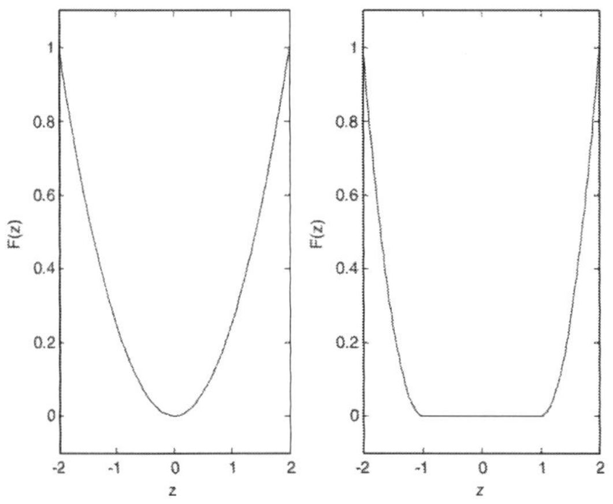

Figure 9 A unique solution (left) and multiple solutions (right).

Multiple solutions can be analyzed by means of a SVD (Singular Value Decomposition) of the Jacobian of the active first order KKT (Karush Kuhn Tucker) conditions around the optimal point. Let J be this Jacobian and Δz be small changes in the decision variables, then multiple solutions exist if:

$$J\Delta z = USV^T \Delta z = 0 \tag{7}$$

In other words one or more singular values should be zero. Figure 10 shows the result for the single tank example. Note that indeed a number of singular values are zero. By the way this analysis provides more information; for example the dimension of the multiple solutions space.

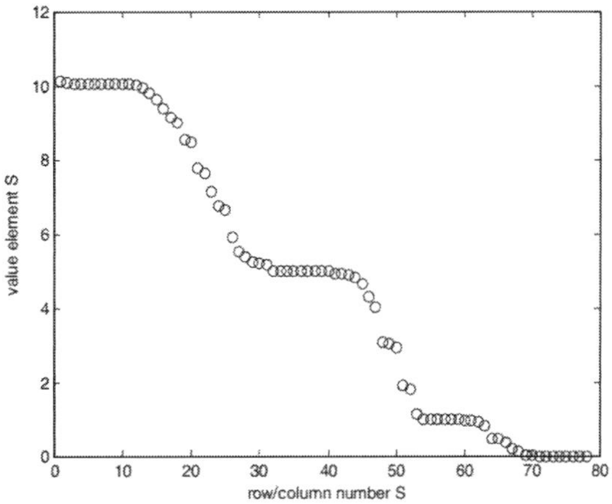

Figure 10 The singular values of the Jacobian J for the single tank problem.

There are two reasons why the final solution should be locally unique:

1. In a closed-loop setting EDO will be performed repeatedly. This can easily lead to "jumping around the multiple solutions space".
2. The existence of multiple solutions implies that there are still DOF left. So there is an opportunity to perform another optimization. Of course this optimization should be based on a different objective and it should not sacrifice economic performance (too much).

The last reason strongly suggests the use of lexicographic optimization; a form of multiobjective optimization, see Miettinen [9]. For two objectives F_1 and F_2 it boils down to:

$$\min F_1(z) \;\; \text{s.t.} \begin{cases} h(z) = 0 \\ g(z) \le 0 \end{cases} \Rightarrow \text{a solution } z^* \qquad \min F_2(z) \;\; \text{s.t.} \begin{cases} h(z) = 0 \\ g(z) \le 0 \\ F_1(z) \le (1+S) \cdot F_1(z^*) \end{cases} \tag{8}$$

So the overall optimization is done in two stages. The 1st stage involves an economic objective F_1 and determines the best economic performance F_1*. The 2nd stage is based on a different objective F_2 and only uses the remaining DOF because of the extra constraint on economic performance. The last constraint can be relaxed with the parameter *S*.

Is lexicographic optimization not somewhat complex and computationally expensive? In order to answer that question lets consider the alternative that is used in MPC; regularization. In MPC regularization is provided by the term $\Delta u^T R \Delta u$, with Δu being the input moves and *R* a weighing matrix. For regularization to work properly the matrix *R* must be tuned carefully in order to balance regularization with the original objective. In the case of lexicographic optimization only one scalar parameter *S* needs to be specified and this is straightforward since its influence is completely transparent. By the way the computational effort of the 2nd stage will be small provided that this stage continues where the 1st stage stopped (hot-start of the 2nd stage). To summarize; compared to regularization lexicographic optimization is less complex and about comparable in computational effort.

Lexicographic optimization was applied to the three experiments previously discussed. For the choice of the 2nd objective and the parameter S see table 1.

Table 1 Details of the 2nd stage optimization of the three experiments.

Experiment	**2nd objective**	***S***
Single tank	Achieve steady operation: $\min \int_0^2 \left((u(t)-5)^2 \right) dt$	0
STR and tank	Realize schedule in minimal time: $\min \int_0^2 \left((VT(t)-4)^2 + (AR(t)-0)^2 \right) dt$	0.03
DC and tank	Achieve steady operation: $\min \int_0^2 \left((F(t)-3)^2 + (V(t)-3)^2 + (L(t)-3)^2 \right) dt$	0.001

The hot start of the 2nd stage was implemented in GAMS by using two model and two solve statements. The results are shown in figure 4b, 6b and 8b. In all cases lexicographic optimization resulted in considerable changes of the optimal trajectories and unique solutions. Note that the remaining DOF were really utilized by the lexicographic approach. For example in the STR and tank experiment the time needed to realize the schedule was halved.

Conclusions and Further Work

A general well-motivated formulation of EDO has been given. It was shown that this formulation does not intrinsically fix all DOF. It was also shown that remaining DOF can be utilized by lexicographic optimization.

Future work will focus on the background of multiple solutions (why and when do they appear?), a general approach for the 2nd objective (what are sensible objectives and how to formulate them?) and extension of the open-loop to the closed-loop case.

References

1. T. Backx, O. Bosgra, and W. Marquardt. Integration of model predictive control and optimization of processes. In *Advanced Control of Chemical Processes*, June 2000.
2. L.T. Biegler. An overview of simultaneous strategies for dynamic optimization. *Chemical Engineering and Processing: Process Intensification*, 46(11): 1043-1053, 2007.
3. D. Brennan. *Process Industry Economics An international perspective*. Institution of Chemical Engineers, Warwickshire, 1998.
4. M. Diehl, H.G. Bock, J.P. Schlöder, R. Findeisen, Z. Nagy, and F. Allgöwer. Real-time optimization and nonlinear model predictive control of processes governed by differential-algebraic equations. *Journal of Process Control*, 12(4):577-585, 2002.

5. A.E.M. Huesman, O.H. Bosgra and P.M.J. Van den Hof. Degrees of Freedom Analysis of Economic Dynamic Optimal Plantwide Operation. Preprints 8th IFAC International Symposium on Dynamics and Control of Process Systems (DYCOPS), Vol. 1, pages 165-170, Mexico, 2007.
6. J. V. Kadam, W. Marquardt, M. Schlegel, T. Backx, O. H. Bosgra, P. J. Brouwer, G. Dünnebier, D. van Hessem, A. Tiagounov, and S. de Wolf. Towards integrated dynamic real-time optimization and control of industrial processes. In *Proceedings Foundations of Computer Aided Process Operations* (FOCAPO2003), pages 593-596, 2003.
7. W. L. Luyben, B. Tyreus, and M. L. Luyben. *Plantwide Process Control*. McGraw-Hill, New York, 1999.
8. T. E. Marlin. *Process Control*. McGraw-Hill, New York, 1995.
9. K.M. Miettinen. *Nonlinear Multiobjective Optimization.* Kluwer Academic Publishers, Boston, 1999.
10. P. A. Rolandi and J. A. Romagnoli. A framework for online full optimizing control of chemical processes. In *Proceedings of European Symposium on Computer Aided Process Engineering* (ESCAPE-15), pages 1315-1320, 2005.
11. M. Schlegel and W. Marquardt. Detection and exploitation of the control switching structure in the solution of dynamic optimization problems. *Journal of Process Control,* 16(3):275-290, 2006.
12. D.C. White. Online Optimization: What have we learned?. *Hydrocarbon Processing,* 77(6):55-69, 1998.

Mixed $\mathcal{H}_2/\mathcal{H}_\infty$ Model Predictive Control for Unstable and Non-Minimum Constrained Processes

P. Aadaleesan and Prabirkumar Saha
Indian Institute of Technology Guwahati, Assam, India.

Abstract

In this paper we propose a newer kind of robust model predictive controller (MPC), namely, Nash game approach based mixed $\mathcal{H}_2/\mathcal{H}_\infty$ model predictive control. Two quadratic objective functions are considered, viz. one for $\mathcal{H}_2$ problem and another for $\mathcal{H}_\infty$ problem. The resulting open-loop control problem reduces to solving a pair of coupled non-symmetric algebraic Riccati equations. The pair of coupled Riccati equations are solved, by finding the invariant subspace spanning eigenvectors of a specially constructed matrix. The novel control algorithm is demonstrated using numerical example for non-minimum unstable system.

1 Introduction

Robust model predictive control (MPC) is one of the very demanding area of research in recent years (Cuzzola et al., 2002; Löfberg. J, 2003; Ramírez et al., 2006; Zhao et al., 2000). Even if there are good models available both for the plant and the disturbances affecting the plant, any unmodelled phenomenon and/or some unknown disturbances causing discrepancies in the system's performance could be taken care by the available feedback in MPC design. However the open-loop controller is not actually meant for that. Eventually, the controller should be robustly designed enough to meet such discrepancies which affects both stability and performance of the process. $\mathcal{H}_\infty$ robust MPC is a well known such kind of controller and its analogy to game theoretic strategy is also well documented in the literature (Başar and Olsder, 1995; Rao, 2000). Moreover, to benefit from the advantage of combining $\mathcal{H}_2$ and $\mathcal{H}_\infty$, a newer hybrid robust controller called mixed $\mathcal{H}_2/\mathcal{H}_\infty$ controller was developed in early 90's (Kaminer et al., 1993; Limebeer et al., 1994). However, this type of robust controller has not be much investigated in the MPC regime, except recently by Orukpe et al. (2007). In this present work, the mixed $\mathcal{H}_2/\mathcal{H}_\infty$ model predictive control problem is approached from a different perspective, i.e., using Nash game approach is proposed. Here, the existence of an optimal saddle point solution is claimed as the solution of the optimal control problem. The coupled non-symmetric algebraic Riccati equation resulting from such problem is solved in a systematic manner, with the construction of a special matrix, as shown in Freiling et al (1999). The efficacy of the present controller is compared against that of Orukpe et al. (2007) for non-minimum phase unstable processes, which are of much interest in process industries.

The paper has been organized in the following manner. Section 2 gives the systematic formulation of the problem. The existence of the optimal saddle point solution is shown in section 3. The procedure of solving the resulting coupled algebraic Riccati equations is shown in section 4. In section 5 the efficiency of the proposed method is showed using numerical example. Finally, in section 6 the conclusions are given.

2 Problem Formulation

Consider the discrete-time linear system affected with process disturbance:

$$\begin{aligned} x_{k+1} &= Ax_k + B_u u_k + B_\omega \omega_k, \quad x(0) = x_0 \\ z_k &= \begin{bmatrix} Cx_k \\ Du_k \end{bmatrix} \end{aligned} \tag{1}$$

where, $x_k = [x_k^1, x_k^2, \cdots, x_k^n]^T \in \mathbb{R}^n$ is the system state, $u_k = [u_k^1, u_k^2, \cdots, u_k^m]^T \in \mathbb{R}^m$ is the control input; $\omega_k \in \mathcal{L}_{2,[0,\infty]}$ is the unknown but bounded disturbance. Also, A, B_u and B_ω are matrices of appropriate dimensions. Also consider $Q = C^T C$ and $R := D^T D$.

The $\mathcal{H}_\infty$ and $\mathcal{H}_2$ control performance are given by equations (2) and (3), respectively, as

$$J_\infty(u, \omega) = \frac{1}{2}\sum_{k=0}^{\infty}\left(x_k^T Q x_k - \gamma^2 \omega_k^T \omega_k\right) \tag{2}$$

$$J_2(u, \omega) = \frac{1}{2}\sum_{k=0}^{\infty} z_k^T z_k = \frac{1}{2}\sum_{k=0}^{\infty}\left(x_k^T Q x_k + u_k^T R u_k\right) \tag{3}$$

where γ^2 is the upper bound of the worst case performance (or the attenuation factor for the disturbance).

The Hamiltonian function (H_2) of $\mathcal{H}_2$ problem is given as,

$$H_2 := J_2(u_k, \omega_k) + \bar{p}_{k+1}^T\left[(Ax_k + B_u u_k + B_\omega \omega_k) - x_{k+1}\right] \tag{4}$$

$$= \frac{1}{2}\left(x_k^T Q x_k + u_k^T R u_k\right) + \bar{p}_{k+1}^T\left[(Ax_k + B_u u_k + B_\omega \omega_k) - x_{k+1}\right] \tag{5}$$

where $\bar{p}_k$ is the Lagrange multiplier. To minimize the function H_2, we need to partially differentiate H_2 w.r.t. its components, say, x_k and u_k and setting them to zero.

$$\frac{\partial H_2}{\partial x_k} = 0: \quad Qx_k + A^T \bar{p}_{k+1} - \bar{p}_k = 0 \tag{6}$$

$$\frac{\partial H_2}{\partial u_k} = 0: \quad Ru_k + B_u^T \bar{p}_{k+1} = 0 \tag{7}$$

From equations (6) and (7) we have

$$\bar{p}_k = Qx_k + A^T \bar{p}_{k+1} \tag{8}$$

$$u_k = -R^{-1} B_u^T \bar{p}_{k+1} \tag{9}$$

Assume that $\bar{p}_k$ can be given in the form,

$$\bar{p}_k = P_k^I x_k \tag{10}$$

where P_k^I is an $n \times n$ real symmetric positive definite (Lyapunov) matrix. Using equation (10) into equation (9) gives

$$u_k = -R^{-1} B_u^T P_{k+1}^I x_{k+1} \tag{11}$$

Now, the $\mathcal{H}_\infty$ problem's Hamiltonian function (H_∞) is given as,

$$\begin{aligned} H_\infty &:= J_\infty(u_k, \omega_k) + \bar{\bar{p}}_{k+1}^T \left[(Ax_k + B_u u_k + B_\omega \omega_k) - x_{k+1}\right] && (12) \\ &= \frac{1}{2}\left(x_k^T Q x_k \; - \; \gamma^2 \omega_k^T \omega_k\right) + \bar{\bar{p}}_{k+1}^T \left[(Ax_k + B_u u_k + B_\omega \omega_k) - x_{k+1}\right] && (13) \end{aligned}$$

where $\bar{\bar{p}}_{k+1}$ is again the Lagrange multiplier. Partially differentiating the Hamiltonian function (H_∞) w.r.t. its components, say, x_k and ω_k and equating them to zero,

$$\begin{aligned} \frac{\partial H_\infty}{\partial x_k} &= 0: \quad Qx_k + A^T \bar{\bar{p}}_{k+1} - \bar{\bar{p}}_k = 0 && (14) \\ \frac{\partial H_\infty}{\partial \omega_k} &= 0: \quad -\gamma^2 \omega_k + B_\omega^T \bar{\bar{p}}_{k+1} = 0 && (15) \end{aligned}$$

From the equations (14) and (15) we have,

$$\begin{aligned} \bar{\bar{p}}_k &= Qx_k + A^T \bar{\bar{p}}_{k+1} && (16) \\ \omega_k &= \gamma^{-2} B_\omega^T \bar{\bar{p}}_{k+1} && (17) \end{aligned}$$

Again considering a similar assumption as in equation (10),

$$\bar{\bar{p}}_k = P_k^{II} x_k \tag{18}$$

Using equation (18) in (15) we get,

$$\omega_k = \gamma^{-2} B_\omega^T P_{k+1}^{II} x_{k+1} \tag{19}$$

Now, from equations (11) and (19) we could write the closed-loop system equation as

$$x_{k+1} = [I + B_u R^{-1} B_u^T P_{k+1}^I - B_\omega \gamma^{-2} B_\omega^T P_{k+1}^{II}]^{-1} A x_k := \Phi_k x_k \tag{20}$$

where Φ is the closed-loop state transition matrix. By substituting equations (18) and (20) in (16), the discrete algebraic Riccati equation ($\mathcal{H}_\infty$-DARE) obtained from $\mathcal{H}_\infty$ problem is,

$$P_k^{II} x_k \;=\; Qx_k + A^T P_{k+1}^{II} [I + B_u R^{-1} B_u^T P_{k+1}^I - B_\omega \gamma^{-2} B_\omega^T P_k^{II}]^{-1} A x_k \tag{21}$$

Similarly, the discrete algebraic Riccati equation ($\mathcal{H}_2$-DARE) obtained from $\mathcal{H}_2$ problem, by substituting equations (10) and (20) in (8),

$$P_k^I x_k = Qx_k + A^T P_{k+1}^I [I + B_u R^{-1} B_u^T P_{k+1}^I - B_\omega \gamma^{-2} B_\omega^T P_k^{II}]^{-1} A x_k \tag{22}$$

3 Minimal values of the $\mathcal{H}_2$ and $\mathcal{H}_\infty$ cost functions

In this section the minimal values of the cost functions are given. Let us rewrite the closed-loop system equation with the feedback gain matrices for the system given in (1) as;

$$x_{k+1} = (A + B_u K_u + B_\omega K_\omega)\, x_k \tag{23}$$

which could be otherwise given as,

$$x_{k+1} = \bar{A}_\omega x_k + B_u u_k \tag{24}$$
$$x_{k+1} = \bar{A}_u x_k + B_\omega \omega_k \tag{25}$$

where, $\bar{A}_\omega := A - B_\omega K_\omega$ and $\bar{A}_u := A - B_u K_u$. Assume that there exists an optimal feedback gain matrix K_ω from solving a $\mathcal{H}_\infty$ cost function and K_u from solving a $\mathcal{H}_2$ cost function.

Before finding the minimal value of the function, let us rewrite some of our earlier expressions in a form that suits with our new format of the closed loop equation as shown in equations (24) and (25). The $\mathcal{H}_2$ problem's Hamiltonian function (H_2) is rewritten as,

$$H_2 := J_2(u_k, \omega_k) + \bar{p}_{k+1}^T \left[(\bar{A}_\omega x_k + B_u u_k) - x_{k+1} \right] \tag{26}$$

which eventually yields,

$$\frac{\partial H_2}{\partial x_k} = 0 : \quad \bar{p}_k = Q x_k + \bar{A}_\omega^T \bar{p}_{k+1} \tag{27}$$

Therefore, the equation (24) from using equations (11) and (10),

$$\begin{aligned} x_{k+1} &= \bar{A}_\omega x_k - B_u B_u^T P_{k+1}^I x_{k+1} \\ \left[I + B_u B_u^T \bar{P}_{k+1}^I \right] x_{k+1} &= \bar{A}_\omega x_k \end{aligned} \tag{28}$$

Likewise, the $\mathcal{H}_\infty$ problem's Hamiltonian function (H_∞) is also rewritten as,

$$H_\infty := J_\infty(u_k, \omega_k) + \bar{\bar{p}}_{k+1}^T \left[(\bar{A}_u x_k + B_\omega \omega_k) - x_{k+1} \right] \tag{29}$$

which yields,

$$\frac{\partial H_\infty}{\partial x_k} = 0 : \quad \bar{\bar{p}}_k = Q x_k + \bar{A}_u^T \bar{\bar{p}}_{k+1} \tag{30}$$

The equation (25) from using equations (19) and (18),

$$\begin{aligned} x_{k+1} &= \bar{A}_u x_k + \gamma^{-2} B_\omega B_\omega^T \bar{P}_{k+1}^I x_{k+1} \\ \left[I - \gamma^{-2} B_\omega B_\omega^T \bar{P}_{k+1}^I \right] x_{k+1} &= \bar{A}_u x_k \end{aligned} \tag{31}$$

3.1 Minimal value of the $\mathcal{H}_2$ cost function

To find the minimum value of the $\mathcal{H}_2$ control performance, premultiply both sides of the equation (27) by x_k^T and using (10), we get,

$$x_k^T P_k^I x_k = x_k^T Q x_k + x_k^T \bar{A}_\omega^T P_{k+1}^I x_{k+1} \tag{32}$$

Substituting equation (28) into the previous equation, we get,

$$x_k^T Q x_k = x_k^T P_k^I x_k - x_{k+1}^T P_{k+1}^I x_{k+1} - x_{k+1}^T P_{k+1}^I B_u B_u^T P_{k+1}^I x_{k+1} \tag{33}$$

From equation (11), we have

$$\begin{aligned} u_k &= -B_u^T P_{k+1}^I x_{k+1} \\ \Rightarrow u_k^T u_k &= \left(-x_{k+1}^T P_{k+1}^I B_u\right)\left(-B_u^T P_{k+1}^I x_{k+1}\right) \\ &= x_{k+1}^T P_{k+1}^I B_u B_u^T P_{k+1}^I x_{k+1} \end{aligned} \tag{34}$$

By adding equations (33) and (34), we have

$$x_k^T Q x_k + u_k^T u_k = x_k^T P_k^I x_k - x_{k+1}^T P_{k+1}^I x_{k+1} \tag{35}$$

By substitution of this equation into the equation of $J_2(u, \omega)$ gives,

$$\begin{aligned} J_2(u^*, \omega^*) &= \frac{1}{2}\sum_{k=0}^{\infty} x_k^T P_k^I x_k - x_{k+1}^T P_{k+1}^I x_{k+1} \\ &= \frac{1}{2} x_0^T P_0^I x_0 \end{aligned} \tag{36}$$

Equation (36) is due to the cancelation of the terms when the summation is expanded over the time horizon $k = [0, \infty)$ and also using $lim_{k\to\infty} x_k = 0$. Note that the above assertion starts and is extended to the control law formulation with the assumption that there exists an optimal feedback gain matrix for the other (maximizing) player ω_k^*. So the $\mathcal{H}_2$ performance index given in equation (3) can be taken as $J_2(u, \omega^*)$. From the above argument and the result shown in equation (36), it could be directly stated that,

$$J_2(u^*, \omega^*) \leq J_2(u, \omega^*) \tag{37}$$

3.2 Minimal value of the $\mathcal{H}_\infty$ cost function

The minimal value of the $\mathcal{H}_\infty$ cost function is found out in the following. Premultiplying equation (30) by x_k^T and using (18) we get

$$x_k^T P_k^{II} x_k = x_k^T Q x_k + x_k^T \bar{A}_u^T P_{k+1}^{II} x_{k+1} \tag{38}$$

Using equation (31) in the above equation, we get

$$\begin{aligned} x_k^T P_k^{II} x_k &= x_k^T Q x_k + x_{k+1}^T [I - \gamma^{-2} P_{k+1}^{II} B_\omega B_\omega^T] P_{k+1}^{II} x_{k+1} \\ x_k^T Q x_k &= x_k^T P_k^{II} x_k - x_{k+1}^T [I - \gamma^{-2} P_{k+1}^{II} B_\omega B_\omega^T] P_{k+1}^{II} x_{k+1} \end{aligned} \tag{39}$$

From equation (19), we can have

$$\omega_k^T \omega_k = \gamma^{-4} x_{k+1}^T P_{k+1}^{II} B_\omega B_\omega^T P_{k+1}^{II} x_{k+1} \tag{40}$$

Multiplying equation (40) by γ^2 and then subtracting it from equation (39) gives,

$$x_k^T Q x_k - \gamma^2 \omega_k^T \omega_k = x_k^T P_k^{II} x_k - x_{k+1}^T P_{k+1}^{II} x_{k+1} \tag{41}$$

Using the previous equation in the equation of $J_\infty(u, \omega)$ i.e., the $\mathcal{H}_\infty$ cost function gives,

$$\begin{aligned} J_\infty(u^*, \omega^*) &= \frac{1}{2} \sum_{k=0}^{\infty} x_k^T P_k^{II} x_k - x_{k+1}^T P_{k+1}^{II} x_{k+1} \\ &= \frac{1}{2} x_0^T P_0^{II} x_0 \end{aligned} \tag{42}$$

Using the similar argument as given in section 3.1, rather, the other player being the minimizing player, u_k^*, with the assumption that there exist an optimal feedback gain matrix (K_u) for the minimizing player, the $\mathcal{H}_\infty$ cost function can be taken as $J_\infty(u^*, \omega)$. Moreover, from the result of the equation (42), the following condition holds good;

$$J_\infty(u^*, \omega^*) \leq J_\infty(u^*, \omega) \tag{43}$$

So the saddle point of the Nash game is ensured once the coupled algebraic Riccati equations (AREs), given in equations (22) and (21), are solved for their solutions.

4 Solution of the coupled AREs

Solving the pair of coupled AREs to get the values of the Lyapunov function matrices, P^I and P^{II} will fetch us the optimal (saddle point) solution of the Nash game approach to robust control of the linear system (1). Freiling et al. (1999), has given the necessary condition(s) that need to be satisfied, so as to get the solution of the AREs. From the AREs in equations (22) and (21) and the closed loop system equation (20), the following theorem can be stated.

Theorem 4.1 *If* $S(P^I, P^{II}) := \mathtt{span}\begin{pmatrix} I \\ P^I \\ P^{II} \end{pmatrix} \subset \mathbb{C}^{3n\times n}$ *is an invariant subspace of* M_{Na} *with* $\mathtt{det}(I + S_1 P^I + S_2 P^{II}) \neq 0$ *then* $\begin{pmatrix} P^I \\ P^{II} \end{pmatrix}$ *is the solution of the AREs, such that* $\begin{pmatrix} x \\ \bar{p} \\ \bar{\bar{p}} \end{pmatrix}(k+1) = M_{Na} \begin{pmatrix} x \\ \bar{p} \\ \bar{\bar{p}} \end{pmatrix}(k)$*, where,* $S1 := B_u R^{-1} B_u^T$, $S2 := B_\omega \gamma^{-2} B_\omega^T$ *and*

$$M_{Na} := \begin{pmatrix} A^{-1} & A^{-1}S_1 & A^{-1}S_2 \\ QA^{-1} & A^T + QA^{-1}S_1 & QA^{-1}S_2 \\ QA^{-1} & QA^{-1}S_1 & A^T + QA^{-1}S_2 \end{pmatrix}.$$

Corollary 4.1 *If* $\texttt{span}(v_1, \cdots, v_n)$ *is an invariant subspace such that* $\texttt{det}X \neq 0$ *for* $\begin{pmatrix} X \\ Y_1 \\ Y_2 \end{pmatrix} := (v_1, \cdots, v_n)$, *then* $\begin{pmatrix} P^I \\ P^{II} \end{pmatrix} := \begin{pmatrix} Y_1 X^{-1} \\ Y_2 X^{-1} \end{pmatrix}$ *is the solution of the AREs, if* $\texttt{det}(I + S_1 P^I + S_2 P^{II}) \neq 0$.

Using the above theorem and the immediate corollary, the coupled algebraic Riccati equations resulting from the open-loop Nash games can be solved, enabling us to obtain the mixed $\mathcal{H}_2/\mathcal{H}_\infty$ MPC using Nash game approach.

Remark 4.1 *In the present study, we have* $Q_1 = Q_2 := Q$.

Remark 4.2 *The solution of the coupled AREs could exist if and only if the system's states are completely controllable and observable. The interested readers are refereed to Freiling et al. (1999) for further details.*

The invariant subspace spanning eigenvectors of M_{Na} are the required solution of the coupled Riccati equations (22) and (22) as per Corollary 4.1, such that the closed loop eigenvalues lies within the unit circle (i.e., $\texttt{eig}(A + B_u K_u) \leq 1$, where $K_u := -R^{-1} B_u^T P_{k+1}^I \Phi_k x_k$ from equation (11) and (20)).

5 Numerical Example

The performance of both the control algorithms is checked for a non-minimum phase unstable system. The discrete time state-space model of such a system with sampling time, $T = 0.1$ is taken

$$A = \begin{bmatrix} 2.3150 & -1.3500 \\ 1.0000 & 0.0000 \end{bmatrix} \quad B_u = \begin{bmatrix} 1 \\ 0 \end{bmatrix} \quad B_\omega = \begin{bmatrix} 1 \\ 1 \end{bmatrix} \quad C = [0.1133 \quad -0.1191]$$

The system is subjected to a constant disturbance, $\omega = 0.1$. The poles and zero of the system are $1.1575 \pm 0.1010i$ and 1.0512, respectively. With $\gamma^2 = 100$ and $x_0 = [1 \;\; 0]^T$, the performance of Nash game approach is compared against Orukpe et al.'s algorithm and the results are furnished in Fig. 1. Although the present approach gives an oscillatory response (which could be improved by decreasing the input weighing matrix), it gives highly appreciable response than the other mixed $\mathcal{H}_2/\mathcal{H}_\infty$ MPC algorithm.

6 Conclusion

A Nash game theory based mixed $\mathcal{H}_2/\mathcal{H}_\infty$ robust MPC is proposed. The coupled algebraic Riccati equations resulting from this approach is solved efficiently by constructing a special matrix, and from finding its generalized eigenvectors that span its invariant subspace (Freiling et al., 1999). The performance of the proposed robust predictive control algorithm is compared against the one proposed by Orukpe et al. (2007). The proposed

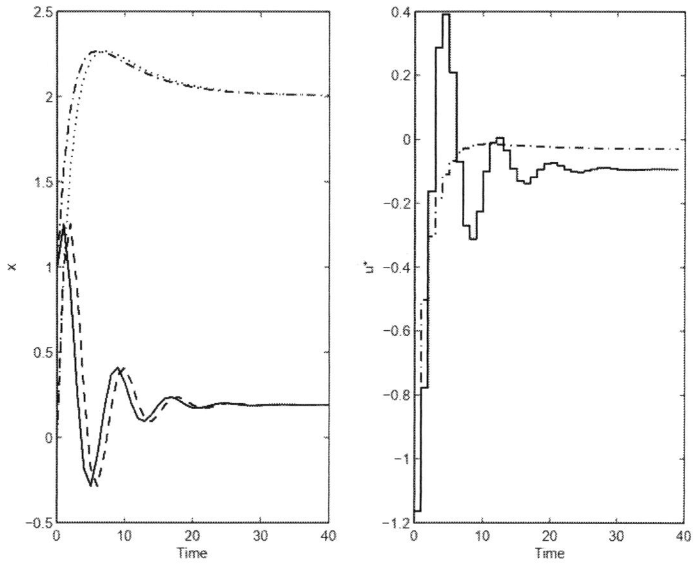

Fig 1 Comparison of Nash game approach mixed $\mathcal{H}_2/\mathcal{H}_\infty$ robust MPC (solid and dashed lines) against the mixed $\mathcal{H}_2/\mathcal{H}_\infty$ MPC proposed in Orukpe et al (dash-dot and dotted lines) for non-minimum unstable system for $\gamma^2 = 100$.

algorithm gives better results when used for non-minimum, unstable systems, which are common in many process industries. It should be noted that the present algorithm could only be used for systems, whose states are both completely controllable and observable.

In the present work, the algorithm has been developed only for unconstrained case. Enhancement of the present work to handle constraints, especially input constraints, is underway.

References

1. Başar, T. and Olsder, G. J. (1995) "Dynamic noncorporative game theory", *SIAM*, Philadelphia.

2. Cuzzola, F. A., Geromel, J. C. and Morari, M. (2002) "An improved approach for constrained robust model predictive control", 38, pp.1183-1189.

3. Freiling, G., Jank, G. and Abou-Kandil, H. (1999) "Discrete-time Riccati equations in open-loop Nash and Stackelberg games", *Euro. J. of Contr.*, 5, pp. 56-66.

4. Kaminer, I., Khargonekar, P. P. and Rotea, M. A. (1993) "Mixed $\mathcal{H}_2/\mathcal{H}_\infty$ control for discrete-time systems via convex optimization", *Automatica*, 29 (1), pp. 57-70.

5. Löfberg, J. (2003) "Minimax approaches to robust model predictive control", *Ph.D. Dissertation*, Linköping Univ., Sweden.

6. Limebeer, D.J.N., Anderson, B. D. O. and Hendel, H. (1994) "A Nash game approach to mixed $\mathcal{H}_2/\mathcal{H}_\infty$ control", *IEEE. Trans. Auto. Contr.*, 39 (1), pp. 69-82.

7. Orukpe, P. E., Jaimoukha, I. M. and El-Zobaidi, H. M. H. (2007) "Model predictive control based on mixed $\mathcal{H}_2/\mathcal{H}_\infty$ control approach", *Amer. Contr. Conf.*, New York, USA.

8. Ramírez, D. R., Alamo, T., Camacho, E. F. and de la Peña, D. M. (2006) "Min-max MPC based on a computationally efficient upper bound of worst case cost", *J. Process Contr.*, 16, pp. 511-519.

9. Rao, C. V. (2000) "Moving horizon strategies for the constrained monitoring and control of nonlinear discrete-time systems", *Ph.D. Dissertation*, Univ. Wisconsin-Madison.

10. Zhao, H., Li, W. and Bentsman, J. (2000) "$\mathcal{H}_\infty$ prediction and unconstrained $\mathcal{H}_\infty$ predictive control:multi-input-multi-output case", *Int. J. Robust and Nonlin. Contr.*, 11(1), pp. 59-86.

OptCon – an Efficient Tool for Rapid Prototyping of Nonlinear Model Predictive Control Applications

Z. K. Nagy

Chemical Engineering Department, Loughborough University, Loughborough, LE11 3TU, United Kingdom, email: z.k.nagy@lboro.ac.uk

Introduction

Nonlinear model predictive control is an optimization-based multivariable constrained control technique that uses a nonlinear dynamic process model for the prediction of the process outputs. At each sampling time the model is updated on the basis of new measurements and state variable estimates. Then the open-loop optimal manipulated variable moves are calculated over a finite prediction horizon with respect to some cost function, and the manipulated variables for the subsequent prediction horizon are implemented. Then the prediction horizon is shifted by usually one sampling time into the future and the previous steps are repeated. NMPC is one of the approaches which inherently can cope with process constraints, nonlinearities and different objectives formulated form economical or environmental considerations. Although the advantages of NMPC have been demonstrated through numerous simulation studies the gap between the number if simulation papers and actual practical implementations is significant. This gap is even more pronounced if consider real industrial applications. The talk will briefly asses the major implementation aspects of NMPC introducing the *integral NMPC design concept*. Figure 1 illustrates the *NMPC pentagon*, which constitutes the basis of the integral NMPC design. All sides of the pentagon have to be considered for successful practical implementation.

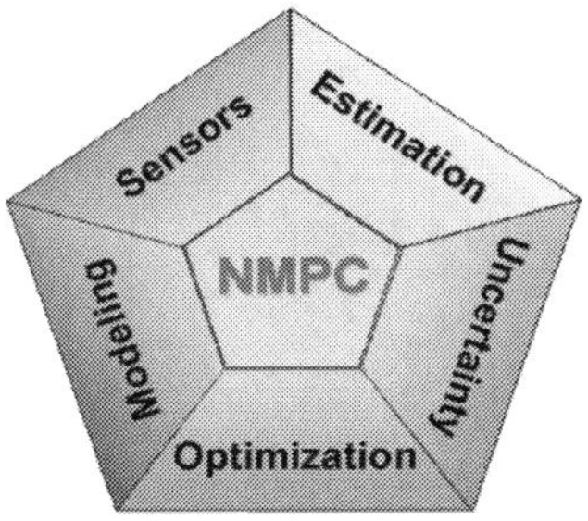

Fig.1. "NMPC Pentagon" that illustrates the main aspects to be considered in an integral design of a practical NMPC.

OPTCON – A Toolbox for rapid NMPC prototyping

A Matlab toolbox is illustrated which has been developed with the aim to provide a rapid prototyping tool for NMPC applications. The NMPC approach presented in this talk is a first principles model based approach that can be used for the solution of moving or shrinking horizon (batch) control problems. The NMPC tool developed includes a number of desirable features. In particular the NMPC setup can be done in Matlab® (The Mathworks Inc.) the most widely used modeling and control environment for control engineers. The model used in the controller has to be developed in the form of Simulink® (The Mathworks Inc.) "mex S-function" using C. The S-function interface of the optimization tool provides convenient and fast connectivity with Matlab®. The NMPC approach is based on a state-of-the-art large-scale nonlinear optimization solver (HQP) [2], which offers one of the most efficient approaches, based on a multiple shooting algorithm [1], that exploits the special structure of optimization problem that arise in NMPC. In this algorithm the optimization horizon of interest is divided into a number of subintervals (stages) with local control parameterizations. The differential equations and cost on these intervals are integrated independently during each optimization iteration, based on the current guess of the control. The continuity/consistency of the final state trajectory at the end of the optimization is enforced by adding consistency constraints to the nonlinear programming problem. This multiple shooting formulation concept is shown in Figure 2.

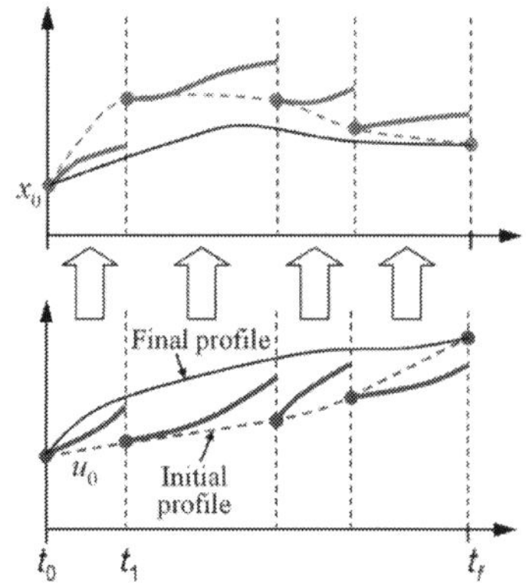

Figure 2. Illustration of the multiple shooting approach.

The optimal control problem to be solved on-line in every sampling time in the NMPC algorithm can be formulated as:

Problem *P1*:

$$\min_{u(t)\in\mathcal{U}} \mathcal{H}(x(t), u(t); \theta), \tag{1}$$

subject to:

$$\dot{x}(t) = f(x(t), u(t); \theta), \tag{2}$$

$$y(t) = g(x(t), u(t); \theta), \tag{3}$$

$$x(t_k) = \hat{x}(t_k), \quad x(t_0) = \hat{x}_0, \tag{4}$$

$$h(x(t), u(t); \theta) \leq 0, \quad t \in [t_k, t_F] \tag{5}$$

where $\mathcal{H}$ is the performance objective, t is the time, t_k is the time at sampling instance k, t_F is the final time at the end of prediction, $x(t) \in \mathbb{R}^{n_x}$ is the n_x vector of states, $u(t) \in \mathcal{U}$ is the n_u set of input vectors, $y(t) \in \mathbb{R}^{n_y}$ is the n_y vector of measured variables used to compute the estimated states $\hat{x}(t_k)$, $\theta \in \Theta \subset \mathbb{R}^{n_\theta}$ is the n_θ vector of possible uncertain parameters, where the set Θ can be either defined by hard bounds or probabilistic, characterized by a multivariate probability density function. The function $f : \mathbb{R}^{n_x} \times \mathcal{U} \times \Theta \to \mathbb{R}^{n_x}$ is the twice continuously differentiable vector function of the dynamic equations of the system, $g : \mathbb{R}^{n_x} \times \mathcal{U} \times \Theta \to \mathbb{R}^{n_y}$ is the measurement equations function, and $h : \mathbb{R}^{n_x} \times \mathcal{U} \times \Theta \to \mathbb{R}^{c}$ is the vector of functions that describe all linear and nonlinear, time-varying or end-time algebraic constraints for the system, where c denotes the number of these constraints. The objective function can have the following general form:

$$\mathcal{H}(x(t),u(t);\theta) = \mathcal{M}(x(t_F);\theta) + \int_{t_k}^{t_F} \mathcal{L}(x(t),u(t);\theta)dt \,. \tag{6}$$

We assume that $\mathcal{H} : \mathbb{R}^{n_x} \times \mathcal{U} \times \Theta \to \mathbb{R}$ is twice continuously differentiable, thus fast optimization algorithms, based on first and second order derivatives may be exploited in the solution of (6). The form of (6) is general enough to express a wide range of objectives encountered in NMPC applications (moving or shrinking horizon approach on regulation and/or setpoint tracking, direct minimization of the operation time, optimal initial conditions, multiple simultaneous objectives, treatment of soft constraints, terminal penalty term for stability, etc.). For batch processes with end-point optimization the objective usually reduces to the Mayer form ($\mathcal{L}(\cdot) = 0$), however the Lagrange term ($\mathcal{L}(\cdot)$) still may be used, e.g. to implement soft constraints on control rate. In NMPC the optimization problem (1)-(5) is solved iteratively on-line, in a moving (receding) horizon ($t_F < t_f$) or shrinking horizon ($t_F = t_f$) approach, where t_f is the batch time.

Considering the discrete nature of the on-line control problem, the continuous time optimization problem involved in the NMPC formulation is solved by formulating a discrete approximation to it, that can be handled by conventional nonlinear programming (NLP) solvers [6]. The time horizon $t \in [t_0, t_f]$ is divided into N equally spaced time intervals Δt (stages), with discrete time steps $t_k = t_0 + k\Delta t$, and $k = 0, 1, \ldots, N$. Model equations are discretized,

$$x_{k+1} = f_k(x_k, u_k; \theta) \tag{7}$$

and added to the optimization problem as equality constraints. Problem *P2* gives the general discrete time formulation of Problem *P1*.

Problem *P2*:

$$\min_{u_k, u_{k+1}, \ldots, u_{k+N_p}} \{\mathcal{M}^{k+N_p}(x_{k+N_p};\theta) + \sum_{j=k}^{k+N_p} \mathcal{L}_j(x_j, u_j;\theta)\}, \tag{8}$$

subject to:

$$G_k(x_k, u_k; \theta) = 0\,, \tag{9}$$

$$H_k(x_k, u_k; \theta) \leq 0\,, \tag{10}$$

where N_p is the number of stages in the prediction horizon $[t_k,\ t_F]$, $G_k:\mathbb{R}^{n_x}\times\mathcal{U}\times\Theta\rightarrow\mathbb{R}^{n_x+n_y}$ corresponds to all equality constraints resulted from (7) and (3), $H_k:\mathbb{R}^{n_x}\times\mathcal{U}\times\Theta\rightarrow\mathbb{R}^{c+2n_u}$ is the vector function of all inequality constraints (5), including the constraints on the inputs considering. We consider here that the set of possible inputs is given as hard bounds. It is assumed that the vector functions G and H are twice continuously differentiable.

Real time iteration NMPC algorithm

In NMPC simulation studies usually immediate feedback is considered, i.e. the optimal feedback control corresponding to the information available up to the moment t_k, is computed, $u^\star(t_k)=[u_{0|t_k}, u_{1|t_k}, \ldots, u_{N_p|t_k}]$, and the first value ($u_{0|t_k}$) is introduced into the process considering no delay. However, the solution of the NLP problem P2 requires a certain, usually not negligible, amount of computation time δ_k, while the system will evolve to a different state, where the solution $u^\star(t_k)$ will no longer be optimal. Computational delay δ_k has to be taken into consideration in real-time applications. In the approach used here, in moment t_k, first the control input from the second stage of the previous optimization problem $u_{1|t_{k-1}}$ (which corresponds to the first stage of the current optimization) is injected into the process, and then the solution of the current optimization problem is started, with fixed $u_{0|t_k}=u_{1|t_{k-1}}$. After completion, the optimization idles for the remaining period of $t\in(t_k+\delta_k, t_{k+1})$, and then at the beginning of the next stage, at moment $t_{k+1}=t_k+\Delta t$, $u_{1|t_k}$ is introduced into the process, and the algorithm is repeated. This approach requires real-time feasibility for the solution of each open-loop optimization problems ($\delta_k\leq\Delta t$).

This multistage formulation is advantageous to the efficient application of the SQP-type solver in HQP, and often leads to faster solution of the optimization problem due to the better approximation of the Lagrangian Hessian of the nonlinear subproblems using low rank updates. HQP uses a sparse interior point algorithm for the efficient treatment of the linear-quadratic subproblems in the nonlinear SQP iterations. The main structure of the OptCon software is shown on Figure 3.

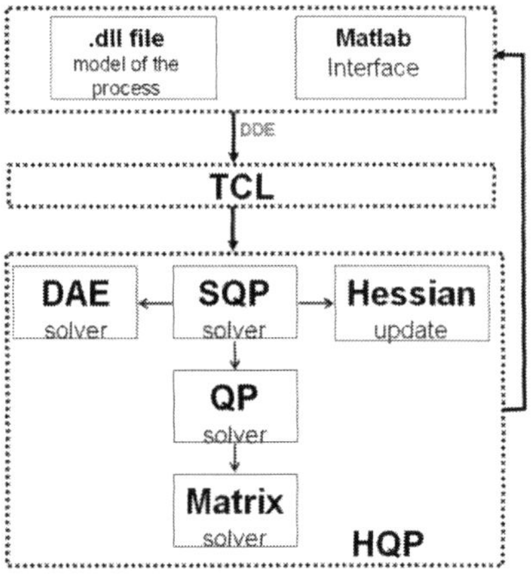

Fig. 3. Structure of the main components of the NMPC toolbox - OptCon

Some fo the features and available options of the toolbox are:

- Lagrangian Hessian approximation
    - BFGS – dense BFGS updates for separate diagonal blocks
    - DScale – numerical approximation with a diagonal matrix
    - Gerschgorin – positive definiteness guaranteed by a diagonal offfset

- QP solver:
    - Use of barrier method
    - Mehrotra – primal-dual predictor-corrector method
    - Franke – based on interior point algorithm from Wright

- QP matrix solver to solve the extended Karush-Kuhn-Tucker conditions:
    - SpBKP – solution of the system
    - RedSpBKP – solution of the reduced system
    - SpSC – sparce Schur complement method
    - LQDOCP – block – wise solution of the multistage problems.

Available DAEA/ODE solvers: RKsuite – explicit ODE solver; Euler – explicit ODE solver, fixed step; RK4, IMP – implicit ODE solver; DASPK – implicit DAE solver; SDIRK – implicit DAE solver; dt – fixed integration step size.

The optimization problems are configured by setting the required weights in the predetermined objective functions. There are two precompiled objective functions included in the toolbox using the SFunctionOpt and SFunctionEst function from HQP.

The optimization time horizon [t0, tf] is split into k = 0,…, KK stages KK = sps K sample periods, (sps = # of samples per stage)

The optimization problem:

$$J=\sum_{kk=0}^{KK}\Delta t^{kk}\sum_{i=1}^{n_u}\left\{u_{weight1}u\left(t^{kk}\right)+u_{weight2}\left[u\left(t^{kk}\right)-\frac{u_{ref}}{u_{nominal}}\right]^2\right\}_i+$$

$$+\sum_{kk=0}^{KK-1}\Delta t^{kk}\sum_{i=1}^{n_u}\left\{dcr\_u_{weight1}\,\dot{u}\left(t^{kk}\right)+dcr\_u_{weight2}\left[\dot{u}\left(t^{kk}\right)-\frac{dcr\_u_{ref}}{u_{nominal}}\right]^2\right\}_i+$$

$$+\sum_{kk=0}^{KK}\Delta t^{kk}\sum_{i=1}^{n_y}\left\{y_{weight1}y\left(t^{kk}\right)+y_{weight2}\left[y\left(t^{kk}\right)-\frac{y_{ref}}{y_{nominal}}\right]^2\right\}_i+$$

$$+\sum_{i=1}^{n_y}\left\{y_{f\_weight1}y\left(t_f\right)+y_{f\_weight2}\left[y\left(t_f\right)-\frac{y_{ref}}{y_{nominal}}\right]^2\right\}_i\rightarrow\min$$

The estimation problem is formulated as

$$J=\sum_{l=1}^{N_{ez}}\sum_{k=k_{l,v}}^{K_l}\sum_{i=1}^{\dim(I)}\left\{\left[\frac{y_{s^{kl}}(I)-y_{s_{ref}^{k_l}}}{y_{nominal}(I)}\right]^2\right\}_i \rightarrow \min,\ I=find(y_{active}),$$

Some of the main functions available in OPTCON are:

- f_optHQP – optimization/optimal control
- f_estHQP – state and parameter estimation
- f_NMPCHQP – NMPC function
- f_MHEHQP – MHE estimation
- f_rtsim – real-time process simulator

Additional functions which provide implementation templates for NMPC and MHE applications are also available.

The use of the estimation function is illustrated in Figure 4

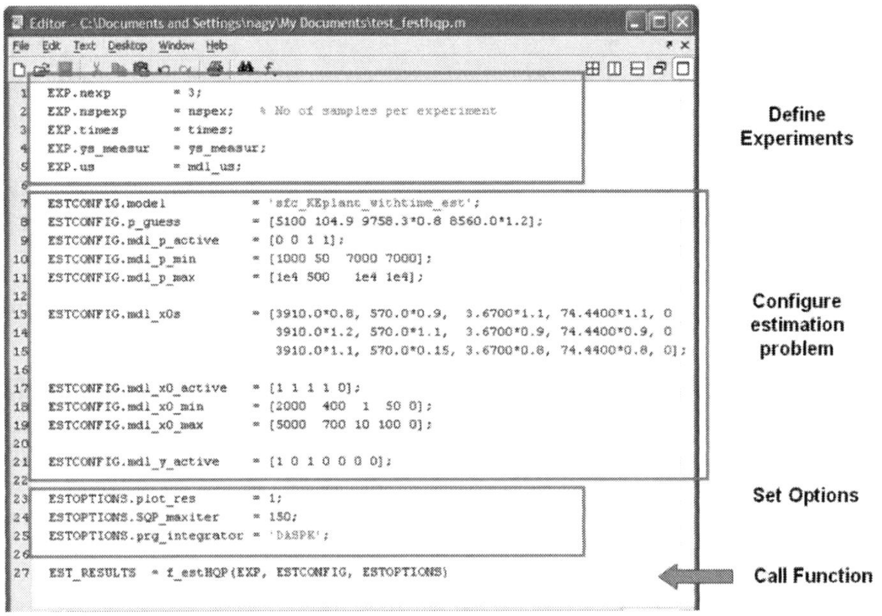

Figure 4. Matlab script illustrating the use of the estimation function on OptCon./

OptCon has been successfully applied in a variety of NMPC applications. The application of OptCon to a laboratory four tank system will be illustrated briefly, which is the first practical application of a theoretically well founded NMPC approach (the so called quasi-infinite horizon NMPC), that guarantees stability. Figure 5 and 6 illustrate these results [4].

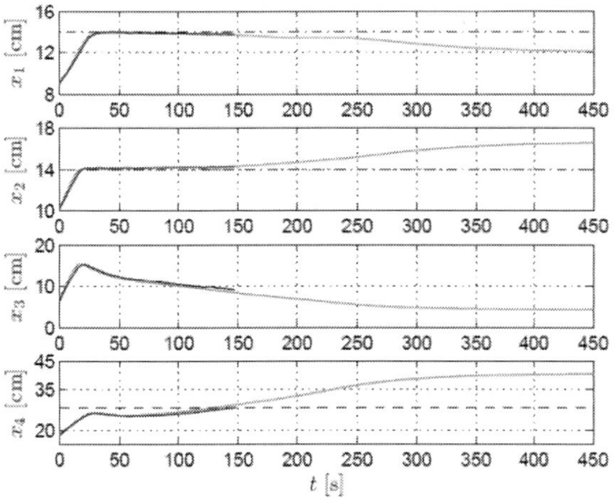

Figure 5. NMPC without stability and state constraints. Setpoint (black dash-dotted), simulation (grey-solid) and experiment (black solid). The figure illustrates that the system cannot be stabilized at the origin without considering stability constraints in the problem formulation.

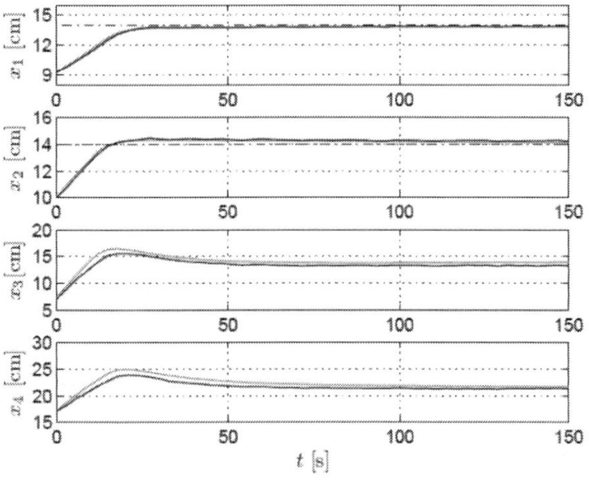

Figure 6. Quasi infinite horizon NMPC with prediction horizon of 60s: set-point (balck dash-dotted), simulation (grey-solid) and experiment (black solid). The figure illustrates that the system is stabilized at the origin using the stability constraints.

Next we illustrate how the same tool can be used for prototyping industrial NMPC applications and go from simulation studies to the actual practical tests within minutes. The quality of the product strongly depends on how tightly the temperature profile is tracked during the batch. Disturbances caused by the effects of exothermic reaction cause overshooting when classical cascade-PI control is used, with an increasing degradation in control performance when batch time is reduced. It is shown that NMPC can provide higher control performance for setpoint tracking. Additionally, simulations show that product quality can significantly decrease in case of certain disturbance scenarios. The NMPC approach proposed here is a hierarchical (two level) NMPC approach, in which a higher level NMPC controller reoptimizes on-line the setpoint, based on the desired end-point properties of the product. The setpoint is sent to a lower level, setpoint-tracking NMPC. The results show that with this approach good temperature tracking is achieved and the product quality at the end of the batch is well controlled. Figure 7 illustrates the temperature tracking results of the NMPC controller implemented to an industrial pilot plant [3].

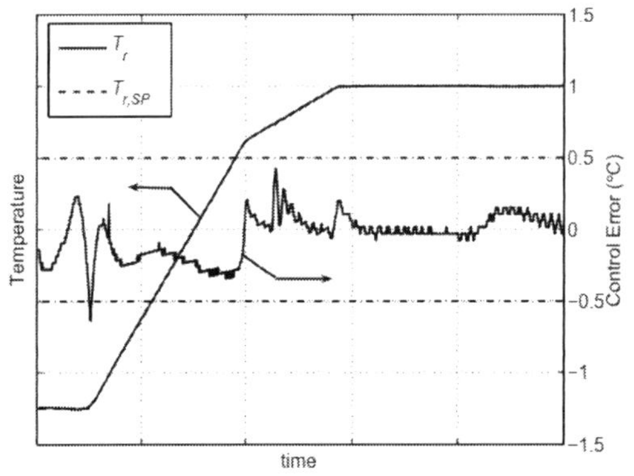

Figure 7. Experimental data with the setpoint tracking performance of the NMPC in the BASF pilot plant. The reactor temperature almost perfectly follows the setpoint (within ±0.5 °C).

Conclusions

The paper described the main features of a new user friendly tool, OptCon, which can provide an efficient environment for rapid prototyping of industrial nonlinear model predictive control applications. The tool provides a Matlab based connectivity between he efficient optimizations package HQP and Matlab. The package is distributed as a freeware application and can be obtained by contacting the author at (z.k.nagy@lboro.ac.uk).

References

[1] Diehl, M., H. G. Bock, J. P. Schlöder, R. Findeisen, Z. Nagy, F. Allgöwer (2002). Real-time optimization and nonlinear model predictive control of processes governed by differential-algebraic equations, *Journal of Process Control*, 12, 577-585.

[2] Franke, R., E. Arnold and H. Linke. HQP: A solver for nonlinearly constrained large-scale optimization. http://hqp.sourceforge.net.

[3] Nagy, Z. K., B. Mahn, R. Franke, F. Allgower, Efficient output feedback nonlinear model predictive control for temperature control of industrial batch reactors, *Control Engineering Practice,*, 15, 839-859, 2007.

[4] Raff, T., S. Huber, Z. K. Nagy, F. Allgower, Nonlinear model predictive control of a four tank system: an experimental stability study, in Proc. of IEEE International Conference on Control Applications, Munich, Germany, 237-242, October 4-6, 2006.

329d Robust Multi-Parametric Model-Based Control

Efstratios N. Pistikopoulos, Konstantinos Kouramas, and Nuno P. Faísca. Centre for Process Systems Engineering, Imperial College London, Roderic Hill Building, South Kensington Campus, London, United Kingdom

In this work we present the fundamental theoretical developments for a unified and general framework for the design of Robust Multi-Parametric Controllers (Pistikopoulos et al. 2007ab, ParOS 2007). Given a dynamic system with model uncertainties both in the right and left hand-sides (of constraints of the underlying uncertain optimisation problem), the proposed framework is realised in three steps:

1.robust reformulation of the constraints: the state/output and input constraints are reformulated to account for the worst-case model uncertainty (Ben-Tal and Nemirovski, 2001, Lim et al., 2004).

2.dynamic programming: the uncertain optimisation problem is reformulated with a dynamic programming multi-stage setting (Bellman, 2003, Faísca et al., 2008).

3.multi-parametric programming: each of the stages of the DP problem are solved as multi-parametric programming problems and the control input is obtained as an explicit function of the incumbent state (Pistikopoulos et al, 2007ab).

The above framework can guarantee the design of multi-parametric MPC (mp-MPC) controllers that can ensure constraints satisfaction for all admissible uncertainties as opposed to the previous non-robust mp-MPC design techniques (Sakizlis et al., 2004). The results are demonstrated with examples that illustrate the proposed framework.

References:

Bellman, R. (2003). Dynamic Programming. Dover Publications

Ben-Tal, A., Nemirovski, A., (2000). Math. Program 88, 411-424

Faísca, N.P., Kouramas, K.I., Saraiva, P.M., Rustem, B., Pistikopoulos, E.N., (2008). Optimization Letters 2, 267-280

Lim, X., Janak, S.L., Floudas, C.A., (2004). Comp. & Chem. Eng. 28, 1069-1085

Parametric Optimisation Solutions (ParOS) Ltd, Improved Process Control, European Patent EP1399784, 2007

Pistikopoulos, E.N., Georgiadis, M. & Dua, V., (2007). Multi-parametric Parametric Programming. Weinheim: Wiley-VCH

Pistikopoulos, E.N., Georgiadis, M. & Dua, V., (2007). Multi-parametric Parametric Model-based Control. Weinheim: Wiley-VCH

Sakizlis, V., Kakalis, N.M.P., Dua, V., Perkins, J.D., Pistikokpoulos, E.N., (2004). Automatica 40, 189-201

329e Economically Optimizing Model Predictive Control

Rishi Amrit, Department of Chemical and Biological Engineering, University of Wisconsin - Madison, 1415 Engineering Dr., Madison, WI 53706 and James B. Rawlings, Department of Chemical and Biological Engineering, University of Wisconsin-Madison, 1415 Engineering Drive, Madison, WI 53706.

The overall objective of a chemical process is to convert raw materials into desired products using available resources, while satisfying operational constraints. Since there are multiple ways of satisfying these objectives, the primary goal of a process plant is to maximize its profit, i.e. operate the process in the most economical way possible at all times subject to uncertainty and fluctuations in operating conditions, which are inevitable in all processes. Hence, in any modern chemical manufacturing facility, automatic feedback control becomes indispensable. Model Predictive Control (MPC) is the most widely used advanced feedback control technology in the process industries [1, 2]. Its ability to handle constraints and satisfy some optimal performance criterion, by solving an online optimization problem, is its major strength [3]. Currently, MPC is implemented as a two layer strategy, which is based on fundamental ideas on hierarchical control introduced in the 1960's [4, 5, 6]. The first layer, known as the Real Time Optimization (RTO) layer [7], solves an economic optimization problem to estimate the best operating steady state of the process subject to the given constraints. These are the set-points. The second layer, which is the dynamic optimization layer, then manipulates the inputs in such a fashion as to “track” these setpoints.

Backx et al. [8] stresses the need for dynamic operations in process industries in an increasingly market-driven economy. Engell [9] reviews the various answers to this need, most of which focus on ways to improve the structure of the RTO layer. Hence there is an opportunity to improve the economic performance of MPC framework. The set-points are essentially the economic objectives translated into process control objectives. This translation results in loss of economic information as the dynamic regulation layer has no information about the original plant economics except for a point or a trajectory which it tries to track. Reducing this loss of information can improve the economic performance of the controller and hence this provides an opportunity to rethink the distribution of functionality between the two layers of MPC hierarchy.

This presentation introduces a new strategy for MPC to reduce this loss of economic information. This strategy uses plant economics directly in the dynamic regulation layer, avoiding the conversion of economic objective to set-points. In such a scenario the function of the RTO layer is combined with the dynamic regulation, allowing a single layer. We consider a simple process, and compare the performance of the new methodology with the traditional two layer technique, demonstrating the potential of the new strategy. First, the economics for the process are adequately modeled taking into account the operating and utility cost. Next, the performances of both the strategies are compared and the economically better performance of the new strategy is shown. The traditional two layer strategy is implemented using the above economic model in the RTO layer. In the new strategy, the dynamic regulator is made to optimize the modeled economics directly. Finally, the example is used to bring out economic modeling issues that arise when implementing the new strategy, and possible solutions are proposed.

References

[1] S. Joe Qin and Thomas A. Badgwell. A survey of industrial model predictive control technology. Control Eng. Prac., 11(7):733–764, 2003.

[2] Robert E. Young, R. Donald Bartusiak, and Robert W. Fontaine. Evolution of an industrial nonlinear model predictive controller. In James B. Rawlings, Babatunde A. Ogunnaike, and John W. Eaton, editors, Chemical Process Control–VI: Sixth International Conference on Chemical Process Control, pages 342–351, Tucson, Arizona, January 2001. AIChE Symposium Series, Volume 98, Number 326.

[3] David Q. Mayne, James B. Rawlings, Christopher V. Rao, and Pierre O. M. Scokaert. Constrained model predictive control: Stability and optimality. Automatica, 36(6):789–814, 2000.

[4] I. Lefkowitz. Multilevel Approach Applied to Control System Design. Trans. ASME, 88:2–10, 1966.

[5] MD Mesarovic, D. Macko, and Y. Takahara. Theory of hierarchical, multilevel systems. Academic Press, New York, 1970.

[6] W Findeisen, F.N. Bailey, M. Brdys, K. Malinowski, P. Wozniak, and A. Woznaik. Control and Coordination in Hierarchical Systems. John Wiley & Sons, New York, N.Y., 1980.

[7] Thomas E. Marlin and Andrew N. Hrymak. Real-time operations optimization of continuous processes. In Jeffrey C. Kantor, Carlos E. Garcia and Brice Carnahan, editors, Chemical Process Control–V, pages 156–164. CACHE, AIChE, 1997.

[8] T. Backx, O. Bosagra, and W. Marquardt. Integration of model predictive control and optimization of processes. In Advanced Control of Chemical Processes, June 2000.

[9] S. Engell. Feedback control for optimal process operation. J. Proc. Cont., 17:203–219, 2007.

329f Estimation of Experimental Gradients for Real-Time Optimization

Alejandro Marchetti, **Benoît Chachuat**, and Dominique Bonvin. Automatic Control Laboratory, Ecole Polytechnique Federale de Lausanne (EPFL), Station 9, Lausanne, CH-1015, Switzerland

Challenges in real-time process optimization arise from the inability to build and adapt accurate models for complex physico-chemical processes. Two main classes of optimization methods are available for handling model uncertainty. In the absence of measurements, a robust optimization approach is typically used, whereby conservatism is introduced to guarantee feasibility for the entire range of expected variations [1]. When measurements are available, adaptive optimization can help adjust to process changes and disturbances, thereby reducing conservatism [2].

This presentation focuses on the latter class of methods. The measurements can be used in different ways to compensate for model uncertainty. A possible classification can be made as follows [3]: (i) *model-adaptation methods* that use the measurements to update the parameters of the process model before repeating the optimization; (ii) *modifier-adaptation methods* that adapt constraint and gradient modifiers; and (iii) *input-adaptation methods* that convert the optimization problem into a feedback control problem. Here, we argue in favor of modifier-adaptation methods, which use a parameterization and measurements that are tailored to the tracking of the necessary conditions of optimality.

Perhaps the key issue in applying modifier-adaptation methods is tied to the fact that the gradients of the plant outputs with respect to the plant inputs, also called *experimental gradients*, must be available. An accurate estimate of the experimental gradients is indeed necessary for the iterates to yield a KKT point (relative to the plant) upon convergence. A novel way of estimating the experimental gradients is described in this presentation. It is based on the rather natural idea that the expected level of noise in the gradient estimates, as induced by the measurement noise, can be kept sufficiently small by ensuring a certain distance between the successive operating points. In particular, a number of theoretical results are presented regarding the choice of such a distance.

These developments are illustrated for an experimental three-tank system. It consists of three cylinders interconnected in series by two pipes; two pumps (driven by DC motors) supply the leftmost and rightmost columns with liquid (water). The real-time optimization problem is formulated so as to minimize the overall pumping energy needed to maintain the height of liquid in the columns between given limits.

References

[1] Monnigmann M. and Marquardt W., "Steady-state process optimization with guaranteed robust stability and feasibility," *AIChE J* **49**(12):3110-3126, 2003.

[2] Marlin T. E. and Hrymak A. N., "Real-time operations optimization of continuous process," *Proc 5th Int Conf on Chemical Process Control (CPC-5)*, Tahoe City NV, 1997.

[3] Chachuat B., Srinivasan B. and Bonvin D., "Model parameterizaton tailored to real-time optimization," *Proc. 18th Eur Symp on Computer Aided Process Engineering (ESCAPE-18)*, Lyon, France, 2008.

329g The Min-Max Robust Design of Advanced-Step Nmpc Controller

Rui Huang, Victor M. Zavala, and Lorenz T. Biegler. Chemical Engineering Department, Carnegie Mellon University, 5000 Forbes Ave, Pittsburgh, PA 15213

Nonlinear model predictive control (NMPC) with first principle models has recently received a great amount of attention from industry. However, the inherited barriers to this control strategy are the large on-line computational cost and resulting feedback delays. To overcome some of these limitations, the so-called advanced-step NMPC controller (AS-NMPC) has been recently proposed. The controller makes use of the dynamic model to predict the future state of the system, use this state to solve a predicted problem in advance and correct on-line using a fast NLP sensitivity approximation. It has been reported in several large-scale applications that the AS-NMPC controller is able to reduce the feedback delay by at least two orders of magnitude compared to the on-line solution of the full NMPC problem and that the sensitivity approximation errors are negligible [1,4]. Moreover, it has been shown that the AS-NMPC controller enjoys the same nominal stability properties of a standard or ideal NMPC controller and its inherent robust stability properties have also been established [3].

If high degrees of uncertainty are present in the system, the inherent robust stability properties of NMPC are not sufficient. In this case, robust design strategies able to account for uncertainty explicitly in the controller formulation are required. Several formulations have been summarized in [2]. It has been shown that the Min-Max NMPC formulation, which computes the best control policy based on the worst expected realization of the uncertainties, is able to guarantee robust stability. This controller formulation is attractive from a theoretical point of view but dramatically increases the computational cost of the on-line NMPC problem. From a computational point of view, more attractive alternatives based on multi-scenario formulations and linearization arguments have been proposed.

In this work, we explore different robust NMPC formulations and embed them into the AS-NMPC controller framework. In particular, we emphasize on the benefits of multi-scenario formulations due to their favorable structure and we explore different strategies to capture the full uncertainty description using a few scenarios. Finally, we explore connections to state estimation that allows a further reduction of the number of scenarios. Simulation examples are presented to demonstrate the concepts.

Reference:

[1] R. Huang, V.M. Zavala, and L.T. Biegler. Advanced step nonlinear model predictive control for air separation units. Submitted for Publication.

[2] L. Magni and R. Scattolini. Robustness and robust design of mpc for nonlinear discrete-time systems. In R. Findeisen, F. Allgöwer, and L.T Biegler, editors, Assessment and Future Directions of Nonlinear Model Predictive Control, pages 239{254. Springer, 2007.

[3] V.M. Zavala and L.T. Biegler. The advanced step nmpc controller: Optimality, stability and robustness. Automatica, Submitted for publication, 2007.

[4] V.M. Zavala, C.D. Laird, and L.T. Biegler. Fast implementations and rigorous models: Can both be accommodated in nmpc? Int. J. Robust Nonlinear Control, 18:800-815, 2008.

343a Global Optimization of Algorithms

Alexander Mitsos, Aices, RWTH Aachen, Pauwelstrasse 12, Aachen, Germany, **Benoît Chachuat**, Automatic Control Laboratory, Ecole Polytechnique Federale de Lausanne (EPFL), Station 9, Lausanne, CH-1015, Switzerland, and Paul I. Barton, Chemical Engineering, Massachusetts Institute of Technology, 77 Massachusetts Ave., RM 66-464, Cambridge, MA 02139.

Theory, examples and an open-source implementation of subgradient propagation in the McCormick relaxations [1] are presented. This development allows the convex relaxation and global optimization of algorithms. In particular, parameter estimation problems with dynamic systems embedded can be solved provided that an algorithm with a fixed number of iterations exists for the corresponding simulation problem.

Deterministic algorithms based on continuous and/or discrete branch-and-bound have facilitated the global optimization of nonconvex programs. These algorithms rely on convex or affine relaxations of the functions participating in the optimization problem. Convex and concave envelopes or tight relaxations are known for a variety of simple nonlinear terms [2] and this allows the construction of convex and concave relaxations for a quite general class of functions through several methods [1,3,4]. The majority of the literature on global optimization considers nonconvex programs for which explicit functions are known for the objective and constraints. A more general case is that the objective function and/or constraints are calculated implicitly by an algorithm, e.g., the solution of a boundary value problem via the finite element method. Similar to automatic differentiation [5], the relaxations based on McCormick's composition theorem [1,6] can be used to construct convex relaxations of algorithms automatically. However, these relaxations are nonsmooth. Here, the systematic propagation of subgradients is described without introducing auxiliary variables. The calculation of subgradients allows the use of a nonsmooth NLP solver to obtain the required lower bound. Alternatively, linearizations via the subgradients can give an affine relaxation. Similar to the convex relaxation, the subgradient propagation relies on the recursive application of a few rules, namely the calculation of subgradients for addition, multiplication and composition operations. Subgradients at interior points can be calculated for any factorable function for which a McCormick relaxation exists, provided that subgradients are known for the relaxations of the univariate intrinsic functions. For boundary points additional assumptions are necessary.

While the proposed subgradient propagation and relaxation of algorithms can be used in a variety of optimization problems, this talk focuses on dynamic optimization problems, and more specifically on parameter estimation problems. Sophisticated global optimization methods exist for initial value problems [7,8]. Similar to the existing methods, the proposal in this talk relies on discretization of the independent variable and on McCormick relaxations. The main difference is that in the existing algorithms, relaxation is performed before the discretization (done by the integrator), whereas in the proposal the relaxation is performed after the discretization. Another difference is that here state bounds and relaxations are obtained simultaneously. In addition to being more flexible (e.g., applicable also to boundary value problems), it is interesting that the proposed relaxation of algorithms provides a very simple alternative to the elaborate theory of the existing methods.

Case studies from both initial and boundary value problems are presented, including the solution of the heat equation via finite differences and the solution of a nonlinear ordinary differential equation (ODE) system via the explicit-Euler method. The commercial code BARON [9] is used as a benchmark for the performance of the proposed algorithmic relaxations. BARON requires explicit functions and therefore the discretized state variables are encoded as optimization variables along with the equality constraints from the discretization of the ODEs relating them. The case studies have an imperfect match between model and measurement and the challenge for global optimization methods is to converge the lower bound. Establishing that the lower bound is higher than the statistical test for the fit proves that the mismatch is due to the model and not to a suboptimal parameter fit [10]. An obvious advantage of the lower bounding scheme proposed here is the significantly smaller number of optimization variables. The computational results demonstrate that this advantage can result in drastically faster convergence of the lower bound. BARON is a state-of-the-art deterministic global NLP solver, and therefore the comparison shows the promise of the proposed method.

[1] G. P. McCormick. Computability of global solutions to factorable nonconvex programs: Part I. Convex underestimating problems. Mathematical Programming, 10(2):147--175, 1976.

[2] M. Tawarmalani and N. V. Sahinidis. Global optimization of mixed-integer nonlinear programs: A theoretical and computational study. Mathematical Programming, 99(3):563--591, 2004.

[3] C. S. Adjiman and C. A. Floudas. Rigorous convex underestimators for general twice-differentiable problems. Journal of Global Optimization, 9(1):23--40, 1996.

[4] E. M. B. Smith and C. C. Pantelides. A symbolic reformulation/spatial branch-and-bound algorithm for the

global optimisation of nonconvex MINLPs. Computers & Chemical Engineering, 23(4-5):457--478, 1999.

[5] A. Griewank. Evaluating Derivatives: Principles and Techniques of Algorithmic Differentiation. Frontiers in Applied Mathematics. SIAM, Philadelphia (PA), 2000.

[6] G. P. McCormick. Nonlinear Programming: Theory, Algorithms and Applications. John Wiley and Sons, New York, 1983.

[7] A. B. Singer. Global Dynamic Optimization, http://yoric.mit.edu/download/Reports/SingerThesis.pdf. PhD thesis, Massachusetts Institute of Technology, 2004.

[8] B. Chachuat, A. B. Singer, and P. I. Barton. Global methods for dynamic optimization and mixed-integer dynamic optimization. Industrial & Engineering Chemistry Research, 45(25):8373--8392, 2006.

[9] N. V. Sahinidis. BARON: A general purpose global optimization software package. Journal of Global Optimization, 8:201--205, 1996.

[10] A. B. Singer, J. W. Taylor, P. I. Barton, and W. H. Green. Global dynamic optimization for parameter estimation in chemical kinetics. Journal of Physical Chemistry A, 110(3):971--976, 2006.

343b An Inexact Trust Region Algorithm for Large-Scale Process Optimization with Dense Jacobians

Sree Rama Raju Vetukuri[1], Lorenz T. Biegler[1], and Andrea Walther[2]. (1) Chemical Engineering, Carnegie Mellon Univeristy, 5000 Forbes Avenue, Pittsburgh, PA 15213, (2) Institute of Scientific Computing, TU Dresden, Dresden, 01062, Germany

Periodic Adsorption Processes (PAPs) have gained increasing commercial acceptance as an efficient separation technique for a wide range of applications. These processes consist of vessels or beds packed with solid adsorbent. The adsorbent is brought in contact with a multi-component feed and the separation of the components is based on the difference in the affinity towards the adsorbent particles. Examples of PAPs include vacuum swing adsorption (VSA) to separate oxygen from air, pressure swing adsorption to separate hydrogen from hydrocarbons and simulated moving bed chromatography to separate two enantiomers in the liquid phase. Examples of PAPs include vacuum swing adsorption to separate oxygen from air, pressure swing adsorption to separate hydrogen from hydrocarbons and simulated moving bed chromatography to separate two enantiomers in the liquid phase. Unlike most processes e.g. distillation which operates at steady state conditions, PAPs operate under periodic transient conditions with each bed repeatedly undergoing a sequence of steps. These processes reach a cyclic steady state (CSS), where the concentration profiles change dynamically and the profiles at the beginning of each cycle match with those at the end of the cycle. This CSS operation results in dense constraint Jacobians, where the time required for the computation of the Jacobian and its factorization dominates the overall optimization process.

This talk presents a trust-region SQP algorithm for the solution of minimization problems with non-linear equality constrained problems. Instead of forming and factoring the dense constraint Jacobian, this algorithm approximates the Jacobian of equality constraints with a specialized quasi-Newton method [1]. Hence it is well suited to solve optimization problems related to PAPs. This algorithm represents a Byrd-Omojokun [2] trust-region approach that takes the inexactness of the Jacobian and its null-space representation into account. The global convergence of the algorithm to first-order critical points is ensured by the theoretical results presented in [3]. The quality of the approximated constraint Jacobian can be adjusted by verifying two conditions that measure the inexactness of the null space representation. The two required conditions on the inexactness can be easily verified during the optimization process.

Furthermore, we will discuss briefly how the derivative information is computed. Here we apply a targeted approach [4] combining automatic differentiation and more sophisticated integration algorithms, for example by CVODES, to evaluate the direct sensitivity equation, the adjoint equation and the second order adjoint equation.

Numerical results for a Simulated Moving Bed system and non-isothermal VSA O2 bulk gas separation processes are presented.

References:

1. Griewank, A; Walther, A. On constrained optimization by adjoint-based quasi-Newton methods. Optim. Methods. Softw. 2002; 17: 869-889.

2. Omojokun, E. Trust Region Algorithms for Optimization with Nonlinear Equality and Inequality Constraints, Ph.D. thesis, Department of Computer Science, University of Colorado, 1989.

3. Walther, A. A First-Order Convergence analyis of Trust-Region methods with inexact Jacobians. SIAM J. of Optim, 2008; 19(1): 307-325.

4. Ozyurt, D.B.; Barton, P.I. Cheap Second Order Directional derivatives of stiff ODE Embedded Fuctionals. SIAM J. Sci. Comput. 2005; 26(5): 1725-1743.

343c Modeling and Dynamic Simulation of An Industrial Natural Gas Processing Unit Considering Environmental and Economic Impacts

Paul N. Sharratt, School of Chemical Engineering and Analytical Science, The University of Manchester, PO Box 88, M60 1QD, Manchester, United Kingdom, **Aurora Hernández Enríquez**, School of Chemical Engineering and Analytical Science, PhD Student, The University of Manchester, Sackville Street, Manchester, M60 1QD, United Kingdom, and Antonio Flores Tlacuahuac, Department of Chemical Engineering and Sciences, Lecturer, Universidad Iberoamericana . Prol. Paseo de la Reforma 880, Lomas de Santa Fe, Mexico DF, 01219, Mexico.

ABSTRACT

This paper uses analysis of the performance of the process as a means of supporting decision making for future investments in order to improve reliability and environmental issues in an existing cryogenic gas processing unit. An integrated model is presented that identifies opportunity areas for improvements in the operation of the unit. Dynamic simulation is discussed and the question of what is a required level of modelling detail in its applications is addressed. The methodology sets out how the problem can be targetted by considering relevant operators comments and analysing operational data. This can provide deep understanding of various interrelationships between multiple plant operating factors including maintenance activities, which could be essential for substantial improvements of process performance. The dynamic model validation should take into account records saved in the control system and look at how the model reproduces them. The model agrees in a satisfactory way with the actual performance of the unit, and it is used to evaluate the economic and environmental impacts that some of the possible failures could bring to the unit. Further work is planned to link the model to a more accurate model of the probability of failure occurrence, thereby supporting enhanced design of maintenance programmes.

343d Optimal Flow Assurance Policies for Hydrate Prevention In Deep-Water Gas Production Systems

Eduardo Luna-Ortiz[1], Praveen Lawrence[2], Costas C. Pantelides[2], Claire S. Adjiman[1], and Charles D. Immanuel[1]. (1) Centre for Process Systems Engineering, Department of Chemical Engineering, Imperial College London, South Kensington Campus, London, SW7 2AZ, United Kingdom, (2) Process Systems Enterprise Ltd., Bridge Studios, 107a Hammersmith Bridge Rd, London, W6 9DA, United Kingdom

The development of flow assurance policies during transient operations in deep-water production systems has been identified as one of the paramount challenges that the oil and gas industry faces today [1]. High hydrostatic pressures and low ambient temperatures promote the formation of gas hydrates that can deposit in the flowlines and risers, interrupting the flow and causing severe financial losses due to lost production and remediation costs. Therefore, a key issue is to design cost-effective strategies that guarantee the continuous flow of gas to the processing offshore platform or onshore facility by steering the system away from hydrate formation conditions. This is particularly difficult during transient situations such as shut-in and re-startup.

In this work , we use a model-based flow assurance computational framework developed previously [2] for the identification of optimal hydrate prevention policies during transient operation. A multi-phase flow model (based on the drift-flux model [3,4]) that describes the transient thermal-hydraulic behaviour of sub-sea pipelines is coupled with a hydrate thermodynamic equilibrium calculation module [5]. This model-based approach allows us to determine whether hydrate formation can occur at any time instant and at any point along the pipe network by comparing the hydrate formation equilibrium temperature and the actual fluid temperature inside the pipe. The injection of thermodynamic hydrate formation inhibitors such as methanol is also included in the model. The framework is implemented in a state-of-the-art dynamic modeling tool (gPROMS®) [6], which allows the easy modular construction of pipe networks with different geometries and the use of a range of models for heat transfer coefficients or drift-flux correlations. In order to demonstrate the capabilities of the proposed framework, we consider a case study based on typical conditions typical in the Gulf of Mexico. We obtain optimal control policies (optimal inhibitor dosages and depreessurisation rates) that ensure hydrate-free operation under a transient scenario in which the pipeline is shut-in at the choke valve (topsides) for 10 hours and then restarted. We investigate the impact of different objective functions, such as the minimisation of lost production or the minimisation of methanol injection costs, on the operating policy. The re-opening rate of the choke valve is a very sensitive parameter due to rapid cooling (Joule-Thomson effect) which may lead to hydrate formation. The control strategies identified via dynamic optimisation take into account the trade-offs that exist between a rapid restart and a hydrate-free restart. Our studies demonstrate the potential benefits of a model-based approach in dealing with the complex dynamics of the flow assurance problem.

[1] Macintosh, N., 2000, Offshore, 60(10).

[2] Luna-Ortiz, E., Lawrence, P, Pantelides, C. C, Adjiman, C. S. and Immanuel, C. D. (2008), 18th European Symposium on Computer Aided Process Engineering – ESCAPE 18

[3] Ishii, M. and Hibiki, T., 2006, Thermo-Fluid Dynamics of Two-phase Flow, Springer

[4] Shi, H., Holmes, J., Durlofsky, L., Aziz, K., Diaz, L., Alkaya, B. and Oddie, G., 2005, SPE J., March, 24-33.

[5] van der Waals, J. H. and Platteeuw, J. C., 1959, Adv. Chem. Phys., 2, 2-57.

[6] PSE, 2007, gPROMS® ModelBuilder 3.0.3, www.psenterprise.com

343e Plantwide Optimizing Control of Bioprocesses: Bio-Ethanol Process Case Study

Silvia Ochoa, Jens-Uwe Repke, and Günter Wozny. Chair of Process Dynamics and Operation, Berlin Institute of Technology, Straße des 17. Juni 135, Berlin, 10623, Germany

The main purpose of any industrial production process is to achieve all the time, despite the presence of disturbances, the highest profitability for the whole plant. For this reason, the purpose of process control should be to attain optimal operation instead of maintaining a predefined set point [1]. Furthermore, taking into account that in real operation, the different operating units are usually highly interconnected (i.e energy and mass recycle loops), the control of the process should be addressed from a plantwide control perspective because optimal operation of the individual units does not necessarily lead to optimal operation for the whole plant. Plantwide control is a large-scale problem which deals with up to hundreds of measurements and manipulated variables [2]. Approaches for plantwide control are usually classified into optimization-based [3-4] or heuristic-based [5-6]. Usually the optimization-based approaches apply nonlinear optimization methods to select the control structure, whereas the heuristic-based approaches use engineering knowledge coupled with simulation analysis to define a suitable control structure for the process.

In this paper the use of a plantwide optimization-based approach, designated as plantwide optimizing control, is proposed for the bio-ethanol process. In this approach, plantwide optimization and control are integrated into a single dynamic large-scale real-time optimization problem, where the objective function is defined by the plant profitability and the manipulated variables are used as decision variables. In order to solve the problem in real time, different strategies are required to increase the efficiency of the method. Therefore, the optimization routine is only re-called on-demand by the activation of an optimization trigger vector depending on the presence of disturbances and on the plant performance. The values of the trigger vector indicate whether each manipulated variable is considered as decision variable during the optimization or not, and determine the relative size of the search region for each decision variable.

The plantwide optimizing approach has been applied to the continuous bio-ethanol production process from starch, taking into account saccharification, fermentation, cells recycle and distillation stages. Although the bio-ethanol process has ten available manipulated variables, only seven of them are taken as decision variables during the optimization problem, because the remaining are used in an internal control loop proposed for biomass control [7]. This internal loop comprises a split-range controller for keeping the biomass concentration in the fermentor at its optimal value and a ratio controller for achieving a suitable viscosity in the biomass recycle slurry. The biomass internal loop is connected in cascade with the optimization layer involved in the plantwide optimizing control, receiving the optimal set points that should be tracked in order to maximize the profitability objective function defined for the whole plant.

The motivation for selecting the bio-ethanol process as case study relays on the fact that, in spite of recent advances in purification technologies and in the modification of microbial strains for making them more resistant to the different stress and inhibition factors, the economical feasibility of the bio-ethanol industry is still questioned. For this reason, the optimization and control of this process remain as challenging problems [8]. The results obtained using the plantwide optimizing approach for the continuous bio-ethanol process will be presented and compared to a two-layer nonlinear model predictive controller (NMPC) with set point optimization, showing that plantwide optimizing control is a very promising alternative for achieving optimal operation in bioprocesses even in the presence of disturbances.

References:

[1] S. Engell. Feedback control for optimal process operation. Journal of Process Control, 17, 203-219, 2007.

[2] A. C. B. de Araújo. Studies on Plantwide Control. Ph. D. Thesis. Norwegian University of Science and Technology, Trondheim, 2007.

[3] S. Skogestad. Plantwide control: the search for the self-optimizing control structure. Journal of Process Control, 10: 487-507, 2000.

[4] G. Stephanopoulos and C. Ng. Perspectives on the synthesis of plant-wide control structures. Journal of Process Control 10: 97–111, 2000.

[5] W. Luyben, L. B. D. Tyréus and M. L. Luyben. Plantwide process control. McGraw-Hill. USA, 1998.

[6] H.R. Lausch, G.Wozny, M.Wutkewicz and H. Wendeler. Plant-wide control of an industrial process. Trans IChemE, 76: 185-192, 1998.

[7] S. Ochoa, J-U. Repke and G. Wozny. Integrating Real-Time Optimization and Control for Optimal Operation. Proceedings of the CHEMPOR 2008 conference, Braga, Portugal, 2008, 151.

[8] S. Ochoa, A. Yoo, J-U. Repke, G. Wozny and D.R. Yang., Modeling and Parameter Identification of the Simultaneous Saccharification-Fermentation Process for Ethanol Production. Biotechnol. Prog., 23: 1454-1462, 2007

Dynamic Optimization of an Evaporator with a Nonlinear Model Predictive Controller for Application at Modular Micro Rectification

R. Knauss, R. Marr, M. Siebenhofer

Department of Chemical Engineering and Environmental Technology,
Graz University of Technology, Inffeldgasse 25 C, A-8010 Graz, Austria
ralf.knauss@tugraz.at

Intensification and Optimization of unit operation distillation and thus in further consequence of rectification process should be carried out as main objective of this work. A combination of Micro Process Engineering and Advanced Process Control (APC) is therefore applied to latter separation process.

Vapour and liquid phases of a simple batch distillation may be separated theoretically at equilibrium at a certain pressure and temperature state. A small separation effect can be achieved by this, because only one equilibrium stage between liquid and vapour phase is thereby realised. Contacting again vapour and liquid flows leaving one single stage in co-current mode, no further separation effect can be achieved. Under real conditions equilibrium will not be reached, which is then expressed in calculation as tray efficiency. Increased mass and heat transport may be obtained by mixing and separation of flows from single distillation vessels working at different operating points. Such a system of several stages may also be denoted rectification, if the streams leaving the micro devices are connected in counter-current operation mode. Thereby micro devices are as individual modules representing unit operations of such a process. Modular Micro Rectification (MMR) for discontinuous phase contact consists of unit operations heating, cooling, mixing and separating. Heat exchangers, mixers and cyclones for phase separation are serially connected to counter-current rectification system with the highest mass and heat transfer efficiency.
Continuous contacting of phases for the purpose of mass and heat transfer at counter-current operation mode with the exception of membrane applications only makes sense up to now for absorption process. Advances were reported by literature for the design of continuous and also discontinuous contacting, counter-current rectification in only one single micro structured apparatus [1,2,3]. Generally continuous phase contacting counter-current micro devices can only be seen as useful without occurrence of mixing effects caused by friction forces on phase interface.
However, a plant for Modular Micro Rectification containing several unit operations offers a correspondingly large number of manipulable variables, which have to be coordinated for reasonable handling of the single devices (Fig. 2). Due to the design of micro scaled devices plants form sophisticated systems, while for a fully optimized control still no common satisfying solution exists.

When operating an electrical powered evaporator for modular rectification purposes, designed by Forschungszentrum Karlsruhe, a strong interaction of mass flow with the vapour fraction and the outlet temperature can be observed [4]. Defined operating point for mass flow, temperature and vapour fraction can only be held with difficulties using traditional methods of linear control technology. When mass flow is increased, temperature at the evaporator outlet is reduced dramatically. Rising then the power input to maintain the temperature leads due to a increase of pressure in the micro-channels, which in turn lowers the mass flow and lifts temperature excessively again.
In a narrow range around operating points for mass flow and outlet temperature a step test was performed. The resulting step responses were used to tune the PID-controllers for the coupled loops mass flow and outlet temperature (Fig. 1: coupled variables).
Time constants of the systems differ strongly from each other. For the effect of power input on the outlet temperature a dead time of ten seconds and a time constant of one hundred seconds could be revealed by the step tests.

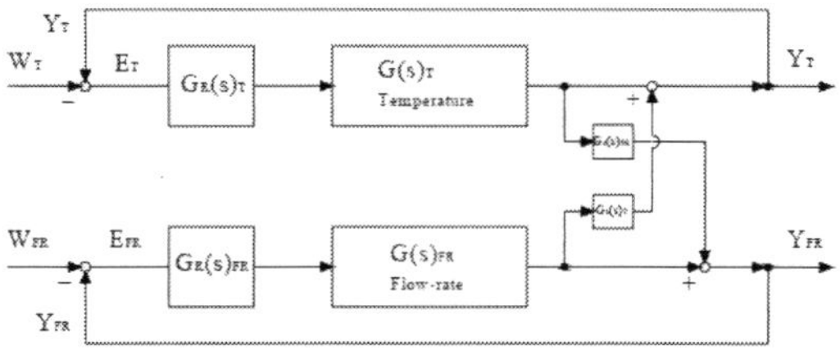

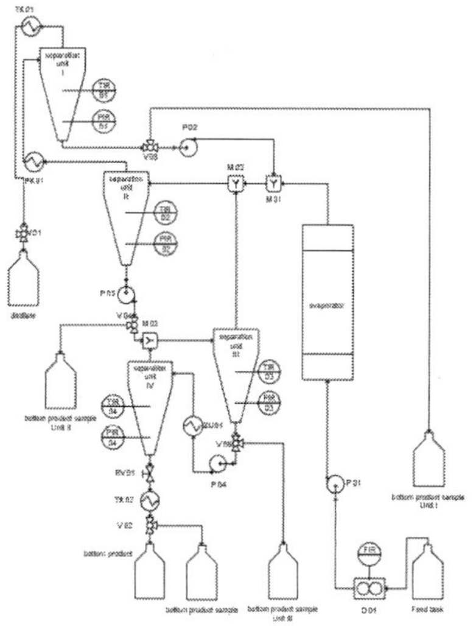

Fig. 1: Coupling of outlet temperature with mass flow displayed as linear transfer functions.
Fig. 2: Four stage Modular Micro Rectification (MMR)

The constants of the influence of pump control signal on the mass flow are determined by a dead time of only 2.6 seconds with a time constant of six seconds.
Sampling intervals of 0.1 seconds for controlled variable mass flow and one second for outlet temperature had to be chosen. Using the before determined PID-gains the system could not be operated fully stable (see Fig. 3). At mass flows or temperatures or combinations of these variables being not positioned within identified range the system performs an undamped oscillation. Due to strong action of the controllers on the system intense stress on the actuators is caused (Fig. 3).

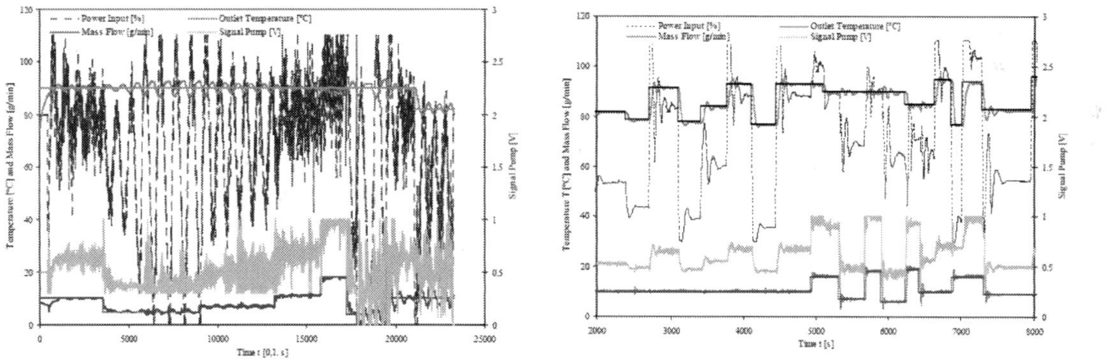

Fig. 3: Before and after Optimization of the Micro Evaporator by NMPC

It could be demonstrated by several experiments, that precise control of the process through the use of PID-control system can be seen as a challenge. For dynamic optimization of this multi-variable evaporation process with nonlinear system dynamics the first time a Nonlinear Model Predictive Controller (NMPC) had to be applied to the field of Micro Process Engineering. MPC generally has an internal model of the complete process by which the historical trend of the process variables (feed forward, manipulated and controlled variables) is monitored and registered. Hence, it is also possible to calculate and predict future trend of controlled variables for a time horizon, when any intervention by the controller is avoided [5].
By use of optimization algorithms similar to a chess computer the best strategy for the manipulated variables is generated, thereby a control is converted to an optimization task.

To calculate an optimal control sequence for the manipulated variables, the MPC performs a minimization of energy consumption and operation costs to obtain a given objective, which is expressed by a minimal difference between set-points to actual value of the controlled variables. The square of the deviations of the controlled variables from their set-points together with action of the manipulated variables are minimized over a certain future time horizon. Such an objective function (J) for a simple **S**[ingle] **I**[nput] **S**[ingle] **O**[utput] system with by definition only one manipulated and one controlled variable may be formulated as followed [6,7,8,9]:

$$J(U,t) = P \cdot \sum_{i=N_1}^{N_2} [y_R(y_R(t+i) - y(t+i)]^2 + \Gamma \cdot \sum_{j=1}^{Nu} [u(t+i-1) - u(t+i-2)]^2$$

This objective function contains also a set of parameters for tuning the controller. Through coefficients Rho (P) and Gamma (Γ) the future control behaviour can be manipulated by penalizing energy consumption or otherwise reducing control efficiency. The resulting constrained cost function is non-convex making detection of relative local optimum and the prevention of suboptimal solutions a difficult task. This obstacle can only be gone around using a combination of heuristic with gradient optimization methods. Optimization of complex states like unsteady operating conditions, changing set-points or constraints, in advance is performed by genetic algorithm.

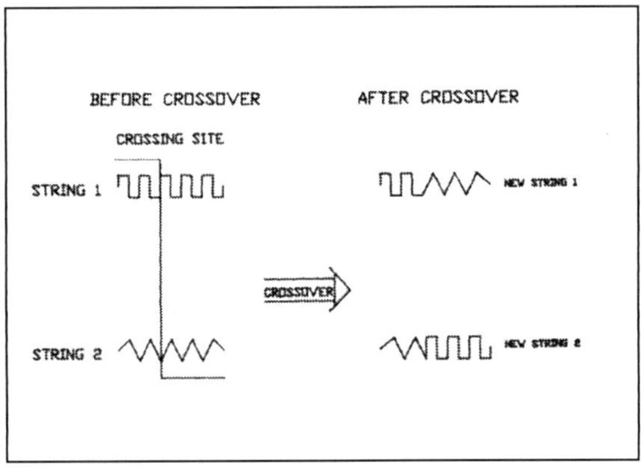

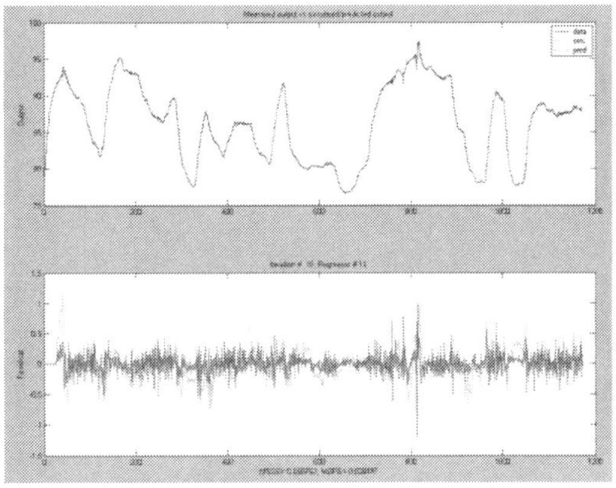

Fig. 4: Simple example for Crossover of parents producing offspring after Goldberg [10]
Fig. 5: NARX-Modelling of the outlet temperature

Genetic algorithms are as stochastic optimization methods based on the principle of natural selection and the evolutionary concept of nature. A natural parameter range of an optimization task is coded as a finite string by a limited set of characters. These search methods base upon a whole population of individuals producing an offspring by the genetic operators variation and selection, which is able to give better solution for an optimization task than their parents generation would have done. These artificial individuals are generated from the strings of a previous generation, additionally by operator mutation entirely new pieces of information are introduced (See Fig. 4). It is an efficient optimization technique that uses historical information for detection of local extrema. At unsteady operation of plant the solution vector is passed in every discrete time step by the heuristic algorithm to an implementation of gradient method of Levenberg-Marquardt, which finally performs the exact detection of the function minimum by sequential step method.

$$(J^T \cdot J + \mu \cdot \mathrm{diag}(J^T \cdot J)) \cdot h_{lm} = -J^T \cdot f \text{ and } u_n^* = u_{n-1}^* + h_{lm} \cdot$$

At completely stationary operating conditions the previously identified solution vector can usually be used for the initial control sequence, what directly leads to a significant reduction of CPU load.

Grouped nonlinear process models in shape of two NARX polynomials were first time implemented in a **M**[ultiple] **I**[nput] **M**[ultiple] **O**[utput] formulation of NMPC (Fig. 5: NARX: **N**[onlinear] **A**[uto]**R**[egressive] [with] [e]**X**[ogenous] [inputs]). Every training procedure a model is evolved from a set of several thousand regressors, which consist of nonlinear combinations of the lagged manipulated with the controlled and feed forward variables of the process. The total number of regressors depends on the highest power of the polynomial terms and the lag time of every variable. High polynomial order or long lag times lead directly to increased computation effort and CPU load. Various methods were proposed for the selection of polynomial structure and regressors. But until now structure selection is still not fully solved. For this work variations of the **F**[orward] **R**[egression] **O**[rthogonal] **E**[stimator] and the **S**[imulation] **E**[rror] **M**[inimization] [with] **P**[runing] methods were considered.
Using FROE algorithm the model structure is iteratively incremented until a specified prediction accuracy is obtained. Thereby parameters are estimated by means of orthogonal least squares, while the structure selection is based on the **E**[rror] **R**[eduction] **R**[atio] criterion:

$$[\mathrm{ERR}]_i = \frac{\mathrm{MSPE}(M_i) - \mathrm{MSPE}(M_{i+1})}{\sum_{t=1}^{N} y^2(t)}$$ and also $$[\mathrm{ERR}]_i = \frac{\hat{g}_i^2 \cdot \sum_{t=1}^{N} w_i^2(t)}{\sum_{t=1}^{N} y^2(t)}$$ with the corresponding model

$$y(t) = w^T(t) \cdot g + \xi(t),$$

where w_i is the auxiliary orthogonal regressor and g_i is an estimated parameter. Because of the orthogonality of regressors $w^T(t)$, the significance of every candidate regressor can be evaluated independently.
A more effective criterion estimates the model structure on the basis of the simulation error, substituting the **M**[ean] **S**[quared] **P**[rediction] **E**[rror] with the **M**[ean] **S**[quared] **S**[imulation] **E**[rror]. The resulting criterion is then denoted **S**[imulation] [Error] **R**[eduction] **R**[atio]:

$$[\mathrm{SRR}]_i = \frac{\mathrm{MSSE}(M_i) - \mathrm{MSSE}(M_{i+1})}{\sum_{t=1}^{N} y^2(t)}.$$

Every iteration step the model M_{i+1} is evaluated and compared to the model M_i one step before. This criterion for selecting regressors is also complemented by a systematic procedure for regressor deletion, included in the training procedure to simplify the regressor selection. Not significant terms are deleted by a so called “pruning procedure“ similar to Neural Nets. A complete iteration of the SEMP algorithm either adds a new regressor to the actual model or substitutes one or more of its terms with it, with the objective to improve the performance by variation of regressors.
Every discrete time step the predictions from these polynomial models of controlled variables mass flow and outlet temperature are then evaluated numerically and integrated into objective function.

The concept was formulated generically in C++ and implemented as a **D**[ynamic] **L**[ink] **L**[ibrary] in the Process Control Software LABVIEW 7 as a MIMO formulation (Fig. 6). For editing the polynomial models also on LABVIEW **G**[raphical] **U**[ser] **I**[nterface] a parser was created.

Based on experimental results it was demonstrated, that NMPC keeps the coupled variables mass flow and temperature with minimal control activity in the entire two-phase region at their set-points. Manipulated variables are calculated in optimal ratio to each other.
As a consequence the mechanical stress on actuators and energy consumption is kept low, which reduces directly investment costs. A decrease of operating cost can easily be achieved by reduction of the amount

of failure output in production process. Model mismatch caused by variation of plant behaviour or inlet composition is compensated by error minimization in closed control loop.

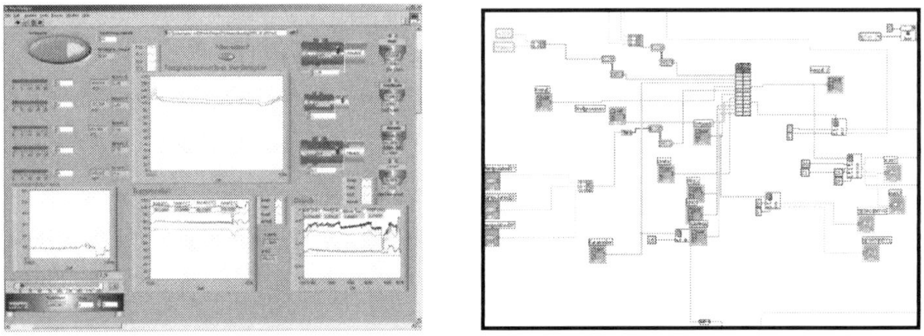

Fig. 6: Programming Interface of LABVIEW with a Library Function Node

The proposed "Nonlinear Model Predictive Control" concept has also already been successfully applied to mechanical transport systems and de-inking plants in the field of paper and pulp industry by the eposC Ltd. in Grambach near Graz (Austria). For the optimisation and production increase of twenty percent with additional reduction of chemical consumption in the paper processing of Norske Skog Bruck (Austria) the project was awarded with the INNOward 2007 of the Austrian Federal Economic Chamber in the category "Energy/Environment".

References:

[1] L. SILVA, R. ARORA, A. TONKOVICH, e. al., Advanced Separations, including Distillation, using Micro Channel Architecture for Process Intensification, Book of Abstracts, European Congress of Chemical Engineering (ECCE-6) Copenhagen, September 2007

[2] L. SILVA, D. WEIDERT, R. LITT, A. TONKOVICH , R ARORA, M. FANELLI, Exergetic Effect Distillation in Microchannel Architecture, Presentation at AIChE Meeting, Houston, April 2007

[3] H. FINK, M.J. HAMPE, Designing and Constructing Microplants, IMRET 3, 1999, pp. 664 - 673.

[4] R. BRANDSTÄTTER, Development of strategies for the realisation of a Modular Micro Distillation, diploma thesis, Graz University of Technology, Graz, 2007.

[5] R. DITTMAR, B.-M. PFEIFFER, Modellbasierte prädiktive Regelung, Eine Einführung für Ingenieure, Oldenbourg Wissenschaftsverlag, München, Wien, 2004

[6] A. BOLLIG, Prädiktive Regelung mit neuronalen Netzen, Theoretische Arbeit, Automatisierungstechnik 51, S. 69-77 , 2003

[7] M. NØRGAARD, L.K. HANSEN et al., Neural networks for Modelling and Control of Dynamic Systems, Springer, London, 2000

[8] F. S. MJALLI, Neural network model-based predictive control of liquid–liquid extraction contactors, Chemical Engineering Science 60, S. 239-253, Chemical Engineering Department, University of Qatar, 2005

[9] P. POTOČNIK, I. GRABEC, Nonlinear Model Predictive control of a Cutting Process, Neurocomputing 43, pp. 107-126, Faculty of Mechanical Engineering - University of Ljubljana, 2002

[10] D. E. GOLDBERG, Genetic Algorithms in Search, Optimization, and Machine Learning, Addison-Wesley Publishing Company Inc., University of Alabama, 1989

343g A Decomposition Approach to Optimal Operation of Fermentation Processes

Harvey Arellano-Garcia, Chair of Process Dynamics and Operation, Berlin Institute of Technology, Str. des 17. Juni 135, 10623, Berlin, Germany

An inherent characteristic of fermentation processes is that a batch phase is required at the beginning when the initial condition of the inhibitory product is very small, i.e. far from its peak which is optimal for the growth of biomass. Furthermore, since the volumetric flow rates are set to zero during the batch period, all the other continuous decision variables have no impact on the process either. On the other hand, the initial concentrations of the substrates and the switching time point from the batch to the continuous phase are crucial decision parameters for the objective function and the constraints in the optimization problem. Thus, the original optimization problem is decomposed into two sub-problems. The first one minimizes the batch time subject to end conditions concerning concentrations bounds, which are supplied by the optimizer of the continuous phase where initial concentrations are used as decision parameters as well. This interaction procedure is finished when the end concentrations of the batch period are equal to the corresponding optimized values of the continuous period. For the fermentation process considered, however, the reformulation of the sub-problems is physically equivalent to the original problem formulation.

The aim of the optimization is the maximization of the product total amount per time while minimizing the start-up period. To keep the production costs at a convenient level, different constraints are included in the optimization problem such as a glucose waste limit, a lower bound for the outlet product concentration and also technical constraints which involve upper bounds for the biomass concentration in the reactor. Previous to the numerical optimization and based on experimental results, several model parameters were adjusted using robust dynamic simulation (discretization by five orthogonal collocation points) in combination with parameter estimation methods. For the simulation and computation of the sensitivities, we propose a new multiple-time-scaling-approach to solving the resulting optimization problem which possesses strong nonlinear properties.

Trends and Strategies for Obtaining Patents; Is the Patent Office Becoming the Rejection Office?

Presented by
Chris Lewis
for
The American Institute of Chemical Engineers
November 18, 2008

RatnerPrestia
WE SPECIALIZE IN THE LAW OF CREATIVITY®

Patent Office = Rejection Office?

The Data

- Between 1961 and 1984
  - Patent applications varied between 70,000 to 100,000
  - Patents issued hovered around 50,000
- Between 1984 and 2005
  - Patents tripled to 150,000
  - Patent applications nearly quadrupled to 380,000
- The Board of Patent Appeals and Interferences
  - Affirmed between 35 and 50% before April, 2007; Affirmed nearly 60% thereafter
  - Reversed between 35 and 55% before April, 2007; Reversed between 25 and 25% thereafter

Correlation or Causation?

RatnerPrestia
WE SPECIALIZE IN THE LAW OF CREATIVITY®

Why is this Happening?

Who's to Blame?

- Silly Patents - The Cat, the Cook, and the Lawyer
- Stifling Competition with Invalid Patents
- Need for Affordable Healthcare
- Blame it on the Press and Politicians
  It all makes for a good story

RatnerPrestia
WE SPECIALIZE IN THE LAW OF CREATIVITY®

Roadmap of Talk

- Basis of patentability
- Supreme Court weighs in
- Federal Circuit's flirtations
- The PTO tries to change the rules
- The "Rules" have changed, so how do we adapt?

RatnerPrestia
WE SPECIALIZE IN THE LAW OF CREATIVITY®

Basis for Patentability

High Level Requirements

- Useful
- New
- Non-Obvious
- Adequately described/"enabled"

RatnerPrestia
WE SPECIALIZE IN THE LAW OF CREATIVITY®

Novelty

Find a Distinction in the Claim

- Claim:
  A composition comprising a zeolite and a catalytic metal comprising platinum supported on the zeolite.
- Reference:
  Discloses rhodium on zeolite
- New?
  Yes, reference does not teach platinum

RatnerPrestia
WE SPECIALIZE IN THE LAW OF CREATIVITY®

Non-Obviousness

The Difference cannot be a Mere Obvious Modification

- Same facts as before
- A second reference teaches that platinum has improved activity over rhodium for the disclosed reaction
- Obvious?
  At least "*prima facie*" obvious

RatnerPrestia
WE SPECIALIZE IN THE LAW OF CREATIVITY®

The Examiner's Approach pre-KSR

The "Prima Facie" Obviousness Case

- References, as combined, disclose all of the claim limitations
- Teaching, suggestion, or motivation to combine the references
- Reasonable expectation of success that, upon combination, the invention would work

RatnerPrestia
WE SPECIALIZE IN THE LAW OF CREATIVITY®

Obviousness

Holdings in *KSR v. Teleflex*

- Do not rigidly apply the "TSM" test
- Any need or problem can provide a reason for combining references in the field
- Not limited to be led to only those elements that solve the "primary purpose" of a patent
- Obvious to try might be enough
- Still important to find a "reason" for combining the references

RatnerPrestia
WE SPECIALIZE IN THE LAW OF CREATIVITY®

Federal Circuit's Flirtations

Pfizer v. Apotex

- Patent: Besylate salt of amlodipine (blood pressure medication)
- Prior Art: Reference listing various amlodipine salts + Reference reciting benzene sulphonate (besylate) of other drugs
- Dist. Ct.: Not obvious; unexpected solubility, stability, processability

RatnerPrestia
WE SPECIALIZE IN THE LAW OF CREATIVITY®

Pfizer v. Apotex

Stop the Press!

- Federal Circuit: Obvious
- Motivation to Combine:
  - ➢ Only 53 anions approved by FDA for pharmaceutical salts, but ...
  - ➢ Other besylate salt drugs also show improved stability, solubility, and pharmacokinetics
  - ➢ Routine to make and test
- Expectation of Success:
  - ➢ New salt derived from finite, known choices
  - ➢ No enhanced therapeutic efficacy or reduced toxicity
- No unexpected properties shown (same as other besylate salt drugs)

The PTO Tries to Change the Rules

GSK Sues the PTO and Wins

- Limits on the Number of Claims
- Limits on the Number of Continuations/RCEs
- Reporting Requirements and the "Presumption"

RatnerPrestia
WE SPECIALIZE IN THE LAW OF CREATIVITY®

So, what do we do now?

Possible Arguments

- Still the best:
  References as combined fail to disclose all of the claim limitations
- No reasonable expectation of success/unpredictability
- Secondary Considerations
- Others:
  - Non-analogous art
  - Modification to primary references changes its basis for operation

RatnerPrestia
WE SPECIALIZE IN THE LAW OF CREATIVITY®

Arguments after KSR

No reasonable expectation of success

- Similar constituents do not work in similar environments
- Reference expressly discounts possibility the claimed invention would work
- Unpredictable technology area

RatnerPrestia
WE SPECIALIZE IN THE LAW OF CREATIVITY®

Arguments after KSR

Unexpected Results

- Prove with evidence, not attorney argument
- Invention has new or unintended use or property (*e.g.*, treats other diseases, better bioavailability, low toxicity, etc.)
- Repeated failures or roadblocks during development
- Materials or methods to produce the invention not well developed or routine

Patent Trends and Conclusions

- Patents
  - Still have their place,
  - Are valued by industry and investors.
- But the trend is, patents
  - Will be harder to obtain,
  - Will be easier to attack,
  - Will have less of a "bang for the buck"
- Nonetheless, protect your worthy assets.

RatnerPrestia
WE SPECIALIZE IN THE LAW OF CREATIVITY®

Is the PTO Becoming the R.O.?

THANK YOU!

RatnerPrestia
WE SPECIALIZE IN THE LAW OF CREATIVITY®

354b The Relationship Between Non-IP Intellectual Assets and IP

Mark Catan, Miles & Stockbridge PC, 1751 Pinnacle Drive, Suite 500, McLean,, VA 22102-3833

Consideration of non-IP intellectual assets is a useful tool in devising how best to "game" the patent system. This notion flows from the idea that patents are *derived from* non-IP intellectual assets. The intellectual asset perspective can help companies develop more efficient and effective internal processes for developing IP, which is what this presentation is about. The ideas that flow from this perspective are consistent with the evolving areas of consensus relating to patent strategy. These includes the concept that patents are a creative outlet in themselves, not a derivative of other design processes; that effective patenting should employ perspectives from multiple disciplines; that knowledge of the patent landscape should feed into the definition of a new product's value proposition; etc. Certain broadly accepted premises flow into the analysis and will also be presented. These include how effective processes for gaming the patent system naturally generates targets for structured innovation processes. A proven analysis tool kit will be presented which is designed to help patent applicants find the most effective tactics for protecting a given technology. The latter is particularly useful for companies who ask "now what do we do?" after learning about the impact of the Supreme Court's KSR decision.

Strategies and Current Trends for Policing, Enforcing and Resolving Disputes about Intellectual Property Assets

Presented by

Harrie Samaras

for

The American Institute of Chemical Engineers

November 18, 2008

RatnerPrestia
WE SPECIALIZE IN THE LAW OF CREATIVITY®

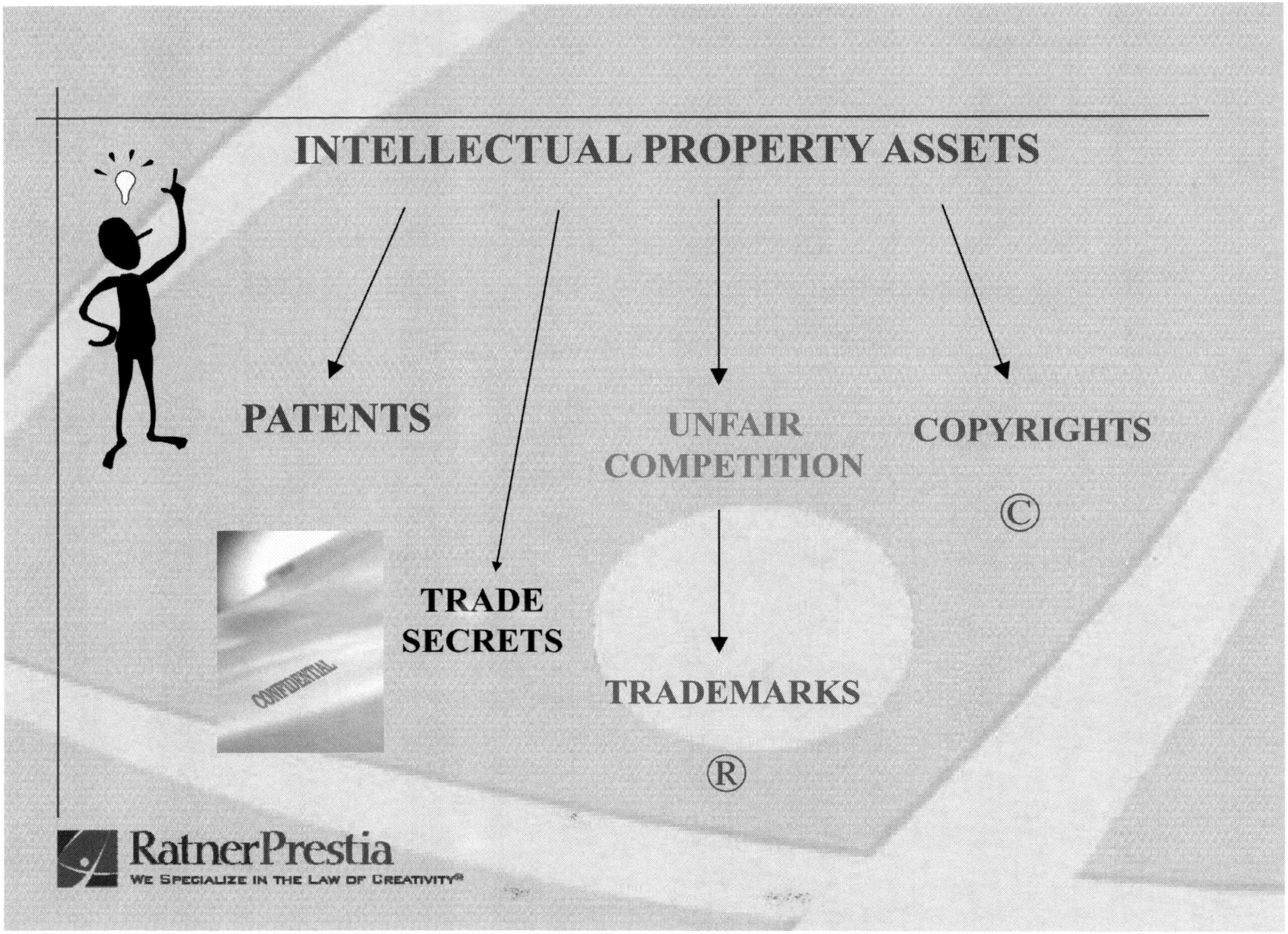
INTELLECTUAL PROPERTY ASSETS
PATENTS
TRADE SECRETS
UNFAIR COMPETITION
TRADEMARKS
®
COPYRIGHTS
©
CONFIDENTIAL
RatnerPrestia
WE SPECIALIZE IN THE LAW OF CREATIVITY®

What is policing, enforcing and resolving?

- **Policing**: monitoring/vigilance to see if others are using your company's IP and having some strategy in mind if they are (e.g., licensing, enjoining, acquiring their rights).

- **Enforcing**: taking some action to prevent the accused infringers from further infringement (e.g., cease and desist letter, court proceeding, administrative proceeding (ITC), UDRP proceeding (domain names)).

- **Resolving**: taking some action to settle the dispute (e.g., jury verdict, court judgment, arbitration, mediation, negotiation)

RatnerPrestia
WE SPECIALIZE IN THE LAW OF CREATIVITY®

Why police and enforce?

- IP is an investment (time and money) for a company.
- IP is an asset, thus the potential is there for a return on that investment.
- How can a company leverage the asset and obtain a worthwhile return?
  - Licensing royalties
  - Damages for infringement
  - Injunction (stop) infringers
  - Cross-licensing
  - Selling/assigning the IP
  - Contribution to a joint venture
  - Enhance the value of a business
  - Leverage in settlements

- You snooze, you lose. Policing and enforcing preserves rights/increases the value of an IP asset.

RatnerPrestia
WE SPECIALIZE IN THE LAW OF CREATIVITY®

Costs to Litigate

- According to the 2005 Economic Survey of the AIPLA, the average cost of bringing a patent infringement suit (regardless of how much is at stake) through trial is about $2.5 million.
  - When the amount at risk is less than $1M, the cost through discovery is about $350,000 and through trial is about $650,000.
  - When the amount at risk is between $1M and $25M, the cost through discovery is about $1.25M and through trial is about $2M.
  - When the amount at risk is more than $25M, the cost through discovery is about $3M and through trial is about $4.5M.
- In 2007, the median time from filing a civil action (including patent cases) in federal district court through disposition by trial was about 24 months, but it can take up to 3 years to get to trial in a patent case.
- E-discovery: Involves ancillary personnel, extensive documents, severe penalties for not complying
- The real costs of a dispute: disruption to the business; loss of personnel resources (corporate employees' ongoing assistance, document gathering, "e-discovery," depositions, settlement conferences, trial); costs of litigation $$$$.

RatnerPrestia
WE SPECIALIZE IN THE LAW OF CREATIVITY®

Time, Money, Resources, Consistency

Policing >> enforcing >> resolving

Policing and Enforcing Trends

- Policing needs support from all sectors of an organization. More companies are getting personnel from all areas involved in monitoring trade shows and trade periodicals, automatic watch services, periodic searches (e.g., internet, TM search firms).
- Assistance of Others (e.g., Internet Service Provider, Customs Agents, Private Investigators, U.S. Marshals)
- Since 1991, the number of patent litigation suits has grown at a rate of 5.8%.*
- The annual median patent damages award, when adjusted for inflation, has remained fairly consistent since 1995 (about $3.9M through 2000) and near $3.8M from 2001-07.*
- Patent damages awards vary sharply by industry. Since 1995: Telecommunications had the highest exposure to damages at a median award of over $31M based on 37 decisions; Software had the next highest exposure, at a median award of $8.5M based on 11 decisions; Pharmaceuticals had a median award of $1.3 M based on 17 decisions.*

*Pricewaterhouse Coopers 2008 Patent Litigation Study

RatnerPrestia
WE SPECIALIZE IN THE LAW OF CREATIVITY®

Policing and Enforcing Trends (cont'd)

- Patent plaintiffs generally enjoy higher success rates before juries.*
- Jury awards have been higher than bench trials in patent cases possibly contributing to the significant increase in jury trials since 1995.*
- Reasonable royalties has supplanted lost profits as the more frequently used basis for patent damages awards (easier to prove, holding company plaintiffs who cannot account for lost sales, patent troll suits).*
- Suits brought by entities owning patents but not making or selling products (patent trolls).*
- Time to trial varies based upon jurisdiction: E.D. Va. and W.D. Wisconsin (under a year); Connecticut (close to 5 years); and Massachusetts (close to 4 years).*

*Pricewaterhouse Coopers 2008 Patent Litigation Study

RatnerPrestia
WE SPECIALIZE IN THE LAW OF CREATIVITY®

Policing and Enforcing Trends

- *MedImmune v. Genentech* (Supreme Court decision), followed by *SanDisk Corporation v. ST Microelectronics, Inc., et al.* and *Sony Electronics, Inc. v. Guardian Media Technologies Ltd.* (letters written to alleged infringers) :
  - ➢ lowered threshold for accused infringers to bring a declaratory judgment action
  - ➢ increased risk of lawsuit against patent (IP) holder seeking to license/enforce
- *eBay, Inc. v. MercExchange* - The Supreme Court rejected the "general rule" of issuing a permanent injunction where a patent is found infringed, holding whether to issue an injunction in a patent infringement case is no different than any other case.
  - ➢ direct competitors + irreparable harm + inadequacy of money damages, more likely to obtain permanent injunction
  - ➢ not direct competitors or only hold patents + no irreparable harm + history of licensing, less likely to obtain a permanent injunction
- Legislative Initiatives – Criminalize use of intellectual property and impose penalties

RatnerPrestia
WE SPECIALIZE IN THE LAW OF CREATIVITY®

Trends in Resolving IP Disputes

- **Patent Rules**
  - Determine claim interpretation sooner
  - Determine infringement and invalidity sooner
- **ITC (International Trade Commission) Proceedings**
  - To block infringing imports coming into the U.S.
  - 14 months from beginning to end
- **Alternatives to Litigation ("ADR")**
  - Growing use in IP cases
  - Being used at all stages of an IP case (enforcement to appeal)
  - ADR is being considered at the policing and enforcement stages
  - U.S. Court of Appeals for the Federal Circuit now has an active mediation program (includes patent appeals)
  - Arbitration and mediation used the most (can combine these)

RatnerPrestia
WE SPECIALIZE IN THE LAW OF CREATIVITY®

Why ADR for IP Cases?

- In 2007, the percentages of IP cases tried: Patent (3.7% - 99 cases); Trademark (1.3% - 49 cases; Copyright (1.1% - 54 cases)
- High costs of litigation (*e.g.*, monetary and non-monetary)
- Shrinking corporate budgets
- Life of technology and/or patent
- Confidentiality
- Changes in corporations (*e.g.*, mergers/acquisitions; change in philosophy; change in leadership; change in product or asset focus)
- Value of IP asset (*e.g.*, cost/benefit)
- Remedies more varied and in accord with the business
- Preserve relationships (*e.g.*, licensor/licensee, companies competing in one area of their business but working together in other areas)

RatnerPrestia
WE SPECIALIZE IN THE LAW OF CREATIVITY®

Trends in Resolving IP Disputes

- ADR Organizations/IP Panels (*e.g.,)*
  - AAA; CPR; WIPO; JAMS; INTA
- ADR Rules for IP Disputes
  - CPR Rules for Non-Administered Arbitration of Patent and Trade Secret Disputes
  - AAA's Commercial Arbitration Rules and Mediation Procedures + Supplementary Rules for the Resolution of Patent Disputes
  - WIPO Arbitration and Mediation Rules
- Cyber-ADR: On-line resolution
- Use of Special Masters
- Court-encouraged ADR
- Voluntary Mediation in the Delaware Court of Chancery
- Maryland's Business and Technology Track
  - 2003, Maryland first state to create specialized trial court
  - Resolves business disputes that involve technology issues
  - Trained ADR Professionals
- Federal legislation: Ten-year pilot program for patent trial judges
- Three-arbitrator appeal panels (*e.g.*,)
  - CPR (former federal judges)
  - JAMS Optional appeal procedure

RatnerPrestia
WE SPECIALIZE IN THE LAW OF CREATIVITY®

It's QUESTION TIME !!
RatnerPrestia
WE SPECIALIZE IN THE LAW OF CREATIVITY®

354d Digitally Networked: Implications for Idea Management and Organizational Capability towards Innovation

Mukund Karanjikar, Chevron Energy Technology Co, 3901 Briarpark Drive, Houston, TX 77042

The rise of digital era brought with it a number of possibilities for people networks. Staying in touch, friends, professional networks, volunteer groups, fan groups and many others are more connected than ever.

Theories proposed at the intersection of ideas, innovation and people networks over the last ten years take the stance that networks are conducive to idea generation and hence innovation. This presentation systematically discusses how organizations can tap into these digital networks and enhance their own capability towards innovation.

Identifying Discontinuities for Strategic Research

Jack Hipple
Innovation-TRIZ
Tampa, FL

Biography

Jack Hipple is Principal with Innovation-TRIZ, a consulting firm focused on organizational problem solving and specializing in the use of the breakthrough problem solving process known as TRIZ ("Theory of Inventive Problem Solving"). He graduated from Carnegie Mellon University with a BSChE and spent 30 years in the chemical industry prior to starting his own consulting business in 1999. His industrial career included responsibility for Dow Chemical's Discovery Research program and its global chemical engineering laboratory, project management responsibility for the National Center for Manufacturing Sciences, and new product development responsibilities for Ansell Edmont and Cabot Corporation

Abstract

R&D and technology planning frequently is done via extrapolation of past businesses and technologies, using the core competencies of the existing organizations. Occasionally, this results in disastrous consequences when a given technology is totally displaced by a technology totally foreign to the organization. An everyday example would be the evolution of "image capturing" from etching to painting to printing to wet chemical photography to electronic photography. Each of these approaches uses different skills and competencies and significantly different technical approaches.

When one studies the evolution of intellectual property, a predictive set of technology development lines can be seen, which are predictive and allows strategic, R&D, and personnel planning that stays one step ahead of discontinuities and anticipates them. These patterns of evolution are a key part of the TRIZ "Inventive Problem Solving" process the principles of which have been derived from the study of millions of the world's most inventive patents. This allows an organization to plan much more effectively in terms of budgets, types of technical disciplines needed, and types of customers and potential customers with which to partner.

etc. the mental process changes in a very sadistic, but productive way. When asked to figure out how to make a process release hazardous materials *all* the time and cause loss of human life, and an environmental catastrophe, a group will not only find many more possibilities, but will have fun doing it. This presentation will review several major incidents from this perspective and demonstrate how this technique could have been used proactively to improve the process design and minimize the potential for the disasters that happened.

HISTORY AND BACKGROUND

In the 1950's an insightful Russian inventor and patent examiner named Genrikh Altshuller, after analyzing hundreds of thousands of breakthrough patents, determined that there was a basic inventive algorithm that could solve virtually any significant engineering problem. He also found there were similar and a limited number of generic inventive principles used across all areas of technology. This work led to "TRIZ" (Russian acronym for "Theory of Solving Inventive Problems"). Altshuller and his colleagues continued their work, continuing to develop these algorithms through the 60's and 70's. TRIZ emigrated from the former Soviet Union to the United States, Europe, and Japan in the late 1980's after Perestroika and is now in use by many Fortune 500 companies to solve difficult problems, improve intellectual property filings, and accelerate the pace of internal innovation. Major users include Motorola, Dow Chemical, Unilever, Siemens, Intel, the US Navy, British Petroleum, and Hewlett Packard. AIChE and ASME currently offer public courses in this methodology[5]. The technology has been extended in its application to not only technology applications, but to business problems, human factors and ergonomics, biological systems, and intellectual property filing and circumvention. Over time, the continued analysis of patents and inventions also showed that there repeatable patterns of technological evolution which could be used to not only predict next steps (sometimes very discontinuous) in technological evolution, but also assist with strategic planning and acquisitions.

TRIZ Lines of Evolution and Examples

In the study of the patent literature, it is possible to identify lines and patterns of evolution that occur repeatedly and are seen across a wide range of technologies. Some of the ones Altshuller and his colleagues identified are as follows:

1 *Systems and products become more ideal over time*.

This implies simpler products and systems, functions which don't require a product or system at all, or the addition of useful complexity to an existing system. It can also mean the elimination of a part of a system and having the function which was being accomplished via that part achieved through another or combination of other elements, eliminating the need for a given part. In the TRIZ world, this is commonly referred to as "trimming". Another approach to a system becoming more idea is the identification and use of resources previously unidentified, especially those that might be inexpensive or normally considered detriments to the system.

Consider the simple example of tank trucks used for chemical transport. With DOT regulations and a simple colored and numbered placard on the truck, an emergency responder immediately knows the general nature of the materials in the truck and has a general feel fore how to respond to a spill.

Example illustrating the "trimming" concept as well as the use of existing resources would be the Black and Decker paint stick as well as traveling toothbrush. Both use the hollow space (unused resource) within a handle to replace, in one case a paint pan, and in the other a toothpaste tube. Note that these two examples, illustrating the same inventive principles, are from two totally different industries and were years apart in their commercial introduction. Note also that using the handles as holders for materials also reduces the use of plastic in the handles' manufacture.

2. *Systems and products become more dynamic with time.*

Dynamism can mean many different things, but in this context, it means more responsive to conditions or needs of the user or system at any given point in time or under any condition.

Examples of this line seen in the chemical industry are variable speed pumps and drives, variable process control systems whose set points vary with process or external conditions, systems that change color to indicate a condition, and alert warning levels which increase with severity of weather or process conditions.

In the everyday world of automobiles, we see brakes whose action are proportional to road conditions such as surface moisture, radio volumes that automatically adjust to external noise and car speed, and seats that adjust automatically to the driver's key. Of course, all of these also make the car more ideal from the driver's viewpoint.

3. *Systems oscillate between simplicity and complexity*

This line of evolution describes a phenomenon often seen as products and systems evolve. A simple device or system is invented that serves a new need or serves it better than another product or system. We then begin to add "useful complexity" in the product or system to make the product more valuable (and potentially raise the price). Then, as we add complexity, the ability to use the product or process becomes problematic. Finally someone finds a simplicity breakthrough and the cycle stars again. A simple everyday example of this is the copier. Additional functionality such as reducing and enlarging as well as faxing are now commonplace. Unfortunately many of these systems are now so complex that only the most experienced office assistant knows how to use, and the exercise of making just a few black and white copies becomes a daunting task. These individuals now have simple black and white copiers in their office to avoid the complexity, or as in the first line cited above, do without copies and traditional storage and mailing via Emailing and electronic storage.

In the chemical industry, we can look at

4. *Evolution with Matching and Mismatching Components*

5. *Evolution Toward Micro-level and Increased Use of Fields*

This is an expression of the fact that products and systems evolve along a field evolution line:

a. Mechanical field

b. Thermal and acoustic fields

c. Chemical fields

d. Electronic fields

e. Electromagnetic fields

Examples

Assuming that we have successfully completed the first three steps of this analysis (ideal result, inverted ideal result, exaggerated inverted ideal result), we look at time, space, fields, field conversion, substances and materials, time, and information as resources that can be used to accomplish this extreme negative situation. Let's look at an example of each.

Time as a Resource for Failure

The CCPS Process Safety Beacon in CEP Magazine (January, 2006) presented a summary entitled "Time Sensitive Chemicals", highlighting the fact that many chemicals have shelf life limits and become unstable or reactive with time in storage. Inhibitors in monomer storage systems have time limits as they are consumed. In principal, this is no different than freshness dates for food in a grocery store. In the sense of this methodology, we would ask, "How much time do we need to allow decomposition? How would I MAKE SURE that this amount of time is available?" The answer may lead us to look closer at process storage, inventory, shipping delays, as well as customer inventory time. Some of these segments may be seen by a checklist review, but if we focus on HOW to create the time, we will tend to see more possibilities, such as changes in the customer's inventory (not just our own). We would more appropriately consider the relationship of time and temperature. Slow rail deliveries in an excessively hot summer may result in consumption of an inhibitor. This has been the case in styrene polymerizations within rail cars.

In a previous example presented at the 2007 CCPS meeting[4], the rupture of a hazardous chemical drum shipment was traced back to a long overseas transport time, allowing a corrosive reaction (iron plus bromine) sufficient time to generate enough heat and pressure to go beyond the pressure rating of the drum. The combination of time, extremely low heat capacity (0.1 vs. water's 1.0), and the heat of corrosion contributed to a dangerous release of bromine.

In an ethylene oxide explosion[3], long standing rust (a polymerization catalyst) in a dead spot (*time!)* had accumulated. How can we *make sure* that we can accumulate reactive materials? How would we design a piping system to do this? Then don't do it!

We frequently recognize time as a potential problem area, but we don't often exaggerate its impact as we do with Predictive Failure Analysis©.

Materials as Resources for Failure

We are generally aware of the potential consequences of mixing the wrong materials together, but we sometimes forget that normally "non-hazardous" materials can create danger, if combined in the wrong way. We also recognize in general the hazards of confined spaces. The combination can be deadly. CCPS reported this past August on the death of a worker using a sheet of black plastic to shield a black light to inspect piping. 150 feet away an open nitrogen line put enough nitrogen into this "closed" space, causing asphyxiation. The checklist approach did not consider checking this possibility. Suppose we had used this type of analysis in this case. We would ask the question, "How can be *make sure* there is no breathable air within this space? Are there any materials nearby that might allow this to happen? How would we *make sure* that nitrogen was introduced into this space? Where is there nitrogen?" The source would probably have been identified and eliminated.

In the previously mentioned bromine release case, the material being shipped and the material from which the drum was constructed were material resources for a runaway reaction beyond the obvious corrosion reaction concern.

In another well publicized case, an explosion within a kerosene tank containing a water layer was traced to bacterial action generating methane which was the source, combined with a welding torch, for an explosion. The checklists used did not catch this, but if we had asked the question, "How would I make sure that a flammable gas was generated in the tank? What would be required? Is it possible to generate the materials?"

Substituting steel and titanium in dry and wet chlorine service is a frequent materials mistake.

Information as a Resource for Failure

Since virtually all of our processes measure and respond to information generated by a variety of sensors and instrumentation, it is important to consider there normally "good" resources and how they can be used in a negative way. In the now famous Texas City fire and explosion, a column level indicator showed level decreasing at the same time the column was overflowing. This occurred because the instrument was submerged and under those conditions, density was being measured, not level. As the temperature rose during column heat up and recirculation, the level "indication" (not the actual level) actually decreased whereas the column was actually overflowing. A back up level indicator failed to respond. Consider how things might be different, if during a safety review we had used this Predictive Failure Analysis© process instead of the normal HAZOP approach. We would have asked, "How could I *make sure* that we were *never* aware of that the level was? That the level indication was *never* correct?" It is probable that someone, during the review, would have suggested that, if the fluid level in the column was higher than the maximum level indication, the level indication would be meaningless and potentially dangerous. It might also have been suggested to find a mechanism to determine the status of any backup instrumentation.

The general question to ask is "How Would I *make sure* that my process instrumentation readings are incorrect *at all times?*"

Unavailable or unknown information causes the same potential for disaster. If explosion potential or explosive ranges are simply not known, the most well intentioned HAZOP team will not take precautions.[3]

Energy as a Resource for Failure

This is an area which we normally consider. We usually think about obvious and visible energy sources such as furnaces, steam, heat transfer fluids, hot water, gas, and electrical sources. In some cases, however, we miss the energy produced by internal processes and the accumulation of energy in ways and places not expected. In a 2005 acetylene cylinder facility explosion in Perth Amboy at Acetylene Services[5], a failed check valve allowed a flow of acetylene backward from a calcium carbide/water reactor into the water supply line and ultimately into a closed space which contained a space heater. Now consider if we had asked, "How can we *make sure* that acetylene flows backwards? How could we make sure that, if this happened, it would concentrate in a confined space? How would I *make sure* that, if this happened, there would be a source of ignition? Why not put a space heater in the space? This was well intentioned to keep operating personal warm in a water area that was never expected to contain acetylene.

The approximately 200 dust explosions in the last decade in the US have been analyzed by the Chemical Safety Board[5] and two key aspects have been pointed out. First, in addition to the traditional fire triangle, it is required to have dust concentrations be elevated and in a confined space. These are mechanisms for concentrating the energy of dusts. Then something disturbs the settled dust is dispersed and ignites. In analyzing our processes, we can ask the simple question (in addition to all the traditional ones), "How can we MAKE SURE that dust concentrates?" Inadequate ventilation? Cleaning? How can we MAKE SURE that we have a confined space for dusts to accumulate? How can we MAKE SURE we have both? How can we *make sure* that there is *always* an ignition source present?

In the 2005 Praxair gas cylinder storage explosion, energy sources came from the sun (it was an unusually hot 97 degrees in late spring) and hot asphalt, the combination of which was sufficient to overwhelm the pressure relief valves in the cylinders. We often think about the sun and asphalt as things that are just "there"—part of the facilities and background, but from a TRIZ analysis perspective, they are resources, and they can be energy sources. If we had asked the question, "How can we *make sure* that there is sufficient heat to heat the cylinders sufficiently to release gas?" it is possible that we might have thought about shielding the storage area or painting the surface with a reflective coating. We might have calculated more thoroughly the heat transfer and relief valve limitations of the system.

Fields as a Source of Failure

Basic fields include thermal, chemical, magnetic, electronic, acoustic, electromagnetic, and biological. The potential impact of recognized fields is normally considered in our check listing, but frequently field conversion is not (see next section). Unrecognized fields are the concern. The simple way to do this is to take this list and ask "is the field present?" If so, what are the consequences? Next, ask how could I generate such a field if it is not recognized as present already? Then, what are the consequences of this field's presence? We can use Table 1 as a template for doing this.

Table 1

Field Source Worksheet

Field	**Present? (Y/N)**	**If No, how to produce?**	**Potential Consequences**
Mechanical			
Thermal			
Chemical			
Electronic			
Electromagnetic			
Acoustical			
Pressure			
Acoustical			
Human Behavior			

Note that we have included "human behavior" as one of these fields. We have many automated chemical plants that we assume are not impacted by human behavior, but as we see frequently (and was the case in the PB fire and explosion) that humans can turn off alarms and ignore them. The question to be asked here is how could we *make sure* that an alarm or warning was ignored? How would we make that happen *all the time?* What resources are required? Are they present? Can they be created with the resources present? In the case of human behavior, we can think in terms of conflicting data presentations, parallel job responsibilities, personal experience and past history of false alarms, etc.

Forcing ourselves to create these various fields is a more comprehensive approach than simply asking ourselves whether they are present or not. Like an ignition source in a fire, we seldom know what it was, but it's always there. This is not to imply that all these fields and sources of energy are everywhere at all time, but to say that we frequently get blindsided by our narrow vision of the process environment.

Field Conversion as a Source of Failure

There is almost an endless list of field conversions. Chemical fields (reactions) generate heat, electrical fields generate both thermal fields and magnetic fields, and magnetic fields generate electrical fields. Chemical fields (reactions) are produced by electrical fields, both producing thermal fields. Mechanical fields (friction, rotating equipment, etc.) all produce friction, heat, and static. Static (an electrical field) is an ignition source for gases and liquid vapors. We know most of these facts. In the use of TRIZ in conventional problem solving, we usually are looking for field conversion in a positive sense, i.e. we are looking for potential resources for problem solving. In failure prediction, we are looking at the opposite: the potential consequences of these field conversions from a negative standpoint. What new fields can be formed and how could these new fields cause harm and damage?

As an example, the corrosion of materials is normally considered a nuisance or quality problem, and a factor that we use to choose materials of construction. However, corrosion is a chemical reaction and as such, has thermal effects. In the aforementioned bromine drum rupture, the concern was about the drums leaking a hazardous, corrosive material and causing personal harm and injury. The drums were properly designed (i.e. wall thickness) to withstand the corrosion rate over the intended time. However, they were not pressure vessels and were not designed to withstand the pressure generated from the liquid turning into a gas from the heat of corrosion. Aggravating this situation was the extraordinarily low heat capacity of liquid bromine (0.1 vs. water's 1.0). The combination of these two things in combination with the drums being overfilled, caused the hazardous has release. Again, if we had asked ourselves, "How could we *make sure* the drums ruptured in transit?" we would have said overfill them, let them sit in a hot Houston ship channel, and not properly calculate the heat balance on the drum during transport. In another case, corrosion of steel in an evaporator waiting to cool prior to vessel entry removed sufficient oxygen to produce a less than breathable atmosphere several hours after a vessel entry check had been performed, indicating safe entry. Several hours later, the maintenance worker entered the vessel without independent air supply and was immediately overcome and died before rescue was possible. This is an example of time, energy, and field conversion all contributing to a fatality.

In the BP raffinator tower explosion mentioned previously, the filled tower's thermal field applied unknowingly to recirculating fluid was changing the density of the fluid and affecting the mechanical force on the level indicator. Basic laws of physics tell us that no field can be used or generated without their being one or more accompanying fields. One of our responsibilities is to make sure that we are aware of all the potential byproduct fields a field, process, system, operation, or movement can produce and then methodically step through the consequences of these additional fields on our processes, equipment, instrumentation, and control. A simple table for doing this is shown in Table 2:

Table 2

Field Conversion Worksheet

Original Field	Byproduct Field 1	Byproduct Field 2	Potential Consequences
Mechanical			
Thermal			
Chemical			
Electronic			
Electromagnetic			
Acoustical			
Pressure			
Optical			
Human Behavior			

There are TRIZ (and other) handbooks[1,2] summarizing these field conversions, some of which are unknown in a practical everyday sense. Usually, in traditional TRIZ problem solving, we are trying to identify fields that can be cleverly used for solving a problem; in this case we are looking for by-product fields in a negative sense.

Some of these field conversion effects can be subtle and difficult to recognize and that is why it is critical to exaggerate our thinking in this process. Compressed air is frequently dried and purified through the use of molecular sieves. Nitrogen is preferentially absorbed and an enriched stream of oxygen is produced initially[3.] If an enriched oxygen stream is a hazard, this needs to be considered. This is a case of a chemical field (adsorption) producing a change in substance concentration, which can be considered as an enriched chemical field in terms of oxygen.

Note that human behavior has been included. This could include stress, long work hours, personal or family concerns, or supervisory/employee friction. The "byproducts" may not be traditional physical items, but certainly could include morale, attention span, and concentration. All of these have the potential to indirectly cause harm in a process through inattention, incorrect data collection and response, and incorrect decision making.

Space as a Resource for Failure

Insufficient space (confined space) is frequently cited as an accident cause. Using Predictive Failure Analysis© will not prevent doing things we already know are wrong or are not in compliance with known safety procedures, such as not using proper confined space procedures. However, space is a much broader term than this. It can be space around a system as well as space within a system (i.e. voids). It can be isolated spaces or dead spots in processing systems that can "hold up" materials for long periods of time (see previous section on time) and allow chemical reactions to proceed that are unanticipated.

A worker trying to clean tank residue was overcome by xylene fumes after a tank cleaning. Solids have voids (a space resource!) which can trap hazardous materials. Table 3 is an outline for use in evaluating space resources as resources for failures.

Table 3

Space as a Hazardous Resource

Type of Space Resource	Is This Space Available?	Potential Consequences	How to Create?
On the surface			
External, surrounding space			
Space inside			
Porous space inside			

Conclusions and Summary

Checklists and diagrams are useful and necessary, but they can be incomplete and the process for completing them can be boring when what is needed for safety thinking is stimulation and completeness. Deliberately trying to make a process, system, or piece of equipment fail produces possibilities that check listing by whatever mechanism, does not.

References

1. Mann, D. et.al. Hands On Systematic Innovation, Creax Press, 2003, pp.236-237, 345-354
2. Salamatov, Y. TRIZ: The Right Solution at the Right Time, Insytec, 1999, pp 245-253
3. Kletz, T. What Went Wrong?: Case Histories of Process Plant Disasters, 4th edition. Gulf Publishing, Houston, TX, 1998
4. Center for Chemical Process Safety Proceedings, 2007 www.aiche.org/ccps
5. Chemical Safety Board Safety Video, March 2007

Resources

1. Mann, D. et.al. Matrix 2003, Creax Press, 2003
2. Hipple, Jack. "Using TRIZ in Reverse to Analyze Failures" Chemical Engineering Progress, May 2005, pp 48-51
3. Altshuller Institute for TRIZ Studies, Worcester, MA: www.aitriz.org
4. Innovation-TRIZ web site: www.innovation-triz.com
5. American Society of Mechanical Engineers web site: www.asme.org
6. On line TRIZ Journal: www.triz-journal.com

367a Lattice Boltzmann Modeling and the Behavior of Droplets

James M. Parker and Goran Jovanovic. Chemical Engineering, Oregon State University, 102 Gleeson Hall, Corvallis, OR 97331

The lattice Boltzmann modeling approach holds the promise of accurate simulations of complex multiphase fluid systems. As new simulation methods based on lattice Boltzmann modeling begin to be used in the research and design of new processes in the field of chemical engineering, benchmarking the performance of these methods becomes necessary and critical. In this work we present a newly developed lattice Boltzmann method for multicomponent-multiphase flow capable of accurately modeling the multiphase mass transfer, chemical equilibrium, momentum transfer, and surface tension effects for various binary systems. To further validate the multiphase lattice Boltzmann model, the behavior of droplets and bubbles is simulated over a range of viscosities, densities, and surface tensions for comparison with experimental data. Good agreement is found between experimental data and simulation based on lattice Boltzmann modeling approach.

367b Development and Application of a Conservative Algorithm for An Adaptive Change of Resolution In Mixed Atomistic / Coarse-Grained Multiscale Simulations

Andreas Heyden and Jong Hyuk Park. Department of Chemical Engineering, University of South Carolina, 301 South Main Street, Columbia, SC 29208

Understanding complex materials often requires investigating multiple, tightly coupled time and length scales. Neither atomistic nor coarse-grained simulations are often able to adequately capture all the relevant scales. To combine the efficiency of coarse-grained models with the accuracy of atomistic models for systems that require atomistic resolution only locally, for example at a reactive group, mixed-resolution models have been developed. These models use a coarse-grained description for the part of the system distant from an active site and atomistic description for the active site and its direct environment. Since the active zone may diffuse during a molecular simulation, the simulation algorithm needs to permit an on-the-fly reclassication of atoms as they transition between the high- and low-resolution regimes. Recently, we derived a Hamiltonian and presented a simulation protocol for mixed-resolution systems that allows for such a change in resolution of selected groups of atoms during a molecular dynamics simulation [1]. A microcanonical simulation protocol conserves energy and angular and linear momentum. In this paper, we briefly present our novel algorithm and compare it to previous mixed resolution algorithms [2,3]. Subsequently, we present simulation results for the structure and diffusion of tetrahedral model molecules and a water molecule in hexane. These model simulations permit identifing challenges in mixed resolution simulations and how they can be overcome.

[1] Heyden, A.; Truhlar, D. G., Conservative algorithm for an adaptive change of resolution in mixed atomistic/coarse-grained multiscale simulations. Journal of Chemical Theory and Computation 2008, 4, (2), 217-221.

[2] Praprotnik, M.; Delle Site, L.; Kremer, K., Adaptive resolution molecular-dynamics simulation: Changing the degrees of freedom on the fly. Journal of Chemical Physics 2005, 123, (22), 224106.

[3] Ensing, B.; Nielsen, S. O.; Moore, P. B.; Klein, M. L.; Parrinello, M., Energy conservation in adaptive hybrid atomistic/coarse-grain molecular dynamics. Journal of Chemical Theory and Computation 2007, 3, (3), 1100-1105.

367c Non-Linear Modeling of the Formation of Activated Carbons: Stochastic Approach

A. Argoti[1], L. T. Fan[1], W. P. Walawender[1], and S. T. Chou[2]. (1) Department of Chemical Engineering, Kansas State University, 1005 Durland Hall, Manhattan, KS 66506, (2) Department of Finance and Banking, Kun Shan University, No. 949, Da-Wan Rd., Yung-Kang City, Taiwan

Activated carbons (ACs) constitute a type of carbon adsorbents, which have long been known to be inordinately effective for the purification of gases and liquids and the separation of mixtures of gases or liquids. The formation of ACs entails the modification of the original internal surfaces of carbonaceous substrates, e.g., coal or biomass, which can be carried out by a variety of chemical and/or physical methods, thereby augmenting the carbonaceous substrates' adsorbing capacities. The formation of ACs tends to proceed randomly or stochastically because of the discrete and mesoscopic nature of the carbonaceous substrates; the random encounters between the activation agent and the carbon on the carbonaceous substrates' internal surfaces; the intricate morphology, or structure, of the carbonaceous substrates' internal surfaces; and the incessant variation of the process with time. Thus, it is highly desirable that the formation of ACs from carbonaceous substrates be analyzed, modeled, and simulated in light of a stochastic paradigm. Herein, the formation of ACs is modeled as a stochastic process, specifically, as a pure-birth process whose intensity of transition is non-linear. This non-linear intensity of transition aims at incorporating the effects of morphological, i.e., structural, characteristics of the carbonaceous substrates' internal surfaces on the formation of ACs. The resultant model gives rise to the process' master equation whose solution renders it possible to compute the mean and higher moments about this mean. These higher moments, of which the variance is of special importance, are collectively a manifestation of the process' inherent fluctuations. Moreover, the master equation is simulated via the Monte Carlo method at the initial stage of the formation of ACs. The results of modeling are validated by comparing them with the available experimental data.

367d Assesment of Transient Responses In Muscle by Comparison of Damkohler Numbers and Consideration of Stretch Response

Richard L. Long, New Mexico State University, Chemical Engineering Department, P.O. Box 30001, MSC 3805, Las Cruces, NM 88003-8001

Fast striated muscle dynamics has been studied by many different authors. The sliding filament model was first introduced by Huxley in 1957. A comprehensive symposium at Cold Springs Harbor, 37 (1973) established many details of muscle response. Long(1973) and later Long and McIntire(1981)formulated a two compartment model of muscle response based on a nonequilibrium thermodynamic approach combined with the kinetics of Lymn and Taylor (1971) and Ashley and Moiscue (1972). This was the first integrated quantitative synthesis of the kinetics and mechanics of muscle action. It overcame the limitation noted by H.E.Huxley in the Cold Spring Harbor Symposium that complete data was not available on a single muscle type in a single specie by making the hypothesis that the same reactions would have roughly similarly rate and equilibrium constants across species in the same muscle types. Long and Phoonsiri(2006)revised the earlier model to employ a crossbridge balance approach to linking the solution chemistry to the protein lattice mechanics. It is a transition state type model which is distinct from the Huxley strain models, but which is similar to some of the so-called "thermal ratchet" models. One of the results of this model is that relative reaction rates of release of ADP from a high strain and low strain state can be computed from physiolgically macroscale mechanical experiment responses. The model gives quantitative results that have since been partially confirmed by the recent qualitative experiments reported by Nabir et al (2007). Mechanical responses depend on Ca+2 transport which depends on diffusion from the sarcoplasmic reticulum as well as reaction. Estimates of the Damkohler number and stretch experiments can be used to look in more detail at this issue. At present it appears that most of the ADP is released post-power stroke but some is release pre-power stroke.

367e Development of a Hierarchical Computational Architecture for Automated Nanomaterial Design

Jie Xiao and **Yinlun Huang**. Department of Chemical Engineering and Materials Science, Wayne State University, 5050 Anthony Wayne Drive, Detroit, MI 48202

Molecular modeling and simulation has been recognized as an effective tool for improving more insightful and comprehensive understanding of nanomaterial structures, properties, which is critical for nanomaterial development. However, computational efficiency has been always a major challenge in molecular simulation, due to the existence of a huge design space which requires a thorough investigation to achieve design optimality. Tremendous efforts have been made on the development of parallel computing focused techniques.[1-3] The developed techniques are very helpful for reducing the computational time of individual in silico experiments. However, nanomaterial development is a much more sophisticated task, involving the design and management of massive runs of simulations and numerous types of subtasks in multiple computational stages. Thus, an effective and efficient nanomaterial development must rely on the appropriate design of a computational architecture as well.

In this work, we will introduce a generic task-distributed hierarchical computational architecture for nanomaterial design, which can be readily constructed on a usual cluster system. In this architecture, the upper layer, designed using the master node of a cluster machine, is for managing the entire material analysis and design process. The lower layer consists of two computational zones, one for material sample development, and the other for material property testing. In each design iteration, nanomaterial formulation candidates are transmitted down to the sample development zone in the lower layer to generate samples by each compute node. The samples generated from this zone will be transmitted back to the master node in the upper layer, and the master node will then manage the sample transmission to the sample testing zone in the lower layer. The test results will be transmitted up to the master node. The material formulation-coating property correlation will be analyzed and the most preferable formulation will be identified for that design iteration. Based on the preset design objective, the Optimal Material Design Algorithm resided in the master node will determine if the next design iteration will be needed or not. If yes, then that algorithm will determine how to modify the formulation and subsequently a new iteration starts; otherwise, the identified material design will be considered the design solution. The developed computational architecture, together with the algorithms for hierarchical computations, can realize design automation, with significant computational efficiency. The introduced computational architecture has been successfully implemented in a Beowulf Cluster system (10 Xeon dual processor nodes @ 2.66 GHz). The efficacy of the architecture will be fully demonstrated through a case study, which will identify an optimal super-scratch-resistant nanopaint formulation.

References:

1. Xiao SP, Ni J, Wang SW. The bridging domain multiscale method and its high performance computing implementation. Journal of Computational and Theoretical Nanoscience. in press, 2008.

2. Ogata S, Lidorikis E, Shimojo F, Nakano A, Vashishta P, Kalia RK. Hybrid finite-element/molecular dynamics/electronic density functional approach to materials simulations on parallel computers. Computer Physics Communications. 2001; 138: 143-154.

3. Liu Z, Chen L, Paghavan P, Du Q, Sofo JO, Langer SA, Wolverton C. An integrated framework for multi-scale materials simulation and design. Journal of Computer Aided Materials Design. 2004; 11: 183-199.

367f "Quorum Quenching" on Pseudomonas Aeruginosa: Optimization of the Dose of 4-Nitropyridine-N-Oxide for the Inhibition of Biofilms Based on Stochastic Kinetics

Virginia Covo[1], Luis Miguel Medina[1], Martha Josefina Vives[2], and Andres Fernando Gonzalez Barrios[1]. (1) Chemical Engineering, Centro de Diseño de Productos y Procesos, Universidad de los Andes, Carrera 1a este No. 19 A 40 7° piso, Bogotá, Colombia, (2) Microbiology, Centro de investigación en Microbiología (CIMIC), Universidad de los Andes, Cra 1E No 18A-10 . Bloque J, Bogotá, Colombia

Discovering that the expression of some phenotypes is regulated by quorum sensing (QS) has broadened the knowledge on the mechanism that several bacteria use for the environmental response for their benefit. Such is the case of *Pseudomonas aeruginosa*, bacteria that are responsible for the development of biofilms as a result of this communication system. Biofilms are sessile communities encountered in many environments and are responsible for the resistance towards antibiotics. It has reported that there exist certain compounds that are able to interrupt this communication system (quorum quenching), and it results in the expression or suppression of certain genes.

This study centers its attention on *4-nitropyridine-n-oxide* (4-NPO) that has already been proven to interrupt *P. aeruginosa* quorum sensing system (Rasmussen et al. 2005). Simulated annealing algorithm was used in order to find a distribution of optimum concentrations of the quencher considering its stochastic behavior modeled by the Gillespie algorithm and minimizing LasR/3O-C12-HSL complex. Distribution of the optimum concentrations was found to behave normally as expected based on the results regarding the distribution of the population at different time points.

On the other hand, optimum concentrations obtained were corroborated by evaluating *P. aeruginosa* biofilm formation capability which has been reported to be significantly controlled by the QS system (Hentzer et al. 2002). Mean optimal concentration was predicted to be 2500 μM while biofilm formation was reduced 75% when adding 2000μM 4-NPO. This approach demonstrates the useful role for stochastic model regarding dose optimization.

372a Synthesis of P-Xylene Crystallization Processes Using Gdp Models

Ricardo M. Lima, Department of Chemical Engineering, Carnegie Mellon University, 5000 Forbes Avenue, Pittsburgh, PA 15213 and Ignacio E. Grossmann, Center for Advanced Process Decision-making, Dept of Chemical Engineering, Carnegie Mellon University, 5000 Forbes Avenue, Pittsburgh, PA 15213.

This paper addresses the synthesis and optimization of crystallization processes for p-xylene recovery from streams of mixed xylenes. Crystallization based processes exploit the large freezing point difference between p-xylene and the remaining components in the mixture. Typical processes consist of one or two crystallization stages operating at different temperature levels, liquid/solid separation devices using different types of centrifuges (imposed by operation ranges of feed solid concentrations and of feed temperature), melting stages with slurry drums and heat exchangers, and a final stage of purification involving centrifuges with wash streams. The units are organized by stages in series and/or parallel configurations giving rise to complex flowsheets. Lima and Grossmann (2008) proposed a novel superstructure, a mixed integer nonlinear programming (MINLP) model, and a decomposition strategy to cope with the complexity of the model. The decomposition strategy developed by these authors allowed them to propose and study alternative flowsheets for the p-xylene separation from streams with high concentration of p-xylene. The size and complexity of their superstructure and respective model prevented the application of global optimization deterministic algorithms to solve the MINLP problems at each level of the decomposition approach proposed.

In this work we propose a generalized disjunctive programming (GDP) model to describe the superstructure proposed by Lima and Grossmann (2008). A specific feature of this superstrucutre that we try to explore is the existence of a network of crystallizers and three sets of centrifuges operating in parallel. The proposed GDP model involves disjunctions associated with the existence or not of each crystallizer and centrifuge. However, in each term of each disjunction, the equations of the downstream splitters and upstream mixers of each unit are also considered. Therefore, whenever a crystallizer is not selected the corresponding splitter and mixer are also not considered. Although, these sets of units in parallel may resemble trays of distillation columns, these sets represent a distinct case, since there are no connections between the units in the same set. A logic-based outer-approximation algorithm is proposed to solve the GDP problem, where the linearizations for the terms of the disjunctions are provided from the solutions of an aggregated model. In this aggregated model, the sets of units in parallel are replaced by a single equivalent unit, decreasing the size and complexity of the associated MINLP problems. Therefore, the solution of the full MINLP problem or the solution of a set-covering problem (Turkay and Grossmann, 1996) to define the sub-structures to provide the linearizations is not necessary. The main idea of this approach is to avoid the solution of large MINLP models where all the units in parallel are present, as it was implemented in the decomposition approach proposed by Lima and Grossmann (2008). The solution of smaller nonlinear programming (NLP) problems at each iteration of the logic-based approach should provide additional solutions in terms of flowsheets and operating conditions.

Optimum flowsheets for a wide range of p-xylene concentrations in the feed stream are presented and compared with flowsheets reported in the literature. In addition, numerical and algorithm performance indicators for comparing the two approaches (two-level decomposition and GDP) are also presented.

References

Lima, R. M. and Grossmann, I. E., (2008), Optimal synthesis of p-xylene separation processes based on crystallization technology, AIChE Journal, submitted.

Turkay, M. and Grossmann I.E., (1996), Logic-based MINLP algorithms for the optimal synthesis of process networks, Computers and Chemical Engineering, 20(8), 959-978.

372b A Graphical Approach to Improving Gasification

David Glasser and Diane Hildebrandt. Centre of Materials and Process Synthesis (COMPS), University of the Witwatersrand, Private Bag 3, WITS 2050, Johannesburg, South Africa

Thermodynamics can be used to describe any process, or system of processes. Of particular interest are the properties of Enthalpy and Gibbs Free Energy. By using these two thermodynamic characteristics together as vectors on a diagram of Free Energy against Enthalpy, it becomes possible to develop optimal process flow sheets that combine the thermodynamics of chemical reactions and the dynamics of physical operations on a single diagram. This concept is also explored in a paper by Baraka Celesin Sempuga (Classification of Chemical Processes: A Graphical Approach using the Heat Engine Concept, Abstract submitted to AICHE in May 2008)

By selecting appropriate operating conditions for a process; such as temperature, pressure and reaction extent; it is possible to find on the G-H diagram the combination of reactions and operations that will allow the process, in its entirety, to be run reversibly, which is to say its change in Entropy will be zero. This would thus allow a flow sheet to be developed that would represent the process performing at its highest efficiency.

Such a flow sheet can provide an important basis for allowing new processes to be developed to minimize its impact on the environment and maximize its productivity and also allow provide a means of identifying the inefficiencies of existing processes and providing a basis for improvement.

The operation of a Gasifier is considered to illustrate the use of the G-H diagram. The primary reaction is the gasification of coal. To provide the heat and work that gasification requires coal is also burned in oxygen. The Water-Gas-Shift reaction is also considered to allow the ratio of Carbon Monoxide and Hydrogen to be adjusted to desired specification. The phase change of liquid water to its vapour phase is also considered which allows for the use of liquid water as a starting feedstock.

Using these ideas one is able to show how to improve the carbon efficiency of a gasification process.

372c A Versatile Simulator for the Analysis and Design of Thermal Batteries

Nir Haimovich[1], Dario R. Dekel[2], and Simon Brandon[1]. (1) Chemical Engineering, Technion, Haifa, 32000, Israel, (2) CellEra, Caesarea Business and Industrial Park, Caesarea, Israel

Thermal Batteries are primary one-shoot high-power energy sources which, due to the molten salt used as an electrolyte within their galvanic cells, are operated at elevated temperatures (300-700 °C). Thanks to the use of a high-temperature molten salt (e.g., LiCl-KCl) rather than an ionic solution, large electrical currents may be drawn from these devices. However, the high operational temperatures require an intelligent management of heat transfer in the system, as electrical current cut-off will follow salt (electrolyte) solidification thereby determining the operation time in which current may be supplied. It is often due to heat transfer limitations that thermal batteries, consisting of serial-connected and thermally-insulated cells, usually function as short time power sources for different autonomous systems. Thermal design is therefore of utmost importance, involving optimization of the geometrical arrangement, cell structure and insulation materials. Meeting performance requirements, while being constrained by size and ambient atmosphere specifications, becomes a time and money consuming process using an experimental trial and error approach. On the other hand, the design process may be significantly enhanced using battery performance simulations based on a reliable mathematical model.

In this contribution, we present our in-house developed thermal simulator, *TThermBat*. The simulator is based on a transient mathematical model of a two-dimensional axis-symmetric geometry, to be used as a versatile and robust simulation tool for thermal battery design and analysis. This thermal model enables a portable and detailed simulation from the level of each cell sub-component (e.g., cathode disc) up to that of the entire product, considering various heat transfer phenomena, including conduction, phase-change (electrolyte salt solidification), heat of reactions and Joule-Heating; heat loss via convection and radiation to an ambient temperature is applied at the model system's external boundaries. In addition, the heat transfer model is supported by an overall (albeit simplified) mass balance, involving the electrical current drawn from the battery and an integral mass transfer resistance coefficient for each cell. Model calculations have been successfully validated through comparison with well-established analytically and numerically solved test-cases, as well as with experiments using a "dummy" (1-component inert-cell) battery.

A leading characteristic of our simulator is its versatility both in terms of geometry and in terms of material systems. For example, the virtual geometrical model is constructed by a MATLAB Graphic User Interface (GUI), allowing the user to define the battery's inner and outer structure with great flexibility; this includes defining the cell ingredients, the number of electrochemical-cells, and details of the insulation layers. The GUI also supports an extensive and upgradable material property library, in which each material is characterized by a set of temperature (and, in some cases, concentration) dependent physical properties, to be applied in each domain of the geometrical model. Model calculations are numerically performed and analyzed via the Finite Element Method (FEM). Finally, the GUI enables various post-processing options: temperature colormaps, dynamic animations and temperature-time plots with their corresponding temperature-mole fraction paths over the binary salt phase-diagram (when relevant). In addition, an approximate electrical cell analysis, based on the narrowing cross-section for electrical charge transfer due to the salt solidification is enabled.

Results of the *TThermBat* simulator have been achieved for both a single cell and a full scale thermal battery, uncovering a number system-level effects. Future plans involve pursuing the development of a new simulator, *EThermBat*, which is based on a rigorous differential mass transfer analysis for the prediction of electrical performance of thermal batteries at both the single cell and the system level.

372d Systematic Screening and Design Evolution for Catalytic Gas Conversion Processes

Patrick Linke, Chemical Engineering Department, Texas A&M University at Qatar, P.O.Box 23874, Doha, Qatar, Daniel Montolio-Rodriguez, Process & Information Systems Engineering, University of Surrey, GU2 7XH, Guildford, United Kingdom, and David Linke, Catalyst Discovery and Reaction Engineering, Leibniz Institute for Catalysis, Richard-Willstaetter-Str. 12, Berlin, 12489, Germany.

The past two decades have seen increased research efforts into how optimal process designs can be systematically achieved for a given chemistry using mathematical optimization of process network superstructures (1). Such methods have recently been highlighted as key technologies that promise to deliver major improvements in process efficiencies (2). However, the methods are rarely applied in industry, mainly as a result of their inability to handle practical process constraints, to robustly process complex kinetic models and to support the design-decision making process. We present the development of advanced optimal process synthesis framework that enables the identification of optimal reaction-separation process structures. The framework enables the exploration of design trends and key design features that maximize process performance in terms of profit for catalytic gas phase reaction systems, taking into account various design options for the reactor, the separation system and the energy systems.

The framework introduces the concepts of design screening and evolution and permits the design engineer to explore the system's performance and complexity trade-offs. In the screening stage, the framework allows the screening of vast numbers of potential design candidates to identify optimal and near-optimal conceptual process design candidates. In the evolution stage, these promising design candidates can be explored further in two directions. In terms of design complexity, the relationship between the design features identified in the screening stage and the performance can be established. Furthermore, detailed reaction process models are employed to explore the potential effects associated with non-ideal behavior of the system. Both design stages employ tailored superstructure formulations that are searched for optimal design options using stochastic techniques in the form of Tabu Search.

The framework has been developed by building upon our previous work (1,4,5). We have developed process superstructure representations specifically tailored to heterogeneously catalyzed gas-phase reacting systems. The superstructures account for detailed representations of the reaction system through combinations of generic units that embed options related to mixing, temperature policies, the mass of catalyst present, as well as constraints related to energy management and composition limits. The separation systems are represented in aggregated form. This is achieved through cost models that are developed using separation systems synthesis models and capture the lowest cost separations possible for given stream compositions and flow rates. The model aggregation allows the decoupling of separation synthesis calculations from superstructure optimization and keeps the problem numerically solvable for problems of industrial complexity. Energy targeting methods are employed to ensure designs require minimum amounts of energy.

The process synthesis is performed in two stages to generate maximum design insight and understanding. Screening stage: the performances of conventional designs are obtained as base cases and the performance limit that can be attained for the system with an innovative process design is obtained from superstructure optimization allowing all possible interactions between feed streams and recycle streams with multiple reaction zones each of which can exhibit different mixing, heat management and catalyst mass. Evolution stage: The designs from step 1 are analyzed and key design features are identified for which targeted superstructures are optimized to study the impact of individual design features as well as the effect of non-ideal behavior of the reaction system.

Applications to a number of industrially relevant reaction systems will be presented, including the partial oxidation of ethane to acetic acid (3) and the production of styrene. The results will show how the framework has produced innovative designs with significantly improved performances over their conventional counterparts.

References

1. P Linke, AC Kokossis, AIChE J 49(2003) 1451.

2. C Tsoka, WR Johns, P Linke, A Kokossis. Green Chem 8(2004), 401.

3. D Linke, D Wolf, S Zeiss, U Dingerdissen, M Baerns, J Catal 205(2002) 32.

4. V Ashley, P Linke. Chem Eng Res Des 82(2004), 952.

5. P Linke, AC Kokossis. Comp Chem Eng 27(2003), 733.

372e Global Optimization of Water Management Problems In Process Plants

Debora C. Faria, Miguel J. Bagajewicz, and David A. Puckett. Chemical, Biological and Materials Engineering, University of Oklahoma, 100 East Boyd Street, T-133, Norman, OK 73019-0628

Over the last few decades, optimization of water network systems in process plants has been analyzed using several simplifying assumptions. Among them one can highlighted the assumptions in which the costs (both capital and operating) associated with the regeneration processes are only a function of flowrate; and, others where the efficiency of the these processes (in some instances given as fixed removal efficiency, and others as fixed outlet concentrations) are independent of operating conditions such as flowrates and concentrations. In reality, these assumptions imply that for the same amount of flowrate treated, capital and operation costs are the same regardless of the concentration (and/or contaminant mass load) of the streams. In other words, this means that a treatment of a diluted stream costs the same as a concentrated one. Moreover, it is assumed these regeneration processes can reach the same efficiency (in terms of fixed removal rate or fixed outlet concentration) for both diluted and concentrated streams.

Most of these simplifying assumptions have been made with the objective of reducing the mathematical difficulties of handling non-convex non-linear models. The optimization of water systems including these assumptions is already a difficulty problem to be solved due to the existence of bilinear terms arisen from the contaminants balances.

We propose a MINLP model to solve water management problems using more elaborated mathematical description of the regeneration processes. To solve this model, a new global optimization methodology that we developed, which is able to handle both bilinear terms and the non-linear function arisen from the more elaborated models for the regeneration processes, is used.

372f Incorporating Inherent Safety Principles In Design Via Property Design Platform

Fadwa T. Eljack, Department of Chemical Engineering, Qatar University, P.O. Box 2713, Doha, Qatar and Mario Richard Eden, Department of Chemical Engineering, Auburn University, 222 Ross Hall, Auburn, AL 36849-5127.

The word safety in design signals accident prevention and protective measures. On a larger scale, safety in processes involves the risk of operating process units and operating conditions, supplied and generated substances as well as a number of other parameters that would affect the safety of people, the environment or the surrounding property (Suardin et al., 2006).

Traditionally there are safety preventive/protective measures, generally applied after implementation of process plant design; and it has been reported to account for 15-30% of capital cost in the petroleum and gas industry (Palaniappan, 2004). Such methods are not all encompassing and hence, failure to identify risks, and according the inability to address or "control" these risks, can result in massive costs, both human and economic. To minimize the risk of hazardous conditions, Kletz (1991) introduced four inherent safer design concepts: minimize, substitute, moderate and simplify. Inherent safer design looks at involving safer concepts at the pre-design stage in order to identify a truly optimal design. Incorporating Kletz's safety design issues into a design algorithm changes the dynamic of what is truly an optimal design. Typically in process design the objective is to identify most favorable operating conditions that maximize productivity. However, considering the safety of those operating conditions (Kletz's second design concept) might alter the optimality of the proposed design.

The objective of the work here is to use the property platform as means of incorporating process safety in the preliminary stages of design. Current application of inherent methods exists, however according to Hendershot (2006), there is still a need to develop specific tools and techniques for the application of inherently safer design. The literature clearly shows that the lack of information in the preliminary phases of design complicates the inclusion of safety in traditional design approaches. Recently, the Property Integration design framework was introduced as a targeting methodology that identifies optimum process performance ahead of design. The clustering method defines and solves the design problem within a property framework. The key feature of this method is that it transforms all of the constraints placed on the design problem into properties, and it uses it as way to identify the feasible solution region. At the same time, principles of inherent safety such as minimization and limiting effects refer to changing the initial conditions/constraints of the design problem. So it would make sense to include such safety principles into the property design methodology.

The work will outline a robust property-based process design methodology using recently developed property clustering technique that will consider the needs of operating process units and the safety risk within the formulation and solution steps. Moreover, the methodology will develop a systematic scheme to transform safety hazards associated with the various process units and streams in terms of properties or functionalities. Safety risks or process hazards will be utilized as another constraint on the design problem. The translated constraints placed on the design allow for the identification of process unit property targets ahead of design. This is a systematic method for placing bounds on the search space without loosing accuracy; hence the approach will be able to enumerate all design alternatives. Thus the presented paper will propose a framework for the incorporation of safety in the preliminary stages of design by infusing Inherent Safety Principles into the Property Clustering Platform.

Keywords: process design, inherent safety, property design

382a Optimal Corticosteroid Treatment for Modulating the Dynamics of An in Vivo Human Response to Endotoxin

Q. Yang, Chemical and Biochemical Engineering, Rutgers - The State University of New Jersey, 98 Brett Road, Piscataway, NJ 08901, P.T. Foteinou, Biomedical Engineering, Rutgers University, Piscataway, NJ 08854, S.E. Calvano, Department of Surgery, UMDNJ-Robert Wood Johnson Medical School, New Brunswick, NJ 08901, S.F. Lowry, Surgery, UMDNJ-Robert Wood Johnson Medical School, New Brunswick, NJ 08901, and Ioannis P. Androulakis, Biomedical and Chemical & Biochemical Engineering, Rutgers - The State University of New Jersey, 599 Taylor Road, Piscataway, NJ 08854.

Researchers have put significant effort in exploring the corticosteroid effect in modulating the dynamics of an inflammatory response 1-6. Although, there is a physiologic rationale for the use of corticosteroids in a systemic inflammatory response syndrome, long-term studies are needed to address optimal drug dosing, duration of treatment and timing of intervention that suffices to reverse the shock of an unresolved inflammatory response that accompanies critical illness 7. The development of mathematical models that capture the complexity of biological systems through the dynamic integration of their elementary components facilitates the rationalization and evaluation of the effectiveness of various treatment interruptions. Designing therapeutic strategies that target the intrinsic dynamics of systemic inflammation is not a trivial task given the fact that a drug can either be administered as an intravenous injection or as an infusion. Each route of drug administration encompasses a set of various parameters such as duration of treatment (for infusion), frequency of injections (for injection), drug dose and it may account for a different biological response. In this study assuming that corticosteroids serve as the putative controllers of an endotoxin-induced inflammatory response a number of questions need to be addressed: "Can we indentify a regime of steroid-based interventions that modulate efficiently the dynamics of an exuberant inflammatory response? Can we distinguish therapeutic strategies that have a critical effect on the response of the system? Do different routes of drug administration account for the manifestation of salient dynamic features of the perturbed state of the system?" It is important to realize that such questions formulate a combinatorial problem that precipitates the development of constraint integer optimization algorithms and motivates the research goals of this study. In the present study we will explore the possibility of formulating a mixed-integer optimization algorithm that opens the research avenues to establish an in silico corticosteroid control policy against an in vivo human response to endotoxin.

The experimental data analyzed in this study were generated as part of the Inflammation and Host Response to Injury Large Scale Collaborative Project funded by the USPHS, U54 GM621119. Human subjects were injected intravenously with endotoxin (CC-RE, lot 2) at a dose of 2-ng/kg body weight (endotoxin treated subjects) and 0.9% sodium chloride (placebo treated subjects). Blood samples were collected and the transcript abundance of leukocytes was measured before endotoxin administration (0hr) and 2, 4, 6, 9 and 24 hrs after endotoxin injection. Parts of these data have been published in 8-10; however the analyses in this article represents an approach that has not been previously published.

Based on our prior work, we developed a NF-kappaB dependent indirect response model of an endotoxin-induced inflammation that integrates mechanistically the opposing effect of two signaling pathways; one associated with the activation of NF-kB that drives downstream the transcriptional activation of inflammatory mediators and one related to the genomic signaling of exogenous corticosteroids; as the putative controllers of inflammation. The development of such a mechanistic model allows us to gain insight about how the system responds to a multitude of external signals through the dynamic interaction of signaling modules. We prompt to formulate a mixed-integer optimization algorithm that takes different routes of drug administration into account via the potential of binary variables. The primary objective is to capture the best corticosteroid intervention strategy under the perturbation of a high concentration of LPS that accounts for the most rapid inflammatory recovery. Therefore, we are interested in identifying an ensemble of steroid interventions that modulate the activity of NF-kappaB so that to rapidly approximate its reference trajectory (baseline) despite the implications of a high LPS stimulus.

Consequently, the formulation of a mixed-integer optimization framework enables us to evaluate the efficacy of particular corticosteroid interventions against the progression of systemic inflammation. Such a framework lays the foundation of an open loop control algorithm that defines research windows shedding invaluable insight on the complex dynamics of the system. Moreover, identifying a regime of optimal intervention strategies may be theoretically tractable but experimentally not practically feasible to validate. However, our approach serves as a critical enabler to improve our understanding about alternative optimal ways of modulating the inflammatory response.

1. Annane, D.; Sebille, V.; Charpentier, C.; Bollaert, P. E.; Francois, B.; Korach, J. M.; Capellier, G.; Cohen, Y.; Azoulay, E.; Troche, G.; Chaumet-Riffaut, P.; Bellissant, E., Effect of treatment with low doses of hydrocortisone and fludrocortisone on mortality in patients with septic shock. Jama 2002, 288, (7), 862-71.

2. Barber, A. E.; Coyle, S. M.; Marano, M. A.; Fischer, E.; Calvano, S. E.; Fong, Y.; Moldawer, L. L.; Lowry, S. F., Glucocorticoid therapy alters hormonal and cytokine responses to endotoxin in man. J Immunol 1993, 150, (5), 1999-2006.

3. Barber, A. E.; Coyle, S. M.; Fischer, E.; Smith, C.; van der Poll, T.; Shires, G. T.; Lowry, S. F., Influence of hypercortisolemia on soluble tumor necrosis factor receptor II and interleukin-1 receptor antagonist responses to endotoxin in human beings. Surgery 1995, 118, (2), 406-10; discussion 410-1.

4. Richardson, R. P.; Rhyne, C. D.; Fong, Y.; Hesse, D. G.; Tracey, K. J.; Marano, M. A.; Lowry, S. F.; Antonacci, A. C.; Calvano, S. E., Peripheral blood leukocyte kinetics following in vivo lipopolysaccharide (LPS) administration to normal human subjects. Influence of elicited hormones and cytokines. Ann Surg 1989, 210, (2), 239-45.

5. Hawes, A. S.; Rock, C. S.; Keogh, C. V.; Lowry, S. F.; Calvano, S. E., In vivo effects of the antiglucocorticoid RU 486 on glucocorticoid and cytokine responses to Escherichia coli endotoxin.

Infect Immun 1992, 60, (7), 2641-7.

6. Calvano, S. E.; Albert, J. D.; Legaspi, A.; Organ, B. C.; Tracey, K. J.; Lowry, S. F.; Shires, G. T.; Antonacci, A. C., Comparison of numerical and phenotypic leukocyte changes during constant hydrocortisone infusion in normal humans with those in thermally injured patients. Surg Gynecol Obstet 1987, 164, (6), 509-20.

7. Jahan, A., Septic shock in the postoperative patient: three important management decisions. Cleve Clin J Med 2006, 73 Suppl 1, S67-71.

8. Calvano, S. E.; Xiao, W.; Richards, D. R.; Felciano, R. M.; Baker, H. V.; Cho, R. J.; Chen, R. O.; Brownstein, B. H.; Cobb, J. P.; Tschoeke, S. K.; Miller-Graziano, C.; Moldawer, L. L.; Mindrinos, M. N.; Davis, R. W.; Tompkins, R. G.; Lowry, S. F., A network-based analysis of systemic inflammation in humans. Nature 2005, 437, (7061), 1032-7.

9. Foteinou, P.; Calvano, S.; Lowry, S.; Androulakis, I., An indirect response model of endotoxin-induced systemic inflammation. Journal of Critical Care 2007, 22, (4), 337-338.

10. Storey, J. D.; Xiao, W.; Leek, J. T.; Tompkins, R. G.; Davis, R. W., Significance analysis of time course microarray experiments. Proc Natl Acad Sci U S A 2005, 102, (36), 12837-42.

11. Ramakrishnan, R.; DuBois, D. C.; Almon, R. R.; Pyszczynski, N. A.; Jusko, W. J., Fifth-generation model for corticosteroid pharmacodynamics: application to steady-state receptor down-regulation and enzyme induction patterns during seven-day continuous infusion of methylprednisolone in rats. J Pharmacokinet Pharmacodyn 2002, 29, (1), 1-24.

382b An Integrated Framework of Transcriptome Analysis and *In Silico* Modeling Leads to Mechanistic Insights on Fti/taxol Synergy

Zeynep H. Gümüs[1], **Ada Gjrezi**[2], Heming Xing[3], Zach Pitluk[3], Ilse Van den Wyngaert[4], William Talloen[4], Hinrich W.H. Göhlmann[5], Iya Khalil[3], Harel Weinstein[1], and Giannakakou Paraskevi[2]. (1) The HRH Prince Alwaleed Bin Talal Bin Abdulaziz Alsaud Institute for Computational Biomedicine, Weill Medical College, Cornell University, 1300 York Ave, New York, NY 10065, (2) Department of Medicine, Division of Hematology and Medical Oncology, Weill Medical College, Cornell University, New York, NY 10065, (3) Gene Network Sciences, 10 Canal Park, Cambridge, MA 02141, (4) Division of Janssen Pharmaceutica, Johnson and Johnson Pharmaceutical Research & Development, Beerse, Belgium, (5) Functional Genomics, Johnson and Johnson Pharmaceutical Research & Development, Turnhoutseweg 30, Beerse, Belgium

Understanding the causal relationships between pathophysiology of the cell and drug effects on biomolecular targets is imperative in the development of new therapeutics. Construction of useful models of drug effects on disease pathophysiology involves (i) the collection of large and diverse data sets and information on functional connections between cellular components; and (ii) the development and application of computational learning methods to integrate the relevant information and enable prediction of cell behavior. We illustrate both the challenges presented by the development of a combined experimental and computational approach that includes functional analysis, data-driven simulation and experimental validation, and its power to identify critical cellular components in the anticancer activity of drug combinations exhibiting therapeutic synergy.

A novel class of experimental anticancer agents, the Farnesyltransferase Inhibitors (FTIs), have exhibited modest activity in the clinic as single agents, but were found to have promising activity in combination with standard chemotherapy drugs. In particular, FTIs synergize with the microtubule-stabilizing drug Taxol in several *in vivo* and *in vitro* models. Furthermore, recent phase I and phase II clinical trials have demonstrated beneficial clinical activity for this combination in chemotherapy of Taxol-resistant cancer patients. We have recently shown that the drug combination inhibits cell growth, enhances tubulin acetylation and induces apoptosis in human cancer cell lines (1). However, the cellular mechanisms underlying the synergistic effects of this promising anticancer drug combination still needs to be understood. To this end, we undertook a collaborative experimental and computational study combining (i) whole-genome transcriptome analysis (ii) measurements of drug effects on biomarkers (iii) utilization of interaction and functional databases for functional and pathway connections, (iv) computational modeling and prediction based on an *in silico* reverse engineering approach.

Transcriptome changes in cells exposed to the FTI Lonafarnib (LNF) (1, 5, 10 μM) and Taxol (Tx) (2, 5, 10 nM) alone or in synergistic combination (LNF 1 μM + Tx 2, 5 or 10nM) were examined using whole-genome Affymetrix HGU133 Plus 2 microarrays. The observed synergistic transcriptome changes were evaluated in the context of detailed protein-protein interaction maps and molecular networks. On this basis, they were grouped into enriched enzymatic, metabolic and signaling pathways and classified according to functional categories utilizing interaction and functional databases. These analyses were complemented by additional experimental measurements of the effects of drugs on tubulin acetylation, to serve as biomarkers for synergistic drug activity. The various data elements were combined and analyzed with a reverse engineering and forward simulation technology in order to develop an *in silico* model predictive of drug synergy.

We report on important insights provided by the results regarding the biology of the combined drug effects. Thus, the dose-dependent effects of Taxol or LNF treatment alone were shown to include expression of genes involved in cell death, cell cycle, DNA replication and repair and p53 signaling pathways (LNF had additional effects on genes involved in cellular assembly, organization, and motility). Genes likely to be involved in synergy belong to signaling networks specific to actin and microtubule cytoskeleton, adherens junctions, and receptor tyrosine kinase signaling, indicating the nature of the synergistic effects.

(1) Marcus, A. I., O'Brate, A. M., Buey, R. M., Zhou, J., Thomas, S., Khuri, F. R., Andreu, J. M., Diaz, F., Giannakakou, P. Farnesyltransferase inhibitors reverse taxane resistance. Cancer Research, 2006, 66 (17), 8838-8846.

382c Rigorous Mathematic Approach for Chemotherapy Design In the Brain

Libin Zhang[1], Jeonghwa Moon[2], Benyamin Grosman[2], and Andreas A. Linninger[3]. (1) University of Illinois at Chicago, 851 S. Morgan St., SEO 218, Chicago, IL 60607, (2) Chemical Engineering, University of Illinois at Chicago, 851 S. Morgan St., Room 218, Chicago, IL 60607, (3) Bioengineering, University of Illinois at Chicago, 851 S. Morgan St., Room 218, Chicago, IL 60607

The malignant glioblastoma is a devastating disease of central nerve system of the brain often lethal in few months time. Combining chemotherapy, radiation therapy, genetic therapy may extend significantly life expectation of patients suffering from this disease, provided that the cancer can be destroyed without harming excessively the patient's brain. Therefore, the challenge for designing effective chemotherapy lies in distributing the chemotherapeutic toxin so that it kills cancer cell while at the same time not causing excessive damage to healthy brain tissue. This problem requires the optimal choice of catheter injection position and design parameters like number of outlet ports and port configuration to best discharge the toxins as well as and design parameters like insertion pressure, flow and dilution rate. It is still the state-of-art treatment of this process to rely on surgeon institution in term of catheter's locations for design parameters and flow rates.

In this presentation, we propose a rigorous mathematical optimization problem on a distributed two-dimensional or three dimensional brain to determine the optimal catheter location, injection pressure and other parameters such as number of hole or outlets. We will use as the objective function the maximal coverage cancerous tissue as well as adjacent white matter tracts where reoccurrence is likely, while at the same time minimizing the concentration of toxins to cortical areas responsible for critical brain function such as smell and sight. We formulate a rigorous distributed optimization problem with partial differential equations constraints representing momentum, specie transport phenomena. The finite volume discretization method using unstructured grid is used to solve the resulting non-linear mathematical program with stochastic mathematical programming techniques.

The methodology of the design of catheter locations integrates interdisciplinary expertise from systems biology and engineering optimization. The approach will allow physicians and scientists to design and optimize chemotherapy in a systematic fashion, thus reducing the need for trial-and-error animal experiments.

382d Dynamic Modeling and Analysis of Glucose and Fatty Acid Interactions In Skeletal Muscle Tissue

Baris U. Agar, Department of Chemical and Biological Engineering, Illinois Institute of Technology, 10 W. 33rd Street, Chicago, IL 60616, Ali Cinar, Chemical and Biological Engineering, Illinois Institute of Technology, 10 W. 33rd Street, Chicago, IL 60616, Emmanuel C. Opara, Department of Biomedical Engineering, Illinois Institute of Technology, 10 W. 32nd Street, Chicago, IL 60616, and Gary Reznik, Microbiology and Immunology Division, IIT Research Institute (IITRI), 10 W. 35th Street, Chicago, IL 60616.

A mathematical model of glucose and free fatty acid (FFA) utilization in skeletal muscle tissue has been developed. The effects of changes in plasma FFA concentration on the utilization of glucose, and the effects of changes in glucose concentration on the utilization of FFA have been investigated via sensitivity analysis tools to illustrate the competing nature of FFA and glucose as fuel sources for skeletal muscle tissue.

Diabetes is estimated to affect 21 million people in the United States and type 2 diabetes accounts for more than 90 percent of all diagnosed cases [1]. Several studies indicate that obesity is associated with an increased risk of type 2 diabetes [2, 3] and FFA levels are elevated in most obese subjects [4]. Increased FFA and intracellular lipid concentrations are the primary suspects for causing insulin resistance in liver and muscle tissue. Physiological elevations of plasma FFA inhibit insulin stimulated glucose uptake in a dose dependent fashion [4].

The effect of elevated FFA concentration on utilization of glucose is described by the glucose-fatty acid cycle hypothesis [5]. According to this hypothesis, glucose and FFA compete with each other to serve as an energy source. Elevation of plasma FFA levels inhibits glucose uptake, which leads to muscle cells with impaired sensitivity to insulin, a suspected cause of Type 2 diabetes. Experimental studies have shown glucose transport as the step at which FFA induces insulin resistance [6, 7]. Similarly, an increase in insulin mediated glucose utilization causes suppression of FFA oxidation [8, 9]. According to experimental studies, increased glucose utilization leads to an increase in cellular malonyl-CoA levels which may inhibit the transportation of fatty acyl-CoA into the mitochondria, the site of FFA oxidation [9].

The kinetic equations governing the rate of formation and disappearance of metabolites involved in skeletal muscle energy metabolism are constructed based on previous work by Dash and co-workers [10]. In addition, the kinetics of FFA transportation, its inhibition by malonyl-CoA formation and the effect of FFA on glucose transportation are formulated and the kinetic constants of the final model are updated based on experimental findings.

Simulation studies indicate that an increase in plasma FFA level alone is sufficient to increase FFA utilization significantly whereas increased level of insulin is necessary for plasma glucose to induce increased glucose utilization. A sensitivity analysis on plasma FFA concentration reveals that glucose disposal may decrease by %20 through a two-fold increase in plasma FFA levels. A bifurcation analysis tool is being developed to observe the stable and unstable steady states, to address steady state multiplicity and to aid in the sensitivity analysis of the system.

REFERENCES

1. Centers for Disease Control and Prevention, National Diabetes Fact Sheet, 2005

2. Ford ES, Williamson DF, and Liu S: Weight change and diabetes incidence: findings from a national cohort of US adults. Am J Epidemiol. 146:214-222, 1997.

3. Resnick HE, Valsania P, Halter JB, and Lin X: Relation of weight gain and weight loss on subsequent diabetes risk in overweight adults. J Epidemiol Community Health. 54:596-602, 2000.

4. Boden G: Free fatty acids (FFA), a link between obesity and insulin resistance. Frontiers in Bioscience 3, 169-175, 1998.

5. Randle PJ, Garland PB, Hales CN, and Newsholme EA: The glucose-fatty acid cycle: its role in insulin sensitivity and the metabolic disturbances of diabetes mellitus, Lancet 1:785-9, 1963.

6. Dresner A, Laurent D, Marcucci M, Griffin ME, Dufour S, Cline GW: Effects of free fatty acids on glucose transport and IRS-1 associated phosphatidylinositol 3-kinase activity, J Clin Invest 103:253-9, 1999.

7. Cline GW, Petersen KF, Krssak M, Shen J, Hundal RS, and Trajanoski Z: Impaired glucose transport as a cause of decreased insulin-stimulated muscle glycogen synthesis in type 2 diabetes, N Engl J Med 341:240-6, 1999.

8. Forsey RG, Reid K, and Brosnan JT: Competition between fatty acids and carbohydrate or ketone bodies as metabolic fuels for the isolated perfused heart. Can J Physiol Pharmacol. 65(3): 401–406, 1987.

9. Awan MM, and Saggerson ED: Malonyl-CoA metabolism in cardiac myocytes and its relevance to the control of fatty acid oxidation. Biochem. J. 295: 61-66, 1993.

10. Dash RK, Dibella II JA, and Cabrera ME. A Computational model of skeletal muscle metabolism linking cellular adaptations induced by altered loading states to metabolic responses during exercise. Biomed Eng Online 6(1): 14, 2007.

382e A Control Engineering Model for Hemostasis Following Blood Vessel Injury

Erik S. Welf, Chemical Engineering, University of Delaware, 150 Academy Street, Colburn Lab, Newark, DE 19716, Ulhas P. Naik, Biological Science, University of Delaware, 150 Academy St., Newark, DE 19716, and Babatunde A. Ogunnaike, Department of Chemical Engineering, University of Delaware, Newark, DE 19716.

Hemostasis, the process by which platelets form a clot to halt blood loss following vessel damage, must be tightly regulated to ensure that a clot forms quickly to prevent excess blood loss without restricting blood flow to the rest of the body. Conceptually, a sophisticated biological control system is responsible for maintaining this tight regulation, realized physiologically by a complex set of interactions between many different molecules both inside and outside the platelet. Precisely how all the individual components work together to maintain this balance has been difficult to understand using traditional experimental methods alone—quantitative models are required.

In this work, we develop an engineering control system-based model of the process of hemostasis, organizing essential components in terms of their respective control system functions, i.e. as sensors, controllers, actuators, and the controlled process. By organizing key molecular interactions according to their function within the hemostatic control system, and using known biological characteristics to model them as inter-connected units, we have created a model that represents the process of hemostasis in terms of it's effectiveness at controlling blood loss following vessel injury. Our model includes mechanistic details within the framework of the overall system properties; this approach enables us to identify the molecular source of system malfunctions and design corrective action based on the desired system properties. We show how different components of the hemostatic system affect overall responses to vessel injury, and how we can use the model to formulate and test hypotheses regarding the causes and effects of different pathological conditions.

382f Long-Term Validation of Inferential Blood Glucose Model Developed from Real Subject In Free-Living Study Using Noninvasive Inputs

Derrick Rollins, Iowa State University, Department of Chemical Engineering, 2114 Sweeney Hall, Ames, IA 50011-2230, Nidhi Bhandari, Chemical and Biological Engineering, Iowa State University, 2114 Sweeney Hall, Ames, IA 50011, Nisarg Vyas, BodyMedia Inc., 4 Smithfield St., 11th floor, Pittsburgh, PA 15222, and Dave Andre, BodyMedia, Inc.

In this work we present the results validating the efficacy and accuracy of a block-oriented modeling approach in modeling blood glucose of a real type 2 diabetic (T2D) subject using real data and its ability to predict over a period of almost two years. The model was developed (trained) from 3 weeks of data collected on a T2D subject in September –October 2006. The model was then tested on five days of data immediately after the training period (October 2006) and showed its excellent ability to model glucose over the test period using only non-invasive food and activity inputs. As a measure of evaluating performance, an average absolute error (AAE) (i.e., the average of the absolute values for the measured glucose minus modeled glucose) was used. The AAE in training was 12.0 mg/dL and 13.6 mg/dL in testing in 2006. While the ability of the model to predict well during testing in 2006, was itself noteworthy and indicative of its ability to give cause-and-effect, an additional testament to the model's capability is its ability to model the data collected on the same subject over a year later in September / October 2007 and then again (over year and a half later) in February/ March 2008 at similar level of accuracy. (For comparison consider that the glucose meter AAE for replicated measurements was 15.3 mg/dL).

The details of the study carried out in 2006 and then repeated in 2007 and 2008 are given below.

Subject: The subject was 50 years old, in good health (except type two diabetes) with a body mass index (BMI) of 27.9 kg/m2. He was not on diabetic medication or insulin.

Measures: To obtain the fast sampling rate necessary for dynamic modeling, the MiniMed Continuous Glucose Monitor, MMT-7102® (Medtronic, Minneapolis, Minnesota) was used to measure glucose at five minute intervals. The activity measurements were obtained using the SenseWear® Pro3 body monitoring system (BodyMedia Inc., Pittsburgh, PA). The glucose monitor requires the subcutaneous insertion of a sensor, typically in the torso, and assesses interstitial glucose at a reported rate of one sample every five minutes. The interstitial glucose measurements are used to infer blood glucose. The sensors were replaced weekly which resulted in one to two hours of no measurements for initialization. This monitor is self-calibrating but is referenced directly to measured blood glucose values obtained four times daily from a glucose meter. The subject in this study obtained these values from his personal One Touch Ultra® blood glucose meter (LifeScan, Inc., Milpitas, CA).

Protocol: Because this was a free-living study, no constraints were placed on diet or lifestyle. The subject recorded food ingested, the approximate serving sizes, and the eating times and durations in a food log. In addition, the subject also obtained at least four daily measurements of blood glucose using finger-stick measurement and entered that into the CGMS for calibration. Using nutrition tables the grams of carbohydrates, fats, and proteins ingested per meal or snack were determined and used in modeling

Model Development: The final Wiener model from this study consisted of the eleven variables and had115 parameters. The selected inputs for this study have nonlinear, highly interactive, and dynamic affects on blood glucose. A theoretical modeling approach is not practical due to inadequate first principle knowledge. An empirical approach is not likely to succeed because, by necessity, data collection is under free-living conditions, and therefore under high input correlation, and thus, inhibiting cause and effect modeling, a critical goal of this research. (By cause and effect modeling we mean model development that determines the independent and specific effect an input has on glucose response over the input space interest.) Therefore, given these limitations, we chose a semi-empirical modeling approach. More specifically, the method was a unique and specific application of Wiener modeling that extends the method developed by Rollins and Bhandari [1] in a novel way to build accurate cause and effect models from free-living data.

References:

1. D.K. Rollins, N. Bhandari Constrained MIMO Dynamic Discrete-Time Modeling Exploiting Optimal Experimental Design. Journal of Process Control: 14(6): 671-683, 2004.

Web Page: public.iastate.edu/~drollins/Publications

407a Consequence of Thermophysical Approximations on Flare and Hydraulic Networks. A Case Study

Srinivasa Rao Karanam, Asuman N. Yilmaz, and Sabyasachi Sen. Invensys SimSci-Esscor, 26561 Rancho Parkway South, Suite 100, Lake Forest, CA 92630

Fluid-flow network analysis simulators used for designing flare and hydraulics systems are comprised of several inter-dependent and layered components. They are:

* The thermophysical properties calculation sub-system,

* The suite of correlations for pressure-drop calculations, and

* The network solver

Of these three layers, the thermophysical properties calculation component represents the core of a simulator and, therefore, plays a critical role in all simulation calculations. It follows, therefore, that the fidelity of this thermodynamics core plays a major role in the accuracy of the predicted relief load that a flare network is designed for. This is an important first step in the avoidance of over-sizing of relief valves which handle relief loads, on account of the significant costs involved. The case for good design is further warranted by the fact that, for safety reasons, uncertainties in thermophysical properties are typically countered by over-designing the system.

Commercially available fluid flow simulators offer a variety of options for calculating thermophysical properties, ranging from non-rigorous methods based on rules-of-thumb and simple bundled correlations, to rigorous equation of state models also involving three-phase flash calculations. Non-rigorous "simplified thermodynamics" approaches yield only rough estimates, and do not, typically, include calculations involving discrete substances. Furthermore, the simplicity often results in inability to describe reasonable vapor-liquid-equilibrium which, in turn, mandates calculations based on single-phase flow assumptions, even if the system experiences multi-phase flow conditions. On the other hand, the solution of flow networks utilizing "simple thermo" can often be an order-of-magnitude less CPU-intensive than those incorporating rigorous component-based thermodynamics and, consequently, a popular choice for analyzing and designing flare networks, even though this approach may lead to over-design.

This case study focuses on several scenarios, comparing and contrasting flare / hydraulic network calculations based on simple versus rigorous thermodynamics. In addition, this investigation seeks to address the over-design penalty, and how savings may be realized with "rigorous thermo" approaches. Several key quantities are compared, including state variables such as density ("microscopic quantities") and design parameters such as relief load ("macroscopic quantities"). Some rules-of-thumb (guidelines) are proposed for flare / hydraulic network calculations based on single-phase flow when, in reality, two-phase flow occurs.

407b Integrated Design and Control Using Adaptive Full State Identification

Benyamin Grosman[1], Jeonghwa Moon[1], Libin Zhang, and Andreas A. Linninger[2]. (1) Chemical Engineering, University of Illinois at Chicago, 851 S. Morgan St., Room 218, Chicago, IL 60607, (2) Bioengineering, University of Illinois at Chicago, 851 S. Morgan St., Room 218, Chicago, IL 60607

High performance processes require design that goes close to design boundaries and specifications while still guaranty robust performances. In order to go tighter to the boundaries of process performances much attention was been devoted to integrated design and control in which decision about the dynamics as well the design are taken by optimal fashion. Rigorous methods solving design and control lead to very challenging mathematical formulation which often becomes intractable. Our group presented a novel approach which embeds control decisions in search for optimal design performances under uncertainty. It does so by working in a reduced work space in which key design variables from the system are determined while embedded control decisions are taken in a close to optimal fashion. The two stage optimization therefore reduce drastically the number of decision variables such as pairing off manipulated and control also reduces the state space and make some problems of industrial interest tractable for computationally analysis that previously were not been able to solve.

In this presentation we will overcome limitations of early approach based on diagonal identification and expand the adaptive identification to a full state representation. This state representation will still show to be computably very efficient using recursive techniques. Case studies will demonstrate the increase performances of the adaptive full state identification and illuminate problem with highly correlated multi variable systems which were previously very difficult to address. We will also present methods to illuminate the offsets and by that improve the quality of previous approach to integrated design and control.

407c A Hybrid Algorithm for 2-Stage Stochastic Programming with Recourse

Li Sun[1], Jiping Pan[1], and Helen H. Lou[2]. (1) Department of Chemical Engineering, Dalian University of Technology, P.O. Box 288, Dalian,Liaoning, 116023, China, (2) Department of Chemical Engineering, Lamar University, P.O.Box 10053, Beaumont, TX 77710

For chemical process design, various uncertain factors appear in the form of manufacturing variations, changes in material properties, market fluctuation, etc. So uncertainty should be considered in the process design and optimization.

2-stage stochastic programming with recourse (2-SPR) is an important approach to process optimization under uncertainty. It can be formulated as followings:

$$\min\ f(x) + E_e\,[Q(u, e)] \qquad (1)$$

$$S.\,t\ \ h(x, e) = 0$$

$$\text{With}\ \ Q(u, e) = \min F(x, e) \qquad (2)$$

$$S.\,t\ \ g(x, u, e) \le 0,$$

Where, $f(x)$ is the first-stage objective function. $Q(u, e)$ is a penalty term, which is interpreted as penalty shortfall or corrective measures against any infeasibility, and it is the second-stage optimization function. $f(x) + E_e[Q(u, e)]$ is the objective function with recourse. h and g are the vectors of the equality and inequality constraints, which are derived from process models. x is structure vector. u is operating vector. e is uncertain factor.

Mathematically, efficient algorithms need to be developed and utilized for the 2-SPR model.

While uncertain factors have discrete probability distribution, the 2-SPR model can be stated as a large-scale equivalent deterministic programming. For the uncertain factors with continuous probability distribution, the decomposition methods and approximation schemes are required to be utilized to bring about large-scale mathematical programs, but there are many difficulties to overcome. For example, the evaluation on the expected value of the second-stage function (2) is an uncertain definition by the interior optimization, and it is a sub-optimization problem embedded in the master optimization (1). It should be integrated in the feasible region, which is unknown at the first stage, but obtained only when the master optimization (1) is available.

Many works have been conducted on this kind of problems. Lerapetritou and Pistikopoulos (1994) proposed a feasible region algorithm. Liu and Sahinidis (1996) used Monte Carlo integration schemes instead of Gauss integration. Wang and Han(2005) proposed Monte Carlo sampling from the entire domain of the distribution and feasibility cuts in Benders decomposition. All these methods have their own characters.

Our work attempts to extend the prior research in this field to propose a hybrid algorithm. Uncertain factors are tackled by Monte Carlo sampling, and improved Benders decomposition is utilized to deal with the problem when the master optimization can not be solved directly. Furthermore, a new approach is derived to overcome the computation difficulty for the model with an extremely large number of constraints. The major steps in this approach are presented below:

Step 1: Monte Carlo sampling.

Step 2: To solve the dual of the second-stage model (2), and one of three cases will arise:

Case 1: if the dual has an unfeasible solution, the solution on the optimization problem (1) is unbounded.

Case 2: if the dual has an optimal solution, the extreme points can be obtained to obtain the expected value of the second-stage model (2) and the upper bound of the optimization model(1).

Case 3: if the dual is unbounded, there is no feasible solution on the optimization model (1).

Step 3 To rewrite the optimization model (1) as the restricted model with more constraints.

Step 4 To solve the restricted model with only a small subset of constraints to obtain the feasible solutions, and then check whether these feasible solutions satisfy any of the non-included constraints.

Step 5 To obtain the optimal solution and the lower bound of the optimization model (1).

Step 6 If the difference between the upper and the lower bounds is small enough for the pre-specified tolerance, the calculation is terminated. Otherwise, return to step 2.

A case study will be presented to demonstrate the efficacy of the algorithm.

Keyword: Benders decomposition; uncertainty; 2-stage stochastic programming; Monte Carlo sampling

407d A Systematic Bi-Level Approach for Efficient Process Optimization Under Uncertainty Using Stochastic Annealing

Athanasios I. Papadopoulos, Chemical Process Engineering Research Institute, Center for Research and Technology Hellas, 6th klm Harillaou-Thermi Road, P.O.Box 60361, Thermi-Thessaloniki, 57001, Greece and Panos Seferlis, Mechanical Engineering Department, Aristotle University of Thessaloniki, P. O. Box 424, Thessaloniki, 54124, Greece.

This work reports on a novel approach for efficient process design under uncertainty using stochastic Simulated Annealing (SA) that employs data mining and model fitting to significantly reduce the associated computational effort. Existing approaches for process design under uncertainty using stochastic SA involve drawing samples of the uncertain parameters from a prespecified probability distribution and the optimization loop of the algorithm requires simulation for each sample separately in order to emulate the effect of the uncertain parameter in the design problem addressed and select the optimum design paths. Apparently, the number of utilized samples is of crucial importance as large numbers are required to maintain a realistic representation of the uncertain parameter distribution but at the expense of reduced algorithmic efficiency due to increased computational effort. In this respect, a major improvement proposed in published work (Kim and Diwekar, 2002) involves the use of an efficient sampling technique in order to achieve an improved representation of the uncertain parameter distribution, hence faster identification of the uncertain space leading to potentially optimum alternatives. A further improvement proposed by Painton and Diwekar (1995) involves the allowance of lower numbers of samples at the initial stages of the optimization search, while the number of samples is forced to increase based on a penalty function as the optimization search is intensified towards the optimum. Although low numbers of samples are allowed at initial search stages, there is no mechanism to prevent the stochastic optimization algorithm from imposing a large number of samples even at initial search stages, since the number of samples is selected randomly by the algorithm in a similar manner to the perturbations imposed in the design variables. Furthermore, increasing the number of samples as the algorithm approaches the optimum still presents an increased computational burden for larger process synthesis problems involving high non-linearities and combinatorial complexities.

In this work, we propose a systematic bi-level approach for process optimization under uncertainty that aims to significantly reduce the number of simulations required by the stochastic optimization algorithm after the sampling procedure. The first level of the proposed method implements a data mining technique in the form of clustering on the samples obtained from the probability distribution of the uncertain parameter. The implementation of clustering in the obtained samples leads to the generation of distinct and coherent groups of similar points representing the uncertain parameter values. A cluster center is then defined for each cluster based on statistical criteria and this central point approximately represents the contents of the cluster. At this stage the central point of each cluster is introduced to process simulation and an objective function value is calculated for each cluster center. The second level of the proposed approach involves using the available cluster center points in conjunction with their corresponding objective function values to fit a continuous model that represents the objective function value as a function of the cluster centers. At this point the objective function values of the remaining sample points are calculated based on the developed model, hence avoiding the time consuming simulations. The fitted model provides objective function value predictions that are either identical or lie within very close proximity to the values calculated through simulations. The fitting of the developed model is facilitated by the beneficial cluster-inherent statistical properties as well as from efficient sampling techniques, such as Hammersley sampling, that allow uniform multi-dimensional representation of the uncertain parameter distribution. The proposed approach enables the use of constantly large uncertain parameter samples regardless of the size of the optimization problem addressed or the stage of the performed optimization search. The number of generated clusters is decided by the utilized clustering algorithm based on statistical criteria, is maintained at low, yet sufficient, sample representation levels and presents very limited variations throughout the optimization search. The implementation of clustering and model fitting are fast and computationally insignificant compared to the numerous, time consuming simulations required.

The proposed approach is illustrated through case studies of industrial interest on the design of process flowsheets involving crystallization reactors followed by separation of the produced crystals. A number of uncertainties in raw material and model related data are accounted for in order to maintain a realistic representation of the industrial reality in the utilized process models. Such uncertainties are simultaneously addressed in the synthesis of simple as well as complex process schemes accounting for multiple recycle and bypass streams, varying reaction and separation topologies as well as varying processing conditions. The employed process synthesis framework under uncertainty allows systematic process flowsheet representation, using generic process modules regardless of the processing task addressed, while facilitating comprehensive interconnectivity options within the flowsheet, based on a synthesis framework presented by Papadopoulos and Linke (2004) and Linke and Kokossis (2003). Results indicate the development of robust solutions and significant time savings in computations as compared to solving the same problems without the use of the proposed approach, while merits and shortcomings of the proposed approach are also discussed.

References cited

Kim, K. J., Diwekar, U (2002). Hammersley stochastic annealing: Efficiency improvement for combinatorial optimization under uncertainty. IIE Transactions, 34, 761-777.

Painton, L., Diwekar, U. (1995). Stochastic annealing for synthesis under uncertainty. European J. of Operational Research, 83, 489-502.

Papadopoulos,A.I, Linke,P.(2004). On the synthesis and optimization of liquid-liquid extraction processes using stochastic search methods, Comp. Chem. Eng., 28, 2391-2406.

Linke,P.,Kokossis,A.C.(2003) Attainable reaction and separation processes from a superstructure-based method, AIChE J., 49, 1451-1470.

407e L-Shaped Bonus Algorithm with Application to Water Pollutant Trading

Yogendra Shastri, Agricultural and Biological Engineering, University of Illinois at Urbana Champaign / Vishwamitra Research Institute (CUSTOM), 1304 W. Pennsylvania avenue, Urbana, IL 61801 and Urmila Diwekar, Vishwamitra Research Institute, Center for Uncertain Systems: Tools for Optimization and Management, Vishwamitra Research Institute, Clarendon Hills, IL 60514.

Applications of optimization in the field of process systems engineering often involve dealing with

nonlinearity and uncertainty, necessitating the solution of stochastic nonlinear programming (SNLP) problems.

However, the existing algorithms to solve such problems suffer from various limitations. L-shaped BONUS

algorithm, an integration of BONUS algorithm and sampling based L-shaped method, has been recently proposed to

overcome some of these problems. It is shown to have desirable computational properties. This work further

investigates the properties of the algorithm by applying it to the environmental problem of pollutant

(nutrient) trading in Christina River Basin. Environmental concerns being heightened, pollution abatement

related decisions are important for the industries, making the efficient solution techniques invaluable. The

results confirm the computational efficiency of the L-shaped BONUS algorithm. Simultaneously, interesting

aspects of the environmental trading problem are also explored, pointing towards the scope of implementing

stochastic programming techniques for better decision making in the field of industrial ecology.

409a Distributed Control System Design Using Lyapunov-Based Model Predictive Control

Panagiotis D. Christofides[1], David Muñoz de la Peña[2], and Jinfeng Liu[2]. (1) Department of Chemical and Biomolecular Engineering, Department of Electrical Engineering, University of California, Los Angeles, Los Angeles, CA 90095, (2) Department of Chemical and Biomolecular Engineering, University of California, Los Angeles, Los Angeles, CA 90095

Modern chemical plants involve highly integrated process units and utilize automatic control systems to meet desired safety, environmental and production specifications. Most control systems are decentralized in nature from a plant point of view, that is, each control system operates on a process using local process information without communicating with control systems operating on adjacent units. Nowadays, with the new developments in communication technologies such as wireless networks, there is an increasing interest in developing distributed control architectures; that is, the various control systems share information in order to take their corresponding decisions autonomously. This distributed control architecture gives rise to many interesting control problems. On one hand, the shared network used by each control system may introduce new dynamics into the closed-loop system such as time-varying delays or data losses. On the other hand, the distributed control algorithms must take into account how local control decisions affect the other control systems and affect the overall closed-loop performance. This is in general a very difficult problem. There are many works in the literature that address distributed control and its applications to different fields such as multi-agents control, sensor networks or internet congestion control, and more recently, chemical processes.

In this work, we introduce a distributed control architecture based on Lyapunov- based Model Predictive Control (LMPC) for nonlinear process systems with both continuous and asynchronous sensing and/or actuation. This class of nonlinear systems arises naturally in the context of process control systems based on hybrid communication networks (i.e, point-to-point wired links integrated with networked wired/wireless communication) and utilizing heterogeneous, asynchronous measurements. The control problem is to coordinate the design of a local (single-process) Lyapunov-based model predictive controller which takes decisions based on synchronous process sensing and actuation, and a supervisory (plantwide) Lyapunov-based model predictive controller that utilizes networked sensors and actuators and accounts for asynchronous sensing, actuation and network behavior, to ensure closed-loop stability and performance. To address this problem, when a new asynchronous measurement arrive, the supervisory controller solves a plantwide MPC optimization problem assuming that the local controller implements an already existing Lyapunov-based control law that guarantees closed-loop stability. The optimal input trajectory is sent to the local controller along a contractive constraint in the form on an upper bound on the value of the Lyapunov function along the predicted closed-loop nominal state trajectory. This contractive constraint is utilized in the MPC optimization problem of the local controller and guarantees that even in the local controller does not bahave as the supervisory controller assumed, stability of the closed-loop system is preserved. The proposed distributed control architecture profits from both the continuous and the asynchronous measurements as well as from additional networked control actuators and preserves the stability properties of the local controller even in the presence of asynchronous communications while improving the closed-loop performance. The theoretical results are demonstrated using chemical process examples.

A parallel algorithm for large-scale dynamic optimization

Arndt Hartwich, Wolfgang Marquardt[1], RWTH Aachen University, Aachen, Germany

Abstract

In nowadays' market conditions in the chemical industry, multi-product facilities, either multi-purpose batch processes or continuous processes operated in production campaigns, become more and more common. Hence, the operation of these processes does not only require optimal stationary but also optimal dynamic modes of operation. Dynamic optimization is considered a powerful tool to determine optimal control trajectories. However, the application of this method is bothered by the very high computational times in case of considering realistic models of complete chemical plants. Therefore, besides continuing algorithmic improvement, parallel computing strategies have to be exploited. Such parallelization strategies are presented in this contribution. Three examples of very different problem sizes are discussed. The speedup for the largest example even obtains a value close to the theoretical limit of the parallelization approach.

1 Introduction

Dynamic optimization can be considered as a powerful tool, where most applications are more or less time-critical. In particular, the applicability of real-time implementations strongly depends on the computational times until the solution to the dynamic optimization problem is obtained. Applying the *sequential approach*, the infinite dimensional optimization problem is transformed into a finite dimensional optimization problem by means of an appropriate discretization, where the computation of the corresponding gradients, also termed sensitivities, of the latter optimization problem results in the largest part of computational times. Thus, the numerical integration of the combined state and sensitivity equation system represents the largest portion of computational effort. Nowadays' developments in computer hardware are not directed towards faster CPUs, but towards parallel computers. Hence, further reductions in computational times can only be accomplished if parallel computing is accounted for. Keeping and Pantelides (1998) and Zhu and Petzold (1999) have exploited BDF-type numerical integration routines in order to allow for distributed memory parallelization. Schlegel *et al.* (2004) have extended the linear-implicit extrapolation routine LIMEX (Deuflhard and Bornemann, 1994; Deuflhard *et al.*, 1987) in order to allow for the solution of sensitivity equation systems. The parallel implementation of this sensitivity integration algorithm is described in this work. Nowak *et al.* (1998) have developed a parallel algorithm of LIMEX. As however the greatest part of the arising computational effort is due to the sensitivity equation system, the parallelization as suggested in this work focuses on the additional computational effort due to the sensitivity equation system similar to the works of Keeping and Pantelides (1998) and Zhu and Petzold (1999). As the operating system Windows Server 2008 has been chosen.

2 Numerical Solution methods of dynamic optimization

The dynamic optimization involves the determination of the optimization variables, control trajectories $\mathbf{u}(t)$ and final time t_f, such that an objective function Φ has to be minimized subject to the model equations and additional path and endpoint constraints. Mathematically this leads to a non-linear optimal control problem and can be stated as

$$\min_{\mathbf{u}(t),t_f} \quad \Phi = \Phi(\mathbf{y}(t_f), \mathbf{u}(t_f)) \qquad \text{(CDYNOPT)}$$

$$\begin{aligned}
s.t.\ \mathbf{B}\dot{\mathbf{y}} &= \mathbf{f}(\mathbf{y}(t), \mathbf{u}(t)), && t \in [t_0, t_f], && \text{(1a)}\\
0 &= \mathbf{y}(t_0) - \mathbf{y}_0, && && \text{(1b)}\\
0 &\geq \mathbf{h}^y(\mathbf{y}(t), \mathbf{u}(t)), && t \in [t_0, t_f], && \text{(1c)}\\
0 &\geq \mathbf{e}(\mathbf{y}(t_f), \mathbf{u}(t_f)), && && \text{(1d)}
\end{aligned}$$

[1]Corresponding author: wolfgang.marquardt@avt.rwth-aachen.de

$$\mathbf{u}(t)^{min} \leq \mathbf{u}(t) \leq \mathbf{u}(t)^{max}, \qquad t \in [t_0, t_f]. \tag{1e}$$

This formulation will be denoted as the continuous dynamic optimization problem (CDYNOPT) in the following. $\mathbf{y}(t) \in \mathcal{R}^{n_y}$ denotes the vector of the state variables, defined on the time horizon $t \in [t_0, t_f]$. The differential-algebraic process model of index one is given by Equation 1a. The time-dependent control variables $\mathbf{u}(t) \in \mathcal{R}^{n_u}$ are degrees of freedom for the optimization problem minimizing the objective function Φ. The state variables are constrained by path constraints $\mathbf{h}^y(\cdot) \in \mathcal{R}^{n_{h^y}}$. Endpoint constraints are denoted by $\mathbf{e}(\cdot) \in \mathcal{R}^{n_e}$.

2.1 Numerical solution

Binder *et al.* (2001) and Srinivasan *et al.* (2003) have discussed in detail the specific advantages and disadvantages of alternative strategies for the solution of dynamic optimization problems. The chosen *direct method* solves the dynamic optimization problem (CDYNOPT) numerically according to the *control vector parameterization approach*, which is also termed the *sequential approach*. This solution method is well-known for its ability to solve very large-scale and stiff problems.

2.1.1 Tranformation into NLP

The dynamic optimization problem (CDYNOPT) is transcribed into a non-linear program (NLP) by time-discretizing the controls $\mathbf{u}(t)$ on the time horizon $t \in [t_0, t_f]$ and using, e.g. piecewise constant or piecewise linear approximations. The discretization and the formulation of the NLP for the piecewise constant approximation reads as $\mathbf{u}_i(t_k) = c_{\mathbf{u}_{i,k}}, k = 1, ..., N, i = 1, ..., n_u$, where N is the number of discretization intervals and n_u the number of control variables. Choosing the discretized controls and the final time $z := [c_{u_{1,1}}, ..., c_{u_{n_u,N}}, t_f]$, as the n_z optimization variables, the dynamic optimization problem can be transcribed into the NLP

$$\min_{\mathbf{z}} \Phi = \Phi_M\left(\mathbf{y}\left(\mathbf{z}, t_f\right)\right) \tag{NLP}$$

$$\text{s.t.} \quad \mathbf{0} \geq \mathbf{h}^{\mathbf{y}}(\mathbf{y}(\mathbf{z})), \tag{2a}$$

$$\mathbf{0} \geq \mathbf{e}\left(\mathbf{y}\left(\mathbf{z}\right)\right), \tag{2b}$$

$$\mathbf{z}^{min} \leq \mathbf{z} \leq \mathbf{z}^{max}. \tag{2c}$$

The approximation and reformulation for piecewise linear intervals is straightforward. The path constraints (1c) of the continuous formulation (CDYNOPT) are relaxed by a pointwise formulation on the discretized grid. The NLP is solved by employing a standard SQP algorithm (Gill *et al.*, 1998). The optimization algorithm requires repetitive function evaluations and gradients.

2.1.2 Numerical methods for the computation of first-order sensitivities

In order to obtain the gradients for the NLP, one can differentiate the DAE system (1a) with respect to each parameter z_i, which yields additional n_z sensitivity equation systems of the form

$$\mathbf{B}\dot{\mathbf{s}}_i^y = \frac{\partial \mathbf{f}}{\partial \mathbf{y}}\mathbf{s}_i^y + \frac{\partial \mathbf{f}}{\partial z_i}, \quad i = 1, ..., n_z, \tag{3}$$

where $\mathbf{s}_i = \frac{d\mathbf{y}}{dz_i}$ denotes the sensitivity of $\mathbf{y}$ with respect to the parameter z_i. The sensitivity systems (3) for each parameter z_i form DAE systems as well, which are independent of each other, but obviously depend on the solution of the state equations (1a). Since for each degree of freedom z_i a sensitivity equation system has to be solved, the computational cost of this approach is rangely proportional to the number of decision variables n_z.

2.1.3 Linearly-implicit Euler extrapolation for sensitivity analysis

This section briefly summarizes the linearly-implicit extrapolation algorithm. The reader is referred to the works of Deuflhard (1983), Deuflhard and Bornemann (1994), Deuflhard *et al.* (1987) and Ehrig *et al.* (1996) for more details on the basic numerical integration routine and to Schlegel *et al.* (2004) for more details on the extensions for simultaneous state and sensitivity integration, that computes the vector $\mathbf{Y} := [\mathbf{y}^T, \mathbf{s}_1^T, ..., \mathbf{s}_{n_z}^T]$.

The key idea of an extrapolation method is a repeated integration over the basic step H for decreasing internal stepsizes

$$h_j = H/m_j, \; j = 1, \ldots, j_{max}, \tag{4}$$

where the harmonic sequence $\{m_j\} = \{1, 2, 3, ...\}$ is utilized. A simplified algorithmic structure is depicted by Algorithm 1, where the extrapolation tableau is computed according

$$T_{j,k} = T_{j,k-1} + \frac{T_{j,k-1} - T_{j-1,k-1}}{m_j/m_{j-k+1} - 1}, \qquad k = 1, \ldots, j. \tag{5}$$

Algorithm 1 Simultaneous state and sensitivity integration (SLIMEX)

```
Compute Jacobian A_n = ∂/∂y (f(y_n, z, t_n) − Bẏ)
for j = 1, ..., j_max while convergence criterion not satisfied do
  h_j = H/m_j
  LU = A_n − B/h_j
  for k = 0, ..., j − 1 do
    y_{k+1} = y_k − (LU)^{-1} f(y_k, z, t_k)
    Γ = (A(y_k) s_{i,k} + ∂f/∂z_i (y_k))
    s_{i,k+1} = s_{i,k} − (LU)^{-1} Γ, i = 1, ..., n_z
  end for
  T_{j,1} = Y_j
  if j > 1 then
    compute T_{j,j} via (5) and check convergence
  end if
end for
Y(H) = T_{j,j}
```

As described by Schlegel *et al.* (2004), a staggered integration by Algorithm 2 has been omitted. Error control is only applied to the states, but not to the sensitivity equation system. A staggered approach first converges the states and afterwards computes the sensitivities on a converged sequence of order $\hat{j}$ and step-sizes H. Hence, eventually redundant extrapolations of the sensitivity system could be avoided by only integrating the sensitivities if convergence in the states has been accomplished. However, housekeeping of the LU decompositions, Jacobians etc. has been considered too much effort for the rather rare occasion of a step-size reduction. For reasons explained below the staggered approach is parallelized.

It should be noted that the first LU decomposition of each step is computed employing the MA28AD routine, whereas every subsequent LU decomposition is computed by the MA28BD routine (Duff, 1979), which reuses the pivot sequence of MA28AD and substantially saves computational times. In some cases, it can be computationally beneficial if even the first LU decomposition on each step is first attempted using MA28BD and afterwards using MA28AD, if the first one has failed. This is done for the third example, which will be discussed in Section 4.

3 Parallel dynamic optimization

According to Algorithm 1, the same Jacobian matrix $\mathbf{A}$ is required for the state system and for the sensitivity equation systems. The numerical integration scheme requires one Jacobian matrix per step for a pure state integration, whereas a sensitivity integration requires significantly more than one Jacobian, as the Jacobian is required for the construction of the right hand side of the sensitivity equation system. The computation of a Jacobian can

Algorithm 2 Staggered state and sensitivity integration (SLIMEX)

Compute Jacobian $\mathbf{A}_n = \frac{\partial}{\partial \mathbf{y}}(\mathbf{f}(\mathbf{y}_n, \mathbf{z}, t_n) - \mathbf{B}\dot{\mathbf{y}})$
for $j = 1, \ldots, j_{max}$ while convergence criterion not satisfied **do**
  $h_j = H/m_j$
  $\mathbf{LU} = \mathbf{A}_n - \frac{\mathbf{B}}{h_j}$
  for $k = 0, \ldots, j-1$ **do**
    $\mathbf{y}_{k+1} = \mathbf{y}_k - (\mathbf{LU})^{-1}\,\mathbf{f}(\mathbf{y}_k, \mathbf{z}, t_k)$
  end for
  $\mathbf{T}^y_{j,1} = \mathbf{Y}_j$
  if $j > 1$ **then**
    compute $\mathbf{T}^y_{j,j}$ via (5) and check convergence
  end if
  $\hat{j} = j$
end for
$\mathbf{y}(H) = \mathbf{T}^y_{\hat{j},\hat{j}}$
for $j = 1, \ldots, \hat{j}$ **do**
  $h_j = H/m_j$
  $\mathbf{LU} = \mathbf{A}_n - \frac{\mathbf{B}}{h_j}$
  for $k = 0, \ldots, j-1$ **do**
    $\Gamma = \left(\mathbf{A}(\mathbf{y}_k)\mathbf{s}_{i,k} + \frac{\partial \mathbf{f}}{\partial z_i}(\mathbf{y}_k)\right)$
    $\mathbf{s}_{i,k+1} = \mathbf{s}_{i,k} - (\mathbf{LU})^{-1}\Gamma, \quad i = 1, \ldots, n_z$
  end for
  $\mathbf{T}^s_{j,1} = \mathbf{s}_{i,j}$
  if $j > 1$ **then**
    compute $\mathbf{T}^s_{j,j}$ via (5)
  end if
end for
$\mathbf{s}_i(H) = \mathbf{T}^s_{\hat{j},\hat{j}}$

be very time consuming - depending on the model under consideration. Furthermore, the additional effort required for sensitivity evaluation when compared to pure state integration comprises n_z additional evaluations of the sensitivity residuals and n_z additional back substitutions with the already available LU decomposition. The error control is only applied to the states. It is important to note that the sensitivity equation system i is independent of the system $j \neq i$. The implementation is done using the Message Passing Interface (MPI). Windows Server 2008 has been chosen as the operating system.

3.1 Parallelization strategies

For a wide range of case studies also wide ranges of the number of sensitivity equation systems n_z and the size of the DAE model n_y have to be accounted for in order to obtain high efficiency for each case study at hand. It is expected, that increasing computational times will allow for more communication between the processes, whereas very small examples might even result in no speedup in a parallel setting. In the following, three different parallelization strategies are discussed, where the extent to which information is communicated, increases. These three different strategies are expected to cover the wide range of possible case studies.

3.1.1 Distributed sensitivities (DS)

The main idea for parallelization is the distributed computation of sensitivities, since the sensitivity equations are independent of each other and only dependent on the state system. All processes are assigned an equal number of sensitivity equation systems. This approach is very simple and requires communication only once per sensitivity integration, i.e. after each integration sensitivity information of each process is broadcasted to all other processes such that each process contains the complete sensitivity information. The processes will be denoted as *distributed sensitivity integrators* (DSI) in the following. All processes have to perform the state integration, the computation of the Jacobians and the LU decomposition.

3.1.2 Distributed sensitivities with centralized Jacobian (CJ)

The fundamental idea of this approach is to decouple the state integration and the sensitivity integration. The state integration requires one Jacobian per integration step H, i.e. $\mathbf{A}_n$, whereas the sensitivity integration requires additional evaluations of the Jacobian $\mathbf{A}(\mathbf{y}_k)$ for each iteration $k = 1, \ldots, j-1$ on the extrapolation tableau. However, the states $\mathbf{y}_k$, from which the Jacobians can be computed, are independent of the sensitivity system and can be obtained by a pure state integration. Hence, a preceding state integration can deliver the states $\mathbf{y}_k$ and the Jacobians $\mathbf{A}(\mathbf{y}_k)$ can be computed by separate processes. Processes which integrate the sensitivity equations follow the state integration with a small time delay and receive the Jacobians from separate processes that compute the Jacobians. Thus, the computational burden to obtain Jacobian matrices for the sensitivity integration can be reduced to the communication overhead. However, additional processes that compute the Jacobians and communicates the entire information, respectively, are required. One of the drawbacks of this approach is the very high frequency of communication, which has to take place after each step of the integrator. This strategy, which has been suggested similarly by Keeping and Pantelides (1998), will require four different types of processes:

- *Communicator* (CO): This process communicates the entire information.
- *State integrator* (STI): This process carries out the sole state integration and communicates the states $\mathbf{y}$, etc. to the communicator.
- *Jacobian calculator* (JC): This process receives the state vectors $\mathbf{y}$ from the Communicator (CO) and returns the computed Jacobians $\mathbf{A}(\mathbf{y})$.
- *Sensitivity integrator* (SEI): Analogously to the distributed sensitivities strategy this process performs a sensitivity integration of a subset of sensitivities and additionally receives the required Jacobians from (CO). In contrast to the distributed sensitivities strategy, the states are not included in the numerical integration, but are communicated to the SEI to allow for the computation of the LU decomposition. As in the original algorithm error control on the sensitivities has been omitted, the order j and the step-size H are sent to the sensitivity integrator from the state integrator via the communicator.

The staggered approach is parallelized and the entire communication of information has to account for the fact, that states $\mathbf{y}_k$, that are computed, have to be omitted, if a step size reduction has to be performed. The staggered approach has been considered too much effort for the serial code since step size reductions are rather rare and housekeeping of LU decompositions and Jacobians, etc. has been considered too much effort. However, the parallel code has to store the LU decompositions and Jacobians anyway and therefore the latter argument is not valid for the parallelization. Hence, the staggered approach is parallelized as it saves eventually redundant basic steps on the sensitivity equations.

3.1.3 Distributed sensitivities with centralized Jacobian and communicated LU decomposition (CJLU)

This approach follows the idea of (CJ), but the LU decompositions, which are computed by the state integrator, are additionally communicated to the sensitivity integrators via the communicator. The size of the LU decomposition is significantly larger compared to the size of the Jacobians. Hence, the communication overhead will increase significantly compared to (CJ). Furthermore, the efficient implementation of LIMEX, which reuses the same pivot sequence over one step, can result in higher computational times for (CJLU) compared to (CJ). MA28AD is employed only once to compute the LU decomposition on one step. All subsequent computations of the LU decompositions are computed by calls to MA28BD, which substantially saves computational effort. Therefore, the overhead due to communication might be equal to the potential savings.

3.2 Load balancing

In order to obtain the optimal degree of speedup, the appropriate numbers of CPUs assigned to each task has to be determined. The state integrator has to be sufficiently fast to allow for computation of the Jacobians in time, when the sensitivity integrators require the Jacobians. The theoretical limit of the reduction in computational time is the pure state integration, which follows from Amdahls law. Hence, there exists an upper limit for the total number of CPUs for the chosen approach of parallelization. Nowak *et al.* (1998) presented a parallel version of the LIMEX state integrator. Merging both approaches would result in a lower theoretical limit.

3.3 Memory issues

During one of the case studies, which will be described in more detail in Section 4, the required memory space for sensitivity integration exceeded the allowed limit of 3GB for 32Bit software on Windows. Due to the fact that parts of the code is commercial and at that point in time not available in 64Bit, the sensitivity integration had to be split into several sequential sensitivity integrations, where each integration computes a part of the entire sensitivity information. Hence, the parallelization using MPI greatly contributes to the efficiency of the software, since all sensitivities can be computed by one parallel sensitivity integration, as the sensitivity load is distributed on several processes, each of which can address up to 3GB. The assessed numerical example features a rather small number of sensitivities, which can be computed by one serial sensitivity integration in order to allow for a comparison of the serial code to the parallel one. However, for the real case study the speedup is not only larger due to the higher overall load of sensitivities and a longer time horizon, the real speedup is even higher, since the serial code would have to compute the sensitivities by repeated sensitivity integrations on a reduced load.

4 Numerical Experiments

In this section the algorithmic performance is investigated by means of three case studies. The efficiency

$$\Theta = \frac{\Psi}{n_p} \tag{6}$$

is of interest not only for very large-scale models, but also when considering small-scale problems. n_p denotes the number of CPUs and Ψ speedup

$$\Psi = \frac{t_s}{t_p}, \tag{7}$$

where t_s denotes the serial computational time and t_p the parallel one. The examples are chosen over a wide range of applications. The chosen approach for parallel computing is targeted at large problems, where the model consists of many equations and many sensitivities have to be computed. The numerical experiments have been conducted on Xeon 2.66 GHz double quad core machines employing 16 GByte of random access memory running on Windows Server 2008. Since only 8 CPUs on one node have been available and the network between the nodes has not been considered to be sufficiently fast, the numerical experiments have not yet assessed the parallelization beyond 8 CPUs.

4.1 Case studies

4.1.1 Example 1: Polymerization process (236 DAEs)

The controls are discretized by 68 parameters. The model is considered very small, which is suggested by the very low computational times for one single sensitivity integration. The applied strategy, computational times required for one sensitivity integration, speedup and efficiency are depicted in Table 1. For strategies (CJ) and (CJLU) also the number of CPUs per task are given. Since strategy (DS) only employs one type of process, all CPUs are used as distributed sensitivity integrators. The number in brackets shows the computational time for one single state integration. This number represents the theoretical limit for the parallelization. Only strategy (DS) has been successfully applied to this example, since the other approaches have resulted in higher computational times compared to the serial code because of large overhead. Even though this example is very small and features only a small number of sensitivities, a speedup in the range of two has still been achieved.

Strat.	CO	STI	JC	SEI	#CPU	Time [sec]	Ψ	Θ
serial	0	0	0	0	1	2.54(0.5)		
DS	0	0	0	0	2	1.67	1.52	0.84
DS	0	0	0	0	3	1.44	1.76	0.48
DS	0	0	0	0	4	1.51	1.68	0.38
DS	0	0	0	0	5	1.34	1.90	0.27
DS	0	0	0	0	6	1.31	1.94	0.28
DS	0	0	0	0	7	1.29	1.97	0.18
DS	0	0	0	0	8	1.26	2.02	0.16

Table 1: Example 1: Strategy, number of processors and computational times

4.1.2 Example 2: Polymerization process (2104 DAEs)

The controls are discretized by 197 parameters. For this example, strategy (CJ) turns out to be the fastest strategy (Tab. 2), whereas DS gives slightly worse results and CJLU is significantly worse compared to the other approaches. In total, a speedup of around 3 is obtained. Only one JC is considered sufficient, since the major computational effort is required for the computation of the sensitivities. Strategy DS does not even show major improvements in the total computational time if the number of employed CPUs is increased beyond six.

Strat.	CO	STI	JC	SEI	#CPU	Time [sec]	Ψ	Θ
serial	0	0	0	0	1	33.27(2.3)		
DS	0	0	0	0	2	17.68	1.88	0.94
DS	0	0	0	0	3	13.04	2.55	0.85
DS	0	0	0	0	4	12.97	2.57	0.64
DS	0	0	0	0	5	12.01	2.77	0.55
DS	0	0	0	0	6	11.57	2.88	0.48
DS	0	0	0	0	7	11.34	2.93	0.42
DS	0	0	0	0	8	11.28	2.95	0.37
CJ	1	1	1	1	4	29.19	1.14	0.28
CJ	1	1	1	2	5	15.73	2.12	0.42
CJ	1	1	1	3	6	12.53	2.66	0.44
CJ	1	1	1	4	7	11.4	2.92	0.42
CJ	1	1	1	5	8	10.74	3.10	0.39
CJLU	1	1	1	1	4	33.29	1.00	0.25
CJLU	1	1	1	2	5	20.98	1.59	0.32
CJLU	1	1	1	3	6	18.83	1.77	0.29
CJLU	1	1	1	4	7	17.95	1.85	0.26
CJLU	1	1	1	5	8	17.55	1.90	0.24

Table 2: Example 2: Strategy, number of processors and computational times

4.1.3 Example 3: Large-scale intermediate chemicals process (13956 DAEs)

The controls are discretized by 75 parameters. This example is considered the most important example for the studied parallelization strategies. One of the most interesting aspects of this case study is, that all physical properties are evaluated by external functions, i.e. so-called physical property packages. Hence, they have a strong impact on the computational time for the Jacobians, whereas however they result in a rather small Jacobian, since the corresponding equations are solved within the property packages and do not show up in the model. These physical properties require many floating point operations, whereas the corresponding LU decomposition is computed comparably fast, since the LU decomposition is only dependent on the size and structure of the Jacobian.

Considering the discussion in Section 3.3 the speedup would be larger for a higher number of parameters. However in this case, the serial performance could not be determined due to memory issues. Since several sequential sensitivity integrations with a reduced load would have to be performed, the real speedup of the parallel code is significantly larger, such that even the efficiency Θ can exceed 100%.

In contrast to the other examples, the parallelization performance almost reaches its theoretical limit, which is the state integration without sensitivities. Strategy CJ with 4 CPUs assigned to JC (489 sec) almost reaches the limit (435 sec). The communication overhead that occurred for this example is comparably small, since the general computational times are large enough. Furthermore, the frequency of communication decreases.

Strat.	CO	STI	JC	SEI	#CPU	Time [sec]	Ψ	Θ
serial	0	0	0	0	1	2080 (435)		
DS	0	0	0	0	2	1997	1.04	0.52
DS	0	0	0	0	3	1914	1.09	0.36
DS	0	0	0	0	4	1866	1.11	0.28
DS	0	0	0	0	5	1832	1.14	0.23
DS	0	0	0	0	6	1755	1.19	0.20
DS	0	0	0	0	7	1738	1.20	0.17
DS	0	0	0	0	8	1701	1.22	0.15
CJ	1	1	1	1	4	1375	1.51	0.38
CJ	1	1	2	2	6	698	2.98	0.50
CJ	1	1	3	3	8	543	3.83	0.48
CJ	1	1	2	4	8	735	2.83	0.35
CJ	1	1	4	2	8	489	4.25	0.53
CJLU	1	1	3	3	8	611	3.40	0.43
CJLU	1	1	2	2	8	776	2.68	0.34
CJLU	1	1	1	1	8	1434.58	1.45	0.18
CJLU	1	1	2	4	8	838.54	2.48	0.31
CJLU	1	1	4	2	8	593.65	3.50	0.44

Table 3: Example 3: Strategy, number of processors and computational times

4.2 Discussion of results

The above examples show that the chosen approaches work very well for large-scale problems and still quite efficient for small and medium-scale problems. Strategy (CJ) performs best for the larger two examples and (DS) for the smallest one, which was expected to a certain extent. The difference between (CJ) and (CJLU) is rather small for the largest example. However, for this example (CJ) works better, since the LU decomposition is rather small compared to the computational effort for each Jacobian.

5 Conclusion

In this paper, the parallelization of the combined state and sensitivity integration of a dynamic optimization approach has been suggested. The one-step extrapolation code of Schlegel *et al.* (2004), that requires by far the largest computational effort in the chosen method of the *sequential approach*, has been parallelized by means of an exploitation of the algorithmic structure. Three numerical examples were examined, which assess the three implemented approaches to parallelize the algorithms. The simple distributed computation of the sensitivities turns out to be the best approach for very small examples, whereas the additional centralized computation of the Jacobians gives considerable extra speedup, even though the overhead increases significantly. For the largest example, the theoretical limit has almost been reached. The additional communication of the LU decomposition turns out to be not efficient for the chosen examples, since the communication overhead of the enormous amount of data is larger than the computational times for the LU decomposition at all. For future work, different implementations for the computation of the LU decomposition will be assessed, but also different possibilities of the parallelization of the LU decomposition. Furthermore, OpenMP can be employed in order to speed up each process of the illustrated algorithm. Since only nodes with 8 CPUs have been employed, the usage of OpenMP has been omitted for this work, since already 4 tasks are required to be divided onto the available CPUs.

References

Binder, T., L. Blank, H. G. Bock, R. Bulirsch, W. Dahmen, M. Diehl, T. Kronseder, W. Marquardt, J. P. Schlöder and O. von Stryk (2001). Introduction to model based optimization of chemical processes on moving horizons. In: *Online Optimization of Large Scale Systems* (M. Grötschel, S. O. Krumke and J. Rambau, Eds.). Springer-Verlag Berlin Heidelberg. pp. 297–339.

Deuflhard, P. (1983). Order and stepsize control in extrapolation methods. *Numerische Mathematik* (41), 399–422.

Deuflhard, P. and F. Bornemann (1994). *Numerische Mathematik II.* De Gruyter. Berlin.

Deuflhard, P., E. Hairer and J. Zugck (1987). One-step and extrapolation methods for differential-algebraic systems. *Numerische Mathematik* (51), 501–516.

Duff, I.S. (1979). Ma28 - a set of fortran subroutines for sparse unsymmetric linear equations. Technical Report AERE-R8730. AERE Harwell.

Ehrig, R., U. Nowak and P. Deuflhard (1996). Highly scalable parallel linearly-implicit extrapolation algorithms. Technical Report TR 96–11. Konrad-Zuse-Zentrum für Informationstechnik. Berlin.

Gill, P.E., W. Murray and M.A. Saunders (1998). SNOPT: An SQP algorithm for large-scale constrained optimization. Technical report. Stanford University. Stanford, USA.

Keeping, B.R. and C.C. Pantelides (1998). A distributed memory parallel algorithm for the efficient computation of sensitivities of differential-algebraic systems. *Math. comp. Sim.* **44**, 545–558.

Nowak, U., R. Ehrig and L. Overdieck (1998). Parallel extrapolation methods and their application in chemical engineering. *High-Performance computing and networking lecture notes in computer science* **1401**, 419–428.

Schlegel, M., W. Marquardt, R. Ehrig and U. Nowak (2004). Sensitivity analysis of linearly-implicit differential-algebraic systems by one-step extrapolation. *Appl. Num. Math.* **48**, 83–102.

Srinivasan, B., S. Palanki and D. Bonvin (2003). Dynamic optimization of batch processes I. Characterization of the nominal solution. *Comp. Chem. Eng.* **27**, 1–26.

Zhu, W.J. and L. Petzold (1999). Parallel sensitivity analysis for daes with many parameters. *Concurrency-practice and experience* **11**, 571–585.

409c Quasi-Decentralized State Estimation and Output Feedback Control of Process Systems Over Communication Networks

Yulei Sun and Nael H. El-Farra. Department of Chemical Engineering & Materials Science, University of California, Davis, One Shields Avenue, Davis, CA 95616

Traditionally, control of chemical plants with multiple geographically-distributed interconnected units has been studied within either the centralized or decentralized control frameworks. In centralized control, all measurements are collected and sent to a central unit for processing, and the resultant control commands are then sent back to the plant. In decentralized control, on the other hand, the plant is decomposed into a number of simpler subsystems (typically based on functional and/or time-scale differences of the unit operations) with interconnections, and a number of local controllers are connected to each subsystem with no signal transfer taking place between different local controllers. An approach that provides a compromise between the complexity of traditional centralized control schemes and the performance limitations of decentralized control approaches is quasi-decentralized control. The term quasi-decentralized control refers to a distributed control structure in which most signals used for control are collected and processed locally -- although some signals (the total number of which is kept to a minimum) still need to be transferred between local units and controllers to adequately account for the interactions between the different units and minimize the propagation of disturbances and process upsets from one unit to another.

One of the key problems that need to be addressed in the design of quasi-decentralized control systems is the coordination between the control and communication tasks and how to account for possible limitations of the communication medium in the formulation and solution of the control problem. The importance of this problem stems from the increased reliance in the process industries in recent years on distributed sensor and control systems that are accessed over communication networks rather than dedicated links, which is motivated in part by the substantial savings in installation and maintenance time and costs as well as the flexibility and enhanced fault-tolerance capabilities of networked control systems. Also, as the trend towards augmenting dedicated control networks with low-cost wireless sensor and actuator networks in the process industry continues to take hold in order to achieve high-density sensing and actuation (see, for example, [1]), the need to account for communication issues in the controller design framework becomes apparent. In this context, communication limitations arise both from the disruptions caused by interference in the field and/or environmental impact, as well as the inherent constraints on the power, computation and communication capabilities of the wireless devices.

The design of a quasi-decentralized control strategy that enforces the desired closed-loop objectives with minimal cross communication between the component subsystems is an appealing goal since it reduces reliance on the communication medium. This is an important consideration particularly when the communication medium is a potentially unreliable wireless sensor network where conserving network resources is key to prolonging the service life of the network. Beyond saving on communication costs, the study of this problem provides an assessment of the robustness of a given networked process control system, and allows designers to identify the fundamental limits on the tolerance of a given control system to communication suspensions. In an effort to address this problem, we developed in [2] a quasi-decentralized networked control architecture that enforces closed-loop stability with minimal cross communication between the constituent subsystems. The main idea was to embed in the local control system of each unit a set of dynamic models that provide the local controller with estimates of the states of the neighboring units, in order to be used when state information is not transmitted over the network. Both the control and communication laws in this case were derived under the assumption that the full state of each unit is available for measurement. In many practical applications, direct measurements of the full state are seldom available, and the lack of full state measurements has important implications that need to be accounted for both at the local control and the plant-wide communication levels.

Motivated by these considerations, we develop in this work a quasi-decentralized output feedback control architecture for multi-unit plants with limited state measurements and tightly interconnected units that exchange information over a shared communication network. We address the problem of designing an integrated state estimation, control and communication policy that requires minimal communication between the units without sacrificing closed-loop stability. To this end, we embed in the local control system of each unit a set of dynamic models that provide an approximation of the interactions between the given unit and its neighbors in the plant when communication is suspended over the network. To deal with the lack of full state measurements, an appropriate state observer is included in the local control system of each unit to generate estimates of the local state variables from the measured outputs. The estimates are used to implement the local state feedback controllers and are also broadcasted at the plant-wide level over the shared communication network to update the state of the corresponding model embedded in the interconnected subsystems when communication is re-established. Using a switched system formulation, an explicit characterization of the maximum allowable update period (i.e., minimum cross communication frequency) between the distributed control systems is obtained in terms of plant-model mismatch, the controller and observer design parameters. The results are illustrated using a chemical plant example and compared with existing networked control strategies.

References:

[1] Song, J., A. K. Mok, D. Chen, and M. Nixon, ``Challenges of wireless control in process industry," Proceedings of Workshop on Research Directions for Security and Networking in Critical Real-Time and Embedded Systems, San Jose, CA, 2006.

[2] Sun, Y. and N. H. El-Farra, ``Quasi-decentralized model-based control of networked process systems," Comp. & Chem. Eng., in press.

409d Cooperative, Distributed Model Predictive Control for Systems with Coupled Input Constraints

Brett Stewart and James B. Rawlings. Department of Chemical and Biological Engineering, University of Wisconsin-Madison, 1415 Engineering Drive, Madison, WI 53706

We consider a plant-wide control problem divided into several subsystems, each of which is controlled by a model predictive controller (MPC). The MPC systems employ two optimization layers. The steady-state target optimization layer calculates feasible steady states, accounting for disturbances in plant operating conditions. The subsystem dynamic controllers move the plant to these targets. Coupled constraints are constraints shared between subsystems within the plant. These constraints arise because of shared subsystem feeds or utilities and are usually represented as the o -diagonal blocks in the plant-wide constraint matrix. Distributed model predictive control, the control strategy considered here, decomposes the plant-wide control problem into a collection of subsystem problems while ensuring stability and optimality [1, 2]. Distributed control is viewed as more flexible and robust because the plant does not rely on a single centralized controller.

The distributed control problem can be viewed as an N-player game in which subsystems minimize their local objective [3]. In communication-based strategies [1], the subsystems iterate to a Nash equilibrium, which is sufficient for stability if the subsystems are weakly coupled dynamically. This strategy is inadequate, however, for the steady-state problem because typically the interactions between subsystems are strong. In this case a Nash equilibrium is not necessarily a stable operating point. Cooperative strategies, however, produce iterates that converge to the Pareto optimal point, a stable point for any strength interactions [2]. Cooperative model predictive control has been shown to work with both centralized and distributed steady-state target calculations but has so far assumed that the decomposed subsystems' constraints are uncoupled in the distributed target calculation. Introduction of coupled constraints into the cooperative steady-state target calculation produces suboptimal plant-wide steady states.

Most previous work for distributed target calculation with coupled constraints uses a coordinator [4]. The coordinator solves a dual problem for the Lagrange multipliers of the coupled constraints. Structurally, however, the coordinator is similar to the centralized target calculation. A completely distributed control system does not have a coordinator.

Using industrial examples, we illustrate how the coupled constraint problem arises. We show that while cooperative control with coupled constraints is suboptimal, it guarantees closed-loop stability for all stationary points of the steady-state target calculation. Finally, we show how performance can be improved using algorithms that account for the constraint coupling without the need for third-party coordination. These methods maintain the attractive stability and optimality properties of cooperative controller while preserving the distributed control topology in plants with coupled constraints.

REFERENCES

[1] Eduardo Camponogara, Dong Jia, Bruce H. Krogh, and Sarosh Talukdar. Distributed model predictive control. IEEE Ctl. Sys. Mag., pages 44–52, February 2002.

[2] Aswin N. Venkat, James B. Rawlings, and Stephen J. Wright. Stability and optimality of distributed, linear MPC. part 1: state feedback. Technical Report 2006–03, TWMCC, Department of Chemical and Biological Engineering, University of Wisconsin–Madison (Available at http://jbrwww.che.wisc.edu/tech-reports.html), October 2006.

[3] James B. Rawlings and Brett T. Stewart. Coordinating multiple optimization-based controllers: new opportunities and challenges. In DYCOPS, Cancun, Mexico, June 2007.

[4] R. Cheng, JF Forbes, and WS Yip. Price-driven coordination method for solving plant-wide MPC problems. J. Proc. Cont., 17(5):429–438, 2007.

409e Large-Scale Optimization Strategies for the Operation of Low-Density Polyethylene Tubular Reactors

Victor M. Zavala and Lorenz T. Biegler. Department of Chemical Engineering, Carnegie Mellon University, 5000 Forbes Ave., Pittsburgh, PA 15213

There is substantial economic interest to optimize the operations of low-density polyethylene (LDPE) tubular reactors. Due to the high complexity of these units, systematic optimization techniques need to be used for this. One of the main limitations associated to this is the high dimensionality and complexity of the multi-zone tubular reactor model which is described by a large set of partial, ordinary and algebraic equations (PDAEs). In this work, we demonstrate that a simultaneous full-discretization approach coupled to a full-space nonlinear programming (NLP) solver results in an efficient strategy to cope with these limitations. We exploit these advantages in the analysis of different scenarios arising in the operation of LDPE reactors. In particular, we propose moving horizon estimation and nonlinear model predictive control strategies to track and compensate for persistent disturbances such as fouling in order to keep the reactor temperature profile and final properties of the polymer at targets. Finally, we show that the controller can easily be extended to incorporate economic decisions in the objective and we illustrate the potential benefits and bottlenecks of this approach.

Web Page: www.andrew.cmu.edu/~vzavala

409f Analytical Results for Tiered Intermediate Product Prices for Apc Using a Refinery Lp

Jeyarathan Arjunan and James R. Whiteley. Chemical Engineering, Oklahoma State University, 423 EN, Stillwater, OK 74078

Intermediate tiered prices are needed for advanced or multivariable process control (APC) and economic optimization in petroleum refineries. The objective function of an APC controller requires economic values for intermediate products such as naphtha, gas oils, resids, etc. These economic values are used directly or indirectly depending on the choice of APC technology. Controlled variable targets and manipulated variable set-points are directly determined from the once-a-minute (typically) solution of the APC LP or QP. The tiered price of refinery products such as aviation gasoline and motor fuel are established by their market value. Intermediate products like naphtha and heavy oils generally do not have a direct market value and their prices are determined implicitly by variations associated with fuel product prices, yield characteristics, operating costs, capacity limitation of equipment and, crude type.

This paper addresses the use of marginal values generated by the Refinery Planning & Economics plant-wide LP for intermediate product prices in APC controllers. Solution of a plant-wide LP produces marginal values or shadow prices for all intermediate products by virtue of material balance equality constraints. As background material, this paper provides a detailed discussion of the unique characteristics, correct interpretation, and theoretical implications of marginal values associated with material balance constraints. The background discussion (geometric and analytic) that is provided extends beyond the surprisingly limited treatment of marginal values for equality constraints found in most LP texts. A solid understanding of the implications of material balance equality constraints is required to correctly interpret the results presented in the remainder of the paper.

There is no consensus regarding the potential use of refinery-wide LP marginal values for APC intermediate product prices. Rigorously speaking, a purist can state two reasons not to use LP marginal values for APC intermediate product prices. The first is that the model used to generate the marginal values is linear while the actual process is nonlinear. The second is that the refinery will normally operate at a point different than the optimum LP solution. In both cases the purist can claim that the LP marginal values will be different from the "true" marginal values. While true, the practical questions are 1) how large is the difference between the LP and "true" marginal values, and 2) is the relative difference between LP and "true" marginal values for different intermediate products constant? The APC controller is more concerned with the latter. While the purist is correct, the APC engineer needs intermediate product prices.

The results reported in this paper assume the process is linear and investigate the change in LP marginal values for intermediate product prices over varying operating conditions. The intent is to understand the applicability of intermediate product price over a range of different operating conditions. The concept of tiered price is introduced to address this issue. We define tiered prices as the marginal values at degenerate points when a maximum capacity constraint is systematically reduced. In this work tiered prices were generated by varying the crude distillation capacity of a refinery LP with 33-decision variables and 39-constraints obtained from literature.

The feasible region in a refinery LP is characterized by the null space defined by the material balance constraints. While tiered prices were generated by varying a particular constraint, the constraint of interest traverses the null space. The feasibility of the LP solution is maintained as long as the constraint movement is restricted within the null space. Analytically, the LP basis changes at the optimum solution associated with each degenerate LP solution or tier. Changes in intermediate product prices were not observed at all the tiers. Instead, the marginal values of intermediate products were found to vary only at specific tiers. The rationale behind this occurrence is correlated to the type and behavior of the active constraints at each tier. A physical and theoretical interpretation of tier prices relative to constraint movement and the activity of decision variables are provided and offer a better appreciation for the values and ranges of intermediate product prices. The discussion also addresses the APC engineers concern on when to change the price in APC objective function. To assist better understating and visualization of these concepts an example 3-D LP problem with algebraic and graphical solution will be provided.

409g Implementation of Coordinator Mpc on a Large-Scale Gas Plant

Elvira M. B Aske, Dept. of Chemical Engineering, Norwegian Univ of Sci & Tech (NTNU), Trondheim, N7491, Norway, Stig Strand, Process Control, Statoil, Trondheim, Norway, and **Sigurd Skogestad**, Chemical Engineering, Norwegian Univeristy of Science and Technology, Sem Sælandsvei 4, Trondheim, 7491, Norway.

In many cases economic optimal operation is the same as maximum plant throughput. This corresponds to maximum flow through the bottleneck(s). This insight may greatly simplify implementation and a real-time optimization (RTO) based on a nonlinear model is not necessary. The key issue is to identify the active "bottleneck" constraint and a coordinator MPC based on a linear model with constraints is proposed (Aske *et.al.*, 2008)

The main objective is to maximize the throughput (feeds) of the plant, subject to achieving feasible operation(satisfying operational constraints in all units). With the assumption that the flow through the network is represented by a set of units with linear flow connections, the maximum throughput problem is then a linear programming (LP) problem. The use of MPC to solve the LP has the benefit of allowing for a coordinated dynamic implementation. The constraints for the coordinator MPC are the maximum flows through the individual units. These may change with time and a key idea is that they can be obtained with almost no extra effort using the models in the existing local MPCs. The coordinator MPC receives a remaining capacity measure from each of the local MPC applications (each unit) and thereby tries to maximize throughput by adjusting feed rates and crossovers (between parallel trains). In this paper, an actual implementation of coordinator MPC at the Kårstø gas plant is described.

The plant

The Kårstø gas processing plant plays a key role in the transport and treatment of gas and condensate from central parts of the Norwegian continental shelf. The plant receives rich gas and unstabilized condensate through pipelines and separates the feeds into its various components. More than 30 fields export its gas through the Kårstø plant, and this set high demands to the plant efficiency and regularity. If the gas plant stops or limits the export, the one or several fields must reduce, or in worst-case, shout down oil- and gas export since the fields can not dispose the gas elsewhere (burning the gas offshore is prohibited).

Calculating remaining capacity

The constraints for the coordinator MPC are non-negative remaining capacities ($R \geq b \geq 0$) in all units, where R is the remaining capacity in the unit and b denotes the back off. The remaining capacity for unit k is given by R_k and is the distance between the current feed and maximum feed, $R_k = F^l_{k,max} - F^l_k$, where F^l_k is the current feed to the local unit and $F^l_{k,max}$ is the maximum feed the unit can receive and still operate within the operation constraints.

Most of the distillation columns at the Kårstø gas plant have already MPC installed with two-point composition control. These local (unit) MPC applications are used to calculate the maximum feed rate each unit can receive, subject to given controlled variable (CV) and manipulated variable (MV) constraints. The maximum feed to the unit k is then obtained by solving the

additional steady-state problem:

$$F^{l}_{k,max} = max\ F_k$$

subject to the linear model equations and constraints of the local MPC, and this is a LP problem. Here u^{l}_{k} is the vector of manipulated variables in the local MPC, and the optimization is subject to satisfying the linear constraints for the unit. To include past MV moves and disturbances, the end predictions of the variables should be used instead of the present values.

Coordinator MPC design

For the total plant, the coordinator MPC consist of approximately 36 CVs, 12 MVs and 3 DVs. However, a sub-application is implemented first, because it is easier and we want to demonstrate the coordinator ability. Another reason is that parts of the plant will be rebuilt shortly and therefore no local MPC application is implemented on this part.

The coordinator MPC is implemented in SEPTIC MPC (Strand and Sagli, 2003). The sub-application consist of 22 CVs (14 remaining capacities, 7 other constraints and total plant feed), 6 MVs (4 feed rates, 1 crossover and 1 feed split) and 5 DVs (2 feed compositions, 1 feed rate, 1 crossover and 1 feed split). The CVs are constraints with low- and/or high limits, except for the total plant feed, which is a set point controlled CV. The total plant feed is

the sum of the feeds rates, which should be maximized. To maximize throughput, the total plant feed set point is a high, unreachable set point with lower priority than the constrained CVs.

All distillation columns in the sub-application (12 in total) have local MPC installed and the remaining capacity is calculated there. The compressors have not MPC control, but here are "dummy" MPC applications to calculate their remaining capacity. The two "dummy" applications consist only of CVs and DVs. The other CV constraints in the coordinator are measurements to avoid emptying or filling pipelines and tanks.

The coordinator MPC receives three values from each of the local MPC: the calculated remaining capacity, quality of the remaining capacity calculation(GOOD/BAD) and status of the local MPC application (ON/OFF). The latter value is used to avoid that the coordinator MPC is active without the local MPC applications is active.

Each variable (CV, MV and DV) is located in one or more sub-groups that will deactivate if one critical variable is deactivated. For instance, if a local MPC application is turned off, the corresponding remaining capacity is deactivated, which leads deactivation of the whole sub-group, including the correspondingfeed rate. The coordinator MPC can then operate even if parts of the plant is no running or not available for throughput maximizing for some reason. For the coordinator MPC, each MV defines a sub-group with corresponding CVs as members.

Coordinator MPC modelling and tuning

The models in the coordinator MPC are all SISO step-response models. The models are obtained from step tests and historical plant data. The steady-state gain in the models from the feed rates is calculated from a typical feed composition to validate step-tests. The prediction and control horizon is set to be 6 hours, whereas the longest models are approximately 4.5 hours. The coordinator has 4 integrating CVs, and for these the control horizon is a tuning parameter that can be set individually. Here they are selected to half the prediction horizon, which is 3 hours.

The tuning of the coordinator MPC is a trade-off between robustness and MV (e.g. feed) variations on one side and keeping the flows through the bottlenecks close to their maximum on the other side. A rational use of the MVs is prioritized to avoid unnecessary variations in throughput.

Experience from implementation

The performance of the local MPC applications are crucial for the coordinator MPC to work as intended. The weakness is the calculation of the remaining capacity in each unit. The goodness of the models and in particular the steady-state gain is important. Our experience is when a larger disturbance occurs, the predicted steady-state values may violate limits and if this violation is large enough, the LP optimization do not find a feasible solution, and the maximum capacity calculation (Flmax) turns bad. The end prediction value is in such cases often not reasonable because the MPC application assumes that the disturbance will persist constant (or with a low-pass filter)throughout the prediction horizon, which is seldom the case. To overcome this problem, several approaches are used:

With a known, measured, short-time disturbance: the maximum capacity (F_max) is held constant throughout the disturbance. This is used for the disturbance that occurs at each dryer exchange.

Include a minimum value of maximum capacity, Flmax, that the unit for sure is able to process.

Include gain scheduling in models to describe typically non-linear behavior.

The coordinator MPC is also vulnerable to model errors in the local MPC applications. The local MPCs consider in general product quality control within some operational constraints. However, due to feed back, acceptable product quality control can be achieved even with some model errors. On the other hand, the coordinator MPC uses the models in the local MPCs directly for estimation of remaining capacity. Thus, model errors leads to errors in the remaining capacity calculation. We have observed variations in the estimated capacity with periods of 1-2 hours. These variations are challenging because this is the bandwidth where we want the coordinator to operate; hence these variations can not be reduced by filter. The variations in the estimated capacities were usually due to model errors especially from the unit feed. A systematic treatment of the inferential models (estimators of product quality) and models in the local MPC applications is necessary to obtain satisfactory performance of the coordinator MPC.

The operators are familiar with the MPC interface after several years of implementing the local MPC applications at the plant and operating those. This is a clear advantage because the coordinator MPC has the same interface. Close dialog with both operator personnel and operator leader is crucial.

The coordinator MPC uses the feed valves (throughput rate) in closed-loop. This directly involves the operator leader because the leader receives guidelines from the booking responsible and operators of the gas pipeline network.

415a Cyberinfrastructure (CI) Funding In the Engineering Directorate (ENG) of the National Science Foundation (NSF)

Maria K. Burka, National Science Foundation, 4201 Wilson Boulevard, Room 525, Arlington, VA 22230

Pursuant to the Atkins report and numerous subsequent workshops, CI has become a major focus of recent initiatives at NSF. This has included funding opportunities in high performance computing, data storage, sharing and management, and the application and uses of CI to all domains in science in engineering. The engineering community has had, and will continue to have in the near future, many opportunities to participate in these funding opportunities. Past successes and failures, future programs, and the funding outlook for research and education in 2009 will be presented.

415b Pharmaceutical Informatics and New Financial Models of the Bio-Pharma Industry

Sangtae Kim, Chemical Engineering, Purdue University, 585 Purdue Mall, West Lafayette, IN 47907

In silico efforts to accelerate drug discovery and development, promoted with great fanfare at the dawn of the genomics era, have largely disappointed the decision makers in the R&D labs of the bio-pharma industry. Academics on the other hand, have greeted successive cyber transitions, e.g., from from Mega- to Giga-, and Giga- to Tera-, with new rounds of hype about accelerating the search for cures or implying implicitly that this time it will be different. Nevertheless, this speaker has reasons to believe that the next transition, from the Giga- to the Peta-scale eras, will really be different. These conclusions lead to some intriguing possibilities of new financial models to propel the R&D supply chain from university labs to the proof of concept, Phase II clinical trials.

Web Page: engineering.purdue.edu/people/sangtae.kim.1/Pharma-essay.htm

415c Challenges in Biological Informatics

Nick Sahinidis, Chemical Engineering, Carnegie Mellon University, 5000 Forbes Avenue, Pittsburgh, PA 15213

Structural biology has witnessed an unprecedented influx of mathematical and computational techniques over the past three decades. These techniques have transformed research in a field that has become increasingly data-driven. As part of this revolution, optimization techniques are now used routinely to solve problems in structure elucidation and the study of structure-function relationships. However, many fundamental problems in this field remain largely unsolved. We present some of the challenges and opportunities in biological informatics, and review recent progress toward the solution of problems in the deciphering of protein structures, as well as the comparison and classification of known protein structures.

SEMANTIC TECHNOLOGIES FOR WORKFLOW AND INFORMATION MANAGEMENT

Wolfgang Marquardt
AVT – Process Systems Engineering & Center for Computational Engineering Science
RWTH Aachen University, Aachen, Germany

Introduction

Product and process design processes (not only but also) in the process industries are of a cooperative nature involving different departments in a single or in more than one enterprise typically at geographically distributed sites. They use and produce a multitude of documents with many interdependencies and overlaps stored in very different electronic formats.

There are many software tools are used to support the project team. Some of them are of a domain specific nature (i.e. process simulators, CAD or CAE systems, etc.) and others are of a generic type and independent of the requirements of chemical engineering design processes (i.e. word processors, project management systems, etc.). An efficient support of chemical engineering design processes requires coherent information technology support across the lifecycle of the project. Existing software tools for dedicated tasks in the project lifecycle have to be enhanced and linked to evolve into a coherent design support system which not only offers user-interface, data and control integration but which rather integrates the distributed, collaborative, and concurrent design process carried out by interdisciplinary teams in different organizations.

To date, no satisfactory solution is available in industry. Therefore, an enormous potential in productivity increase of design teams and consequently in cost reduction and quality improvement exists. Leveraging such potential constitutes a tremendous economical opportunity in particular for enterprises in high wage countries such as Europe, Japan, the US and Canada.

This problem has been subject of the Collaborative Research Center IMPROVE [1], a long-term collaborative research project at RWTH Aachen University, Germany, which has been carried out since 1997 for a period of about 12 years in close collaboration between computer science and chemical engineering research groups. The findings of this project have been summarized in a recently published research monograph [2].

This brief paper addresses the problem from the perspective of semantic technologies. It gives an overview on our recent research which aims at applying and benchmarking ontological modeling and associated semantic technologies to selected key problems arising during chemical engineering design processes. We do extensively refer to our own papers which contain a lot of references to the work of other research groups in the field for the sake of brevity.

Understanding the Work Process

Real industrial work processes are complex and consequently difficult to plan, document, improve and reuse. A fundamental understanding of these work processes is considered to be a prerequisite for their reengineering and for the development of effective information technology based support systems. To this end we have developed a simple and intuitive graphical notation, called C3, which is used to develop a coarse-grained representation of a collaborative work process including the various activities, the actors and their roles, the information used and produced during the activities, and the resources (software tools, experiments etc.) used. An overview on the modeling language and its application to a number of industrial case studies is presented in [3, 4]. The modeling procedure has shown its power in such diverse applications as modeling of product design processes, conceptual process design processes, work processes in detail engineering as well as operational processes.

While the C3 modeling methodology is very useful for the participative modeling of work processes by the design team itself, it lacks the detail and degree of formality to be useful for monitoring the design process, for reasoning on its status, for the detection of inconsistencies of incomplete tasks or for the conceptual design and construction of software tools to support the design process. Therefore, formal information modeling by means of ontologies [5] and their processing with semantic software technologies is a logical next step.

Information Modeling

Information modeling has a long tradition in engineering design processes. Most notably, the STEP [6] modeling activity has been carried out for many years, not only but also in the process engineering domain. ISO 15026 [7] is a prominent example. These formats are intended to be used as data exchange formats in order to integrate software tools used during the engineering design process, to build data warehouses for the integration of existing data stores, or to even implement comprehensive engineering design databases which hold all project-relevant information on the designed process. Ontologies have been used just recently to represent the core concepts of ISO 15026 as an upper ontology [8].

These formats have not gained sufficient attention. Hence, the most prevalent problems of information integration, such as consistency or a desired degree of inconsistency along the project lifecycle, the exchange of information between tools and teams during the design process, the dependencies between documents, or more precisely, between the contents of the documents on different levels of granularity, or even the homogenization of different proprietary data formats has not been successfully addressed by this approach.

One reason of the lack of acceptance seems to be an insufficient capturing of the semantics of the information modeled. Hence, we are exploring the use of more expressive modeling approaches to facilitate the capturing of the detailed semantics of the modeled information items. Ontologies are employed to not only represent the result of the design process ("the product") but also the work process leading this result ("the actions") including the resources employed ("the tools"). The requirements in and the different strategies for application domain modeling are thoroughly discussed in [9], where our own work is also put into the perspective of existing approaches.

OntoCAPE – a Comprehensive Process Engineering Ontology

Object-oriented information modeling has a long tradition in our research group dating back to the early 1990 [10] where information modeling has been addressed in the context of developing advanced software environments for mathematical modeling [11]. OntoCAPE is based on the concepts of general systems theory. While its focus has been largely on the representation of the products of the design process, it also includes concepts to integrate the design process knowledge with the product. Its architectural design is based on abstraction and modularization which guarantees transparency and extensibility. OntoCAPE is represented in OWL [12]. A detailed description of the design principles or even a presentation of its major concepts is beyond the scope of this brief paper. We rather refer to the literature. In particular, an introduction and an overview on OntoCAPE is given in [13, 14], while a series of technical reports available from the author [15-19] present a detailed exposition of the ontology. A comprehensive presentation of the ontology and its design principles will be available in a forth-coming book which will be published early 2009 [20]. OntoCAPE is available under an open source license model at the website of the author (http://www.avt.rwth-aachen.de/AVT/index.php?id=486).

We want to stress that OntoCAPE is a full-fledged ontology which is in contrast to current practice where the term ontology is used in case ontological languages are used for knowledge representation but where all other requirements are not fulfilled [20]. Two types of ontology-like structures can be identified, which can be referred to as *pseudo ontologies* and *lightweight ontologies*. A pseudo-ontology is understood to be a part of a software system formulated in an ontology language, but has not been designed for reuse across various applications which we consider mandatory for a full-fledged ontology. Another differentiating factor is the semantic richness. Often, the so-called lightweight ontologies do not define the semantics of the concepts through axiomatic definitions, but rather utilize only a subset of the available modeling elements such as classes and instances but not relation properties, local constraints, or other forms of axioms. In this sense, OntoCAPE is considered a full-fledged ontology, to our knowledge the most comprehensive if not the only heavyweight ontology in the area of process engineering.

Recent Applications of OntoCAPE

OntoCAPE has been used in a number of large-scale academic and industrial case studies to validate the modeling approach and to demonstrate the applicability and the potential of semantic technologies in process engineering applications.

In the COGents project, funded by the European Union, an early version of OntoCAPE has been used to implement an agent-based system to search and select mathematical models on the web to match a given specification [21]. The ontology has been used to formalize the communication between the agents of the COGents system and to support the match making between a specification and the properties of a certain model. A process data warehouse has been designed and prototypically implemented to demonstrate the use of ontologies for information management in design processes [22] including the capturing of the activities of the designers, the resources used and the design products developed during the design process. Furthermore, an experimental mathematical modeling system ModKit+ has been designed and prototypically implemented using OntoCAPE [23, 24].

An industrial case study is currently in progress where OntoCAPE is used to integrate heterogeneous data in an industrial development process spanning different major tasks. In this project OntoCAPE is used to design a homogenization layer on top of XML-based representations of the content of a number of software tools. OntoStudio, a product of Ontoprise, is used to implement reasoning functionality. The case study is provided by Evonik Degussa. A more detailed description of the scope of this project is given in [25] while first results are included in [26].

Extending OntoCAPE Towards Work Process Modeling

While ontological technologies have been mainly used to represent, integrate and retrieve information on the result of the design process which is used or produced during its various steps, the methodology can be carried over to the formal representation of design products (or documents), the design processes and its resources. First steps towards the development of formal models using semantic modeling and ontological technologies are presented in [27].

This work is currently extended in order to extend OntoCAPE with process modeling capability. In particular, the processes to be modeled are not only confined to product and process design processes but also to operational processes to be applied to the plant. The objective of a current collaborative project aims at the development of such an extension of the ontology and on software tools which support the modeler and which allow an export of the ontological model on either the product and process design or the operational process to another software application which interprets or enacts the process model. Examples include discrete-event simulators to compute performance measures of the design process or a control software based on sequential function charts which implements the operational process of a certain unit. This project is carried out in close cooperation between our research group, Air Products, BASF SE, Bayer Technology Services and Siemens AG. A more detailed description of the project can be found in [28].

Final Comments

Our research on the use of semantic technologies in process engineering has clearly revealed the enormous potential of ontology-based information modeling and of the reasoning capabilities of semantic software technologies which are developed in the context of the semantic web. A properly chosen architecture of the ontology empowers the chemical engineer to extend and modify the ontology by means of high level modeling tools himself. The direct access to the knowledge facilitates maintenance and extension of the information model and the knowledge bases. The reasoning capabilities of modern semantic tools facilitate checking the logical consistency of the information model and the implemented knowledge. We not only have validated the concepts on academic toy problems, but we have implemented prototypical systems of future design support functionality and we are currently validating the methodology, OntoCAPE and the semantic technologies in industrial projects.

There are however also some drawbacks. In particular, the development of an extensible and widely usable ontology is by no means straightforward. Though we believe that OntoCPAE constitutes a very good foundation for a generic and widely usable ontology for process engineering applications, we still see a lot of room for improvement and for extension of the information model and the knowledge bases. Furthermore, the semantic technologies, in particular the ontology editors and reasoning tools are still under development. Performance is an issue for the latter, while usability by application-domain rather than knowledge engineering experts is the key issue for the former. The most important bottleneck at the moment is the lack of few best practice applications and demonstrators in an industrial setting and the lack of detailed application models to be used for the variety of tasks during the lifecycle of a design project or even of the plant itself.

Due to the immense workload required to come up with useful and comprehensive ontologies for process engineering applications, it is highly desirable, that academics and industry join forces in order to develop, maintain and gradually extend a process engineering ontology for process engineering. OntoCAPE is definitely a perfect starting point for such an undertaken.

Acknowledgements

The continuing financial support of Deutsche Forschungsgemeinschaft (DFG) is gratefully appreciated. The author would like to thank his colleagues in computer science, in particular M. Nagl, for many stimulating and fruitful discussions, and his students, in particular R. Hai, J. Morbach, M. Theißen, and A. Wiesner for their contributions to the success of this project and for their support in compiling this paper and its presentation in the conference.

References

1. Marquardt, W., and M. Nagl, (2004), "Workflow and information centered support of design processes - the IMPROVE perspective", *Comp. Chem. Engg.* 29, pp. 65-82.
2. Nagl, M. and W. Marquardt, eds. (2008): *Collaborative and Distributed Chemical Engineering Design Processes: From Understanding to Substantial Support.* Springer, Berlin.
3. Theißen, M., Hai R. and W. Marquardt (2008), "Design process modeling in chemical engineering", *Journal of Computing and Information Science in Engineering* 8, 011007.
4. Theißen, M., Hai, R., Morbach, J., Schneider, R. and W. Marquardt (2008), *"Scenario-based analysis of industrial work processes"*, in M. Nagl and W. Marquardt (2008), Chapt. 5.1.
5. Uschold, M. and M. Grüninger (1996), "Ontologies: principles, methods and applications," *Knowl. Eng. Rev.* 11 (2), pp. 93–136.
6. ISO10303, STandard for the Exchange of Product data
7. ISO15926, Industrial Automation Systems and Integration – Integration of Life-Cycle Data for Process Plants Including Oil and Gas Production Facilities
8. Batres, R., West, M., Leal, D., Price, D., Masaki, K., Shimada, Y., Fuchino, T. and Y. Naka (2007), "An upper ontology based on ISO 15926", Comp. Chem. Eng. 31 (5/6), 519–534.
9. Morbach, J., Theißen, M., Hai, R. and W. Marquardt (2008), "An Introduction to Application Domain Modeling," in M. Nagl and W. Marquardt (2008).
10. Marquardt, W., Gerstlauer, A., Gilles, E.D. (1993), „Modeling and Representation of Complex Objects: a Chemical Engineering Perspective", 6th Int. Conf. on Industrial and Engineering Applications to Artificial Intelligence and Expert Systems, Edinburgh, Scotland, pp. 219-228.
11. Bogusch, R., Lohmann, B. and W. Marquardt (2001), "Computer-Aided Process Modeling with MODKIT, *Comput. Chem. Engg.* 25, pp. 963-995.
12. W3C (2002), *"The OWL representation ontology"*, Online available at www.w3.org/2002/07/owl, accessed October 2007.
13. Morbach, J., Yang, A. and W. Marquardt (2007), "OntoCAPE – a large-scale ontology for chemical process engineering," *Engineering Applications of Artificial Intelligence* 20:2, pp. 147–161.
14. Morbach, J., Bayer, B., Yang, A., and W. Marquardt (2008), "Product Data Models", in M. Nagl and W. Marquardt (2008), Chapter 2.2.
15. Morbach, J., Bayer, B., Wiesner, A., Yang, A. and W. Marquardt (2008), *„OntoCAPE 2.0 – The Upper Level",* Technical Report (LPT-2008-25), Lehrstuhl für Prozesstechnik, RWTH Aachen University.
16. Morbach, J., Wiesner, A. and W. Marquardt (2008), *"OntoCAPE 2.0 – The Meta Model",* Technical Report (LPT-2008-24), Lehrstuhl für Prozesstechnik, RWTH Aachen University.
17. Morbach, J., Yang, A., Wiesner, A. and W. Marquardt (2008), *„OntoCAPE 2.0 – Supporting Concepts"*, Technical Report (LPT-2008-26), Lehrstuhl für Prozesstechnik, RWTH Aachen University.
18. Morbach, J., Yang, A. and W. Marquardt (2008), *"OntoCAPE 2.0 – Materials*", Technical Report (LPT-2008-27), Lehrstuhl für Prozesstechnik, RWTH Aachen University.
19. Morbach, J., Yang, A. and W. Marquardt (2008), *"OntoCAPE 2.0 – Mathematical Models",*

Technical Report (LPT-2008-28), Lehrstuhl für Prozesstechnik, RWTH Aachen University.
20. Marquardt, W., Morbach, J., Wiesner, A. and A. Yang (2009), *"OntoCAPE – A (Re)Usable Ontology for Chemical Process Engineering",* Springer, Berlin, in preparation.
21. Yang, A., Braunschweig, B., Fraga, E.S., Guessoum, Z., Marquardt, W., Nadjemi, O., Paen, D., Piñol, D., Roux, P., Sama, S., Serra, M., and I. Stalker (2007), "A multi-agent system to facilitate component-based process modelling and design", *Comp. Chem. Engg.* 32, pp. 2290-2305.
22. Brandt, S.C., Morbach, J., Miatidis, M., Theißen, M., Jarke, M., and W. Marquardt (2008), "An ontology-based approach to knowledge management in design processes", *Comp. Chem. Engg.* 32, pp. 320–342.
23. Yang, A., Morbach, J. and W. Marquardt (2004*), "From conceptualization to model generation: the roles of ontologies in process modeling",* Proc. Sixth International Conference on Foundations of Computer-Aided Process Design, pp. 591-594.
24. Yang, A. and W. Marquardt (2004), *"An ontology-based approach to conceptual process modelling",* Proc. of ESCAPE-14 (European Symposium on Computer Aided Process Engineering), Lisbon, Portugal, pp. 1159-1164.
25. Morbach, J. and W. Marquardt (2008), "Ontology-based integration and management of distributed design data", in M. Nagl and W. Marquardt (2008), pp. 647–655.
26. Wiesner, A., Morbach, J. and W. Marquardt (2008), "Semantic data integration for process engineering design data", Proceedings of the 10th International Conference on Enterprise Information System - ICEIS, pp. 190-195.
27. Eggersmann, M., Kausch, B., Luczak, H, Marquardt, W., Schlick, C., Schneider, N., Schneider, R. and M. Theißen (2008), „Work process models", in M. Nagl and W. Marquardt (2008), Chapt. 2.4.
28. Theißen, M., Hai, R. and W. Marquardt (2008), "Computer-assisted work process modeling in chemical engineering", in Nagl, M. and W. Marquardt (2008), pp. 656–666.

415e Informatics Needs, Challenges and Opportunities in Chemical Engineering: Roundtable Panel Discussion

Larry Megan, Praxair, 175 East Park Drive, PO Box 44, Tonawanda, NY 14150 and **Venkat Venkatasubramanian**, Chemical Engineering, Purdue University, 480 Stadium Mall Drive, West Lafayette, IN 47906.

10E Plenary Roundtable - The session co-chairs will lead a discussion on this theme with the speakers with participation from the audience.

424a Model-Based Product and Process Development for Polymers in An Industrial Environment

Klaus-Dieter Hungenberg, Polymer Process Development and Modeling, BASF SE, GKE/M - B001, D-67056, Ludwigshafen, Germany

Polymers are complex molecules where the properties strongly depend on the production process. Moreover, their properties themselves in turn have a strong impact on the process. For example, the molecular weight or the molecular weight distribution may change during the process and so the viscosity, which leads to changes in mixing behaviour, flow properties etc. Such interdependencies are captured by models, which not only take into account mass and energy balances, but also the polymerization kinetics and thermodynamics. By this, these models link process conditions to the molecular structure and to product properties. This facilitates the development of new products and processes and the optimization of existing processes. The process of model-building and the use of such models is demonstrated.

Development of On-line Optimization-Based Control Strategies for a Starved-Feed Semi-batch Copolymerization Reactor

Onyinye E Okorafo, Luis Fernando Torres, Robin A Hutchinson and Martin Guay. Chemical Engineering, Queen's University, 19 Division St, Dupuis Hall, Kingston, ON K7L 3N6, Canada.

Introduction

Acrylic resins for automotive coatings are of great interest to the polymer industry. Most automotive coating resins are complex copolymers primarily composed of styrene, acrylate, and methacrylate monomers. A "starved-feed" feeding policy is implemented that involves adding the initiator and monomer to the reactor at a constant rate over several hours. This feeding strategy is employed at high temperature conditions to ensure uniform average molecular weight and copolymer composition while operating at relatively low solvent levels. At these conditions of low monomer concentrations, high instantaneous conversion is achieved which results in polymer with relatively constant properties requiring minimal on-line measurement.

Operation of the robust reactor is at the cost of long batch time as well as drift in both polymer molecular weight and composition in the early stages of the batch. Therefore, it is important to improve reactor efficiency of quality monitoring and operation which will enhance product quality, production rates, safer operation, and subsequently improve profits. A dynamic model is developed to test the on-line optimization algorithm for the starved-feed free-radical copolymerization of butyl methacrylate and styrene (BMA/ STY) system, which reduces batch time while improving product uniformity. Three case studies are discussed to show the result of the optimization technique.

I. Description of System

Solution free-radical BMA/STY copolymerization was performed under starved-feed condition at 138°C. Monomer is added to the reactor at a constant rate over 6 hours at different mass ratio (feed compositions) with tert-butyl peroxyacetate initiator fed at a constant mass ratio of 2.0 wt% relative to monomer feed. The process is illustrated in Figure 1.

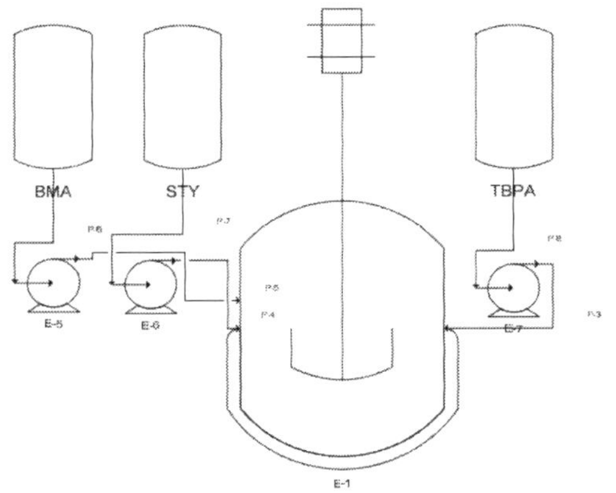

Figure 1: Schematic of semibatch reactor system. The flow rates of butyl methacrylate (BMA) and styrene (STY) monomers and tert-butyl peroxyacetate (TBPA) initiator can be independently manipulated.

The monomer and polymer compositions were shown [1] to be relatively uniform throughout the reaction in the semi-batch reactor as the free BMA and STY monomer levels were low for the whole experimentation.

II. Process Model

A full model [1] was developed for off-line optimization and product development of the present starved-feed feeding method containing a defined set of kinetic mechanisms for semi-batch high temperature BMA/ STY copolymerization. For this study, a mathematical model was formulated based on a reduced set of kinetic mechanisms embedded into the overall material balances in order to capture the basic process performance to design an efficient and robust system to control polymer quality. Methacrylate depropagation, styrene thermal initiation, and penultimate propagation kinetics were not incorporated in the mechanistic set.

Under standard assumptions such as perfect mixing, constant physical properties, quasi steady-state assumption (radical stationarity) and long chain hypothesis, mass balances on the initiator, solvent and monomers, and moment balances on the radical (live) and dead polymer chains yield the isothermal mathematical model for the semibatch system summarized below. The resulting set of differential equations describing the semibatch isothermal copolymerization is:

$$\begin{aligned}
\frac{d(x_1(t))}{dt} &= u_1(t) - x_1(t)(k_{p_{AA}}\lambda_o^A(t) + k_{p_{BA}}\lambda_o^B(t)) \\
\frac{d(x_2(t))}{dt} &= u_2(t) - x_2(t)(k_{p_{BB}}\lambda_o^B(t) + k_{p_{AB}}\lambda_o^A(t)) \\
\frac{d(x_3(t))}{dt} &= u_3(t) - k_d x_3(t) \\
\frac{d(x_4(t))}{dt} &= x_1(t)(k_{p_{AA}}\lambda_o^A(t) + k_{p_{BA}}\lambda_o^B(t)) \\
\frac{d(x_5(t))}{dt} &= x_2(t)(k_{p_{BB}}\lambda_o^B(t) + k_{p_{AB}}\lambda_o^A(t)) \\
\frac{d(x_6(t))}{dt} &= V(t)(R_{\mu_0}(t))
\end{aligned} \qquad (1)$$

where $x_1(t)$ and $x_2(t)$ represent the total mass of unreacted BMA and STY in the reactor, while $x_3(t)$ is the mass of unreacted initiator. States $x_4(t)$ and $x_5(t)$ represent the mass of BMA and STY contained in the polymer chains, and the number of moles of polymer chains is represented by $x_6(t)$.

The reduced model derived in this work capture the essential dynamics of the full model and the experimental data [1]. The primary references for these coefficients and parameters are contained in the work of Li and Hutchinson [1]. Various quantities of interest can be calculated from the states. For the semibatch reactor, monomer conversion is defined as the mass of polymer in the reactor divided by the mass of monomer and polymer:

$$\chi_m(t) = \frac{x_4(t) + x_5(t)}{x_1(t) + x_4(t) + x_2(t) + x_5(t)}$$

Number-average molecular weight and copolymer composition (BMA mole fraction), denoted by M_n and F, are calculated as follows:

$$M_n(t) = \frac{x_4(t) + x_5(t)}{x_6(t)}, \quad F(t) = \frac{x_4(t)}{M_A} \Bigg/ \left(\frac{x_4(t)}{M_A} + \frac{x_5(t)}{M_B} \right)$$

An objective of the starved-feed strategy is to keep these latter two quantities uniform throughout the course of the semibatch reaction, while maintaining high conversion in the reactor. Thus, the objective function is defined so as to minimize the batch time and maintain the copolymer molecular weight (M_n) and composition (F) at their targets values throughout the batch while also satisfying path and terminal constrains on the states and/or the controls.

III. Real-Time Optimization Mechanism

An optimization technique [2] is utilized to design a control structure that satisfies the specific objectives of the system under study discussed in the previous section. The feedback measurements are incorporated directly into the optimization procedure rather than into a low level tracking controller making it an online optimization instead of an online re-optimization, therefore eliminating the need for controller design and related difficulties. The technique used in this study is different from existing online optimization techniques ([3],[4],[5],[6]) as it involves minimal computational requirements, provides stable closed-loop performance and does not require knowledge of a nominal optimal solution.

The control problem undertaken in this study is to find the input trajectory that solves the dynamic optimization problem for some user defined cost function *J*. A dynamic optimization problem is formulated describing the feed flow rates parameters and the time interval lengths as decision or optimizing variables. For tracking purposes the objectives for M_n and F have been aggregated into a single scalar function while defining the batch time as a terminal objective function:

$$J = \int_0^{t_f} (\omega_1 (\frac{M_n(t)}{M_{sp}} - 1)^2 + \omega_2 (\frac{F(t)}{F_{sp}} - 1)^2) dt + \omega_3 t_f \qquad (2)$$

where J is the objective function to be minimized, and M_{sp} and F_{sp} are the target values for molecular weight and composition respectively. The set of weights to scale the objectives are defined by $\omega_1, \omega_2, \omega_3$. The optimization is subject to process dynamics (1) and the constraints described below.

A suitable means to limit potential heat release is to place constraints on unreacted monomer in the reactor. The monomer and initiator levels in the batch must not exceed certain maximum limits to meet safety constraints related to heat release. These constraints are described by:

$$0 \le x_k(t) \le x_k^{\max}, \; x_{k+3}(t) \ge 0 \quad k \in \{1,2,3\} \qquad (3)$$

Actuator limitations often dictate control constraints, whereas non-negativity of flow rates is a common input constraint.

$$0 \le u_k(t) \le u_k^{\max}, \quad k \in \{1,2,3\} \qquad (4)$$

Terminal constraints are commonly related to safety or productivity considerations. With regard to the latter, the desired mass of polymer at the batch end is specified as an endpoint constraint:

$$x_4(t_f)+x_5(t_f)=m_{pol}(t_f) \quad (5)$$

The control functions are parameterized in three distinct strategies. The first (Case 1) and simplest strategy considers using a single constant: $u_k(t)=u_k$ ($k=1,2,3$) through the entire batch time horizon subject to a starved-feed operation. The corresponding control parameterization is given by:

$$u_1(t)=\theta_1,\ u_2(t)=\theta_2\theta_1,\ u_3(t)=\theta_3\theta_1 \quad (6)$$

The second parameterization (Case 2) considered in this work allows for exponential feeding in a starved-feed operation. This parameterization is written as follows:

$$u_1(t)=\theta_1 e^{\theta_2 t},\ u_2(t)=\theta_3 u_1(t),\ u_3(t)=\theta_4 u_1(t) \quad (7)$$

The final parameterization (Case 3) allows for independent exponential feeding strategy. The objective is to verify the cost of utilizing a starved-feed strategy in practice. This slightly more complex parameterization is given by:

$$u_1(t)=\theta_1 e^{\theta_2 t},\ u_2(t)=\theta_3 e^{\theta_4 t},\ u_3(t)=\theta_5 e^{\theta_6 t} \quad (8)$$

Note that eq.(8) can be reduced to either eqs.(7) or (6) for specific choices of the parameters $(\theta_1,\theta_2,\theta_3,\theta_4,\theta_5,\theta_6)$.

The objective of the optimization technique is to steer the system to the local optimum of the user-defined cost eq.(2) subject to a specific parameterization (*i.e.* from eqs. (6), (7) and (8)) while responding in real-time to changes in plant operation. The technique cannot be used to recover the actual optimal control but provides suboptimal control strategies and allows the adjustment of the parameters of the input parameterization in a way that provides continuous improvement with respect to the user-defined cost function. Constraint satisfaction is also satisfied.

The constrained problem is transformed to the minimization of the unconstrained cost:

$$J_{ip}=\int_{\tau}^{1}\left[t_f(\omega_1(\frac{M_n(\sigma)}{M_{sp}}-1)^2+\omega_2(\frac{F(\sigma)}{F_{sp}}-1)^2)-\mu_1\sum_{j=1}^{z=m}\log(w_j+\varepsilon)\right]d\sigma+\omega_3 t_f+M(w_f)^2 \quad (9)$$

where the m path constraints w_j are given by:

$$w_k=-x_k-x_k^{\max}+\varepsilon,\ w_{k+3}=x_k+\varepsilon\ ,\quad k=1,2,3$$

control constraints by:

$$w_{k+6}=-u_k+u_k^{\max}+\varepsilon,\ w_{k+9}=u_k+\varepsilon\ ,\quad k=1,2,3$$

and end point constraints by:

$$w_f=x_4+x_5-m_{pol}(t_f)$$

Numerical values for the bounds of the path constraints are given with the cases studies. The integral part of the modified cost is considered as an additional state such that:

$$\dot{x}_7 = \frac{dx_7(\tau)}{d\tau} = t_f(\omega_1(\frac{M_n(\tau)}{M_{sp}} - 1)^2 + \omega_2(\frac{F(\tau)}{F_{sp}} - 1)^2) - \mu_1 \sum_{j=1}^{z=m} \log(w_j + \varepsilon) \quad (10)$$

Finding a locally optimal solution can often be achieved through appropriate initial guesses to the optimizer and the use of realistic constraints. However, to find a feasible set of initial guesses is sometimes not a trivial task due to the interior feasible point method (barrier function) used. The barrier method requires that the initial guess of the solution to be strictly feasible. In addition, ill conditioning inherent to the nonlinearity of the reactor system may play a role in the sensitivity of the solution to initial conditions.

The constrained parameter set $\Omega = \{\theta \in \Re^N \mid w(x^p(t), \varphi(\theta, t)) \geq 0\}$ describes a convex subset of $\Re^N$. It is assumed that the parameters evolve on a compact subset γ of $\Re^N$, and that the cost functional $J: \Re^N \rightarrow \Re^N$ is convex and continuously differentiable on γ. The assumptions made up to this point guarantee a local optimization of the constrained problem exists and that the gradient can be used to achieve that minimization [2].

An interior point method with penalty function is used to include the constraint costs. The interior point method incorporating a log barrier function enforces the state and input constraints (essentially converting the constrained optimization problem into an unconstrained one) while the end-point constraints are incorporated through a terminal penalty function. In the remaining equations obvious notation has been omitted. Thus, let the path cost with barrier function be stated as follows:

$$L(x, u) = q(x, u) - \sum_{i=1}^{nc} \mu_1 \log(w_i(x, u) + \varepsilon) \quad (11)$$

To emphasize that the optimization is based on the current conditions, the new cost functional with interior point method and penalty function is stated as:

$$J_{ip} = \int_t^{t_f} L(x^p(\tau), \varphi(\theta, \tau)) d\tau + M \| x^p(t_f) - x_f \|^2 \quad (12)$$

where $\tau \in [t, t_f]$ is the integration variable and nc is the number of inequality constraints. The parameter $\mu_1 > 0$ is the barrier parameter for the logarithm term, $M > 0$ is the penalty term and $\varepsilon > 0$ is a constraint relaxation factor (back-off) that prevents the barrier term from singularity.

The formulation uses a gradient-based method for the solution of the dynamic optimization problems in real-time, with a straightforward diagonally scaled steepest descent parameter update law of the form:

$$\dot{\theta} = -\text{Proj}(\Gamma \nabla_\theta J_{ip}, \Omega_w) \quad (13)$$

where $\Gamma > 0$ is the scaling matrix (referred as a gain matrix) to be defined later. To avoid divergence of the update law a projection algorithm is used. This guarantees that the parameters remain in a convex set $\Omega_w = \{\theta \in \Re^N \mid \|\theta\| \leq \eta\}$ for some $\eta > 0$ while assuring that the cost decreases until the optimal profile parameters are determined.

A Lyapunov-based method was used to show convergence to the local minimizers of a user-defined cost functional. The projection algorithm is given by:

$$\dot{\theta} = \begin{cases} \gamma, & \text{if } \|\theta\| < \eta \text{ or } (\|\theta\| = \eta \text{ and } \nabla P(\theta) \le 0) \\ \psi, & \text{otherwise} \end{cases}$$

where $\psi = \gamma - \gamma \frac{\nabla P(\theta)\nabla P(\theta)^T}{\|\nabla P(\theta)\|^2}$, γ is a compact subset of $\Re^n$ where the parameters are assumed to evolve, $P(\theta) = \theta^T\theta - \eta$, θ is the vector of input parameters and η is chosen such that $\|\theta\| \le \eta$.

IV. Case Studies

Cases 1-3 illustrate the optimization of the BMA/STY system relative to the conventional starved feed strategy for the 50:50 BMA: STY mass ratio. Cases 1 to 3 follow experimental initial conditions (monomer and initiator feed rates) used in the experimental study of [1] and summarized in Table 1. In Table 1, m_{TBPA}^{fed}, m_{BMA}^{fed}, and m_{STY}^{fed} represent the total mass of initiator, BMA and STY fed to the reactor to the final batch time of t_f. The mass of polymer produced at the final batch time is $m_{pol}(t_f)$, taken as the terminal (end-point) constraint in the optimization routine and set so that the final weight fraction of polymer in solution is 0.7 at the end of the batch.

Table 1: Conventional starved feed strategey used in [1]

Results	50:50 BMA:STY (wt ratio)	75:25 BMA STY (wt ratio)
m_{TBPA}^{fed}, Kg	9.7E-03	9.7E-03
m_{BMA}^{fed}, Kg	2.46E-01	3.6E-01
m_{STY}^{fed}, Kg,	2.46E-01	1.23E-01
t_f, min	360	360
$m_{pol}(t_f)$	0.474	0.483

For all three cases, the target values of F_A and M_n (g/mol) are 0.43 and 11 respectively. The constraints for the states and input variables are set to: $x_1(t) \in [0, 0.02]$, $x_2(t) \in [0, 0.02]$, $x_3(t) \in [0, 5.0\times10^{-4}]$ and $u_i(t) \in [0, 0.01]$, $i = 1,2,3$. The algorithm parameters for all three cases are as follows. The adaptive gain (k) in the steepest descent law used by the algorithm is 0.0001. The end constraint parameter M is set to 10^8. This large value ensures that endpoint values are strictly enforced. The barrier function parameters, μ and ε, are set to 0.1 and 10^{-7}, respectively. The initial conditions for the system are $x_1(0) = x_2(0) = x_3(0) = x_4(0) = x_5(0) = x_6(0) = 0$.

Table 2 Performance metrics for the three cases.

Case	$m_{pol}(t_f)$ (kg)	t_f (sec)	$J_{ip}(t_f)$
1	0.47386	15335	1.8988×10^5
2	0.47383	15479	1.8565×10^5
3	0.47383	15234	1.7706×10^5

As summarized in Table 2, the more general parameterization from Case 3 provides some improvements in performance over the simpler Cases 1 and 2. All cases provide a significant improvement over the experimental batch time of 21600 seconds. In all three cases, the optimization routine successfully implements a feasible locally optimal strategy for the given parameterization that 1) improves the nominal performance of the fed-batch system and 2) enforces constraint satisfaction in the monomer feed. Figure 2 compares the results obtained using the real-time optimization technique for Case 3 to the nominal results from the feed-rate used in [1] that resulted in a batch time of 21600 seconds. These curves are labeled "simulation" on the plots.

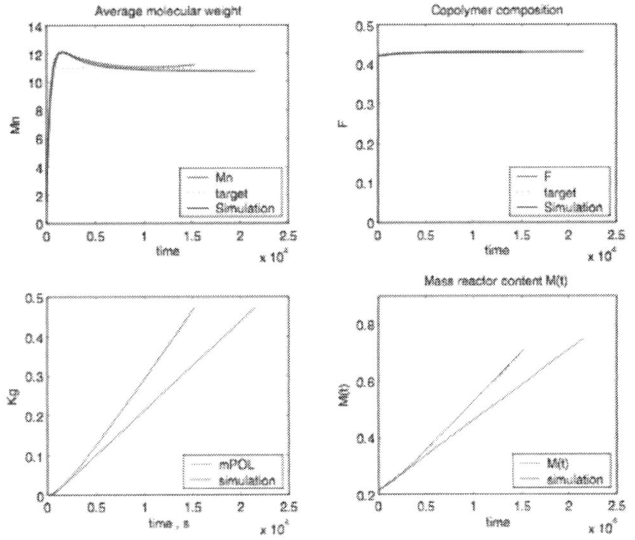

Figure 2: Performance of the real-time optimization control system and the nominal fed-batch strategy.

Figure 3 shows the state trajectories for Case 3. It can be seen that the real-time optimizing controller pushes $x_3(t)$ to its upper limit. Relaxation of this constraint should result in further reduction in batch time. Figure 4 shows the corresponding input trajectories. As expected, the effect of the constraint in $x_3(t)$ is shown in the trajectory for $u_3(t)$. On average the real-time optimization routine requires 1 to 10 msec of computation for every second of operation. The corresponding controller is therefore viable for implementation.

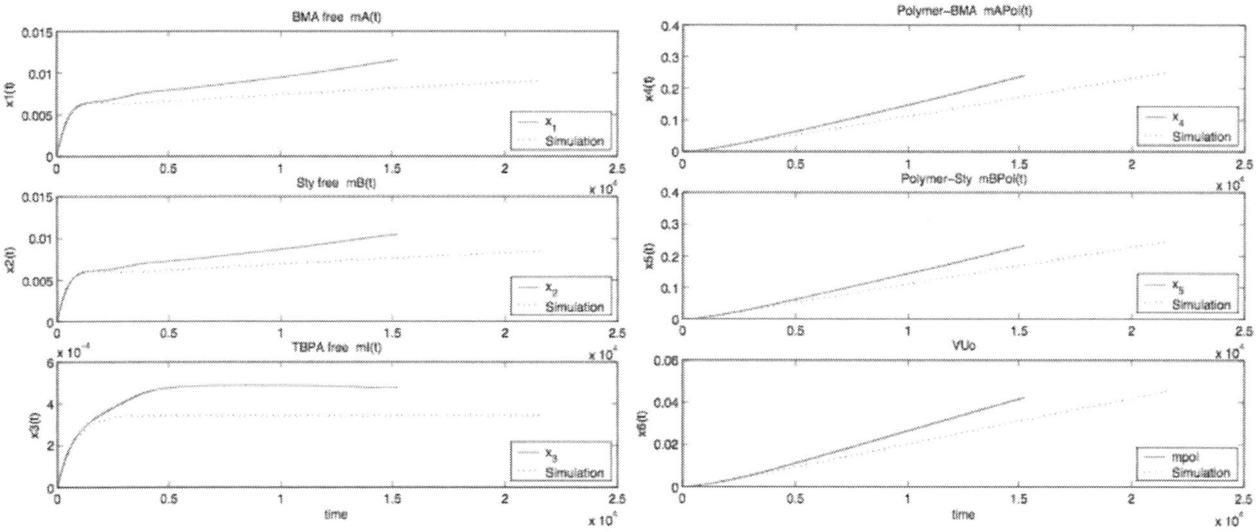

Figure 3: State trajectories for Case 3. The nominal simulation based on the experimental conditions in [1] are also shown.

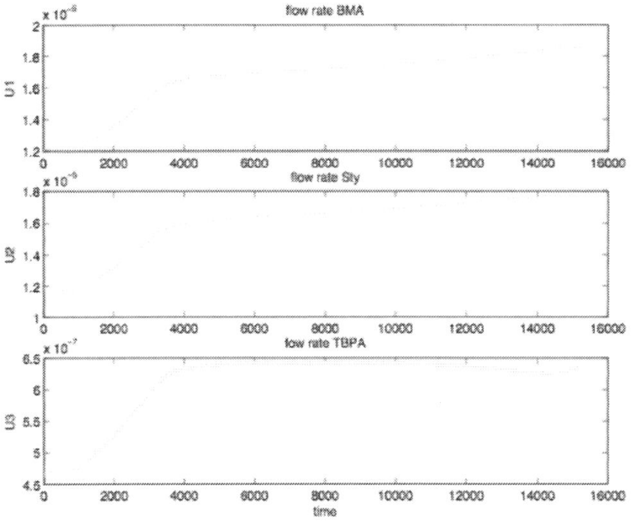

Figure 4: Input trajectory for Case 3.

V. Conclusions

Considerable improvements in polymer quality (less drift in molecular weight and copolymer composition) and productivity (batch time reduction) can be obtained while satisfying the constraints imposed to the system. Moreover, this novel technique and the use of a reduced order model result in low computational effort that can be further exploited through more complex input parameterization and/or the use of a more complex model to improve the accuracy of the model predictions. The actual implementation will be the main focus of further experimental study. Ongoing research in this area has demonstrated the promise that a viable estimation routine can be developed and implemented.

Acknowledgement

The authors acknowledge the support of the Natural Sciences and Engineering Research Council of Canada and E.I. du Pont de Nemours and Co.

References

[1] Li, D. and R.A. Hutchinson, *High temperature semibatch free radical copolymerization of butyl methacrylate and styrene.* Macromolecular Symposium, 2006. **243**: p. 24-34.

[2] Peters, N., D. Dehaan and M. Guay, *Real-Time Dynamic Optimization of Nonlinear Batch Systems.* Journal of Process Control, 2007. **17**(3): p. 261-271.

[3] Abel, O.and W. Marquardt, *Scenario-integrated on-line optimization of batch reactors.* Journal of Process Control, 2003. **13**: p. 703-715.

[4] Chen, H. and. F. Allgöwer,, *Nonlinear model predictive control schemes with guaranteed stability*, in *Nonlinear Model Based Process Control.* 1998, Kluwer: Dordrecht, The Netherlands. p. 465-494.

[5] Mayne, D.Q., Rawlings, J.B., Rao, C.V., Scokaert, P., *Constrained model predictive control: stability and optimality.* Automatica, 2000. **36**(6): p. 789-814.

[6] Richards, J.R. and J.P.Congalidis, *Measurement and control of polymerization reactors.* Computers and Chemical Engineering, 2006. **30**: p. 1447-1463.

424c Determination of the Dominant Nucleation Mechanism for Oil-Soluble Initiators In Miniemulsion Polymerization Using Kinetic Monte Carlo Simulations

Jonathan A. Rawlston, School of Chemical and Biomolecular Engineering, Georgia Institute of Technology, 311 Ferst Drive N.W., Atlanta, GA 30332-0100, Juchen Guo, Department of Chemical and Biomolecular Engineering, University of Maryland, College Park, MD 20742, F. Joseph Schork, Department of Chemical & Biomolecular Engineering, University of Maryland, 2113D Chemical Engineering Bldg 090, College Park, MD 20742, and **Martha Grover Gallivan**, School of Chemical and Biomolecular Engineering, Georgia Institute of Technology, 311 Ferst Drive, Atlanta, GA 30332-0100.

The use of oil-soluble initiators in free-radical miniemulsion polymerization has increased due to their ability to generate radicals primarily within monomer droplets. Existing theories concerning the nucleation mechanism for oil-soluble initiators suggest that a single radical must be formed within a particle for propagation to occur, despite the fact that an oil-soluble initiator molecule decomposes to form two radicals. According to existing theories, the primary source of nucleating radicals may be either within the particle [1], or from the small amount of initiator in the aqueous phase [2].

At the nanometer size of miniemulsion particles, concentration variables do not adequately describe the reactions in a particle, and use of such variables may lead to the inaccurate description of mechanisms within a particle. Application of the kinetic Monte Carlo algorithm [3] allows for the simulation of individual reactive events, such as propagation or termination, while tracking macroscopic observables, such as conversion and molecular weight. Simulation results are qualitatively compared to the conversion rate and the peak of the molecular weight distribution for two sets of experimental data for styrene miniemulsion polymerization, one taken from Alduncin and Asua [4], and another collected by ourselves.

Comparison of the simulation results to Asua's experimental data shows that desorption is insignificant at his experimental conditions. Our experimental data were collected at a smaller particle size and a lower temperature, and comparison of the simulation results to our data indicates that desorption of radicals is significant at our experimental conditions. Desorption of radicals is shown to be more significant at lower temperatures, when the propagation rate is lower, and at smaller particle sizes, where a radical may reach the particle surface more quickly. Absorption of radicals is shown not to be necessary for qualitative agreement with either set of experimental data.

[1] Asua, J.M., Rodriguez, V.S., Sudol, E.D., El-Aasser, M. S. "The Free-Radical Distribution in Emulsion Polymerization Using Oil-Soluble Initiators." Journal of Polymer Science Part A: Polymer Chemistry, 1989, 27, 3569 – 3587.

[2] Nomura, M. and Suzuki, K. "The Kinetic and Mechanistic Role of Oil-Soluble Iinitiators in Micro-

and Macroemulsion Polymerizations." Industrial and Engineering Chemistry Research, 2005, 44, 2561 – 2567.

[3] Fichthorn, K.A. and Weinberg, W.H. "Theoretical Foundations of Dynamic Monte-Carlo Simulations." Journal of Chemical Physics, 1991, 95, 1090 – 1096.

[4] Alduncin, J.A. and Asua, J.M. "Molecular-Weight Distributions in the Miniemulsion Polymerization of Styrene Initiated by Oil-Soluble Initiators." Polymer, 1994, 35, 3758 – 3765.

Modular Optimization and Optimal Control of Polymerization Reactions

Michael Wulkow, CiT GmbH,
Oldenburger Str. 200, Rastede, 26180, Germany, m.wulkow[at]cit-wulkow.de

Modeling and simulation of polymerization reactions range from studies of elemental kinetic phenomena to real-time plant simulations. Where formerly monomer conversion and property averages had been studied, now product properties like full molecular weight distributions, copolymer composition, branching degree and network densities are under consideration.

Because of the numerical complexity of the underlying mathematical equations, even off-line optimization and optimal control of such properties is still challenging. An optimal control problem for polymerization might consist of several objectives and controls requiring the solution of high-dimensional population balances. For example, a typical task is to produce a certain shape of the molecular weight distribution in optimal time and under some constraints regarding feed strategy and maximal temperature. If various controls like initial mixture, time-dependent feed rates, reactor cooling function and variable process time are to be identified in one single problem set-up, the number of variables can easily increase to several dozens. Apart from the pure technical and mathematical aspects to solve such a problem, in practice one has to follow a typical course of action.

- Start with a well-tested, predictive model in a reliable simulation environment
- Define appropriate objective functions and their relative weighting. For example, if one wants to minimize the difference between a property A of the model and the requirement A*, an integral condition $\int_0^T (A(t)-A^*)^2$ will not suffice. Instead one should use a relative measure and a weighting $\frac{\omega}{T}\left(\int_0^T\left(\frac{A(t)-A^*}{A^*}\right)^2\right)^{1/2}$. This term will describe the relative weighted deviation of state A per time.
- Decide about the controls (number, constraints, further parameters)
- Solve a nonlinear optimization problem
- Analyze the results (optimality, uniqueness, problems)
- Use the optimized model.

It can always happen, that at intermediate steps the model has to be reformulated, requirements change etc. As a consequence the optimized model drifts apart from the original model and a new optimization has to be performed. At the same time, quite new optimization tasks arise such that a fixed user-interface will be at its limits soon. Here the basic technical aspects mix with ergonomic and software issues. It is required to have a modularized approach, where the simulation part is separated from the optimization tasks. Therefore a modular framework has to be developed, where simulation, preparations and parameter identification are separated.

A simplified version of a classical optimal control problem often has the following structure:

$$
\begin{aligned}
&\min_u F(u) = g(x_0, x_T) + \int_0^T h(x(t), u(t))dt \\
&\frac{dx}{dt} = f(x(t), u(t)) \qquad\qquad (1) \\
&b(x(t), u(t)) \le 0 \\
&r(x_0, x_T) = 0
\end{aligned}
$$

Here the state variables are denoted by x(t) and the controls by u(t). For polymer systems describing a chain-length distribution, x(t) is a vector of formally infinite dimension, in practice including ten thousands or even more than one million single components. The control vector u(t) might consist of time-.dependent functions like feed rate per time and/or cooling temperature, but also of single parameters like initial mass of initiator. The right-hand side f(x(t),u(t)) summarizes all reactions and additional balances of a kinetic system. The setup of such systems is described in [1]. Among the state constraints there are conditions like a maximal temperature or heat production or an upper limit of feed rate. Moreover the process time T may be left open as another optimization variable in order to reach a "time optimal" control.

The functions g and h in the objective function formally describe the deviation of the actual states and controls from some required values. For polymer systems the evaluation of these expressions can be very difficult. If the shape of a molecular weight distribution is prescribed (usually at the end of the process), its deviation compared to a simulated distribution has to be defined in a reasonable way. Scaling and weighting play a particular role in this context.

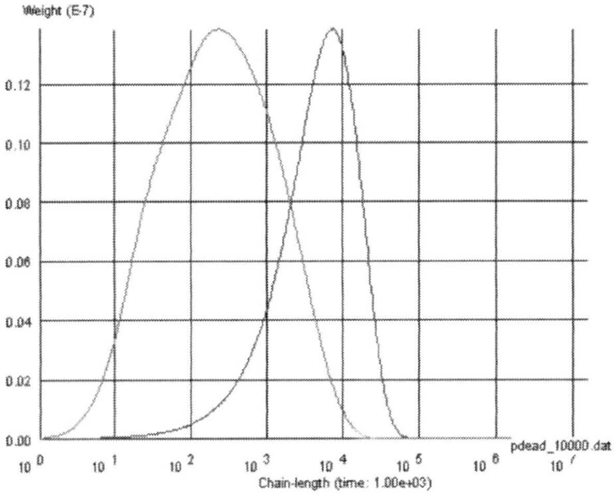

Figure 1: Normalized comparison of shapes of molecular weight distributions, distribution from GPC (red) and from actual simulation (blue).

A general approach to optimal control problems related to differential equations is to apply the Maximum principle to the Hamilton function of the system leading to a boundary value problem and implicit expressions for the controls. The related analytical preparations and distinction of cases are more or less prohibitive in the context if infinite-dimensional population balances. Therefore we pursue the so-called direct approach, where all controls are parameterized using certain basis functions (e.g. piecewise constant/linear functions, splines, polynomials etc.). By that the optimal control problem can be transferred into a large scale parameter optimization problem with constraints on the parameters, the resulting control functions and the state variables. The direct approach has the advantage that it can be applied to a basic differential equation system without much preparation. Furthermore constrains on controls and states can be set directly. A disadvantage might be the fact, that dependent on the chosen parameterization, large numbers of optimization variables can arise and that the obtained results are usually not unique and not optimal in the strict sense of the Maximum principle. Fortunately and in contrast to classical parameter identification of kinetic rate constants, for control purposes parameters and controls have not to be unique - but (sub-) optimal „only". This requires special numerical algorithms efficient enough to ensure fast convergence, but capable of dealing with underestimated least-square problems.

Therefore we sketch some brief details of a new algorithm, originally developed for parameter estimation problems, which turned out to be particularly helpful for parameterized control problems too. The basic idea is to solve the optimization problem by use of a Gauss-Newton method and perform an advanced analysis of the related Jacobian matrix (containing the derivatives of all objectives with respect to all parameters) aiming at the number of so-called "essential degrees of freedom", i.e. the number of parameters which can be uniquely identified in the present model in view of the present information (i.e. data for classical parameter estimation or constraints and requirements for optimal control problems).

If the number of essential parameters is not maximal, there will be correlations or total insensitivities among the parameters. Technically we define the essential directions of the Jacobian matrix as those directions which belong (under singular value decomposition) to *large* singular values, hoping that large singular values hold more characteristic of the matrix than small ones and that neglecting small singular values will not imply loss of too much information. It has to be mentioned, that the obtained number of essential directions in parameter space (leading to number of essential parameters) depends on a threshold, which cannot be given as a fixed theoretical value, but has to be chosen as parameter of the algorithm itself. However, this threshold could be identified to be in a small range for many tested examples (as a rule of thumb: the ratio of singular values of the remaining lower-dimensional sub problem should be smaller than 100).

The program package Predici [1] has been used for more than 12 years for the modeling of polymerization kinetics. The core feature of Predici and its underlying Galerkin h-p-method - the computation of full chain-length distribution - has been extended to additional distributed properties such as mentioned above. The question is how to make use of this complexity when it comes to optimization without change of model structure and definition (since models are permanently changed and improved, an optimization procedure should leave the model itself unchanged). Whereas for classical optimization problems special algorithms are available - mostly based on first and second derivatives of a differentiable objective function - in the context of full chain-length distributions a computation of higher order derivatives is prohibitive. Therefore in Predici the basic parameter estimation tool has been extended by additional functions to allow for optimal control problems.

Principally all input for a control problem (1) can be done in any typical Predici version. For example, inequality conditions of the form $b(x_i) \leq c_i$ can be reformulated in terms of a slack variable L_i to $b(x_i) + L_i^2 = c_i$. The new parameter L_i can be added to the list of model parameters and included in the estimation process. Control functions can also be formulated by use of commands of Predici's script language (polygon, polynomial). However, after first successful tests with the general approach, it turned out, that for the daily work the permanent redefinition of parameters and script function would be error prone and time consuming.

Instead a preprocessor has been implemented and equipped with a full user interface, called Priamoz. The idea is to take the unchanged model set-up, define optimization based requirements in the user-interface and automatically generate all functions, parameters and weightings by an automatic procedure leading to an augmented system. The subsequent parameter estimation – applying the essential directions approach – leads to results which can be checked and confirmed and are finally transferred back to a basic model without traces of the intermediate solution process.

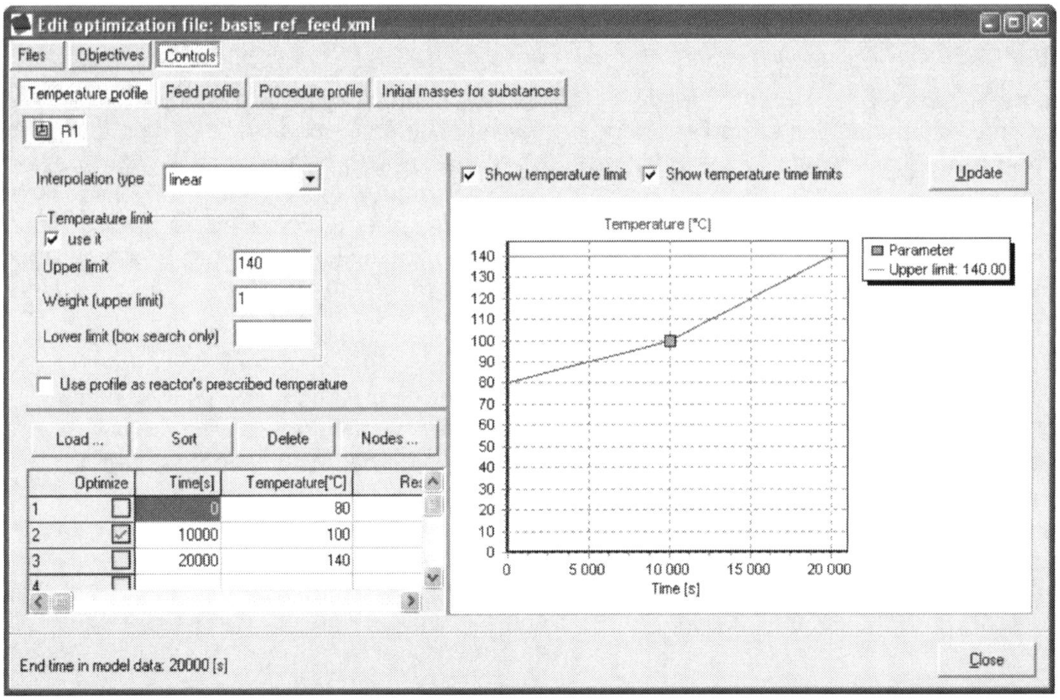

Figure 2: Selection and initialization of a temperature profile as control function in Priamoz. The input is transformed to Predici model input.

Regarding the objectives all model outputs can be selected. Furthermore also a time-optimal control can be added. This requires a variable end time of the process which in turn leads to transformation of the controls to be identified. In many theoretical papers thus a transformation [0,T] → [0,1] is applied to the differential equation system. Such a transformation, based on an optimization parameter T_{end}, has also been introduced in Predici. Instead of using the unit interval, however, a reasonable process time T_p is used to keep all other parts of the dynamic system nearly unchanged and to allow direct suggestive graphical output.

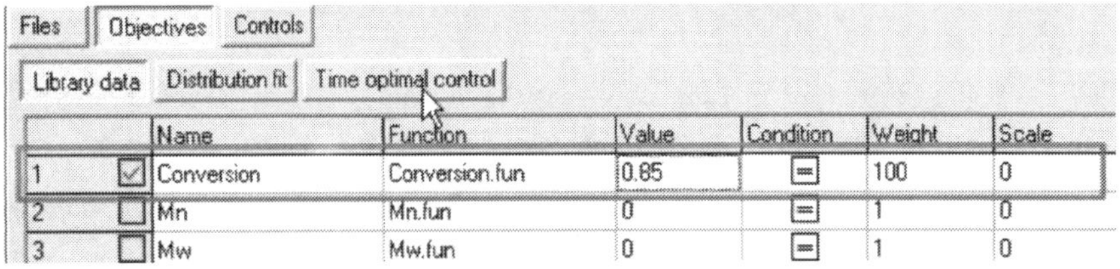

Figure 3: Selection of objectives in Priamoz. The reported functions for conversion and polymer mean values are part of a Predici model and defined by the user.

This approach might look like an aspect of pure software architecture, but actually provides an efficient and comfortable way to extent the current status of the used tools to new problems without programming. It can be used for many kinds of objectives, even in connection to direct rheology computations (for mostly linear polymer chains) which are available employing the full information of the molecular weight distribution. The whole framework could also be a way to use a common interface for the generation of model files of different simulation tools.

References:

[1] M. Wulkow, *Computer Aided Modeling of Polymer Reaction Engineering - The Status of Predici, 1 – Simulation*, Macromol. React. Eng. 2008, 2 (Issue 6), in print; published online: 04.09.2008, DOI: 10.1002/mren.200800024

424e Integrated Modeling and Control of Polymeric Coating Curing Process and Product Under Uncertainty

Zheng Liu[1], Yinlun Huang[1], and Zhen Wu[2]. (1) Department of Chemical Engineering and Materials Science, Wayne State University, 5050 Anthony Wayne Drive, Engineering Building 2404, Detroit, MI 48202, (2) Chemical Engineering and Materials Science, Wayne State University, 5050 Anthony Wayne Drive, Detroit, MI 48202

One of the most challenging issues in automotive surface coating manufacturing is how to ensure the polymer coating quality developed on the vehicle surface. In production, vehicle bodies, before entering each baking oven, are covered by a thin layer of wet film of waterborne or solvent-borne paint on the panel surface. When a vehicle body travels in the oven, its panels are heated, which leads to the corsslinking reaction between the precursor polymer chains and crosslinkers within the polymer coating film. Within a pre-set operational period, the curing process should be accomplished and the surface coating is expected to achieve a desired coating quality in terms of its corsslinking network structures. [1], [2] However, due to a variety of disturbances on both the polymer curing process and the polymer coating product in operation, the final coating quality of the vehicle surface film is always a challenge in industry.

In this paper, both the polymer curing process and the polymer coating product are deeply analyzed and carefully modeled. [1]-[3] The reference models of the process and the product are also introduced to compensate the model uncertainty due to the lack of sufficient knowledge and information about the curing process and more critically the coating product. With the model set developed, an internal model control (IMC) based integrated process and product control (IPPC) methodology is introduced for the control of both the polymer coating curing process and coating quality. By this methodology, the control system design is capable of realizing simultaneous dynamic control of both the operational performance of the curing process and the manufacturing quality of the coating product.

The proposed IMC-IPPC design methodology is featured by the following advantages. First, the control of the polymer coating corsslinking quality can be ensured through all-time, on-aim proactive quality control (QC). This allows prediction of potential quality problems and derive early actions to prevent the occurrence of certain types of quality problems in the earliest manufacturing steps. Second, the energy supplied to the curing process can be balanced with the coating film temperature and the polymer crosslinking conversion rate, which can make the process operation much more efficient. Third, the process constraints on the curing hot air temperature can be practically imposed in controller design. Thus, the integrated control scheme can be more realistically implemented in real applications.

The introduced IMC-IPPC design methodology has been successfully tested with industrial problems. It has shown that although there exist a number of model inaccuracies for both the polymer curing process and the polymer coating product, and although the automotive polymer coating curing system experiences various disturbances, the crosslinking quality of the polymer product can be effectively controlled to achieve a desired conversion rate, and the polymer curing process performance has been fully acceptable as well with a high efficiency of the process control energy, and a satisfaction of the hot air temperature constraints. The introduced methodology is general enough for application in many other types of polymeric coating manufacturing processes.

Reference: [1] Lou HH, Huang YL. Integrated Modeling and Simulation for Improved Reactive Drying of Clear Coat. Ind Eng Chem Res. 2000;38:500–507.

[2] Dickie RA, Bauer DR, Ward SM, Wagner DA. Modeling Paint and Adhesive Cure in Automotive Applications. Prog Org Coat. 1997;31:209–216.

[3] Xiao J, Huang YL, Qian Y, Lou HH. Integrated product and process control of single-input-single-output systems. AIChE J. 2007; 53: 891-901.

Experiences in the Building of Dynamic Flowsheet Models for Embedded Nonlinear Control of Polymer Processes

Kenneth V Allsford*, Jeffrey G Renfro and John Goodman

Abstract — **Controllers that directly incorporate an embedded first principles dynamic flowsheet model have been applied to a range of transition metal and free radical polymerization kinetics industrial processes. These controllers handle grade runs, transitions and constraint pushing. This paper discusses lessons learned from these applications, with reference to industrial examples, in the contexts of (i) scoping the flowsheet model; (ii) offline parameter estimation to match the model to process steady state; and iii) online state estimation for continuous and discontinuous process measurements. An emphasis is placed on (ii). The discussion in this paper is of a general nature, although the experiences were gained with specific tools not in the public domain.**

I. INTRODUCTION

TABLE 1 lists some generic challenges inherent to polymer process control.

TABLE 1: POLYMER PROCESS CONTROL CHALLENGES

Concept	Notes
Core challenge	Complex catalytic reaction.
No in-situ quality measurements	Quality properties such as melt index, density, % solubles, etc measured outside the reactor (often in the laboratory) with some time delay.
Process capability issues	Quality properties set tight relative to process sampling, laboratory measurement and consistent process operation capabilities given typical measured and unmeasured disturbances.
Transition control	Implementation of significant changes to property targets during which other operating conditions such as production rate may also be changed significantly to maximize process unit profitability.
Constraint pushing and turndown ratio	Controlling to unit constraints and/or through 60%+ turndown as apparent dynamics and gains change with different properties and/or production rate.
Adapting to process change	For example, handling new catalysts or quality properties.

Xie et al [1] were among the first to discuss nonlinear reactor modeling for gas phase polymerization processes with emphasis on control application. Subsequently, controllers that directly incorporate an embedded first principles dynamic flowsheet model have been applied for handling grade runs, transitions and constraint pushing to a range of transition metal and free radical polymerization kinetics industrial polymer processes. This paper focuses on issues associated with the development and implementation of the embedded dynamic flowsheet model.

* ken.allsford@honeywell.com, +1-832-252-3344 (corresponding author), jeffrey.renfro@honeywell.com and john.goodman2@honeywell.com are associated with Honeywell Process Solutions, 1250 W Sam Houston Pkwy S, Houston, TX 77042.

II. CONCEPT OF PROCESS MODEL DESCRIPTION

By definition models both simplify a real system and also synthesize state-of-the-art knowledge. The bridging of these potentially contradictory definitions is the model description, the specific set of questions about the real world for which the model is required to provide an answer. Through a review of the model description, the model detail that can be sacrificed without reduction in relevant predictive capability can be identified. The process model description is influenced through the use of established tools in the problem realm and by applying generic model-building concepts during use of these tools.

Established problem-solving approaches within the process industries are the concepts of flowsheeting for modeling and online state estimation for control model update ... and the evolution of tools that apply these concepts. For example, flowsheeting has evolved to support the generation of dynamic models through entry of actual equipment volumes and geometry. Also within the context of flowsheeting for polymer processes the framework has evolved to allow the engineer configuring the application to:

- Select the reactions to use from a library of reaction schemes (free radical or transition metal kinetics).
- Predict end-use properties based on the most common correlations proposed and validated in the open technical literature.

Online state estimation has evolved from constant or integrating output disturbance value through Extended Kalman Filter (EKF) to methodologies that have a more direct connection with a flowsheet model – such as Implicit Dynamic Feedback (IDF) and Moving Horizon Estimation (MHE) – see Hedengren et al [2].

Generic model-building concepts to be applied in conjunction with these established tools include:

- Clarify model objectives
- Start small
- Verify (test/debug) and validate (does the model adequately mimic reality?)
- Improve incrementally (prioritizing additions in terms of the greatest anticipated improvement in the model)
- Refactor frequently (backtrack and/or simplify)

III. DEFINING THE PROCESS MODEL DESCRIPTION FOR THE APPLICATION CONTEXT

For a process control application, the embedded flowsheet model must be both robust (provide a usable solution) and solve relatively quickly. Within the context of these generic constraints, the core requirement is to predict the effect of the manipulated (MV) and feed forward (FF) variables in the controller on the controlled variables (CV). This definition implies the exact objective is a function of controller design. However, for the general range of polymer production processes and reaction mechanisms, we can practically think of the CVs as falling into three groups:

- Polymer properties
- Polymer production rate
- Polymer production rate constraints

Real-time process values are typically available (measured directly or inferred through relatively simple calculation of direct measurements) for the latter two groups of CVs. Although this blurs defining the process model description, continuous state estimation algorithms can play a role in the overall prediction capability of the model for these CV groups. In actual practice, this becomes iterative in the sense of applying the generic model-building concepts of "Start Small", "Validate" and "Improve Incrementally" until a model that is judged satisfactory for control objectives is obtained – section VIII reviews some selected examples.

However, the set of CVs in the first group are not typically continuously measured in real time and as such cannot be updated by continuous state estimation. As indicated in Table 1, these CVs are also the CVs associated with process capability issues i.e. constraints on these CVs are tight given typical measured and unmeasured disturbances. As such, prediction capability similar to the measurement capability is sought for these CVs – and this becomes part of the process model description. Measurement capability can be viewed from the context of examining the cumulative effect of sampling issues, sample pre-treatment variation, measurement system repeatability, etc. on the measurement variability. More practically, measurement capability can be looked at from the context of actual variability when the process is nominally steady. From experience, this is typically less than 5% (in terms of the actual measurement for properties such as polymer melt index or across the range of measurement for properties such as polymer density). As well as such base prediction capability, the prediction "error" should show no significant bias across the range of operating conditions e.g. not biased in terms of poor prediction for one product type, or when operating at high production rates, etc.

IV. Implication of Process Model Description

As implied earlier, the basis for prediction of polymer properties within a flowsheet model is as outputs of the "reactor" unit operation. Flowsheeting packages usually provide a continuous stirred tank reactor (CSTR) and, if appropriate, combinations of these CSTR are used to represent the reality. For example, a tubular reactor can be represented as a set of CSTR in series.

Quality properties of interest are correlated as functions of more fundamental properties such as polymer molecular weight distribution. These more fundamental properties are predicted subsequent to the solution of polymer kinetic expressions. The general rate expression is of the form:

$$r = kc_A^{\gamma} c_B$$

where:

C_A is the monomer, modifier, solvent, etc.
C_B is the live polymer site or radical concentration
γ is the reaction exponent
k is the rate constant and is of the form:

$$k = k_0 e^{\frac{-(EA+VA\,P)}{RT}}$$

where:

P is pressure
T is temperature
R is the gas constant
k_j is the rate constant for reaction j
$k0_j$ is the frequency factor for reaction j
EA_j is the activation energy for reaction j
VA_j is the activation volume for reaction j

For transition metal kinetics such rate expressions are applied to the following reaction classes:

- Site activation – monomer, hydrogen, cocatalyst, electron-donor, spontaneous, etc.
- Initiation – vacant active site to monomer with active site
- Propagation – monomer addition
- Chain transfer – monomer addition terminated (by monomer, hydrogen, cocatalyst, electon donor, spontaneous, etc.) and new vacant active site created.
- Site transformation – conversion of active site types by monomer, hydrogen, cocatalyst, electron donor, spontaneous, etc.
- Site deactivation – monomer, hydrogen, cocatalyst, electron donor, spontaneous, etc

For free radical kinetics such rate expressions are applied to the following reaction classes:

- Initiator decomposition
- Propagation – monomer addition
- Chain transfer – monomer, modifier, polymer
- Chain termination – combination, disproportionation
- Beta scission
- Backbiting

The relative rates of the different reactions determine the model predictions. For example, the relative rates of chain transfer and propagation reactions determine the length of the polymer chain. In turn, subsequent to transformation through the method of moments [3] to predict polymer molecular weight distribution, this is typically the primary correlating input for melt index prediction.

As each reaction class applies to a number of explicitly defined reactions, the implication of the process model description is that parameter determination approaches are required for the kinetic and additional correlating parameters. For multiple catalyst site types – maybe also multiple monomer systems – the number of different relevant parameters in the model may approach 100. Irrespective of the operation of the polymer process, there is insufficient information in the available process data to enable all these parameters to be treated as independent. As such the process of parameter determination requires application of engineering judgment in the context of the available body of knowledge in the open literature. A core objective of this paper is to provide an overview of the application of such knowledge for some specific examples.

V. PREDICTING QUALITY PROPERTIES FOR UNIPOL™ POLYETHYLENE (PE) PROCESS

Gas phase reactors are often discussed in the polymers technical literature. An example is UNIPOL™ technology, licensed through Univation – a process description is given in references [4] and [5] (interactive format). Figure 1 shows the main process equipment.

FIGURE 1: PROCESS FLOW DIAGRAM

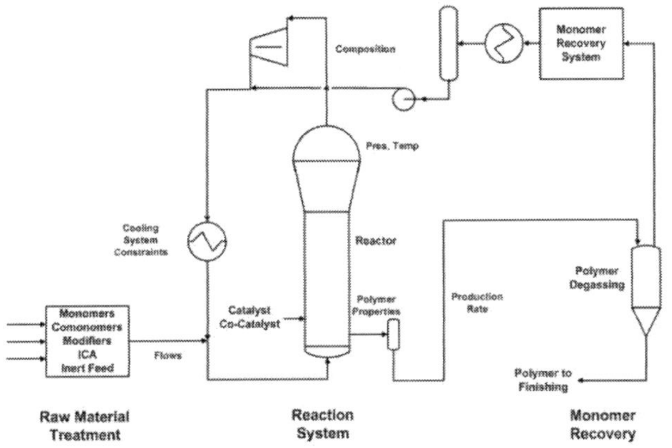

Different product types (grades) are produced by selecting reaction conditions, comonomers and (transition metal) catalyst system. Each grade represents some balance in properties for different end uses. Grade specifications are typically quoted in terms of melt index and density as defined by ASTM measuring standards [6].

Steps involved in "identifying" the model are listed in Table 2 (next page). In practice, simultaneous manipulation of parameters that involves an obvious but difficult to define trade-off occurs only in step 8 (chain transfer to monomer as compared to chain transfer to hydrogen). In particular for such trade-off – but of general use especially for the latter steps in Table 2 – a multiple data-set reconciliation problem is auto-configured. The objective function is specified as:

$$\phi(\mathbf{y},\mathbf{p},\mathbf{w}) = \sum_{i=1}^{n_y}\left(\frac{w_i^y(y_i - y_i^{meas})}{y_i^{scale}}\right)^2 + \sum_{i=1}^{n_p}\left(\frac{w_i^p(p_i - p_i^{start})}{p_i^{scale}}\right)^2$$

where:

n_y is the number of measured variables

n_p is the number of parameter variables

$\mathbf{y}$ is the model predicted variable values

$\mathbf{y}^{meas}$ is the measured values

$\mathbf{y}^{scale}$ is the measurement scaling factor values

$\mathbf{w}^y$ is the measurement weights

$\mathbf{p}$ is the model parameter variable values

$\mathbf{p}^{start}$ is the initial model parameter variable values

$\mathbf{p}^{scale}$ is the parameter scaling factor values

$\mathbf{w}^p$ is the user supplied parameter weights

The weighting factors for the measured values and parameters in this objective function are defined by the user. Normally the $\mathbf{w}^p$ weights are zeroed out, but when non-zero they can help the regularization of the problem where weak observability may result in ill-conditioning (or solution non-uniqueness). This objective function is an L2-norm that penalizes "outliers". Some alternate objective function definitions have been utilized. An example is the automatic zeroing out of objective function contributions for predictions that fall within "tolerance" of the measurement: however from a mathematical perspective this is a more complicated implementation (non-smooth objective function changes for variation of parameters in certain ranges).

Generally, a valid model is not obtained just by matching process data. The use of the tool has to be "engineering smart". Engineering knowledge of the process gains are often incorporated into the estimation process – either directly through constraining the gains or indirectly by gain analysis – combined with parameter limits and data set adjustments. Also, unmeasured disturbances present in the process data can result in poor model sensitivities even though the model matches the process data quite well. If the structure and presence of unmeasured disturbances are known it can be appropriate to include simultaneous disturbance estimation in the parameter estimation problem through model specification changes to address this issue.

Typically the input multiple data-sets are recipes – typical operating data covering different grades (maybe production rates) – adjusted to be consistent with recent operating data. The change in operating conditions necessary to implement transition from grade to grade is of more importance than absolute values. Some of the key inputs for this example are cycle gas composition and reactor temperature.

An approach and tools to develop a kinetic model focused on prediction of polymer properties have been described. How good are such kinetic models? In our experience predictions for 70-80% of **consistent** data-sets are within the 5% criterion defined above; the remaining 20-30% are within 10%. Why not all predictions within the 5% criterion?

TABLE 2: STEPS TO DETERMINE KINETIC PARAMETERS FOR UNIPOL™ PE PROCESS

Step	Activity	Observations	Outcomes
1	Literature search: tabulate reaction schemes and associated kinetic parameters.	Success reported based on modeling fluidized bed as single CSTR. Polydispersity not a typical input for correlations to predict bulk melt index and polymer density. Moderate, relatively constant pressure.	Fluidized bed modeled as CSTR. Only a single catalyst site type is considered (as a consequence, also no site transformation reactions). Activation volume kinetic parameters zeroed out.
2	General process observations	Single reactor. Site activation and initiation are "fast" (after adjusting catalyst flow to bed, the effect is seen in a few minutes).	Deactivation kinetics are zeroed out. Site activation and initiation reactions are made dependent on reactants always present in the model. Kinetic parameters for these reactions set to values that impart small sensitivity to production rate (made "fast").
3	From literature search, identify base activation energies.	Differences in activation energies (not absolute values) are important for predicting sensitivity of polymer properties to reactor temperature. For polymer density, differences for propagation of comonomer as compared to ethylene are important. For melt index, differences between chain transfer and propagation reactions are important.	Rough check is made to confirm that the predicted sensitivities of polymer properties to reactor temperature obtained with the selected activation energies are consistent with experience/observation.
4	Identify base propagation frequency factors	Match the general polymer production rate versus catalyst usage data for the process.	Initially the propagation frequency factors for the different reactions are all set equal.
5	Determine polymer density correlation	Several density correlation forms (usually relatively simple) available in the literature; primary correlated variable is percent comonomer incorporated. Some approximate data across operating range are available e.g. comonomer wt% incorporation is known from historical plant performance.	Several forms are fitted – the form and parameters with best fit and extrapolation relationship are selected. Note: willing to accept different correlation parameters for different comonomers.
6	Identify comonomer propagation kinetic parameters.	Recipes for different grades are input data. For fluidized bed reactor the recipes typically include cycle gas composition ratios, e.g. comonomer/ethylene ratio. Frequency factors and reaction order exponents are set equal for reactions with the same monomer with active reaction site in the polymer chain – to reduce the number of different parameters.	The propagation frequency factor for ethylene is held and the propagation frequency factors and (possibly) reaction order exponents for comonomer(s) are adjusted to improve the polymer density prediction. A check is made to determine if errors of the fit for polymer density are correlated with reactor temperature: if this is the case the activation energies for comonomer propagation are adjusted to eliminate this bias.
7	Select melt index correlation function	In the literature, melt index is normally correlated as a function of polymer weight average molecular weight. A few laboratory data may be available.	A correlation form and parameters are selected from the literature (form and parameter values independent of comonomer type). Parameters are fitted if sufficient lab data available.
8	Identify chain transfer kinetic parameters	Recipes for different grades are input data. For fluidized bed reactor the recipes typically include cycle gas composition ratios, e.g. comonomer/ethylene, hydrogen/ethylene ratios. Frequency factors and reaction order exponents are set equal for reactions with the same monomer with active reaction site in the polymer chain – to reduce the number of different parameters.	The propagation kinetic parameters are held and the chain transfer to hydrogen and chain transfer to monomer kinetics are adjusted. A check is made to determine if errors of the fit for polymer melt index are correlated with reactor temperature: if this is the case the activation energies for comonomer propagation are adjusted to eliminate this bias.
9	Check overall quality of model prediction and iterate to earlier step	Most common to iterate steps 4, 7 and 8.	Require fit of polymer melt index within 5% of target values and fit of polymer density to be within 5% of range of polymer density with no identified biases that are functions of operating conditions.

The model is an approximation: even though there are many available parameters, prediction limitations still exist. Are prediction discrepancies greater than 5% just accepted? The answer is "it depends". Generally, if the predictions that are out by more than 5% also bias the model predictive capability, then additional modeling and parameter fitting of a different form will be implemented. Sometimes the form has a known process basis that is not covered by the detail of the model as generally described above. For example polymer density may be known to be a (weak) function of polymer melt index - in which case this relationship is added and correlating parameters identified.

Similarly, several mechanisms not initially in the model may be proposed for effects on production rate. Again, the model can be enhanced – although engineering judgment is needed regarding accepting the implementation costs, as online state estimation can also be utilized to "improve" the model prediction.

Otherwise, a purely statistical model can be added and parameterized that addresses the model prediction "deficiencies". In our experience, this is not typically required but is better than catching model prediction deficiencies on feedback of irregular and delayed plant data.

Model-fitting as described above is abstracted from the

actual process. If sufficient relevant historical data are available these data can be utilized relatively easily to provide a "feel-good" factor that connects with the actual process – the kinetic parameter optimization tool can be used to generate input data for identifying back-test success rate on predicting historical transitions. (As a steady state model is being used, some subjectivity is involved in identifying the pre- and post-transition "steady states".) Although it is a function of unmeasured disturbances (usually particularly associated with the catalyst) a success rate of higher than 75% is generally experienced.

What is the engineering effort involved for this model-fitting? On the basis of a good understanding of the engineering concepts and tools, the effort related to the process described herein is akin to model identification effort for traditional linear multivariable controllers. Generally, the major activity involves collating historical plant operation to provide the objectives of the model-fitting and ensure "coverage" is consistent with actual plant operations.

VI. Predicting Quality Properties for Single Reactor Free Radical Polymer Processes

Although many of the reaction classes for free radical reactions have a parallel to the reaction classes for transition metals, there are some differences that impact the detail of the procedure for determining kinetic parameter values. For example, initiator decomposition kinetic parameters may be available – from open literature, or the manufacturer. As another example, usually deactivation reactions need to be active to enable reasonable prediction of production rate. Understanding the predictive effect of the different termination reactions (disproportionation and combination) may be difficult. One approach is to initially fix the ratio of these reaction rates at some literature value; then complete the overall model fit (in much the same way as reviewed in Table 2). Then the ratio is varied; the model re-fitted and the effects on the model prediction capability studied (i.e. an overall iterative step). As before, updated values for kinetic parameters are only accepted after analysis deems the steady state gain matrix remains reasonable.

VII. Predicting Quality Properties for Multiple Reactor Polymer Processes

Multiple reactor polymer processes bring challenges to controller design. However, a model-fitting advantage exists as there are effectively additional measurement data due to inference of production rates from each reactor. These data can be used to determine some of the kinetic parameters for reactions previously set primarily from engineering assumptions. For transition metal kinetics, this includes the deactivation and site activation or initiation reactions.

VIII. Expanding the Process Model Description

Sections IV to VII focused on achieving the requirements for prediction of quality properties CVs. However, section III also listed the CV "groups" of polymer production rate and production rate constraints. Reasonable prediction for polymer production rate effectively falls out of the approach discussed for kinetic parameter fitting. In the online environment, model-based state estimation such as IDF or MHE can use the instantaneous production rate, as determined by energy and/or material balance, to update an input disturbance of catalyst activity. This is particularly effective for multiple reactors in series. The CV group remaining for discussion is therefore polymer production rate constraints. In reality each CV within this group has unique aspects for a particular application. Some examples for a Unipol™ PE process controller are discussed below.

An important operability constraint is maintaining the cooled cycle gas returned with fresh monomer to the fluidized bed distributor as dry (no hydrocarbon condensation) or wet (including condensed hydrocarbon). If returned wet, then there is typically an upper limit on the liquid fraction. Polymer production rate is typically maximized at the intersection of maximum cooling constraint – such as valve position at output limit – and maximum liquid fraction of cycle gas (or minimum approach of cycle gas temperature to dew point for dry operation).

First, predicting the valve position constraint is discussed. The flowsheet approach for the process model conserves energy and material balances. Use of an exchanger model enables prediction of required cooling water flow, which has a relatively simple, direct relationship with cooling water valve position that can be determined in a spreadsheet using traditional linear regression on plant data. (This is akin to determining the "measured" valve characteristic.) Online state estimation can be used to update an output bias for the predicted valve position. It can also be noted that online state estimation can be used to update the input disturbance of exchanger UA thus facilitating use of a "simple" exchanger UA model within the flowsheet.

Now we turn to the cycle gas cooler exit liquid fraction constraint. An online measurement is not available so, as was the case for polymer properties, a more detailed flowsheet modeling approach is required. Given that the process may be operated in a high hydrogen concentration regime, for offline studies specialized thermodynamic relationships can be used to predict the process-side flash. However, there are concerns to incorporate such relationships within the control flowsheet model due to expectation of multiple roots from the relevant equations that may affect the solution robustness. Resolution of this issue is to emulate the specialized thermodynamic relationships by determining and using "k-value" multipliers that adjust base k-values predicted from a more basic equation of state such that the model predictions with these "k-value" multipliers result in a reasonably closed energy balance for various sets of process data. These "k-value" multipliers are determined using the offline tool discussed for determining kinetic parameters.

IX. MATCHING THE CONTROL MODEL TO THE PROCESS: ONLINE STATE ESTIMATION – CONTINUOUS MEASUREMENTS

Unmodeled features of a physical process pose a limitation, in particular, for model-based control. Before optimized manipulated variable (MV) moves can be computed, the current model states must reasonably match the measurements from the actual process. Several approaches are discussed in the technical literature for implanting this match (termed state estimation). We prefer more direct optimization approaches such as MHE because the real world nonlinear differential algebraic equation (DAE) model forms, including partial DAEs (PDAE), are easily accommodated in this formulation. This allows for the exploitation of sparsity, which is critical to success in using flowsheet models for process control/optimization applications. Constraints are also naturally handled in MHE approaches. Our focus is to use state estimation approaches that facilitate the use of less sophisticated models for predictive analysis as discussed in [2].

We have found that EKF is not appropriate for large sparse models such as those generated by flowsheet models as elimination of algebraic equations during the linearization process is required for the computation of Kalman gains. This typically destroys any sparsity present in the model. In addition, flowsheet models generally contain a large number of algebraic equations, usually precluding state reduction to reduce the problem size.

X. MATCHING THE CONTROL MODEL TO THE PROCESS: ONLINE STATE ESTIMATION – DISCONTINUOUS MEASUREMENTS

Section III discussed that measurements for the actual quality CVs are often only available on a delayed and discrete basis (for example as laboratory measurements). This was the argument for initially focusing model description on prediction of the quality property CVs. However, the CV measurement needs to be used to update the control model as it is this measurement (not model prediction) that determines whether the product is within quality specifications.

In the absence of an automated system the board operator uses human judgment to respond to a new laboratory measurement and the objective of automation is to match or exceed the judgment of the "best operator". Application of traditional Statistical Process Control (SPC) to the laboratory measurement tends to be insufficient as:

- The time to obtain the measurement subsequent to sampling is relatively long compared to reactor residence time.
- Inhibitors/poisons in small quantities can cause properties to drift at steady state.

Our experience has been to apply additive or multiplicative output biases to the underlying model-predicted values for polymer properties to generate the controller predicted value. On receipt of a new laboratory measurement, a new additive bias is calculated through:

$$Aerror = CV_{meas} - CV_{model}$$
$$Abias = Abias_{old} + Gain * Aerror$$

And a new multiplicative bias is calculated through:

$$Merror = \text{Ln}\,(CV_{meas} - \text{Offset}) - \text{Ln}\,(CV_{model} - \text{Offset})$$
$$Mbias = \text{Exp}\,(\text{Ln}\,(Mbias_{old}) + Gain * Merror)$$

where:

CV_{meas}	New laboratory measurement value
CV_{model}	Controller unbiased model prediction value corresponding to the process time (sample time – process/sample dead time) the sample was taken.
Gain	Bias feedback gain term
Offset	Fixed offset value (typically zero)

Aerror, Merror	Error
Abias, Mbias	New bias
$Abias_{old}$, $Mbias_{old}$	Previous value of bias

for which engineering judgment applies to:

- Selection of the model variable to be updated
- Choice of additive or multiplicative output bias
- Setting values for Gain

If the model prediction change for a step input is a fairly constant absolute delta, independent of the underlying model prediction value, an additive output bias is most appropriate. A multiplicative output bias is most appropriate when the model prediction change for a step input is generally a fairly constant fraction of the current measurement.

Melt index is known to be consistent with the expectations for using a multiplicative output bias – this can also be inferred from the relatively common practice described in the literature of using a logarithm transformation on the melt index to generate a variable more suitable to be handled with a more traditional linear multivariable controller. Thus model predicted melt index is updated using the multiplicative output bias approach (and this same bias is also applied to the instantaneous melt index which is a constraint controlled variable in the controller matrix).

For polymer density, the approach to update the model from the laboratory measurement has been varied – and the choice between additive and multiplicative biases has been application-specific. Multiplicative output biases have been applied on variables transformed from the base laboratory measurement, such as percent comonomer incorporated.

Our practice also uses a variable Gain set by a heuristic algorithm, nominally emulating the judgment of the "best operator". Inputs to the algorithm include the trend of the discrepancy between unbiased model prediction and actual laboratory measurement: if drift is detected the Gain is increased in order to catch up with the process. In the absence of drift the Gain is set lower. If the occasional "outlier" is sometimes present in the laboratory measurements, then the algorithm is set to use a low value for Gain on the first occasion of a significant discrepancy between model prediction and laboratory measurement.

XI. CONCLUSIONS

In the context of development and implementation of dynamic flowsheet models embedded within nonlinear controllers, this paper has discussed:

- Factors affecting choice of flowsheet modeling approach
- Determination of kinetic and other parameters for the flowsheet model
- Coupling use of continuous state estimation with less fidelity in the flowsheet model
- Use of delayed discrete measurements for state estimation

Much of the discussion has been based around engineering "common sense". As a very broad generalization, an argument has been made for a rigorous first principles approach for the reactor(s) coupled with simpler flowsheeting modeling that maintains energy and material balances for the remainder of the process. Practically some exceptions will occur to this generalization, as was discussed for the process side cooler for UNIPOL™ PE process.

Our experience is that the effort to develop the flowsheet model embedded within the controller is akin to the effort required to develop empirical models for traditional linear multivariable controllers. As such, this approach to nonlinear control bears serious consideration as it brings the potential benefits listed in Table 3, some of which were discussed in [7].

TABLE 3: CONTROLLER EMBEDDED MODEL ADVANTAGES

Issue	Notes
Offline study	Dynamic chemical engineering control model from the online controller available for offline study use.
Regression validity concerns	Reduced engineering focus for regression evaluation due to the underlying nature of the control model.
Handling new grades	Validation exercise rather than a re-engineering exercise due to the nature of the control model.
Inferred measurements	All states in the control model are available, including intrinsic instantaneous polymer properties.

REFERENCES

[1] Xie, T., K. B. McAuley, J. C. C. Hsu and D. W. Bacon, "Gas Phase Ethylene Polymerization: Production Processes, Polymer Properties and Reactor Modeling", Ind. Eng. Chem. Res., 33, pp. 449-479, 1994.

[2] J.D. Hedengren, K.V. Allsford and J. Ramlal, "Moving horizon estimation and control for an Industrial Gas Phase Polymerization Reactor", Paper WeB19.4, *Proc of American Control Conference (ACC)*, New York, NY, Jul. 2007

[3] Ray, W.H. "On the mathematical modeling of polymerization reactors", J. Macromol. Sci.-Revs. Macromol. Chem., Vol. C8(1), pp. 1-56, 1972.

[4] M McCown Kaus, Ch 14.9 (pp 14.113-14.130) in "Handbook of Petrochemicals Production Processes" (Ed R. A. Meyers) ISBN-13: 9780071410427, McGraw-Hill, New York, NY, Sept 2004

[5] www.univation.com/unipol.animation.html, Mar 2008

[6] ASTM, Procedures D1238-04c and D1505-03, www.astm.org, 2007.

[7] K. V. Allsford, J. Goodman, J. Ramlal and R. Beigley, "Nonlinear Multivariable Control and Optimization of a Polyethylene Process Based on Embedded Dynamic Chemical Engineering Model", Proc of 3rd International Symposium on Advanced Control of Industrial Processes (ADCONIP), Banff, Canada, May 4-7 2008

ADVANCED PROCESS CONTROL FOR CHEMICAL BATCH REACTORS APPLIED TO A POLYMERIZATION PROCESS

Wim Van Brempt, Bert Pluymers, Bjorn Vandecraen, IPCOS, Technologielaan 11/0101 3001 Heverlee, Belgium

Jan De Wilde, Cytec Surface Specialties, Anderlechtstraat 33, 1620 Drogenbos, Belgium

Introduction

Advanced Process Control ('APC') has become widespread and even a commodity tool for certain continuous processes. APC encompasses model based control solutions, where control and on-line optimization go hand in hand, and a Model Predictive Controller is often used for this purpose.

The advantage of this technology is straightforward: while delivering a significantly improved process control result, it enables to optimize your process on-line taking operational constraints into account.

However, until recently these tools could not be applied to chemical reaction batch processes. The reasons for this are twofold: on the one hand the general APC tools as applied in continuous processes cannot deal with the more complex nature of batch processes (changing dynamics, nonlinear responses). On the other hand the modeling effort and related cost would also be prohibitive to implement an APC project on batch processes with a reasonable return on investment. This would even be more true in a multi-product multi-reactor environment.

Therefore a new strategy has been developed for chemical batch processes. A hybrid modeling solution is proposed which needs a limited engineering input, and delivers a high precision representation of the process. The hybrid model contains both a rigorous mechanistic process description as well as a parametric part. The mechanistic part is used to model often encountered devices in a batch process, like half tube jackets, coils, external heat exchangers, condensers and an overall mass and heat balance. The parametric part is used to describe the reaction kinetics.

This non-linear model is then used within the Model Predictive Control solution. This solution contains an EKF based observer, as well as an interbatch control and observer solution.

Objective of the control solution is to deliver improved control performance, however in a lot of applications batch cycle time optimization is the prime objective. As such the reaction phase of the batch is minimized while keeping the process within the allowable constraints (available heat exchange duty, adiabatic temperature, quality aspects...).

The above mentioned APC methodology has now been applied to several industrial processes. An application case at CYTEC Industries Inc. is shown on a polyacrylate polymerization process.

Some observations on batch control challenges

Batch reactor processes are commonly praised for their flexibility. Indeed, with limited variety in hardware equipment, a wide range of products can be produced, ranging from performance polymers to specialty chemicals, from pharmaceutical intermediates to food ingredients. This flexibility however results in much more complex operation. Recipes must be defined and implemented, while production scheduling is a challenge in a multi-reactor multi-product environment. Also the control task can be more complex since the process can no longer be characterized having a single process gain, but the changing product characteristics throughout the batch cause the gains to vary dynamically. This nonlinear behavior has made traditional linear MPC technology inadequate for solving the problem. One can only conclude that a different type of model is required.

Different types of nonlinear models come to mind. On the one hand, one could choose very descriptive mechanistic models based on phase descriptions and heat and mass balances of all the components in the reactor. These models can give a very accurate description of all phenomena occurring within the reactor. To be able to be used within an APC application these models should have a guaranteed short calculation time, be stable at all times, be validated and tuned on the process, and have a limited numbers of states. Even more importantly, although not always trivial, they should contain a proper input-output description for control. One has to be sure that all relevant control degrees of freedom and interactions are well modeled. If all these criteria are met, a mechanistic model approach may be used.

The financial picture however looks a bit different if the rigorous model has to be developed from scratch for a specific APC application. This is especially true in a multi-product reactor, where a given recipe is only utilized a limited time per year. If the APC application is commissioned for that recipe, the payback should be generated on that recipe as well. To cut project costs, modeling efforts must be reduced as much as possible. A simple modeling technology requiring limited implementation time is needed.

APC for Batch applications

Standard techniques which are applied in continuous APC applications offer no solution for batch processes. In order to enable online optimization of batch processes, novel modeling and control techniques must be developed.

Dedicated Batch APC Models: Hybrid Modeling

Optimization is implicitly embedded in a Model Predictive Control solution. A process model is required that features a short design time and sufficient accuracy to be used within a control and on-line optimization application. For reasons explained above, a purely mechanistic model does not address these requirements. A black box approach, which represents the opposite extreme, is not straightforward either. Linear black box models do not work in a highly nonlinear application as a batch process. General nonlinear black box functions (e.g. neural nets) possess insufficient extrapolation capability. It is the assertion of the authors that a hybrid combination of mechanistic and black box modeling delivers the best of both worlds.

The envisaged model should give a good representation of the heat balance of the process. This will automatically include information on conversion, build up of unreacted feed material, and adiabatic run-away temperature and pressure.

In fact, most batch reactors share common principles. All of them use heat exchange devices, like jacketed reactors (half pipes or other design), coils or external heat exchangers. In some cases vapor builds up and a condenser is used. In a fed-batch or semi-batch reactor, mass is accumulated in the reactor while the reactor is being filled. All these principles can be modeled in a general mechanistic way that requires only a few parameters to be tuned to the actual process. This general mechanistic model can be applied to most batch reactors and drastically decreases the engineering time since one can take the advantage of general pre-built modules.

On the other hand the reaction scheme and kinetics, the phase behavior, etc. differ from process to process. Mechanistic modeling of these phenomena is the most time consuming part. Therefore a specific black box approach is chosen which uses a surface map to relate overall conversion and temperature to reaction energy or conversion speed (Figure 1). Some a priori knowledge is built into the map as well.

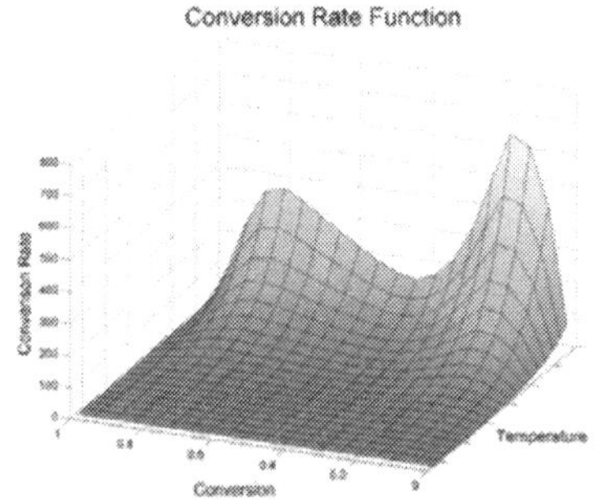

Figure 1. Reaction Surface Map

Experience with this method has revealed that combining the rigorous model part with the black box reaction part, results in a very accurate process model demonstrating nice extrapolation capability. All heat balance related issues such as reaction energy, conversion, reactant excess, and adiabatic temperature are taken into account. Design times are short as most of the information can be extracted from historical data and a limited number of tests on the process. Once a model has been validated on the process, generating extra models for new products on the same reactor is only a small effort.

Nonlinear MPC (NL-MPC)

Given the fact that the proposed model format is not standard and certainly nonlinear, traditional linear MPC principles, calculations, and procedures do not hold. A nonlinear MPC (INCA® for BATCH) solution is required.

The proposed strategy splits the NL-MPC problem into two stages: a prediction step and an optimization step. First a nonlinear prediction is made with the aid of the hybrid model. Based on the previous process inputs and the previous optimization result, a prediction emerges of the process behavior over a horizon into the future. Along this predicted batch trajectory a set of linear dynamic models is automatically provided. These models are combined into a linear time variant model (LTV model).

In the second stage this linear model is used to optimize the process. Based on the nonlinear prediction, the optimization observes the gap between where the process is heading and where the process is desired to be. The optimizer minimizes this gap by taking into account operating constraints on process inputs and process variables such as adiabatic temperature, and reactant excess. The model used during the optimization is the LTV model, enabling fast calculation speed since a QP algorithm can be applied. An Extended Kalman

Filter technology based observer modifies model states such that the hybrid model stays on track with the actual running batch.

In addition to the aforementioned optimization and observer, additional modules are optionally executed in between subsequent batches. Two of these modules are an inter-batch observer and an inter-batch controller. Both deal with phenomena which are only visible over longer time periods than one batch. The interbatch observer adapts the hybrid model in case of fouling or catalyst deactivation which only become tangible after several batch runs. The interbatch controller modifies the optimization targets and constraints from batch to batch to maintain the product at or within quality specifications. These specifications may be affected due to slowly varying feedstock quality and become visible only after a lab sample is analyzed after the batch has ended.

Application Example

INCA® for BATCH was applied on a water based polyacrylate process at CYTEC Drogenbos, Belgium. The site contains multiple workshops, amongst others a water based and solvent based polyacrylate process. It is a typical multi-product multi-reactor environment. The solvent based polyacrylate workshop counts 5 reactors covering 60 product grades, while the water based polyacrylate workshop produces 70 different product grades using 5 reactors.

The water based polyacrylate process is an exothermal fed batch process taking place under atmospheric conditions. A process lay-out is shown in Figure 2. Heat is exchanged by the means of a half-tubed jacket. The jacket temperature is controlled making use of a frequently applied configuration: a circulation loop with a fresh cooling water inlet to decrease the circulation water temperature and a heat exchanger to increase the circulation water temperature. Raw materials are dosed making use of premix vessels. An agitator is used to achieve proper mixing.

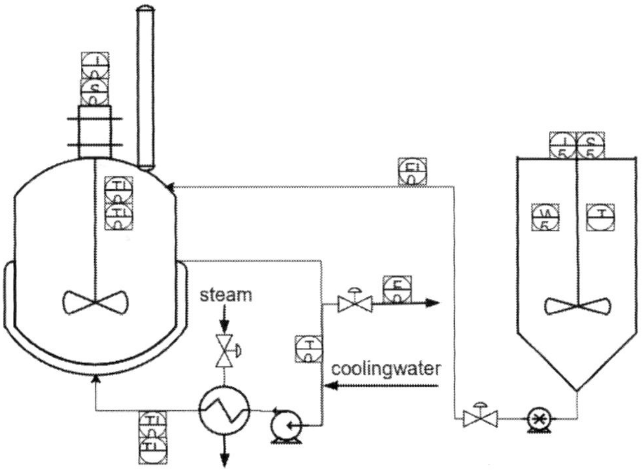

Figure 2. Batch reactor Layout

The project was implemented via a number of phases. In order to gather data for model development, a logging environment was set up to enable data logging at high sample rates without data compression. The logging software was installed on a PC and connected to the plant DCS via OPC. The next stage demonstrated significant benefits of the hybrid model approach. If using traditional model identification techniques, tests spanning many batches would have been required but the hybrid model approach only required tests to be executed during two batches. During a first batch a slightly elevated temperature profile was used. In a second batch a step disturbance was applied to the reactor feed flow (Figure 3). Both experiments were discussed thoroughly with key process engineers. The amplitudes of the applied changes were selected such that they would have minimal impact on the process operation and especially the batch-end quality, while the effect would still be observable in the measurements.

In a next step the hybrid model was identified. This was largely done based on historical data, enriched by the data delivered by the above described tests. The resulting reactor temperature profiles (hybrid model versus measurements) for two different product grades are shown in Figure 4. Hybrid model derivation for extra product grades becomes a smaller task, since large parts of the models can be reused.

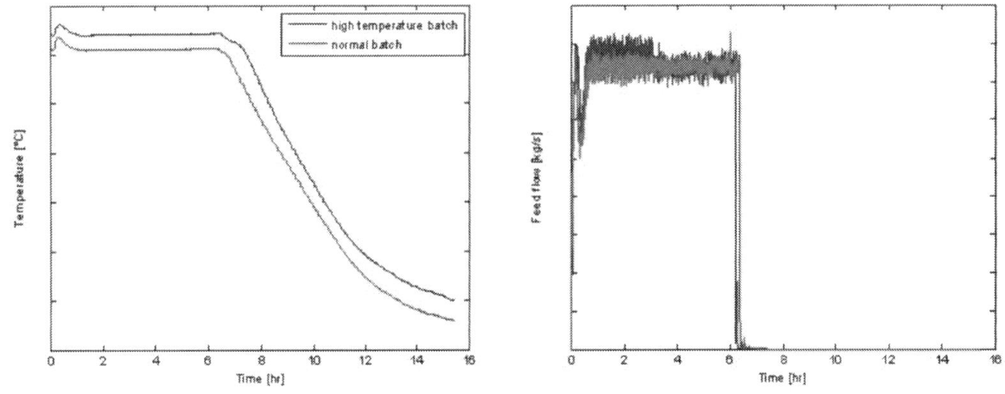

Figure 3. Applied tests to identify hybrid model

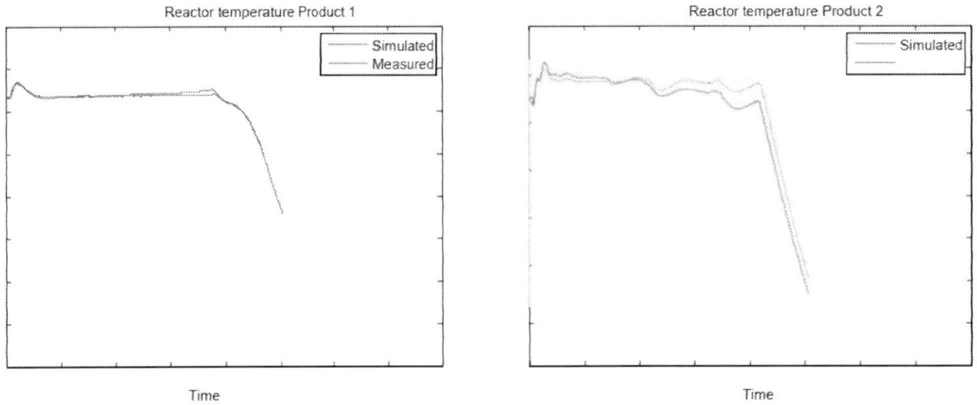

Figure 4. Modeling Results

Comparing to the measured signal, one can notice a slow drift on the model simulation for the second product shown in Figure 4. To improve model predictions, an Extended Kalman Filter based observer is tuned. The observer updates model states such that the model stays in track with the process despite unmodeled disturbances and model shortcomings. Next, a complete offline simulator was set up, consisting of a controller, an observer, and a modified hybrid model to represent the process-model mismatch. Based on this environment, controller tuning was performed and several process scenarios were analyzed. This simulator was then used for the Factory Acceptance Test where satisfactory performance for multiple case studies and sensitivity analyses was demonstrated. The control environment supports a broad set of different product grades, each represented by its dedicated hybrid model.

Due to the computationally intensive calculations required by the technology described in this paper, the controller is implemented in a separate, dedicated PC. This PC is typically connected to the process DCS or PLC via OPC. A separate interface is designed on the control system operator screen to permit the operator to interact with the application. Under normal operation, batch parameters are defined when the batch recipe is scheduled. It is especially worth noting that this is also when the choice to utilize MPC is defined. The recipe ensures that DCS/PLS controllers are switched into the proper input mode (local, remote, supervisory) and the MPC controller is then automatically started at the proper moment. At any given moment the operator can decide to switch off the MPC controller, and then a fallback is initiated towards a safe PID control solution. In rare cases of PC hardware, communication, or software malfunction the fallback is also triggered by means of a watchdog timer.

The results for two different product grades are discussed next. In Figure 5 the implementation results are shown for a first product grade. In this case limited operational freedom was available due to product quality related constraints. Therefore the focus was geared on batch reproducibility and product consistency. It is shown that a hybrid model predictive approach enables a tight process control, reducing the variance dramatically compared to conventional PID control. Mainly the overshoot at the start of the introduction phase was recognized as being critical to control the desired product end-quality parameters. Elimination of this overshoot was therefore essential to improve product consistency.

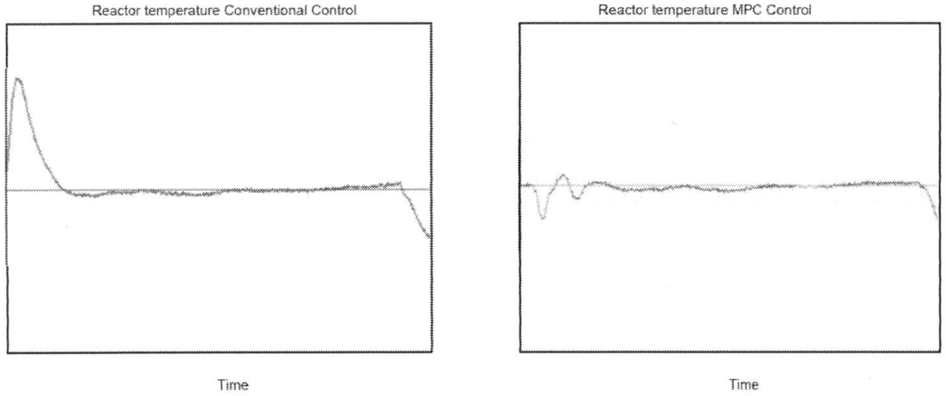

Figure 5. MPC results for polyacrylate product grade 1

In Figure 6 the results for another product grade are shown. In this case the focus was directed upon batch cycle time reduction. The MPC controller is configured such that the feed flow is maximized, while guaranteeing that at the current time instant and in the future the available cooling power will not be exceeded. The eventual control result shows an operation switching between the cooling constraint and the feed flow constraint, while keeping the process temperature on the setpoint. As a result 60 minutes batch cycle time is eliminated.

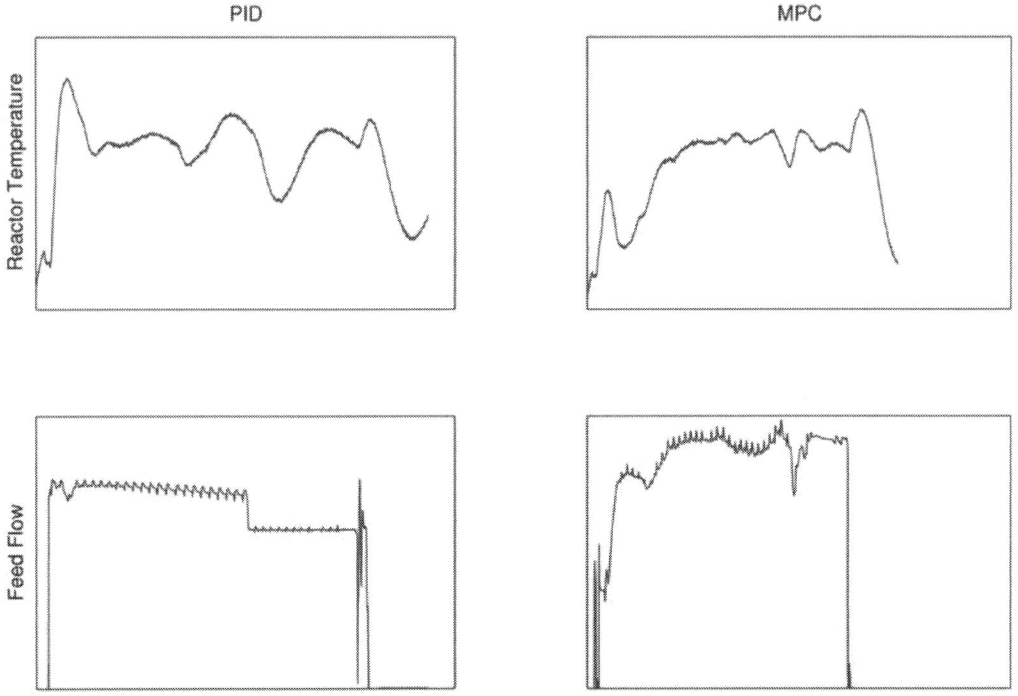

Figure 6. MPC results for polyacrylate product grade 2

Conclusion

In this paper an industrial application is shown of a hybrid non-linear model predictive control technology. The solution is specifically geared towards batch reactor systems, and aims at achieving two goals. At first a technical solution is provided to address the non-linear dynamical complexity of a batch reactor, encompassing a modeling and optimizing control solution. Secondly, the technology is set up to minimize modeling and configuration time, in order to achieve acceptable payback time in an industrial multi-reactor multi-product setting. Industrial results for different polyacrylate product grades show enhanced batch reproducibility and significant cycle time reduction.

433a Electromechanically-Driven Complex Void Dynamics In Metallic Thin Films

Vivek Tomar, M. Rauf Gungor, and Dimitrios Maroudas. Department of Chemical Engineering, University of Massachusetts, 159 Goessmann Laboratory, North Pleasant Street, Amherst, MA 01003

Electromigration-driven nonlinear void dynamics in metallic thin films has been studied widely in order to fundamentally understand metallic interconnect failure that is responsible for serious materials reliability problems in integrated circuits. Recent theoretical work in this area has demonstrated extremely rich nonlinear dynamics for the electromigration-induced morphological evolution of void surfaces due to surface diffusional anisotropy and pronounced current crowding effects. In this presentation, we report results from the analysis of the nonlinear dynamics of void surface morphological response in metallic thin films under the combined action of electric fields and mechanical stresses. We place emphasis on the role of the symmetry of surface diffusional anisotropy in determining electromechanically driven complex void dynamics. Most importantly, we report complex shape evolution that sets the material system on a route to chaos characterized by a sequence of period-doubling bifurcations.

Our nonlinear analysis is based on self-consistent numerical simulations of current-induced and stress-induced migration and morphological evolution of void surfaces in metallic thin films. The simulations account rigorously for current crowding and stress concentration effects that become particularly important in narrow metallic films, as well as surface curvature effects that are particularly strong due to the strong anisotropy of adatom diffusion on void surfaces. The mass transport problem on the void surface is solved coupled with the electrostatic and mechanical deformation problems in the conducting film that contains the morphologically evolving void, assuming that the metallic material responds to stress elastically. A Galerkin boundary-integral method is used for the computation of the electrostatic potential and the elastic displacement field in conjunction with a front tracking method for monitoring the evolution of the void surface morphology.

In the absence of stress, the electromigration-induced void surface morphological response is studied for single-crystalline films with twofold and fourfold symmetry of surface diffusional anisotropy, characteristic of <110>- and <100>-oriented film planes in face-centered cubic (fcc) metals. Variation of the electric-field strength, the surface diffusional anisotropy strength, and the void size past critical values is found to cause transitions from stable steady to stable oscillatory morphological responses on voids migrating along the films at constant speed through Hopf bifurcations at the corresponding critical points. The nature of the Hopf bifurcation is determined by the symmetry of the surface diffusional anisotropy on the film plane: the bifurcation is supercritical for fourfold symmetry and subcritical for twofold symmetry. Our simulations reveal hysteresis phenomena and bistability for both twofold and fourfold symmetry. For twofold symmetry, the transitions are between a stable steady state and a stable time-periodic state. For fourfold symmetry, the transitions may also be between two stable steady states, one of which bifurcates to a time-periodic state at the corresponding Hopf point.

For twofold symmetry, under the simultaneous action of an electric field and mechanical (biaxial tensile) stress, our analysis predicts that increasing the applied stress level leads to morphological transition from a steady state to a time-periodic state through a subcritical Hopf bifurcation. Further increase of the applied stress level leads to a second critical point where a period-doubling bifurcation occurs, resulting in more complex surface wave propagation. The bifurcation sequence continues with increasing stress level, setting the system on a route to chaos. With further increase in the stress level, the system exits from the chaotic regime to a complex time-periodic state, characterized by three periods. Further increase in stress pushes the system into another chaotic regime and leads to film failure beyond a certain stress level. The complex shape evolution is characterized in detail over the range of stress and electric field levels examined and the features of the resulting chaotic states (strange attractors) are reported. An analogous study for the effects on the nonlinear void dynamics of the simultaneous action of an electric field and mechanical stress also is presented for <100>-oriented thin films with fourfold symmetry of surface diffusional anisotropy.

433b Elastic Stability and Structural Response of Body-Centered Cubic Crystals Under Uniaxial and Hydrostatic Loading

Hadrian Djohari, Department of Chemical Engineering, University of Massachusetts, Amherst, 686 N Pleasant St, Amherst, MA 01375, Frederick Milstein, Departments of Materials and Mechanical Engineering, University of California, Santa Barbara, Engineering II Bldg, Room 2355, Santa Barbara, CA 93106-5070, and Dimitrios Maroudas, Department of Chemical Engineering, University of Massachusetts, 686 N Pleasant Street, Amherst, MA 01003.

Understanding large-strain mechanical deformation and failure of crystalline solids requires analyses of elastic stability in order to determine the crystal's mechanical strength. At given temperature, the structural response of a crystal to applied mechanical loading under a certain loading mode becomes unstable beyond a critical stress level. A major challenge in materials mechanics is to predict the theoretical strength of crystals under various mechanical loading modes and to analyze and fundamentally understand the underlying atomic-scale mechanisms of the crystals' structural response beyond the onset of stress-induced instability.

In this presentation, we report results of systematic elastic stability analyses in body-centered cubic (bcc) crystals under uniaxial and hydrostatic loading. The bcc structure is not a close-packed one and, therefore, it tends to be less stable and with higher energy than that of the face-centered cubic (fcc) and hexagonal close-packed (hcp) structures. Stress-induced bcc-to-fcc and bcc-to-hcp transitions have been observed experimentally in K, Rb, Cs, Ca, Sr, Tl, and Fe crystals and in Be, Mg, Ba, Tl, Ti, Zr, and Fe crystals, respectively. Parameters that affect the bcc crystal lattice stability under mechanical loading include the loading mode, the direction of loading, temperature, and the presence of defects (such as nanovoids and other lattice imperfections) in the crystal. We employ isostress-isothermal molecular-dynamics (MD) simulation to analyze systematically the structural response of the crystal, placing special emphasis on the atomic pattern formation characteristics during the structural transformation. Subsequently, we conduct isothermal-isostrain MD simulations in order to compute the elastic moduli through canonical strain fluctuation formulae and use them to assess the crystal's elastic stability rigorously, according to proper stability criteria.

Our analysis emphasizes the similarities and differences in the atomic pattern formation during the stress-induced structural transformations of bcc crystals and the bifurcations in the crystal structural response that are exhibited when the applied load and/or the temperature are varied. We find that a bcc crystal would transform readily to a more stable structure, rather than fail, under both uniaxial and hydrostatic loading. In most of the cases that we examined, the loss of structural stability is consistent with Born's criterion. The bcc structure transforms to a fcc one, when it follows the tetragonal path of Bain deformation, and when it is under very high uniaxial compressive stress. However, the bcc structure transforms to a hcp one at the points of bifurcation from the tetragonal path and when a shear-modulus instability occurs. The bcc-to-hcp transformation mechanism is governed by the shearing of alternate planes of the body-centered tetragonal (bct) structure; this produces the stacking sequence of the {0001} basal planes of the hcp phase. Finally, we find that the presence of nanovoids within the crystal generates heterogeneous nucleation sites for the shearing transformation mechanisms and enhances the likelihood of structural transformation; this heterogeneous nucleation of the new phase at the defect site lowers the overall critical stress levels for the transformation.

433c Engineering Complexity In Chemically-Reacting Dynamical Systems: Studies on Electrode Arrays

Craig G. Rusin[1], Istvan Z. Kiss[1], Hiroshi Kori[2], and John L. Hudson[1]. (1) Chemical Engineering, University of Virginia, Charlottesville, VA 22904, (2) Mathematics, Hokkaido University, Japan

Collective organization, such as rhythmic patterns and synchronized states, is commonly found in complex systems composed of a large number of interacting oscillatory elements. Such behavior is important in biological functions such as circadian rhythm and memory as well as in epilepsy and essential tremors. A major question of both theoretical and practical importance is how to bring these dynamical systems to a desired state or, equivalently, how to avoid a deleterious state without destroying the inherent rhythmic behavior of the individual elements.

We have developed a general methodology for engineering the collective emergent behavior of rhythmic systems by using a weak non-linear feedback signal to gently steer the system to a desired state. Flexibility is obtained through the use of experiment-based reduced models in designing the feedback. Such models capture the essential dynamical behavior of the system without requiring detailed chemical or biological descriptions of the individual rhythmic units.

We have experimentally demonstrated the effectiveness of the method by engineering the collective behavior of a population of oscillatory electrochemical elements. Application is made to the design of a nonlinear anti-pacemaker for the destruction of pathological synchronization of a population of interacting oscillators and to the generation of sequentially-visited dynamic cluster patterns similar to reproducible sequences seen in biological systems.

433d Dynamical Behaviour of the Phenylacetylene Carbonyation Reaction in a Calorimeter

Ankur Mukherjee[1], Katarina Novakovic[2], Dominic P. Searson[2], and Mark Willis[2]. (1) School of Chemical Engineering and Advanced Materials, Newcastle University, Merz Court, Claremont Road, Newcastle upon Tyne, United Kingdom, (2) School of Chemical Engineering & Advanced Materials, Newcastle University, Merz Court, Claremont Road, Newcastle upon Tyne, United Kingdom

This paper reports dynamical behaviour of the palladium catalysed oxidative carbonylation reaction of phenylacetylene in a homogeneous catalytic system (PdI_2–KI–O_2–NaOAc in methanol). The dynamics of this system is studied through the simultaneous oscillations in pH and reaction exotherm (Q_r) in a calorimeter. The objective of the study is to contribute towards understanding the reaction kinetics underlying the oscillation generation. To achieve this goal, the reaction is perturbed by methanol and phenylacetylene at different phases and the dynamics is characterised according to the phases of the pH oscillations. In addition, concentration profiles of reactants and products in a larger period pH oscillation at lower reactor temperature are reported. Other observations, such as quasiperiodic pH oscillations in response to changes in air-flow rate and pH oscillations after the incidence of white light are presented.

The dynamical behaviour of the reaction system is analysed by nonlinear dynamical tools. The pH oscillations in response to single pulse perturbations by methanol and phenylacetylene are represented by phase transition curves (PTC's). The dynamical characteristics of the system under periodic perturbations are simulated from iterations of the PTC's for a range of forcing time intervals. Resonance regions are identified for specific forcing periods and are illustrated by respective firing number plots. The firing number sequences denote devil's staircase like dependence on the perturbation time intervals.

Web Page: www.ncl.ac.uk/htt/

433f Front Velocity and Criterion of Transversal Patterns In Packed-Bed Reactors

Moshe Sheintuch and Olga Nekhamkina. Chemical Engineering, Technion, Israel Institute of Technology, Technion City, Haifa, 32000, Israel

1. Introduction. Predicting hot zones in catalytic reactors is the focal point of design criteria aimed to avoid severe safety hazards. Thermal fronts are known to exist in such reactors and some experiments indicate that they may undergo symmetry breaking in the transversal direction under certain conditions. Note, that the linear stability analysis of the 1-D front solution is a rather complex problem as the basic (steady or moving) front state is not known, a priori. Reported studies employ numerical simulations[1] which are very time consuming and are difficult to analyze. To simplify the full 3-D problem analysis several reduced models yielding a 2-D presentation have been proposed[2-3]. Linear stability analysis can be applied then after a single, or multiple steady state (SS) solutions have been determined. The only analytical result obtained with such reduced models, in the case of a single stable SS is that the necessary condition for emergence of stationary transversal patterns is: $Pe_{Ctrans} < Pe_{Ttrans}$[2]. This is exactly a condition of the Turing activator-inhibitor mechanism and can be applied for a stationary front, but it is considered to be unrealistic for fixed beds. A new criterion of transversal symmetry breaking of moving fronts has been recently proposed by us[4]. Using both a learning two-tube model exchanging heat and mass, and a thin cylindrical annular 2-D bed model we obtained for a first order activated reaction the critical condition to be: $\alpha=\Delta T_{ad}Pe_T/\Delta T_m Pe_C>1$, (1) where ΔT_{ad} and ΔT_m are the adiabatic and the maximal temperature rises. This condition coincides with the previous condition for stationary fronts ($\Delta T_{ad}=\Delta T_m$) when ratios of transversal to axial dispersions of heat and mass are consistent ($Pe_{Ctrans}/Pe_C=Pe_{Ttrans}/Pe_T$). The novel condition (1) can be satisfied in an upstream propagating front ($\Delta T_{ad}/\Delta T_m>1$) for feasible operating conditions ($Pe_C>Pe_T$) as was verified by direct numerical simulations of a 2-D shell PBR model[4]. To implement condition (1) we need to know relation $\Delta T_m=\Delta T_m(\Delta T_{ad},Pe_C,Pe_T)$, which to the best of our knowledge is not available yet. The purpose of this study is to derive the approximate relations for the propagation velocity and the maximal temperature rise of a planar front in a PBR, accounting for a finite mass dispersion and to analyze transversal symmetry breaking of moving and stationary fronts using the relation between the velocity (Vf) and the local curvature (K) of the front.
2. Analysis. As a first stage we derived for a planar front two algebraic equations with two unknowns (ΔT_m,Vf), which in the limiting case of $Pe_C\rightarrow\infty$ can be reduced to the approximate relations known for a 1-D "ideal" front in the case of negligible mass dispersion[5]. Numerical simulations of a 1-D PBR show a good agreement between the simulated and approximated values for moderate and large Pe_C. At the next stage we considered a curvilinear front propagation in a local polar coordinate system and derived approximate relations for the propagation velocity (Vf^c) and the maximal temperature rise for the case of a finite curvature. Applying the condition $dVf^c/dK|_{K=0}=0$ we obtained a necessary condition for

symmetry breaking in the following form: $C(\alpha,\Delta T_{ad},Da,Pe_T,Pe_C,\Delta T_m{}^c,Vf^c,\gamma)[1-\alpha] <0$, (2) where C is a certain algebraic expression. In the limiting case $\alpha \to 0$ ($Pe_C \to \infty$) we have C>0 and the planar 1-D front is stable. Analysis of simplified Eq. (2) with $\alpha>1-\delta$ ($\delta>0$, small) shows that C>0 at least for $1< \alpha <2.4$ and thus a planar front is unstable within this domain. Note, that condition $\alpha>1$ coincides with the previously obtained criterion (1), while refined condition (2) can define the upper α boundary of transversal patterns to emerge. 3. Simulations. The derived criterion was verified by direct numerical simulations of the 2-D cylindrical shell and a full 3-D PBR models. With both models simulations revealed formation of transversal patterns in upstream propagating fronts. On the cylindrical shell the sustained patterns include "frozen" multi-wave, rotating and aperiodic structures. In the 3-D PBR model rotating spirals are generic forms of transversal patterns. The wave structure becomes more complex with increasing the ratio Pe_{trans}/Pe. In conclusion we would like to point that transversal three-dimensional patterns were predicted and simulated in PBR's with a first order activated kinetics within the feasible domain of parameters for the first time. Figure 1.Typical front transformation on an unfolded cylindrical shell leading to formation of upstream propagating transversal patterns (moving coordinates). Figure 2. Typical rotating spiral structure of a front propagating in a 3-D PBR showing the temperature pattern in a cross-section normal to the flow (behind the front). References

1. V.Balakotaiah, E.L.Christaforatou, & D.H.West, Chem. Eng. Sci., 54, 1725-1734 (1999). 2. G.Viswanathan, A. Bindal, J. Khinast, & D. Luss, D, AIChE J., 51, 3028-3038 (2005). 3. V.Balakotaiah, N. Gupta, & D.H.West, Chem. Eng. Sci., 57, 435-448 (2002). 4. O. Nekhamkina, & M. Sheintuch, Chem. Eng. Sci, in press. 5. O.V. Kiselev, & Yu. Sh. Matros, Combustion, Explos. and Shock Waves, 16, 152 (1980).

(1)

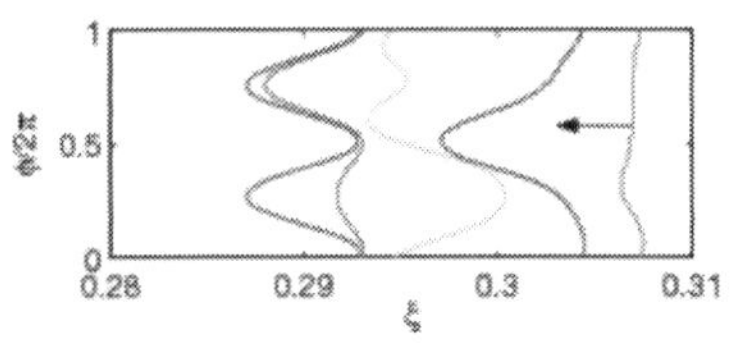

(2)

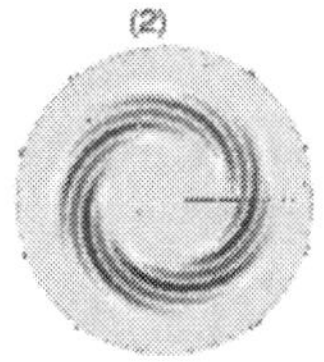

433g Pattern Formation and Chaotic Dynamics In a Model of Simultaneous Oxidation and Reduction In Three-Way Catalytic Converters

Martin Kohout[1], Milan Kubicek[2], Milos Marek[3], and **Igor Schreiber**[1]. (1) Department of Chemical Engineering, Institute of Chemical Technology, Prague, Technicka 5, Prague 6, 16628, Czech Republic, (2) Department of Mathematics, Prague Institute of Chemical Technology, Technicka 5, Prague 6, 16628, Czech Republic, (3) Department of Chemical Engineering, Prague Institute of Chemical Technology, Technicka 5, Prague 6, 16628, Czech Republic

Dynamical behavior of reaction-transport processes in a detailed model of three-way catalytic converter, where catalytic oxidation of carbon monoxide and hydrocarbons occurs simultaneously with reduction of nitrogen oxides possesses several sources of kinetic rather than thermokinetic autocatalysis and can support sustained oscillations and patterns.

The instabilities are provided by kinetics of certain subnetworks found by a reaction network analysis. First, bifurcation analysis is applied to identify basic dynamical modes in a spatially homogeneous reactor. Then the kinetic subnetwork is coupled with transport by diffusion and convection. Reaction-transport processes are assumed to occur in a tubular cross-flow reactor, where reactants are supplied at the inlet as well as along the reactor. We examine whether and how spatiotemporal structures such as spatially nonhomogeneous steady states and travelling waves are produced. Furthermore we find that the system displays a wide variety of chaotic spatiotemporal patterns, which are examined in detail.

References

1. M. Marek, M. Schejbal, P. Ko•í, V. Nevoral, M. Kubí•ek, O. Hada• and I. Schreiber, Chaos, 16, 037107 (2006).

2. M. Kohout, I. Schreiber and M. Kubí•ek, Comp. & Chem. Eng., 26, 517 (2002).

439a A Statistics Pattern Based Framwork for Process Monitoring

Qinghua (Peter) He, Department of Chemical Engineering, Tuskegee University, Tuskegee, AL 36088 and Jin Wang, Department of Chemical Engineering, Auburn University, Auburn, AL 36849.

In the semiconductor industry, process monitoring or fault detection and classification (FDC) has become a critical component of the manufacturing system in order to meet the challenges posed by ever decreasing tolerance of process variation and error. Because process variations often depend on the specific product being produced as well as the specific manufacturing tools being used, to address the dependence of product quality on manufacturing context, currently deployed process monitoring techniques are "threaded" methods [1]. In the last decade, high-mix is becoming the standard manufacturing paradigm in response to rapid and frequent changes in factory plans and process recipes. This trend makes it increasingly difficult for existing FDC techniques to effectively meet manufacturers' requirements [2]. This is because current practice of FDC, i.e., "threaded" FDC, utilizes only part of the historical data, i.e., the ones from the same tool and the same product to build monitoring models. Threaded FDC addresses tool and product variability explicitly, however, in high-mix manufacturing, threaded FDC results in: (1) significant number of models that increases in proportion to the number of product-tool combinations; (2) insufficient data to build models for low-running threads. Although non-threaded (i.e., using data across tools and products) state estimation has drawn some research interest, little attention has been given to non-threaded process monitoring.

In this work, a novel correlation pattern based non-threaded FDC framework is developed to address the challenges posed by high-mix manufacturing. The basic idea of the non-threaded FDC framework is based on the following factor: although normal process data usually show considerable variation across different tools, different products and over time, the correlation patterns (i.e., covariance structures) among different variables are much more stable compared to process mean, due to physical and chemical principles governing the process operation, such as mass and energy balances, transport, kinetics and thermodynamics. Therefore, instead of monitoring the process mean that changes from batch to batch, in the developed approach, we monitor the correlation patterns that are stable from batch to batch to detect and classify process abnormalities.

Because different threads of a same process share a single recipe, they are governed by the same physical and chemical principles, and the data correlation patterns are independent of threads. Therefore, the developed fault detection framework is a true non-threaded fault detection method and only one model is needed for one process/recipe. For similar reasons, the proposed fault detection framework does not suffer from multimodal batch trajectories. In addition, the correlation pattern based monitoring approach does not require data unfolding and there is no need to equalize the unequal batch and step lengths, because data in each step of each batch will generate a correlation pattern matrix whose dimension is determined by the number of the variables instead of step durations.

Both fault detection and classification approaches are developed within the correlation based framework.

The effectiveness of the developed fault detection approach is demonstrated using both simulation examples, and industrial data (TI etch data set [3]). Analysis results show the data correlation pattern of multimodal, unequal step length batches are well conserved.

Reference

[1] J. Wang, Q.P. He and T.F. Edgar (2008), "On state estimation of high-mix semiconductor manufacturing using a singular Gauss-Markov model", submitted to Journal of Process Control

[2] Q.P. He and J. Wang (2007), "Fault detection using K-nearest neighbor rule for semiconductor manufacturing", IEEE transaction on semiconductor manufacturing, Vol. 20 (4) 345-354.

[3] B.M. Wise, N.B. Gallagher, S.W. Butler, D.D. White JR and G.G. Barna (1999), "A comparison of principal component analysis, multiway principal component analysis, trilinear decomposition and parallel factor analysis for fault detection in a semiconductor etch process", J. Chemometrics, Vol. 13: 379-396

439b Agent-Based Distributed Monitoring and Fault Detection

Sinem Perk, Fouad Teymour, and Ali Cinar. Chemical and Biological Engineering, Illinois Institute of Technology, 10 West 33rd St., Chicago, IL 60616

In spatially distributed, large chemical processes, holistic monitoring and fault detection using the highly correlated, nonlinear process data is difficult and may lead to too many false and missed alarms that would render the monitoring process as unreliable. The performance of the monitoring method is also dependent on the process operating region, the fit of the statistical model to the process data and on the performances of the monitoring statistics for that method. Ideally, a monitoring method should explain the process data very well and its monitoring statistics should neither be too strict nor loose.

The aim of this study is to provide an agent-based adaptive distributed monitoring and fault detection framework, which calculates the performance of each agent under different process conditions, keeps this performance history for future reference, and enables the betterment of each agent through adaptation and coordination. In this scheme, using object oriented programming, monitoring and fault detection are embedded as two interdependent layers, where multiple local monitoring and fault detection agents are located. In this paper, an agent-based distributed monitoring and fault detection scheme is proposed for a spatially distributed reactor network.

To increase the performance of the monitoring method multiblock principal component analysis (PCA) has been proposed for large processes, where meaningful division of the process into separate monitoring blocks is possible. Since, multiblock PCA provides statistical process information both at the local and holistic level, fast fault detection and fault localization is provided. Building multiple principal component analysis (PCA) and dynamic PCA in a decentralized manner for each of the above mentioned monitoring blocks provides extra flexibility for each model, such as different number of principal components can be employed if necessary.

Here, the monitoring layer consists of a series of agents that simultaneously perform decentralized PCA, dynamic PCA for each reactor in the network and an agent that perform multiblock PCA for the whole network, and monitoring organizers that track and store the performances of each of the agents for different parts of the network, and for different operation states. The fault detection layer consists of a series of fault detection agents, namely statistical prediction error (SPE) and T2 agents, associated with each monitoring method for each reactor, and a series of fault detection organizers, that keep track of the performances of each of the local fault detection agents and provide a consensus fault flag for the parts of the network they are located. The fault detection agents' performances are updated online as a weighted moving average in time, where the instantaneous performances are based on the payoffs specified for their actions. A missed alarm is considered as the worst performance, and is followed by false alarm and correct detection, latter being the best performance. The overall performance is also dependent on the performances of other coordinating fault detection methods working on the same part of the network. If a monitoring method's fault detection agents are doing well, this implies that the method itself is also performing well. If the monitoring methods performance is degrading in time because its fault detection methods are performing badly, this requires either an update of the monitoring method or update of the fault detection statistic's limits. This approach helps identify if the fault detection agents are irresponsive or highly aggressive or if the monitoring method is not sufficient for the data provided.

The reactor network model for multiple interconnected CSTRs, each having feed and exit streams, as well as multiple connections to their neighbors, is written in C and linked with Repast Simphony via the Java Native Interface (JNI). The framework is built in Java using Repast Simphony, which is an open source agent modeling and simulation environment, and using COLT distribution. Repast Simphony is a java-based framework for agent simulation and provides features such as an event scheduler and 2D, 3D visualization tools. The agents created with Repast Simphony interact with virtual representations of the physical reactor network. The virtual network objects map the states of the physical system to objects that can be manipulated by the agents in the framework. COLT distribution is used to implement the statistical model building and statistical testing methods and is connected to the agent simulation for online monitoring using Repast Simphony event scheduler. The COLT distribution consists of several free Java libraries, for user convenience bundled up under a single name and provides an infrastructure for scalable scientific and technical computing in Java. It contains, among others, efficient and usable data structures and algorithms for off-line and on-line data analysis, linear algebra, multi-dimensional arrays, statistics, histogram computation and Monte Carlo simulation and parallel and concurrent programming.

A case study is performed in the network of reactors by introducing a disturbance to the network. The effectiveness of the methods to detect the fault, the performance evaluation of the methods, and the cooperation and adaptation of the agents is demonstrated.

439c Real-Time Kinetic Hard-Modelling for the Optimisation of Reaction Conditions and the Detection of Process Upset in Semi-Batch Reactors

Julien Billeter[1], Yorck-Michael Neuhold[1], Graeme D. Puxty[2], and Konrad Hungerbühler[1]. (1) Institute for Chemical and Bioengineering, Swiss Federal Institute of Technology (ETH) Zürich, Wolfgang-Pauli-Str. 10, Zürich, 8093, Switzerland, (2) Division of Energy Technology, CSIRO, PO Box 330, Newcastle NSW 2300, Australia

Process Analytical Technology (PAT) has greatly evolved in the last decades due to the development of multivariate online sensors that are able to monitor the properties of industrial processes in real time [1, 2]. The online monitoring of product quality and the detection of process upsets are important for the pharmaceutical and fine chemical industry in order to maintain their product specifications and their commitments regarding safety, health and environment. Most frequent sources of deviations from normal operating conditions in semi-batch processes are due to slightly imprecise initial conditions or impurities in the initial reactants causing unexpected side reactions [3].

Online monitoring of industrial processes usually relies on calibration methods, such Principal Component Regression (PCR), Partial Least Squares (PLS) or Black Box models (e.g. Neural Networks) [4, 5]. A drawback of these calibration methods is their poor behaviour regarding extrapolation, requiring a constant effort for the operator to maintain the calibration conditions. Kinetic modelling techniques [6] do not suffer from this drawback as they are based on first principal models and can also be adapted for the monitoring of highly fluctuating processes, e.g. semi-batch processes.

In this contribution, we propose a method for the online monitoring of semi-batch processes based on kinetic hard-modelling. The proposed method assumes that the kinetic model and the associated rate constants have already been determined at an earlier stage in R&D. In a first phase, the algorithm corrects estimates for the initial concentrations from dosing a small amount of reagent and fitting the kinetic model to the measured signals, e.g. mid-IR, UV-vis or released/consumed heat. In a second phase, if no process upset is detected, the corrected initial concentrations are fed back into the kinetic model and the algorithm optimises the dosing rate of the reagent or the operating temperature by maximising a property of the process, e.g. yield, selectivity or conversion. When optimum operating conditions are found, the algorithm forces the reactor to work under these improved conditions and the process is continuously re-optimised to detect possible process upsets.

The method will be discussed based on simulated and experimental data taken from our high performance small scale reaction calorimeter coupled to in-situ mid-IR and UV-vis ATR-spectroscopy [7].

<u>References:</u>

[1] P. Gemperline, G. Puxty, M. Maeder, D. Walker, F. Tarczynski, M. Bosserman, Analytical

Chemistry 76 (2004) 2575-2582.

[2] J. Workman, M. Koch, D. Veltkamp, Analytical Chemistry 77 (2005) 3789-3806.

[3] E.N.M. van Sprang, H.J. Ramaker, H.F.M. Boelens, J.A. Westerhuis, D. Whiteman, D. Baines, I. Weaver, Analyst 128 (2003) 98-102.

[4] M. Spear, Chemical Processing 70 (2007) 20-26.

[5] T.J. Thurston, R.G. Brereton, D.J. Foord, R.E.A. Escott, Journal of Chemometrics 17 (2003) 313-322.

[6] M. Maeder, Y.M. Neuhold, Practical Data Analysis in Chemistry, Elsevier, Amsterdam NL, 2007.

[7] F. Visentin, S.I. Gianoli, A. Zogg, O.M. Kut, K. Hungerbühler, Process Research & Development 8 (2004) 725-737.

439d Industrial Implementation of on-Line Multivariate Quality Control (Part III: Relationship to molecular weights)

Leo H. Chiang, Kathy Walsh, Shawn Maynard, Bart Neff, and Lloyd Colegrove. Core R&D, The Dow Chemical Company, 2301 Brazosport Blvd., B1225, Freeport, TX 77584

Development and implementation of on-line multivariate quality control was discussed at the 2005 AIChE meeting (Chiang and Colegrove, 2005) and long-term performance, model maintenance, and model leveraging was discussed at the 2006 AIChE meeting (Chiang and Colegrove, 2006)

To further satisfy customer application requirements a new analyzer was installed to monitor molecular weights for two key products. Representative samples from 7 products were collected for both molecular weight and product quality variable measurements. A partial least squares (PLS) model was built to predict product quality based on molecule weights. The PLS model provided accurate prediction of product quality for most time periods. Model discrepancy was observed for some time periods due to analyzer calibration problems. The new analyzer provided additional product quality information, but it was not feasible to build a PLS model to predict molecular weights based on product quality variables.

The project confirms that both multivariate quality control and molecular weights are needed for an effective product quality management program for these two key products.

Reference:

L. Chiang and L. Colegrove, Industrial implementation of on-line multivariate quality control, AIChE Annual Meeting, Cincinnati, OH, 2005. Paper 320a.

L. Chiang and L. Colegrove, Industrial implementation of on-line multivariate quality control, AIChE Annual Meeting, San Francisco, 2006. Paper 92d.

439f Fault-Detection and Isolation and Fault-Tolerant Control of Nonlinear Process Systems Using Asynchronous Measurements

Panagiotis D. Christofides[1], Charles McFall[2], James F. Davis[3], and Ben Ohran[2]. (1) Department of Chemical and Biomolecular Engineering, Department of Electrical Engineering, University of California, Los Angeles, Los Angeles, CA 90095, (2) Department of Chemical and Biomolecular Engineering, University of California, Los Angeles, Los Angeles, CA 90095, (3) University of California - Los Angeles, Los Angeles, CA 90095

Chemical process operation is relying extensively on automated control systems in order to deal with the stringent requirements of safety, environmental sustainability, and profitability. Automation, however, tends to increase the vulnerability of a chemical process to faults (e.g., failures in the actuators, sensors or in the controllers) potentially causing a host of safety, environmental and economic problems. Management of abnormal situations is a major challenge in the chemical and process industries since abnormal situations account annually for at least 10 billion USD in lost revenue in the US alone. An important consideration of many process measurements is that they are often asynchronous, subject to time-varying delays, and vulnerable to data losses. This can be due to the nature of the measurement it self (i.e., difficult concentration measurements) or due to the communication network (wired or wireless) that may introduce such dynamics.

This works addresses the problem of fault-detection and isolation and fault-tolerant control when several system measurements are not available synchronously. First, a fault-detection and isolation (FDI) scheme that employs model-based techniques is proposed that allows for the isolation of faults. This scheme employs model-based FDI filters similar to those found in [1] in addition to observers that estimate the fault free evolution of asynchronously measured states during times when they are unmeasured. Specifically, the proposed FDI scheme provides detection and isolation of faults that are relative degree one to the set of synchronously measured states, and grouping of faults that relative degree one to the set of asynchronously measured states. The detection occurs immediately after a fault takes place, and the isolation, limited by the arrival of asynchronous measurements, occurs as asynchronous measurements come in. Once the FDI methodology has provided the system supervisor with a fault diagnosis, the supervisor takes appropriate action to seamlessly reconfigure the system to an alternative control configuration that will enforce the desired operation. We will present applications of the proposed asynchronous FDI and FTC framework to a polyethylene reactor simulation.

1. Mhaskar, P., C. McFall, A. Gani, P. D. Christofides and J. F. Davis, "Isolation and Handling of Actuator Faults in Nonlinear Systems," Automatica, 77: 83-62, 2008.

439g Fault Tolerant Controller Design for Transport-Reaction Processes Employing Spatial Controllability

Antonios Armaou, The Pennsylvania State University, University Park, PA 16802 and Michael A. Demetriou, Mechanical Engineering, Worcester Polytechnic Institute, Worcester, MA 01609.

Actuator failures have too often plagued chemical processes, often leading to deteriorating product quality and potentially dangerous process operation, such as runaway conditions. Motivated by the importance of the aforementioned problem, the issue of fault tolerant and fault accommodating controller design has been an active research topic in the chemical engineering community. However, while there has been extensive research from the control community on fault detection and diagnosis of finite dimensional systems using model-based robust and adaptive control techniques, the issue of fault tolerant control for distributed processes has been investigated only recently [1,2].

To address the issue of fault tolerance we embark on a completely new direction; we explicitly consider the spatial variability that transport-reaction processes naturally enjoy, and explicitly employ a new concept of identifying actuator locations that while being physically apart they have the same authority on the process states [3]. This artifact is only applicable to distributed parameter systems wherein two or more different locations within the spatial domain of definition can provide the same level of controllability and at the same time the same feedback gain can be applicable to all such locations. We capitalize on this property of spatially distributed parameter systems to find locations for different groups of actuators with similar controllability levels, and when a given actuator group fails, then simply deactivating the faulty actuator group and activating another actuator group constitutes the fault accommodation policy. The control signal in this case is not changed, a situation that cannot be implemented in lumped parameter systems.

Therefore the major conceptual contribution of this work is to employ ``optimal'' actuator locations that have the same level of controllability and design a single feedback controller for the archetypal actuator group. An on-line supervisor is subsequently built that includes a fault detection scheme that monitors for possible faults by comparing the plants performance with that of the nominal performance. When a deviation is observed, then a fault is declared. The fault accommodation scheme in this special case does not require a control reconfiguration, but simply the deactivation of the current actuator group and the activation of another healthy actuator group. The profound simplicity of the fault accommodation significantly reduces the costs associated with process supervision with obvious economic and performance savings.

[1] M. A. Demetriou, "Utilization of LMI methods for fault tolerant control of a flexible cable with faulty actuators," Proceedings of the 2001 40th IEEE Conference on Decision and Control, pp. 1885-1890, vol. 2, Orlando, Florida, December 4-7, 2001.

[2] A. Armaou and M. A. Demetriou, "Robust Detection and Accommodation of Incipient Component and Actuator Faults in Nonlinear Distributed Processes," AIChE J., in press, 2008.

[3] A. Armaou and M. A. Demetriou, "Optimal Actuator/Sensor Placement for Linear Parabolic PDEs Using Spatial H2 Norm," Chem. Eng. Sci., 61, 7351-7367, 2006.

448a Centralized - Decentralized Optimization for Refinery Scheduling

Nikisha Shah[1], Zhenya Jia[1], Georgios Saharidis[1], and M.G. Ierapetritou[2]. (1) Dept. of Chemical and Biochemical Engineering, Rutgers University, 98 Brett Rd., Piscataway, NJ 08854, (2) Chemical & Biochemical Engineering Department, Rutgers University, the State University of New Jersey, 98 Brett Road, Piscataway, NJ 08854

In the area of process operations, as the scale of the production increases, the mathematical complexity of the corresponding mathematical model and the time it takes to find its optimal solution increase exponentially. There are relatively few papers that have addressed the planning and scheduling problems using decentralized optimization strategies. Kelly and Zyngier (2007) presented a procedure to find a suitable way to decompose large decision-making problems and compared different decentralized approaches using hierarchical decomposition heuristics. Saharidis et al. (2006, 2008) studied the problem of production planning in deterministic and stochastic environments and compared centralized and decentralized optimization for an enterprise consisting of two production plants in series producing many different outputs with subcontracting options.

In this work, a novel decentralized strategy is proposed for large scale refinery scheduling problems. Two operating scenarios are studied - the first one corresponds to the case where simultaneous input and output flows are not allowed to occur in the storage tanks and the second one to the case where this restriction is not applied.

The proposed decentralized strategy contains two main steps. In the first step, the centralized system is decomposed following a general decomposing rule and optimal schedules are determined for each sub-problem. The main idea of the decomposition approach is that the system is decomposed at intermediate storage tanks such that the inlet and outlet stream of the tanks belong to different sub-problems. The decomposition starts with the final products or product storage tanks, and continues to include the reactors/units that are connected to them and stops when the storage tanks are reached. The products, units and storage tanks are part of the one sub-problem. Then following the input stream of each storage tank, the same procedure is used to determine the next sub-problem. If input and output stream of the storage tank are included at the same local problem then the storage tank also belongs to that local problem. The second step of the proposed approach concerns the integration of the optimal solutions of the sub-problems to achieve an optimal schedule for the entire system. The integration is based on the demand and mass balance constraints at connecting storage tanks. Since the storage tanks are connecting the sub-problems, the outflow of each of the storage tank becomes the demand for the preceding sub-problem. The material coming into the tank and the material already present in the tank should be at least equal to the material flowing out of the tank to satisfy the mass balance constraints. The mathematical formulation is based on continuous time representation and involves material balance constraints, capacity constraints, sequence constraints, assignment constraints, demand constraints and an objective function which corresponds to the minimization of makespan.

Different case studies of refinery scheduling problems are presented to illustrate the applicability and effectiveness of the decentralized strategy and the optimal solutions are compared with the ones obtained using centralized models. It is observed and proved analytically that if our decomposing rule is applied for the cases presented above, the local and global optimization give the same optimal makespan. Finally, it is important to notice that the computational effort of the proposed decentralized strategy is significantly less than the centralized approach.

Reference:

[1] J.D. Kelly and D. Zyngier (2007). Hierarchical decomposition heuristic for scheduling: Coordinated reasoning for decentralized and distributed decision-making problems. Comput. Chem. Eng.

[2] G.K. Saharidis, Y. Dallery and F. Karaesmen (2006). Centralized versus decentralized production planning. RAIRO Operations Research, 40, 113.

[3] G.K. Saharidis, V. Kouikoglou and Y. Dallery (2008). Analysis of centralized and decentralized control policies for a two-stage stochastic supply chain with subcontracting options. Submitted

448b Dynamic Modeling of a Global Specialty Chemicals Supply Chain

Charles Wei Kang Wong, 1Department of Chemical and Biomolecular Engineering, National University of Singapore, 10 Kent Ridge Crescent, Singapore 119260, Singapore, Singapore, Arief Adhitya, Process Science and Modeling, Institute of Chemical and Engineering Sciences, 1 Pesek Road, Jurong Island, Singapore 627833, Singapore, Singapore, and **Rajagopalan Srinivasan**, Chemical and Biomolecular Engineering, National University of Singapore, 10 Kent Ridge Crescent, Singapore 119260, Singapore, Singapore.

In today's competitive business environment, enterprises consider supply chain management (SCM) to be a key factor for achieving cost efficiency and better profitability. A hierarchy of decisions with important economic implications has to be made in SCM – strategic, tactical, operational, and ad hoc. However, as attested by the large body of research work in this area, SCM is not a straightforward task. Supply chains (SCs) span across continents, involve numerous entities with different interests, and contend with various uncertainties. The complex maze of the network results in complex dynamics, which in turn could lead to unforeseen domino effects. These motivate the development of simulation models of the SC, which could capture the behavior of the entities, their interactions, the resulting dynamics, and the various uncertainties. These models provide decision support for SCM by allowing the user to evaluate the impact of a particular decision on the SC performance, analyze different SC policies, identify the consequences of a disruption, and determine remedial actions, through simulation.

In this paper, we present a dynamic model of a global specialty chemicals SC dealing with lubricant additive products. A typical global lubricant additive SC comprises raw material suppliers, 3rd party logistics (3PL) providers, shippers, the lubricant additive company, and customers. The lubricant additive company comprises a global sales department which directly interacts with customers and a number of lubricant additive plants in various geographical locations. In the lubricant additive SC operation, decision-making is distributed across various departments in the enterprise. Each department has its own tasks, for example the procurement department buys raw materials and the operations department makes products. Each operates based on certain rules and policies. The integrated, overall SC performance and its dynamics emerge from the combined effects of the individual actions of each department.

The entities and their interactions have been modeled as a dynamic block-based simulation model, implemented in MATLAB/Simulink. The model explicitly considers the various supply chain entities: upstream raw material suppliers, downstream customers as well as the internal entities of the lubricant additive company. These internal entities include a centralized sales department and a number of production sites located at different regions in the world. Each production site has its own procurement, scheduling, storage, operations, and packaging departments. The model simulates various supply chain activities such as raw material procurement, order assignment, job scheduling, storage, and production. Stochastic variations in transportation, production, prices, and demands are considered in the proposed model. SC infrastructure is modeled using blocks in Simulink while policies and decision-making algorithms are implemented as MATLAB m-files. This provides the user with the flexibility of simulating different policies in a plug-and-play manner. This paper describes the model and illustrates its capabilities for decision support using several case studies.

Optimization of Discrete Time Supply Chain Models with Guaranteed Robust Stability and Feasibility

Darya Kastsian and Martin Mönnigmann,
Technische Universität Braunschweig, Germany

Abstract

This contribution presents a method for the simultaneous economic optimization and optimal design for disturbance rejection of discrete time systems. The method is an extension of a previously presented approach for the optimization of discrete time systems with constraints on robust stability [10]. In this contribution we extend the method to a new set of constraints on the location of poles of discrete time systems.

1 Introduction

Supply chain optimization has recently received a lot of attention. When seeking optimal supply chain designs it is important to consider the stability of the system, since supply chains are known to permit unstable behavior as witnessed, for example, in the bullwhip effect. Beyond mere stability, it is also important to design the system such that disturbances decay quickly enough. In this contribution we discuss a method for the economic optimization of discrete time systems which simultaneously ensures robust disturbance rejection.

The proposed method is an extension of the normal vector methods for the design of nonlinear dynamical systems with guaranteed robust stability and feasibility [4, 7–11]. The normal vector method was originally developed for parametric uncertainty with respect to stability properties in optimization problems. However, it naturally applies to the robust treatment of more general dynamical properties as well as robust feasibility. In Section 2 we discuss the normal vector method and its extension to ensuring disturbance rejection. The proposed approach is applied in Section 3 to the optimization of the supply chain model of an automatic pipeline feedback compensated inventory and order-based production control system (APIOBPCS) embedded within a vendor managed inventory (VMI) supply chain. Finally, in Section 4 we give a conclusion.

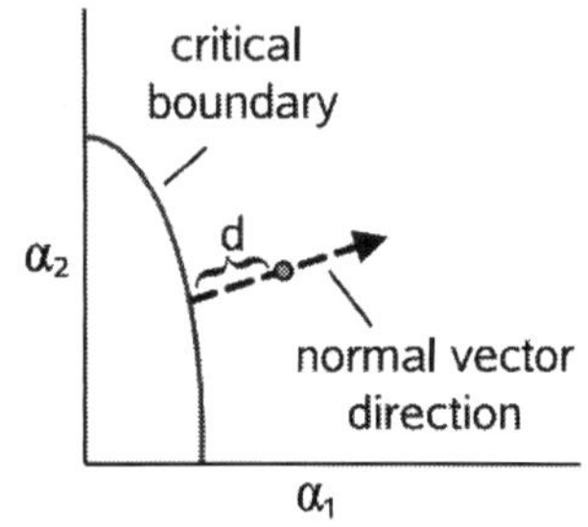

Figure 1: Optimal point of a hypothetical system with a critical boundary. The parametric distance d to the locally closest point on the critical boundary can be measured along the normal vector direction. The symbols α_1 and α_2 denote model parameters. If we ensure the distance between the optimal point and the nearest critical point to be larger than zero, then the optimal point is guaranteed to stay away from the critical boundary.

2 Normal vector method and rejection of disturbances

The normal vector method is based on the fact that equilibrium solutions of dynamical systems can be characterized by their parametric distance to the manifolds of critical points [7]. Typical critical points of interest are bifurcation points or points at which state variable constraints are violated. Normal vectors to the critical manifolds can be used to measure the distance from the nominal point of operation to stability and feasibility boundaries in the space of the system design parameters α_i. In order to keep the optimal point away from the critical boundary we constrain the parametric distance between the critical manifold and critical boundary to be larger than zero. This condition can be added to the optimization problem by posing an appropriate nonlinear inequality constraint. In Figure 1 the critical boundary and the normal direction to it are sketched for a hypothetical system. In Figure 1 we show a case with two parameters, but for higher dimensional cases the idea remains the same. In general more than one critical boundary can exist. By staying sufficiently far away from all critical manifolds we can guarantee robust stability and feasibility of the system.

For discrete time systems three types of bifurcation points exist, namely the fold, Neimark-Sacker and flip bifurcations [6]. These bifurcation points can be characterized by eigenvalues on the unit circle in the complex plane. In order to ensure robust disturbance rejection we introduce a simple but new type of critical point that is an extension and generalization of fold, Neimark-Sacker and flip bifurcations. Applying the normal vector method to this type of critical point amounts to forcing all eigenvalues of the linearized discrete time system into a circle of radius $R < 1$ on the complex plane.

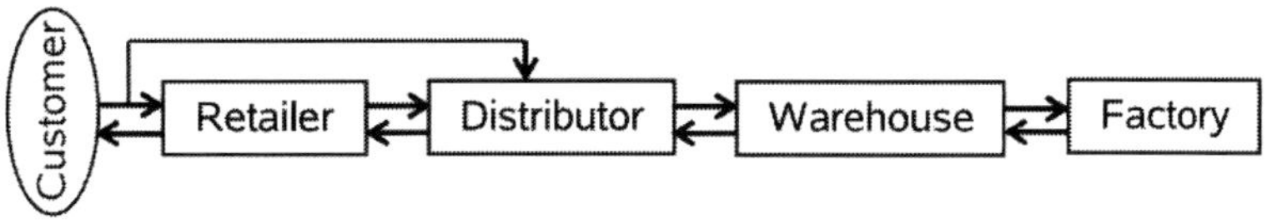

Figure 2: Flow of information in a VMI supply chain.

By specifying the radius R, the user of the method can specify the decay rate by which disturbances are rejected. We stress that while the normal vector method makes use of the linearized system and its eigenvalues, the decay rate constraint holds for the nonlinear system in a finite ball around the nominal optimal system parameters [5].

3 Applications

We present the result of the optimization of a VMI-APIOBPCS supply chain model with constraints on the decay rate for robust disturbance rejection. In contrast to traditional supply chains the vendor managed inventory supply chain distributors share inventory information and/or point of sales data rather than orders with their manufacturers. Alternatively, the manufacturer can determine the inventory from deliveries and sales if the distributor does not have the ability to share inventory information or is hesitant to do so, but is willing to share end sales data. A simple schematic VMI supply chain is illustrated in Figure 2 [3]. The system belongs to the class of automatic pipeline feedback compensated inventory and order-based production control systems (APIOBPCS). In the APIOBPCS the ordering rule is based upon forecast demand and the difference between a fixed target level of inventory and the actual level. The ordering rule also takes into account the work in progress, comparing actual levels with a target value. The pipeline and inventory levels information in the APIOBPCS are incorporated into the production order rate by adding a fraction of the difference between desired work in progress and actual work in progress, and a fraction of the difference between distributors inventory holding and actual inventory level, respectively. A full description of the VMI-APIOBPCS is given in [1]. Figure 3 shows the stability conditions for VMI-APIOBPCS in its parameter space. The line $Ti = 0.5$ shows the stability boundary due to flip bifurcations. All other curves in Figure 3 are sets of Neimark-Sacker bifurcation points. Three sample fixed points are used to illustrate the system's dynamical response. Figures 3 (a), (b) and (c) show stable, critically stable and unstable system response, respectively. In order to ensure the robust disturbance rejection we guarantee that all eigenvalues of the VMI-APIOBPCS lay inside the circle of radius $R = 0.8$ on the complex plane. Figure 4 illustrates

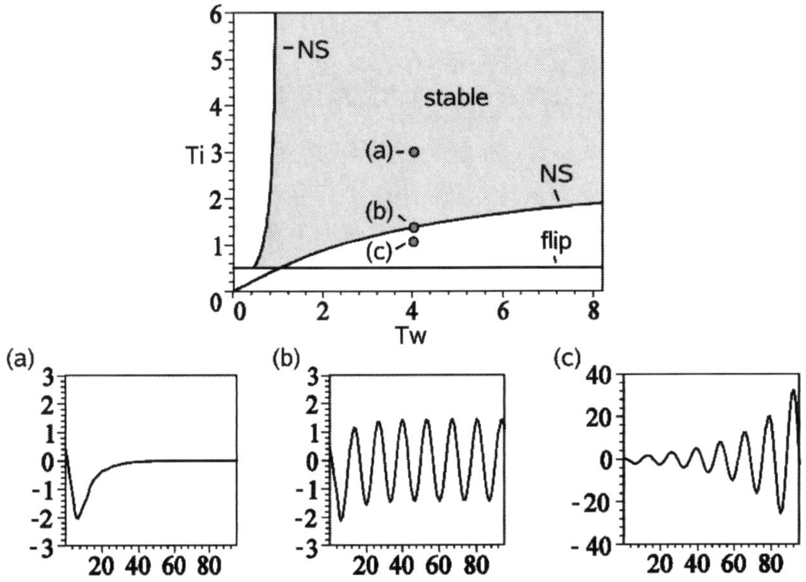

Figure 3: Bifurcation picture of the VMI-APIOBPCS. Diagram (a), (b) and (c) show unstable, stable, and critically stable dynamic response of the actual inventory, respectively.

the modified critical boundaries where the region with radius less than 0.8 is marked by gray color. The dynamic response of the actual inventory is shown in Figure 4 (a).

We will optimize the VMI-APIOBPCS supply chain model according to the following three criteria. The motivation and detailed description for these criteria is given in [2]. The first criterion is the noise of bandwidth, which is traditionally a useful measure to characterize the frequency response of a system and hence the production adaptation costs. The second and third criterion are scaled quadratic errors in inventory and in virtual consumption. The normal vector method is used to ensure robust disturbance rejection such that all eigenvalues lay inside the circle of radius 0.8. The result of the optimization procedure with the normal vector method is shown in Figure 5.

4 Conclusion

We showed that the concept of normal vector method can be applied to the simultaneous economic optimization and optimal design for disturbance rejection of discrete time systems. This concept was successfully demonstrated on the optimization of the VMI-APIOBPCS supply chain model.

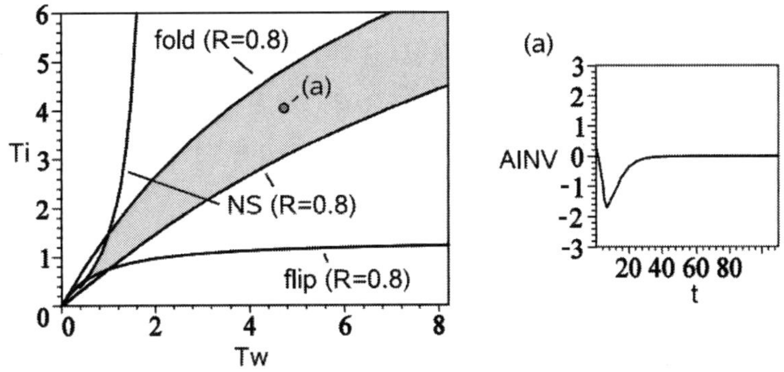

Figure 4: Modified critical boundaries with $R = 0.8$ of the VMI-APIOBPCS. Diagram (a) shows the dynamic response of the actual inventory.

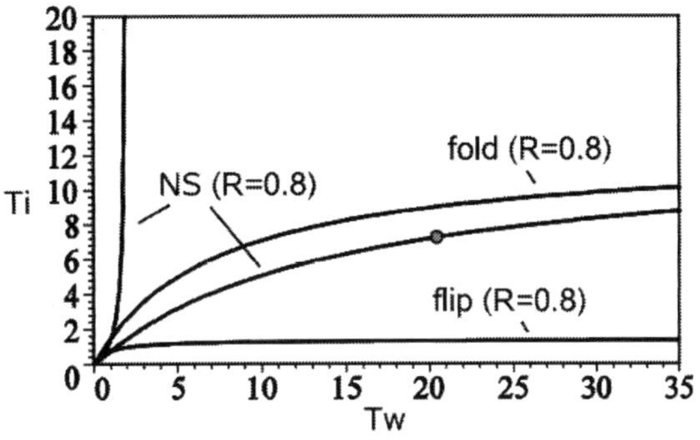

Figure 5: The optimal point of the VMI-APIOBPCS obtained with the normal vector method and modified critical boundaries with $R = 0.8$.

References

[1] S. M. Disney, and D. R. Towill, *A discrete transfer function model to determine the dynamic stability of a managed inventory supply chain*, Int. J. Prod. Res., 40 (2002), pp. 179–204.

[2] S. M. Disney, and D. R. Towill, *A procedure for the optimization of the dynamic response of a Vendor managed inventory system*, Comput. Ind. Eng., 43 (2002), pp. 27–58.

[3] S. M. Disney, A. T. Potter, and B. M. Gardner, *The impact of vendor managed inventory on transport operations*, Transportation Research, Part E 39 (2003), pp. 363–380.

[4] J. Gerhard, W. Marquardt, and M. Mönnigmann, *Normal vectors on critical manifolds for robust design of transient processes in the presence of fast disturbances*, SIAM J. Appl. Dyn. Syst., 7 pp. (2008),461–490.

[5] D. Hinrichsen, and A. J. Pritchard, *Mathematical systems theory I. Modelling, state space analysis, stability and robustness*, Springer-Verlag, Berlin, 2005.

[6] Yu. A. Kuznetsov, *Elements of Applied Bifurcation Theory*, Springer, New York, 1998.

[7] M. Mönnigmann, and W. Marquardt, *Normal vectors on manifolds of critical points for parametric robustness of equilibrium solutions of ODE systems*, J. of Nonlinear Sc., 12 (2002), pp. 85–112.

[8] M. Mönnigmann, and W. Marquardt, *Steady state process optimization with guaranteed robust stability and feasibility*, AIChE J., 49 (2003), pp. 3110–3126.

[9] M. Mönnigmann, and W. Marquardt, *Steady-state process optimization with guaranteed robust stability and flexibility: Application to HDA reaction section*, Ind. Eng. Chem. Res., 44 (2005), pp. 2737–2753.

[10] M. Mönnigmann, J. Gerhard, and W. Marquardt, *Design of uncertain discrete time systems with constructive nonlinear dynamics methods*, AIChE Annual Meeting, San Francisco, 2006.

[11] M. Mönnigmann, W. Marquardt, C. H. Bischof, T. Beelitz, B. Lang, and P. Willems, *A hybrid approach for efficient robust design of dynamic systems*, SIAM Review, 49 (2007), pp. 236–254.

448d A Process Attainable Region Approach for Supply Chain Management

Charles Sung and Christos T. Maravelias. Chemical and Biological Engineering, University of Wisconsin - Madison, 1415 Engineering Drive, Madison, WI 53706

Despite advances in computer-aided optimization, a continuing obstacle in supply chains coordination is effective integration of disparate models having differing scope, operating on different time scales, and containing different levels of detail [1]. This is apparent even when integrating two closely related problems such as production planning and scheduling. Several general strategies may be employed when optimizing over two problems: the combined problem may be dealt as one problem, the two problems may be treated as separate and solved iteratively, or the two problems may be treated as separate and solved hierarchically/sequentially [2]. The last approach is most straightforward and generally requires the least of amount of online computational effort. A number of authors have suggested that it is sufficient if the upper level problem is provided with an accurate, low-complexity surrogate for approximating the lower level problem's feasibility and effect on overall optimality [3, 4].

We previously [5] proposed a method that uses off-line computations based on a detailed scheduling model to generate the convex hull of feasible production targets, the Process Attainable Region (PAR), of a given production facility. The PAR of a facility is expressed only in terms of the planning variables (i.e. the scheduling variables are projected out), yet provides all the necessary information to solve the production planning problem effectively. Here we extend that work by proposing a framework for the development of non-convex process attainable regions, via the identification and subtraction of convex infeasible regions [6]. A major advantage of this line of research is that it does not require that special knowledge be known about the model to be approximated. Instead, the geometry of the projected feasible region is intelligently investigated and the approximation constructed from such. The end result is a surrogate model that provides all the necessary information for the solution of operational planning problems in supply chain management (SCM).

The proposed method constructs the surrogate model in four phases. In the first phase, points on the boundary of the feasible region are identified. Line searches look inward thus allowing investigation of a disjoint feasible region. In the second phase, points found in the first phase are used to construct many convex infeasible regions. These are optimally merged using the Multi-Parametric Toolbox [7] to reduce the total number of convex infeasible regions. The third phase begins by checking to see if any of the infeasible regions can be expanded. Then recognizing that inequalities defining convex infeasible regions are later flipped to become disjunctive inequalities defining the non-convex feasible region, unnecessary regions and inequalities are removed. Remaining inequalities are "cleaned-up" as necessary. In the fourth phase, big-M coefficients are calculated, and the surrogate model is formally posed. An alternative approach where the non-convex region is expressed as a union of convex polytopes is also presented.

Given the large discrepancy between information provided by the scheduling problem and the information needed to solve SCM problems, a great opportunity exists for reducing the scheduling problem to a compact surrogate model. We show how the proposed framework can be used to provide information for complex process networks via compact surrogate models. We present reductions for different types of process facilities (e.g. network, sequential) and classes of scheduling formulations (e. g. discrete-time, continuous-time). Examples are also presented for integrated operational planning problems for SCM with many production facilities, and intermediate and customer-facing nodes, that were previously intractable.

[1] Stadtler, H. (2005). Supply Chain Management and Advanced Planning – Basics, Overview and Challenges. Eur. J. Oper. Res., 163, 575.

[2] Maravelias, C. T., Sung, C. (2008). Integration of Production Planning and Scheduling: Overview, Challenges and Opportunities. In Proceedings: Foundations of Computer-aided Process Operations, Boston, MA, June 29 – July 2, 2008.

[3] Bassett, M. H., Pekny, J. F., Reklaitis, G. V. (1996). Perspectives on Model-based Integration of Process Operations. Comput. Chem. Engr, 20, 821-844.

[4] Grossmann, I.E. (2005). Enterprise-wide Optimization: A New Frontier in Process Systems Engineering, AIChE, 1846-1857.

[5] Sung, C., Maravelias, C.T. (2007). An Attainable Region Approach for Production Planning of Multi-Product Processes. AIChE, 53, 1298-1315.

[6] Goyal, V, Ierapetritou, M.G. (2003). Framework for Evaluating the Feasibility/Operability of Nonconvex Processes. AIChE, 49,1233-1240.

[7] Kvasnica, M., Grieder, P., Baoti, M. (2004). Multi-Parametric Toolbox (MPT). http://control.ee.ethz.ch/~mpt/

448e Selection of Third-Party Service Contracts for Chemical Logistics

Mukta Bansal[1], I.a. Karimi[2], and Raj Srinivasan[2]. (1) Department of Chemical & Biomolecular Engineering, National University of Singapore, 4 Engineering Drive 4, Singapore, 117576, Singapore, (2) Chemical and Biomolecular Engineering, National University of Singapore, 4 Engineering Drive 4, Singapore, 117576, Singapore

"It is remarkable that an industry that is built around the management of process and materials flows in plants has been slow to grasp the principle of organizational processes and logistics flows. But based on the success of such methods in other sectors, this is the future; …" – Braithwaite, 2002. This clearly indicates the state of the chemical industry in paying attention to supply chain logistics. Logistics is the glue that binds the entities of a supply chain (Karimi et al., 2005). According to the Council of Logistics Management (Lambert, 2001), "Logistics is that part of the supply chain process that plans, implements, and controls the efficient, effective flow and storage of goods, services, and related information from the point-of-origin to the point-of-consumption in order to meet customer requirements".

The definition of logistics – "the flow of material, information, and money between consumers and suppliers" (Frazelle, 2002) emphasizes the link between logistics performance and customer satisfaction. Today, logistics is much more than transportation alone; it includes many other services. Lieb and Randall (1996) showed that the most frequently outsourced logistics services are warehouse management/operations, shipment consolidation, carrier selection, logistics information systems, rate negotiation, and fleet management/operations with the significant expansion of services including product assembly/installation, product returns, and customer spare parts.

Whether it is a chemical company that manages its own logistics, or a third party logistics provider (3PL) that manages it for the chemical company, the ultimate cost of logistics directly affects the cost effectiveness of global chemical supply chains (Jetlund et al., 2004). According to Karimi et al. (2002), "Often an overlooked component of the chemical business, a critical examination of logistics practices can result in substantial savings". While logistics is an issue of increasing importance to almost all industries, it is of most relevance to the chemical industry, as various types of chemical and related industries have some of the highest logistics costs. Logistics costs can vary from 3.6% of the purchase price for a best-in-class (BIC) site to 20% at the other extreme (Karimi et al., 2002).

Logistics costs in the chemical and related industries are among the highest in asset-intensive supply chains. Having managed the intra-plant logistics well for years, the companies are now looking for ways to lower the costs of enterprise-wide logistics by increasingly outsourcing a variety of logistics services to third-party logistics (3PL) firms globally. We present a systematic and quantitative decision-making formalism to address the integrated logistics needs of a MNC in a global business environment. Although our goal is to address the logistics in chemicals and related industries in particular, the methodology is general and applicable to other supply chains as well. The formalism involves a novel representation of logistics activities in terms of a recipe superstructure and a static MILP model based on that to select the optimal contracts (or/and in-house execution) that minimize the total logistics cost. The contracts are designed to fulfill partial or full bundles of various logistics needs, tasks, and services, which can be performed at globally distributed sites. We account for product bundling and transport expenses between various sites in bulk or packaged or container forms and to various customers. Our goal is to identify the contracts (hence the providers) and the location of each service, which serve the total needs of a company in an integrated and most cost-effective manner.

We address this problem from the perspective of a chemical company who signs one/multiple contracts with logistics companies. We first develop a powerful framework for representing this complex problem in a comprehensive and general manner, and then use that to develop a multi-period mixed integer linear programming (MILP) formulation. We use several examples to highlight the advantages of the proposed approach. It allows the flexibility of selecting partial contracts, which reduces the combinatorial complexity and computation time considerably, along with some reduction in costs under certain assumptions. The model not only accommodates existing contracts, but also allows new contracts to extend beyond the horizon. Thus, it is able to address in a reactive manner the various dynamic disruptions that normally arise in chemical supply chains. We also found that the full selection of contracts are not only costly, it is also computationally expensive. The CPU time is much larger in the case of full selection of contracts.

Key words: Logistics; 3PLs; contracts; supply chain; chemical

Braithwaite, A. Achieving world class supply chain and logistics in the chemical industry. EPCA Logistics Meeting Monaco, November 2002.

Karimi, I.A., Sharafali, M., Mahalingam, H. Scheduling tank container movements for chemical logistics. AICHE, 2005, 51, 178-197.

Lambert, D.M. The Supply Chain Management and Logistics Controversy. In: Brewer, A.M., Button, K. J., Hensher, D.A., (Eds.), Handbook of Logistics and Supply-Chain Management. Pergamon Publishers, Netherland, pp 99 – 126, 2001.

Frazelle, E. Supply chain strategy: the logistics of supply chain management, New York, McGraw Hill Pub Co, 2002.

Lieb, R., and Randall, H. A comparison of the use of third party logistics services by large American manufacturers, 1991, 1994, and 1995. Journal of Business Logistics, 1996, 17(1), pp 305-320.

Jetlund, A.S., Karimi, I.A. Improving the logistics of multi-compartment chemical tankers. Computers & Chemical Engineering, 2004, 28, 1267-1283.

Karimi, I.A., Srinivasan, R., Han, P.L. Unlock supply chain improvements through effective logistics. Chemical Engineering Progress, 2002, 5, 32-38.

448f Optimal Supply Chain Design with Multi-Echelon Stochastic Inventory and Risk-Pooling

Fengqi You, Dept of Chemical Engineering, Carnegie Mellon University, 5000 Forbes Ave, Pittsburgh, PA 15213 and Ignacio E. Grossmann, Center for Advanced Process Decision-making, Dept of Chemical Engineering, Carnegie Mellon University, 5000 Forbes Avenue, Pittsburgh, PA 15213.

Due to the increasing pressure for remaining competitive in the global market place, optimizing inventories across the supply chain has become an emerging challenge for the process industries to reduce costs and improve the customer service.[1,2] This challenge requires integrating inventory management with supply chain network design, so that decisions on the locations to stock the inventory and the associated amount of inventory in each stocking location can be determined simultaneously for lower costs and higher customer service level. However, the integration is usually nontrivial for multi-echelon supply chains and their associated inventory systems in the presence of uncertain customer demands.[3]

In this work we develop a joint multi-echelon supply chain design and inventory model for simultaneously optimizing the transportation, inventory and network structure in a supply chain under demand uncertainty. By assuming that each node in the supply chain network operates with a base-stock policy for a bounded normally distributed demand, [4] we capture the stochastic nature of the problem and develop an equivalent deterministic optimization model. The model determines the supply chain design decisions such as the locations of distribution centers (DCs), assignments of customers to DCs, assignments of DCs to plants, shipment levels from plants to the DCs and from DCs to customers, and inventory decisions such as the working inventory, pipeline inventory and safety stock in each node of the supply chain network. The model also captures risk-pooling effects [5] by consolidating the safety stock inventory of downstream nodes to the upstream nodes in the multi-echelon supply chain.

We formulate this problem as a mixed-integer nonlinear program (MINLP) with a nonconvex objective function including bilinear, trilinear and square root terms. By exploiting the properties of the basic model, we reformulate the problem as a separable concave minimization program. A tailored hierarchical decomposition algorithm based on Lagrangean relaxation is developed to obtain near global optimal solutions with reasonable computational expense. Three heuristics based on the model properties are also proposed to speed up the computation and are incorporated into the Lagrangean relaxation and decomposition algorithm. Computational examples for chemical and industrial gases supply chains with up to 20 plants, 200 potential distribution centers and 250 customers are presented to illustrate the application of the model and the performance of the algorithm.

[1] Grossmann, I. E., Enterprise-wide Optimization: A New Frontier in Process Systems Engineering. AIChE Journal, 2005, 51, 1846.

[2] Chopra, S.; Meindl, P., Supply Chain Management: Strategy, Planning and Operation. Prentice Hall:

Saddle River, NJ, 2003.

[3] Zipkin, P. H., Foundations of Inventory Management. McGraw-Hill: Boston, MA, 2000.

[4] Graves, S. C.; Willem, S. P., Optimizing Strategic Safety Stock Placement in Supply Chains. Manufacturing & Service Operations Management, 2000, 2, 68.

[5] Eppen, G., Effects of centralization on expected costs in a multi-echelon newsboy problem. Management Science, 1979, 25, (5), 498.

458a Design of An on-Farm Ethanol Dewatering System

Anuradha Mukherjee[1], James R. Whiteley[1], and Danielle Bellmer[2]. (1) Chemical Engineering, Oklahoma State University, 423 EN, Stillwater, OK 74078, (2) Biosystems and Agricultural Engineering, Oklahoma State University, 108 FAPC, Stillwater, OK 74078

The use of ethanol as a bio-fuel has gained momentum in recent years to meet national and international energy needs. Most bio-ethanol produced in the United States is derived from corn (starch). Among other raw materials, sweet sorghum is an alternative feedstock which gives high ethanol yields per acre and is a low input, drought tolerant high carbohydrate producer. These desirable characteristics have spurred great interest world-wide in the potential viability of sweet sorghum as bio-fuel feedstock.

Production of ethanol from sweet sorghum is more direct compared to starch and cellulosic feedstock as the carbohydrates produced by sweet sorghum are sugars which can be directly fermented to ethanol. This benefit of sweet sorghum is partially offset by the need to initiate fermentation immediately (on-farm) to minimize the loss of microbial conversion of the sugars to non-ethanol products. Researchers in the School of Biosystems & Agricultural Engineering at Oklahoma State University have demonstrated the ability to harvest, express, and ferment sweet sorghum on-farm. The fermentation product is an aqueous 6 – 10 wt % ethanol product. The resulting challenge is to devise a means to economically deliver fuel-grade ethanol to the marketplace.

The high concentration of water (90+ wt %) in the fermentation product provides a strong incentive to perform on-farm dewatering in lieu of costly transport to a central processing facility. This paper describes the unique challenges associated with design of an on-farm distillation system capable of producing a near azeotropic (95 vol%) ethanol product. Final on-farm processing using a membrane pervaporation process to produce the final fuel-grade ethanol is under investigation but not addressed in this paper. The on-farm or small scale production of ethanol presents a set of novel technological challenges and opportunities from a design and engineering stand point, the success of which lies in it being economical and easy to operate by on-farm personnel.

Existing centralized bio-ethanol separation systems are large scale process plants capable of producing 100+ gpm of fuel grade ethanol. The design of the beer still, rectifier, and side stripper columns used to produce the azeotropic ethanol intermediate product for this service are well established. Unfortunately, these designs cannot be applied without significant modification to the on-farm case. The two main factors that drive the design of an on-arm operation are its scale and operational environment. Unlike industrial scale facilities, on-farm units must operate at low feed rates of about 250 gallons per hour. The corresponding ethanol production rate of less than 1 gpm is more than two orders of magnitude less than a commercial facility. It is also unlikely that an on-farm facility will have access to traditional utilities such as steam or cooling water. Therefore, the design of an on farm unit will differ from its industrial counterpart.

This paper focuses on the design and simulation of a distillation system for ethanol separation which is representative of an 'on-farm' scenario. A two column distillation configuration with associated ancillary equipment was designed for the above purpose. The column configuration, feed stream ethanol concentration, product ethanol concentration and recovery ethanol quantities were chosen to realistically represent on-farm operation. The first column, generally referred to as the beer still was designed to handle a heavy solid content. The second column known as the rectifying column can use either trays or packing for separation. Propane or natural gas is assumed to be the only available source of heat, requiring non-standard design of the column reboilers. An air-cooled condenser is employed due to the lack of availability of cooling water. Simulation results, preliminary mechanical design, and costs are presented along with special maintenance and operational requirements. A preliminary costing analysis of the on-farm ethanol water separation unit was also undertaken and compared to industrially successful large scale units.

458b A Process Integration Approach to Strategic Design and Scheduling of Multi-Feedstock Biodiesel Plants

René D. Elms[1], Grace Nworie[2], and Mahmoud M. El-Halwagi[1]. (1) Texas A&M University, 3122 TAMU, College Station, TX 77843, (2) Chemical Engineering, Texas A&M University, College Station, TX 77845

Various feedstocks may be utilized to produce biodiesel. These feedstocks include soybean oil, waste oil from the food industry (e.g., waste cooking oil), as well as a variety of other oils such as palm, sunflower, jatropha, olive, rapeseed, cottonseed, safflower, and seasame oils. With changing prices, supply, and demand of these feedstocks, there is a need to design flexible plants that can process various inputs. The objective of this work is to develop a systematic procedure for the design and operation of flexible biodiesel plants that can accommodate a variety of feedstocks. In particular, the proposed project will employ a holistic approach and a combination of process simulation, synthesis, and integration techniques that will result in the:

1. Synthesis of innovative processes for the cost-effective production of biodiesel from various feedstocks

2. Integration of energy and mass resources within the process

3. Optimization of process design and scheduling for the flexible plant

4. Techno-economic and environmental assessment and sensitivity analysis of the proposed schemes

The developed simulation, integration, and optimization models along with the integrated experiments will provide a new platform that can systematically provide a complete market-sensitive strategy for innovative pathways for producing biodiesel. These new techniques will lay the foundations for fundamental research in the emerging area of sustainable biofuels. A case study is solved to demonstrate the merits and applicability of the devised procedure.

458c Online State Estimation In Enzymatic Synthesis of Ampicillin Process: Experimental and Simulation Study

Marcelo Perencin de Arruda Ribeiro[1], Roberto de Campos Giordano[1], and Sridhar Ungarala[2]. (1) Department of Chemical Engineering, Federal University of São Carlos, Via Washington Luiz, km 235, C.P. 676, São Carlos, 13564-210, Brazil, (2) Cleveland State University, Department of Chemical Engineering, 2121 Euclid Avenue, SH 455, Cleveland, OH 44115

Semi-synthetic penicillins such as ampicillin and amoxicillin are among the most widely used antibiotics worldwide. The common industrial process, carried out at low temperatures (-30°C) using organochloride solvents, has been criticized for generating considerable amounts of non-recyclable waste (Bruggink and Roy, 2001). Alternatively, the synthesis of beta-lactam antibiotics through an enzyme-catalyzed process (using penicillin G acylase) may be carried out in aqueous medium, at room temperature (25°C) and almost neutral pH (Ospina et al., 1996). The enzymatic process known as kinetically-controlled synthesis (KCS) has its yields driven by a maximum transient and overcomes the necessity of organic (co-)solvents (Kasche, 1986). This approach has the potential to be a "green process" alternative to the common industrial process.

The use of a semi-continuous integrated reactor, where substrates are fed and product is crystallized (Kasche and Galunsky, 1995), may overcome the obstacles of KCS and it is a promising alternative for large-scale production based on an enzymatic process. Optimization of this reactor requires knowledge of the state variable transients (pH, temperature and concentration of substrates), such that selectivity and productivity may be maximized (Ribeiro and Giordano, 2005). An experimental monitoring and control system for ampicillin synthesis has been built including UV spectrophotometry and multivariate calibration (Ribeiro et al., 2008). The measurements need to be filtered and estimators must be designed to infer variables that are not directly measured such as the mass of antibiotic produced (in solid phase).

The present work focuses on online state estimation in enzymatic synthesis of ampicillin in a semi-continuous integrated reactor. Bayesian estimation via Sequential Monte Carlo sampling (SMC), such as Sampling Importance Resampling (SIR) has been shown to be feasible for online estimation in chemical engineering problems with highly non-linear but low dimensional systems (Chen et al., 2004).

The ampicillin process is nonlinear, comprising 7 states: total concentration of D(-)phenylglycine methyl ester, PGME, and 6-aminopenicillanic acid, 6-APA (substrates); total concentration of ampicillin and D (-)phenylglycine (desired and undesired products); volume of the aqueous phase; pH; and apparent enzyme activity. The system has 3 control variables (inputs): flow rates of EMFG, 6-APA and sodium hydroxide. Inputs also carry uncertainties and are dynamically driven by a proportional controller. The measured data (outputs) are: concentration of substrate and products in aqueous phase, and pH throughout time.

First, a simplified kinetic model from the literature (Ferreira et al., 2000) was fitted against experimental data generated in a batch reactor. Since the simplified model cannot fit the data for all range of concentration of substrate, only a specific region with more relevance to the process was used to fit the model. The aqueous phase plus the insoluble biocatalyst was assumed to be one pseudo-homogenous phase. Possible effects of gradients of concentration and pH within the biocatalyst particle and uncertainties of activity of the enzyme are lumped in the noise of the apparent enzymatic activity, which can be related to the dynamic effectiveness of reaction. Crystallization of products was assumed as soon as they reached their limit of solubility. Therefore, a possible metastable zone was neglected. Covariance matrices of the measurement noise were estimated from results of the fitting procedure. Some terms of the covariance matrices of the noises of the system were estimated and others were assumed based on prior information of the real system.

Secondly, evaluation of the SIR on the experimental data used to fit the model and a validation batch showed good agreement between predicted states and the states calculated by the fitted model. On the other hand, this approach also exhibited high sensitivity to modeling errors. Incorporation of initial condition uncertainties and noise term in the enzymatic activity appeared to overcome error modeling and gave better filtering results. In other words, it permitted to track a parameter related to the global effectiveness of reaction.

In a simulated semi-continuous integrated reactor, the SIR approach produced state estimates close to the simulated states and allowed good prediction of the mass of antibiotic crystallized. pH and concentrations were measured every 0.5 and 5 minutes respectively. The Monte Carlo state estimation method used 500 samples for each variable and the computation time was less than 3 seconds for each 30 seconds of real batch. Therefore, this approach may be used for online state estimation. Average results obtained after 35 runs showed a mean squared error (mse) between estimated and simulated values for concentration was 2.3x10-5 (mol.L-1)2 and an mse for pH was 1.4x10-4. Results obtained by using Extended Kalman Filter (EKF) for the same runs showed an mse value for the concentrations of 5.0x10-5 (mol.L-1)2 and an mse for pH of 16.8x10-4. However, the computation time for the SIR approach was 35 times greater than the time for EKF. In contrast to EKF, the SIR approach does not require simplifying assumptions about the probability density function of state variables or the dynamic models.

References:

Bruggink, A. & Roy, P. D. (2001), Synthesis of beta-Lactam antibiotics (Chapter I), Kluwer Academic Publishers, Netherlands.

Chen, W. S.; Bakshi, B. R.; Goel, P. K. & Ungarala, S. (2004), 'Bayesian estimation via sequential Monte Carlo sampling: Unconstrained nonlinear dynamic systems', Ind. Eng. Chem. Res. 43(14), 4012-4025.

Ferreira, A. L. O.; Goncalves, L. R. B.; Giordano, R. C. & Giordano, R. L. C. (2000), 'A simplified kinetic model for the side reactions occurring during the enzymatic synthesis of ampicillin', Braz. J.

Chem. Eng. 17(4-7), 835-839.

Kasche, V. (1986), 'Mechanism And Yields In Enzyme Catalyzed Equilibrium And Kinetically Controlled Synthesis Of Beta-Lactam Antibiotics, Peptides And Other Condensation Products', Enzyme Microb. Technol. 8(1), 4-16.

Kasche, V. & Galunsky, B. (1995), 'Enzyme-Catalyzed Biotransformations In Aqueous 2-Phase Systems With Precipitated Substrate And/Or Product', Biotechnol. Bioeng. 45(3), 261--267.

Ospina, S.; Barzana, E.; Ramirez, O. T. & LopezMunguia, A. (1996), 'Effect of pH in the synthesis of ampicillin by penicillin acylase', Enzyme Microb. Technol. 19(6), 462--469.

Ribeiro, M. P. A. & Giordano, R. C. (2005), 'Variational calculus (optimal control) applied to the optimization of the enzymatic synthesis of ampicillin', Braz. Arch. Biol. Tech. 48, 19--28.

Ribeiro, M. P. A.; Padua, T. F.; Leite, O. D.; Giordano, R. L. C. & Giordano, R. C. (2008), 'Multivariate calibration methods applied to the monitoring of the enzymatic synthesis of amipicilin', Chemom. Intell. Lab. Syst. 90, 169--177.

458d Non-Linear Parameter Estimation for a Dynamic Model of Mammalian Cell Culture

Adam C. Baughman, Chemical & Biological Engineering, Rensselaer Polytechnic Institute, 110 8th Street, Ricketts Building, Troy, NY 12180, Susan Sharfstein, Department of Chemical and Biological Engineering, Rensselaer Polytechnic Institute, 110 8th Street, Biotech 2nd Floor, Troy, NY 12180, and Lealon L. Martin, Chemical and Biological Engineering, Rensselaer Polytechnic Institute, 110 8th Street, Troy, NY 12180.

To appropriately capture the behavior of intricate biological processes, dynamic models of mammalian cell culture systems may be formulated using non-linear kinetic expressions or other complex mathematical constructions. Such models can potentially incorporate a large number of parameters, only a few of which may be experimentally measurable. As a result, it is diffcult to obtain or estimate parameter values that are both mathematically suitable and physically relevant. In the present work, we describe a systematic, optimization-based, technique for the estimation of unknown parameter values in dynamic models. The approach is general, and applicable to any model intended to describe a fundamentally smooth process (i.e. a process described by continuous functions). Explicit Euler discretization is used to convert the given (continuous) differential model into an equivalent series of differential-algebraic equations. The discretized expressions are then used to pose a constrained non-linear optimization problem whose solution is the set of ideal parameter estimates. We demonstrate efficacy of this approach using the example of a dynamic model of mammalian cell culture proposed by Gao, et al. [2007] Our parameter estimation technique is shown to provide parameters that are both consistent with established biological behavior in the cell culture and that also provide noticeably improved agreement between the experimental data and model proposed by the original authors.

458e Decision-Making In Biopharmaceutical Process Design

Charles Siletti and Demetri Petrides. INTELLIGEN, INC., 700 Walton Ave., Mount Laurel, NJ 08054

Objectives and constraints in biopharmaceutical design often differ from those associated with the design of large scale chemical processes. The choices of unit operations are more limited and operating conditions must frequently be determined experimentally. Time to market may outweigh process optimization as an objective. This presentation will explore case studies for several key design decisions including: fed-batch vs. continuous perfusion for cell culture, disposable vs. fixed storage vessels, and process scale and batch size.

Web Page: www.intelligen.com/literature

458f Formulation of a Herbal Medicinal Product - Sheng Mai San

Elizabeth J. Zhang, Bioengineering Graduate Program, Department of Chemical Engineering, Hong Kong University of Science and Technology, Clear Water Bay, Kowloon, Hong Kong, China, Kam M. Ko, Department of Biochemistry, Hong Kong University of Science and Technology, Clear Water Bay, Kowloon, Hong Kong, China, and **Ka M. Ng**, Department of Chemical Engineering, Hong Kong University of Science and Technology, Clear Water Bay, Kowloon, Hong Kong, China.

The objectives of this study were to explore the effective components of Sheng Mai San (SMS), a traditional Chinese herbal recipe, and to develop medicinal products based on this recipe. The effective components were characterized by HPLC analysis. The effect of different ratios between the three herbs of SMS - Radix Ginseng, Radix Ophiopogonis and Fructus Schisandrae was studied. The origin of 5-HMF, an effective component, was identified. Then MTT assay using H9c2 cells was employed to evaluate the anti-oxidant activity of SMS decoctions. Moreover, both ethanol precipitation and chitosan precipitation were used to purify the effective components to develop high quality medicinal products. The two methods were primarily identified to be effective for conserving most of the effective components and the anti-oxidant activity of SMS, with ethanol precipitation being more efficient and the chitosan method being lower cost.

466a A Framework for Assessing the Biocomplexity of Sustainable Material Use

Jun-Ki Choi, Center for Resilience, Department of Chemical and Biomolecular Engineering, The Ohio State University, 140 West 19th Avenue, Columbus, OH 43235, Bhavik R. Bakshi, Department of Chemical and Biomolecular Engineering, The Ohio State Unversity, Columbus, OH 43210, Timothy C. Haab, Department of Agricultural, Environmental and Development Economics, The Ohio State University, Columbus, OH 43235, and Prem K. Goel, Department of Statistics, The Ohio State University, 404 Cockins Hall 1958 Neil Avenue, Columbus, OH 43210.

With the rapidly growing level of the degradation of our natural capitals, there are strong needs for analyzing the biocomplexity of our current systems in economic, social, and environmental aspects. Biocomplexity is the dynamic web of often surprising interrelationship that arises when components of various disciplines interact in multiple temporal, spatial scales. There are various modeling techniques available for the analysis of engineering, environmental, and economic systems however most of the methodologies usually ignore the multiscale aspect of different disciplines and the importance of the synthesis of each methodology. The goal of this research is to develop a systematic framework for assessing the biocomplexity of material use and modeling the lifecycle of industrial material in multiple aspects in order to utilize the framework for improving decision making in various disciplines. Special emphasis will be placed on the importance of the multiscale modeling of the complex systems by comparing the result of single scale modeling. And the broader implication of the expanding the system boundary of the study to the ecosystem goods and services will be discussed.

A framework is developed for the integration of the different scales of modeling scheme in the context of economic input-output (IO) modeling and engineering optimization problem in order to provide a broader implication of the life cycle aspect of our current industrial system. The major benefit of input-output analysis is that it provides a snap-shot of an economy at specific period with readily available empirical data and many researchers still rely on the monetary transaction. However, such a procedure may introduce inconsistencies in the resultant accounting of physical consumption of commodities, necessitating adjustments in the procedure to insure reasonable results. Physical accounts become more important in the study of the biocomplexity of material use and there are strong needs for enhancing the current states of the physical accounting in the IO context. In addition, fuller incorporation of the demand side is vital for analyzing alternative scenarios such as the interaction of consumer behavior, commodity price, and technology alternatives. The proposed framework use both physical and monetary input output tables in every step of modeling process to provide the insight of the significant difference of using both accountings. Leontief price model and cost-push system (Dietzenbacher 1997) are used to capture the change of the price caused by the certain shock (such as carbon tax) in the economy (coarser scale). Ghosh model (Miller and Blair 1985) has been used to balance out the monetary and physical change at the short period time after the disturbance to the system. And then engineering (finer scale) optimization problem should be solved to cope with the price change of resource inputs to the industry production process. In the little longer period time, change of price of industry production (outputs) will affect the consumer behavior in purchasing the necessary goods and services. Therefore, the concept of the price elasticity of demand is integrated in the input-output framework to capture the amount of the physical demand change accrued by the commodity price change. Finally, Leontief physical model and demand-pull system (Wei, et al. 1979; Leontief 1986) is used to capture the total amount of change in production (outputs) of each sector after consumer demand change. Different environmental policy scenarios are tested for the analysis of different level of the economy wide pollution reduction and the enhancement of the use of natural capital as a raw material for the engineering process is analyzed through the proposed framework.

As a case study, the effect of carbon tax will be illustrated in the economic, environmental, and social aspect of the use of material in the electricity generation process. Different level of disaggregated level electricity generation sectors are embedded to the readily available detailed level of economic transaction data from the U.S. Bureau of Economic Analysis. As the period of projection gets longer, the consideration of the changes in production technology and consumer demand have to be incorporated. Therefore, we analyze both short-run and long-run effect of economic, ecological disruption to the current system in the IO framework. The short-run impact analysis is employed when the exogenous changes occur in the short term or the changes occur because of the actions of only a small number of agents. While the long-run forecasting is used when longer-term and broader changes are examined.

Reference

[1] Dietzenbacher, E. (1997). "In Vindication of the Ghosh Model: A Reinterpretation as a Price Model." Journal of Regional Science 37(4): 629.

[2] Leontief, W. (1986). Input-Output Economics. Oxford, Oxford University Press.

[3] Miller, R. E. and P. D. Blair (1985). Input-Output Analysis: Foundations and Extensions. NJ, Prentice-Hall.

[4] Wei, J., T. W. F. Russel, et al. (1979). The Structure of the Chemical Industries: Function and Economics. New York, McGraw Hill.

466b Safeparking of Multi-Unit Process Systems

Rahul Gandhi and Prashant Mhaskar. Department of Chemical Engineering, McMaster University, Hamilton, ON L8S 4L7, Canada

The operation of chemical processes often encounters faults in process equipments such as actuators and sensors. Equipment failure can have a serious impact on the product quality, lead to undesirable pollutant emission and impact the overall plant productivity and economy negatively. To overcome these problems without interrupting nominal process operation, significant research efforts have focussed on devising automated methods for online diagnosis and isolation of faults and in developing online strategies for performance recovery from the degraded conditions. The extensive research in fault-tolerant control (FTC) is based on the assumption of availability of sufficient control effort or redundant control configurations to preserve operation at the nominal equilibrium point in the presence of faults (see, e.g., [1,2,3]). In contrast, the problem of faults that do not allow continuation of operation at the nominal operating point has not received sufficient attention. In the absence of a framework for handling such faults, ad-hoc approaches could result in temporarily shutting down the process which can have substantially negative economic ramifications.

In [4], a safe-parking framework was developed to address the problem of determining how to run the process during fault-rectification to prevent onset of hazardous situations and enable smooth transition to nominal operation upon fault recovery. In [5], the safe-parking framework is extended to handle uncertainty and unavailability of measurements. The results in [4,5], however, consider safe-parking in the context of single unit. The connected nature of chemical processes via material, energy or communication lines can have significant impact on the ability of safe-parking the process. In particular, a change in operating condition of one unit naturally acts as disturbance to the downstream units and hence large changes in operating conditions of one unit, while possibly enabling safe-parking of the unit in question, can jeopardize the operation of the downstream units, and therefore of the whole plant. This necessitates that the safe-park point for a unit in multi-unit process be chosen with adequate consideration of its effect on downstream processes.

Motivated by above consideration, this work addresses the problem of handling faults in the context of multi-unit networked processes. We consider a multi-unit nonlinear process system subject to input constraints and faults in one unit that preclude the possibility of operating the unit at its nominal equilibrium point. We first consider the case where there exists a safe-park point for the faulty unit such that its effect can be completely absorbed (via changing the nominal values of the manipulated inputs) in the downstream unit. Steady-state as well as dynamic considerations (including the presence of input constraints) are used in determining the necessary conditions for safe-parking the multi-unit system. We next consider the problem where no viable safe-park point for the faulty unit exists such that its effect can be completely absorbed in the subsequent unit. A methodology is developed that allows simultaneous safe-parking of the consecutive units. Finally, we incorporate performance considerations in the safe-parking framework for multi units systems and illustrate the implementation of the safe-parking framework using a multi-unit chemical reactor system.

References

[1] Mhaskar, P., A. Gani, N. H. El-Farra, C. McFall, P. D. Christofides and J. F. Davis, "Integrated Fault Detection and Fault-Tolerant Control of Nonlinear Process Systems," *AIChE J., 52,* 2129-2148, 2006.

[2] P. Mhaskar, ``Robust Model Predictive Control Design for Fault-Tolerant Control of Process Systems," *Ind. Eng. & Che. Res., 45,* 8565-8574, 2006.

[3] Mhaskar, P., C. McFall, A. Gani, P. D. Christofides and J. F. Davis, "Isolation and Handling of Actuator Faults in Nonlinear Systems," *Automatica, 44,* 53-62, 2008.

[4] Gandhi, R. and P. Mhaskar. Safe-parking of nonlinear process systems. *Comp. & Chem. Eng.,* accepted for publication.

[5] Mahmood, M., Gandhi, R. and P. Mhaskar. Safe-parking of nonlinear process systems: Handling uncertainty and unavailability of measurements, *Chem. Eng. Sci.*, submitted.

466c Model-Based Sensor Scheduling for Power Management In Wireless Sensor Networks

Yulei Sun and Nael H. El-Farra. Department of Chemical Engineering & Materials Science, University of California, Davis, One Shields Avenue, Davis, CA 95616

The convergence of recent advances in sensor manufacturing, wireless communications and digital electronics has produced low-cost wireless sensor networks (WSNs) that can be installed for a fraction of the cost of wired devices (see, for example, [1],[3],[4]). WSNs offer unprecedented flexibility ranging from high density sensing capabilities to deployment in areas where wired devices may be difficult or impossible to deploy (such as inside waterways and high-temperature areas in oil refineries). Augmenting existing process control and monitoring systems with additional WSNs has the potential to expand the capabilities of the existing control technology beyond what is feasible with the wired networked architectures alone. Specifically, deploying additional WSNs throughout the plant and interfacing those devices with the existing control systems permit collecting and broadly disseminating additional real-time information about the state of the plant units which in turn can be used to enhance the performance and robustness of the plant operations. The extra information, together with the increased levels of sensor redundancy achieved with WSNs, also enable achieving proactive fault-tolerance and real-time plant reconfiguration based on anticipated market demand changes. These are appealing goals that coincide with the recent calls over the last few years for expanding the traditional process control and operations paradigm in the direction of smart plant operations [2],[6].

To harness the full potential of WSNs in process control, there is a need to address the fundamental challenges introduced by this technology from a control point of view. One of the main challenges to be addressed when deploying a low-cost WSN for control is that of handling the inherent constraints on network resources, including the limitations on the computation, processing and communication capabilities. Other constraints such as limited power (battery energy) are also important when the WSN is deployed in harsh or inaccessible environments where a continuous power supply is not feasible and the wireless devices have to rely on battery power instead [4]. A tradeoff exists between the achievable control performance and the extent of network resource utilization. Specifically, maximizing the control performance requires continuous (or at least frequent) collection of data and disseminating it broadly to the target control systems. On the other hand, the limited resources of a WSN, together with the difficulty of frequent battery replacement in a plant environment, suggest that sensing and communication should be reduced in order to aggressively conserve resources and extend the lifetime of the network as much as possible. Realizing the potential of WSNs to improve process control requires the development of effective strategies for characterizing and managing this tradeoff.

An effort to address this problem was initiated in [5] where a quasi-decentralized networked control architecture was developed for plants with distributed interconnected units that exchange information over a shared communication network. The main idea is to reduce the exchange of information between the local control systems as much as possible without sacrificing stability of the individual units and the overall plant. To this end, dynamic models of the interconnected units are embedded in the local control system of each unit to provide it with an estimate of the evolution of its neighbors when measurements are not transmitted through the network. The use of a model to recreate the interactions of a given unit with one of its neighbors allows the sensor suite of the neighboring unit to send its data in a discrete fashion since the model can provide an approximation of the unit's dynamics. The state of each model is then updated using the actual state of the corresponding unit provided by its sensor suite at discrete time instances to compensate for possible model uncertainty. Finally, the minimum allowable communication frequency is characterized for the case when the update period is constant and the same for all the units. Under this scenario, all sensor suites are given access to the network and can successfully transmit their data simultaneously.

In addition to controlling the transmission frequencies of individual sensors in the network, another important way of conserving the network's energy resources is to select and activate only a subset of the deployed sensor suites at any given time to communicate with a given unit. Under this restriction, the performance of each unit in the plant becomes dependent not only on the controller design but also on the selection of the scheduling strategy that, at any time, determines the order in which the sensor suites of the neighboring units transmit. The scheduling problem is also important in cases where access constraints in the wireless communications medium limit the number of available channels so that at any one time only some of the sensors and actuators can exchange information, while others must wait.

To address this problem, we present in this work a model-based sensor scheduling approach for power resource management in wireless sensor networks. To illustrate the main ideas, we consider as an example the configuration where only one wireless sensor suite is allowed to transmit its measurement updates to the appropriate units at any one time, while the others remain dormant for some time before the next suite is allowed to transmit its data. The time intervals between the transmissions of the different suites can vary but the update period -- i.e., the time interval between two consecutive transmissions of the same sensor suite -- remains the same for all sensor suites. The objective is to find an optimal sequence for establishing and terminating communication between the sensors suites and the target controllers that maximizes the update period (and thus minimizes the overall network utilization) without jeopardizing the required stability and performance characteristics. To this end, we consider a large-scale distributed plant with interconnected processing units, and initially design a quasi-decentralized control structure where each unit has a local control system with its sensors and actuators connected to the local controller through a dedicated communication network, while the local control systems communicate with one another through a wireless sensor network to minimize the propagation of disturbances and meet the overall plant objectives. Embedded in the local control system of each unit is a set of dynamic models that provide an approximation of the interactions between the given unit and its neighbors in the plant when measurements from the neighboring sensor suites are not transmitted through the network. The state of each model is updated using actual measurements from the corresponding unit when communication is re-established. Since the sensor suites of the different units are forced to transmit their data at different times, the models within the local control system of a given unit will be updated at different times according to the chosen transmission schedule.

Exploiting the periodic nature of the transmission schedule, the networked closed-loop plant is formulated as a switched system where the process states evolve continuously while the model forecasting errors are reset to zero at transmission times. This formulation allows obtaining and explicit characterization of the maximum allowable update period in terms of the sensor transmission schedule, the update times of the different sensor suites, the uncertainty in the models as well as the controller design parameters. The results can be used to compare a set of possible transmission schedules to determine the one that leads to the maximum update period and thus results in minimal network resource utilization. The theoretical results are illustrated using a chemical plant example.

References:

[1] Akyildiz, I. F., W. Su, Y. Sankarasubramaniam, and E. Cayirci, ``Wireless sensor networks: a survey," Computer Networks, 38:102--114, 2002.

[2] Christofides, P. D., J. F. Davis, N. H. El-Farra, K. Harris, and J. N. Gibson, ``Smart plant operations: Vision, progress and challenges," AIChE J., 53: 2734-2741, 2007.

[3] Kumar, P. R., ``New technological vistas for systems and control: The example of wireless networks," IEEE Contr. Syst. Mag., 21:24--37, 2001.

[4] Song, J., A. K. Mok, D. Chen, and M. Nixon, ``Challenges of wireless control in process industry," Proceedings of Workshop on Research Directions for Security and Networking in Critical Real-Time and Embedded Systems, San Jose, CA, 2006.

[5] Sun, Y. and N. H. El-Farra, ``Quasi-decentralized model-based control of networked process systems," Comp. \& Chem. Eng., in press.

[6] Ydstie, E. B., ``New vistas for process control: Integrating physics and communication networks," AIChE J., 48:422--426, 2002.

466d Modeling Technology Transitions and Risks Using Input-Output Framework

Vikas Khanna, Department of Chemical and Biomolecular Engineering, The Ohio State University, Koffolt Laboratories, 140W, 19th Avenue, Columbus, OH 43210 and Bhavik R. Bakshi, Department of Chemical and Biomolecular Engineering, The Ohio State Unversity, Columbus, OH 43210.

Several recent events in the U.S. have highlighted the criticality and vulnerability of infrastructure systems to sudden shocks such as natural disasters, terrorist attacks, and food shortages. Proper understanding of such disruptive scenarios and their impact using holistic and integrated systems modeling techniques is crucial for effective resource allocation and disaster management. For example, the disruptive ability of emerging technologies such as biofuels, nanotechnology, and newer materials has been anticipated by many. Development and in turn impact of such radically different technologies on the economy and the associated infrastructure systems will be fueled by environmental and economic concerns and changing human preferences. However, attempts to study their long term economic-environmental impact and future trajectories pose both methodological and empirical challenges for engineers and policy makers alike. This is because emerging technologies and the associated infrastructure systems are part of complex hierarchical dynamic industrial systems with highly interconnected processes and technologies that makes system characterization difficult. Understanding their dynamics and the associated economic-environmental implications require data at both the micro and the macro level which is not readily available. Another challenge is associated with the technological change over time and hence the associated uncertainties with respect to the resource consumption and its impact. These issues ranging from the lack of data for evaluating emerging technologies to the development of models that could provide a better understanding of complex hierarchical systems and its implications for long-term sustainability assessments of emerging technologies are the focus of this work.

The present work focuses on exploring the utility of Input-Output (I/O) models for studying the effect of sudden shocks and quantifying the associated risks. Economic Input-Output (EIO) models have been used for various purposes including exploring the effect of changes in final demand, taxes and other economic changes [1,2]. EIO model uses information contained in the detailed I/O tables of the economy. It uses a matrix representation of interindustry relations in a nation's (or a region's) economy. The matrix contains data about the monetary transactions between each pair of industrial sectors. Models based on the EIO model have also been developed for life cycle assessment by combining the EIO model with data about emissions and resource use. Economic Input-Output life cycle assessment (EIOLCA) mainly focuses on the impact of emissions and energy use by combining economic data with the emissions details for specific sectors and is mainly an output-side approach [3]. More recently, Ecologically based life cycle assessment (Eco-LCA) has been developed which is novel and unique in its ability to quantify the contribution of ecosystem goods and services. [4,5]

We are using the Eco-LCA model to understand the impact of changes in the availability of natural resources including natural capital. This includes understanding the potential impact of loss of services such as pollination, water scarcities, and soil fertility. Such information is used to determine sectors that are likely to face maximum risk due to sudden disruptions. Since the EIO model considers a static and linear state of the economy, it is not able to simulate long-term effects of such disruptions. However, it is appropriate for gaining insight into the short-term effects of disruptions before any adaptation due to market forces or policies. Such simulation is relevant to understanding the effect of environmental changes as well as human-induced changes such as terrorism and natural disasters.

The utility of the proposed approach will be illustrated using case studies involving energy system transitions. The case studies involve the use of I/O models coupled with scenarios in a prospective dynamic analysis and with explicit treatment of uncertainty to evaluate the broader impact of technology transitions. The implications of the present work are to enhance the fundamental understanding and modeling and provide insights for resilient and sustainable technology transitions. Methodologically, the goal is to enhance and extend presently available tools and develop new modeling approach that explains the dynamics of complex systems with specific focus on emerging technologies. In the long term, this work is expected to complement the traditional static biophysical models and methods by including the dynamic behavior of complex adaptive systems, various uncertainties, risks, and study their implications for decision-making and sustainability.

References

1. R. E. Miller and P. D. Blair. Input-Output analysis. Foundations and extensions. Prentice-Hall, Inc., Englewood Cliffs, New Jersey, 1985.

2. W. W. Leontief. Input-Output Economics. Oxford University Press, 1985.

3. EIOLCA. www.eiolca.net, accessed on May 10, 2008.

4. N. U. Ukidwe and B. R. Bakshi. Industrial and ecological cumulative exergy consumption of the united states via the 1997 input-output benchmark model. Energy, 32(9):1560–1592, 2007.

5. N.U. Ukidwe and B.R. Bakshi. Thermodynamic Accounting of Ecosystem Contribution to Economic Sectors with Application to 1992 U.S. Economy. Environmental Science and Technology, 38(18):4810–4827, 2004.

466e A Reaction Route Network Approach to the Thermodynamic Consistency of Microkinetic Models

Michael Morin[1], Saurabh Vilekar[1], Ravi Datta[2], and Ilie Fishtik[1]. (1) Chemical Engineering, Worcester Polytechnic Institute, 100 Institute Rd, Worcester, MA 01609, (2) Chemical Engineering, WPI, 100 Institute Rd, Worcester, MA 01609

Complex reactions proceed via a large number of elementary reaction steps involving reactants, reaction intermediates and products. The kinetic characteristics of the reaction steps are normally subject to rigorous thermodynamic constraints such as Wegschieder conditions. The enumeration and implementation of these constraints in complex reaction mechanisms may be tedious. In this presentation we show that the problem of thermodynamic consistency of reaction mechanisms may be naturally treated in terms of reaction route networks. Moreover, a reaction route networks cannot be assembled in principle if all of the thermodynamic constraints placed on a reaction mechanism are not satisfied. The approach is illustrated with examples of various microkinetic models.

466f Graph-Theoretic Approach for Discovering Alternative Synthetic Routes Forming Complex Networks

L. T. Fan[1], **Tengyan Zhang**[1], A. Argoti[1], Jiahong Liu[1], Botond Bertok[2], and Ferenc Friedler[2]. (1) Department of Chemical Engineering, Kansas State University, 1005 Durland Hall, Manhattan, KS 66506, (2) Department of Computer Science, University of Veszprém, Egyetem u. 10., Veszprém, H-8200, Hungary

The discovery of alternative synthetic routes, or alternative reaction paths, for synthesizing a chemical compound is a subject of vital theoretical importance in synthetic chemistry. It is also a subject of practical importance for developing a chemical process by taking into account such alternative synthetic routes.

Any available set of alternative synthetic routes invariably gives rise to a highly complex reaction network that almost always contains numerous additional synthetic routes resulting from the feasible combinations of the available set; nevertheless, it is extremely difficult, if not impossible, to exhaustively recover the alternative synthetic routes from the complex network. The mathematically exact and computationally efficient approach presented herein resorts to the graph-theoretic method for process-network synthesis based on process graphs (P-graphs). The efficacy of the approach is illustrated with examples.

466g Integrating Networked Process Systems to Solve Dynamic Optimization Problems

B. Erik Ydstie and Michael R. Wartmann. Department of Chemical Engineering, Carnegie Mellon University, 5000 Forbes Ave, Pittsburgh, PA 15213

We introduce a new framework for distributed control and optimization of complex networks based on variational calculus. Conservation laws for extensive quantities and the second law of thermodynamics lead to conditions for stability and optimality of the network. We derive a general way of describing interconnections in networks through matrix representations that capture a network's topology using basic principles from electrical engineering methodologies. This shows how the dynamics of independent entities in a network define the objective function of the optimization problem that is simultaneously solved. A generalized version of Tellegen's theorem from electrical circuit theory plays a central role in developing the objective function of the regarded dynamic networks. These results indicate that we can solve optimization problems using dynamical systems, and how the objective function depends on the choice of feedback control and strategies. We highlight the concept of duality between the extensive variables and intensive variables and their role for the process network. Several examples are presented to illustrate these principles for different types of network connections, for continuous and discrete and linear and nonlinear flow connections. The boundary or terminal connections of the network are investigated in connection with the resulting primal and dual optimization problem and the effect on the dynamic behavior of the system. We illustrate how distributed control and optimization can be implemented in network structures using the proposed framework.

471a Explicit Real-Time Optimization

Johannes E. P. Jäschke[1], Sridharakumar Narasimhan[2], and Sigurd Skogestad[2]. (1) Chemical Engineering, Norwegian University of Science and Technology, Sem Sælandsvei 4, Trondheim, 7491, Norway, (2) Chemical Engineering, Norwegian Univeristy of Science and Technology, Sem Sælands vei 4, Trondheim, 7491, Norway

Rising competition, energy prices and environmental demands make it increasingly necessary to operate chemical processes as close to optimality as possible. In order to remain close to optimality in spite of disturbances, two approaches may be considered [1]. The first approach is to obtain optimal operation via conventional real-time optimization. This implies that the optimal setpoints of the controlled variables are computed online and are updated at certain time intervals based on the last available measurements. Setting up, solving and maintaining an RTO system can be a very time-consuming and complex task, as the uncertainty in the model and parameters can have a severe impact and the new setpoints have to be available at a given time instant.

A different approach is to move all considerations and calculations off-line, in order to find self-optimizing control variables (certain measurement combinations). Controlling these variables keeps the process at or close to the optimal operating point in presence of disturbances without the need to re-optimize. Once these variables are obtained and the control structure is set up, the conventional RTO optimization problem can be reduced significantly and may even be completely replaced by the new Explicit-RTO layer.

To the authors knowledge, these optimal measurement invariants have only been considered systematically only for linear plants with quadratic cost functions. This work extends the ideas of self-optimizing control, particularly the concept of the null-space method [2] to systems described by rational functions.

Using polynomial elimination techniques based on Grobner bases it is possible to determine polynomials in the measured variables, which are used as self-optimizing control variables. Keeping these variables constant yields zero loss from optimality for arbitrary large disturbances, as long as the corresponding active set does not change.

The elimination method for non-linear polynomial models makes it not only applicable for small disturbance ranges, where the process can be approximated linearly, but also for large disturbances, when non-linearities are taken into account. As every non-linear process may be approximated by a higher order Taylor series, this method may also be used to find optimal invariants for other processes, described by more general equations.

The method was tested for a model of a four component system in an isothermal CSTR with two reactions, taken from [3]. The process in consideration has three regions defined by the set of active constraints, one unknown reaction rate as a disturbance and four measurements: The two feed flow rates, one outlet concentration and the heat applied to the jacket. Using the Computer Algebra package 'Singular' [4], a polynomial self-optimizing variable is found for each of the regions. E.g. in the first region, where only the applied heat is at a constraint, a polynomial of total degree 4 in the outlet concentration and one feed rate is found. Controlling this variable yields zero loss in the entire region. Moreover in this case-study it is found that the self-optimizing variables of the neighbouring regions can be used for determining the switching point between the active sets. This shows that it is unnecessary to track the NCO for this system by more complicated methods.

Designing a self-optimizing control structure might require more efforts in advance, but is straightforward to implement and to maintain in practice. Once the control structure is obtained and implemented, there is no longer need to invest in expensive real-time equipment in order to obtain optimal operation.

[1] S. Skogestad. Control structure design for complete chemical plants, Comp. Chem. Engng., Vol. 28 (1-2), (2004) 219-234 .

[2] V. Alstad, S. Skogestad, E.S. Hori. Optimal measurement combinations as controlled variables. Journal of Process Control, Article in press doi:10.1016/j.jprocont.2008.01.002

[3] B. Srinivasan, L.T. Biegler and D. Bonvin, Tracking the necessary conditions of optimality with changing set of active constraints using a barrier-penalty function, Computers and Chemical Engineering, Vol 32, (2008) 572-279

[4] G.-M. Greuel, G. Pfister, and H. Schönemann. Singular 3.0. A Computer Algebra System for Polynomial Computations. Centre for Computer Algebra, University of Kaiserslautern (2005). http://www.singular.uni-kl.de.

471b Identification and Control of a Tubular Solid Oxide Fuel Cell (SOFC)

Debangsu Bhattacharyya, Chemical Engg. Dept., Clarkson University, 8 Clarkson Avenue, Potsdam, NY 13699, Raghunathan Rengaswamy, Chemical Engineering, Clarkson University, Box 5705, Clarkson University, Potsdam, NY 13699-5705, and Finnerty Caine, NanoDynamics,Inc., 901 Fuhrmann Blvd., Buffalo, NY 14203.

In view of its use in a grid-connected system, SOFC is expected to face large variations in power demand. Under varying load, a SOFC has to satisfy three requirements. First, it has to meet the stringent requirements of the external load (such as power, voltage etc.). At the same time, it has to maintain the efficiency of the system. Also no structural or material damage can be afforded. Coupled with significant temperature gradients and fast transients, this gives rise to a very challenging control problem.

Model predictive control (MPC) is a very efficient and robust control technique for such challenging control problems. For implementation of MPC, a model needs to be identified which can be solved in real time. A data-driven model is identified in this study. Data for model identification are generated by a model which has been validated with industrial data for steps of different magnitude and directionality in different variables for wide operating conditions. So the process nonlinearity is expected to be reflected in the data and needs to be captured by the identified model. The input signal is designed maintaining linear persistence of excitation. Several linear (such as FIR, ARX etc.) and nonlinear models (such as NAARX) with increasing complexity are compared and the best model is chosen according to the AIC values of the models. Both SISO and MIMO models are identified. Several SISO models such as Voltage-power, H2 flow-power, voltage-H2 Utilization Factor are identified. For the MIMO model, voltage and H2 flow are considered as inputs. Power and utilization factors are considered as outputs. A linear model is found to be satisfactory for most SISO cases. However, a nonlinear model such as NAARX model with more cross terms is found to improve the model performance significantly for the MIMO case. All through this work, efforts have been made to come up with the simplest, yet representative model that can be used for real-time applications.

The identified model is used for designing the controller for both the SISO and MIMO systems. A PID controller and a nonlinear Model Predictive Controller are implemented. In the presence of strong interactions between the loops, the implementation of the controllers is very challenging. For the SISO control, the PID control is found to work satisfactorily. However, large overshoots in PID performance are observed for the MIMO problem. The performance of NMPC is very good for all the criteria tested. Reasons for large overshoots in the PID performance are found by comparing the control moves of the PID in comparison to that of the NMPC. Issues related to real-time implementation and tuning of the NMPC controller will also be presented in this talk.

471c Passivity Based Modeling and Control of Coal Power Plants

B. Erik Ydstie and Chengtao Wen. Department of Chemical Engineering, Carnegie Mellon University, 5000 Forbes Ave, Pittsburgh, PA 15213

The main advantage of passivity based control is that stabilizing, high performance control systems with optimal bandwidth can be developed without detailed process modeling [1-2]. In this paper, we show that the passivity theory can also be used to guide modeling complex, interconnected process systems, like a coal fired, Rankine cycle power plant.

Coal fired power plants currently account for more than 50% of the electricity generation in the US. Coal is abundant and cheap relative to other sources of fossil energy. Unfortunately the carbon foot-print is large. In the long run it is desirable to phase out coal based power generation as nuclear and renewable energy systems are developed. In the short run it is imperative that the coal based fleet is operated as efficiently as possible to minimize the emission of greenhouse gases and other pollutants like fly-ash, NOX and SOX.

Advanced process control and optimization have not made significant inroads into power plant control. Most power plants are operated using old control system structures using PI controllers that may be poorly tuned and some feed-forward. These control system structures were adequate for steady state operation. In the current, deregulated market power plants are subject to frequent load changes. Shortcomings in the classical control system designs have become evident and there is a need to implement advanced, model based control systems which can be used for stabilization and setpoint tracking.

Power generation systems are highly integrated internally and the dynamics change considerably due to strong integration with the power grid. The dynamics display a wide range of time-scales ranging from millisecond in the turbine and electrical systems to seconds and minutes in the steam temperature, combustion and emission control systems. All the time-scales must be modeled and controlled to ensure stable and safe power plant operation.

Classical approaches to power plant modeling use intensive variables like pressure and temperature to represent the system state [3-4]. Unfortunately, several approximations have to be made to implement such models, thermodynamic calculations often become cumbersome and the model structure becomes complex and non-intuitive. Nonlinearities are present in the state equations and it is difficult to develop systematic methods for modeling the network interconnections. Control system design can becomes cumbersome since inherent passivity properties, which facilitate simple control system design, may be lost.

In the passivity based approach we choose the inventories (the so-called Gibbs ensemble $\{U,V,N\}$) to represent the system state. The conservation laws are represented in the control affine form which can be connected with the passive systems theory. Network structures are easy to develop since the state space consisting of the extensive variables, can be described using concepts from affine geometry [5] [6]. Thermodynamics, chemical reaction and transport are described using algebraic constraints. The second law of thermodynamics can be used to show that the resulting DAE system has index 1, which means that open and closed loop dynamics can be solved using well tested integration methods suitable for stiff systems like ODE15s in MATLAB.

In the paper we use the passivity based approach to develop a dynamic model and multivariable control systems for a Rankine cycle power plant. The model components include the water-wall, riser and drum-boiler complex. We include multiple super-heaters, turbines with steam extraction, condenser, furnace, feed and air heaters. Each component is modeled as a system of DAEs with standard interconnects which allow accurate representations of the material and energy flows. The resulting model system maintains the invariant (Hamlitonian) structures which can be derived from the mass and energy conservation laws. The resulting model has been verified against data from a 600MW power plant system in the Pittsburgh region.

We furthermore prove that the state space spanned by the intensive variables such as temperature and pressure is isomorphic to the space spanned by extensive variables such as the mass and energy inventories for a single component system. This result allows us to develop a passivity-based inventory control scheme the power plant system. The asymptotic stability of the closed-loop system is proved using Lyapunov stability theory using a thermodynamic based dissipation function.

The proposed inventory controller has been tested in two typical tests: step response test and area regulation test. In the step response test, the power plant model is subjected to a set-point change in the mass or energy inventory of a unit operation. The area regulation (AR) test is a standard test and frequently used in the daily operation of a power station. The AR test signals consist of several ramps, which fluctuate around the base loading with a maximum deviation of ±10%. The numerical simulations suggest the performance and efficiency of the inventory controllers in power plant systems.

[1] A. van der Schaft. L2-Gain and Passivity Techniques in Nonlinear Control. Springer-Verlag, New York, second edition, 2000.

[2] K. R. Jillson and B. Erik Ydstie. "Process networks with decentralized inventory and flow control" Journal of Process Control, vol. 17:399-413, 2007.

[3] K. J. Astrom and R. D. Bell. "Drum-boiler dynamics, " Automatica, 36:363-378, 2000

[4] M. E. Flynn , M. J. Oapos Malley. "A drum boiler model for long term power system dynamic simulation, " IEEE Transactions on Power Systems, vol. 14:209-217, 1999

[5] E.D. Gilles. Network theory for chemical processes. Chem. Eng. Technol., 21(2):121–132, 1998.

[6] C. A. Farschman, K. P. Viswanath, B. Erik Ydstie. "Process systems and inventory control," AIChE Journal, Vol. 44(8):1841-1857, 1998

[7] K. L. Chien, E. I. Ergin , C. Ling and A. Lee. "Dynamic analysis of a boiler," ASME Transactions 80:1809-1819, 1958.

471d Cfd Simulation of Urea-Based Hybrid Sncr-Scr in a Pilot Scale Flow Reactor

Thanh D. B. Nguyen[1], Tae-Ho Kang[1], Young-Il Lim[1], Seong-Joon Kim[2], and Kyung-Seun Yoo[2]. (1) Department of Chemical Engineering, Hankyong National University, Jungangno 167, Anseong-si, Gyonggi-do, 456-749, South Korea, (2) Department of Environmental Engineering, Kwangwoon University, 447-1, Wolgye-dong, Nowon-gu, Seoul, 139-701 Korea

Abstract

Controlling the emissions of nitrogen oxides (NO_x) from the flue gas has been developed and innovated over the world to meet the needs of the stringent regulations. The post-combustion techniques include selective catalytic reduction (SCR), selective non-catalytic reduction (SNCR), and the hybridization of SNCR-SCR. The hybrid SNCR-SCR technology, as applied for NO_X control from the stationary combustion systems, combines SNCR and SCR with the intention of optimizing the NO_X reduction performance, minimizing the operating cost, and reducing the ammonia slip from the process.

In this study, the urea-based hybrid SNCR-SCR process in a pilot-scale flow reactor is investigated through the experimentation and computational fluid dynamics (CFD) simulation. The pilot system is installed with a 150kW LPG burner. The V_2O_5-WO_3/TiO_2 monolith honeycomb is used as a catalyst in the SCR zone.

A reduced chemical kinetics for the SNCR process using urea-water solution is cooperated with a two phase CFD model to account for the coupling between gas phase and droplet phase. The turbulent reacting flow CFD model is used to predict the turbulent chemistry of NO reduction, ammonia slip and N_2O evolution. A Arrhenius-type reduced model is applied to predict the surface chemistry at the SCR reactor.

The full scale two-dimensional turbulent volumetric and surface reacting CFD model is validated with the experiments in the pilot scale hybrid SNCR-SCR. The CFD simulation results agree well with the experimental data.

Keywords: NO_X reduction, Selective non-catalytic reduction (SNCR), Selective catalytic reduction (SCR), Hybrid SNCR-SCR, Urea solution, Computational fluid dynamics (CFD), Pilot-scale flow reactor.

471e Fast Nonlinear Predictive Control of Autothermal Hydrogen Generation Subject to Economic Constraints

Michael Baldea, Praxair, Inc., 175 East Park Drive, Tonawanda, NY 14150, Prodromos Daoutidis, Chemical Engineering and Materials Science, University of Minnesota, 151 Amundson Hall, Minneapolis, MN 55455, and Zoltan K. Nagy, Chemical Engineering Department, Loughborough University, Loughborough, LE11 3TU, England.

Environmental and political pressure to mitigate the effects of global warming has fostered a growing interest in renewable and clean energy resources, particularly in the development of Ethanol- and Hydrogen-based transportation and energy generation solutions. While Ethanol can be derived from US-based resources such as corn via well known processes, the quest for efficient and economical Hydrogen sources continues. Autothermal reactors that combine, in a single unit, endothermic Hydrogen generation and exothermic reactions that create the thermal energy necessary to support it, represent a very promising technology in this field. These devices are compact and efficient, lending themselves naturally to on-board or on-site hydrogen production for transportation/power applications.

The above applications typically assume that hydrogen can be produced at a variable rate that can satisfy the changing demands of a downstream consumer. Consequently, steady-state performance and safety criteria notwithstanding, autothermal reactors need to meet stringent transient performance requirements. Prior work [1,5] has considered the dynamics of unidirectional autothermal reactors featuring hydrogen generation via endothermic methane steam reforming and water gas shift, supported by the catalytic combustion of methane. Using singular perturbation arguments, it was demonstrated that such systems exhibit a two time scale behavior. Also, it was shown that, ceteris paribus, counter-current designs, whereby the reforming mixture and the fuel flow in opposite directions in dedicated channels of a monolith reactor, exhibit are more robust to disturbances and production rate changes than their co-current counterparts. It was also shown that, in the absence of a control system, autothermal reactors--regardless of flow configuration--suffer a significant performance degradation (measured in terms of energy efficiency and conversion rates) when operated in a transient manner.

Owing to its ability to handle non-square systems with constraints, model predictive control (MPC) constitutes an obvious choice for tackling the stringent operating requirements and limited degrees of freedom of autothermal reactors. Traditional linear MPC lends itself well to slow-varying systems that are operated in the vicinity of a steady-state, being, however, is ill-suited to highly nonlinear systems operated in a transient manner. In view of the above, the present work addresses the control of the transient operation of autothermal reactors within the framework of Nonlinear Model Predictive Control (NMPC). Initially, we define a performance index based on the energy efficiency of the reactor. Subsequently, the index is embedded in the formulation of the nonlinear control problem, thereby addressing, in a unified manner, the operational, dynamic and economic constraints typically posed by autothermal reactors for hydrogen production.

NMPC implementations are oftentimes hindered by the dimensionality of the underlying optimization problems. In the second part of the work, we present the use of modern, fast optimization strategies [2] to alleviate these difficulties. Exploiting the special structure of the optimization problems that arise in NMPC [3], the OptCon package [4] (based on the HQP[2] solver) provides an extremely efficient solution strategy based on a multiple shooting algorithm. Within this approach, the control interval is divided into a series of grid points. Using local control parameterizations, a shooting method is employed to transition between successive points. The differential equations and the cost function on these time intervals are integrated independently during each optimization iteration, based on the current guess of the controller's output, while the continuity/consistency of the final state trajectory is enforced with additional constraints. This multistage formulation is advantageous to the efficient application of the sequential quadratic programming (SQP)-type solver in HQP, leading to very short solution times. A sparse interior point algorithm is used for the efficient treatment of the linear-quadratic subproblems in the nonlinear SQP iterations.

The proposed control strategy is demonstrated on a detailed, first principles reactor model [5]. Our simulation results exhibit a significant increase (compared to the open loop case) in the dynamic performance of the reactor, as well as improved operation economics, while the short control execution times represent an incentive for real-time implementation.

References

[1] Baldea, M. and Daoutidis, P., Dynamics and Control of Autothermal Reactors for the Production of Hydrogen, Chem. Eng. Sci., 64, 2007, 3218-3230

[2] Franke, R., Arnold, E., and Linke, H., HQP: A solver for nonlinearly constrained large-scale optimization. http://hqp.source¬forge.net.

[3] Diehl, M., Bock, H. G., Schloeder, J. P., Findeisen, R., Nagy, Z. K., and Allgoewer, F., Real-time optimization and nonlinear model predictive control of processes governed by differential algebraic equations, Journal of Process Control, 12, 577-585, 2002.

[4] Nagy, Z. K., Mahn, B., Franke, R., and Allgower, F., Efficient output feedback nonlinear model predictive control for temperature control of industrial batch reactors, Control Engineering Practice, 15, 839-859, 2007.

[5] Baldea, M., Zanfir, M. and Daoutidis, P. Counter-Current Autothermal Reactors For Hydrogen Generation: Modeling And Dynamic Analysis, In Preparation

471f Stochastic Prediction and Control of Cyclic Combustion Variation in Internal Combustion Engines

James C. Peyton Jones and **Kenneth R. Muske**. Electrical and Computer Engineering, Villanova University, 800 Lancaster Ave., Villanova, PA 19085-1681

Cyclic variability of the combustion process in internal combustion gasoline engines has long been known to produce an undesirable increase in emission levels, decrease in engine efficiency, and reduced driveability. These variations are closely reflected in the cyclic variation of cylinder pressure and/or ionization measurements. However, the majority of the information from these sensors has traditionally been discarded, either by considering only the ensemble mean or by analyzing only a few scalar quantities such as peak pressure, peak pressure rise, or mean effective pressure. This loss of information, and the lack of a more complete description of the combustion process, significantly limits the generality and capability of monitoring and control of the combustion process at the individual cylinder event level.

In this work, a control-relevant stochastic combustion model is developed to describe the cyclically varying time history of the combustion process during the combustion period within the cylinder. The resulting set of model parameters, which can be efficiently estimated from sensor data, provide a much more complete description of the cyclic combustion process and are capable of characterizing not just point values, but variations in the entire evolution of the combustion period.

This work describes how the generality of the stochastic combustion model can be exploited. Examples include simulation of cyclic data sets with similar statistical properties to actual engine data and derivation of the traditional monitoring statistics, such as standard deviation of maximum pressure and start of burn angle, directly from the parameters of the model. The same approach also provides insight into the phasing of the cyclic process because a number of these monitoring statistics can be computed directly as a function of crank angle. It is also straightforward to evaluate the contribution from each of the physically meaningful model parameters to these quantities and, thereby, obtain some physical understanding of the mechanism involved.

The presentation begins with the framework of the stochastic combustion model and the variety of possible parameterizations. These parametric forms are fitted to engine data and the model validated by comparing the model simulations to verification data sets. The model is then used to estimate the stochastic properties of undesirable engine events such as knock through an evaluation of residuals. Finally, a stochastic control algorithm is proposed to maximize combustion efficiency.

487a Estimation of Spatial Coupling and Temporal Dynamics In the Spread of Childhood Infectious Disease

Carl D. Laird, Chemical Engineering, Texas A&M University, 3122 TAMU, College Station, TX 77843, Derek A. Cummings, Epidemiology, Johns Hopkins Bloomberg School of Public Health, Baltimore, MD 21205, and James Young, Industrial Engineering, Texas A&M University, 3122 TAMU, College Station, TX 77843.

Infectious diseases continue to be a significant public health problem, especially in developing nations where limited resources and other factors prevent sustained administration of effective public health programs. The need to develop reliable mechanistic models for the spread of infectious disease is two-fold. From a public health perspective, it is clear that reliable models will aid in program decision making and control of infectious disease spread. From a scientific perspective, the identification of an appropriate mechanistic model can help improve our understanding of the important factors affecting the spread of infectious disease.

While the spread of infectious disease between individuals is an inherently stochastic process, it is remarkable to observe that distinct patterns, both temporal and spatial, emerge in aggregate with sufficiently large populations. For example, measles incidence appears to be correlated with school term holidays, while also displaying annual and biennial patterns in time[1]. Furthermore, spatial patterns of synchrony and traveling waves have been observed in data for dengue[2] and influenza[3]. The emergence of these patterns implies that there are fundamental temporal and spatial influences that need to be further understood. In previous work, we have demonstrated a nonlinear programming approach for estimating seasonal drivers using models for a single community, and showed that this seasonality reflects school term holiday schedules.

Here, we extend this approach to the estimation of spatially-coupled models and seek to understand both temporal and spatial drivers influencing the spread of infectious disease. We present a nonlinear programming formulation for reliable and efficient parameter estimation of these models, including the spread within communities and the spread between communities. These estimation problems include very large-scale models and a large parameter space. Therefore, we require a scalable optimization approach and a sufficiently large data set. We present an interior-point internal decomposition strategy [4] that allows efficient solution of these large estimation problems in parallel, exploiting both spatial and temporal structure in the model. Estimation is performed over two extensive sets of data. The first, from England and Wales, includes weekly reported cases of scarlet fever, whooping cough, measles, and others diseases for hundreds of cities. The second set of data, from Thailand, includes monthly case counts and annual age distributions of various infectious diseases (including measles, chickenpox, influenza, and dengue) from 1972 to present. In addition to advancing our understanding of the fundamental drivers through estimation from case data, these improved models open the door for optimal planning of control programs using rigorous dynamic models in a mathematical programming framework.

[1] B.F. Finkenstadt and B.T. Grenfell. Time series modelling of childhood diseases: a dynamical systems approach. Journal of the Royal Statistical Society, Series C, 49:187–205, 2000.

[2] D.A.T. Cummings, R.A. Irizarry, N.E. Huang, T.P. Endy, A. Nisalak, K. Ungchusak, and D.S. Burke. Travelling waves in the occurrence of dengue haemorrhagic fever in thailand. Nature, 427 (6972):344–347, 2004.

[3] C. Viboud, O. N. Bjornstad, D. L. Smith, L. Simonsen, M. A. Miller, and B.T. Grenfell. Synchrony, Waves, and Spatial Hierarchies in the Spread of Influenza . Science, 12:447 – 451, 5772.

[4] Laird, C. D., Biegler, L. T., "Large-Scale Nonlinear Programming for Multi-scenario Optimization", accepted for publication in proceedings of the International Conference on High Performance Scientific Computing, Hanoi, Vietnam, 2006.

487b Verified Solution of Epidemiological Models with Probabilistic Uncertainty

Joshua A. Enszer and Mark A. Stadtherr. Chemical and Biomolecular Engineering, University of Notre Dame, 182 Fitzpatrick Hall, Notre Dame, IN 46556

Epidemiological models can be used to study the impact and spread of disease through a population. The continuous-time version of these models takes on the form of a system of ordinary differential equations (ODEs) in which the state variables are the population or population fraction of members belonging to different states relative to the disease, such as susceptible or infected. Often these models are coarse and the precise number of individuals in a given class, or the numerical parameters explaining the interactions and transitions among the different classes, are not known with certainty. These quantities may be bounded with intervals or probability distributions.

In this presentation, we demonstrate a method for the verified solution of nonlinear epidemiological models, thus computing rigorous bounds on the predicted population of classes over a given time frame, based on the specified ranges of the uncertain parameters and/or initial conditions. The method is based on the general approach described by Lin and Stadtherr [1], which uses an interval Taylor series to represent dependence on time, and uses Taylor models to represent dependence on uncertain values. We also demonstrate an approach for the propagation of uncertain probability distributions in one or more model parameter and/or initial condition. Assuming an uncertain probability distribution for each parameter and/or initial condition of interest in the epidemiological model, we use a method, based on Taylor models and probability boxes (p-boxes) and recently described by Enszer et al. [2], that propagates these distributions through the dynamic model. As a result, we obtain a p-box describing the probability distribution for each state variable at any given time of interest. The traditional Monte Carlo simulation approach may not bound all possible population trajectories because it is impossible to sample the complete parameter space with a finite number of trials, but this Taylor model method provides completely rigorous results that fully capture all possible dynamics of the system under uncertain conditions.

The use of this method in epidemiology is demonstrated using the SIRS model, and other variations of Kermack-McKendrick models, including a case with time-dependent transmission [3]. The results of Monte Carlo simulation are compared to those of this method.

[1] Lin, Y., Stadtherr, M.A. Validated Solutions of Initial Value Problems for Parametric ODEs. Applied Numerical Mathematics, 57: pp. 1145--1162, 2007.

[2] Enszer, J.A., Lin, Y., Ferson, S., Corliss, G.F., Stadtherr, M.A. Propagating Uncertainties in Modeling Nonlinear Dynamic Systems. In Proceedings of the 3rd International Workshop on Reliable Engineering Computing, Georgia Institute of Technology, Savannah, GA: pp. 89--105, 2008.

[3] J. Dushoff et. al. Dynamical Resonance can Account for Seasonality of Influenza Epidemics. Proceedings of the National Academy of Sciences, 101: pp. 16915--16916, 2004.

487c Estimation of Infectious Disease Parameters from Initial-Time Asymptotic Analysis

Kenneth R. Muske[1], Hashem Ashrafiuon[2], and Alan M. Whitman[2]. (1) Villanova University, Department of Chemical Engineering, 800 Lancaster ave., Villanova, PA 19085-1681, (2) Mechanical Engineering, Villanova University, 800 Lancaster Ave., Villanova, PA 19085

For many infectiuos diseases, the short infection period relative to the life span of the host generally causes acute pathologies in the host. When the pathogen first establishes an infection site, the immune response in the host may be inactive for some initial period during which time the pathogen multiplies. This initial delay period in the stimulation of the immune response can be a critical factor in host mortality. For example, low levels of sporozites can lead to clinically significant malaria parasitization while higher levels can cause death in a number of days. Studies with animals intravenously injected with low doses of HIV revealed a production of neutralizing antibodies and high survival rates while those inoculated with high doses rapidly developed clinical disease. Similar observations have been made for measles where large inoculums, which may ocasionally result from airborne transmission, have been suggested to increase the risk of vaccine failure.

For acute infections, the estimation of the time and extent of the pathogenic load can be of great importance in developing effective intervention treatment programs for those individuals who have been exposed to a pathogen. Mathematical models of the timing and extent of pathogen infection, and possible reinfection, processes in an individual provide information on the early dynamics of acute infections, the maximum expected pathogenic load, and the host immune response. Deterministic mathematical models of

infectious diseases are typically expressed as a series of coupled, nonlinear, first-order, initial value ODEs [1]. Because these models are nonlinear, numerical solutions are typically obtained. Approximate analytical solutions based on linearization have restricted utility because this approach eliminates all of the interesting nonlinear effects incorporated in the model. Parameter perturbation is an alternative method for generating approximate analytical solutions.

Asymptotic theory is used to develop an approximate solution to a nonlinear coupled ODE model of an infectious disease in which a pathogen attacks a host whose immune system responds defensively. This asymptotic model is then used to estimate the original model parameters based on initial-time measurements

of the pathogen or immune level. The model under consideration is that presented by Moshtashemi and Levins

[2]. There is a short time scale that characterizes the initial growth of the invading pathogen, referred to as the initial layer, and a longer time scale that characterizes the immune system behavior after it has reacted to the invasion, referred to as the remission layer. This process repeats with alternate periods of short-time pathogen growth and long-time immune system dominance. Using singular perturbation theory, it is

possible to derive very accurate approximate analytical solutions in each of these regions starting from the

initial layer. The method of matched asymptotic expansions is used to describe the transient dynamics of the

interaction between the invading pathogen and the host immune system [3]. The advantage of using the asymptotic model is the simplification that arises in the parameter estimation scheme when only the initial time is considered. Uncertainty in the assay associated with pathogen loading or immune response measurements

is taken into consideration by the use of confidence interval estimation for the parameters of interest. The

result is a point estimate with uncertainty bounds of the maximum predicted pathogen loading and immune system

response based on information obtained early in the infection/immune response cycle that can be of great

utility in evaluating treatment options.

References

[1] Hethcote, The mathematics of infectious diseases, SIAM Rev., 42, 599, 2000.

[2] Mohtashemi and Levins, Transient dynamics and early diagnosis in infectious disease, J. Math. Biol.,

43, 446, 2001.

[3] Whitman and Ashrafiuon, Asymptotic theory of an infectiuos disease model, J. Math. Biol., 53, 287,

2006.

487d Engineering Control Approaches for the Design and Analysis of Adaptive Behavioral Interventions

Daniel E. Rivera, Arizona State University, Dept. of Chemical Engineering, Mail Stop 876006, Tempe, AZ 85287-6006 and Linda M. Collins, Human Development and Family Studies, Penn State University, 204 E. Calder Way, Suite 400, State College, PA 16801.

Control engineering examines how to transform the behavior of systems over time from undesirable conditions to desirable ones. Cruise control in automobiles, the home thermostat, and the insulin pump are just some examples of control systems at work in our daily lives. The last half-century has seen the extensive application of control engineering concepts to physical systems; however, control engineering has yet to substantially impact the field of behavioral health. An increasing interest by government and community agencies for developing comprehensive systems solutions to the prevention and treatment of chronic disorders (among them drug and alcohol abuse, HIV/AIDS, cancer, mental health, diabetes, obesity, and cardiovascular health) has created new opportunities for novel approaches to these important public health problems that rely on control engineering principles.

The presentation will provide a perspective on the role that control engineering principles can play in behavioral health by examining the problem of adaptive, time-varying interventions. Adaptive interventions systematically individualize therapy through the use of decision rules that determine intervention dosages and forms of treatment by relying on measurements of tailoring variables over time. Adaptive interventions constitute a form of feedback control system in the context of behavioral health. Consequently, drawing from control engineering has the potential to significantly inform the analysis, design, and implementation of adaptive interventions, leading to improved adherence, better management of limited resources, a reduction of negative effects, and overall more effective interventions.

The conceptual framework underlying this work comes from Collins et al. (2004) and Rivera et al. (2007). A simulation study of a hypothetical adaptive intervention inspired by the Fast Track program for prevention of conduct disorders in at-risk children is presented. The results of explicit decision rules (similar to those proposed by Collins et al. (2004)) are compared to both Proportional-Integral-Derivative (PID)-type and Model Predictive Control (MPC)-based decision policies designed on the basis of model-based engineering control principles. In light of this analysis and simulation study, a series of systems technologies that will impact future research on this problem are presented.

[1] Collins, L.M., S.A. Murphy and K.L. Bierman (2004). "A conceptual framework for adaptive preventive interventions," Prevention Science, 5 (3), 185-196.

[2] Rivera, D.E., M.D. Pew, and L.M. Collins (2007). "Using engineering control principles to inform the design of adaptive interventions: a conceptual introduction," Drug and Alcohol Dependence, Special Issue on Adaptive Treatment Strategies, 88, Supplement 2, pgs. S31-S40.

487e Modeling the Control of An Excitatory Neurotransmitter Receptor during Alcoholism and Alcohol Withdrawal

Mary K. McDonald[1], James S. Schwaber[2], and Babatunde A. Ogunnaike[1]. (1) Chemical Engineering, University of Delaware, 150 Academy Street, Newark, DE 19716, (2) Daniel Baugh Institute for Functional Genomics and Computational Biology, Thomas Jefferson University, 1020 Locust St Room 381, Philadelphia, PA 19107

In 1998, it was estimated that 14 million Americans were affected by alcohol abuse and the annual economic impact of the associated healthcare and productivity losses totaled $100 billion dollars [1]. Over the last decade, there has been no improvement in the epidemic, according to a Centers for Disease Control report indicating an average of 35,915 chronic alcohol-related deaths per year between 2001 and 2005, along with a steady rise in the number of alcohol-related morbidity hospital discharges [2]. Despite its prevalence and many decades of alcohol research, however, we do not completely understand the neurological basis, epidemiology, optimal treatment regimen, and progression of alcoholism.

Previous alcohol research has elucidated the prominent role of the brain during alcoholism and alcohol withdrawal. Furthermore, numerous studies have suggested that many of the physiological and emotional consequences of alcoholism and alcohol withdrawal may be due to the deregulation of neurotransmitter systems in the brain. Ethanol molecules block a particular type of glutamate receptor (glutamate is the brain's primary excitatory neurotransmitter), the N-methyl-D-aspartate (NMDA) receptor, and thereby abolish its glutamatergic response [3]. Clinical studies with NMDA receptor agonists and antagonists have solidified the importance of NMDA receptors in alcoholism and alcohol withdrawal: NMDA receptor antagonists (like ethanol itself) lessen alcohol withdrawal symptoms while NMDA receptor agonists increase the severity of these symptoms [4, 5]. Additional clinical studies have shown an increase in glutamatergic excitation during withdrawal [6], which is indicative of a larger than normal NMDA receptor response.

We hypothesize that the level of glutamatergic response from NMDA receptors is regulated during chronic alcoholism via the control of available, non-ethanol-blocked NMDA receptors. While blood alcohol levels are non-zero, ethanol molecules bind and block NMDA receptors, thereby decreasing the glutamatergic response. Experimental results suggest that there is compensatory control action to maintain steady glutamatergic response achieved either via generation of new NMDA receptors or via recruitment of NMDA receptors to the synapse (where neurotransmission occurs) from a nearby reserve pool. Upon alcohol withdrawal, however, the ethanol molecules dissociate from blocked NMDA receptors, producing additional unblocked receptors at the synapse, and resulting in a marked increase in glutamatergic excitation, which could explain many of the physiological and emotional effects observed clinically during alcohol withdrawal.

We have developed three separate computational models of this isolated hypothesized control system. All three models examine the effect of a typical chronic alcohol intake pattern on both NMDA receptor levels and glutamatergic response during alcoholism and alcohol withdrawal. In the first model, we include only new NMDA receptor generation via protein synthesis as the control action. In the second model, we include only receptor relocalization from a reserve pool to the synapse as the means of control action. Finally, in the third model, we include both mechanisms. In all three models, we are able to capture both the expected NMDA receptor dynamics during alcoholism and the expected glutamatergic increase during withdrawal. In this presentation, we will discuss the model development and parameterization, as well as the differences in sensitivity and stability characteristics of the three models. While we do not expect that NMDA receptor control alone completely describes the organism-wide effects of alcoholism and alcohol withdrawal, the widespread effects of glutamatergic regulation in the brain could explain many of the clinical symptoms, and future work with these models will include the incorporation of additional molecular details to obtain a more comprehensive model that can then be utilized to postulate alternative alcohol withdrawal treatments.

1. Rich, B., Ed. (1998). *The Dana Brain Daybook: What's New in Neuroscience.* 2(1).

2. Centers for Disease Control, Alcohol-Attributable Deaths Report, Average for United States 2001-2005, http://www.cdc.gov/Alcohol

3. Grant, K.A. and D.M. Lovinger (1995). Cellular and behavioral neurobiology of alcohol: receptor-mediated neuronal processes. *Clinical Neuroscience*, 3, 155-164.

4. Krupitsky, E.M., A.A. Rudenko, A.M. Burakov, T.Y. Slavina, A.A. Grinenko, B. Pittman, R. Gueorguieva, I.L. Petrakis, E.E. Zvartau, J.H. Krystal (2007). Antiglutamatergic strategies for ethanol detoxification: comparison with placebo and Diazepam. *Alcoholism: Clinical and Experimental Research*, 31(4), 604-611.

5. Krystal, J.H., I.L. Petrakis, G. Mason, L. Trevisan, D.C. D'Souza (2003). N-methyl-D-aspartate glutamate receptors and alcoholism: reward, dependence, treatment, and vulnerability. *Pharmacology & Therapeutics*, 99, 79-94.

6. Rosetti, Z.L., S. Carboni, F. Fadda (1999). Glutamate induced increase of extracellular glutamate through N-methyl-D-aspartate receptors in ethanol withdrawal. *Neuroscience*, 93(3), 1135-1140.

487f Blood Glucose Regulation with An Adaptive Model-Based Control Algorithm

Meriyan Eren-Oruklu[1], Ali Cinar[1], Lauretta Quinn[2], and Don Smith[2]. (1) Chemical and Biological Engineering, Illinois Institute of Technology, 10 West 33rd St., Chicago, IL 60616, (2) College of Nursing, University of Illinois at Chicago, 845 South Damen Avenue MC 802, Chicago, IL 60612

Diabetes is a disease characterized by degradation of insulin secretion from pancreas and consequently failure in blood glucose regulation in the body. All patients with type 1 diabetes and some patients with type 2 diabetes depend on exogenous insulin. The main objective of diabetes management is to reduce glucose variability and to keep the blood glucose concentration within tight control (normoglycemic range). Currently, insulin-dependent patients deal with this life-long problem by taking 3-7 daily fingerstick blood glucose measurements and 3-5 daily insulin injections or administering insulin with a manually controlled pump. Patients generally experience prolonged hyperglycemia or hypoglycemic episodes as it is a difficult task to decide on the required insulin amount/rate and correct timing of injection or bolus insulin infusion, with changing daily life conditions (e.g. diet, exercise, stress or illness). Closing the glucose control loop with a fully automated artificial pancreas will definitely improve quality of life for insulin-dependent patients. In this research, we propose an adaptive model-based control algorithm that keeps glucose concentrations within normoglycemic range and dynamically responds to disturbances like physiological or external glucose perturbations. The algorithm handles delays associated with insulin absorption, transient difference between plasma and interstitial glucose, and variations in inter/intra-subject glucose-insulin dynamics. The adaptability of the controller is further assured with subject-specific linear models developed from patient's continuous glucose monitoring (CGM) device data.

The performance of model-based control algorithms is highly dependent on the accuracy of the process-model used to represent the true dynamics of the system. In literature, model-based control strategies employing physiological glucose-insulin dynamics models have been previously proposed for closing the loop for insulin-dependent patients. Differently from the physiological models that are generally representative of only an average subject under specific disturbance-free conditions, we make use of frequently sampled glucose concentration data for development of patient-specific models that can dynamically adapt to inter-/intra-subject variations or disturbances like meal consumption or exercise. Time-series analysis is utilized to develop linear and low-order autoregressive moving-average models (ARMA) from patient's own continuous glucose monitoring data. At each sampling step, the model parameters are recursively tuned using the weighted recursive least squares (RLS) method. The RLS algorithm is further integrated with a change detection method to dynamically capture the unpredicted glucose fluctuations due to physiological or external perturbations and to provide a faster response under such conditions. The proposed glucose prediction strategy is then integrated with adaptive model-based control methods for closing the glucose control loop for patients with diabetes.

Depending on the site of glucose measurement and/or insulin delivery, one of the major challenges of the closed-loop glucose regulation is the variable time-delay introduced to the system due to insulin absorption into the bloodstream and time-lag between subcutaneous glucose and blood glucose concentrations. Subcutaneous space appears to be the minimally invasive site and therefore is the most preferred site by CGM devices and insulin pumps currently available in the market. Differently from many of the proposed glucose control algorithms in the literature that assume the minimally delayed intravenous site for either glucose sensing or insulin delivery (or both), we focus on the most delayed site (subcutaneous) for both glucose measurements and insulin administration. To cope with delays associated with insulin absorption, we introduce a Smith predictor like structure to the closed-loop algorithm. Even tough almost all of the available CGM sensors monitor glucose in the subcutaneous interstitial fluid rather than blood, management of diabetes is still defined in terms of blood glucose (not subcutaneous) by clinicians and therefore the standard choice for control performance is still the blood glucose regulation. Therefore, we also introduce a lag-filter to account for the time-lag between subcutaneous and blood glucose to the closed-loop structure.

In summary, variations in glucose-insulin dynamics are monitored by online identification of the model. At each step with the new coming subcutaneous glucose measurement, the current blood glucose concentration is estimated using the lag-filter. Model parameters are updated, and estimation of future time course of blood glucose is performed. These parameters are then used in the model-based control algorithm for calculation of the appropriate control action (insulin infusion rate). The required insulin infusion rate is determined by solving an optimization problem that minimizes the deviation of the predicted glucose values from a reference glucose trajectory. Two well-known model-based control strategies, generalized predictive control (GPC) and linear quadratic control (LQC) are used for control-law calculations, and are modified to include a delay compensator (Smith predictor like structure) associated with insulin absorption and a time-varying reference glucose trajectory.

Assuming subcutaneous glucose measurements, the closed-loop performance is evaluated on a simulated patient with type 1 diabetes and results are reported for subcutaneous delivery of rapid-acting insulin. The algorithm is able to keep blood glucose concentrations within normoglycemic range in presence of multiple-meal consumptions with simultaneous physiological glucose variations (e.g. changes in insulin sensitivity). The proposed closed-loop algorithm for blood glucose regulation provides robust control to a wide range of inter-/intra-subject variability of glucose-insulin dynamics and it further accounts for delays associated with insulin absorption or time-lag between plasma and interstitial glucose concentrations.

487g Human-Friendly and Control-Relevant Modeling for a Closed-Loop Artificial Pancreas

Hyunjin Lee and B. Wayne Bequette. Chemical and Biological Engineering, Rensselaer Polytechnic Institute, 110 8th st, Ricketts #201, Troy, NY 12180

A closed-loop artificial pancreas composed of a continuous glucose sensor, insulin infusion pump, and a control algorithm, has been the "holy grail" of engineering solutions for automated regulation of blood glucose in individuals with type 1 diabetes mellitus ([1],[2],[5],[6],[7],[9]). Model predictive control algorithms have the potential for improved performance compared to PID-based schemes, particularly if the insulin-glucose model is tailored to each individual. In this paper we address limitations to the current methods of model identification for model-based controller design, in both ambulatory subjects and those undergoing in-hospital clinical studies.

There is no shortage of mathematical models relating insulin and glucose dynamics (see [3],[6],[8], for example), but because of a large number of parameters these can be difficult to tailor to a specific individual and are hence more useful for in silico simulations of closed-loop behavior. As in the chemical process industries, the development of a good empirical model inherently requires a trade-off between keeping sufficient input excitation and low output variable deviations [7]. More frequent and large magnitude changes in the input (insulin delivery rate) result in more deviations of the output (glucose) away from the desired value. A methodology of "human-friendly" identification that accounts for desirable limits in changes in the insulin infusion rate and acceptable deviations from the glucose setpoint with shorter test duration is desirable.

In this paper we show that the meal consumption, insulin infusion and glucose sensor data from typical ambulatory care or clinical studies is generally not adequate for model identification. Type 1 diabetic subjects normally administer an insulin bolus at or near a meal intake in to avoid high postprandial glucose excursions, resulting in two input signals (meal carbohydrate content and magnitude of insulin bolus) that are highly correlated since they are injected at very close times. This often yields an insulin-glucose gain that is estimated to be positive even though it is clearly negative (insulin is administered for the purpose of reducing the glucose concentration). The strong correlation between input signals makes it impossible to estimate accurately the dynamics of insulin and meal to glucose under normal living conditions.

Introducing a time-delay possible between the meal and an insulin bolus can reduce the input correlation and yield improved models for controller design. A meal generates an impulse response in glucose concentration with a shape that depends on the meal content and size. A meal insulin bolus produces an impulse response, and a change in basal rate results in a step response. The main goal of this paper is to present how to design human-friendly and control-relevant modeling experiments, leading to suitable models for a closed-loop artificial pancreas. A series of single-input single-output experiments are designed to reduce the correlation by perturbing either meal or insulin, one at a time. We illustrate the model identification procedures by using simulation studies based on two common compartmental models for insulin-glucose dynamics [3],[6].

It is clear that using meal knowledge to provide rapid feedforward control action yields significantly better glucose control performance compared to feedback-only control [10], however it is known that over 65% of adolescents fail to provide meal-time insulin boluses two or more times each week, resulting in higher mean glucose concentrations and the increased probability of long-term complications. In addition, it is common for individuals to underestimate meal size by the order of 30%, resulting in insulin boluses that are too low and yielding higher than necessary postprandial glucose values. We have developed a meal detection [4], and meal size estimation strategy to provide the benefits of feedforward-type control without requiring individuals to specify that they are consuming a meal or providing the meal size. We compare the closed-loop performance of our estimation-based strategy with feedback-only and feedforward-feedback where the feedforward portion is missing on two meals per week. Our estimation-based strategy results in lower mean and maximum glucose values than the other two strategies.

Literature Cited

1. B. W. Bequette, "A critical assessment of algorithms and challenges in the development of a closed-loop artificial pancreas," Diabetes Technology & Therapeutics, vol. 7, no. 1, pp. 28–47, 2005.

2. R. Bellazzi, G. Nucci, and C. Cobelli, "The subcutaneous route to insulin dependent diabetes therapy," Engineering in Medicine and Biology Magazine, IEEE, vol. 20, no. 1, pp. 54–64, 2001.

3. C. Dalla Man, D.M. Raimondo, R.A. Rizza and C. Cobelli "GIM, Simulation Software of Meal Glucose-Insulin Model," J. Diabetes Science and Technology, vol. 1, no. 3, 323-330, 2007.

4. E. Dassau, B. W. Bequette, B. A. Buckingham, and F. J. Doyle III, "Detection of a meal using continuous glucose monitoring: Implications for an artificial beta-cell," Diabetes Care, vol. 31, no. 2, pp. 295–300, 2008.

5. F. J. Doyle III, L. Jovanovic, and D. Seborg, "A Tutorial on Biomedical Process Control. I. glucose control strategies for treating type 1 diabetes mellitus," Journal of Process Control, vol. 17, no. 7, pp. 572–576, 2007.

6. R. Hovorka, V. Canonico, L. J. Chassin, U. Haueter,M.Massi-Benedetti,M. O. Federici, T. R. Pieber, H. C. Schaller, L. Schaupp, T. Vering, and M. E. Willinska, "Nonlinear model predictive control of glucose concentration in subjects with type 1 diabetes," Physiological Measurement, vol. 25, pp. 905–920, 2004.

7. H. Lee and B.W. Bequette, "A closed-loop artificial pancreas based on MPC: human-friendly identification and automatic meal disturbance rejection," in Proceedings of the 17th IFAC World

Congress, Seoul, South Korea, 2008.

8. R.S. Parker and F.J. Doyle III: Control-relevant modeling in drug delivery. Adv Drug Deliv Rev 2001;48:211–228.

9. R.S. Parker, F.J. Doyle FJ III and N.A. Peppas: The intravenous route to blood glucose control. IEEE Eng Med Biol Mag 2001;20:65–73.

10. S. A. Weinzimer, G. M. Steil, K. L. Swan, J. Dziura, N. Kurtz, and W. V. Tamborlane, "Fully automated closed-loop insulin delivery vs. semi-automated hybrid control in pediatric patients with type 1 diabetes using an artificial pancreas," Diabetes Care, vol. 31, pp. 934–939, 2008.

498a A New Methodology for Integrated Planning and Scheduling Using Multi-Level Optimization

Zukui Li, Department of Chemical and Biochemical Engineering, Rutgers University, Piscataway, NJ 08854 and M.G. Ierapetritou, Chemical & Biochemical Engineering Department, Rutgers University, the State University of New Jersey, 98 Brett Road, Piscataway, NJ 08854.

The objective of integrated planning and scheduling is to coordinate the planning and scheduling decision levels to avoid infeasible operations and guarantee consistent plant operations. The main challenges in the integration of planning and scheduling is the fact that the problems are at different time scales and that the resulted optimization model includes a very large number of variables and constraints. Although significant effort has been made towards the integration of planning and scheduling problems, development of consistent planning and scheduling model and efficient solution strategy is still being identified as one of the challenges in enterprise-wide optimization (Grossmann 2005).

Two main methodologies have been proposed in the literature for the integrated planning and scheduling. The first type of methods follows the simultaneous approach where the detailed scheduling problem is solved for the entire time horizon. The scheduling model is incorporated into the planning problem, which leads to very large scale multiperiod optimization problem. Different solution strategies based on aggregation and decomposition has been studied in the literature (Schilling and Pantelides 1999, Grossmann, Van den Heever et al. 2002). The other type of methods follow the hierarchical approach where the overall problem is decomposed into upper level planning and lower level scheduling problem that can be decoupled. In this method the upper level aggregate model is solved to set the production targets resulting in an upper bound to the original problem followed by a detailed scheduling problem solution whish is individually optimized with fixed targets, thus decreasing the computational effort. The challenge lies in developing an aggregated model that yields tight bounds to reduce the number of upper and lower level problems (Wilkinson, Shah et al. 1995, Orcun, Altinel et al. 2001, Zhu and Majozi 2001).

In this work, we propose a new formulation of the integrated planning and scheduling problem which is a bilevel hierarchical problem with multiple lower level problems (scheduling subproblems in different periods and same time horizon). Based on this formulation we then develop an alternative solution methodology which combines the application of parametric programming and convexification technique to improve the computational efficiency but still maintain the quality of the solution for the planning and scheduling integration problem. It should be noticed that each lower level scheduling subproblem is independent from the other subproblems and they are only linked through the upper lever planning decision variables (production target, inventory level). Thus by treating the upper level decision variables in the lower level scheduling problem as parameters, all the lower level problems correspond exactly to the same multiparametric problems and give rise to the same parametric solution. Thus solving only one parametric scheduling problem, the characteristics of different periods are captured through the parametric solution composed by a set of parametric objective functions and critical regions (Li and Ierapetritou 2007), which are further used to reformulate the integrated planning and scheduling problem.

Different strategies for solving the reformulated integrated planning and scheduling problem are further studied in this work. The first strategy adopts a direct reformulation of the problem through the use of big-M constraints, which transform the logic constraints representing the critical regions through introducing additional binary variables and consequently a new MILP problem is formulated. The main disadvantage of this method is brought by the number of critical regions in the parametric solution. It is shown that the number of binary variables is equal to the product of planning periods times the critical regions. Comparing with original planning and scheduling integration model, the problem complexity will be reduced only if the critical regions are less than the binary variables in the scheduling model. To avoid this limitation, a second strategy based on convexification of the parametric solutions (Sun, Luo et al. 2007) is proposed given that the parametric objective function of the scheduling subproblem is nonsmooth monotone function and the boundary of the feasible region can also be described with a nonsmooth monotone function (Li and Ierapetritou 2007). Thus the supporting hyperplane method for convex problems (Slyke and Wets 1969) can be applied and a new formulation is derived which contains only continuous variables of the planning level decision variables. The scale of the original bilevel integrated problem is greatly reduced because the scheduling subproblems are substituted with a set of support hyperplane constraints to describe the objective function and the feasible region. Furthermore, the method ensures the consistency between planning and scheduling problem without sacrificing the solution accuracy. Applications of the proposed methodology on several examples illustrate the effectiveness of the proposed methodology.

[1] Grossmann, I. (2005). "Enterprise-wide optimization: A new frontier in process systems engineering." AIChE J. 51: 1846-1857.

[2] Grossmann, I. E., S. A. Van den Heever, et al. (2002). "Discrete optimization methods and their role in the integration of planning and scheduling." AIChE Symp. Ser. 98: 150.

[3] Li, Z. and M. G. Ierapetritou (2007). "Process scheduling under uncertainty using multiparametric programming." AIChE J. 53: 3183-3203.

[4] Orcun, S., I. K. Altinel, et al. (2001). "General continuous-time models for production planning and scheduling of batch processing plants: Mixed Integer linear program formulations and computational issues." Comput. Chem. Eng. 25: 371-389.

[5] Schilling, G. and C. C. Pantelides (1999). "Optimal periodic scheduling of multipurpose plants." Comput. Chem. Eng. 23: 635-655.

[6] Sun, X. L., H. Z. Luo, et al. (2007). "Convexification of Nonsmooth Monotone Functions." J. Opt. Theor. Appl. 132: 339-351.

[7] Wilkinson, S. J., N. Shah, et al. (1995). "Aggregate modeling of multipurpose plant operation." Comput. Chem. Eng. 19: S583-S588.

[8] Zhu, X. X. and T. Majozi (2001). "Novel continuous-time MILP formulation for multipurpose batch plants. 2. Integrated planning and scheduling." Ind. Eng. Chem. Res. 40: 5621-5634.

[9] Slyke R. V. and R. J. Wets (1969). "L-shaped linear programs with applications to optimal control and stochastic linear programming", SIAM J. Appl. Math.,17: 638-663.

498b Integrated Operational Planning and Medium-Term Scheduling of a Large-Scale Industrial Batch Plant Under Uncertainty

Peter M. Verderame and Christodoulos A. Floudas. Department of Chemical Engineering, Princeton University, Dept of Chemical Engineering; A215 Engineering Quadrangle, Princeton, NJ 08544

The operational planning and the medium-term scheduling of a chemical plant are interrelated activities dealing with the allocation of plant resources. Operational planning typically occurs over a time horizon of between one to three months and determines the daily production targets for the plant in question. Medium-term scheduling occurs over a time horizon of between two to four weeks and deals in greater detail with the allocation of plant resources such as process units. Given the production targets supplied by the operational planning model, the medium-term scheduling model attempts to determine when certain units should be utilized and the sequencing of these units. In order to ultimately generate a final production profile which is immune to uncertainty in demand, price, and processing time, a planning and scheduling framework needs to explicitly model uncertainty.

The effective integration of operational planning and medium-term scheduling under uncertainty has proven to be a formidable task. Shah [1] and Kallrath [2] both presented excellent reviews highlighting the challenges surrounding the integration of planning and scheduling. Floudas and Lin [3,4] outlined the various objectives and modeling aspects surrounding the scheduling of chemical plants, and Li and Ierapetritou [5] have presented a review of process scheduling under uncertainty. The main approaches for optimizing under uncertainty are stochastic programming, fuzzy programming, stochastic dynamic programming, and robust optimization. The work of Lin et al. [6] and Janak et al. [7] has laid the foundation for a robust optimization approach toward addressing uncertainty within the field of planning and scheduling.

In order to address the objective of providing a production profile which is immune to various forms of uncertainty, the integrated operational planning and medium-term scheduling framework developed by Verderame and Floudas [8] has been extended in order to take into account uncertainty at both the planning and scheduling level using in part the techniques presented by Lin et al. [6] and Janak et al. [7]. The novel Planning with Production Disaggregation Model, as well as the medium-term scheduling model developed by Janak et al. [9] have been reformulated into their respective robust counterparts for both bounded uncertainty and uncertainty with known distribution. The rolling horizon framework, which allows for the two-way interaction between the planning and scheduling levels, also has been modified so as to take into account the presence of uncertainty. It will be demonstrated through an industrial case study of a large-scale, multiproduct and multipurpose batch plant capable of producing hundreds of different products that the extended planning and scheduling framework is capable of providing a daily production profile which is immune to various forms of uncertainty.

[1] Shah, N. Process Industry Supply Chains: Advances and Challenges. Comp. Chem. Eng. 2005, 29, 1225.

[2] Kallrath, J. Planning and Scheduling in the Process Industry. OR Spectrum, 2002, 24, 219.

[3] Floudas, C.A.; Lin, X. Continuous-time versus Discrete-time Approaches for Scheduling of Chemical Processes: A review. Comp. Chem. Eng. 2004, 28, 2109.

[4] Floudas, C.A.; Lin, X. Mixed Integer Linear Programming in Process Scheduling: Modeling, Algorithms, and Applications. Annals of Operations Research. 2005, 139, 131.

[5] Li, Zukui; Ierapetritou, M. Process scheduling under uncertainty: Review and Challenges. Comp. Chem. Eng. 2008, 32, 715.

[6] Lin, X.; Janak, S.L.; Floudas, C.A. A new robust optimization approach for scheduling under uncertainty: I. Bounded uncertainty. Comp. Chem. Eng. 2004, 28, 1069.

[7] Janak, S.L.; Lin, X.; Floudas, C.A. A new robust optimization approach for scheduling under uncertainty: II. Uncertainty with known probability distribution. Comp. Chem. Eng. 2007, 31, 171.

[8] Verderame, P.M.; Floudas, CA. Integrated Operational Planning and Medium-term Scheduling of Large-Scale Industrial Batch Plants. Ind. Eng. Chem. Res. Accepted for Publication.

[9] Janak, S.L.; Floudas, C.A.; Vormbrock, N. Production Scheduling of a Large-Scale Industrial Batch Plant. I. Short-Term and Medium-Term Scheduling. Ind. Eng. Chem. Res. 2006, 25, 8234.

498c Bi-Level Planning for Profit Increment In Petroleum Industry

Jian Zhang and Qiang Xu. Department of Chemical Engineering, Lamar University, Beaumont, TX TX 77710

Petroleum refining and petrochemical industries are facing more challenges than ever before. Crude oil price is continuously rising while the quality is becoming low. Although the products' price can be increased to some extent, however, the major petroleum products, such as gasoline and diesel, have to meet more rigorous quality specifications at the same time. To efficiently make profits under these challenges, faster and more accurate planning and scheduling models for both incorporation and plant levels are urgently needed. In this paper, an integrated planning model for maximum profit making has been developed. The internal interactions between incorporation and plant wide models are also discussed.

A plant-level planning model forecasts detailed raw material requirement, process and product management in the next few months. It is a multi-period mixed-integer nonlinear model. Multi-period means future plan is divided into several time intervals. The intervals in the near future describe detailed plans in a short time period to cooperate with scheduling. The intervals in the far future describe brief plans in a long time period. Mixed-integer features discrete information including unit minimal capacity requirements, crude oil shipping constraints and utility constraints. Nonlinearity suggests quality related information, which is including swing cut, whole plant quality tracking and nonlinear blending. The corporation-level planning model is the integration of over 20 plant-level models plus distribution network. It predicts corporation's crude oil trading, distribution of raw materials, intermediate products and final products.

Both corporation-level and plant-level models should be solved very quickly. Usually, users want to get 6 solution scenarios within one day, which suggests each scenario should be completed within 1 hour. Given averagely 5 times tuning for one scenario, a corporation-level model should be solved in 5~7 minutes, which is too stringent. Thus, plant-level planning models should be simplified before the integration. To avoid the possible discrepancy between corporate and plant level models, a simplified methodology imbedded with many methods and heuristic rules is presented in this paper.

The proposed bi-level planning model system has been applied in a large petroleum corporation in Asia and has brought tremendous profit.

498d Cutting Planes and Decomposition Techniques for Solving Large-Scale Milp Model for Petroleum Supply Operations

Roger Rocha, Logistic Division - Research and Development Center, PETROBRAS, Av. Horácio de Macedo 950, Ilha do Fundão, Rio de Janeiro, Brazil, Ignacio E. Grossmann, Center for Advanced Process Decision-making, Dept of Chemical Engineering, Carnegie Mellon University, 5000 Forbes Avenue, Pittsburgh, PA 15213, and Marcus V. S. Poggi de Aragão, Dept. of Informatics, Pontific Catholic University – PUC-Rio, R. Marques de São Vicente 225, Rio de Janeiro, Brazil.

We consider a problem faced by PETROBRAS. This problem represents an important link in its petroleum supply chain optimization as it is responsible for the integration between the strategic and the operational decision levels. A brief description of the petroleum supply chain at PETROBRAS is as follows: Petroleum can either be locally produced or imported from aboard. Domestic petroleum production comes mostly from offshore platform, and it is shipped to maritime terminals either by tankers or pipelines depending on the infrastructure available. After reaching terminals, petroleum can be transferred to refineries by pipelines or exported via tankers. At the refineries, petroleum is processed in crude distillation units on daily scheduled production campaigns.

The problem tackled in this work consists of: given a set of production platforms; a set of terminals; a set of refineries; a set of tanker classes; the daily oil production of each platform; the consumption campaign of each crude distillation unit in refineries; the logistic infrastructure restrictions, namely, the storage capacity of production sites, terminals, and refineries, maximum pipeline flow rate between terminals and refineries, number and transportation capacity of tankers available in each tanker class, number of tankers that can operate simultaneously in each terminal. Determine, for a horizon of two months, a minimum cost offloading scheduling of each platform in order to supply each refinery with the correct type and amount of oil indicated by the strategic planning.

Our first approach to solve this problem was to build a straightforward fixed-charge network model relying on a daily discretization of the time horizon [1]. This model was important to point out the part of the model that deserves a special attention in terms of formulation, specifically, the offloading scheduling of platforms. We propose a reformulation of the inventory constraint at platforms that basically consist of projecting this constraint onto all variables but the stock variables. Even tough the original constraint and the reformulation are equivalent in terms of their Linear Programming relaxation, the latter is more suitable in an integer programming framework as it embodies a cascade knapsack constraint structure and can be better exploited by the solvers commercially available. In addition, we propose a splitting of offloading variables in order to impose a hierarchical branching on variables.

Despite the effort to tight the formulation, we observe that the lower bound obtained is quite weak due to the economies of scale, i.e., the cost per volume decreases as the ship capacities increase, hence the linear relaxation tends to use fraction of a possible bigger ship available instead of trying a smaller ship or a whole bigger ship. Studying the structure of this part of the problem, we devise some valid inequalities reasoning on the minimum and maximum storage capacity of the each platform. The results motivate us to take a step further and develop some tighter inequalities. Firstly, we can show that Chvatal inequalities rank 1 [2], which can be derived directly from the inventory constraint reformulated, dominate the valid inequalities. Moreover, using PORTA[3] we observe that the facet of the polyhedron of integer points associated with the inventory constraint at the platforms can be obtained by lifting the former valid inequalities proposed. In order to obtain lifted valid inequalities, we develop an approximate algorithm that combine inequalities derived for each class of ship. This last idea helps to speed up the solution process at the cost of the addition of a few more inequalities.

Another solution approach followed in this work is the decomposition algorithm. This problem can be viewed as the interaction between of an offloading scheduling platform problem and a refinery supply scheduling problem. The literature has been reported in the last two decades successful results using decomposition algorithms to solve large scale problems, among the most cited technique we can mention Lagrangean and Benders Decompositions [4,5]. In this work we propose a new decomposition algorithm that resembles in some aspects the Benders decomposition algorithm. It is well known in the literature that generate Benders cuts from an integer sub-problem remains a challenge and recently some promising ideas have been presented to overcome this difficult. Our decomposition algorithm relies in the idea of decomposing the problem by copying variables, and instead of generating cuts using some straightforward ideas, e.g., Combinatorial cuts (no-good constraints) or Benders cuts, we apply disjunctive programming theory to derive them. We prove that this approach of generating cuts is most effective than the Combinatorial cuts and preliminaries results show that this method is a viable alternative especially when the instance size grows.

Our algorithm was tested on test instances of the problem involving 10 platforms, 9 tanker types, 8 maritime terminals, 11 refineries, and 20 distillation units over a time horizon of 72 discretized intervals. The typical instance size is approximately 27,000 binary variables, 56,000 continuous variables, and 42,000 constraints.

References

[1] Aires, M., Lucena, A., Rocha, R., Santiago, C., Simonetti, L. (2004). Optimizing the Petroleum Supply Chain at PETROBRAS. In Proceedings of the ESCAPE 14. Lisbon, Portugal.

[2] Chvatal, V. (1973). Edmonds polytopes and a hierarchy of combinatorial problems. Discrete Mathematics 4, 305-337

[3] T. Christof, A. LÄobel, PORTA - POlyhedron Representation Transformation Algorithm,

http://www.zib.de/Optimization/Software/Porta/.

[4] Yan H. (1996). Solving some difficult mixed-integer programming problems in production and forest mmanagement. PhD. Thesis Dissertation, University of Pennsylvania.

[5] Binato S., Pereira M.V.F., Granville S. (2001). A new Benders Decomposition approach to solve power transmission network design problems. IEEE Transactions on Power Systems, 16, 235-40.

498e Simultaneous Batching and Scheduling In Multi-Stage Processes with Storage Constraints

Arul Sundaramoorthy and **Christos T. Maravelias**. Chemical and Biological Engineering, University of Wisconsin - Madison, 1415 Engineering Drive, Madison, WI 53706

Existing methods for the optimization of multi-stage batch processes consider the batching and schduling problems independently. Orders are first divided into batches based on the unit capacities, and these batches are used as input in the scheduling model. This sequential approach can lead to sub-optimal solutions. Furthermore, storage constraints are important in multi-stage batch processing: the existence of storage vessels decouples processing steps and enables better utilization of resources, e.g. storage before and after a bottleneck stage leads to higher production because it allows uninterrupted utilization of the most scarce resources. However, despite its importance, the problem of simultaneous batching and scheduling of multi-stage processes with storage constraints has not been addressed in the literature. The goal of this work is the development of a general framework for this problem generalized problem.

First, we study the different storage policies in the context of multi-stage processes. Interestingly, the modeling of storage constraints in multi-stage processes, where batch mixing and splitting is not allowed, is more challenging than in network-based formulations, where batches can be mixed and the number of states is fixed. In particular, storage in multi-stage processes depends not only on the size, but also on the number of storage vessels and batches. The number of batches, however, is an optimization decision which means that we cannot always determine a priori the storage policies of a given process. To address this challenge, we propose a general classification of storage policies in multi-stage processes based on the storage capacity (number and size of vessels) and timing constraints (limited/ unlimited/zero-wait). We use this classification to develop a formulation that accounts for all policies and their combinations.

Second, we develop a mixed-integer programming (MIP) formulation for the generalized problem we consider. Storage vessels are treated as pseudo-units for which assignment and sequencing constraints are expressed. While a batch has to be assigned to one of the processing units in a given stage, a batch can be transferred to a unit in the following stage directly. We use a precedence-based method to achieve sequencing of batches in units/vessels. Also, we sequence the batches assigned to the same vessel in intermediate stages, but allow the batches of the same order to share a storage vessel in the last stage as a finished product. Further, we account for storage time (in storage vessels) and waiting time (in processing units) through bounding constraints. We present how the proposed MIP model can be used to address all the classes of storage policies identified in our classification.

Third, we present several extensions. We extend our model to address sequence-dependent changeover times and costs. We discuss how the model can be simplified to address problems with fixed number and sizes of batches. We develop a class of tightening inequalities based on time windows.

Finally, we illustrate the significance of storage constraints in batching and scheduling of multi-stage processes using several examples. We show how storage constraints can impact both the objective and the batching decisions and illustrate how storage considerations for finished products affect the solution. We also show how standalone scheduling problems can be solved. We close with a computational study that shows that the addition of the tightening constraints reduces the integrality gap and enhances the solution process.

498f Refinery Planning Considering Financial Risk Management Using Crude Oil Derivatives

Jeongho Park[1], Sunwon Park[1], Young Kim[2], and Choamun Yun[2]. (1) Department of Chemical and Biomolecular Engineering, KAIST, 373-1 Guseongdong, Yuseonggu, Daejeon, 305-701, South Korea, (2) Department of Chemical and Biomolecular Engineering, Korea Advanced Institute of Science and Technology, Daejeon, South Korea

The profit of refinery sensitively fluctuates as crude oil price changes. Uncertain prices and demands of products also affect the profit. The refinery cannot control these exogenous market conditions. So the refinery must actively manage the financial risk. Crude oil derivatives are used for financial risk management. Futures or options help the refinery reduce the financial risk because they can buy the crude oil at the pre-determined price.

In this research we make a risk management model using contracts in the planning of refinery. First, we maximize the profit of a hypothetical refinery based on two-stage stochastic programming. For Monte Carlo simulation, various scenarios are generated with the assumption that prices and demands follow geometric Brownian motion. Then prices of futures and options are modeled with the stochastic process. The optimal number of contracts varies depending on decision maker's objectives. Lastly we draw cumulative risk curves to compare risk profiles of three different objectives.

534a Enterprise-Wide Data Management Connects People

Gregory M. Banik, Infomatics Division, Bio-Rad Laboratories, Two Penn Center Plaza, Suite 800, 1500 John F. Kennedy Boulevard, Philadelphia, PA 19102

Optimizing enterprise operations by deploying system wide infomatics.

Rapid, Global Communication of Company Policies and Standards

Jeremiah O'Brien
Process Data Control Corp
Arlington, Texas

Richard Wiand
Process Data Control Corp
Arlington, Texas

Introduction

The challenge of "getting everyone on the same page" is a significant one for companies that operate globally, due to such factors as distributed office locations, language differences, and technology constraints. At the same time, studies show that successful companies adhere to a common vision and core set of values enterprise-wide, from "branding" of the company's image, to key business processes and procedures. Many companies are challenged to find new ways with which they can galvanize the creativity and productivity of their workforce to pursue company goals that employees share and support.

Company standards and policies have traditionally served the function of communicating new corporate programs and initiatives to the staff. To be effective, standards and policies rely on a shared culture in the workforce. Information management (IM) systems are becoming an integral part of this cultural landscape, partly because they provide the physical means to exchange information between management and staff, and partly because they are a ubiquitous part of the shared experience of company workers. IM systems also cross boundaries and hierarchies of management that makes them uniquely suited to serving purposes beyond the objectives for which they were designed.

This paper discusses an approach to developing IM systems whose mission is related to task management. In the experience of the authors, this approach promotes common vision, understanding, and action, in addition to improving efficiency in both the initial development and ongoing maintenance of task management systems.

Reasons for Dysfunctional or Sub-optimal IM Systems

Deploying new IM systems that meet requirements and support an organization's goals has proven to be an elusive objective, and the likelihood of success does not correlate positively with company size. Some systems fail because of strategic or tactical deficiencies, but others are held back by disconnects between work processes and procedures, and the supporting elements of the system. Because computers have transitioned over the past two decades from being a helpful resource, to being an indispensible member of most work teams, our ability to interact effectively with them is mission-critical.

IM system developers can improve user acceptability by following two suggestions. First, user needs must be defined along with user requirements. Requirements can be reduced to "punch lists" and prioritized as to importance to the system and to users. Needs are more complex, and are likely to defy simplification or reduction, such as to a list of criteria. Second, both user needs and requirements must be integrated with IM processes in a fundamental and intuitive way.

The problem appears to be that organizations, and in many cases, the IM developers themselves, fail to see the importance of understanding the needs of users for a "common language." Consequently, both the systems and data are predicated on meeting functional requirements, rather than addressing user needs for working with recognizable processes and procedures. Both Windows and web page development tools such as ASP.NET have greatly improved our ability to create colorful and interesting computer screens, menus, and other functional components. However, the system as a whole – including data -- must be intuitive to earn the reputation of being "user-friendly," and most IM systems fall well short of this goal.

IM systems that are not viewed as being intuitive almost never succeed, at least not for very long, because they invariably give rise to unsatisfactory ratings and frustration on the part of the users. Users simply have too many examples of highly intuitive systems in their everyday use of the Internet to be content with anything less, and the excuse that business systems deal with more complex subjects than common websites tends to fall on deaf ears.

Purpose of Organizing Tasks

This section discusses why intuitive IM systems and data have been replaced with much more efficient, but often less intuitive and difficult to use systems. An alternate approach to organizing data is proposed, which is based on the concept of an operational activity, or "Ops Activity," as it is more commonly known, to communicate common work processes across a spectrum of IM systems functional components.

Functionality and Data

From a user's perspective, systems comprise both Functionality (i.e., automated work processes, such as filling in a form or searching in a list), and Data. It would be difficult to conceive of a business IM system that does not comprise both of these elements. Programmers often add a third layer, "Business Logic," when describing their system model; however, business rules and logic appear to be embedded in the Functionality of systems from the user's point of view.

In the parlance of knowledge management (KM), the functionality of an IM system exists to enable users to derive information from data, and, if sufficiently rich context exists, to evolve wisdom from this information. Information alone can fulfill numerous mission requirements; however, most companies find that expanding wisdom is the most reliable means to improving decision-making.

Using Intuitive Labels and Concepts to Organize Data

Ops Activities represent work processes and use the common sense identification to communicate meaning to users through the data itself, as opposed to using system features for this communication. The reason why intuitive labeling of data may not "come naturally" to system developers is because standard practices often lead them in the opposite direction.

For example, Equipment Identification Numbers (EINs) are used in many task management systems to uniquely identify plant equipment. In the simpler and less data-intensive IM systems used in the past, EINs could be descriptive of the items to which they applied while also serving the main purpose of providing a unique identifier to facilitate data searches. A storage tank, for example, may have been identified as "Tank-101," and a cooling tower as "CT-002." In current data management systems, such as SAP, equipment are often represented by a 20-digit or longer EIN whose sole function is to uniquely index the equipment list. Users must rely on the functionality of the systems or an external cross-reference list to deliver even the most basic information about the equipment.

Why have we moved from intuitive ways of identifying equipment to extremely precise, but non-intuitive approaches? The answer is that older database management systems were less efficient at indexing alphanumeric data than indexing pure numeric data. Today's computer systems are sufficiently advanced to handle indexing and searches in lists that contain mixed data types, and can therefore handle other methods of "packaging" data to be more intuitive to users. Self-explanatory labels for data elements such as Ops Activities can facilitate communication of knowledge management (KM) elements across the enterprise. These elements may then be leveraged to integrate other work processes and enhance IM systems.

Ops Activities are self-explanatory and capable of leveraging work across a broad cross-section of business practices. They are, in a word, "inclusive" of a spectrum of data management procedures and concerns, as opposed to being defined for use within the boundaries of an immediate use of the system. An analogy may be found in the field of sustainable design for buildings, a subject area that was spearheaded by popular author and practicing architect William McDonough.

In his groundbreaking book entitled, Cradle to Cradle: Remaking the Way We Make Things, Mr. McDonough argues convincingly for the benefits from a fundamental shift in design philosophy. In a recently published interview that appeared in Discover Magazine (October, 2008), he described his design process as "starting from a principles perspective [instead of] starting with goals or with tactics or with conventional practices." He goes on to discuss how this philosophy led his team to build the largest "living roof" in the world (10.4 acres) atop the Ford Motor Company assembly plant in Dearborn, Michigan. "We started from the idea of a healthy workplace," he explained, "and from that principle grew the question of how you would have healthy water... healthy soil... [and] healthy air."

IM systems are not typically conceived or designed with an organization's principles in mind; they are more commonly proposed to solve specific problems associated with processing data. However, until Mr. McDonough explored sustainability in his design practice, the same could be said of designing buildings. Ops Activities are responsive to a core principle in the IM systems that manage tasks and document task results; namely, an intuitive and flexible organization of work processes. The basic ideas of this approach may be transferrable to other types of IM systems as well.

Ops Activities Development Process

Ops Activities are predicated on the "common language" of users in the workplace and a universal understanding of what the terms in that language represent. The beginning point is an analysis of all business processes that occupy the time or attention of users – not the purpose of the activities, but the activities themselves.

Ops Activities Analysis attempts to define the functional elements of work process supply chains, and then define the most appropriate and widely understood terminology for data associated with those elements. Where "layers" of activities are found, as one frequently finds in complex work processes, the Ops Activities Analysis must also "nest" sub-activities under and within higher-level ones in the manner of an outline.

Step 1– Define Work Elements

Table 1 contains a list of the unique regulatory requirements that apply to designing, installing, operating and maintaining a closed vent system (CVS). This list is a subset of federal CVS requirements, which constitutes only a subset of all rules that apply to a CVS. However, as an illustration of how an Ops Activity Analysis is performed, this "subset of a subset" of rules will suffice.

Table 1 consists of two columns, one showing the citation of the rule that contains the requirement, and the other showing the regulatory text of each rule. A listing such as this is the first step in developing an Ops Activity Analysis, however, unlike Table 1, the scope of the tasks listed should be as broad as possible. The idea is to capture all requirements and explain what is known about the responsibilities of company personnel to meet those requirements.

Step 2 – Identify Tasks

Table 2 consists of an abbreviated version of the CVS requirements shown in Table 1, plus two new columns. One column shows the description of a task which will satisfy the requirement when performed as scheduled. The other column consists of a brief description of the Ops Activity that best describes the action(s) of all tasks that share the same Ops Activity (note: these are examples only).

The level of detail in defining tasks and Ops Activities depends on the scope and complexity of the business areas being studied. For example, a company that operates a chain of general automotive repair shops might conduct an Ops Activity Analysis to improve its business practices, in which the major business components likely to be defined would include oil changes, front end alignments, and brake jobs. A company that specializes in providing oil changes, on the other hand, may consider all major business components to be oil change related, such as removing the oil drain plug, replacing the oil filter, and adding new oil. In the case of a company specializing in analyzing used lubricating oil to assess engine condition, a common practice with aircraft and certain industrial machinery, major work elements would be categorized differently from either of the first two companies, and so on.

With respect to CVS compliance, the work elements correlate with actions that personnel must take to properly design, operate, maintain, inspect, and keep records for, a CVS. Because the requirements are conveniently explained in the applicable rules, we can move quickly past the requirements analysis and move directly to task definitions, i.e., what tasks need to be performed by plant personnel to comply with CVS requirements? Experience has shown that task descriptions communicate most effectively when they begin with a verb and are followed as closely as possible by the subject of the verb. Additional modifiers and supporting data should follow this verb-subject pairing as needed, as shown in Table 2.

Step 3 – Assign Tasks to Ops Activities

The last column of Table 2 contains the Ops Activity that was assigned to each of the listed tasks. More than a subject matter categorization, the three Ops Activity definitions shown in Table 2 summarize, in operational terms, all of the tasks to which they have been associated. A 30-character label has been found to be the ideal length of an Ops Activity, i.e., long enough to differentiate Ops Activities even in extremely long lists of tasks, yet concise enough to provide a name, as opposed to a description.

Ops Activities are not detailed enough to inform plant personnel of all steps needed to complete a work assignment. They are intended only to provide a mechanism for grouping tasks into higher assemblies. An average-sized process industry plant may have tens of thousands of equipment-specific tasks involving monitoring and inspections, reporting, testing, recordkeeping, and other types of work processes. Hence, the very act of categorization creates opportunities for more efficient management and reporting of data.

By way of example, the plant whose data was used to develop Table 2 is required to perform about 22,000 equipment-specific compliance tasks under its applicable requirements. These tasks represents 3,600 unique (i.e., not equipment-specific) tasks, attesting to the fact that some rules require action at all equipment of a certain type (e.g., storage tank inspections). The Ops Activity Analysis for this plant categorized 3,600 unique tasks into 46 Ops Activities.

Ops Activities are not inscrutable identification numbers or obscure index labels. They are self-explanatory labels that represent and summarize a large number of tasks. The utility of developing Ops Activities is discussed in the next section.

Table 1 Selected Federal Requirements (40 CFR Part 61) for Closed Vent System Requirements	
Citation No.	Regulatory Text
61.349	For each closed-vent system and control device used to comply with standards in accordance with §61.343 through §61.348 of this subpart, the owner or operator shall properly design, install, operate, and maintain the closed-vent system and control device
61.349	Vent systems that contain any bypass line that could divert the vent stream away from a control device used to comply with the provisions of this subpart shall install, maintain, and operate according to the manufacturer's specifications a flow indicator
61.349	Repair shall be completed no later than 15 calendar days after the emissions are detected or the visible defect is observed.
61.349	The visual inspection shall include inspection of ductwork and piping and connections to covers and control devices for evidence of visible defects such as holes in ductwork or piping and loose connections.
61.349	Each closed-vent system and control device shall be visually inspected initially and quarterly thereafter.
61.356	For each closed-vent system complying with paragraph (a) of this section, one or more devices which vent directly to the atmosphere may be used on the closed-vent system provided each device remains in a closed, sealed position during normal operations ex
61.356	Specifications, drawings, schematics, and piping and instrumentation diagrams prepared by the owner or operator, or the control device manufacturer or vendor that describe the control device design based on acceptable engineering texts.
61.356	A description of the operating parameter (or parameters) to be monitored to ensure that the control device will be operated in conformance with these standards and the control device's design specifications and an explanation of the criteria used for selection.
61.356	Dates of startup and shutdown of the closed-vent system and control device.
61.356	A statement signed and dated by the owner or operator certifying that the closed-vent system and control device is designed to operate at the documented performance level when the waste management unit vented to the control device is or would be operating

Table 2 Tasks and Ops Activities for Selected for Closed Vent System Requirements			
Citation No.	Basis of Task (i.e., Rule Text)	Task Description	Ops Activity
61.349	For each closed-vent system and control device used to comply with standards...	Design, install, operate, and maintain CVS & control device in accordance with following requirements if used to comply with standards in accordance with 61.363-61.348.	CVS Installation & Maintenance
61.349	Vent systems that contain any bypass line that could divert the vent stream...	Install, maintain, and operate according to manufacturer's specifications flow indicator that provides record of vent stream flow away from control device at least once every 15 minutes.	CVS Installation & Maintenance
61.356	For each closed-vent system complying with paragraph (a) of this section...	May use one or more devices which vent to atmosphere provided each device remains closed, sealed during normal ops except when needs to open to prevent physical damage.	CVS Operations and Records
61.356	Specifications, drawings, schematics, and piping and instrumentation...	Maintain design analysis that includes specifications, drawings, schematics, and piping & instrumentation diagrams, prepared by owner; or control device manufacturer or vendor that describe control device design.	CVS Operations and Records
61.356	A description of the operating parameter (or parameters) to be monitored...	Maintain record of description of operating parameter(s) to be monitored to ensure that control device will be operated in conformance with these standards & control device's design specifications.	CVS Operations and Records
61.356	Dates of startup and shutdown of the closed-vent system...	Maintain records of dates of startup & shutdown of closed-vent system & control device.	CVS Operations and Records
61.356	A statement signed and dated by the owner or operator certifying that the closed-vent system and control device is designed to operate ...	Maintain statement signed and dated by owner/operator certifying that CVS & control device is designed to operate at documented performance level when waste manage unit vented to control device is op at highest capacity expected to occur.	CVS Operations and Records
61.349	Repair shall be completed no later than 15 calendar days after...	Complete repair of CVS/control device no later than 15 calendar days after emissions are detected or visible defect is observed.	CVS Tests, Inspection & Repair
61.349	The visual inspection shall include inspection of ductwork and...	Visually inspect CVS ductwork, piping and connections to covers & control device, for evidence of visible defects such as holes in ductwork and loose connections.	CVS Tests, Inspection & Repair
61.349	Each closed-vent system and control device shall...	Visually inspect each closed-vent system and control device initially and quarterly thereafter.	CVS Tests, Inspection & Repair

Step 4 – Define the Primary Tasks for Ops Activities

Ultimately, human effort (or, in some cases, machine effort) will be needed to take action on the tasks that have been defined and categorized by Ops Activity. The need to document the fact that prescribed actions were taken often leads compliance managers to using task groups to streamline recordkeeping as well as gain acceptance from users. Process industry personnel, especially operators, do not have enough hours in their work day to validate each and every compliance task on an equipment-specific basis. Tasks must be summarized to reduce the number of required confirmations of task completion by users.

In the case of compliance assurance systems, the level of detail of compliance records, regulatory sufficiency of those records, and user acceptability, are all major objectives. Ops Activities exist to provide a definition for the tasks summarization level. "Primary Tasks" is the term that identifies operational deployment of Ops Activities for recordkeeping. All tasks grouped under an Ops Activity are members of a task group whose Primary Task is identified by the same Ops Activity. Although the grouping of tasks may be changed from time to time, or in certain situations, Ops Activities are useful because they define a default schema for organizing tasks. This is the reason why Ops Activities are assigned to unique tasks instead of equipment-specific tasks.

Table 3 shows the same tasks and Ops Activities that were shown in Table 2, but it also includes Primary Task definitions in the columns that have been added on the right side of the chart. Specific attributes of Primary Tasks that are recommended to be considered in an Ops Activity Analysis include:

- Primary Task Description – a long description of the kinds of actions that the task group comprises
- Task Frequency – how often task "history" data is planned to be updated
- Responsible Person – who will be expected to update task "history" data
- Task Group Level –the level at which task groups will be created for recordkeeping, i.e., equipment-level, process unit, or site-wide.

Step 5 – Review Use and Maintainability of Ops Activities

One principle of an organization that is pertinent to Ops Activities is increasing staff efficiency and effectiveness. Employees generally react favorably to empowerment, and the bonds of trust that are formed with management can lead to a more productive corporate culture. Task management systems are usually deployed at least in part to increase accountability; any factors that empower staff to accept greater accountability for their work are likely to have a positive impact on the corporate culture.

In addition to enabling intuitive recordkeeping and analysis of task completion records, Ops Activities can also empower staff to efficiently maintain their task lists. The IM system must provide capabilities for adding new Ops Activities, rearranging task members of existing Ops Activities, and modifying Primary Task information. To ensure that future maintenance is efficient, all adds and changes to the "official" Ops Activity list should be made centrally and deployed with little or no effort to all users.

<table>
<tr><th colspan="7">Table 3
Tasks and Ops Activities and Primary Tasks for Closed Vent System Requirements</th></tr>
<tr><th colspan="2">Citation No. and Task Description</th><th>Ops Activity</th><th>Primary Task Description</th><th>Task Fre-quency</th><th>Respon-sible Person</th><th>Task Group Level</th></tr>
<tr><td>61.349</td><td>Design, install, operate, and maintain CVS & CD in accordance with following requirements if used to comply with standards….</td><td rowspan="2">CVS Design and Maint-enance</td><td rowspan="2">Design, Install, and Maintain Closed Vent Systems as required by NESHAPS.</td><td rowspan="2">Semi-annual</td><td rowspan="2">Plant Engineer</td><td rowspan="2">Unit</td></tr>
<tr><td>61.349</td><td>Install, maintain, & operate according to manufacturer's specifications flow indicator that provides record of vent stream flow away from…</td></tr>
<tr><td>61.356</td><td>May use one or more devices which vent to atmos provided each device remains closed…</td><td rowspan="5">CVS Operations and Records</td><td rowspan="5">Comply with NESHAPS requirements concerning the design and operation of closed vent systems and control devices.</td><td rowspan="5">Monthly</td><td rowspan="5">Unit Super-visor</td><td rowspan="5">Equip-ment</td></tr>
<tr><td>61.356</td><td>Maintain design analysis that includes specifications, drawings, schematics, & piping & instrumentation diagrams prepared by….</td></tr>
<tr><td>61.356</td><td>Maintain record of description of operating parameter(s) to be monitored to ensure that CD will be operated in...</td></tr>
<tr><td>61.356</td><td>Maintain records of dates of startup & shutdown of closed-vent system &…</td></tr>
<tr><td>61.356</td><td>Maintain statement signed and dated by owner/operator certifying that CVS & control device is designed to operate….</td></tr>
<tr><td>61.349</td><td>Complete repair of CVS/control device no later than 15 calendar days after emissions are detected…</td><td rowspan="3">CVS Tests, Inspection & Repair</td><td rowspan="3">Inspect and operate closed vent systems and control devices as required, maintain records, and make timely repairs of malfunctions in accordance with applicable rules.</td><td rowspan="3">Quar-terly</td><td rowspan="3">Maint-enance Super-visor</td><td rowspan="3">Site-wide</td></tr>
<tr><td>61.349</td><td>Visually inspect CVS ductwork, piping and connections to covers & control device, for….</td></tr>
<tr><td>61.349</td><td>Visually inspect each closed-vent system and control device initially…</td></tr>
</table>

The IM system must also "broadcast" changes that were made to a central database to their facility and unit-specific data domains. When updating is done, detailed change logs are essential, especially when system records are being created to confirm compliance with regulatory requirements.

Summary

Tasks and task groups have become more common in recent years as the use of task management systems has grown. Several reasons are behind this trend, including new government rules, plant safety concerns, and quality improvement initiatives. As discussed in this paper, recordkeeping for environmental compliance has relied on task management systems for compliance assurance for many years. Other examples of corporate initiatives that involve task management systems include:

ISA/ANSI 84.00 Standards - Compliance with OSHA rules require plant engineers to maintain records of interlock trips, investigations, and follow-up actions, to demonstrate a company's adherence to prescriptive (e.g., NFPA 85) and performance (e.g., IEC) standards.

Operations Logs and "Routine Duties" - Many plants now require operators to log their completion of checklists, inspections, etc., often using mobile computing devices to make the transition to paperless a system.

Pharmaceuticals Industry Procedures - New federal rules require every step that is taken to produce an FDA-controlled product to be fully documented and literally signed-off on by managers.

Plant Maintenance - Task lists have been used to describe maintenance procedures for many years, but companies have only recently started keeping actual-performed maintenance records. Regulatory requirements associated with air emissions from maintenance, start-up and shut-down events have generated interest in actual-performed maintenance records at many process industry plants.

Sarbanes-Oxley - This federal law requires extensive documentation of corporate actions, procedures, and decisions, to ensure that shareholders and creditors received fair and balanced reporting of a company's potential liabilities.

Task management will continue to grow in many businesses because it is associated with numerous corporate goals and objectives, such as increased accountability and efficient oversight of operations. Ops Activities provide a method for organizing tasks into intuitively identifiable blocks to support tracking of results, staff training, and communication of responsibilities to company personnel. They have also been shown to enhance performance reporting and analysis, and simplify the maintenance of databases following changes in tasks and recordkeeping policies.

534c Net-Centric Integration of Manufacturing, Supply Chain, and Enterprise Management – Delivering the Process Industry's Equivalent to Sensor Fusion and the Battlefield of Tomorrow

Michael R. Basila, 17130 Deer Run Dr, BP Chemicals Company, Orland Park, IL 60467-8801

The Process Industry has led the application of digital manufacturing technology and its integration with Enterprise Management Systems since the early days of the computer. However, the promises of benefits from the full integration of manufacturing, supply chain, and enterprise management information have been elusive as the PC vapor ware applications of the 1980's. Boeing is currently implementing the Department of Defense Sensor Fusion and Battlefield of Tomorrow net-centric warfare initiatives, which give every soldier from the Theater Commander down to the individual grunts in the fields a real-time, 360o view of the battlefield and the tools necessary to manage the engagements. DOD is solving management and IT issues that closely parallel the manufacturing, supply chain, and enterprise management issues of the chemical manufacturing and refining industries. There are many examples of how companies have solved pieces of the enterprise data fusion puzzle from which we can extend a more holistic enterprise solution framed in terms of the DOD experience and current / near-term IT technologies.

534d Achieving Energy Reductions Via a Global Energy Management System

Flynn M. Collins, Energy CoE, ExxonMobil Chemical, 13501 Katy Freeway, Houston, TX 77079

For many years, ExxonMobil has undertaken voluntary actions to improve efficiency in our operations and in customer use of our products. Our Global Energy Management System (GEMS) is an important initiative that is having a positive impact at each of our refineries and chemical plants. The system builds on international best practices and benchmarking to identify energy efficiencies. Launched in 2000, it utilizes a common methodology to identify performance gaps, implement closure plans, sustain progress, and drive energy efficiency toward leading-edge performance.

534e Roundtable Discussion - Implementing Information Technology Applications to Support Global Organizations

Gary K. Stenerson, ExxonMobil Chemical, 4500 Bayway Drive, Baytown, TX 77520-9728

Authors and meeting participants discuss considerations in the design and implementation of information technology systems used to support global organizations. Discussion offers opportunity to ask questions and share experiences in how to effectively use information technology.

AUTHOR INDEX

AUTHOR INDEX

AUTHOR INDEX

AUTHOR INDEX

AUTHOR INDEX

AUTHOR INDEX

AUTHOR INDEX

AUTHOR INDEX

AUTHOR INDEX

AUTHOR INDEX

AUTHOR INDEX

AUTHOR INDEX

American Institute of Chemical Engineers
3 Park Avenue
New York, New York 10016-5991

CURRAN ASSOCIATES INC
proceedings.com

Computing and Systems Technology Division

Presentations at the 2008 AIChE Annual Meeting

Philadelphia, Pennsylvania
16 - 21 November 2008

Volume 2 of 2

ISBN: 978-1-61567-225-7

Printed from e-media with permission by:

Curran Associates, Inc.
57 Morehouse Lane
Red Hook, NY 12571

Some format issues inherent in the e-media version may also appear in this print version.

Printed by Curran Associates, Inc. (2009)

For permission requests, please contact AIChE
at the address below.

AIChE
3 Park Avenue
New York, NY 10016-5991

Phone: (203) 702-7660
Fax: (203) 775-5177

www.aiche.org

Additional copies of this publication are available from:

Curran Associates, Inc.
57 Morehouse Lane
Red Hook, NY 12571 USA
Phone: 845-758-0400
Fax: 845-758-2634
Email: curran@proceedings.com
Web: www.proceedings.com

TABLE OF CONTENTS

Volume 1

Volume 2

545a A Modified Network Component Analysis (NCA) Methodology for the Decomposition of X-Ray Scattering Signatures from Polymers

Ian Tolle[1], Yvonne A. Akpalu[2], and Lealon L. Martin[1]. (1) Department of Chemical and Biological Engineering, Rensselaer Polytechnic Institute, 110 8th St., Troy, NY 12180, (2) Department of Chemistry and Chemical Biology, Rensselaer Polytechnic Institute, 110 8th St., Troy, NY 12180

In this work we present an optimization-based methodology to identify network structure-process relationships in polymeric materials in order to provide a robust understanding of how processing conditions affect their structural domains. The approach is largely inspired by Network Component Analysis (NCA), originally developed by Liao and co-workers (2003) to infer unobservable phenomena using large multivariate data sets. The fundamental assumption in NCA is that the connective relationship between desired unobservable phenomena and measured data is based on a largely known bipartite network topology. However in many such analyses, this topology is either partially or completely unknown. To address this issue, the original NCA problem is reformulated as a mixed integer nonlinear program (MINLP). Optimal solutions of the MINLP formulation provide network topologies and physically meaningful component signatures that correspond to the best possible data reconstruction. We demonstrate this approach for the analysis of wide angle X-ray scattering (WAXS) data from a branched copolymer system. The copolymer samples vary in side chain length and isothermal crystallization temperature. Employing our analysis we isolate crystalline and amorphous components and observe their temperature-dependent variation. Reasonable agreement is achieved between the degree of crystallinity calculated by this NCA-based decomposition and the experimentally reported values.

545b Multi-Scale Modelling of Chinese Hamster Ovary Cell Cultures

Srinivas Karra, Chemical Engineering, Texas Tech University, Lubbock, TX 79409-3121 and **M. Nazmul Karim**, Chemical Engineering Department, Texas Tech University, Lubbock, TX 79409-3121.

Recombinant DNA technology has made possible the generation of high-valued proteins with therapeutic potentials. Sales of US$ 16 billion are expected for monoclonal antibodies alone in 2010. The complex structural nature of these therapeutic proteins and their requirement for post-translational modifications make mammalian cells a good host for complete expression of the proteins.

Mammalian cell culture reproducibility and optimal operation depends on various environmental and physiological factors; fragile and susceptible to shear stresses, inhibition of growth rate by several products of cell metabolism, sensitivity to process parameter (pH, temperature and dissolved oxygen tension) changes on cell metabolism, apoptosis, and variable nutrient consumption and energy requirements. Heterogeneous nature of the culture system due to the apportioning of cell cycle phases, further worsen this problem.

At this juncture, models of mammalian cell culture systems have a wide range of potential applications, such as analysis and prediction of experimental results, optimization of culture conditions, and perhaps most importantly, the investigation of fundamental metabolic processes and their subsequent elucidation, which will result in the design of more efficient mammalian cell culture systems both from a genetic engineering and process control perspective.

In this work, most of the predominant physiological phenomena that determines the intrinsic state of Chinese Hamster Ovary cells are modeled in an effort to unify them to estimate the onset of metabolism shift from available on-line and off-line measurements which may be noisy. Multi-level modeling is adapted to characterize the intra-cellular and inter-cellular processes in mammalian cell cultures. A hybrid single cell model is developed to quantify cell growth, death, lysis, nutrient uptake, metabolite and protein production and their dependency on various environmental and physiological factors. Population balance models are combined with this hybrid model to characterize various phases in the cell cycle and in turn account for the heterogeneity in the cell populations.

545c A Systematic Multiscale Modelling Framework for Product-Process Design and Development

Ricardo Morales-Rodríguez and Rafiqul Gani. CAPEC, Department of Chemical and Biochemical Engineering, Technical University of Denmark, Søltofts Plads, Building 229, 2800 Kgs. Lyngby, Denmark

The design, development and manufacture of a product and its process need to be consistent with the end-use characteristics of the desired product. Also, in many cases involving structured formulated chemical products, the end-use properties are a function of the microscopic properties that defines the structure of the product and/or process. Therefore, multiscale analysis of properties of the product-process is necessary. One of the common ways to achieve the desired characteristics of the product-process is through trial and error based experiments, which can be expensive and time consuming, but is also reliable. An alternative approach is the introduction of a systematic modelling framework where some of the time consuming and/or repetitive experimental steps are replaced by reliable model-based computational tools. To achieve this objective, it is necessary to have a flexible modelling tool as part of a model-based framework that allows systematic generation and analysis of product-process alternatives. For example, multi-scale features are needed when the properties defining the chemical structure and end-use characteristics of the product are dependent on parameters of different size and time scale. The advantages of the use of multi-scale modelling approach is that the design, development and/or manufacturing of a product-process can be described at different scales of length and time, providing thereby the knowledge of the applied phenomena at diverse degrees of abstractions and details.

In this paper, we propose the use of a systematic multiscale modelling framework for chemical product-process design and development. This framework allows the design of the chemical product-process assisted by computer-aided modelling tools to not only guide the user through the various stages of the desired product-process (related to the structured formulated chemical products), but also provides the different models (and their parameters) needed for the various model based calculations. The framework combines the product design and process methods & tools with modelling tools and integrates them with design templates (work-flow) for guiding the user through the design steps. For each step of the work-flow, data corresponding to the use of the models need also to be available. Therefore, the framework also includes a library of models and knowledge-base of pure component data, model parameters and product-process data. The work-flow starts with a specification of the characteristics and properties of the product to be designed. Based on this information, alternatives are generated (product design tool), which are then tested and evaluated to identify the chemicals and/or their mixtures that satisfy the desired product specifications (property models at various scales combined with database and product design tool). Next, modelling tools are employed to generate the appropriate performance models for simulation of the product behaviour and end-use characteristics.

The presentation will highlight the modelling framework in terms of the modelling test-bed (ICAS-MoT), the model library, the work-flow for product-process design and the corresponding data-flow. The performance of the systematic modelling framework for product-process design and development will be illustrated through a case study embedding the modelling and design issues related to the uptake and release of formulated pesticide and pharmaceutical products. With respect to process design and modelling, the simulation of the uptake/release of the active ingredient is performed, highlighting the use of multi-scale models. Through this case study, the advantages of the use of the model-based framework in formulated chemical product design will be illustrated.

545d Ab Initio Crystal Structure Determination Using Global Optimization

Angelo Lucia and Rajeswar R. Gattupalli. University of Rhode Island, Dept of Chem. Engineering, Kingston, RI 02881

Crystal structure determination is important to the understanding of the structure-function relationship in many branches of theoretical and applied chemistry. In this presentation, an ab initio global optimization method for crystal structure determination based on the terrain/funneling method is described. The global optimization formulation is comprised of a large constrained optimization problem, in which the objective function is an all-atom description of the potential energy and expressed in terms of the fractional coordinates of all atoms in the crystal and lattice parameters, and where the constraints remove translational and rotational symmetries and enforce bounds on variables. The key features of the proposed approach are that it is truly ab initio since random starting molecular and crystal configurations are used and the terrain/funneling method simultaneously adjusts both the conformation of each molecule as well as the packing of all molecules in the crystal. Numerical results for several crystal structure determination examples are presented that clearly show that the terrain/funneling method is a robust, reliable and efficient methodology for crystal structure determination, can handle polymorphism, solve large dimensional problems, and does not get trapped at local minima. Many geometric illustrations are used to elucidate key points.

545e Processmodeller. A Modelling Environment for Multi-Scale, Large-Scale Modelling

Heinz Preisig, Chemical Engineering, Norwegian University of Science and Technology, Trondheim, 7491, Norway

Scope

Modelling is not an art, it is absolutely possible to identify patterns, which can be generalised yielding a highly structured modelling methodology. Over the last years we have proven the effectiveness of the approach in industrial application and education. Since the last version of the modelling software Modeller has been publicised on the basis of which the high-tech company Mobatech has been established in the Netherlands, the project has progressed further and expands now into model reduction, thus enabling multi-scale modelling.

Presentation

The basic approach is presented extending on the handling of large-scale systems. This is then augmented with the associated mathematical representation of the plant as it is represented in the current theory definition component of the program environment. The discussion is extended to the mathematical background of the model assumption tool, which enables three types of order of magnitude assumptions all of which have an effect on the structure of the problem. Changing the structure changes the structure of the mathematical representation, which in turn often implies mathematical problems in the resulting equation structure. These problems have been identified and will be briefly discusses as well as the resolution of the problem. In all cases the result is a reduced model, thus the developed methodologies can be used for model reduction thus resulting in models characterized by a larger granularity. Change of granularity also implies a different time scale.

Recent Project-Related References:

1. Preisig, H A;Modelling: Compartmental Networks and Topologies - A Comparison with Bond Graphs;ESCAPE 14 (2004), Lisabon, Portugal, 16-19/05/2004, pp 1111-1116;2004

2. Preisig, H A;A Topology Approach to Modelling;SIMS 45, Copenhagen, Danmark, 24-25/09/2004; pp 413-420;2004

3. Preisig, H A & Westerweele, M R;Modeller--An interactive model editor for physical-chemical-biological models;FOCAPO 2003, Coral Springs, Florida, USA, 12-15 Jan 2003, paper 92; pp 531-534;2004

4. Preisig, H A;Gymnastic Exercises with Topologies Relating to Time-Scale Assumptions;ESCAPE 14 (2004), Lisabon, Portugal, 16-19/05/2004, pp 1105-1110;2004;

5. Preisig, H A;Computer-aided modelling: A study on the dynamic flash;ESCAPE 15 (2005), Barcelona, Spain, 29/6 - 1/6;2005

6. Preisig, H A & Haug-Warberg, T & Loevall, B T;On model portability;ESCAPE 16 / PSE 9, Garmisch-Partenkirchen, Germany; pp 483;2006

7. Preisig, H A;A Graph-Theory-Based Approach to the Analysis of Large-Scale Plants;ESCAPE 17, Bucharest, Rumenia, 2007/05/27-30; pp 81-86;2007

8. Preisig H A;Three Principle Model Reductions Based on Time-Scale Considerations;ESCAPE 18; Lyon, France; 2008/06/2-4;2008

Parameter set selection via clustering of parameters into pair-wise indistinguishable groups of parameters

Yunfei Chu and Juergen Hahn*
Artie McFerrin Department of Chemical Engineering
Texas A&M University
College Station, TX 77843-3122
hahn@tamu.edu

1. Introduction

Mathematical modeling has become an essential approach to understand the mechanisms and dynamics in a system. With advent of computing power and experimental techniques, modeling of large-scale complex systems is accessible and it has become a key component in various research areas. For example, in systems biology mathematical modeling has been a keystone (Kitano 2002). To study signal transduction pathways a number of large scale models have been developed (e.g. Schoeberl *et al.*, 2002; Yamada *et al.*, 2003; Singh *et al.*, 2006) which provide a fresh insight into the biological systems. The structure of a model is often built by mechanism analysis while the parameters are often required to be updated from experimental data. The usefulness of a model is highly dependent on the quality of the model. However accurate estimation of the large scale systems is still a challenge.

A great number of parameters contrasted with the limited experimental data results in the over-parameterized model and not all the parameters are identifiable in practice. If parameters are not practically identifiable then a small amount of noise in the data will result in large variations of the estimated value of the parameters and the parameters can not be estimated accurately (Walter and Pronzato, 1990). Estimation of an over-parameterized model is an ill-posed inverse problem and some regularization methods are required to guarantee the uniqueness and stability of the solution (Aster *et al.*, 2005). A widely used regularization is to select a subset of parameters to be estimated while all other parameters are fixed at a constant value. The question then becomes how to select the identifiable parameters to be estimated.

The identifiability of a set of parameters depends on the effect that changes in their values have on the output and the effect can be in turn characterized by the sensitivity vectors. Several methods for parameter selection based on sensitivity vectors have been proposed in the literature. These include, but are not limited to, a collinearity index method (Brun et al., 2001), a column pivoting method (Velez-Reyes and Verghese, 1995), an extension of the relative gain array (Sandink et al., 2001), a Gram-Schmidt orthogonalization method (Yao et al., 2003), a recursive approach based upon principal component analysis (Li et al., 2004) and a combination of Hankel singular value and singular value decomposition (Sun and Hahn, 2006). A systematic approach for parameter selection is based on optimality criteria computed from the Fisher information matrix. The inverse of Fisher information matrix provides a lower bound for the covariance matrix of parameter estimators (Walter and Pronzato, 1990) and it can serve as a measure for the quality of a parameter set. A subset of identifiable parameters can be selected based upon optimizing some criteria such as the D-optimality or the modified E-optimality criterion (Weijers et al., 1997; Brun et al., 2002). Parameter selection has been used in a variety of applications, ranging from ecological systems (Anh et al., 2006), power systems (Hiskens, 2001), production systems (Bastogne et al., 2007), chemical reactions (Kou et al., 2005), biochemical networks (Gadkar et al., 2005) to wastewater treatment processes (Sin and Vanrolleghem, 2007).

Generally by defining an objective function to measure the quality of a parameter set parameter selection can be formulated as a combinatorial optimization problem, however, it is nontrivial to find the solution. For small scale system the exhaustive search (Weijers et al., 1997) can provide the exact solution. However, even for medial scale system the total number of possible combination of parameters is too large to be enumerated. Stochastic search such as the genetic algorithm (Chu and Hahn, 2007) provides a way to select parameters for medial and large scale systems. Another approach is the stepwise methods which select parameters one by one and in each step a parameter is selected to optimize the criterion function. The orthogonalization method including householder transformation (Velez-Reyes and Verghese, 1995) and Gram-Schmidt procedure (Yao et al., 2003) can be regarded as an approach to maximize *D*-optimality criterion stepwisely (Chu and Hahn, 2007).

However, the existing methods all concentrate on search in identifiable parameter sets. This work provides a different avenue for parameter selection. The pairwise indistinguishable parameter sets are determined first, in which any two parameters can not be distinguished. Since the parameters belonging to the same indistinguishable set can not be estimated simultaneously the search of all combinations of parameters can be reduced to search of parameters from different indistinguishable sets. The number of possible combinations will be reduced dramatically.

2. Background

2.1 Identifiability

Parameter identifiability is an important issue in estimation and basically it is cast into two category: analytical (or structural, priori) identifiabilty and numerical (or practical, posteriori) identifiability (Walter 1987; Walter and Pronzato, 1997; Ljung, 1999). The analytical identifiabilty studies the uniqueness of solution in estimation while the numerical identifiability investigates the stability of the solution. The analytical identifiabilty can be also divided into two kinds: the global identifiability which guarantees the unique solution in a large range of parameter space and the local identifiability which guarantees the unique solution in a neighborhood of a given parameter value. It is easy to see the global identifiability includes the local identifiability however it is much more difficult to test. The common methods including differential algebra (Ljung and Glad, 1994), Taylor series approach and similarity transformation (Chappell *et al.*, 1990) are restricted to small scale systems.

The local identifiability is relatively easy to check. The condition is based on the rank of the sensitivity matrix. The local identifiability is related with this work and to make it clear the definition and the condition from Rothenbe (1971) are used.

Definition 1: A parameter point $\boldsymbol{\theta}_0$ is said to be locally identifiable if there exists an open neighborhood of $\boldsymbol{\theta}_0$ containing no other $\boldsymbol{\theta}$ which produces the identical observations $\mathbf{y}$.

Condition 1: Let $\boldsymbol{\theta}_0$ be a parameter point and the sensitivity matrix $\mathbf{S}(\boldsymbol{\theta}) = \partial \mathbf{y} / \partial \boldsymbol{\theta}^{\mathrm{T}}$ has constant rank in a neighborhood of $\boldsymbol{\theta}_0$. Then $\boldsymbol{\theta}_0$ is locally identifiable if and only if $\mathbf{S}(\boldsymbol{\theta}_0)$ is nonsingular.

It should be noted the condition that the sensitivity matrix has constant rank is necessary. If this condition is removed then the nonsingularity of the sensitivity matrix is just a sufficient condition for local identifiability, or, in another word, a rank deficient sensitivity matrix does not imply that the parameters are not locally identifiable. The condition of constant rank has to be checked analytically and only the value at a parameter point is not enough.

The analytical identifiability guarantees the existence of the unique solution at least in a small range. However the analytically identifiable parameters may not be estimated accurately in practice. If sensitivity matrix is not singular but is close to singularity, the noise in data will be amplified in the procedure of estimation and result in large variations in estimated parameter value. To obtain an accurate estimation it is also required that the sensitivity matrix should be far from singularity. However the term of 'far from singularity' has no unified definition and various functions are used to measure the distance to singularity of the sensitivity matrix. A set of them are the experimental optimality criteria presented in the next subsection.

2.2 Optimality criteria of the Fisher Information Matrix

The numerical identifiability can be measured by the parameter covariance matrix. If the covariance matrix is large then the parameters are not numerical identifiable. However, the covariance matrix can only be computed after the parameters have been estimated and it is also affected by the estimation algorithm. Alternatively, the Fisher information matrix can be used as its inverse provides the Cramer-Rao lower bound for the covariance matrix (Ljung, 1999).

Assume the measured output is a function of the parameters affected by measurement noise

$$\tilde{\mathbf{y}} = \mathbf{y}(\boldsymbol{\theta}) + \boldsymbol{\varepsilon}, \tag{1}$$

where $\tilde{\mathbf{y}} = [\tilde{y}(t_1), \cdots, \tilde{y}(t_n)]^{\mathrm{T}}$ is the observation of the output, $\mathbf{y}(\boldsymbol{\theta}) = [y(t_1, \boldsymbol{\theta}), \ldots, y(t_n, \boldsymbol{\theta})]^{\mathrm{T}}$ is the true value and $\varepsilon = [\varepsilon(t_1), \ldots, \varepsilon(t_n)]^{\mathrm{T}}$ is the measurement noise. In practice the measurement noise is often assumed to be normally distributed with zero mean and a covariance given by the matrix $\boldsymbol{\Sigma}=\sigma^2\mathbf{I}$. The Fisher information matrix, $\mathbf{F}$, is given by

$$\mathbf{F}(\boldsymbol{\theta}) = \left(\frac{\partial \mathbf{y}}{\partial \boldsymbol{\theta}^{\mathrm{T}}}\right)^{\mathrm{T}} \frac{\partial \mathbf{y}}{\partial \boldsymbol{\theta}^{\mathrm{T}}}. \tag{2}$$

Without loss of generality for the procedure, it can be assumed that $\sigma^2 = 1$.

To reduce the variations in estimated parameters it is desired to maximize the Fisher information matrix in some sense in order to reduce its inverse. To measure the size of the Fisher information matrix precisely a set of real function is defined. Such functions are called the experimental optimality criteria and named alphabetically (Kiefer, 1959). The most popular experimental optimality criterion is the *D*-optimality criterion which maximizes the logarithm of the determinant of the Fisher information matrix:

$$\varphi_D^* = \max \varphi_D(\mathbf{F}) = \max \log \det(\mathbf{F}). \tag{3}$$

This criterion minimizes the volume of the confidence ellipsoid with an arbitrary fixed confidence level for a least square estimator. Other common criteria (Kiefer, 1959; Atkinson et al., 2007) includes the *E*-optimality which maximizes the least eigenvalue of Fisher information matrxi

$$\varphi_E^* = \max \varphi_E(\mathbf{F}) = \max \lambda_n(\mathbf{F}). \tag{4}$$

and modified *E*-optimality which minimizes the condition number

$$\varphi_{ME}^* = \min \varphi_{ME}(\mathbf{F}) = \min \kappa(\mathbf{F}). \tag{5}$$

To optimize these criteria, the Fisher information matrix should be far from singularity. And from relation between the Fisher information matrix and the sensitivity matrix (Eq.2) it is equivalent to that the sensitivity matrix is far from singularity. The optimality criteria provide a set of functions to measure the distance to singularity of the sensitivity matrix for numerical identifiability test.

If the parameters are not all numerical identifiable a set of identifiable parameters are often selected for estimation. When a set of parameters selected the Fisher information matrix become

$$\mathbf{F}_{\mathbf{L}} = \mathbf{L}^{\mathrm{T}}\mathbf{F}\mathbf{L} = (\mathbf{S}\mathbf{L})^{\mathrm{T}}(\mathbf{S}\mathbf{L}). \tag{6}$$

where the selection matrix $\mathbf{L}$ is given by

$$\mathbf{L} = \begin{bmatrix} \mathbf{e}_{i_1} & \mathbf{e}_{i_2} & \cdots & \mathbf{e}_{i_m} \end{bmatrix}. \tag{7}$$

The set $\{i_1, i_2, \cdots, i_m\}$ denotes the selected parameters and $\mathbf{e}_i$ is the *i*-th column of the identity matrix. Then the problem of parameter selection becomes search of the selection matrix to let the Fisher information matrix far from singularity by optimizing some criterion function.

However, this is a combinatorial problem and not trivial to solve. A way to obtain an adequately good solution is the forward selection, which adds a new parameter into the selected set each step and the parameter added is determined by optimizing the criterion. One kind of forward selection is the Gram-Schmidt orthogonalization method (Yao et al., 2003). Using this technique, the first parameter selected has the largest norm of the sensitivity vector. In a second step the sensitivity vectors of other parameters are projected on to the space orthogonal to the first sensitivity vector. From the projected sensitivity vectors, the longest one is chosen and the corresponding parameter is selected as the second parameter in the set. The same procedure is repeated to select the remaining parameters until the number of parameters to be estimated is reached or until the length of the projected sensitivity vectors decreases below a certain threshold. Another version of orthogonalization method uses the Householder transformation for orthogonalization (Velez-Reyes and Verghese, 1995) which extends the linear version (Golub and van Load, 1989) to the nonlinear systems. The basic procedure is the same as the Gram-Schmidt procedure except there is a Householder mapping in each step. It has been shown that the orthogonalization method is in fact a stepwise method to maximize *D*-optimality criterion (Chu and Hahn, 2007). The forward selection can not guarantee to find the optimal solution however it can quickly find a suboptimal solution close to the optimal one. Because of it simplicity, the orthogonalization method has attracted a wide attention (e.g., Burth, 1999; Gadkar et al. 2005; Kou et al., 2005; Yue et al., 2006; Jaqaman and Danuser, 2006).

3 Investigation of pairwise indistinguishable parameter sets

If the sensitivity vectors of two parameters are not parallel then the parameters can not be reduced to one and they are distinguishable. However, if the sensitivity vectors are close to paralleling the two parameters will have very similar effects and in practice the difference in effects to distinguish them may be so tiny that they are overwhelmed by the noise in the data. In this case the parameters are said to be numerically pairwise distinguishable.

Before search for the numerically distinguishable sets a value to measure how close a sensitivity vector parallel to one another is required. Naturally the angle between two sensitivity vectors provides a way to measure the closeness of the vectors. A similarity measure of two parameters can be defined by the angle as

$$\cos\phi_{ik} = \frac{\left|\mathbf{s}_i^{\mathrm{T}}\mathbf{s}_k\right|}{\left\|\mathbf{s}_i\right\|_2 \left\|\mathbf{s}_k\right\|_2}, \tag{8}$$

where $\phi_{ik} \in [0, \pi/2]$ is the angle between sensitivity vector $\mathbf{s}_i$ and $\mathbf{s}_k$. The range of the similarity measure is from 0 to 1. When it equals 1, the two parameters have equivalent effect and the case returns to the analytical indistinguishability. When it equals 0, the sensitivity vectors of the parameters are orthogonal and the parameters have distinct effects. It is should be noted that the similarity measure is the absolute value of Pearson's correlation coefficient. The absolute value is because the parameters will have equivalent effects when their sensitivity vectors parallel to each other no matter if they are in the identical direction or in the opposite direction.

Based on the similarity measure, the parameters can be grouped by the clustering algorithms. Clustering is an unsupervised classification which partition a data set into subsets (groups), so that the data in each subset share some common trait. In this work the parameter set is partitioned by the parameter effects and the parameters in the same group have similar effects. The groups clustered are indeed the numerical indistinguishable required. There are a great number of clustering algorithms (Duda et al., 2001; Theodoridis and Koutroumbas, 2006) and the agglomerative hierarchical clustering is used.

This method builds the hierarchy from the individual parameter to the whole set by progressively merging groups. In the initial stage, each parameter is clustered into a one parameter group. At each middle stage two groups are merged into a new group and the two groups selected have the largest similarity measure. There are several approaches to define the similarity between groups based on the similarity between individual parameters. The approach of complete linkage is used which uses the least similarity value between the parameters in two groups as the similarity measure of the two groups. By this approach the least similarity value of parameters in a group can be controlled by choosing the number of groups. The fewer groups partitioned the smaller the least similarity value in a group.

Investigation of the pairwise indistinguishable sets reveals a relationship among parameters. The information of which parameters have the similar effects is very helpful to study the mechanism of a system. An important application is to select the identifiable parameter set. Parameter selection can be formulated as a combinatorial optimization problem to maximize some criterion function which measures the degree of identifiability of a set of parameters. A commonly used criterion function is the D-optimality criterion which is formulated as

$$\begin{aligned} & \mathbf{z}^* = \arg\max_{\mathbf{z}} \ \log\det\left(\mathbf{F}(\mathbf{z})\right) \\ \text{s.t. } & \mathbf{F}(\mathbf{z}) = \mathbf{FIM}^{(i_1,\cdots i_{n_s})}_{(i_1,\cdots i_{n_s})} \text{ with } i_j \text{ that } z_{i_j} = 1,\ j = 1\cdots n_s \\ & z_1 + z_2 + \cdots + z_{n_\theta} = n_s \\ & z_i \in \{0,1\},\ i = 1\cdots n_\theta \end{aligned} \quad (9)$$

The decision vector $\mathbf{z} \in \{0,1\}^{n_\theta}$ denotes whether a parameter is included in the selected parameter subset. If z_i=1 then θ_i belongs to the selected subset with the size of n_s. The matrix **FIM** is the Fisher information matrix of all parameters. $\mathbf{F}(\mathbf{z})$ is the Fisher information matrix of the parameters included in the selected subset and it is equal to the principal submatrix of **FIM** with the indices of the non-zero decision variables (the entries of column i_j and row i_k, j, $k = 1\ldots n_s$).

However, the solution to the combinatorial problem is not trivial. Even for medial scale system (e.g., 10 selected from 50) the total number of possible combination of parameters is too large to be searched exhaustively. Clustering parameters provides a new approach to solve the combinatorial problem for parameter selection. Search of which parameters can not be estimated together is performed before search of which parameters can be estimated simultaneously. Since the parameters in a pairwise indistinguishable set can not be estimated together there is no need to try their combinations in the search. Identification of indistinguishable sets inserts the new constraints to the optimization problem to reduce the possible number of combination

$$\sum_i z_{n_i} \leq 1, \quad (10)$$

where z_{n_i} represents if the i-th parameter in the n-th indistinguishable set is selected or not and in an indistinguishable set at most one parameter can be selected for estimation. Further, a representative parameter can be selected from each indistinguishable set and the parameters for estimation can be searched from the combinations of the representative parameters. Clustering parameters can be regarded as a 'dimensional reduction' method. It can reduce the number of possible combination dramatically and make it possible for an exhaustive search.

Algorithm of parameter selection base on parameter clustering

Step 1. Calculate the sensitivity vectors of the output with respect to the parameters.

Step 2. Determine n_s, the number of parameters per set, by singular value decomposition of the sensitivity matrix or the methods of forward selection.

Step 3. Wipe out the insignificant parameters whose sensitivity vectors have small length (e.g., less than 5% of the largest one).

Step 4. Cluster the parameters into n_g ($n_g \geq n_s$) groups with the similarity measure (Eq. 8) by hierarchical clustering.

Step 5. Select the parameter which has the largest sensitivity vector in a group as the representative of the group.

Step 6. Select n_s parameters from n_g representatives to optimize the criterion function by exhaustive search.

There are different approaches to calculate the parameter sensitivity in Step 1 and one commonly used is to calculate the sensitivity value $\partial \mathbf{y}(t)/\partial \boldsymbol{\theta}^{\mathrm{T}}$ by solving the system equations

$$\begin{aligned} \frac{d\mathbf{x}}{dt} &= \mathbf{f}(\mathbf{x},\mathbf{u},\boldsymbol{\theta}) \\ \mathbf{y} &= \mathbf{h}(\mathbf{x},\mathbf{u},\boldsymbol{\theta}) \end{aligned}, \tag{11}$$

and the sensitivity equations simultaneously

$$\begin{aligned} \frac{d}{dt}\frac{\partial \mathbf{x}}{\partial \boldsymbol{\theta}^{\mathrm{T}}} &= \frac{\partial \mathbf{f}}{\partial \mathbf{x}^{\mathrm{T}}}\frac{\partial \mathbf{x}}{\partial \boldsymbol{\theta}^{\mathrm{T}}} + \frac{\partial \mathbf{f}}{\partial \boldsymbol{\theta}^{\mathrm{T}}} \\ \frac{\partial \mathbf{y}}{\partial \boldsymbol{\theta}^{\mathrm{T}}} &= \frac{\partial \mathbf{h}}{\partial \mathbf{x}^{\mathrm{T}}}\frac{\partial \mathbf{x}}{\partial \boldsymbol{\theta}^{\mathrm{T}}} + \frac{\partial \mathbf{h}}{\partial \boldsymbol{\theta}^{\mathrm{T}}} \end{aligned}, \tag{12}$$

where **x**, **u**, **y**, **θ** are states, inputs, outputs and parameters respectively. The sensitivity vector is consisted of the sensitivity value at different time points

$$\mathbf{s}_i = \left[\partial \mathbf{y}^{\mathrm{T}}(t_1)/\partial \theta_i \quad \partial \mathbf{y}^{\mathrm{T}}(t_2)/\partial \theta_i \quad \cdots \quad \partial \mathbf{y}^{\mathrm{T}}(t_n)/\partial \theta_i\right]^{\mathrm{T}}, \tag{13}$$

where $\mathbf{s}_i$ is the sensitivity vector with respect to parameter θ_i.

The number of parameters per set can be determined by the numerical rank of the sensitivity matrix. Each column of the sensitivity matrix is a sensitivity vector of a parameter. The number of columns is equal to the number of parameters. However, due to the parameter correlation sensitivity matrix will be rank-deficient or close to singularity. In this case some singular values of the sensitivity matrix are zero or close to zero. The numerical rank of the sensitivity matrix is the number of the largest singular values greater than a threshold. The forward selection as the orthogonalization method can be also used to determine the number of parameters per set and it will have the similar results with singular value decomposition of the sensitivity matrix.

Step 3 is a screening of parameters based on their sensitivity values only. Science the parameters have small sensitivity value are difficult to be estimated no matter what combinations they are in, these parameters can be discarded in further analysis. The number of groups is a key variable in the algorithm which controls the discrepancy value between the original function and the reduced one and it also have a significant effect on the optimal solution searched. Using the hierarchical clustering the hierarchical tree is obtained. From the tree it is easy to see the relation between the least similarity value and the number of groups. So the number of groups can be determined by the least similarity value required.

A parameter in an indistinguishable set can compensate the effects of other parameters in the set. However, if the sensitivity value of the parameter is small a large change of the value is required for the compensation and if the sensitivity value is large a small change is required. A parameter with a large sensitivity value is easy to be estimated from the noise data than a parameter with a small sensitivity value. So the representative of a set is selected as the one has the largest sensitivity value in Step 5. The number of combination of representatives is much smaller than the number of all parameters and an exhaustive search is feasible.

4 Case study

Modeling of biochemical reactions in a cell has been a keystone in systems biology. However a difficulty in the modeling is that the models often contain even hundreds of parameters while the experimental data gathered is still scarce. So not all the parameters are identifiable and a set of identifiable parameters is often selected for estimation. Parameter selection, such as the Gram-Schmidt orthogonalization method, has attracted a wide attention in modeling of large scale biochemical networks (Gadkar et al. 2005; Yue et al., 2006; Jaqaman and Danuser, 2006).

To illustrate the technique a model of a signal transduction network (Huang *et al.*, 2007) is used. The model, shown in Figure 1, contains two pathways: Janus-associated kinases & signal transducers and transcription factors are activated in one pathway while the other pathway involves the activation of mitogen-activated protein kinases and it is updated from the model of Singh *et al.* 2006. This model consists of 72 nonlinear ordinary differential equations which include 124 parameters. The details description of the model and value of the parameters can be seen in the references (Singh *et al.*, 2006; Chu *et al.*, 2007; Huang *et al.*, 2007)

From the 124 parameters 7 is selected for estimation. The number of parameters selected is determined by singular value decomposition of the sensitivity matrix. The singular values beyond the largest 7 are close to zero. Use of the orthogonalization method will have the same result. The 74 insignificant parameters the sensitivity value of which are less than 5% of the largest one are removed before selection. The problem then becomes a selection of 7 parameters from 50 to optimize the *D*-optimality criterion. The total number of possible combinations is near 10^7 and there is an intensive computation burden for exhaustive search. To compare, the orthogonal method, a genetic algorithm and the clustering method are all applied.

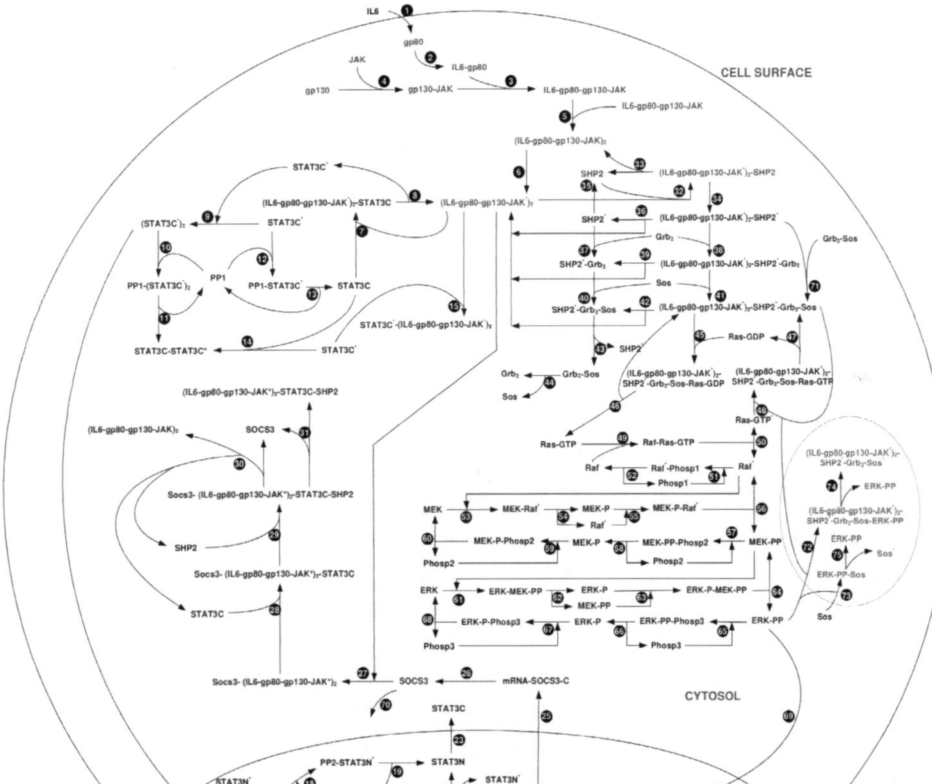

Figure 1. The model of IL-6 signaling pathway

Figure 2 shows the dendrogram of hierarchical clustering of parameters. It is obvious to see that some parameters have very high similarity value. The sensitivity vectors of those parameters nearly parallel and it is difficult to distinguish the effects of any two parameters. The least similarity value can be also read from the figure. For example, the dash line shows the least similarity when the parameters are clustered into 12 groups. As increase of the number of groups the dash line will move down and the least similarity value will increase. Table 1 lists the least similarity value changing with the number of groups.

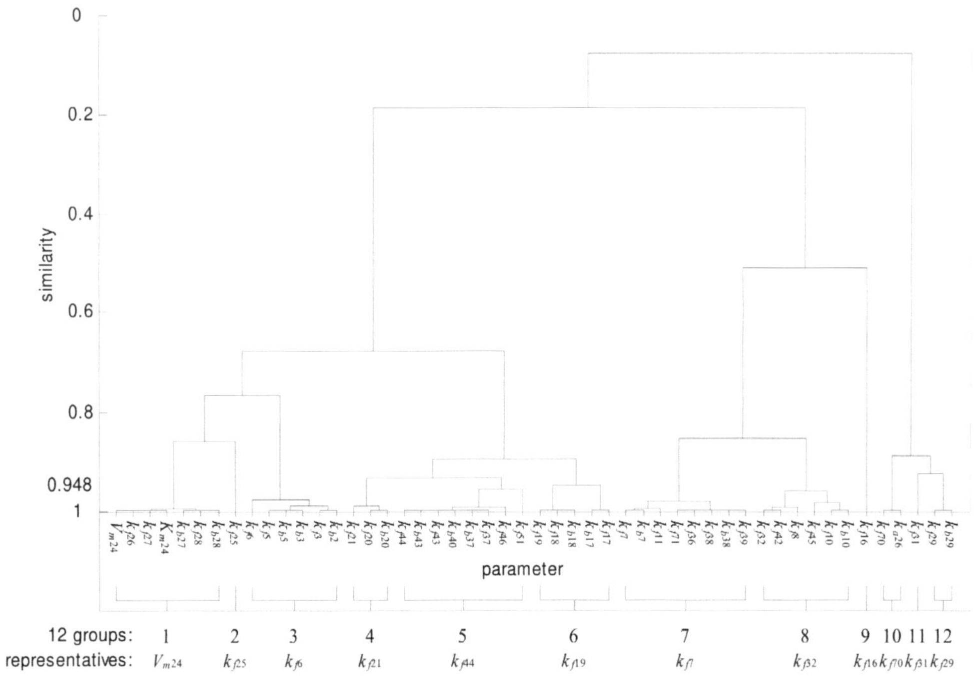

Figure 2. The dendrogram of hierarchical clustering of parameters

When the parameters are grouped, the parameter which has the longest sensitivity vector in a group is chosen as the representative of the group. Then the parameters for estimation are selected from the representatives by an exhaustive search to maximize the *D*-optimality criterion. The largest criterion value searched for different number of groups is shown in Table 1. The criterion value is small when the parameters are clustered into 7 groups. This is because the similarity of parameters in a group is not large enough that the representative is able to cover the total effects of a group. There is also a large discrepancy between the original model and the reduced one. However, as increase of the number of groups the largest criterion value rises. When the parameters are clustered into 12 groups the criterion value increases to the highest one and it stays at the value as increase of the number of groups. It gives a sign that an adequately good solution is found. The parameters in the 12 groups and the representatives are shown in Figure 2.

Table 1. The results by clustering method for different number of groups

No. of groups	7	8	9	10	11	12	13	14	15
Least similarity	0.862	0.889	0.894	0.924	0.935	0.948	0.959	0.962	0.977
Criterion	-4.139	-0.346	2.471	3.680	3.680	4.113	4.113	4.113	4.113
Discrepancy	0.430	0.305	0.272	0.074	0.031	0.029	0.014	0.009	0.008

To have an idea of how good the subset searched by clustering method, the optimal subsets searched by the orthogonalization method and the genetic algorithm are listed in Table 2. Rather than the optimal one the genetic algorithm can also find some suboptimal sets listed in Table 2 as well. The optimal set searched by clustering method is identical with that searched by genetic algorithm. The set has the criterion value larger than the one searched by the orthogonalization method. The orthogonalization method is a forward selection approach to maximize the criterion stepwise. If the number of parameters per set is one then the *D*-optimality criterion is reduced to the length of the sensitivity vector. So in the first step the method of forward selection will select the parameter which has the longest sensitivity value. The parameter in this case is k_{f7}. However the optimal set searched by clustering method and genetic algorithm does not include this parameter. This indicates that the forward selection may fail to find the optimal solution.

Ability to search a collection of (sub-)optimal solutions rather than the only optimal one is an advantage for the genetic algorithm. From the solutions some relationship among the parameters can be found. For example, the second set searched by genetic algorithm has only one parameter, k_{a26}, different from the one, k_{f70}, in the first set and the difference in the criterion value between the two sets is minor. This means that in practice the two parameters are interchangeable. However, the clustering method shows it more clearly. The two parameters are clustered in the same group (the 10th group) shown in Figure 2 so one can be used to replace another. The same situation happens between the 3rd set and the 1st one. The different parameters k_{f18} and k_{f19} are in the same group (the 6th group). Rather than providing the optimal set for parameter selection clustering method can also reveal the relationship among parameters.

Table 2. The results by the three methods

	Parameters selected	Criterion
Clustering	$k_{f32}, k_{f70}, k_{f16}, k_{f21}, k_{f31}, k_{f44}, k_{f19}$	4.113
Forward selection	$k_{f7}, k_{f6}, k_{f21}, k_{f70}, k_{f31}, k_{f19}, k_{f44}$	3.850
Genetic algorithm	$k_{f32}, k_{f70}, k_{f16}, k_{f21}, k_{f31}, k_{f44}, k_{f19}$	4.113
	$k_{f32}, k_{a26}, k_{f16}, k_{f21}, k_{f31}, k_{f44}, k_{f19}$	4.106
	$k_{f32}, k_{f70}, k_{f16}, k_{f21}, k_{f31}, k_{f44}, k_{f18}$	4.088
	$k_{f32}, k_{a26}, k_{f16}, k_{f21}, k_{f31}, k_{f44}, k_{f18}$	4.071
	$k_{f32}, k_{f70}, k_{f6}, k_{f21}, k_{f31}, k_{f44}, k_{f18}$	4.064

5 Conclusion

A great number of parameters are involved in a large scale system. However, the effects of some parameters are every similar and these parameters can not be estimated simultaneously. Investigation of pairwise indistinguishable set is to uncover when the effect on the output produced by variations of parameters in a set can be compensated by change of one parameter in the set. Identification of the pairwise indistinguishable sets is very useful in parameter selection. Parameter selection is a combinatorial problem to optimize a criterion function of the set. Identification of the indistinguishable sets provides a 'dimensional reduction' technique to simplify the optimization problem. The search of all possible combinations of parameters is reduced to search of combinations of the representatives and the number of sets for enumeration is reduced so dramatically that an exhaustive search is feasible to find the solution.

References

1. Anh DT, Bonnet MP, Vachaud G, Minh CV, Prieur N, Duc LV, Anh LL. Biochemical modeling of thc Nhue River (Hanoi, Vietnam): Practical identifiability analysis and parameters estimation. Ecological Modelling 193 (3-4): 182-204 2006.
2. Aster RC, Borchers B, Thurber CH. Parameter estimation and inverse problems. Boston: Elsevier Academic Press, 2005.
3. Atkinson A, Donev A, Tobias R. Optimum experimental designs, with SAS. Oxford: Oxford University Press, 2007.

4. Bastogne T, Thomassin M, Masse J. Selection and identification of physical parameters from passive observation. Application to a winding process. Control Engineering Practice 15 (9): 1051-1061 2007.
5. Brun R, Reichert P, Kunsch HR. Practical identifiability analysis of large environmental simulation models. Water Resources Research 37 (4): 1015-1030 2001.
6. Brun R, Kuhni M, Siegrist H, Gujer W, Reichert P. Practical identifiability of ASM2d parameters - systematic selection and tuning of parameter subsets. Water Research 36 (16): 4113-4127 2002.
7. Burth M, Verghese GC, Velez-Reyes M. Subset selection for improved parameter estimation in on-line identification of a synchronous generator. IEEE Transactions on Power Systems. 14 (1): 218-225 1999.
8. Chappell MJ, Godfrey KR, Vajda S. Global identifiability of the Parameters of Nonlinear-systems with specified inputs-A comparision of methods. Mathematical Biosciences 102 (1): 41-73 1990.
9. Chu Y, Jayaraman A, Hahn J. Parameter Sensitivity Analysis of IL-6 Signaling Pathways. IET Systems Biology 1 (6): 342-352 2007.
10. Chu Y, Hahn J. Parameter Set Selection for Estimation for Nonlinear Dynamic Systems. AIChE Journal 53 (11): 2858-2870 2007.
11. Duda RO, Hart PE, Stork DG. Pattern classification (2nd edition). New York: Wiley, 2001.
12. Gadkar KG, Varner J, Doyle FJ. Model identification of signal transduction networks from data using a state regulator problem. IEE Proceedings Systems Biology 2 (1): 17-30 2005.
13. Golub GH, van Loan CF. Matrix computations (2nd edition). Baltimore: Johns Hopkins University Press, 1989.
14. Hiskens IA. Nonlinear dynamic model evaluation from disturbance measurements. IEEE Transactions on Power Systems 16 (4): 702-710 2001.
15. Huang Z, Chu Y, Senocak F, Jayaraman A, Hahn J. Model Update of Signal Transduction Pathways in Hepatocytes based upon Sensitivity Analysis. Proceedings Foundations of Systems Biology. 45-50 Stuttgart, Germany 2007
16. Jaqaman K, Danuser G. Linking data to models: data regression. Nature Reviews Molecular Cell Biology 7 (11): 813-819 2006.
17. Kiefer J. Optimum experimental designs. Journal of the Royal Statistical Society Series B - Statistical Methodology 21 (2): 272-319 1959.
18. Kitano H. Systems biology: A brief overview. Science 295 (5560): 1662-1664, 2002
19. Kou B, McAuley KB, Hsu CC, Bacon DW, Yao KZ. Mathematical model and parameter estimation for gas-phase ethylene homopolymerization with supported metallocene catalyst. Industrial & Engineering Chemistry Research 44 (8): 2428-2442 2005.
20. Li RJ, Henson MA, Kurtz MJ. Selection of model parameters for off-line parameter estimation. IEEE Transactions on control systems technology 12 (3): 402-412 2004.
21. Ljung L, Glad T. On global identifiability for arbitrary model parametrizations. Automatica 30 (2): 265-276 1994.
22. Ljung L. System identification: theory for the user (2nd edition). Upper Saddle River: Prentice Hall PTR, 1999.
23. Rothenbe TJ. Identification in parametric models. Econometrica 39 (3): 577-591 1971.
24. Sandink CA, McAuley KB, McLellan PJ. Selection of parameters for updating in on-line models. Industrial & Engineering Chemistry Research 40 (18): 3936-3950 2001.
25. Singh A, Jayaraman A, Hahn J. Modeling regulatory mechanisms in IL-6 signal transduction in hepatocytes. Biotechnology and Bioengineering 95 (5): 850-862, 2006.
26. Schoeberl B, Eichler-Jonsson C, Gilles ED, Muller G. Computational modeling of the dynamics of the MAP kinase cascade activated by surface and internalized EGF receptors. Nature Biotechnology 20 (4): 370-375, 2002.

27. Sin G, Vanrolleghem PA. Extensions to modeling aerobic carbon degradation using combined respirometric-titrimetric measurements in view of activated sludge model calibration. Water Research 41 (15): 3345-3358, 2007.
28. Sun CL, Hahn J. Parameter reduction for stable dynamical systems based on Hankel singular values and sensitivity analysis. Chemical Egnineering Science 61 (16): 5393-5403, 2006.
29. Theodoridis S, Koutroumbas K. Pattern recognition (3rd edition). Boston: Elsevier/Academic Press, 2006.
30. Velez-Reyes M, Verghese GC. Subset selection in identification, and application to speed and parameter estimation for induction machines. Proceedings of the 4th IEEE Conference on Control Applications. 991 – 997, 1995.
31. Walter E. Identifiability of parametric models. New York: Pergamon Press, 1987.
32. Walter E, Pronzato L. Qualitative and quantitative experiment design for phenomenological models - A survey. Automatica 26 (2): 195-213 1990.
33. Walter E, Pronzato L. Identification of parametric models from experimental data. Berlin: Springer, 1997.
34. Weijers SR, Vanrolleghem PA. A procedure for selecting best identifiable parameters in calibrating activated sludge model no.1 to full-scale plant data. Water Science and Technology 36 (5): 69-79 1997.
35. Yamada S, Shiono S, Joo A, Yoshimura A. Control mechanism of JAK/STAT signal transduction pathway. FEBS LETTERS 534 (1-3): 190-196, 2003.
36. Yao KZ, Shaw BM, Kou B, McAuley KB, Bacon DW. Modeling ethylene/butene copolymerization with multi-site catalysts: Parameter estimability and experimental design. Polymer Reaction Engineering 11 (3): 563-588 2003.
37. Yue H, Brown M, Knowles J, Wang H, Broomhead DS, Kell DB. Insights into the behaviour of systems biology models from dynamic sensitivty and identifiability analysis: a case study of an NF-kappa B signalling pathway. Molecular Biosystems. 2 (12): 640-649, 2006.

546a Construction of Stochastic Differential Equation Models for Porous Thin Film Deposition Processes

Gangshi Hu[1], Xinyu Zhang[1], Gerassimos Orkoulas[1], and Panagiotis D. Christofides[2]. (1) Department of Chemical and Biomolecular Engineering, University of California, Los Angeles, Los Angeles, CA 90095, (2) Department of Chemical and Biomolecular Engineering, Department of Electrical Engineering, University of California, Los Angeles, Los Angeles, CA 90095

In many industrial applications, structural defects in thin film growth processes, usually expressed in terms of thin film porosity, play an important role in determining electrical and mechanical properties of the thin films. For example, low-k dielectric films of high porosity are being used in current interconnect technologies to meet resistive capacitive (RC) delay goals and minimize cross-talk. However, increased porosity negatively affects the mechanical properties of dielectric films, increasing the risk of thermo-mechanical failures [1]. Furthermore, in the case of gate dielectrics, it is important to reduce thin film porosity as much as possible and eliminate the development of holes close to the interface.

Kinetic Monte Carlo (kMC) methods are widely employed in the modeling of thin film growth and porosity. However, kMC methods are not suitable for the design of model-based control systems owing to their computational demand and the need to carry out calculations in real-time in feedback control systems. Stochastic differential equation (SDE) model provide an attractive alternative, as the basis for control system design, owing to their potential to capture key process behavior from a control point of view and their computational efficiency [2,3]. SDE models can be derived from the corresponding master equations and their coefficients can be estimated from kMC data. Successful applications of SDE modeling of the surface height profile have been done in the context of thin film growth processes, such as SDE modeling of surface height profile in random deposition with surface relaxation deposition processes, ballistic deposition processes and sputtering processes [4]. However, there are no works that have focused on SDE modeling of thin film porosity.

In this work, a method is presented for the construction of lumped and spatially distributed SDE models that describe the evolution of the film porosity in thin film deposition process. Film density is introduced into the SDE model to represent porosity and is defined as the ratio between the numbers of occupied sites and total bulk sites. The data for model identification are generated from a lattice-based kMC simulator of the deposition process, which allows porosity during film growth. Two microscopic events are considered in the process: random deposition and thermal diffusion. Vacancies and overhangs are allowed in the kMC model to introduce porosity, which is mathematically represented by the thin film density. Various lumped and spatially distributed SDE models are constructed. The coefficients of the SDE models are estimated by least-square fitting method to match the profiles of expected values and variances of the film density of the SDE model to that of the kMC data. The dependence of the model coefficients of the SDE on the process parameters, i.e., substrate temperature and deposition rate, is investigated and expressed in terms of explicit functions. Simulation results demonstrate that the constructed SDE models reproduce well the kMC model behavior for a variety of operating conditions.

1. Kloster, G., Scherban, T., Xu, G., Blaine, J., Sun, B. and Zhou, Y. Porosity effects on low-k dielectric film strength and interfacial adhesion. Proceedings of the IEEE 2002 International; 242-244.

2. Edwards, S. F. and Wilkinson, D. R. The surface statistics of a granular aggregate. Proceedings of the Royal Society of London. Series A, Mathematical and Physical Sciences 1982;381:17-31.

3. Lauritsen, K. B., Cuerno, R. and Makse, H. A. Noisy Kuramoto-Sivashinsky equation for an erosion model. Physical Review E 1996; 54:3577-3580.

4. Hu, G., Lou, Y. and Christofides, P. D. Model Parameter Estimation and Feedback Control of Surface Roughness in a Sputtering Process. Chemical Engineering Science 2008;63:1800-1816.

546b Optimal Operation with Respect to Multiscale Objectives during Layered Heterostructure Growth

Christopher Behrens and Antonios Armaou. Chemical Engineering, The Pennsylvania State University, University Park, PA 16802

Spatially distributed multiscale systems have been recently developed to describe transport-reaction processes for which phenomena occur across length scales that differ by several orders of magnitude. These descriptions can range from the smaller regimes (microscale, quantum and atomistic) to the intermediate (mesoscale) and larger (macroscale/continuum) ones. The level of detail required to accurately describe the dynamic behavior of the system at smaller length scales cannot be provided by solely using a description from larger length-scales. On the other hand, describing larger length-scale behavior using descriptions from smaller length-scale models would be infeasible due to the required computational overhead of this procedure. Motivated by the above, multiscale models, which combine larger and smaller-length scale models that are intimately connected, have been developed.

Such multiscale models could prove very useful in optimization, as continuous pressure on profit margins has resulted in the desire to improve efficiency in order to reduce costs. Multiscale optimization of such processes, however, is a problem with three levels of hierarchy in computation. In the lower hierarchy, potentially computationally intensive simulations may need to be performed to model the properties of interest at both smaller and larger length-scales. If bi-directional information (conditions at a larger length-scale affect the conditions at a smaller length-scale and vice-versa) is required, both these smaller and larger length-scale simulations must be performed iteratively at each process condition in order to achieve convergence of the multiscale description, which is the intermediate hierarchy. Finally, gradient information might not be available in closed form, necessitating the use of black-box optimization algorithms. Black-box optimization can require several function evaluations, which in this case is at the top of the hierarchy, and therefore all of these evaluations must include reevaluation of the process behavior for all length scales involved. With the need to perform numerous simulations, solving a multiscale optimization problem could quickly become intractable. Many different methods have been proposed to circumvent this computational intractability. One promising idea is to obtain the necessary information from linear interpolations whenever possible instead of using costly simulations. This is the basic idea behind *in situ* adaptive tabulation (ISAT). Initially used in solving combustion chemistry problems [1], ISAT has been subsequently extended to stochastic systems and combined with black-box optimization for maximizing uniformity and minimizing surface roughness in a gallium nitride thin-film [2]. Additionally, if lattice-based atomistic simulations are used, lattice reconstruction techniques based on the low-order properties of interest can be employed [3].

The implementation and further refinement of these methods is investigated by modeling the deposition of a thin-film consisting of alternating gallium arsenide and aluminum arsenide layers (GaAs/AlAs.) This example uses macroscale "inputs" from the reactor description [2,4] to determine properties such as temperature, flow rates, and concentrations, and mesoscale kinetic Monte Carlo (kMC) simulations to measure the previously characterized interfacial properties of the film. The objectives of this problem are to minimize the interfacial step-densities between GaAs and AlAs layers, while also minimizing the temperature and the time spent in-between depositing species (termed annealing time,) as well the macroscopic objective of reducing spatial non-uniformity. This problem only involves adsorption and therefore only requires uni-directional flow of information from the macroscale level to the mesoscale level. Based on this model, we explore the methodology needed to accelerate the computational process by using techniques such as ISAT. Using this methodology, a multiscale model for a GaAs/AlAs thin-film deposition process was developed for optimization of the interfacial properties, and used within an efficient optimization framework to identify optimal-time varying process conditions for the process.

1. S.B. Pope, "Computationally efficient implementation of combustion chemistry using in situ adaptive tabulation," *Combust. Theory Modelling*, **1**, 41-63, 1997.

2. A. Varshney and A. Armaou, "Multiscale optimization using hybrid PDE/kMC process systems with application to thin film growth," *Chem. Eng. Sci.*, **60**, 6780-6794, 2005.

3. A. Varshney and A. Armaou, "Identification of macroscopic variables for low-order modeling of thin-film growth," *Ind. Eng. Chem. Res.*, **45**, 8290-8298, 2006.

4. A. Varshney and A. Armaou, "Reduced order modeling and dynamic optimization of multiscale PDE/kMC process systems," *Comp. & Chem. Eng.*, in press, 2008.

546c Multiscale Model Development for Microstructure-Based Product Quality Control In Polymeric Coating Curing

Jie Xiao and **Yinlun Huang**. Department of Chemical Engineering and Materials Science, Wayne State University, 5050 Anthony Wayne Drive, Detroit, MI 48202

Polymeric surface coating has numerous applications in the automotive, aerospace, electronics, steel, chemical and paper-making industries. It is widely recognized that the coating quality in terms of corrosion prevention and scratch resistance is directly determined by the coating microstructure (i.e., the 3D polymer network structure) that is formed during the curing process. Thus microstructure-based real-time quality control is the key to achieving superior coating performance, for which characterizing the 3D polymer network formation during the curing process becomes a crucial basis. Numerous efforts have been devoted to either microscopic modeling of the polymer network formation[1-3] or macroscopic modeling of the curing process.[4] The monoscale modeling efforts can be hardly applied for realizing microstructure-based product quality control.

In this work, we will introduce a unified multiscale modeling method for characterizing polymeric coating curing. The macroscopic dynamic curing condition is modeled using the computational fluid dynamics (CFD), where different heating mechanisms (e.g., convection and radiation) can be readily applied. The microscopic crosslinked network structure is created through a Lattice Monte Carlo (LMC) simulation, which has unique capability for dealing with polymeric materials with any distributions of molecular weight and functional groups on the polymer chains. The macroscopic and microscopic models are seamlessly coupled by a special bi-directional coupling method. The curing temperature at a specific time instant obtained from the CFD model is sent to the LMC model for evolving the coating microstructure. On the other hand, the crosslinking conversion derived from the microstructure is sent back to the CFD model, which determines the next time instant for sending a temperature to the LMC model. Consequently, the microstructure evolution throughout a complete curing process under any curing condition can be revealed. Furthermore, a microstructure analysis method is introduced to correlate the micostructure (at a length scale of 10^1-10^2 nm) with the macroscopic coating quality (at a length scale of 10^{-2}-10^1 m). The introduced methodology is applied to an automotive coating curing case study. The attractiveness of revealing on-line all-time information about the coating microstructure and performance evolution will be fully demonstrated.

References:

1. Gina N, Cohen C, Panagiotopoulos AZ. A Monte Carlo study of the structural properties of end-linked polymer networks. Journal of Chemical Physics. 2000; 112: 6910-6916.

2. Balabanyan AG, Kramarenko EY, Ronova IA, Khokhlov AR. Monte Carlo study of structure and kinetics of formation of end-linked polymer networks. Polymer. 2005; 46, 4248-4257.

3. Hagn C, Wittkop M, Kreitmeier S, Trautenberg HL, Holzl T, Goritz D. The creation and spatial structure of end-linked bimodal polymer networks: a Monte Carlo study. Polymer Gels and Networks. 1995; 5: 327-337.

4. Lou HH, Huang YL. Integrated modeling and simulation for improved reactive drying of clearcoat. Ind. Eng. Chem. Res. 2000; 39: 500-507.

546d Efficient and Robust Multi-Scale Models of Conjugate Heat Transfer and Melt Crystal Growth

Andrew Yeckel[1], Lisa Lun[1], Thomas Jung[2], Jochen Friedrich[2], and Jeffrey J. Derby[1]. (1) Chemical Engineering and Materials Science, University of Minnesota, 151 Amundson Hall, Minneapolis, MN 55455-0132, (2) Crystal Growth Laboratory, Fraunhofer Institute IISB, Erlangen, Germany

The modeling of melt crystal growth processes is an inherently multi-scale challenge, with relevant length scales ranging from furnace dimensions to atomic-sized features in the grown crystal. A faithful depiction of the interaction of these scales is needed to advance our scientific understanding of crystal growth processes, as well as to make our process models quantitatively predictive and technologically useful. However, efficient and robust techniques are needed to couple strongly nonlinear conjugate heat transfer with models that depict phase change and crystal growth.

A strategy that simultaneously represents all chosen phenomena at all scales in a single, large mathematical model is referred to as a monolithic, analytic, or direct-coupling approach. From the points of view of mathematical self- consistency and algorithmic robustness, this course of action is preferred and is the chosen strategy for many global- scale models for melt crystal growth processes. However, such approaches require intensive and coordinated programming efforts, are typically system-specific, and are often difficult to maintain and modify. Due to these challenges, there is an increasing desire for innovative ways to couple existing software that have been developed to solve specific problems, especially for modeling multi-scale and multi-physics problems. Such alternative approaches are modular, partitioned, or synthetic, in which different models and solvers are used together. In this manner, software tools that have desired features can be exploited to enhance existing capabilities.

We discuss the mathematical and algorithmic challenges for the modular coupling of global-scale furnace heat transfer models and local-scale models for melt crystal growth based on an innovative Block-Newton approach implemented using a Jacobian-free Newton-Krylov algorithm. To clarify some of the underlying issues, we present initial studies of this and other approaches using simple models. Then we present our experience with the coupling of the global model, CrysMAS developed by the Crystal Growth Laboratory, Erlangen, Germany, that computes high-temperature, furnace heat transfer, with the local model, Cats2D and Cats3D, developed by the Derby group to solve for heat transfer, incompressible melt flow, and melt-crystal interface shape.

546e Computational Studies of Semiconductor Nanocrystal Synthesis In Templating Media

Sreekumar Kuriyedath[1], Yannis G. Kevrekidis[2], and T. J. Mountziaris[1]. (1) Department of Chemical Engineering, University of Massachusetts, Amherst, MA 01003, (2) Department of Chemical Engineering, Princeton University, Princeton, NJ 08540

Semiconductor nanocrystals exhibit quantum confinement effects below a certain size and their optical and electronic properties become size-dependent. Applications of these materials include biosensors, solar cells, and nanoelectronics. The use of templates in the synthesis of semiconductor nanocrystals allows precise control of their shape and size. It also enables easy scale-up for commercial production.

We have developed mesoscopic models that describe the synthesis of zinc selenide nanocrystals in microemulsions and liquid crystals [1-3]. The templates are formed by self assembly of a ternary system that includes a polar solvent, a non-polar solvent, and an amphiphilic block copolymer. The nanocrystals are grown inside the dispersed nanodomains of the templates, e.g. inside the nanodroplets of a microemulsion or the cylindical nanodomains of a liquid crystal. The initial nuclei are formed by a spontaneous and irreversible reaction between a zinc precursor that is dissolved in the dispersed nanodomains of the template and a selenium precursors that is dissolved in the continuous phase and diffuses through the block copolymer layer. The mesoscopic models are based on kinetic Monte Carlo simulations that track the diffusion of the zinc precursor molecules inside a dispersed nanodomain, the formation of zinc selenide clusters (nuclei) by reaction at the interface with the selenium precursor, and the formation of a single nanocrystal in each nanodomain by cluster-cluster coalescence. Parametric studies have been performed to elucidate the effects of particle coalescence efficiency and interfacial flux of the selenium precursor on the evolution of nanoparticle populations and the time required for formation of a single nanocrystal in each domain.

The mesoscopic models are coupled through the interfacial flux of the selenium precursor to macroscopic (PDE-based)reactor-level models that treat the two phases of the template as interpenetrating continua. This multi-scale modeling approach enables optimal reactor design by linking reactor-level operating parameters with particle formation processes taking place inside the dispersed nanodomains of each template.

References

1. G. N. Karanikolos, P. Alexandridis, G. Itskos, A. Petrou, and T. J. Mountziaris, Langmuir 20, 550 (2004).

2. G.N. Karanikolos, P. Alexandridis, R. Mallory, A. Petrou, and T. J. Mountziaris, Nanotechnology, 16, 2372-2380 (2005)

3. G.N. Karanikolos, N.-L. Law, R. Mallory, A. Petrou, P. Alexandridis, and T. J. Mountziaris, Nanotechnology, 17, 3121-3128 (2006).

546f Multi-Scale Modelling and Design of Catalytic Microreactor Systems

Bostjan Hari and Constantinos Theodoropoulos. School of Chemical Engineering and Analytical Science, University of Manchester, Sackville St, Manchester, M60 1QD, United Kingdom

Microreactor systems are receiving increasing attention nowadays due to their enhanced operation characteristics, such as increased mass and heat transfer capabilities, flow uniformity, inherent safety and potential for high throughput through the construction of array configurations. They also ensure a smaller plant size, lower cost of production and more flexible response to market demand [1].

In this work we construct multi-scale models of continuous microreactors with catalytic walls as an alternative to the conventional packed-bed configuration. The bulk (gaseous) phase is treated macroscopically using Computational Fluid Dynamics (CFD), including transport phenomena and gas-phase reactions, while the catalytic activity is modeled at the micro/mesoscopic level using kinetic Monte Carlo (kMC) simulations including effects such as adsorption, desorption surface reactions and surface diffusion. A number of kMC lattices depending on the discretisation of the computational domain are used to effectively represent the reactive surfaces. Massively parallel computations both for the gas-phase using SANDIA's code for reaction transport processes (codename: MPSALSA) [2] and for the surface are used to accelerate the tedious computational tasks.

The efficient coupling of time and length scales is investigated through coarse-graining methodologies and comparisons with fully macroscopic mean field approaches are made in order to assess the window of parameters where the multi-scale approach becomes important. The effect of both vertical interactions (between the macroscopic gas-phase and the microscopic surface) and lateral interactions between neighbouring kMC lattices is studied with the use of efficient numerical techniques. The relative importance of these interactions is evaluated for a wide range of conditions, through sensitivity studies. The methodology is demonstrated through the use of two illustrative reaction systems with well-known kinetics: CO oxidation [3] and the simultaneous NO reduction with CO oxidation [4].

References

1. Ehrfeld W, Hessel V, Lowe H, Microreactors: New Technology for Modern Chemistry, Wiley-VCH (2000).

2. Shadid J, Hutchinson S, Hennigan G, Moffat H, Devine K, Salinger AG, Parallel Computing 23, (1997), 1307-1325.

3. Zhdanov VP and Kasemo B 20 (1994) 111-189.

4. Fink Th, Dath JP, Imbihl R, Ertl G, J. of Chem. Phys. 95 (1991) 2109-2126.

546g Temperature Accelerated Dynamics Based Kinetic Monte Carlo Method

Abhijit Chatterjee[1], Blas P. Uberuaga[2], and Arthur F. Voter[1]. (1) Theoretical Division, Los Alamos National Laboratory, T-12 MS B268, Los Alamos, NM 87545, (2) Material Science and Technology Division, Los Alamos National Laboratory, Los Alamos, NM 87545

The temperature accelerated dynamics (TAD) method is a powerful tool for studying rare event dynamics in materials and provides significant computational speed-up compared to molecular dynamics simulations. TAD simulations can easily access large time scales, such as milliseconds to seconds. However, modeling of large length scales with TAD is challenging. Kinetic Monte Carlo (KMC) is an ideal tool for studying dynamics and steady state of noise-controlled phenomena at large length and time scales. However, KMC requires a list of atomic processes, i.e., a process catalog, for system evolution, which are typically not available a priori. As a result, several assumptions are introduced into the process catalog about possible atomic mechanisms. In addition, often atoms are assumed to reside on a lattice. This leads to unrealistic kinetics and thermodynamic behavior in multicomponent and strained material systems.

To overcome these shortcomings of the two methods, we introduce a TAD-based KMC method that generates an on-the-fly process catalog with a prescribed confidence level. The process catalog provides rate information regarding the possible escape pathways from visited potential energy basins to neighboring basins. For the first time confidence measures associated with the TAD process catalogs are derived, which guarantees that the process catalog contains all the relevant escape pathways for the confidence level chosen. The system evolution is studied by employing the KMC method. Computational requirement of the TAD-based KMC method is discussed. In addition, the local TAD-based KMC is discussed, which exploits past visit to similar local atomic environments and provides additional computational speed-up while retaining accuracy of molecular dynamics. Finally, domain decomposition strategies based on the TAD-based KMC method, that can access even larger length and time scales, are discussed. The application of TAD-based KMC method to metal alloys and oxide materials for electronic, catalytic and energy applications is discussed.

554a Optimal Scheduling of Multistage Batch Plants. A Comparative Study

Pedro Castro, Department of Process Modeling and Simulation, INETI, Lisbon, Portugal and Augusto Q. Novais, Dms, INETI, Lisbon, Portugal.

Scheduling in the process industry has received considerable attention during the last 15 years and extensive reviews can be found in the literature. Because of the great variety of aspects that need to be taken into consideration, many approaches have been proposed to handle specific problem types. However, the main developments have resulted from more general approaches, those that can effectively handle a large number of problem features. Important landmarks are the unified frameworks for process representation: the state-task (STN) and the resource task networks (RTN).

Scheduling models are first characterized by the representation of time. Discrete-time approaches can sometimes be a good option but, most of the times, the characteristics of the problem compels us to employ a continuous-time model, which can be either time grid or sequence based. Nowadays, it is clear that single time grid, continuous-time models based on unified frameworks are the most general since they can consider resource constraints other than equipment, together with various storage policies. However, it is also true that they are usually not the most efficient for some classes of problems. Recent detailed computational studies have shown that unit-specific formulations can be orders of magnitude faster in problems arising from multipurpose batch plants (Shaik and Floudas, Ind. Eng. Chem. Res. 2006, 45, 6190 and Comput. Chem. Eng. 2008, 32, 260) and also from multistage multiproduct batch plants (Castro and Grossmann, Ind. Eng. Chem. Res. 2005, 44, 9175 and Liu and Karimi, Chem. Eng. Sci. 2007, 62, 1549), under an unlimited intermediate storage policy.

Traditional approaches for multistage multiproduct batch plants consider that the number and size of batches is known a priori. Different batches of the same product are treated as distinct entities that maintain their identity throughout the processing stages, similarly to batches belonging to different products. Hence, they can be viewed as single batch approaches. Decoupling batching from scheduling decisions may not be an easy task in processes with parallel units of dissimilar capacities. Furthermore, it often leads to suboptimal solutions since one cannot take advantage of batch mixing/splitting to capitalize on the different capacities from one stage to the next, as noted by Sundaramoorthy and Maravelias (Ind. Eng. Chem. Res. 2008, 47, 1546).

In this paper, we propose a new multiple time grid (a.k.a. unit-specific) formulation that can simultaneously determine the optimal set of batches (number and size) together with unit assignment and sequencing of batches. It can be viewed as an extension of our previous work (Castro et al. Ind. Eng. Chem Res. 2006, 45, 6210), which dealt with problems involving a single product batch per stage. The main novelty is the definition of a single intermediate storage unit per stage to which an explicit storage unit is associated. The events belonging to the time grid of a particular storage unit are then related to the corresponding event points of those processing units feeding to and receiving material from it. Such timing constraints, together with the material balances, guarantee that a product can only start to be produced in stage k+1 after enough intermediate has been produced in stage k. For problems with sequence dependent changeovers, the model uses 4-index binary variables to identify the execution of combined processing and changeover tasks. Wherever changeovers are not an issue, it can be simplified to a 3-index model.

The new formulation is thoroughly compared to the unit-specific approaches of our previous work (Castro and co-workers, 2005 and 2006) and that of Shaik and Floudas (2008). A variety of scenarios are considered involving single and multiple batch cases, alternative objective functions and problems with and without sequence-dependent changeovers. The results show that its higher generality, when compared to that of Castro et al. (2005, 2006), comes at the expense of a small increase in the computational cost for single batch problems. However, when compared to the one of Shaik and Floudas (2008) it can be up to 3 or 4 orders of magnitude faster due to lower integrality gap, to the use of a smaller number of event points to find the global optimal solution, and to fewer constraints. On the other hand, the latter has the advantage of being suitable for multipurpose plants and other storage policies.

554b CP-Based Symmetry-Breaking Branching for Crude-Oil Operations Scheduling

Sylvain Mouret, Dept of Chemical Engineering, Carnegie Mellon University, 5000 Forbes Ave, Pittsburgh, PA 15213, Ignacio E. Grossmann, Center for Advanced Process Decision-making, Dept of Chemical Engineering, Carnegie Mellon University, 5000 Forbes Avenue, Pittsburgh, PA 15213, and Pierre Pestiaux, Centre de Recherche de Gonfréville, Total France, Le Havre, 76700, France.

Scheduling formulations often involve, depending on the objective, many symmetrical solutions. For instance, a discrete-time representation may involve symmetrical solutions when the objective value is not affected by moving a solution's task one time interval to the left (resp. to the right). Therefore, it is important to detect and circumvent the symmetry challenge and trying to solve a scheduling problem. This paper presents a methodology using a CP-based branching rule in order to eliminate irrelevant symmetric solutions from the search. The main idea is to select a representative for each set of symmetrical discrete solutions and use logic constraint propagation to restrict the branch and bound search to theses solutions.

Many representations have been used to solve various crude-oil operations problems. Lee et al. (1996) developed a discrete-formulation to solve a relaxation of a crude-oil operations problem where the composition constraints are dropped. This approach can lead to many symmetrical solutions especially when defining a large number of time intervals to improve the quality of the solution. Jia et al. (2003) and Furman et al. (2007) used the event-point representation introduced by Ierapetritou and Floudas (1998) to solve the same problem. The computational expenses reported seem to show that this type of formulation would benefit from detecting and breaking possible symmetries present in the problem.

In this paper, the multi-operations time-slots MINLP model introduced by Mouret et al. (2008a) is used. This representation consists in postulating a number of time-slots and assigning exactly one operation to each time-slot. As shown by Mouret et al. (2008b), this representation leads to many symmetrical solutions that have been identified. Linear symmetry-breaking constraints based on regular languages (Côté et al., 2007) have been added to the model reducing the computational expense by several orders of magnitude. The drawback of this approach is the large number of variables and constraints added to the model, thus increasing the solution time of the relaxation node.

The aim of this work is to further reduce the overall expense by using Constraint Programming (CP) to handle more efficiently the logic symmetry-breaking constraints. A CP model is defined including the symmetry-breaking constraint and other logic constraints involved in the problem formulation. The MILP branch and bound search is then improved by adding a constraint propagation step extending the branching decisions at each node. Similar ideas have been applied to Integer Programming models by Margot (2002) and Ostrowski (2007). The main difference here is that CP is used as the symmetry-breaking algorithm, thus benefiting from its flexible modeling capabilities.

The CP-based branching algorithm extends the branching decisions made by the MILP solver by fixing additional binary variables from the model on each branch initially created. Therefore, it is possible to use any pre-defined branching rule (using pseudo costs, most fractional variable, strong branching, branching on SOS1 variables, etc).

Computational results are presented for the four crude-oil operations problems introduced by Lee et al. (1996) showing an average 60% improvement compared to the extended MILP model presented by Mouret et al. (2008b). This is due to the low cost of constraint propagation compared to solving the node relaxation when the symmetry-breaking constraints are included in the MILP model.

[1] Lee, H., Pinto, J.M., Grossmann, I.E., and Park, S. (1996). "Mixed-integer linear programming model for refinery short-term scheduling of crude oil unloading with inventory management." Ind. Eng. Chem. Res. 35(5): 1630-1641

[2] Jia, Z., Ierapetritou, M.G., and Kelly, J.D. (2003). "Refinery Short-Term Scheduling Using Continuous Time Formulation: Crude-Oil Operations." Ind. Eng. Chem. Res. 42(13): 3085-3097

[3] Furman, K.C., Jia, Z., and Ierapetritou, M.G. (2007). "A Robust Event-Based Continuous Time Formulation for Tank Transfer Scheduling." Ind. Eng. Chem. Res. 46(26): 9126-9136

[4] Ierapetritou, M.G. and Floudas, C.A. (1998). "Effective Continuous-Time Formulation for Short-Term Scheduling. 1. Multipurpose Batch Processes." Ind. Eng. Chem. Res. 37(11): 4341-4359

[5] Mouret, S., Grossmann, I.E., and Pestiaux, P. (2008) "Multi-Operations Time-Slots Model for Crude-Oil Operations Scheduling." ESCAPE 18. To appear.

[6] Mouret, S., Grossmann, I.E., and Pestiaux, P. (2008) "A Deterministic Finite Automaton Approach Generating Symmetry-Breaking Constraints for Crude-Oil Scheduling Problems." FOCAPO 2008. To appear.

[7] Côté, M.C., Gendron, B., and Rousseau, L.M. (2007). "Modeling the Regular Constraint with Integer Programming." CPAIOR 2007. LNCS 4510: 29-43

[8] Margot, F. (2002). "Pruning by Isomorphism in Branch-and-Cut." Mathematical Programming. 94 (1): 71-90

[9] Ostrowski, J., Linderoth, J., Rossi, F., and Smriglio, S. (2007). "Orbital Branching." IPCO XII. LNCS 4513: 104-118

554c A Convex Hull Discretization Approach to the Global Optimization of Pooling Problems

Viet Pham, Carl D. Laird, and Mahmoud El-Halwagi. Chemical Engineering, Texas A&M University, College Station, TX 77843

The pooling problem is an important optimization problem that is encountered in process operation and scheduling. Because of the presence of bilinear terms, the traditional formulation is nonconvex. Consequently, there is a need to develop computationally-efficient and easy-to-implement global-optimization techniques. In this paper, a new approach is proposed based on three concepts: linearization by discretizing nonlinear variables, pre-processing using implicit enumeration of the discretization to form a convex-hull which limits the size of the search space, and application of integer cuts to ensure compatibility between the original problem and the discretized formulation. The continuous quality variables contributing to the bilinear terms are first discretized. The discretized problem is a mixed integer linear program (MILP) which is globally solvable in a computationally effective manner using branch and bound method. The merits of the proposed approach are illustrated with case studies from literature and comparison with published results.

554d The Unit-Specific General Precedence Scheduling Model: Addressing Sequence-Dependent Setup Times and Costs Issues

Georgios Kopanos and Luis Puigjaner. Chemical Engineering Department, Universitat Politècnica de Catalunya - ETSEIB, Diagonal, 647, Barcelona, E-08028, Spain

Setup times appear often into industrial scheduling problems and are one of the most frequent additional complications in production scheduling. Setup time examples for the process industry are processes such as cleaning of processing units, inspecting of material (analytic and quality control), preheating, mixing and etc. Scheduling problems involving setup times can be mainly categorized in sequence-independent or sequence-dependent setup. Setup is sequence-dependent if its duration depends on both the current and the immediately preceding task, and is sequence-independent if its duration depends only on the current processing task. Sequence-dependent setups scheduling problems constitute the more complicated case.

Some industrial examples of sequence-dependent setups include: (i) chemical compounds manufacturing, where the extent of the cleansing depends on both the chemical most recently processed and the chemical about to be processed and (ii) the printing industry, where the cleaning and setting of the press for processing the next job depend on its difference from the colour of ink, size of paper and types used in the previous job. The case of sequence-dependent setups can be found in numerous other industrial systems, such as the stamping operation in plastic manufacturing, die changing a metal processing shop, and roll slitting in the paper industry. In many cases sequence-dependent setup costs are significant, thus cannot be omitted during the scheduling/planning optimization procedure. It worth mentioning that in some cases could be convenient incorporate process operations into setup times in order to simplify the scheduling model.

Nowadays, the trend in manufacturing of the production of small batches or unit products to satisfy demand and avoid inventory has made more relevant the scheduling problems with sequence-dependent setup times between all jobs, and not only between batches.[1] Despite of the fact that sequence-dependent setups are a reality for the majority of process industries little work has been done towards the development of an efficient scheduling mathematical model addressing the aforementioned aspects. It is pointed out that the consideration of sequence-dependent setups drastically increases the complexity of scheduling models and limits their efficiency and their practical application, since yields to computational expensive mathematical models.

In this work, the unit-specific general precedence (USGP) scheduling MILP model is presented. The concept of the proposed scheduling model has been inspired by the general precedence framework first introduced by Méndez and Cerdá[2]. In the USGP the general precedence concept is applied only for each unit permitting the consideration of sequence-dependent setup times and costs, an issue that the general precedence framework is not able to tackle. The novel USGP framework results into mathematical models that have extremely low computational times for complicated scheduling problems for the already existed scheduling mathematical models. Thus, complex scheduling problems including sequence-dependent setup costs can be solved efficiently. Some case studies are addressed here in order to shed light on the advantages and the special features of the USGP scheduling model. The proposed model efficiently overcomes the deficiencies of the general precedence framework when tackling sequence-dependent setup times and/or costs.

References

[1] Noivo, J. A.; Ramalhinho-Lourenço, H., Solving two production scheduling problems with sequence-dependent set-up times, J. of Economic Literature Classification: C61, M11, L60. 1998.

[2] Méndez, C.; Cerdá, J., Dynamic scheduling in multiproduct batch plants. Comps & Chem. Eng. 2003, 27, 1247–1259.

Acknowledgements

Financial support received from the Spanish Ministry of Education and Science (FPU grants) is fully appreciated. Besides, financial support from PRISM-MRTN-CT-2004-512233 and DPI2006-05673 projects is gratefully acknowledged.

554e a New Network-Based Representation for Process Scheduling

Diego M. Giménez[1], Gabriela P. Henning[1], and **Christos T. Maravelias**[2]. (1) Universidad Nacional del Litoral, CONICET, Santa Fe, Argentina, (2) University of Wisconsin - Madison, 1415 Engineering Drive, Madison, WI 53706

In this paper we present a new mixed-integer programming (MIP) formulation for the scheduling of general (multi-purpose) process networks. Current models are based on the state-task network (STN) or the resource-task network (RTN) representation (Kondili et al., 1994; Pantelides, 1994). In most existing formulations, it is implicitly assumed that:

a) material transfer between units is always possible, i.e. all process units are connected to all the vessels that are used for the storage of the corresponding input and output states;

b) all input (output) states of a task are transferred simultaneously to (from) the process unit when the task starts (ends);

c) stable output states can be temporarily stored in a process unit after a task is completed, but input states cannot be temporarily stored before a task starts, i.e. in continuous time representations, the beginning of a task must coincide with a time point and at such point all the materials should be available.

However, these assumptions do not always hold. For example, if an intermediate chemical is produced in multiple processes, then several vessels may be used for its storage and each vessel may not be connected to all downstream units. Also, in many processes, input (output) states are not transferred simultaneously to (from) the processing unit. For example, in recovery and purification processes the solvent can be drained earlier. Similarly, in certain chemical reactions reactants are fed before the beginning of the task which actually occurs when the catalyst is added. It is interesting to note here that although the scheduling of multipurpose facilities has received considerable attention, most approaches focus on the development of alternative MIP formulations for the existing STN and RTN representations, but there are very few attempts to address the above limitations. The issue of unit connectivity was tackled by Barbosa-Póvoa and Macchietto (1994). However, to our knowledge none of the existing approaches deals with the shortcomings due to assumptions (b) and (c).

The goal of this paper is the development of a MIP formulation that overcomes these limitations. In achieving this, we develop four key new ideas. First, we use a time representation that does not require tasks to start exactly at a time point, thus probably reducing the number of time points necessary to represent a solution. Second, we allow input (output) states to be transferred to (from) a process unit before (after) a task begins (ends) via the introduction of a "storage" state for each unit. Third, we explicitly model material transfer via "flow" variables, thus explicitly taking into account forbidden connections between units. Fourth, we introduce new variables to denote the idle and storage time in a process unit, as well as the time elapsed before (after) a task starts (finishes). These new variables allow us to enforce matching between tasks and time points without using big-M constraints.

The proposed representation introduces several novel modeling concepts that turn it into a powerful tool. The advantages are considerable in comparison with previous approaches. In spite of its generality, the resulting model complexity is not greater than in many other continuous-time formulations. Also, the proposed approach can be extended to model utility consumption, changeovers and continuous processes. We present several examples where the proposed approach yields solutions that cannot be represented by existing models.

Kondili, E., Pantelides, C.C., Sargent, R. (1993). Comput. Chem. Eng., 17, 211-227.

Pantelides, C.C. (1994). Proceedings on the 2nd FOCAPO Conference, 253–274.

Barbosa-Póvoa, A.P., Macchietto, S. (1994). Comput. Chem. Eng., 18, 1013-1042.

554f Short-Term Scheduling of Multipurpose Batch Processes Using Unit Slots

Naresh Susarla, Jie Li, and **Karimi I. A.**. Department of Chemical and Biomolecular Engineering, National University of Singapore, 4 Engineering Drive 4, Singapore, 117576, Singapore

Scheduling of multipurpose batch plants is challenging and has received considerable attention so far. Several formulations for this problem exist in the literature. The early attempts using Mixed Integer Linear Programming (MILP) formulations were with discrete-time representations (Kondili et al., 1993; Shah et al., 1993; Mockus and Reklaitis, 1994; Lee et al., 2001). However, recent focus has been on continuous-time representation (Maravelias and Grossmann, 2003a; Janak et al., 2004; Sundaramoorthy and Karimi, 2005; Shaik et al., 2006; Mendez et al., 2006) using slot-based, event-based, and sequence-based modeling techniques. For problem/process representation, these models employ various approaches such as State Task Network (STN), Resource-task network (RTN), and Generalized Recipe Diagram (GRD). Sundaramoorthy & Karimi (2005) used process slots (Liu & Karimi, 2007) in their work, but no model has so far used the concept of unit slots (Liu & Karimi, 2007) for scheduling multipurpose plants.

In this work, we use the GRD approach to focus on the generic problem of short-term scheduling in batch processes. For this, we present a novel continuous-time mixed integer linear programming (MILP) model that uses unit slots (Liu & Karimi, 2007). As expected, this leads to a significant reduction in the numbers of slots and binary variables in the formulation. Our model allows variable batch sizes and processing times, various storage configurations (Classes: UIS, LIS, and FIS with policies: UW, LW, and NW), different scheduling objectives (such as profit maximization and makespan minimization), and limited resources and utilities. To demonstrate the effectiveness of our model, we evaluate its performance with some recent models from the literature. Through our extensive numerical evaluation, we further shed light on various considerations that affect solution time. Our comparisons show that our model outperforms the best formulations existing in the literature, and it also uses fewer binary variables, continuous variables, and constraints.

References:

1. Kondili, E., Pantelides, C.C., Sargent, R.W.H., 1993. A general algorithm for short-term scheduling of batch operations-I. MILP formulation. Computer and Chemical Engineering 17, 211-227.

2. Mockus, L., Reklaitis, G.V., 1994. Mathematical programming formulation for scheduling of batch operations based on non-uniform time discretization. Computers and Chemical Engineering 21, 1147.

3. Maravelias, C.T., Grossmann, I.E., 2003a. New general continuous time state-task network formulation for short-term scheduling of multipurpose batch plants. Industrial and Engineering Chemistry Research 42, 3056-3074.

4. Karimi, I.A., McDonald, C.M., 1997. Planning and scheduling of parallel semi continuous processes. 2. Short-term scheduling. Industrial and Engineering Chemistry Research 36, 2701-2714.

5. Ierapetritou, M.G., Floudas, C.A., 1998. Effective continuous time formulation for short term scheduling. 1. Multipurpose batch processes. Industrial and Engineering Chemistry Research 37, 4341-4359.

6. Arul, S., Karimi, I.A., 2005. A simpler better slot-based continuous-time formulation for short term scheduling in multipurpose batch plants. Chemical Engineering Science 60, 2679-2702.

7. Janak, S. L., Lin, X., & Floudas, C. A. (2004). Enhanced continuous-time unit-specific event-based formulation for short-term scheduling of multipurpose batch processes: Resource constraints and mixed storage policies. Industrial and Engineering Chemistry Research, 43, 2516-2533.

8. Pantelides, C. C. Unified frameworks for optimal process planning and scheduling. In Proceedings of the Second International Conference on Foundations of Conputer-Aided Process operations, Crested Butte, CO, 1993; Rippin, D. W. T., Hale, J. C., Davis, J., Eds.; CACHE: Austin, TX, 1993; pp 253-274.

9. Shaik, M. A., Janak, S. L., Floudas, C. A., 2006. Continuous-time models for short-term scheduling of multipurpose batch plants: A comparative study. Industrial and Engineering Chemistry Research, 45, 6190.

10. Mendez, C. A., Jaime C., Grosmann, I. E., Harjunkoski, I., Fahl, M., 2006. State-of-the-art review of optimization methods for short-term scheduling of batch processes. Computers and Chemical Engineering, 30, 913-946.

11. Liu Yu, Karimi, I. A., 2007. Novel continuous-time formulations for scheduling multi-stage batch plants with identical parallel units. Computers and Chemical Engineering, 31, 1671-1693.

565a Fermentative Poly(3-hydroxybutyrate) Production In Alcaligenes Latus: A Combined Metabolic/kinetic Modelling Approach

Giannis Penloglou[1], Avraam Roussos[2], Christos Chatzidoukas[2], and **Costas Kiparissides**[1]. (1) Department of Chemical Engineering, Aristotle University of Thessaloniki & Chemical Process Engineering Research Institute, P.O. Box 472, Thessaloniki, 54124, Greece, (2) Chemical Process Engineering Research Institute, P.O. Box 60361, Thessaloniki, 57001, Greece

Introduction

Polyhydroxyalkanoates (PHAs) are microbial thermoplastics produced in a variety of micro-organisms as intracellular carbon and energy storage compounds. These compounds exhibit significant advantages over conventional polymeric materials as they are non-toxic, they come from renewable sources and they are 100% biodegradable. Polyhydroxybutyrate (PHB) constitutes the most important and widely studied representative of the PHAs. It is a biopolymer with mechanical properties similar to conventional commercial polymers like polypropylene or polyethylene. Despite the promise of these new materials, their introduction to the worldwide market is inhibited by a series of economic and engineering considerations. Commercially available biopolymers are at present, significantly more expensive than their synthetic alternatives. Therefore, there is an emerging need to reduce the overall cost of PHB production by designing novel processes and product separation/purification procedures in order to maximize its yield and productivity. Towards this direction, advanced mathematical models with simulation capabilities of the complex biochemical phenomena occurring in cell factories within a bioreactor, are of decisive importance.

The objective of the present work is to develop a structured, segregated modeling framework for the process design and optimization of the microbial PHB production process. The framework will comprise individual models at different length/time scales namely, metabolic and kinetic models at the single-cell level. At the present stage our focus is on the development of models at the cell-metabolism and polymerization-kinetics level in order to be able to predict both the concentration gradients of intracellular metabolites including PHB, as well as to account for the dynamics of the molecular build-up of polymer chains in terms of their molecular weight distribution (MWD), which in turn reflects the final mechanical properties of the produced biopolymer. Despite the fact that there is a large number of studies based on (structured and unstructured) metabolic models of biopolymer-producing bacteria, none of them has focused on the prediction of the MW properties of the produced polymer in microbial production systems.

Model Development

In the present study a mathematical model of different scales is developed for the description of the cell metabolism, in terms of substrate utilization, biomass growth and PHB accumulation, and at the same time, for the prediction of the molecular weight distribution of the produced biopolymer in Alcaligenes latus.

Sucrose, provided as carbon source, is assimilated by the cells by means of active transport and metabolized either for PHB accumulation and/or for biomass growth (synthesis of new cells) and maintenance.

The pathway of the central aerobic carbon metabolism that leads to the production of PHB and biomass growth in this bacterium consists of a large network of biochemical reactions. For the needs of this study part of the complex metabolic network is simplified into a single reaction step, since we are only interested in specific an not all the produced metabolites.

Initially, sucrose is converted into acetyl-coenzyme A (AcCoA) through the Entner-Doudoroff pathway. The main biosynthetic pathway of P(3HB) consists of three enzymic reactions. In the first reaction, two AcCoA molecules are condensed by 3-ketothiolase (PhaA) into one molecule of aceto-acetyl-coenzyme A (Ac-AcCoA). In the second reaction, the Ac-AcCoA is reduced by reductase (PhaB) to hydroxybutyryl-CoA (3-HBCoA) at the expense of NADH. In the final stage, the 3-HBCoA monomers are polymerized by synthase (PhaC) to form P(3HB).

Alcaligenes latus is a growth associated bacterium that has the ability to accumulate PHB even when biomass growth is not ceased. As long as the above three reactions take place, biomass growth from AcCoA occurs in parallel, in the presence of a nitrogen source, while catabolism of AcCoA also occurs through the Krebs cycle, where it is converted to CO2. The necessary energy for the synthesis of every molecule of residual biomass from AcCoA, is considered constant. The formula of the produced non-polymer biomass (residual) is assumed to be CH1.76O0.47N0.19, as determined by Yamane et al. (1995), through elemental analysis of A. latus cells containing PHB up to 45% of dry cell weight.

Finally, the accumulated P(3HB) can be hydrolysed into its building blocks (3-HB monomers) by the action of depolymerase (PhaZ) and then utilised by the bacteria as carbon and energy source, when carbon-limited conditions occur in the cells environment (i.e., depletion of carbon source).

The third stage of the enzymic reactions described above comprises a set of comprehensive, elementary reactions that constitute the enzymic polymerization mechanism of the monomer unit (3-HBCoA) into polymer P(3HB), which can be used as carbon source. The polymerization-depolymerization mechanism considered in this study consists of initiation, propagation, chain transfer to water and depolymerization reactions (Kawaguchi and Doi, 1992). The P(3HB) synthesis on the active site of polymerase is initiated by the reaction with the hydroxybutyryl-CoA monomer and is followed by the chain propagation reaction. The chain transfer reaction transforms the live (propagating) polymer chains into dead (inactive) polymer chains. Subsequently, under conditions of carbon starvation the produced (PHB) polymer chains degrade in order to be used as carbon source.

The rate of monomer production is assumed to be a constant fraction of the rate of substrate consumption, thereby associating the polymerization kinetic model with the metabolism of the cell.

Notice that the connection point between the polymerization kinetic model and the central metabolism of A. latus bacteria is the metabolite 3-HBCoA. 3-HBCoA is produced from sucrose and at the same time is consumed for the formation and propagation of the PHB polymer chains. The dynamics of these two competing rates designate the accumulation of polymer within the bacterial cells. In order to calculate the rate of substrate consumption, a metabolically unstructured (lumped) model is employed, accounting for the dynamic production/consumption rates of biomass, carbon and nitrogen substrates and polyhydroxybutyrate.

Model Application

Initially, the metabolic and polymerization kinetic models are individually developed and verified. The unknown parameters (i.e., kinetic constants, monomer production rate, etc.) are estimated based on experimental data. At the first stage a uniform cell population is assumed and we focus on the single cell carbon metabolism (metabolic level) and the polymerization mechanism (kinetic level), with respect to the prediction of the molecular weight distribution of the produced biopolymer. In this framework and for the accurate development of the kinetic/polymerization model it is necessary to obtain information about the synthesis of monomer directly from the uptake rate of the carbon source and then its consumption rate in the polymerization process. The operation of the central carbon metabolism in A. latus for biomass growth and PHB accumulation is utilized, in order to obtain this kind of information. Furthermore the time evolution of the uptake rate of the carbon source derives from a lumped model of cell metabolism, comprising simple differential equations.

The developed mathematical model is able to predict the consumption rate of carbon and nitrogen sources, the dynamic evolution of biomass growth and polymer product accumulation and also the molecular weight distribution of the produced polymer, along the fermentation of A. latus bacteria.

References

• Beun J.J., Dircks K., van Loosdrecht M.C.M., Heijnen J.J., "Poly-b-hydroxybutyrate metabolism in dynamically fed mixed microbial cultures", 2002, Wat. Res., vol. 36, pp. 1167–1180.

• Kawaguchi Y., Doi Y., "Kinetics and mechanism of synthesis and degradation of Poly(3-hydroxyburate) in Alcaligenes eutrophus", Macromolecules, 1992, vol. 25, pp. 2324-2329.

• Yamane T., Fukunaga M., Lee Y.W., "Increased PHB productivity by high-cell-density fed-batch culture of Alcaligenes latus, a growth-associated PHB producer", Biotechnol. Bioeng., 1995, vol. 50, pp. 197–202.

565b Indirect Response Model of Arsenic Exposure on Gene Expression

Meric A. Ovacik[1], M.G. Ierapetritou[2], P.G. Georgopoulos[3], W. Welsh[3], and I.P. Androulakis[4]. (1) Chemical and Biochemical Engineering, Rutgers University, Piscataway, NJ 08854, (2) Chemical & Biochemical Engineering Department, Rutgers University, the State University of New Jersey, 98 Brett Road, Piscataway, NJ 08854, (3) University of Medicine and Dentistry of New Jersey (UMDNJ)/Robert Wood Johnson Medical School (RWJMS), Piscataway, NJ 08854, (4) Biomedical Engineering, Rutgers University, Piscataway, NJ 08854

Arsenic is an environmental toxicant and the exposure to it is a worldwide public health issue. Arsenic is categorized as a human carcinogen based on epidemiological studies in human liver. Aside from cancer, arsenic exposure is associated with liver injury, developmental abnormalities and genotoxicity[1]. However, the mechanism of action to arsenic exposure, especially at the molecular level, is not fairly understood.

A simple conceptual model for arsenic metabolism in a cell includes incorporating components for influx, efflux of inorganic arsenic, its metabolites and molecular targets that refer to any intracellular molecule , which inorganic arsenic or its metabolites interact with[2]. Intracellular concentrations of arsenicals, which are responsible for the arsenic toxicity, depend on the uptake rate and the loss of the species from the cell. In this study, we used of gene expression data as molecular targets from zebrafish liver, which as a toxicological model has a potential to propose new perspectives into chemical toxicity, to construct an indirect response model of arsenic exposure. The experimental data was obtained from Gene Expression Omnibus (GEO). Adult zebrafish were treated with 15ppm As(V) for 96 hr and fish livers were sampled at 8, 24, 48 and 96 hours for the microarray study, where the histological studies confirmed liver injury[1]. For the arsenic metabolism, we employed gene expression of specific ABC transporters, MRP1 and MRP2, which are the critical factors in removal of arsenic metabolites from cells [3]. The expression profiles of these genes were used to characterize arsenic removal activity in the cells.

We employed a shape-based microclustering algorithm [4] to reveal important profiles of the gene expression over time and the overpopulated profiles. The functional categorizations of the extracted intrinsic responses of arsenic exposure were stress response, DNA damage, and increased metabolism due to stress response and cytoskeletal changes. Based on the adaptive response model of the zebrafish liver, we proposed an indirect response model of arsenic exposure on transcriptional regulation that can be used for understanding liver injury. Arsenic metabolism intermediates were known to directly apply oxidative stress damage to protein and DNA methylation[5]. Heat shock proteins and DNA repair protein could help to stabilize the arsenic effect on DNA and protein. Meanwhile, central liver metabolism was increased due to repair processes. The hypermetabolism and repair processes taken together required active transportation and many of the transport proteins are coupled with large movements within the cell. Therefore, genes coding for cytoskeletal changes were altered and suggesting large-scale reshaping and restructuring of cell that were the reasons of the histological changes in liver tissues.

The indirect response model, as it is, describes the liver injury of steady arsenic exposure in zebrafish for 96 hour. The intrinsic responses showed a constant decrease between 48 hour and 96 hour interval suggesting that the regulation of liver injury initiated in this time interval. The indirect response model aims to simulate and to analyze the responses where the liver injure can be interfered. Moreover, the indirect response model includes arsenic metabolism intermediates implicitly, which allows the model to be integrated in multiscale modeling of arsenic. Incorporation of available pharmacodynamic models of arsenic metabolism for several organisms, such as mouse and rats, would allow the indirect response model to analyze also different dose effects on liver injury. Additionally, since arsenic trioxide is a competent cancer medication, the model also can be used in drug administration complications such as over dose or understanding the side effects.

1. Lam SH, Winata CL, Tong Y, Korzh S, Lim WS, Korzh V, Spitsbergen J, Mathavan S, Miller LD, Liu ET, and Gong Z. Transcriptome kinetics of arsenic-induced adaptive response in zebrafish liver. Physiol Genomics, 2006. 27:351-61.

2. Thomas DJ. Molecular processes in cellular arsenic metabolism. Toxicol Appl Pharmacol, 2007. 222:365-73.

3. Liu J, Kadiiska MB, Liu Y, Lu T, Qu W and Waalkes MP. Stress-related gene expression in mice treated with inorganic arsenicals. Toxicol Sci, 2001. 61:314-20.

4. Yang E, Maguire T, Yarmush ML, Berthiaume F and Androulakis IP. Bioinformatics analysis of the early inflammatory response in a rat thermal injury model. BMC Bioinformatics, 2007. 8:10.

5. Zhao CQ, Young MR, Diwan BA, Coogan TP and Waalkes MP. Association of arsenic-induced malignant transformation with DNA hypomethylation and aberrant gene expression. Proc Natl Acad Sci U S A, 1997. 94:10907-12.

565c Design and Optimization of An RNAi Based Biomolecular Logic Circuit

Leonidas Bleris[1], Keller Rinaudo[1], Rohan Maddamsetti[1], Sairam Subramanian[2], Ron Weiss[2], and Yaakov Benenson[1]. (1) FAS Center for Systems Biology, Harvard University, 7 Divinity Avenue, Cambridge, MA 02138, (2) Electrical Engineering, Princeton University, E-Quad B-312, Olden St., Princeton, NJ 08544

Components of a living organism, from organs and tissues to single cells and subcellular compartments, exchange and process numerous molecular signals in order to coordinate their activity. When these components fail, they generate characteristic signals that often trigger self-repair processes but can also cause disease when left unchecked. In the not-so-distant future, engineered biomolecular circuits will process information in human cells monitoring in parallel multiple inputs, detecting minute changes, rapidly assessing a patient's condition, and responding in infinitesimal time. Such systems will be used for diagnosing, preventing, treating, and monitoring disease in ways that achieve optimal and highly specific individual health-care, redefining personalized medicine and opening the path to new technologies.

Today, scientists in the cross-sections of disciplines such as biology, chemistry, mathematics, and engineering strive to produce molecular circuits with novel and useful functionalities, in a strikingly similar manner to physicists and engineers that built the fundamental building blocks of computers and modern electronic devices during the 19th century. Similar to a transistor, the basic component of an electrical circuit, with the voltage indicating binary high and low output, in cells a gene can have a binary high and low state depending on the protein concentration. Towards this direction, there are several prototype "biodevices" [1-5] that operate in cells, such as oscillators, toggle switches, and circuits implementing basic Boolean operations (i.e. AND, OR, NOT logic gates). These devices comprise of genetic and biochemical components such as RNA, DNA fragments, proteins, and inducer molecules.

In [6] we presented an experimental implementation of modules that can evaluate logic expressions in human kidney cells, using disjunctive and conjunctive normal forms (DNF and CNF). Since any Boolean expression can be represented in CNF and in DNF form these modules allow for the evaluation of any arbitrary logic or condition in vivo. The CNF and DNF modules use a combination of transcriptional and post-transcriptional regulation pathways as the underlying molecular "hardware". The logic expressions are encoded in a multigene network as the "software", and the inputs, i.e. the truth values of the variables, are encoded by the presence or absence of small interfering RNAs (siRNAs) utilizing the RNA interference pathway. The result of the evaluation is read out using a fluorescent reporter protein. We conducted experiments to prove the feasibility of the computation framework using transient cotransfections of the software segment genes and siRNA molecules. We present experimental results of subsequent generations of the evaluator, starting from two-input logical AND and OR operations to more complex five variable cases. We highlight the approach used for the optimization of the performance of this circuit and discuss extensions and limitations.

[1]. Weiss, R., Homsy, G.E., Knight, T.F. Toward in vivo digital circuits. DIMACS workshop on evolution in computation (1999)

[2]. Benenson, Y., Gil, B., Ben-Dor, U., Adar, R. & Shapiro, E. An autonomous molecular computer for logical control of gene expression. Nature 429, 423-429 (2004)

[3]. Stojanovic, M. N., Stefanovic, D. A deoxyribozyme-based molecular automaton. Nature Biotechnology 21, 1069-1074 (2003)

[4]. Hasty J, McMillen D and Collins JJ. Engineered gene circuits. Nature 420: 224-230 (2002).

[5]. Elowitz MB, Leibler S. A Synthetic Oscillatory Network of Transcriptional Regulators. Nature 403, 335-8 (2000)

[6]. K. Rinaudo*, L. G. Bleris*, R. Maddamsetti, S. Subramanian, R. Weiss, Y. Benenson. A Universal RNAi-based Logic Evaluator that Operates in Mammalian Cells. Nature Biotechnology 25, 795-801 (2007) (* Equal contribution)

Web Page: www.sysbio.harvard.edu/csb/benenson/bleris.html

Analysis procedure for signal transduction pathways by clustering parameters according to their sensitivity profiles

Yunfei Chu, Zuyi Huang and Juergen Hahn
Artie McFerrin Department of Chemical Engineering
Texas A&M University
College Station, TX 77843-3122

1. Introduction

Mathematical modeling and analysis of signal transduction networks plays an important role in systems biology. Models of signal transduction networks generally consist of nonlinear differential equations (Schoeberl et al., 2002; Singh et al., 2006) with a large number of parameters whose values are not precisely known and only a portion of which can be estimated from experimental data.

Sensitivity analysis techniques are commonly used to determine the key parameters of a model. Sensitivity values are used to rank the importance of the effect of changes in parameter values on the output. However, for a dynamic system the sensitivity is a function of time and lumping the effect over time into a scalar value only provides partial information about the importance of a parameter. One reason for this is that parameters which have similar cumulative effects may still cause very different dynamic changes of the outputs. It has been recognized that distinct temporal activation profiles of the same signaling proteins result in diverse physiological responses (Marshall, 1995; Hoffmann et al., 2002; Kholodenko 2006) and, therefore, categorizing the dynamic effects of parameters is of great importance.

The aim of this work is to use the entire time-dependent sensitivity profile of parameters for ranking the importance of parameters of signal transduction networks but also to determine which parameters have similar dynamic sensitivity profiles that cannot be distinguished from one another. A technique based on parameter clustering of the sensitivity profiles is developed to rank the parameters. A similarity measure is defined to quantify correlations among the effects that changes in parameters have on the measurements. If the similarity measure has a value of unity then the effects of two parameters cannot be distinguished from one another, i.e., the effect caused by changes in one of the parameters can be compensated by changing the other one. If the similarity is zero then the two parameters have distinct effects. The parameters can be grouped by a clustering algorithm based upon their sensitivity profile. Since parameters in a group have correlated effects, the magnitude of the sensitivity vector can be used to rank the parameters in a group. The parameter with the longest sensitivity vector in each group can be selected as the representative parameter of that group. Thus, the technique identifies the important parameters of the signal transduction network as the representative parameters for each of the groups.

The advantage of this technique over conventional methods is that the parameters are ranked by their dynamic effects rather than the cumulative effects only. This allows to not only determine a set of parameters that are important for the signal transduction network but also to characterize the effect that changes in parameters have on the output. As a result it is possible to view each cluster of parameters as a set of parameters where uncertainty in the value of any of the parameters can be compensated by the values of other parameters. For illustration purposes this technique is applied to the Jak/STAT and MAPK/NF-IL-6 signal transduction network stimulated by interleukin 6.

2. Background

Successful parameter estimation depends, among other things, on parameter identifiability. Parameter identifiability can be determined either analytically, also called structural identifiability, or numerically, also called practical identifiability (Walter 1987; Walter and Pronzato, 1997; Ljung, 1999). Analytical identifiabilty investigates uniqueness of the solution derived from parameter estimation while the numerical identifiability focuses on the stability of the solution. Additionally, analytical identifiabilty can be either global or local. While global identifiability includes local identifiability as a special case, it is significantly more difficult to determine global identifiability (Ljung and Glad, 1994) as approaches based upon Taylor series approximations and similarity transformations (Chappell *et al.*, 1990) are restricted to small systems. Local identifiability on the other hand is relatively straightforward to test as the rank of the parameter output sensitivity matrix determines local identifiability. As the techniques introduced in this work are based upon these concepts, the definitions of identifiability as presented in Rothenbe (1971) are briefly reviewed next.

Definition 1: A parameter point $\boldsymbol{\theta}_0$ is said to be locally identifiable if there exists an open neighborhood of $\boldsymbol{\theta}_0$ containing no other $\boldsymbol{\theta}$ which produces the identical observations $\mathbf{y}$.

Condition 1: Let $\boldsymbol{\theta}_0$ be a parameter point and the sensitivity matrix $\mathbf{S}(\boldsymbol{\theta}) = \partial \mathbf{y} / \partial \boldsymbol{\theta}^{\mathrm{T}}$ has constant rank in a neighborhood of $\boldsymbol{\theta}_0$. Then $\boldsymbol{\theta}_0$ is locally identifiable if and only if $\mathbf{S}(\boldsymbol{\theta}_0)$ is nonsingular.

It should be noted that the sensitivity matrix having constant rank is a necessary condition. If this condition is removed then the nonsingularity of the sensitivity matrix is just a sufficient condition for local identifiability, or, in other words, a rank-deficient sensitivity matrix does not imply that the parameters are not locally identifiable. The condition of constant rank has to be checked analytically and evaluating this condition for one nominal value of the parameters is insufficient.

Analytical identifiability guarantees the existence of a unique solution in at least a small neighborhood of the nominal point. However, analytically identifiable of parameters does not guarantee accurate estimation in practice. If sensitivity matrix is not singular but is ill-conditioned, then noise in the data will result in large variations of the estimated parameter values. To obtain an accurate estimate it is also required that the sensitivity matrix should not be ill-conditioned.

3 Investigation of pairwise indistinguishable parameter sets

As the number of parameters in many fundamental models far exceeds the number of parameters that can be accurately estimated from available data, it is necessary to determine a subset of parameters which can be estimated. Parameter selection can be viewed as a special case of model reduction as only the values for some parameters are determined from parameter estimation while all other parameters are assumed to remain at their constant value.

This section presents two techniques for determining subsets of parameters to be estimated. The first technique is an analytical approach which derives the condition where the output of a system with fewer parameters is identical to the one for the entire parameter set. The second method is a numerical approach which does not focus on the outputs being identical but instead investigates the error bound that results from including fewer parameters whose values can change in a model.

3.1 Analytical approach

Determining which parameters can be lumped together in a model is a problem that is related to parameter identifiability. Each time a parameter is not locally identifiable, it is possible to reduce the parameter by setting it to a constant value. If the effects that two parameters have on the output are identical then each parameter may be individually identifiable, however, only one of the two parameters needs to be considered and the other can be set to a constant value. The following definitions and propositions define this situation.

Definition 2: A parameter set is said to be a pairwise indistinguishable set when any two parameters in the set are not locally identifiable.

From Condition 1 it can be seen that if the two parameters are not locally identifiable then their sensitivity matrix is rank deficient. This implies that sensitivity vectors of two parameters are parallel and that the two parameters can be replaced by a linear combination of them, e.g., by setting one parameter to its nominal value and using only the other parameter.

Proposition 1.
Assume the output $\mathbf{y} \in R^m$ is an analytical function of the parameter $\boldsymbol{\theta} \in R^n$

$$\mathbf{y} = \mathbf{f}(\boldsymbol{\theta}), \tag{1}$$

and $\overline{\boldsymbol{\theta}}$ is a nominal value of the parameter. If the sensitivity value of the output with respect to θ_i (i=1,...,r) is nonzero

$$\frac{\partial f_s}{\partial \theta_i} \neq 0, \quad i = 1,\cdots,r \text{ and } s = 1,\cdots,m. \tag{2}$$

then for any $\boldsymbol{\theta}$ in a neighborhood of $\overline{\boldsymbol{\theta}}$, there exist a function $\psi(\boldsymbol{\theta})$ such that

$$\mathbf{f}(\theta_1,\cdots,\theta_{r-1},\theta_r,\theta_{r+1},\cdots,\theta_n) = \mathbf{f}(\overline{\theta}_1,\cdots,\overline{\theta}_{r-1},\psi(\theta_1,\cdots,\theta_n),\theta_{r+1},\cdots,\theta_n) \tag{3}$$

if and only if the sensitivity vectors of the output with respect to θ_i (i=1,...,r) are parallel to each other

$$\frac{\partial \mathbf{f}}{\partial \theta_i} + \alpha_i \frac{\partial \mathbf{f}}{\partial \theta_r} = 0, \qquad i = 1,\cdots,r-1 \tag{4}$$

where α_i is a function of $\boldsymbol{\theta}$.

Proposition 1 states that the effects that changes of the values of parameters in a pairwise indistinguishable set have on the outputs can be lumped together and the same changes can be expressed by any one parameter of the set. As a result of this, all other parameters can be fixed at their nominal values. Accordingly, the output function can be re-parameterized to have fewer parameters, where one possible re-parameterization is shown in Proposition 1.

Illustrative example
A simple nonlinear regression model is used to illustrate the presented analytical procedure.
Let

$$\mathbf{f}(\theta_1,\theta_2,\theta_3) = \begin{bmatrix} \theta_1\theta_2 + \theta_3 \\ \theta_1\theta_2\theta_3 \end{bmatrix} = \mathbf{g}(\psi,\theta_3), \tag{5}$$

where it can easily be seen that a substitution $\psi = \theta_1\theta_2$ can be made. However, this result is derived here using the procedure presented above.
The sensitivity vectors for θ_1 and θ_2 are computed to be

$$\frac{\partial \mathbf{f}}{\partial \theta_1} = \begin{bmatrix} \theta_2 \\ \theta_2\theta_3 \end{bmatrix} \text{ and } \frac{\partial \mathbf{f}}{\partial \theta_2} = \begin{bmatrix} \theta_1 \\ \theta_1\theta_3 \end{bmatrix}. \tag{6}$$

These two sensitivity vectors are parallel and are related by the following differential equation:

$$\frac{\partial \mathbf{f}}{\partial \theta_1} - \frac{\theta_2}{\theta_1}\frac{\partial \mathbf{f}}{\partial \theta_2} = 0. \tag{7}$$

Equation (7) can be used to compute the re-parameterization of ψ. The characteristic ordinary differential equation is given by

$$\frac{d\theta_1}{1} = \frac{d\theta_2}{-\theta_2/\theta_1}. \tag{8}$$

which can be solved by separation of variables and the solution is

$$\theta_1\theta_2 = C, \tag{9}$$

where C is a constant and the first integral is $\psi = \theta_1\theta_2$ which is the new variable to re-parameterize the model.

3.2 Numerical approach

The procedure presented in the last section results in a set of characteristic ordinary differential equations which needs to be solved. As it is rarely the case that this expression can be solved analytically, a numerical approach is presented here. This numerical approach does not require the sensitivity vectors to be parallel, however, the angle between the sensitivity vectors should be small. In this case the parameters can be viewed as being pairwise indistinguishable with a certain numerical precision.

A similarity measure of the effect of two parameters on the output can be defined by

$$\cos\phi_{ik} = \frac{\left|\mathbf{s}_i^{\mathrm{T}}\mathbf{s}_k\right|}{\left\|\mathbf{s}_i\right\|_2 \left\|\mathbf{s}_k\right\|_2}, \tag{10}$$

where $\phi_{ik} \in [0, \pi/2]$ is the angle between sensitivity vector $\mathbf{s}_i$ and $\mathbf{s}_k$. The value of the similarity measure ranges from 0 to 1 where a value of one indicates that the two vectors are parallel to one another and that the two parameters cannot be distinguished. A value of zero, on the other hand, refers to the sensitivity vectors of the parameters being orthogonal and that the parameters have distinct effects on the output. It is should be noted that the absolute value is used for the similarity measure as it is of little important which direction the sensitivity vectors have.

Based on the similarity measure, the parameters can be grouped by a clustering algorithms. Agglomerative hierarchical clustering is used here since from the hierarchical tree it is easy to determine how many groups the parameters should be clustered into and how the least similarity value is changed as the number of groups changed. However, other clustering algorithms (Duda et al., 2001; Theodoridis and Koutroumbas, 2006) could be used with little significant change in the outcome.

This method forms groups by repeatedly merging different groups of parameters. Initially, each parameter is in a group by itself. In a second step, the two groups with the largest similarity measure are merged into a new group. The similarity within a group can be controlled by the number of groups that one chooses to have.

Since the parameters in a numerical pairwise distinguishable set have similar effects on the output, a parameter in the set can be selected as the representative for the group. Parameter estimation is simplified due to this procedure as the number of parameters can be reduced to the number of groups. However, since the sensitivity vectors for the parameter in a group are not perfectly parallel, it has to be taken into account that there will be a discrepancy between the parameter-output effect of the original system and the one with a reduced number of parameters.

This discrepancy can be measured by the prediction gap between the two functions

$$d(\boldsymbol{\theta}) = \min_{\boldsymbol{\psi}} \left\|\mathbf{f}(\boldsymbol{\theta}) - \mathbf{g}(\boldsymbol{\psi})\right\|_2, \tag{11}$$

where **f** is the original output function and **g** is derived from **f** when only one parameter per group is considered and all other ones are fixed at their nominal values. The individual parameter of $\boldsymbol{\psi}$ are the representative parameters for each group and are a subset of $\boldsymbol{\theta}$. The prediction gap indicates how well the model with a reduced parameter set can approximate the behavior of the original model.

It is non-trivial to compute this discrepancy for general nonlinear functions. Due to this an approximation of d based upon linearization is used in this work. The truncated Taylor series approximation of original function, **f**, with respect to the parameters is given by

$$\mathbf{f}(\boldsymbol{\theta}) \approx \mathbf{f}(\bar{\boldsymbol{\theta}}) + \frac{\partial \mathbf{f}}{\partial \boldsymbol{\theta}^{\mathrm{T}}}(\boldsymbol{\theta} - \bar{\boldsymbol{\theta}}), \tag{12}$$

and the approximated Taylor expression of the function **g** is

$$\begin{aligned} \mathbf{g}(\boldsymbol{\psi}) &\approx \mathbf{g}(\bar{\boldsymbol{\psi}}) + \frac{\partial \mathbf{g}}{\partial \boldsymbol{\psi}^{\mathrm{T}}}(\boldsymbol{\psi} - \bar{\boldsymbol{\psi}}) \\ &= \mathbf{f}(\bar{\boldsymbol{\theta}}) + \sum_i \frac{\partial \mathbf{f}}{\partial \theta_{s_i}}\left(\psi_i - \bar{\theta}_{s_i}\right) \end{aligned}, \tag{13}$$

where s_i is the index for the unfixed parameters. Then the discrepancy becomes

$$\mathbf{f}(\boldsymbol{\theta}) - \mathbf{g}(\boldsymbol{\psi}) = \mathbf{S}\mathbf{x} - \mathbf{T}\mathbf{y}, \tag{14}$$

where $\mathbf{S} = \left.\frac{\partial \mathbf{f}}{\partial \boldsymbol{\theta}^{\mathrm{T}}}\right|_{\bar{\boldsymbol{\theta}}}$, $\mathbf{T} = \left.\frac{\partial \mathbf{f}}{\partial \boldsymbol{\theta}_{\mathbf{s}}^{\mathrm{T}}}\right|_{\bar{\boldsymbol{\theta}}}$, $\mathbf{x} = \boldsymbol{\theta} - \bar{\boldsymbol{\theta}}$ and $\mathbf{y} = \boldsymbol{\psi} - \bar{\boldsymbol{\theta}}_{\mathbf{s}}$. The discrepancy is dependent on the value of parameters and the worst case can be considered

$$d = \max_{\|\mathbf{x}\|_2 = 1} \min_{\mathbf{y}} \|\mathbf{S}\mathbf{x} - \mathbf{T}\mathbf{y}\|_2 . \tag{15}$$

Since the discrepancy may increase unbounded with increased of length of **x,** a constraint is placed on the length of **x**.

The similarity of parameters within a group can be controlled by determining the number of groups for the parameter set. In the extreme case where each group only contains one parameter, the discrepancy is between the original function and the one with a reduced parameter set is zero as the two parameter sets are equal. However, the discrepancy will increase as fewer groups are used and the similarity within groups decreases. It will be shown in the following that the discrepancy can be bounded by a decreasing function of the least similarity value in a set.

Proposition 2. Let $\mathbf{s}_k$ be the *k*-th column vector of the matrix **S**. Then the discrepancy

$$d = \max_{\|x\|_2 = 1} \min_{y} \|\mathbf{S}\mathbf{x} - \mathbf{s}_k y\|_2 \tag{16}$$

can be bounded by

$$d \le \sqrt{1 - \cos^2 \phi_m} \sqrt{\sum_{i \ne k} \|\mathbf{s}_i\|_2^2}, \tag{17}$$

where $\cos\phi_m$ is the smallest similarity value in a group.

Proposition 3. Let $\mathbf{s}_{l_i}$ be the l_i-th column of the matrix **S** where *l* is the index of groups and *i* is the index of the sensitivity vector in a group. Then the discrepancy (Eq. 15) can be bounded by

$$d \le \sqrt{\sum_l d_l^2}, \tag{18}$$

where d_l is the discrepancy of the *l*-th group.

3.3 Application of determining pairwise indistinguishable parameter set for parameter estimation

One important step for parameter estimation is to select the set of parameters to be estimated. It is possible to formulate the parameter set selection procedure as an optimization problem, such as

$$\begin{aligned} &\mathbf{z}^* = \arg\max_{\mathbf{z}} \ \log\det\left(\mathbf{F}(\mathbf{z})\right) \\ \text{s.t. } &\mathbf{F}(\mathbf{z}) = \mathbf{FIM}_{(i_1,\cdots i_{n_s})}^{(i_1,\cdots i_{n_s})} \text{ with } i_j \text{ that } z_{i_j} = 1,\ j = 1\cdots n_s \\ &z_1 + z_2 + \cdots + z_{n_\theta} = n_s \\ &z_i \in \{0,1\},\ i = 1\cdots n_\theta \end{aligned} \quad . \tag{19}$$

where the decision vector $\mathbf{z} \in \{0,1\}^{n_\theta}$ denotes whether a parameter is included in the selected parameter subset. If z_i=1 then θ_i belongs to the selected subset with the size of n_s. The matrix **FIM** is the Fisher information matrix of all parameters. **F(z)** is the Fisher information matrix of the parameters included in the selected subset and it is equal to the principal submatrix of **FIM** with the indices of the non-zero decision variables (the entries of column i_j and row i_k, j, k = 1…n_s).

The optimization problem given by equation (19) is nontrivial to solve as the number of possible combinations of parameters grows exponentially with the number of parameters in the problem. Reducing the number of parameters to be considered can significantly reduce the computational burden. The parameter clustering algorithm can be used as described in the previous subsection. Since only one parameter per can be reliable estimated from data, it is unnecessary to consider all parameters for the parameter set selection problem.

Algorithm of parameter selection base on parameter clustering

Step 1. Calculate the sensitivity vectors of the output with respect to the parameters.

Step 2. Determine n_s, the number of parameters per set, by singular value decomposition of the sensitivity matrix or the methods of forward selection.

Step 3. Set parameters whose sensitivity vectors have small length (e.g., less than 5% of the largest one) to their nominal values.

Step 4. Cluster the parameters into n_g ($n_g \geq n_s$) groups with the similarity measure (Eq. 10) by hierarchical clustering.

Step 5. Select the parameter which has the largest sensitivity vector in a group as the representative of the group.

Step 6. Select n_s parameters from n_g representatives to optimize the criterion function by solving the optimization problem given by equation (19).

Different approaches exist for calculating the parameter sensitivity in Step 1. One commonly used technique is to calculate the sensitivity value $\partial\mathbf{y}(t)/\partial\boldsymbol{\theta}^{\mathrm{T}}$ by solving the system equations

$$\begin{aligned} \frac{d\mathbf{x}}{dt} &= \mathbf{f}\left(\mathbf{x},\mathbf{u},\boldsymbol{\theta}\right) \\ \mathbf{y} &= \mathbf{h}\left(\mathbf{x},\mathbf{u},\boldsymbol{\theta}\right) \end{aligned} , \tag{20}$$

and the sensitivity equations simultaneously

$$\begin{aligned} \frac{d}{dt}\frac{\partial\mathbf{x}}{\partial\boldsymbol{\theta}^{\mathrm{T}}} &= \frac{\partial\mathbf{f}}{\partial\mathbf{x}^{\mathrm{T}}}\frac{\partial\mathbf{x}}{\partial\boldsymbol{\theta}^{\mathrm{T}}} + \frac{\partial\mathbf{f}}{\partial\boldsymbol{\theta}^{\mathrm{T}}} \\ \frac{\partial\mathbf{y}}{\partial\boldsymbol{\theta}^{\mathrm{T}}} &= \frac{\partial\mathbf{h}}{\partial\mathbf{x}^{\mathrm{T}}}\frac{\partial\mathbf{x}}{\partial\boldsymbol{\theta}^{\mathrm{T}}} + \frac{\partial\mathbf{h}}{\partial\boldsymbol{\theta}^{\mathrm{T}}} \end{aligned} , \tag{21}$$

where **x**, **u**, **y**, **θ** are states, inputs, outputs, and parameters, respectively. The sensitivity vector consists of the sensitivity values at different time points

$$\mathbf{s}_i = \left[\partial \mathbf{y}^{\mathrm{T}}(t_1)/\partial \theta_i \quad \partial \mathbf{y}^{\mathrm{T}}(t_2)/\partial \theta_i \quad \cdots \quad \partial \mathbf{y}^{\mathrm{T}}(t_n)/\partial \theta_i \right]^{\mathrm{T}}, \tag{22}$$

where $\mathbf{s}_i$ is the sensitivity vector with respect to the parameter θ_i.

The number of parameters per set from Step 2 can be determined by the rank of the sensitivity matrix. Each column of the sensitivity matrix is a sensitivity vector of a parameter. The number of columns is equal to the number of parameters. However, due to correlation between parameters, the sensitivity matrix may be ill-conditioned. The rank of the sensitivity matrix can be determined by the number of singular values greater than a certain threshold. Step 3 represents a simple methodology for reducing the parameter set as no parameter with a small length of the sensitivity vector needs to be considered. Step 4 performs clustering of the remaining parameters into groups and the parameter with the largest sensitivity vector for each group is chosen as the representative of this group in Step 5. Step 6 selects the parameters to be considered for solution of the optimization problem from equation (19) by taking one parameter per cluster as described in Step 5.

The presented technique can significantly reduce the computational burden for solving the optimization problem given by equation (19) as the computational effort for solution of this problem grows exponentially with the number of parameters to be considered.

4 Case study

Modeling signal transduction networks is one important component of systems biology. Signal transduction networks are biochemical reaction networks that can contain a large number of proteins and the number of parameters in a model usually exceeds the number of proteins. Since it is not possible to determine the majority of parameters from experimental data, some form of parameter set selection has to take place before parameter estimation is performed (Gadkar et al., 2005; Yue et al., 2006; Jaqaman and Danuser 2006).

To illustrate the technique presented in this work, a model of a signal transduction pathway for hepatocytes stimulated by Interleukin-6 is used (Singh et al., 2006). The model, shown in Figure 1, contains two pathways: Janus-associated kinases & signal transducers and transcription factors are activated in one pathway while the other pathway involves the activation of mitogen-activated protein kinases. This model consists of 66 nonlinear ordinary differential equations and includes 115 parameters. The state variables are the concentrations of the proteins in the pathway and the input variable is the concentration of Interleukin-6 outside of the cell that initiates signal transduction. The output variable is the concentration of $(STAT3N^*)_2$ (dimer of activated STAT3 in the nucleus) as this transcription factor can be indirectly measured using a green fluorescent protein (GFP) reporter system. A detailed description of the original version of the model and the nominal values of the parameters can be found in the literature (Singh et al., 2006; Chu et al., 2007), however, the model has been updated to describe the mechanism that SOCS3 and SHP2 compete for the same binding site on the receptor (Huang et al., 2007).

The Fisher information matrix is computed in a first step. The sensitivity value is sampled every minute during the time interval from 0 to 12 hr to form the sensitivity vector. Singular value decomposition of the sensitivity matrix determines that the 9th through 115th singular values are close to zero. Accordingly, a parameter set consisting of 8 parameters will be selected. In a second step, the lengths of the sensitivity vectors are analyzed. 70 of the parameters have sensitivity vectors with a length that is less than 5% of the length of the largest sensitivity vector. These 70 parameters will be set to their nominal values and not considered further. The problem to be solved turns into a problem where a combination of 8 parameters needs to be chosen from a set of 45 parameters such that the *D*-optimality criterion is maximized. If an exhaustive search were to be performed then the number of possible parameter sets that would have to be evaluated would be $\sim 2\times10^{8}$. For the purpose of comparison, the forward selection (the orthogonalization method), a solution of the optimization problem via genetic algorithm, and the clustering method introduced in this paper are applied and discussed in the following.

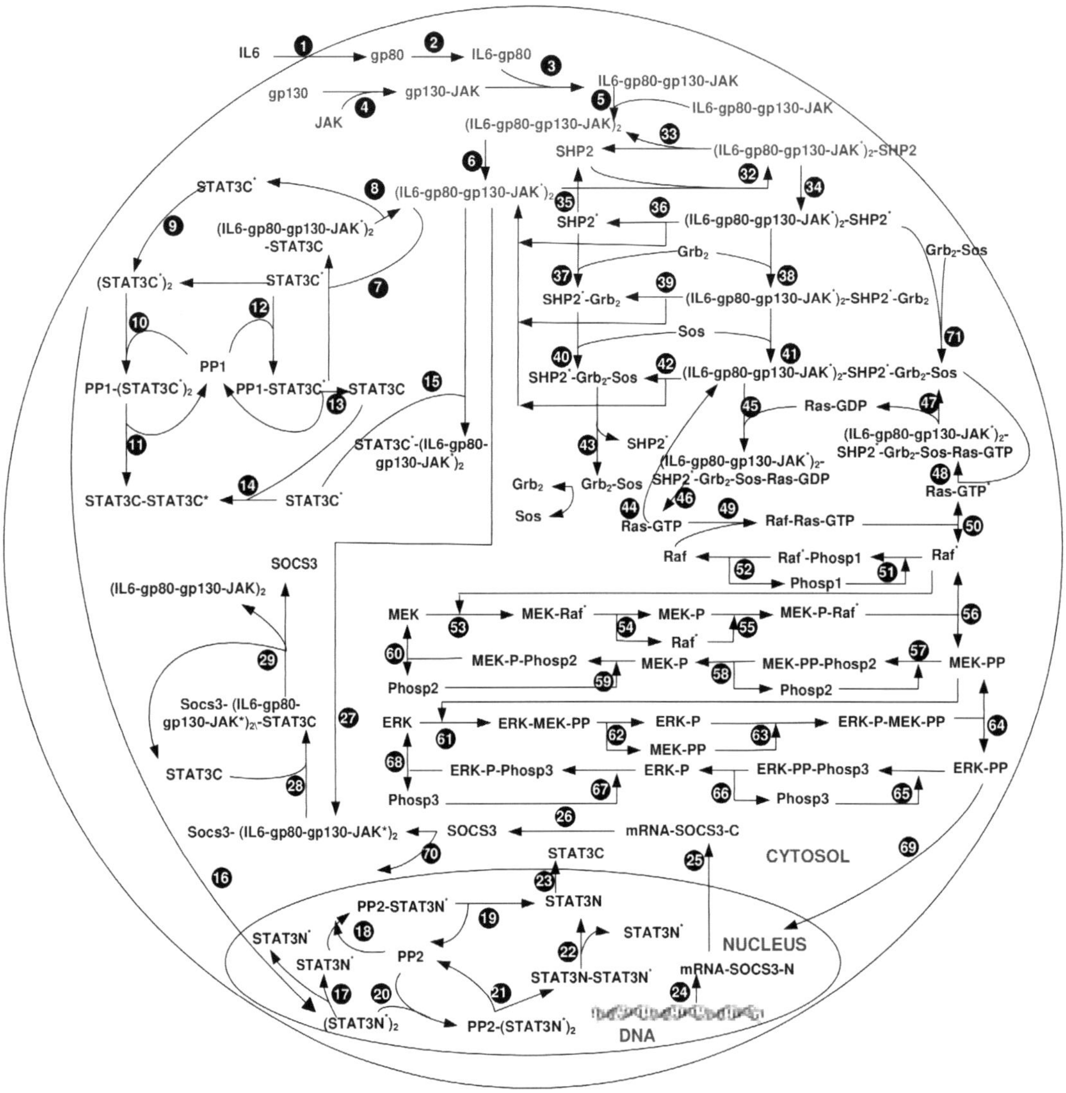

Figure 1. Model of the Interleukin-6 signaling pathway

Figure 2 shows the dendrogram of hierarchical clustering of parameters. It can be concluded that the similarity values between some of the parameters is very high as their sensitivity vectors are almost parallel. The diagram also illustrates how the selection of the similarity value influences the number of group. For example, if 11 groups are used then the smallest similarity value is equal to 0.941 which is illustrated by the dashed line. An increase in the number of groups leads to an increase of the lowest similarity value of the system.

Reducing the parameter space via parameter clustering can be viewed as one type of model reduction. The discrepancy between the original model and the reduced model is important as it indicated how many groups need to be selected to appropriately represent the original model. As discussed in Proposition 2 and Proposition 3, the model discrepancy can be bounded by the least similarity measure. At the same time, the smallest similarity measure can be determined by choosing the number of groups from the dendrogram in Figure 2. Therefore the number of groups can be determined by assuming the discrepancy to be less than a certain threshold value. Table 1 lists the least similarity measure and the discrepancy value for different number of groups. If the parameters are clustered into more groups, then the least similarity measure is increased and the discrepancy value is decreased. In this case the number of groups is determined to be equal to 11 as the discrepancy value drops below 0.05.

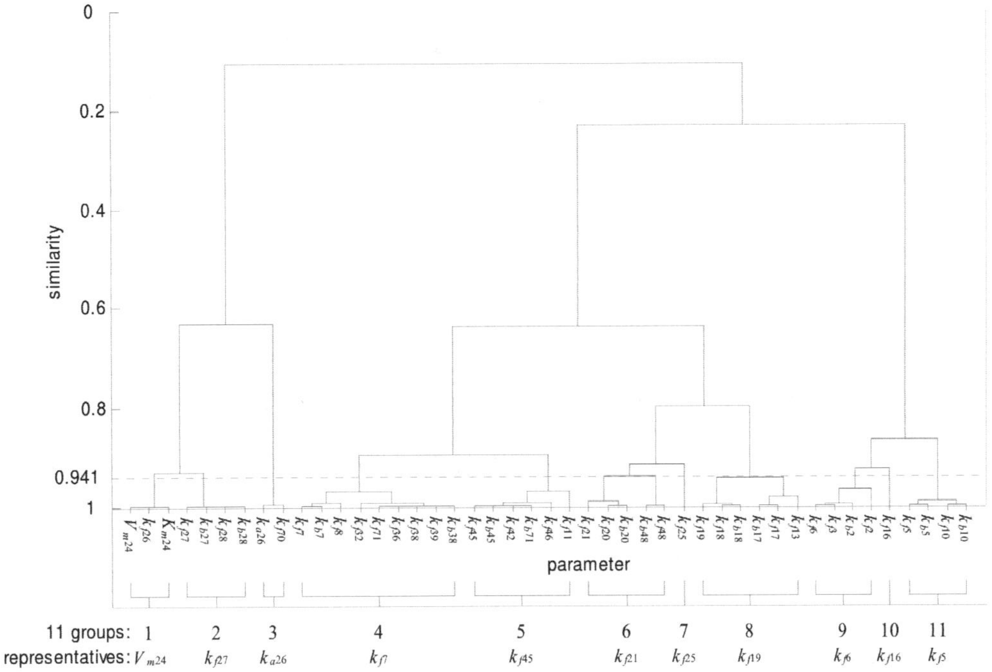

Figure 2. Dendrogram of hierarchical clustering of parameters

The parameter with the largest sensitivity vector in each group is chosen as the representative parameters for the group. The parameters selected for estimation are now chosen from the set of 11 representative parameters instead of the original 45 parameters. The optimization problem has reduced to determining a set of 8 parameters out of 11 possible parameters to maximize the *D*-optimality criterion of the Fisher information matrix, as compared to the original problem that involved choosing a set of 8 parameters out of 45 parameters. The computational effort decreases significantly, from $\sim 2\times10^{8}$ possible combinations to 165, due to this reduction in the number of parameters that need to be considered.

Discussions

The phenomenon of highly correlated parameters is an important feature of the complex biochemical network. It has been demonstrated that the phenomenon is universe in the existing models in systems biology via analysis of the eigenvalue of the Hessian matrix (Gutenkunst et al., 2007). Similar inferences were also made by global inversion of a metabolic model (Piazza et al., 2008). The parameter clustering method developed and applied to the signal transduction network in this work supports the inference from another perspective. Some parameters are highly correlated parameters and have very high value of the similarity measure.

The effect of the correlated parameters on estimation is two folds. On one side the correlated parameters result in an ill-conditioned problem for finding the optimal parameter value. Over-fitting occurs and the estimation which is sensitivity to the noise deteriorates the predictability of the model. On the other side, correlated parameters have the overlapping effects on the output and it is unnecessary to estimate all parameters accurately to generate an accurate prediction. Parameter selection is both a procedure to overcome the over-fitting by controlling the model complexity and an approach to simplify the parameter optimization by reducing the dimension of the decision variables.

Analysis of the highly correlated parameters is essential to not only estimation of the dynamic model but also investigation of the underlying mechanism in a biochemical network. The hierarchy tree clustered provides a clear map of the parameter relationship. Sensitivity analysis is a widely used technique to identify the important parameters. For dynamic system the parametric sensitivity is a function of time and is recorded as a sensitivity vector by sampling at specified time points. The length of the sensitivity vector which represents the accumulative effect of a parameter is often used to rank a parameter. However, it is the whole sensitivity profile that records all the information of dynamic effects of a parameter rather than the accumulative sensitivity value. The parameters which have the distinct sensitivity profiles may have the similar accumulative sensitivity value (Chu and Hahn, 2007). It has been recognized that distinct temporal activation profiles of the same signaling proteins result in diverse physiological responses (Marshall, 1995; Hoffmann et al., 2002; Kholodenko 2006) and, therefore, discriminating the dynamic effects of the parameters is of great importance.

Clustering of parameters provide a new approach to investigate the importance of a parameter taking the dynamic information into account. Since the importance of a parameter will change with time not all parameters are comparable. However, for the correlated parameters since their sensitivity profiles are similar the parameter which has large accumulative sensitivity value is more important than the parameter which has small sensitivity value. So in a group the representative parameter selected for estimation which has the largest sensitivity value is also the most important parameter in the group. Parameters from different groups have to be investigated separately since their sensitivity profiles are different.

It should be noted that some groups only contain one parameter, such as k_{f16}, k_{f25} in the shown example. These parameters have distinct sensitivity profiles and their effects cannot be compensated for by other parameters. These parameters reveal some key parts in the signal transduction pathway. The parameter k_{f16} is in the reaction where the transcription factor enters the nucleus. The parameter k_{f25} controls the reaction where the mRNA leaves the nucleus and the mRNA initiates the down regulating via activating the suppressor SOCS3.

For groups containing multiple parameters it is interesting to know if the parameters in a group are close to each other or distributed around the pathway. From the results in the case study, the answer is that both situations exist. The parameters represented by k_{f21} are all in the down regulating reactions associated with suppressor PP2 in the nucleus while the parameters represented by k_{f27} are all in the down regulating reactions associated with suppressor SOCS3. However, the parameters in the group represented by k_{f7} are distributed more widely. The group contains parameters from both JAK/STAT pathways and MAPK pathway. This provides the insight into the coupling of the two pathways and presents the clue for further study.

Investigation of correlated parameters is also very helpful to analyze the robustness with variations of parameter value. It is well known that robustness is a common property of biological networks, e.g., resilience to perturbations in current conditions (Stelling et al., 2004). The kinetic parameters in a biochemical network can change due to alteration of enzyme activity caused by a mutation or a disease. However, biological networks have a certain tolerance regarding variations in kinetic parameters while still being able to maintain their functions. One important reason for robustness is redundancy built into a network (Chen et al., 2005). As some components of a network may fail, there are other components that have a similar effect which allows the network to function properly. Parameter clustering can reveal some of these redundancies inherent to a system. The variations of some parameters by the change of the environment can be compensated by change of other parameters in the group via self regularization of the cell.

Conclusion

The great number of correlated free parameters contrasted with the limited number of noisy experimental data is the main difficulty in estimation of a complex network. The model is too complex to fit the data and over-fitting will occur. A method of parameter selection via parameter clustering to improve model predictability is developed. The hierarchical clustering method is used to group parameters according to their effects on the output. The parameters in a group have similar effects on the output and the effect of a parameter can be compensated by another. The representative parameters from every group are the parameters selected for estimation. The parameter selection circumvents the over-fitting. It results in slightly larger value in the loss function, however, it produces a smaller mean squared error and also a much more accurate prediction outside the experimental conditions for estimation. Parameter clustering also sheds insights into the relationship among parameters and provides a very useful tool for analysis of the biochemical network.

References

1. Chappell MJ, Godfrey KR, Vajda S. Global identifiability of the Parameters of Nonlinear-systems with specified inputs-A comparision of methods. Mathematical Biosciences 102 (1): 41-73 1990.
2. Chen BS, Wang YC, Wu WS, Li WH. A new measure of the robustness of biochemical networks. Bioinformatics 21 (11): 2698-2705 2005.
3. Chu Y, Jayaraman A, Hahn J. Parameter Sensitivity Analysis of IL-6 Signaling Pathways. IET Systems Biology 1 (6): 342-352 2007.
4. Chu Y, Hahn J. Parameter Set Selection for Estimation for Nonlinear Dynamic Systems. AIChE Journal 53 (11): 2858-2870 2007.
5. Duda RO, Hart PE, Stork DG. Pattern classification (2nd edition). New York: Wiley, 2001.
6. Gadkar KG, Varner J, Doyle FJ. Model identification of signal transduction networks from data using a state regulator problem. IEE Proceedings Systems Biology 2 (1): 17-30 2005.
7. Gutenkunst RN, Waterfall JJ, Casey FP, Brown KS, Myers CR, Sethna JP. Universally sloppy parameter sensitivities in systems biology models. PLOS Computational Biology 3 (10) 1871-1878 2007.
8. Hoffmann A, Levchenko A, Scott ML, Baltimore D. The IκB-NF-κB signaling module: temporal control and selective gene activation. Science 298 (5596): 1241-1245, 2002.
9. Huang Z, Chu Y, Senocak F, Jayaraman A, Hahn J. Model Update of Signal Transduction Pathways in Hepatocytes based upon Sensitivity Analysis. Proceedings Foundations of Systems Biology. 45-50 Stuttgart, Germany 2007
10. Kholodenko BN. Cell-signalling dynamics in time and space. Nature Reviews Molecular Cell Biology 7 (3): 165-176 2006.
11. Ljung L, Glad T. On global identifiability for arbitrary model parametrizations. Automatica 30 (2): 265-276 1994.
12. Ljung L. System identification: theory for the user (2nd edition). Upper Saddle River: Prentice Hall PTR, 1999.
13. Marshall CJ. Specificity of receptor tyrosine kinase signaling: Transient versus sustained extracellular signal-regulated activation. Cell 80 (2): 179-185 1995.
14. Piazza M (Piazza, Matthew), Feng XJ (Feng, Xiao-Jiang), Rabinowitz JD (Rabinowitz, Joshua D.), Rabitz H. Diverse metabolic model parameters generate similar methionine cycle dynamics. Journal of Theoretical Biology 251 (4) 628-639 2008.
15. Rothenbe TJ. Identification in parametric models. Econometrica 39 (3): 577-591 1971.

16. Singh A, Jayaraman A, Hahn J. Modeling regulatory mechanisms in IL-6 signal transduction in hepatocytes. Biotechnology and Bioengineering 95 (5): 850-862, 2006.
17. Schoeberl B, Eichler-Jonsson C, Gilles ED, Muller G. Computational modeling of the dynamics of the MAP kinase cascade activated by surface and internalized EGF receptors. Nature Biotechnology 20 (4): 370-375, 2002.
18. Stelling J, Sauer U, Szallasi Z, Doyle FJ, Doyle J. Robustness of cellular functions. Cell 118 (6): 675-685 2004.
19. Theodoridis S, Koutroumbas K. Pattern recognition (3rd edition). Boston: Elsevier/Academic Press, 2006.
20. Walter E. Identifiability of parametric models. New York: Pergamon Press, 1987.
21. Walter E, Pronzato L. Identification of parametric models from experimental data. Berlin: Springer, 1997.
22. Yue H, Brown M, Knowles J, Wang H, Broomhead DS, Kell DB. Insights into the behaviour of systems biology models from dynamic sensitivty and identifiability analysis: a case study of an NF-kappa B signalling pathway. Molecular Biosystems. 2 (12): 640-649, 2006.

565e The Effect of Biological Variability on the Angiotensin II Gene Regulatory Network In the Central Regulation of Blood Pressure

Gregory M. Miller[1], Rajanikanth Vadigepalli[2], James S. Schwaber[2], and Babatunde A. Ogunnaike[1]. (1) Chemical Engineering, University of Delaware, 150 Academy St., Newark, DE 19716, (2) Daniel Baugh Institute for Functional Genomics and Computational Biology, Thomas Jefferson University, 1020 Locust St Room 575, Philadelphia, PA 19107

Neurons in the nucleus tractus solitarius (NTS) communicate through chemical messengers to play a major role in blood pressure regulation, and aberrant communications within this brain nucleus cause serious diseases such as hypertension. To fulfill their role in blood pressure regulation, NTS neurons must receive chemical messengers and process these signals through biochemical networks comprising a large number of interacting proteins and genes; but because the population of NTS neurons is heterogeneous, there is significant variability in the abundance and activity of these network components. The effect of this neuron-to-neuron variability on blood pressure regulation is not well-understood and an improvement in understanding is necessary for developing effective treatments of hypertension. Our overall objective, therefore, is to understand how NTS neurons function in the presence of biological variability.

In this work, we investigate the effect of biological variability on a critical mechanism for the central regulation of blood pressure: angiotensin II type 1 receptor (AT1R) activation of tyrosine hydroxylase in the brain. Using the regulatory mechanisms established in the literature (Veerasingham and Raizada, 2003), we construct a mechanistic, ordinary differential equation model for the induction of tyrosine hydroxylase gene expression modulated by the gene regulatory network activated by AT1R. This model allows us to explore the effect of biological variability on neuron function by performing model simulations with variations in reaction rate constants and species initial concentrations, and then comparing the predicted response of tyrosine hydroxylase from AT1R gene networks with variations in model parameters.

We present simulation results showing AT1R activation induces tyrosine hydroxylase robustly in the presence of biological variability, and discuss properties of the AT1R gene regulatory network that ensure this robust performance. By finding this critical mechanism for the central regulation of blood pressure to be robust to neuron-to-neuron variability, our results may lead to the development of improved treatments of hypertension.

Reference

Veerasingham S.J. & Raizada M.K. (2003) **Brain renin–angiotensin system dysfunction in hypertension: recent advances and perspectives**. *Br. J. Pharmacol.* **139**(2):191-202.

565f Optimal Cancer Treatment Using Radiovirotherapy

Ricardo Dunia, LabVIEW R&D, National Instruments, 11500 N. Mopac Expwy, Austin, TX 78759 and **Thomas Edison**, Chemical Engineering, University of Texas in Austin, 1 University Station C0400, Austin, TX 78712.

Protein engineering has enabled the production of viruses that infect tumor cells [1]. Models of tumor and virus interactions have been recently developed and validated [2][3]. This new and effective cancer therapy has been modeled at the cell level by three state variables, which include the population of free virus particles controlled by the virotherapy dosage. A three dimensional simulations of a solid tumor at the macroscopic level has shown that the tumor eradication requires widespread distribution within the tumor at the time of infection [3]. It has been suggested that virotherapy alone may failed to eliminate a tumor in cases where the tumor vascular irrigation is not homogenously widespread [4]. The use of radiovirotherapy has been also modeled at the macroscopic level and relevant therapeutic scenarios have been proposed for its optimal treatment. This research work develops an efficient control strategy to eliminate a localized tumor subject to the maximum free virus concentrations and radiotherapy frequency attainable by the human body. The controller adapts to the different stages of the therapy by reducing the dosage as the tumor reduces its size. Although the core therapy model is developed at the cellular level, the feedback information provided to the nonlinear controller is at a macroscopic level, making the implementation of the closed loop therapy feasible to real applications.

[1] Antibody Targeted Cell Fusion ; Nakamura, T., Peng, K. W., Vongpunsawad, S., Harvey, M., Mizuguchi, H., Hayakawa, T., Cattaneo, R., Russel, S., 2004. Nat. Biotechnology. 22 (3) 331-336.

[2] Modeling of cancer virotherapy with recombinant measles viruses ; Zeljko Bajzer, Thomas Carr, Kresimir Josic, Stephen J. Russel David Dingli, 2008, Journal of Theoretical Biology, 252, 109-122.

[3]Validation and Analysis of a Mathematical Model of a Replication-competent Oncolytic Virus for Cancer Treatment: Implication for Virus Design and Delivery; L.M. Wein, J. Wu and D. Kirn, 2003, Cancer Research, 63, 1317-1324.

[4] Mathematical Modeling of Cancer Radiovirotherapy; D. Dingli, M. Cascino, K. Josic, S. Russell, Z. Bajzer, 2006, Mathematical Biosciences, 199, 55-78.

565g Combined Statistical and Genome-Scale Analysis of Mammalian Cell Lines Producing Bio-Therapeutics

Suresh Selvarasu, **Dong-Yup Lee**, and **I. A. Karimi**. Department of Chemical & Biomolecular Engineering, National University of Singapore, 4, Engineering Drive 4, Singapore, 117576, Singapore

Mammalian cell cultures constitute a major source for producing high-value bio-therapeutics such as monoclonal antibodies, recombinant proteins, viral vaccines, etc. The increasing demand for these commercially valuable compounds can be satisfied by developing high-yielding mammalian cell culture processes (1). The development of such cultures requires a good understanding of the metabolism and cellular response of the production host. However, the host's structural, functional, and dynamic complexities make it difficult to elucidate the behavior of the mammalian cellular system. In order to alleviate this difficulty, statistical techniques have been recently adopted for investigating correlations among the available experimental profiles (2). These techniques do not require the development of structured models and they use only a minimal set of experimental data, which makes them applicable to cell cultures with different operating modes. However, they still have limitations in explaining the effect of culture conditions on the internal cell metabolism. In this sense, in silico analyses using genome-scale models of organisms are invaluable in characterizing the cellular physiology and internal metabolism. A combined approach involving statistical and genome-scale analyses can be crucial in elucidating the effect of environmental conditions and its impact on internal cell metabolism. However, such techniques have received limited attention in the literature. They offer promise for identifying optimal feeding strategies and potential cell engineering targets as well as facilitating media optimization for the enhanced production of bio-therapeutics in mammalian cells. Furthermore, using such a systematic approach may also reduce the number of wet experiments.

In this work, we present a combined statistical and in silico genome-scale analysis for mouse cell lines producing mAb to elucidate their physiological and metabolic states under different environmental conditions and genetic manipulations. First, we applied multivariate statistical data analysis techniques on experimental fermentation data to identify the relationships among nutrient uptakes and desired products. The analysis identified significant correlations between mAb production and certain amino acids. Subsequently, we expanded and curated an existing genome-scale metabolic model (3) of mouse cells using updated biochemical and genomic data. The internal metabolic activities were then explored to elucidate the effect of positive correlation obtained from statistical analysis. The effect of culture media on intracellular metabolism was also explored further (4). The in silico simulation results revealed interesting results on ATP utilization, metabolic shifting and nutrient utilization. Thus we see that a combined approach involving statistical analysis and in silico genome-scale analysis can be very effective for analyzing the cellular metabolism.

References:

1)Wurm FM. 2004. Production of recombinant protein therapeutics in cultivated mammalian cells. Nat

Biotechnol 32:1393-1398.

2)Alwis DMD, Dutton RL, Scharer J, Moo-Young M. 2007. Statistical methods in media optimization for batch and fed-batch animal cell culture. Bioprocess Biosyst Engg, 30:107-113.

3)Sheikh K, Forster J, Nielson LK. 2005. Modeling hybridoma cell metabolism using a generic genome-scale metabolic model of Mus musculus. Biotech. Prog. 21:112-121.

4)Selvarasu. S., Lee, D-Y., Wong, V. V. T., Karimi, I. A. Elucidation of metabolism in hybridoma cells grown in fed-batch culture by genome-scale modeling. Submitted

574b Direct PID Tuning from Closed-Loop Data and Its Application to Unstable Polymerization Process

Kenichi Tasaka[1], Manabu Kano[1], Morimasa Ogawa[2], Shiro Masuda[3], and Toru Yamamoto[4]. (1) Chemical Engineering, Kyoto University, Nishikyo-ku, Kyoto, 615-8510, Japan, (2) Yamatake Corporation, 2-7-27, Kanda, Kurashiki, 712-8061, Japan, (3) Tokyo Metropolitan University, 6-6, Asahigaoka, Hino, 191-0065, Japan, (4) Hiroshima University, Higashi-Hiroshima, 739-8524, Japan

In the present work, a new method for direct tuning of PID controllers using operation data under feedback control is proposed, and its application to unstable polymerization processes is investigated.

PID control has been dominant as a control algorithm in process industry. To achieve high profitability, the tuning of PID controllers is still an important issue in practice. PID controllers are tuned usually on the basis of process models identified from operation data and empirical tuning rules. However, good identification is a difficult task. In addition, closed-loop identification is often required, especially when a process is unstable.

To avoid identification and optimize PID control parameters directly from closed-loop data, several techniques such as virtual reference feedback tuning (VRFT) and iterative feedback tuning (IFT) have been proposed. Without a process model, these methods can determine the PID control parameters so that the closed-loop response corresponds closely with that of the desired reference model. More recently, fictitious reference iterative tuning (FRIT) was proposed. FRIT can optimize the PID control parameters from a set of input-output data like VRFT, but its objective function is different from that of VRFT.

The objective of this research is to extend FRIT for improving its practicability. The problems of FRIT are: 1) there was a risk of obtaining a local minimum because the Gauss-Newton method is commonly used in FRIT and 2) it is difficult to properly determine a reference model without information on a process. The use of an inappropriate reference model deteriorates the control performance achieved. At the worst, the control system becomes unstable. The second problem is serious not only in FRIT but also in VRFT and IFT.

To solve these two problems, FRIT is extended. First, particle swarm optimization (PSO) is used to find a global minimum. Second, the objective function is modified to include the penalty for changes of the input variable. Third, the parameter in the reference model is optimized together with PID control parameters. The second and the third extensions are very useful to realize peaceful control responses, especially when the process is unstable. The proposed method is referred to as extended FRIT (E-FRIT).

E-FRIT is applied to two types of unstable polymerization processes, and its performance is compared with that of the conventional methods. The results clearly show that each extension plays an important role to derive better PID control parameters. In fact, conventional methods tend to make the control system oscillating or even unstable. The direct PID tuning via E-FRIT is promising for chemical process control.

Inverse Model of the Glass pH Sensor

Jietae Lee[1]*, Su Whan Sung[1], Seungjae Lee[2] and Dae Ryook Yang[3]
1 Department of Chemical Engineering, Kyungpook National University, Taegu, 702-701, KOREA
2 Daewon Foreign Language High School, Seoul, KOREA
3 Department of Chemical Engineering, Seoul, KOREA
* Corresponding author. Tel: +82-53-950-5620, Fax: +82-53-950-6615, E-mail: jtlee@knu.ac.kr

Keywords: Glass pH Electrode, Transient Response, Time Constant

Introduction

The concentration of ion H^+ (pH) in solution plays very important role in chemical, biological and physical processes. Since Cremer (1906) first reported the pH response of oxide glasses, many theoretical researches about the glass electrode and its applications have been carried out. The glass electrode is stable mechanically and chemically and can be used in the wide range of solvents. In spite of new developments of semiconductor sensors and optical sensors for pH measurements (Bakker, 1998), the glass electrode is still of great practical importance.

The glass electrode is relatively fast and can be applied effectively in on-line measurements. However, for some applications such as identification tanks (Sung et al., 1995; Obut and Ozgen, 2008), end-point estimations and auto-titrators, faster responses will be preferable. The detailed mechanism of glass electrode is still questionable. Roughly, the dynamic response of glass electrode consists of quick response to a quasi-equilibrium and slow drift due to ion permeation through the glass membrane (Kohler et al., 2005). The former, response to the quasi-equilibrium, can be explained through the equivalent circuit having a very high internal resistance and a capacitance (Buck, 1968). Time constant of the glass electrode is the time that the capacitance is charged through the very large internal resistance. Here a measuring method that reduces this time by an order of magnitude is proposed. External voltage is introduced to charge the capacitance. It is compared and controlled to trace the glass pH electrode voltage by the integral controller.

Glass pH Sensor

A typical pH combination electrode consists of glass and reference electrodes in one body. It has a thin glass bulb which is sensitive on the pH of solution. This pH-sensitive glass membrane has a typical resistance of $10^8\Omega$ and capacitance of 10^2pF. Fig. 1 shows a rough equivalent circuit of the glass pH electrode. Because capacitances are due to electrodes and electrical double layer of solutions and are distributed throughout elements of the pH sensor body, the equivalent circuit of Fig. 1 is approximate and will be effective for low frequency operations.

The voltage of the ideal pH electrode changes by 59.16 mV per each pH unit change at 25°C (Harris, 2007). It can be measured through operational amplifiers with input resistance over $10^{12}\Omega$ and very low bias current. For a step change of pH, the voltage output of glass pH sensor shows responses as shown in Fig. 2. This dynamical response is due to times to charge the capacitance through the resistance of the glass membrane. Times to reach the steady state pH values can be as large as 5 sec for some pH sensors and they can be too large for some applications. A method to reduce these times is investigated.

Responses for the External Voltage Forcing

Here an external voltage forcing as shown in Fig. 3 is considered. Dynamic model for this system can be expressed as follows:

$$E_b(s) = \frac{R/(R+R_1)}{\tau s+1}E_a(s) + \frac{R_1/(R+R_1)}{\tau s+1}E(s)$$
$$\tau = R_1 C \Big/ \left(1+\frac{R_1}{R}\right) \qquad (1)$$

When E_b is not connected (R_1 is infinite), the time constant τ in Eq. (1) becomes RC. That is, the time constant for the step change of E (the voltage of glass pH sensor) is RC (multiplication of the internal resistance and the capacitance). Responses will be like Fig. 2.

When R_1 is much smaller than R, the time constant τ in Eq. (1) becomes R_1C. Smaller time constant will make responses for voltage changes of E and Ea be faster. Fig. 4 shows this. However, the pH sensor output E_b at the steady state is $R/(R+R_1)E_a + R_1/(R+R_1)E$, which is dependent of R. Because R is not known in advance and time varying, especially for the change of temperature (Buck and Krull, 1968), a simple measurement of the voltage E_b at the steady state for a constant E_a cannot be used to estimate E. To solve this problem, methods to find the internal resistance R and the sensor voltage E simultaneously with two or more experiments for different forcing voltages of E_a are invented (Koluvek, 1995). A square wave with sufficient period to ensure steady states of E_b can be used for measuring E and R continuously.

Proposed Measurement Circuit

Methods with external alternating-current voltages make electric currents flow through the glass pH electrodes. It is much different from the conventional zero current measurement. Here, a method which is similar to the conventional zero current measurement of pH is proposed. A circuit for this is shown in Fig. 5. Q_1 is an instrumental amplifier with very low input bias current and very high input resistance. Its gain is adjusted by the resistance R_2 and measures the current flowing to the glass pH electrode. The circuit around Q_2 is an integrator. This feedback system is an integral control system which will make the current to the glass electrode be zero. It will make E_a (Output) trace E continuously. Changing feedback loop gain, time constant for the step change of the glass pH electrode E can be made small.

Conclusion

A simple method to make time constant of the glass pH electrode be small is proposed. It can trace the voltage of glass pH electrode and measure it at the condition of near zero current.

Literature Cited

Bakker, E., *Chem. Rev.*, 98, 1593, 1998.

Buck, R. P. and Krull, I., "Transient Electrical Behavior of Glass Membranes Part. III. Experimental," *J. Electroanal. Chem.*, 18, 387-399, 1968.

Daniel, D. C. *Quantitative Chemical Analysis*, 7th ed., W.H. Freeman and Company, N.Y., 2007.

Kohler, H. H., Haider, C., and Woelki, S., "Selectivity and Dynamic Behavior of Glass Electrodes," *Advances in Colloid and Interface Science*, 114, -115, 281-290, 2005.

Koluvek, R. H., "Circuit for Measuring Source Resistance of a Sensor," US Patent 5469070, 1995.

Obut, S. and Ozgen, C., "On-line Identification and Control of pH in a Neutralization System," *Ind. Eng. Chem. Res.*, in press, 2008.

Sung, S. W., Lee, I. B. and Yang, D. R., "pH Control Using an Identification Reactor," *Ind. Eng. Chem. Res.*, 34, 2418-2426, 1995.

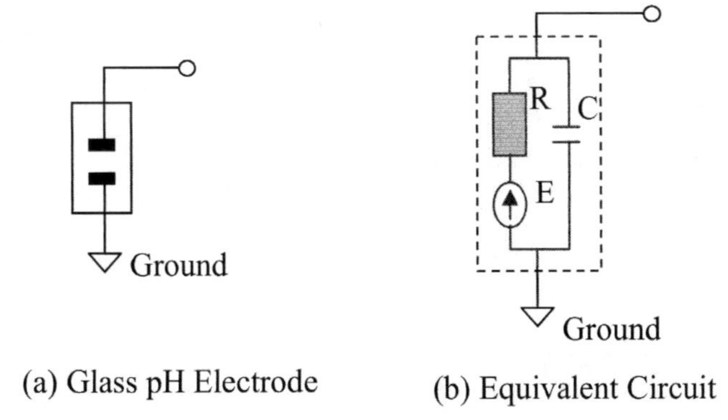

Figure 1. An equivalent circuit for the glass pH electrode.

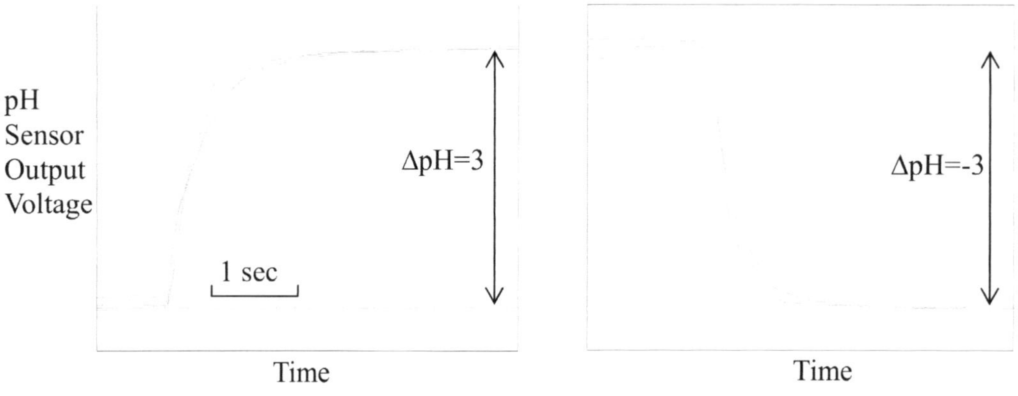

(a) pH Sensor Responses for Positive Step Changes of pH

(b) pH Sensor Responses for Negative Step Changes of pH

Figure 2. Responses of step changes in solution pH (Experimental results).

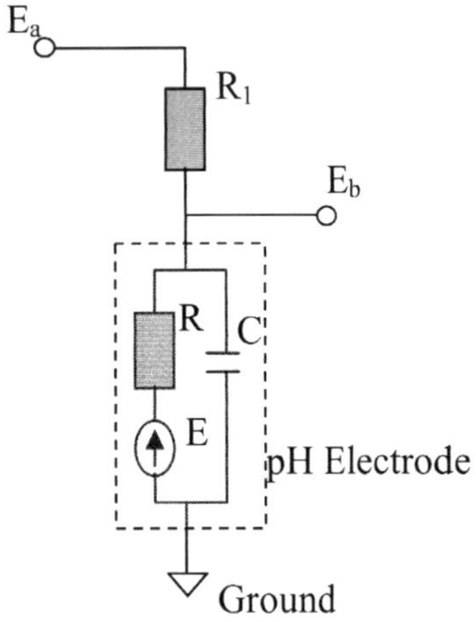

Figure 3. A model for the external voltage forcing.

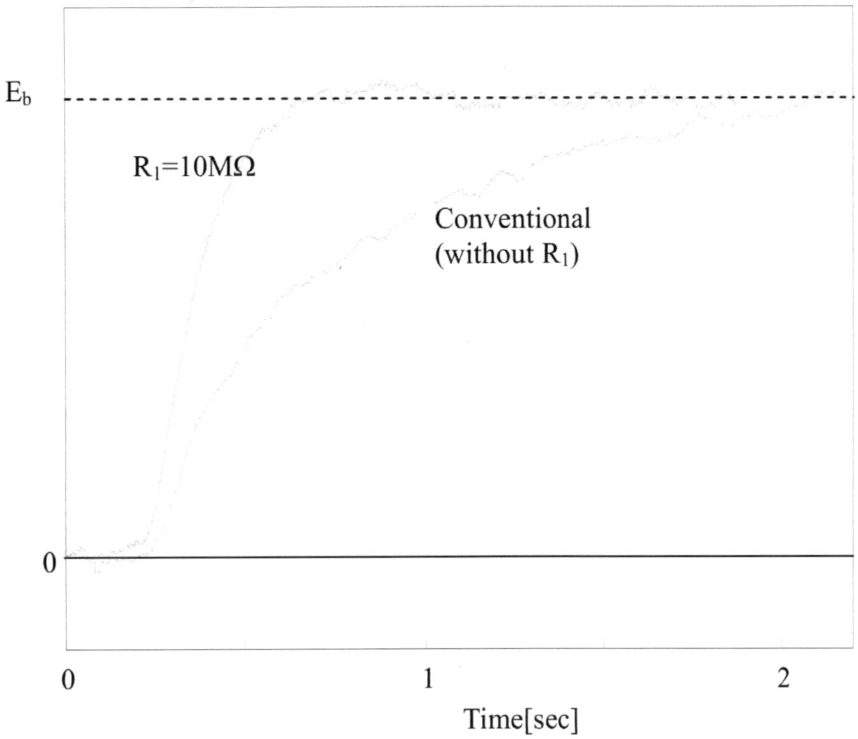

Figure 4. Responses of step changes in solution pH (Experimental results).

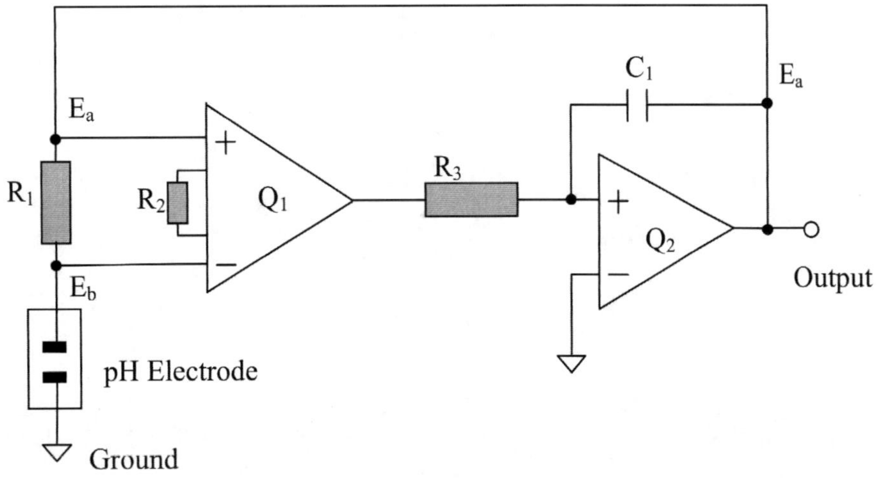

Figure 5. A proposed circuit for improved transient response of the glass pH electrode.

574d A Dynamic Systems Approach to the Synthesis and Analysis of Mammalian Metabolic Networks

Adam C. Baughman, Lealon L. Martin, and Susan T. Sharfstein. Chemical & Biological Engineering, Rensselaer Polytechnic Institute, 110 8th Street, Ricketts Building, Troy, NY 12180

We present a novel, multi-level, computational framework for the synthesis and analysis of mammalian metabolism from the point of nutrient uptake, through a complex network of catabolic (energetic) and anabolic (biosynthetic) reactions, and to the point at which a large protein product is synthesized and possibly secreted. This framework is predicated upon dynamic systems theory, and adopts a perspective of input and output relationships as determined through a unique network state. This network state is, in turn, determined through the simultaneous optimization of a universal linear network flux-balance subject to certain unique non-linear constraints. Further, we will demonstrate the ability of this framework to identify valid, yet non-intuitive, metabolic chemistry when tasked with the optimization of cellular processes relative to the maximum yield of a protein product. We use this framework to probe for bottlenecks in synthesis of large non-native proteins (i.e. monoclonal antibodies) from mammalian cell hosts. Finally, to integrate this computational effort with the experimental laboratory, we explore specific challenges in the engineering of cell lines to adopt this non-intuitive metabolic chemistry. It is our hope that such engineered cell lines will demonstrate the improved productivity predicted by our model.

574e Alternative Model Structure with Simplistic Noise Modelling for Linear Processes Subjected to Non-Stationary Distrubances

Srinivas Karra, Chemical Engineering, Texas Tech University, Lubbock, TX 79409-3121 and M. Nazmul Karim, Chemical Engineering Department, Texas Tech University, Lubbock, TX 79409-3121.

A simple system identification technique to identify the linear processes affected by non-stationary disturbances is proposed. This uses a time varying bias term, a representative of the additive non-stationary external disturbance entering the process, in addition to the output predictions in an ARMAX or OE model framework. Decoupled loss function and covariance update with different forgetting factors for linear time invariant input-output dynamics part and time varying part (bias term) of the model ensured the unbiased estimation of true process dynamics along with disturbance dynamics.

Practical issues such as time delay estimation, model order selection are discussed. Extensions for time varying processes and MIMO processes are also proposed. Efficacy of the proposed identification technique in both open loop and closed loop cases against OE, ARMAX and BJ identification techniques is performed using various simulation studies.

Passive Heat Exchanger Temperature Control Using Variable Conductance Heat Pipes

Richard W. Bonner III
Advanced Cooling Technologies, Inc., Lancaster, Pa

Abstract

Heat exchangers are important unit operations for a variety of industrial processes. They are often necessary to recuperate waste heat and improve process efficiencies. The heat exchanger's outlet streams often feed processes that are very sensitive to temperature (chemical reactors, separation membranes, fuel cells, etc.) where robust temperature control is a must. Most heat exchangers maintain outlet temperatures by using active means such as temperature controllers with actuated valves and inefficient by-pass streams to maintain a set point. These active control schemes have several drawbacks including the low reliability of moving parts, susceptibility to operator error, and the complicated systems level modeling required to validate control schemes.

Presented in this paper is a passive temperature control technology that utilizes variable conductance heat pipes (VCHP) to control outlet temperature from a counter current heat exchanger. VCHP's are heat pipes filled with a prescribed amount of non-condensable gas. The non-condensable gas expands and contracts in the condenser of the heat pipe as the saturated working fluid's vapor pressure and temperature changes. As a result, the evaporator temperature of the VCHP is nearly constant, regardless of operating conditions (flow rates, inlet temperatures, etc.). A heat exchanger using a tube bank of VCHP's has the unique property that the hot side fluid exits at nearly constant temperature regardless of heat exchanger operating conditions (hot and cold side inlet temperatures and flow rates.) A transient model for a VCHP heat exchanger is presented along with experimental data from a VCHP heat exchanger designed to control the outlet temperature of a hydrogen stream for a fuel cell application. The passive temperature control feature of VCHP heat exchangers is ideal for applications where reliability and safety are of paramount concern.

Background

Heat pipes passively transfer heat at low thermal resistances by capillary circulation of fluid at saturated conditions[1]. Shown in Figure 1, a heat pipe consists of a vacuum leak tight envelope enclosing a working fluid and wick structure. Heat input vaporizes the saturated liquid contained within the wick in the evaporator section of the heat pipe. The vapor, carrying the latent heat of vaporization, flows towards the cooler condenser section. In the condenser, the vapor condenses within the wick, giving up its latent heat. The condensed liquid returns to the evaporator through the wick structure by capillary action. Fluid circulation continues as long as the temperature gradient between the evaporator and condenser is maintained. The heat transfer process occurs with minimal temperature gradient between the evaporator and condenser because the saturated fluid within the heat pipe circulates with small pressure differences. Since the fluid is at saturated conditions, small pressure differences correlates to small temperature differences.

A variable conductance heat pipe (VCHP) is a variation of a heat pipe possessing some inherent temperature control abilities[2]. Variable conductance heat pipes get their temperature control features from the expansion and contraction of non-condensable gas in the condenser of the heat pipe. At low temperatures, the heat pipe working fluid exerts a low pressure on non-condensable gas in the condenser of the heat pipe, causing the non-condensable gas to expand. A conceptual diagram of a variable conductance heat pipe operating at a low temperature condition is shown in Figure 2. At this low temperature condition most of the condenser is blocked off by the non-condensable gas, which prevents the heat pipe working fluid from cooling itself down further. As the thermal load on the heat pipe is increased, the heat pipe working fluid temperature increases, along with pressure. As shown in Figure 3, the non-condensable gas contracts to expose more of the condenser area for cooling. As the gas expands, the conductance (inverse of thermal resistance) of the heat pipe is increased. The increase in conductance with temperature gives the variable conductance heat pipe the ability to maintain a temperature around a given set point. The set point is chosen by charging the heat pipe with an exact amount of non-condensable gas.

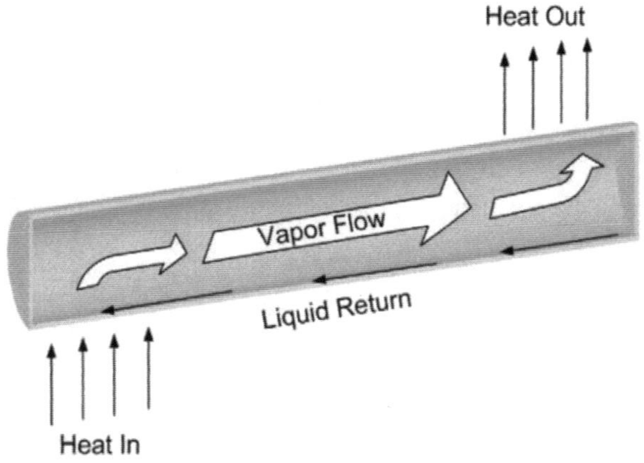

Figure 1 A schematic of a heat pipe is shown. Heat entering the evaporator of a heat pipe boils fluid out of a wick structure. Vapor flows axially down the inner diameter of the heat pipes and condenses. The wick structure then returns fluid back to the evaporator.

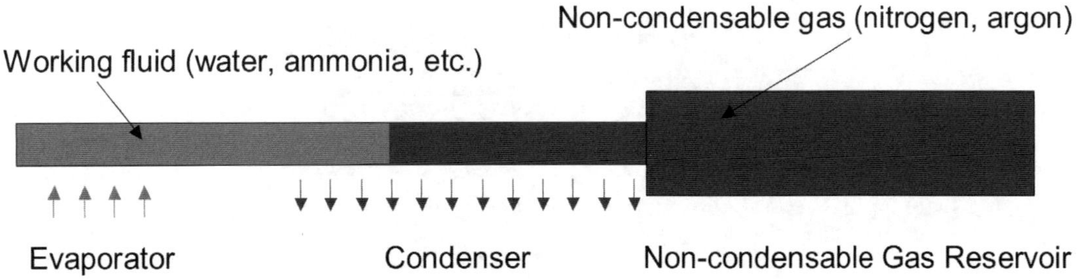

Figure 2 A diagram of a variable conductance heat pipe operating at a low temperature condition is shown. With the saturated working fluid at low saturated temperature (and saturated pressure) the non-condensable gas present in the reservoir expands to occupy most of the condenser section.

Variable Conductance Heat Pipe Heat Exchanger

A variable conductance heat pipe heat exchanger (VCHP HX) is an array of variable conductance heat pipes in a heat exchanger configuration[3-4]. A diagram of a VCHP HX with operating conditions for a Navy fuel cell application is shown in Figure 4 [5-6]. The hot side fluid (hydrogen in this example) enters the heat exchanger and passes over the evaporator side of the variable conductance heat pipe elements. Coolant flows counter currently to this stream and passes over the condenser end of the heat pipes. Nearest to the hot side inlet, the heat pipes are the hottest. The non-condensable gas is compressed, leaving most of the condenser available to efficiently transfer heat to the coolant side. In this region of the heat exchanger the highest heat transfer rates occur. As the hot side fluid flows toward the exit of the heat exchanger it cold down. The local heat pipes also decreases in temperature, expanding non-condensable gas and "blanketing" the condenser of the heat pipe. As the hot side fluid approaches the exit, the condenser is almost completely blocked by non-condensable gas, and heat transfer rates approach zero. The hot side outlet temperature then exits the heat exchanger at approximately the set point of the variable conductance heat pipe elements.

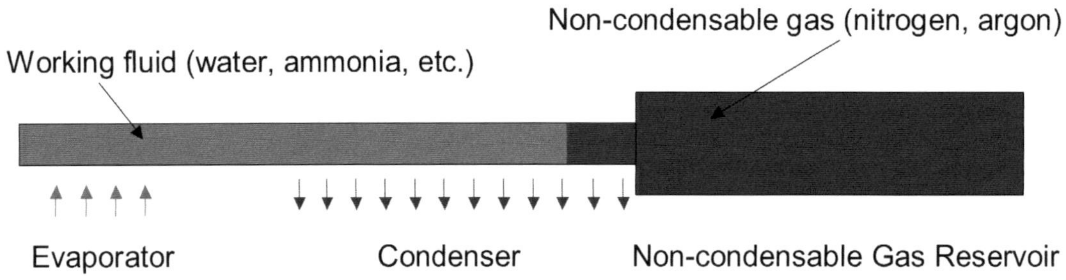

Figure 3 A diagram of a variable conductance heat pipe operating at a high temperature condition is shown. With the saturated working fluid at high saturated temperature (and saturated pressure) the non-condensable gas present in the reservoir compresses to occupy most of the condenser section.

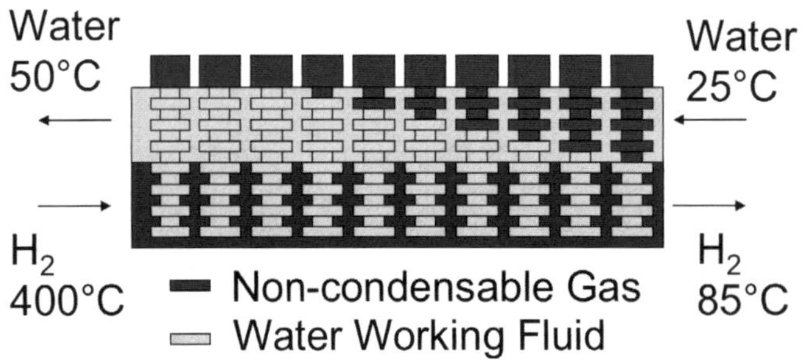

Figure 4 A variable conductance heat pipe (VCHP) heat exchanger is shown. Essentially an array of individual heat pipes, the VCHP heat exchanger passively responds to changes in inlet process temperatures by altering available condenser area on the cold fluid side. As a result, hot side outlet temperatures remain nearly constant. Operating conditions for a hydrogen fuel cell application are shown.

The advantage of the VCHP HX is that outlet hot side temperature is passively controlled. As the hot side inlet temperature or flow rate changes, the outlet of the hot side temperature side remains constant due to the expansion and contraction of non-condensable gas in the condensers of the heat pipes. If an operating condition requiring more thermal duty (such as increased inlet temperature or higher flow rate) is encountered the heat pipes in the heat exchange will rise slightly in temperature. Non-condensable gas in the heat pipes will contract, exposing more condenser and increasing the thermal duty of the heat exchanger. The result is a relatively constant outlet hot side temperature.

VCHP Heat Exchanger Modeling

The following summarizes a model developed to transiently model the VCHP HX. Nomenclature is provided after the conclusion section. Equations 1 and 2 describe the time and position dependent energy balances for the process fluid and steam in the heat exchanger. The process fluid exchanges heat with the local heat pipe at its current temperature. The necessary heat transfer coefficient is not that of the heat transfer coefficient on the heat pipe surface, but the effective heat transfer coefficient observed across the footprint of the heat exchanger. An equation for the effective heat transfer coefficient is shown in Equation 3. Equation 4 describes the time and position dependent heat balance around the heat pipe, accounting for the mass of the heat pipe. The heat pipe exchanges heat with the local process streams. The area of the heat pipe on the condenser side is sensitive to local process conditions and heat pipe properties such as geometry and non-condensable gas charging. The area for heat transfer on the condenser side is a function of the non-blanketed condenser length and amount of heat conducted up the heat pipe wall. The equation describing the length of non-condensable gas covered condenser is displayed in Equation 5. Equation 6 describes the effective length of open condenser accounting for the conduction up the heat pipe wall in the non-condensable gas covered condenser length. Equation 8 describes a non-dimensional fin efficiency parameter for conduction through a pin (in this case a hollow pin). This model is capable of predicting both transient and steady state performance for the VCHP HX. Steady-state predictions are compared to experimental data in Figures 5 and 6.

$$\frac{\partial T_1}{\partial t} = -v_1 \frac{\partial T_1}{\partial x} + \frac{h_{1,eff}}{(\rho CpH)_1}(T_{HP} - T_1) \quad (1)$$

$$\frac{\partial T_2}{\partial t} = -v_2 \frac{\partial T_2}{\partial x} + \frac{h_{2,eff}}{(\rho CpH)_2}(T_{HP} - T_2) \quad (2)$$

$$h_{n,eff} = h_n \frac{A_n}{W^2} \quad (3)$$

$$\frac{\partial T_{HP}}{\partial t} = \frac{h_1 A_1 (T_1 - T_{HP}) + h_2 A_2 (T_2 - T_{HP})}{(mCp)_{HP}} \quad (4)$$

$$L_{2,ncg} = \frac{\dfrac{n_{ncg} T_{ncg} R}{P_{Sat,HP}} - V_R}{\pi R_{HP}^2} \quad (5)$$

$$L_2 = L_{cond} - L_{2,ncg} + \frac{\tanh(mL_{2,ncg})}{m} \quad (6)$$

$$L_1 = L_{evap} \quad (7)$$

$$m^2 = \frac{h_2 P}{kA_c} \quad (8)$$

Experimental Data

A VCHP HX heat exchanger was successfully fabricated and tested. The heat exchanger was tested over a range of operating conditions to verify thermal control. The hydrogen flow rate was varied from 2.5kg/hr down to 0.5kg/hr. The hydrogen inlet temperature was also varied from 400°C down to 120°C. These particular design requirements were supplied from a Navy fuel cell reforming application. The water flow rate was kept constant at 800ml/min. A plot of hydrogen outlet temperature versus hydrogen inlet temperature at the maximum flow rate of 2.5kg/hr is shown in Figure 5. The outlet temperature varied 11°C over the change in inlet temperature from 95°C to 84°C. As expected the heat exchanger operated hotter than the set point under these conditions since the most thermal loading on the heat exchanger is encountered at the highest flow rate. The VCHP heat exchanger model predicted the measured data well. Also shown on the graph is the predicted performance for a typical counter current heat exchanger (non-VCHP, constant conductance) without temperature control. The predicted change in temperature for a typical heat exchanger under these operating conditions would be 56°C.

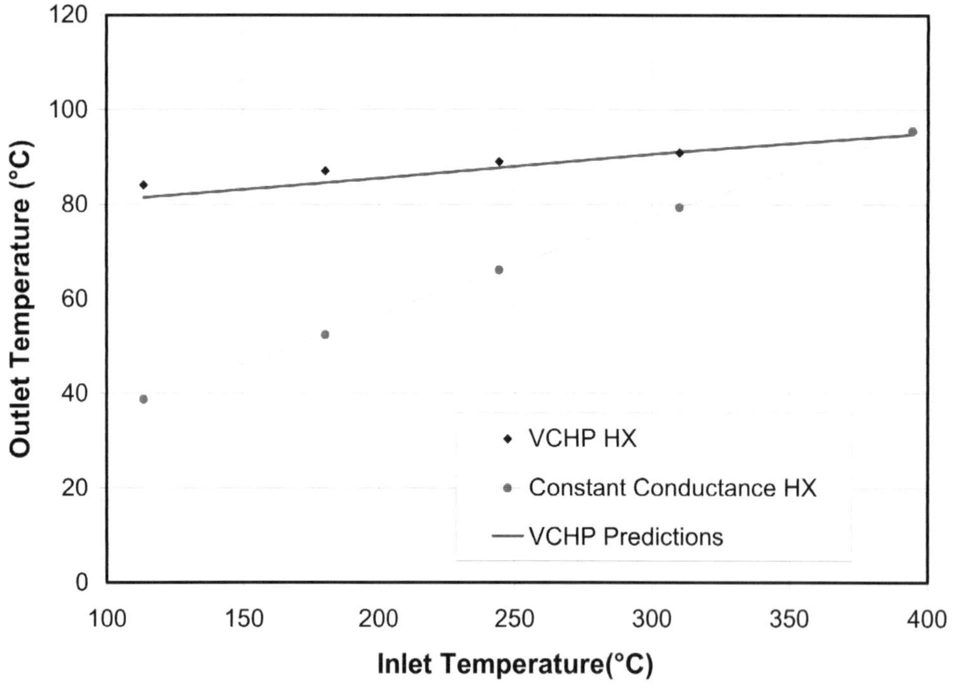

Figure 5 The measured outlet hydrogen temperature is plotted against inlet hydrogen temperature for the finless heat exchanger at a mass flow rate of 2.5kg/hr. The predicted outlet temperature from the VCHP heat exchanger model is also shown. The calculated performance for a constant conductance heat exchanger (standard heat exchanger) is also shown.

The results for the 0.5kg/hr testing are shown in Figure 6. The comparison to the typical heat exchanger is also shown. Overall, the VCHP is able to keep the outlet hydrogen temperature within a 22°C temperature range over a 5:1 turndown ratio in flow rate and 280°C change in inlet temperature. The size of the heat exchanger used to make the prediction is the same size used to make the predictions in Figure 5. In other words the size of the heat exchanger used in the prediction was large enough to remove enough heat to maintain the hydrogen outlet temperature of 95°C at 400°C inlet temperature and 2.5kg/hr. For the prediction made in Figure 6, the only variable that needed to be accounted for was the change in heat transfer coefficient due to the lower flow rate.

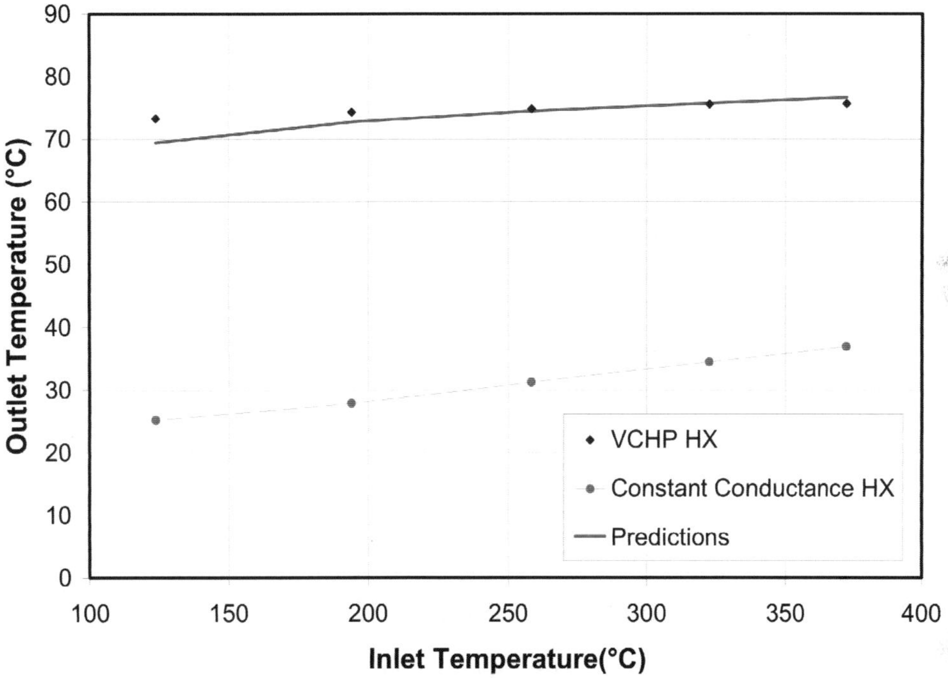

Figure 6 The measured outlet hydrogen temperature is plotted against inlet hydrogen temperature for the finless heat exchanger at a mass flow rate of 0.5kg/hr. The predicted outlet temperature from the VCHP heat exchanger model is also shown. The calculated performance for a constant conductance heat exchanger (standard heat exchanger) is also shown.

Conclusions

A variable conductance heat pipe heat exchanger has been successfully designed, fabricated and tested for a Navy fuel cell application. A performance prediction model for the VCHP HX was verified using experimental data. The experimental data also verified the overall ability of the VCHP HX to passively maintain temperature control under real world design requirements. The VCHP HX is also expected to have benefits in other applications, such as temperature control in chemical reactors where local temperature control is desired.

Acknowledgements

This material is based upon work sponsored by the Office of Naval Research under Naval Surface Warfare Center contract number N00014-06-C-0651. Mr. Donald Hoffman, Denis Colahan and Ed House were the technical monitors. Any opinions, findings, and conclusions and recommendations expressed in this material are those of the authors and do not necessarily reflect the views of the Naval Surface Warfare Center or the Office of Naval Research.

Nomenclature

A_1	Area of heat pipe evaporator
A_2	Area of heat pipe exposed to working fluid
Cp_1, Cp_2	Heat capacity of hot and cold streams
Cp_{HP}	Heat capacity of heat pipe (envelope and working fluid)
h_1, h_2	Local heat transfer coefficient of hot and cold streams
$h_{1,eff}, h_{2,eff}$	Local effective heat transfer coefficient, normalized by heat pipe footprint
H_1, H_2	Height of flow passage on hot and cold side
m_{HP}	Mass of heat pipe (envelope and working fluid)
P_{sat}	Saturation pressure of working fluid
R	Ideal gas constant
R_{HP}	Heat pipe outer radius
R_R	Radius of non-condensable gas reservoir
t	Time
T_1, T_2	Temperature of hot and cold streams
T_{HP}	Heat pipe temperature
T_{ncg}	Non-condensable gas temperature
T_∞	Ambient temperature around reservoir
v_1, v_2	Velocity of hot and cold streams
V_c	Volume of non-condensable gas in condenser
V_r	Volume of reservoir
x	Position along heat exchanger
ρ_1, ρ_2	Density of hot and cold streams

References

[1] Dunn, P. D. and Reay, D. A., "Heat Pipes", 4th ed. 1994. ISBN 0-08-041903-8

[2] Faghri, A., 1995, Heat Pipe Science and Technology, Taylor & Francis, Washington, DC.

[3] Marcus, B. D., Heat Pipes: Control Techniques, Report 2, NASA Contract No. NAS2-5503, 7/1971.

[4] Bienert, W., Heat Pipes for Temperature Control, Proceedings of the Fourth Intersociety Energy Conversion Conference, Wash., DC, 9/1969, pp. 1033-1041.

[5] Naval Shipboard Fuel Cell Program, Donald Hoffman, Naval Sea Systems Command; Anthony Nickens, ONR 33X; Harry Skruch, NAVSEA 05Z, Jan 16-17, 2003 , Shipbuilding Technologies 2003, Biloxi, MS

[6] Sarraf, D., Bonner, R., Colahan, D. "Passive Thermal Management for a Fuel Cell Reforming Process," International Energy Conversion Engineering Conference. San Diego, CA, June 2006.

Multi-scale Analysis and Control Structure Design for Biotechnological Processes

Jobrun Nandong[1], Yudi Samyudia[1] and Moses O. Tadé[2]
[1]Department of Chemical Engineering, Curtin University of Technology Sarawak Malaysia;
[2]Department of Chemical Engineering, Curtin University of Technology, Perth WA, Australia

Abstract

In this paper, a novel methodology of multi-scale analysis and control structure design is proposed for typical multi-scale biotechnological processes. This methodology combines the concepts of factorial design, multiprojection analysis and control relevant metric. The effectiveness of this methodology in handling multi-scale variables including micro-scale phenomena is demonstrated using two examples, which are two-stage distributed feed (TSDF) extractive alcoholic fermentation and recycled micro-aerobic fermentation processes. In both examples, the control structure designs resulting from the proposed methodology shows superior robustness/performance than the arbitrarily chosen control structures.

Keywords: multi-scale, fermentation, PCA, v-gap, control structure design

1. INTRODUCTION

Most of the fundamentally important processes are multi-scale in nature where one of the most striking examples which have served the humanity since prebiblical time is biotechnological process (Knorr & Sinskey, 1985). Recent progress in biotechnological field has created tremendous amount of disparate data and information about many biological systems of interests. This disparate data and information are multi-scale in nature spanning across different time and length scales. Despite this progress, the realization of the potential benefits from this multi-scale data particularly in practical aspects remains a daunting task. This provides the challenges and opportunities which have become the subject of intensive research especially in the last few years. A unified framework which is capable to integrate all this data and information across different time and length scales is a vital aspect towards producing new insights and meaningful knowledge which then, could subsequently be used to improve bioreactor performance – the heart of biotechnological processes (Cooney, 1983). In this respect, one of the key enabling tools which could serve as platform for integrating this disparate data and knowledge is multi-scale modeling where recently, its application to complex biological processes has shown significant progress (Ayton et al., 2007).

Although the applications of multi-scale modeling and analysis have recently shown some progress in the performance improvement of biochemical processes, for examples (Teixeira et al. 2007; Bideaux et al., 2006; Galaktionov et al., 2002; Charbon & Swaminarayan, 1998; Schwarzer & Peukert, 2005), the most prevalence approach currently adopted in the bioprocess improvement remains relying on the macroscopic models. These models are generally based on the unstructured kinetics i.e. *formal* macro-scale approach. In view of the recently growing competition in biochemical industry worldwide, the development of efficient process monitoring and control are considered as one of the key aspects in bioprocess improvement (Henson, 2006; Alford, 2006; Schügerl, 2001). Not withstanding this significance, the applications of advanced control methods which have been practiced extensively in chemical industries are still scarcely implemented in biological processes (Komives & Parker, 2003).

Moreover, the extension of the advanced controller techniques to bioprocesses might not be straight forward owing to the different natures between chemical and biotechnological processes. One of the key differences between chemical and biochemical processes is that the former consists of large numbers of functionally diverse, and frequently multifunctional, sets of elements that interact selectively and nonlinearly, which produce coherent rather than complex behaviours as in the later (Kitano, 2002). In other words biological systems are finely tuned across multiple resolutions e.g. from genomic to metabolic resolution. Another important aspect in bioprocess control adopted so far is that, most of the research efforts have only been dedicated in the development of controller algorithms and only scarce reports are available on the control structure design (Nandong et al., 2008a).

In chemical industries, the choice of control structure design has been recognized as a factor that has far more important impact than the choice of controller algorithms on the plant dynamic performance (Arbel et al., 1996; Hovd & Skogestad, 1993; Morari et al., 1980). The tasks of designing control structure or strategy involve the selection of and pairings of suitable manipulated and measured variables. To date in bioprocess control design, the choice of controlled outputs and manipulated variables are generally predetermined from process knowledge and experience, for examples, substrate and dissolved oxygen concentrations are normally controlled owing to their well-known influences on the cellular metabolisms. The reason why this approach works well in bioprocess could be due to the relatively small number of available inputs and outputs (e.g. frequently only one bioreactor involved) as compared with the traditional chemical processes. Additionally, in bioprocess the control system is normally focused on bioreactor.

But credible studies indicate that many of the bioprocess improvements including the finding of innovative operations could be achieved by using multiple bioreactors rather than single large bioreactor (Dourado et al. 1987; Bayrock & Ingledew, 2001; Chaabane et al., 2006; Xiu et al., 2002). Although increasing the number of bioreactors would not increase dramatically the number of inputs, the number of total bioreactors outputs generally multiplied by the number of bioreactors i.e. two bioreactors could have twice the number of outputs as single bioreactor. Thus, whenever multiple bioreactors are used, the engineers would then have to face more challenges in term of selecting the most suitable controlled outputs. Furthermore, an interesting result from recent studies (Nandong et al., 2007a; Nandong et al., 2008a) show that the implementation of appropriate control structure based on multi-scale analysis could enable the control of nonlinear system using only the traditional linear PID-type controllers. Thus, this reveals that by adopting an appropriate control structure design based on multi-scale concept could lead to an interesting, and yet yields practical alternative to applying complex advanced control strategies. Additionally, application of simple PID controllers is advantageous over that of complex advanced controller algorithms owing to its robustness and ease of maintenance and operations.

Presently, there is no control structure design methodology which has been reported in literature which is suitable for the multi-scale processes that include micro-scale phenomena i.e. most of the control structure methodologies presented to date are suitable for the macro-scale variables. In this paper, the key aim is to present a proposed methodology of multi-scale analysis and control structure design for multi-scale biotechnological processes (see Nandong et al., 2008a). The effectiveness of this proposed methodology is demonstrated on two examples, which are (1) two-stage distributed feed (TSDF) continuous extractive alcoholic fermentation, and (2) recycled micro-aerobic fermentation system. It would be shown that based on the proposed methodology, one could find a control strategy that could be efficiently used to control a nonlinear multi-scale system using only simple linear PID controllers. Furthermore, the simulation study shows that the control structure design based on the proposed methodology has superior robustness and performance than that of the arbitrary one.

2. METHDOLOGY

Due to the space limitation, only brief description of the proposed methodology is presented. The details about this methodology are presented in Nandong et al. (2008a). The key idea underlying the proposed methodology is to identify the *critical* variables which are strongly correlated with the specified plant performance measures. The foremost performance measure which is to be achieved by the resulting control structure design is to minimize the nonlinear excitation in the face of external disturbance occurrence. If the nonlinear excitation is small then this should justify the use of simple linear PID-type controller. And of course other important performance measures such as yield and productivity must also be achieved or at least be maintained closed at the desired values. But, it is expected that to be able to control a system such that steady-sate performance measures (i.e. yield and productivity) could be maintained closed at desired values (or minimized variability), it is crucial that the system is easily control (i.e. has favorable dynamic operability). Thus this is the reason why finding the control structure which could minimize the nonlinear excitation when external disturbance occurs is the foremost objective to be fulfilled (i.e. dynamic performance measure) before meeting steady-state performance measures.

Figure 1 shows the mapping concept from the key or *critical* variables to the performance measures. One way to do this mapping is by adopting the technique of multiprojection analysis e.g. Principal Component Analysis (PCA). In the PCA, the objective is to reduce the large number of interacting variables in a dataset into smaller group of key variables i.e. dimensional reduction. As applied to the control structure analysis as illustrated by Figure 1 below, the PCA analysis is used to identify the *critical* variables that strongly correlate with the performance measures. This mapping could lead to three general cases as:

1. **Single Variable - Single Performance (SVSP) Mapping** – One variable/parameter is strongly correlated with a certain performance measure.
2. **Single Variable-Multiple Performances (SVMP) Mapping** – One variable/parameter is strongly correlated with more than one performance measures.
3. **Multiple Variables-Multiple Performance (MVMP) Mapping** – Multiple variables/parameters are strongly correlated with multiple performance measures – cross mapping exists.

The complexity of the control structure design problem depends on the cases above, for example, if the mapping results in case 1, then the control structure problem is expected to be relatively simple as compared with case 2 or 3. But in most practical applications, case 2 and 3 are expected to be common. Additionally, in view that the control structure also depends on the operating level, thus it is possible that there is no proper control structure exists at all i.e. there is no critical variables that map strongly to the specified performance measures as reported in Nandong et al. (2007b). The methodology for finding the operating conditions with favorable dynamic operability could be found in this paper. The interested readers could refer to Kourti and MacGregor (1995) for background on PCA and Vinnicombe (2001) for v-gap metric concept used in this methodology.

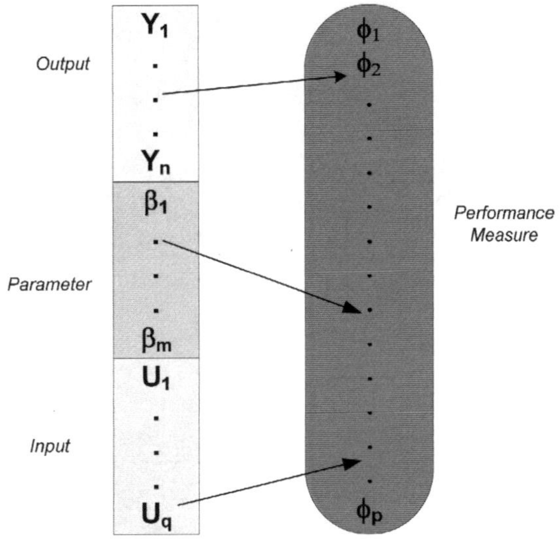

Figure 1 - Schematic of mapping from critical variables to performance measures

Algorithm

The proposed control structure design methodology in this paper could be summarized as following:

Step 1 – Generation of Operating Regimes: Select a set of nominal operating levels, and for each operating level apply step inputs perturbations using a factorial design concept.

Step 2 – Generation of Linear Model Sets: Linearize the nonlinear process model at each nominal operating level and at the corresponding perturbed levels.

Step 3 – Generation of Quality Variables: Determine the performance measures, for examples, productivity, yield, etc at nominal and perturbed operating levels. Also, compute the nonlinearity excitation as measured by *v-gap* value between the nominal and perturbed levels.

Step 4 – Mapping of Critical Variables to Performance Measures: Gather and combine all the generated data on the process parameters, input-output variables and the computed performance measures (or quality variables) in **Step 3** to form a dataset, which is to be used in the PCA (multiprojection) analysis.

Step 5 – Control Structure Design: There are three main tasks involved which are the selection of controlled variables, manipulated variables and controller pairings.

3. RESULTS AND DISCUSSIONS

The application of the proposed methodology to multi-scale biotechnological processes is demonstrated using two fermentation examples. The first example which is on TSDF alcoholic fermentation process is taken from the work reported in Nandong et al. (2008a). Meanwhile the multi-scale model describing the recycled micro-aerobic fermentation process is taken from Nandong et al. (2008b).

Also note that, in these examples only linear PI controllers are used. As for the case of multi-scale control strategy, the micro-scale controller is not specified but it is assumed that this controller could maintain the micro-scale output within certain range.

3.1 Example 1: Two-stage Distributed Feed (TSDF) Alcoholic Fermentation Process

Figure 2 shows the schematic of TSDF continuous alcoholic fermentation processes. The key feature of this design is that the fresh feed is distributed between the two bioreactors that are in series (R1 and R2) according to ratio SR (fraction of fresh feed that enters the R1). The details about this process could be found in Nandong et al. (2008). Table 1 shows the variables forming the dataset. Note that there are 16 perturbed levels (P1, P2...P16) around the nominal level P0 based on the four inputs (r, R, SR and So).

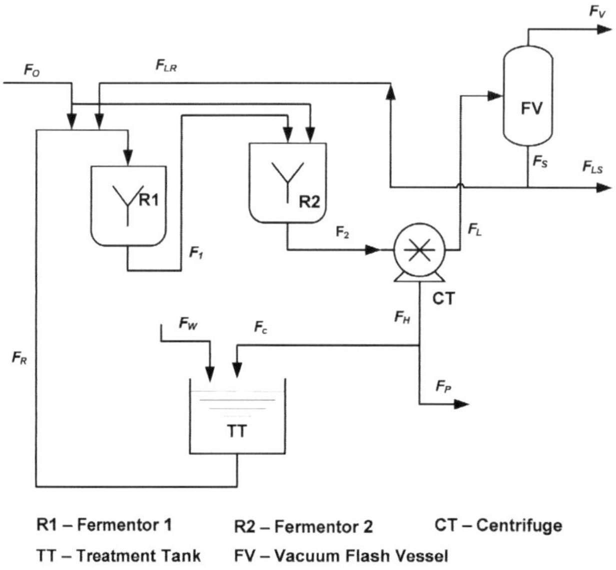

Figure 2 – Schematic of TSDF continuous alcoholic fermentation process

Table 1 – Variables forming the dataset for PCA analysis

Variable	Description	Variable	Description
Yield	*Ethanol yield*	S_1	*Substrate concentration in R1*
Prod	*Ethanol productivity*	S_2	*Substrate concentration in R2*
Conv	*Substrate conversion*	T_1	*Temperature in R1*
r	*Flash recycle ratio*	T_2	*Temperature in R2*
R	*Cell recycle ratio*	P_1	*Ethanol concentration in R1*
SR	*Split feed ratio*	P_2	*Ethanol concentration in R2*
So	*Fresh substrate concentration*	Xv_1	*Viable cell concentration in R1*
L_1	*Liquid level in R1*	Xv_2	*Viable cell concentration in R2*
L_2	*Liquid level in R2*	v-gap$_1$	*v-gap of R1*
Rp_1	*Rate ethanol production in R1*	v-gap$_2$	*v-gap of combined R1 and R2*
Rp_2	*Rate ethanol production in R2*	Rx_1	*Rate cell production in R1*
Rs_1	*Rate substrate consumption in R1*	Rx_2	*Rate cell production in R2*
Rs_2	*Rate substrate consumption in R2*		

There are 26 variables in total which are considered in the PCA analysis as following:

$$\phi = [v-gap_1 \quad v-gap_2 \quad Yield \quad Prod \quad Conv]^T$$
$$U = [r \quad R \quad SR \quad So]^T$$
$$Y = [L_1 \quad L_2 \quad S_1 \quad S_2 \quad T_1 \quad T_2 \quad P_1 \quad P_2 \quad Xv_1 \quad Xv_2]^T \quad (1)$$
$$\beta = [Rp_1 \quad Rp_2 \quad Rs_1 \quad R_2 \quad Rx_1 \quad Rx_2]^T$$

In this case, there are 5 performance measures, 4 macro-scale inputs, 10 macro-scale outputs and 6 macro-scale parameters. These parameters could be viewed as the outputs of micro-scale inputs *(see Nandong et al., 2007a)*. Thus in the dataset there is no direct micro-scale variables being considered i.e. only indirectly via the parameters. Table 2 shows some of the selected values from the original dataset for few observations. Notice that for observation #3 at P2, the value of v-gap$_2$ is the largest. Thus it is interesting to know whether this will lead to outlier or not in the PCA analysis latter. And if this becomes outlier, then the question is whether v-gap$_2$ has any contribution to this outlier.

Table 2 – Selected values from the complete dataset

No	Level	v-gap$_2$	S_1 (kg/m^3)	S_2 (kg/m^3)	Yield (%)	Prod (kg/m^3.hr)	Conv (%)
1	P0	0	17.35	1.60	89.90	14.83	98.49
2	P1	0.980	43.18	20.31	75.09	18.51	79.39
3	**P2**	**0.995**	**99.17**	**79.58**	**45.42**	**16.73**	**45.87**
5	P4	0.980	73.85	38.85	67.41	21.15	75.44

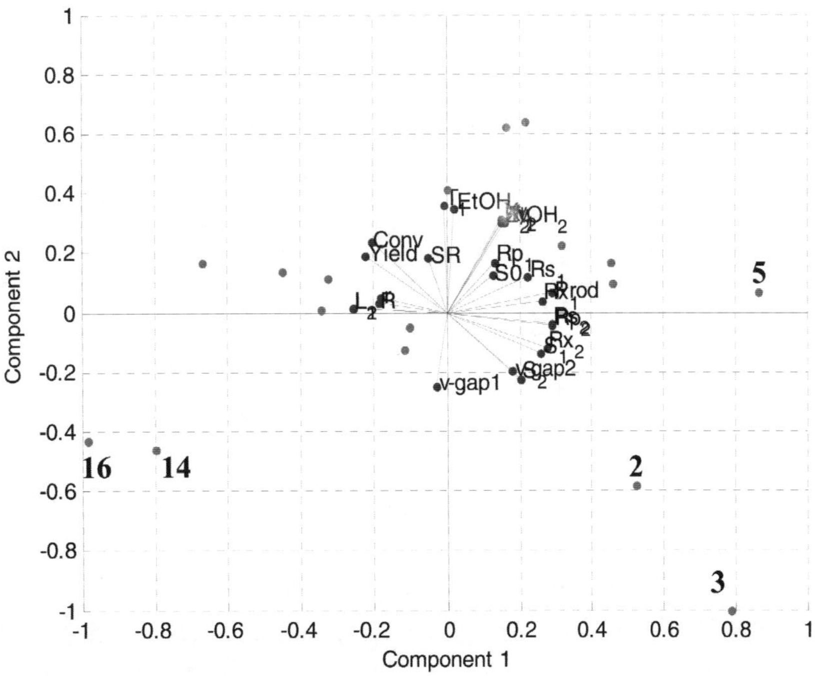

Figure 3- Plot of PC for two principal components PC1 and PC2 of the complete dataset

From Figure 3, the outliers #2 and #3 (at P1 and P2 respectively) could be due to 14 variables which are *v-gap*$_2$, T_1, L_1, L_2, r, R, *SR*, *Yield*, *Conv*, S_1, S_2, R_{x2}, R_{p2}, R_{s2} (i.e. these variables occupy 2nd and 4th quadrants. Other variables which occupy 1st and 3rd quadrants could be responsible for other outliers i.e. outlier #14, #15 and #16. Since the main objective in this example is to obtain the critical variables that strongly correlate with nonlinear excitation (i.e. v-gap), thus further analysis will be focused on the variables that responsible for outlier #3 where v-gap$_2$ could be one of the contributor to this outlier. Therefore, to identify which variables that really have strong correlation with the outlier #3, the original dataset is reduced to 14 variables and another PCA analysis is performed on this first reduced dataset.

The PCA analysis performed on this first reduced dataset indicates that there are 7 variables that may have strong correlations with outlier #3, which are *v-gap*$_2$, S_1, S_2, *SR*, T_1, *Yield* and *Conv*. However, it is still rather unclear at this stage which variable/s actually has dominant contribution to the outlier #3. More importantly, could it be that v-gap$_2$ has strong correlation with this outlier. Thus, the dataset is further reduced to only 7 variables (*v-gap*$_2$, S_1, S_2, *SR*, T_1, *Yield* and *Conv*). Another PCA analysis is performed on this second reduced dataset.

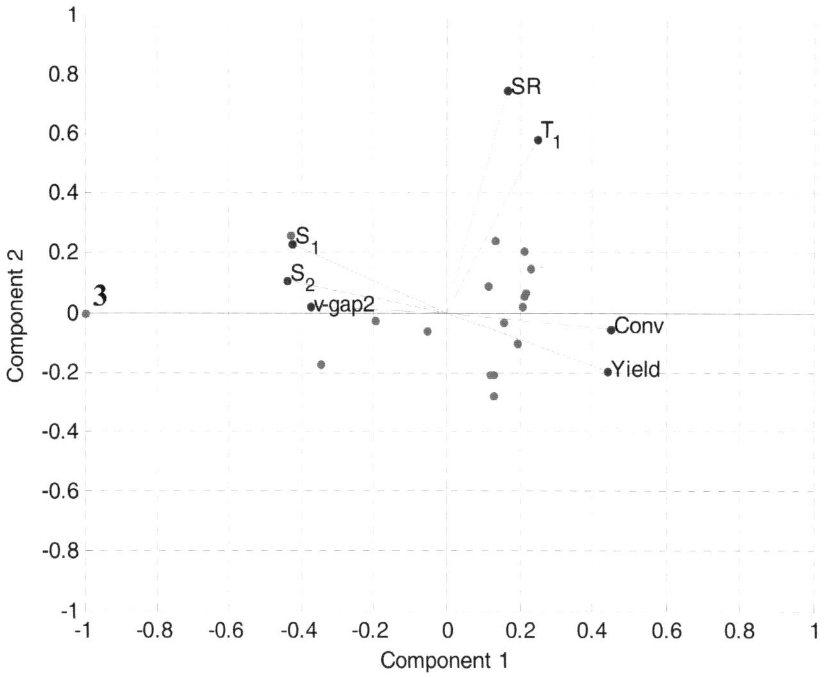

Figure 4 - Plot of PC score and loading for PC1 and PC2 of the second reduced dataset

Figure 4 shows the plot of two principal components from this second reduced dataset. This plot indicates that the outlier #3 is positively correlated with S_1, S_2 and *v-gap*$_2$ and is negatively correlated with *Conv* and *Yield*. Moreover, this mapping analysis reveals two critical variables which are S_1 and S_2 that strongly correlate with the performance measures, which are v-gap$_2$, Yield and Conv. Because S_1 and S_2 are positively correlated with v-gap$_2$ (in the same quadrant), thus to reduce the nonlinear excitation (i.e. as measure by v-gap) the values of S_1 and S_2 must be reduced. Also in this example the mapping results in case 3 i.e. Multiple Variables-Multiple Performance (MVMP).

Now the next question is whether it is necessary to control both S_1 and S_2? Because they are positively correlated, this implies that if one of these variables is controlled then this would indirectly control another variable. So in this case S_2 is selected as the controlled variable. Another variable to be controlled is liquid level in bioreactor 2 (i.e. L_2). Two inputs which are chosen as manipulated variables are r and R because these inputs have strong *cause*-and-*effect* relationship with the outputs i.e. high process gains. This control structure is called S-L structure. As a benchmark, another arbitrary control structure uses viable cell concentration in bioreactor 2 (Xv_2) and ethanol concentration in bioreactor 2 (P_2) as the controlled variables. In this control structure, r and R are also used as manipulated variables. This arbitrary control structure is called Xv-P structure. Note that in both structures, only PI controllers are used and tunings are based on linearized model at nominal operating level (P0) using Ziegler-Nichols formula.

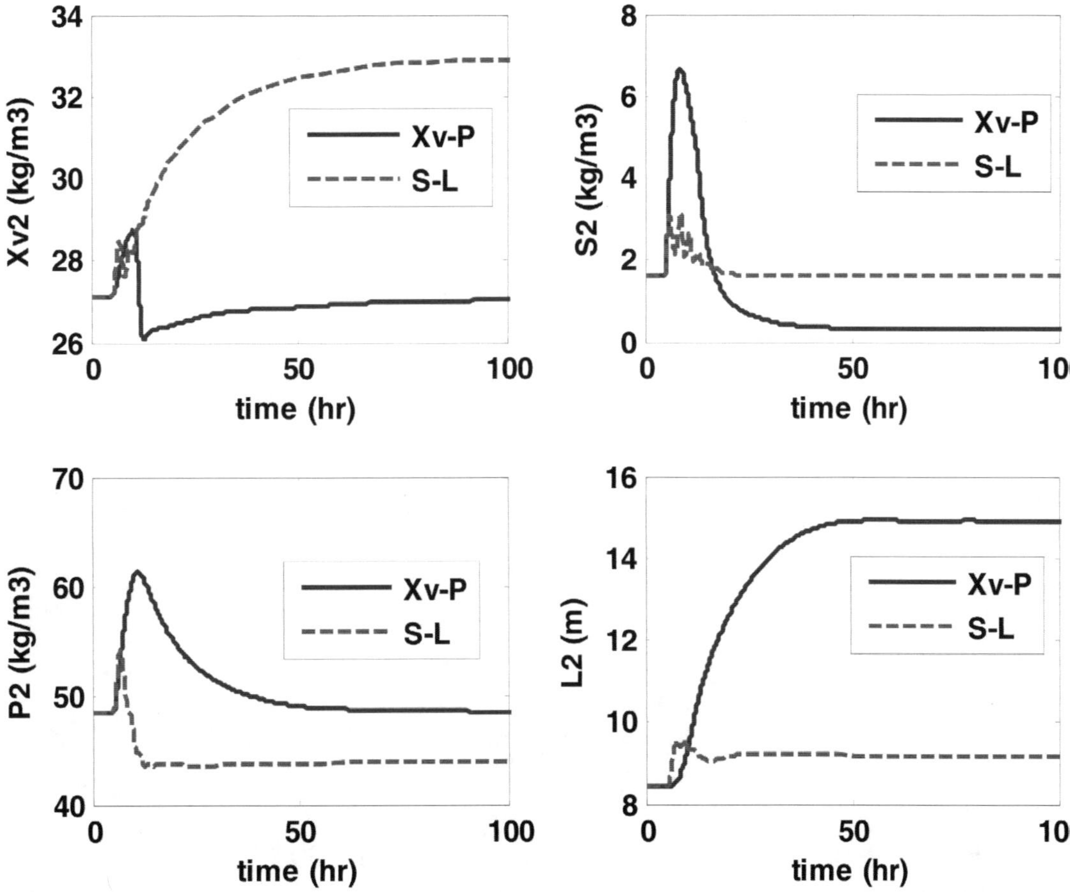

Figure 5 – Response to step change ΔS_o = 60 kg/m^3: nominal S_o = 170 kg/m^3

Figure 5 shows the closed-loop responses of the two control structures when subject to step disturbance in fresh substrate concentration (So) of 60 kg/m^3. For the S-L control structure, note that the uncontrolled Xv_2 and P_2 show significant deviations from the nominal value and yet the closed-loop responses in S_2 and L_2 are still stable and faster than that of arbitrary Xv-P control structure. It is interesting to note that the Xv-P strategy is unstable when subject to step decrease in So by 60 kg/m^3. However S-L strategy remains stable. For Xv-P structure, under this disturbance it seems that S_2 increases sharply (result not shown). Because S_2 is positively correlated with v-gap$_2$, sharp increase in S_2 would lead to increase in nonlinear excitation thereby causing severe performance degradation to the linear control system – hence leading to instability. Meanwhile, under S-L control strategy the value of S_2 is prevented from drifting from the nominal value. Consequently this prevents the severe nonlinear excitation.

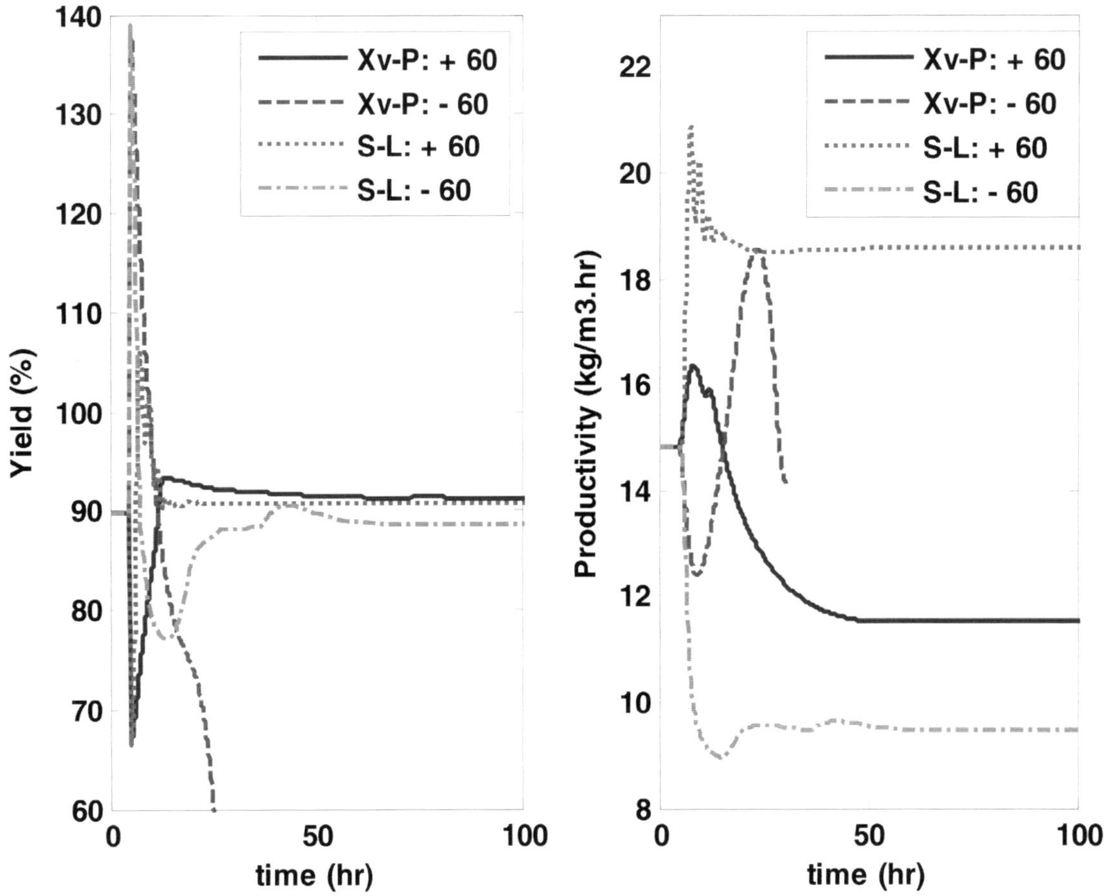

Figure 6 – Impact of ΔS_o = ±60 kg/m^3 on yield and productivity: nominal S_o = 170 kg/m^3

Figure 6 displays the impact of step disturbance in So of ±60 kg/m^3 on the steady state performances (yield and productivity). Notice that the S-L control strategy is able to maintain the yield closed to the nominal value (small steady-state offset). On the other hand, Xv-P strategy fails to remains stable when subject to -60kg/m^3 in So. For S-L strategy the ability of the control structure to have good impact on yield is expected since the S_2 is negatively correlated with this performance measure. Thus controlling S_2 is not only essential to prevent severe nonlinear excitation but also necessary to keep yield at nominal value. However, there is no clear advantage of S-L structure (other than it is stable) over Xv-P strategy for the productivity i.e. large offset of yield for S-L structure. Again this could be expected because neither S_2 nor L_2 is strongly correlated with productivity. So, to control this productivity measure, one needs to find other critical variable/s that correlates well with this performance but this is not done in this work.

3.2 Example 2 – Recycled Micro-aerobic Fermentation System

The multi-scale model of the recycled micro-aerobic fermentation system used in this example is taken from Nandong et al. (2008b). Table 3 shows the variables forming the dataset used in the PCA analysis. There are 20 variables in total which are considered in this PCA analysis as following:

$$\begin{aligned} &\phi = Magap \\ &U = [So \;\; ADP \;\; ATP \;\; NADH]^T \\ &Y = [GlcE \;\; Gly \;\; Ace \;\; EtOH \;\; Cx \;\; Xa \;\; Glc \;\; G6P \;\; F6P \;\; FBP \;\; DHAP \;\; GAP \;\; PEP \;\; PYR \;\; ALDE]^T \end{aligned} \tag{2}$$

Note that U and Y are multi-scale in nature i.e. GlcE is macro-scale output while G6P, PYR and ALDE are micro-scale outputs or intracellular metabolite concentrations. It is important to point out in this dataset there is no parameter i.e. both macro- and micro-scale variables are treated explicitly. Of course, one could also include the parameters such as specific growth and product formation rates but these parameters are calculated directly by the micro-scale system.

Table 3 – Variables forming the dataset used in PCA analysis

Variable	Description	Variable	Description
Ma gap	*v-gap metric*	Xa	*Fraction of active cells*
So	*Inlet glucose concentration*	GlcC	*Intracellular glucose concentration*
ADP	*Adenosine diphosphate*	G6P	*Glucose-6-phosphate*
ATP	*Adenosine triphosphate*	F6P	*Fructose-6-phosphate*
NADH	*Nicotinamide adenine dinucleotide*	FBP	*Fructose-1,6-phosphate*
GlcE	*Extracellular glucose concentration*	DHAP	*Dihydroxyacetone phosphate*
Gly	*Extracellular glycerol concentration*	GAP	*Glyceraldehyde-3-phosphate*
Ace	*Extracellular acetate concentration*	PEP	*Phosphoenol pyruvate*
EtOH	*Extracellular ethanol concentration*	PYR	*Pyruvate*
Cx	*Total cells concentration*	ALDE	*Acetaldehyde*

Table 4 – Selected values from complete dataset

No	Level	Ma gap	Ace (g/m^3)	G6P (mM)	F6P (mM)	PYR (mM)	ALDE (mM)
1	P0	0	85.62	3.6883	0.6949	6.5833	0.0358
11	P10	0.0002	87.69	3.5608	0.6680	8.7191	0.0369
14	**P13**	**0.7072**	**104.93**	**3.5907**	**0.6744**	**8.4752**	**0.0473**
15	P14	0.0002	87.20	3.8174	0.7220	5.1704	0.0367

Table 4 displays selected values from the complete dataset for 4 observations only. An interesting observation is that at P13 the value of v-gap (Ma gap) is abnormally large equals to 0.707 compared with at other perturbed levels (very small). PCA analysis is performed and the analysis shows that there are 9 variables which could significantly correlate with the outlier #14 (at P13) that are Glc, PYR, Ace, ALDE, Ma gap, G6P, F6P, ATP and PEP. To clearly identify the key variables, the original dataset is reduced to these 9 variables and second PCA analysis is performed. The plot of the two principal components which describe about 90% of the total variances in the reduced dataset is shown in Figure 7. From this plot, one can conclude that there are two critical variables which positively correlate with Ma gap which are Ace and ALDE.

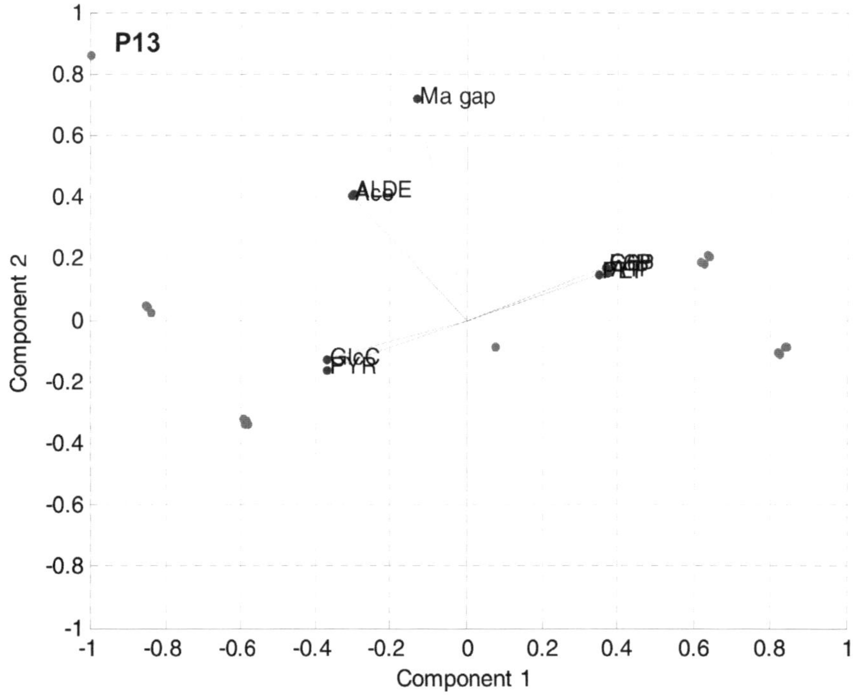

Figure 7 - PCA plot of reduced dataset for two major principal components

Both variables positively correlate with Ace and ALDE. So, the implication to control structure design is that whether both of these critical variables need to be controlled or not in order to prevent large nonlinear excitation in the face of disturbance occurrence. Recall in the previous example, if two critical variables are positively correlated then only one need to be controlled because this also lead to indirectly control another critical variable. But this strategy seems to work if both critical variables are of similar scale i.e. previously both variables are macro-scale. The question now, does this strategy works if the critical variables are of different scales as in this example.

To evaluate the effectiveness of applying the strategy which is adopted in example 1, two MIMO control structures would be proposed:

Single-scale Control (SSC) Strategy 1 – Only Ace is controlled while the micro-scale variable ALDE is left uncontrolled. In addition the EtOH is selected as another controlled variable.

Multi-scale Control (MSC) Strategy – Both of the critical variables are controlled in addition to controlling the EtOH.

Another macro-scale control structure which adopts GlcE and EtOH as controlled variables is proposed as a benchmark – called *SSC strategy 2*. Also, note that all macro-scale controllers are PI-type where the tuning is based on the linearized model at the nominal operating level using the Ziegler-Nichols open-loop formula. For the MSC strategy, the micro-scale controller is simply assumed to be able to keep the ALDE within ±0.0002 mM. The performance is assessed based on the EtOH closed-loop response when subject to unit step disturbance in cellular ATP concentration.

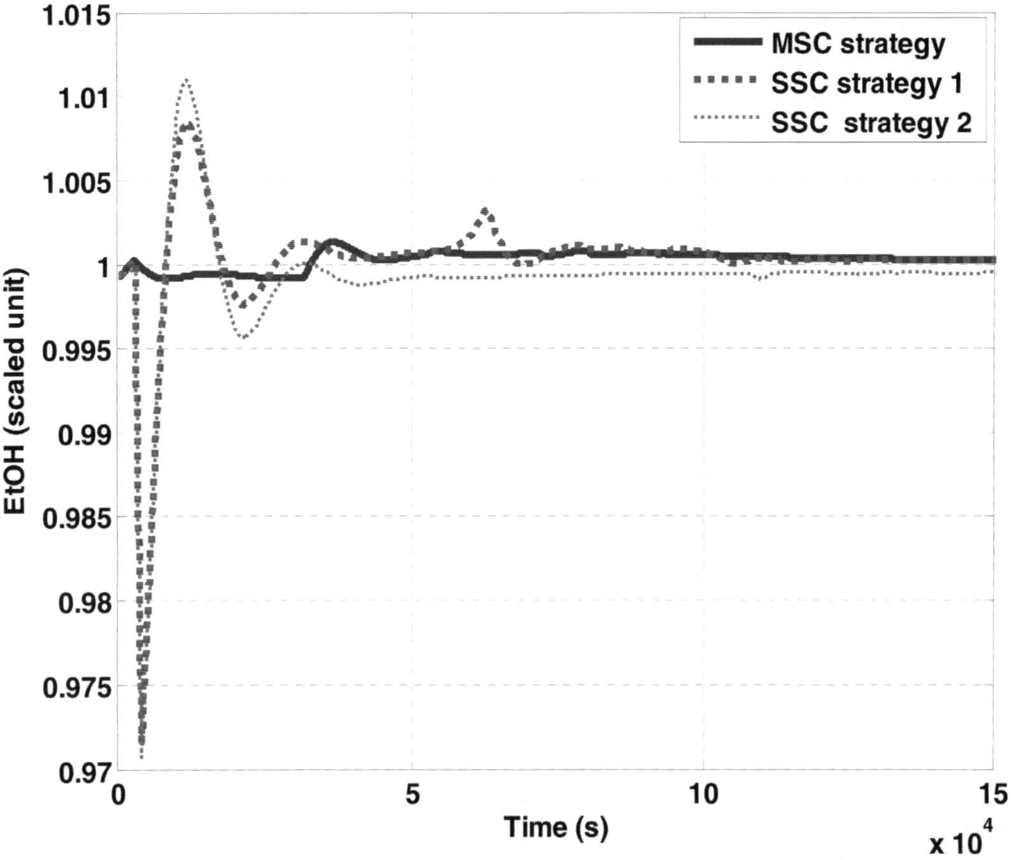

Figure 8 – Closed loop response of EtOH to pulse disturbance in ATP at time 2000s (duration 1000s, amplitude 1) for three different control strategies

From Figure 8, it shows that multi-scale control (MSC) strategy outperforms the two macro-scale (or single-scale) strategies i.e. much smaller overshoot and faster settling time. For the case of single-scale control (SSC) strategy 1, it seems that the advantage to control Ace (i.e. one of the critical variables) is small as compared to the arbitrary control structure that controls GlcE and EtOH (SSC strategy 2).

Thus in contrary to the previous example, it is important to control both *critical* variables which are actually multi-scale in terms of length and time resolution i.e. multi-scale control strategy required in this example to have significance improvement in control performance. Note that, the offset in MSC strategy and SSC strategy 1 could be due to the lack of integral action and no controller retuning has been considered in this example.

CONCLUSIONS

The effectiveness of proposed methodology in handling multi-scale processes has been confirmed using extractive fermentation and recycled micro-aerobic fermentation systems. The key novelty in this approach is its ability to find which variables among the multi-scale outputs which directly or strongly correlated with the specified performance measures and must be controlled. Essential to this methodology is to find which variables that have strong impacts on the nonlinear excitation i.e. v-gap metric. Controlling these variables is vital to ensure good dynamic operability so that it can justify the use of simple linear PID-type controllers for strongly nonlinear system. If two critical variables exist for certain performance measure/s, then only one of these variables need to be controlled if they are both positively correlated and if both are also of similar macro-scale type. On the other hand, if two critical variables are multi-scale than both variables must be controlled i.e. as shown in the example 2. Thus this would lead to full multi-scale control (MSC) strategy.

The key limitation of adopting MSC strategy is the lack of suitable micro-scale manipulated variables that could be used to control the micro-scale outputs directly. Additionally, the lack of currently suitable sensors to measure the micro-scale outputs online is another major hindrance in the adoption of MSC strategy. However, these obstacles would soon be resolved by the rapid advances in (1) cellular design methodology (see Kaznessis, 2006), (2) measurement and sampling technologies, and (3) multi-scale modeling and computation.

Acknowledgement

This work is part of the project that is supported by Malaysian Ministry of Science, Technology and Innovation (grant no. 02-02-07-SF0001).

References

Alford, J.S. (2006). Bioprocess control: advances and challenges. *Computers Chemical Engineering*, **Vol. 30**, 1464-1475.

Arbel, A., Rinard, I.H. and Shinnar, R. (1996). Dynamics and control of fluidized catalytics crackers. 3. Designing the control system: choice of manipulated and measured variables for partial control. *Industrial Engineering Chemistry Research*, **Vol. 35**, 2215-2233.

Ayton, G.S., Noid, W.G. and Voth, G.A. (2007). Multiscale modelling of biomolecular systems: in serial and in parallel. *Current Opinion in Structural Biology*, **Vol. 17**, 192-198

Bayrock, D.P. and Ingledew, W.M. (2001). Application of multistage continuous fermentation for production of fuel alcohol by very-high-gravity fermentation technology. *J. Industrial Microbiology Biotechnology*, **Vol. 27**, 87-93.

Bideaux, C., Alfenore, S., Cameleyre, X., Jouve, C.M., Uribelarrea, J.L. and Guillouet, S.E. (2006). Minimization of glycerol production during the high-performance fed-batch ethanolic fermentation process in Saccharomyces cerevisiae, using a metabolic model as a predictive tool. *Applied Environmental Microbiology*, **Vol. 72, No. 3**, 2134-2140.

Chaabane, F.B., Aldiguier, A.S., Alfenore, S. and et al. (2006). Very high ethanol productivity in an innovative continuous two-stage bioreactor with cell recycle. *Bioprocess Biosystem Engineering*, **Vol. 29**, 49-57.

Charbon, C. and Swaminarayan, S. (1998). A multiscale model for polymer crystallization. II: Solidification of a macroscopic part. *Polymer Engineering Science*, **Vol. 38, No. 4**, 644-656.

Cooney, C.L. (1983). Biorecators: design and operation. *SCIENCE*, **Vol. 219**, 728-733.
Dourado, A., Calvet, J.L., Sevely, Y. and Goma, G. Modeling and static optimization of the ethanol production in a cascade reactor. II. Static optimization. *Biotechnology Bioengineering*, **Vol. XXIX**, 195-203.
Galaktionov, O.S., Anderson, P.D., Peters, G.W.M. and Tucker III, C.L. (2002). A global, multi-scale simulation of laminar fluid mixing: the extended mapping method. *Int. J. Multiphase Flow*, **Vol. 28**, 497-523.
Henson, M.A. (2006). Exploiting cellular biology to manufacture high-value products. *IEEE Control Systems Magazine*, August, 54-62.
Hovd, M. and Skogestad, S. (1993). Procedure for regulatory control structure selection with application to the FCC process. *AIChE*, **Vol. 39, No. 12**, 1938-1953.
Kaznessis, Y.N. (2006). Multi-scale models for gene network engineering. *Chemical Engineering Science*, **Vol. 61**, 940-953.
Kitano, H. (2002). Computational systems biology. *NATURE*, **Vol. 420**, 206-210.
Knorr, D. and Sinskey, A.J. (1985). Biotechnology in Food Production and Processing. *SCIENCE*, **Vol. 229**, 1224-1229.
Komives, C. and Parker, R.S. (2003). Bioreactor state estimation and control. *Current Opinion in Biotechnology*, **Vol. 14**, 468-474.
Kourti, T., & MacGregor, J.F. (1995). Process analysis, monitoring diagnosis multivariate methods, *Chemo. Intel. Lab. Syst.* **Vol. 28**, 3-21.
Morari, M., Arkun, Y. and Stephanopoulos, G. (1980). Studies in the synthesis of control structures for chemical processes. *AIChE*, **Vol. 26, No. 2**, 220-232.
Nandong, J., Samyudia, Y. and Tade, M.O. (2007b). Control structure analysis and design for nonlinear multivariable systems, *8th Proceedings DYCOPS,* Cancun, Mexico, 2007.
Nandong, J., Samyudia, Y. and Tadé, M.O. (2008a). Control structure design for multi-scale biotechnological processes. *Industrial Engineering Chemistry Research*, (unpublished).
Nandong. J., Samyudia, Y. and Tade, M.O. (2007a). Control of multi-scale dynamics system. *16th IEEE International Conference on Control Applications*, Singapore, 1-3 October 2007.
Nandong. J., Samyudia, Y. and Tade, M.O. (2008b). Multi-scale framework for modeling and control of fermentation processes. *17th IFAC World Congress*, Seoul, Korea, 6-11 July 2008.
Schügerl, K. (2001). Progress in monitoring, modeling and control of bioprocesses during the last 20 years. *J. Biotechnology*, ***Vol. 85***, 149-173.
Schwarzer, H.C. and Peukert, W. (2005). Prediction of aggregation kinetics based on surface properties of nanoparticles. *Chemical Engineering Science*, **Vol. 60**, 11-25.
Teixeira, A.P., Alves, C., Alves, P.M, Carrondo, M.J.T. and Oliveira, R. (2007). Hybrid elementary flux analysis/nonparametric modeling: application for bioprocess control. *BMC Bioinformatics*, 8:30.
Vinnicombe, G. (2001). *Uncertainty and feedback*. Imperial College Press, London.
Xiu, Z.L., Song, B.H., Sun, L.H. and Zeng, A.P. (2002). Theoretical analysis of effects of metabolic overflow and time delay on the performance and dynamic behavior of a two-stage fermentation process. *Biochemical Engineering J.*, **Vol. 11**, 101-109.

574h Robust Nonlinear Model Predictive Control of An Augmented-Reality Lab-Scale Plant

Udo Schubert, Harvey Arellano-Garcia, and Guenter Wozny. Department of Process Dynamics and Operation, Berlin Institute of Technology, Straße des 17. Juni 135, 10623 Berlin, Berlin, Germany

Model predictive control (MPC) turns out to become the current standard in latest industrial applications. The achieved improvements arising from replacement of outdated PID-control-schemes with MPC-systems confirm this evolution. However, the common approach by now is to utilize linear models to predict the future process behavior. This method has some well known drawbacks e.g. in case of strong nonlinearity of the process, or a large region of operation [1]. These deficiencies become an issue, because the need for matching higher and stringent quality standards is continuously increasing. Therefore current research focuses on nonlinear model predictive control (NMPC) to overcome limitations of linear MPC.

The objective of this work is the application of Robust NMPC to a lab-scale plant of an industrial continuously stirred tank reactor (CSTR) process. The existing assets are extended with a simulation layer, which results in an augmented reality plan. The chemical reaction (reactor hold-up) is the virtual part, represented by a simulation that contributes to the heat exchange process (reactor jacket and heat exchanger) in the real world. Finally, the NMPC-controller is used to control the temperature in the virtual reactor hold-up using the heat exchange process of the real world. With this setup, it is easy to introduce different levels of uncertainty, disturbances and failures into the system. However, excluding the part of the chemical reaction still leaves enough non-artificial phenomena and nonlinearity in the system. It is known that a heat exchanger process possesses strong dependency of the coolant temperature and flow, as well as on the product temperature and flow.

The design of the augmented-reality plant is illustrated by reviewing the design of the robust NMPC-scheme. For the explicit consideration of uncertainty a stochastic optimization framework is used [2]. The results are discussed concerning robustness against model parameter uncertainty, process uncertainty and overall control performance.

[1] HENSON, M.A.: Nonlinear model predictive control: current status and future directions. Comp. & Chem. Eng., Vol. 23, No. 2, pp 187-202 (1998.

[2] ARELLANO-GARCIA, H.: Chance constrained optimization of process systems under uncertainty. PhD Thesis (2006)

Relay with a Negative Hysteresis for Autotuning of Control Systems Under Noisy Environments

Jietae Lee[1], Su Whan Sung[1] and Thomas F. Edgar[2]
[1] Department of Chemical Engineering, Kyungpook National University, Taegu 702-701, KOREA
Tel: +82-53-950-5620, Fax: +82-53-950-6615, E-mail: jtlee@knu.ac.kr
[2] Department of Chemical Engineering, University of Texas, Austin, TX78712, U.S.A.

When measurements contain noises, to avoid chattering, relays with positive hysteresis are often used for the relay feedback autotuning of control systems. However, their switching times fluctuate and average switching periods are not those at a given phase angle. Their phase angles are dependent on the oscillation periods and can be found after the relay feedback experiments are done. To find amplitudes of relay feedback oscillations, average values are needed.

Recently, a method which reduces errors in the ultimate gain identification by using integrals of relay feedback responses has been proposed [1]. This integral which reduces the effects of noises can be put in the feedback loop, but its oscillation becomes at the phase angle of $-\pi/2$. The frequency response data at the phase angle of $-\pi/2$ is usually not recommended for tuning of PID controllers. To find frequency responses of phase angles between $-\pi/2$ and $-\pi$, relay with a negative hysteresis [2] is introduced here.

First, we investigate examples showing phase angles appropriate to tune PID controllers [3]. Then we propose a method to obtain the frequency response at the phase angle. For this, relay with a negative hysteresis and the integral element are used. The integral element reduces the effect of noises and the relay with negative hysteresis compensates the phase lag due to the integral element. Typical responses of the proposed relay feedback method are shown as follows.

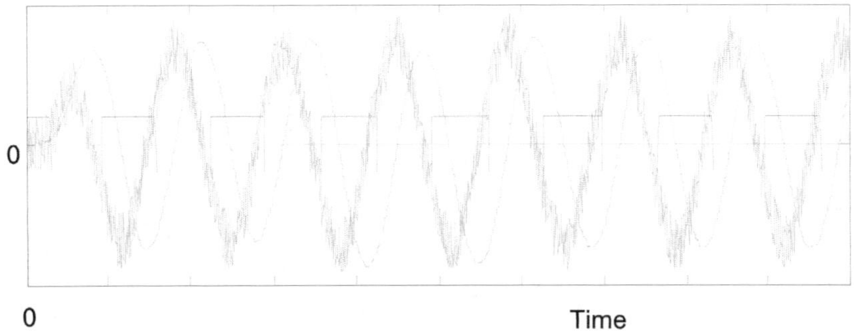

Figure 1. Typical responses of the proposed relay feedback identification.

References
[1] J. Lee, S.W. Sung and T.F. Edgar, AIChE J., 53, 2329 (2007).
[2] D.P. Atherton, Nonlinear Control Engineering, Van Nostrand Reinhold, N.Y. (1975).
[3] M. Friman and K.V. Waller, I&EC Research, 36, 2662 (1997).

574j Dynamic Optimization of *Mma* and *Vac* Copolymerization Process

Nádson M. N. Lima[1], Lamia Zuñiga Liñan[1], Rubens Maciel Filho[1], **Flavio Manenti**[2], Davide Manca[2], and Marcelo Embiruçu[3]. (1) School of Chemical Engineering, University of Campinas, UNICAMP, P. O. Box 6066, 13083-970, Campinas-SP, Brazil, (2) CMIC Dept. "Giulio Natta", Politecnico di Milano, Piazza Leonardo da Vinci, I-20133, Milano, Italy, (3) Polytechnique Institute, Federal University of Bahia, UFBA, 40210-630, Salvador-BA, Brazil

Increasing worldwide market competitiveness and reduced profit margins are pressing chemical and process industries to move towards a predictive control approach, based on first-principles mathematical models, as well as plant dynamic optimization.

In this perspective, the paper focuses on the development of a nonlinear model predictive control (NMPC) to manage the copolymerization process of methyl methacrylate (MMA) and vinyl acetate (VAc), consisting of a jacketed continuous stirred tank reactor, a separator, and a recycle loop. This system presents a highly complex behavior, thus making difficult the success the controllers based on linear models.

A detailed differential and algebraic mathematical model as well as process operating conditions were obtained from the literature (Congalidis et al., 1989). Also, improvements in kinetic parameters proposed by Maner and Doyle (1997) have been considered. The mathematical model, which consists of 53 differential and algebraic equations, is implemented in Fortran 90/95, to simulate the plant and setup the NMPC, which is fundamental for those systems characterized by long process transients and strong nonlinear dynamics, such as polymerization plants. The numerical solution is performed by using IMSL library.

Keywords: Nonlinear Model Predictive Control; dynamic optimization problems; copolymerization.

574k Modeling and Control of a Packed Distillation Column

Carolina Ramirez[1], Iván Dario Gil Chaves[1], and Nicanor Quijano[2]. (1) Department of Chemical Engineering, Universidad de los Andes, Calle 19A 1-37 Este office ML-729, Bogotá, Colombia, (2) Electric and Electronic Engineering, Universidad de los Andes, Carrera 1 #18A-10, Bogotá, Colombia

The vast majority of industrial processes have as final operation separation stages. Distillation represents 95% of the operations in chemical industries having a major impact on product quality and energy use. The control of distillation columns presents a major challenge because of its non-linearities, its non-stationary behavior, the relationship between the control of two compositions, and the large amount of disturbances. One way to simulate the control of a unit operation is by developing a theoretical model that presents the performance of this operation. This work presents a model based on the equilibrium for the distillation of the solution water – ethanol in the packed column of the Chemical Engineering Laboratory at the Universidad de los Andes, a sensitivity analysis a PI control for the compositions of bottoms and distillate, and the comparison of such control with the Ideal Free Distribution (IFD) model.

First of all, it is necessary to simulate the packed column in a stable state. For that, we assume that the column is in thermodynamic equilibrium in each stage, a partial condenser, a liquid feed in its saturation temperature, constant pressure, negligible heat losses with the surroundings and a five columns stage with its feed in the third theoretical plate. With these assumptions, the model is formulated, using the MESH equations and the Wilson model for the equilibrium of the solution, and the solution is provided based on the model of the tridiagonal matrix. This model solves in a rigorous way the distillation column, but it is necessary to divide the column in theoretical plates. Five theoretical equilibrium stages are found, and a total condenser. Then, a sensitivity analysis is performed in order to find which and in what magnitude the disturbances in the feed alter the stable state response. This result shows that the disturbance that generates the biggest modification in the response in the stable condition is the feed composition. Further, this simulation provides the needed information to find some indispensable values for the dynamic simulation and some constants needed to find a transfer function of the system.

With the obtained results, a dynamic state model is developed primarily in the mass balances at each theoretical stage using the Wilson thermodynamic model. The results give the temperature and composition profile for a given period of time. This model assumes a binary non-ideal mixture, constant pressure along the column, constant molar flows without steam retention, linear liquid dynamic, thermodynamic equilibrium at every stage, and a total condenser. With this model, it is found the required time to obtain the stable state and the possibility to make a proper simulation control. Besides, we can get a transfer function for the system that relates every composition or temperature in each stage with two inlets, specifically the distillate and reboil reflux.

Using the dynamic state model we develop two PI controllers tuned in order to control the distillate and bottoms temperature with L-V configuration giving very good results with pulse and step signals in the feed properties (i.e., the feed concentration, flow and the state in which it is entering to the column). The control gives stabilization in around ten minutes but it generates great changes in the manipulated variables and in high frequencies it presents an oscillation with large amplitude generating enormous energy losses and very distant concentration with the set point. Outside these disadvantages, the controllers yield a stable response in a wide range of specified set points and the columns acquire high molar concentrations of ethanol in the distillate with values around the 78%. In addition, the model gives the possibility to control every single stage if it is necessary.

Even though we get a reasonably good performance with the PI controllers, the amount of energy lost is big. In order to improve this, a behavioral ecology model is developed to keep the enthalpy loss in the entire column constant using the concept of the minimum thermodynamic condition (MTC). Using the bioinspired model based on the Ideal Free Distribution (IFD) concept, and the replicator dynamics, we guarantee that the performance is better than for a linear controller (i.e., the energy lost is improved, and the response is faster).

574l Challenges to Using Mechanisms of Action to Quantify Efficacy of Nucleoside-Analogue Reverse Transcriptase Inhibitors

James Monaco[1], **Samira Khalili**[2], and Antonios Armaou[2]. (1) Undergraduate student, Pennsylvania State University, University Park, PA 16802, (2) Chemical Engineering, The Pennsylvania State University, University Park, PA 16802

In this work, the mechanisms by which nucleoside-analogue reverse transcriptase inhibitors (NRTIs), the most common class of drugs used in the treatment of HIV-1 exerts their antiviral effects are thoroughly analyzed. We subsequently seek to identify ways in which these known mechanisms could be employed to generate mathematical models for drug efficacy in terms of measurable physical values.

NRTIs are congeners of natural nucleotides that cannot form 3'-5'linkages, and function by competing with natural nucleotides for incorporation into the HIV genome during its reverse transcription. It is demonstrated that the probability of an RTI inclusion instead of its natural nucleotide can be expressed in terms of intracellular drug concentrations, natural nucleotide concentrations, and relevant rate constants derived from Reverse Transcriptase's mechanism of nucleotide addition. Unfortunately the probability of NTRI inclusion is only half of the puzzle. In order to determine the ultimate effect, the resistance of the NTRI to removal from the genome must be considered.

It is demonstrated that under certain idealized conditions, this mechanism is reducible to a simple mathematical expression: If each RTI incorporation might be assumed to carry with it some probability of permanently disabling the developing HIV genome, then efficacy and drug concentration can be shown to have a power relationship, while efficacy and natural nucleotide concentration have an inverse power relationship. However, detailed modeling of the entire process will require well informed in sillico simulations. While technically feasible, such simulation appears to require the collection of more physical data than is currently available. Under certain assumptions for the values of physical parameters, the use of stochastic simulations is demonstrated to quantify the relationship between NRTI efficacy and concentration

574m Mpc-Based Integrated Product and Process Control

Zhen Wu[1], Zheng Liu[2], Yinlun Huang[2], and Yu Qian[3]. (1) Chemical Engineering and Materials Science, Wayne State University, 5050 Anthony Wayne Drive, Detroit, MI 48202, (2) Department of Chemical Engineering and Materials Science, Wayne State University, 5050 Anthony Wayne Drive, Engineering Building 2404, Detroit, MI 48202, (3) School of Chemical Engineering, South China University of Technology, Guangzhou, 510640, China

Model Predictive Control (MPC) is a very attractive advanced process control technique for difficult multivariable control problems. One of the major advantages of MPC is its capability of predicting future system behavior based on the model, if accurate, and current measurements, and providing early warnings of potential problems [1]. This technique, offering a general control scheme, should be valid for product quality control as well, if the quality indicator(s) can be properly defined and measured (or estimated), and the product manufacturing dynamics can be adequately modeled. Note that in most industrial cases, product quality control must be coordinated with process efficiency control. This renders a challenging problem of how to achieve simultaneous control of both the process efficiency (in terms of energy and material use) and the product quality by means of MPC technique.

Integrated Product and Process Control (IPPC) was formally introduced by Xiao et al. [2]. The essence of IPPC design methodology is the coordination of the control of process efficiency and product quality using a classical cascade control scheme. Nevertheless, the control effectiveness by that methodology is restricted by the use of conventional PID control laws and the inability of dealing with model inaccuracy. Furthermore, the methodology does not take process capability issues into account during control design.

In this paper, we will introduce an enhanced IPPC design methodology where MPC is employed. In the resulting cascade control scheme, the inner-loop is designed to realize MPC-based process efficiency control under certain process capability constraints. The control objective of the inner loop is to ensure that the controlled process variable(s) provide the needed actions for product manufacturing. The outer loop, where MPC is also utilized, is designed to control the manufacturing of the product in a dynamic domain, which is driven by the "means" provided by the process system. By using this methodology, the product quality during the manufacturing process can be predicted more realistically, not just based on the measurement (either directly or indirectly obtained) of the product behavior, but also on the information about the process under capability constraints. In this work, a complete mathematical system will be presented with necessary derivations. The efficacy of the design methodology will be demonstrated through a case study which is about the control of polymeric coating quality and process energy efficiency under various process operational constraints.

Reference

1. Seboge, D. E., Edgar, T. F., and Mellichamp, D. A.. Process Dynamics and Control. John Wiley & Sons, 1989.

2. Xiao J, Huang YL. Integrated product and process control of single-input-single-output systems. AIChE J., 2007; 53:891-901.

DEVELOPMENT OF NONLINEAR PREDICTIVE MODEL-BASED FEEDFORWARD CONTROL FROM CLOSED-LOOP FREELY-EXISTING REAL DATA

Derrick K. Rollins, Stephanie D. Loveland, and Nidhi Bhandari
Department of Chemical and Biological Engineering
Iowa State University, Ames, IA, USA

Introduction

The control of chemical processes in industry is a very important aspect of everyday operations. The ability to maintain control has an impact on process safety, product quality and plant profitability [1]. In recent years, many different advanced control techniques have been developed. Many chemical and biological processes exhibit nonlinear behavior, but model-based control schemes have often used linear models, which can be sufficient if the process is operated over a small range of inputs [2, 3]. Real processes also often exhibit complicated dynamic responses to changes in the process inputs, including nonlinear dynamics.

The performance of any model-based controller is highly dependent on the model that is used to predict process behavior. The procedures for developing the model can be time-consuming and costly, requiring the process to be perturbed in order to determine cause-and-effect behavior between the process inputs and outputs. It is desirable to be able to identify the process model without causing significant upsets to everyday operations. Ideally, historical data from the plant database could be used to develop the models to be used for prediction of process output response to changes in the inputs. The advantages of using this historical data are numerous. The data is readily available, is collected frequently, covers the "typical" operating space of the process, and does not require specific perturbations of the process inputs. However, several problems can be encountered if plant historical data is used. The process inputs are likely to be highly correlated, and the range of the inputs may not be very broad.

Distillation is an example of a process that would benefit from a model-based controller that can be built using plant historical data. The high degree of interaction between process variables makes it a complicated problem for control. There have been many applications of feedforward control to distillation processes. Among these, several have attempted to address the process response of product composition to various input disturbances such as feed flow and feed composition [4-6], but none have attempted to compensate for multiple input disturbances occurring simultaneously.

The purpose of this work is to demonstrate a method of developing a nonlinear process model under highly correlated inputs that can be used for a feedforward controller that will compensate for multiple input disturbances simultaneously. The model can be developed using historical plant data collected under closed-loop conditions and still effectively determine cause-effect behavior between the inputs and output of the process.

Methodology

Block-oriented models have been used to accurately describe many types of nonlinear process behavior. The two most common block-oriented models are the Hammerstein and Wiener models. In this paper, we will discuss the LNL block-oriented model which has a linear dynamic block followed by a nonlinear static block and then another linear dynamic block.

Closed-form solutions to the Hammerstein and Wiener processes have been proposed by Rollins et al. [7] and Bhandari and Rollins [8] for continuous-time modeling (CTM), and are called the H-BEST and W-BEST methodologies, respectively. A discrete-time modeling (DTM) method has also been developed by Rollins and Bhandari [9] that follows the CTM procedures for H-BEST and W-BEST. This work is an extension of the DTM method for the W-BEST methodology. This work also extends the developments of Rollins, et al. [10] in modeling diabetic subjects under free-living data collection to the modeling of real process data under freely existing conditions such as plant data. In addition, it applies the principles used by Rollins et al. [11] to address inputs that are serially correlated.

The Wiener model

Procedure for Model-Building

The following is the procedure we developed for model building under freely existing data, and is a modification of the experimental design method in Rollins et al. [11]:

1. Select the dynamic model form for Eq. 1, and estimate the static and dynamic model parameters under Model 1. This is repeated until an acceptable model form is found.
2. Using the residuals from step 1, determine the ARMA (p,q) form of Eq. 7 and estimates of the θ_p and ϕ_q parameters.
3. Simultaneously refit the dynamic, ultimate response, and ARMA parameters under Model 2 if necessary.
4. Check the residuals in step 3 for compliance to white noise

The form of the ARMA (p,q) model is found using the autocorrelation function (ACF) and partial autocorrelation function (PACF). Note that in Rollins et al. [11], statistical design of experiments (SDOE) was used to determine input changes to the process. In this case, we will be using historical plant data so no additional experiments need to be determined.

Feedforward Control Methodology

General Feedforward Controller Methods

The concept of feedforward control allows for theoretically perfect control of a process system. By measuring the disturbances (loads) of a process and modeling how these disturbances affect the process outputs, corrective action can be taken before the process outputs deviate from their desired values [1]. However, because there are many process disturbances that cannot be measured in a timely or efficient manner, feedforward control is typically used in conjunction with standard feedback control, which compensates for any deviation of the output variable from its setpoint, regardless of what caused the deviation.

The feedforward controller, G_f, is generally approximated by a linear model [12], but nonlinear process models can also be used [13, 14]. The model is found by determining how the process output responds to a given input disturbance.

In order for the feedforward/feedback controller to provide perfect control of the output variable Y, we must first find the closed-loop response to an arbitrary change in the disturbance variable X_i.

Some ideal feedforward controllers are physically unrealizable, in these cases, the feedforward controller is often approximated by a lead-lag unit [1]. We will discuss how this problem is addressed with the feedforward controller of the proposed approach in the next section.

Proposed Feedforward Controller Methodology

The block diagram for the feedforward/feedback controller under Wiener modeling has not been shown for space considerations. The first step in implementing feedforward/feedback control under Wiener modeling is to identify the model parameters for the process being studied. Once the model has been identified, the feedforward controller is designed.

As in the case of the traditional feedforward controller, we can find the closed-loop output response to a change in the input disturbance X_i.

The Distillation Process

The distillation process used was a pilot-scale methanol/water distillation column consisting of 12 trays, with an inside diameter of 6 inches. Feed was introduced at Tray 4 and had a concentration of 15% (mol) methanol. The column is connected to a DeltaV distributed control system from Emerson Process Management.

Open-Loop Model Building

Training Phase

In order to first determine the feasibility of developing a Wiener model under highly correlated inputs for the distillation process, training and test data were collected in the open-loop mode [15]. The input variables chosen for the experimental tests that were conducted included feed flow rate, feed temperature set point, reboiler level and reboiler steam pressure. Other variables that were measured on-line and used as input variables included reflux flow rate, column pressure, bottoms product flow rate, distillate product flow rate and overhead condensate temperature. The output response of the process was the top tray (Tray 12) temperature, which was also measured on-line.

During the experiment the feed temperature set point was the input variable, and the feed temperature displayed significant dynamic response to the set point changes and also to the feed flow changes. Because of this, we included dynamics on the feed temperature set point which relate the observed feed temperature with the feed temperature set point and feed flow changes. The other input changes displayed very little dynamic response to set point changes, so the measured values of those input variables were used directly in the model building process.

Testing Phase

In order to test the model that was built during the training phase, another series of input changes was made to the distillation process. In this case, a Box-Behnken design with four factors and three center points was run, but this time the correlation between the feed flow rate and feed temperature set point was zero. This was done to demonstrate that the model built under high correlation was able to determine accurate cause-and-effect behavior between the input and output variables. As in the training phase, only four inputs were deliberately changed, but an additional five inputs were measured on-line and used in the prediction of the output response.

PI controller vs. FFPI controller process responses

The feedforward/feedback controller that we developed was tested and compared to the feedback controller alone.

In order to quantitatively compare the results of implementing the feedforward controller, the average absolute error (AAE) term is used.

In the case of the PI controller, AAE=0.192, while in the case of the FFPI controller, AAE = 0.134. This represents a 42.8% reduction in the variability of the Tray 12 temperature.

Conclusions

In this paper, we have presented a methodology for developing a Wiener block-oriented model from plant data that accurately predicts process response behavior to multiple input disturbances that are occurring simultaneously. The model was implemented into a feedforward/feedback control scheme and demonstrated marked improvements over traditional feedback control. The ability to develop the model with plant historical data under closed-loop conditions represents a significant advantage over traditional model-building techniques, which require specific perturbations of the process that can affect plant operations.

This work will be extended to other types of chemical and biological process systems for further investigation. Specifically, the work done by Rollins et al. [24] to predict glucose response in type 2 diabetics will be extended to close the loop on glucose concentration. The implications of this for persons with type 1 and type 2 diabetes are tremendous, as the ability to control glucose levels at a desired level will greatly affect their longevity and quality of life.

References

[1] Seborg, D.E., T.F. Edgar and D.A. Mellichamp, *Process Dynamics and Control*, 2nd edition, John Wiley and Sons, 2003.

[2] Clarke, D.W., C. Mohtadi and P.S. Tuffs, "Generalized Predictive Control- Part 1: The Basic Algorithm," Automatica, Vol. 23, pp. 137-148, 1987.

[3] Muske, K.R. and J.B. Rawlins, "Model Predictive Control with Linear Models," AIChE Journal, Vol. 39, pp. 262-287, 1993.

[4] Rix, A., K. Loewe and H. Gelbe, "Feedforward Control of a Binary High Purity Distillation Column," Chemical Engineering Communications, Vol. 159, pp. 105-118, 1997.

[5] Castellanos-Sahagun, E., J. Alvarez-Ramirez and J. Alvarez, "Two-Point temperature Control Structure and Algorithm Design for Binary Distillation Columns," Industrial and Engineering Chemistry Research, Vol. 44, pp. 142-152, 2005.

[6] Baratti, R., S. Corti and A. Servida, "Feedforward Control Strategy for Distillation Columns," Artificial Intelligence in Engineering, vol. 11, No. 4, pp. 405-412, 1997.

[7] Rollins, D.K., N. Bhandari, A.M. Bassily, G.M. Colver and S. Chin, "A Continuous-Time Nonlinear Dynamic Predictive Modeling Method for Hammerstein Processes," Industrial and Engineering Chemistry Research, Vol. 42, No. 4, pp. 861-872, 2003.

[8] Bhandari, N. and D.K. Rollins, "A Continuous-Time MIMO Wiener Modeling Method," Industrial and Engineering Chemistry Research, Vol. 42, No. 22, pp. 5583-5595, 2003.

[9] Rollins, D.K. and N. Bhandari, "Constrained MIMO dynamic Discrete-Time Modeling Exploiting Optimal Experimental Design," Journal of Process Control, Vol. 14, No. 6, pp. 671-683, 2004.

[10] Rollins, D.K., J. Kleinedler, A. Strohbehn, L. Boland, M. Murphy, D. Andre, D. Wolf and W.E. Franke, "Modeling Glucose Noninvasively Using Wiener Simulation Modeling for Type 2 Diabetic Patients Under Free-living Conditions," submitted to *IEEE Transactions on Biomedical Engineering*, in review.

[11] Rollins, D.K., N. Bhandari, S. Chin, T. Junge and K. Roosa, "Optimal Deterministic Transfer Function Modeling in the Presence of Serially Correlated Noise," Chemical Engineering Research and Design, Vol. 84(A1), pp. 9-21, 2006.

[12] Zhang, J. and R. Agustriyanto, "Inferential Feedforward Control of a Distillation Column," Proceedings of the American Control Conference, pp. 2555-2560, Arlington, VA, June 25-27, 2001.

[13] Smith, C.A. and A.B. Corripio, *Principles and Practice of Automatic Process Control*, Wiley, New York, 1985.

[14] Luyben, W.L., *Process Modeling, Simulation and Control for Chemical Engineers*, 2nd ed., McGraw-Hill, New York, 1990.

[15] Loveland, S.D. and L. dela Rosa, "Distillation – A Unit Operations Laboratory Manual," Department of Chemical and Biological Engineering, Iowa State University, Ames, Iowa, 2005.

574o Blood Glucose Regulation with Stochastic Optimal Control for Insulin-Dependent Diabetic Patients

Saadet Ulas Acikgoz, UOP LLC, Des Plaines, IL 60016 and **Urmila Diwekar**, Vishwamitra Research Institute, Center for Uncertain Systems: Tools for Optimization and Management, Vishwamitra Research Institute, Clarendon Hills, IL 60514.

20.8 million people in the U.S. suffer from diabetes, which has many complications such as heart disease and stroke, high blood pressure, kidney disease, nervous system disease and amputations. The hormone insulin has many functions in the body; most importantly it influences the entry of glucose into cells. The lack of insulin prevents glucose from entering the cells and be utilized, which leads to excess blood sugar and excretion of large volumes of urine, dehydration and thirst. The current treatment methods for insulin-dependent diabetes include subcutaneous insulin injection or continuous infusion of insulin via an insulin pump. The former treatment requires patients to inject insulin four to five times a day. The amount of injection is usually determined by a glucose measurement, an approximation of the glucose content of the upcoming meal and estimated insulin release kinetics. The continuous insulin infusion pump allows for more predictable delivery due to its constant infusion rate into a subcutaneous delivery site. Keeping the blood glucose levels as close to normal (non-diabetic) as possible is essential for preventing diabetes related complications. Ideally this level is between 90 and 130 mg/dl before meals and less than 180 two hours after starting a meal. The Diabetes Control and Complications Trial (DCCT Research Group, 1993) followed 1441 people with diabetes for several years. This trial concluded that the patients who followed a tight glucose control program were less likely to develop complications such as eye disease, kidney disease and nerve disease, than the ones who followed the standard treatment, because the former group had kept the blood glucose levels lower.

The ideal treatment for controlling blood glucose levels in insulin dependent diabetic patients would be the use of an artificial pancreas which would have the following components: (a) a glucose sensor which monitors the blood glucose continuously with sufficient reliability and precision; (b) a computer which could calculate the necessary insulin infusion rates by an appropriate feedback algorithm; (c) an insulin infusion pump which would release the required amount of insulin into the blood. Safe delivery of insulin in this way requires reliable glucose sensors. Two types of sensors have been developed during the last 30 years, minimal invasive and non-invasive (Koschinsky and Heinemann, 2001). The non-invasive approaches are carried out using optical glucose sensors. These sensors work by directing a light beam through intact skin and measuring the properties of the reflected light that are altered either as a result of direct interaction with glucose (spectroscopic approach) or due to the indirect effects of glucose by inducing changes in the physical properties of skin (scattering approach). However, these optical sensors are not able to measure glucose with sufficient precision. On the other hand, minimally invasive sensors measure the glucose concentration in the interstitial fluid of the skin or in the subcutis. There is a free and rapid exchange of glucose molecules and interstitial fluid. Therefore, changes in blood glucose and interstitial glucose are correlated. However, there is a time delay between these changes varying from a few seconds to 15 minutes; which complicates the interpretation of measurement results (Roe and Smoller, 1998). The magnitude of this delay depends on factors such as the absolute glucose concentrations and direction of change. This delay shows intra and inter-individual variability. Furthermore, it has been found that the absolute values of interstitial glucose concentrations vary between 50 and 100% of the intravasal value.

The wide-spread use of these glucose sensors is also complicated by biocompatibility issues and skin reactions. Also these sensors should be available at a reasonable cost in order to be applicable to insulin-dependent diabetic patients. The implementation of a closed loop system in daily life conditions requires these reliability, compatibility, cost and safety issues to be resolved.

The aim of this paper is to develop an optimal control system. Optimal control is different from a closed loop feedback control where the desired operating point is compared with an actual operating point and knowledge of the difference is fed back to the system. Optimal control problems are defined in their time domain, and their solution requires establishing an index of performance for the system and designing the course (future) of action so as to optimize a performance index. Therefore optimal control allows us to make future decisions. Using optimal control theory we can minimize the deviations of blood glucose from non-diabetic levels, while penalizing the use of large amounts of infused insulin for safety. Swan (1982), Fisher and Teo (1989), Ollerton (1989), Fisher (1991) and Parker et al. (1999) applied optimal control theory to this problem. However, uncertainties in model parameters and variability among different individuals were not considered in these papers.

The success of optimal control method depends on the accuracy of the model; therefore, the inherent uncertainties in the patient need to be addressed. If the uncertainties are omitted and if the model cannot accurately represent the glucose and insulin dynamics, this can lead to significant performance degradation. Significant variability of relevant parameters among patients and within a given patient during the course of the day or week has been reported in literature (Simon et al., 1987; Bremer and Gough, 1999). Meals and exercise, the age and weight of the patient also affect the insulin/ glucose dynamics. These daily and hourly fluctuations of patient parameters can create difficulties in continuous glucose control. These dynamic uncertainties affect the optimal insulin infusion profiles.

The aim of this paper is to model these uncertainties by a novel approach and incorporating them into formulations of optimal control. Time-dependent uncertainties are commonly encountered in finance literature. Dixit and Pindyck (1994) and Merton and Samuelson (1990) described optimal investment rules developed for pricing options in financial markets, and Ito's Lemma (Ito, 1951; 1974) to generalize the Bellman equation or the fundamental equation of optimality for the stochastic case. This new equation constitutes the base of the so called Real Options Theory. Although such a theory was developed in the field of economics, it was recently applied to optimal control problems encountered in other branches of science. For example, in chemical engineering literature, time-dependent uncertainties in batch processing and pharmaceutical separations were represented by Ito processes and time-dependent stochastic optimal control profiles were obtained. Using this approach, the performance of separation processes where stochastic optimal control was applied, has increased significantly as high as 69%. Using Ito processes, ideal and non-ideal systems were represented and thermodynamic parameter uncertainties associated with locally optimal parameter estimates as a result of nonlinear regression were addressed (Ulas and Diwekar, 2004; Ulas et al., 2005).

This approach could also be extended to optimal glucose control in insulin dependent diabetic patients. The blood glucose profiles can be represented using Ito processes and stochastic optimal control profiles could be derived to achieve better treatment for diabetes. The results show that the hourly and daily variations of blood glucose in response to meals and insulin action can be modeled using this methodology and using stochastic maximum principle; optimal insulin infusion profiles can be computed. The stochastic optimal control profile results in fewer variations from the reference blood glucose value of 4.5 mmol/L as compared to the deterministic profile and could potentially be useful in preventing the complications of diabetes.

References:

Fisher M.E. and Teo K.L. (1989) 'Optimal insulin infusion resulting from a mathematical model of blood glucose dynamics', *IEEE Transactions in Biomedical Engineering* 36:479-486.

Ito K. (1951) 'On stochastic differential equations', *Memoirs of American Mathematical Society* 4(1).

Ito K. (1974) 'On stochastic differentials', *Applied Mathematics and Optimization*, 4, 374

Koschinsky T. and Heinemann L. (2001) 'Sensors for glucose monitoring: technical and clinical aspects', *Diabetes Metab. Res. Rev.* 17(2):113-23

Merton R.C., and Samuelson P.A. (1990), *Continuous-time Finance*, Blackwell Publishing, Cambridge MA

Ollerton R.L. (1989) 'Application of optimal control theory to diabetes mellitus'. *International Journal of Control* 50: 2503-2522.

Parker R.S., Doyle III F.J. and Peppas N.A. (1999), 'A model-based algorithm for blood-glucose control in type I diabetic patients', *IEEE Transactions in Biomedical Engineering* 46(2): 148-157

Roe J.N. and Smoller B.R. (1998) 'Bloodless glucose measurements', *Crit Rev Ther Drug Carrier Syst.* 15(3):199-241.

Simon G., Brandenberger G., and Follenius M. (1987), 'Ultradian oscillations of plasma glucose, insulin and c-peptide in man during continuous enteral nutrition', *Journal of Clinical Endocrinology & Metabolism* 64: 669-674.

Swan G.W. (1982), 'An optimal control model of diabetes mellitus', *Bulletin of Mathematical Biology* 44: 793-808.

The Diabetes Control and Complications Trial Research Group (1993) 'The effect of intensive treatment of diabetes on the development and progression of long-term complications in insulin-dependent diabetes mellitus', *New England Journal of Medicine* 329: 977-986.

Ulas S. and Diwekar U.M. (2004), 'Thermodynamic uncertainties in batch processing and optimal control', *Computers and Chemical Engineering* 28(11): 2245-2258.

Ulas S., Diwekar U.M., and Stadtherr M.A. (2005), 'Uncertainties in parameter estimation and optimal control in batch distillation', *Computers and Chemical Engineering* 29(8): 1805-1814.

574p Data Analysis and Parameter Estimation of Bioprocesses

Harvey Arellano-Garcia, Chair of Process Dynamics and Operation, Berlin Institute of Technology, Str. des 17. Juni 135, 10623, Berlin, Germany

The estimation of state variables and model parameters based on a rigorous process model together with measurement data is a key step in model-based process applications. Models which describe fermentation processes show high non- linearity and time variant kinetic parameters. However, a good quality of parameter estimation enables the realization of well posed and effective control strategies. Thus, the potential to evaluate the quality of the parameters can be increased based on practical identifiability analysis i.e. using real noise-corrupted experimental data series. In this work, the dynamical behaviour of the parameter system is taken into account. One requirement for a successful state or parameter estimation is that the system is in general identifiable or observable. Therefore, in this work an approach is applied using the analysis of the quantitative identifiability, since measurement errors are considered. Moreover, in order to determine the quantitative identifiability, the confidence region of the parameter estimates has been analyzed. Here, an asymptotic lower bound of this matrix can be given by the inverse of the Fischer Information matrix (FIM). To scale the different values in the objective function and to consider the accuracy of the measurement devices, the inverse of the variance-covariance matrix of the measurement errors is used as a weighting in the objective function. The proposed optimization-based approach has been applied to improve the predictivity of kinetic models based on available measurements together with a process model. In this study, experimental cultivation data are used for model-based parameter identification.

574q Population-Based Colorectal Cancer Care Control

Selen Aydogan-Cremaschi[1], Tapas K. Das[2], Chaitra Gopalappa[2], and Seza Orcun[1]. (1) e-Enterprise Center, Discovery Park, Purdue University, 203 S. Martin Jischke Drive, West Lafayette, IN 47907, (2) Department of Industrial & Management Systems Engineering, University of South Florida, 4202 E. Fowler Avenue, Tampa, FL 33620

Colorectal cancer (CRC) is the second most common cause of death in the United States. Survival depends on the stage of the cancer when diagnosed, around 90% for early stage diagnosis and declining to 15% for advanced stages [1]. Therefore, it is crucial to have effective and efficient screening policies, high compliance to screening, and adequate access to necessary healthcare resources for controlling CRC in a population. This requires understanding of the current CRC care system data and the impact of different elements of the system on each other.

The CRC care system is highly complex and its performance depends on many decision makers, such as patients, service providers, and insurers/payers, often with conflicting objectives and constraints. For example, controlling cost for services, an incentive for insurers/payers, has inherent conflict with a patient's objective of receiving abundant services. Potential gaps in the quality of care exist in terms of both underutilization of services, which disproportionately affects vulnerable populations, and over-utilization resulting in increased costs with no matching benefit. Some problems in the CRC care system are due to the distributed and fragmented nature of the knowledge requiring collaborative efforts of autonomous and diverse practitioners like nurses who counsel prevention, specialists who conduct screening, and oncologists who treat patients with CRC.

Agent-based simulation modeling is a growing area with successful applications in manufacturing, consumer markets, supply chains, anti-terrorism work force, etc [2]. An agent is defined as an identity that interacts with its environment, learns and makes independent decisions with a defined goal [3]. Hence, multi-agent systems are suitable for analyzing the healthcare system and they have been applied to various health care problems [4-6].

In this work, we propose a robust and scalable multi-agent framework where characteristics of CRC care constituents (patients, hospitals, doctors, nurses, insurers/payers) and their interactions are configured to study the system behavior and identify effective control policies. The framework is composed of three major agent classes, as shown in Figure 1. *Persona Agents* focus on modeling behaviors of individuals involved (patients, physicians, nurses, social workers, patient-families) in CRC care by expanding MISCAN modeling [7, 8]. Some examples of the behaviors we study are degree of patient engagement, patient demographics, and physicians' evidence-based practice compliance. *Organizational Agents* define the environment in which persona agents interact. For example, adoption and exchange level of best practices by provider networks (hospitals, clinics, insurers/payers, medical analysis laboratories, employers, pharmacies) might influence physicians' performance. *Regulatory Agents* (local, state and federal government, FDA) set the healthcare system rules under which organizational agents function.

They define the type of allowable insurance plans, pricing and settlement rules like cap on malpractice insurance and pay for performance, and patient driven healthcare incentives (taxes and patient tariffs).

In this presentation, patient and physician agents, and their interactions in the developed framework are discussed. Indiana is used as the pilot site and the agents are populated with the Indiana CRC care system data. The preliminary analyses show that we can approximate the current state of Indiana CRC incidence rates. Further validation and verification of the model is in progress.

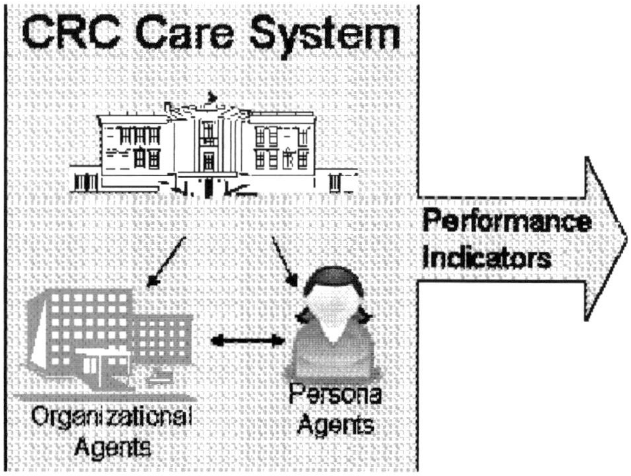

Figure 1. CRC Care System

* Authors are listed in alphabetical order.

References

1. Mitchell, E., Macdonald, S., Campbell, N.C., Weller, D., and U. Macleod. Influences on pre-hospital delay in the diagnosis of colorectal cancer: a systematic review. British Journal of Cancer, 2008. 98: p. 60-70.
2. Drogoul, A., D. Vanbergue, and T. Meurisse. Multi-Agent Based Simulation: Where are the Agents? in Multi-Agent-Based Simulation Conference (MABS'02). 2002. Bologna, Italy: LNCS, Springer-Verlag.
3. Macal, C.M. and M.J. North. Tutorial on agent-based modeling and simulation. In Proceedings of 37th Winter Simulation Conference. 2005. Orlando, FL, United States: Institute of Electrical and Electronics Engineers Inc., New York, NY 10016-5997, United States.
4. Nealon, J.L. and A. Moreno, Agent-based applications in health care, in Applications of Software

Technology in the Health Care Domain, J.L. Nealon and A. Moreno, Editors. 2003, Whitestein Series Software Agent Technologies: Basel, Switserland. p. 3-18.
5. Beer, M., et al., An agent-based architecture for managing the provision of community care - The INCA (Intelligent Community Alarm) experience. AI Communications, 2003. 16(3): p. 179-192.
6. Kanagarajah, A.K., et al. An exploration into the uses of agent-based modeling to improve quality of health care. in 6th International Conference on Complex Systems. 2006. Boston, MA.
7. Habbema, J.D., et al., The MISCAN simulation program for the evaluation of screening for disease. Comput Methods Programs Biomed, 1985. 20(1): p. 79-93.
8. Loeve, F., et al., The MISCAN-COLON simulation model for the evaluation of colorectal cancer screening. Comput Biomed Res, 1999. 32(1): p. 13-33.

574r The Physical Structure of Intracellular Feedback Loops

Christopher V. Rao, Department of Chemical Engineering, University of Illinois, 600 South Matthews Ave., 211 Roger Adams Laboratory, Box C-3, Urbana, IL 61801

The parallels between the feedback control mechanisms used in cells and manufacturing processes have long been noted. Both need to function robustly in response to dynamics changes in supply/demand and also in response to unforeseen disturbances. To deal with these challenges, natural and synthetic systems employ feedback control. As a general control theory for synthetic systems has already been developed, one can imagine that the same tools may be applicable for the analysis and design of natural systems. However, one challenge in applying concepts from synthetic systems to natural one is that with the latter a natural decoupling between the process and control system often does not exist. Moreover, natural systems physical encode their control systems in a number of different ways. Understanding this encoding is often necessary for elucidating the mechanism and associated degrees of freedom. In this work, we explore two biological control examples, one involved in sugar metabolism and the other in antibiotic resistance, where the relative location and orientation of the regulators encoded within the genome plays a critical role in the feedback mechanism. We demonstrate experimentally that perturbations to the genetic encoding of the feedback dramatically alter its behaviors. These results suggest that the physical instantiation of the feedback needs to be accounted for in any theory of biological control.

576a Assigning Physical Significance to the Diffuse Interface Between Terraces In Phase-Field Modeling of Steps on Crystal Surfaces

Igal G. Rasin and **Simon Brandon**. Chemical Engineering, Technion, Haifa, 32000, Israel

We apply a modified phase field modeling approach to the analysis of steps on a crystalline surface. Specifically we are interested in capturing phenomena associated with non-mass transport related interaction between steps. To this end we assign a physical significance to the form of the interfacial region between terraces (i.e. steps), inherent in the phase field approach, by tuning the double well potential to produce long range interaction energies varying as $1/L^{2}$, where L is half the distance between steps. Resultant repulsive interactions between adjacent steps of the same sign are shown to affect step-flow kinetics in a manner consistent with Gibbs-Thomson related supersaturation. This phenomenon is further demonstrated to cause dislocation-driven (spiral) crystal growth kinetics to deviate, for large supersaturation values, from the classical quadratic growth law. Attractive interactions between adjacent steps of opposite sign, also resulting from the finite interfacial width, are briefly explored particularly with respect to their possible impact on two-dimensional nucleation

576b Catalytic-Pathway Identification for Methanol Decomposition on Pt (1 1 1) Via the Graph-Theoretic Method Based on P-Graphs

Yu-Chuan Lin[1], L. T. Fan[1], Shahram R. Shafie[1], Botond Bertok[2], and Ferenc Friedler[2]. (1) Department of Chemical Engineering, Kansas State University, 1005 Durland Hall, Manhattan, KS 66506, (2) Department of Computer Science, University of Veszprém, Egyetem u. 10., Veszprém, H-8200, Hungary

Catalytic decomposition of methanol (MD) plays a significant role in the production of hydrogen, which is deemed to be one of the most, if not the most, desirable clean fuels. Thus, the catalytic mechanisms, or pathways, of MD have lately been the focus of intense research interest. Recently, the feasible independent pathways (IP_i's) have been reported on the basis of a set of highly plausible elementary reactions. Nevertheless, no feasible acyclic combined pathways (AP_i's) comprising IP_i's have been concomitantly reported. Such AP_i's can not be ignored in identifying dominant pathways, as demonstrated in the current work.

576c Coarse-Grained Kinetic Monte Carlo Models: Complex Lattices, Multicomponent Systems, and Homogenization at the Stochastic Level

Stuart D. Collins, Chemical Engineering, University of Delaware, Newark, DE 19716, Abhijit Chatterjee, Theoretical Division, Los Alamos National Laboratory, T-12 MS B268, Los Alamos, NM 87545, and Dion Vlachos, Director of Center for Catalytic Science and Technology (CCST), University of Delaware, Newark, DE 19716.

On-lattice kinetic Monte Carlo (KMC) simulations have extensively been applied to numerous systems. However, their applicability is severely limited to relatively short time and length scales. Recently, the Coarse-Grained MC (CGMC) method was introduced to greatly expand the reach of the lattice KMC technique. Herein, we extend the previous spatial CGMC methods to multicomponent species and/or site types. The underlying theory is derived and numerical examples are presented to demonstrate the method. Furthermore, we introduce the concept of homogenization at the stochastic level over all site types of a spatially coarse grained cell. Homogenization provides a novel coarsening of the number of processes, an important aspect for complex problems plagued by numerous microscopic processes (combinatorial complexity). As expected, the homogenized CGMC method outperforms the traditional KMC method on computational cost while retaining good accuracy.

A three-dimensional interfacial spectral boundary element algorithm for capsules and erythrocytes in strong flows

Walter R. Dodson III
Fischell Department of Bioengineering
University of Maryland
College Park, MD, USA

Panagiotis Dimitrakopoulos
Department of Chemical and Biomolecular Engineering,
Bioengineering Graduate Program
University of Maryland
College Park, MD, USA

Introduction

The study of the interfacial dynamics of artificial or physiological capsules (i.e. membrane-enclosed fluid volumes) in Stokes flow has seen an increased interest during the last few decades due to their numerous engineering and biomedical applications. In this work we restrict our interest to elastic membranes with shearing and area-dilatation resistance but negligible bending resistance. This class represents a wide range of artificial capsules as well as the human red blood cells. To solve this problem, several computational methodologies based on low-order interpolation schemes have been employed (e.g. immersed boundary and boundary integral algorithms). While these methodologies are able to study the capsule deformation at low and moderate flow rates, they are unable to determine the capsule dynamics at strong flows where the interfacial deformation is large.

Results

To overcome this, we have developed a spectral boundary element algorithm for interfacial dynamics of three-dimensional capsules in Stokes flow. Our methodology preserves the main characteristic of the spectral methods, i.e. the exponential convergence in the interfacial accuracy as the number of spectral points increases, but without creating denser systems as spectral methods used in volume discretization do. Owing to its spectral nature, our interfacial algorithm has the significant advantage of the accurate determination of any interfacial property, including geometric derivatives and membrane tensions. We believe that this is an important issue for the correct and accurate determination of very deformed capsule shapes made from membranes obeying non-linear elastic laws such as the Skalak and neo-Hookean (or Mooney-Rivlin) laws.

Conclusions

In contrast to the existing low-order methodologies which are unable to study interfacial dynamics at high flow rates, our spectral algorithm predicts stable transient and steady-state capsule shapes in these flows. The pointed profiles of these shapes are in qualitative agreement with experimental findings. Our results for erythrocytes in strong shear flows are also in excellent agreement with experimental findings from ektacytometry.

577b A Software Tool for Design of Process Monitoring and Analysis Systems

Ravendra Singh[1], Krist V. Gernaey[2], and Rafiqul Gani[1]. (1) CAPEC, Department of Chemical and Biochemical Engineering, Technical University of Denmark, Søltofts Plads, Building 227, 2800 Kgs. Lyngby, Denmark, (2) BioEng, Department of Chemical and Biochemical Engineering, Technical University of Denmark, DTU - Building 227/250, Søltofts Plads, 2800 Kgs. Lyngby, Denmark

Abstract

The use of systematic methods and tools for design of process monitoring and analysis system for product quality control can be very useful in the manufacturing industry, particularly the pharmaceutical industry. The design of process monitoring and analysis system involves the identification of the critical quality parameters, selection of economical & reliable data measurement methods and integration with the control system.

A model and data (knowledge) based computer-aided framework including the methods and tools through which the design of a process monitoring and analysis system for product quality control can be generated, analyzed and/or validated, has been developed. The developed framework consists of three main parts: a knowledge base of methods and tools for monitoring/analysis, a model library to supplement the gaps in the knowledge base and an algorithm for design and analysis of process-product specific monitoring and analysis systems. The knowledge base provides the necessary information/data during the design of the process monitoring & analysis system. It contains information for typical unit processes in terms of the type of operation they perform, the process variables involved with those operations, the corresponding actuator candidates, and the equipments used for on-line measurement of data (type of equipment, accuracy, precision, operating range, response time, drift, resolution, cost etc.). The model library contains a set of mathematical models for different types of unit processes, sensors and controllers. They are used to generate data (when measured data are not available) as well as for verification of the performance of the closed-loop controlled process. The design-analysis algorithm identifies the optimal monitoring & analysis system in terms of cost and/or time subject to the product-process constraints. It consists of mainly four analysis steps: process analysis to identify the process points and process variables, sensitivity analysis to identify the critical process points and critical process variables, interdependency analysis to couple the critical process variables with the actuators and performance analysis of monitoring tools to select the measuring equipments. A software prototype for design of process monitoring and analysis systems, based on the mentioned framework has been developed.

The objective of this presentation is two-fold, first to highlight the design software together with the design methodology and then to demonstrate the application of the developed software. The developed software is able to design new process monitoring & analysis systems as well as to analyze existing process monitoring & analysis systems. The application of the software for design of process monitoring & analysis systems will be illustrated through two practical case studies involving pharmaceutical products

577c Data-Based Fault Detection and Isolation Using Feedback Control: Output Feedback, Optimality and Asynchronous Measurements

Panagiotis D. Christofides[1], Ben Ohran[2], David Muñoz de la Peña[2], and James F. Davis[3]. (1) Department of Chemical and Biomolecular Engineering, Department of Electrical Engineering, University of California, Los Angeles, Los Angeles, CA 90095, (2) Department of Chemical and Biomolecular Engineering, University of California, Los Angeles, Los Angeles, CA 90095, (3) University of California - Los Angeles, Los Angeles, CA 90095

With billions in lost revenue each year in U.S. industries [1] due to abnormal events, the need for better methods for handling abnormal situations in chemical processes is critical. Fault-tolerant control (FTC) minimizes the economic, environmental, health and safety impacts in abnormal situations encountered in chemical processes by maintaining system stability in the presence of a fault. Fault detection and isolation (FDI) is a key element of successful fault-tolerant control structures. In FTC, information obtained from an FDI scheme allows systems with redundant control and/or measurement devices to be reconfigured in order to maintain stability, thereby averting potential disaster (see, for example, [2]). As one of the most complex elements in fault-tolerant control, fault detection and isolation is an area of great importance that has received a lot of attention over the past years (see [3, 4] for reviews).

Recently, we introduced a method of fault detection and isolation that combines aspects of both model-based FDI and data-based FDI by using model-based control laws to induce an appropriate closed-loop system structure that allows fault detection and isolation based on available measured process data [5]. The present work focuses on extending these results to consider three new aspects. First, the case of output feedback control is considered. A continuously stirred tank reactor example is used to demonstrate how controller-enhanced FDI can be performed without full state information through the use of an appropriate state estimator. Second, the same model is used to demonstrate the use of model predictive control (MPC) as the external control law to optimize system performance. Third, the case where some of the state measurements are received asynchronously is considered and the cases in which it is possible or not possible to perform fault detection and isolation is explored. These results relax the requirement of full-state feedback requirement and demonstrate the general applicability of the controller-enhanced FDI method.

References

[1] Nimmo I.. Adequately Address Abnormal Operations /Chem. Eng. Prog.. /1995;91:36-45.

[2] Mhaskar P., Gani A., El-Farra N. H., McFall C., Christofides P. D., Davis J. F.. Integrated Fault Detection and Fault-Tolerant Control of Process Systems /AIChE Journal. /2006;52:2129-2148.

[3] Venkatasubramanian V., Rengaswamy R., Kavuri S.N., Yin K.. A review of process fault detection and diagnosis Part III: Process history based methods /Computers and Chemical

Engineering. /2003;27:327-346.

[4] Venkatasubramanian V., Rengaswamy R., Yin K., Kavuri S.N.. A review of process fault detection and diagnosis Part I: Quantitative model-based methods /Computers and Chemical Engineering. /2003;27:293-311.

[5] Ohran B. J., Muñoz de la Peña D., Christofides P. D., Davis J. F.. Enhancing Data-based Fault Isolation Through Nonlinear Control /AIChE Journal. /2008;54:223-241.

577d Novel Method for Global Optimization of Nonconvex Minlp Problems

Debora C. Faria, School of Chemical Engineering and Material Science, University of Oklahoma, 100 East Boyd Street, T-133, Norman, OK 73019-0628 and Miguel J. Bagajewicz, Chemical, Biological and Materials Engineering, University of Oklahoma, 100 E. Boyd St., T-335, Norman, OK 73019.

Most of chemical engineering problems rely on the use of non-linear models, which are frequently non-convex. Global optimization methods are desired, especially because they are able to generate lower bounds that allow us to know how far we are from the global optimum solution. Although in some cases there is no strict need of finding the global optimum solution (gap zero), it is very important to have at least an idea of how much better a solution could be. Another important advantage of global optimization methods is that initial starting points for the problem are not required.

We present two global optimization methods. One of them relies on the discretization of a critical function, which is part of a non-linear non-convex constraints (or a variable for bilinear cases). This method requires the use of binary variables that increase in number as the model increases. In a second approach, the critical function is strategically rewritten as parameters and a set of equations based on lower and upper values of the function values, which replaces the non-convex non-linear constraints to generate lower bounds.

In both cases the lower bound model is also used to eliminate portions of the intervals of different variables when it is identified that these cannot be part of the global optimum solution. Based on this ability, an elimination procedure is built. The advantages of both methods will be discussed and examples will be presented to show and compare the efficiency of the methods.

A Nearest In Control Neighbour Based Method to Estimate Variable Contributions to the Hotelling's Statistic

Carlos R. Alvarez[a,b], Adriana Brandolín[a], Mabel C. Sánchez[a], Luis Puigjaner[b]

[a]*Planta Piloto de Ingeniería Química,BahiaBlanca 8000, Argentina*
[b]Universitat Politècnica de Catalunya, Barcelona 08029, Spain

Abstract

Hotelling's statistic, also called T^2-statistic, is widely used in statistical process control as an extension of the univariate student's chart to reliably detect out of control status in multivariate processes. Although it is a very efficient tool for detection purposes, by itself, it offers no assistance about the origin of the declared faulty status. Several different approaches have been proposed to estimate the variable values' effect on the overall statistic's value. Some of these strategies work in the original measurement space while others interpret the results coming from the analysis in latent variable spaces using for example principal component analysis (PCA) or independent component analysis (ICA). In this work we present a novel strategy based on finding the nearest "in-control" neighbour of the observation point.

Introduction

Statistical process monitoring involves three activities: detection of the out-of-control or faulty status, identification of the variable or variables that signal such condition, and diagnosis of the source cause for the abnormal behaviour. Although Hotelling's statistic is widely used to reliably detect out of control status it offers no assistance in the identification stage. A number of strategies have been proposed to assign variable-contribution values to the T^2-statistic taking into account the multivariate nature of process data.

Mason *et al.* [1,2] proposed to decompose the T^2-statistic value as a summation of J independent parts (where J is the number of measured variables). The first term is calculated squaring a univariate t statistic for one variable. The jth term (j=2, ..., J) of the sum is the jth measurement adjusted by using estimates of the mean and standard deviation of its conditional probability distribution given the (j−1) previously considered variables. Since there exists no fixed order variables, J! different but non-independent partitions can be obtained. As a possible solution for this problem, authors suggested to focus the interest in only two of those terms for each partition: the one corresponding to the unadjusted contribution of a single selected variable and, the term containing the adjusted contributions of this variable after the adjustment of the (J−1) remaining ones. Nevertheless, when the inspection of this reduced set of terms is not enough to come to a clear conclusion, all significant conditional terms should be compared to a critical value, increasing the complexity of the identification of the source fault. An alternative straightforward method to decompose the T^2-statistic as a unique sum of variable contributions was recently presented by Alvarez *et al.* [3]. This method also provides a clear understanding of positive and negative contributions which often results from the techniques mentioned here and, estimates a bound for the negative ones.

Among the methods that work in latent variable-spaces, it was Jackson [4] who first proposed the decomposition of the T^2-statistic into a sum of principal components and perform the identification in terms of the weight of each variable in the of out-of-control component. However, in most of the industrial applications it results very difficult to associate a physical meaning to each principal component and, the variables associated with out-of-control signals cannot be determined easily. Miller *et al.* [5] and MacGregor *et al.* [6] proposed to evaluate the contributions of each process variable to the scores that are outside of their confidence limits. Nomikos [7] presented an approach to calculate the contributions of each process variable to the T^2-statistic instead of to the scores, when latent variables cannot be associated to a meaningful group of process variables. Westerhuis *et al.* [8] extended the theory of contribution plots to latent variable models with correlated scores and, introduced control limits for the contributions that help in finding the variables which behaviour are different with respect to those contained in the reference data set.

In all the above mentioned methods, the contribution to the T^2-statistic for each variable is estimated considering the remaining *J-1* variables fixed at their measured values. As a result of this, there is a sole "parametric curve" defining all the possible values of the T^2-statistic as function of only the analyzed variable, as it was pointed out by Alvarez *et al.* [3].

Variable contributions and contribution plots

Let's consider a chemical process in which J variables are measured and monitored over the time, and let $\mathbf{x}$ be a process observation vector containing all of those measurements for a given time instant t. The value of the Hotelling's statistic for $\mathbf{x}$ can be estimated by:

$$T^2 = (\mathbf{x} - \overline{\mathbf{x}})^{\mathrm{T}} \mathbf{S}^{-1} (\mathbf{x} - \overline{\mathbf{x}}), \tag{1}$$

where $\overline{\mathbf{x}}$ is the estimation of the population mean ($\boldsymbol{\mu}$) and $\mathbf{S}$ is the estimation for the variance-covariance matrix $\boldsymbol{\Sigma}$. If it is possible to assume that $\mathbf{x}$ follows a normal multivariate distribution ($\mathbf{x} = N(\boldsymbol{\mu}, \boldsymbol{\Sigma})$), then T^2 follows a $[J(I^2-1)/(I^2-IJ)]F_{J,I-J,\alpha}$ distribution, where, $F_{J,I-J,\alpha}$ is the value of the F distribution for a level of significance α, with J and $(I-J)$ degrees of freedom and I is the number of observations of the reference population.

As can be seen in equation (2), the Hotelling statistic has a squared form with the minimum at $\mathbf{x} = \overline{\mathbf{x}}$. Since matrix $\mathbf{S}$ is positive semidefinite, all the possible values for $\mathbf{x}$ will generate statistic's values that are greater than or equal to zero. For the sake of simplicity in nomenclature, the observation vectors $\mathbf{x}$ will be considered to be standardized so that $\mathbf{x} \approx N(\mathbf{0}, \mathbf{R})$, where $\mathbf{R}$ is the correlation matrix.

$$T^2 = \sum_{i=1}^{J}\sum_{j=1}^{J} a_{i,j}x_i x_j \tag{2}$$

$$\mathbf{R}^{-1} = \begin{pmatrix} a_{1,1} & \cdots & a_{1,N} \\ \vdots & \ddots & \vdots \\ a_{N,1} & \cdots & a_{N,N} \end{pmatrix} \tag{3}$$

This particular structure can be exploited in order to estimate the influence of each measurement in the final statistic's value.

Mason *et al.* [1,2] proposed to decompose the T^2-statistic value as a summation of J independent parts:

$$\begin{aligned} T^2 &= t_1^2 + T_{2\bullet 1}^2 + T_{3\bullet 1,2}^2 + T_{4\bullet 1,2,3}^2 + T_{p\bullet 1,2,\ldots,p-1}^2 \\ &= t_1^2 + \sum_{n=1}^{p-1} T_{n+1\bullet 1,\ldots,n}^2 \end{aligned} , \tag{4}$$

where t_1^2 is the student value of the first variable and $T_{n+1\bullet 1,\ldots,n}^2$ is the contribution of the jth measurement adjusted by using estimates of the mean and standard deviation of its conditional probability distribution given the $(j-1)$ previously considered variables. Since there exists no fixed order variables, it results possible to obtain $J!$ different (but non-independent) partitions for T^2.

Alvarez *et al.* [3] have recently presented an alternative straightforward method to decompose the T^2-statistic as a unique sum of variable contributions as:

$$T^2 = \sum_{j=1}^{J} a_{j,j}\left(x_j^2 - x_j^{*2}\right) = \sum_{j=1}^{J} c_j \tag{5}$$

$$x_j^* = -\frac{\sum_{\substack{j=1 \\ j\neq k}}^{J} a_{i,j}x_i}{a_{j,j}} , \tag{6}$$

where $x_j^*/2$ is the x_j value that minimizes the value of T^2 given the remaining $J\text{-}1$ variables values. This decomposition of the Hotelling's statistic also allows to understand the meaning of a negative variable-contribution and to estimate a bound for it. The variable contribution will take negative values if $0 \leq x_j \leq x_j^*$. The minimum contribution value is c_k^{min} that is located at $x_j = x_j^*/2$. If x_j is out of $0 \leq x_j \leq x_j^*$ the T^2-statistic is positive and it increases with $|x_j|$. The value of variable x_j contradicts the correlation structure if $x_j \leq 0$. On the other hand, a value of $x_j > x_j^*$ represents a large positive deviation with respect to the mean, in the direction indicated by the correlation matrix.

Several methods to calculate the variables' contributions when latent variables projection methods are used have bee presented as well. Nomikos [7] proposed an approach to calculate the contributions of each process variable to the Hotellings' statistic (called D when projection methods are used) instead of to the scores when latent variables cannot be associated to a meaningful group of process variables. Westerhuis *et al.* [8] presented an extended approach of the contribution plots proposed by Nomikos [7] to be applicable also to latent variable models with correlated scores. They proposed to calculate the contribution of the jth variable to the inflated statistic value (c_j) as:

$$D^2 = \mathbf{t}^{\mathrm{T}}\mathbf{S}_{\mathrm{L}}^{-1}\mathbf{t} = \mathbf{t}^{\mathrm{T}}\mathbf{S}_{\mathrm{L}}^{-1}[\hat{\mathbf{x}}^{\mathrm{T}}\mathbf{P}]^{\mathrm{T}} \tag{7}$$

$$D^2 = \mathbf{t}^{\mathrm{T}}\mathbf{S}_{\mathrm{L}}^{-1}\sum_{j=1}^{J}[\hat{x}_j\mathbf{p}^{\mathrm{T}}]^{\mathrm{T}} \tag{8}$$

$$D^2 = \sum_{j=1}^{J}\mathbf{t}^{\mathrm{T}}\mathbf{S}_{\mathrm{L}}^{-1}[\hat{x}_j\mathbf{p}^{\mathrm{T}}]^{\mathrm{T}} = \sum_{n=1}^{N}c_j\,, \tag{9}$$

where $\mathbf{t}$ and $\mathbf{S}_{\mathrm{L}}$ are the coordinate of $\hat{\mathbf{x}}$ in the considered latent space and variance-covariance matrix respectively. The main drawback of this technique is that it associates no interpretation to the negative values that can result from the calculations.

A Nearest In Control Neighbour Based Method (NICN)

Without assuming any gross error in measurements it still being possible to obtain an observation point whose statistic value exceeds the critical distribution value for a given significance level. This is possible if one or more variables in the observation vector does not behave as the observations in the reference population do (i.e. an anomalous event is really happening). Since one or more variables can be causing high values in the Hotelling's statistic, the question of which are the faulty variables has to be answered. One possible answer to that question could be given based in the knowledge of the nearest neighbour of the observation point that is in control. This information gives us an idea of how far from an in control allocation the faulty observation is and, which the minimum distance and direction that should be explained by an anomalous situation is. The problem of finding this nearest neighbour can be stated as an optimization problem, where the objective is to find an alternative point that minimizes a distance function to the measured point, subject to the constrain that the corresponding T^2-value is equal to or less than T_C^2.

$$\begin{aligned}&\min \quad (\mathbf{x}-\hat{\mathbf{x}})^{\mathrm{T}}\boldsymbol{\Psi}^{-1}(\mathbf{x}-\hat{\mathbf{x}})\\ &s.t.\end{aligned} \tag{10}$$

$$\mathbf{x}^{\mathrm{T}}\boldsymbol{\Sigma}^{-1}\mathbf{x} \le T_C^2\,, \tag{11}$$

where $\boldsymbol{\Psi}$ is the matrix that defines the type of distance chosen to measure the proximity to the faulty observation and, $\boldsymbol{\Sigma}$ is the correlation matrix estimated from the reference population.

The first order optimality conditions for the constrained optimization problem stated in equation (10) result:

$$L = (\mathbf{x} - \hat{\mathbf{x}})^{\mathrm{T}} \boldsymbol{\Psi}^{-1} (\mathbf{x} - \hat{\mathbf{x}}) - \lambda (\mathbf{x}^{\mathrm{T}} \boldsymbol{\Sigma}^{-1} \mathbf{x} - T_C^2) \tag{12}$$

$$L = \mathbf{x}^{\mathrm{T}} \boldsymbol{\Psi}^{-1} \mathbf{x} - \hat{\mathbf{x}}^{\mathrm{T}} \boldsymbol{\Psi}^{-1} \hat{\mathbf{x}} + 2 \hat{\mathbf{x}}^{\mathrm{T}} \boldsymbol{\Psi}^{-1} \mathbf{x} - \lambda (\mathbf{x}^{\mathrm{T}} \boldsymbol{\Sigma}^{-1} \mathbf{x} - T_C^2) \tag{13}$$

$$\frac{\partial L}{\partial \mathbf{x}} = 2 \boldsymbol{\Psi}^{-1} \mathbf{x} + 2 \lambda \boldsymbol{\Sigma}^{-1} \mathbf{x} - 2 \boldsymbol{\Psi}^{-1} \hat{\mathbf{x}} \tag{14}$$

$$\frac{\partial L}{\partial \lambda} = \mathbf{x}^{\mathrm{T}} \boldsymbol{\Sigma}^{-1} \mathbf{x} - T_C^2 \tag{15}$$

If we choose the distance measure to be minimized as the Mahalanobis distance between the observation and its nearest in control neighbour (i.e. $\boldsymbol{\Psi} = \boldsymbol{\Sigma}$) equations (14) and (15) can be written as:

$$\frac{\partial L}{\partial \mathbf{x}} = 2(1 + \lambda) \boldsymbol{\Psi}^{-1} \mathbf{x} \tag{16}$$

$$\frac{\partial L}{\partial \lambda} = \mathbf{x}^{\mathrm{T}} \boldsymbol{\Psi}^{-1} \mathbf{x} - T_C^2 \tag{17}$$

and the feasible points can be obtained by solving the equation system below.

$$\frac{\partial L}{\partial \mathbf{x}} = 0 \Rightarrow (1 + \lambda) \boldsymbol{\Psi}^{-1} \mathbf{x} = \boldsymbol{\Psi}^{-1} \hat{\mathbf{x}} \tag{18}$$

$$\frac{\partial L}{\partial \lambda} = 0 \Rightarrow \mathbf{x}^{\mathrm{T}} \boldsymbol{\Psi}^{-1} \mathbf{x} = T_C^2 \tag{19}$$

$$(1 + \lambda) \underbrace{\mathbf{x}^{\mathrm{T}} \boldsymbol{\Psi}^{-1} \mathbf{x}}_{T_C^2} = \mathbf{x}^{\mathrm{T}} \boldsymbol{\Psi}^{-1} \hat{\mathbf{x}} \tag{20}$$

$$\lambda = \pm \left(\frac{T_C^2}{T_{Obs}^2} \right)^{1/2} \tag{21}$$

$$\mathbf{x} = \lambda \hat{\mathbf{x}} \tag{22}$$

Considering that there are only two possible solutions; the comparison of the objective function at both solutions results much easier than evaluating the second order optimality conditions to decide which one is the corresponding nearest neighbour.

The solution for the more general situation, when $\boldsymbol{\Psi} \neq \boldsymbol{\Sigma}$ is more difficult to be obtained in an analytical way. Nevertheless the computational cost of considering different distances' metrics is not much bigger.

Figure 1 shows the level curves for the two variables' case. It shows the shape and intersections between different T^2-value curves and the lines defined by the Euclidean distance (green full lines) and the Mahalanobis distance (blue dashed lines) to the out of control observation. Two different points are considered to be out of control $\mathbf{p}_1$=[-1,2] and $\mathbf{p}_2$=[2,0] and the reference distribution is supposed to have its mean in [0,0]. It may be noticed that the line defined for the neighbours when $\mathbf{\Psi}=\mathbf{\Sigma}$ forms a straight line. On the other hand, when the Euclidean distance is considered the intersections follows a non-linear curve.

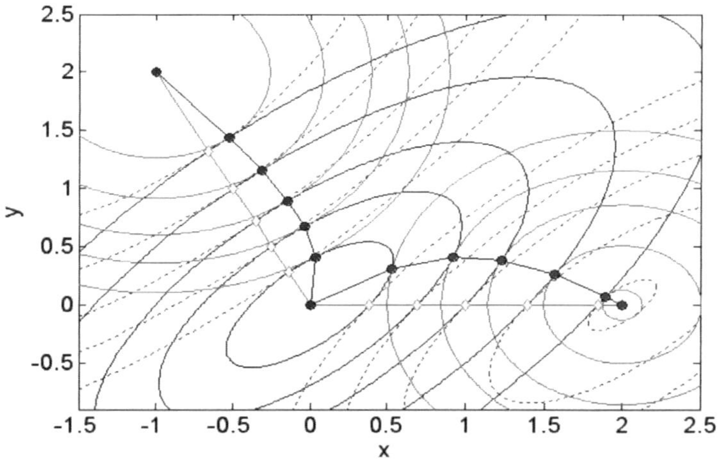

Figure 1- Different positions for the nearest in control neighbour depending on the selection of the distance metric.

The contribution of each variable to the inflated statistic value (c_j) can be estimated in several ways using this information. The easiest way is to compare the distance in which each variable should be modified to reach the nearest in control point against the average value of these distances for that observation point. Since all the variables should have been previously standardized (to be dimensionless) the comparison of the resulting movements in each direction can be used as estimates of the deviation degree of each variable, and can be used in a similar way as classically used in contribution plots. Then the directions whose changes are greater than the single threshold value (τ), are considered as suspicious variables:

$$\tau = \frac{1}{J}\sum_{j=1}^{J} c_j \tag{23}$$

Other alternatives can be used, for example, evaluating the contribution of each variable at the out of control observation as Alvarez *et al.* [3] proposed. After that, the same can be done at the resulting neighbour point. The change in each variable's contribution between these two points is an estimation of the contribution of each variable to the out of control state.

Case Studies

In this section, variable contributions to the T^2-statistic are obtained by applying the proposed strategy and compared with those estimated by using the original space strategy (OSS) (Alvarez *et al.* [3]) and the ones calculated by using equations (7) to (9). due to Westerhuis *et al.* [8]. Results for two case studies are provided.

Case Study I

Let us consider the data set formed by 20 observations of four variables presented by De Maesschalck *et al.* [9] as a reference data set. In addition, seven test observations are proposed to show how their T^2-statistic's values are interpreted by each of the three strategies (Table 1). The pairs of measurements $TEST_1/TEST_2$ and $TEST_3/TEST_4$ have the same Euclidean distance from the mean vector but some variables present deviations of different sign and magnitude.

Table 1 Test observations

	Observation	Euclidean Distance to the Mean
$TEST_1$	[1.000 5.350 3.125 3.245]	5.000
$TEST_2$	[11.00 5.350 3.125 3.245]	5.000
$TEST_3$	[1.000 7.000 3.125 3.245]	5.265
$TEST_4$	[11.00 7.000 3.125 3.245]	5.265
$TEST_5$	[8.000 7.000 11.00 5.000]	8.475
$TEST_6$	[2.000 8.000 8.000 7.000]	7.803
$TEST_7$	[2.100 3.100 7.900 4.900]	6.769

The same statistic value is obtained for $TEST_1$ and $TEST_2$. It is independent of the deviation sign because the three remaining variables are at their mean values. For $TEST_3$ and $TEST_4$ two variables deviate with respect to their means. The deviation of variable 2 is the same for both observations. In contrast, the deviation of variable 1 has the same magnitude but different sign. Note that $T^2_{TEST_3} > T^2_{TEST_4}$ and it is also greater than the critical value. The difference between $TEST_3$ and $TEST_4$ arises because the sign of $TEST_3$'s deviation contradicts more the correlation structure. For the observation $TEST_5$, $T^2_{\mathrm{TEST}_5}$ is greater than the statistic critical value for α=0.05 but not for α=0.01. Regarding $TEST_6$, $T^2_{\mathrm{TEST}_6}$ is larger than the critical value of the statistic for both levels of significance. The value of the variable contributions to the T^2 obtained by using the NICN and the OSS approaches are shown in Table 2 and Table 3 respectively. Contributions which are above their corresponding limit are underlined and in bold.

Table 2. OSS variable contributions

Observation	$c_1^{T^2}$	$c_2^{T^2}$	$c_3^{T^2}$	$c_4^{T^2}$	T^2
$TEST_1$, $TEST_2$	**<u>11.92</u>**	0.000	0.000	0.000	11.92
$TEST_3$	**<u>16.59</u>**	**<u>7.906</u>**	0.000	0.000	24.49
$TEST_4$	**<u>7.256</u>**	-1.425	0.000	0.000	5.832
$TEST_5$	1.024	-0.233	**<u>14.97</u>**	-0.402	15.36
$TEST_6$	**<u>9.872</u>**	**<u>7.986</u>**	1.292	**<u>8.266</u>**	27.42
$TEST_7$	0.582	3.290	3.905	3.105	10.88

Table 3. NICN variable contributions

Observation	$c_1^{T^2}$	$c_2^{T^2}$	$c_3^{T^2}$	$c_4^{T^2}$
$TEST_1$, $TEST_2$	**0.1762** **2.0337**	0.000 0.000	0.000 0.000	0.000 0.000
$TEST_3$	**0.7847** **0.0033**	**0.3740** **0.0016**	0.000 0.000	0.000 0.000
$TEST_4$	**2.6438** **7.5539**	0.5190 1.4829	0.000 0.000	0.000 0.000
$TEST_5$	0.0001 0.0614	0.0000 0.0140	**0.0021** **0.8977**	0.0001 0.0241
$TEST_6$	**0.6693** **0.0459**	**0.5414** **0.0371**	0.0876 0.0060	**0.5604** **0.0385**
$TEST_7$	0.0176 0.1336	**0.0996** **0.7549**	**0.1182** **0.8959**	**0.0940** **0.7124**

α=0.05 (above), α=0.01 (bellow)

It can be easily noticed in the tables above that the identification capabilities of OSS and NICN are similar, only $TEST_7$ presents differences in the suspicious set of variables since no contribution exceeds its limits when OSS strategy is applied.

Since the strategy developed by Westerhuis *et al.* [8] uses a latent variable model, a PCA model of the reference data was performed considering three retained PCs, reaching a variance reconstruction over 95%. Detection capabilities of the Hotelling's statistic are strongly affected by the dimension reduction. Only $TEST_5$ and $TEST_6$ are detected as faulty observations by *D*-statistic. Consequently, only values for those will be analyzed for comparison purposes. In addition, it should be pointed out that neither $TEST_4$ nor $TEST_7$ were indicated as faulty observations for any statistic. Table 4 shows the corresponding values for *D* and *SPE* statistics. Values in bold indicate that the corresponding statistic detects a fault.

Table 4. *D* and *SPE* statistics for Case Study I when PCA is applied

	D	*SPE*
$TEST_1$	2.85	**1.82**
$TEST_2$	2.85	**1.82**
$TEST_3$	2.91	**4.48**
$TEST_4$	4.138	0.34
$TEST_5$	**15.31**	0.01
$TEST_6$	**20.34**	**1.42**
$TEST_7$	10.12	0.15

$D_{C,0.05}=11.25$ $\quad SPE_{C,0.05}=0.81$

$D_{C,0.01}=18.25$ $\quad SPE_{C,0.01}=1.48$

Table 5 shows the values for the contributions for these observations by using equations (7) to (9). For $TEST_5$, variable x_3 is pointed out as the faulty variable in concordance with the results obtained before. In contrast, when $TEST_6$ is analysed only variables x_1 and x_4 are highlighted leaving x_2 as an in-control variable.

Table 5. Variable contributions according to Westerhuis

Observation	c_1^D	c_2^D	c_3^D	c_4^D	D
$TEST_5$	0.7743	0.1206	**15.1038**	-0.6818	15.31
$TEST_6$	**3.4648**	0.6809	0.2392	**15.9576**	20.34

Case Study II

The second case study is a tubular reactor where the reaction A+B→3C takes place. The set of measured variables is composed by ten observations: the inlet composition of A, B and C compounds, inlet reactor and refrigerant temperatures, inlet flowrate, reactor temperature at axial positions of 10 m and 20 m, outlet reactor temperature and outlet composition of C compound, which are identified as variables 1 to 10, respectively. The reference population is formed by thirty seven observations and, four additional runs are considered to perform the same comparisons as in the previous case. For run (*R1*) an increment in the composition of component C in the feed is simulated. An increase in the outlet temperature is considered in run (*R2*). The third run, (*R3*), shows a reduction in both the refrigerant temperature and the reactor temperature at 10 m as well as a high value on the C concentration at the reactor outlet. The last run, (*R4*), corresponds to a reduction in the outlet temperature and an increment in the inlet concentration of component C.

Table 6 OSS variable contributions for Case Study II

Run	$c_1^{T^2}$	$c_2^{T^2}$	$c_3^{T^2}$	$c_4^{T^2}$	$c_5^{T^2}$	$c_6^{T^2}$	$c_7^{T^2}$	$c_8^{T^2}$	$c_9^{T^2}$	$c_{10}^{T^2}$
R1	3.040	-0.020	**336.8**	3.770	39.85	0.540	21.59	2.770	-59.70	-3.290
R2	-356.0	-33.10	-0.100	-207.0	-1486	5.300	74.70	4.100	**1294**	**805.3**
R3	-523.0	35.70	1.000	-273.0	**955.8**	-0.900	**504.0**	16.20	-1287	**680.0**
R4	11.75	-111.0	2.890	34.64	-912.0	1.320	-83.80	3.120	**736.6**	**358.4**

Variable contributions to the T^2-statistic, calculated using OSS strategy, are presented in Table 6 for each run in which the D-statistic is greater than the critical values. Critical values for the OSS-based contributions were calculated as the mean value plus three standard deviations of those obtained from the reference data set.

Table 7 NICN variable contributions for Case Study II

Run	$c_1^{T^2}$	$c_2^{T^2}$	$c_3^{T^2}$	$c_4^{T^2}$	$c_5^{T^2}$	$c_6^{T^2}$	$c_7^{T^2}$	$c_8^{T^2}$	$c_9^{T^2}$	$c_{10}^{T^2}$
R1	13.037	0.9767	**166.81**	13.366	43.323	0.7645	53.699	3.4070	71.829	21.995
R2	72.776	6.7550	0.0149	42.257	**303.41**	1.0724	15.178	0.8274	**264.21**	164.41
R3	117.28	7.9861	0.2316	61.078	**214.39**	0.2058	113.11	3.6351	**288.37**	152.42
R4	0.2924	2.7674	0.0722	0.8654	**22.877**	0.0328	2.0993	0.0778	**18.488**	8.9806

Table 7 shows the variables' contributions when the NICN approach is used. In this case the threshold value for contributions was set at 2 τ. Some differences appear when comparing against

Table 6. This is due to the fact that all the directions are modified when NICN approach is applied in contrast to the "fixed curve" approach given by the OSS which produce very high negative values for some contributions.

A PCA of the same data have been carried out giving a total variance reconstruction of 75.4% when three P.C.s are retained (Cattell's criterion [4] has been used to choose the number of retained PCs). When the PCA model is used to evaluate these observations, only *R1* is pointed out as an out of control observation and the variables' contributions are shown in Table 8.

Table 8 Variable contributions according to Westerhuis

Run	c_1^D	c_2^D	c_3^D	c_4^D	c_5^D	c_6^D	c_7^D	c_8^D	c_9^D	c_{10}^D
R1	0.4363	-0.0088	**113.03**	0.1250	-0.0361	0.0840	-0.1359	-0.1765	0.0191	0.2039

As can be seen in Table 8, for (*R1*) the main contribution corresponds to the third variable (inlet C concentration), which is consistent with the actual simulated deviation. The same result is obtained using the other strategies.

The proposed technique has also been applied to industrial case studies, and to interpret results in latent variable spaces. More congruous results between the D-statistic variable contributions and the SPE-statistic variable contributions are obtained when the first ones are calculated using the proposed method

Conclusions

When projection techniques are used to reduce the space dimension it is possible to have either some missing or false alarms when compared with the original space results as it was shown by Alvarez *et al* [3]. This discrepancy could be due to an incorrect selection of the number of retained PCs and therefore, a strategy working in the original space can help to overcome this problem. In this work, a novel method to estimate the variable contribution to the T^2-statistic is presented. Given a measured point which T^2-value exceeds the critical value T_C^2, the contribution of each variable is determined in terms of the minimum distance between the measured point and its closer neighbour with a T^2-value equal to T^2_C. Results have shown a good performance when this technique is applied in the original variable space and they are similar to those obtained using the OSS strategy proposed by Alvarez *et al* [3]. Moreover, in other case studies it was noticed that results of NICN approach are more congruent with *SPE* contributions than those obtained using the classical approaches when PCA models are employed.

Acknowledgments

The authors wish to thank the financial support of CONICET (National Research Council of Argentina), ANPCyT (National Agency for the Science and Technological Promotion), UNS (Universidad Nacional del Sur, Bahía Blanca, Argentina) and the European Community project MRTN-CT-2004-512233

References

1. Mason, R. L., N. D. Tracy and J. C. Young (1995), "Decomposition of T2 for multivariate control chart interpretation", *Journal of Quality Technology*, 27, 99-108.

2. Mason, R. L., N. D. Tracy and J. C. Young (1997), "A practical approach for interpreting multivariate T2 control chart signals", *Journal of Quality Technology*, 29, 396-406.

3. Alvarez, C. R., A. Brandolin and M. C. Sanchez (2007), "On the variable contributions to the D-statistic", *Chemometrics and Intelligent Laboratory Systems*, 88, 189-196.

4. Jackson, J. E. (1991), "A User's Guide to Principal Components", John Wiley & Sons, Inc., United States of America.

5. Miller, P., R. E. Swanson and C. F. Heckler (1999), "Contribution plots: The missing link in multivariative quality control", *Applied Mathematics Computer Science*, 8, 775-792.

6. Macgregor, J. F., C. Jaeckle, C. Kiparissides and M. Koutoudi (1994), "Process monitoring and diagnosis by multiblock PLS methods", *AIChE Journal*, 40, 826-838.

7. Nomikos, P. and J. F. Macgregor (1994), "Monitoring of Batch Process Using Multiway Principal Component Analysis", *AIChE Journal*, 40, 1361-1375.

8. Westerhuis, J. A., S. P. Gurden and A. K. Smilde (2000), "Generalized contribution plots in multivariate statistical process monitoring", *Chemometrics and Intelligent Laboratory Systems*, 51, 95-114.

9. De Maesschalck, R., D. Jouan-Rimbaud and D. L. Massart (2000), "The Mahalanobis distance", *Chemometrics and Intelligent Laboratory Systems*, 50, 1-18.

577g A Functional Approach to Optimization of Supply Chain Networks with Both Efficiency and Robustness Objectives

Aviral Shukla[1], Vishal Agarwal[2], and Venkat Venkatasubramanian[1]. (1) School of Chemical Engineering, Purdue University, 480 Stadium Mall Drive, West Lafayette, IN 47907, (2) School of industrial Engineering, Purdue University, 480 Stadium Mall Drive, West Lafayette, IN 47907

As an event driven system, a supply chain network will face not only the uncertainties in the supply chain, but also the unexpected events in the supply chain network such as contingency, disruption, disaster and terrorism. These uncertainties and unexpected events have a negative impact on the survival and performance of the supply chain network (Dong 2006). Some work has been done in this field based on a graph-theoretic approach to define robustness indices for networks (Dong 2006; Venkatasubramanian et al. 2004). Since most firms are interested in analyzing the effect of network disruptions on their bottom-line, a graph-theoretic approach might not always be the best way to tackle this problem. Snyder and Daskin have looked at this problem for a case when risk pooling is possible by supply from the unaffected nodes, in case of failure of some nodes in the network (Lawrence and Mark 2005). In several cases high transportation cost or absence of a link between two nodes might render risk pooling from other nodes impractical. We have used a simple supply chain facility location model (Meepetchdee and Shah 2007) for determining the location of warehouses. Scenario planning is a valuable tool for analyzing the behavior of the network due to disruptions (Gaonkar and Vishwanadham 2004). We have incorporated this approach in our model to calculate the extra cost incurred to the network due to anticipated failures. Since network disruption can occur, both due to failure of nodes and the links connecting them, separate models were developed for both of these cases. The networks that emerge show interesting insights into designing optimal supply chain networks with least long-term costs.

References

Dong, M. (2006). "Development of supply chain network robustness index." International Journal of Services Operations and Informatics, 1, 54-66.

Gaonkar, R., and Vishwanadham, N. (2004). "A conceptual and analytical framework for the management of risk in supply chains." IEEE International Conference on Robotics and Automation, 26, 2699-2704.

Lawrence, V. S., and Mark, S. D. (2005). "Reliability Models for Facility Location: The Expected Failure Cost Case." Transportation Science, 39(3), 400-416.

Meepetchdee, Y., and Shah, N. (2007). "Logistical network design with robustness and complexity considerations." International Journal of Physical Distribution & Logistics Management, 37(3), 201-222.

Venkatasubramanian, V., Katare, S., Patkar, P. R., and Mu, F.-P. (2004). "Spontaneous emergence of complex optimal networks through evolutionary adaptation." Computers and Chemical Engineering, 28 (9), 1789-1798.

Web Page: cobweb.ecn.purdue.edu/~lips/

577h Minimization Under Uncertainty of Fresh Water Consumption for Pulverized Carbon (PC) and Integrated Gasification Combined-Cycle (IGCC) Power Plants

Juan M. Salazar and Urmila Diwekar. Vishwamitra Research Institute, Center for Uncertain Systems: Tools for Optimization and Management, 368 56th Street, Clarendon Hills, IL 60514

Conventional coal-fired power plants are significant consumers of fresh water for cooling and production purposes. However, recent power demand increments and environmental concerns have constrained the fresh water usage and consumption. These constraints affect the efficiency and productivity of the power plants due to the close relationship between power generation and water consumption. Additionally, new technologies (including carbon sequestration units) entail considerable augmentation of the already high water consumption of thermoelectric processes. Water loss is strongly affected by uncertain variables like atmospheric conditions and novel technological configurations of the cooling systems. Thus, optimal operating conditions and parameters, subjected to uncertain variations, are necessary to minimize the water consumption while fulfilling the required generation levels and the environmental regulations. Pulverized Carbon (PC) and Integrated Gasification Combined-Cycle (IGCC) plants including and excluding carbon sequestration units are studied. A conventional stochastic nonlinear programming approach for this optimization problem is computationally expensive due to the complexity of the required models for the processes. Comprehensive Aspen Plus simulation models are necessary to ensure appropriate closure of the mass and energy balances and to consider all the cooling system losses (makeups and blowdowns) and process consumption losses (humidification and reactions) to be minimized. Uncertainties are determined, characterized, and included in the Aspen Plus model generating the stochastic simulation of the processes. Employment of the better optimization of nonlinear uncertain systems (BONUS) algorithm decreases the computational requirements of the stochastic optimization.

577i Enhancing Supply Chain Network Design by Considering Financial Analysis Issues

Georgios Kopanos, José Miguel Laínez, Mariana Badell, and Luis Puigjaner. Chemical Engineering Department, Universitat Politècnica de Catalunya - ETSEIB, Diagonal, 647, Barcelona, E-08028, Spain

One of the challenges for the new millennium in the chemical industry (and not only) are the development of new products to achieve value growth, the optimization of the Supply Chain (SC) to achieve value preservation, and the improvement of the global life cycle to maintain living viability. Financial analysis constitutes the major measure of the profitability, the solvency, the liquidity, the stability and, thus, the viability of an enterprise or a project.

Financial analysts usually employ techniques such as 'funds flow analysis' and financial ratios to understand the problems and opportunities inherent in an investment or financing decision. Afterwards, their reports are usually presented to the Chief Financial Officer (CFO) as one of their main yardsticks in making business strategic decisions.

Despite being the chain of the most expensive resource, the financial chain is the less planned (e.g., budgeted), and the less supported by literature and Advance Planning Software (APS) computer-aided programs. Nowadays, useful SC models and APS will be only those that are capable of preserving the firms' liquidity while providing a quick and optimal response to orders, through the optimization of delivery dates and price–time trade-off solutions. In SC design-planning literature can be found some important contributions considering cash flow and budget management.[1,2] However, the SC design literature neglects the consideration of financial ratios despite the fact that constitutes a crucial part of the enterprise financial analysis.

In the current work, financial analysis aspects are taken into account in the optimization of supply chain network design problems. Several financial ratios addressing different economic enterprise aspects have been examined and incorporated into the problem representation. Illustrative SC design-planning examples will be presented in order to shed light on the special characteristics and the benefits of this strategy.

To end up with, our belief is that financial analysis issues have to be taken into account during the SC design-planning optimization procedure in order to generate and propose strong financial standing projects increasing the possibility of their acceptance from the CFO while reducing the probability of the company's financial analysts disapproval (with the latter being a frequent fact now days).

References

[1] G. Guillén-Gosálbez; Badell, M.; Espuña, A.; Puigjaner L., Computers & Chemical Engineering, Volume 30, 2006, Pages 421-436.

[2] Laínez, J. M. ; G. Guillén-Gosálbez; Badell, M.; Espuña, A.; Puigjaner L., Computer Aided Chemical Engineering, Volume 24, 2007, Pages 721-726.

Acknowledgements

Financial support received from the Spanish Ministry of Education and Science (FPU grants) and Generalitat de Catalunya (FI grants) is fully appreciated. Besides, financial support from PRISM-MRTN-CT-2004-512233 and DPI2006-05673 projects is gratefully acknowledged.

577j Global Optimization of Noisy Black Box Functions Using the Artificial Chemical Process and Evolutionary Algorithms

Roberto Irizarry, Electronic Technologies, DuPont, 14 TW Alexander Drive, Research Triangle Park, Raleigh, NC 27709

Many chemical processes are modeled using stochastic simulations (MC, KMC, SSA, etc). The optimization of these processes is very challenging due to: (a) The model is in a form of simulation, (b) The computational cost of the simulation is high and (c) There is a statistical noise associated with the simulations. The objective of this work is to accelerate the optimization by using a very crude model in most part of the optimization procedure. Evolutionary algorithm (EA) and artificial chemical process (ACP) algorithm [1] are modified using concepts of optimal comparison and stochastic ruler during the search. The modified algorithms accelerate the optimization tremendously. Dynamic and hybrid dynamics optimization of a complex nano-particle formation is solved using this methodology ([2], [3]). The algorithms are compared in terms of robustness and speed in finding the global optimum as a function of the model noise (from accurate simulations to very crude simulations with very large noise).

[1] R. Irizarry, LARES: An Artificial Chemical Process Approach for Optimization (2004) Evolutionary Computation Journal, 12 (4), 435-460.

[2] R. Irizarry, Fast Monte Carlo Methodology for Multivariate Particulate Systems-I: Point Ensemble Monte Carlo (2008) Chemical Engineering Science 63, 95-110.

[3] R. Irizarry, Hybrid Dynamic Optimization using Artificial Chemical Process: Extended LARES-PR (2006) Industrial & Engineering Chemistry Research, 45, 8400-8412.

577k A Novel Continuous-Time Formulation for Short-Term Scheduling of Continuous Processes

Jie Li[1], Naresh Susarla, and **I. a. Karimi**[2]. (1) Department of Chemical and Biomolecular Engineering, National University of Singapore, 4 Engineering Drive 4, Singapore 117576, Singapore, Singapore, (2) Department of Chemical & Biomolecular Engineering, National University of Singapore, 4, Engineering Drive 4, Singapore, 117576, Singapore

Scheduling of continuous plants has received less attention compared to batch plants in the literature although it is substantially important in chemical industries such as petroleum industry. Because continuous plants have some key differences from batch plants, for instance, the processing times in continuous plants depend on unit-dependent processing rates, final product demand, and storage capacities, separate models are required for scheduling continuous processes.

So far, several works (Karimi and Mcdonald, 1997; Mendez and Cerda, 2002; Reddy et al. 2004; Castro et al. 2004; Shaik and Floudas, 2007) in the literature have addressed this problem and developed various formulations based on different time representation such as unit slot-based (McDonald & Karimi, 1997), process slot-based (Reddy et al., 2004), global event-based (Castro et al., 2004), unit-specific event-based (Shaik and Floudas, 2007) and sequence-based (Mendez and Cerda, 2002). All these models incorporated different storage configurations such as unlimited, finite, dedicated, flexible, and no intermediate storage.

In this paper, we develop a novel continuous-time formulation for short-term scheduling of continuous processes based on unit-slots (Pinto & Grossmann, 1995; Lim & Karimi, 2003). We incorporate various storage configurations such as unlimited, finite, dedicate, flexible and no intermediate storage policies. We solve several examples including the benchmark example of a fast moving consumer goods manufacturing plant, which is extensively studied in the literature, to evaluate the performance of our proposed model. The results show that our model requires fewer discrete variables and constraints and gets the same or better solutions with less computation times compared to existing models. We also use our model to investigate the effect of different modeling and computing considerations such as computer configurations, MIP solvers and software versions, etc.

Key words: Asynchronous; slot-based; continuous-time model; continuous processes; optimality

References

Castro, P. M., Barbosa-Povoa, A. P., Matos, H. A., Novais, A. Q., Simple Continuous-time Formulation for Short-Term Scheduling of Batch and Continuous Processes, Ind. Eng. Chem. Res., 2004, 43, 105-118.

Karimi, I. A., Mcdonald, C. M., Planning and Scheduling of Parallel Semi-Continuous Processes: 2.

Short-Term Scheduling, Ind. Eng. Chem. Res., 1997, 36, 2701-2714.

Lim, M. F., Karimi, I. A., A slot-Based Formulation for Single-Stage Multiproduct Batch Plants with Multiple Orders per Product, Ind. Eng. Chem. Res., 42 (9), 1914-1924.

Mendez, C. A., Cerda, J., An Efficient MILP Continuous-Time Formulation for Short-Term Scheduling in Multiproduct Continuous Facilities, Comput. Chem. Eng., 2002, 26, 687-695.

Pinto, J. M., Grossmann, I. E., A Continuous Time Mixed Integer Linear Programming Model for Short-Term Scheduling of Multi-Stage Batch Plants, Ind. Eng. Chem. Res., 34, 3037-3051.

Reddy, P. C. P., Karimi, I. A., Srinivasan, R., A Novel Continuous-Time Formulation for Scheduling of Crude Oil Operations, Chemical Engineering Science, 2004a, 59, 1325-1341.

Shaik, M. A., Floudas, C. A., Improved Unit-Specific Event-Based Continuous-Time Model for Short-Term Scheduling of Continuous Processes: Rigorous Treatment of Storage Requirements, Ind. Eng. Chem. Res., 2007, 46, 1764-1779.

577l Eliminating Gaps Between Advanced Planning and Scheduling Models

Yanqin Wen, Department of Chemical Engineering, Lamer University, Beaumont, TX TX 77710 and Qiang Xu, Department of Chemical Engineering, Lamar University, Beaumont, TX TX 77710.

Advanced planning and scheduling systems are widely used in oil refining and petrochemical plants all over the world. A planning system is focused on corporation-wide or plant-wide optimization, and gets production plan that is relatively preliminary. A scheduling system is focused on plant-wide or unit-wide realization, and gets detailed schedule schemes based on the plan. A plan and the corresponding schedule should be consistent, otherwise a gap between them appears and either the plan or the schedule should be rebuilt. This paper discussed when will the gap between planning and scheduling be generated and how to eliminate the gap.

Primary reason for the gap is the difference of periods. A production plan is normally involved in one or more periods of year, month or week. A schedule is involved in periods of day or hour. A plan is "feasible" means it is feasible in monthly or weekly average, but it may not feasible in a specific day or hour. Many cases were reported that plans caused nonsensical scheduling requirements. For example, two streams may be pooled in a plan, but the two streams are produced in different time and no inventory available. To avoid such situation, common model structures were presented, which could be adopted by planning and scheduling models. For the planning model, special mixed-integer features were introduced to avoid periodic inconsistent between planning and scheduling. For the scheduling model, multi-period predictive features are introduced, which enable the scheduling model simultaneously to generate "near future" schedule and to predict "far future" plan. The "far future" plan produced by the scheduling system should be aligned with the one produced by the planning system, which will reduce the gap.

The methods have been used in several refineries. In a case study, plant-wide planning and scheduling models were built to deal with diesel transition problem through summer to winter. The planning model had 6 periods to optimize production plans from September to the next February, and the scheduling models were 35 periods rolling models. In this case, gap between planning and scheduling was effectively eliminated.

Building a High-Fidelity Dynamic Simulation and Control System Fast

David C. Zhou, Ph.D.
President, 2States Technology
100 Cowing Street, West Roxbury, MA 02132
David.zhou@2stastes.com

KEYWORDS

Sizing and Selection, Modeling, Steady-State and Dynamic Simulation, Control, Tuning

ABSTRACT

Process operations are dynamic in nature. It is the ultimate goal to own a high-fidelity virtual dynamic system that simulates behaviors and performance not only under normal conditions but also during startup and shutdown as well as under abnormal conditions. Such a system could be used to exam designs, analyze operations, obtain simplified models for model-based controllers, tune controllers, train operators and find optimum solutions. However, building such a system remains a challenging task.

Integrating high-fidelity equipment blocks, with built-in sizing, rating and testing methods and environment into a simulation tool would help a user achieve such a goal. Furthermore, adding product databases and selection methods to obtain equipment parameters automatically simplifies the building process for the user. Examples are included to demonstrate the application of such a tool to process industries.

INTRODUCTION

Steady-state simulation is widely used in process design. However, a high-fidelity dynamic simulation is particularly useful to design control systems, tune controllers and train operators. Furthermore, dynamic simulation may be an effective way to resolve the conflicts between optimum design from a steady-state standpoint and optimum dynamic control and operation.

It is natural to build a virtual dynamic system the same way to build a real plant, using pre-built equipment, instrument and controller blocks. You can size, select and connect these blocks together, set up initial conditions and push a button to run. These virtual blocks interact among themselves and with operators via fluid and signal to simulate an existing process or the process to be built.

A simulation platform with these blocks and an integrated environment is a useful tool for users to build high-fidelity dynamic systems fast. With the unified platform for both steady-state and dynamic simulation, process and control engineers can work together to optimize the process and control design, size and select equipment, and tune the virtual plant to ensure the real plant can be started up quickly and run as expected.

VIRTUAL BUILDING BLOCKS

To quickly build a dynamic system, it is critical that a simulation tool is armed with pre-built model block libraries. Libraries contain blocks ranging from math, signal, system and control to process instrument and equipment such as piping, tanks, valves, pumps, heat exchangers and distillation columns for process industries.

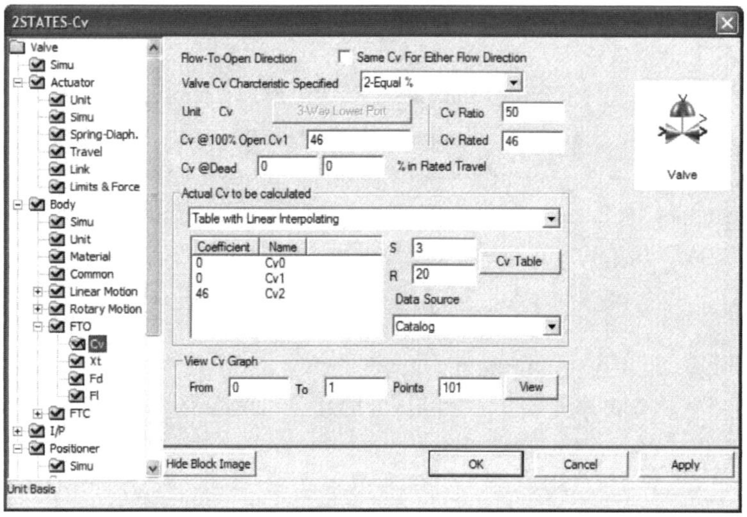

Figure 1. Components of a Pneumatic Control Valve System

For the virtual dynamic system to be accurate in a wide operation scope, building blocks should be accurately modeled, based on physical and chemical laws and rigorous math and engineering relationships that apply. However, extremely rigorous modeling should be avoided to save development work and make simulation time reasonable by a personal computer.

A general equipment block may require hundreds of parameters to be modeled accurately for most possible cases. A pneumatic control valve system shown in figure 1 consists of a valve body, an actuator and position controller with optional airset and air booster. All these components are required to be modeled accurately and interact among them and with process fluid passing through valve body to exhibit its function and behaviors.

Figure 2 shows dynamic response on 2% input step change of a control valve with high packing friction. High friction in a valve may cause stick-slip movement of valve stem and limit cycling of its control loop. Furthermore, it may prevent a valve from fully opening or closing if its actuator is not sized properly. If the control valve is oversized in capacity, high friction may significantly degrade the performance of its control system. Simulation tests with such a high-fidelity control valve modeling can reveal these problems and help find the solutions.

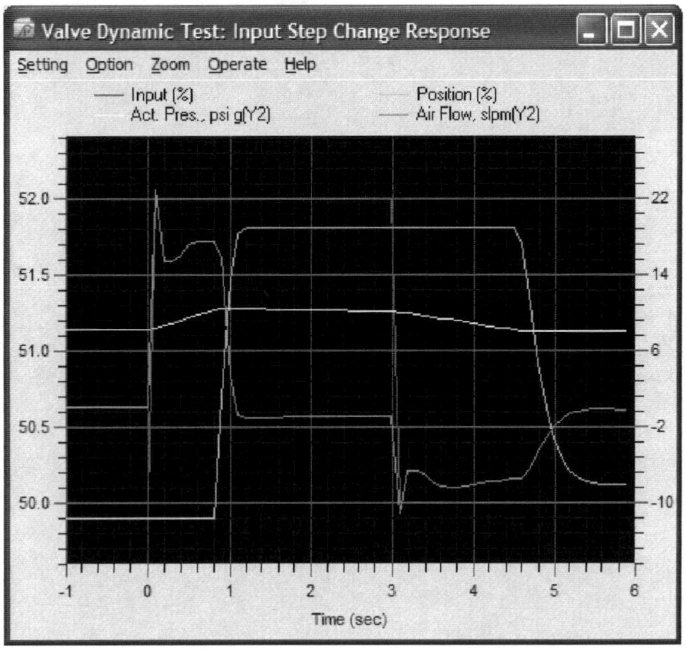

Figure 2. Dynamic Response of a High Friction Control Valve

A user may make simplification for a customized block in modeling and calculation based on his or her judgment and the task undertaken, so their impact on calculation is either negligible or unimportant for the particular task.

OBTAIN DATA BY SIZING AND SELECTING EQUIPMENT

Equipment may vary in types and sizes. For a specific application, there may be many equipment products to choose from. Sizing and selection functions in an equipment block can be used to size and select equipment and to obtain the equipment parameters from databases automatically. This alleviates burdens on users for providing data for the block and reduces development time for building steady-state and dynamic simulation systems.

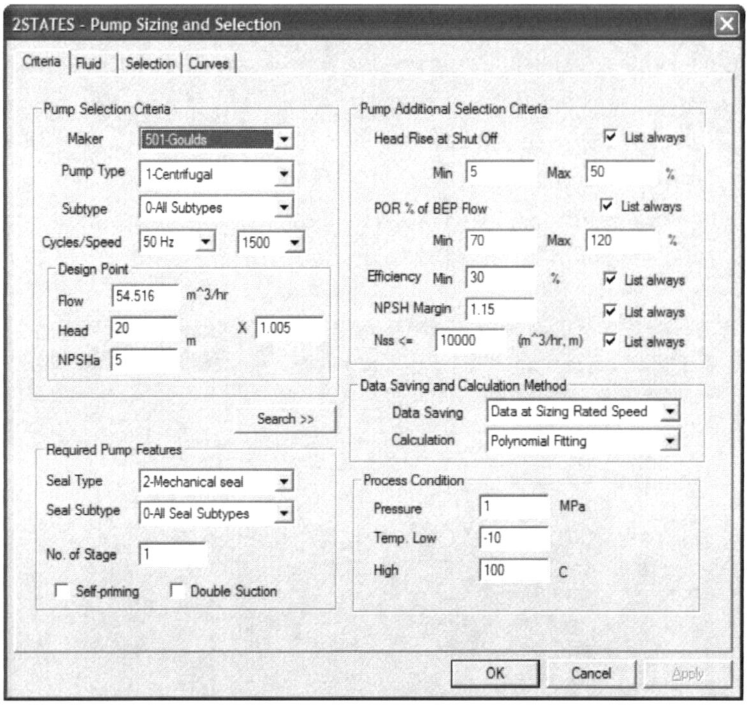

Figure 3. Pump Sizing and Selection Interface

Passing process data from upstream or related equipment for sizing downstream or auxiliary equipment may reduce data entry errors and increase work efficiency. The same applies for rating and testing. For example, after sizing a distillation column, its reboiler and condenser can be sized, based on the results of a distillation column sizing, such as stream flow, temperature, pressure and compositions as well as thermal duties required. Similarly, pumps, control valves, flow measuring elements and connecting pipes in the column system can also be sized consequently in the process.

Equipment sizing may produce multiple options, as illustrated in Figure 4 for heat exchanger sizing. Built-in rating and testing methods in a block can be used to evaluate the options. Various steady-state and dynamic tests can be conducted and comparisons can be made to make an informed choice. Figure 5 shows dynamic responses of three heat exchangers under consideration. Of course, important equipment should be further evaluated from steady-state and dynamic simulation tests of its loop-wide, unit-wide even plant-wide systems to determine if the selection is the best.

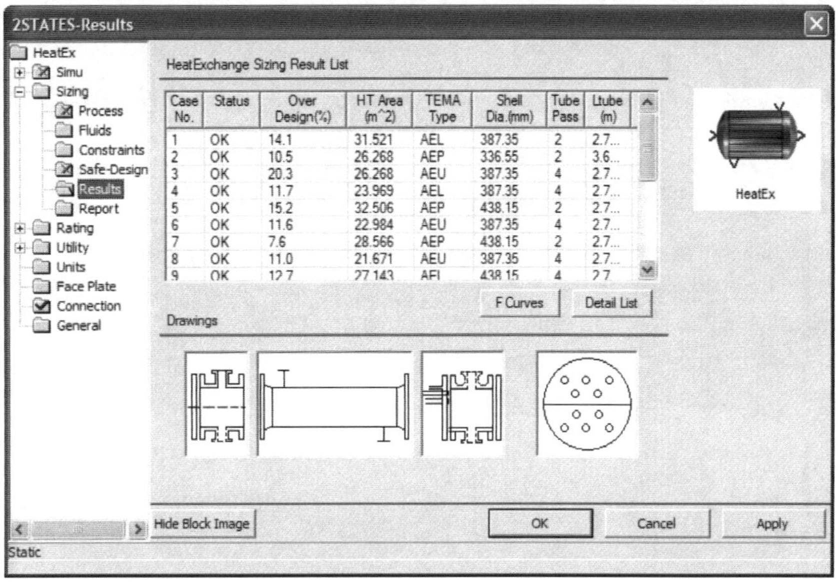

Figure 4. Multiple Options after Heat Exchanger Sizing

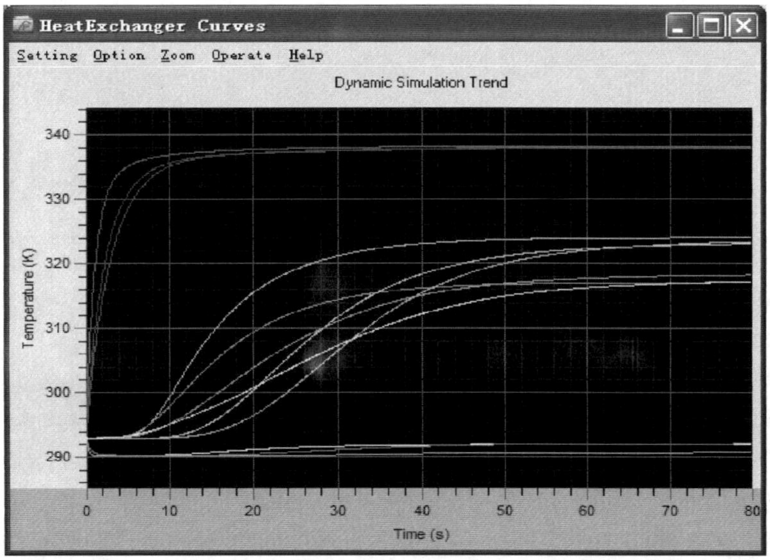

Figure 5. Dynamic Responses of Three Heat Exchangers

FLUID PROPERTY CALCULATION

Process equipment is typically used to process fluids. Setting up fluids in applicable blocks and choosing the appropriate fluid property models and methods are important in equipment sizing and rating as well as steady-state and dynamic simulation. Fluid property models and methods are used to calculate thermodynamic and transport properties such as density, viscosity, thermal conductivity, fugacity, enthalpy and entropy. Therefore they contribute to the accuracy of calculation results as long as fluid is involved.

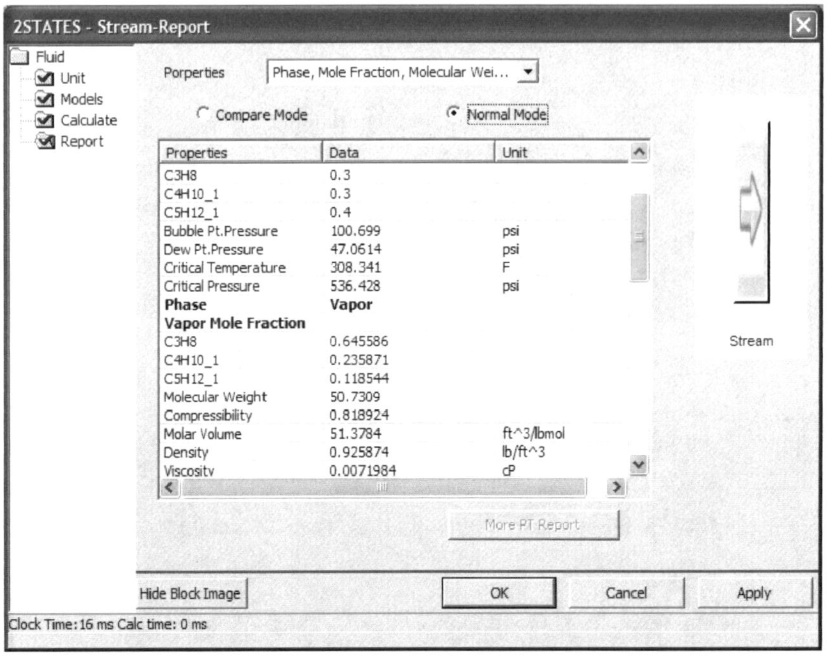

Figure 6. Property Calculation Results of a Fluid Mixture

BUILDING AND SIMULATING A PROCESS

Simulating a process is useful in research, design, operation and training. The block based simulation structure and graphic interface make the building and modify a process easy. To build a process, select the blocks needed and connect the blocks with wires. Rigorous modeling of individual blocks and fluid property calculation combining integration methods allow simulating behaviors and performance of a process not only under normal operating conditions but also during startup, shutdown, and under abnormal conditions. That would help find optimum process and control structures and strategies, minimize energy consumption and obtain controller tuning parameters that applicable to various operating conditions.

With such a simulation tool, you can interactively modify equipment data, operating conditions such as temperature and pressure, feed compositions and controller parameters during the simulation or to run new cases and evaluate alternatives. You can monitor the operation through faceplates, data views, trends, graphs, tables and other interfaces that are configurable and adjustable and can be opened or closed at any time.

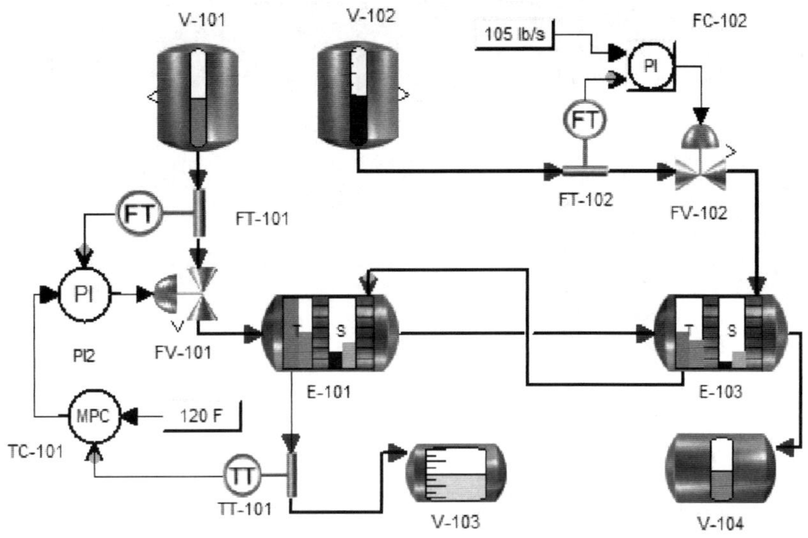

Figure 7. Temperature Control Simulation of Heat Exchangers in Series

Figure 7 gives a simulation diagram for two heat exchangers in series. It may demonstrate how different dynamic behaviors become when flow load changes. You can tune controller parameters to see its impact on dynamic performance under different operation conditions. You can compare model predictive control (MPC) against PID control side by side for this particular application to arrive at your conclusion.

A distillation column (Figure 8) is an important process and a good example for multivariable control system design and simulation. High-fidelity dynamic column simulation allows a user to develop its control system and test it under a wide operating condition. You can identify a simplified model for model predictive control and evaluate its performance under manipulated operating conditions.

After the control method is finalized, the system may be used for operator training. A high-fidelity simulation system is much more valuable then a simulator with simplified models, because it can train operators to rum the process under much wider operating conditions, especially abnormal ones.

If there are operating data available, you can refine data such as fouling resistances for heat transfer and column tray efficiencies for separation to match simulation results with actual process data. The virtual system can be used to trouble-shoot existing process and control problems, optimize plant operation and study modification of processes and procedures.

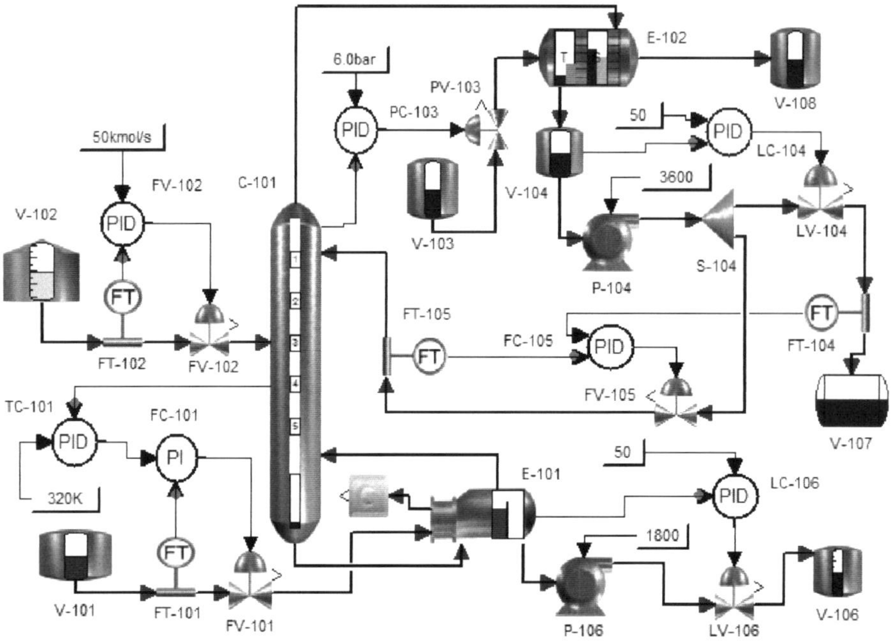

Figure 8. Distillation Column Dynamic Simulation

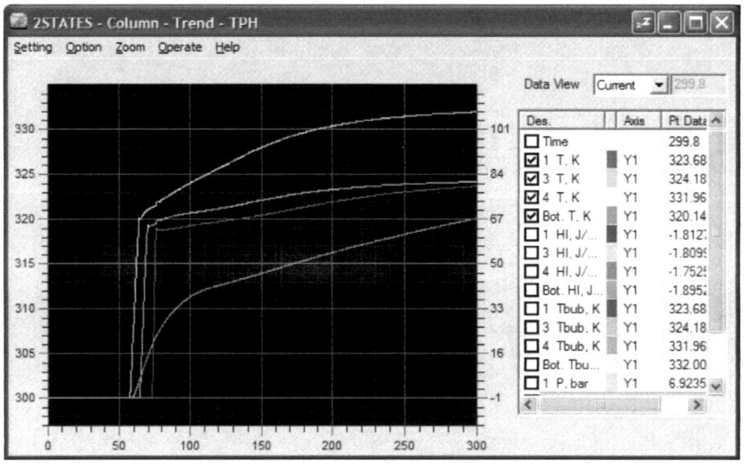

Figure 9. Temperature Response during Distillation Column Startup

SUMMARY

It is possible to quickly build a high-fidelity dynamic simulation system for wide operating ranges with a simulation tool. It is essential such a tool is created with various accurate blocks and sizing and selecting functions to obtain data from built-in databases. In the near future, high-fidelity dynamic plant simulation would be widely used in design, troubleshooting and training, thus become a valuable tool to own.

577n Identification of the Attainable Region for Batch Reactor Networks

Benjamin J. Davis, UCLA Chemical Engineering Department, 5531 Boelter Hall, Los Angeles, CA 90095 and Vasilios I. Manousiouthakis, Chemical Engineering Department, UCLA,, 5531 Boelter Hall, Los Angeles, CA 90095-1592.

Quantification of performance targets is essential to identifying the overall merit of a particular process design or operating chemical plant. In order to determine where improvements can be made in terms of energy efficiency, cost-effectiveness, emissions reduction, and overall process sustainability, one must identify realistic benchmarks to compare to. This work here focuses on identifying performance limits for any system of discretely-fed batch reactors. By evaluation of the best possible performance of a batch reactor network under known physical constraints, a designer or decision maker can quantify exactly how good a job their flowsheet or actual reactor network is doing.

I will describe a method [Davis et al. "Identification of the Attainable Region for Batch Reactor Networks" Ind. Eng. Chem. Res., in press] for automatically identifying the set of all points in concentration space that represent outlet compositions of some network of discretely fed batch reactors for a given reaction set with known kinetics. This so-called batch attainable region (BAR) is dependent on the batch network's feed and total operating time, and it is shown to be quantifiable using the Infinite DimEnsionAl State-space (IDEAS) framework. Because a simple batch reactor model possesses the properties that allow application of the IDEAS framework, we can formulate the resulting IDEAS Infinite Linear Program (ILP) whose solution is guaranteed to identify the globally optimal network of batch reactors. A simple transformation of this IDEAS ILP leads to two algorithms that can be used to construct an approximation to the BAR. The first is a "Shrink-Wrap"-like algorithm which creates increasingly accurate approximations of a set guaranteed to contain the true BAR for all network operating times by eliminating infeasible points in concentration space. The second is a breadth-first algorithm that creates increasingly accurate inner approximations to the BAR for a given network operating time. I have implemented these two algorithms and will outline an example where their output converges to the analytically identified BAR.

577o Wastewater Treatment: New Perspectives Provided by Interactive Multiobjective Optimization

Jussi Hakanen[1], Kaisa Miettinen[1], and Kristian Sahlstedt[2]. (1) Dept. of Mathematical Information Technology, University of Jyväskylä, P.O. Box 35 (Agora), FI-40014 University of Jyväskylä, Jyväskylä, Finland, (2) Pöyry Environment Oy, P.O. Box 50, FI-01621 Vantaa, Vantaa, Finland

Operational requirements of wastewater treatment plants, notably the effluent limits of nitrogen and phosphorus, are becoming more and more strict. Consequently, more complex wastewater treatment processes are gaining ground. At the same time, the requirements for economical efficiency (for example, minimizing plant footprint and the consumption of chemicals and energy) as well as for operational reliability are also tightening. This makes the design of a wastewater treatment plant a complex process involving trade-offs between a number of conflicting economical and operational criteria.

The overwhelming majority of wastewater treatment modelling considers the activated sludge process (ASP), globally the most common method of wastewater treatment. In this process, biomass (which is called activated sludge) suspended in the wastewater to be treated is cultivated and maintained in an aerated bioreactor. The wastewater is purified, i.e. organic carbon, nitrogen and phosphorus are removed, during its retention in the bioreactor. The bioreactor is followed by a clarifier basin, in which the biomass is separated by gravitational settling and returned to the bioreactor, and the treated wastewater is directed as overflow to further treatment or to discharge. Excess activated sludge is removed from the process and treated separately.

The wastewater treatment plant design has been previously considered by optimizing only one objective function, which in one way or another describes the costs of the process to be minimized. However, practical real-world optimization problems, like the ones in wastewater treatment, often have to be considered from many different perspectives. This gives rise to several conflicting evaluation criteria. When the optimization problem in question needs to be solved with respect to several conflicting criteria simultaneously, the concept of optimality needs to be redefined. When dealing with multiple conflicting objective functions, the solution can be seen as optimal when no objective function value can be improved without impairing any other objective. These optimal solutions are called Pareto optimal solutions or compromise solutions. Typical to these optimal solutions is that there usually are (infinitely) many of them and they all are mathematically equivalent. Solving these types of problems requires using the methods of multiobjective optimization.

To guarantee a final design which takes into account all the relevant criteria related to wastewater treatment, we propose an interactive design strategy that utilizes numerical simulation of wastewater treatment processes combined with an efficient interactive multiobjective optimization method. This enables the designer to simultaneously consider the process from different perspectives and optimally balance the final design between different conflicting design criteria.

In this paper, we concentrate on utilizing modern optimization techniques to provide decision support for the designer which helps him/her to locate the best trade-offs between different competing design alternatives. By utilizing interactive multiobjective optimization in the design process, the designer is able to learn about the problem and about the interdependences between the conflicting design criteria. In addition, (s)he can concentrate only on those compromises that are of interest to him/her.

In our approach, novel aspects to wastewater treatment plant design are the consideration of multiple evaluation criteria and interactive nature of the design process. So far, we have found only one paper in this field that deals with multiple objectives and there the idea is to produce an approximation of all the compromise solutions to the multiobjective optimization problem considered. When compared to this approach, our interactive approach is more computationally efficient because we do not try to approximate all the compromise solutions of which many can be uninteresting to the designer. To our knowledge, there are no papers on utilizing interactive multiobjective optimization in wastewater treatment plant design which makes this kind of approach an entirely novel in wastewater treatment plant design, although such interactive tools are succesfully utilized in other fields.

We have conducted a preliminary study on combining an interactive multiobjective optimization tool IND-NIMBUS (http://ind-nimbus.it.jyu.fi/) with the GPS-X simulator (http://www.hydromantis.com/software02.html) designed for simulating wastewater treatment processes. The approach was tested with a case study of wastewater treatment plant design following the activated sludge process described above. The case study included three conflicting objectives, namely, the residual ammonium nitrogen concentration, the dose of alkalinity chemical and the consumption of energy by aeration. In this paper, we extend the case study to be more realistic by considering more conflicting objectives. The possibilities of this interactive design strategy are illustrated by reporting results from this case study and analyzing their significance and the potential of this novel point of view.

577p Dynamic Simulation for Best Performance of Multi-Stage Compression System during Plant Startup

Xiongtao Yang, Kuyen Li, and Qiang Xu. Department of Chemical Engineering, Lamar University, Beaumont, TX 77710

Chemical plant startup is a plant-wide dynamic operation that usually involves discontinuous and/or parallel operating procedures, as well as the wide change of controllers' set points. During the plant startup, the operation for multi-stage compression system (MSCS) especially needs sufficient care. This is not only because the compression equipments are the most valuable facilities in the plant, but also its failure operation will cause dangerous situations to the whole process and thereby cause tremendous raw material and energy loss. Thus, the startup procedure for MSCS must be virtually evaluated prior to the real implementation.

This paper employs rigorous dynamic simulation to test both safety and economic performance for MSCS startup operation. For safety consideration, anti-surge process design and control strategies are dynamically evaluated along with startup procedures. Operating point trajectory for each compressor and their potential surge risks will be identified. For economic consideration, on the other hand, feasible and energy-efficient operating (economic) zone for each compressor will also be identified. The difference between the operating trajectory and the economic zone is thus available, which will be utilized to guide the performance optimization for MSCS design and control.

The rigorous dynamic simulation is a cost-effective and general way for MSCS system performance evaluation. The best-performance oriented rigorous dynamic simulation can be extended in future to any critical chemical process systems to improve their turnaround or routine operations. A case study for performance improvement of an MSCS during an ethylene plant startup will be demonstrated in this paper.

577q Production Planning, Scheduling, and Debottlenecking Practices In the Biopharmaceutical Industries

Charles Siletti and Demetri Petrides. INTELLIGEN, INC., 700 Walton Ave., Mount Laurel, NJ 08054

This paper presents industrial experience with a resource-constrained batch process scheduling tool. The batch process representation is partially based on the ISA S88 batch process standard. This representation allows the import of batch process information from other software, e.g. batch process simulators. The scheduling algorithm is a non-optimization approach that proceeds in two steps. First a bottleneck analysis is done to determine a lower bound on the process cycle time, and all the batches are scheduled accordingly. Second, if conflicts remain, they are resolved by applying progressively aggressive modifications to the schedule. This approach to scheduling was tested on several biotech processes. These processes consist of a sequence of batch steps performed with dedicated equipment. The scheduling challenges in biotech processes lie in the ancillary operations: media and buffer preparation, vessel and line cleaning, and chromatography column preparation. Such operations may use shared resources that can to couple process suites with otherwise dedicated equipment. These considerations are further complicated by variability in process durations. Three case studies, which are based on a process for the manufacture of monoclonal antibodies (MABs), illustrate the value of a constrained-resource scheduling tool for biotech processes. In the first case study, the scheduling tool shows that auxiliary cleaning equipment can limit batch production. A second case study shows how scheduling tools can be used to identify and eliminate bottlenecks associated with utilities. A third case study illustrates how to use scheduling tools to mitigate the effects of process variability.

Web Page: www.intelligen.com/literature

577r Monitoring, Detection and Diagnosis of Leaks from Gas Pipes Based on Real-Time Data Collected Over the Ubiquitous Sensor Network

Dongil Peter Shin, Dept. of Chemical Engineering, Myongji University, Nam-Dong, Yongin, Kyunggido, 449-728, South Korea and Jae Wook Ko, Dept. of Chemical Engineering, Kwangwoon University, Wolgye-Dong, Nowon-Ku, Seoul, 139-701, South Korea.

As the world's second largest LNG importer, South Korea imports most of its LNG from Indonesia, Qatar, and Oman with smaller volumes from Malaysia, Brunei, the UAE, and Australia. South Korea has three regasification terminals owned and operated by state-owned Korea Gas Corporation (KOGAS) at Pyeongtaek, Incheon, and Tongyeong. Natural gas is mainly supplied to city gas companies and power generators. With three terminals and eight district offices nationwide, KOGAS is supplying natural gas to every corner of the country.

LNG is stored in storage tanks at three receiving terminals. After going through a gasification process, it is distributed nationwide through an extensive pipeline network. The operating branches, which receive natural gas from the terminals, supply it stably to 17 power stations of 10 power companies and 30 city gas firms. As natural gas is supplied to end users through pipelines, effective operation of a network of pipelines is crucial. By completing the pipeline to Gangwon province in 2002, after starting construction of the natural gas pipeline network in the Seoul metropolitan area in 1983, KOGAS had constructed a ring-shaped 2,520km supply network linking the whole nation. This network allows for safe and convenient use of natural gas around the clock anywhere on the peninsula.

For many years, Korea has proven to be a trailblazer in the information-technology sector. And the country won't likely stop until it attains its ultimate vision of creating a "ubiquitous" society where people can connect to the web, television and other services anytime, anywhere. Korea's central and local governments, as well as businesses, have begun to test out the possibilities of creating that visionary state through "u-city" or ubiquitous city projects. U-city is intended to give its residents not only a more convenient lifestyle but also more secure, safe, environmentally-friendly and humane way of life. This will be achieved by integrating the latest IT infrastructures and information services into urban space. Technologies that are likely to be featured in u-cities include broadband convergence network, radio frequency identification, ubiquitous sensor network, home networking, WiBro, digital multimedia broadcasting, telematics, geographic information system, location-based system, smart card system and video conference technologies.

By using its advanced IT, Korea is trying to improve the security of its energy infrastructure and the safety in everyday's energy use around the country. In this presentation, we will give an overview of a government-sponsored research on the monitoring, detection and diagnosis of releases from gas pipes, based on real-time data collected over the ubiquitous sensor network. A ubiquitous sensor network covering the city gas pipe network is to be installed for the sensor-network based source tracking. Fixed sensors and mobile sensors, if necessary, send operational information at designated points over the network and measured concentrations of gases in real-time.

From the real-time sensor data including flow rates, acoustic emission signals, and gas concentrations, we monitor, detect and analyze the suspected location and magnitude of gas releases. The type and amount of the release and its location on the GIS map are tracked relying on a combination of diagnostic methods and massive computation of dispersion models. Warnings are generated and corresponding emergency responses are executed based on the result of the simulation of the expected dispersion consequences, by combining the sensor data, meteorological information, GIS data and the diagnostic result of the source term. A prototype system of the proposed framework is being tested against the simulated sensor data. In addition to the enhanced security of the energy infrastructure, the result of this study will help improving the confidence of residents living next to storage, handling and transporting facilities of hazardous materials, by supplying the improved real-time emergency response plan and monitoring system for various hazardous gas releases.

577s Energy Optimization through Non-Linear Modeling and Control of Fuel Gas Process at Hovensa

Manish Misra[1], Victor M. Marrero[1], Kevin S. Brooks[2], and Hannes de Beer[2]. (1) HOVENSA, LLC, 1 Estate Hope, Christiansted, 00820, US Virgin Islands, (2) BluESP (Pty.) Ltd., 53 Platina St, Jukskei Park, Randburg, 2188, South Africa

Increases in prices of raw crude and utility costs, and stricter environmental regulations are some of the primary drivers for efforts towards energy optimization in refineries. An efficient optimization of a refinery's fuel gas system can provide stability to the fuel gas header pressure and account for variations in fuel gas quality, thereby leading to optimal control of charge heaters and boilers and optimization of the LPG inventories.

Hovensa LLC, with a refining capacity of 495,000 BPD of raw crude, has recently implemented a Fuel Gas Optimizer (FGO^{TM}). The FGO^{TM} is an APC platform, developed by BluESP on AspenTech's InfoPlus21 database and DMCplus controllers. FGO is a non-linear dynamic optimizer that simultaneously controls the fuel gas header pressure and the fuel gas quality, in real time. It does this my continuously monitoring the flow and quality of the fuel gas suppliers, as well as the demand from consumers. The models in the DMCPlus controller are updated online in order to mange the system non-linearities. The optimization objective of the system is adjusted to the needs of the refinery – typically either maximize or minimize fuel gas usage.

As part of the FGO^{TM} implementation at Hovensa, fuel gas producers (off-gases from the FCC and Coker units, excess Hydrogen, enrichment gases, etc.) were balanced with the consumers of fuel gas (charge heaters, boilers, gas turbines, etc.). FGOTM provided a mechanism to manipulate the fuel gas production from the LPG and propane vaporizers. Additionally, by specifying the amount of fuel gas that can be consumed in each boiler, FGO effected a mechanism for controlling the fuel gas:fuel oil ratio in the fire boilers. Towards providing a real-time measure of the BTU quality of fuel gases, a multivariate model, accounting for the BTU of each producer and consumer stream, was developed by using InfoPlus21. FGO then predicted the quality of the fuel gas, and provided an on-line advisory on manipulating the number of burners on the crude heaters.

An efficient implementation of FGO^{TM} at Hovensa has brought stability to the fuel gas header pressure, and provided a handle to predict and account for the variations in fuel gas quality, which has led to desirable control of the operations of fire boilers and charge heaters and yielded economically attractive returns on the sale of some of the products.

Keywords: fuel gas, non-linear optimization, pressure stability, BTU quality.

577t Environmental Evaluation of Cogeneration Scenarios In Sugar Industry

Mauricio Colombo[1], Fernando Daniel Mele[1], Maria Rosa Hernandez[1], Jorge E. Gatica[2], and José Luz Silveira Sr.[3]. (1) Ingenieria de Procesos y Gestion Industrial, Universidad Nacional de Tucuman, Avda. Independencia 1800, Tucuman, Argentina, (2) Chemical and Biomedical Engineering, Cleveland State University, 455 Stilwell Hall, 2121 Euclid Avenue, Cleveland, OH 44115-2214, (3) Energy, UNESP (São Paulo State University), Av Ariberto Pereira da Cunha, 333, Guaratinguetá, 12514-410, Brazil

Keywords: Optimization, Cogeneration, Environmental impact, Sugar Industry.

Abstract

The development of new techniques that allow the analysis and optimization of energy systems bearing in mind environmental issues is indispensable in a world with finite natural resources and growing demand of energy.

Among the energy systems that deserve special attention, cogeneration in the sugar industry must be pointed out, because it uses efficiently a common fuel for generation of useful heat and power. Within this frame, thermoeconomical optimization - 2nd Law of Thermodynamics analysis by exergy function and economic evaluation of the thermal system - gradually is taking importance as a powerful tool to assist to the decision making process.

Also, the explicit consideration of environmental issues offers a better way to explore trade-offs between different aspects to support the decisions that must be made. In this work it is used the technique of Life Cycle Analysis (LCA) which allows to consider environmental matters as an integral part of the problem, in opposite to most of the environmental approaches that only reduce residuals generation , without taking into account impacts associated to other related processes.

On the other hand, the consideration of environmental issues in optimization of energy systems is a novel and promissory contribution in the state of the art of energy optimization and LCA.

The system under study is a sugar plant of Tucumán (Argentina) given the particular importance that this industry had inside the regional economy of the Argentinean Northwest.

Although cogeneration comes being used a while ago in sugar industry, being the main objective the generation of heat and as secondary objective the electric power generation and mechanic power to cover several needs of working machineries, to the date it is no available a versatile tool that allows to analyze economical feasible alternatives bearing in mind environmental issues.

At sugar plants, steam is generated in boilers using as fuel bagasse - cellulosic fiber waste obtained crushing the sugar cane- and it is used to give useful heat and shaft work to the plant, but it can also be used to generate electricity with export opportunities to the electrical network. The great number of process alternatives outlines a serious decision making problem in order to take advantage of the resources.

Although the problem turns out to be a mixed non-linear problem (MINLP), the main contribution of this work is the development of a hybrid strategy to evaluate cogeneration alternatives that combines optimization approaches with environmental indicators.

This powerful tool for its versatility and robustness to analyze cogeneration systems, will be of great help in the decision making process, because of their easy implementation to analyze the kind of problems presented in the sugar industry.

577u Industrial Biosystems Engineering: Application of Systems Engineering In Biorefinery Systems

Shulin Chen, Department of Biological Systems Engineering, Washington State University, 258 LJ Smith Hall, Pullman, WA 99164

The concept of "Industrial Biosystems Engineering (IBE)" was proposed by the author as a framework of scientific knowledge and methodology of a new branch of engineering serving the needs for the upcoming bioeconomy that requires a biorefinery and biomanufaturing of commodity fuel and chemicals at industrial production scale. The key disciplinary support for this new engineering branch includes basic systems biology, bioprocess engineering, and systems engineering and systems analysis. This presentation discusses the relationship between the IBsE and existing disciplines and provides examples of the IBE application to biofuel and biochemical production. Developing such a framework offers a great opportunity for applying systems engineering theory and approach to biorefinery analysis and design for supporting the development of the bioeconomy.

577v Model-Based Fault Detection and Monitoring of Uncertain Hybrid Process Systems

Nael H. El-Farra, Department of Chemical Engineering & Materials Science, University of California, Davis, One Shields Avenue, Davis, CA 95616

The problems of on-line fault detection and handling in dynamic process systems have been the subject of considerable research interest over the past few decades in both the academic and industrial circles. Traditionally, most of the research work in this area has focused on the monitoring of purely continuous dynamic processes described by systems of ordinary differential equations using data-based or model-based approaches. Yet, there are many examples in the chemical process industries where the dynamical properties of the process depend rather on an intricate interaction between discrete and continuous variables, and are more appropriately modeled by intervals of piecewise continuous dynamics interspersed by discrete transitions. Process systems wherein the overall behavior is shaped by the interactions of continuous dynamics with discrete events are referred to as hybrid systems, and are characterized by a switched multi-modal structure in which discrete events trigger autonomous or controlled transitions between a collection of continuous dynamical subsystems or modes. The continuous dynamics in each mode typically arise from the underlying physical laws such as mass, momentum, and energy conservation, and are typically modeled by continuous-time differential equations. The discrete events, on the other hand, can be the result of inherent physico-chemical discontinuities in the continuous dynamics, transitions between different operating regimes, the use of discrete actuators and sensors in the control system, or the use of logic-based switching for supervisory and safety control.

Over the past two decades, the challenges posed by the combined discrete-continuous interactions in hybrid systems, together with the abundance of practical applications where such interactions arise, have motivated significant research work on the design of supervisory and control schemes for several classes of hybrid systems (see, for example, [1], [2], [3]). More recently, the fusion of hybrid system tools with advances in nonlinear process control has led to the formulation and solution of several practical control problems for nonlinear hybrid systems [4]. Compared with the efforts on control of hybrid systems, the problems of fault detection and monitoring have received limited attention. Existing results in this direction have focused mainly on the related problems of state estimation and observer design for switched linear systems (see, for example, [5]). For the majority of chemical processes, however, the underlying continuous dynamics are inherently nonlinear and cannot be monitored effectively using linear approaches. The lack of systematic methods for model-based fault detection in nonlinear hybrid process systems can significantly limit the achievable process performance and control quality in those processes.

One of the key issues in the design of model-based fault detection schemes for hybrid systems is handling the switched dynamics which requires the design of a hybrid observer that reconstructs both the continuous and discrete variables and ensures residual convergence under switching. Another key consideration is the ability of the fault detection scheme to discriminate effectively between process and/ or control system faults on the one hand, and the discrete transitions that take place between the continuous modes on the other. Failure to distinguish between faults and mode transitions can lead to false or missed alarms and subsequent instability or deterioration in the overall process performance. The monitoring problem for a hybrid system is further complicated by the presence of uncertainty in both the continuous dynamics and the discrete events governing the transitions between them. Uncertainty in the continuous dynamics arises typically due to the presence of unknown or partially known process parameters as well as time-varying exogenous disturbances which if not properly accounted for can adversely affect the implementation of the monitoring and control systems. Uncertainty in the mode transitions, on the other hand, stems from the lack of a priori knowledge of either the timing or the sequence of transitions between the constituent modes. For example, in processes undergoing autonomous transitions due to intrinsic physico-chemical discontinuities (e.g., phase changes, flow reversals and transitions, saturation of control valves), the timing and sequence of switches are determined by the evolution of the process and cannot be determined beforehand. An effective hybrid monitoring scheme must therefore be robust with respect to model uncertainty and have the ability to decouple the effects of faults from mode transitions from disturbances.

Motivated by these considerations, we focus in this contribution on the problem of fault detection in a class of hybrid process systems modeled by switched nonlinear systems with uncertain continuous dynamics and uncertain mode transitions. A robust hybrid monitoring scheme that discriminates between faults, mode transitions and uncertainty is developed using tools from unknown input observer theory and results from switched system stability. The monitoring scheme consists of (1) a family of robust fault detection filters that detect the faults within the continuous modes, (2) a family of mode observers that locate the active operating mode at any given time, and (3) a supervisor that switches synchronously between the fault detection filters as the process transitions from one mode to another.

Each fault detection filter in this architecture relies on the available process measurements to estimate the fault-free behavior of a given mode and uses the discrepancy from the actual behavior as a residual signal. The filters are designed using the unknown input observer principle to ensure that the estimation errors, which are used as residual signals, are decoupled from the model uncertainty and disturbances. Conditions for the convergence of the fault-free residuals under switching are also derived to ensure that the residuals are insensitive to the mode transitions. In this manner, the residual of each filter becomes sensitive only to the faults within a given mode. The ability of the switched fault detection scheme to detect faults depends on the ability of the supervisor to correctly identify which mode is active at any given time to ensure that the corresponding filter is activated. For this purpose, a set of robust state observers that recreate the expected dynamic behavior of each mode are constructed. The observers are designed to ensure that the observation errors, which are used as residual signals, are decoupled from the model uncertainty and disturbances, as well as the faults within each mode, and are thus sensitive only to the mode location. By running the observers in parallel with the process, a unique pattern of residuals is obtained where at any given time only the observer estimating the active mode's states returns a converging residual, while the rest do not. Once a mode transition is detected and the active mode is identified, the supervisor switches to the corresponding fault detection filter to monitor the active mode. Finally, the design and implementation of the monitoring scheme are demonstrated using a chemical process example.

References:

[1] Yamalidou, E.C. and J. Kantor, ``Modeling and optimal control of discrete-event chemical processes using Petri nets," Comp. Chem. Eng., 15:503-519, 1990.

[2] Bemporad, A. and M. Morari, ``Control of systems integrating logic, dynamics and constraints," Automatica, 35:407-427, 1999.

[3] Engell, S., S. Kowalewski, C. Schulz and O. Stursberg, ``Continuous-Discrete Interactions in Chemical Processing Plants," Proc. IEEE, 88:1050-1068, 2000.

[4] Christofides, P. D. and N. H. El-Farra. Control of Nonlinear and Hybrid Process Systems: Designs for Uncertainty, Constraints and Time-Delays, Springer-Verlag, Berlin, Germany, 2005.

[5] Alessandri, A., M. Baglietto, G. Battistelli, ``Luenberger observers for switching discrete-time linear systems," Int. J. Contr., 80:1931-1943, 2007.

578a Operation and/or Design of Sustainable Processes Using Multi-Objective Optimization

Elaine S. Q. Lee and G. P. Rangaiah. Chemical and Biomolecular Engineering, National University of Singapore, Faculty of Engineering, Block E5, #B-04, 4 Engineering Drive 4, Singapore 117576, Singapore, Singapore

Almost every article advocating sustainability (e.g. Azapagic and Perdan, 2000; Sikdar, 2003a and 2003b; Heinzle et al., 2006; Darton, 2007) provides this quote: "Sustainable development is development that meets the needs of the present, without compromising the ability of future generations to meet their own needs" from the report of the World Commission on Environment and Development (1987). It is not possible to journey towards sustainability if it only depended on environmentalists and/or the government; it requires the awareness of every individual and their efforts to realize this laudable objective. Chemical process industries (CPI) are known to process and produce environmentally unfriendly compounds (e.g. benzene, chlorofluorocarbons, SOx, particulate matter). While chemical engineers are implementing pollution prevention solutions, they can also learn to further optimize their processes by modifying operating or design variables, making the processes more benign to the environment.

There are three spheres of sustainability: economic development, environmental stewardship and societal equity (Azapagic and Perdan, 2000; Sikdar, 2003a and 2003b; Heinzle et al., 2006). This is often touted as the "triple bottom line". Economic indicators measure the profitability of a chemical process. These usually are the first criteria for companies; if the process is not economically feasible, the project would be aborted. Environmental indicators are concerned with efficient use of raw materials and energy in the process as well as the environmental impacts caused by emissions. The latter include human toxicology, ecotoxicity, global warming, ozone depletion, acidification, eutrophication and photochemical oxidation (Young and Cabezas, 1999; Jolliet et al., 2003; Bare et al., 2006). Societal indicators measure different aspects of working conditions and regulations. As these indicators are based on company and governmental policies, it is not appropriate to include them in the analysis and optimization of a chemical process.

In the literature, studies which considered process optimization for sustainability can be divided into two classes. The first comprises of studies where the various aspects of sustainability have been aggregated into a single indicator for single objective optimization (e.g., Dantus and High, 1999; Chen et al., 2002; Ramzan and Witt, 2006). The use of weighted method or analytical hierarchy process (AHP) to aggregate the indicators has their drawbacks; the most important drawback is its disability to find the non-convex portion of the Pareto optimal solutions (Deb, 2001). This calls for the use of multi-objective optimization. The second class comprises of studies that have performed multi-objective optimization on chemical processes (e.g., Lim et al., 1999; Kheawhom and Hirao, 2002; Hoffmann et al., 2004; Hugo and Pistikpoulos, 2005; Flores et al., 2007). These studies, however, have either chosen one environment impact category or a combined environmental index, and then optimized it along with an economic indicator (i.e. bi-objective optimization). There are many contributing factors for the environmental aspect as mentioned above. Minimizing the weighted sum of the various factors does not necessarily minimize the contribution of each individual factor. Hence, it would be more appropriate to categorize various impacts factors into groups and minimize the impact groups individually. An example of impact factor categorization is where 'human toxicity by ingestion' and 'human toxicity by inhalation' are grouped as 'impact on humans'. This leads to a number of environmental impact groups. Optimization can be carried out with some/all environmental impact groups and with economic objectives simultaneously. Hence, for the first time, multi-objective optimization would be employed to optimize a basket of environmental and economic indicators, and not just two.

Two case studies have been chosen for the present study; they are a VOC (volatile organic component) recovery system (Chen et al., 2003) and a solvent recovery system (Chakraborty and Linninger, 2002). Six groups were identified for environmental impacts: (1) impact on humans, (2) impact on ecosystem – terrestrial and aquatic, (3) impact on local/global temperatures – global warming and ozone depletion, (4) photochemical oxidation, (5) acid rain, and (6) eutrophication. In addition, economic aspects would be considered in the optimization. Potential economic criteria are profit before taxes (PBT), payback period (PBP) and net present worth (NPW). Some or all these objectives will be optimized simultaneously using a multi-objective optimization tool, namely, non-dominated sorting genetic algorithm (NSGA-II), which was successfully used for many chemical engineering applications (Masuduzzaman and Rangaiah, 2008). As the objectives may be partially or totally conflicting, Pareto optimal solutions will be obtained. This will elucidate the trade-offs present and the decision maker would be better equipped in choosing the best solution. Net flow method and/or rough set method could be used as a tool in identifying the best Pareto optimal solution, whereby the decision maker's preference has to be declared (Thibault, 2008). While the choice of a solution may be subjective, the generation of the Pareto optimal solutions provides a quantitative foundation in reducing the biasness of the decision maker (Deb, 2001). Pareto optimal solutions and the best Pareto optimal solution for the two case studies will be presented and discussed. Insights gained from considering several objectives instead of just two will be highlighted.

References

Azapagic, A. and Perdan, S. (2000). Indicators of Sustainable Development for Industry: A General Framework, Process Safety and Environmental Protection: Transactions of the Institution of Chemical Engineers, 78, Part B, pp. 243-261.

Bare, J., Gloria, T. and Norris, G. (2006). Development of the Method and U.S. Normalization Database for Life Cycle Impact Assessment and Sustainability Metrics, Environmental Science & Technology, 40, pp. 5108-5115.

Chakraborty, A. and Linninger, A. A. (2002). Plant-Wide Waste Management. 1. Synthesis and Multiobjective Design, Industrial and Engineering Chemistry Research, 41, pp. 4591-4604.

Chen, H., Wen, Y., Waters, M. D. and Shonnard, D. R. (2002). Design Guidance for Chemical Processes Using Environmental and Economic Assessments, Industrial and Engineering Chemistry Research, 41, pp. 4503-4513.

Chen, H., Rogers, T. N., Barna, B. A. and Shonnard, D. R. (2003). Automating Hierarchical Environmentally-Conscious Design Using Integrated Software: VOC Recovery Case Study, Environmental Progress, Vol. 22, No. 3, pp. 147-160.

Dantus, M. M. and High, K. A. (1999). Evaluation of Waste Minimization Alternatives under Uncertainty: a Multiobjective Optimization Approach, Computers and Chemical Engineering, 23, pp. 1493-1508.

Darton, R. (2006). Sustainable Development – A Particular Challenge for Engineers, 11th APCChE, Kuala Lumpur, August 27-30, 2006.

Deb, K. (2001). Multi-objective Optimization using Evolutionary Algorithms, John Wiley & Sons, New York.

Flores, X., Rodriguez-Roda, I. and Poch, M. (2007). Systematic Procedure to Handle Critical Decisions during the Conceptual Design of Activated Sludge Plants, Industrial and Engineering Chemistry Research, 46, pp. 5600-5613.

Heinzle, E., Biwer, A. and Cooney, C. (2006). Development of Sustainable Bioprocesses: Modeling and Assessment, John Wiley & Sons Ltd, England.

Hoffmann, V. H., McRae, G. J. and Hungerbühler, K. (2004). Methodology for Early-Stage Technology Assessment and Decision Making under Uncertainty: Application to the Selection of Chemical Processes, Industrial and Engineering Chemistry Research, 43, pp. 4337-4349.

Hugo, A. and Pistikopoulos, E. N. (2005). Environmentally Conscious Long-range Planning and Design of Supply Chain Networks, Journal of Cleaner Production, 13, pp. 1471-1491.

Jolliet, O., Margni, M., Charles, R., Humbert, S., Payet, J., Rebitzer, G. and Rosenbaum, R. (2003). IMPACT 2002+: A New Life Cycle Impact Assessment Methodology, The International Journal of Life Cycle Assessment, Vol. 8, No. 6, pp. 324-330.

Kheawhom, S. and Hirao, M. (2002). Decision Support Tools for Process Design and Selection, Computers and Chemical Engineering, 26, pp. 747-755.

Lim, Y. I., Floquet, P. And Joulia, X. (1999). Multiobjective Optimization in Terms of Economics and Potential Environment Impact for Process Design and Analysis in a Chemical Process Simulator, Industrial and Engineering Chemistry Research, 38, pp. 4729-4741.

Masuduzzaman and Rangaiah, G.P. (2008). Multi-objective Optimization Applications in Chemical Engineering. In Rangaiah, G. P. (Ed.), Multi-objective Optimization: Techniques and Applications in Chemical Engineering, World Scientific, Singapore, in press.

Ramzan, N. and Witt, W. (2006). Multi-objective Optimization in Distillation Unit: A Case Study, The Canadian Journal of Chemical Engineering, 84, pp. 604-613.

Sikdar, S. K. (2003a). Sustainable Development and Sustainability Metrics, AIChE Journal, Vol. 49, No. 8, pp. 1928-1932.

Sikdar, S. K. (2003b). Journey Towards Sustainable Development: A Role for Chemical Engineers, Environmental Progress, Vol. 22, No. 4, pp. 227-232.

Thibault, J. (2008). Net Flow and Rough Sets: Two Methods for Ranking the Pareto Domain. In Rangaiah, G. P. (Ed.), Multi-objective Optimization: Techniques and Applications in Chemical Engineering, World Scientific, Singapore, in press.

World Commission on Environment and Development (1987). Our Common Future, Oxford University Press, Oxford.

Young, D. M. and Cabezas, H. (1999). Designing Sustainable Processes with Simulation: The Waste Reduction (WAR) Algorithm, Computers and Chemical Engineering, 23, pp. 1477-1491.

578b A C.F.D.-Guided Mixed Integer Nonlinear Programming (MINLP) Model for Conceptual Design of a Carbothermic Aluminium Reactor

Dimitrios I. Gerogiorgis, Department of Chemical Engineering, Massachusetts Institute of Technology (M.I.T.), 77 Massachusetts Avenue, Cambridge, MA 02139

Mixed Integer Nonlinear Programming (MINLP) theory and applications have pervaded the field of computer-aided design and optimization of many traditional, well-studied chemical processes. Exemplary processes studied include many common petrochemical unit operations, such as multicomponent, azeotropic, reactive and isotope distillation, heat and mass exchanger networks, reactor networks, solvent and supercritical fluid extraction, and membrane separation systems. Petrochemical industries have certainly benefited a lot from MINLP problem formulations. Metallurgical industries, however, are historically characterized by a high degree of complexity (high-temperature reactions, multiphase flow) resulting in lack of accurate and detailed models. Advances can provide design recommendations for cost savings and profitability improvements.

Carbothermic reduction of alumina for the production of pure aluminium is a high-yield, economically favorable and environmentally benign process with identified industrial potential. This method is identified as a feasible alternative to the prevalent electrochemical reduction by feasibility studies, but its complexity poses remarkable technical obstacles to implementation. The conceptual multistage electrothermic reactor recently proposed (Johansen, Aune et al., 2000) is a quite attractive idea for achieving reactor scaleup but entails significant design challenges. The most interesting one is the distributed nature of this process, since the electrothermic heating necessary for this endothermic reaction is achieved by using independent AC electrode pairs. Both placement as well as electrode voltage tuning are nontrivial problems given the complexity: optimal design should maximize productivity, minimize superheating and avoid short-circuiting.

The paper elaborates on a previous coarse MINLP model for carbothermic aluminium production we presented in the last ESCAPE-13 at Lappeenranta, Finland (Gerogiorgis and Ydstie, 2003). The present study employs a suitably simplified finite volume model of the carbothermic reactor: a detailed two-dimensional structured discretization of the domain is considered and a steady state CSTR reactor model is used at each of the resulting rectangular finite volumes to probe accurately mass, heat and component balances using temperature-dependent physical properties. The CSTR connectivity is decided by analyzing the slag flow calculated by a 2D CFD model. The presence of electrode pairs (inert DC current conductors) is modeled using binary variables, with piecewise constant voltage and field intensity profiles considered along the horizontal axis. Detailed thermodynamic equilibrium calculations are performed as part of the MINLP model in order to reliably elucidate chemical species' concentrations within each of the finite volumes. The goal is to (a) perform electrode placing and voltage optimization under balance constraints in order to achieve maximization of liquid Al product and minimization of inevitable gas losses, (b) conduct sensitivity analyses illustrating the effect of possible designs on reactor productivity.

578c Effectiveness of Coatings to Reduce Fouling In Shell and Tube Heat Exchangers

Gabriel T. Fernandez and Talea A. Rothan. Process & Control Systems Department, Saudi Aramco, 12921, Dhahran 31311, Saudi Arabia

This paper studies the effectiveness of applying a coating on the tube side to resist fouling in reboilers on crude oil stabilizers. In this process, crude flows in the tubes of vertical reboilers heated by 60 psig (300 °F) steam to stabilize crude oil to meet the True Vapor Pressure (TVP) and H_2S specifications. Due to the high temperature, reboilers are subject to fouling predominantly on the crude side, resulting in low heat transfer and shut down for cleaning. Based on vendor claims on the effectiveness of coatings to reduce fouling and increase the interval between cleanings, a study was undertaken to compare the operation of one set of reboilers coated on the crude side (inside) with a protective coating on Column A, with another set of similar reboilers and similar operating conditions with no coating, on the adjacent Column B. The study was conducted for an interval of about 11 months and the results of this comparative study are presented in this paper.

578d Cfd Predictions of Bod Removal In Wastewater Treatment Lagoons

Binxin Wu, R&D, Philadelphia Mixing Solutions LTD, 1221 East Main Street, Palmyra, PA 17078

An integrated model on mixing flow, heat transfer, and BOD removal in a large scale wastewater treatment lagoon is presented. The flow model will be developed with multiphase flow theory, in which the multiple mixers are treated as momentum sources, and the effect of wind speed on the water surface is included. The heat exchange between the lagoon walls and surrounding ground, weather conditions, and solar radiations will be included into the heat boundary conditions for the heat transfer model. The flow and temperature fields will be simulated by Fluent CFD (computational fluid dynamics) commercial software. RTD (residence time distribution) in the lagoon will be predicted by solving scalar transport equation based upon the converged flow fields. BOD (biochemical oxygen demand) removal in the lagoon will be developed on the basis of first-order kinetics, and simulation of BOD removal will be conducted through coupling temperature and RTD.

578e Heat Integrated Reactive Distillation Designs for Hydrolysis of Methyl Acetate

Hao-Yeh Lee, Hsiao-Ping Huang, and Yi-Chen Lee. Chemical Engineering, National Taiwan University, No. 1, Sec. 4, Roosevelt Road, Taipei, Taiwan

In this paper, two designs of heat-integrated reactive distillation for the hydrolysis of methyl acetate are studied. The first one employs internal heat-integrated reactive distillation column (r-HIDIC) and the other utilizes feed-split reactive distillation column (r-FS). The savings in energy consumption and total annual costs are focused on the reactive column only. The reference base case for comparison is the RD column in the work of Lin et al. (2007). In both cases, part of the reactive distillation utilizes a higher pressure. This is because a higher pressure makes heat transfers in the system possible. In fact, this higher pressure benefits the reaction taking place. Because, the hydrolysis reaction of methyl acetate is slightly endothermic, a higher operating pressure implies higher temperatures, faster reaction rates, and better equilibrium constants in the reaction part. The energy conservation studies show that HIDIC can reduce energy saving up to 27%, and FS had 38% saving when the process conditions are both optimized. Nevertheless, as far as economic aspect is concerned, HIDIC has no benefit, because of 33% more total annual cost (TAC) than the original RD column is required. This is due to an expensive gas compressor which takes 61.9% total capital cost and 31.3% utility cost. On the other hand, FS has a better economical efficiency with 6.4% TAC and 15.2% operating cost savings. FS has two smaller distillation columns but the capital cost is almost same as the original one. Raising the operating pressure in one part of the RD columns increases the temperature and the reaction rate which, in turn, results in smaller equipments and catalyst requirements.

Modeling and Simulation of a pilot-scale fixed bed reactor for dimethyl ether (DME) synthesis

Daesung Song[a], Dongil Shin[c], Wonjun Cho[b], Dal Keun Park[a] and En Sup Yoon[*a]

[a]School of Chemical Engineering, Seoul National University, San 56-1 Sillim-dong, Gwanak-gu, Seoul, Korea
[b]LNG Technology Research Center, KOGAS, Dongchun-dong, Yeonsu-gu, Incheon, Korea
[c]Dept. of Chemical Engineering, Myongji University Yongin, Kyunggido 449-728, Korea

ABSTRACT Dimethyl Ether (DME, CH_3OCH_3) is the simplest ether and is considered as one of the leading candidates in the quest for a substitute for petroleum-based fuels. In this work, we analyzed the one-step synthesis of DME in a shell and tube type fixed bed reactor and carried out a simulation with a one-dimensional, steady state model of a heterogeneous catalyst bed, while taking into consideration the heat and mass transfer between the catalyst pellets and reactants gas and the effectiveness factor of the catalysts, together with the reactor cooling through the reactor tube wall. The reactor simulation was carried out under steady state condition and we compared the simulation results with the experimental data obtained from operations of a pilot-scale reactor and found good agreement between them.

Keywords Reactor, Modelling, DME

1. INTRODUCTION

Dimethyl ether (DME), the simplest ether, is considered as a substitute fuel that could potentially replace petroleum-based fuels [1]. Its physical properties are similar to those of liquefied petroleum gas (LPG), so it can exploit the existing land-based and ocean-based LPG infrastructures with minor modifications. DME with its cetane number of 55 to 60 is also considered as a substitute for diesel fuel (cetane number 55)[2-5]. Compared with diesel fuel, the combustion of DME produced much less pollutants such as hydrocarbons, carbon monoxide, nitrogen oxides and particulates in CIDI engine tests [5, 6-9]. DME has been traditionally produced by the dehydration of methanol, which is produced from syngas, a product of natural gas reforming. This traditional process is thus called the two-step method of preparing DME. However, DME can also be prepared directly from syngas (single-step) [10-12]. The single-step method needs only one reactor for the synthesis of DME, instead of two for the two-step process. It can also alleviate the thermodynamic limitations associated with the synthesis of methanol, by converting the produced methanol into DME, thereby potentially enhancing the overall conversion of syngas into DME. In a fixed bed reactor, the main difficulty would be the prevention of the occurrence of hot spots, since the reactions involved in the synthesis of DME are highly exothermic. Catalysts can be irreversibly deactivated when they are exposed to certain temperatures. Therefore, it is necessary to understand and predict the behavior of the reactor under various conditions for the design and scale-up of the DME synthesis process. However, it is not feasible to gather all the data experimentally and, therefore, numerical simulations would be highly valuable in the development of such a process. In this study, we developed a mathematical model to simulate a pilot-plant scale shell and tube type DME reactor and analyzed the one-step synthesis of DME, and carried out the simulation of a pilot scale fixed bed reactor with a one-dimensional heterogeneous reactor model under steady state conditions, while considering the heat and mass transfer between the catalyst pellets and reactant gas and the effectiveness factor of the catalysts, together with the cooling of the reactor through the reactor wall. We used the hybrid catalyst which is, according to our definition, catalyst pellets prepared from a mixture of a fine powdered commercial methanol synthesis catalyst, $CuO/ZnO/Al_2O_3$, and a fine powdered methanol dehydration catalyst, γ-alumina. Thus, the active sites for the synthesis of methanol and those for its dehydration are intermixed within the catalyst pellets. Using the simulator developed in this study, we compared the experimental data from operations of a pilot-scale DME reactor with the simulation results, such as the temperature profile, CO conversion, DME yield and out stream composition.

2. REACTOR MODEL DEVELOPMENT

2.1 Reaction kinetics

Preparation of DME from syngas can be represented by three catalytic reactions as shown below. [13]

$$CO_2 + 3H_2 \longleftrightarrow CH_3OH + H_2O \qquad (1a)$$

$$CO + H_2O \longleftrightarrow CO_2 + H_2 \qquad (1b)$$

$$2CH_3OH \longleftrightarrow CH_3OCH_3 + H_2O \qquad (1c)$$

Reaction (1a) corresponds to the synthesis of methanol from carbon dioxide and hydrogen. Reaction (1b) is the water gas shift reaction. These two reactions are catalyzed by the methanol synthesis catalyst ($CuO/ZnO/Al_2O_3$). Reaction (1c) is the methanol dehydration reaction, i.e. the DME synthesis reaction catalyzed by an acidic catalyst (γ-alumina). Combinations of these three reactions can explain the other schemes of preparation of DME from syngas. We also use the reaction rate equations as follows [14, 15].

$$r_{CO_2 hydrogenation} = \frac{k_1(p_{H_2}p_{co_2})[1-(1/K_{eqm1})(p_{CH_3OH}p_{H_2O})/(p_{CO_2}p_{H_2}^3)]}{(1+K_2(p_{H_2O}/p_{H_2})+\sqrt{K_3p_{H_2}}+K_4p_{H_2O})^3}$$

$$\Delta Hr = -49.47kJ/mol \qquad (2a)$$

$$r_{RWGS} = \frac{k_5p_{CO_2}[1-K_{eqm2}(p_{CO}p_{H_2O}/p_{CO_2}p_{H_2})]}{1+K_2(p_{H_2O}/p_{H_2})+\sqrt{K_3p_{H_2}}+K_4p_{H_2O}}$$

$$\Delta Hr = 41.17kJ/mol \qquad (2b)$$

$$r_{MeOHdehydration} = k_6K_{CH_3OH}^2[\frac{C_{CH_3OH}^2-(C_{H_2O}C_{DME}/K_{eqm3})}{(1+2\sqrt{K_{CH_3OH}C_{CH_3OH}}+K_{H_2O}C_{H_2O})^4}]$$

$$\Delta Hr = -21.003kJ/mol \qquad (2c)$$

These three reactions are all exothermic. Therefore, the handling of the reaction heat is very important for the control of the reactions. The values of the kinetic parameters in the kinetic expressions are summarized in Table 2.1. The equilibrium constant of each reaction is taken from the literature [16, 17].

Table 2.1. kinetic parameters

	A(i)	B(i)	parameter
k_1	1.65	36696	4846.93
K_2	3610	0	3610
K_3	0.37	17197	15.61
K_4	7.14×10^{-11}	124119	38.34
k_5	1.09×10^{10}	-94765	12.07
K_{ch3oh}	0.00079	70500	3633.13
k_6	3.7×10^{10}	-105000	4.417
K_{H20}	0.084	41100	643.376

$parameter = A(i)\exp(B(i)/RT)$

$$Log_{10}K_{eqm1} = \frac{3066}{T} - 10.592 \qquad (3a)$$

$$Log_{10}\frac{1}{K_{eqm2}} = -\frac{2073}{T} + 2.029 \qquad (3b)$$

$$Log_{10}K_{eqm3} = \frac{10194}{T} - 13.91 \qquad (3c)$$

2.2 Heat and Mass Transfer on the catalyst surface

Due to the highly exothermic nature of the reactions, the temperature of the catalyst pellets can differ from that of the bulk stream of the reactants. Likewise, there could be a difference in the concentration of the reactants between the bulk and catalyst pellets surface. The heat and mass transport rate correlations on the surface of the catalyst are summarized as shown below.

$$N_{nu} = \frac{h_p d_p}{k_g} = 2 + 0.6 N_{\mathrm{Re}\,sph}^{0.5} N_{pr}^{1/3} \qquad (4)$$

$$N_{\mathrm{Re}\,sph} = \frac{d_p v \rho}{\mu}$$

$$N_{pr} = \frac{C_p \mu}{k_g}$$

$$\sum_i r_i \Delta H_r g_{cat} = h_p A_p (T_{p1} - T_f) \qquad (5)$$

Since both diffusion and reaction steps are involved in the reactor, the relative rates of each step are important. When the reaction rate is relatively rapid compared to the rate of diffusion, the process is said to be diffusion-controlled. In contrast the process is said to be reaction-controlled when the rate of mass transfer is limited by the reaction step. Therefore, we investigated mass transfer on the catalyst and employed conventional expressions (6) for the diffusion rate of the diffusing components and mass transfer coefficient.

$$W_{Ar.i} = k_{ci}(C_{A0.i} - C_{As.i})$$

$$N_{Sh} = \frac{k_{ci} d_p}{D_{i,j}} = 1.17 N_{\mathrm{Re}}^{0.585} N_{Sc}^{1/3}$$

$$N_{Re} = \frac{d_p v \rho}{\mu} \qquad (6)$$

$$N_{Sc} = \frac{\mu}{\rho D_{i,j}}$$

Then, we decided whether the process is diffusion-controlled or reaction-controlled.

2.3 Effectiveness factor

The intrinsic reaction rate can differ from the global reaction rate, because of pore diffusion in the catalyst pellets when their diameter is of the order of several millimeters. We assume that the temperature of each catalyst pellet is uniform and equal to the temperature on its surface and consider only the mass balance within it. For spherical catalyst pellets the concentration of the reactants and products can be obtained by solving the following partial different equations.[21].

$$\frac{d^2 C_{H_2}}{dr^2} + \frac{2}{r}\frac{dC_{H_2}}{dr} = \frac{3\rho_{P_1}}{D_{eff.H_2}} r_{CO_2} + \frac{\rho_{P_1}}{D_{eff.H_2}} r_{RWGS} \qquad (7a)$$

$$\frac{d^2 C_{CO}}{dr^2} + \frac{2}{r}\frac{dC_{CO}}{dr} = -\frac{\rho_{P_1}}{D_{eff.CO}} r_{RWGS} \qquad (7b)$$

$$\frac{d^2 C_{CO_2}}{dr^2} + \frac{2}{r}\frac{dC_{CO_2}}{dr} = \frac{\rho_{P_1}}{D_{eff.CO_2}} r_{CO_2} + \frac{\rho_{P_1}}{D_{eff.CO_2}} r_{RWGS} \qquad (7c)$$

$$\frac{d^2 C_{H_2O}}{dr^2} + \frac{2}{r}\frac{dC_{H_2O}}{dr} =$$

$$-\frac{\rho_{P_1}}{D_{eff.H_2O}} r_{CO_2} - \frac{\rho_{P_1}}{D_{eff.H_2O}} r_{RWGS} - \frac{\rho_{P_2}}{2D_{eff.H_2O}} r_{MeOH} \qquad (7d)$$

$$\frac{d^2 C_{DME}}{dr^2} + \frac{2}{r}\frac{dC_{DME}}{dr} = -\frac{\rho_{P_2}}{2D_{eff.DME}} r_{MeOH} \qquad (7e)$$

$$\frac{d^2 C_{MeOH}}{dr^2} + \frac{2}{r}\frac{dC_{MeOH}}{dr} = -\frac{\rho_{P_1}}{D_{eff.MeOH}} r_{CO_2} + \frac{\rho_{P_2}}{D_{eff.MeOH}} r_{MeOH} \qquad (7f)$$

Boundary conditions

$$r = R, C = C_o$$

$$r = 0, \frac{dC}{dr} = 0$$

The effective diffusivity of the gases through the porous catalyst pellets was estimated from the molecular diffusivity, porosity and tortuosity of the pores using equation (8). The solutions of these equations provide the concentration profiles of the gas species in the catalyst pellets together with the global reaction rates.

$$D_{eff.i} = D_i \frac{\varepsilon}{\tau} \qquad (8)$$

$$\varepsilon = void\ fraction(porosity)$$

$$\tau = tortuosity$$

2.4 Heat transfer between tubes and shell

The heat generated from the chemical reactions in the catalyst is transferred to the gases flowing through the reactor tubes, and then to the cooling medium on the shell side of the reactor. Equation (9) holds for the whole reactor, as well as for any given section of it. It calculates the temperature rise of the reactants by subtracting the heat transferred to the coolant from the heat generated by the chemical reactions.

$$(\sum_i r_i \Delta H_r g_{cat} - UA_i(T_f - T_o)) = C_{p,fluid} m_f \Delta T_f \qquad (9)$$

In order to estimate the heat transfer coefficient (10a), U, we need to know both the shell side and tube side heat transfer coefficients. The common method of calculating the heat transfer coefficient can be employed for the estimation of the shell side heat transfer coefficient (10b), as heat transferring liquid or boiling water is used as the cooling medium in the shell [22].

$$U_o = \frac{1}{\frac{1}{h_i}(\frac{D_o}{D_i}) + \frac{x_w}{k_m}(\frac{D_o}{D_L}) + \frac{1}{h_o} + \frac{D_o}{D_i}\frac{1}{h_{di}} + \frac{1}{h_{do}}} \qquad (10a)$$

$$h_o = \frac{0.023 C_p v \rho}{N_{pr}^{2/3} N_{re}^{0.2}} \qquad (10b)$$

However, the presence of the catalyst pellets makes the estimation of the heat transfer coefficient on the tube side much more complicated. Although several correlations are available in the literature, their reliability should be experimentally verified. As we do not have experimental values for the tube side heat transfer coefficient, we selected the correlation (11) for the estimation of the tube side heat transfer coefficient [18].

$$h = (\frac{1}{\frac{1}{h_r + 2k_{ew}^0 / d_p + a_w C_{pg} \rho_g v} + \frac{1}{h_{packet}}}) \qquad (11)$$

$$h_{packet} = 1.13(\frac{k_e^0 \rho_p (1-\varepsilon_{mf}) C_{ps}}{\tau_t})^{0.5}$$

$$k_{ew}^0 = \varepsilon_w k_g + (1-\varepsilon_w) k_s (\frac{1}{\phi_w (k_s / k_g) + 1/3})$$

$$k_e^0 = \varepsilon_{mf} k_g + (1-\varepsilon_{mf}) k_s (\frac{1}{\phi_b (k_s / k_g) + 2/3})$$

$$a_w = 0.05, h_r(radiation) = 0$$

2.5 Simulation of the reactor

Equation (12) is used to calculate the mass flowrate profile in the tube of the reactor [23].

$$\frac{dF_i}{dz} = \rho_b \frac{\pi D_i^2}{4} \sum_{j=1}^{n_r} \gamma_{i,j} r_j \quad i = 1,2,..,n_c \quad (12)$$

In order to solve the above ODE, the axial length of the reactor tube is divided into 1000 sections. First, the reaction rates are calculated with the use of the correlations listed in section 2.1. For this calculation, the heat and mass transfer on the surface of the catalyst described in section 2.2 and the effectiveness factor described in section 2.3 are incorporated in order to reflect their effects on the reaction rates. Then, the correlations of section 2.4 are used for the calculation of the reactor temperature in each section. Iterative procedures for determining the temperature and reaction rates in each section are necessary, as the temperature and reaction rates are inter-dependent. After determining the temperature in the section of the tube, the mass balance in each section is calculated using Equation (12) and then the simulation is allowed to proceed to the next step.
We used the following correlations to evaluate the reactor performance, such as the CO conversion and DME Yield:

$$COconversion = \frac{COin - COout}{COin} \times 100 \quad (13a)$$

$$DMEyield = \frac{DMEout \times 2}{COin - COout} \times 100 \quad (13b)$$

3. RESULTS AND DISCUSSION

Using the simulator we developed, the behavior of a pilot-scale DME reactor was studied. The base condition of the reactor for the simulation is as follows: feed composition (yH2 = 0.653, yCO = 0.326, yCO2 = 0.003, yCH4 = 0.0025, yH2O = 0.0155, yCH3OCH3 = 0, yCH3OCH3 = 0), GHSV 2000 hr^{-1}, feed temperature 220 ℃, reactor pressure 50 bar, H_2 : CO ratio 2:1, inlet temperature of cooling water 210 ℃. The hybrid catalyst consists of the methanol synthesis and methanol dehydration components, whose ratio is 8:2. We compared the experimental data obtained from the pilot scale fixed bed reactor using the hybrid catalyst with the simulation results.

3.1 Comparison of experimental data and simulations result

The properties of the catalyst and the in formation concerning the pilot-plant scale reactor are shown in Table 3.1. These data were used for the simulation of the reactor.

Table3.1. Property of catalyst and the pilot-plant scale reactor

Catalyst	Density (kg/m_3)	1783.5
	Porosity (%)	45.53
	Pore tortuosity	1.69
	Mass (kg)	7.85
	Pellet diameter (m)	0.006
Reactor	Tube diameter (m)	0.03
	Length (m)	1.6
	Tube number	7

Fig. 3.1 shows the simulated temperature profile and the reactor temperature at 5 points (0.16 m, 0.608 m, 1.08 m). Although the simulated temperature profile has an error range of within ±8.8°C, this simulator can predict the temperature profile of the reactor well, thus helping to control the temperature of the reactor and preventing the irreversible deactivation of the catalysts.

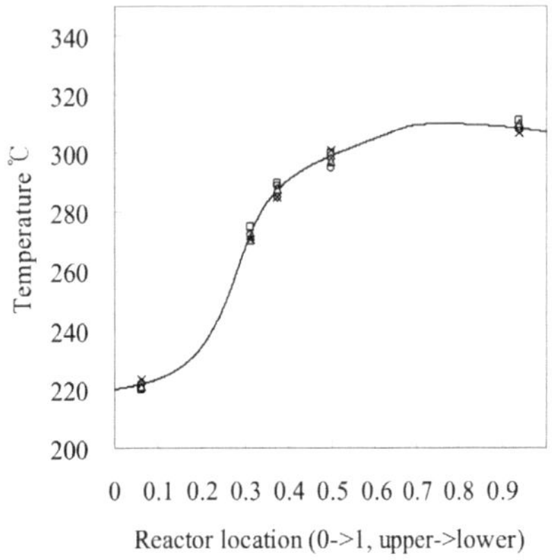

Fig. 3.1. Simulated temperature profile and the experimental data of the reactor filled with only hybrid catalyst

Table 3.2 shows the simulation result with corresponding pilot plant reactor data. As can be seen there is a good agreement between the simulations result and pilot plant reactor data.

Table3.2. Comparison of the simulation result and pilot plant reactor data

Property	Simulation result	Pilot plant result 1	Pilot plant result 2	Pilot plant result 3	Pilot plant result 4	Pilot plant result 5
CO conversion	39.85	39.4	38.93	37.92	37.39	36.27
DME yield	50.58	50.02	49.41	48.01	47.20	45.39
Exit temp.(℃)	307	312	309	308	308	309
Mole %						
yH_2	54.96	54.74	56	54	57	53
yCO	29.6	29.8	29.9	30.3	28.4	32.61
yCO_2	6.79	6.68	5.42	7.31	6.17	5.88
yCH_4	0.31	0.31	0.30	0.24	0.31	0.31
yH_2O	0.43	0.41	0.39	0.36	0.34	0.31
yCH_3OCH_3	4.97	4.78	4.71	4.44	4.29	3.98
yCH_3OH	2.94	3.02	3.12	3.33	3.45	3.71
Byproduct		0.26	0.16	0.02	0.04	0.2

4. CONCLUSION

In this work, we developed a simulator of a shell and tube type pilot-scale fixed bed reactor for the single-step synthesis of DME from syngas by applying the one-dimensional heterogeneous model under the steady state. We examined the behavior of the reactor in the hybrid catalyst. We found that complex reactions coupled with pore diffusion within the catalyst pellets can result in unusual values of the effectiveness factor. Therefore, the reactor showed higher performance. However, more careful cooling is needed due to reaction heat. We applied this model to a pilot-scale reactor and found that there was good agreement between the simulation result and experimental data. We were also able to predict the behavior of the reactor using this simulator. In this way, we were able to predict the temperature profile of the reactor and prevent the irreversible deactivation of the catalyst, thus allowing stable and optimized operation conditions to be obtained and the scale up of the DME synthesis reactor to be achieved.

ACKNOWLEDGEMENTS

The support provided for this study by KOGAS (Korea Gas Corporation) and BK21 (Brain Korea21) is gratefully acknowledged.

REFERENCES

1 Troy A. Semelsberger, Rodney L. Borup, Howard L. Greene., "Dimethyl ether (DME) as an alternative fuel" *J Power Sources* **156,** 497-511(2006).

2 Alam M, Fujita O, Ito K., "Performance of NOx reduction catalysts with simulated dimethyl ether diesel engine exhaust gas" *Proc. Inst. Mech. Eng. Part A: J Power Energy* **218**, 89-96(2004).

3 Sorenson SC., "Dimethyl ether in diesel engines: Progress and perspectives." *J Eng Gas Turbines Power-Trans ASME* **123**, 652-658(2001).

4 Rouhi AM. AMOCO, "HALDOR-TOPSOE Develop Dimethyl Ether As Alternative Diesel Fuel" *Chem Eng News* **73**, 37-39(1995).

5 Fleisch TH, Basu A, Gradassi MJ, Masin JG., "Dimethyl ether: A fuel for the 21st century" *Natural Gas Conversion IV Studies in Surface Science and Catalysis* **107**, 117-125(1997).

6 E.W. Kaiser, T.J. Wailington, M.D. Hurley, J. Platz, H.J. Curran, W.J. PITZ, C.K. Westerbrook., "Experimental and modelling study of premixed atmospheric-pressure dimethyl ether-air flames" *J Phys Chem A* **104**, 8194-8206(2000).

7 Song J, Huang Z, Qiao XQ, Wang WL. "Performance of a controllable premixed combustion engine fuelled with dimethyl ether" *Energy Convers Manage* **45**, 2223-2232(2004).

8 Wang HW, Zhou LB., "Performance of a direct injection diesel engine fuelled with a dimethyl ether/diesel blend" *Proc. Inst Mech EngPart D: J Automobile Eng* 217, 819-824(2003).

9 Zannis TC, Hountalas DT., "DI diesel engine performance and emissions from the oxygen enrichment of fuels with various aromatic content" *Energy Fuels* **18**, 659-666(2004).

10 Peng XD, Wang AW, Toseland BA, Tijm PJA., "Single-Step Syngas-to Dimethyl Ether Processes for Optimal Productivity Minimal Emissions, and Natural Gas-Derived Syngas" *Ind Eng Chem Res* 38, 4381-4288(1999).

11 Kunpeng Sun, Weiwei Lu, Fengyan Qiu, Shuwen Liu, Xianlun Xu., "Direct synthesis of DME over bifunctional catalyst: surface properties and catalytic performance" *Applied Catalysis A: General* **252**, 243-249(2003).

12 Kohji Omata, Yuhsuke Watanabe, Tetsuo Umegaki, Gunji Ishiguro, Muneyoshi Yamada., "Low-pressure DME synthesis with Cu-based hybrid catalysts using temperature-gradient reactor" *Fuel* **81**, 1605-1609(2002).

13 Ng KL, Chadwick D, Toseland BA., "Kinetics and modelling of dimethyl ether synthesis from synthesis gas" *Chem Eng Sci* **54**, 3587-3592(1999).

14 Vanden Busshe KM, Froment GF., "A steady-state kinetic model for methanol synthesis and the water gas shift reaction on a commercial $Cu/ZnO/Al_2O_3$ catalyst" *J Catalyst* **161**, 1-10(1996).

15 Gorazd Bercic, Janez Levec., "Intrinsic and Global reaction rate of methanol dehydration over γ-Al_2O_3 pellets" *Ind Eng Chem Res* **31**, 1035-1040(1992).

16 Martyn V., Twigg. Catalyst Handbook, 2nd ed, Wolfe, London (1986).

17 Daniel R. Stull, Edgar F. westrum. Jr, Gerard C. Sinke., The Chemical Thermodynamics of Organic Compounds, John Wiley , New York(1969).

18 Daizo Kunii, Octave Levenspiel., Fluidization Engineering, 2nd ed, Butterworth-Heinemann , Boston(1990).

19 James R. Welty, Charles E. Wicks, Robert E. Wilson., Fundamentals of momentum, heat and mass transfer, 3rd ed, Wiley , New York(1976).

20 Cengel., Heat transfer a practical approach, 2nd ed, McGrawHill , Boston(1998).

21 Charles N. Satterfield., Mass Transfer in Heterogeneous Catalysis, M.I.T. Press , Cambridge Mass(1970).

22 Warren L.McCabe, Julian C.Smith, Peter Harriott., Unit Operations of Chemical Engineering, 5th ed, McGrawHill , New York(1995).

23 Riggs, J.M., An introduction to numerical methods for chemical engineers, Texas Tech University Press , Lubbock(1994).

578g Energy Analysis of Chemical Batch Plant through Advanced Integration of Energy Conversion, Production Processes and Waste Management

Claude Rerat, Stavros Papadokonstantakis, and Konrad Hungerbühler. Institute for Chemical and Bioengineering, Swiss Federal Institute of Technology Zurich, Wolfgang-Pauli-Str.10, HCI G138, Zurich, 8093, Switzerland

This study presents the development of a model for a chemical batch plant which is used in the integration of energy demanding unit operations, energy conversion systems and waste management during process design. If energy conversion and waste management systems are only considered as "slaves to the production plant", this can lead to oversizing of installations or alternatively to problems with peak loads and generally to operating with low efficiency due to variations of operating conditions from initial design conditions. The latter scenario has a high importance in the case of a batch plant, where the product portfolio can change quickly leading to changes in utility requirements both quantitatively and qualitatively. In order to avoid the problems mentioned above and to look for a global optimum from financial and environmental point of view, an integrated design of the chemical production, the energy conversion and the waste management system must be achieved.

For this purpose, models for energy conversion systems and waste treatments based on real-life applications should be built to predict the energy consumption/production and the emissions as a function of process demands. Initially, a monoproduct plant has been chosen to represent the simplest case for the modeling of utility consumption (steam, brine, water, electricity). The Bottom-Up approach (Bieler et al 2004) was applied to calculate the energy consumption of a whole building with the help of process data and results were compared with the real consumption. Direct measurements of unit operations were performed to develop empirical parametric models of thermal losses and heat exchange efficiencies were calculated. These values will be useful during tasks scheduling in order to choose the most appropriate equipment for operation depending on their energy requirement.

Then, the energy profiles that are predicted by the models will serve as data for the design of a heat exchanger network specially built for batch processes in order to estimate the potential heat recoveries by applying direct heat exchange or by using an energy storage structure. The overall result has to be assessed with financial and environmental indicators in order for Pareto optimal alternatives to be identified. Only such an approach will ensure that an improvement of one part of the process is actually beneficial for the overall system and therefore it has to be undertaken.

Reference

1. Bieler PS, Fischer U, Hungerbuhler K. 2004. Modeling the Energy Consumption of Chemical Batch Plants: Bottom-Up Approach. Ind. Eng. Chem. Res. 43:7785-95

578h Superstructure-Based Nlp Formulation for Design of Batch Water Networks

Cheng-Liang Chen, Chia-Yuan Chang, Jui-Yuan Lee, Ying-Jyuan Ciou, and Chih-Yao Lin. Department of Chemical Engineering, National Taiwan University, #1, Sec 4, Roosevolt Rd., Taipei, 10617, Taiwan

The rapidly industrial development causes a considerable increase of water requirement in process plants. For that reason, water minimization has received more and more attention in the past decades and becomes a popular research topic. The efficiency of water utilization can be improved with a proper conservation strategy which not only reduces the operating cost but also lowers the environmental impact. Up to now, most of works on water minimization are concerned with continuous processes (Bagajewicz, 2000; Yoo et al., 2007), whereas the works for batch processes are rather less because of the fact that water usually constitutes a small part of the operating cost in batch operations. However, from the aspect of environmental protection and resource saving, water integration is still important and necessary to be considered in batch processes.

Batch processes are practically common in various industrial sectors, such as the production of food, pharmaceuticals and agrochemicals, etc. In this work, a general mathematical formulation is developed for the design of water-using network in batch plants, which is capable to deal with the cases of multiple contaminants and multiple storage tanks. For modelling the design problem, a set of superstructures are proposed to incorporate all possible connections in the batch water-reuse system, where a number of water-using tasks are proceeded and a number of storage tanks are placed for temporary storage to enhance the opportunities of water reuse/recycle. Besides the exploration of ultimate water recovery, the forbidden matches between specific water-using tasks in terms of water reuse are further taken into consideration as the practical restraint and an effective technique is presented to meet this special request. Representative examples from literature (Majozi 2005a, 2005b, 2006) are provided to demonstrate the adequacy of presented formulation.

References

1. Bagajewicz, M. A Review of Recent Design Procedures for Water Networks in Refineries and Process Plants. Comput. Chem. Eng. 2000, 24, 2093-2113.

2. Yoo C.K., Lee T.Y., Kim J., Moon I., Jung J.H., Han C., Oh J.M. and Lee I.B. Integrated Water Resource Management through Water Reuse Network Design for Clean Production Technology: State of the Art. Korean J. Chem. Eng. 2007, 24, 567-576.

3. Majozi, T. Wastewater Minimisation Using Central Reusable Water Storage in Batch Plants. Comput. Chem. Eng. 2005a, 29, 1631-1646.

4. Majozi, T. An Effective Technique for Wastewater Minimisation in Batch Processes. J. Clean. Prod. 2005b, 13, 1374-1380.

5. Majozi, T. Storage Design for Maximum Wastewater Reuse in Multipurpose Batch Plants. Ind. Eng. Chem. Res. 2006, 45, 5936-5943.

WAYS OF INCREASING TO EFFICIENCY OF THE REFINERIES

Nickolay V. Kuzichkin, Nickolay V. Lisitsyn, Vladimir I. Fedorov

Rapidly growing demand and high oil prices force oil companies to search potential efficiency improvements of the refineries. Global Shell Solution technical audit has shown possibility of increasing to efficiency of the refineries without essential investment. How can companies define priority challenges, allowing perfecting technical-economic factors of the production?

The criteria (safety factors of quality, heat recovery, engineering standards deviation and others) to allow defining the bottleneck problems in sphere of management and technical usage of the plant are examined in this paper. The analysis of the plant was organized by means of criterion. The improved processing characteristics of the reformer, CDU, xylene plant and blending of gasoline and diesel oil are shown.

The heat recovery reduction results [1,2] to account of the partial reconstruction of the heat exchanger networks of CDU and the reduction quality reserve in the diesel and gasoline blending were presented. The technological process rate analysis is considered here.

*Case study 1 (well-*founded reduction of the quality norms of the intermediate products).

In existing scheme of the process (Fig. 1) summary xelenes get from catalysate fractions 105-127°C on distillation block, herewith-heavy aromatic hydrocarbons are separated in column K-6. Hereinafter ortho-xelene is separated by the method clear rectification, para-xelene is separated by the cryogenic crystallization while meta-xelene and ethyl benzene enters on isomerization for the reason increase fission yield. At the same times a result side reaction of disproportionation and trans-alkylation is newly formed heavy aromatic hydrocarbons. It is separated in column T-103.

The power-consuming is increased in this unit because of double separation of heavy aromatic hydrocarbons in column K-6 and T-103, and what is more process is conducted under raised reflux ratio since contents target component C8 in column distillate is specified (99.3 and 99.2% masses).

Results of executed modeling have shown that column K-6 can be excluded from scheme without reduction quality to marketable products; herewith heat load will fall on 20 GJ/hour.

The experiment was given for acknowledgement of these conclusions

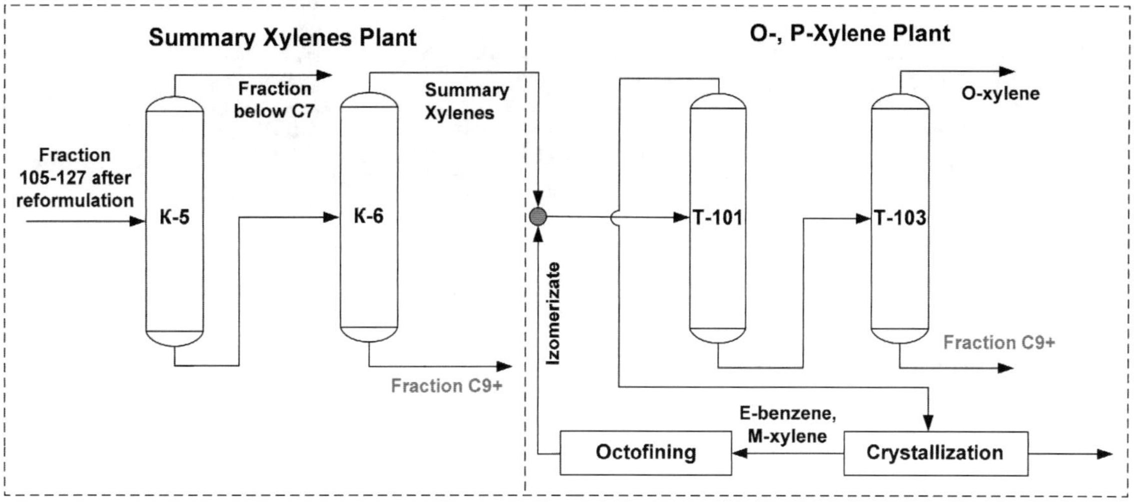

Fig 1. The goods xelenes separation.

On the first stage warm-up profile column K-6 was changed so (Fig. 2) to enlarge the an aromatic hydrocarbons concentration C9+ in distillate before 2% masses. Since quality to marketable products did not grow worse (Fig. 3), the absence of the column K-6 directly is imitated on second stage by means of mixing its selections in reservoir.

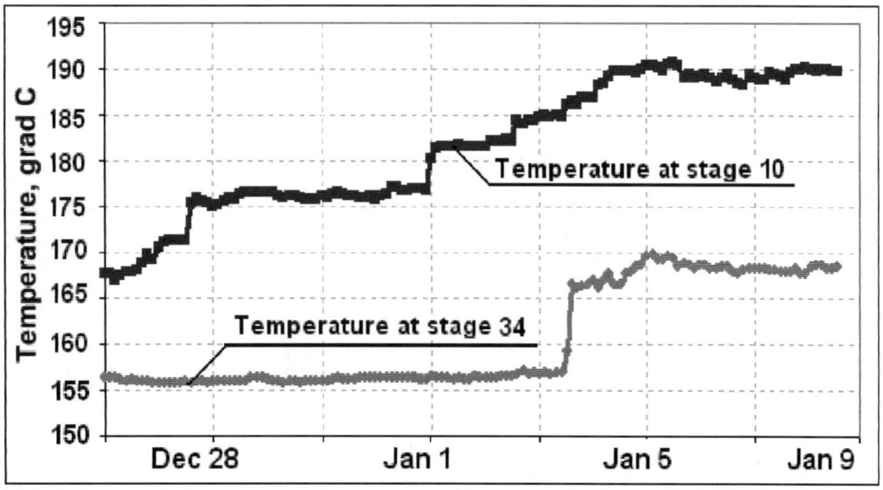

Fig.2 The column K-6 warm-up profile .

The results of the laboratory checking have shown (Fig. 3) that quality to product herewith did not grow worse. Thereby, the experiment has confirmed that column K-6 reasonable to exclude from scheme (the economic effect is approximately $1.000.000/year).

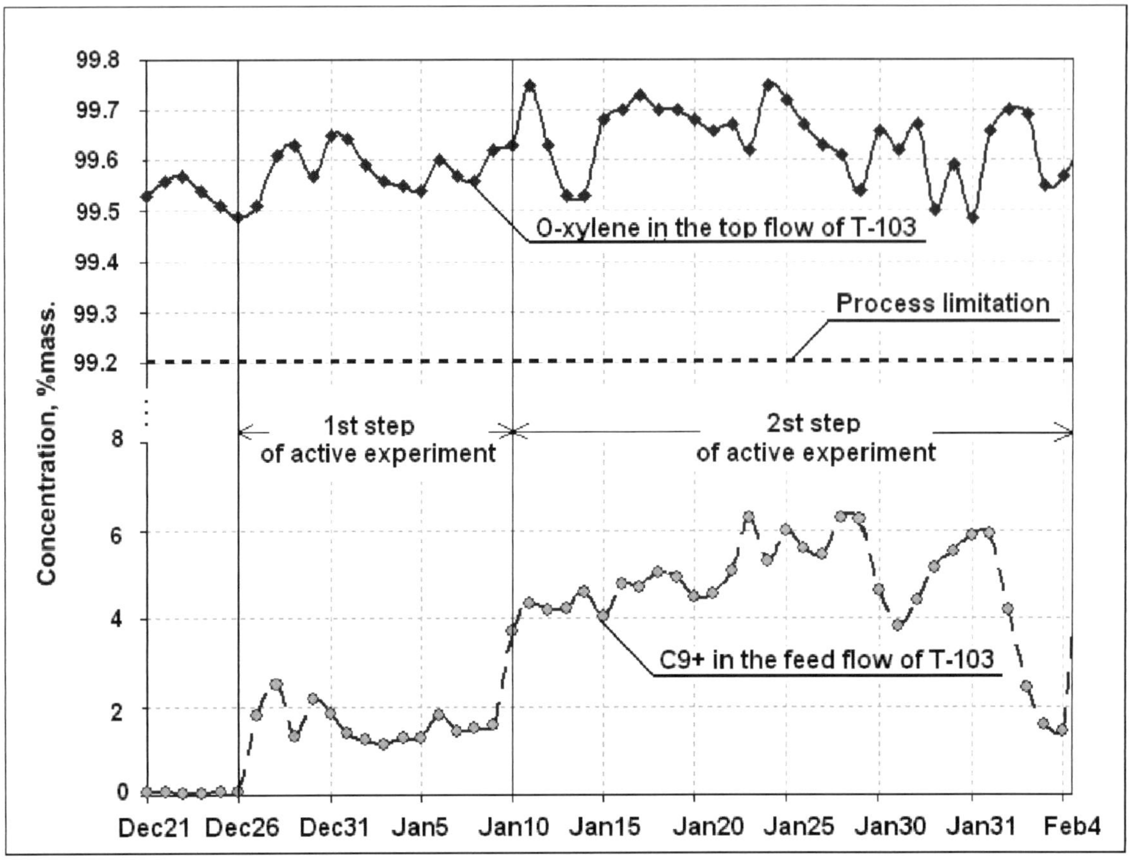

Fig. 3. The experiment results.

Case study 2 (preservation of quality to product at the excess to design loading)

In benzene production scheme (Fig. 4) heat is going from the reactor block in stabilizations in fixed amount. Presenting additional cool feedstock in the reactor block has brought about reduction available heat resource, but loading the stabilization block increased.

This has brought about reduction reflux ratio of the column K-4 and are growing of the contents hydrocarbon C6 in reflux with 5% before 23% masses.

Modeling of the process has shown the regulation quality reflux possible reach to account of the partial redistribution heat resource is between columns K-3 and K-4.

Called on industrial check (Fig. 5) has confirmed relevancy of the recommendation. The Block to stabilizations translated in new mode, economic effect is $1.000.000/year

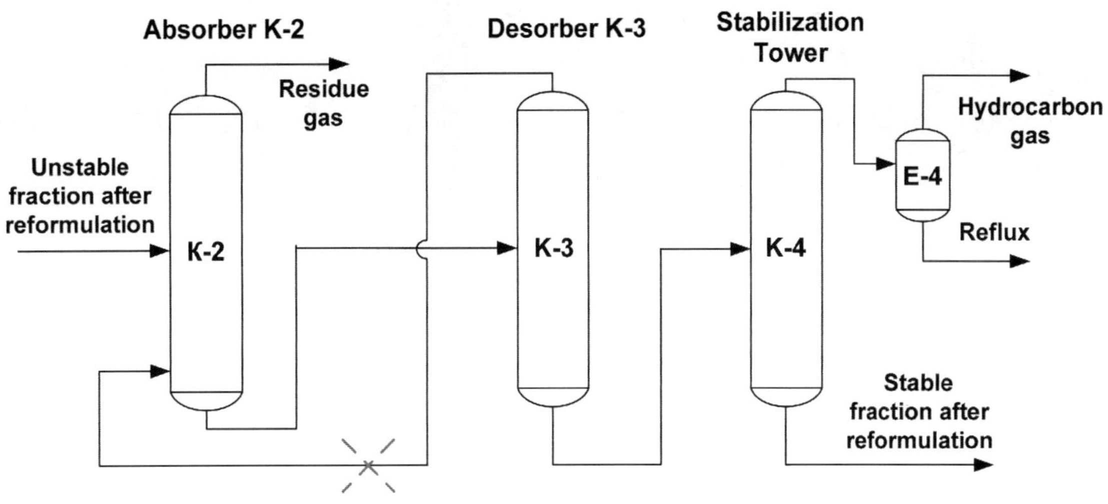

Fig 4. Principle schemes of the block to stabilizations of the benzene reformer

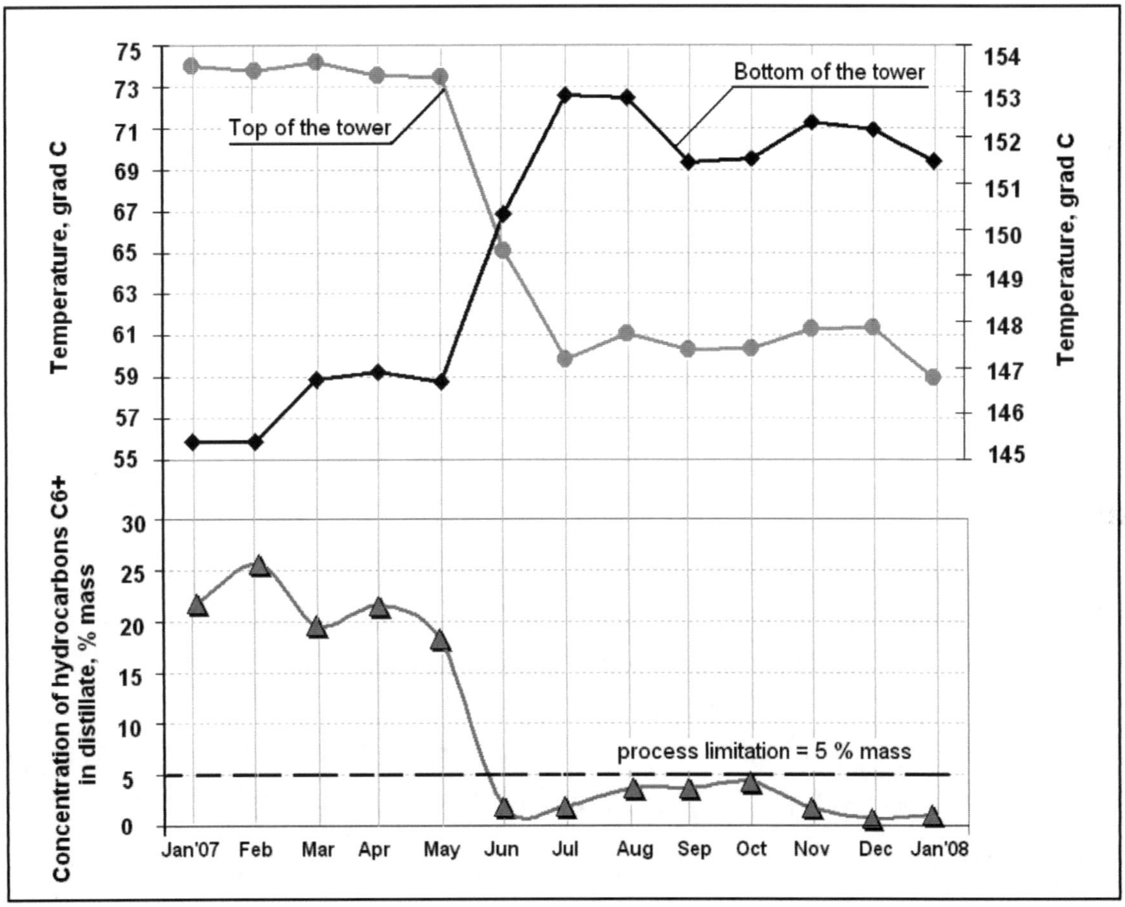

Fig. 5 Changes of the mode of the block to stabilizations of the benzene reformer

Case study 3 (change the mode for production of the product to more high category)

The block secondary rectification of the benzene reformer (Fig. 6) is intended for separation of the benzene and toluene of the high purity.

The process is complicated from ratio "benzene: toluene" in raw material is changed over a wide range (Fig 7).

The goods benzene satisfies the Russian standard a quality with heavy stocks, but does not correspond to the European rate.

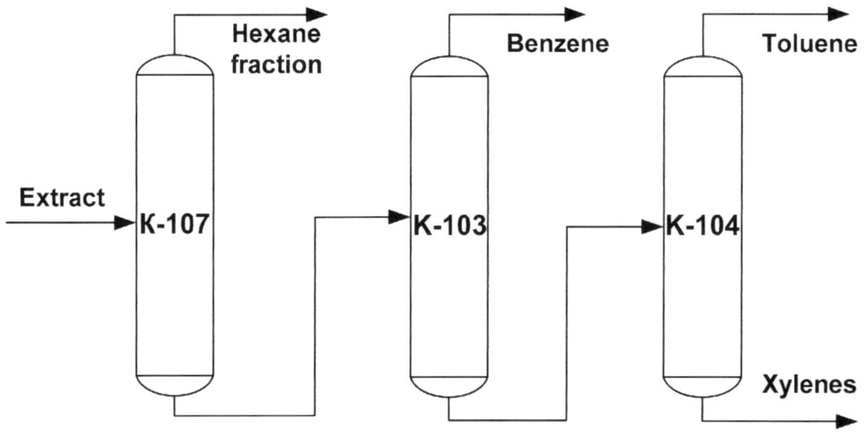

Fig. 6. The block secondary rectification of the benzene reformer

The modeling of the process has shown that possible to provide firm functioning the block in broad range of the change the composition feedstock with reception more clean benzene. The active experiment on industrial installation has confirmed these results (Fig. 7).

Thereby, the correspondence to of the goods benzene specification europen company can be provided when change the technological mode. The potential economic effect is \$1.500.000/year.

Case study 4 (the *well*-founded reduction of spare quality in products)

The production of butane and i-butane are realized on gas fractionation plant in the column with clear rectification. In existing mode products has a significant spare on contents of the admixtures (Fig. 8).

The modeling has shown that under relatively soft change the mode possible to reduce the consumption of the heat on 5-10% (depending on composition of the feeding), having saved fair supplies on quality of the marketable products (Fig 8). The profit is \$130.000 when change the mode are carried out.

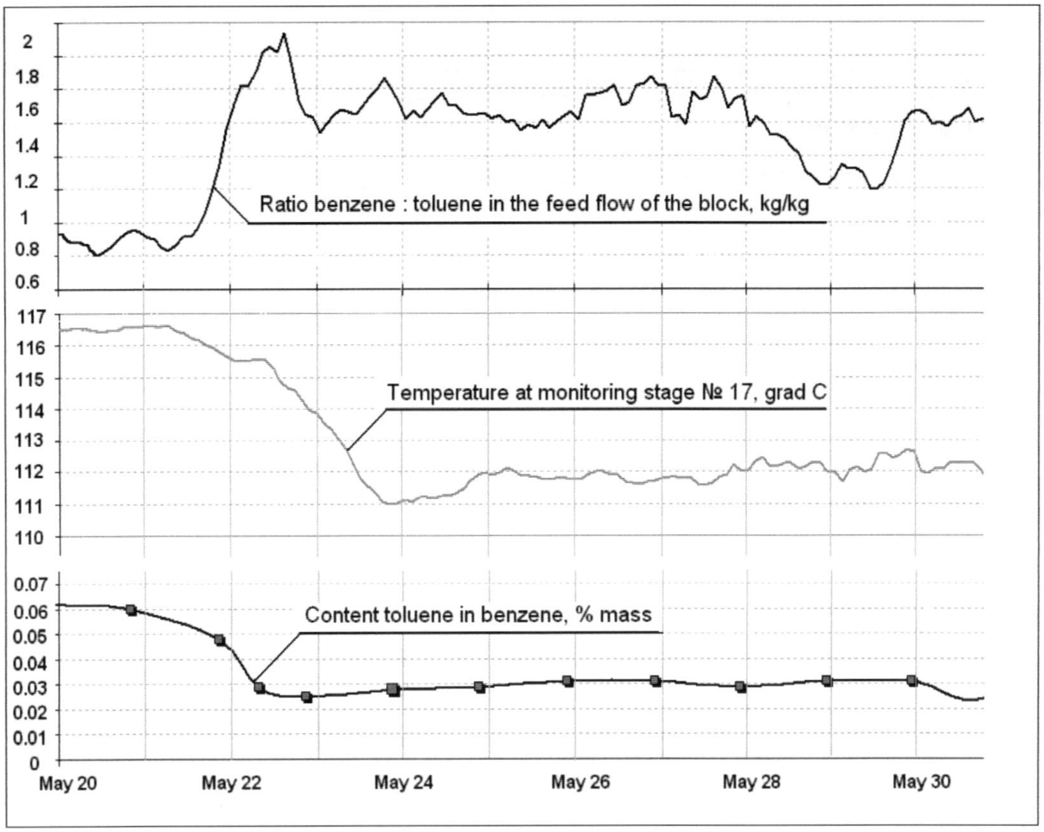

Рис. 7 Results of the experiment on block secondary rectification

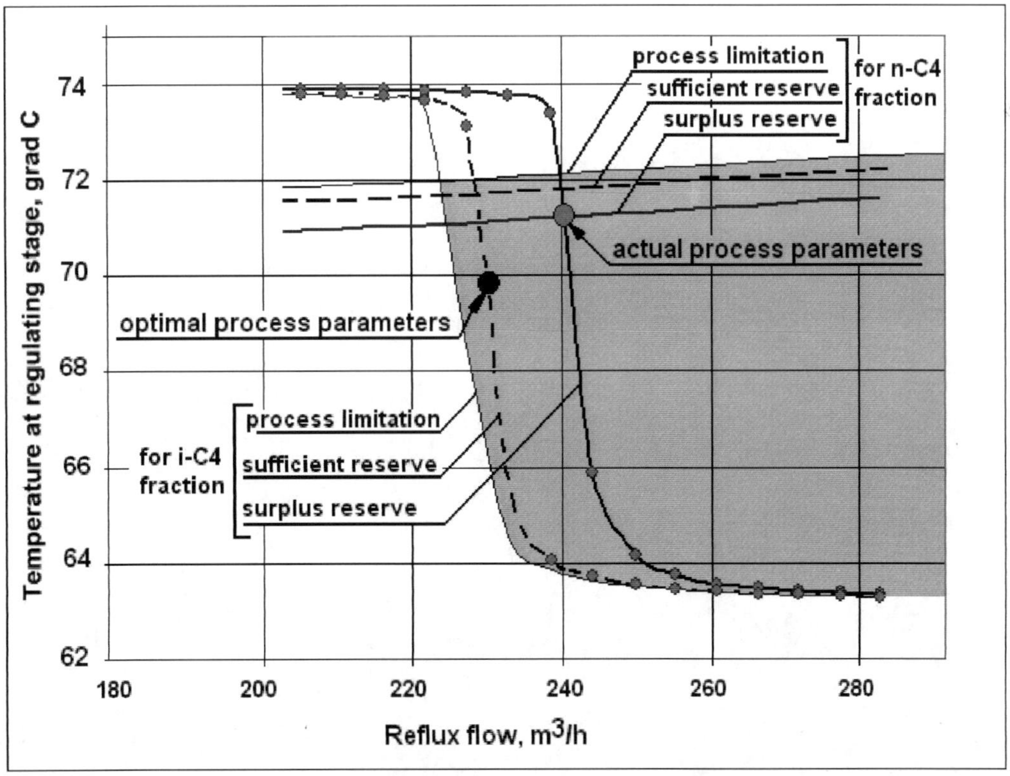

Fig 8 The temperature as function of reflux flow.

Thereby we can find the bottlenecks in production, analysing and revising the technological mode rates.

The significant economic effect is possible to get due to optimization technological mode on units with bottlenecks

REFERENCES

1.Nickolay V. Lisitsyn, Nickolay V. Kuzichkin, Vladimir I. Fedorov. Increasing the Efficiency of Mass and Heat Exchange Networks. 2006 Annual Meeting Abstract ID# 61619 *Submitted to* COMPUTING AND SYSTEMS TECHNOLOGY DIVISION.

2. Nickolay V. Lisitsyn, Nickolay V. Kuzichkin. Optimization of Benzine and Diesel Fuel Blending. 2006 Spring National Meeting *Abstract ID# 41637 Submitted to* TOPICAL A: 9TH TOPICAL CONFERENCE ON REFINERY PROCESSING

578j Design of Divided Wall Reactive Distillation for Hydrolysis of Methyl Acetate

Hao-Yeh Lee, Hsiao-Ping Huang, and Chih-Chun Hsu. Chemical Engineering, National Taiwan University, No. 1, Sec. 4, Roosevelt Road, Taipei, Taiwan

Because the price of the crude oil is arising constantly, the exploitation of energy saving becomes one of the recent focuses in research. Utilizing integrated multi-functional unit to replace some single function units is one effective approach toward this goal. Reactive distillation column (RD) and divided wall column (DWC) are examples of such integration. A further integration of RD and DWC is thus motivated and is called reactive divided wall column (RDWC). The methyl acetate hydrolysis process with RDWC design was first proposed by Sander (2007). However, their work was aimed to illustrate that the concept of RDWC could be physically feasible for industry. From their research, the major products (i.e. methanol and acetic acid) are not as pure as the industrial grade. To overcome this deficiency, the design of RDWC aimed to high quality products and the possible energy savings achievable are studied.

In this paper, a traditional RD configuration which contains one RD column and one sidedraw distillation column for methyl acetate hydrolysis is studied. In this configuration, RD top condenser contains 10 times catalyst than reactive tray and has total reflux to improve the methyl acetate conversion. Except the methyl acetate, the acetic acid, methanol, and excess reactant water are drawn from the RD bottom and fed into the second column. In the second column, top distillate with high pure methanol and bottom product with high purity acetic acid are obtained. The sidedraw of second column contains rich water which is fed as a recycle stream back to the RD column. It is found that although the above-mentioned configuration can achieve high product specifications, it needs much more reboiler duty and total annual cost. To decrease the energy consumption and capital cost reduction, the above process is integrated to a RDWC configuration. By simulation, it proves that this RDWC can be an effective design. Compared with the original configuration, the original RD reboiler is removed. From this RDWC design, all energy required is provided by the reboiler of the second column. The resulting RDWC can reduce 40 % energy demand and 25 % total annual cost. It is observed that, to make this major saving possible, this RDWC avoids the remixing effect that used to take place is the original RD bottom.

578k Investigation of Isobutene Dimerization Process In Reactive Distillation Using Rigorous Three-Phase Non-Equilibrium Stage Model

Amit Katariya[1], Mariyana Chalakova[2], Hannsjörg Freund[1], Sanjay M. Mahajani[3], and Kai Sundmacher[4]. (1) Physical and Chemical Process Engineering, Max Planck Institute for Dynamics of Complex Technical Systems, Sandtorstr. 1, Magdeburg, 39106, Germany, (2) Process System Engineering, Otto-von-Guericke-University, Universitatplatz 2, Magdeburg, 39106, Germany, (3) Department of Chemical Engineering, Indian Institute of Technology, Bombay, I.I.T. Bombay, Powai, Mumbai 400076, India, (4) Process Systems Engineering, Otto-von-Guericke-Univ. Magdeburg, Universitaetsplatz 2, Magdeburg, 39016, Germany

Dimerization of isobutene is an important step in the production of isooctane, which is a potential gasoline fuel additive used for enhancing the octane number. The process for the production of isooctane involves dimerization of isobutene followed by subsequent hydrogenation of the formed dimer to isooctane. In the conventional scheme there are serious selectivity issues involved due to the presence of subsequent oligomerization reactions. Recently, dimerization has been successfully carried out in a reactive distillation column, and it was possible to significantly improve the yield of the desired dimer.

The present study involves the investigation of the process performance in the presence of mass and heat transfer limitations. It has also been found that mass transfer inside the catalyst pores is significant and has to be considered while modeling the process. Hence a rigorous three phase non-equilibrium (NEQ) stage model has been developed. The modeling of the mass transfer inside the catalyst pores in the presence of multiple non-linear reactions, and incorporating such model in a rigorous NEQ model is computationally intensive and challenging.

Design of ethanol dehydration process: heterogeneous azeotropic distillation vs. extractive distillation

Gunawan and I-Lung Chien, Department of Chemical Engineering, National Taiwan University of Science and Technology, Taipei 106, Taiwan

Ethanol dehydration process is important in producing high-purity ethanol from renewable biomass resources. In a diluted stream from fermentation after filtration or centrifuge to remove heavy phase materials, the ethanol content is typically less than 10 wt%. Taking a feed stream to the ethanol dehydration system with 6.4 wt% of ethanol as an example, this paper will investigate the optimal design flowsheet to produce high-purity ethanol via heterogeneous azeotropic distillation and to compare the result versus extractive distillation. A classical example of ethanol dehydration system is to use benzene as entrainer in a heterogeneous azeotropic distillation configuration to aid the separation. However, because benzene is a known carcinogen, this work will demonstrate the design concept using other possible alternative entrainer. As for the extractive distillation configuration, ethylene glycol will be used as extractive agent to aid the separation.

For the heterogeneous azeotropic distillation configuration, a column will be designed to act as both the pre-concentrator column and the recovery column in the process. The fresh feed together with the aqueous outlet stream from decanter is fed into this combined column with the bottom product of near pure water and the top product is fed into the heterogeneous azeotropic column. The bottom product of this heterogeneous azeotropic column is pure ethanol and the top vapor of this column will be close to the lowest boiling point of the ternary system (ternary azeotrope). This top vapor stream after condensation can naturally separated into two liquid phases with the organic phase containing entrainer which can be recycled back to this heterogeneous azeotropic column to aid the separation and the aqueous outlet stream is fed into the combined column as mentioned previously. The Total Annual Cost (TAC) will be used as the objective function to be minimized for the above two-column design. Possible alternative entrainers will be compared to obtain the optimized flowseet with the lowest TAC.

An alternative design is to use ethylene glycol as extractive agent in the ethanol dehydration system. The overall process with the extractive distillation configuration will contain three columns. The first column is the pre-concentrator column with the purpose of taking out most of the water in the fresh feed. The top product of this pre-concentrator column (near ethanol-water azeotrope) is fed into an extractive distillation column with the extractive agent also entering this column. The presence of the extractive agent alters the relative volatility between ethanol and water causing ethanol to move toward the top part and water to move toward the bottom part of this column together with the heavy extractive agent. The bottom product of the extractive distillation column is fed into the recovery column to produce pure water in the distillate and pure extractive agent in the column bottoms. This bottom stream is recycled back to the extractive distillation column for continue usage. The TAC of this optimized flowsheet via extractive distillation configuration will be compared to the optimized flowsheet via heterogeneous azeotropic distillation configuration to find out which configuration will have the lowest TAC and the lowest operating energy consumption.

Keywords: Ethanol Dehydration, Heterogeneous Azeotropic Distillation, Extractive Distillation, Process Design.

A Non-linear Programming Formulation for Optimal Design and Operation of a Multitubular Packed-Bed Reactor Using a One Stage Solution Strategy (simultaneous solution): Case Study of *o*-xylene Oxidation to Phthalic Anhydride

Carlos A. Llinás Vargas
Chemical Engineering graduate student, Universidad de los Andes, Bogotá, Colombia

Jorge M. Gómez Ramírez
Chemical Engineer, M.Sc, PhD, Professor at Universidad de los Andes, Bogotá, Colombia

ABSTRACT: A one-dimensional pseudo-homogeneous model is used for simulation and subsequent optimization of a MPBR. The reactive system chosen for the present study is Finlayson extended *o*-xylene oxidation to phthalic anhydride production in the presence of undesired side reactions. The case study is actually taken from Froment and Bischoff and in the present article the optimal design and operation of the reactor is searched upon. All programmed simulations and optimizations were achieved by means of Wolfram Mathematica 6.0®. As is usual for the present case, sensitivity analyses are done for *o*-xylene inlet concentration (molar fraction) and reactor inlet temperature. The problem for optimization with differential equation constraints is also discussed and several objective functions, technical and economical ones, are maximized.

1 INTRODUCTION

O-xylene oxidation to phthalic has been studied widely and rigorous two-dimensional pseudo-homogenous models have been applied by many [1, 2]. Yet optimal design and operation of such reactor has never been discussed. Though the present article studies the latter the main goal of the case study is actually to formulate a one stage solution strategy for optimization with differential equations as constraints. For such reasons the author admonishes *a priori* any attempt to search for actual optimal design of the reactor at industrial scale in the present article. One has to remember that actual optimal reactor design and operation must be accompanied with pilot plant studies and thoroughly experimental designs. On the other hand kinetic rate law determination through rigorous experimentation and statistics [3] becomes essential when evaluating chemical reactor performance and design even though literature may provide some information on reaction kinetics.

It is then by these means that an already proposed kinetic rate law of pseudo-first order is used in all simulations and no attempt is done for experimental corroboration of the mechanism proposed. Yet simulations with typical system's parameters are run in order to confirm that the results are physically coherent. Likewise reactor profiles are exhibited for different reactor lengths in such a way that becomes suitable for analysis.

Finally, all simulations and optimizations are done working with a steady state model. All modeling, simulation and optimization was programmed in Wolfram Mathematica 6.0® which becomes a convenient software package for solving complex system of differential equations (ODE, PDE, DAE) and optimization problems with already "Built-In Functions". As will be seen the optimization problem with differential equation constraints doesn't become a trivial problem when analytical solutions to such are lacked.

2 MATHEMATICAL MODELING OF FIXED BED REACTORS

When it comes to generalizing the mathematical models for fixed bed reactors, "...it is not possible to concentrate on specific cases and processes. Instead [one must] discuss general models and principles involved in the design and analysis of any type of fixed bed reactor, no matter what process" [1]. The required degree of sophistication in the mathematical modeling is subject in the first place on the process, more specifically on the sensitivity to perturbations in the operating conditions. Yet of equal importance is the degree of accuracy with which the kinetic and transport phenomena parameters are known.

When it comes to modeling fixed bed reactors probably the most important characteristic of the model is whether it considers the presence of the solid catalyst or not, that is, if a pseudo-homogeneous or heterogeneous model is used. These are usually considered as the two broad categories in fixed bed reactor design. Moreover axial and radial dispersion of mass and energy in either model may be inserted. Yet in the present case axial and radial effects were omitted because the main goal of the one stage solution required the already complex discretization of several differential equations and the solution of these in the optimization problem. Having not neglected such effects would have given rise to new complexities which fall beyond the scope of the present work.

Therefore in the present work a pseudo-homogeneous model without either axial or radial dispersion but variable coolant agent temperature is used. The case study involved 3 reactions, the desired one and two undesired reactions which take place one parallel and the other in series. Solutions to the optimization problems were studied under two scenarios; one in which the desired reaction was only considered and one in which all reactions were taken into account. For such reason the resulting ODE for both cases are shown subsequently.

For the model in which only the desired reaction is studied the resulting ODE's arise from mass balance of the limiting reactant expressed in terms of conversion and energy balance over the reactor and coolant agent. Thus three dependent variables result:

[illegible] (1)

[illegible] (2)

[illegible] (3)

For the model in which all reactions are taken into account the resulting ODE's consist of the mass balance from the 5 species present in the reacting system and the energy balance over the reactor and coolant agent:

$$[illegible] = (r_1' + r_3')\rho_b A_c \quad (4)$$

$$[illegible] = (3r_1' + [illegible]\, r_2' + [illegible]\, r_3')\rho_b A_c \quad (5)$$

$$[illegible] = (r_1' \; [illegible] \; r_2')\rho_b A_c \quad (6)$$

$$[illegible] = (-3r_1' - 2r_2' - 5r_3')\rho_b A_c \quad (7)$$

$$[illegible] = (-8r_2' - 8r_3')\rho_b A_c \quad (8)$$

[illegible] (9)

[illegible] (10)

3 THE OPTIMIZATION PROBLEM

The formulation of an optimization problem consists mainly of three elements: the objective function, the constraints, and the optimization variables or decision variables.

3.1 *Objective functions*

Three objective functions were maximized in the present article: (1) conversion of o-xylene, (2) production of phthalic anhydride and (3) reactor profitability.

For the optimizations where only optimal operation of the reactor was studied the following objective functions were formulated:

Maximization of o-xylene conversion in the absence of undesired reactions:

$$(1)\, Maximice\ X_{[illegible]}$$

Maximization of phthalic anhydride production in the presence of undesired reactions:

$$(2)\, Maximice\ FC_{[illegible]}$$

Maximization of reactor profitability in the presence of undesired reactions:

$(3)\ Maximice\left(\frac{[illegible]}{[illegible]} - ser\left(\frac{(FE_{1000ae})(m.w.FE)(0.41)(P)}{1000}\right) [illegible]\right)$

On the other hand for the optimization problems where optimal operation and design is search upon simultaneously, the following objective functions were formulated:

Maximization of phthalic anhydride production in the presence of undesired reactions:

$(1)\ Maximice\ FC_{maeshatne}$

Maximization of reactor profitability in the presence of undesired reactions:

$(2)\ Maximice\left(\frac{[illegible]}{[illegible]} - ser\left(\frac{(FE_{1000ae})(m.w.FE)(0.41)(P)}{1000}\right) [illegible]\right)$

3.2 Optimization constraints

The optimization constraints are made up of the model ODE's which lack an analytical solution. Furthermore the following inequality constraints are involved in the optimization problems formulation:

$$0 \leq FA \leq FA_0 \quad (11)$$
$$0 \leq FB \leq FB_0 \quad (12)$$
$$FC \geq 0 \quad (13)$$
$$FD \geq 0 \quad (14)$$
$$FE \geq 0 \quad (15)$$
$$0 \leq X \leq 1 \quad (16)$$
$$T \leq 668 \quad (17)$$
$$T_0, Ta_0 \geq 0 \quad (18)$$
$$Ta \leq T \quad (19)$$
$$l \geq 0 \quad (20)$$

3.3 Optimization variables

The resulting optimization variables are the dependent variables of the system of ODE's; the o-xylene conversion profile, the molar profiles of the species involved and the reactor and coolant agent temperature profiles. Likewise in order to study the optimal design and operation of the reactor, the reactor length *l* and the inlet temperatures of the reactor and coolant agent were taken as optimization variables.

In the present work results convenient then to divide the optimization problems into two sub-groups: (1) one degree of freedom optimization and (2) two degrees of freedom optimization. The one degree of freedom optimization problems studied uniquely the optimal operation of the reactor, where the degree of freedom resulted in the inlet temperature of the reactor and coolant agent, taken always as equal (i.e. $T_0 = Ta_0$). On the other hand the two degrees of freedom optimization problems studied the optimal operation and design of the reactor simultaneously. In such case the two degrees of freedom turn out to be the inlet temperatures mentioned previously and the reactor length.

3.4 Optimization method: One-stage solution strategy

First of all it has to be recognized that the already known optimization techniques are unable to work with ODE as constraints of the optimization problem formulation, that is if it is to be solved simultaneously (one-stage). As the main goal of the present work is to solve the optimization problem in one stage, discretization of the mathematical model turns out to be the first step to take in order to enable a one stage solution. There have been recent proposals mainly in the area of reaction engineer that actually have already encountered the need to use such technique [4].

Most optimization methods work by using first and second derivatives of a Lagrangian function in order to meet the necessary and sufficient conditions of optimality [5]. Accordingly, such local methods of optimization cannot work with differential equations as constraints because their implementation would lead to second and third derivatives which clearly do not represent coherence in optimization techniques. Likewise, global optimization methods, though not using derivatives of the constraints to search for optimum, cannot work with differential equations as constraints. Hence a way must be devised by which the pseudo-homogeneous model can be described in analytical terms. Furthermore, an analytical solution of such model is lacked and one must turn to numerical methods to solve it. Clearly optimization with ODE, PDE, DAE constraints is not a trivial problem to solve.

Discretization of mathematical models in order to apply optimization techniques is encountered widely in reaction engineering problems due to the mathematical complexity of these. Numerous methods have been devised to meet such need: Finite differences, finite elements, orthogonal collocation, Rayleigh-Ritz method, etc [6, 7]. Centered finite difference method was used in the case study. In a summarized manner the method simply uses discretization of the dependent variables in n arbitrary collocation points or "mesh". Depending on the mesh size, the number of variables and equations in the model will vary. Clearly as the number of points in the mesh increases computation time increases and the mathematical model becomes bigger. In the present study some optimizations were run with a 50 points mesh and others with a 100 points mesh. Yet is important to mention that results were independent of such parameter and that computation time didn't exceeded 2 minutes in all cases. As an example, the resulting model from the discretization of the reactor temperature can be represented in the following manner:

$$\frac{[illegible]}{[illegible]} = f(F_A, F_B, F_C, F_E, F_R, T, T_a)$$

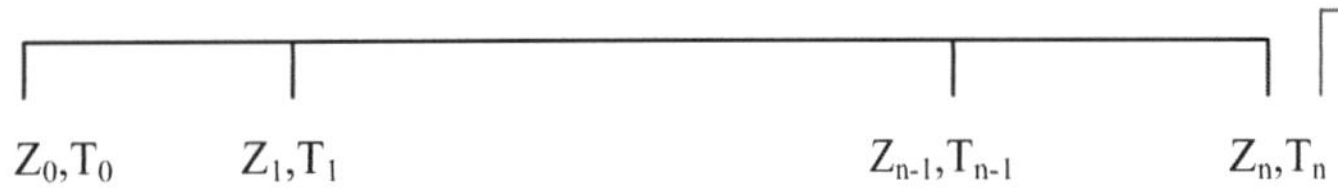

Figure1. Discretization of reactor temperature

$$For\ i = \{1, n-1\} \rightarrow \frac{[illegible]_{i+1} - [illegible]_{i-1}}{[illegible]} = f(F_{Ai}, ..., T_{Ai})$$

$$For\ i = n \rightarrow \frac{[illegible]}{[illegible]} = f(F_{Ai}, ..., T_{ai})$$

$$where: \Delta h = \frac{[illegible]}{[illegible]}$$

From the discretization process it arises that both the number of equality constraints (model) and optimization variables are dependent of the mesh size taken. For the one degree of freedom optimization 100 points were used whereas for the two degrees of freedom optimization a 50 points mesh size was used.

Hence one has from the discretization process mentioned that the actual optimization constraints are the resulting highly nonlinear equations from the discretization of the n conversions, n reactor temperatures, n coolant agent temperatures and n molar flows, n been the number of collocation points (mesh size). Likewise, the resulting optimization variables are each of the mesh points from the discretization of the dependent variables (i.e. FA_i, FB_i, FC_i, FD_i, FE_i, X_i, T_i, Ta_i) which exhibit up as will be seen as discrete reactor profiles, plus the degrees of freedom. In tables 1 and 2 the resulting optimization problems formulation are summarized.

Table 1. One degree of freedom optimizations

Objective function	Mesh size	Number of equations	Number of variables	Number of inequality constraints
Maximization o-xylene conversion	100	300	301	301
Maximization phthalic anhydride production	100	700	701	701
Maximization profitability	100	700	701	701

Table 2. Two degrees of freedom optimizations

Objective function	Mesh size	Number of equations	Number of variables	Number of inequality constraints
Maximization phthalic anhydride production	50	350	352	352
Maximization profitability	50	350	352	352

3.5 Optimization technique and software

Interior point method was used for all optimizations. The algorithm for such optimization technique is already programmed as a "Built-In Function" in the simulation software Wolfram Mathematica 6.0®. That means that the optimization technique, either local or global, can be easily called for by a simple programming line. In the present study for the one stage solution strategy only the interior point method was used because global optimization techniques led to difficulties in computation time. Though a trickier programming was needed for global optimization, global optimization techniques (i.e. Nelder-Mead, Differential Evolution, and Genetic Algorithm) were easily called for in a two level solution strategy. Furthermore it is important to remember that the main purpose of the present study was to formulate a simultaneous solution (one stage solution strategy). Results by both means led to the same solution.

4 CASE STUDY: *O*-XYLENE OXIDATION FOR PHTHALIC ANHYDRIDE PRODUCTION

The reaction system studied is the oxidation of *o*-xylene in the presence of undesired reactions [2]:

$$C_8H_{10} + 3O_2 \rightarrow C_8H_4O_3 + 3H_2O$$

$C_8H_4O_3$ + [illegible] $O_2 \rightarrow 8CO_2 + 2H_2O$

C_8H_{10} + [illegible] $O_2 \rightarrow 8CO_2 + 5H_2O$

The reactor usually consists of 2500 tubes up to 10,000 tubes. The length of the tubes is 3 m and their internal diameter is 1 inch. The reaction is carried out over V_2O_5 catalyst and due to selectivity and deactivation problems reactor temperature should not exceed 668K (415°C). A heat transfer medium consisting of sodium nitrite-potassium nitrate fused salts circulates outside the tubes through the shell to remove the heat of reaction [8]. In the present study as in others the reactor is considered to operate at 1.0 atm, that is, pressure drop is neglected.

Due to the large excess of oxygen at reactor entrance (mole fraction>0.99), the rate law is considered pseudo first order yet in the present study variations for the oxygen partial pressure was considered. The reaction kinetics and specific reaction rate are taken from [2]:

$$r_1' = -k_1 P_{oxy} P_{O2} \quad (21)$$

$$r_2' = -k_2 P_{pa} P_{O2} \quad (22)$$

$$r_3' = -k_3 P_{oxy} P_{O2} \quad (23)$$

$k_1 = exp(19.837 -$ [illegible]$)$ (24)

$k_2 = exp(20.86 -$ [illegible]$)$ (25)

$k_3 = exp(18.97 -$ [illegible]$)$ (26)

Where partial pressures are in bars and the subscripts *oxy*, O_2, *pa* corresponds to the partial pressures of o-xylene, oxygen and phthalic anhydride respectively.

Given that the undesired reactions are favored at higher temperatures, the inlet reactor and coolant agent temperatures become interesting optimization variables. Similarly, due to the presence of a series reaction the reactor length becomes another interesting optimization variable.

5 RESULTS

Initial results from simulations with typical parameters are exhibited first. Sensitivity analysis is done for reactor inlet conditions, mainly reactor inlet temperature and inlet *o*-

xylene molar fraction. Afterwards results from the one stage optimizations are shown as discrete reactor profiles.

5.1 Simulation results

For initial simulation of the model certain parameters were taken from the literature [1, 2], others were calculated (i.e. heat capacities), and yet others were changed arbitrarily. Integration of the set of ordinary differential equations was done for a reactor length of 4 m for a better understanding of the reactor profiles. The simulation involved all of three reactions. The study of the system in the absence of undesired reactions was only taken under the optimization section. All parameters used in the study correspond to those of a 2500 tube reactor

Table 3. Parameters for simulation

Parameter	Value
r (m)	0.0127
FT_0(kmol/s)	0.0006057
yA_0	0.00224
yB_0	0.99776
T_0(K)	628
T_{a0}(K)	628
P,(Pa)	101325
Φ	0.45
ρ_c(kg/m^3)	2800
U(Kj/m^2 s K)	0.0845
R,(m^3 Pa/ mol K)	8.314
Cp_A (Kj/kmol K)	244.53
Cp_B(Kj/kmol K)	32.6
Cp_C(Kj/kmol K)	238.17
Cp_D(Kj/kmol K)	36.837
Cp_E(Kj/kmol K)	48.144
Cp_a(Kj/kmol K)	1527.769
m(kg/s)	0.04
ΔH_{RX1}(Kj/kmol)	-1285000
ΔH_{RX2}(Kj/kmol)	-3279000
ΔH_{RX3}(Kj/kmol)	-4564000

Illustrations 1, 2 and 3 show the reactor temperature profile, molar profiles and molar profile of the desired product respectively for a 4 m reactor.

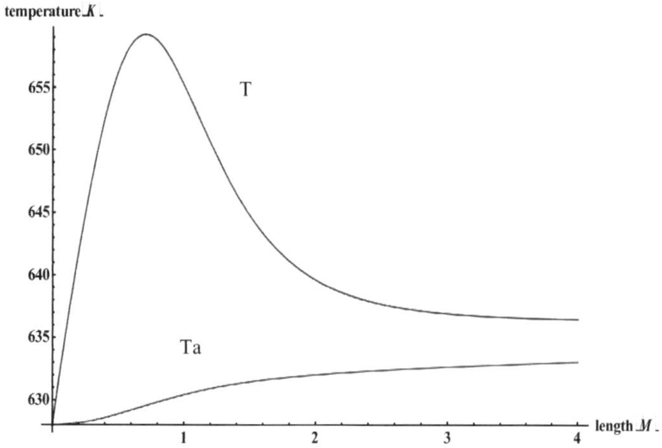

Illustration 1. Reactor temperature profiles

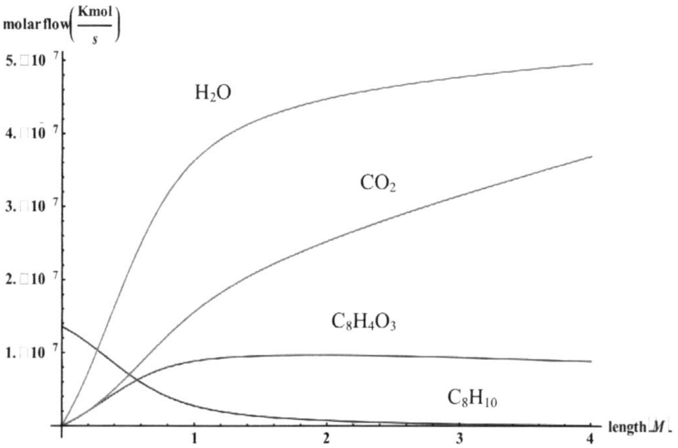

Illustration 2. Reactor molar flow profiles

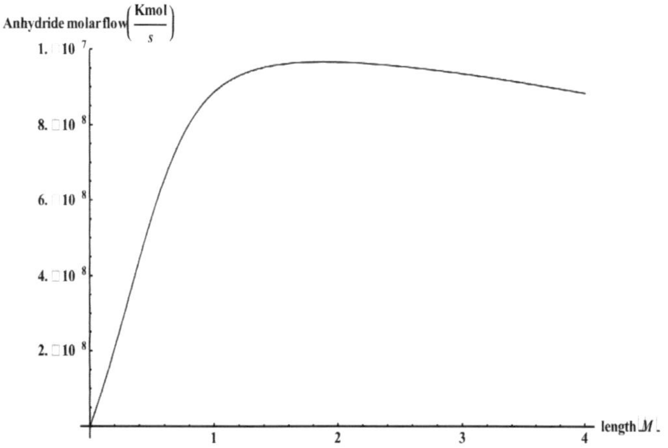

Illustration 3. Phthalic Anhydride molar flow profile

As one may see the reactor exhibits its typical hot spot whose magnitude is function mainly of *o*-xylene inlet concentration and reactor inlet temperature and its location mainly on flow velocity. For the parameters in the table one may appreciate likewise that side reactions strongly affect reactor performance beyond 2 meters. Thus reactor length becomes an important decision variable to optimize as side reactions are wanted to be kept at a minimum.

5.2 Results for sensitivity to inlet conditions: Sensitivity analysis

For the sensitivity analysis all parameters from the simulation remain the same except that one for which analysis is been done. In the present case sensitivity to changes in reactor inlet temperature and o-xylene inlet molar fraction was performed.

As mentioned previously for the case study, reactor performance is strongly dependent on reactor inlet temperature and concentration. As shown in illustration 4 reactor hot spot and run away ignition starts approximately when inlet temperature is 655K. Likewise as argued initially the idea of operating such reactor over 668K becomes unacceptable thus reactor inlet temperature clearly becomes an optimization variable.

Similarly, sensitivity to *o*-xylene inlet molar fraction is show in illustration 5. Run away reaction is also evident when inlet molar fraction reaches up to 0.0032. Such strong hot spots clearly affect catalyst performance and lifetime. In the present case study is clear then that a thermal runaway limit exists and its prediction becomes important if one desires to avoid reactor sensitivity. Molar fraction value for the first entrance condition is now shown on illustration 5. Its corresponding value is 0.001.

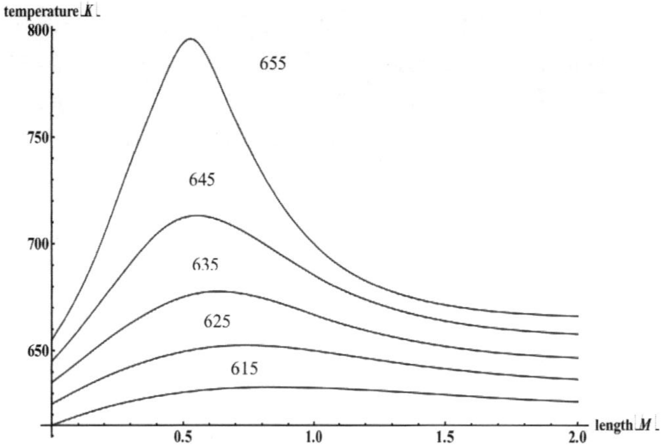

Illustration 4. Sensitivity to inlet reactor temperature

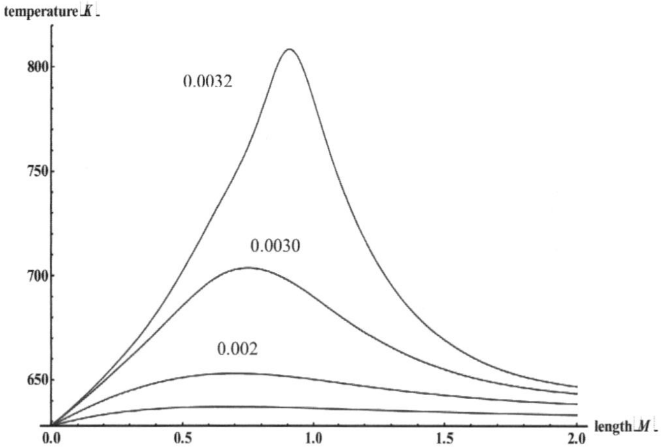

Illustration 5. Sensitivity to o-xylene inlet molar fraction

Thermal runaway limit actually may be defined at the inflection point of the temperature profile. Such problem may be visualized quantitatively under a strict mathematical perspective. It is this what actually Shahid Bashir et al. do as they prove that to avoid thermal runaway in cooled reactors the second derivative of the reaction temperature along the tube length must remain less than zero[9]:

$$\frac{d^2T}{dz^2} < 0 \qquad (27)$$

Moreover they emphasize that the cause for a temperature runaway is that the $d^2T/dz^2=0$ limit is exceeded. It is evident for the present reactive system under study that such limit has already been exceeded when the inlet molar fraction is 0.0032 (i.e. a change in concavity has already occurred).

On the other hand, for the inlet temperature sensitivity, it seems as if the threshold value for reaction thermal runaway were slightly less than 655K. A small change in concavity is weakly appreciated at the beginning of the temperature profile for the latter inlet temperature indicating that the second derivative criterion has already been violated.

5.3 Optimization results

For the optimization problems all parameters from table 3 remained the same except the *o*-xylene inlet molar fraction which was parameterized to 0.003 and the inlet temperatures (i.e. T_0, Ta_0) which were taken always as optimization variables and equalized. For the optimizations which studied only the optimal operation of the reactor (i.e. one degree of freedom), that is, where the length of the reactor is a parameter, a 2 meter reactor was assumed. Tables 4 and 5 summarize the optimization results:

Table 4. One degree of freedom optimization results: Inlet temperature

Objective function	Optimal inlet temperature(K)	Maximum reactor temperature attained(K)	Objective function value
Maximization o-xylene conversion	624.83	668	0.941475
Maximization phthalic anhydride production	617.88	657	0.46738 mol/hr
Maximization profitability	613.97	643	0.468689 $/hr

Table 5. Two degrees of freedom optimization results: Inlet temperature and reactor length

Objective function	Optimal reactor length(m)	Optimal inlet temperature(K)	Maximum reactor temperature attained(K)	Objective function value
Maximization phthalic anhydride production	3	613.874	645	0.473131 mol/hr
Maximization profitability	2.4	611.119	637.8	0.447479 $/hr

In what follows sub-optimal temperature profiles result from simulations where system constraints are intentionally violated, mainly the maximum allowable temperature, are shown. After each sub-optimal profile optimal profiles result from the optimizations are illustrated. It remains important to emphasize that different optimal criterions led to different outcomes in the optimal design and operation of the reactor.

Maximization o-xylene conversion:

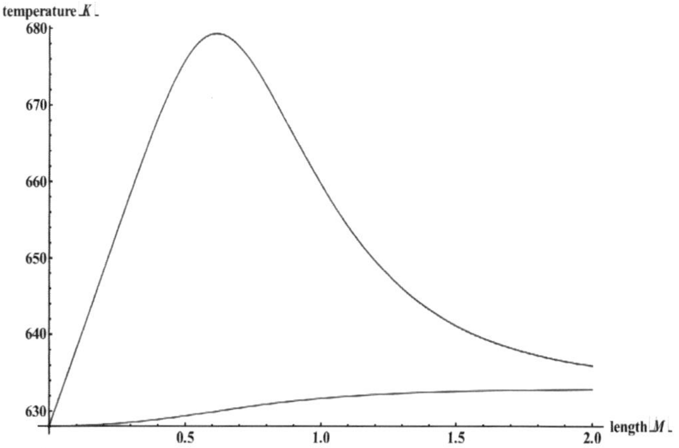

Illustration 6. Initial sub-optimal temperature profile

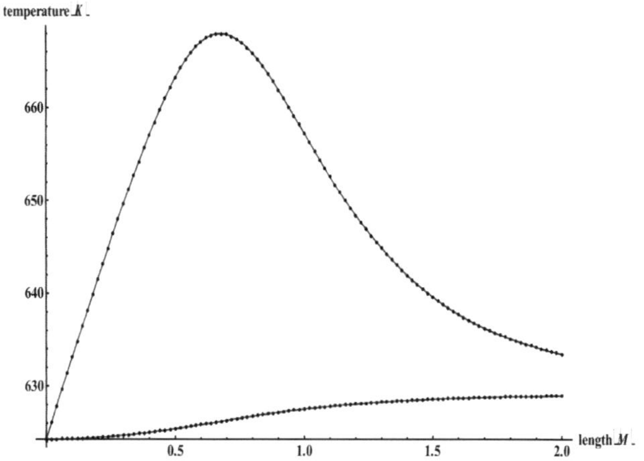

Illustration 7. Final optimal temperature profile

Maximization phthalic anhydride production (1 degree of freedom):

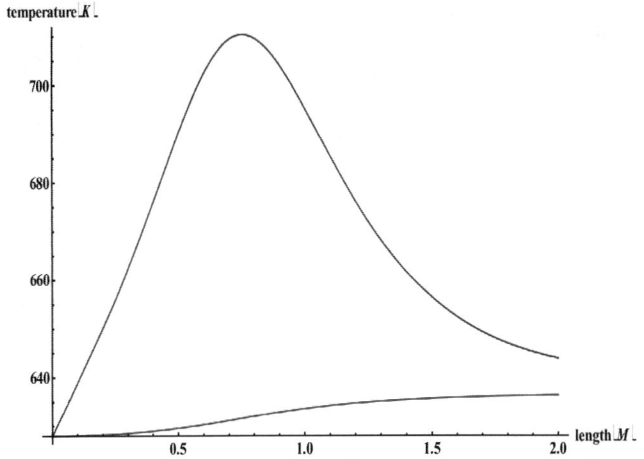

Illustration 8. Initial sub-optimal temperature profile

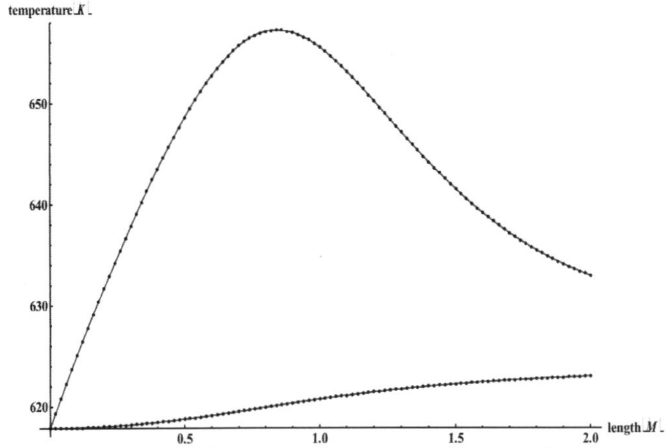

Illustration 9. Final optimal temperature profile

Maximization reactor profitability (1 degree of freedom):

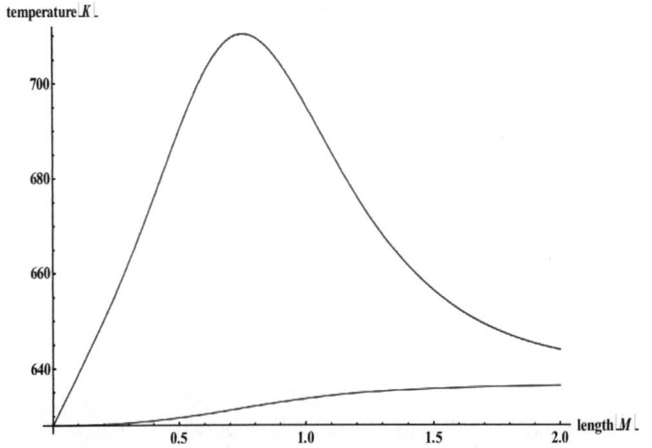

Illustration 10. Initial sub-optimal temperature profile

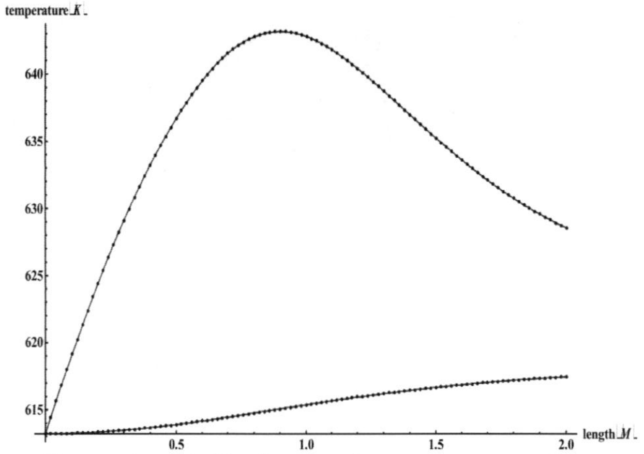

Illustration 11. Final optimal temperature profile

The molar flow profiles for the production of the desired product (i.e. phthalic anhydride) behaved quite the same in all optimizations. Illustrations 12 and 13 show the sub-optimal and optimal molar flow profile for phthalic anhydride respectively. The illustrations were taken from the maximization of the reactor profitability with one degree of freedom:

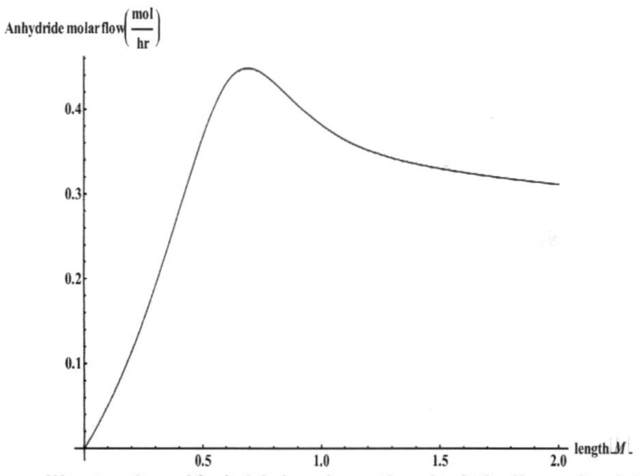

Illustration 12. Initial sub-optimal phthalic anhydride molar flow profile

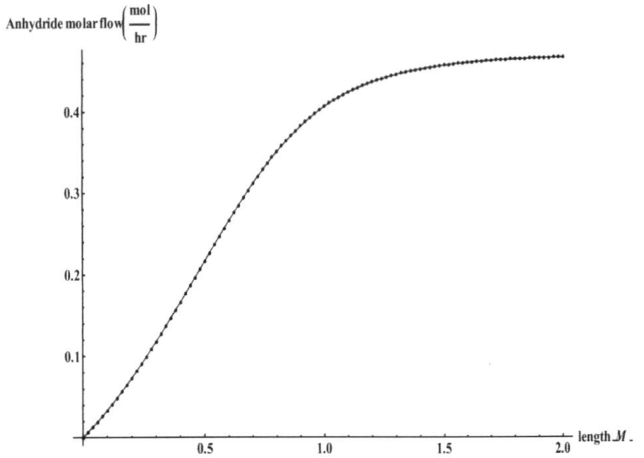

Illustration 13. Final optimal phthalic anhydride molar flow profile

Maximization phthalic anhydride production (2 degrees of freedom):

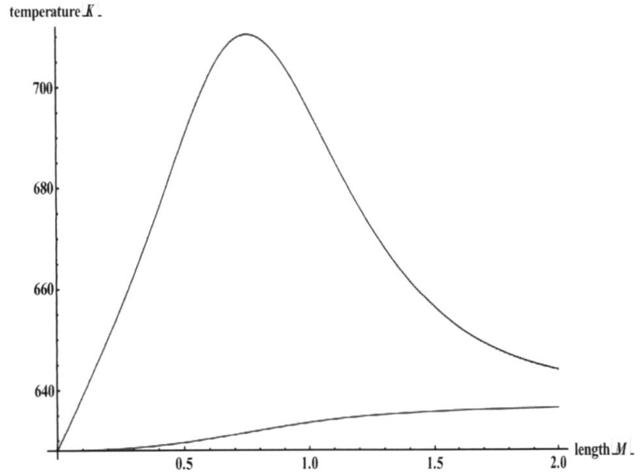

Illustration 14. Initial sub-optimal temperature profile

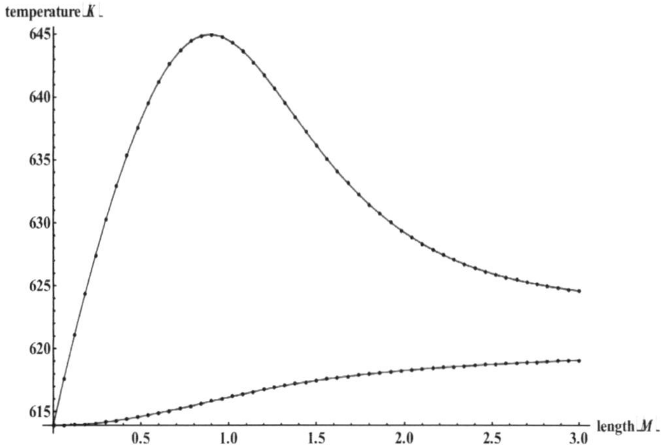

Illustration 15. Final optimal temperature profile

Maximization reactor profitability (2 degrees of freedom):

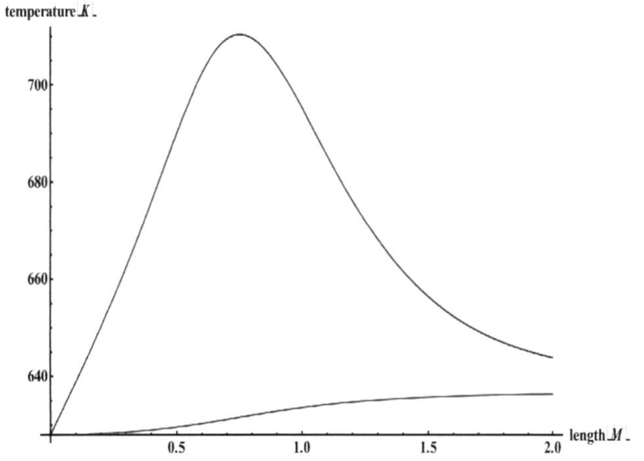

Illustration 16. Initial sub-optimal temperature profile

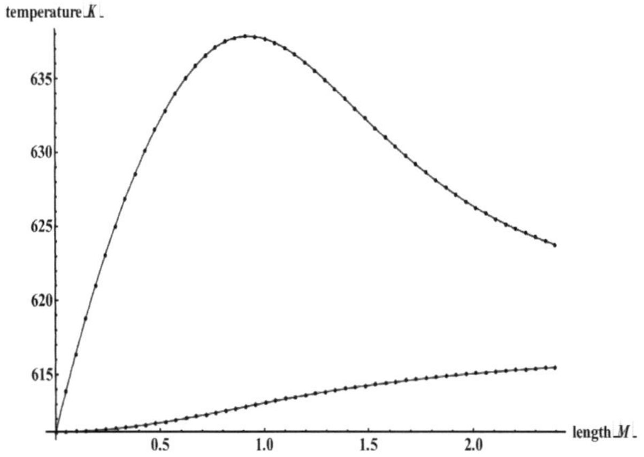

Illustration 17. Final optimal temperature profile

The following illustrations exhibit the initial and final molar flow profiles for the case where maximization the reactor profitability with two degrees of freedom is studied.

Clearly optimal reactor design and operation leaded to optimal molar flow profiles as phthalic anhydride production is kept at its maximum whilst CO_2 from the undesired reactions is kept at a minimum.

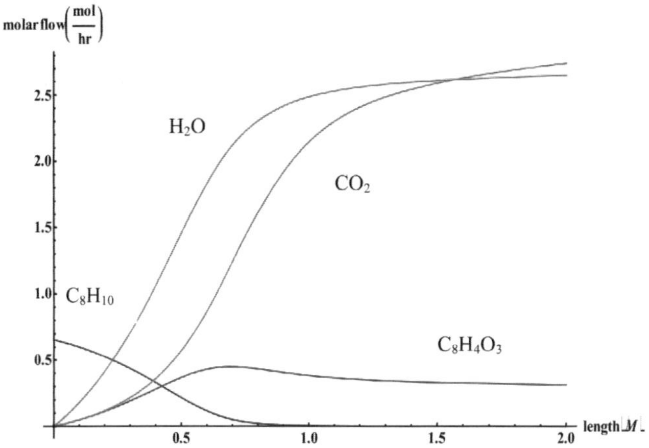

Illustration 18. Initial sub-optimal molar flow profiles

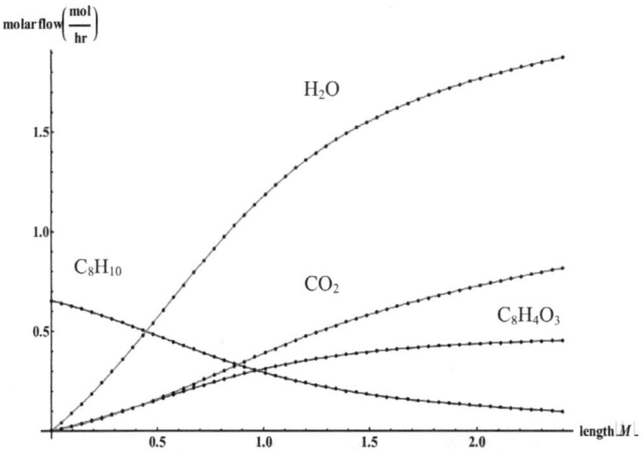

Illustration 19. Final optimal molar flow profiles

Finally a graph of the objective function for the reactor profitability with two degrees of freedom is shown. By using the results from the profit objective function, one may insert the result for inlet optimal temperature (i.e. 611.119K) in a one degree of freedom optimization and obtain a graph of profit as a function of reactor length. Notice that a maximum is attained when reactor length is approximately 2.4m. Hence by reducing the reactor profit optimization from two degrees to a one degree of freedom optimization and inserting the optimal inlet temperature from the two degrees of freedom case, the graph bolsters the result obtained from the two degree of freedom optimization.

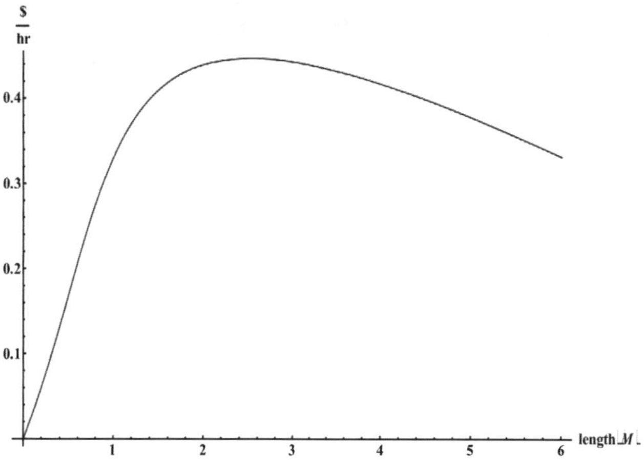

Figure 18.Objective function maximum

6 CONCLUSIONS.

Reactor simulation for phthalic anhydride production was done for some typical parameters. Reactor sensitivity analysis for inlet temperature and *o*-xylene inlet concentration was done in order to pinpoint reaction thermal runaway limit. The concavity criterion showed to be right for the present case as an unacceptable temperature of 800K exhibited when change in concavity of the profiles took place.

Optimization led to minimization of the problems of selectivity in the reactive system, mainly the presence of the side reactions. Moreover such minimization of side reactions was accomplished by maintaining the temperature below its constrained value, 668K. While the one degree of freedom problems solved the optimal operation of the reactor, two degrees of freedom optimizations were able to solve the problem of optimal operation and design simultaneously.

All optimizations kept phthalic anhydride production at a maximum and CO_2 production at a minimum as presented in illustrations 19. The one degree of freedom maximization of the *o*-xylene conversion was the only case where undesired side reactions were not considered thus leading to the highest optimal inlet temperature. Likewise this case reached the highest hot spot, up to the limit imposed by the inequality constraint (i.e. T≤ 668K).

The problem of optimization with differential equations that lack an analytical solution was solved by discretization of the mathematical model and using a one level solution strategy. Centered finite difference was used to discrete the mathematical model. Solution of the optimization problem was seen to be independent of the mesh size used yet computation time did show dependent of such parameter. Finally optimization results were obtained for reactor inlet temperature and length. Such optimizations led to optimal temperature profiles and molar profiles by keeping production of undesired products to a minimum.

NOTATION

$r_1' =$ RX 1 kinetic rate law [illegible]
$r_2' =$ RX 2 kinetic rate law [illegible]
$r_3' =$ RX 3 kinetic rate law [illegible]
$k_1 =$ RX 1 specific rate [illegible]
$k_2 =$ RX 2 specific rate [illegible]
$k_3 =$ RX 3 specific rate [illegible]
$X =$ o − xylene conversion
$T =$ reactor temperature $[K]$
$T_a =$ coolant agent temperature $[K]$
$F_A =$ o − xylene molar flow [illegible]
$F_B = O_2$ molar flow [illegible]
$F_C =$ phthalic anhydride molar flow [illegible]
$F_D = H_2O$ molar flow [illegible]
$F_E = CO_2$ molar flow [illegible]
$r =$ reactor radius $[m]$
$A_c =$ reactor cross sectional area $[m^2]$
$a =$ heat exchange area per unit reactor volume $[m^{-1}]$
$FT_0 =$ total inlet molar flow [illegible]
$yA_0 =$ o − xylene inlet molar fraction
$yB_0 = O_2$ inlet molar fraction
$T_0 =$ reactor inlet temperature $[K]$
$T_{a0} =$ coolant agent inlet temperature $[K]$
$P =$ system pressure $[Pa]$
$\phi =$ bed porosity
$\rho_c =$ catalyst density [illegible]
$\rho_b =$ bulk density [illegible]
$U =$ overall heat transfer coefficient [illegible]
$R =$ ideal gas constant [illegible]
$Cp_A =$ o − xylene heat capacity [illegible]
$Cp_B = O_2$ heat capacity [illegible]
$Cp_C =$ phthalic anhydride heat capacity [illegible]
$Cp_D = H_2O$ heat capacity [illegible]
$Cp_E = CO_2$ heat capacity [illegible]
$Cp_a =$ coolant agent heat capacity [illegible]
$m =$ coolant agent mass flow [illegible]
$\Delta H_{RX1} =$ RX1 heat of reaction [illegible]
$\Delta H_{RX2} =$ RX2 heat of reaction [illegible]
$\Delta H_{RX3} =$ RX3 heat of reaction [illegible]
$z =$ reactor axial coordinate $[m]$

OPTIMIZATION CONSTANTS NOTATION

$m.w.FC =$ phthalic anhydride molecular weight
$m.w.FE = CO_2$ molecular weight
$p =$ anhydride price, 9 [illegible]
$pr =$ reactor cost, 50 [illegible]
$par, ser =$ weight contribution to lost incomes, 0.5
$w =$ repayment multiplier, 0.624 $[year^{-1}]$
$l =$ reactor length $[m]$
0.301 = parallel reaction stoichiometry [illegible]
1.36 = desired reaction stoichiometry [illegible]
0.41 = series reaction stoichiometry [illegible]

REFERENCES

[1] Froment, G.F. and K.B. Bischoff, *Chemical Reactor Analysis and Design*, John Wiley & Sons, New York (1990)

[2] B. Rosendall and B.A. Finlayson, *Transport Effects in Packed-Bed Oxidation Reactors*, Computers chem. Engng Vol. 19 No. 11 (1995)

[3] H. Scott Fogler, *Elements of Chemical Reaction Engineering*, Prentice Hall, Fourth Edition (2006)

[4] Arturo M. Cervantes, Andres Wächter, Reha H. Tütüncü, Lorenz T. Biegler, "A reduced space interior point strategy for optimization of differential algebraic systems",

Computers and Chemical Engineering 24 (2000)
[5] Thomas F. Edgar, David M. Himmelblau, “Optimization of chemical processes”, McGraw-Hill, Second Edition, (2001)
[6] Finlayson, B. “Nonlinear Analysis of Chemical Engineering”, McGraw-Hill, New York, (1980)
[7] Richard L. Burden. “Numerical Analysis”, Thomson-Learning, Seventh edition, (2002)
[8] Warren D. Seider, J.D. Seader, Daniel R. Lewin, “Product and Process Design Principles”, John Wiley & Sons, Second Edition, (2004)
[9] Shahid Bashir, Tibor Chován, Bassam J. Masri, Ananya Mukherjee, Alok Pant, Sumit Sen, and P. Vijayaraghavan, “Thermal Runaway Limit of Tubular Reactors, Defined at the Inflection Point of the Temperature Profile”, Ind. Eng. Chem. Res, Vol 31, No.9, (1992)

ANALYSIS OF MINIMUM VAPOR FLOWRATE REQUIREMENTS FOR THE FULLY THERMALLY COUPLED DISTILLATION SYSTEM WITH POSTFRACTIONATOR

Lorena E. Ruíz-Marín [a], *Angel Castro-Agüero* [a], *Nelly Ramírez-Corona* [b], *Arturo Jiménez-Gutiérrez* [c]

[a] Facultad de Ciencias Básicas, Ingeniería y Tecnología,Universidad Autónoma de Tlaxcala, Apizaco, Tlaxcala, Méx. C.P. 90300
[b] Departamento de Ingeniería Química y Alimentos, Universidad de las Américas-Puebla, San Andrés Cholula, Puebla, Méx. , C.P. 72820
[c] Departamento de Ingeniería Química, Instituto Tecnológico de Celaya, Celaya, Gto, Mex. 38010, México

Abstract

In recent years, thermally coupled distillation schemes, especially the fully thermally coupled distillation system (Petlyuk column), have been studied because of their important energy savings with respect to the conventional distillation sequences. A disadvantage of the Petlyuk column is that the savings decrease when a high purity of the middle component is required. A Petlyuk column with postfractionator has been recently proposed, which can correct the differences between the interlinking trays to increase the thermodynamic efficiency.

However, because of the extra recycle streams, design and simulation of this new system is not trivial. This work presents a degrees of freedom analysis and a set of suggested design variables for this system. Then, as a basis for a shortcut design method, Underwood´s equation is developed to calculate the minimum vapor flowrate.

One important problem for the solution of Underwood's equation is the selection of values for the flowrates and compositions for the postfractionator interlinking streams. We show an alternative solution for this problem using a graphical McCabe-Thiele analysis.

1. Introduction

Energy saving has been the focus of many studies in distillation systems. In recent years, thermally coupled distillation schemes, especially the fully thermally coupled distillation system (Petlyuk column), have been studied because of their potential energy savings with respect to the conventional distillation sequences (Tedder and Rudd, 1978, Glinos and Malone, 1985, Fidkowski y Krolikowski, 1986, 1987 and 1990, Triantafyllou and Smith, 1992, Wolff and Skogestad,1995 Halvorsen and Skogestad, 2004). Petlyuk columns are known to have the lowest energy demand for the separation of ternary mixtures, although for certain values of relative volatilities and feed compositions, they can show considerably lower values of thermodynamic efficiencies than the thermally coupled and conventional configurations (Agrawal and Fidkowski, 1998). A disadvantage of the Petlyuk column is that the energy savings decrease when a high purity of the intermediate component is required.

Degrees of freedom for Petlyuk columns have been reported (Chavez-Contreras, 1985; Fidkowski and Krolikowski, 1986). If one assumes adiabatic stages and constant pressure, there exist C+9 degrees of freedom. Chavez-Contreras (1985) established two possible sets of variables that can meet the Petlyuk columns' degrees of freedom.

Shortcut methods have been proposed for the design of Petlyuk and Kaibel columns (Triantafyllou and Smith, 1992, Muralikrishna et. al., 2002). Shortcut methods can be used to quickly evaluate the applicability of a Peltyuk system for a given problem (Halvorsen and Skogestad, 2004). Underwood's equation (Underwood, 1948) has also been used to compare energy requirements of Petlyuk arrangements with other coupled and non coupled distillation systems (Fidkowski and Agrawal, 2001; Agrawal and Fidkowski, 1999).

The thermodynamic efficiency of conventional distillation systems is affected by the remixing of the middle component (Triantafyllou and Smith,1992). For a Petlyuk column, the mismatch between the feed stream and feed tray composition, and between the interlinking streams and interlinking trays compositions are the main reasons for the reduction of its thermodynamic efficiency (Kim, 2002a). The composition differences at the interlinking trays for a Kaibel column have also been studied for quaternary systems (Kim, 2002b).

A new thermally coupled distillation column with postfractionator (FTCDCP) was proposed by Kim (2006). The main purpose of the postfractionator is to solve the mismatch in composition in the interlinking trays of the Petlyuk column, therefore raising its thermodynamic efficiency. Petlyuk and FTCDCP designs have been made using Fenske's equation to calculate the minimum stages in each section, and then using twice that value for a practical column. The interlinking trays between the prefractionactor and the main column are selected so that the compositions match as close as possible. The postfractionator is taken from the middle section of the main column, which has a high composition of intermediate component.

This work presents a degrees of freedom (DOF) analysis for the FTCDCP. A set of design variables is proposed, and Underwood´s equation is developed to calculate the minimum vapor flowrate for FTCDCP. Finally, a McCabe-Thiele analysis is used to fix the values for the flowrates and compositions for the postfractionator interlinking streams.

2. Degrees of Freedom Analysis

The analysis of DOF for the FTCDCP system was made using a similar methodology as the one used by Chavez-Contreras (1985) for the Petlyuk column. FTCDCP can be divided into subsystems (prefractionator, main column and postfrationator), which are formed by basic elements (stages, condenser, reboiler, etc), Figure 1. DOF of each subsystem is computed adding the DOF for its basic elements, and then subtracting (C+2) DOF for each redundant stream. Finally, the DOF for the FTCDCP system is obtained adding the DOF for each subsystem and then subtracting the DOF of the corresponding interlinking streams.

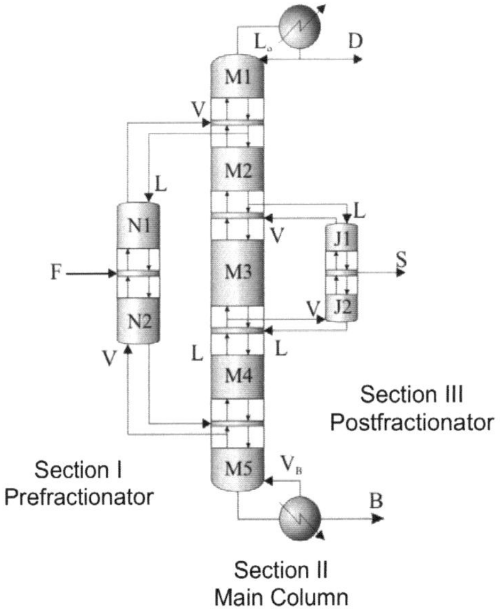

Figure 1. Subsystems of FTCDCP column

A summary of the DOF computation is shown in Table 1. All stages are considered non-adiabatic; feed stage and interlinking stages are included. The DOF for Section I are 2N+3C+8, which are the same as those obtained by Chavez-Contreras (1985). Section II has 2M + 4C + 20 DOF. This section has 2C+3 additional DOF than the Petlyuk main column because of the two additional feeds at the interlinking stages with postfrationator and the flow rates of side streams of the interlinking stages. Finally, the overall system has 2(N + M + J) + C + 19 DOF.

Table 1. DOF of the FTCDCP system

SUBSYSTEM	DOF
SECTION I	**2N + 3C + 8**
SECTION II	**2M + 4C + 20**
SECTION III	**2J + 2C + 7**
Sum	2(N + M + J) + 9C + 35
Redundant streams= 8(C+2)	-8C – 16
FTCDCP's DOF	**2(N + M + J) + C + 19**

Two convenient initial sets of specified variables are shown in Table 2. The first 2(N+M+J)+C+3 variables are quite common. Sixteen additional degrees of freedom remain. Case I of Table 2 is to be used for design problems, and Case II for simulation cases.

Table 2. Proposed set of specified variables

Heat transfer to/from all stages, including divider and excluding condenser and reboiler	M+N+J-1
Pressure of each equilibrium stage, including condenser, reboiler and divider	M+N+J+1
Condenser degree of subcooling	1
Feed stream completely specified	C+2
Sub Total	**2(N + M + J) + C + 3**

CASE I Component recoveries specified		**CASE II Number of equilibrium stages specified**	
Total flow rate of two product streams	2	Number of stages of each column	3
Reflux ratio (> minimum) of main column	1	Feed location	1
Recovery of light, middle and heavy key component in distillate, sidestream and bottom respectively	3	Location of the interlinking stage and stage with sidestream	5
Purity of light or heavy key in sidestream	1	External reflux ratio	1
Recovery of light and heavy component in section I	2	Flow rate of the interlinking vapor streams	2
Reflux ratio (L/V > minimum) of section I	1	Flow rate of the interlinking liquid strems	2
Net flow rate from main column to section VI or two interlinking flow rates	2	Total flow rate of two product streams	2

3. Minimum flows of FTCDCP

Underwood`s equation is developed to obtain the minimum vapor flowrate for the FTCDCP system, which is divided as shown in Figure 2.

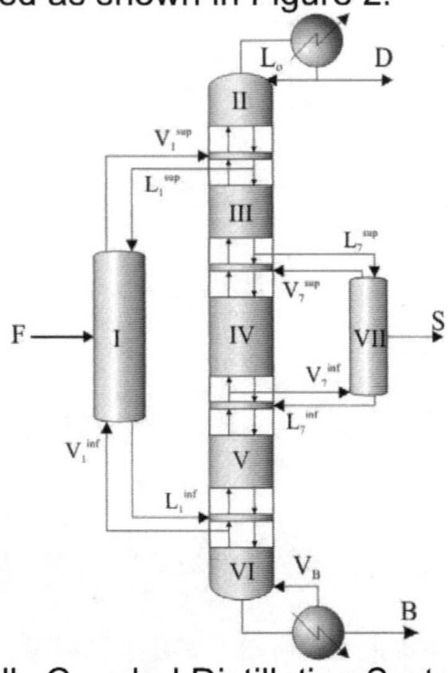

Figure 2. Sections of Thermally Coupled Distillation System for Underwood's procedure

Section I

Section I can be treated as a conventional column, with a given distribution of the middle component in the top and bottoms products; the minimum vapor flowrate is determined using a conventional Underwood's procedure (Triantafyllou and Smith, 1992, Muralikrishna et. al., 2002).

Sections II, III, IV, V and VI

The main column of FTCDCP is divided into 5 sections, Figure 4. Each of these sections must operate above minimum flow rate conditions, and we must estimate which one of them is the controlling section. Minimum vapor flow rates for complex columns arrangements (as the main column of FTCDCP) can be determined from extensions of Underwood´s equation (Kister, 1992; Nikolaides and Malone, 1987). Equations 1-4 show the minimum flow rate required for each Section of the main column.

$$V_{2\min} = \sum_{i=1}^{3} \frac{\alpha_i \left(D x_{Di} - L_1^{\sup} x_{1i}^{\sup} \right)}{\alpha_i - \phi^{II}} \qquad (1)$$

$$V_{3\min} = \sum_{i=1}^{3} \frac{\alpha_i (D x_{Di} - D^1 x_{Di}^1)}{\alpha_i - \phi^{III}} \qquad (2)$$

$$V_{4\min} = \sum_{i=1}^{3} \frac{\alpha_i (D x_{Di} - D^1 x_{Di}^1 + D^7 x_{Di}^7)}{\alpha_i - \phi^{IV}} \qquad (3)$$

$$V_{5\min} = \sum_{i=1}^{3} \frac{\alpha_i (D x_{Di} - D^1 x_{Di}^1 + D^7 x_{Di}^7 + B^7 x_{Bi}^7)}{\alpha_i - \phi^{V}} \qquad (4)$$

The postfractionator can be divided into two sections, as shown in Figure 3; the upper section is taken as a stripping section, while the lower section is considered as a rectifying section. Side stream flow rate, S, will be the sum of bottoms flow rate S1 and distillate flow rate S2.

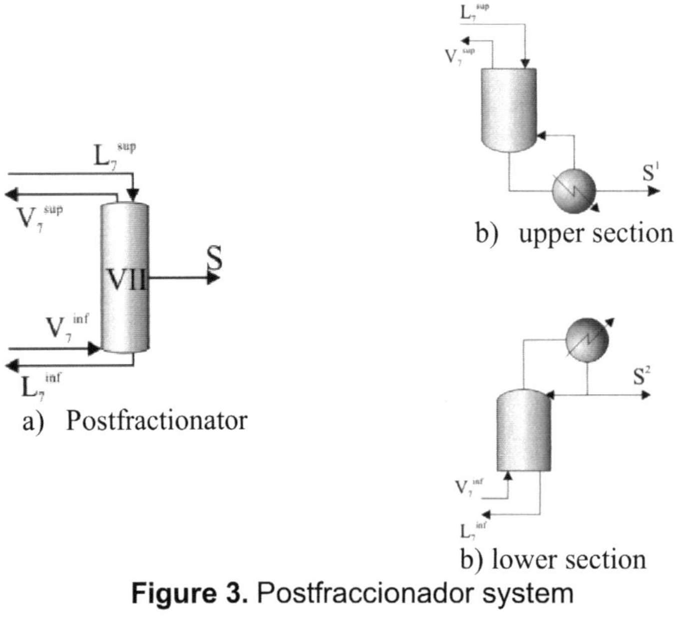

Figure 3. Postfraccionador system

Equation 5 estimates the upper section minimum flow rate, while Equation 6 estimates the lower section minimum flow rate.

$$-\frac{V_7^{\sup}}{S^1} = \sum_{i=1}^{3} \frac{\alpha_i x_{Si}^1}{\alpha_i - \phi_{\sup}^{VII}} \qquad (5)$$

$$\frac{V_7^{\inf}}{S^2} = \sum_{i=1}^{3} \frac{\alpha_i x_{Si}^2}{\alpha_i - \phi_{\inf}^{VII}} \qquad (6)$$

At this point, minimum vapor flow rates for each section of FTCDCP could be determined if all the values of ϕ were known. However, the interlinking feed stage of section III-IV is the same that for the upper section of the postfractionator; these sections therefore have the same minimum vapor flow rate and the same absorption factor ($\phi^{III} = \phi_{\inf}^{VII}$). A similar analysis can be done with the interlinking feed stage of sections IV-V and the lower section of the postfractionator. Section IV should be a simultaneous rectifying and stripping section. If this observation holds, the structure could be rearranged as shown in Figure 4, with section IV operating at total reflux. In the modified structure, the minimum flow rate in section IV is not calculated, and flow rates of sections VII (upper and lower) are set equal. In addition, we need to find the optimum values of liquid and vapor interlinking streams for the postfractionator.

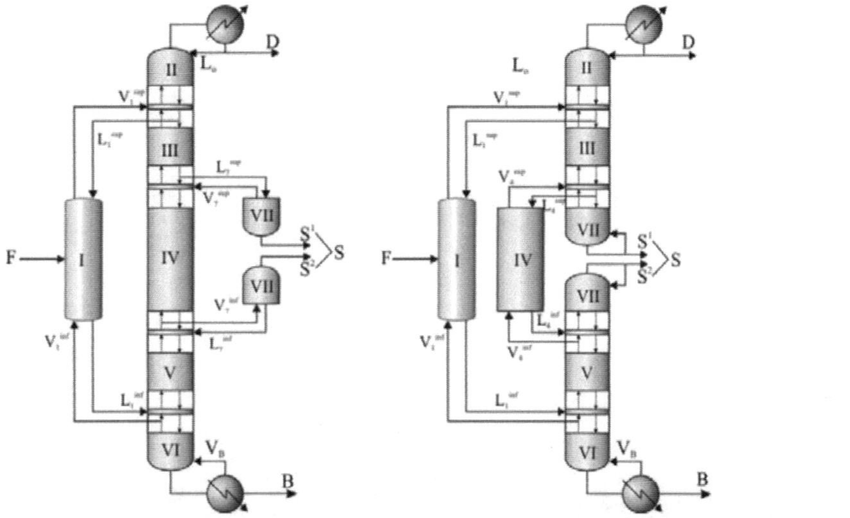

a) Initial Structure b) Modified Structure

Figure 4. Thermally coupled distillation system with post-fractionator

The Underwood's equations for the modified structure are:

<u>Section between D – S[1]</u>

$$R_{I\,\mathrm{Im}\,in} = \sum_{i=1}^{3} \frac{\alpha_i x_{Di}}{\alpha_i - \phi^{II}} - 1 \quad where \quad R_{II\,\mathrm{Im}\,in} = \frac{L_{I\,\mathrm{Im}\,in}}{D}$$

$$\phi^{II} \text{ is determined by equation: } \sum_{i=1}^{3} \frac{\alpha_i x_D^1}{\alpha_i - \phi^{II}} = 1 + \frac{L_1^{\sup}}{D^1} \qquad (7)$$

$$R_{3\min} = \frac{1}{\left(L_1^{\sup} + D\right)} \left[\sum_{i=1}^{3} \frac{\alpha_i (D x_{Di} - D^1 x_{Di}^1)}{\alpha_i - \phi^{III}} + D^1 - D \right] \quad where \quad R_{3\min} = \frac{L_{3\min}^{\sup}}{L_1^{\sup} + D}$$

$$\phi^{II} \text{ is determined by equation: } \sum_{i=1}^{3} \frac{\alpha_i x_{4i}^{\sup}}{\alpha_i - \phi^{III}} = 0 \qquad (8)$$

$$R_{3\min} = \frac{1}{\left(L_1^{\sup} + D\right)}\left[\sum_{i=1}^{3} \frac{\alpha_i (Dx_{Di} - D^1 x_{Di}^1)}{\alpha_i - \phi^{III}} + D^1 - D\right] \quad where \quad R_{3\min} = \frac{L_{3\min}^{\sup}}{L_1^{\sup} + D}$$

$$\phi^{II} \text{ is determined by equation: } \sum_{i=1}^{3} \frac{\alpha_i y_{4i}^{\sup}}{\alpha_i - \phi^{III}} = 1 \tag{9}$$

One should notice that the value of D_4 is 0 because section IV operates at total reflux conditions (the absorption factor is infinite). Once the reflux ratios for the top zone have been calculated, we establish the dominant reflux as follows.

$$R_{\min}^{up} = \max\begin{cases} \dfrac{L_{2\min}^{\sup}}{D} \\ R_{3\min} + \dfrac{L_1^{\sup}}{D}\left(R_{3\min} + 1\right) + \dfrac{L_4^{\sup}}{D} \\ R_{3\min} + \dfrac{L_1^{\sup}}{D}\left(R_{3\min} + 1\right) + \dfrac{L_4^{\sup}}{D} \end{cases} \tag{10}$$

<u>Section between S^2-B</u>

$$R_{7\inf\min} = \sum_{i=1}^{3} \frac{\alpha_i x_{Si}^2}{\alpha_i - \phi^{VII\inf}} - 1 \quad where \quad R_{7\inf\min} = \frac{L_{7\inf\min}^{\sup}}{S^2}$$

$$\phi^{VII} \text{ is determined by equation: } \sum_{i=1}^{3} \frac{\alpha_i y_{4i}^{\inf}}{\alpha_i - \phi^{VII\inf}} = 1 \tag{11}$$

$$R_{7\inf\min} = \sum_{i=1}^{3} \frac{\alpha_i x_{Si}^2}{\alpha_i - \phi^{VII\inf}} - 1 \quad where \quad R_{7\inf\min} = \frac{L_{7\inf\min}^{\sup}}{S^2}$$

$$\phi^{VII} \text{ is determined by equation: } \sum_{i=1}^{3} \frac{\alpha_i x_{4i}^{\inf}}{\alpha_i - \phi^{VII\inf}} = 0 \tag{12}$$

$$R_{5\min} = \sum_{i=1}^{3} \frac{\alpha_i x_{Si}^2}{\alpha_i - \phi^{V}} - 1 \quad donde \quad R_{5\min} = \frac{L_{5\min}^{\sup}}{S^2}$$

$$\phi^{V} \text{ is determined by equation: } \sum_{i=1}^{3} \frac{\alpha_i x_{Bi}^1}{\alpha_i - \phi^{V}} = 1 - \frac{L_1^{\inf}}{B^1} \tag{13}$$

The net flow rate B4 is also 0 because section IV operates at total reflux conditions. The absorption factor has to be calculated with the compositions the interlingking streams. The controlling minimum reflux is calculated using Equation 14.

$$R_{\min}^{low} = \max\begin{cases} \dfrac{L_{7\inf\min}^{\sup}}{S^2} \\ R_{5\min} - \dfrac{L_4^{\inf}}{S^2} \\ R_{5\min} + \dfrac{L_4^{\inf}}{S^2} \end{cases} \tag{14}$$

Finally, the minimum reflux ratio for the whole system can be set as the highest value from each section.

4. McCabe-Thiele analysis

In the FTCDCP arrangement for ternary mixtures, sections II, III and VII can be viewed as a column that separates a binary mixture (AB) if the amount of the heavy component (C) is low. The same can be said for sections VII, V and VI, which practically separate the binary mixture BC if the amount of A is not significant. McCabe-Thiele diagrams for both cases are shown in Figures 5 and 6.

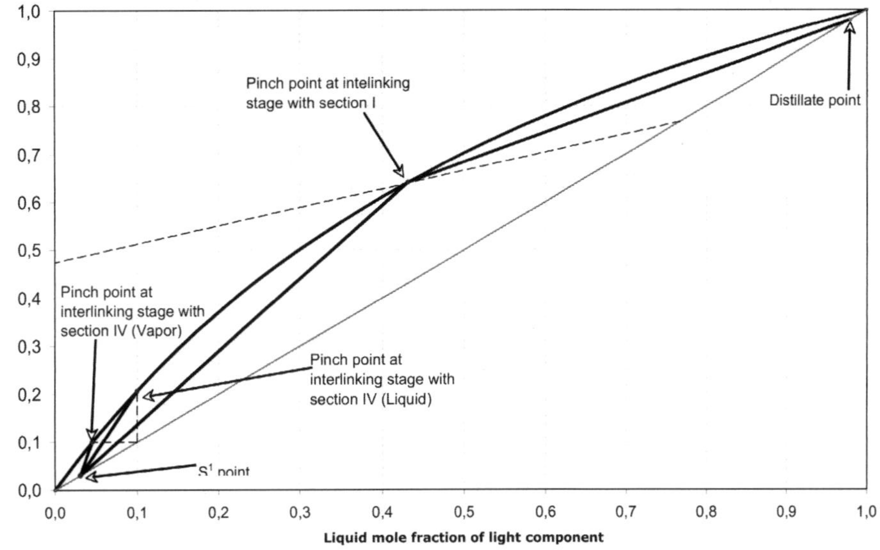

Figure 5. McCabe-Thiele diagram (minimum conditions) of the binary column that represents Sections II, III and

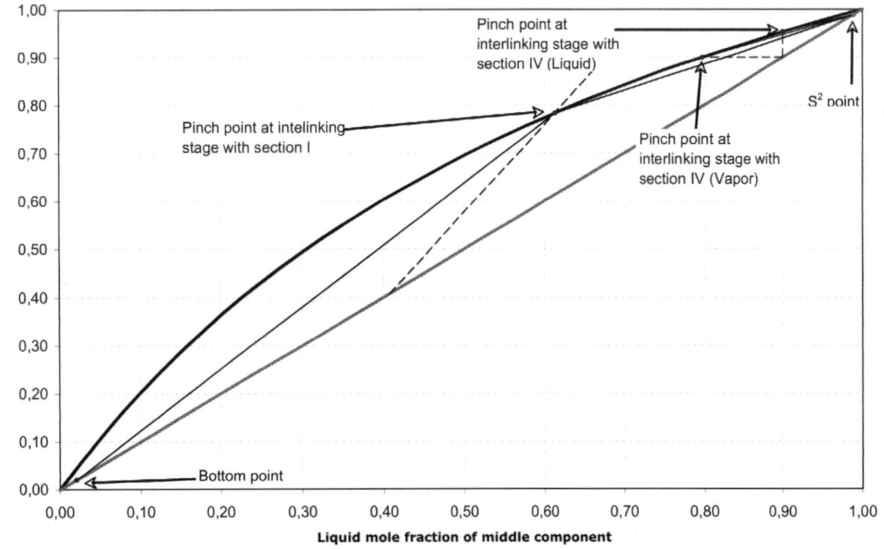

VII

Figure 6. McCabe-Thiele diagram (minimum conditions) of the binary column that represents Sections VII, V and VI

Pinch points, which are equivalent to Equations 7, 8 and 9 (Figure 5) and Equations 11, 12 and 13 (Figure 6), are shown in both figures. Also, one can find the compositions of the interlinking stages (at minimum reflux conditions). Compositions of interlinking streams between sections IV and VII are part of the specifications (Table 2). Therefore, pinch points compositions at interlinking stages with section VII are a consequence of those specifications. On the other hand, compositions of the interlinking stream between sections I and II, and I and VI, which are not specified, can be obtained directly with a McCabe-Thiele diagram. Interlinking streams can be represented by an equivalent net flow with a superheated or subcooled thermal condition (Glinos and Malone, 1985); therefore, distillate and bottom lines of section I can be easily drawn on the McCabe-Thiele diagram.

A limiting case occurs if the interlinking flow rate between Sections IV and VII is cero. In this case, the operating lines of the postfractionator and the main column overlap each other; the FTCDCP becomes equivalent to a Petlyuk column, and only pinch points with Section I can exist.

5. Conclusions

Degrees of freedom have been determined for the FTCDCP system, and a set of design variables have been proposed to meet those degrees of freedom. Underwood's equation was developed to find the minimum flow rate for this complex arrangement. There are several variables (interlinking streams between the Prefractionator and the main column, interlinking streams between the Postfractionator and the main column, and compositions or recoveries of such interlinking streams) that can be part to an optimization problem for this distillation system.

6. Nomenclature

q	Feed thermal condition
B	Bottom flow rate of FTCDCP, Kg-mol/hr
D	Distillate flow rate of FTCDCP, Kg-mol/hr
S	Sidestream flow rate of FTCDCP, Kg-mol/hr
F	Feed flow rate of FTCDCP, Kg-mol/hr
L	Liquid flow rate above feed, Kg-mol/hr
L'	Liquid flow rate below feed, Kg-mol/hr
R	Reflux ratio
V	Vapor flow rate above feed, Kg-mol/hr
V'	Vapor flow rate below feed, Kg-mol/hr
x	Mole fraction
Greek symbols	
α	Relative volatilities
ϕ	Underwood`s equation roots
subindices	
B	Refers to Bottom of column
D	Refers to Dome of column
S	Refers to Sidestream of column
F	Refers to Feed
i	Refers to the *i* component
min	Minimum
(1,2,3,4,5,6,7)	Refers to the section number of FTCDCP
(I,II,III)	Refers to the section number of FTCDCP to obtain the DOF
superindices	
inf	Inferior
min	Minimum
sup	Superior
(1,7)	Sección del sistema acoplado térmicamente con postfraccionador

7. References

Agrawal, R. and Fidkowski, Z.T., 1998, Are Thermally Coupled Distillation Columns Always Thermodynamizally More Efficient for Ternary Distillations?, *Ind. Eng. Chem. Res.*, **37**, 3444-3454.

Agrawal, R. and Fidkowski, Z.T., 1999, New Thermally Coupled Schemes for Ternary Distillation, *AIChE Journal*, **45**, 485-496.

Chavez-Contreras, R., 1985, Multiple Solutions for Interlinked Distillation Systems, PhD Thesis, University of Utah, Department of Chemical Engineering.

Fidkowski, Z.T. and Krolikowski, L., 1986, Thermally Coupled System of Distillation Columns: Optimization Procedure, *AIChE Journal*, **32**, 537-546.

Fidkowski, Z.T. and Krolikowski, L., 1987, Minimum Energy Requirements of Thremally Coupled Distillation Systems, *AIChE Journal*, **32**, 643-653.

Fidkowski, Z.T. and Krolikowski, L., 1990, Energy Requirements of Nonconventional Distillation Systems, *AIChE Journal*, **36**, 1275-1278.

Fidkowski, Z.T. and Agrawal, R., 2001, Multicomponent Thermally Coupled Systems of Distillation Columns at Minimum Reflux, *AIChE Journal*, **47**, 2713-2724

Glinos, K. and Malone, M.F., 1985, Minimum Vapor Flows in a Distillation Column with a Sidestream Stripper, *Ind. Eng. Chem. Process Des. Dev.,* **24**, 1087-1090.

Halvonsen, I.J. and Skogestad, S., 2004, *Shortcut Analysis of Optimal Operation of Petlyuk Distillation, Ind. Eng. Chem. Res.*, **43**, 3994-3999.

Kim, Y.H., 2002a, Structural design of a fully thermally coupled distillation column, *Chemical Engineering Journal*, **85**, 289-301.

Kim, Y.H., 2002b, Rigorous Design of Extended Fully Thermally Coupled Distillation Columns, *Chemical Engineering Journal*, **89**, 89-99

Kim, Y. H., 2006. A New Fully Thermally Coupled Distillation Column with Postfractionator, *Chemical Engineering and Processing*, **45**, 254-263.

Kister, H.Z., 1992, Distillation Design, *McGraw Hill, Inc.*

Muralikrishna, K., Madhavan, K.P. and Shah, S.S., 2002, Development of Dividing Wall Distillation Column Design Space for a Specified Sepration, *Trans IchemE, Part A, Chem Eng. Res. Des.*, **80**, 155-166.

Nikolaides, I.P. and Malone, F.M., 1987, Approximate Design of Multiple-Feed/Side-Stream Distillation Systems, *Ind. Eng. Chem. Res.*, **26**, 1839-1845.

Tedder, D.W. and Rudd, D.F., 1978, Parametric studies in industrial distillation, *AIChE Journal*, **24**, 303-334.

Triantafyllou, C and Smith, R, 1992, The Design and Optimisation of Fully Thermally Coupled Distillation Columns, *Trans IchemE, Part A, Chem Eng. Res. Des.*, **70**, 118-132.

578o Composition Estimation of Reactive Distillation Systems and Its Implication to Feasibility

Jae W. Lee, Department of Chemical Engineering, The City College of New York, 140th street and Convent Ave, New York, NY 10031 and James Chin, Chem. Eng., Department of Chemical Engineering, The City College of New York, 140th street and Convent Ave, New York, NY 10031.

This work presents a new mathematical procedure for estimating the liquid still composition trajectory of batch reactive distillation systems. Simple information such as material balance and reaction equilibrium will be utilized to estimate liquid trajectory of simultaneous reaction and V-L separation. The trajectory information is essential to determining whether products are reachable when several distillation boundaries are present, initial feed compositions can vary, and the number of components exceeds our visualization capability. For a given feed to product flow ratio with constant feed charge and product compositions, the still pot composition trajectory is mathematically confined to the intersection between the reaction equilibrium manifolds and a "material balance plane" that is the union of stoichiometric lines and material balance rays connecting still and product compositions. Starting from this estimated still pot trajectory, we can evaluate the feasibility of various batch reactive distillation configurations by constructing the residue curves for non-extractive cases and the feasible region formed by upper and lower bounds of reflux ratios for extractive cases.

The Study of Process Design for Production of the Clean Fuel – Dimethyl Ether (DME)

You-Syuan Chen, Jyun-Da Lin, Jhen-Ren Lee, An-Li Shieu, Jing-Jie Lan, and Gow-Bin Wang[*]

Department of Chemical and Materials Engineering / Green Technology Research Center, Chang Gung University, Kwei-Shan, Tao-Yuan 333, Taiwan

Abstract

It is known that dimethyl ether (DME) has wide application, such as solvent, propellant, chemical intermediate, refrigerant replacement, LPG substitute and transportation fuel. Furthermore, DME has currently attracted considerable attention as an alternative clean fuel for diesel engines because it can emit much less pollutants such as CO, NO_x and particulates. Traditionally, in industrial processes, DME is produced by dehydration of methanol using acidic porous catalyst. Methanol is commercially synthesized from CO/CO_2 hydrogenation over Cu-based catalyst. Thus one can prepare DME from synthesis gas (syngas) in a two-step process of syngas to methanol and further to DME, or in a single-step process of synthesis gas direct to DME.

Due to attracting attention of the DME synthesis research, this work first uses commercial process simulators to analyze different DME synthesized methods presented in the literature. The optimal operating conditions for DME direct synthesis are then clearly evaluated. Finally, a feasible process flow diagram for the direct DME process is developed to meet the desired production specifications. These steady-state simulation results are demonstrated to fit with those shown in the literature.

Introduction

Dimethyl ether (DME) has wide application, such as solvent, propellant, chemical intermediate, refrigerant replacement, LPG substitute and transportation fuel. Recently, dimethyl ether has attracted attention as an alternative fuel for diesel engines and LPG substitute. Moreover, because it can emit much less pollutants such as CO, NO_x and particulates, dimethyl ether really meets the standards for clean fuel (Ng et al., 1999). In the future, DME will be much used in fuel industries.

It is known DME is commercially prepared by dehydration of methanol using acidic porous catalyst. Some of the proposed reaction rates for this traditional DME synthesis are presented in Table 1 (Chen, 2008). In industrial process, methanol is synthesized from CO/CO_2 hydrogenation over Cu-based catalyst. Therefore, one can prepare DME from synthesis gas (syngas) in a two-step (or indirect) process of syngas to methanol and further to DME. Recently, it has been reported that DME can be synthesized from syngas in a single-step (or direct) process, which is more thermodynamically and economically favorable than the indirect process (Lu et al., 2004). Recently, Lee et al. (2006) have developed a simulator of a shell and tube type fixed bed reactor for DME direct synthesis. Ohno et al. (2007) have demonstrated their slurry phase DME direct synthesis technology by long term operation of 100 tons/day plant.

Due to powerful development of the commercial process simulators and attracting attention of the DME synthesis research, this work tries to apply the most widely used commercial process design simulation software to constructing process flowsheets for the direct DME process. First, commercial process simulators such as AspenPlus and AspenDynamics are used to analyze different DME synthesized methods presented in the literature. The optimal operating conditions for DME direct synthesis are then clearly evaluated. Finally, a feasible process flow diagram for the direct DME process is developed to meet the desired production specifications. These steady-state simulation results are demonstrated to fit with those shown in the literature.

Steady-state Simulation for DME Synthesis

In industrial processes, DME is traditionally produced by dehydration of methanol using acidic porous catalyst. Bondiera and Naccache (1991) have provided a simple kinetic expression in power law for this reaction. Later, Bercic and Levec (1992) propose the following rate equation

$$-r_{CH_3OH} = \frac{k_s K_{CH_3OH}^2 (C_{CH_3OH}^2 - C_{H_2O} C_{DME} / K_{eq})}{(1 + 2(K_{CH_3OH} C_{CH_3OH})^{1/2} + K_{H_2O} C_{H_2O})^4} \qquad \Delta H_r = -21.003 \text{ kJ/mol.}$$

By using this more complicated expression of reaction rate, Figure 1 shows the temperature distribution of the adiabatic fixed bed reactor for traditional DME synthesis. Simulation results for traditional DME synthesis in adiabatic fixed bed reactor are shown as Table 2. In this case, methanol conversion is about 9% higher.

It is known methanol is commercially synthesized from CO/CO_2 hydrogenation over Cu-based catalyst. Thus one can prepare DME from synthesis gas (syngas) in a single-step (or direct) process. In this study, the following reaction rate equations given by Bussche and Froment (1996) and Bercic and Levec (1992) are utilized to find the optimal operating conditions of the fixed bed reactor for DME direct synthesis.

$$CO_2 + 3H_2 \leftrightarrow CH_3OH + H_2O \qquad r_{CH_3OH} = \frac{k_1 (P_{H_2} P_{CO_2}) [1 - (1/K^*)(P_{CH_3OH} P_{H_2O})/(P_{CO_2} P_{H_2}^3)]}{(1 + K_2 (P_{H_2O}/P_{H_2}) + \sqrt{K_3 P_{H_2}} + K_4 P_{H_2O})^3} \qquad \Delta H_r = -49.47 \text{ kJ/mol}$$

$$CO + H_2O \leftrightarrow H_2 + CO_2 \qquad r_{RWGS} = \frac{k_5 P_{CO_2} [1 - K_3^* (P_{CO} P_{H_2O} / P_{CO_2} P_{H_2})]}{1 + K_2 (P_{H_2O}/P_{H_2}) + \sqrt{K_3 P_{H_2}} + K_4 P_{H_2O}} \qquad \Delta H_r = -41.17 \text{ kJ/mol}$$

$$2CH_3OH \leftrightarrow CH_3OCH_3 + H_2O \qquad -r_{CH_3OH} = \frac{k_s K_{CH_3OH}^2 (C_{CH_3OH}^2 - C_{H_2O} C_{DME} / K_{eq})}{(1 + 2(K_{CH_3OH} C_{CH_3OH})^{1/2} + K_{H_2O} C_{H_2O})^4} \qquad \Delta H_r = -21.003 \text{ kJ/mol}$$

By using this more complicated expression of rate, Figure 2 shows the temperature distribution of the adiabatic fixed bed reactor for DME direct synthesis. Simulation results for DME direct synthesis in adiabatic fixed bed reactor shown as Table 3 are consistent with those provided by Lee et al. (2006).

Optimal Operating Conditions for DME Direct Synthesis

Due to facts of the going up prices of feedstock and thermodynamic and economical advantages of DME direct synthesis from syngas, this study tries to determine the optimal operating conditions for DME direct synthesis. The effects of important process variables on DME reactor have been clearly tested by use of the reaction rates presented in previous section (Chen, 2008). Then, it is found the following suitable ranges of variables: inlet temperature 450~470 K, inlet pressure 70~80 atm, $H_2/(CO_2+CO)$ flow ratio 3.5~4.5, $CO_2/(CO+CO_2)$ flow ratio 0~0.1, and exit temperature 470~478 K.

Table 4 shows the determined optimal operating conditions of the fixed bed reactor for DME direct synthesis. Figure 3 presents the temperature distribution of the optimal fixed bed reactor for DME direct synthesis. It is clear that within the middle-section reactor temperature range of 500~520 K, the better DME productivity could be obtained (Lee et al., 2006). The CO conversion is about 99% in this case.

Design Flowsheet for Direct DME Process

According to the obtained optimal operating conditions of the fixed bed reactor for DME direct synthesis shown in Table 4, a feasible process flow diagram for the direct DME process is then developed to meet the desired production specifications. The designed steady-state process flowsheet is shown as Figure 4. This direct DME process flowsheet involves following major unit operations: a fixed bed chemical reactor, a vapor-liquid separator, a product stripper, and two distillation towers.

Table 5 shows the simulated operating conditions of the fixed bed reactor for overall direct DME process. The obtained reactor conditions are consistent with those shown in Table 4. Table 6 illustrates all information of the recycle and purge streams for this proposed direct DME process. It is easy to find the DME production rate of 54.5 tons/day under the proposed design flowsheet for direct DME process.

Conclusions

Due to powerful development of the commercial process simulators and attracting attention of the DME synthesis research, this work tries to apply the most widely used commercial process design simulation software to constructing process flowsheets for the direct DME process. First, commercial process simulators such as AspenPlus and AspenDynamics are used to analyze different DME synthesized methods presented in the literature. The optimal operating conditions for DME direct synthesis are then clearly evaluated. Finally, a feasible process flow diagram for the direct DME process is developed to meet the desired production specifications. These steady-state simulation results are demonstrated to fit with those shown in the literature.

Acknowledgements

Financial support from Chang Gung University (BMRP 573) is gratefully acknowledged. We also greatly acknowledge Aspen Technology for having granted special permission for the use of the ASPEN PLUS under the condition of the academic licensing agreement.

References

1. Bercic, G. and J. Levec (1992), "Intrinsic and Global Reacrion Rate of Methanol Dehydration over γ-Al_2O_3 Pellets," Ind. Eng. Chem. Res., 31, 1035-1040.
2. Bondiera, J. and C. Naccache (1991), "Kinetics of Methanol Dehydration in Dealuminated H-Mordenite: Model with Acid and Basic Active Centres," Applied Catalysis, 69, 139-148.
3. Bussche, K. M. V. and G. F. Froment (1996), "A Steady-State Kinetic Model for Methanol Synthesis and the Water Gas Shift Reaction on a Commercial $Cu/ZnO/Al_2O_3$ Catalyst," Journal of Catalysis, 161, 1-10.
4. Chen, Y. S. (2008), Steady-State and Dynamic Modeling of Dimethyl Ether (DME) Synthesis from Syngas, M.S. Thesis, Dept. of Chemical and Materials Eng., Chang Gung University, Tao-Yuan, Taiwan.

5. Lee, S. B., W. Cho, D. K. Park, and E. S. Yoon (2006), "Simulation of Fixed Bed Reactor for Dimethyl Ether Synthesis," Korean J. Chem. Eng., 23, 522-530.
6. Lu, W. Z., L. H. Teng, and W. D. Xiao (2004), "Simulation and Experiment Study of Dimethyl Ether Synthesis from Syngas in a Fluized-Bed Reactor," Chemical Engineering Science, 59, 5455-5464.
7. Ng, K. L., D. Chadwick, and B. A. Toseland (1999), "Kinetics and Modeling of Dimethyl Ether Synthesis from Synthesis Gas," Chemical Engineering Science, 54, 3587-3592.
8. Ohno, Y., H., Yagi, N. Inoue, K. Okuyama, and S. Aoki (2007), "Slurry Phase DME Direct Synthesis Technology – 100 tons/day Demonstration Plant Operation and Scale up Study," Natural Gas Conversion VIII, 403-408.

Table 1. Some published rate equations for traditional DME synthesis (Chen, 2008).

Author (Year)	Rate equation
Kallo and Knözinger (1967)	$-r_{CH_3OH} = k\dfrac{C_{CH_3OH}^{1/2}}{C_{CH_3OH}^{1/2} + k_2 C_{H_2O}}$
Figueras et al. (1971)	$-r_{CH_3OH} = \dfrac{kK_{CH_3OH}C_{CH_3OH}^{1/2}}{1 + K_{CH_3OH}C_{CH_3OH}^{1/2} + K_{H_2O}C_{H_2O}}$
Than et al. (1972)	$-r_{CH_3OH} = \dfrac{kK_{CH_3OH}C_{CH_3OH}}{(1 + K_{CH_3OH}C_{CH_3OH})^2}$
Schmitz (1978)	$-r_{CH_3OH} = k_1 + k_2 C_{CH_3OH}$
Klusaček and Schneider (1982)	$-r_{CH_3OH} = \dfrac{kK_{CH_3OH}C_{CH_3OH}}{[1 + 2(K_{CH_3OH}C_{CH_3OH})^{1/2} + K_{H_2O}C_{H_2O}]^2}$ $-r_{CH_3OH} = \dfrac{kK_{CH_3OH}C_{CH_3OH}}{[1 + 2(K_{CH_3OH}C_{CH_3OH})^{1/2} + K_{H_2O}C_{H_2O}]^3}$ $-r_{CH_3OH} = \dfrac{kK_{CH_3OH}C_{CH_3OH}}{[1 + 2(K_{CH_3OH}C_{CH_3OH})^{1/2} + K_{H_2O}C_{H_2O}]^4}$
Berčič and Levec (1992)	$-r_{CH_3OH} = \dfrac{k_s K_{CH_3OH}^2 (C_{CH_3OH}^2 - C_{H_2O}C_{DME}/K_{eq})}{[1 + 2(K_{CH_3OH}C_{CH_3OH})^{1/2} + K_{H_2O}C_{H_2O}]^4}$

Table 2. Simulation results for traditional DME synthesis in adiabatic fixed bed reactor.

	Reactor inlet	Reactor outlet
DME (kmol/hr)	0	113.86
Methanol (kmol/hr)	262	34028
Water (kmol/hr)	0	113.86
Total flow (kmol/hr)	262	262
Temperture (K)	551.45	681.71
Pressure (atm)	14.9	14.4

Table 3. Simulation results for DME direct synthesis in adiabatic fixed bed reactor.

	Reactor inlet	Reactor outlet
DME (kmol/hr)	0	4.39
CO (kmol/hr)	110.55	94.64
CO_2 (kmol/hr)	0	6.35
H_2 (kmol/hr)	221.1	208.35
Methanol (kmol/hr)	0	0.78
Water (kmol/hr)	3.35	1.38
Total flow (kmol/hr)	335	315.9
Temperture (K)	555	683.5
Pressure (atm)	49.35	48.26

Table 4. Optimal operating conditions of the fixed bed reactor for DME direct synthesis.

	Reactor inlet	Reactor outlet
Total flow (kmol/hr)	335	238.84
DME	0	23.64
CO	54.27	0.52
CO_2	6.03	11.7
H_2	271.35	180.86
Menthanol	0	0.81
Water	3.35	21.31
Temperaure (K)	468	480
Pressure (atm)	78.95	77.97

Table 5. Simulated operating conditions of the fixed bed reactor for direct DME process.

	Reactor inlet	Reactor outlet
DME (kmol/hr)	0.59	50.36
CO (kmol/hr)	100.4	0.47
CO_2 (kmol/hr)	21.58	2.68
H_2 (kmol/hr)	452.23	252.59
Methanol (kmol/hr)	0	0.33
Water (kmol/hr)	10	59.67
Total flow (kmol/hr)	584.83	385.1
Temperture (K)	468	480
Pressure (atm)	78.95	77.97

Table 6. Information of recycle and purge streams for direct DME process.

	Recycle	Purge
DME (kmol/hr)	0.59	0.50
CO (kmol/hr)	0.44	0.03
CO_2 (kmol/hr)	21.58	20.39
H_2 (kmol/hr)	252.22	0.36
Methanol (kmol/hr)	0	0
Water (kmol/hr)	0	0
Total flow (kmol/hr)	274.83	21.29
Temperture (K)	450	295.28
Pressure (atm)	79.15	37.5

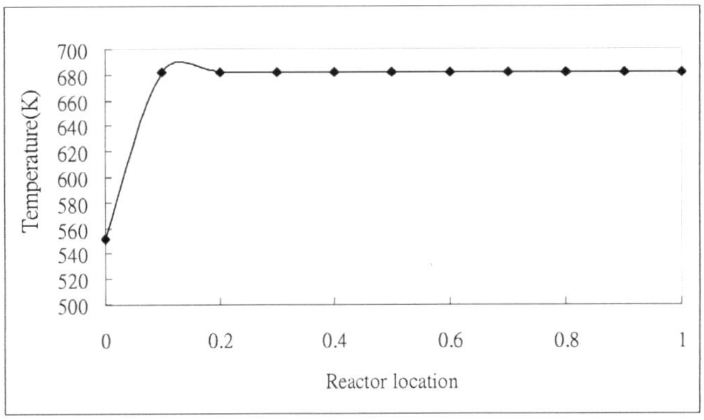

Figure 1. Temperature distribution of the adiabatic fixed bed reactor for traditional DME synthesis.

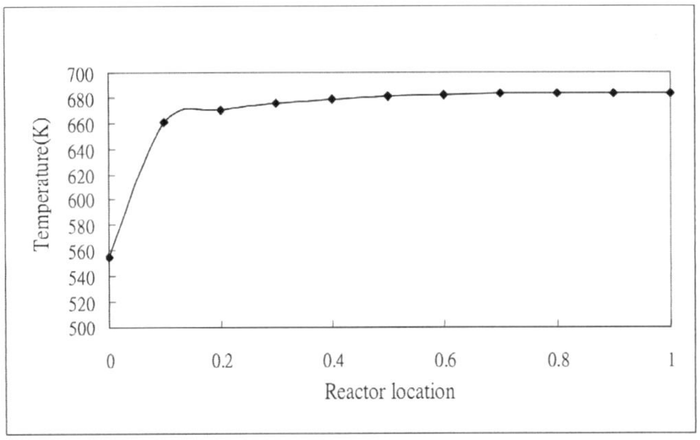

Figure 2. Temperature distribution of the adiabatic fixed bed reactor for DME direct synthesis.

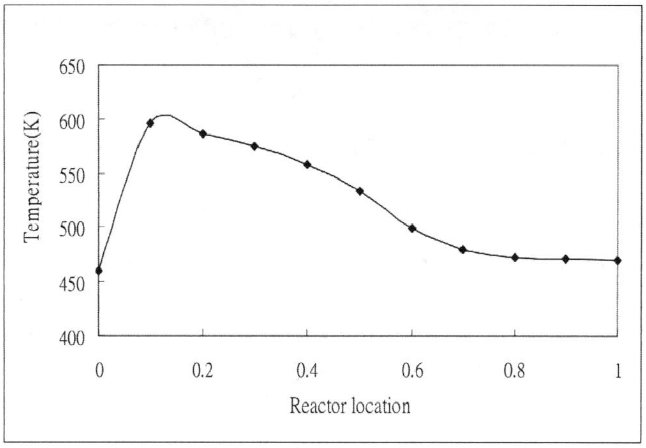

Figure 3. Temperature distribution of the optimal fixed bed reactor for DME direct synthesis.

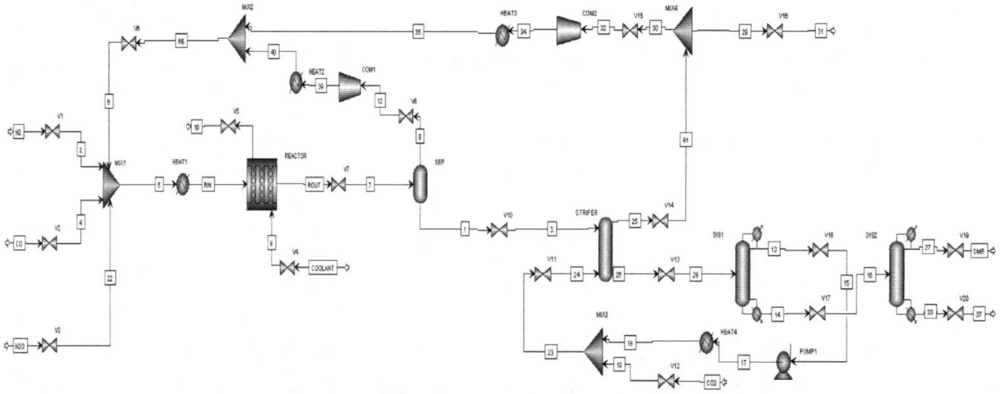

Figure 4. Overall flowsheet of the direct DME process.

578q Web-Based Object-Oriented Modelling Environment for the Hierarchical Design and Simulation of Chemical Processes

Harvey Arellano-Garcia, Walter Martini, Stefan Kuntsche, and Günter Wozny. Department of Process Dynamics and Operation, Berlin Institute of Technology, Str. des 17. Juni 135, 10623 Berlin, Berlin, Germany

An advanced generation of modeling support systems for efficient interdisciplinary research as well as multi-scale modeling and simulation is being developed based on the tool MOSAIC. This work is concerned with the ongoing development of a modeling environment for large and modular equation systems. The basic concept is modularity and interconnectivity of equations and equation systems in different levels. The modeling environment shall be controlled by standard internet browsers to avoid the necessity of installing additional software on local computers. While there are many commercial and well established simulation and modeling tools for chemical engineering on the market, the tool developed in this work follows clear didactic guidelines about how to work with mathematical models of processes in chemical engineering. The project is thus aimed to provide both a learning tool for chemical engineer students and to help better understand modeling. The tool will encourage working in groups via internet and it will address certain cooperation problems such as documentation, model library organization and reusability of both models and supplementary knowledge.

The modeling in the tool developed is based on the indication of equation systems as it is known from tools such as gPROMS. The equations and their respective comments will be kept in a formal specification sheet using MathML and other web standards. This enables the implementation and presentation of the models in a form known from books, papers and handwritten mathematic formulations. Thus, the students or researchers can focus on the meaning of the model equations without having to work hard on the numeric part of the problem. Since complicated computer code is avoided, the communication between co-workers is made easier. While this principle is already known from tools like Mathcad, the software developed in this work will have the additional characteristic of allowing models and parts of models to be reused, exchanged and interconnected. This modularity is aimed to facilitate working in groups on complex projects where specialization and exchange of information are important.

Currently, the main focus lays on different process alternatives for oxidative coupling of methane, which is being investigated in a cooperation project within the framework of unicat (www.unicat.tu-berlin.de). Here, the evolution of catalytic properties at different levels of complexity is studied by comparing catalytic model systems.

Use of Commercial Process Simulator to Mode Transition Control of the Tennessee Eastman Challenge Problem

*Fei-Yu Lin, You-Syuan Chen, and Gow-Bin Wang**

Chang Gung University, Tao-Yuan, Taiwan

Abstract

The Tennessee Eastman (TE) benchmark process provided by Downs and Vogel in 1993 is a realistic simulation of an actual commercial process. Due to the confidential considerations, real components, some physical properties, and kinetic data in this TE program are modified for academic research and industrial training. It is known that TE plant has seven possible operating modes for three different production mass ratios. Since then, several researches have utilized this TE plant to evaluate optimal operating conditions and performance of the proposed process flowsheets and control schemes.

By studying the characteristics of different TE control structures presented in the literature, this work first apply commercial process design simulators to setting reasonable TE process flowsheets. Different operating modes and control performance are then tested by steady-state and dynamic simulations of this designed TE flowsheet. Finally, mode transition control strategies proposed by Tu (2006) are then used to transfer the TE plant from one operating mode to other modes.

Introduction

In 1993, Downs and Vogel proposed the TE plantwide industrial process control problem as a study of alternative control and optimization strategies. In next year, McAvoy and Ye presented a multiple single-input single-output control loop structure to handle this plantwide control problem. They used NI and RGA rules to analyze the pairings between manipulated variables and controlled variables. This systematic base control structure for TE process was tested by use of four design stages. McAvoy et al. (1995) then modified the base control structure to improve the disturbance rejection ability. The main change was to choose the more suitable pairing for reactor pressure. Therefore, it could manage the disturbances in TE process more effectively. It has been reported that the TE process is highly nonlinear and is open-loop unstable (Downs and Vogel, 1993). Ricker and Lee (1995) developed a model predictive control algorithm by use of a simplified nonlinear model to improve the TE process control performance.

It is known TE plant has seven possible operating modes for three different production mass ratios. The nominal operating condition is set as base case mode. In this direction, Tian and Hoo (2005) proposed two model predictive control approaches which differ in the type of models used to deal with regulation and transition control problems. They claimed better closed-loop regulation and mode transition performance especially when the worse disturbance is presented could be obtained by use of their control strategies. On the other hand, Tu (2006) treated the mentioned TE mode transition control under the case of multiple single-input single-output control loops.

By studying the characteristics of different TE control structures presented in the literature, this work first apply commercial process design simulators to setting reasonable TE process flowsheets. Different operating modes and control performance are then tested by steady-state and dynamic simulations of this designed TE flowsheet. Finally, mode transition control strategies proposed by Tu (2006) are then used to transfer the TE plant from one operating mode to other modes.

Control Schemes for Tennessee Eastman Process

The Tennessee Eastman process, as shown in Figure 1, involves five major unit operations: a two-phase chemical reactor, a partial condenser, a vapor-liquid separator, a product stripper, and a recycle compressor (Downs and Vogel, 1993). Four gaseous reactants A, C, D, and E and an inert B are fed to the exothermic reactor where two liquid products G and H are formed. A liquid byproduct F is also produced by two side reactions.

Table 1 gives seven modes of process operation corresponding to different products G/H mass ratios (50/50, 10/90, and 90/10) and production rates for TE plant. The base operating mode is a 50/50 G/H mass ratio and a production rate of 14,072 lb/hr. The plant simulation gives 12 manipulated variables, 41 measurements, and 50 state variables. Four setpoint changes, 20 different types of disturbances, and some process constraints are also specified by Downs and Vogel (1993).

Since 1993, several researches have utilized the TE for evaluating process control technology. Early work on this problem was reported by McAvoy and Ye (1994), who developed a decentralized plantwide control system design based on multiple single-input single-output control loops. They used steady-state tools such as the RGA, NI, and disturbance analysis and dynamic simulation to determine the most promising pairings of the TE control loops. Figure 2 shows the alternative decentralized control structures. Ricker and Lee (1995) tested this base control strategy and found that the reactor pressure loop cannot handle a large disturbance in the A feed (IDV(6)) and the compressor power loop may cause saturation problem for large setpoint changes. Therefore, McAvoy et al. (1995) proposed an improved base control scheme shown in Figure 3 which uses the A feed to control the A/C ratio in the reactor feed. Moreover, it is clear that the three-level cascade control for the TE reactor with cooling jacket was designed to enhance the regulatory control performance of the reactor pressure loop.

The Proposed Flowsheet for Tennessee Eastman Process

The chemical reactions occurred in TE reactor are expressed as below

$$A(g) + C(g) + D(g) \rightarrow G(liq)$$
$$A(g) + C(g) + E(g) \rightarrow H(liq)$$
$$A(g) + E(g) \rightarrow F(liq)$$
$$3D(g) \rightarrow 2F(liq)$$

and assumed that all the reactions are irreversible and exothermic. It can be found that basic physical properties of these components are shown by Downs and Vogel (1993) but the true chemical formulas of these components are not given. In this study, by trying any possible reacting cases, the following reactions are determined to represent the original TE reactions.

$$H_2 + CO + CH_3OH \rightarrow C_2H_6O_2$$
$$H_2 + CO + C_2H_5OH \rightarrow C_3H_8O_2$$
$$H_2 + CO + CH_3OH \rightarrow C_2H_6O_2$$
$$H_2 + CO + C_2H_5OH \rightarrow C_3H_8O_2$$

As shown in Figure 4, reasonable TE process flowsheet can be constructed by using commercial process simulator such as Aspen Plus. Lin (2008) has presented the fact that the obtained steady-state operation conditions under base case are consistent with those shown in the literature. Figure 5 shows the different disturbance responses for step change in D feed temperature in stream 2. It is apparent that the corresponding response of reactor pressure in the designed dynamic flowsheet is consistent with those shown in the literature.

Mode Transition Control Performance

It is well known that Tennessee Eastman process has seven possible operating modes for three different production mass ratios (50/50, 10/90, and 90/10) and production rates. How to smoothly transfer the TE plant from one operating mode to other modes is an important issue for practical industry. In the work, according to the mode transition control strategies proposed by Tu (2006), one can complete the corresponding flowsheet of plandwide control structure for TE plant as shown in Figure 6.

It is known that the performance testing and evaluation of various TE process control strategies can be done with four setpoint changes and/or 20 different types of disturbances specified by Downs and Vogel (1993). Here, mode transition control performance from base case to mode 2 for TE plant is shown in Figure 7. The responses of reactor temperature and reactor pressures are quite fast. Figure 7 also shows the smooth but slow response of the G/H mass ratio. In fact, this hysteresis characteristics can be improved by fine tune the controller parameters (Tu, 2006).

Conclusions

By studying the characteristics of different TE control structures presented in the literature, this work first apply commercial process design simulators to setting reasonable TE process flowsheets. Different operating modes and control performance are then tested by steady-state and dynamic simulations of this designed TE flowsheet. The obtained steady-state operation conditions and closed-loop control performance under base case are consistent with those shown in the literature. Finally, mode transition control strategies proposed by Tu (2006) are then used to transfer the TE plant from one operating mode to other modes. The smooth but slow response of the G/H mass ratio should be further improved.

Acknowledgements

Financial support from Chang Gung University (BMRP 573) is gratefully acknowledged. We also greatly acknowledge Aspen Technology for having granted special permission for the use of the ASPEN PLUS under the condition of the academic licensing agreement.

References

1. Downs, J. J. and E. F. Vogel (1993), "A Plant-wide Industrial Process Control Problem," Computers and Chemical Engineering, 17, 245-255.
2. Lin, F. Y. (2008), The Application of Commercial Simulators for Mode Transition Control of the Tennessee Eastman Benchmark Process, M.S. Thesis, Dept. of Chemical and Materials Eng., Chang Gung University, Tao-Yuan, Taiwan.

3. McAvoy, T. J. and N. Ye (1994), "Base Control for the Tennessee Eastman Problem," Computers and Chemical Engineering, 18, 383-413.
4. McAvoy, T. J. N. Ye. and C. Gang (1995), "An Improved Base Control for the Tennessee Eastman Problem," Proceedings of the American Control Conference, Seattle, WA, 240-244.
5. Ricker, N. L. and J. H. Lee (1995), "Nonlinear Model Predictive Control of the Tennessee Eastman Challenge Process," Computers and Chemical Engineering, 19, 961-981.
6. Tian, Z. and K. A. Hoo (2005), "Multiple Model-Based Control of the Tennessee-Eastman Process," Industrial Engineering and Chemistry Research, 44, 3187-3202.
7. Tu, H. G. (2006), Mode Transition and Plantwide Control of the Tennessee Eastman Process, M.S. Thesis, Dept. of Chemical and Materials Eng., Chang Gung University, Tao-Yuan, Taiwan.

Table 1. Several operating modes for TE process (Downs and Vogel, 1993).

Mode	G/H mass ratio	Production rate (Stream 11)
1	50/50	7,038kgh^{-1} G and 7,038kgh^{-1} H (Base case)
2	10/90	1,408 kgh^{-1} G and 12,669 kgh^{-1} H
3	90/10	10,000 kgh^{-1} G and 1,111 kgh^{-1} H
4	50/50	maximum production rate
5	10/90	maximum production rate
6	90/10	maximum production rate

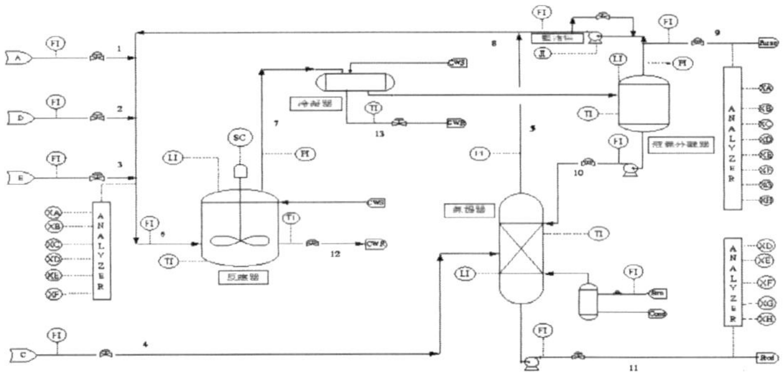

Figure 1. TE process flowsheet (Downs and Vogel, 1993).

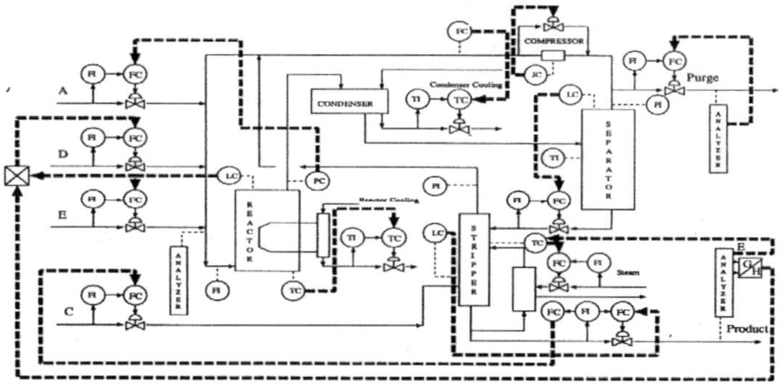

Figure 2. Base control structure by McAvoy and Ye (1994).

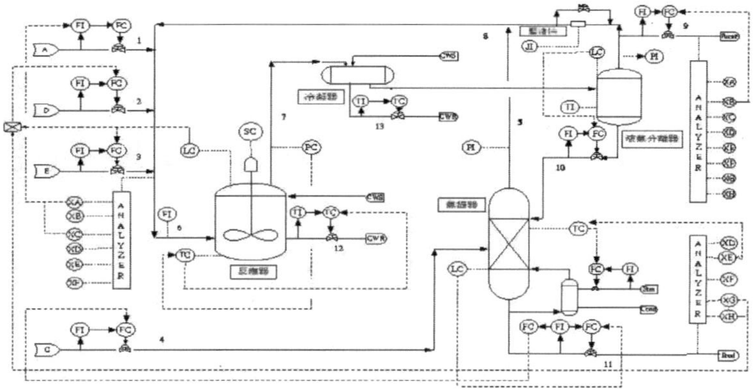

Figure 3. Improved control structure by McAvoy et al. (1995).

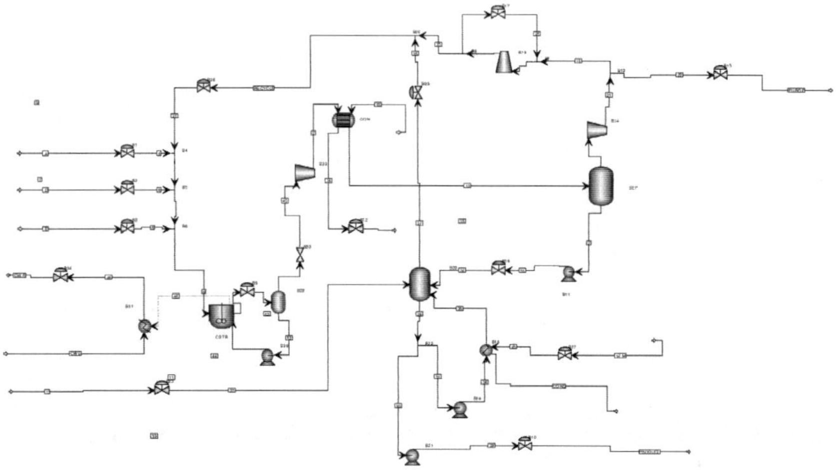

Figure 4. Steady-state process flowsheet for TE plant.

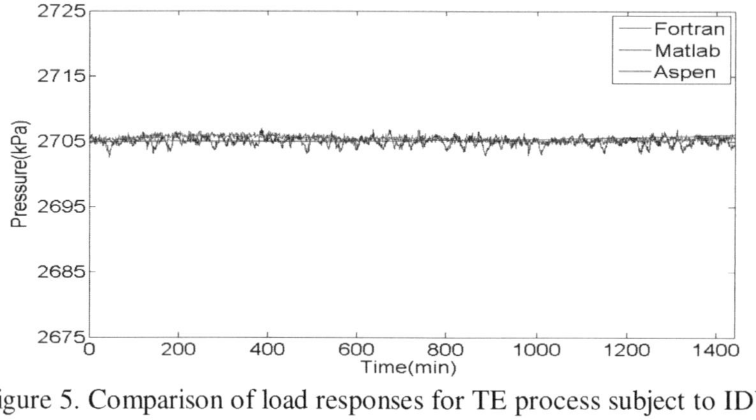

Figure 5. Comparison of load responses for TE process subject to IDV(3).

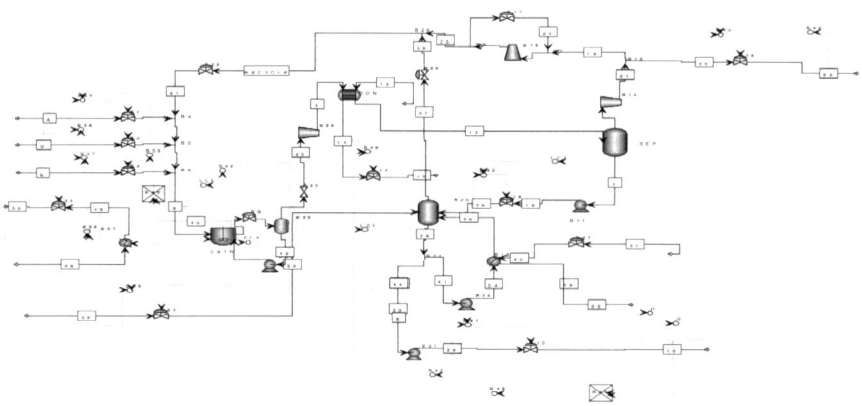

Figure 6. Proposed flowsheet of plandwide control structure for TE plant (Lin, 2008).

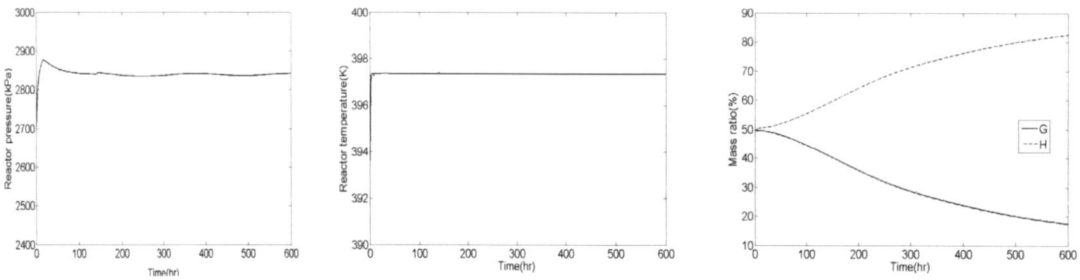

Figure 7. Mode transition control performance from base case to mode 2 for TE plant.

Experimental Studies and Optimal Design for a Small-Scale Autonomous Power System Based on Methanol Reforming and a PEM Fuel Cell

Martha Ouzounidou[1], Dimitris Ipsakis[1,2], Spyros Voutetakis[1], Simira Papadopoulou[1,3] and Panos Seferlis[1,4]

[1] *Chemical Process Engineering Research Institute (C.P.E.R.I.), Centre for Research and Technology – Hellas (CE.R.TH.), 6th km Charilaou-Thermi Road, P.O. Box 60361, 57001 Thermi-Thessaloniki, Greece*
[2] *Department of Chemical Engineering, Aristotle University of Thessaloniki, P.O. Box 1517, 54124 Thessaloniki, Greece*
[3] *Department of Automation, Alexander Technological Educational Institute of Thessaloniki, P.O. Box 14561, 54101 Thessaloniki, Greece*
[4] *Department of Mechanical Engineering, Aristotle University of Thessaloniki, P.O. Box 424, 54124 Thessaloniki, Greece*

Abstract

An experimental and preliminary design study on the production and purification of hydrogen derived from autothermal reforming of methanol is examined in this paper. Methanol, air and water are co-fed in the reformer reactor to produce hydrogen that is intended to be used in a polymer electrolyte membrane (PEM) fuel cell. However, due to the presence of CO at levels that are poisonous to the anode (Pt) of the PEM fuel cell, the effluent of the reformer is treated at a preferential oxidation (PROX) reactor for the minimization of CO at acceptable levels (<50ppm). The present study, deals with the examination of the catalysts performance at both reactors for various temperatures, flowrates (reformer) and ratios of reactants (PROX). The PEM fuel cell is also examined for various operation temperatures. Furthermore, the design study investigates three possible configurations (modeled at ASPEN Plus®), as far as heat management is of concern, in order to identify which could be the optimal solution for an autonomous power unit with minimal auxilliary supplies.

Introduction

Greenhouse gases, mainly produced by power plants that use conventional fuels such as lignite, oil and natural gas, have increased the global warming effect the past decades. Hydrogen is considered as an energy carrier for the future and can be totally non-polluting when used in fuel cells (water is the exhaust), while it becomes economically competitive with gasoline or diesel. Moreover, hydrogen can prevent the depletion of fossil fuel reserves and can result in the reduction of greenhouse gases.
Despite the many benefits that hydrogen carries, there are also constraints on its direct use to the various applications [1]. Consequently, hydrogen must be produced by fuels that could directly fed it to fuel cells. Nowadays, hydrogen is mainly produced from non-renewable sources, such as natural gas and petroleum fractions. These fuels however, lead to high CO_2 emissions that further contributes to the greenhouse effect. On the other hand, methanol is a liquid which is easily available in the market, it does not require special conditions of storage, while it is also free from high reforming temperatures and sulphur oxides that are usually met in natural gas reforming. Moreover, methanol has a high

hydrogen-to-carbon ratio and no carbon-to-carbon ratio and thus, prevents the soot formation [2], while biomass resources can be used to produce methanol (bio-methanol).
Currently, there are three main processes for producing hydrogen from methanol: (i) steam reforming of methanol, (ii) partial oxidation of methanol and (iii) autothermal reforming of methanol [3]. Although steam reforming produces a high level of hydrogen at the outlet stream (~75%), it is a reaction that is endothermic and requires heat to be provided by an external source (e.g. burner) [1, 3]. Partial oxidation is highly exothermic, which leads to the formation of hot spots in the catalyst and resulting in sintering and deterioration of the overall catalytic process, while the produced hydrogen remains at low levels (~45%). In the autothermal steam reforming, oxygen, water and methanol are co-fed to the reformer and the reactants ratios are chosen in such a way in order to have a slightly exothermic or thermally neutral reaction [3]. Therefore, autothermal reforming contains most of the advantages of the above procedures and is currently the most reliable process for hydrogen production at an autonomous power unit.
A drawback of reforming hydrocarbons like methanol, is the production of CO at levels that are very poisonous to the electrocatalyst at the Pt-anode of the fuel cells [4]. CO is preferentially adsorbed onto the surface of the catalyst, blocking its access to H_2 and degrading the electrochemical performance of fuel cells operating at temperature below 100°C [5, 6]. Several processes used for the minimization of CO content at acceptable levels (less than 50ppm) have been discussed in the past where among them preferential oxidation is considered to be the simplest and the least expensive method [7].
The present study will focus on the experimental study of the reformer and PROX reactor and on the analysis of various configurations schemes that deal with the heat integration on important sections of the unit. The results will form the basis of enhancing the already developed mathematical model of the integrated system [8].

Description of the Integrated System

In order to obtain realistic data for an integrated system, a pilot plant was constructed (Fig.1) and the main parts of the pilot plant are the following:

- The system of the fuel processing that includes the preheater, the methanol steam reformer reactor and the PROX reactor.
- The PEM fuel cell.
- The electronic system for the automation of the pilot plant.

Within two diaphragm pumps, methanol and water (liquid phase) with N_2 as carrier gas, are preheated and evaporated before they are introduced to the reformer simulating the operation of a catalytic burner. The effluent line of the preheater is mixed with air, creating a new stream for the autothermal reforming of methanol. The methanol steam reformer used in this work is a fixed bed reactor and was constructed from stainless steel with an inside diameter of 10 cm and a length of 15 cm. The catalyst used for methanol reforming is a copper-based catalyst in the form of CuO-MnO and is able to process a hydrogen–rich stream composed of 55-65% H_2, 15-25% CO_2, 12-18% N_2, 1-3% CO in dry basis [11]. The outlet stream is analyzed by a gas chromatograph (HP-6890 Series GC System).
The reactions that take place at the reformer are a combination of endothermic and exothermic reactions. Partial oxidation of methanol is allowed to take place due to the highly endothermic reaction of steam reforming in order to provide the necessary heat to the system. With a proper selection of the oxygen/methanol (O_2/CH_3OH) ratio the whole

process at the reformer can be considered to be adiabatic [3, 8, 9]. The other two reactions that take place are the water gas shift reaction and the decomposition of methanol [8].
As mentioned in the introduction section, the high levels of CO at the reformer outlet, render it unsuitable for using the hydrogen stream straight to the anode of the PEM fuel cell. Therefore, the PROX reactor becomes a necessity to the system in order to minimize CO with oxidation. The PROX reactor constructed from the same material as the reformer, has an inside diameter of 10 cm and a length of 10 cm. The catalyst used at the PROX reactor is an oxide catalyst in the form of $CuO\text{-}CeO_2$ [12]. CO is analyzed by a CO Analyzer (Horiba VA-3000 Gas Analyzer). The reactions that take place are the CO oxidation and the inevitable H_2 oxidation. The main objective of the analysis of the PROX reactor is to keep the CO concentration at a maximum limit of less than 50 ppm.
The pilot plant unit is fully automated and based on the continuous supervisory control and data acquisition system (SCADA) with the industrial software iFIX by GE Fanuc. For ease of indentification and operation, a status display is provided on the screen of a computer where the flow diagram of the system is graphically represented. Pumps, heaters, valves, power e.t.c., are controlled through the computer system by on/of commands or by pre-programmed start-up commands. The computer control system acts to automatically maintain the system parameters, to allow programmed keyboard operations, to provide safety management and to perform parametric retrieval.

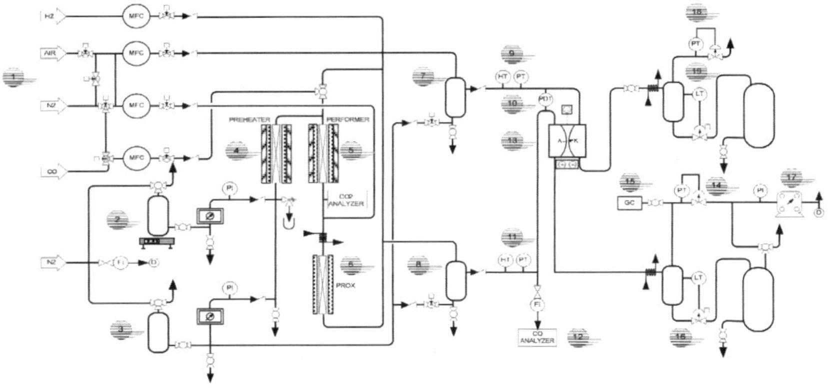

Fig. 1: Schematic diagram of the controlled pilot plant unit for hydrogen production and power generation

1. Feedstock lines: H_2, O_2, N_2, CO 2. Feedstock vessel of methanol 3. Feedstock vessel of water 4. Preheater 5. Reformer reactor 6. PROX reactor 7,8 Saturators 9,11 Humidity and Pressure Indicators	10. Pressure indication among anode and cathode of the fuel cell 12. CO Analyzer 13. PEM Fuel Cell 14,18. Pressure indicators 15. Gas Chromatograph Analyzer of the reformate gas 16,19. Level transmitters 17. Wet Test Meter (WTM)

Results of the Integrated System

Experimental Studies on the Integrated Power System

The catalyst loading in the reformer was 90 gr and methanol conversion in various methanol flowrates and temperature was studied. Fig. 2 presents the experimental results performed in the range of W/F_{CH3OH} ratio 43.4-214.1 gr/mmol·s. The ratios of steam/methanol (H_2O/CH_3OH) and of oxygen/methanol (O_2/CH_3OH) ratios are 1.5 and 0.1 respectively, unless stated otherwise. Furthermore, Fig.3 presents the effect of temperature at the methanol conversion and consequently hydrogen production.

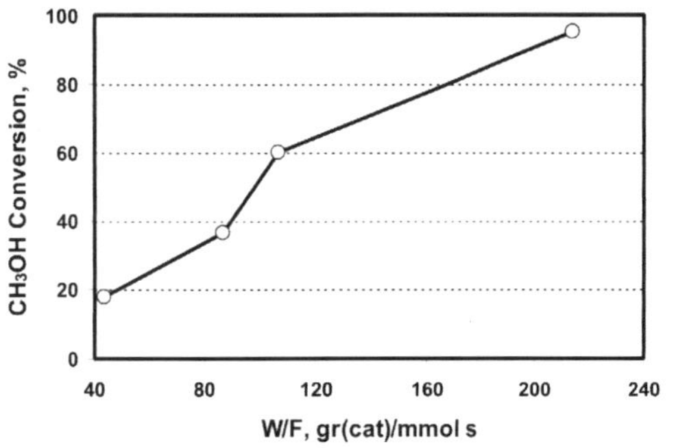

(2)

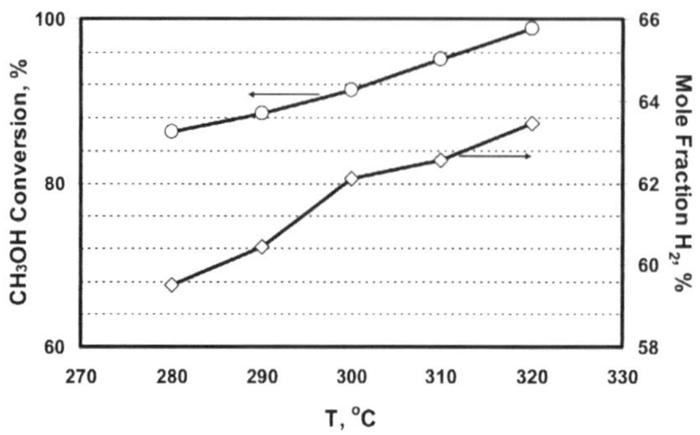

(3)

Fig. 2: CH_3OH conversion as a function of W/F_{CH3OH} at the temperature of T=300 °C.
Fig. 3: CH_3OH conversion and hydrogen production (% mole) as a function of temperature at F_{CH3OH}= 1.013 ml/min.

From the above figures, it is obvious that with the decrease of the methanol feed, methanol conversion is increased and a methanol conversion of 95% was achieved at the temperature of 300 °C and W/F_{CH3OH}=214.1 gr/mmol·s. Moreover, based on the results presented in Fig. 3, a maximum methanol conversion of 98.9% was obtained at the temperature of 320 °C for a methanol flowrate feed of 1.013 ml/min. The hydrogen production reached 63% mole, which is sufficient for a power generation of 140 W. Such a power generation is considered very low for the design of the system, since the system is capable of providing power up to 10kW based on the inlet methanol flowrate. Thus, a better analysis and study of the catalyst performance is needed in order to achieve >90% of methanol conversion for much higher flowrates than 1.013ml/min. Fig. 4 presents the dependence of CO formation on temperature where it was found that higher temperatures favors CO production and this needs to be taken into account at PROX reactor analysis.

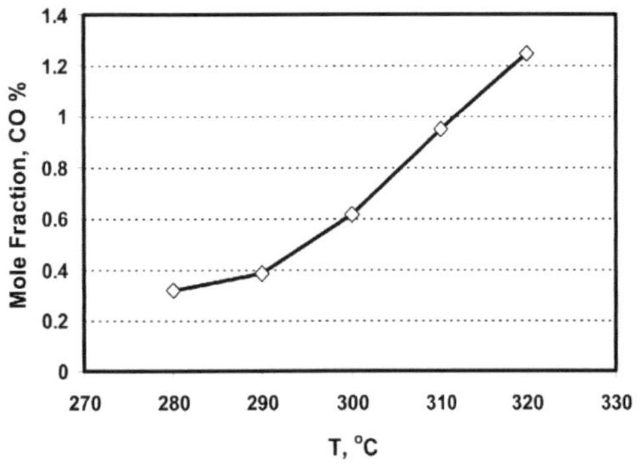

(4)

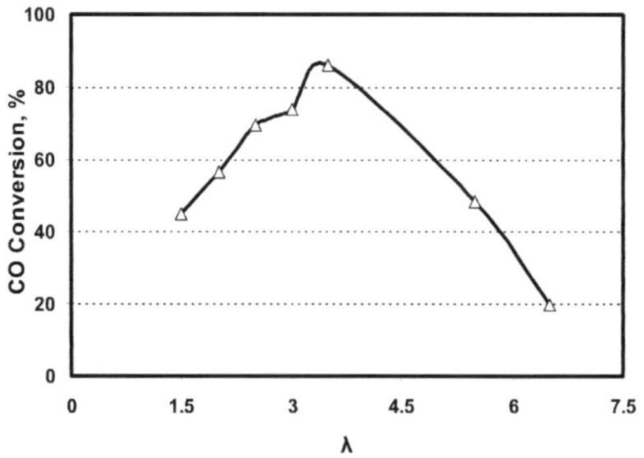

(5)

Fig. 4: CO production as a function of temperature (280-320 °C).
Fig. 5: CO conversion as a function of λ= $2[O_2]/[CO]$ (F_{CH3OH}=0.5 ml/min, T=211 °C)

Experiments in order to minimize the CO concentration of the reformer effluent were done in the PROX reactor at a temperature range of 180-220 °C. In Fig. 5 the conversion of CO is presented as a function on the ratio λ= $2[O_2]/[CO]$. At the temperature of T=211 °C [12] and λ=3.5 a conversion of 86% was achieved, but with hydrogen losses up to 6.5%. As can be seen in Fig.5, the increase of λ causes an increase in CO conversion. However, after a certain value of λ, the conversion decreases because in hydrogen oxidation water is produced and blocks active sites of the catalyst and thus, CO cannot reach them.

Finally, the performance of a small-scale PEM fuel cell stack consisted of 10 cell was investigated. Pure H_2 was fed to the anode and air to the cathode. In order to examine the hysterisis phenomena the experiments were conducted both by increasing and by decreasing the current density. As shown in Fig. 6, no hysterisis phenomena occur and increased temperatures favors the operation of the fuel cell because the overvoltages are decreased. At the temperature of T=75 °C a maximum power density was observed at 0.48 W/cm^2 at a cell voltage of 4.5 V and current density 0.106 A/cm^2.

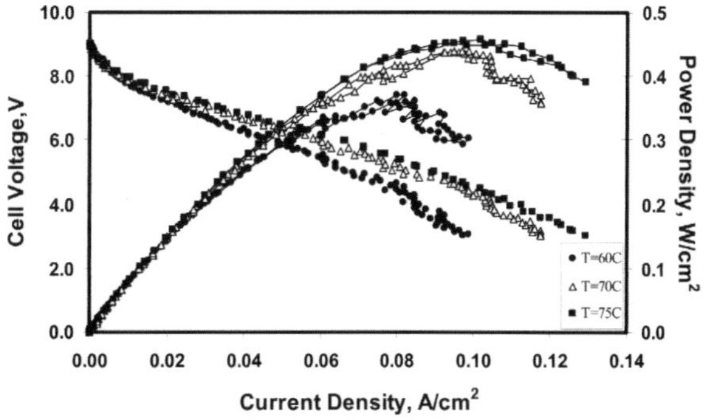

Fig. 6: Power density and cell voltage as a function of current density at T= 60,70,75 °C.

Optimal design based on proposed process flow diagrams

The production of hydrogen via autothermal reforming of methanol, reveals that besides the aforementioned basic components of the two reactors and the fuel cell, there are numerous other parameters that need to be considered. Two of the most important are the start-up of the unit, where the evaporation of the reactants should take place, and the cooling on the outlet of the reformer and PROX reactors. For the current study, we will discuss on three configurations where heat integration is needed for the evaporation of methanol and water and for the cooling of the outlet streams of the reformer and PROX reactors. The proposed schemes were studied with the use of the simulation tool ASPEN Plus® for a power generation of 1kW and design specifications were used in order to ensure that the reactants ratios at the reformer and PROX reactors produced the desired conditions (adiabatic conditions at reformer, minimal CO at PROX reactor outlet, methanol and air addition for the burner e.t.c.).

Desciption of the Process Flow Diagrams (PFDs)

In general, it can be said for all the configurations that methanol and water are mixed and the product stream is being evaporated by the exhaust gases from the burner (H1 (1)). Then, the hot stream (gaseous form of methanol and water) is mixed with air and enters the reformer for the hydrogen production. The outlet stream from the reformer is cooled and mixed with air for hydrogen purification (removal of CO) at the PROX reactor. Then, the outlet stream from the PROX reactor (after water is removed) is further cooled and enters the anode of the fuel cell. Finally, all the hydrogen that has not reacted along with the other components (CH_3OH, CO_2, N_2 etc.) is fed at the burner where additional methanol and air are used for providing the required energy (exhaust gases) for evaporating the reactants methanol and water. The configurations for the current study will be based on minimizing the methanol and air addition at the burner by using possible integration and/or recycling of process streams.

Fig.1 presents the first configuration. The outlet stream of the reformer and the PROX reactors are cooled using water that is regarded as an auxilliary unit in HI (2) and HI (3), respectively. With this configuration, the results for the used water and the additional methanol and air at the burner are shown in table 1 and discussed later.

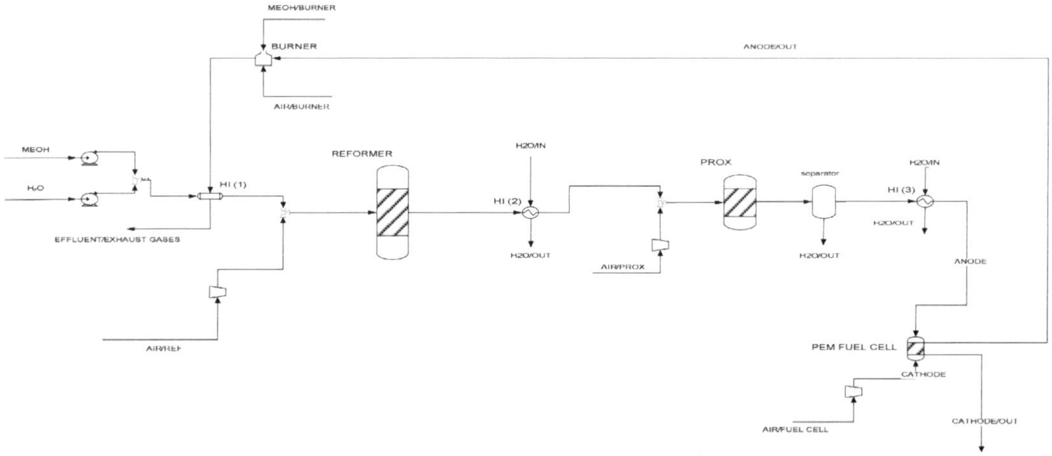

Fig.7: Description of the 1st process flow diagram (PFD1)

Trying to improve the first operation scheme, some modifications were made and shown in Fig.8. Specifically, the water hot streams after HI(2) and HI (3) are mixed and used to heat up (1st stage) the inlet methanol and water in HI (4). After the 1st stage, methanol and water are further heat by the exhaust gases at the burner to meet a temperature around the reaction temperature (HI (1)). With this configuration, the methanol and air in the burner can be reduced as shown in table 1.

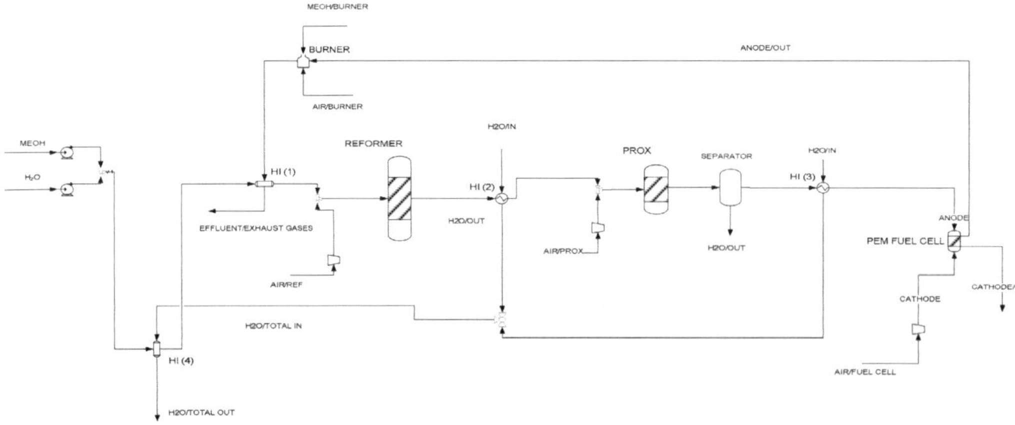

Fig. 8: Description of the 2nd process flow diagram (PFD2)

For the third configuration (Fig.9), the inlet methanol and water stream is heated up and evaporated at three stages in order to avoid using water as a coolant medium. In the first stage, the outlet stream of the reformer is used for heating up the reactants in HI (2) which are then further heated up by the outlet stream of the PROX reactor in HI (3). In the 3rd stage, the reactants are being heated by the exhaust gases from the burner in HI (1).

Despite, the fact that this configuration eliminates the use of water as coolant, it lacks at the start-up of the unit where external heat is needed for the system to start operating. Thus, for a more detailed PFD at the start-up, heat exchangers will need to be present and when necessary, the involved streams will first pass (with a by-pass subunit) from these heat exchangers in order for the unit to start its operation. The same by-pass can be applied for PFD1 and PFD2 when necessary. The results for PFD3 are also shown in table 1 and discussed below.

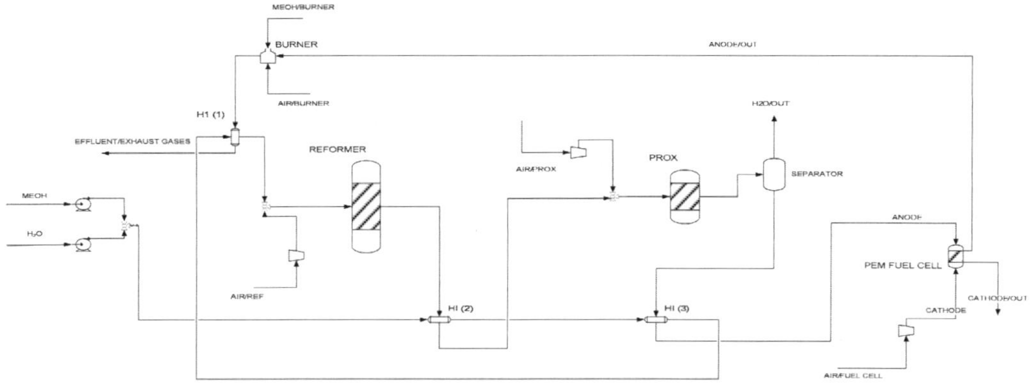

Fig. 9: Description of the 3rd process flow diagram (PFD3)

Results and discussion on the results of process flow diagrams

As can be seen from table 1, PFD2 and PFD3 results in the reduction in the methanol and air at the burner up to 32-34% as compared to PFD1. However, in PFD2 the cooling of the outlet streams from the two reactors takes place with the use of water and this favors the process in case of unexcpected start-up and shut-down actions. On the other hand, PFD3 totally eliminates the use of water, but cannot be used as shown in Fig.9, because problems at the start-up of the unit might arise and by-pass units should be present. PFD1 can be considered an unsatisfactory solution since no heat or mass recycle is present and this results to higher methanol and air mixtures at the burner which can be detrimental for an autonomous power system. Overall, it can be said that a combination of PFD2 and PFD3 will be the best solution in the system because, at the start-up cooling will take place with water (PFD2) and after steady-state is reached, heat integration with streams will be involved (PFD3).

Table 1.Results for the various configurations

	BURNER		**HEAT EXCHANGERS**	
	Methanol, mol/min	Air , mol/min	Water (HI-2), mol/min	Water (H1-3), mol/min
PFD1	0.137	1.182	0.506	0.702
PFD2	0.089	0.804	0.506	0.702
PFD3	0.09	0.816	-	-

Conclusions

In this study, an integrated and fully automated pilot plant for the production of hydrogen was studied both from an experimental and design point of view. As far as the experimental part is of concern, it was found that hydrogen production increases with temperature and reaches 63 (% mole) in the effluent of the reformer at T=320 °C. A maximum CH_3OH conversion (98.9%) was also obtained at T=320 °C and W/F_{CH3OH}=214.1 gr/mmol·s. The experimental results from the investigation of the performance of the PROX reactor showed a CO conversion of 86% with hydrogen losses up to 6.5% but increased O_2/CO ratios can deteriorate the performance of the catalyst due to the inevitable hydrogen oxidation. The maximum power density at the PEM fuel cell was 0.48 W/cm^2 and increased for higher temperatures. The preliminary design study, showed many possibilities for using process streams in the unit. The cooling of the reactors outlet is of major concern, while efforts on methanol and air minimization at the burner showed that recycling and heat integration is necessary for the operation of an autonomous power system. All the above results are going to be used for the enhancing the already developed dynamic mathematical model of the power unit.

References

[1] S.K. Kamarudin, W.R.W. Daud, A.Md. Som, M.S. Takriff, A.W. Mohammad, Y.K. Loke, Design of a fuel processor unit for PEM fuel cell via shortcut design method, Chemical Engineering Journal, Volume 104, (2004), 7-17

[2] B. Lindström and L.J. Petterson, Hydrogen generation by steam reforming of methanol over copper-based catalysts for fuel cell applications, International Journal of Hydrogen Energy, Volume 26, (2001), 923-933.

[3] B. Lindström and L.J. Petterson, Development of a methanol fuelled reformer for fuel cell applications, Journal of Power Sources, Volume 118, (2003) 71-78.

[4] L. Gubler, G.G. Scherer, A. Wokaun, Effects of cell and electrode design on the CO tolerance of polymer electrolyte fuel cells, Phys. Chem. Chem. Phys., Volume 3, (2003), 325-329.

[5] T.E. Springer, T. Rockwald, T.A. Zawodzinski, S. Gottesfeld, Model for polymer electrolyte fuel cell operation on reformate feed-Effects of CO, H-2 dilution, and high fuel utilization, Journal of the Electrochemical Society, Volume 148 (2001), 11-23.

[6] R. K. Ahluwalia, E. D. Doss, R. Kumar, Performance of high-temperature polymer electrolyte fuel cell systems, Journal of Power Sources, Volume 117, Issues 1-2, 9(2003), 45-60

[7] F. Mariño, C. Descorme, D. Duprez, Noble metal catalysts for the preferential oxidation of carbon monoxide in the presence of hydrogen (PROX), Applied Catalysis B: Environmental, Volume 54, Issue 1, (2004), 59-66

[8] D. Ipsakis, P. Kechagiopoulos, C. Martavaltzi, S. Voutetakis, P. Seferlis, P. Daoutidis, F. Stergiopoulos, Study of an integrated system for the production of hydrogen by autothermal reforming of methanol, Computer Aided Chemical Engineering, Volume 24, (2007), 913-918

[9] S. P. Asprey, B. W. Wojciechowski, B.A. Peppley, Kinetic studies using temperature-scanning: the steam-reforming of methanol, Applied Catalysis A: General, Volume 179, Issues 1-2, (1999), 51-70

[10] G.Sedmak, S. Hočevar, J. Levec, Kinetics of selective CO oxidation in excess of H_2 over the nanostructured $Cu_{0.1}Ce_{0.9}O_{2-y}$ catalyst, Journal of Catalysis, Volume 213, Issue 2, (2003),135-150

[11] J. Papavasiliou, G. Avgouropoulos, T.Ioannides, Combined steam reforming of methanol over Cu–Mn spinel oxide catalysts, Journal of Catalysis, Volume 251, Issue 1, (2007), 7-20

[12] J. Papavasiliou, G. Avgouropoulos, Th. Ioannides, Production of hydrogen via combined steam reforming of methanol over $CuO–CeO_2$ catalysts, Catalysis Communications, Volume 5, Issue 5, (2004), 231-235

578t Preliminary Study on the Inclusion of Inherent Safety Principles Using Property Design Framework

Fadwa T. Eljack, Department of Chemical Engineering, Qatar University, P.O. Box 2713, Doha, Qatar and Mario Richard Eden, Department of Chemical Engineering, Auburn University, 222 Ross Hall, Auburn, AL 36849-5127.

The word safety in design signals accident prevention and protective measures. On a larger scale, safety in processes involves the risk of operating process units and operating conditions, supplied and generated substances as well as a number of other parameters that would affect the safety of people, the environment or the surrounding property (Suardin et al., 2006).

Traditionally there are safety preventive/protective measures, generally applied after implementation of process plant design; and it has been reported to account for 15-30% of capital cost in the petroleum and gas industry (Palaniappan, 2004). Such methods are not all encompassing and hence, failure to identify risks, and according the inability to address or "control" these risks, can result in massive costs, both human and economic. To minimize the risk of hazardous conditions, Kletz (1991) introduced four inherent safer design concepts: minimize, substitute, moderate and simplify. Inherent safer design looks at involving safer concepts at the pre-design stage in order to identify a truly optimal design. Incorporating Kletz's safety design issues into a design algorithm changes the dynamic of what is truly an optimal design. Typically in process design the objective is to identify most favorable operating conditions that maximize productivity. However, considering the safety of those operating conditions (Kletz's second design concept) might alter the optimality of the proposed design.

The objective of the work here is to use the property platform as means of incorporating process safety in the preliminary stages of design. Current application of inherent methods exists, however according to Hendershot (2006), there is still a need to develop specific tools and techniques for the application of inherently safer design. The literature clearly shows that the lack of information in the preliminary phases of design complicates the inclusion of safety in traditional design approaches. Recently, the Property Integration design framework was introduced as a targeting methodology that identifies optimum process performance ahead of design. The clustering method defines and solves the design problem within a property framework. The key feature of this method is that it transforms all of the constraints placed on the design problem into properties, and it uses it as way to identify the feasible solution region. At the same time, principles of inherent safety such as minimization and limiting effects refer to changing the initial conditions/constraints of the design problem. So it would make sense to include such safety principles into the property design methodology.

The work will outline a robust property-based process design methodology using recently developed property clustering technique that will consider the needs of operating process units and the safety risk within the formulation and solution steps. Moreover, the methodology will develop a systematic scheme to transform safety hazards associated with the various process units and streams in terms of properties or functionalities. Safety risks or process hazards will be utilized as another constraint on the design problem. The translated constraints placed on the design allow for the identification of process unit property targets ahead of design. This is a systematic method for placing bounds on the search space without loosing accuracy; hence the approach will be able to enumerate all design alternatives. Thus the presented paper will propose a framework for the incorporation of safety in the preliminary stages of design by infusing Inherent Safety Principles into the Property Clustering Platform.

Keywords: process design, inherent safety, property design

578u A Molecular Signature Technique for Process and Product Design

Nishanth Chemmangattuvalappil, Charles C. Solvason, and Mario R. Eden. Department of Chemical Engineering, Auburn University, 230 Ross Hall, Auburn University, AL 36849-5127

Recent developments in the area of integrated process and product design have shown that products can be designed in terms of their properties without committing to any specific components a priori. However, the identification of suitable candidate molecules that satisfy a set of property targets is still a challenging issue. While current techniques make use of group contribution methods (GCM) to design molecules, there are many properties which can not be estimated by GCM. Furthermore, not all possible atomic arrangements are represented in GCM. Hence, there is a need of an efficient methodology for the design of molecules that can handle the multitude of non-GCM described properties. In this work, an algorithm is presented that integrates the concept of molecular signatures into the reverse problem formulation framework. A signature is a systematic coding system of atom types. A molecular signature is a linear combination of its atomic signatures. It has been shown that any of the topological indices (TI) of molecules can be represented in terms of molecular signatures and that it is possible to correlate the TI to the actual properties. Using this information, the new algorithm utilizes the property clustering technique in a reverse problem formulation to obtain the molecular structure based on the property requirements estimated in the process design step. The accuracy of this method depends only on two factors: (1) how well the property-TI relationships are estimated and (2) the level of atomic signatures used to describe the TI. Since, many TI can be used to describe each property, this algorithm generally provide reliable results. This contribution will illustrate the developed methods and demonstrate their use in a case study.

578v Systematic Integration of Product Attributes and Molecular Synthesis

Charles C. Solvason, Nishanth Chemmangattuvalappil, and Mario R. Eden. Department of Chemical Engineering, Auburn University, 230 Ross Hall, Auburn University, AL 36849-5127

The complete and efficient solution to the enumeration of candidate compounds and mixtures that meet specified consumer attributes is often a difficult mathematical programming problem. Most approaches to this problem involve the solution of a mixed integer nonlinear program (MINLP) which may achieve only local optima solutions. In this paper a proof-of-concept study is presented to show that empirical models can be used in a reverse problem formulation to ensure a complete set of candidate compounds and mixtures are found subject to the predictive power of the models. The method utilizes a transformation of consumer attributes to properties described by the group contribution method and solves the reverse problem formulation using the property clustering technique. The design of an environmentally benign refrigerant for the replacement of 1,1,1,2-tetrafluoroethane (R-134a) in a typical refrigeration cycle is used to highlight the method.

Byproduct Hydrogen Recycling Networks Design of Petrochemical Complex

Changhyun Jeong, Sangho Lee and Chonghun Han

School of Chemical and Biological Engineering, Seoul National University
599 Gwanangno, Gwanak-gu, Seoul, 151-744, South Korea,

Abstract— In a petrochemical complex, large amount of hydrogen is produced as a by-product and used as a fuel in petrochemical and oil refinery plants. By recycling this byproduct hydrogen as a raw material, the value of hydrogen can be greatly improved. This study proposes a design methodology for byproduct hydrogen recycling network between plants in petrochemical complex by analyzing the hydrogen pinch, required cost and constraints.

INTRODUCTION

This study describes the application of mass integration principles to the problem of a petrochemical complex hydrogen management. Mass integration is a methodology that provides an understanding of the flow and allocation of hydrogen offgas within a process.

In general, hydrogen recycling network design determines the optimal routing and allocation between the sources and the sinks. Sometimes, sources can be intercepted and treated in order to change flowrates and purities. This study considers a specific type of process which is a petrochemical complex hydrogen network.

BACKGROUND

The superstructure shown in Figure 1 has been built to consider trade-offs. Hydrogen suppliers and hydrogen consumers are decomposed as the sources and the sinks. Hydrogen purification processes are installed between these sources and sinks. In order to integrate different purification processes, they are assumed to be able to feed each other. Currently, only the integration of PSA is considered because it is the most commonly used purification processes in petrochemical complex.

The superstructure contains all the possible network configurations with purification processes. The optimal design may need one purification unit(PSA or membrane) or no purification process at all. The possibility of streams fed to purification units is fully covered by the superstructure. The installation of new compressors and pipelines is also included in this superstructure.

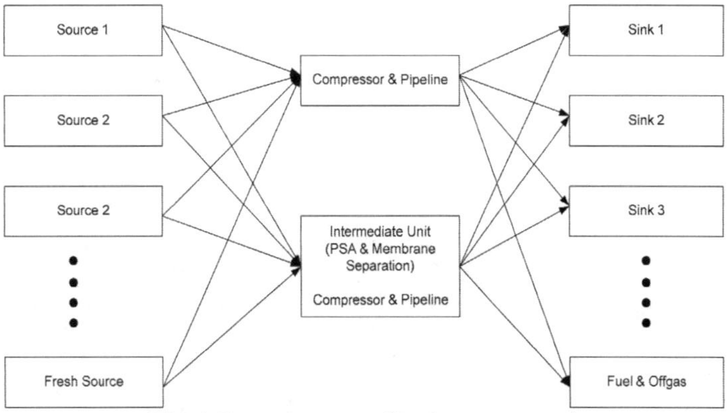

Fig. 1 Superstructure of hydrogen network

THEORY

The hydrogen recovery network of petrochemical complex is formulated as a mathematical model based on the assumption above. The objective is to maximize the profit of the hydrogen recovery, considering the fresh resource, interception device, and pipeline.

To investigate the distribution of profits between industries in Petrochemical complex, the optimization problem is reformulated based on the assumptions as follows.

1.The hydrogen from each company is reused by another company, or purified by separation units of hydrogen network or thrown away outside.

2.Each company uses 'high purity hydrogen' from other companies directly, or 'purified hydrogen' by separation units of hydrogen network, or purchased raw materials from outside.

3.Separation units of hydrogen network treat 'low purity hydrogen' from each company and transfer them to each company or fuel.

CONCLUSION

In this study, the economic effectiveness of hydrogen recovery network is represented by proposing a network optimization model. The method is based on setting up a superstructure that includes the feasible networks. The objective function consists of the average total cost saving in hydrogen network.. And this method can be applied to solve real scale problems.

REFERENCES

1. Towler, G.P., Mann, R. Serriere, A.J-L. and Gabaude, C.M.D., Refinery hydrogen Management: Cost Analysis of Chemically-Integrated Facilities, Ind.Eng.Chem.Res., 35(7), 2378-2388,1996
2. Alves, J.J. and Towler, G.P., Analysis of Refinery Hydrogen Distribution Systems, Ind.Eng.Chem.Res., 41(23), 5759-5769,2002
3. Hallale, N. and Liu, F., Refinery Hydrogen Management for Clean Fuels Production, Adv.Enc.Res.,6, 81-98,2001
4. Liu, F. and Zhang, F,, Strategy of Purifier Selection and Integration in hydrogen networks, Chem.Eng.Res,Des.,82(20), 1315-1330,2004

578x Development of Quantitative Structure-Property Relations for Ionic Liquids with Refrigerant Gases

Samantha E. McLeese, Aaron M. Scurto, and Kyle V. Camarda. Department of Chemical and Petroleum Engineering, The University of Kansas, 1530 W. 15th St., 4132 Learned Hall, Lawrence, KS 66045

Ionic liquids (ILs) are currently being considered for novel uses in hydrofluorocarbon (refrigerant) gas separations and as solvents in absorption refrigeration systems. Due to the low to immeasurable volatility of ILs, these compounds may be well-suited for such applications, and could potentially be considered as environmentally-benign solvents. Their molecularly-tunable nature yields an extraordinary number of possible cation and anion combinations, the majority of which have never been synthesized. This project has developed quantitative structure-property relations (QSPRs) to allow prediction of such important properties of ionic liquids as solubility in a given refrigerant, diffusivity, viscosity, toxicity, molar volume, melting and decomposition temperatures, as well as many excess thermodynamic properties directly from structure. The electronic structure of ionic liquids is quantified using modified connectivity indices, which describe bonding environments, charge distribution, orbital hybridization and other interactions within and between ions, and between the ions and the refrigerant molecules present. These topological descriptors provide a unique set of values which identify a specific ionic liquid structurally. QSPR relations have been developed using linear regression between these descriptors and experimentally measured physical and chemical properties. Both the quality of fit and the predictive capability of these expressions will be presented. These relations provide a methodology for quick estimation of the most relevant properties of ionic liquids, and can be used within a computational molecular design framework to provide candidate structures for use as environmentally benign solvents with refrigerant gases.

578y A Novel Integrated Design of Wastewater and Heat Exchange Networks Based on Cost Estimation

Jiyong Kim, Jinkyung Kim, Younghee Lee, Eunyong Kim, and Il Moon. Department of Chemical Engineering, Yonsei University, 134 shinchon-dong, Seodaemun-gu, Seoul, 120-749, South Korea

A new systematic methodology for designing wastewater and heat exchange networks for process industries involving effluent streams containing multiple contaminants is proposed. A simultaneous optimization approach mathematically combines the problems of wastewater and heat exchanger network optimization into a single step. This process includes two global iterations of a similar two-stage approach and optimizes networks for water and heat exchange simultaneously based on cost estimates. The objective function is to minimize the total annual cost of the wastewater and heat exchange network design which is subject to the mass and energy balance constraints on all the pertinent flows and to constraints related to the concentrations of contaminants. The proposed method employs a strategy to address a mixed integer non-linear programming (MILNP) formulation. This approach was applied to optimize the wastewater and heat exchanger networks connecting oil refinery processes, which included CDU, HDS, and desalter units, and three contaminants: hydrocarbons, H2S, and salt. In order to validate the proposed method, the solutions from four different optimization problems were compared. Simultaneous optimization reduced total costs by 76.4% compared to the base case (optimized solution without water reusing or heat exchange), in which water recycling and heat exchange were not employed, and by 3.9% compared to case 3 (wastewater reusing and heat exchanger networks optimized using the two-stage optimization approach). The novel method of simultaneous optimization was not only simpler than the established sequential design method but also less expensive than the sequential method. The simultaneous optimization method proposed in this study provides useful guideline for designing wastewater and heat exchanger networks with greater cost efficiencies and environmental performances.

578z Global Sensitivity Analysis: Estimation of Sensitivity Indexes in Metabolic Network Dynamic Models

Jimena A. Di Maggio, Chemical Engineering, Planta Piloto de Ingenieria Quimica (PLAPIQUI), Universidad Nacional del Sur, Camino La carrindanga Km 7, Bahia Blanca, 8000, Argentina, Juan C. Diaz Ricci, Instituto Superior de Investigaciones Biológicas (INSIBIO), Universidad Nacional de Tucuman, Chacabuco 461, Tucuman, 4000, and Maria Soledad Diaz, Chemical Engineering, Planta Piloto de Ingenieria Quimica (PLAPIQUI), Universidad Nacional del Sur - CONICET, Camino La Carrindanga Km 7, Bahia Blanca, 8000, Argentina.

Dynamic models for metabolic pathways that can predict the microbial behavior constitute important tools in metabolic engineering. During the past few years, there has been a great increase in the availability of high-throughput data that characterize the status of cells at the genomic, proteomic, metabolic and physiological levels. It is now possible to use the same methods to record the status of cells over time, which gives a large amount of information on the dynamics of functioning cells. Dynamic models comprise a nonlinear differential algebraic system of equations, which arise from mass balances for metabolites and have a large number of kinetic parameters that require tuning for an specific growth condition. The first step in the solution of the inverse problem is the sensitivity analysis. Global sensitivity approaches estimate the effect on the output of a factor when all the others are varying, enabling the identification of interactions in nonlinear and/or nonadditive models. Generally, global approaches allow the use of model-independent methods as they do not require assumptions of additivity or linearity. Model responses are then analyzed statistically to yield the empirical distribution function/probability distribution of the model outputs.

In this work, a global sensistivity analysis has been performed through variance-based techniques to identify which parameters have the largest impact on the model output and which of them account for most of the uncertainty in that output. Sensitivity indexes have been calculated for each parameter, based on Sobol's method (1993); in which the first-order sensitivity index for each parameters is calculated as the ratio between the variance of the expected valued of the mean of the output (Vi) for the corresponding parameter and the total variance (V).

The global sensitivity analysis has been performed on a dynamic model for the Embden-Meyerhof-Parnas pathway, the phosphotransferase system and the pentose phosphate pathway of Escherichia coli K-12 strain W3110 (Chassagnole et al., 2002). The model comprises nineteen dynamic mass balance equations for extracellular glucose and intracellular metabolites, twenty nine kinetic rate expressions and seven additional algebraic equations to represent the concentration of co-metabolites. The model involves around one hundred parameters. Each parameter has been considered to have a normal probability distribution centered on its nominal value and sample sizes of one thousand scenarios have been considered. The preceding analysis has allowed identification of less than thirty parameters as the most influential ones on the complex metabolic network under study.

References

Chassagnole, C., Noisommit-Rizzi, N., Schmid, J.W., Mauch, K., Reuss, M., Dynamic modeling of the central carbon metabolism of Escherichia coli. Biotechnol. Bioeng. 79, 53–73, 2002.

Sobol, I.M., Sensitivity estimates for nonlinear mathematical models. Math. Model. Comput. Exp. 1, 407–414, 1993.

578aa Numerical Modeling of Horizontal Wafering of Silicon Melts on a Molten Substrate

B. Erik Ydstie, Sukumar Balaji, and Sudhir Ranjan. Department of Chemical Engineering, Carnegie Mellon University, 5000 Forbes Ave, Pittsburgh, PA 15213

In the current study, we model the variations in geometric topology and the interactions between two immiscible fluids using Finite Element Method. The well known technique, Level Set Method is employed to track the changes in shapes, moving boundaries and interactions between fluids. The Multiphysics modeling software COMSOL is used for this purpose. To test the predictions, the modeling work is coordinated with experimental efforts.

In order to obtain a silicon film of varying thickness at the surface of molten substrate, a drift of silicon melt in a known volume of molten substrate such as slag was modeled. Based on the slag properties, the moving boundary problem is simulated. Experiments were carried out at 1500°C to melt silicon in slag, which is comprised of various metal oxides and whose melting point is around 1000-1200°C. The set-up is then slowly cooled to study the stability of the melt at room temperature. Both the experimental and the simulated results confirm the formation of stabilized shapes with varying sizes based on the physical properties of both the slag and the silicon melt.

The simulated model is then used to carry out sensitivity analysis for the two phase system. The most important physical parameters that characterize the final form of the silicon film are density, surface tension, viscosity and the film thickness. The sensitivity of each parameter is studied individually by keeping other parameters constant. Based on the influence of each parameter, outstretching of silicon melts on molten slag system and ways of continuous floating of the thin layer on the slag surface are being devised.

Keywords: Level set method, two phase flow, moving boundary, sensitivity analysis, COMSOL Multiphysics.

578ab

OPTIMAL DESIGN OF BATCH-STORAGE NETWORK UNDER JOINT UNCERTAINTIES

Gyeongbeom Yi[*]

[*] *Department of Chemical Engineering*
Pukyong National University
San 100, Yongdang-Dong, Nam-Ku
Pusan, Korea 608-739
(Tel) 82-51-629-6439
(Fax) 82-51-629-6429
(E-mail) gbyi@pknu.ac.kr

Key words : Optimal Lot Size, Process-Buffer Network, Joint Uncertainty, Waste Treatment Process, Analytical Solution.

Prepared for Presentation at the 2008 Annual Meeting, Philadelphia, PA, Nov. 16-Nov. 21
Unpublished

ABSTRACT

This study sought to find analytic solutions to the problem of determining the optimal capacity of a batch-storage network to meet demand for finished products in a system undergoing joint random variations of operating time and batch quantity. The raw material purchasing flow and final product demand flow are susceptible to joint random variations in the order cycle time and batch size. The production processes also have joint random variations in production cycle time and product quantity. Waste regeneration or disposal processes are included into the network to treat the spoiled materials from failed batches. The objective function of the optimization is minimizing the expected total cost, which is composed of setup and inventory holding costs as well as the capital costs of constructing processes and storage units. A production and inventory method, the PSW (Periodic Square Wave) model, provides a unique graphical method to find the upper/lower bounds and average of random flows, which are used to construct terms of the objective function and constraints of the optimization model. The advantage of this model is that it provides a set of simple analytic solutions while also maintaining a realistic description of the random material flows between processes and storage units; as a consequence of these analytic solutions, the computation burden is significantly reduced. The proposed method has the potential to rapidly provide very useful data on which to base investment decisions during the early plant design stage. It should be particularly useful when these decisions must be made in a highly uncertain business environment..

Introduction

The production and inventory analysis method known as the periodic square wave (PSW) method was recently developed and used to determine the optimal design of a parallel batch-storage system.[1] Subsequently, the method was extended to handle a sequential multistage batch-storage network (BSN),[2] and further modified to handle a non-sequential network structure that can deal with recycle material flows in the network.[3] The key advantage of the PSW model over other models lies in its simple analytical sizing and timing equations. This advantageous characteristic has been exploited in an analysis of an integrated financial and production system.[4,5] The range of physical process structures for which the PSW model can be used has expanded from batch processes to multitasking batch or semi-continuous processes[6] and a multi-site batch production/transportation network (i.e., a large-scale supply chain system)[7]. A BSN basically resembles a state task network (STN) in which batch corresponds to task and storage corresponds to state; thus, BSNs provide an effective representation of supply chains such as purchasing, production, transportation and demand processes.

One of the challenging problems in supply chain optimization is dealing with uncertainty. Major business uncertainties arise from product demand forecasting, production equipment malfunctions, off-spec materials, changes in the prices of raw materials or finished products, and raw material supply shortages. A promising approach to accounting for such uncertainties is to include a computational mechanism to mitigate such uncertainty effects into the supply chain optimization model. However, the models developed to date are unsuitable for use in real applications due to severe computational complexity arising from the treatment of uncertainty as well as unknown probabilistic parameters. Given this situation, the present study aims to develop a compact analytical solution to optimize the design and/or operation of large-scale supply chains with batch processes. In this study, a novel optimization model resulting in simple analytical solutions with negligible computational burden is introduced.

A previous study using the PSW model on a BSN developed analytical solutions of supply chain optimization to deal with uncertainty[8]. The sources of uncertainty were batch size and cycle time variations in raw material purchasing, batch production, transportation and finished product demand processes. The processes were classified into three types according to their random characteristics: (1) processes that possess uncertainty only in the cycle time; (2) processes that possess uncertainty only in the batch size; and (3) processes that possess joint uncertainty in both the cycle time and batch size. In modern society, batch material losses associated with raw material purchasing and transportation processes occur infrequently; therefore, these processes were exclusively considered as type 1 processes in the previous study. Production processes may be either type 1 or type 2 processes. Mixing or blending processes do not usually involve batch material loss, and thus were considered to type 1 processes. Batch material losses do, however, occur in many reaction processes; hence these processes were considered type 2 processes. Most processes had joint uncertainties in cycle time and batch size (type 3 process); however, such joint uncertainties were excluded in the previous study due to the complexity of handling them. In the present study, an approach is developed that overcomes the complexity of handling joint uncertainties of cycle time and batch size. Here it is considered that all processes are subject to joint uncertainties of cycle time and batch size; therefore all processes are type 3 processes. Under this scheme, type 1 and 2 processes are subsystems of type 3 and hence do not need to be considered separately. In spite of the increased problem complexity associated with this approach, analytical solutions for the optimization problem are still available. These analytical solutions greatly reduce the computation time.

When a batch production process is susceptible to random failures, the volume of on-spec product material fluctuates randomly for a given feed volume. In other words, a random amount of waste material is produced, as the amount of waste material equals feed volume minus on-spec product volume. Therefore, the system that deals with material quantity uncertainty should include waste material storage units and waste disposal processes. In this study, the BSN is modified to include a waste material stream in a batch process connected with a waste material storage unit and a waste disposal process connected with a waste material storage.

A multi-period formulation to treat long-term trends of variables and parameters that was introduced in the previous study[8] is omitted in this study and, therefore, the derivation corresponds to single-period formulation that accounts for short-term variations of variables and parameters. The single-period formulation presented here can be easily expanded to the multi-period formulation, as was done in the previous study.[8]

Upper and Lower Bounds of a Flow with Joint Uncertainties

Figure 1 shows three types of uncertainties: (a) uncertainty only in cycle time, (b) uncertainty only in batch size and (c) joint uncertainties in both cycle time and batch size.

The focus of the present study is joint uncertainties. The random properties of joint uncertainties are characterized by two random variables, $\mathbf{B}_{(l)}$ and $\boldsymbol{\omega}_{(l)}$, as shown in Figure 1(c), where subscript **(l)** represents the sequence of batch occurrence. It is not necessary to know the exact distribution functions of $\mathbf{B}_{(l)}$ and $\boldsymbol{\omega}_{(l)}$. It is assumed that $\mathbf{B}_{(l)}$ has a symmetrical distribution function with $\underline{\underline{B}} \le \mathbf{B}_{(l)} \le \overline{\overline{B}}$ and $\boldsymbol{\omega}_{(l)}$ has a non-symmetrical distribution function with $\underline{\underline{\omega}} \le \boldsymbol{\omega}_{(l)}$. The mean values of $\mathbf{B}_{(l)}$ and $\boldsymbol{\omega}_{(l)}$ are $\bar{B} \equiv \frac{\overline{\overline{B}} + \underline{\underline{B}}}{2}$ and $\bar{\omega}$ respectively. Two design parameters, time availability α and size availability β, are introduced such that $\alpha \equiv \frac{\underline{\underline{\omega}}}{\bar{\omega}}$ and $\beta \equiv \frac{\underline{\underline{B}}}{\bar{B}} \equiv 2 - \frac{\overline{\overline{B}}}{\bar{B}}$, where $0 < \alpha \le 1$ and $0 \le \beta \le 1$. As α and/or β approach 1, the process becomes more deterministic. The concept of availability, defined as minimum value without failure divided by average value with failure, comes from failure modes and effects analysis (FMEA). Note that the process with $\beta = 1$ and $0 < \alpha < 1$ is type 1, and the process with $\alpha = 1$ and $0 < \beta < 1$ is type 2 in the previous study.[8]

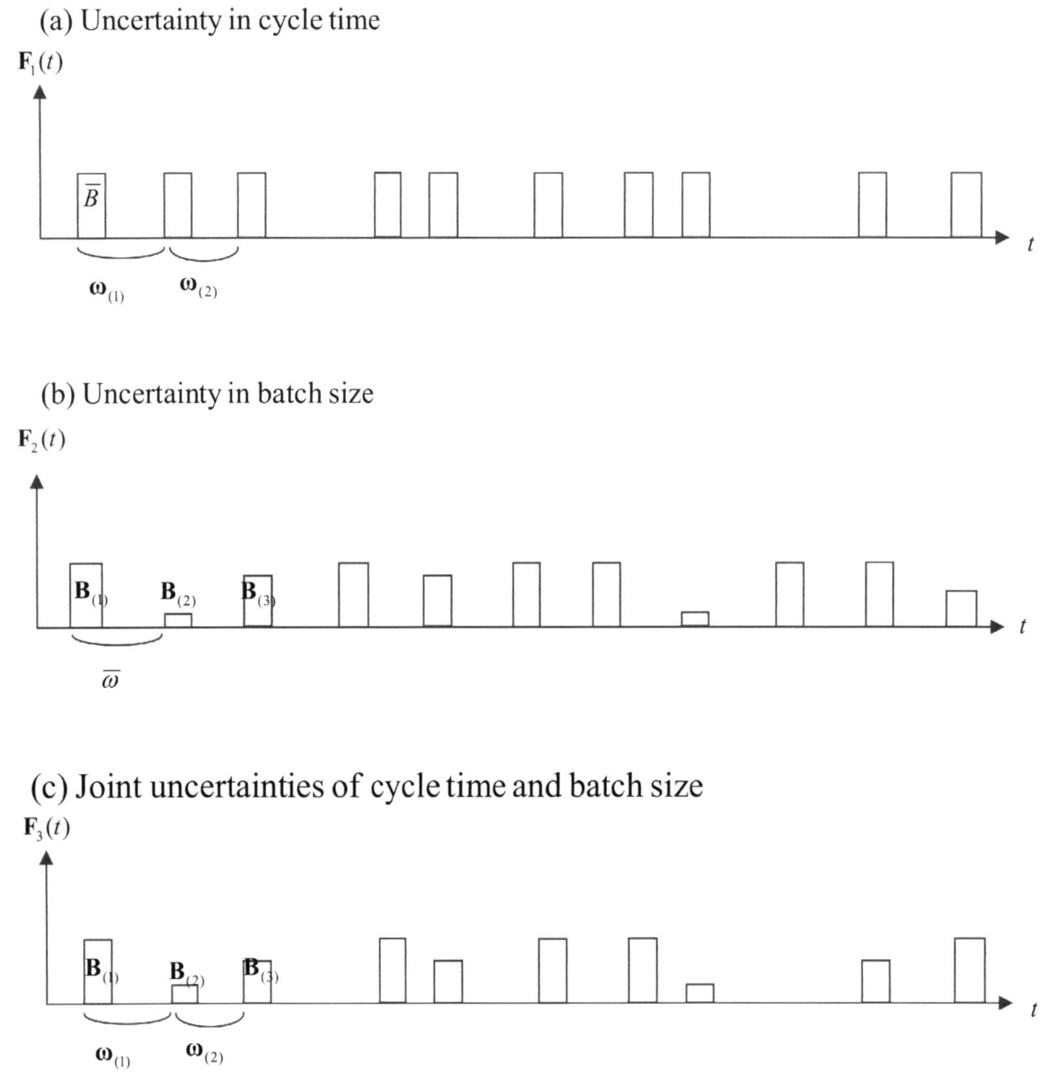

Figure 1. Types of Uncertainty

Suppose that $\mathbf{B}_{(l)}$ and $\boldsymbol{\omega}_{(l)}$ have identical independent distribution functions with respect to **(l)**. For given convergence limits $0 < \varepsilon_1, \varepsilon_2 << 1$ and confidence levels $0 < \delta_1, \delta_2 << 1$, the weak law of large numbers says that there exists an integer η such that $P\left\{\left|\frac{1}{\eta}\sum_{l=1}^{\eta}\mathbf{B}_{(l)} - \overline{B}\right| < \varepsilon_1\right\} \geq 1 - \delta_1$ and $P\left\{\left|\frac{1}{\eta}\sum_{l=1}^{\eta}\boldsymbol{\omega}_{(l)} - \overline{\omega}\right| < \varepsilon_2\right\} \geq 1 - \delta_2$. From Tchebycheff's inequality, $\eta \geq \frac{Var(\mathbf{B}_{(l)})}{\delta_1 \varepsilon_1^{\,2}}$ and $\eta \geq \frac{Var(\boldsymbol{\omega}_{(l)})}{\delta_2 \varepsilon_2^{\,2}}$, that is, $\eta = \max\left\{\mathrm{int}\left[\frac{Var(\mathbf{B}_{(l)})}{\delta_1 \varepsilon_1^{\,2}}\right], \mathrm{int}\left[\frac{Var(\boldsymbol{\omega}_{(l)})}{\delta_2 \varepsilon_2^{\,2}}\right]\right\} + 1$ if the least integer is chosen.[8] Where $Var(.)$ is variance operator and int[.] is a truncation function to make integer. The parameter η, called the occurrence number, should be an even number in order for 0.5η to be an integer. The time interval during which η batches occur is defined as the long cycle time $\tilde{\omega}$. Because the sample means of $\mathbf{B}_{(l)}$ and $\boldsymbol{\omega}_{(l)}$ converge to their mean values during the long cycle time according to the weak law of large numbers, $\tilde{\omega} = \eta\bar{\omega} = \frac{\eta}{\alpha}\underline{\underline{\omega}}$. Here, the long cycle time corresponds to the least period within which all random effects diminish with a given confidence level. Two more parameters are introduced for convenience, the average flow rate $D \equiv \frac{\bar{B}}{\bar{\omega}} = \frac{\alpha\bar{B}}{\underline{\underline{\omega}}}$ and the total dead time within a long cycle time $d \equiv \tilde{\omega} - \eta\underline{\underline{\omega}} = \left(\frac{1}{\alpha} - 1\right)\eta\underline{\underline{\omega}}$.

To generate the optimization formulation, the upper/lower bounds and average inventory level of storage units under joint uncertainties are needed. The upper bound of the inventory level will be used to compute the storage size; the lower bound of the inventory level will be used in the optimization constraint that ensures that the inventory level is always nonnegative; and the average of the inventory level will be used to compute the inventory holding cost of the optimization problem. If the upper/lower bounds and average of all flows coming into and going out of the storage units are known, the upper/lower bounds and average of the inventory level of the storage units can be easily identified. Note that the flow has a constant average flow rate D measured during a long cycle time. This means that in spite of the randomness, the total quantity processed during a long cycle time is constant. Two extreme cases of the flow with joint uncertainties exist-(a) the Upper bound case and (b) the Lower bound case-as shown in Figure 2 for the case of $\eta = 4$.

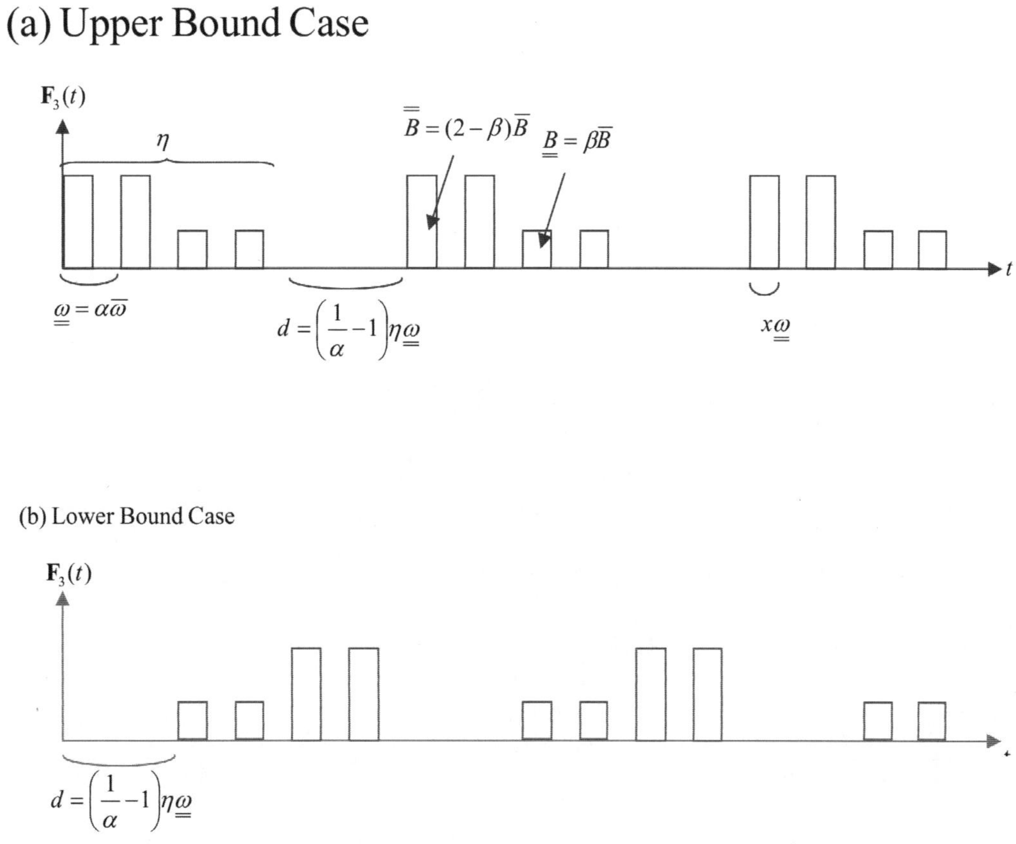

Figure 2. Two Extreme Cases of Flow

The upper bound case has 0.5η times of maximum batch size $\overline{\overline{B}}$ with minimum cycle time $\underline{\underline{\omega}}$, 0.5η times of minimum batch size $\underline{\underline{B}}$ with minimum cycle time $\underline{\underline{\omega}}$, and a total dead time d within repeated long cycle times. The lower bound case has total dead time d, 0.5η times of minimum batch size $\underline{\underline{B}}$ with minimum cycle time $\underline{\underline{\omega}}$, and 0.5η times of maximum batch size $\overline{\overline{B}}$ with minimum cycle time $\underline{\underline{\omega}}$ within repeated long cycle times.

Note that in spite of the greater difference between these two cases, the total quantity processed during a long cycle time of both cases is a constant, $D\tilde{\omega}$. Figure 3 shows the cumulative flow functions of the two cases.

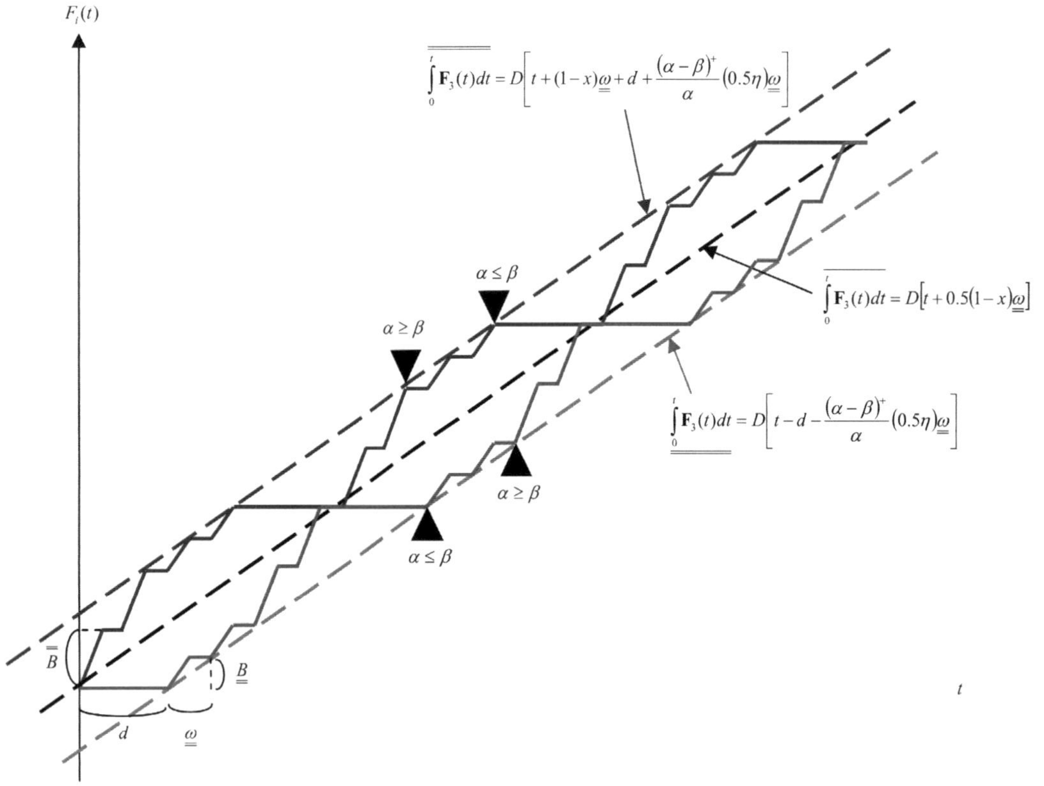

Figure 3. Cumulative Flow Functions for Two Extreme Cases.

The dotted lines are the upper and lower bounds of the two extreme cases. Note that there are two contacting points depending on the values of α and β. The integral of all flows with joint uncertainties exists between the dotted lines, that is, $\underline{\underline{UPSW}} \le \int_0^t \mathbf{F}_3(t)dt \le \overline{\overline{UPSW}}$, where

$$\underline{\underline{UPSW}}(t; D, \underline{\underline{\omega}}, t', x, \theta) = D\left[t - t' - \theta\underline{\underline{\omega}}\right] \tag{1}$$

$$\overline{\overline{UPSW}}(t; D, \underline{\underline{\omega}}, t', x, \theta) = D\left[t - t' + (1-x)\underline{\underline{\omega}} + \theta\underline{\underline{\omega}}\right] \tag{2}$$

Here

$$\theta = \left(\frac{1}{\alpha} - 1\right)\eta + \frac{(\alpha - \beta)^{+}}{\alpha}(0.5\eta) \tag{3}$$

where $(X)^{+} \equiv \max\{0, X\}$, x is called storage operation time fraction and t' is initial start-up time. The average inventory level is highly dependent on the random properties of failures. The exact value of the average inventory level cannot be obtained without defining the probability distribution functions of all random variables, which is a nontrivial task. In this study, an intuitive approach is taken. Specifically, the average of flow $\overline{UPSW} \equiv \overline{\int_0^t \mathbf{F(t)}dt}$ is selected as the line equidistant from the upper and lower bounds[8]:

$$\overline{UPSW}(t; D, \underline{\underline{\omega}}, t', x) = D\left[t - t' + 0.5(1 - x)\underline{\underline{\omega}}\right] \tag{4}$$

This selection could be the most probable[8].

The objective function for the design of the batch-storage network is to minimize the annualized expectation of total cost, which consists of the setup cost of processes, the inventory holding cost of storage units, and the capital cost of the processes and storage units for a given time availability, size availability and occurrence number in a long cycle time of each process. The optimization constraints are no depletion of all storage units

Solving Kuhn-Tucker conditions gives optimal cycle times:

$$\underline{\underline{\omega}}_k^j = \sqrt{\frac{\alpha_k^j A_k^j}{D_k^j \Psi_k^j}} \tag{5}$$

$$\underline{\underline{\omega}}_n^j = \sqrt{\frac{\alpha_n^j A_n^j}{D_n^j \Psi_n^j}} \tag{6}$$

$$\underline{\underline{\omega}}_i = \sqrt{\frac{\alpha_i A_i}{D_i \Psi_i}} \tag{7}$$

where

$$\Psi_k^j = \left[0.5H^j + b^j\right](1 - x_k^j) + \frac{(2 - \beta_k^j)a_k^j}{\alpha_k^j} + \left(H^j + 2b^j\right)\theta_k^j \tag{8}$$

$$\Psi_n^j = \left[0.5H^j + b^j\right](1 - x_n^j) + \frac{(2 - \beta_n^j)a_n^j}{\alpha_n^j} + \left(H^j + 2b^j\right)\theta_n^j \tag{9}$$

$$\Psi_i = \frac{(2-\beta_i)a_i}{\alpha_i} + (1-`x_i)\sum_{j=1}^{|J|}\left(\frac{H^j}{2} + b^j\right)f_i^j$$
$$+(1-x_i')\sum_{j=1}^{|J|}\left(\frac{H^j}{2} + b^j\right)\frac{1}{2-\beta_i}g_i^j + (1-x_i')\sum_{j=1}^{|J|}\left(\frac{H^j}{2} + b^j\right)\frac{1-\beta_i}{2-\beta_i}\hat{g}_i^j$$

$$+\sum_{j=1}^{|J|}\left(H^j+2b^j\right)\left(f_i^{j}{'}\theta_i+\frac{1}{2-\beta_i}g_i^j\theta_i'+\frac{1-\beta_i}{2-\beta_i}\hat{g}_i^j\hat{\theta}_i\right) \tag{10}$$

Example Plant Design

Suppose the plant that produces 3 finished products from 4 raw materials as shown at Figure 4. Figure 4 also includes most input data for computation. A design problem without process I7, storage J10 and J11 was studied in the previous study[3]. The waste materials of failed batches are collected in Storage J10. The waste material J10 can be disposed through the waste disposal process J10 or can be regenerated through the Process I7 to raw materials J2 and J4. The waste material of process I7 goes to storage J11. The waste material J11 is disposed through the waste disposal process J11.

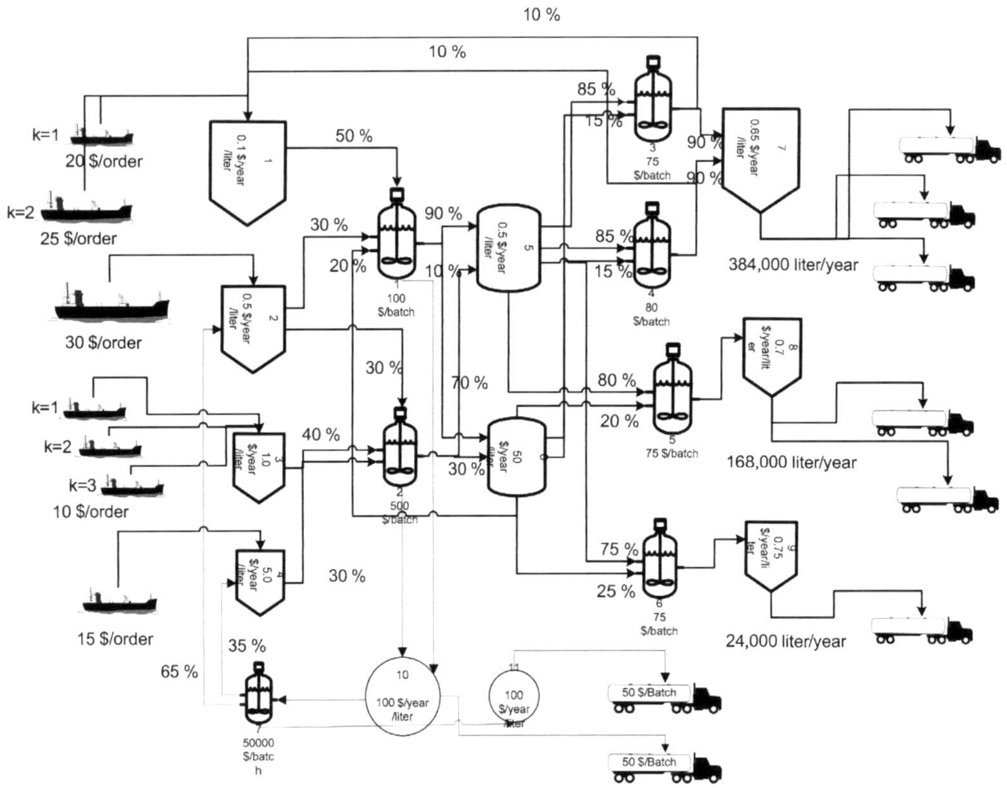

Figure 4. An Example Plant Design Problem

Conclusion

This study deals with determining the optimal sizes of batch processes and storage units interconnected in a general network structure when the processes are subject to joint uncertainties of operating time and batch quantity. Waste regeneration and disposal processes were included into the network to treat materials from failed batches. An unique graphical method was used to find the upper and lower bounds and average of material flows susceptible to short-term joint random variations in the cycle time and batch size. In the definition of the random properties, availabilities and occurrence number were introduced as input parameters instead of more widely used parameters such as the mean and variance. The availability is commonly used in process reliability analysis methods such as failure modes and effects analysis, and the occurrence number is proportional to the variance. These parameters were chosen as they are more practical and easier to estimate based on human perception. The optimization problem consisted of minimizing the expected sum of the setup cost, capital cost of processes/storage units, and inventory holding cost under the constraints of meeting random product demand and no depletion of storage materials. The stochastic version of the PSW model with unique graphical analysis provided analytical solutions of the optimization problem. These analytical solutions greatly reduce the computational burden, which is the major achievement of this study. The analytical optimal solutions made it possible to conduct sensitivity analysis with respect to the input parameters, time availability and size availability with two fixed values of occurrence number.

Acknowledgement

"This work was supported by the Korea Research Foundation Grant funded by the Korean Government (MOEHRD)" (KRF-2007-313-D00132)."

Literature Cited

1. Yi G, Reklaitis GV. Optimal design of multiple batch units with feedstock/product storages. *Chem. Eng. Comm.* 2000;181:79-106.
2. Yi G, Reklaitis GV. Optimal design of batch-storage network using periodic square model. *AIChE J.* 2002;48:1737-1753.
3. Yi G, Reklaitis GV. Optimal design of batch-storage network with recycle streams. *AIChE J.* 2003;49: 3084-3094.
4. Yi G, Reklaitis GV. Optimal design of batch-storage network with financial transactions and cash flows. *AIChE J.* 2004;50:2849-2865.
5. Yi G, Reklaitis GV. Optimal design of batch-storage network considering exchange rates and taxes. *AIChE J.* 2007;53:1211-1231.
6. Yi G, Reklaitis GV. Optimal design of batch-storage network with multitasking semi-continuous processes. *AIChE J.* 2006;52:269-281.
7. Yi G, Kim Y, Lee E. Optimal design of multisite batch-storage network under scenario-based demand uncertainty. *Chem. Eng. Comm.* 2007;194:1297-1327.
8. Yi G, Reklaitis GV. Optimal design of batch-storage network with uncertainty and waste treatments. *AIChE J.* 2006;52:3473-3490.
9. Yi G. Optimal design of batch-storage network under joint uncertainties. *AIChE J.* 2008;54:2567-2580.

578ac Modeling Association in the Saft Equation Using Ab Initio Quantum Chemistry

Angelo Lucia, University of Rhode Island, Dept of Chem. Engineering, Kingston, RI 02881, Leah Octavio, Chemical Engineering, University of Rhode Island, Dept of Chem. Engineering, Kingston, RI 02881, Donald P. Visco Jr., Tennessee Technological University, Box 5013, Cookeville, TN 38505, Eudes E. Fileti, Universidade Federal do ABC, Santo Andre SP, Brazil, and Sylvio Canuto, Instituto de Fisica, Universidade de Sao Paulo, Caixa Postal 66318, San Paulo, Brazil.

Recently, ab initio quantum chemistry calculations (specifically Moller-Plesset and Coupled-Cluster methods in Gaussian 98) by Fileti et al. (2004) have been used to show that hydrogen bonding strengths in alcohol-water mixtures depend on which molecule serves as the proton donor and which molecule serves as the proton acceptor. In particular, gas phase potential energies at 0 K for both self association and cross association have been determined and results clearly show that cross association is stronger if water serves as the proton donor while hydrogen bonds in which alcohol is the proton donor are weaker. Similar results hold for the liquid phase as well.

The original Statistical Associating Fluid Theory (SAFT) equation of state (Chapman et al., 1990; Huang and Radosz, 1990), and its many variations (e.g., simplified SAFT, PC-SAFT, and SAFT-VR), are well established as reliable mathematical tools for modeling the phase behavior of fluid mixtures. The primary advantages of the SAFT approach are that it takes into account the shape and size differences in fluid molecules and incorporates the effects of any association between molecules (e.g., hydrogen bonding). Unfortunately, relative hydrogen bonding effects modeled by SAFT are usually very poor and do not properly describe association when multiple hydrogen bonding sites are available.

The purpose of this poster is to

1) Provide further evidence that the association term in the SAFT formalism often predicts relative hydrogen bonding energies for mixtures that are grossly in error when the assumption of equal association strengths is used.

2) Describe a new methodology for determining unequal site-site association strengths and hydrogen bonding energies by fitting the SAFT association strengths directly to binding energies calculated from ab initio quantum chemistry.

In SAFT, in order to reduce the number of independent parameters, it is frequently assumed that the cross association bonding volumes are equal and that cross association bonding energies are equal. These assumptions lead directly to equal association strengths. Numerical examples are presented that clearly illustrate that using equal association strengths in the SAFT formalism often predicts relative hydrogen bonding energies that are completely different than those predicted by ab initio quantum chemistry calculations (e.g., Coupled Cluster methods and density functional theory). This, in turn, can lead to incorrect values for mole fractions of un-bounded sites, erroneous compressibility factors, and poor predictions of phase equilibrium. To resolve these difficulties, a new global optimization-quantum chemistry methodology for separating the determination of association strengths from non-associating SAFT parameters is proposed such that the resulting hydrogen bonding energies match more closely, in a relative sense, those predicted by quantum chemistry. Thus SAFT association parameters are not correlated to experimental vapor pressure and liquid density data but rather are fit to first principles information (i.e., using hydrogen bonding measurements to determine hydrogen bonding parameters). Numerical results for various associating mixtures, including ethanol and water and hydrogen fluoride and water, are presented to illustrate the efficacy of the proposed global optimization-quantum chemistry approach and clearly show that there are significant differences between the calculated mole fractions of un-bonded sites for equal and unequal association strengths.

MINLP Optimization of Cooling Towers

Medardo Serna-González [a], José M. Ponce-Ortega [a,b] and Arturo Jiménez-Gutiérrez [b]

[a] Universidad Michoacana de San Nicolás de Hidalgo, Morelia, México
[b] Instituto Tecnológico de Celaya, Celaya, México

Abstract

This paper presents a mixed-integer nonlinear programming (MINLP) model for the optimal design of mechanical counter-flow cooling towers, subjected to standard design constraints. The objective function consists of the minimization of the total annual cost, which includes the capital cost of the cooling tower (that depends on the filling material and the air flowrate) and the operating cost (that includes the costs for the water makeup and for the power consumed by the tower fan). Merkel's method is used to set the dimensions of the tower, and empirical correlations are used to estimate mass transfer coefficients for the packed section of the tower. The design variables to be optimized are the water-to-air flow ratio, the water flow rate, the water inlet and outlet temperatures, the operating temperature approach, the type of packing, the type of draft, the volume of packing material, the total pressure drop due to the air flow rate, the fan power, and the water consumption.

The MINLP model was solved with the GAMS software package using the DICOPT solver. The performance of the proposed model is shown with six examples. From the results of the examples, it can be observed that cooling towers with low temperature approaches are not suitable because driving forces become a limiting factor. On the other hand, dry-bulb temperature variations have a negligible effect on the tower size. From the set of design parameters analyzed, the temperature approach is shown to be a critical factor for the optimal design of cooling towers.

Introduction

Closed-cycle cooling water systems are widely used to dissipate the low grade heat of chemical and petrochemical process industries, electric-power generating stations, refrigeration and air conditioning plants. In closed-cycle systems, water cools the hot process streams, then it is cooled by evaporation and direct contact with air in a wet-cooling tower and recycled to the cooling network.

The wet or evaporative cooling towers may be classified as natural draft and mechanical draft types. In both, warm water from the cooling network of a plant enters the top of the tower and always flows downward over an internal labyrinth-like packing, called fill, and air can flow upward (counter flow) or horizontally (cross flow). The fill distributes the water flow uniformly and provides a large air-water interface area for the simultaneous heat and mass transfer processes. As a result of the direct contact between the water and the air in the packing region, part of the water is vaporized and the water temperature is reduced while the air enthalpy is increased. The cooled water is then collected in a cold water basin below the fill from which it is returned to the cooling network. The blow-down water and the water lost through evaporation and drift are replaced with fresh make-up water. The moist air leaves the top of the tower.

In the mechanical draft cooling tower, air is circulated through the tower by means of electrically driven fans. On the other hand, the natural draft cooling tower uses the natural buoyancy of the warmed air to circulate it through the tower. Mechanical draft towers can be either induced draft (fan located at the top of the tower) or forced draft (fan located at the bottom of the tower).

In practice, large natural draft cooling towers are used in power plants for cooling the water supply to the condenser. Mechanical draft cooling towers are preferred for oil refineries and other process industries, as well as for central air-conditioning systems, due to they cover a much wider range of sizes, can be made more compact, give more uniform cooling and have lower water loss than natural draft towers.

Though the fundamentals of mechanical draft counter flow cooling towers are presented in a number of references, very few studies are available for the optimal economic design of this type of cooling towers. In this paper, an MINLP formulation is developed for optimizing mechanical draft counter flow cooling towers, which takes into consideration the above mentioned group of independent variables in conjunction with the constraints imposed on the problem. The objective is to minimize the total annualized cost of the tower. The Merkel's method is used for estimating the size and performance of cooling towers. The design of cooling towers also requires the optimal selection of the type of packing and, in this work, the choice is limited to film, splash, and tickle type fills. The mass transfer and pressure drop characteristics of these types of packing are presented in the form of empirical correlations given by Kloppers and Kröger.[1, 2] In addition, mechanical cooling towers can be either induced or forced. Disjunctive programming is used to formulate the discrete choices of type of packing and type of draft. In thermodynamic property modeling, the enthalpy of saturated air-water vapor mixture is expressed as an exponential function of the local bulk water temperature. For developing this correlation, the ASHRAE[3] property table was used as the source of the enthalpy of saturated air data from 8ºC to 55ºC for standard atmospheric pressure. The fitted equation reproduces the data from ASHRAE[3] with 0.047% of average absolute deviation. The optimization problem is solved with GAMS, using the solver DICOPT.[4]

Cooling tower model

Figure 1 shows the general arrangement and variables employed for the representation of a counter flow cooling tower.

The temperature difference between the water outlet temperature and the wet-bulb temperature of the air entering the tower is called the tower approach.

$$Approach = TW_{out} - TWB_{in} \tag{1}$$

The temperature difference between the cooling tower water inlet and outlet temperatures is termed the range of the tower.

$$Range = TW_{in} - TW_{out} \tag{2}$$

The integrated form the Merkel's equation is presented as:

$$Me = \int_{TW_{out}}^{TW_{in}} \frac{c_{pw} dTW}{(hsa - ha)} \tag{3a}$$

where the Merkel number, *Me*, is given by

$$Me = \frac{h_d a_{fi} A_{fr} L_{fi}}{m_w} = \frac{h_d a_{fi} L_{fi}}{G_w} \tag{3b}$$

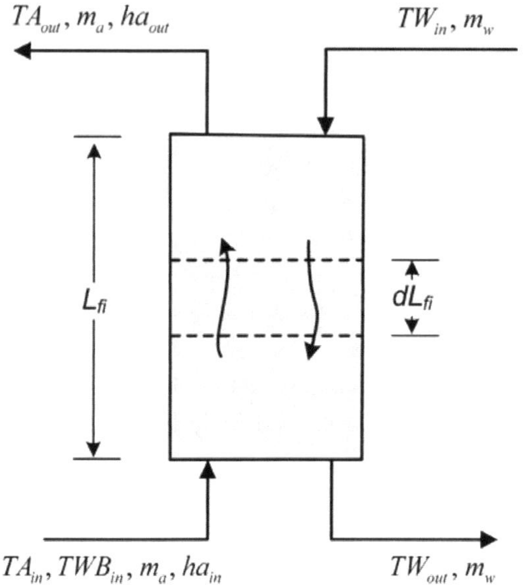

Figure 1. A differential control volume of a counter flow cooling tower

The operating line, which results from an overall energy balance on the fluid streams entering and leaving the tower is given as follows,

$$c_{pw}m_w(TW_{in} - TW_{out}) = m_a(ha_{out} - ha_{in}) \tag{4}$$

MINLP model formulation

Using the Merkel method, the optimal mechanical draft counter flow cooling tower problem is formulated as an MINLP optimization problem as follows.

Heat load

$$Q = c_{pw}m_w(TW_{in} - TW_{out}) \tag{5}$$

Required Merkel number

$$Me = \int_{TW_{out}}^{TW_{in}} \frac{c_{pw}dTW}{(hsa - ha)} = 0.25c_{pw}(TW_{in} - TW_{out})\sum_{i=1}^{4} 1/\Delta h_i \tag{6}$$

For each temperature increment, the local enthalpy difference (Δh_i) is calculated from

$$\Delta h_i = hsa_i - ha_i, \qquad i = 1,...,4 \tag{7}$$

and the algebraic equations for calculating the water temperature and air enthalpies corresponding to each Chebyshev point are given by

$$TW_i = TW_{out} + TCH_i(TW_{in} - TW_{out}), \qquad i = 1,...,4 \tag{8}$$

$$ha_i = ha_{in} + \frac{c_{pw}m_w}{m_a}(TW_i - TW_{out}), \qquad i = 1,...,4 \tag{9}$$

$$hsa_i = -6.38887667 + 0.86581791 * TW_i + 15.7153617 \exp(0.05439778 * TW_i), \qquad i = 1,...4 \tag{10}$$

Packing performance: Transfer and loss coefficients.

The measured transfer coefficients are correlated in terms of the *available Merkel number* for a particular fill type as,

$$Me = c_1\left(\frac{m_w}{A_{fr}}\right)^{c_2}\left(\frac{m_a}{A_{fr}}\right)^{c_3}(L_{fi})^{1+c_4}(TW_{in})^{c_5} \tag{11}$$

The following disjunction is the one associated with the discrete choice of the type of packing. This is accomplished with the Boolean variable Y^k, which can be true or false depending on whether type of packing k is selected or not. The disjunction is as follows:

$$\begin{bmatrix} Y^1 \\ (\text{splash fill}) \\ c_j = c_j^1, \quad j=1,\ldots,5 \end{bmatrix} \vee \begin{bmatrix} Y^2 \\ (\text{trickle fill}) \\ c_j = c_j^2, \quad j=1,\ldots,5 \end{bmatrix} \vee \begin{bmatrix} Y^3 \\ (\text{film fill}) \\ c_j = c_j^3, \quad j=1,\ldots,5 \end{bmatrix}$$

This disjunction is transformed using the convex hull reformulation as following:

$$y^1 + y^2 + y^3 = 1 \tag{12}$$

$$c_j = c_j^1 + c_j^2 + c_j^3, \quad j=1,\ldots,5 \tag{13}$$

$$c_j^k = a_j^k\, y^k, \quad k=1,\ldots,3. \quad j=1,\ldots,5 \tag{14}$$

where y^k is the set of binary variables that decides the type of packing.

The tower fill performance is described not only by the transfer coefficient, but also by the loss coefficient per meter depth of fill, K_{fi}, that is required to calculate the pressure drop through a fill. For the splash, trickle, and film type fills, the loss coefficient correlations can be expressed in the following form[1]

$$K_{fi} = \left[d_1 \left(\frac{m_w}{A_{fr}}\right)^{d_2} \left(\frac{m_a}{A_{fr}}\right)^{d_3} + d_4 \left(\frac{m_w}{A_{fr}}\right)^{d_5} \left(\frac{m_a}{A_{fr}}\right)^{d_6} \right] L_{fi} \tag{15}$$

The corresponding disjunction is given by,

$$\begin{bmatrix} Y^1 \\ (\text{splash fill}) \\ d_l = d_l^1, \quad l=1,\ldots,6 \end{bmatrix} \vee \begin{bmatrix} Y^2 \\ (\text{trickle fill}) \\ d_l = d_l^2, \quad l=1,\ldots,6 \end{bmatrix} \vee \begin{bmatrix} Y^3 \\ (\text{film fill}) \\ d_l = d_l^3, \quad l=1,\ldots,6 \end{bmatrix}$$

This disjunction using the convex hull reformulation is expressed as follows

$$d_l = d_l^1 + d_l^2 + d_l^3, \quad l=1,\ldots,6 \tag{16}$$

$$d_l^k = b_l^k\, y^k, \quad k=1,\ldots,3. \quad l=1,\ldots,6 \tag{17}$$

Tower pressure drop

In mechanical draft cooling towers, the total pressure drop along the air path, ΔP_t, is the sum of the static pressure drop and the velocity pressure.[5] The static pressure drop includes the pressure drop through the fill, ΔP_{fi}, and the sum of miscellaneous pressure losses, ΔP_{misc}, e.g., in drift eliminator, air inlet, water distribution piping, etc.

The pressure drop through the fill matrix is coupled to the loss coefficient by the following relation[1]

$$\Delta P_{fi} = K_{fi} L_{fi} \frac{mav_m^2}{2\rho_m A_{fr}^2} \tag{18}$$

where

$$mav_m = \frac{mav_{in} + mav_{out}}{2} \tag{19}$$

and

$$\rho_m = 1/\left(1/\rho_{in} + 1/\rho_{out}\right) \tag{20}$$

The air-vapor flow at the fill inlet and outlet mav_{in} and mav_{out} are calculated as follows

$$mav_{in} = m_a + w_{in} m_a \tag{21}$$

$$mav_{out} = m_a + w_{out} m_a \tag{22}$$

As pointed out by Mills[6], the miscellaneous pressure losses in components such drift eliminator, air inlet, water distribution piping, column supports, etc., are calculated by using a velocity head equation:

$$\Delta P_{mcl} = K_{mcl} \frac{\rho_{mcl} v_{mcl}^2}{2}$$

Some information on typical values for K_{mcl} is given by Li and Priddy[5] and Mills[6].

Total miscellaneous pressure losses can be represented by

$$\Delta P_{misc} = 6.5 \frac{\rho_m v_m^2}{2} = 6.5 \frac{mav_m^2}{2 \rho_m A_{fr}^2} \tag{23}$$

Another source of the total pressure drop in cooling towers is the velocity pressure, ΔP_{vp}, which is calculated as follows

$$\Delta P_{vp} = (2/3)(\Delta P_{fi} + \Delta P_{misc}) \tag{24}$$

By summing equations (22), (23), and (24), one can express the total pressure drop of the air stream for counter flow cooling towers as

$$\Delta P_t = 1.667(\Delta P_{fi} + \Delta P_{misc}) \tag{25}$$

Power consumption

The required power for a cooling tower fan can be determined by multiplying the total pressure drop with the air volume flow rate at the fan location. Hence, the power consumption is in terms of the air volume flow rate at the packing inlet for forced draft towers or at the packing outlet for induced draft towers. Consequently, the location of the air fan has effect on the power requirement. According to this feature, the following disjunction is required for the discrete decision associated with the optimal selection of draft type:

$$\begin{bmatrix} Y_{CT}^{mfd} \\ (\text{mecanical forced draft}) \\ P = \dfrac{mav_{in} \Delta P_t}{\rho_{in} \eta_f} \end{bmatrix} \vee \begin{bmatrix} Y_{CT}^{mid} \\ (\text{mecanical induced draft}) \\ P = \dfrac{mav_{out} \Delta P_t}{\rho_{out} \eta_f} \end{bmatrix}$$

This disjunction can be described by the following set of equations:

$$P \le \frac{mav_{in} \Delta P_t}{\rho_{in} \eta_f} + M_P^{\max} (1 - y_{CT}^{mfd}) \tag{26}$$

$$P \ge \frac{mav_{in} \Delta P_t}{\rho_{in} \eta_f} - M_P^{\max} (1 - y_{CT}^{mfd}) \tag{27}$$

$$P \le \frac{mav_{out} \Delta P_t}{\rho_{out} \eta_f} + M_P^{\max} (1 - y_{CT}^{mid}) \tag{28}$$

$$P \ge \frac{mav_{out} \Delta P_t}{\rho_{out} \eta_f} - M_P^{\max} (1 - y_{CT}^{mid}) \tag{29}$$

$$y_{CT}^{mfd} + y_{CT}^{mid} = 1 \tag{30}$$

Water Consumption

Makeup water (m_{mw}) is constantly added to the cooling tower basin to compensate for the loss of water to evaporation, drift, and blowdown.

From the conservation of mass, the rate of water evaporated (m_{wev}) into the air stream is related to the dry air flow rate and the air-inlet and outlet water vapor contents as follows

$$m_{wev} = m_a (w_{out} - w_{in}) \tag{31}$$

Since the evaporation water is essentially free of dissolved solids and impurities, all those introduced in the makeup water must be equal to that in blowdown and drift losses. Thus, the amount of blowdown can be expressed as

$$m_{bw} = \frac{m_{mw}}{n_{cycles}} - m_{wd} \tag{32}$$

In a well designed tower the drift loss is no more than 0.2 percent of the total water circulating flow[7]

$$m_{wd} = 0.002 m_w \tag{33}$$

The makeup water must equal the sum of the water losses to maintain a steady circulating water flow, that is, $m_{mw} = m_{mev} + m_{bw} + m_{wd}$,

$$m_{mw} = \frac{n_{cycles} m_{wev}}{n_{cycles} - 1} \tag{34}$$

Physical properties

For total pressure of 1 atm, the enthalpy of the air entering the tower, ha_{in}, is equal to the enthalpy of saturated air at a temperature equal to the inlet air wet-bulb temperature, TWB_{in} (Singham, 1983). Thus, ha_{in} can be calculated as follows

$$ha_{in} = -6.38887667 + 0.86581791 * TWB_{in} + 15.7153617 \exp(0.05439778 * TWB_{in}) \tag{35}$$

The mass fraction humidity of the air stream at tower inlet, w_{in}, is a function of the inlet air dry- and wet-bulb temperatures and is given by[8]

$$w_{in} = \left(\frac{2501.6 - 2.3263\,(TWB_{in})}{2506 + 1.8577\,(TA_{in}) - 4.184\,(TWB_{in})} \right) \left(\frac{0.62509\,(PV_{wbin})}{P_{tot} - 1.005\,(PV_{wbin})} \right) - \left(\frac{1.00416\,(TA_{in} - TWB_{in})}{2506 + 1.8577\,(TA_{in}) - 4.184\,(TWB_{in})} \right) \tag{36}$$

Eq. (4) can be rearranged to give

$$hsa_{out} = ha_{in} + \frac{c_{pw} m_w}{m_a} (TW_{in} - TW_{out}) \tag{37}$$

The outlet air temperature can be represented by the following functional relationship

$$TA_{out} = f_1(hsa_{out}) \tag{38}$$

The mass fraction humidity of the saturated air stream at tower outlet is given by (Kröger, 2004)

$$w_{out} = \frac{0.62509\, PV_{out}}{P_{tot} - 1.05\, PV_{out}} \tag{39}$$

The vapor pressure of water corresponding to a specified temperature is calculated from the correlation given by Hyland and Wexler[9],

$$\ln(PV) = \sum_{n=-1}^{3} c_n T^n + 6.5459673 \ln(T) \tag{40}$$

The density of air-water vapor mixtures can be calculated from the ideal gas law as

$$\rho = \frac{P_{tot}}{287.08\, T} \left[1 - \frac{w}{w + 0.62198} \right] [1 + w] \tag{41}$$

Feasibility constraints

In this section, inequality constraints are written to define the feasible region of the optimization problem as well as feasible operation of the optimal cooling tower as follows,

$$TW_{out} - TWB_{in} \geq 2.8 \tag{42}$$

$$TW_{out} \leq TMPO - \Delta T_{min} \tag{43}$$

$$TW_{in} \le TMPI - \Delta T_{min} \tag{44}$$

$$TW_{in} \le 50^o C \tag{45}$$

$$TW_{in} > TW_{out} \tag{46}$$

$$TA_{out} > TA_{in} \tag{47}$$

$$hsa_i - ha_i > 0 \quad i = 1,...,4 \qquad (47)$$

$$0.5 \le \frac{m_w}{m_a} \le 2.5 \tag{48}$$

$$2.90 \le \frac{m_w}{A_{fr}} \le 5.96 \tag{49}$$

$$1.20 \le \frac{m_a}{A_{fr}} \le 4.25 \tag{50}$$

$$m_w > 0 \tag{51}$$

$$m_a > 0 \tag{52}$$

Objective function

The objective function involves the minimization of the total annualized cost, *TAC*, which is the sum of the annualized capital cost of the cooling tower and annual operating costs C_{op}:

$$minimize\ TAC = K_f C_{cap,CT} + C_{op} \tag{53}$$

The annual operating costs are determined by the makeup water consumed and the fan power required for a given application:

$$C_{op} = H_Y c_w m_{mw} + H_Y c_e P \tag{54}$$

The formula for the installed capital cost of cooling towers, given below, is from Kintner-Meyer and Emery[10]:

$$C_{cap,CT} = C_{CTF} + C_{CTV} A_{fr} L_{fi} + C_{CTMA} m_a \tag{55}$$

The cost coefficient C_{CTV} depends on the type of packing and is represented by the following disjunction:

$$\begin{bmatrix} Y^1 \\ (splash\ fill) \\ C_{CT,V} = C^1_{CT,V} \end{bmatrix} \vee \begin{bmatrix} Y^2 \\ (trickle\ fill) \\ C_{CT,V} = C^2_{CT,V} \end{bmatrix} \vee \begin{bmatrix} Y^3 \\ (film\ fill) \\ C_{CT,V} = C^3_{CT,V} \end{bmatrix}$$

This disjunction can be described by the following equations:

$$C_{CT,V} = C^1_{CT,V} + C^2_{CT,V} + C^3_{CT,V} \tag{56}$$

$$C^k_{CT,V} = e^k y^k, \quad k = 1,...,3 \tag{57}$$

Numerical examples

The software DICOPT included in the GAMS[4] optimization package was used to solve the proposed MINLP model given by Eqs. (5) to (57). In Table 1, the design specifications and process constraints for six examples serving as test problems are presented. Example 1 is the base case. In the other examples, only one input variable or design constraint varies while the other input variables or design constraints maintain at the base values. For all examples, the cooling tower is required to remove 3400 kW at an ambient air pressure of 101 325 Pa. The c_{pw} for the water is taken as 4.187 kJ/kg°C, the H_Y parameter is assumed as 2.934×10^7 s/year, and the annualizing factor for the capital cost as 0.2983/year. The values of c_w, c_e, C_{CTF}, C_{CTMA}, η_f and n_{cycles}, are taken as US$5.283x10^{-4}/kg, US$0.085/kWh, US$31,185, US$1,097.5/(kg of dry air/s), 0.75 and 4, respectively.

Table 1. Design specifications and process Constraints for the examples

	Examples					
	1	2	3	4	5	6
Q (kW)	3400	3400	3400	3400	3400	3400
TA_{in} (°C)	22	17	22	22	22	22
TWB_{in} (°C)	12	12	7	12	12	12
$TMPI$ (°C)	65	65	65	55	65	65
$TMPO$ (°C)	30	30	30	30	25	30
ΔT_{min} (°C)	10	10	10	10	10	5

The results of the total annualized cost minimization for the six examples are presented in Table 2. For each example, the selected type of filling material is the film packing that offers the best combination of heat transfer and pressure losses, enabling the lowest total annualized cost. Also, the selected type of draft is the forced draft in all examples. The water and air loadings of all examples are in the ranges given by constraints (49) and (50), respectively. It should be noted that the total annualized cost is dominated by the operating costs for all examples. In general, the cost of makeup water is greater than the cost of electric power required for the fan performance.

The effect of variation of the inlet air dry-bulb temperature, TA_{in}, on the total annualized cost can be established from the Examples 1 and 2. As TA_{in} is decreased from 22 to 17°C, the optimum total annualized cost shifts from US$66,065.14/yr to US$64,604.64/yr. It causes a 2.21% decrease in the total annualized cost. Thus, the total annualized cost is fairly insensitive to the inlet air dry-bulb temperature for the range of conditions tried, and decreases slightly with TA_{in}.

However, as shown in Table 2, decreasing the tower approach has a pronounced effect on the optimum value of total annualized cost. By taking the optimum solution of Example 1 as reference, we see that, when the optimal tower approach is decreased from 8 to 3 (Example 5), the total annualized cost is increased by 77.52%. In general, as the approach is reduced, tower size increases exponentially if other conditions remain unchanged, and it follows that total annualized cost of the tower also increases exponentially. This is because the driving forces become more limiting as the tower approach decreases and so a larger Me or tower size is required for rejecting the same heat load. In contrast to this behavior, Examples 3 and 6 show that as the tower approach increases the total annualized cost decreases significantly. Thus, it is clear that the tower approach is an important optimization variable.

Examples 3 and 6 show that optimal tower approach increases when TWB_{in} and ΔT_{min} decrease. On the other hand, Example 5 shows that, as $TMPO$ decreases, the optimal tower approach also decreases. Since TWB_{in} is site-specific and $TMPO$ is limited by the temperature conditions of the process, these variables are taken as input parameters. Thus, only the value of ΔT_{min} could be determined by optimization. This decision variable could be minimized to maximize the tower approach and, therefore, minimize the size of the cooling tower. However, this is offset by increases in the size of the coolers in the cooling network due to lower log-mean temperature differences. Thus, this optimization problem must take into account the interaction between tower performance and cooling network performance to select the optimum value of ΔT_{min} in the context of the entire system.

The effect of the cooling range on the cooling tower cost is illustrated in Examples 3 and 6 for a constant tower approach and inlet water temperature. It was found that a reduced cooling range results in a higher water outlet temperature, with the corresponding increase in driving forces for cooling. Thus, as the cooling range is decreased, the tower size (or Merkel number) and tower cost become smaller.

Table 2. Optimization Results for the Examples

	Examples					
	1	2	3	4	5	6
m_w (kg/s)	25.720	25.794	25.700	30.973	22.127	30.749
m_a (kg/s)	31.014	31.443	28.199	36.950	32.428	27.205
m_w/m_a	0.829	0.820	0.911	0.838	0.682	1.130
m_{mw} (kg/s)	1.541	1.456	1.564	1.547	1.542	1.540
m_{wev} (kg/s)	1.156	1.092	1.173	1.160	1.157	1.155
m_{bw} (kg/s)	0.334	0.312	0.340	0.325	0.341	0.323
m_{wd} (kg/s)	0.051	0.052	0.051	0.062	0.044	0.061
TW_{in} (°C)	50	50	50	45	50	50
TW_{out} (°C)	20	20	20	20	15	25
TA_{out} (°C)	37.077	36.871	36.998	34.511	36.411	39.083
Range (°C)	30	30	30	25	35	25
Approach (°C)	8	8	13	8	3	13
L_{fi} (m)	2.294	2.239	1.858	2.154	6.299	1.480
A_{fr} (m^2)	8.869	8.894	8.862	10.680	7.630	9.296
K_{fi}	21.946	21.950	21.926	21.942	22.066	22.639
ΔP_{fi} (Pa)	280.331	277.727	186.621	254.540	1139.529	131.908
ΔP_{misc} (Pa)	36.189	36.740	29.782	35.011	53.288	25.596
ΔP_t (Pa)	527.640	524.216	360.744	482.683	1988.425	262.560
P (hp)	24.637	24.474	15.205	26.852	97.077	10.754
Type of packing	Film	Film	Film	Film	Film	Film
Type of draft	Forced	Forced	Forced	Forced	Forced	Forced
Me	3.083	3.055	2.466	2.923	7.335	1.858
C_{mw}	23885.109	22566.366	24239.785	23983.449	23901.657	23865.877
C_{power}	12737.595	12653.677	7861.037	13882.754	50190.495	5559.875
C_{op}	36622.703	35220.043	32100.822	37866.203	74092.153	29425.752
$KC_{cap,CT}$	29442.436	29384.597	26615.995	32667.705	43186.526	25030.274
TAC	66065.139	64604.640	58716.817	70533.909	117278.68	54456.026

In addition, Examples 2 and 4 (or 1 and 4) show the effect of cooling range for fixed tower approach and fixed water outlet temperature. In this case, we see that as cooling range increases, the inlet water temperature also increases. This results in decreased water and air mass flow rate, which gives smaller operating costs and capital cost.

It should be noted that the best optimal solution (Example 6) has a range and an approach temperature of 25°C and 13°C, respectively. This result is in agreement with above discussion. The associated operating and capital costs are US$29,425.752/yr and US$25,030.274/yr. This constitutes a cost reduction of 14.98% for the capital cost and 19.65% for the operating cost when compared to the base case.

Conclusions

This paper presents an MINLP formulation for the optimal design of mechanical counter flow cooling towers. The Merkel method is used for sizing cooling towers due to it is usually adopted in the practice to design these industrial units.[2] This method relates quantities associated with the specified cooling requirement to quantities associated with the heat transfer performance of a given packing.[11] The *required Merkel number* is calculated using the Chebyshev integration technique. The film, splash, and trickle type fills are the three different types of packing arrangements considered in this paper. Thus, the MINLP model has different empirical correlations for calculating the *available Merkel number* and packing pressure drop of air depending on the type of packing. Also, the model considers that mechanical cooling towers could be induced or forced. These design decisions are represented by disjunctive programming models.

For a given heat load of the tower and inlet air conditions, the solution yields the optimal geometric dimensions of fill height, and cross sectional area of the tower and the optimal operating parameters of water mass flow rate, air mass flow rate, water consumption, power consumption, water outlet temperature, and water inlet temperature, as well as the optimal type of packing and type of draft. As shown in the examples, cooling towers with low approaches are more expensive since driving forces become more limiting. The reverse is also true. On the other hand, dry-bulb temperature variations have negligible effect on tower size. The conclusion is that approach is critical for the optimal design of cooling towers.

References

1. Kloppers, J.C., and Kröger, D.G. (2003). Loss coefficient correlation for wet-cooling tower fills. *Applied Thermal Engineering,* 23(17): 2201-2211.
2. Kloppers, J.C., and Kröger, D.G. (2005). Refinement of the transfer characteristic correlation of wet-cooling tower fills. *Heat Transfer Engineering,* 26(4): 35-41.
3. ASHRAE. (2001). *ASHRAE Handbook – Fundamentals* (American Society of Heating, refrigerating and Air-Conditioning Engineers, Inc., USA).
4. Brooke, A., Kendrick, D., and Meeraus, A. (2006) *GAMS User's Guide* (The Scientific Press, USA).
5. Li, K.W., and Priddy, A.P. (1985). *Power Plant System Design* (John Wiley & Sons, USA).
6. Mills, A.E. (1999). *Basic Heat and Mass Transfer* (Prentice Hall, USA).
7. Kemmer, F.N. (1988). *The NALCO water handbook (*McGraw-Hill, USA).
8. Kröger, D.G. (2004). *Air-Cooled Heat Exchangers and Cooling Towers* (PenWell Corp., USA).
9. Hyland, R.W. and Wexler, A. (1983). Formulation for the thermodynamic properties of the saturated phases of H O from 173.15 K and 473.15 K, *ASHRAE Transactions*, 89(2A): 500-519.
10. Kintner-Meyer, M. and Emery, A.F. (1995). Cost-optimal design for cooling towers, *ASHRAE Journal*, 4: 46-55.
11. Singham, J.R. (1983). *Heat Exchanger Design Handbook* (Hemisphere Publishing Corporation, USA).

Optimal detailed design for integral cooling water systems

José M. Ponce-Ortega [a,b], Medardo Serna-González [a] and Arturo Jiménez-Gutiérrez [b]

[a] *Universidad Michoacana de San Nicolás de Hidalgo, Morelia, Mich. 58060, México*
[b] *Instituto Tecnológico de Celaya, Celaya, Gto. 38010, México*

Abstract

A cooling water system is used in industrial processes to cool down hot process streams from their supply conditions to target temperatures. A cooling water system is comprised of three major components, namely a cooling tower, a heat exchanger network (HEN), and a pumping system; these elements have strong interactions with each other and therefore must be optimized simultaneously to yield an optimal cooling water system.

This paper presents a mixed-integer non linear programming model for the optimal design of integral cooling water systems. The model includes a superstructure for the HEN component of the system that allows arrangements of coolers in series and/or in parallel, and also allows the bypass of fresh water and/or previously-used water to improve the performance of the cooling network. The optimization of HEN coolers is made simultaneously to take into consideration from the synthesis stage the interactions between the pumping effects and the heat transfer coefficients that determine their transfer areas. The model considers as objective function the minimization of the capital costs for the cooling tower, HEN coolers and pumps, as well as the operating costs due to the fan of the tower, pumping requirements and water makeup. One example problem is presented to show the application of the proposed method.

Introduction

Cooling water systems are used to treat hot process streams in the process industries. The cooling water system is constituted by three major components, namely the cooling tower, the heat exchanger network and the pumps; these elements have strong interactions with each other and therefore must be optimized simultaneously to yield an optimal cooling water system (see Figure 1).

Published works on cooling water systems have not carried out a simultaneous optimization that considers all of its components.[1,2,3,4,5,6,7,8,9,10,11] In addition, all previous methodologies have ignored the optimization for the total cooling water system considering the capital and operating cost simultaneously. This paper proposes a new mathematical model for the synthesis of cooling water systems that simultaneously considers the cooling tower, the cooling network and the pumping system. The model takes into account the capital and operating costs simultaneously, as well as the geometrical and operational constraints for the cooling tower and coolers. The methodology considers pumping effects in the cooling system through the heat exchangers detailed design. The objective function for the model consists in the minimization of the capital costs for the cooling tower, coolers and pumps, and the operating costs for the fan of the tower and the pumps, as well as the cost for the makeup cooling water.

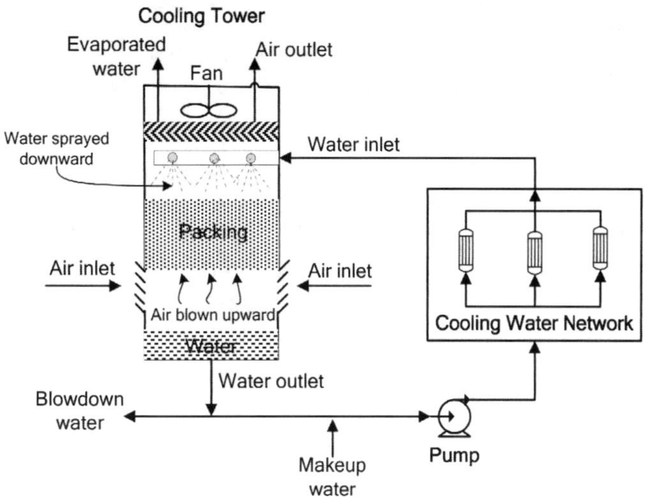

Figure 1. Cooling water system

Model formulation

A superstructure is proposed that considers the interactions between the cooling tower, the cooling water network and the pumping effects (Figure 2). Only one cooling tower is needed, and the cooling water network allows arrangements in parallel, in series and their combinations to get an optimum cooling water network. Also, the superstructure considers the pumping effects in the heat exchangers design. The cooling tower, the heat exchanger units and the pumping design are optimized simultaneously to synthesize the overall cooling water system.

Model for the cooling water network. The model for the cooling water network is based on the one previously reported by Ponce-Ortega et al.[11]

Coolers design equations. When a heat exchanger is needed in the cooling network, a detailed design must be considered to take into account the pumping effects simultaneously. To model this situation, a generalized disjunctive programming model is developed. When the exchanger exists, the set of equations for the detailed design is applied rigorously; otherwise the optimization variables involved are set as zero, as follows:

$$\begin{bmatrix} z_{i,k} \\ \text{Apply exchanger design equations} \end{bmatrix} \vee \begin{bmatrix} \neg z_{i,k} \\ \text{Set designvariables as zero} \end{bmatrix} \quad i \in HPS, k \in ST$$

The equations used for the detailed heat exchangers design include two compact relationships that relate the pressure drops for the heat transfer area and the film heat transfer coefficients for the shell side and the tube side. The compact expression for the shell side is based on the Kern method.[12] Also, the model includes equations for the calculation of the velocities of the fluids, the log-mean temperature difference using Chen's approximation[13], the global heat transfer coefficient and the heat transfer area. The big-M formulation is used to model this disjunction.

Layout and equivalent diameter. The equivalent diameter for the heat exchanger units is needed for the calculation of the shell side heat transfer coefficient, and it depends on the arrangement of the tubes. To model this situation the following disjunction is proposed,

$$\left[\begin{array}{c} z_{i,k} \\ De_{i,k}Dt_{i,k} = \dfrac{4\left(a_{i,k}^{layout1}Lpt_{i,k}^{2} - a_{i,k}^{layout2}\pi Dti_{i,k}^{2}/4\right)}{\pi} a_{i,k}^{layout3} 1000 \\ \left[\begin{array}{c} Y_{TRI_{i,k}}^{layout} \\ (\text{triangular}) \\ C_{AT_{i,k}} = 0.866 \\ a_{i,k}^{layout1} = 1 \\ a_{i,k}^{layout2} = 1 \\ a_{i,k}^{layout3} = 1 \end{array}\right] \vee \left[\begin{array}{c} Y_{SQU_{i,k}}^{layout} \\ (\text{square}) \\ C_{AT_{i,k}} = 1 \\ a_{i,k}^{layout1} = 0.435 \\ a_{i,k}^{layout2} = 0.5 \\ a_{i,k}^{layout3} = 2 \end{array}\right] \end{array}\right] \vee \left[\begin{array}{c} \neg z_{i,k} \\ De_{i,k} = 0 \\ C_{AT_{i,k}} = 0 \\ a_{i,k}^{layout1} = 0 \\ a_{i,k}^{layout2} = 0 \\ a_{i,k}^{layout3} = 0 \end{array}\right] \quad i \in HPS, k \in ST$$

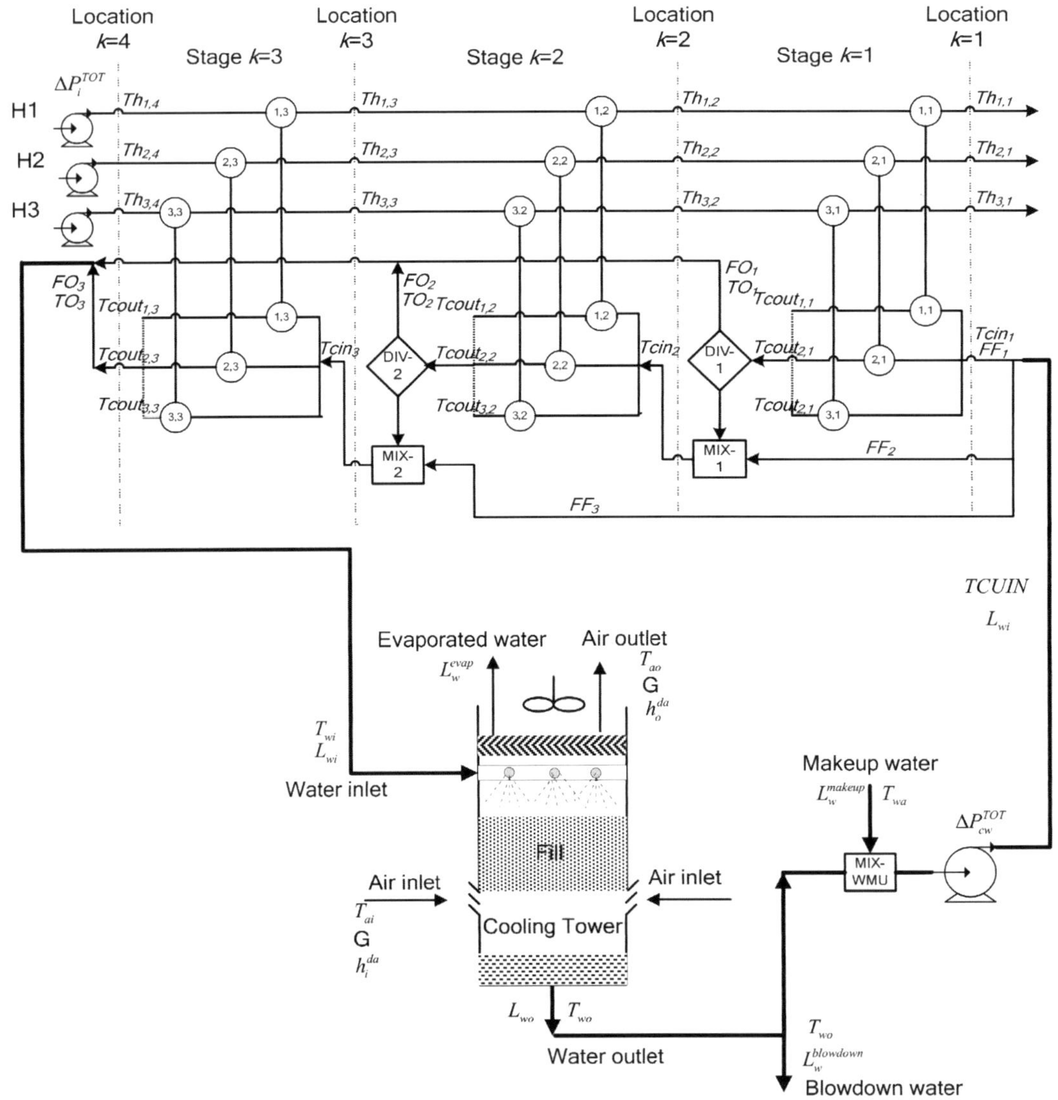

Figure 2. Superstructure for the cooling system.

Tube size. The model must select the optimum tube size from standard dimensions when the heat exchanger exists. Therefore, the following disjunction is proposed,

$$\left[\begin{array}{c} z_{i,k} \\ \begin{bmatrix} Y^{tube}_{1_{i,k}} \\ Dt_{i,k} = 6.350 \\ Dti_{i,k} = 4.928 \\ Lpt_{i,k} = 7.938 \end{bmatrix} \vee \ldots \vee \begin{bmatrix} Y^{tube}_{it_{i,k}} \\ Dt_{i,k} = \ldots \\ Dti_{i,k} = \ldots \\ Lpt_{i,k} = \ldots \end{bmatrix} \vee \ldots \vee \begin{bmatrix} Y^{tube}_{IT_{i,k}} \\ Dt_{i,k} = 50.8 \\ Dti_{i,k} = 45.26 \\ Lpt_{i,k} = 63.5 \end{bmatrix} \end{array} \right] \vee \begin{bmatrix} \neg z_{i,k} \\ Dt_{i,k} = 0 \\ Dti_{i,k} = 0 \\ Lpt_{i,k} = 0 \end{bmatrix} \quad i \in HPS, k \in ST$$

Localization of the fluids. The model must select the optimal localization of the fluids inside of the heat exchangers in the shell or tubes sides. This situation is modeled using the following disjunction,

$$\left[\begin{array}{c} z_{i,k} \\ \begin{bmatrix} Y^{loc}_{1_{i,k}} \\ \text{hot fluid in the tubes} \\ \text{cold water in the shell} \end{bmatrix} \vee \begin{bmatrix} Y^{loc}_{2_{i,k}} \\ \text{hot fluid in the shell} \\ \text{cold water in the tubes} \end{bmatrix} \end{array} \right] \vee \begin{bmatrix} \neg z_{i,k} \\ \text{set all variables as zero} \end{bmatrix} \quad i \in HPS, k \in ST$$

Pressure drops constraints. Because the heat exchanger units for the hot process streams are placed in the superstructure in series, the total pressure drop for each hot process stream is calculated as follows,

$$\Delta P^{TOT}_i = \sum_{k \in ST} \Delta P^h_{i,k}, \quad i \in HPS$$

For the cold water there are combinations of arrangements in series and in parallel in the superstructure; therefore, we need to identify the biggest pressure drop in each stage of the superstructure because this is the one that must be consumed in each stage of the superstructure,

$$\Delta P^{STAGE}_{cu_k} \geq \Delta P^c_{i,k}, \quad i \in HPS, k \in ST$$

Multipass exchangers. Because of space limitation, the use of multipass heat exchangers is sometimes needed. In this case it is necessary to consider the selection of multipass or single pass exchangers, as shown in the following disjunction,

$$\left[\begin{array}{c} z_{i,k} \\ \begin{bmatrix} Y^{Npass1}_{i,k} \\ \text{Single pass exchanger} \end{bmatrix} \vee \begin{bmatrix} Y^{Npass2}_{i,k} \\ \text{Multipass exchanger} \end{bmatrix} \end{array} \right] \vee \begin{bmatrix} \neg z_{i,k} \\ \text{Set variables as zero} \end{bmatrix}, \quad i \in HPS, k \in ST$$

Interconnection between cooling network and the cooling tower. The following balances between the cooling network and the tower are needed.

Fresh cold water,

$$L_{wi} = \sum_{k \in ST} FO_k$$

Energy balance for the mixer in the last stage of the superstructure,

$$\sum_{k \in ST} \left(FO_k TO_k \right) = L_{wi} T_{wi}$$

The heat load for the cooling tower,

$$Q_{CT} = Cp_{cw} \left(L_{wi} T_{wi} - L_{wo} T_{wo} \right)$$

Energy balance in the makeup mixer,

$$L^{makeup}_w \left(T_{wa} - T_{wo} \right) = L_{wi} \left(TCUIN - T_{wo} \right)$$

Cooling tower. This work uses the equations proposed by Serna-Gonzalez et al.[17] to model the cooling tower performance.

Water Consumption. Makeup water is constantly added to the cooling tower basin to compensate the loss of water for evaporation, drift, and blowdown. The total makeup water is given by,

$$L_w^{makeup} = \frac{n_{cycles} L_w^{evap}}{n_{cycles} - 1}$$

where n_{cycles} is the number of cycles of concentration. A cycle of concentration is the ratio of the solid concentration in the circulating water to the solid concentration in the makeup water.

Feasible constraints. There are a set of geometrical and operational constraints for the heat exchanger units and the cooling tower.

Feasible constraints for the coolers. To get a practical cooler, several works have proposed a set of geometrical and operational constraints. These constraints must be applied when the cooler exists in the network, according to the following disjunction,

$$\begin{bmatrix} Z_{i,k} \\ \text{Feasible constraints for exchangers} \end{bmatrix} \vee \begin{bmatrix} \neg Z_{i,k} \\ \text{Set variables as zero} \end{bmatrix}$$

Feasible constraints for the cooling tower. There is a set of constraints to allow an efficient operation of the cooling tower, and another set of constraints to model the cooling tower.[14]

Objective function. The objective function is taken as the minimization of the total yearly cost for the cooling water system, given by the capital cost for the cooling tower, the capital cost for the heat exchangers that considers the number of shells needed when multipass exchangers are selected, the capital cost for the pumps required for the hot process streams and for the cooling water, the operational costs for the electricity needed to operate the fan of the cooling tower, the costs due to the makeup water, and the electricity costs needed to operate the pumps of the system. The objective function is then written as,

$$\begin{aligned}
\min TAC = {} & k_F \left[C_V^{CT} A_{CT} L_{fi} + C_G^{CT} ma_{mean} + C_0^{CT} \right] \\
& + k_F \left[\sum_{i \in HPS} \sum_{k \in ST} CF_i^{exc} z_{i,k} + \sum_{i \in HPS} \sum_{k \in ST} Ns_{i,k} C_i^{exc} \left(\frac{A_{i,k}}{Ns_{i,k}} \right)^{\beta} \right] \\
& + k_F \left[\sum_{i \in HPS} CF_i^{pump} + \sum_{i \in HPS} C_i^{pump} \left(\frac{\Delta P_i^{TOT} F_i^h}{\rho_i} \right)^{\gamma} \right] \\
& + k_F \left[CF_{cu}^{pump} + C_{cu}^{pump} \left(\frac{\sum_{k \in ST} \left(\Delta P_{cu_k}^{STAGE} \right) \sum_{k \in ST} \left(FF_k \right)}{\rho_{cu}} \right)^{\gamma} \right] \\
& + C_{pow} H_Y 3600 \Delta P_{CT} \\
& + C_{pow} \frac{H_Y}{\eta_{pump}} \left[\sum_{i \in HPS} \left(\frac{\Delta P_i^{TOT} F_i^h}{\rho_i} \right) + \frac{\sum_{k \in ST} \left(\Delta P_{cu_k}^{STAGE} \right) \sum_{k \in ST} \left(FF_k \right)}{\rho_{cu}} \right] \\
& + C_{water} H_Y 3600 L_w^{makeup}
\end{aligned}$$

where C_V^{CT}, C_G^{CT} and C_o^{CT} are constants for the capital cost of the tower, CF_i^{exc}, C_i^{exc} and β are parameters for the capital cost of the heat exchangers, CF_i^{pump}, C_i^{pump} and γ are parameters for the capital cost of the pump, C_{pow} is the unit cost for electricity and C_{water} is the unit cost for the makeup water.

Results and discussion

One example is used to show the application of the proposed model. The parameter H_Y was assumed as 8,500 hr/year, with an annualization factor 0.23 year^{-1}, and the coefficients for the capital cost function for the coolers as follows, CF_i^{exc} of \$30,800, C_i^{exc} of \$1,650 (where A is in m^2) and β of 0.65. The capital costs for the pumps are estimated as follows, CF_i^{pump} is set as \$2,000, C_i^{pump} as \$5 (where ΔP , F and ρ are in Pa, kg/s and kg/m^3, respectively) and γ as 0.68. The coefficients for the capital cost of the cooling tower are \$1,097.5 and \$31,185 for C_G^{CT} and C_o^{CT}, respectively. The electricity cost, C_{pow}, is \$0.00005/W-hr and the efficiency for the pumps is 70%. The unit cost for the cold makeup water was taken as \$1.5949 x 10^{-5}/kg. The efficiency for the fan of the cooling tower is 75%, and the number of cycles for the cooling water system is set as 4.

The following conditions were taken as a basis. The maximum allowable temperature for the cooling water is 50°C, the total pressure of the system is 101,712.27 Pa, the temperature for the air at the inlet conditions is 9.7°C, and the wet bulb temperature for the air at the inlet conditions is 8.23°C. The temperature for the makeup water was taken as 12°C.

The physical properties for the cooling water in the network are assumed as follows. The heat capacity is 4.187 kJ/(kg K), the viscosity 1x10^{-3} kg/(m s), the thermal conductivity 0.58 W/(m K), the density 998 kg/m^3 and the individual fouling factor 0.00015 (m^2 K)/W.

To solve the model, the solver DICOPT, implemented in the general algebraic modeling system (GAMS) was used.[15]

The stream data for the hot process streams are given in Table 1, and their physical properties are given in Table 2.

Table 1. Stream data for the Example

Stream	TIN [°C]	TOUT [°C]	FCP [kW/(m^2 K)]
H1	76.60	40.00	100.00
H2	82.00	60.00	60.00
H3	108.85	45.00	400.00
CU	10.00	-	-

Table 2. Physical properties for streams of the Example

Property	*Cp* [J/(kg K)]	μ [kg/(m s)]	*k* [W/(m K)]	ρ [kg/m^3]	*Rd* [(m^2 K)/W]
H1	2,454	2.4 x 10^{-4}	0.114	634	0.00015
H2	1,670	2.3 x 10^{-4}	0.23	780	0.00017
H3	2,680	2.1 x 10^{-4}	0.14	890	0.00016

After the application of the proposed methodology, the water system design shown in Figure 3 was obtained, with the coolers designs given in Table 3. In the optimal system design, the inlet temperature to the cooling network is 17.893°C; for the cooling network there is a parallel arrangement for coolers 1 and 2, foolowed by cooler 3 in series. The outlet temperature of cooling water for cooler 3 is the maximum allowable temperature. The cold water is then sent to the cooling tower.

Table 4 reports a summary of the costs obtained for the optimal design for the cooling water system. Notice that the main contribution to the total annual cost is the capital cost for the cooling tower, which accounts for 50.7%. The second contribution is the cost associated with the operation of the fan for the cooling tower (27.0%); the capital cost for the exchanger contributes with 17.6% of the total.

Table 4 also shows the results for the case when only a classical arrangement of coolers in parallel is allowed for the cooling network (which is achieved by setting the number of stages in the superstructure as one). In this case the total annual cost is 2.22% higher than the in the optimal configuration; this result is mainly due to an increase in the capital costs for the coolers.

Table 4 also reports the results for the case when the cooling tower and the cooling network are optimized sequentially. The optimal design for the cooling water system shows savings by 7.1% with respect to the sequential optimization process, mainly due to the reduction in the water makeup consumption.

Table 3. Designs for the coolers

Concept	Cooler 1	Cooler 2	Cooler 3
Q [kW]	3,660	1,320	25,540
Hot stream allocation	Tubes	Tubes	Tubes
LMTD [K]	35.481	50.019	37.311
F_T	1	1	1
ΔP_T [kPa]	0.175	0.165	0.377
ΔP_S [kPa]	7.691	7.691	16.316
A [m^2]	52.643	16.563	304.885
Dt [mm]	10.211	18.008	4.928
Dti [mm]	12.700	22.225	6.350
Lpt [mm]	15.875	27.781	7.938
Ds [m]	0.914	0.549	1.086
Lbc [m]	0.914	0.549	1.086
L_{TT} [m]	3.657	2.196	4.346
U [kW/(m^2 K)]	1.960	1.593	2.245
Ns	1	1	1
Ntp	1	1	1
Tube arrangement	Square	Square	Square
v_T [m/s]	2.176	1.674	2.500
v_S [m/s]	1.000	1.000	0.963

Table 4. Results comparison for Example 1

Cost	Simultaneous optimal design	Simultaneous design with in parallel cooling water network	Sequential design
Cooling tower capital cost [$/year]	127,462.914	127,462.914	138,210.023
Coolers capital cost [$/year]	44,222.850	48,748.490	46,909.650
Pumps capital cost [$/year]	2,269.802	2,464.469	2,187.345
Electricity fan cost [$/year]	67,811.467	67,811.467	66,224.223
Electricity pumps cost [$/year]	3,365.536	4,312.879	2,432.673
Makeup water cost [$/year]	5,989.807	5,989.807	13,066.278
Total annual cost [$/year]	251,122.376	256,709.206	269,030.193

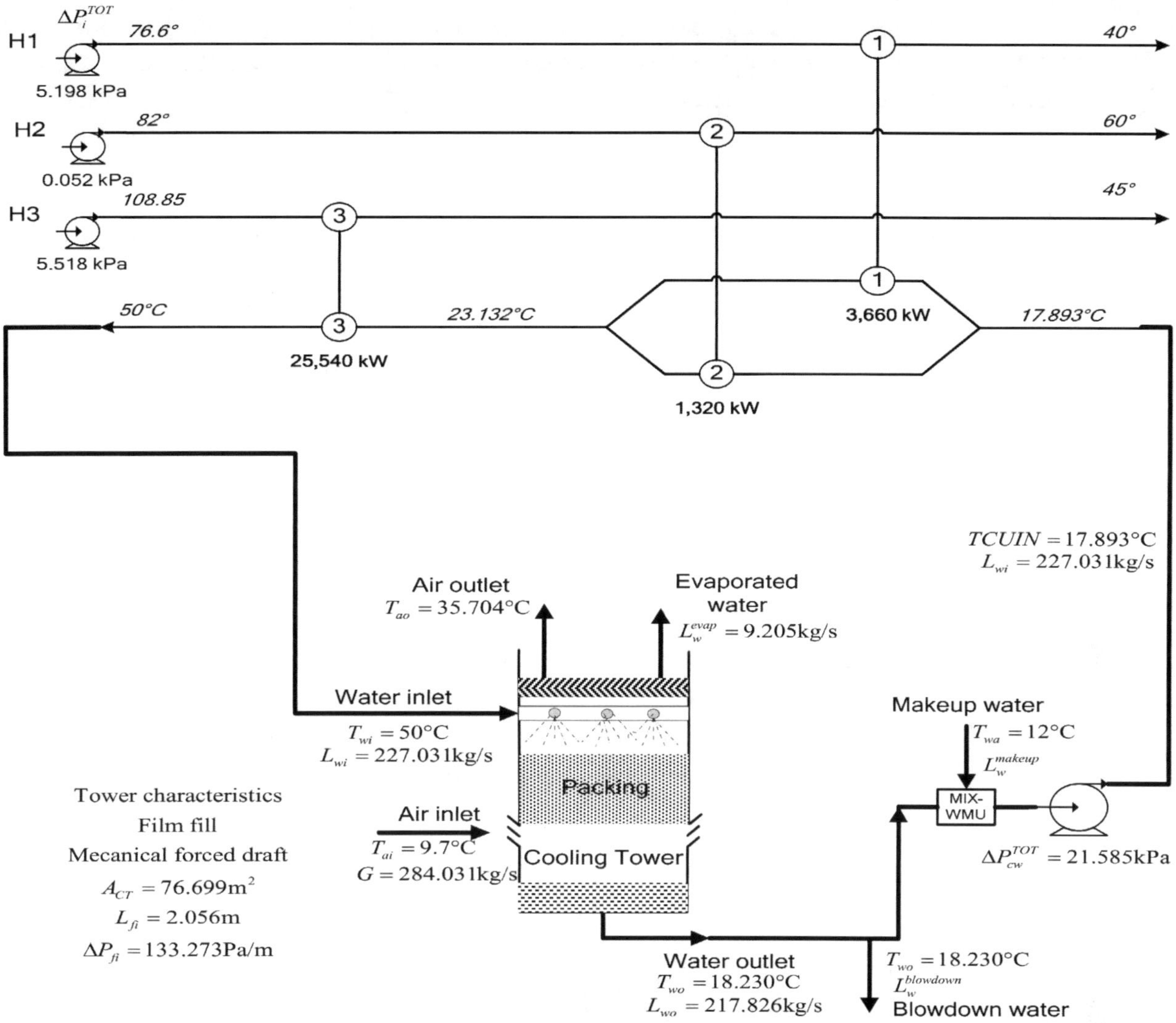

Figure 3. Optimal network for Example 1

Conclusions

An MINLP model for the optimal design of cooling water systems has been presented. The model takes into account the interactions between the components of the system, namely the cooling tower, the cooling network and the pumping system. A simultaneous optimization of the continuous and discrete variables associated with the overall cooling water system is carried out to get a minimum total annual cost for the system. The model considers the operational and geometrical constraints given by standard codes to get feasible designs for the units. A disjunctive programming model is formulated to model the discrete decisions for the system, which is then solved as an MINLP problem. This problem is highly nonlinear and nonconvex, so proper initialization process is needed to obtain a good solution.

The model has been applied to solve several example problems. In all of those problems, the maximum allowable temperature for the cooling water was reached; this result helps the cooling tower performance. In addition, the outlet temperature for the cooling water from the cooling tower is determined by the ambient conditions and the targets temperatures for the hot process streams. In the example problem here presented, the optimal network structure was relatively simple; no bypass for the cooling water was needed, and one cooler for each hot process stream was required. This solution differed form the ones obtained using simplified formulations that do not consider simultaneously the capital and operating costs.

References

1. Castro, MM, Song TW and Pinto JM. (2000). Minimization of operational costs in cooling water systems. *Chemical Engineering Research and Design. 78* (A2): 192-201.
2. Kim JK and Smith R. (2001). Cooling water system design. *Chemical Engineering Science. 56* (12): 3641-3658.
3. Kim JK, Lee GC, Zhu F and Smith R. (2002). Cooling system design. *Heat Transfer Engineering. 23* (6): 49-61.
4. Kim, JK and Smith R. (2003). Automated retrofit design of cooling-water systems. *American Institute of Chemical Engineers Journal. 49* (7): 1712-1730.
5. Kim JK and Smith R. (2004). Cooling system design for water and wastewater minimization. *Industrial and Engineering Chemistry Research. 43* (2): 608-613.
6. Feng X, Shen RJ and Wang B. (2005). Recirculating cooling-water network with an intermediate cooling-water main. *Energy & Fuels. 19* (4): 1723-1728.
7. Picon-Nuñez M, Nila-Gasca C and Morales-Fuentes A. (2007). Simplified model for the determination of the steady state response of cooling systems. *Applied Thermal Engineering. 27* (7): 1173-1181.
8. Majozi T and Moodley A. (2008). Simultaneous targeting and design for cooling water systems with multiples cooling water supplies. *Computers and Chemical Engineering.* In press.
9. Picon-Nuñez M, Morales-Fuentes A and Vazquez-Ramirez EE. (2007). Effect of network arrangement on the heat transfer area of cooling networks. *Applied Thermal Engineering. 27* (16): 2650-2656.
10. Chen CL, Lin LF, Ciou YJ and Chen WC. (2007). Superstructure-based MINLP formulation for synthesis of re-circulating cooling-water networks with intermediate mains. *Journal of Chemical Engineering of Japan. 40* (3): 235-243.
11. Ponce-Ortega JM, Serna-Gonzalez M and Jimenez-Gutierrez. A. (2007). MINLP synthesis of optimal cooling networks. *Chemical Engineering Science. 62* (21): 5728-5735.
12. Kern DQ. (1950). *Process Heat Transfer.* New York: McGraw Hill.
13. Chen JJJ. (1987). Letter to the Editors: Comments on improvements on a replacement for the logarithmic mean. *Chemical Engineering Science. 42* (10): 2488-2489.
14. Serna-González M, Ponce-Ortega JM and Jiménez-Gutiérrez A. (2008). MINLP optimization of mechanical draft counter flor wet-cooling towers. Under review.
15. Brooke A, Kendrick D, Meeraus A and Raman R. (2006). *GAMS-Language guide.* Washington, DC: GAMS Development Corporation.

GTL Process optimization Simulation Study

Yong Heon Kim[1,2], Heoung Yeoun Lee[1], Ki-Won Jun[3], Hyunku Joo[4], Chonghun Han[1] and In Kyu Song[1*], (1)School of Chemical and Biological Engineering, Seoul National University, Seoul National University, Shinlim-dong, Kwanak-ku, Seoul, 151-744, South Korea , (2)Petroleum Technology Insititute, Korea National Oil Corporation, 1588-14 Gwanyang Dong, An Yang, 431-711, South Korea, (3)Alternative Chemicals/Fuel Research Center, Korea Research Institute of Chemical Technology, 100 Jang-dong, Yuseong-gu, Daejeon, 305-343, South Korea, (4)Air Pollution Research Center, Korea Institute of Energy Research, 71-2 Jang-dong, Yusong-ku, Daejeon, 305-343, South Korea,

* Corresponding author
Email address : inksong@snu.ac.kr
Tel : + 82-2-880-9227

ABSTRACT

GTL(Gas to Liquids), especially natural gas to FT(Fisher-Tropsch) synthetic fuel, process optimization study was performed in order to find optimum parameters for optimum production. The minimum required energy and optimum operating conditions of GTL process are determined through the changes of system variables, such as temperature, H_2/CO ratio, etc. During the simulation, the overall synthetic processes are assumed to proceed with steady-state reaction of GTL, and catalyst and the physical properties of reacting medium are governed by Redlich-Kwong-Soave (RKS) equation. The reaction model of GTL process are assumed to be composed of ATR syngas(synthesis gas) reforming unit and FTS(Fisher-Tropsch Synthesis) reactor unit.

The simulation procedures are carried out to determine optimum condition of GTL process while varying syngas composition and operating temperature of the reactor. According to simulation results, optimum operating parameters of GTL process are temperature of 255℃ and H_2/CO ratio of 2. The simulation results will be compared to the actual experimental data from the lab-scale GTL process which is composed of ATR(Auto Thermal Reforming) reaction and FT slurry phase reaction.

Key Words: GTL, Synthetic Fuel, Synthesis gas, FTS, ATR, Simulation

1. Introduction

There is increasing need about energy conversion process as high oil price lasts. GTL process is one of the most promising way to convert other energy than oil to synthetic fuel. The interest in GTL processes as a way of utilizing associated and remote natural gas is growing. Furthermore, natural gas is a relatively evenly distributed resource. Proven world natural gas reserves, which currently exceed 5,000 trillion cubic feet (TCF), have been growing at a faster rate than proven oil reserves. These gas reserves currently represent over 83% of the energy equivalence of proven oil reserves, of which about 75% is considered to be the less desirable heavy crude. However, for using natural gas pipelines must be constructed and LNG infrastructure which has manufacturing, shipping, receiving and transporting is indispensable. Therefore, there are many the stranded gas resources not for development by economic reason. About 3,000 TCF of such gas reserves is considered to be stranded; i.e., accessible by drilling but located too far from potential markets for economical transportation to those markets.

GTL process is to convert natural gas to environmental friendly naphtha, kerosene and light oils. GTL synthetic fuel has been suggested recently as a clean alternative fuel. GTL synthetic fuel is a well-known alternative fuel for diesel, gasoline engines due to its good properties, such as environmental advantage. The GTL diesel fuel has near zero sulfur content and aromatic components and a very high cetane number. Thus, this technology contributes to environment and diversification of oil oriented energy usage.

GTL technology generally entails the chemical conversion of natural gas into readily transportable liquids such as methanol or conventional petroleum refinery type distillate fuels. More recently the term GTL has also been more loosely applied to physical conversion processes such as that for liquefied natural gas (LNG) as well as for chemical conversion processes that produce products which may not be in a liquid state under ambient conditions, such as dimethylether (DME). In this paper, we focus on GTL process optimization based on the production of diesel fuel via syngas production, which is consisting primarily of a mixture of CO + H_2, and FTS.

FTS generally involves the synthesis of hydrocarbons and oxygenates from syngas. The types of hydrocarbons produced can include olefins such as ethylene and propylene, and an extremely wide range of saturated hydrocarbons ranging from methane and ethane to long straight chain paraffinic waxes. The oxygenated materials that are produced consist primarily of alcohols such as methanol and ketones such as acetone. When configured to maximize the production of paraffinic hydrocarbons, the resulting intermediate product mix is often described as synthetic crude oil(syncrude). Such syncrudes can be readily refined into desirable distillate fuel fractions such as kerosene, naphtha, and heating oil using conventional petroleum refining techniques. The kerosene can be further refined or blended into high quality diesel or jet fuel products while the naphtha can be further refined into gasoline or used as a thermal cracking feedstock for olefins production.

Catalysts for FTS were first developed in the early 1900s. Following the discovery by Sabatier and Senderens in 1902 that CO could be hydrogenated over Co, Fe, and Ni catalysts to methane, BASF reported the production of liquids over Co catalysts in 1913. Fischer and Tropsch later reported production of hydrocarbons over alkalized iron in 1923. Much of the early catalyst and process development continued in Germany during the 1930s and 1940s. During World War II, coal based FT production played an important role providing the transportation fuel requirements of the German war effort because of insufficient access to crude oil resources. This production was discontinued when the war ended.

Cobalt catalysts as used in the FTS(Fischer-Tropsch Synthesis) are relatively expensive and need to have a high metal dispersion and long life to be able to offer a good balance between cost and performance. The oxidation of nano-sized metallic cobalt oxide during FTS has long been postulated as a major deactivation mechanism.

Supported cobalt catalysts are receiving widespread attention as the preferred FTS catalyst for the GTL process[1-4]. The development of economically attractive cobalt based FTS catalysts with a high stability requires detailed fundamental understanding of the deactivation mechanisms at play for supported nano-sized cobalt crystallites.

In this study, ATR based synthesis gas process and slurry bed reactor based on cobalt catalyst FTS process is considered. A process simulation with Aspen HYSYS software is conducted to demonstrate the possibly available effects of temperature, H_2/CO ratio, etc. The results will be the baseline for the design and operation of GTL pilot scale process.

2. Technical approach and process simulation

A GTL plant consists of three main process units: a reforming unit where syngas is produced, a FT unit where syngas is converted into liquids(syncrude), and a product upgrading unit where syncrude is upgraded into synthetic fuel. Product upgrading unit is not included in this study. ATR is applied to reforming unit and slurry bed reactor is used for FT synthesis process simulation. A schematic process flowsheets of ATR unit and FTS unit are shown in Figure 1, 2.

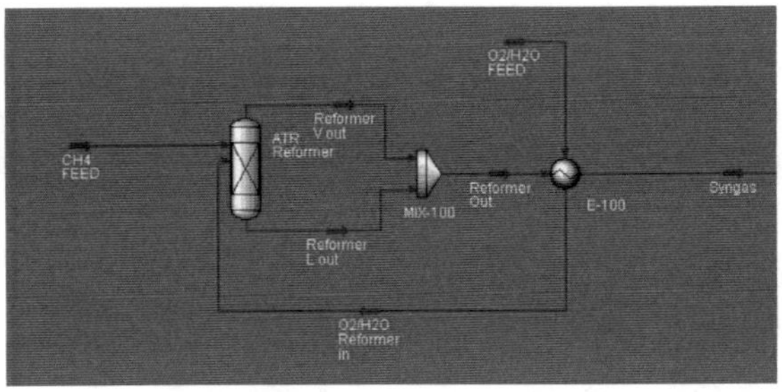

Fig.1. ATR Syngas Production unit simulation PFD

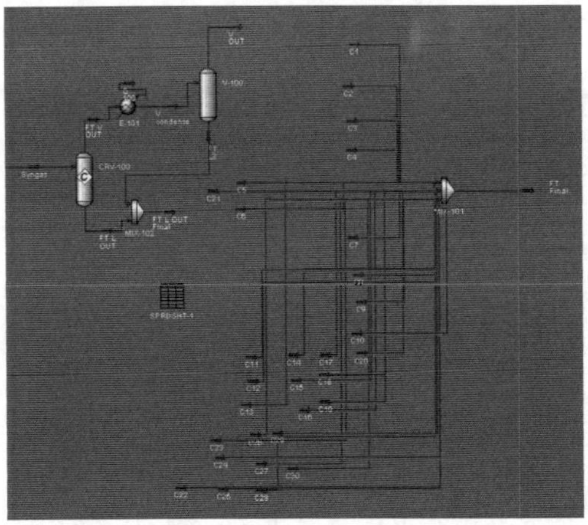

Fig.2. FT FBR Reaction unit simulation PFD

There are some kinds of methods for producing syngas. As a general flow, there is a move toward ATR that uses oxygen. However, because it requires an oxygen plant, thus incurring enormous costs, the issue of excess equipment costs becomes a problem.

Slurry bed is more efficient in removing the heat of the reaction in the FT synthesis, and because the equipment can be more compact. However, this method still has many unanswered technical issues, such as establishing a method for designing scaled-up systems.

Ni catalyst and Co catalyst are often used in FTS. Due to the high selectivity of Co-based catalysts, a lot of effort has been spent on developing and improving catalysts for use in industrial applications.

2.1 Reaction Mechanism of GTL Synthesis

2.1.1 ATR process

ATR consists of steam methane reforming and partial oxidation. Steam methane reforming, the most commonly used process to produce hydrogen and syngas, can be represented as[13]:

$$CH_4 + H_2O \rightarrow CO + 3H_2, \quad \Delta H^R=206 \text{ kJ/mol}$$
$$CO + H_2O \rightarrow H_2 + CO_2, \quad \Delta H^R=-41.2 \text{ kJ/mol}$$

The partial oxidation of methane consists of sub-stoichiometric oxidation written as[14] :

$$CH_4 + 0.5O_2 \rightarrow CO + 2H_2, \quad \Delta H^R=-36 \text{ kJ/mol}$$

The active catalysts for this process are the same as those used for steam reforming and for total oxidation of methane written as[15] :

$$CH_4 + 2O_2 \rightarrow CO_2 + 2\,H_2O, \quad \Delta H^R=-802 \text{ kJ/mol}$$

ATR technology is the preferred and most cost effective technology for natural gas conversion into synthesis gas to GTL plants. The Oryx GTL plant in Qatar is based on ATR technology at low steam/carbon ratio (S/C=0.6). ATR process produces a synthesis gas with a H2/CO ratio equal to 2, which is suitable for the subsequent use in the FTS[16].

2.1.2 FT Synthesis process

The FTS is a catalytic process that converts CO and H2 into a mixture of linear gaseous, liquid and solid hydrocarbons. The FTS is in principle a carbon chain building process, where CH_2 groups are attached to the carbon chain. Which reactions exactly taking place and how, is a matter of controversy, as it has been the last century since 1930's [9]. The main reaction involved in this process can be schematically written as [10]:

$$nCO + 2nH_2 \rightarrow -(CH_2)n- + nH_2O, \quad \Delta H^R=-165kJ/mol.$$

There is also other reactions taking place in the reactor, but the detailed behavior of the reactions is not known. The reactions reported are [11]:

Table 1: Reactions taking place in the FT reactor [11]

Reaction	Reaction enthalpy $\Delta H_{300\ K}$ [kJ/mol]
$CO + 2H_2 \rightarrow -CH_2- + H_2O$	- 165.0
$2CO + H_2 \rightarrow -CH_2- + CO_2$	-204.7
$CO + H_2O \rightarrow H_2 + CO_2$	-39.8
$3CO + H_2 \rightarrow -CH_2- + 2CO_2$	-244.5
$CO_2 + 3H_2 \rightarrow -CH_2- + 2H_2O$	-125.2

These reactions are highly exothermic, and to avoid an increase in temperature, which results in lighter hydrocarbons, it is important to have sufficient cooling, to secure stable reaction conditions [10]. The total heat of reaction amounts to 25 % of the heat of combustion of the synthesis gas [11], and lays thereby a theoretical limit on the maximal efficiency of the FT process.

The reaction is dependent of a catalyst, mostly an iron or cobalt catalyst where the reaction takes place. There is either a low or high temperature process (LTFT, HTFT), with temperatures ranging between 200-240 °C for LTFT and 300-350 °C for HTFT [12]. The HTFT uses an iron catalyst, and the LTFT either an iron or a cobalt catalyst.

The different catalysts include also nickel and Ru based catalysts, which also have enough activity for commercial use of FT. But the availability of Ru is limited, thus forcing a high price. The nickel based catalyst has high activity but produces too much methane, and additionally the performance at high pressure is poor, due to production of volatile carbonyls. This leaves only cobalt and iron as practical catalysts, and this study will only consider these two. Iron is cheap, but cobalt has the advantage of higher activity and longer life, though it is on a metal basis 1000 times more expensive than iron catalyst [9].

Cobalt-based catalysts have been successfully applied in the industrial processes due to their high FT activity and their low oxygenates selectivity, which makes this catalysts suitable for the H2-rich syngas obtained from natural gas.

2.2 Reaction Kinetics of GTL Process

The Aspen HYSYS reactor was used to simulate the ATR process. In this step, methane is converted to syngas, CO and H_2.

Dry [6] and Huff and Statterfield [7] showed that when hydrogen conversion is below 60% the first-order FT kinetics is a good approximation. Except in a few more detailed approaches [8] linear kinetics is used in most of the previous FT slurry models. Given the small solids particle size (50μm and smaller) usually encountered in industrial slurry systems, intra-particle temperature and concentration profiles are most often negligible. Therefore, solids intra-particle mass and heat transfer resistences are presently neglected.

There are a lot of efforts that have been made on FT reaction mechanism and kinetics, which is undoubtedly complex. Table 2 shows the studies about Co catalyst based FT kinetic model.

Table 2: Co catalyst based FT kinetic model

Reactor Type	Kinetic expression	Study
Fixed Bed Reactor	$-r_{CO}=\dfrac{aP_{H_2}P_{CO}^{0.5}}{(1+bP_{CO}^{0.5})^3}$	[18]
Fixed Bed Reactor	$-r_{CO}=\dfrac{aP_{H_2}^{0.6}P_{CO}^{0.65}}{(1+bP_{CO})}$	[19]
Slurry Phase Reactor	$-r_{CO}=\dfrac{aP_{H_2}P_{CO}}{(1+bP_{CO})^2}$	[20]

This kinetic model was initially coupled through a user-supplied subroutine to the Aspen HYSYS reactor model in order to simulate the slurry phase FT reactor and validate the models using the experimental data[21].

Slurry phase kinetic study was applied to simulate FT reaction. It was used for calculating CO conversion.

Modeling FT-product distributions

There have been many attempts to model the product distributions of the FT process. The hydrocarbon synthesis reactions involved in FTS may be regarded as analogous to a polymerization reaction. An essential characteristic of catalysts applied to FT diesel production therefore is the ability to catalyze chain propagation versus chain termination steps. The specific selectivity for a particular hydrocarbon and the overall product distribution may generally be described by a chain polymerization kinetics model involving the stepwise addition of one carbon to another on the growing chain.

There are deviations between inspected product distributions and the different models when the conditions are changed. But there is an agreement in literature that the distributions follow some sort of exponential function, with the probability of chain growth as an important factor. The methodology is mainly divided into a kinetic approach and a thermodynamic approach. The best known model of the latter kind was developed in the early years of FT-synthesis, where a constant probability of chain growth was assumed. This model has been ascribed to Anderson, Schultz, and Flory and is commonly referred to as the Anderson-Schultz-Flory (ASF) model.

$$W_n/n=(1-\alpha)^2\alpha^{(n-1)}$$

(The W_n is equal the weight fraction)

Chain growth probability(α) is calculated by experiment based on Co catalyst[21].

2.3 Simulation Methodology

It is required to utilize thermodynamic parameters which can be applied to fundamental equation of state for simulating a GTL process by Aspen HYSYS. Many equations of state of varying complexity have been developed. No equation is sufficiently accurate to represent all real gases under all conditions. In this simulation study, RKS (Redlich Kwong Soave) equation is utilized for calculating thermodynamic parameters in the model. RK (Redlich-Kwong) equation of state is an extension of the more familiar van der Waal's equation. The RK equation is not suitable for use near the critical pressure, or for liquids. The RK equation is widely applied to binary components. It has good accuracy in volumetric and thermal properties between pure components and mixture, but it is tend to lower accuracy of VLE (Vapor Liquid Equilibrium) calculation in multi-components. Giorgio Soave (1972) modified the RK equation to extend its usefulness to the critical region and for use with liquids in order to make up for the weak point of RK state equation. This equation is written as[17] :

$$P = \frac{RT}{v-b} - \frac{a(T)}{v(v+b)}$$

where, $a(T) = \{1+m(1-T_r^{0.5})\}^2$, $m = 0.480 + 1.574\omega - 0.17\omega^2$

Because GTL process is composed of multi-components accompanied with vapor-liquid phase, RKS equation is selected as governing equation for simulating of GTL process.

2.4 Simulation Assumptions

The following assumptions were considered in modeling GTL process:

- Process is steady state and isothermal
- Constant input flow rate of natural gas
- GTL synthesis catalyst is composed of homogeneous catalyst and charged with constant void fraction of catalyst bed in the GTL reactor
- Catalytic poisoning effected by H_2S neglected

3. Results and discussion

Simulation RESULTS

3.1 Comparison of experimental and simulation results in FT product distribution

The experimental carbon distribution with Co based alumina support catalyst and simulation results are shown in Fig.3. As non-linear form, $Log(W_n / n)$ decreases while carbon number increases. In simulation study, $Log(W_n / n)$ increases linearly while carbon number increases.

Mass fraction FT product distribution by simulation study is shown in Fig.4. Mass fraction reached maximum for carbon number region around 10.

There is a slight difference in FT product distribution between simulation output and actual FT product behavior. The difference between experimental value and simulation value is explained by actual discrepancy of ASF distribution in FTS. But the heavier the carbon is, the smaller the difference between actual distribution and simulation output. Therefore, in heavy carbon region, FT product distribution can be predicted by this simulation model.

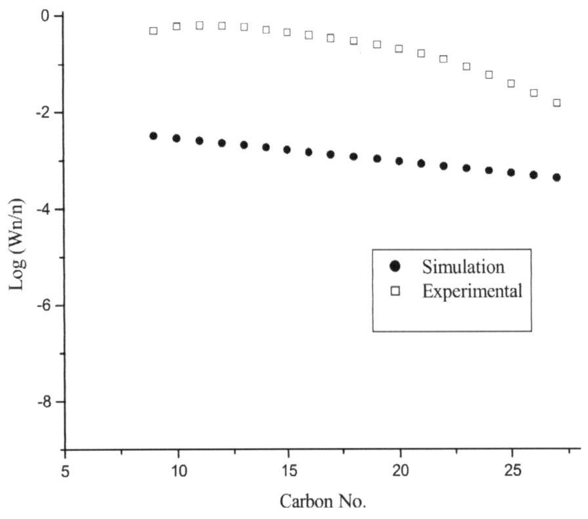

Fig.3. Co based catalyst FT Product carbon number distribution

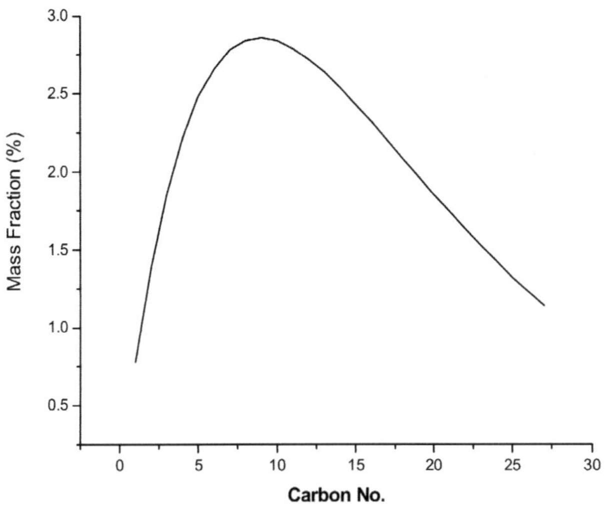

Fig.4. Product distribution by simulation

3.2 Effect of H_2/CO ratio

The influence of the H_2/CO ratio on CO conversion was also studied. The effect of the H_2/CO ratio on CO conversion is shown in Fig.4. CO conversion reached maximum for a H_2/CO ratio of 2.0 as indicated in Fig.4.

Simulation shows good agreement with experimental data. Fig.4 demonstrates good prediction of FTS reaction condition in H_2/CO ratio.

In syngas process, H_2/CO ratio produces 3:1 in SMR (Steam Methane Reforming). In ATR process, H_2/CO ratio is around the desired level of about 2:1. So, ATR has advantage for FTS in H_2/CO ratio.

In actual plant, for the low-temperature FT (LTFT, 200-240℃) process, the H_2/CO ratio is about 1.8 and so is suitable as feed gas to the wax producing LTFR reactors.

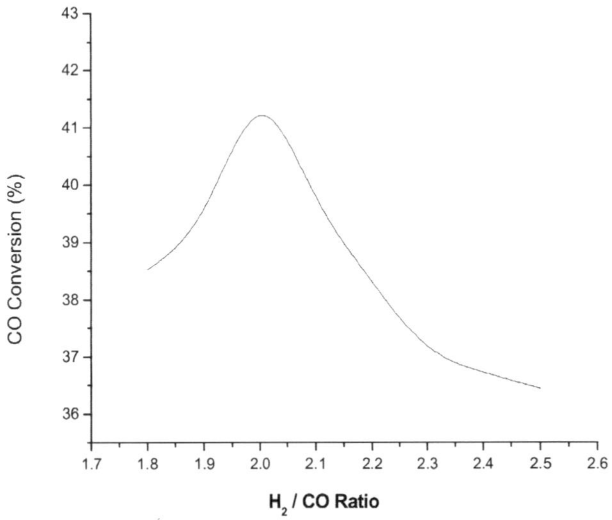

Fig.5. Effect of H_2/CO ratio on CO conversion

3.3 Effect of Temperature on CO conversion and C5+ selectivity simulation case vs experimental case

The simulation results varying reaction temperature of FTS are shown in Fig.6. CO conversion reached maximum for 255℃ of reaction temperature both experimental case and simulation case.

In low temperature region (below 240℃), CO conversion is relatively lower. Therefore, the optimum operating temperature for high CO conversion is 255℃ in this study by experimental and simulation output.

The C5+ selectivity drops rather sharply in experiment case than in simulation case. These results are shown in Fig.7. The C5+ selectivity decreases slightly while temperature is increased in simulation study. In experiment, The C5+ selectivity decreases remarkably while temperature is increased.

In simulation study, optimum temperature considering CO conversion and C5+ selectivity is around 250℃ because of the slight decrease of C5+ selectivity in this temperature region. But actual optimum temperature considering both CO conversion and C5+ selectivity is around 220~230℃.

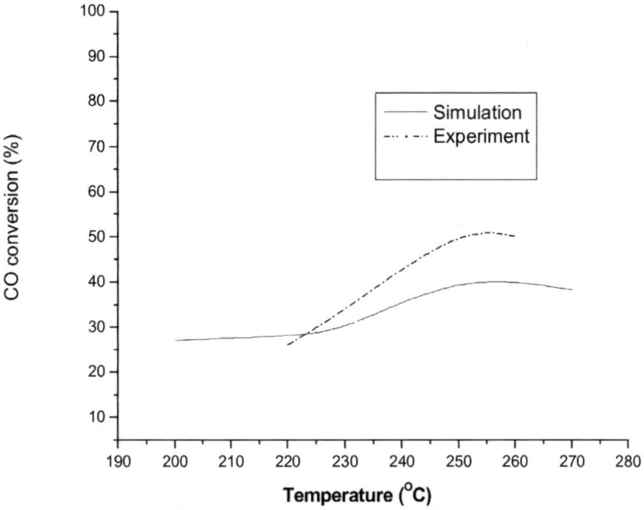

Fig.6. Effect of Temperature on CO conversion simulation case vs experimental case

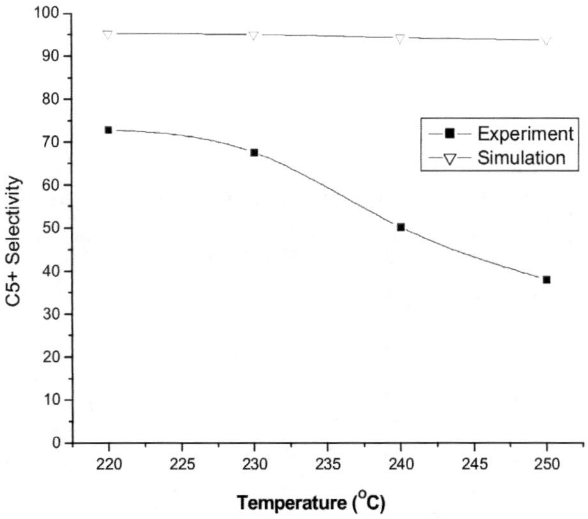

Fig.7. Effect of Temperature on C5+ Selectivity

5. Conclusions

The GTL process simulation can be simulated by the flow sheeting program in Aspen HYSYS. The results of this simulation and the actual results obtained in experiment are practically in reasonable agreement with each other.

Simulation of the processes, including chemical reactions and heat/mass balance, was carried out with Aspen HYSYS software. The effects of temperature on CO conversion, C5+ selectivity and the effects of H_2/CO ratio on CO conversion were discussed. According to simulation results, optimum operating parameters of GTL process are temperature of 255℃ and H_2/CO ratio of 2.

Acknowledgement

The authors express their gratitude to Korea GTL Research Consortium (SK enery, Daelim Cor., KNOC, Doosan Mecatec). Special thanks are also extended to Dr.Bae for his valuable comments provided toward accomplishing this study.

Nomenclature

ω : acentric factor
W_n : weight fraction
α : chain growth probability
T_r : reduced temperature

References

[1] P. van Berge, S. Barradas, J. van de Loosdrecht, J.L Visagie, Erdol Erdgas Kohle 117 (3) (2001) 138.
[2] The Catalyst Review Newsletter, Gas-to-Liquids : Peering into the Crystal Ball, 2005, p. 4.
[3] M.M.G. Senden, A.D. Punt, A. Hoek, Stud. Surf. Sci. Catal. 199 (1998) 943.
[4] A.P. Steynberg, M.E. Dry, Fischer-Tropsch technology, Stud. Surf. Sci. Catal. 152 (2004) 441.
[5] The Catalyst Review Newsletter, Gas-to-Liquids : Peering into the Crystal Ball, 2005, p. 4.
[6] M.E. Dry, Advances in Fischer-Tropsch chemisty, Ind. Eng. Chem. Prod. Res. Dev. 15 (1976) 282.
[7] G.A. Huff Jr., C.N. Satterfield, Intrinsic kinetics of the Fischer-Tropsch synthesis on a reduced fused-magnetite catalyst, Ind. Eng. Chem. Prod. Res. Dev. 23 (1984) 696.
[8] G.P. van der Laan, A.A.C.M. Beenackers, R. Krishna, Multicomponent reaction engineering model for Fe-catalyzed Fischer-Tropsch synthesis in commercial scale slurry bubble column reactors, Chem. Eng. Sci. 54 (1999) 5013.

[9] Dry, Mark E, High quality diesel via the FS process - a review, Journal of Chemical Technology and Biotechnology, 77/2001, 43-50.
[10] Fürnsinn, Stefan, Ripfel, Karl, Rauch, Reinhard, Hofbauer, Hermann, Diesel aus Holz - Die FS Synthese als zukunftsweisende Technologie zur Gewinnung flüssiger Brennstoffe aus Biomasse, 4. Internationale Energiewirtschaftstagung an der TU Wien, 2005.
[11] Rauch, Reinhard, Biomass Gasification to produce Synthesis Gas for Fuel Cells, Liquid Fuels and Chemicals, IEA Bioenergy Agreement, Task 33: Thermal Gasification of Biomass (2001-2003)
[12] Dry, Mark E., The Fischer-Tropsch process: 1950-2000, Catalyst today, 71/2002, 227-241.
[13] J. Sehested, Catalyst today 111 (2006), 103.
[14] Y.S Seo, A. Shirley, S.T. Kovalaczkowski, J. Power Sources 108 (2002), 213-225
[15] T.S.Christensen, I.I.Primdahl, Hydrogen Processing, March 1994, p39-46
[16] K. Aasberg-Petersen, T.S. Christensen, C.S. Nielsen, I. Dybkjaer, Fuel Process. Technol. 83 (2003), 253
[17] Giorgio Soave, Chemical Engineering Science, 27, 1197 (1972)
[18] A.O.I. Rautavuoma, H.S. van der Baan, Apply Catalysis, 1 (1981), 247-272
[19] E. Iglesia, S.C. Reyes, S.L. Soled, in: E.R Becker, C.J. Pereira (Eds.), Computer-Aided Design of Catalysts, Dekker, New York, 1993
[20] I.C. Yates, C.N. Satterfield, Energy and Fuels 5 (1991), 168-173
[21] S-H Song, S-B Lee, J-W Bae, P.S Sai Prasad, K-W Jun, Catalysis Communications (2008), 2284

Retrofit of Heat Exchanger Networks Including Process Modifications

José M. Ponce-Ortega [a,b], Arturo Jiménez-Gutiérrez [a], and Ignacio E. Grossmann [c]

[a] Instituto Tecnológico de Celaya, Celaya, Gto. 38010, México.
[b]Universidad Michoacana de San Nicolás de Hidalgo, Morelia, Mich. 58060, México.
[c] Carnegie Mellon University, Pittsburgh, PA 15213, USA.

Abstract

This paper presents a new formulation for the retrofit of heat exchanger networks that takes into account process modifications. The method accounts for the interactions between the process conditions and the heat integration options to provide an optimal structure for a redesigned heat exchanger network. The formulation is based on a superstructure that considers explicitly the plant layout and the piping arrangement, which yields a mixed-integer nonlinear programming model. The model presented here includes the treatment of isothermal process streams that exchange their latent heats, in addition to the streams commonly considered with sensible heat loads. The objective function consists of the maximization of the total annual profit for the retrofit process, which includes the income from products sales and the expenses due to raw materials, capital cost for new units, utility costs and the piping modification costs. One case of study is presented, and the results show that a significant improvement in the process profitability can be obtained with the simultaneous approach presented in this work for process retrofit with respect to the sole consideration of the heat exchanger networks.

Introduction

Heat exchanger networks (HEN) have been widely applied in industrial projects over the past decades because they provide significant energy and economic savings. A good number of methodologies have been proposed for the HEN synthesis problem; for retrofit problems, however, the available methods are more limited, as has been noted in the review paper by Furman and Sahinidis[1].

Only few works have attempted the simultaneous treatment of process modifications as part of HEN problems. Some ideas along these lines have been considered in the works by Duran and Grossmann[2], Lang et al.[3], and Grossmann et al.[4], in which the optimal flowsheet for the process is obtained by enforcing the minimum utility target. Zhang and Zhu[5] addressed the problem of HEN retrofit considering process changes. However, these authors only considered the effects on utility consumption and did not take into account the capital cost associated with the retrofit process.

Most of the methods described above for HEN retrofit assume that the process conditions (inlet and outlet temperatures, stream flowrates) are fixed, so that no interactions with process modifications are considered. Allowing for potential adjustments in the operating conditions should provide the basis for more cost-effective heat integration. We should also note that none of the methodologies reported for the HEN retrofit has included the explicit treatment of isothermal streams, which are very common in the chemical industry, for instance in the operation of distillation columns. A fairly common approach to the treatment of isothermal streams in HEN synthesis problems consists in assuming a one degree change with a suitable adjustment of a pseudo-heat capacity value to equal the enthalpy change. This approach, however, is prone to scaling problems during the problem numerical solution.

In this paper an MINLP model for the HEN retrofit that considers simultaneously the HEN structure and process modifications is presented. The proposed model considers the plant layout and modification costs through the superstructure by Yee and Grossmann[6] in which operational and structural modifications of the process are added and considered simultaneously. In addition, the model considers explicitly the utility and the capital cost of the units, and takes into account the isothermal process streams that may appear in the process using the approach reported recently by Ponce-Ortega et al.[7]

Model formulation

The proposed mathematical formulation applies to the generalization of the superstructure of Figure 1 for arbitrary number of process streams, existing exchangers and potential new exchangers. Rules to establish the number of potential new exchangers are given in Yee and Grossmann[6]. The following sets are used for the model development. *HPS1* contains the hot process streams that exchange sensible heat in the network, *HPS2* contains the hot process isothermal streams (i.e., they exchange only latent heat and their temperatures remain constant), *HPS* contains all the hot process streams ($HPS = HPS1 \cup HPS2$), *HU* corresponds to the hot utility streams and *HS* contains all the hot streams ($HS = HPS \cup HU$). *HS1* is a set that contains the hot streams excluding isothermal streams ($HS1 = HPS1 \cup HU$). Similarly, *CPS1* is the set for the cold process streams that exchange sensible heat and *CPS2* includes the cold process isothermal streams, whereas *CPS*, *CU* and *CS* represent the total cold process streams ($CPS = CPS1 \cup CPS2$), the cold utility stream, and all the cold streams ($CS = CPS \cup CU$), respectively. *CS1* contains the cold non-isothermal streams ($CS1 = CPS1 \cup CU$). *HCPS* contains the hot and cold process streams ($HCPS = HPS \cup CPS$), and the set *HCTS* contains all hot and cold process and utility streams ($HCTS = HS \cup CS$). The set *E* represents all the exchangers in the superstructure, whereas the sets *EE* and *NE* represent the existing and the new exchangers, respectively. The detailed description of the symbols used in the model formulation is given in the nomenclature section.

Logical assignment constraints. For convenience, the following sets of binary variables are defined. w_s^k denotes that stream *s* is assigned to exchanger *k*. The variable yi_s^k is used when the inlet of stream *s* is assigned to exchanger *k*, whereas ye_s^k is used when stream *s* exits the HEN from exchanger *k*. $x_s^{k,l}$ denotes the interconnection between exchangers *k* and *l* for stream *s*. The binary variables $z_{k,l}^h$ and $z_{k,l}^c$ are used for hot or cold streams if piping segments between exchangers *k* and *l* exist, while ze_k is used for piping segments between exchanger *k* and the exit of the HEN. The variable v^k is used for new exchangers in the retrofitted HEN.

Selection of streams for heat units.

$$\sum_{i \in HS} w_i^k = 1 \qquad k \in EE \tag{1}$$

$$\sum_{j \in CS} w_j^k = 1 \qquad k \in EE \tag{2}$$

$$\sum_{i \in HS} w_i^k \leq 1 \qquad k \in NE \tag{3}$$

$$\sum_{j \in CS} w_j^k \leq 1 \qquad k \in NE \tag{4}$$

$$\sum_{k \in E} yi_s^k \geq 1 \quad s \in HCPS \tag{5}$$

$$\sum_{s \in HS} yi_s^k \leq 1 \quad k \in E \tag{6}$$

$$\sum_{s \in CS} yi_s^k \leq 1 \quad k \in E \tag{7}$$

Figure 1. Superstructure for the HEN retrofit for two hot and two cold process streams

Fresh streams constraints.

$$yi_s^k - w_s^k \leq 0 \qquad s \in HCTS \tag{8}$$

Logical interconnection constraints. Consistency constraints are necessary to allow a connection between two exchangers only if both exchangers service the same process streams. By definition, the variable $x_s^{k,l}$ denotes that stream *s* is assigned to both exchangers *k* and *l*, and the variables $z_{k,l}^h$ and $z_{k,l}^c$ denote the existence of piping sections.

$$x_s^{k,l} - w_s^k \leq 0 \qquad s \in HCPS, k = 1, 2, \ldots, K-1, \text{ and } l = k+1, \ldots, K \tag{9}$$

$$x_s^{k,l} - w_s^l \le 0 \qquad s \in HCPS, k = 1,2,\ldots,K-1, \text{ and } l = k+1,\ldots,K \tag{10}$$

$$z_{k,l}^h - \sum_{i \in HPS} x_i^{k,l} \le 0 \qquad k = 1,2,\ldots,K-1, \text{ and } l = k+1,\ldots,K \tag{11}$$

$$z_{k,l}^c - \sum_{j \in CPS} x_j^{k,l} \le 0 \qquad k = 1,2,\ldots,K-1, \text{ and } l = k+1,\ldots,K \tag{12}$$

$$z_{l,k}^h - \sum_{i \in HPS} x_i^{k,l} \le 0 \qquad k = 1,2,\ldots,K-1, \text{ and } l = k+1,\ldots,K \tag{13}$$

$$z_{l,k}^c - \sum_{j \in CPS} x_j^{k,l} \le 0 \qquad k = 1,2,\ldots,K-1, \text{ and } l = k+1,\ldots,K \tag{14}$$

Definition of an exit stream from a heat exchanger.

$$ze_k^h + w_i^k - ye_i^k \le 1 \quad i \in HS, k \in EE \tag{15}$$

$$ze_k^c + w_j^k - ye_j^k \le 1 \quad j \in CS, k \in EE \tag{16}$$

Definition of new heat exchangers. If any pair of hot and cold streams are serviced by a new heat exchanger, then $v^k = 1, k \in NE$, and the following constraint must be satisfied,

$$\sum_{i \in HS} w_i^k + \sum_{j \in CS} w_j^k - v^k \le 1 \quad k \in NE \tag{17}$$

Definition of new units for isothermal process streams. If the new exchanger *k* processes an isothermal hot process stream *i*, then the binary variable μ_i^k must be equal to one,

$$v^k + w_i^k - \mu_i^k \le 1 \quad i \in HPS2, k \in NE \tag{18}$$

whereas for the isothermal cold process stream *j*, the binary variable μ_j^k must be one,

$$v^k + w_j^k - \mu_j^k \le 1 \quad j \in CPS2, k \in NE \tag{19}$$

Relocation of Heat Exchangers. The assignment variable $\kappa_{l,k}$ is used to denote the relocation of exchanger *l* with area EAM_l to a new position *k*.

$$\sum_{l=1}^{K} \kappa_{k,l} = 1 \qquad k \in E \tag{20}$$

$$\sum_{k=1}^{K} \kappa_{k,l} = 1 \qquad l \in E \tag{21}$$

Mass balance for initial splitters.

$$FCp_s^{IN} = \sum_{k \in E} fCp_s^k \qquad s \in HCPS \tag{22}$$

Mass balance at inlet mixers for each exchanger.

$$fCp_k^{h,in} = \sum_{i \in HPS} fCp_i^k + \sum_{\substack{l \in E \\ l \ne k}} fCp_{l,k}^h + FCp_{HU} w_{HU}^k \qquad k \in E \tag{23}$$

$$fCp_k^{c,in} = \sum_{j \in CPS} fCp_j^k + \sum_{\substack{l \in E \\ l \ne k}} fCp_{l,k}^c + FCp_{CU} w_{CU}^k \qquad k \in E \tag{24}$$

Energy balance at inlet mixers for each exchanger.

$$fCp_k^{h,in} t_k^{h,in} = \sum_{i \in HPS} \left(fCp_i^k T_i^{IN} \right) + \sum_{\substack{l \in E \\ l \ne k}} \left(fCp_{l,k}^h t_l^{h,out} \right) + FCp_{HU} T_{HU}^{IN} w_{HU}^k \qquad k \in E \tag{25}$$

$$fCp_k^{c,in} t_k^{c,in} = \sum_{j \in CPS} \left(fCp_j^k T_j^{IN} \right) + \sum_{\substack{l \in E \\ l \ne k}} \left(fCp_{l,k}^c t_l^{c,out} \right) + FCp_{CU} T_{CU}^{IN} w_{CU}^k \qquad k \in E \tag{26}$$

Existing piping segment.

$$fCp_s^k \le FCp^{UP} yi_s^k \qquad k \in E, s \in HCPS \tag{27}$$

Mass balance at outlet splitters for each exchanger.

$$fCp_k^{h,in} = \sum_{\substack{l \in E \\ l \ne k}} fCp_{k,l}^h + fCp_{k,exit}^h \qquad k \in E \tag{28}$$

$$fCp_k^{c,in} = \sum_{\substack{l \in E \\ l \ne k}} fCp_{k,l}^c + fCp_{k,exit}^c \qquad k \in E \tag{29}$$

Overall heat balance for each process stream.

$$FCp_i^{IN}\left(T_i^{IN} - T_i^{OUT}\right) = \sum_{k \in E}\sum_{j \in CS} Q_{i,j,k} \qquad i \in HPS1 \tag{30a}$$

$$F\lambda_i^{cond} = \sum_{k \in E}\sum_{j \in CS} Q_{i,j,k} \qquad i \in HPS2 \tag{30b}$$

$$FCp_j^{IN}\left(T_j^{OUT} - T_j^{IN}\right) = \sum_{k \in E}\sum_{i \in HS} Q_{i,j,k} \qquad j \in CPS1 \tag{31a}$$

$$F\lambda_j^{evap} = \sum_{k \in E}\sum_{i \in HS} Q_{i,j,k} \qquad j \in CPS2 \tag{31b}$$

Equations for heat exchangers.

$$\sum_{i \in HS1}\sum_{j \in CS} Q_{i,j,k} - fCp_k^{h,in}\left(t_k^{h,in} - t_k^{h,out}\right) \le 0 \qquad k \in EE \tag{32}$$

$$\sum_{i \in HS}\sum_{j \in CS1} Q_{i,j,k} - fCp_k^{c,in}\left(t_k^{c,out} - t_k^{c,in}\right) \le 0 \qquad k \in EE \tag{33}$$

$$t_k^{h,in} - t_k^{h,out} \le M^1\left(1 - w_i^k\right) \qquad i \in HPS2, k \in EE \tag{34}$$

$$t_k^{c,out} - t_k^{c,in} \le M^1\left(1 - w_j^k\right) \qquad j \in CPS2, k \in EE \tag{35}$$

$$DT_{HU} w_{HU}^k - \left(t_k^{h,in} - t_k^{h,out}\right) \le 0 \qquad k \in EE \tag{36}$$

$$DT_{CU} w_{CU}^k - \left(t_k^{c,out} - t_k^{c,in}\right) \le 0 \qquad k \in EE \tag{37}$$

$$\sum_{i \in HS}\sum_{j \in CS}\left(Q_{i,j,k}\left(\frac{1}{h_i} + \frac{1}{h_j}\right)\right) - \left(EA_k + AA_k\right) LMTD_k \le 0 \qquad k \in EE \tag{38}$$

$$LMTD_k - \left(\left(t_k^{h,in} - t_k^{c,out}\right)\left(t_k^{h,out} - t_k^{c,in}\right)\left(\frac{\left(t_k^{h,in} - t_k^{c,out}\right)\left(t_k^{h,out} - t_k^{c,in}\right)}{2}\right)\right)^{1/3} \le 0 \qquad k \in EE \tag{39}$$

$$\Delta T_{MIN} - \left(t_k^{h,out} - t_k^{c,in}\right) \le 0 \qquad k \in EE \tag{40}$$

$$\Delta T_{MIN} - \left(t_k^{h,in} - t_k^{c,out}\right) \le 0 \qquad k \in EE \tag{41}$$

$$t_k^{h,out} \le t_k^{h,in} \qquad k \in EE \tag{42}$$

$$t_k^{c,in} \le t_k^{c,out} \qquad k \in EE \tag{43}$$

For new exchanger units, the set of equations is applied only if the unit is selected for the retrofit network as part of the optimization process.

Heat loads for heat exchangers.

$$Q_{i,j,k} - Q^{UP} w_i^k \le 0 \qquad i \in HS, k \in E \tag{63}$$

$$Q_{i,j,k} - Q^{UP} w_j^k \le 0 \qquad j \in CS, k \in E \tag{64}$$

Definition of flows in piping segments.

$$fCp_{k,l}^{h} - FCp^{UP} z_{k,l}^{h} \le 0 \quad k,l \in E, k \ne l \tag{65}$$

$$fCp_{k,l}^{c} - FCp^{UP} z_{k,l}^{c} \le 0 \quad k,l \in E, k \ne l \tag{66}$$

$$fCp_{k,exit}^{h} - FCp^{UP} ze_{k}^{h} \le 0 \quad k \in E \tag{67}$$

$$fCp_{k,exit}^{c} - FCp^{UP} ze_{k}^{c} \le 0 \quad k \in E \tag{68}$$

Feasibility constraint for isothermal streams.

$$fCp_{k}^{h,in} \ge FCp^{LO} w_{i}^{k} \quad i \in HPS2, k \in EE \tag{69}$$

$$fCp_{k}^{c,in} \ge FCp^{LO} w_{j}^{k} \quad j \in CPS2, k \in EE \tag{70}$$

Existing exchangers moved to a different location.

$$EA_k = \sum_{l=1}^{K} \left(EAM_l \kappa_{k,l} \right) \qquad k \in E \tag{71}$$

Process modeling constraints. The types of process modifications depend upon each particular case. It is necessary to identify the types of process modifications allowed for each particular case. We identify two types of process modifications. The first type is associated to process conditions, for example conversion, pressure, and temperature of reactors, temperature and pressure of separation units, flow rates of purge, feed and product streams, and so forth. These variables are clearly restricted by feasibility conditions. A second type is related to structural modifications of the process, for example the addition or replacement of equipment. Process modifications may affect the temperatures and flow rates of the process streams and as a consequence the utility requirements.

In this way, the constraints *h* and *g* represent material and energy balances, design specifications and structural relationships as follows,

$$h(\mathbf{x},\mathbf{z},\mathbf{y}) = 0 \tag{72}$$

$$g(\mathbf{x},\mathbf{z},\mathbf{y}) \le 0 \tag{73}$$

where **x** represents the continuous variables of the process that are involved in the HEN model (i.e., fCp_s^{IN} and T_s^{IN} for all the process streams), **z** corresponds to the continuous variables that are not used in the HEN model (i.e., pressures and temperatures of the equipments, equipments sizes, etc) and **y** represents the binary variables for structural modifications in the flowsheet.

Objective function. The objective function maximizes the total annual profit of the process. The income in the objective function depends on the sales of the products. The expenses depend on the raw materials, costs due to the modification in the process conditions and the HEN retrofit annual cost. Additional expenses may be considered for the purchase of new process equipment. The HEN retrofit costs include the hot and cold utilities cost, the annualized capital costs and fixed charges for new heat exchangers, the fixed charge for the new piping segments (which depends on the distance between two exchangers *k* and *l*), the variable charge for new piping segments (which depends on the distance and the amount of flow) and finally the cost to relocate a heat exchanger from one position to another.

$$
\begin{aligned}
\max\ profit = & \sum_{s \in produc} CO_{m_s} f_s - \sum_{s \in rowmat} CO_{m_s} f_s - r(\mathbf{x}, \mathbf{z}, \mathbf{y}) \\
& - \sum_{i \in HU} \sum_{j \in CPS} \sum_{k \in E} CO_{HU} \left(Q_{i,j,k} \right) \\
& - \sum_{i \in HPS} \sum_{j \in CU} \sum_{k \in E} CO_{CU} \left(Q_{i,j,k} \right) \\
& - CO_{AA} \sum_{k \in E} AA_k^{\beta} \\
& - CO_{NE} \sum_{k \in NE} \left(v^k \right) \\
& - \sum_{k \in E} \sum_{l \in E} \left(CO^F_{NPH_{k,l}} z^h_{k,l} + CO^F_{NPC_{k,l}} z^c_{k,l} \right) - \sum_{k \in E} \sum_{l \in E} CO^V_{NP_{k,l}} \left(f^h_{k,l} + f^c_{k,l} \right) \\
& - \sum_{s \in HCS} \sum_{k \in E} CO^F_{NPE_{s,k}} \left(ye^k_s \right) - \sum_{s \in HCS} \sum_{k \in E} C^V_{NPE_{s,k}} \\
& - \sum_{s \in HCS} \sum_{k \in E} CO^F_{NPI_{s,k}} \left(yi^k_s \right) - \sum_{s \in HCS} \sum_{k \in E} CO^V_{NPI_{s,k}} \left(f^k_s \right) \\
& - \sum_{k \in E} \sum_{l \in E} CO^{MOV}_{k,l} \left(\kappa_{k,l} \right)
\end{aligned} \qquad (74)
$$

Example

This example takes a retrofit problem from an extension of a process discussed by Duran and Grossmann[2] (see Figure 2). The feed involves three chemical species A, B and C, where C is an inert component. The feed mixture is treated with a two-stage compressor with intermediate cooling to raise its pressure, and then mixed with a recycle stream. The resulting stream is preheated with the reactor outlet stream and fed to the reactor where the components A and B react in an exothermic reaction to produce D. The effluent of the reactor is cooled and sent to a flash unit to recover the product D in the liquid stream. The product stream (liquid from the flash) is heated to deliver the product as saturated vapor. A fraction of the resulting vapor stream of the flash is purged to avoid the accumulation of inert C in the process. The purge stream is heated to deliver it at a required temperature.

The data for the process specifications are given in Table 1. The phase equilibrium in the flash is predicted with an ideal model, while isentropic compression corrected by efficiency factors is assumed for the compressors. For the process streams, heat capacities are assumed to be linear functions of composition.

A superstructure with five existing heat exchangers and three new units was used for this problem. The resulting MINLP model contained 1094 constraints with 852 variables, which included 296 binary variables. After solving this problem with GAMS/DICOPT[8] software in 240.35 secs CPU time, we obtained the retrofit process shown in Figure 3. The new conditions in the retrofitted process are such that no hot utilities are required. In addition, new piping segments for the hot stream effluent to the reactor (*H1* stream) are needed to obtain a better heat integration with the cold process streams *C1*, *C3* and *C2* in exchangers 5, 4 and 2, respectively, and a cold utility is used in exchanger 3 to provide the temperature needed at the inlet of the flash. None of the new exchangers formulated as part of the superstructure were selected. Only additional area in exchanger 1 of 14.95 m^2 was needed. The total annual profit obtained from the simultaneous optimization of the process and the HEN retrofit is \$101,568,482/yr. Table 2 reports the results from the economics for the retrofitted process and for the original process. Notice how the retrofitted process provides a noticeable improvement in the annual profit.

When the retrofit problem was formulated without process modifications, the solution has a total annual profit of \$84,556,338/yr. The simultaneous retrofit optimization has a total profit 32.6% higher than the original process, which is significantly better that the improvement of 10.4% provided when only the HEN retrofit was considered.

Table 1. Specifications for Example 2

Design basis	Cost
Product D	\$3.81/kmol
Feed	\$0.65/kmol
Purge gas	\$0.55/yr
Generated steam	\$1.8537E-5/kJ
Working time	8,376 hr/yr
Payout factor	0.3 /yr
Utilities	
Cooling water [320-290] °K	\$2.4642E-6/kJ
Hot utility [690-690] °K	\$5.5613E-5/kJ
Purchased electric power	\$0.025/kW hr
Demineralized water	\$2.34E-3/kmol
Fixed cost for new heat exchangers	\$3000
Capital cost for new heat exchanger area [*A* in m^2]	\$1650 *A*
Compressor efficiency	γ=1.4, η_m=0.9, η_c=0.8
Reactor conversion	$x = 0.5\exp(-0.002T)\left(\frac{80}{90}\right)P\left[\frac{y_A y_B}{(1+y_C+y_D)}\right]$

Antoine constants [*P* in mm Hg]

Component	*A*	*b*	*C*
A	13.6333	164.90	3.19
B	14.3686	530.22	-13.15
C	15.2243	897.84	-7.16
D	18.5875	3,626.55	-34.29

Film heat transfer coefficients

Stream	*h* [kW/(m^2 °K)]
H1	2.31
H2	0.85
CU	2.50
C1	0.75
C2	0.93
C3	2.18
CU	1.00

Constraints

Reactor	Other
$T_{outlet} \geq T_{inlet}$	$320°K \leq T_{flash} \leq 380°K$
$T_{outlet} \leq 690°K$	$0 \leq \text{purge} \leq 100\%$
$450°K \leq T_{inlet} \leq 670°K$	$\text{product } D \leq \text{product}(0.96)$
$9\text{atm} \leq \text{Pressure} \leq 29\text{atm}$	
$0 \leq \text{conversion} \leq 100\%$	

Conclusions

This paper has presented an MINLP formulation for the retrofit of chemical processes considering simultaneously process modifications and heat integration. The model considers the plant layout and complex piping configurations. The superstructure used for the heat exchanger networks configuration is general, and does not require imposing constraints such as no bypass or no splitting of streams. Also, the model formulation includes the treatment of isothermal process streams. For the economic assessment of alternatives, a simultaneous consideration is included for the capital cost of the new exchanger units, the additional area required and the new piping segments, as well as the operating cost for hot and cold utilities. Fixed and variable piping costs can be considered in the model formulation, as well as the relocation of heat exchangers within the process.

The examples presented here show that significant earnings can be obtained in the retrofit process when the process modifications and the heat integration retrofit are considered simultaneously, as opposed to the solution given by the consideration of heat integration restricted with unchanged process conditions.

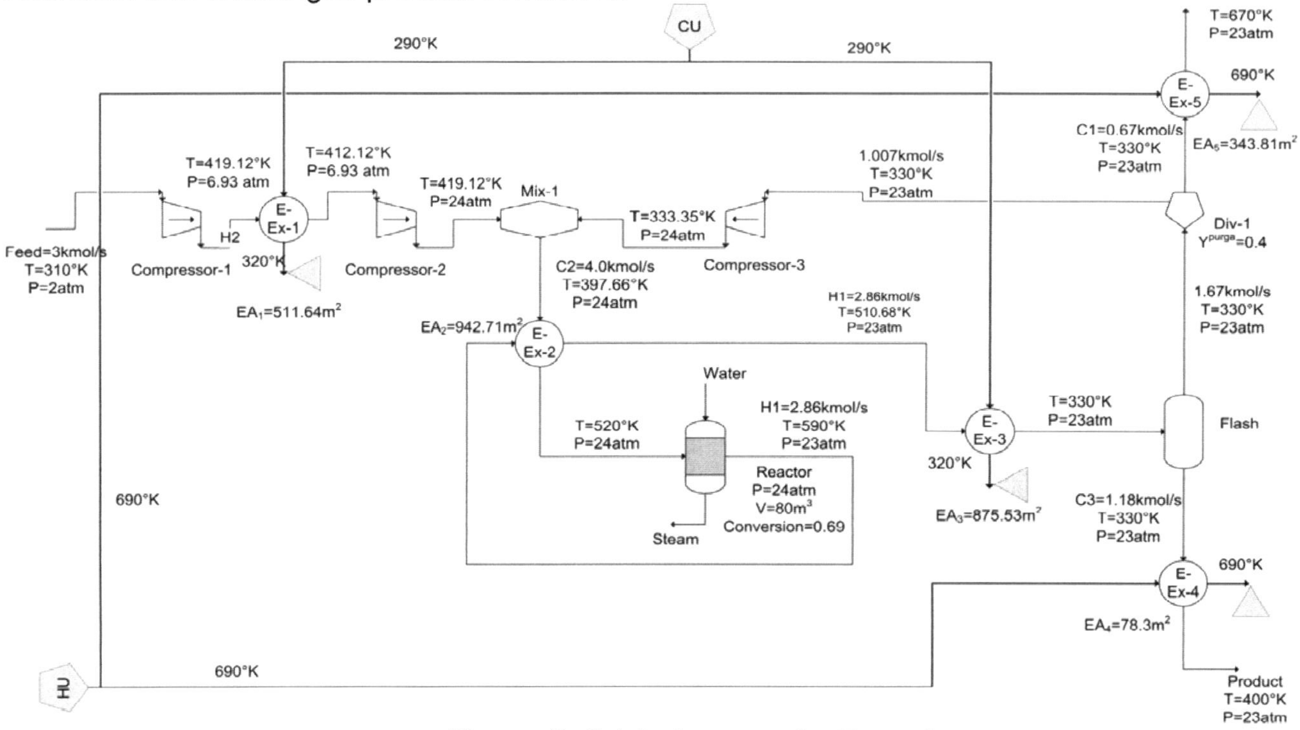

Figure 2. Original process for Example

Table 2. Results comparison for the Example

Concept	Original process	Original process with HEN retrofit	Simultaneous retrofitted process
Costs [$/yr]			
Raw materials	58,799,520.00	58,799,520.00	58,799,520.00
Hot utility	8,033,410.62	233,851.98	0.00
Cold utility	700,799.85	453,231.67	756,332.01
Electricity	2,897,998.30	2,897,998.30	2,278,500.00
Demineralized water	9,303,820.08	9,303,820.08	3,165,612.90
Capital exchangers	0	76,083.69	7,399.73
Piping	0	1,905.00	849.00
Total costs	79,735,548.85	71,766,410.72	65,008,213.64
Earnings [$/yr]			
Product	135,942,969.55	135,942,969.55	158,950,000.00
Purge	11,138,588.39	11,138,588.39	4,482,500.00
Generated vapor	9,240,876.40	9,240,876.40	3,144,196.39
Total earnings	156,322,434.34	156,322,434.34	166,576,696.39
Annual profit	76,586,885.49	84,556,023.62	101,568,482.75

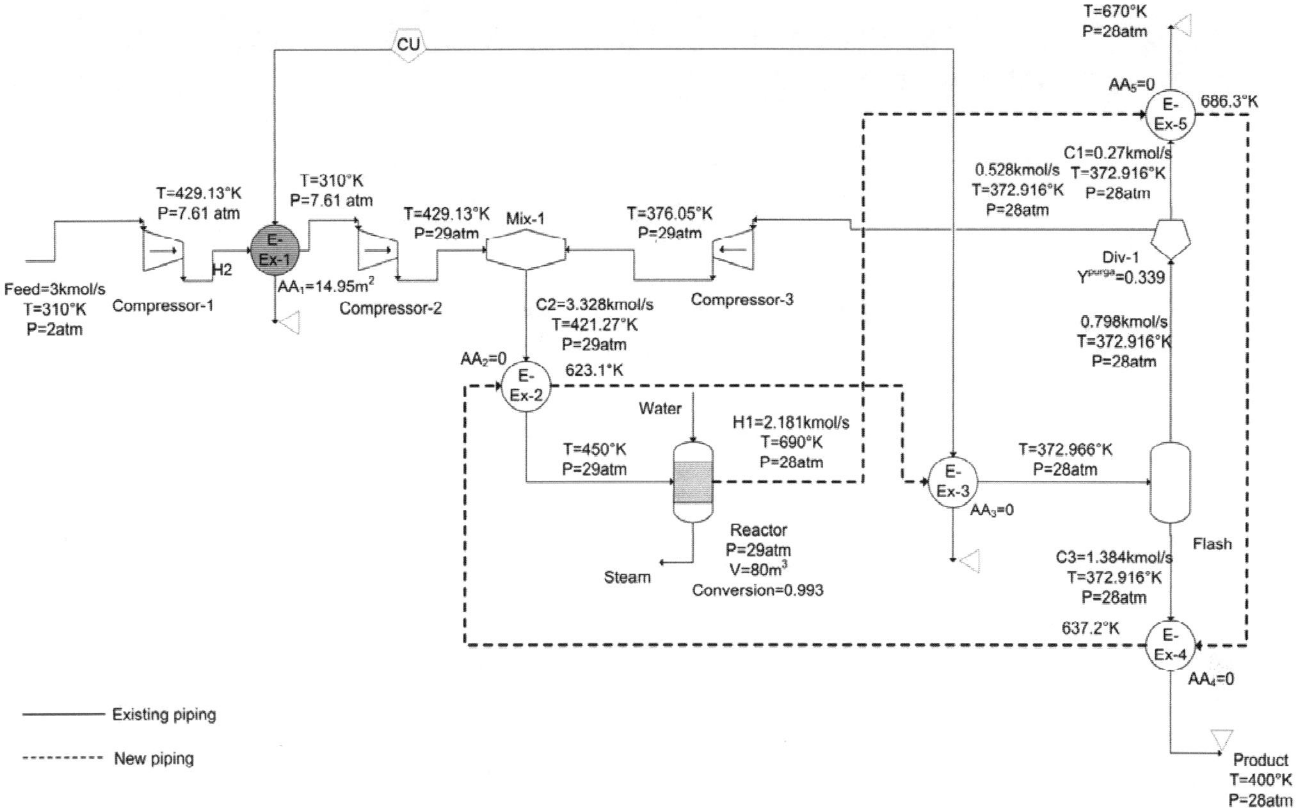

Figure 3. Retrofit process for the Example

Literature cited

1. Furman, K. C. and Sahinidis, N. V. (2002). A critical review and annotated bibliography for heat exchanger network synthesis in the 20th century. *Ind. Eng. Chem. Res. 41* (10), 2335-2370.
2. Duran, M. A. and Grossmann, I. E. (1986). Simultaneous optimization and heat integration of chemical processes. *AIChE J. 32* (1), 123-138.
3. Lang, Y. D., Biegler, L. T. and Grossmann, I. E. (1988). Simultaneous optimization and heat integration with process simulators. *Comput. Chem. Eng. 12* (4), 311-327.
4. Grossmann, I. E., Yeomans, H. and Kravanja, Z. (1998). A rigorous disjunctive optimization model for simultaneous flowsheet optimization and heat integration. *Comput. Chem. Eng. 22*, S157-S164.
5. Zhang, J. and Zhu, X. X. (2000). Simultaneous optimization approach for heat exchanger network retrofit with process changes. *Ind. Eng. Chem. Res. 39* (12), 4963-4973.
6. Yee, T. F. and Grossmann, I. E. (1991). A screening and optimization approach for the retrofit of heat exchanger networks. *Ind. Eng. Chem. Res. 30* (1), 146-162.
7. Ponce-Ortega, J. M., Jiménez-Gutiérrez, A. and Grossmann, I. E. (2008). Optimal synthesis of heat exchanger networks involving isothermal process streams. *Comput. Chem. Eng.* In press **2008**.
8. Brooke, A., Kendrick, D., Meeraus, A. and Raman, R. (2006). GAMS-Language guide. GAMS Development Corporation: Washington, D.C.

590a How Much Is the Pharmaceutical Industry Potentially Losing Due to Lack of Information Interoperability?

Leaelaf M. Hailemariam[1], Venkat Venkatasubramanian[2], and **Arun Giridhar**[2]. (1) School of Chemical Engineering, Purdue University, 480 Stadium Mall Drive, West Lafayette, IN 47907, (2) Chemical Engineering, Purdue University, 480 Stadium Mall Drive, West Lafayette, IN 47907-2100

A looming challenge facing the pharmaceutical industry is the management and integration of the vast amount of data, information and knowledge generated during drug product development [1]. Documentation for a new drug application to the FDA includes over 500,000 pages [2] and is expected to double every 5 years [3]. Information is stored in hundreds to thousands of repositories [4], making integration and efficient use a significant problem.

While there had been multiple efforts to identify the scope of the challenge, there is little work done in quantifying the costs associated with the lack of information interoperability. In addition few attempts have been made to quantify the indirect effects of lack of efficient information processing. These issues are addressed in this work. The cost of information interoperability is computed by adaptation of an approach used in the US Capital Facilities Industry [5]. In addition, the indirect effects of more efficient information processing on the industry's revenue, through improvement of the likelihood of drug substances to get to market, is discussed. Solutions being developed in our group at Purdue to address these challenges are described.

References

1. Venkatasubramanian V., C. Zhao, G. Joglekar, A. Jain, L. Hailemariam, P. Suresh, V. Akkisetty, K. Morris and G.V. Reklaitis (2006) Ontological Informatics Infrastructure for chemical product design and process development. Computers and Chemical Engineering. CPC 7 Special Issue, 30(10-12) 1482-1496

2. Fitzmartin R. (1998) The challenge of Global Electronic Submission Standards in the Biopharmaceutical Industry Drug Information Journal 32 745–756, 1998

3. Morris K, Venugopal S and Eckstut M (2005) Making the most of drug development data Pharmaceutical Engineering 11 399-402

4. Kataria A. (2007) Information Management: Challenges and Opportunities in Achieving Qbd. Presentation at the AIChE Annual Conference, Salt Lake City UT

5. Gallaher M.P., O'Connor A.C., Dettbarn J.L and Gilday L.T. (2004) Cost Analysis of Inadequate Interoperability in the U.S. Capital Facilities Industry. NIST GCR 04-867

590b a Problem Solving Environment for Integrated Solvent and Process Design on the Grid

Athanasios I. Papadopoulos, Chemical Process Engineering Research Institute, Center for Research and Technology Hellas, 6th klm Harillaou-Thermi Road, P.O.Box 60361, Thermi-Thessaloniki, 57001, Greece and **Patrick Linke**, Chemical Engineering Department, Texas A&M University at Qatar, P.O. Box 23874, Doha, Qatar.

Grid technologies have received a lot of attention by the scientific community recently; however, developments are at their infancy and very few engineering applications have been reported to date. A typical application in the domain of process engineering that would benefit from grid computing is the integrated computer-aided solvent and process design. The design problem makes an ideal candidate for grid computing as it integrates a number of software resources such as computer-aided molecular design (CAMD) tools, data mining tools for the analysis of molecular design information, process synthesis tools and web mining tools with data repositories such as in-house and web databases required in the different design stages. At the same time all the above tools may be parts of different design stages spanning from conceptual design to detailed simulation models and may belong to various research groups or companies spread around the world. Furthermore, the optimization problems involved require vast screening of design options, which in its turn requires high computational effort. Grid technologies enable the development of an integrated solution environment for such a problem, where software and hardware heterogeneity is addressed through the use of appropriate tools and vast computing and storage resources become available.

In this context we develop a problem solving environment that interfaces and manages a number of available heterogeneous resources to be used in the integrated solvent and process design. The proposed architecture of the developed problem solving environment involves the following; The available in-house design and optimization tools exist in the form of modules and we develop a web-portal to render them available to the end user. The developed web portal further serves as a connecting node between remote, geographically distributed computing resources and external databases available through web channels. Such resources involve exploitation of computer networks for distribution of the design calculations and available data repositories for storage of design related information. The utilized computing networks include over 3000 processors while the data repositories have a storage capacity of over 100TB. The geographically distributed computing resources are managed by middleware available through a third party organization, which provides generic functionalities, is hard to use and lacks the automation levels matching particular requirements of a problem such as the integrated solvent and process design. In this respect, we develop a set of services that allow user friendly access of the service functions available at the third party middleware through the web portal, focus specifically to the requirements of the integrated solvent and process design problem, while also maintaining functionalities that facilitate the use of the existing services. One major service developed is the workflow generator, which allows the user to select from a number of developed solvent and process design workflows that render the design and optimization tools available to the user, depending on the design problem at hand. Following preparation of the workflow, the job scheduling service exploits a collection of services available at the third party middleware to launch the workflow in the distributed resources for execution, while it dynamically manages the selection of appropriate execution resources, the storage of generated data, and the termination of the workflow execution. Data management and resource allocation services are also available independently of the job scheduling services that allow treatment of the generated data (i.e. storage, retrieval etc.) in the available storage resources as well as querying of the status of the available storage and computing resources. Furthermore, we develop appropriate service functions that allow querying and retrieval of results in third party web-based databases through the web portal. The functionalities of the developed problem solving environment are illustrated through the solution of an integrated solvent and process design problem, while usability issues of the developed portal, computational characteristics and time performance of the workflows execution in the geographically distributed resources are also discussed.

590c a Data Model Supporting Intelligent Search for Materials Research

Stephen D. Stamatis[1], Balachandra B. Krishnamurthy[1], Amr Shehab[2], Tanu Malik[3], Leif Delgass[1], Steven R. Dunlop[4], and James M. Caruthers[1]. (1) School of Chemical Engineering, Purdue University, Forney Hall of Chemical Engineering, 480 Stadium Mall Drive, West Lafayette, IN 47907-2100, (2) Department of Computer Science, Purdue University, (3) Cyber Center, Purdue University, (4) ITaP, Purdue University

High-Throughput Experimentation (HTE) enables researchers with the ability to generate massive amounts of data. HTE is typically used to indentify 'hot' material candidates, where the information from the 'cold' candidates is discarded. However, if experiments are properly designed all data is useful in developing an understanding of the material or process under investigation. In order to gain maximal value from HTE data, the data must be managed in a manner that facilitates knowledge extraction by integrating data from multiple sources, including both lab experiments and computer simulations. Traditional techniques of managing data using nested file folders on local computer network resources are capable of storing the data, but searching data that is stored in this manner is nearly impossible because the data are typically incomplete, have insufficient provenance, are not cataloged correctly and cannot be queried intelligently using domain specific language with complex syntax.

We have developed a general data model as part of the SciAether™ (www.SciAether.org) software system that is capable of serving the data needs of multiple chemistry domains in a flexible and searchable manner. The data model for experimental information distinguishes two classes of materials: (1) an ideal material, which could be a mixture of ideal materials; (2) the actual material in a given batch including all impurities, which is characterized by both its composition and the sequences of 'unit operations' used in its synthesis. Materials are involved in synthesis, characterization and performance experiments. *In silico* experiments are also included in this data model, because they involve ideal materials. The intelligent searching is enabled by a project specific ontology and similarity searching tools. The ontology enables the researcher to use terms that were not originally defined when data was entered into the database, e.g. find all α-olefins, where the term α-olefin was never entered with the data rather the specific instances of α-olefin like propylene, 1-hexene, etc. A user defined rule based engine is employed to classify ideal materials and this classification is stored in the database to enhance query speed. The implementation of similarity search uses an information retrieval approach to give a rank ordered list of structures that are similar to a target molecule. The method consists of the following steps: (1) the molecules of interest are broken into a series of individual fragments, which provide a 'fingerprint' of each molecule; (2) an inverse document list is prepared which indicates all molecules that contain a given fragment; and, (3) the frequency at which specific fragments occur in the molecular population of interest is computed and stored in the database. Using this frequency table, the similarity between related molecules can be determined using heuristics detected by the user's preferences in previous searches. This organization of data along with domain specific ontologies and similarity searching capabilities allows for high level querying of the database, e.g. find all data for batch polymerization (specific unit operation) of a-olefins (class of ideal materials) with metallocene catalysts (class of ideal materials) that contain an aryloxide ligand (class of ideal materials) similar to a 2,6 di-isopropylphenoxy (ideal material) which will then return all actual materials that meet this criterion. The data model described above will be illustrated with case studies in the area of single-site olefin polymerization, which is a homogeneous catalytic process, and the water gas shift reaction, which is a heterogeneous catalytic process.

Web Page: www.SciAether.org

590d An Ontology Approach to Represent and Extract Process Synthesis Knowledge in High-Throughput Optimization

Antonis Kokossis, Franjo Cecelja, Claudia Labrador-Darder, and Patrick Linke. Chemical and Process Engineering, University of Surrey, Guildford, Surrey, GU2 7XH, United Kingdom

The paper presents a systematic approach to represent and extract process synthesis knowledge in high-throughput optimization. Means of analysis include dynamic ontologies populated by computer experiments and continuously upgraded in the course of optimization.

Process synthesis makes extensive use of superstructure optimization to address the systematic development of process design options through generic representations integrating all available designs. Although results can be produced in a single stage, practice dictates the employment of multiple stages - first with the deployment of simple conceptual models and then with the use of more detailed formulations – as intermediate solutions can be reviewed, analyzed and understood. Even in cases where the development of solutions is rigorous and proven, synthesis solutions can prove rather inconclusive and difficult to translate. Indeed, process synthesis typically yields a multitude of solutions with (very) similar design performance but different structural and operational characteristics. In the same vane, minor changes in the design parameters often yield major changes in the selected designs (types of units, connections employed) which are difficult to interpret, comprehend or summarize. The systematic interpretation of the solution can yield not only to a better understanding of the solution space but also to a systematic reduction of the superstructure employed (both in size and complexity) and, thus, to much simpler experiments.

The work combines optimization, semantic models (in the form of ontologies) and analytical tools. The research is presented against a background of stochastic optimisation techniques that enable the development of solution clusters for chemical reactors synthesis. The work explains the systematic extraction of information with a purpose to simplify and interpret design results. The simplification is achieved with a gradual evolution of the superstructure and corresponding adjustments of the optimization search. The interpretation is accomplished with the use of analytical tools to translate data into descriptive terms understood by users. Both stages make use of ontology models bridged with analytical tools to extract information patterns from solution data. Data mining and clustering techniques are employed in order to identify design trends (units, connectivities) and to build the components required for the knowledge models. Ontologies feature concepts and relationships processed by external analysis. They can be interfaced with synthesis workers and can be used to enable solution summaries or the visualization of the results. In the implementation, the work combines the optimization models with an ontology environment and a reasoner. Experiments are registered on the ontology that in turn guides the search of the analysis. The ontologies feature: (i) fixed concepts and relationships directly related to the structural and operational components of the superstructure and (ii) floating concepts upgraded with new associations and relationships in the course of analysis. Dynamic ontologies were developed using Protégé-OWL (supports OWL). They were supported by the RACER reasoner that was functioning as a logical classifier and an ontology verification tool. Visualization was partly enabled through the GrOWL editor.

Results are presented in several examples of reactor network synthesis but are not essentially restricted to a particular type of application.

590e Cyberinfrastructure Roundtable

Larry Megan, Praxair, 175 East Park Drive, PO Box 44, Tonawanda, NY 14150

Roundtable of session participants to discuss key issues in cyberinfrastructure

Comparison of Coal Based Hydrogen and Electricity Co-Generation Processes under a Carbon Constrained Scenario

Liang Zeng, Fanxing Li, Deepak Sridhar, Fei Wang, Hyung Rae Kim, and L. -S. Fan[1]

Abstract

Hydrogen and electricity produced from coal will represent a significant share in the energy mix within the foreseeable future. Various novel processes are developed or currently under development for hydrogen and electricity production from coal derived syngas. The commercial coal to hydrogen and coal to electricity processes include the coal gasification-water gas shift process and integrated gasification combined cycle (IGCC) process. More advanced approaches include enhanced water-gas-shift membrane technology and chemical looping technology.

This paper compares these technologies on the same baseline via Aspen Plus simulation with common assumptions. Firstly, performance of the key reactors, e.g. the reducer/oxidizer of the chemical looping system and the hydrogen-/carbon dioxide-selective water-gas-shift membrane reactor of the membrane system, are investigated and modeled based on thermodynamic principles. Operating conditions such as feed rate, operating temperature and pressure are then optimized. The optimized reactor models are integrated into a common coal gasification system to obtain the overall performance and energy conversion efficiencies of different process configurations. The results are helpful in opting and optimizing the design of future coal to hydrogen/electricity processes.

[1] To whom the correspondence should be addressed

1. Introduction

In view of the volatile crude oil and natural gas prices, coal, which accounts for 40% of the electricity generated world wide at present, is expected to remain as an important fossil energy resource within the foreseeable future [1].

The green house effect of CO_2 has led to increased concerns over CO_2 emissions from coal conversion plants. To mitigate such emissions, future coal conversion plants should have the capability to convert coal into carbon free energy carriers such as hydrogen and electricity while sequestrating at least 90% of the CO_2 generated during the energy conversion scheme.

Traditional pulverized coal combustion (PCC) technology converts approximately a third of the energy in coal into electricity. Such an energy conversion efficiency can be further reduced by up to 40% when the envisaged CO_2 regulation is implemented [2]. Coal gasification techniques such as integrated gasification combined cycle (IGCC) and gasification – water gas shift (WGS) processes can potentially increase the energy conversion efficiency with reduced CO_2 capture penalty; however, CO_2 capture in coal gasification plants will nevertheless reduce the plant efficiency by up to 24%. More advanced approaches such as enhanced WGS membrane technology and chemical looping technology has the potential to further improve the performance of the gasification based coal conversion processes. The enhanced WGS membrane technology integrates membrane into a regular WGS reactor to break the equilibrium by simultaneous removal of the WGS reaction product. Chemical looping technology, on the other hand, utilizes a chemical looping medium such as iron oxide to convert syngas into a concentrated hydrogen stream, avoiding the CO_2 separation step. In order to evaluate the potential of the various coal gasification schemes, process simulations using common assumptions are highly desirable.

2. Thermodynamic Analysis

Thermodynamic model can predict the conversion of reactants when thermodynamic equilibrium is achieved under certain conditions. Although the conversions predicted by thermodynamic modeling may not be identical to that obtained in actual reactor operations, it can be used as a good indicator to evaluate the potential of a process configuration. In the following sections, ASPEN plus® software is used to evaluate three key components of gasification based coal conversion processes. These components include water gas shift reactor in conventional gasification-WGS process, membrane enhanced WGS reactor in the membrane based coal gasification process, and the reducer and the oxidizer in the SCL process.

2.1 water gas shift reaction

Using Aspen Plus, a thermodynamic model for water gas shift reaction can be constructed with ease using one RGibbs block. Throughout this section, the feedstock is limited in H_2, CO and steam. The ratio between H_2 and CO is fixed at 1:2, which is similar to that from a dry feed, oxygen blown gasifier. The property method is set to be PR-BM, which is used in all the models in this paper. The variables of interest in this case study include the operating temperature and steam to CO ratio since pressure will not have significant effect on the equal molar WGS reaction.

In order to determine the effect of temperature, steam and CO molar ratios of 1, 1.5, 2 and 2.5 are investigated in the sensitivity analysis. As shown in Figure 1, at temperature below 250 °C, most CO can be converted into H_2 in all cases. The CO conversion decreases with increased water gas shift reactor temperature. Moreover, higher steam to CO ratio leads to higher CO conversion. To ensure a high CO conversion, steam to CO ratio should exceed 2 and the WGS reactor operating temperature should be less than 250 °C.

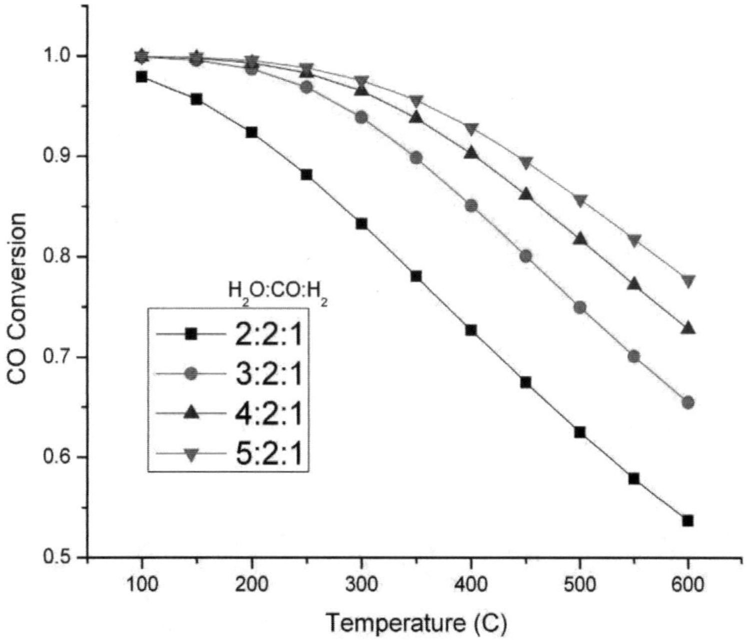

Figure 1 Temperature influence on CO conversion in WGS at 30 bar

2.2 Enhanced WGS Membrane Reactor

The basic thermodynamic principles dictate that simultaneous removal of reaction products enhances the conversion of the reactants. Since the products include both CO_2 and H_2, removal of either specie can enhance the reaction. Therefore, both H_2-slective and CO_2-selective membranes can be used in the enhanced WGS reactor. In the simplified thermodynamic simulation, the selectivity and permeation flux for both types of membranes are assumed to be infinity, which leads to similar performance of the two types of membranes. Therefore, only H_2-selective membrane is investigated.

Although there is no built-in model for membrane reactor in Aspen Plus, a series of RGibbs and Sep blocks can be stringed to represent permeate side and retentate side respectively. Similar model has been reported in previous study on steam reforming of methane [3-5]. As mentioned previously, two configurations for the flow direction between permeate side and retentate side, i.e. co-current and counter-current, will be investigated. The models for both co-current and counter-current design are illustrated in Figure 2.

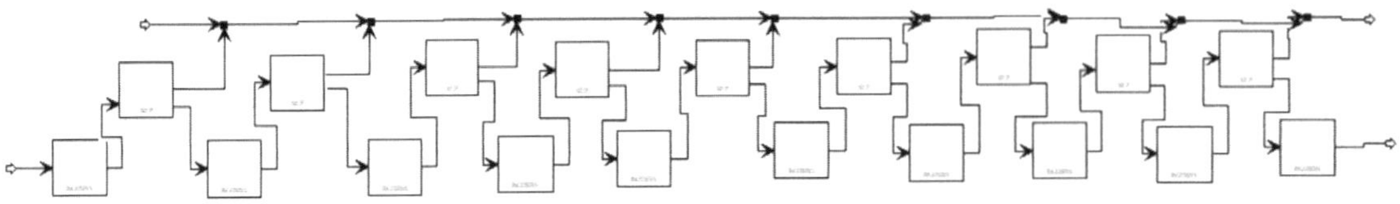

a

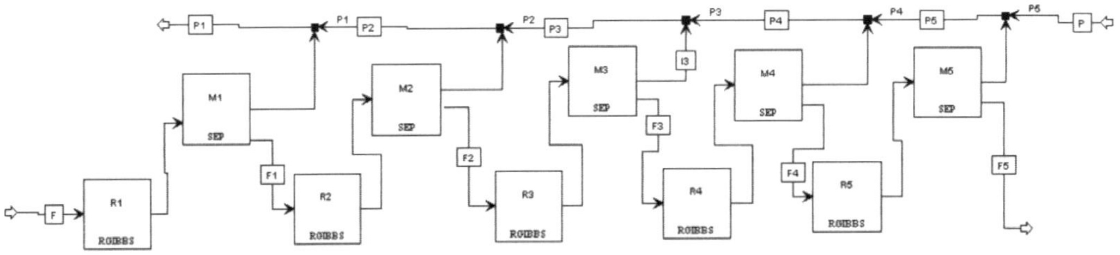

b

Figure 2 a. Cocurrent layout with 9 stages b. Countercurrent layout with 5 stages

As shown in the Figure 2, the inferior retentate gas goes through all the RGibbs blocks one by one, and permeating gas (H_2) together with sweeping gas (steam or N_2) goes through all the Mix blocks. After each stage of reaction equilibrium, the hydrogen is separated by a Sep block to ensure that the partial pressures of hydrogen at both sides are equal:

$$P_{H_2}^{Fn} = P_{H_2}^{Pn} \qquad 1$$

Where P_{H2} represent the partial pressure of hydrogen, and Fn is for retentate side while Pn is for the permeate side

The separation step that satisfies equation 1 can be carried out using the Design Spec function in Aspen Plus. After the separation, the hydrogen is mixed with the permeating stream. The remaining gas in the retentate side will enter the next stage for further conversions. The simulation result converges to the thermodynamic equilibrium as the number of blocks increases. Primary analysis is discussed in the following section which evaluates the key variables for membrane reactors.

2.2.1 Cocurrent Design

In this case, both the streams flow in the same direction. For simplicity, the membrane reactor is assumed to operate isothermally at 400 °C. The total pressure of the permeate side is assumed to be 1 bar.

In one case, with steam: CO equal to 1 in the feedstock at 30 bar and same amount of steam injected as sweeping gas, a 9-stage membrane reactor can convert nearly 95.9% of the CO into H_2. The corresponding H_2 recovery ratio, which is defined as the percentage of hydrogen recovered at permeate side relative to the total amount of hydrogen after fulfilling the reaction, is 98.6%. The trends of H_2 partial pressure and CO conversion along with these stages can be examined in Figure 3. The limitation can be also calculated from a black box model [6] based on overall hydrogen balance and chemical reaction equilibrium.

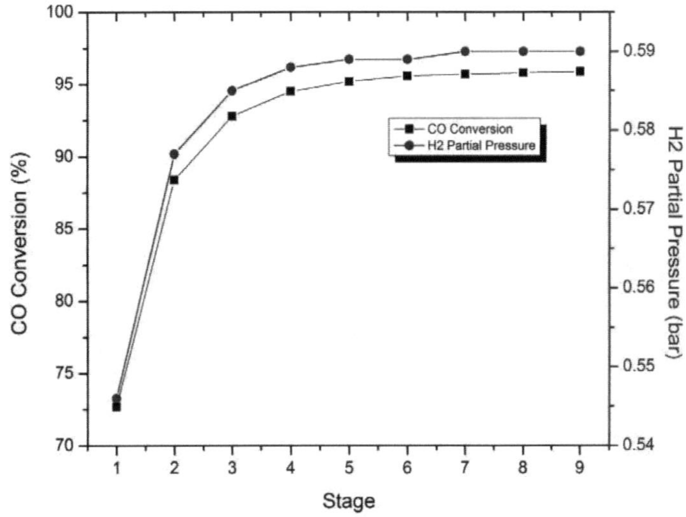

Figure 3 CO Conversion and H_2 partial pressure in a 9-stage cocurrent membrane reactor with a steam: CO ratio of 1.

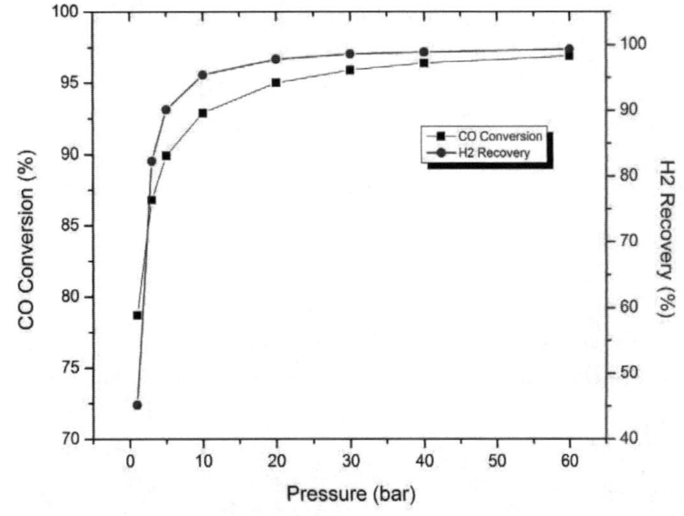

Figure 4 CO conversion and H_2 recovery ratio at different feed pressures

Unlike previous WGS reaction, pressure becomes a factor in the thermodynamic equilibrium. By changing the pressure in feedstock within last case, the CO conversion and H_2 yield vary significantly. As shown in Figure 4, higher pressure in reactant stream will increase the partial pressure of hydrogen at the same side. Therefore, more hydrogen will be driven to the permeate side, which enhances H_2 production. At above 20 bar, the influence of feed pressure becomes minor. Similar results can be expected for the sweeping gas flow rate. Higher sweeping gas flow rate leads to lower hydrogen partial pressure in the permeate side, resulting in increased hydrogen yield.

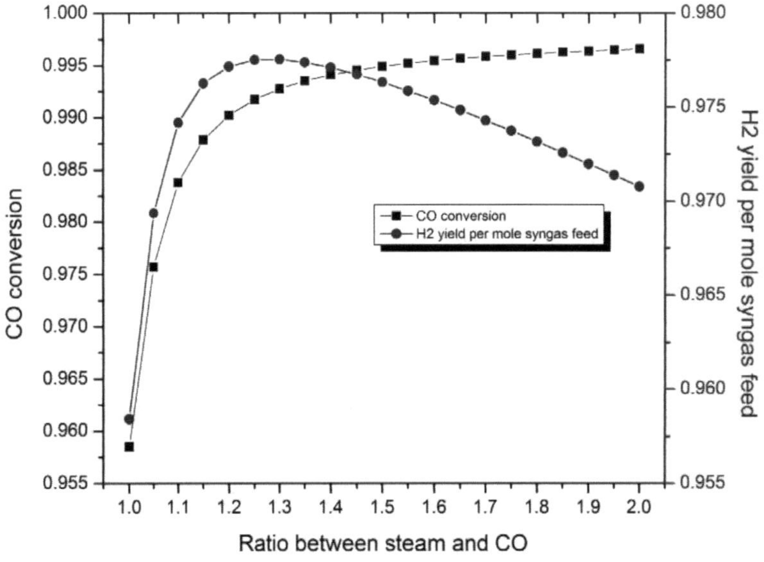

Figure 5 Effect of steam ratio in feedstock

The steam to CO ratio in the reactant stream can also affect hydrogen yield. As shown in Figure 5, on one hand, high steam ratio can promote the WGS reaction. On the other hand, high steam ratio decreases the partial pressure of hydrogen in the reaction side, which limits the hydrogen permeation via membrane. Therefore, there is a peak value in the trend of hydrogen yield. In this case, the highest hydrogen yield appears at a ratio between 1.2 and 1.3.

2.2.2 Countercurrent Design

In this configuration, sweeping gas (steam or nitrogen) together with permeated hydrogen will flow in an opposite direction to the flow of reactant gas. Figure 6 shows the simulation results from a 10-stage membrane reactor model, with steam: CO = 1:1 in the feedstock at variational pressure and sweep gas (1 bar): CO = 1:1. The reaction is also kept at 400 °C.

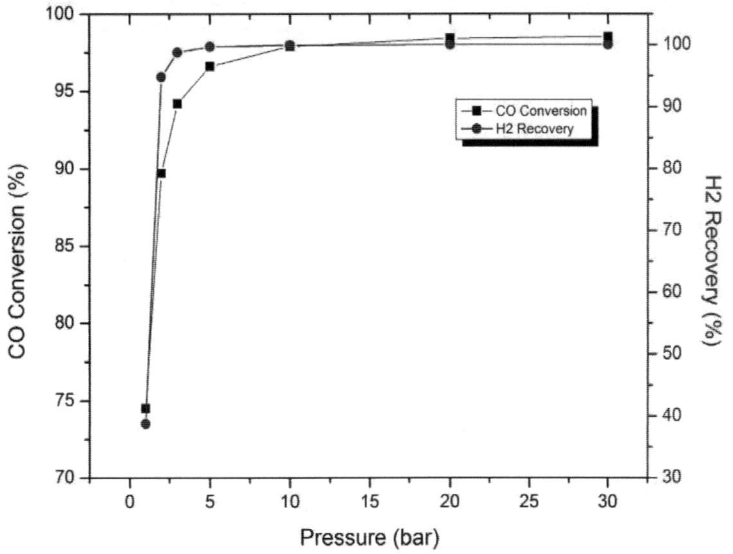

Figure 6 CO conversion and H_2 recovery ratio at different feed pressure

In this configuration and interested operating condition (pressure>3 bar), CO can almost be fully converted to H_2 and transferred to the permeate side. However, the full conversion may require huge membrane surface area and reliable membrane performance.

2.3 Syngas Chemical Looping System

Syngas chemical looping (SCL) system can convert syngas into hydrogen via an oxygen intermediate – iron oxides. The conversion can be completed within a loop of three reactors: the reducer, the oxidizer and the combustor. The reducer utilizes iron oxides to fully oxidize syngas into steam and CO_2. Consequently, CO_2 can be extracted by condensing all the steam without extra energy penalty. The solid product, reduced iron oxide, can be reoxidized to Fe_3O_4 by low temperature steam in oxidizer together with hydrogen production. The regeneration of iron oxides from Fe_3O_4 can be achieved in combustor by direct oxidation using air. The heat released in this step can be used to compensate the heat required in the reducer or to generate steam or power.

Analogously to membrane reactor designs, there are mainly two solid gas contacting patterns, i.e. cocurrent and countercurrent. For cocurrent design, one block of Rgibbs can be used to mimic the thermodynamic situation. However, to fully convert syngas in reducer, the result shows that huge amount of iron oxides are in need, which leads the effluent solids are Fe_3O_4. Due to thermodynamic restrictions, the consequence is that no hydrogen can be generated by steam in oxidizer. Thus, cocurrent fluidized bed reactor design is excluded for hydrogen production.

For countercurrent moving bed reactor design, novel multistage models for both reducer and oxidizer have been developed. As shown in Figure 7, Solid reactants are introduced from the top down, whereas gas stream flows antrorsely through the reactor, represented as a series of Rgibbs reactors. Oxygen transfers between gas and solid streams during each stage by chemical reaction. The opposite operating flow direction ensures that reductive stream and oxidative stream get in touch and react to the largest extent.

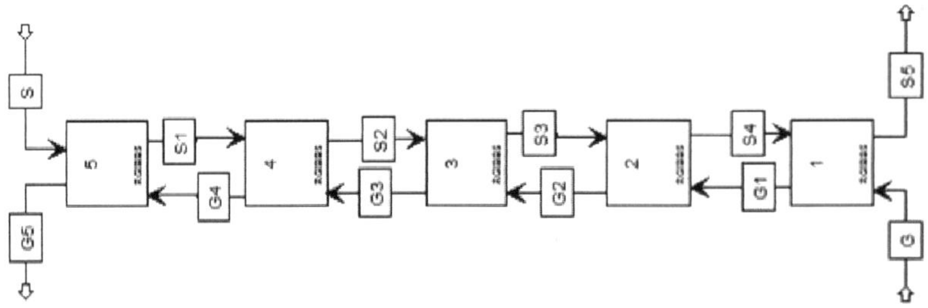

Figure 7 5-stage moving bed reactor model for SCL system

Figure 8 illustrates both gas (CO: H_2 = 2:1) and solid (pure Fe_2O_3) conversions in a reducer working at 900 °C, 30 bar by varying the molar ratio between solid and gas. As shown in Figure 8, to fully convert syngas, the minimum solid gas molar ratio is 0.913, which corresponds to a mixture of 90.7% FeO and 9.3% Fe. These reduced particles can then react with steam to produce hydrogen. Figure 9 concludes the minimum steam consumption per mole CO input for completed hydrogen production at different temperatures. The results suggest that hydrogen production favors low temperature, which is consistent to the exothermic nature of steam iron reaction.

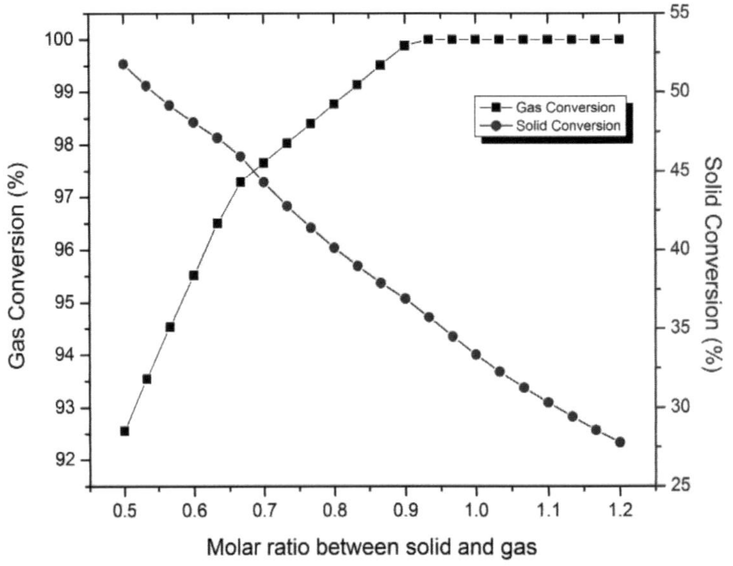

Figure 8 Influence of feed ratio in reducer

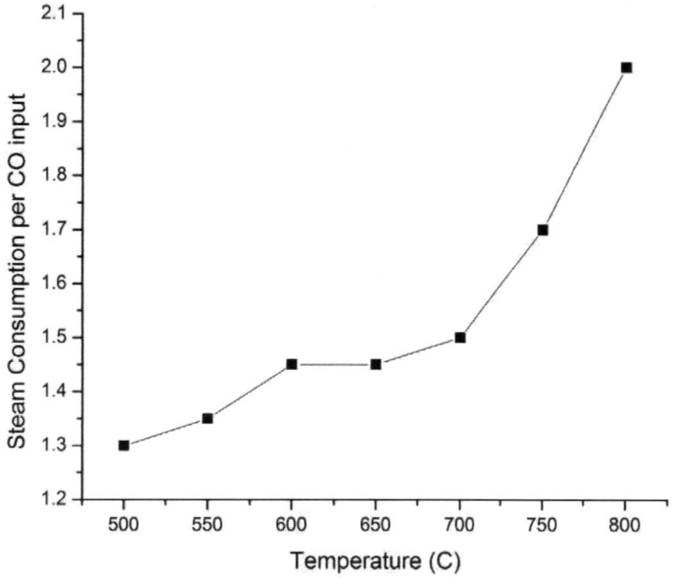

Figure 9 Influence of temperature in oxidizer

3. Process Simulation for Advanced Hydrogen Production

To evaluate the performance of the conventional gasification-WGS process, the enhanced WGS membrane process, and the SCL process, a set of common coal gasification and syngas cleanup system is used for comparison based on identical assumptions.

All the process simulations are based on a Shell dry feed oxygen blown gasifier with 99% carbon conversion and 0.5% heat loss. The gasifier utilizes dried (5% moisture) Illinois #6 coal and pure oxygen (from Air Separation Unit, ASU) to generate syngas with a composition shown in Table 1. Syngas is cooled through gas quench to 500 °C, during which ash and fine particles have been cleaned off. Afterwards, the syngas is routed to high temperature cleanup units to remove sulfur. The clean syngas will then be introduced to the following systems for hydrogen and power production.

Table 1 Summary for Shell Gasifier

Mass Flow ton/hr	Feedstocks	Products
H_2O		4.18
N_2	4.75	6.42
O_2	103.1	0
S		4.81e-005
H_2		5.98
HCL		0.40
CO		190.8

CO_2		6.80
NH_3		0.0048
COS		0.438
H_2S		3.29
CH_4		0.069
Illionis #6 COAL (5% moisture)	124.3	0
ASH		12.88
C		0.847
Temperature °C	133	1233
Pressure bar	35	30

3.1 Traditional WGS Process

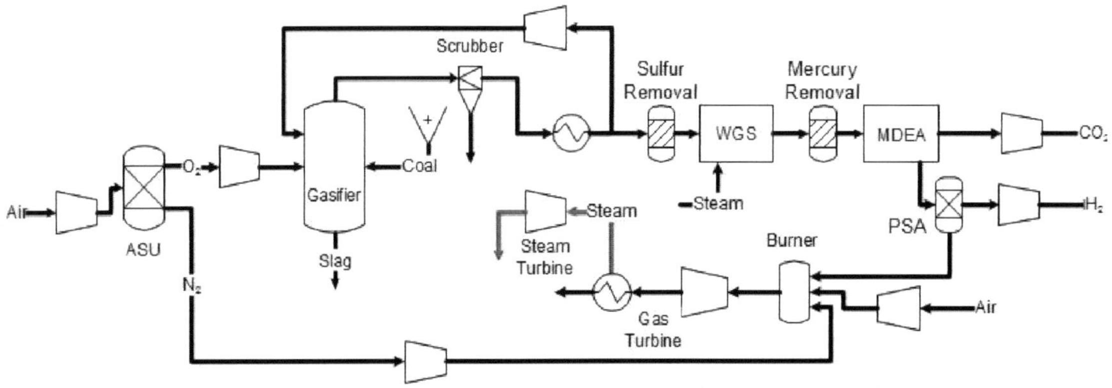

Figure 10 Traditional WGS process for hydrogen production

As shown in Figure 10, clean syngas goes through WGS unit together with sufficient excessive steam (steam: CO = 2.1). The WGS unit is operated at 230 °C with 95% CO conversion. The effluent gas products are then cooled for mercury removal. MDEA unit is used to capture CO_2, and then pressure swing absorption (PSA) unit is used to purify H_2. The tailgas from PSA, including 23% of the total H_2 product and unconverted combustible gas, is to be burned in a gas turbine. For all the cases in this section, the pressure specifications for CO_2 and H_2 are 150 bar and 60 bar, respectively.

3.2 CO_2 selective membrane reactor

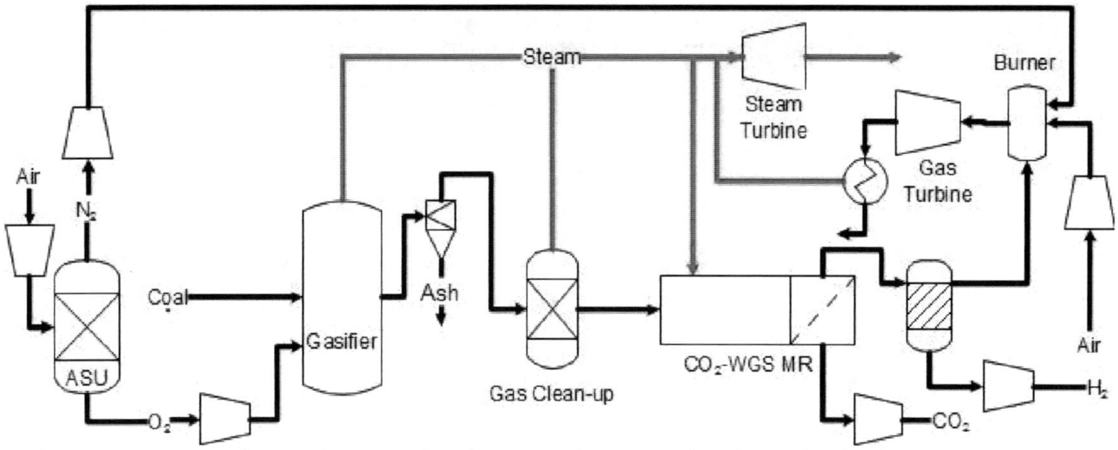

Figure 11 Enhanced CO_2 selective membrane technology for hydrogen production

The CO_2 selective membrane reactor can be integrated into hydrogen production process as shown in Figure 11. For membrane reactor simulations both in 3.1 and 3.2, the ratio between steam and CO is adjusted to 1.3 to achieve full conversion of CO, and no sweeping steam is considered at present. Also, the pressure is set to be 1 bar at permeate side and the whole reactor is kept at 28 bar and 400 °C. To eliminate the impurities in raw hydrogen product, PSA is required for purification. 24% of the total H_2 product goes together with the tailgas for combustion.

3.3 H_2 selective membrane reactor

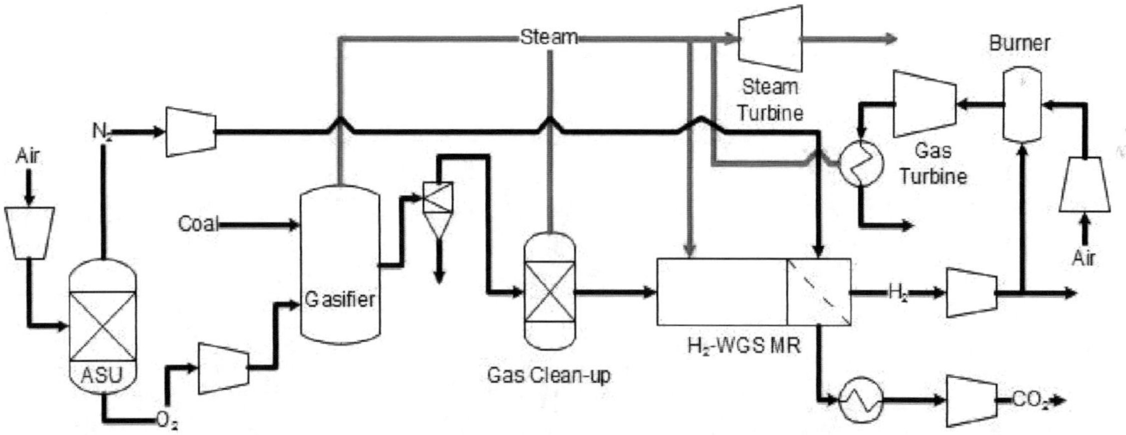

Figure 12 Enhanced H_2 selective membrane technology for hydrogen production

Figure 12 shows the process based on enhanced H_2 selective membrane technology. The pure hydrogen stream is separated after membrane reactor, with 23.5% for power generation and 76.5% for further compression. The CO_2 rich stream is to be cooled and cleaned before sequestration.

3.4 SCL process

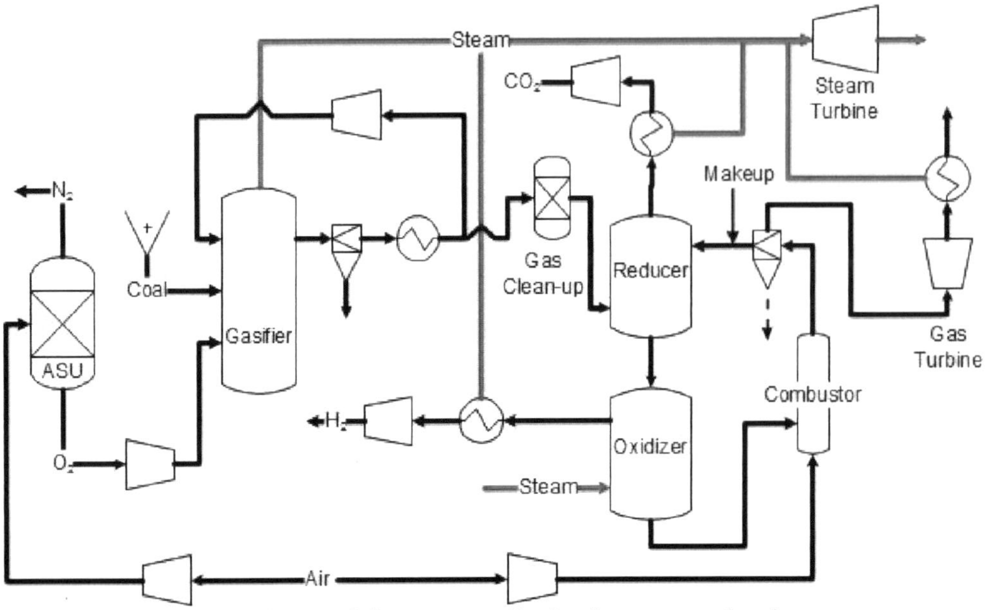

Figure 13 SCL process for hydrogen production

SCL process (shown in Figure 13) adopts chemical looping strategy to produce CO_2 and H_2 in two different reactors. An iron oxide based looping medium is first reduced by syngas and then partially regenerated with steam to produce electricity. The partially regenerated iron oxide is then further combusted in the combustor with air. The heat generated in the combustor is used to generate electricity.

The results for all the four processes are summarized in Table 2. As shown in Table 2, the SCL process has the highest thermal efficiency for hydrogen production. There is little difference between two kinds of enhanced WGS membrane reactor. Conventional WGS process gets the lowest efficiency. However, it is better than most current coal gasification-WGS processes due to the incorporation of the hot/warm gas cleanup unit.

Table 2 Summary of process simulation results

	WGS	CO_2-WGS MR	H_2-WGS MR	SCL
Thermal Input (MW)	1000	1000	1000	1000
CO2 Capture (%)	90	100	100	100
Hydrogen production (kg/hr)	14.65	15.0	15.1	15.3
Net power (MW)	59.7	62	59.8	61.2
Overall Efficiency (%)	63.71	65.32	65.42	66.58

4. Conclusions

Present work studies four advanced hydrogen production schemes using Aspen Plus for both thermodynamic analysis and process evaluation. Optimum operating conditions for water gas shift reaction have been decided based on thermodynamic restrictions. Novel models for both cocurrent and countercurrent design have been developed membrane based WGS reactor and syngas chemical looping reactors. Sensitivity analyses have also been performed. Based on preliminary process simulations using thermodynamic modeling results, the SCL process can potentially deliver the highest thermal efficiency among the advanced hydrogen production technologies. Enhanced WGS membrane technology can also improve the overall efficiency of the coal to hydrogen process.

References

1. EIA, *Annual Energy Outlook 2008*. 2008, Energy Information Administration.
2. Li, F. and L.S. Fan, *Clean Coal Conversion Processes - Progress and Challenges*. Energy and Environmental Science, 2008. **1**: p. 248 - 267.
3. Hara, S., G. Barbieri, and E. Drioli, *Limit conversion of a palladium membrane reactor using counter-current sweep gas on methane steam reforming*. Desalination FIELD Full Journal Title:Desalination, 2006. **200**(1-3): p. 708-709.
4. Marigliano, G., G. Barbieri, and E. Drioli, *Equilibrium conversion for a Pd-based membrane reactor. Dependence on the temperature and pressure*. Chem. Eng. Process. FIELD Full Journal Title:Chemical Engineering and Processing, 2003. **42**(3): p. 231-236.
5. Barbieri, G., et al., *Conversion-Temperature Diagram for a Palladium Membrane Reactor. Analysis of an Endothermic Reaction: Methane Steam Reforming*. Ind. Eng. Chem. Res. FIELD Full Journal Title:Industrial & Engineering Chemistry Research, 2001. **40**(9): p. 2017-2026.
6. Rezai, S.A.S. and Y. Traa, *Equilibrium shift in membrane reactors: A thermodynamic analysis of the dehydrogenative conversion of alkanes*. J. Membr. Sci. FIELD Full Journal Title:Journal of Membrane Science, 2008. **319**(1+2): p. 279-285.

592b Modeling of Fouling in Crude Oil Pre-Heat Trains to Reduce Energy Losses and Carbon Emissions

Francesco Coletti and Sandro Macchietto. Chemical Engineering, Imperial College London, South Kensington Campus, SW7 2AZ, London, United Kingdom

A large amount of energy in oil refineries is used to heat the crude oil in furnaces, following a pre-heat train (PHT), a network of heat exchangers which maximise energy recovery from the hot products. The crude oil flowing in the exchangers deposits unwanted layers of material (fouling) which impede the energy recovery and can obstruct the flow of oil in the pipes. This results in severely decreased thermal and hydraulic efficiencies, large production losses (due to throughput reduction and need for periodic cleaning shut-downs). Energy recovery losses in the PHT must be compensated by substantial increase in fuel burned in the furnace, with large CO2 emission. Avoiding or minimising fouling is therefore important.

Fouling deposition is not well understood, and current exchanger designs (based on empirical fouling factors) and mitigation solutions (ranging from use of chemical additives to mechanical cleaning devices) do not prevent efficiency losses. A US$ 5.4M project (CROF, for CRude Oil Fouling), involving a consortium of universities supported EPSRC and industry, is presently underway, aiming at producing a step change in understanding, data, predictive models and methods for the improved design and operation of PHT.

In this paper, a detailed dynamic, distributed mathematical model for a single tubular heat exchanger with fouling is proposed. By including local effects of process variables (temperature and velocity) and deposition mechanisms, the model accounts for interactions between fouling layer and fluid flow and reproduces well overall experimental behaviour.

Various retrofit options for individual heat exchangers are assessed and their energy and CO2 release impact established. For the future, it is envisaged to use the approach to simulate and optimise the design and operation of the overall energy recovery network, its cleaning schedules and energy efficiency strategies. Results so far indicate the return on such work can be substantial.

Web Page: www.imperial.ac.uk/crudeoilfouling

592c Modeling Multi-Stream Heat Exchangers with or without Phase Change for Simultaneous Optimization and Heat Integration

Ravindra S. Kamath[1], Ignacio E. Grossmann[2], and Lorenz T. Biegler[1]. (1) Dept of Chemical Engineering, Carnegie Mellon University, 5000 Forbes Avenue, Pittsburgh, PA 15213, (2) Center for Advanced Process Decision-making, Dept of Chemical Engineering, Carnegie Mellon University, 5000 Forbes Avenue, Pittsburgh, PA 15213

A multi-stream heat exchanger (MHEX) is a single process unit in which multiple hot and cold streams exchange heat simultaneously. MHEX are more popular in low temperature (cryogenic) applications like air separation and LNG production where the focus is on minimizing energy costs (or maximizing heat recovery from streams) using very small temperature driving forces. Unlike conventional heat exchangers which involve heat exchange between two streams, modeling MHEX is not trivial because of two reasons: a) Matches between hot and cold streams are not known *a priori* b) Since the matches are not known, it is not clear how to apply the criterion of minimum temperature driving force. Moreover, MHEX have complex hardware designs and the streams involved typically undergo phase change during heat transfer.

It is to be noted that a simulation-based process model for MHEX can have exactly one degree of freedom (e.g. temperature of one of the outlet hot or cold streams is unknown). This variable is determined by overall enthalpy balance (assuming MHEX does not use any hot or cold utilities and behaves like an adiabatic black box). However, the method to calculate the enthalpy of the corresponding stream depends on its phase (liquid, vapor or two phase) which again depends on its outlet temperature and hence is not known *a priori*. Also, there is no guarantee that the solution obtained does not involve temperature crossovers or violates the minimum temperature driving force criterion and to avoid these problems, a feasible set of input parameters needs to be specified. There are hardly any simulation or optimization based process models for MHEX available in the open literature which take care of all these issues. Optimization of flowsheets containing one or more MHEXs can be regarded as a case of simultaneous optimization and heat integration where the inlet and outlet streams conditions of MHEXs are optimized simultaneously along with the rest of the process variables in order to minimize overall cost while satisfying the constraints imposed by external process as well as feasible heat transfer constraints inherent for MHEXs.

In this paper, we propose a general nonlinear equation-oriented model for MHEX which addresses all the problems mentioned above. The proposed model for MHEX can be easily connected to models of other process units and is suitable for use in simulation and optimization of flowsheets containing MHEXs. Different models for heat integration can be incorporated depending on the objective (e.g. minimize energy cost only or simultaneously determine matches in the MHEX with or without including capital cost of MHEX).

When energy cost is the dominant factor in the objective, we propose the use of an aggregate model which is based on pinch technology for heat integration. The model equations can be considered as the solution to the following problem statement which is inverse of that used in pinch technology (Papoulias and Grossmann, 1983): *Determine feasible inlet and outlet temperatures and heat capacity flowrates for a given set of hot and cold streams such that there are no utility requirements.* As the concept of pinch ensures that the hot and cold composite curves are separated by HRAT (heat recovery approach temperature), the issues of temperature crossover and minimum temperature driving force are taken care of automatically without knowing the matches between hot and cold streams.

To deal with phase changes, we need to address the following two issues: a) procedure for enthalpy calculation changes with phase which is not known *a priori*. b) Treating heat capacity-flowrate as a constant is not a good assumption for a stream changing phase. We propose to define *a priori* a set of candidate streams which are capable of phase change. The streams belonging to this set are split into three substreams corresponding to superheated (SUP), two phase (2P) and subcooled (SUB) regions. This splitting is based on dew point and bubble point temperatures of the stream that may change during the course of the optimization as pressure and composition of the stream are treated as process variables and can be optimized. From the point of view of heat integration, each of the above substreams can be treated as an independent stream with an associated heat load and inlet and outlet temperatures. The inlet and outlet temperatures of substreams are assigned appropriate values using a disjunctive representation involving Boolean variables (Raman and Grossmann, 1994) where the Boolean variables are associated with the phase of parent stream at the inlet and outlet conditions. The disjunctions can be formulated either as a discrete-continuous model involving binary variables (Lee and Grossmann, 2000) or as a continuous model by solving an inner minimization problem with complementarity constraints (Raghunathan, 2004). Also, when a candidate stream does not change its phase, the inlet and outlet temperatures of irrelevant substreams are manipulated in such a way that the associated heat loads are set to zero. It is to be noted that this representation assumes that the enthalpy of the streams can be approximated as a piecewise linear function of temperature in each of three regions. If necessary, the two phase region can be split further into more segments to improve this approximation Isothermal streams can also be handled easily through an implementation strategy that decouples temperature information used for process calculations from that used for heat integration.

The capability of model is demonstrated using the PRICO (Poly Refrigerant Integrated Cycle Operations) process for LNG production (Lee et al., 2002). Although the PRICO process is a relatively simple process involving a single mixed refrigerant circulating in a simple cycle, it incorporates most of the components present in more complex LNG liquefaction processes. The proposed model is used within a mathematical programming formulation to determine the optimal operating conditions and composition of mixed refrigerant that minimizes the shaft work required for vapor compression.

References

Papoulias, S. A. and Grossmann, I. E. (1983) A Structural Optimization Approach in Proceas Synthesis. II: Heat Recovery Networks, *Computers and Chemical Engineering*, 7, 707-721.

Raman, R. and Grossmann, I. E. (1994). Modeling and computational techniques for logic based integer programming. *Computers and Chemical Engineering*, 18(7), 563-578.

Sangbum L. and Grossmann I. E. (2000) New algorithms for nonlinear generalized disjunctive programming. *Computers and Chemical Engineering*, 24, 2125-2141.

Raghunathan, A. (2004) PhD Thesis, Department of Chemical Engineering, Carnegie Mellon University.

Lee, G. C., Smith, R. and Zhu, X. X. (2002) Optimal Synthesis of Mixed Refrigerant Systems for Low-Temperature Processes. *Industrial and Engineering Chemistry Research, 41, 5016-5028.*

592d Mixed-Integer Bi-Objective Optimization Algorithm for Solving Energy Supply Chain Problems

Ali Ozturk and Metin Turkay. Industrial Engineering, Koc University, RUMELIFENERI YOLU, SARIYER, Istanbul, 34450, Turkey

Energy supply chains consist of raw materials, production facilities and demands for end products, i.e. energy. According to the Energy Information Administration (EIA)'s Report, over one-third of all primary energy consumption goes into producing and delivering electricity and most of the world's primary energy consumption comes from fossil fuels, such as coal, natural gas and oil. Burning fossil fuels release emissions that are harmful to the environment. These emissions can be classified into two main groups: SOx and CO2 equivalent emissions. SOx equivalent emissions are the particles that are quantified strictly and limited with certain regulations. The case with CO2 equivalent emissions is different; these are gases that have effect on global warming, i.e. greenhouse gases (GHG).

Due to increasing environmental impact and the central role of energy in every industrial activity, the improvement of economical and environmental performance of energy production systems continues to be a major issue. The Kyoto Protocol demands for reductions in greenhouse gas emissions by the industrialized countries. The energy sector will be seriously affected by the Kyoto Protocol since it is one of the leading industries that release harmful substances to the environment. Therefore, the energy production companies and the researchers are in the search for recipe solutions for reducing emissions without sacrificing from the amount of energy production and low prices of production.

In a typical energy production optimization model, the objective is to minimize production cost while satisfying energy requirements with given operational constraints. To extend the model, we introduce a new objective function aimed to minimize GHG emissions as well as production cost. It is essential to include the option to use renewable energy technologies in the existing energy production systems in order to improve environmental performance. For that purpose, bio diesel is added to the model with its own limitations and constraints. The last extension we introduced to our model is the carbon capturing and sequestration (CCS) system. The CCS technology involves capturing CO2 from the gas streams generated by burning fossil fuel and then injecting it underground. The CCS system is not modeled as a new detailed process unit in the system, its existence is modeled by new constraints and its operating cost is model as an energy penalty. As a result, the extended model is an MILP problem with two objectives.

Due to the conflicting nature of these two objectives, it is not possible to find a solution that optimizes both objectives simultaneously. Therefore, the main objective in this bi-objective problem is to find efficient or Pareto optimal solutions. The Pareto optimal is the solution where none of the objective functions can be improved without worsening at least one of the others. In order to solve this challenging problem, a mixed-integer bi-objective optimization algorithm is developed. The developed algorithm is designed to present all efficient solutions to the decision maker and it is based on the combination of Tchebycheff scalarizing and weighted sum methods. In this algorithm, the bi-objective problem is converted into single objective problem and efficient solutions are obtained by solving single objective problem iteratively.

In this work, detailed models and suggestions are developed for energy production systems within emission reduction and cost minimization objectives by using different analysis and modeling techniques. A model is developed for future emission reduction targets of energy production companies. As an advance modeling technique, a new objective function aimed to minimize GHG emissions as well as production cost is introduced. These two contradicting objectives are solved by the developed mixed-integer bi-objective algorithm. Efficient solutions that can be used as production planning tools by decision makers are obtained. In order to improve environmental performance, bio diesel and CCS system are added to the model with their own limitations and constraints. The results showed that up to a 26% reduction of CO2 emission can be achieved by fuel balancing strategy which leads to the 35% increase in production cost. When bio diesel is added to the model, up to a 30% reduction of CO2 emission can be achieved with the same amount of increase in production cost. When both bio diesel and CCS system are added to the model, 90% reduction of CO2 emission is achieved which results in 55% increase in production cost.

Minimum Total Annualized Cost for a Heat Exchange Network using the IDEAS Framework

Jorge Pena-Lopez and Vasilios Manousiouthakis,*
Chemical and Biomolecular Engineering Department,
Hydrogen Engineering Research Consortium,
University of California, Los Angeles, CA 90095, USA
jpenalopez@ucla.edu, vasilios@ucla.edu

* to whom correspondence should be addressed

A classic problem that has occupied the area of process engineering for over four decades is the synthesis of heat exchanger networks (HEN). Several approaches have been employed to address this problem. They can be generally categorized into sequential synthesis[1-4] or simultaneous synthesis methods[5-9] and are often based on evolutionary or optimization principles. Neither of these method categories are able to guarantee global optimality; the former because they are based on ad-hoc rules and the latter because they are based on nonlinear programming techniques that can only guarantee local optimality.

The *Infinite DimEnsionAl State-space* (IDEAS) method is a novel conceptual framework for the synthesis of general process networks. With this approach, all possible process networks are considered and the resulting mathematical formulation gives rise to infinite dimensional optimization problems with feasible regions defined by linear constraints thus guaranteeing global optimality of the obtained solution[10].

In this work we discuss the application of the *MultiFamily IDEAS* based computer software developed at UCLA to a problem first proposed by Lee, Masso and Rudd[1]: 4sp1. This is a classical case study for HEN synthesis and one of the milestone tests for HEN synthesis methods. The data for this particular problem are summarized in Table 1, since they are often misquoted in the literature.

Table 1. 4sp1 Design Problem Data[1]

Stream No.	Flow Rate [lb/h]	Input Temp. [°F]	Output Temp. [°F]	Heat Capacity [BTU/lb/°F]
1 (C1)	20,643	140	320	0.70
2 (H1)	27,278	320	200	0.60
3 (C2)	23,060	240	500	0.50
4 (H2)	25,000	480	280	0.80

Saturated steam pressure	962.5	psia
Cooling water temperature	100	°F
Maximum water output temperature	180	°F
ΔT_{min} ; Heat Transfer Coefficient		
Heat exchanger, ΔT_{min} ; U_{HE}	20 ; 150	°F ; BTU/h/ft^2/°F
Water cooler, ΔT_{min} ; U_C	20 ; 150	°F ; BTU/h/ft^2/°F
Steam heater, ΔT_{min} ; U_H	25 ; 200	°F ; BTU/h/ft^2/°F
Equipment downtime	260	h/y
Heat exchanger cost parameters α,b	350, 0.6	
Cooling water cost, C_w	5×10^{-5}	$/lb
Steam cost, C_s	1×10^{-3}	$/lb
Annualizing factor, φ	0.1	1/y
Heat exchanger cost function, E	$E\,[\$] = \alpha A^b,\ A \equiv [ft^2]$	
Total Annualized Cost[a]	$TAC\,[\$/y] = \varphi\,E + CU + HU$	

[a] CU ≡ Cold Utility cost, HU ≡ Hot Utility cost

The Rudd and co-worker approach to solution of the HEN synthesis problem[1] consisted of a sequential algorithm that considered all cold stream/ hot stream matches that met thermodynamic/design constraints. For each such feasible match either one or both of the streams involved were exhausted and the match's contribution to the overall cost was evaluated. When a stream could not be matched, it was sent to the corresponding utility. After all the matches were completed, an algorithm was followed abiding to established heuristic guidelines in order to generate the feasible network with the minimal cost. Since the maximum number of possible process networks that could be generated using this methodology for this particular problem was 4200, the authors employed a brand and bound methodology to simplify the computational burden. Using a constraint of ΔTmin = 20 °F, and the total annualized cost function defined in Table 1., the optimal network identified had a TAC of $13,481. This network is presented in **Error! Reference source not found.**.

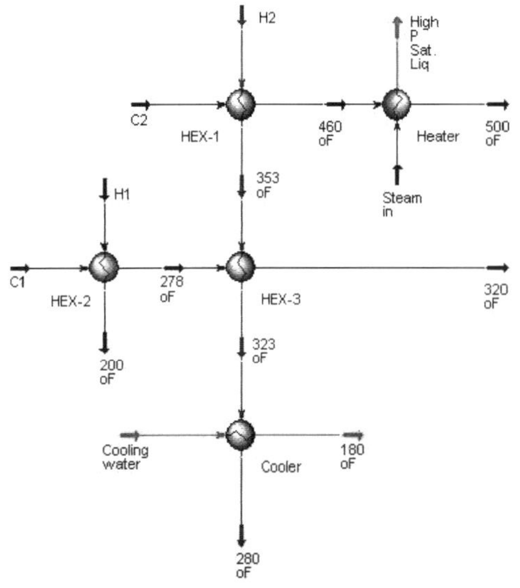

Figure 1. Network for 4sp1 Problem[1] simulated in UniSim Design.

The capital cost of a heat exchanger can be expressed not only using the cost function employed by Lee, Masso, and Rudd, but can also be expressed by a function with linear dependence on the heat exchanger's area and an incremental cost corresponding to an infinitesimally small heat exchanger area. . Figure 2 illustrates how both these cost functions may be realistic representations of a heat exchanger's capital cost.

The incremental cost function is defined as follows:

$$c + \hat{c}A = Inc \tag{1}$$

where *Inc* is the incremental cost function, c is the incremental cost, $\hat{c}$ is the area cost coefficient and A is the heat exchanger area. The obtained values for c and $\hat{c}$ are 2,632.2 and 25.965 respectively.

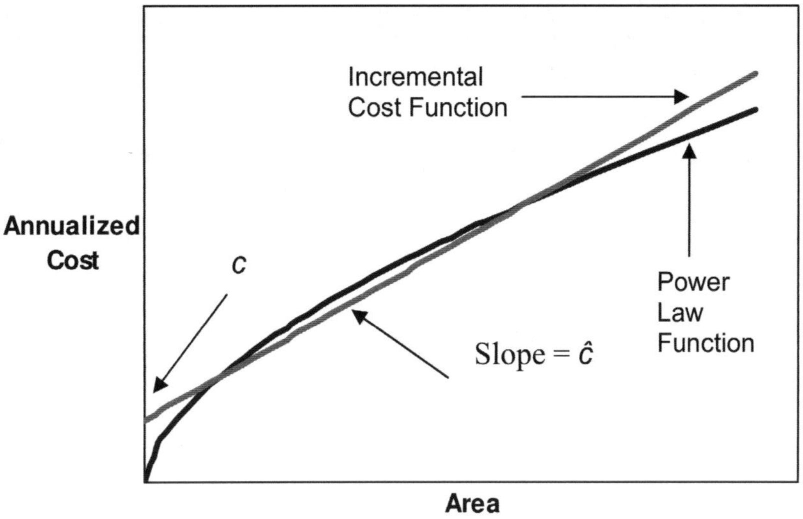

Figure 2. Incremental and power law capital cost functions for a heat exchanger

Using the above incremental cost function, a ΔT_{min} = 20°F and the Lee, Masso, and Rudd synthesis method gives rise to exactly the same optimal HEN and an optimum objective function value that is also equal to $13,481. The 4sp1 problem has been solved by numerous previous investigators. Bagajewicz et al.[11] using the power law cost function, a ΔT_{min} = 0°F and the state-space approach identified a network whose TAC cost (using the original Lee, Masso, and Rudd data) is $10,457 and ΔT_{min} = 1.3°F. The 4sp1 problem has also been approached before by Wilson[12] using the IDEAS framework, the power law objective function, and a ΔT_{min} = 0°F. The identified optimal network has a TAC cost (using the original Lee, Masso, and Rudd data) of $10,453 and a ΔT_{min} = 1.3°F. For this approach the optimization was performed in the ΔT_{min} space as well and yield a network with a cost of $10,453 and a ΔT_{min} = 1.3°F. The same network structure optimized with the incremental cost objective function results in a cost of $10,828 and a ΔT_{min} = 2.3°F. The two networks from references 11 and 12 are presented in Figure 3b, 3a respectively.

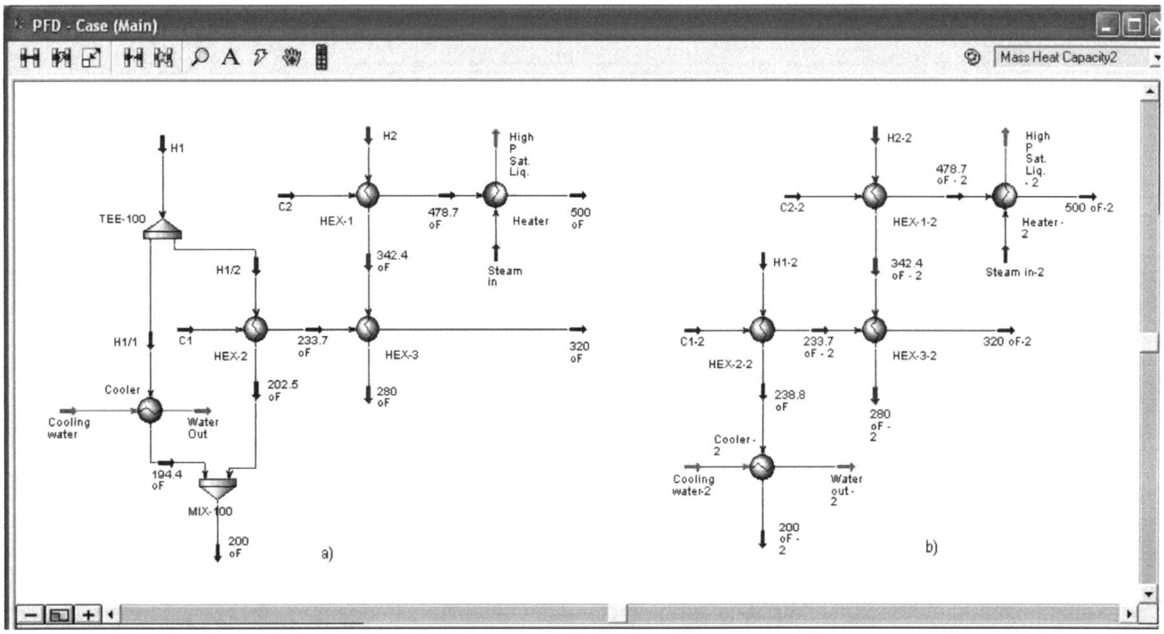

Figure 3. Optimal networks for problem 4sp1. a)IDEAS approach. b) State-space approach.

Multifamily IDEAS automatically generates HENs once the following conditions are specified: definition of fluids used, operating pressures, high and low temperature ranges, heat transfer coefficients, minimum/maximum in/out flow rates (and/or flow cost), inlet and outlet temperatures, heat exchanger cost parameters (slope and ordinate intercept) and ΔT_{min}. In addition to this, an interval discretization must be provided in order for the software to approximate the infinite generation of possible operating units. A large number of discretization points provides a better approximation but it also requires a larger amount of memory allocation and computational time.

To test MF IDEAS the original Lee, Masso, and Rudd 4sp1 data are used with the incremental cost function and ΔT_{min} = 20°F. Using a conventional PC and a coarse interval discretization (6 intervals), the network illustrated in Figure 4 was automatically synthesized by MF-IDEAS in 695.14 CPU seconds. The identified network features a TAC of $15,405 and a ΔT_{min} = 20°F and is predicted to keep improving with a more refined interval discretization.

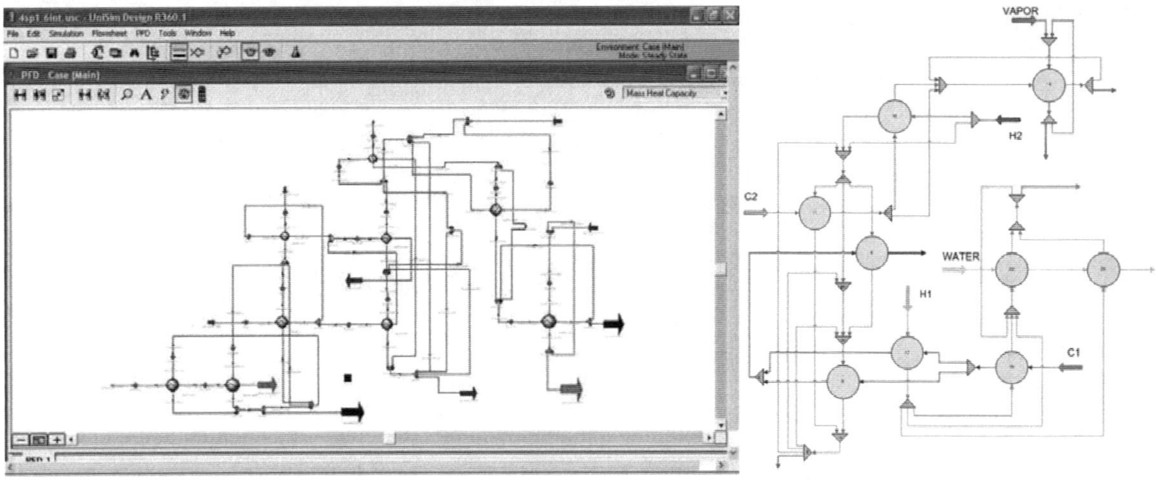

Figure 4. IDEAS-generated flowsheet for 4sp1.

Close examination of the obtained network reveals a significant number of heat exchange, splitting and mixing units. This is attributable to the coarse grid discretization, since the software will split and mix stream flowrates so as to recreate exactly the temperatures corresponding to the employed interval discretization. Increasing the number of intervals provides the software with a better 'pool' of matching temperatures and ultimately, as the interval size becomes small, will be able to generate the flowsheet with global optimality (3a). Further work is being developed to carry out MF-IDEAS HENS with more powerful computational resources and the implementation of parallel distribution of the operations to improve the performance and operational time of the software and hence obtain a better approximation of the optimal network.

References:

[1] Lee, K. F.; Masso, A. H.; Rudd, D. F. Branch and Bound Synthesis of Integrated Process Designs. *Ind. Eng. Chem. Fundam.* **1970**, *9* (1), 48-58.
[2] Linnhoff, B.; Hindmarsh, E. The Pinch Design Method for Heat Exchanger Networks. *Chem. Eng. Sci.* **1983**, *38* (5), 745-763.
[3] Ahmad, S.; Linnhoff, B.; Smith, R. Cost Optimum Heat Exchanger Networks-2. Targets and Design for Detailed Capital Cost Models. *Comput. Chem. Eng.* **1990**, *14* (7), 751-767.
[4] Linnhoff, B. Pinch Analysis-A State-of-the-Art Overview. *Chem. Eng. Res. Des.* **1993**, *71* (A), 503-522.
[5] Papoulias, S. A.; Grossmann, I. E. A Structural Optimization Approach in Process Synthesis-II. Heat Recovery Networks. *Comput. Chem. Eng.* **1983**, *7* (6), 707-721.
[6] Yuan, X.; Pibouleau, L.; Domenech, S. Experiments in Process Synthesis via Mixed-Integer Programming. *Chem. Eng. Process.* **1989**, *25* (2), 99-116.
[7] Yee, T. F.; Grossmann, I. E. Simultaneous Optimization Models for Heat Integration-II. Heat Exchanger Network Synthesis. *Comput. Chem. Eng.* **1990**, *14* (10), 1165-1184.
[8] Daichendt, M. M.; Grossmann, I. E. A Preliminary Screening Procedure for MINLP Heat Exchanger Network Synthesis Using Aggregated Models. *Chem. Eng. Res. Des.* **1994**, *72* (A), 357-363.
[9] Furman, K.C.; Sahinidis, N.V. A critical review and annotated bibliography for heat exchanger network synthesis in the 20th century, *Ind. and Eng. Chem. Res.* **2002**, *41* (10), 2335–2370.
[10] Wilson, S.; Manousiouthakis, V.I. IDEAS Approach to Process Network Synthesis: Application to Multicomponent MEN. *AIChE journal*, **2000**, *46* (12), 2408-2416.
[11] Bagajewicz, M.J.; Pham, R.; Manousiouthakis, V.I. On the state space approach to mass/heat exchanger network design. *Chem. Eng. Sci.*, **1998**, *53*(14), 2595-2621.
[12] Wilson, S. D. Methods for globally optimal mass and heat exchange network synthesis. *Thesis (Ph. D.)-- University of California*, Los Angeles, **2001**.

592f Design of Hydrogen-on-Demand System from a Sodium Borohydride Hydrolysis Reaction

Ai-Jen Hung[1], Cheng-Ching Yu[1], Yih-Hang Chen Sr.[2], Shing-Fen Tsai[3], Ya-Yi Hsu[3], and Jie-Ren Ku[3]. (1) Chemical Engineering, National Taiwan University, Taipei, 106-17, Taiwan, (2) Process Engineering, CTCI Corporation, Taipei, 106, Taiwan, (3) Energy and Environment Research Laboratories, Industrial Technology Research Institute, Hsinchu, 310, Taiwan

Hydrogen has become one of the most promising future energy resources due to concerns about global warming and the depletion of fossil fuels. Hydrogen generation from the hydrolysis reaction of an alkaline sodium borohydride solution ($NaBH_4$) has drawn much attention due to its theoretically high hydrogen storage capacity (10.8 wt%). Because hydrogen can only be generated when an alkaline $NaBH_4$ solution contacts with a catalyst, it is easy to control the generation rate. Therefore, it is preferable to apply the hydrolysis reaction as a hydrogen-on-demand system for proton exchange membrane (PEM) fuel cell applications. Some research has been carried out, for example, Millennium Cell Inc. has developed a 1.2 kW hydrogen-on-demand™ prototype using NaBH4 [1]. Italian National Agency for New Technologies (ENEA) has designed a 6 Wh prototype for cellular phones via $NaBH_4$ hydrolysis reaction [2]. Korean Fuel Cell Research Center has also developed a 2 W PEM fuel cell to power a cellular phone [3]. The objective of this work is to design a procedure to supply the optimum $NaBH_4$ feed flowrate needed to generate sufficient hydrogen in response to process demands. In chemical engineering nomenclature, this is equivalent to design a semi-batch reactor to meet the production demand.

Based on the first-order kinetics and isothermal condition, an analytical expression between the feed and effluent of a semibatch reactor can be derived. The relationship between the input NaBH4 flowrate and the output hydrogen generation is simply a first-order transfer function, where the time constant is the inverse of the reaction rate constant and the process gain follows the reaction stoichiometry, 4. Once the process characteristic of the hydrolysis reaction is determined, the $NaBH_4$ feed flowrate can be determined analytically as the combination of the impulse and step functions to obtain a constant hydrogen generation rate as shown in Fig. 1.

In the implementation phase, two scenarios are explored. One is to design a disposable hydrogen-on-demand system and the other is to design a hydrogen-on-demand system for variable power requirements. For the disposable design, a given amount of NaBH4 solution (which corresponds to the area of the impulse function) is initially placed in the reaction chamber. The reaction is initiated by adding the catalyst, followed by a constant NaBH4 feed flowrate (which corresponds to the step function). As for different power changes, the process inputs also adjusted to meet the demand. Figure 2 gives the response of the hydrogen generation rate according to this scenario. Finally, an experiment is setup to validate the design.

Keywords: sodium borohydride, hydrolysis reaction, hydrogen-on-demand

(A)

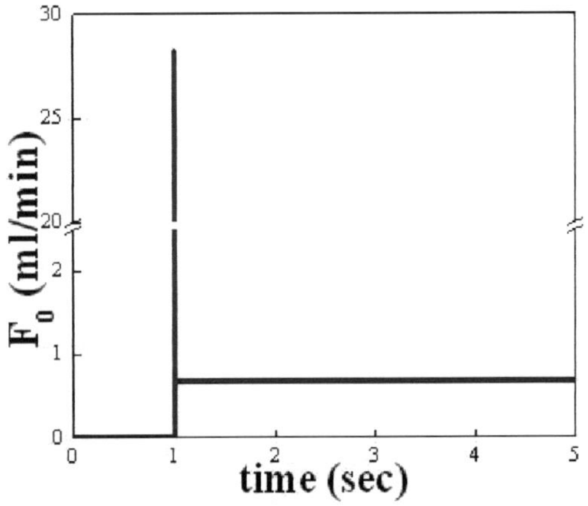

(B)

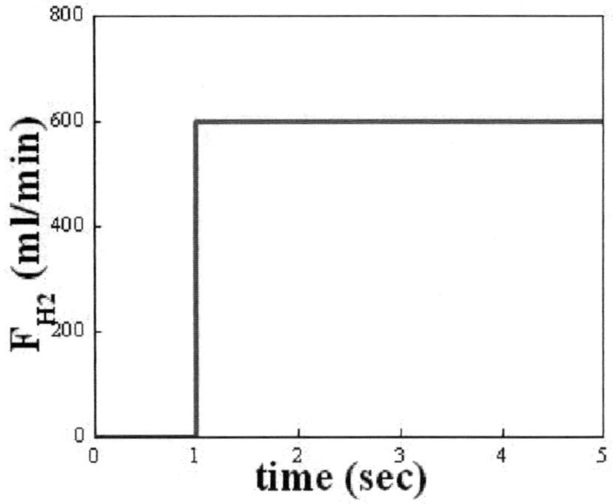

Fig. 1 (A) Combination of impulse and step functions for NaBH4 feed flowrate (B) response of hydrogen generation rate to (A)

(A)

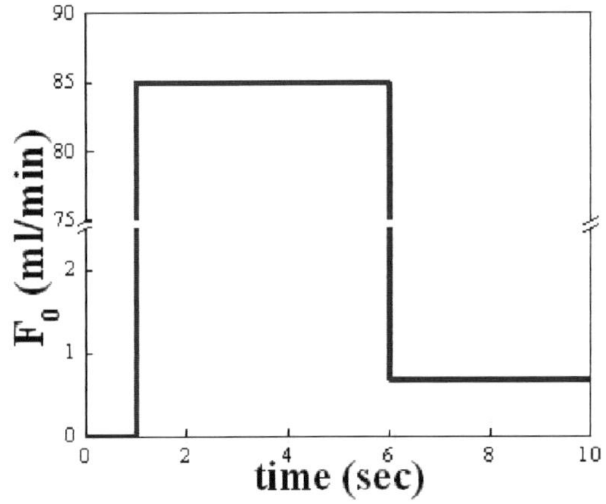

(B)

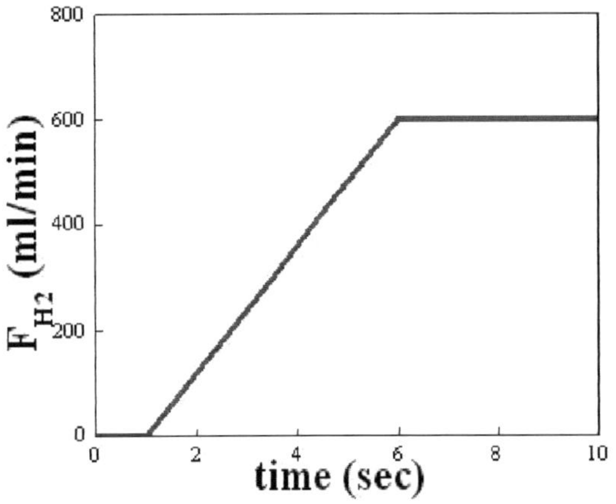

Fig. 2 (A) Combination of pulse and step functions for NaBH4 feed flowrate (B)

response of hydrogen generation rates to (A)

Reference

1. http://gcep.stanford.edu/pdfs/hydrogen_workshop/Wu.pdf

2. http://www.ichet.org/ihec2005/files/manuscripts/Prosini%20P.1-Italy.pdf

3. S. U. Jeong, R. K. Kim, E. A. Cho, H. J. Kim, S. W. Nam, I. H. Oh, S. A. Hong, S. H. Kim, A study on hydrogen generation from NaBH4 solution using the high-performance Co-B catalyst. J. Power Sources, 144(1) (2005) 129-134.

592g Detailed Steady State Modeling of An Anode-Supported Solid Oxide Fuel Cell (SOFC)

Sang Youp Chae[1], Debangsu Bhattacharyya[1], and Raghunathan Rengaswamy[2]. (1) Chemical Engg. Dept., Clarkson University, 8 Clarkson Avenue, Potsdam, NY 13699, (2) Chemical Engineering, Clarkson University, Box 5705, Clarkson University, Potsdam, NY 13699-5705

High reactivity, comparatively inexpensive anode and cathode material, high power density, high level of usable heat and the possibility of using flexible fuels have made SOFC an attractive choice for utility and industrial applications. A detailed steady state model of a SOFC can be a useful tool to evaluate the effects of the operating and design parameters and thus can be used to design new cells. The model can also be used for optimization studies of the cell dimensional and operating parameters. In this talk, the development of an optimization-oriented detailed steady state model will be presented.

To attain this goal, a detailed two dimensional dynamic model for an anode-supported tubular SOFC is developed. The model includes: (i) mass and momentum transport phenomena in the anode and cathode gas flow channels for the reactants and the products, (ii) diffusion from the gas flow channels through the porous electrodes to the reaction sites, (iii) activation overpotential through the Butler-Volmer equation with concentration and temperature dependent expression for exchange current density, and (iv) ohmic resistances. The ohmic resistances of the electrodes are modeled by an equivalent circuit model considering the current path length from the current collectors to the TPB. The enhanced area for electrochemical reactions, supported by experimental evidence, is considered by a cylindrical characterization of the TPB interface. For this industrial cell, the cell temperature is maintained by a furnace. However spatial variation in temperature cannot be ruled out. To consider the effects of these variations, energy conservation equations are written. Due to the unknown heat input from the furnace, consideration of the overall energy balance becomes a challenging task. A MAPLE-MATLAB environment is used to solve the steady state model. The computational issues in solving the model will be discussed. The model is validated using data from a commercial SOFC over a wide range of cell temperatures, reactant flow rates, and DC polarizations. Significant observations made during the validation process will also be presented.

592h Modeling and Optimization of Polymer Electrolyte Fuel Cells

Parag Jain, Lorenz T. Biegler, and Myung S. Jhon. Chemical Engineering, Carnegie Mellon University, 5000 Forbes Ave., Pittsburgh, PA 15213

We develop a generalized modeling and optimization framework that includes rigorous computational fluid dynamics and catalytic reaction phenomena, suitable for analysis of polymer electrolyte fuel cell (PEFC) systems.

Very recently [1], we constructed a multi-dimensional, multi-physics PEFC model that accounted for the transport processes within the gas channels (GC), gas diffusion layer (GDL), and the polymer electrolyte membrane (PEM). These set of partial differential algebraic equations were linked to our in house state-of-the-art interior point optimization algorithm, IPOPT [2] to solve challenging parameter estimation problems possessing multiple experimental operating conditions.

In this paper, we improved our previous integrated PEFC model substantially by incorporating a cathode catalyst layer (CL) model in depth. We adopted an agglomerate CL model, and recasted it to a compact form that is easily convertible to a pseudo-homogenous CL model, for optimization. As a consequence, our new PEFC model captures the essential transport processes and accounts for species, proton, and electron transport within the GC/GDL/PEM as well as reaction/transport processes within the CL. The resulting nonlinear partial differential algebraic equations are fully discretized using a finite difference method and the nonlinear program is linked to IPOPT.

Before solving integrated PEFC model along with optimization, we analyzed our CL model. In our preliminary study, we performed Platinum (Pt) minimization for a specified cell performance as well as, obtained optimal Pt distribution along CL width, which is obtained by solving multi-zone partial differential equation-constrained optimization problem. In this preliminary work, CL was coupled with nearby regions, e.g., GDL and PEM. We intend to optimize the performance of fully integrated PEFC model within both CL and PEM regions. Specifically we will examine Pt and Nafion distributions within CL, and also obtain optimal chemical structures for PEM.

Our proposed framework provides a robust and fast solution methodology, and is general and straightforward for modeling extensions, for addressing other critical issues in the fuel cell technology that require large-scale optimization methodologies.

References:

1. Jain P, Biegler LT, Jhon MS. Parametric Study and Estimation in CFD-based PEM Fuel Cell Models. AICHE J, to appear.

2. Wächter A, Biegler LT. On the implementation of an interior-point filter line-search algorithm for large-scale nonlinear programming. Math Program. 2006;106:25-57.

592i Rib-Channel Optimization for Flowfield Design of PEM Fuel Cells

Shivakumar Kameswaran[1], Arun Pandy[2], Mallika Gummalla[2], and Robert Darling[3]. (1) United Technologies Research Center, East Hartford, CT 06042, (2) Systems Department, United Technologies Research Center, 411 Silver Lane, MS 129-15, East Hartford, CT 06108, (3) UTC Power, South Windsor, CT 06074

Channel design affects ohmic losses, pressure drop and transport losses associated with PEM fuel cells. The objective of this work is to optimally design flow-fields using a combination of modeling tools and limited experiments to improve performance. More specifically, we focus on optimizing rib-channel dimensions and for this purpose we have developed a detailed and distributed steady-state model of a small section of a PEM fuel cell. Two-Dimensional gas and charge transport equations are solved simultaneously, while water transport is modeled using a response-surface obtained from higher fidelity simulations. Limited experimental data are used to tune the model. The model is then employed to determine optimal-rib channel dimensions subject to machining constraints. We present results for different gas-diffusion layer thicknesses and properties. We also demonstrate that it is possible to obtain improvements in power density by optimizing rib-channel dimensions.

602a Bacterial Aerosol Neutralization by Shock Waves Using a Novel Impactor: Design, Computation and Experiments

Patrick R. Sislian[1], Lutz Madler[2], David Pham[1], Mingheng Li[3], Xinyu Zhangh[1], and Panagiotis D. Christofides[4]. (1) Chemical and Biomolecular Engineering, University of California, Los Angeles, 450 Hilgard Ave, Los Angeles, CA 90024, (2) Production Engineering, IWT Foundation Institute of Materials Science, University of Bremen, Badgasteiner Str. 3, 28359 Bremen, Germany, (3) Department of Chemical and Materials Engineering, California State Polytechnic University, 3801 West Temple Ave., Pomona, CA 91768, (4) Department of Chemical and Biomolecular Engineering, Department of Electrical Engineering, University of California, Los Angeles, Los Angeles, CA 90095

Neutralization of bacterial aerosol releases is critical in countering bio-terrorism. As a possible method that avoids the use of chemicals, we investigate the mechanical instabilities of the cell envelope in air as the bacteria pass through aerodynamic shocks. A novel impactor (the device consists of a converging nozzle through which aerosol flows perpendicular to the collection surface) is designed to simultaneously create a controlled and measured shock and to collect the bacteria. Both experimental measurements and computational fluid dynamics simulations are used to characterize the shock wave created. Using the impactor dimensions corresponding to our experimental system we could achieve a shock Mach number of 2.36. The shock presents a discontinuity in the fluid properties (i.e. velocity, pressure, temperature). Due to its inertia, the bacterial particle can not follow the changes in velocity, pressure and temperature. The bacterium therefore experiences strong deceleration as it passes through the shock. Rayleigh-Taylor instabilities are created due to this deceleration on the bacterium surface. The bacterial envelope becomes unstable at a critical deceleration which depends on the surface tension, density and diameter of the bacterium. Spores have a higher surface tension than vegetative cells, and hence, they are more difficult to damage mechanically. The particle motion through the impactor is calculated based on the gas flow simulations. The data indicates that the bacterium (assumed to be spherical of dp=1 micron) experiences decelerations of up to 4.4×10^{9} m/s^2 in the novel impactor system. *E. coli*, a vegetative gram-negative bacterium, requires an acceleration of 3.0×10^{8}m/s^2 and we therefore predict that the novel impactor is capable of neutralizing this aerosol. On the other hand *M. hugatei*, an archaea, which we assume has properties close to *B. atropheus*, a spore forming gram-positive bacterium, requires an acceleration of 1.0×10^{12} m/s^2 which can not be achieved by the current setup. We therefore predict that the impactor as currently designed will not be able to neutralize spores. The computational predictions are corroborated by experimental studies. Specifically,the experimental results indicate that spores of *B. atropheus* retain their viability for shocks with a Mach number of 2.36. Vegetative cells of *B. atropheus*, however, do not retain their viability after passing through the system both under shock and non-shock conditions. The nebulization of both types of cells is tested with the same dye and we conclude that this step has no effect on the viability of the cells.

602b Prediction of the Focal Sites of Tissue Injury Caused by Reactive Gas Uptake In the Respiratory Tract

Banafsheh Keshavarzi, The Pennsylvania State University, Department of Chemical Engineering, University Park, PA 16802, James Ultman, Chemical Engineering, The Pennsylvania State University, University Park, PA 16802, and Ali Borhan, Chemical Engineering, Pennsylvania State University, 122 Fenske Laboratory, University Park, PA 16802.

Inhalation of reactive gases such as ozone results in a reproducible pattern of tissue injury in the respiratory tract which is believed to depend on the local dose delivered to the walls. To predict the focal sites of tissue injury, we perform numerical simulations of reactive gas transport and uptake in anatomically-accurate models of the respiratory tract. The model geometry is created through three-dimensional reconstruction of MRI images of the respiratory tract, including the nasal passages, larynx and tracheobronchial tree. Three-dimensional flow and concentration distributions within the resulting structure are subsequently obtained through numerical solution of the Navier-Stokes, continuity, and species convection-diffusion equations. A quasi-steady diffusion-reaction model is used to account for the interaction between the reactive gas and endogenous substrates in the respiratory tract lining fluid (RTLF). The total rate of reactive gas uptake within different sections of the respiratory tract is determined, and hot spots of reactive gas flux on the respiratory tract wall are identified. For steady inspiratory flow under quiet breathing conditions, high flux regions are predicted to appear at the nasal atrium, middle turbinates, inferior meatus, laryngopharynx, rima glottidis, and downstream of bifurcations. These predictions are qualitatively consistent with experimental observations of formaldehyde-induced nasal lesions in rhesus monkeys [Monticello et al. 1989], and focal sites of epithelial damage in rat lungs exposed to ozone [Postlethwait et al., 2000].

Eulerian Particle Tracking In Bifurcating Biofluids

Clint W Wininger, Arizona State University, Tempe, AZ
Jeff J Heys, Montana State University, Bozeman, MT

Abstract

Biofluid flows in bifurcating geometries has become paramount to the medical community. When addressing inhaled particulate matter in the lungs, it is essential to have an efficient computational model. High-order elements offer the opportunity to model nanoparticle deposition in a case by case basis with moderate computer processor times. This adds insight into cancer precipitation as well as an advantageous route to designing patient specific therapeutic aerosols. This study focuses on developing high-order methods for particle tracking in the weakly turbulent flows that are common in the biological setting. High-order elements are used for direct numerical simulation in the incompressible fluid flow to fully resolve the transporting field. The advection-diffusion is then solved with similar methods to predict a deposition of passive and dilute particles. Eulerian-Eulerian particle tracking has shown a route to developing drug delivery on an individual basis with moderate computing times. The quick adaptation and interpolation between meshes offers a promising perspective that lends towards the utilization in the medical community. The scientific code, Nek5000, and the computer cluster at Arizona State University is used to demonstrate the effectiveness of the field method. Experiments at University of Arizona lend validation to the models. Extensions to Eulerian-Eulerian particle tracking methods with variable approximation orders are explored.

Introduction

Particle deposition has been explored experimentally and numerically since it represents an important problem encountered in various everyday situations. A particular problem of interest is deposition of particles in the human airway. Particles are inhaled through the extrathoracic and tracheobronchial airways and then are transported into the lower lungs. The airways are coated with moist mucus which provides a favorable surface for deposition and interaction with pulmonary tissue [1]. The inhaled particles can range from ultrafine with a diameter less than1 μm to dust particles with a diameter of several hundred microns. Many of the inhaled particles, like second-hand cigarette smoke, can be toxic when they interact with the pulmonary tissue and lead to lung injuries [2]. Several studies have also indicated that the ultrafine particles may be more toxic than their larger counterparts of the same material [3, 4]. Thus, it is important to be able to accurately model the dynamics of nano-particles in complex geometries in order to properly assess the deposition location of the toxic elements as well as efficiently design therapeutic aerosols to target the injured region. Recent work has been completed in modeling nano-particles in the human airway by Choi et al. [5] using a two bifurcation realistic geometry and Zhang et al [6] with an idealized three bifurcation geometry and oral airway. Prior studies have used turbulence models for the fluid velocity field. This study focuses on using high-order spectral elements to fully resolve the velocity field and solve the convection-diffusion equations. In a domain of E spectral elements each with polynomials of degree N, the computational workload is on order EN^4 for EN degrees of freedom, and an exponential convergence rate. The high-order methods are also able to resolve a larger fraction of the resolvable modes for the resolution [7]. These benefits lead to the opportunity resolve the velocity field through direct numeric simulations can provide a more accurate field in a complex geometry of bifurcating tubes as well as novel Eulerian-Eulerian particle tracking using variable approximation orders.

Theory

Governing Equations

The use of high-order spectral methods allows for direct numerical simulation to fully resolve the velocity field. The velocity field is calculated from the unsteady three-dimensional incompressible Navier-Stokes equations:

$$\frac{D\mathbf{v}}{Dt} = -\nabla p + (\mathrm{Re})^{-1}\Delta\mathbf{v} + \mathbf{f}, \quad \nabla\cdot\mathbf{v} = 0 \tag{1}$$

The **v** is a three-dimensional velocity vector, Re is the Reynolds number, *t* is the nondimensional time, *p* is a scaled pressure term, and **f** is a forcing function that is set to zero for this problem. The *div* **v** = 0 is the continuity constraint that enforces the incompressibility. After solving for the velocity field, the passive scalar solution that represents the particle concentration is obtained with the convection-diffusion equation:

$$\frac{\partial\mathbf{C}}{\partial t} + \mathbf{v}\cdot\nabla\mathbf{C} = (Di)\Delta\mathbf{v} \tag{2}$$

C is a scalar concentration field and *Di* is the diffusion coefficient that is assumed to be uniform throughout the domain. The particle concentration is very dilute and therefore can be treated as a passive scalar that does not affect the velocity field. Hence, the convection-diffusion equation can be treated as post-processing step to the fluid solution, and it can be solved multiple times for varying values of *Di* after the velocity field is known. Each different diffusion coefficient represents a particle with a different diameter.

Geometry and Boundary Conditions

For this work, multiple meshes were generated of the same geometry using the in-house program, Bifmesh. The geometry in figure 1 is based on the G0-G1 generations of Weibel's model A of the human lungs as used by Kleinsteuer et al. [8, 9]. The simulations are run at several different approximation orders to examine how the results converge towards a well-resolved solution of the fluid field and particle concentration, which used 1968 elements and 11th-order basis functions (~2.6 million degrees of freedom).

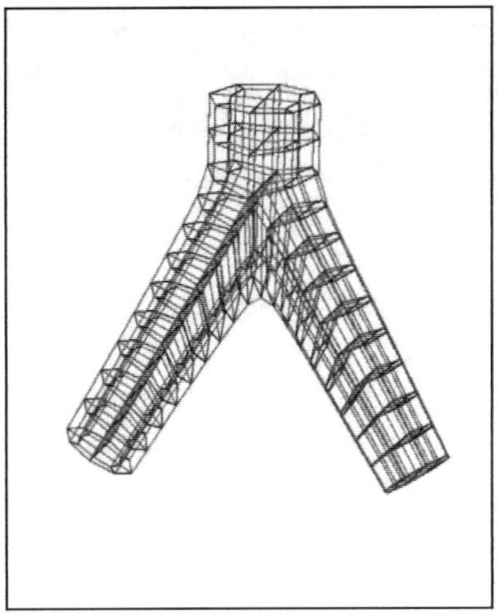

Figure 1: Single bifurcation mesh with 540 elements

The model parameters were set to approximately represent the transport of nanoparticles in the human airways. A Reynolds number of 3000 is reasonable for flow in human airways and is weakly turbulent. The diffusion coefficients range from 1.0×10^{-4} to 5.0×10^{-6} and are representative of small particles in the 1 to 10 nanometer range [6]. The inlet is defined at the top of the geometry with a velocity that is initially zero and follows a sine wave to its maximum value of 1.25 with a parabolic profile. The concentration at the inlet is constant in time defined by a parabolic cross-sectional profile with a maximum of 1.33. The outlet is at the bottom of each daughter branch and has a normal boundary condition of $p = 0$ and $du/dn = 0$. The walls of the tubes have a no-slip boundary condition in the fluid solution and zero concentration for the particle fields. The zero concentration condition allows for particles to diffuse to the walls and represents a perfect sink for particle deposition.

Numerical Methods

The Navier-Stokes equations and the convection-diffusion equation were solved using the spectral element scientific code Nek5000. Nek5000 uses Legendre spectral elements and an overlapping Schwarz preconditioner as developed by Fischer [10]. The Legendre elements use high-order polynomials to approximate the solution. The polynomials of order N introduce N-1 secondary nodes into each element (E) edge. The total nodal count for each mesh is equal to $E*N^3$ for a 3D domain. The polynomials used in this study range from 3rd to 11th order and provide great flexibility by allowing the user to change the accuracy and resolution from one trial to the next without generating a new mesh. The program allows for a variable time step that is based on a CFL condition of 0.25. The method has been used with success for determining velocity fields in similar single bifurcation biological models such as vascular flows [7] and particle fields in [11].

Each computational mesh was run at several different resolutions on the Saguaro computing cluster at Arizona State University. Using 16 Intel 64-bit Xeon EM64T CPUs connected with Infiniband high speed interconnect, the completed simulations ranged in time from 16 to 75 hours per processor depending on the number of degrees of freedom and polynomial order. For each time step, the velocity field is determined, and then the convection-diffusion equation is solved four times to represent particles of various diameters. Since the Nek5000 code scales nearly linearly, the computation times are reported as the total number of processor hours (# of CPUs * wall clock time) necessary for task completion and is relatively independent of the number of processors used [12].

The results are examined by looking at fluid velocity/particle concentration profile across three planes through the domain (inlet, prior to bifurcation, post bifurcation). The three profiles provide a snapshot in time of the chosen quantity as it enters the domain, prior to the most computationally difficult realm, and post that same disturbance. The profiles are compared in the root-mean square error with regards to a highly resolved simulation, which uses 11th order elements and contains nearly 2.6 million degrees of freedom. Results are reported as the percentage difference between the test solution and the resolved solution.

Results

In order to determine the accuracy of the Eulerian particle solution, it is imperative that the velocity field be well resolved and divergence free a priori. The initial mesh was independently refined in the order of approximation and the total number of elements. The order of approximation was allowed to range from 5 to 13, and the elements in the mesh were increased 540 up to 5952. Fast convergence of the velocity field to under 1% error is achieved for both h-and p-refinement.

Since the velocity field can be considered consistently resolved, the particle solutions can be examined independently. The solution is refined in using both p and h. The degrees of freedom range from 130,000 to 1.2 million. The nondimesionalized diffusion coefficient ranges from 10^{-4} to $5x10^{-6}$ (~1-10 nanometer particles). The solution shows good convergence to under 1% error for the largest diffusion coefficient case i.e. smallest particles, and the field with the second largest coefficient also converges but at a slightly slower rate. When the diffusion coefficient is an order of magnitude smaller (10^{-5}), the errors remain over 20% for the h-refinement but show fast convergence with p-refinement.

Since p-refinement increases the amount of interconnectedness between the computational nodes, it creates both greater accuracy and increased computational cost. Figure 2 shows total computation time for all of the simulation described previously. The error is compared for the two most advective particle solutions ($\mathfrak{D} = 1.0 \times 10^{-5}$ & 5.0×10^{-6}) since they best illustrate the accuracy differences between the simulations. Even though h-refinement results in more nodes than the p-refinement, the higher-order (p-refinement) demonstrate improved accuracy for equal simulations times of around 100 hours.

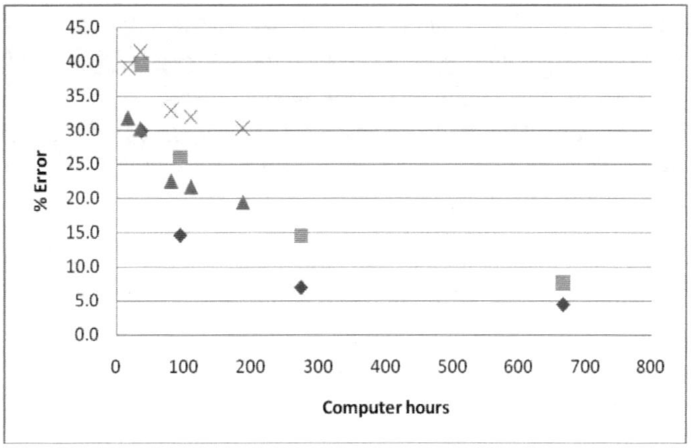

Figure 2: P-refinement demonstrates lower errors in the particle fields, $\mathfrak{D} = 1.0 \times 10^{-5}$ (◆) & $\mathfrak{D} = 5.0 \times 10^{-6}$ (■), than h-refinement, $\mathfrak{D} = 1.0 \times 10^{-5}$ (▲) & $\mathfrak{D} = 5.0 \times 10^{-6}$ (X), does at equal total computation times.

Computational Validation

Collaborators at the University of Arizona have formalized a methodology to link technetium-labeled diethylenetriamine pentaacetate (^{99m}Tc-DTPA) to second hand cigarette smoke. The research cigarettes (1R4 – University of Kentucky Tobacco and Health Research Institute) are injected with ^{99m}Tc-DTPA in four locations, and then allowed to dry for 5 minutes. Normal burn is then initiated with the smoke being drawn into a funnel in order to direct it towards a Kent Scientific small animal ventilator. The animals were prepared by anesthetizing them with a mixture of ketamine HCL (50 mg/kg body weight), xylazine (8 mg/kg body weight), and acepromazine maleate (1 mg/kg body weight). They are then placed in the ventilator with an endotracheal tube placed in the trachea 2 cm above the carina and explosed to a tidal volume of 1.5-2 mls at a respiratory rate of 60 breathes per minute [13, 14].

After the deposition is completed, each rat is placed under a gamma scanner system and observed for 10 minutes with gamma counts being recorded per 30 seconds. This verifies that the particles are securely deposited within the system, and it provides a more accurate assessment of the overall deposition in the different parts of the animal's lungs. The counts were added separately for the trachea, first bifurcation, primary airway (G2-3), and terminal airways. The deposition fraction is provided in figure 3 for the trachea (0), the first branch (1), the primary airways (2), and the terminal airways (3).

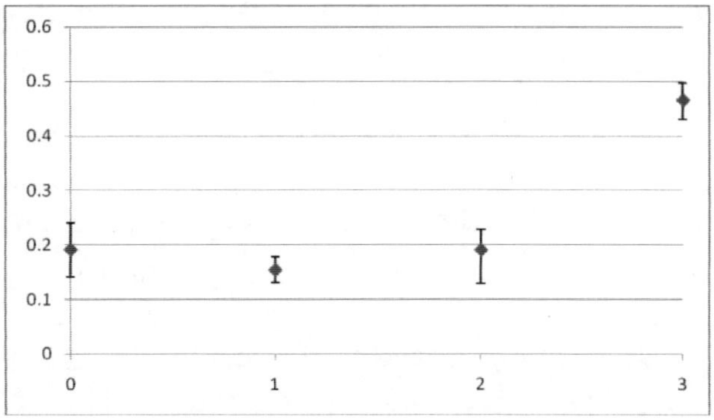

Figure 3: The deposition fraction in the trachea (0), first branch (1), primary airway (2), and the terminal airway (3) for smoke deposition in rats.

The animals in the experiments were euthanized and dissected after the gamma count observation was competed. High resolution images were taken of the entire respiratory system as well as each individual part of interest. An approximate geometry was developed by first overlying the image of first branch onto of the entire system as shown in the first frame of figure 4. This provides a very detailed image of the trachea and first generation which are most important for obtaining accurate results. A skeleton created from the cast of the lungs of a mouse was then overlaid on the image to approximate the primary and terminal airways (frame 2) [15]. Bifmesh was used similar to the prior work to develop the final three bifurcation geometry shown in figure 4.

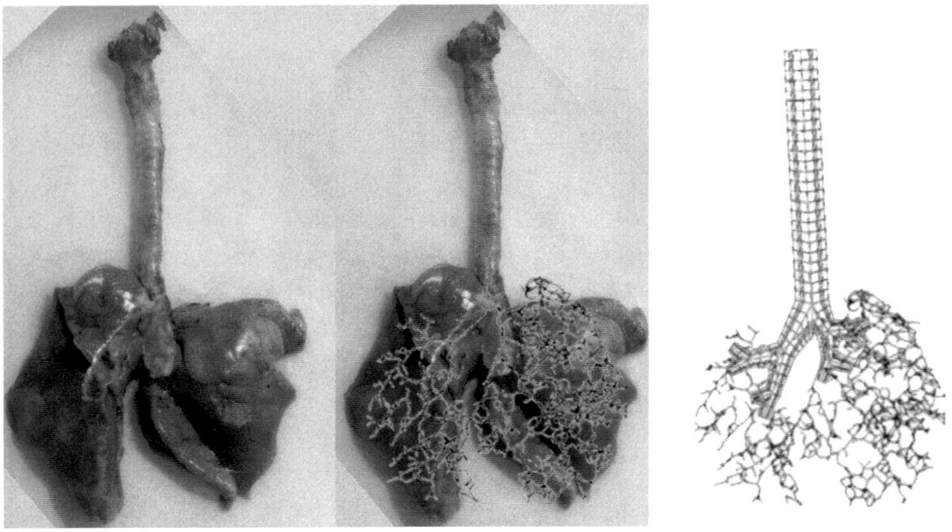

Figure 4: The computational mesh is generated based on a combination of actual rat lung images and published respiratory data [15].

Variable Approximation Orders

One of the main problems demonstrated by the previous work is that to achieve adequate resolution of an advective particle concentration, the fluid solution must be over resolved. This has no impact on the overall accuracy of the simulation, but it creates an unnecessary sacrifice of computational efficiency. This problem is caused by using the same approximation orders within the simulation for the fluid and particle fields. Interpolation can be used to generate a higher resolution solution for the particle concentration than the fluid field without generating multiple meshes or requiring multiple simulations.

The basis functions and parametric spaces used in SEM make it very inexpensive to interpolate data from one basis to another with a higher or lower approximation order. Therefore the expensive fluid problem can be solved on coarser mesh, and then the velocity data is interpolated onto finer mesh to fully resolve the less expensive particle solution. The current method utilizes $\mathbb{P}_N - \mathbb{P}_{N-2}$ spectral elements which means that the pressure is solved on a mesh that is of order N-2 while the velocity is of order N. The values of velocity and pressure are interpolated back and forth between the Gauss-Lobatto-Legendre velocity grid and the discontinuous Gauss-Legendre pressure grid [16].

This same technique can be used to interpolate the final velocity solution onto a higher-order continuous Gauss-Lobatto-Legendre grid for solving the advection-diffusion equation. Figure 5 shows that the fluid solution requires far more computation time than the particle solution. The fluid solution requires solving the Helmholtz operator in each spatial direction as well as a very costly PCG solve of a consistent Poisson operator while the particle solution only requires a single solution to the Helmholtz operator. Therefore in the simulations shown in figure 5 (four particle sizes), the fluid solution requires 50% of the overall simulation time, and each particle solution requires less than 10%.

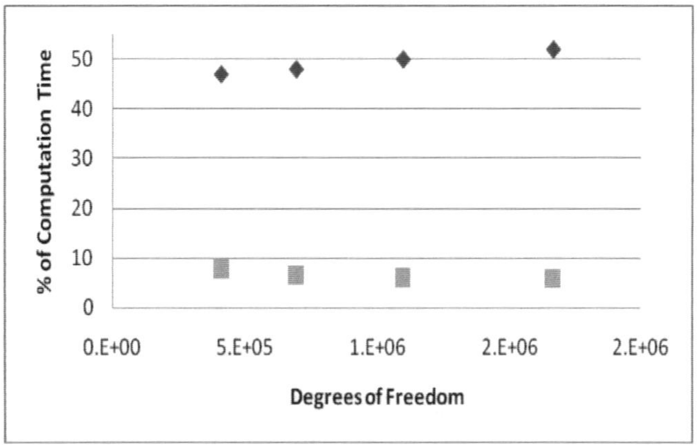

Figure 5: Converging to a solution for the consistent Poisson operator and three Helmholtz operators in the fluid solution (◆) requires over half of overall simulation time, while each of the four solves associated with Helmholtz operator in the advection-diffusion equation (■) requires less than 10% of the time.

This disparity becomes even more pronounced as the particle field becomes more advective because extremely high resolution is required to capture the high frequency waves. Solving the Navier-Stokes equations on a coarser grid becomes very appealing. Also, the stiffness of the Helmholtz operator in the particle solution is proportional to the diffusion coefficient. Therefore, as the diffusion coefficient is decreased, the solution to the advection-diffusion equation can be obtained with fewer PCG loops. The excess time for the highest resolution solutions should be somewhat mitigated by this feature.

Further, the stability of the simulation is based on the CFL condition, $\max_{\Omega}(U\Delta t/\Delta x)$. The axial node spacing can be exploited to maintain a manageable time step while still providing adequate resolution in the particle solution to prevent dispersion error oscillations. The CFL condition could still become unstable do to the decreasing Δx in the axial direction of the particle grid, but the problem can be partly circumvented using variable filter strengths for each field or subcycling [17].

The Nek5000 program will first solve the Navier-Stokes equation exactly as it did before, but then the velocity solution is interpolated onto a set of higher-order Gauss-Lobatto nodes. Two sets of basis functions are demonstrated in figure 6. The AD equation can then be solved on the finely resolved domain. For the output steps, the particle solution is interpolated back onto the initial velocity grid. A less smooth filter can aid in stabilizing the particle solutions.

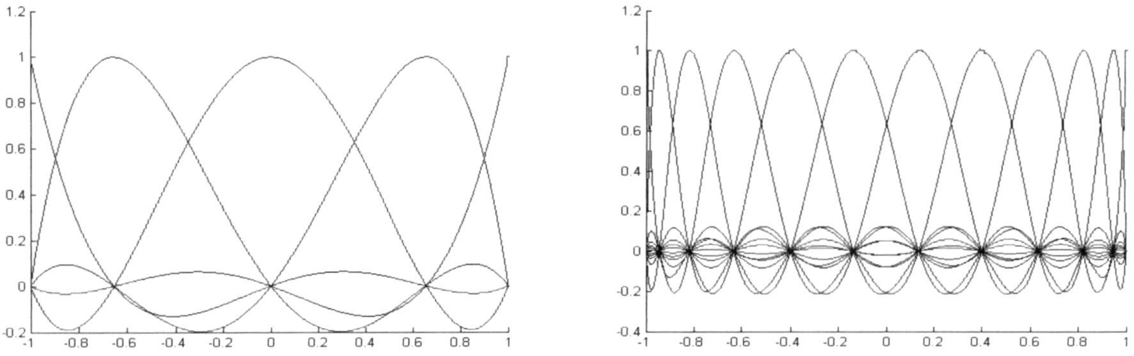

Figure 6: One-dimensional spectral basis functions on a single element for N = 4 (left) and N = 11 (right)

This Eulerian-Eulerian particle tracking method can be functionally implemented into the Nek5000 code to provide a tool for direct numerical simulations of particle fields in many applications. Particle sizes can range to at least 100 nm and are only limited by the physical limitations of the AD equation [18]. These tools can lead towards more accurate and efficient determination of particle deposition patterns for unique geometries with an emphasis on toxic and therapeutic particles in the human respiratory system.

References

1. Gehr, P., F. Blank, and B.M. Rothen-Rutishauser, *Fate of inhaled particles after interaction with the lung surface.* Paediatric Respiratory Reviews, 2006. **7**: p. S73-S75.
2. Singh, N. and G.S. Davis, *Review: occupational and environmental lung disease.* Current Opinion in Pulmonary Medicine, 2002. **8**(2): p. 117-125.
3. Donaldson, K., et al., *The pulmonary toxicology of ultrafine particles.* Journal of Aerosol Medicine-Deposition Clearance and Effects in the Lung, 2002. **15**(2): p. 213-220.
4. Donaldson, K., et al., *Ultrafine particles: mechanisms of lung injury.* Philosophical Transactions of the Royal Society of London Series a-Mathematical Physical and Engineering Sciences, 2000. **358**(1775): p. 2741-2748.
5. Choi, L.T., et al., *Flow and particle deposition patterns in a realistic human double bifurcation airway model.* Inhalation Toxicology, 2007. **19**(2): p. 117-131.
6. Zhang, Z. and C. Kleinstreuer, *Airflow structures and nano-particle deposition in a human upper airway model.* Journal of Computational Physics, 2004. **198**(1): p. 178-210.
7. Fischer, P.F., et al., *Simulation of high-Reynolds number vascular flows.* Computer Methods in Applied Mechanics and Engineering, 2007. **196**(31-32): p. 3049-3060.
8. Zhang, Z. and C. Kleinstreuer, *Transient airflow structures and particle transport in a sequentially branching lung airway model.* Physics of Fluids, 2002. **14**(2): p. 862-880.
9. Weibel, E.R., *Morphometry of the Human Lung.* 1963, New York: Academic Press.
10. Fischer, P.F., *An overlapping Schwarz method for spectral element solution of the incompressible Navier-Stokes equations.* Journal of Computational Physics, 1997. **133**(1): p. 84-101.
11. Chiam, K.H., et al., *Enhanced tracer transport by the spiral defect chaos state of a convecting fluid.* Physical Review E, 2005. **71**(3): p. -.
12. Tufo, H.M. and P.F. Fischer, *Fast parallel direct solvers for coarse grid problems.* Journal of Parallel and

Distributed Computing, 2001. **61**(2): p. 151-177.
13. Witten, M.L., et al., *Acute Cigarette-Smoke Exposure Increases Alveolar Permeability in Rabbits.* American Review of Respiratory Disease, 1985. **132**(2): p. 321-325.
14. Witten, M.L., et al., *New Developments in the Pathogenesis of Smoke Inhalation-Induced Pulmonary-Edema.* Western Journal of Medicine, 1988. **148**(1): p. 33-36.
15. Chaturvedi, A. and Z. Lee, *Three-dimensional segmentation and skeletonization to build an airway tree data structure for small animals.* Physics in Medicine and Biology, 2005. **50**(7): p. 1405-1419.
16. Deville, M.O., P.F. Fischer, and E.H. Mund, *High-Order Methods for Incompressible Fluid Flow*. Cambridge Monographs on Applied and Computational Mathematics, ed. A.I. P.G. Ciarlet, R.V. Kohn, M.H. Wright. 2002: Cambridge University Press.
17. Fischer, P. and J. Mullen, *Filter-based stabilization of spectral element methods.* Comptes Rendus De L Academie Des Sciences Serie I-Mathematique, 2001. **332**(3): p. 265-270.
18. Zhang, Z., et al., *Comparison of micro- and nano-size particle depositions in a human upper airway model.* Journal of Aerosol Science, 2005. **36**(2): p. 211-233.

602d Rheology of Adult Stem Cells and Modelling of Flow Induced Deformation

Alexandre J.S. Ribeiro, Department of Biomedical Engineering, Carnegie Mellon University, Pittsburgh, PA 15213, Ender A. Finol, Institute for Complex Engineered Systems, Carnegie Mellon University, Pittsburgh, PA 15213, and **Kris Noel Dahl**, Department of Chemical Engineering, Carnegie Mellon University, 5000 Forbes Avenue, Pittsburgh, PA 15213.

Injection of stem cells for regenerative therapies requires optimal pressure to efficiently deliver cells, but inadequate designs of devices for injection can generate elevated stresses on cells resulting in altered cell response and cell death. However, few studies have examined rheological properties of stem cells to see how these cells deform and rupture under strain. In this study we measure the viscoelastic properties of different types of adult stem cells: mouse and human hematopoietic stem cells, mouse neural stem cell spheres, and human mesenchymal stem cells. We use a combination of fluorescence microscopy and micropipette aspiration to image the deformation of live cells and subcellular components, such as the nucleus. In stem cells, more so than differentiated cells, the nucleus contributes to the overall mechanics due to the high nuclear to cytoplasmic ratio and poorly developed cytoskeletal structure. We quantify elastic, viscoelastic and rupture strain from micropipette aspiration for these cells. We find variation among cells and among cell types (about 60% difference among cell types), but the adult stem cells studied are much softer than reported values for differentiated cells, as expected. We are incorporating this experimental data into 2D and 3D fluid-structure interaction models of flow-induced stem cell deformation to simulate injection systems, which we hope to then use to optimize injection pressures and possibly system design. Also, we will examine the long-term effects of injection shear stress on stem cell differentiation to better determine long-term cell effects of shear stress.

602e Dynamics and Bifurcation Analysis of the Heat-Shock Response In Eukaryotic Cells

Dimitrios I. Gerogiorgis, Department of Chemical Engineering, Massachusetts Institute of Technology (M.I.T.), 77 Massachusetts Avenue, Cambridge, MA 02139

The heat-shock response mechanism in eukaryotic cells is a vital and therefore ubiquitous molecular reaction to proteotoxicity resulting from the appearance of various classes of non-native (misfolded and damaged) proteins. The accumulation of such non-native protein species can result in the dangerous generation of protein aggregates. To handle this build-up of abnormal proteins, cells employ a complicated machinery of molecular chaperones: these species facilitate the refolding or degradation of misfolded polypeptides, prevent protein aggregation and play a role in formation of aggresome, a centrosome-associated body to which small cytoplasmic aggregates are transported. Protein folding and aggregates have been studied in the domain of neurochemistry and molecular neurobiology, and they are increasingly recognized as responsible for the progression of serious neurodegenerative diseases, which have a severe impact on progressively larger populations. These include Alzheimer's, Parkinson's and Huntington's diseases, and in addition Amyotrophic Lateral Sclerosis (Meriin and Sherman, 2005; Szilagyi, Kardos et al., 2007).

The heat-shock adaptation and survival mechanism is aimed at ensuring protein quality control and homeostasis. The transcriptional induction of genes encoding molecular chaperones and protein degradation mechanisms is a means of efficient management of misfolded and damaged proteins, against the likelihood of the latter persisting as proteotoxic species, protein aggregates, and inclusions. Computational analysis of a mathematical model of the heat-shock network, presented here, led to the identification of the elementary steps that may represent key determinants of network performance.

The central elements of this process are the heat-shock proteins (HSPs) that function as molecular chaperones. Upon sensing a stress signal, such as elevated temperatures, small toxic molecules, oxidants, or heavy metals, cells resort to transient molecular chaperone overexpression whose target is to meet the stress demand by high levels. Chaperones recognize and associate with exposed hydrophobic patches on unfolded polypeptides and conformational intermediates and sequester them uintil they reach their native confirmation by providing an environment for proper refolding, or act as an escort to the proteosomes for orderly degradation.

The importance and function of the heat-shock transcription factor-1 (HSF-1), which is central to the heat-shock response mechanism, has been studied extensively, as documented in numerous publications. Detailed network representations and dynamic mathematical models have been derived and tested in order to analyze quantitatively all species and their concentrations in the heat-shock response cycle. Furthermore, parameter estimation and sensitivity analysis have provided additional insight regarding the relative importance of a large number of kinetic parameters in the respective expression networks. This paper focuses on analyzing the importance of an uncertain, temperature-dependent parameter (βm,

k) and conducting a bifurcation analysis with respect to an uncertain parameter, in order to evaluate multiplicity.

REFERENCES

1. Rieger, T. R., Morimoto, R. I., Hatzimanikatis, V. (2005). Mathematical Modeling of the Eukaryotic Heat-Shock Response: Dynamics of the hsp 70 Promoter. Biophysical Journal, 88, 1646-1658.

2. Szilagyi, A., Kardos, J., Osvath, S. et al. (2007), Protein Folding, in: Handbook of Neurochemistry and Molecular Neurobiology ¨C Neural Protein Metabolism and Function (A. Lajtha & N. Banik, eds.), Springer, USA .

3. Meriin, A. B., Sherman, M. Y. (2005). Role of molecular chaperones in neurodegenerative disorders. International Journal of Hyperthermia, 21(5): 403-419

4. Morimoto, R. I. (1993). Cells in Stress: Transcriptional Activation of Heat-Shock Genes. Science, 259: 1409-1410.

5. MATCONT Software (2007). Gent, Belgium.

602f Quantitative Structure-Property Relationship Modeling of Skin Irritation

S. Golla, K. Yerramsetty, B. J. Neely, S. V. Madihally, R. L. Robinson Jr., and K. a. M. Gasem. School of Chemical Engineering, Oklahoma State University, 423 Engineering North, Stillwater, OK 74078

The importance of developing non-experimental procedures for estimating skin irritation potential of chemicals has been increasing as a result of concerns regarding animal welfare and laboratory expenses. A number of expert systems and quantitative structure-activity relationship (QSAR) models have been proposed in literature for predicting skin irritation potential of compounds. However, the application of these models to predict skin irritation require a priori estimates of several physiochemical properties such as, octanol/water partition coefficient, melting point, lipid solubility, aqueous solubility, surface tension and vapor pressure. Thus, prediction of skin irritation potential using these models often requires other models capable of estimating the physiochemical properties. Further, the literature QSAR models are limited to a particular class of compounds and thus, lack in universal applicability.

In this work we have developed a skin irritation QSPR model using rabbit Draize test data consisting of over two hundred compounds from various chemical classes. The effectiveness of using a combination of traditional, functional group and statistical descriptors to improve the predictive capability has been studied. Our QSPR model is capable of predicting the skin irritation potential of chemical compounds with an R2 of 0.83. External validation of the QSPR model was performed using data from the human-patch test.

602g Multisite Phosphorylation: An Ultrasensitive Mechanism for Protein Degradation

S. Marjan Varedi K., Chemical Engineering, University of Michigan, 2300 Hayward St., 3074 HH Dow Bldg., Ann Arbor, MI 48109 and Xiaoxia (Nina) Lin, Chemical Engineering, The University of Michigan at Ann Arbor, 2300 Hayward St., 3074 HH Dow Bldg., Ann Arbor, MI 48109.

Phosphorylation is one of the most frequent protein modifications in living organisms. Phosphorylation profile has been interpreted as a "molecular barcode" to trigger protein activation, inactivation, translocation, interaction with other proteins and degradation. Phosphorylation-triggered degradation process is a common strategy for elimination of regulatory proteins including cyclin-dependent kinase inhibitors such as Sic1 in yeast and p27 in mammalian cells. Reduced p27 expression is observed in up to 60% of primary human breast cancers, which has been correlated with increased activity of a kinase family.

In this work, phosphorylation-triggered protein degradation has been modeled and analyzed in a systematic approach. Our simulation results show that while a single-site protein does not degrade in an ultrasensitive manner, multisite phosphorylation can lead to thresholding for degradation and causes a switch-like elimination. Switch-like responses are required for binary decisions in the cell cycle transitions and apoptosis. The effects of different parameters and ordering on ultrasensitivity of protein elimination in response to kinase have also been investigated. The ultrasensitivity of the response curve is at maximum when protein degrades after becoming phosphorylated on half of its total sites. With regard to ordering effect, the elimination curve of a multisite protein phosphorylated sequentially is more switch-like than that of a protein phosphorylated randomly.

Other kinds of modification such as ubiquitination, methylation and glycosilation which are involved in tuning macromolecules' stability, activity and translocation can also be explained by the model presented in this work. Generally, multi-site modification provides a precise mechanism for controlling the speed and sharpness of the output in response to an input of stimulus. By using multisite modification of a macromolecule, a gradually changing input signal can be converted into an ultrasensitive output signal. Such single-molecule based switches could also provide a powerful tool for tuning gene expression and protein activity.

614a Operational Planning of a Large-Scale Industrial Batch Plant Under Uncertainty

Peter M. Verderame and Christodoulos A. Floudas. Department of Chemical Engineering, Princeton University, Dept of Chemical Engineering; A215 Engineering Quadrangle, Princeton, NJ 08544

The operational planning of a large-scale industrial batch plant typically occurs over a time horizon of several months with the goal of providing daily production targets and raw material requirements for the plant in question. Kallrath [1] and Shah [2] each provide an overview of the various aspects of planning. Due to the length of the time horizon, multiple forms of uncertainty should be taken into account in order to ensure that the operational planning model does not provide unrealizable production targets and/or raw material requirements. For example, demand, price, and processing time uncertainty should not be overlooked at the operational planning level. The main approaches for optimizing under uncertainty are stochastic programming, fuzzy programming, stochastic dynamic programming, and robust optimization, and the work of Lin et al. [3] and Janak et al. [4] has laid the foundation for a robust optimization approach toward addressing uncertainty within the field of planning.

Extending upon the work of Lin et al. [3] and Janak et al. [4], a novel operational planning model has been developed in order to address the objective of providing a reliable daily production profile (e.g., how much of each product should be produced at a facility on a daily basis) which is immune to various forms of uncertainty. The proposed formulation represents the robust counterpart of the novel Production with Demand Disaggregation Model (PPDM), a discrete-time mixed-integer linear programming (MILP) planning model developed by Verderame and Floudas [5] which still captures the continuous-time nature of the multipurpose and multiproduct batch plant under consideration. The given planning model is a unit aggregation model having the explicit objective of providing a daily production profile for a batch chemical plant. The PPDM has been extended in order to take into account uncertainty associated with the demand, price, and processing time parameters for both the bounded and known probability distribution cases. The ability of the robust PPDM to address the aforementioned objectives of an operational planning model has been validated through an industrial case study of a large-scale, multiproduct and multipurpose batch plant having the capability of producing hundreds of different products over a time horizon of three months.

[1] Kallrath, J. Planning and Scheduling in the Process Industry. OR Spectrum, 2002, 24, 219.

[2] Shah, N. Process Industry Supply Chains: Advances and Challenges. Comp. Chem. Eng. 2005, 29, 1225.

[3] Lin, X.; Janak, S.L.; Floudas, C.A. A new robust optimization approach for scheduling under uncertainty: I. Bounded uncertainty. Comp. Chem. Eng. 2004, 28, 1069.

[4] Janak, S.L.; Lin, X.; Floudas, C.A. A new robust optimization approach for scheduling under uncertainty: II. Uncertainty with known probability distribution. Comp. Chem. Eng. 2007, 31, 171.

[5] Verderame, P.M.; Floudas, CA. Integrated Operational Planning and Medium-term Scheduling of Large-Scale Industrial Batch Plants. Ind. Eng. Chem. Res. Accepted for Publication.

614b Verified Uncertainty Analysis In Modeling Nonlinear Bioreactor Dynamics

Joshua A. Enszer and Mark A. Stadtherr. Chemical and Biomolecular Engineering, University of Notre Dame, 182 Fitzpatrick Hall, Notre Dame, IN 46556

Bioreactors are used in a wide variety of processes and industries, from food processing to waste treatment. Typically, in these reactors, one or more organisms act on one or more substrates, creating more organisms together with desired chemical products. For the analysis, design and operation of bioreactor systems, one may need to consider their dynamics over a wide variety of conditions. Since performing physical experiments is likely to be impractical, numerical models are commonly used in the analysis of such systems. These models typically involve some degree of uncertainty, as some parameters and/or initial conditions may be known only to within some interval. In some cases, information on the probability distribution of the uncertain values may also be available. For analysis of the impact of such uncertainties on modeling results, a Monte Carlo simulation approach is widely used. However, since it is not possible to sample the complete space of the uncertain quantities in a finite number of simulations, the results of Monte Carlo analysis may fail to capture some system behaviors, especially in the case of nonlinear systems.

In this presentation, we demonstrate a method for the verified solution of nonlinear bioreactor models. This method computes rigorous bounds on the concentration profiles in the reactor over a given time horizon, based on specified ranges for uncertain parameters and/or initial conditions. The method is based on the general approach described by Lin and Stadtherr [1], which uses an interval Taylor series to represent dependence on time, and uses Taylor models to represent dependence on uncertain quantities. We also demonstrate an approach for the propagation of uncertain probability distributions in one or more model parameter and/or initial condition through the bioreactor model. Assuming an uncertain probability distribution for each parameter and/or initial condition of interest, we use a method, based on Taylor models and probability boxes (p-boxes) and recently described by Enszer et al. [2], that propagates these distributions through the dynamic model. As a result, we obtain a p-box describing the probability distribution for each state variable at any given time of interest. As opposed to the traditional Monte Carlo simulation approach, which may not properly bound all possible system outputs, these Taylor model methods for verified uncertainty analysis provide completely rigorous results that fully capture all possible system behaviors under the uncertain conditions. These methods are tested using a variety of nonlinear bioreactor models, with uncertainties in some parameters and/or initial conditions. Comparisons are made to the results of Monte Carlo simulations.

[1] Lin, Y., Stadtherr, M.A. Validated Solutions of Initial Value Problems for Parametric ODEs. Applied Numerical Mathematics, 57: pp. 1145-1162, 2007.

[2] Enszer, J.A., Lin, Y., Ferson, S., Corliss, G.F., Stadtherr, M.A. Propagating Uncertainties in Modeling Nonlinear Dynamic Systems. In Proceedings of the 3rd International Workshop on Reliable Engineering Computing, Georgia Institute of Technology, Savannah, GA: pp. 89-105, 2008.

614c Global Optimization of Multi-Parametric Milp Problems

Nuno P. Faísca, Konstantinos I. Kouramas, and **Efstratios N. Pistikopoulos**. Imperial College London, London, sw7 2az, United Kingdom

Multi-parametric programming has recently received considerable attention in the open literature (Pistikopoulos et al, 2007a,b), with important applications to pro-active scheduling and planning under uncertainty problems (Ryu and Pistikopoulos). However, despite these advances, certain important classes of multi-parametric programs have not been fully addressed in the open literature.

In this presentation, we present a global optimisation approach for the general solution of multi-parametric mixed integer linear programs (mp-MILPs), which substantially extends and generalises our earlier work on mp-MILP with only right-hand side (RHS) uncertainty (Dua and Pistikopoulos, 2000). In particular, we describe an optimisation procedure which iterates between a (master) mixed integer non-linear program and a (slave) multi-parametric program. Moreover, we explain how to overcome the presence of bi-linearities, responsible for non-convexity of the multi-parametric program, in two classes of mp-MILPs, with (i) varying parameters in the objective function and (ii) simultaneous presence of varying parameters in the objective function and the right-hand side of the constraints. Planning examples are provided to illustrate the solution steps.

References

Dua, V. and E.N. Pistikopoulos (2000). An algorithm for the solution of multi-parametric mixed integer linear programming problems. Annals of Operations Research 99, pp. 123-139.

Pistikopoulos, E.N., M.C. Georgiadis and V. Dua (2007a). Multiparametric programming: theory, algorithms, and applications, vol 1, Wiley-VCH. Weinheim.

Pistikopoulos, E.N., M.C. Georgiadis and V. Dua (2007b). Multiparametric model-based control: theory and applications, vol 2, Wiley-VCH. Weinheim.

Ryu, JH and E.N. Pistikopoulos (2007). A novel approach to scheduling of zero-wait batch processes under processing time variations. COmputers and Chemical Engineering 31(3), pp. 101-106.

614d Computational Strategies for the Planning of Deepwater Oilfield Infrastructure Under Decision Dependent Uncertainty

Bora Tarhan and Ignacio E. Grossmann. Department of Chemical Engineering, Carnegie Mellon University, Pittsburgh, PA 15213

Oilfield exploration and production operations consist of four major steps: exploration, appraisal, development and production. In each step, many decisions have to be made that affect the overall performance of the operation. In the beginning of exploration phase many uncertainties exist, and depending on the decisions made at each step, the uncertainty reduces gradually. Unfortunately, many crucial decisions related to planning of offshore oilfield infrastructure have to be made in the early steps where the uncertainty level is high. The quality of these decisions affects the overall profitability of the operation. Most of the available literature for exploration and planning of petroleum fields is based on deterministic models (Haugland et al. (1988), Iyer & Grossmann (1998), Van den Heever & Grossmann (2000), Lin & Floudas (2000), Carvalho et al. (2006)), while there are only few papers that deal with uncertainty (Jonsbraten (1998), Lund (2000), Begg et al. (2002), Cullick, Heath, Narayanan (2003), Aseeri, Gorman & Bagajewicz (2004), Goel & Grossmann (2004), Ulstein, Nygreen & Sagli (2007), Ozdogan & Horne (2006)).

In this paper, we consider a multistage stochastic programming approach for the design and planning of an oilfield infrastructure over a planning horizon under gradual uncertainty resolution. The oilfield infrastructure can be composed of Floating Production Storage and Offloading (FPSO) and/or Tension Leg Platform (TLP) facilities, sub-sea and/or TLP wells. The main uncertainties considered are in the initial maximum oil flowrate, size and water breakthrough time of the reservoir, which are represented by discrete distributions. Furthermore, it is assumed that these uncertainties are not immediately realized, but are gradually revealed as a function of design and operation decisions.

In order to account for these decision-dependent uncertainties, we propose a multistage stochastic programming model that captures the complex economic objectives and nonlinear reservoir behavior, and simultaneously optimizes the investment and operation decisions over the entire planning horizon. We propose special computational strategies for a solution algorithm that relies on a duality based branch and bound method (Tarhan and Grossmann, 2007) involving nonconvex mixed-integer nonlinear programming subproblems. The subgradient method has been used for solving the dual problem. During subgradient optimization, it is necessary to solve these nonconvex problems using global optimization in order to generate valid bounds. A variant of the algorithm in which a combination of global optimization and outer-approximation approach used during subgradient iterations is presented. The new approach yields great reductions in solution time and not necessarily sacrificing the quality of the solution. We describe results on several test problems involving nonlinear reservoir models to illustrate the capabilities of the proposed model. Also, we present comparisons of results obtained by using expected values to demonstrate the hedging effect of the proposed multistage programming approach.

614e Optimal Planning and Operation of Cryogenic Air Separation Columns Considering Uncertain Demands and Product Transitions

Yu Zhu and Carl D. Laird. Chemical Engineering, Texas A&M University, 3122 TAMU, College Station, TX 77843-3122

Cryogenic air separation columns, which are widely used in industry, consume a large amount of energy producing significant quantities of high-purity nitrogen, oxygen and argon. In order to meet the requirement of different customers, the production rate and product purities have to be changed rather frequently. Rising material and energy costs, as well as uncertain market and load demands are pushing increased research into the flexible design and operation of cryogenic air separation columns[1-6].

Existing research has focused on the implementation of suitable control strategies to guarantee effective transition between different operating conditions[1-4, 6]. However, a second area to consider is the optimal planning and operation of these air separation systems in order to avoid large or frequent transitions. This problem is difficult for two main reasons. First, it requires optimization formulations that consider correlated uncertainties in the demands and product transitions. Second, the coupled nature of cryogenic air separation systems gives rise to an extremely complex and highly nonlinear model. Nonlinear programming formulations and advanced large-scale algorithms provide a rigorous framework for addressing both of these problems.

Adopting a purely multi-scenario approach that requires the satisfaction of customer demands and constraints on controlled transitions over all the scenarios can lead to solutions that are far too conservative (and expensive). Instead, we combine the multi-scenario approach with Chance-Constrained Programming to obtain more cost effective solutions that are acceptable with regards to the uncertainty. This study presents a multi-scenario mathematical formulation and advanced solution approach for optimal planning and operation with uncertain demands and transitions using chance constraints. A rigorous nonlinear process model is used for the coupled cryogenic air separation system, which considers mass and energy balances, phase equilibrium, and system hydraulics. To effectively solve such large-scale nonlinear problem, we present a parallel internal decomposition approach[6-8] based on the existing primal-dual interior-point NLP solver, IPOPT [9]. Optimal solutions with different inventory and uncertainty distribution assumptions will be discussed, along with the impact of different customer satisfactory levels. Furthermore, as demonstrated by this problem, the parallel interior-point approach is shown to be scalable to many processors.

References:

[1] White, V., Perkins, JD., Espie, DM., (1996). Switchability analysis. *Computers chem. Eng., 2*0, 469.

[2] Roffel, B., Betlem, BHL., de Ruijter, JAF. (2000). First principles dynamic modeling and

multivariable control of a cryogenic distillation process. *Computers chem. Eng.*, 24, 111.

[3] Bian, S., Henson, M., Belanger P., Megan L., (2005) Nonlinear state estimation and model predictive control of nitrogen purification columns. *Ind. Eng. Chem. Res.*, 44, 153.

[4] Seliger B., et al., (2006) Modeling and dynamics of an air separation rectification column as part of an IGCC power plant. *Sep. Purif. Technol.* 49, 136.

[5] Miller, J., Luyben, WL., M., Belanger P., Blouin, S., Megan L., (2008) Improving agility of cryogenic air seapration plants, *Ind. Eng. Chem. Res.*, 47, 394

[6] Zhu, Y., Laird, CD., A parallel algorithm for structured nonlinear programming, *foundations of computer-aided process operations (FOCAP-O)*, Boston, Massachusetts, USA, July 2008.

[7] Laird, CD., Biegler, LT. (2006). Large-scale nonlinear programming for multi-scenario optimization. *proceedings of the int. conference on high performance scientific computing*, Hanoi, Vietnam.

[8] Zavala, VM., Laird, CD., Biegler LT. (2007) Interior-point decomposition approaches for parallel solution of large-scale nonlinear parameter estimation problems. in press, *Chem. Eng. Sci.*

[9] Wächter A., Biegler LT. (2006). On the implementation of an interior-point filter line-search algorithm for large-scale nonlinear programming. *Math. Programm.*, 106, 25

614f Management of the Pharmaceutical R&D Using Stochastic Programming

Matthew Colvin and **Christos T. Maravelias**. Chemical and Biological Engineering, University of Wisconsin - Madison, 1415 Engineering Drive, Madison, WI 53706

Due to the changing circumstances in the managed-health-care environment, the productivity of pharmaceutical research and development (R&D) pipelines in terms of new drugs have started to decline, while the cost of developing a new product has increased. Therefore, systematic methods that aid the management of the pharmaceutical R&D can have substantial impact. The development of such methods, however, is a challenging task due to the stochastic nature and the large number of decisions involved in the R&D pipeline. Specifically:

a) Sources of uncertainty include the cost, duration, resource requirements and outcome of clinical trials (technical uncertainty) as well as the revenues from sales (market uncertainty). While market uncertainty is clearly very important, the uncertainty in the outcome of clinical trials is the most significant source in the development process.

b) Decisions include the selection and prioritization of potential drugs; the timing and resource allocation; resource planning related decisions (hiring, outsourcing, etc.); the out-licensing of drugs.

In this work, we develop a stochastic programming (SP) approach that simultaneously accounts for the selection of drugs, the scheduling of clinical trials, and the resource planning and out-licensing decisions. Given are a) a portfolio of drugs with the corresponding clinical trials and the associated data (duration, cost, resource requirements, probability of success), b) the available resources and associated costs, and c) financial data for the revenues from sales, out-licensing deals, etc. The goal is to determine the decisions that maximize the expected net present value of the R&D pipeline over a fixed time horizon.

The uncertainty in the outcome of clinical trials is modeled via a reduced set of outcomes that leads to a smaller set of scenarios. The planning horizon is divided into a number of periods (stages) resulting into a multi-stage SP mixed-integer programming (MIP) formulation, where decisions are made for each scenario and each time period (stage). The endogenous uncertainty of the R&D process is modeled via non-anticipativity inequalities. To reduce the number of these constraints we develop a number of properties by exploiting the characteristics of the problem at hand (i.e. gradual realization of clinical trial uncertainty). We also develop results that allow us to replace some of the these inequalities with equalities. These results serve the dual purpose of reducing the model and tightening the linear programming formulation. For example, for a three-drug instance, the number of variables was reduced from 19,281 to 4,512 and the number of constraints was reduced from 50,238 to 17,850, with larger improvements for larger problems.

Finally, we develop a branch-and-cut algorithm where we start our search based on a reduced formulation and add only the non-anticipativity constraints that are violated at each node of the search tree. Thus, non-anticipativity constraints do not impede problem generation, allowing larger problems to be solved. Note that unlike conventional branch-and-cut algorithms, we start with a subset of the necessary constraints but we add non-redundant constraints. To correctly find incumbents and upper bounds, we modify the procedure for the pruning of nodes. A number of different implementations were examined. The least computationally intensive approach turned out to be the one where we start from a formulation with no non-anticipativity inequalities and all equalities, while we check for violations at every node. Using the proposed methods, we are able to solve realistic instances with up to seven potential drugs.

614g Optimal Intermediate Storage Level Design for Continuous Production Systems Under Unit Failure Scenarios

Kristin M. Davies, McMaster University, 1280 Main Street West, Hamilton, ON L8S 4L7, Canada and **Christopher L. E. Swartz**, Chemical Engineering, McMaster University, 1280 Main Street West, Hamilton, ON L8S 4L7, Canada.

Process unit shutdowns within a multi-unit process commonly induce additional unit shutdowns and, furthermore, they invariably result in economic consequences. Possibilities of such consequences include the loss of production as well as the creation of off-specification product. Mitigation of these effects may be possible if sufficient buffer capacity is available to introduce a level of independence between the upstream and downstream processes that allow the remainder of the plant to continue operation until the affected units are back in operation. Maximizing the potential of buffer storage, however, is not trivial. A question that arises is what the optimal levels of the buffer storage units should be for use in normal operation, with insight as to what abnormal operation conditions may occur. As demonstrated by Balthazaar (2005), this is a function of the expected unit failure frequencies, failure lengths and recoveries, and process dynamics. The optimization goal is to maximize plant economics by accounting for profit from production as well as costs incurred by process unit shutdowns.

Research into the determination of optimal nominal buffer tank levels is relatively sparse. Related work is that of Tolio at al. (2002) in which a similar process is studied, but with the goal of determining average buffer levels, and without the possibility of control or corrective action to the processing units. In addition, this work focused on discrete part manufacturing, as opposed to continuous process operation. The work by Lee and Reklaitis (1989a,b) while related, focuses on batch systems, which results in a different problem formulation and consequently different results.

Multiple key failure scenarios are simultaneously considered through multi-period dynamic optimization in order to determine a single set of optimal nominal buffer levels common to all scenarios. These scenarios refer to the combination of failure characteristics including failure location, failure length, recovery time, and failure frequency. The use of mixed-integer linear programming (MILP) captures the continuous and discrete decisions that exist within this mathematical model. Discrete decisions are present due to highly constrained processing rates, the triggering of a shutdown when throughput falls beneath a specified minimum value, and the need to count shutdowns.

A case study of two processors in series separated by one intermediate storage unit is first examined. Initially, two failure scenarios (one for the upstream unit, one for the downstream unit) are considered and results studied in detail to gain insight into the influence of various model parameters on the optimization results. Next, a configuration of three processors in series separated by two buffer tanks is examined. Here, three failure scenarios are considered (one for each process unit), but with the added complication of the possibility of sending intermediate product to a purge stream in order to avoid an induced unit shutdown when economically logical. The two processor configuration was then re-examined, but now with ten failure scenarios - five each for the upstream and downstream unit. The purpose of including a large number of failure scenarios is to simulate one of two realistic conditions. The first condition is of several possible failure "modes" for each process unit. For example, different failure durations may be associated with a mechanical failure, electrical disruption, and off-specification product. Alternatively, a distribution in failure durations can simulate uncertainty in the failure duration. Finally, the problem formulation is applied to a typical pulp mill process that includes a recycle stream.

Solution non-uniqueness is a computational issue that arose during this work. A multi-tiered optimization combined with a grid-search approach was used to uncover all optimal solutions for the smaller test cases (two and three processors in series). A soft-constraint problem formulation approach, which avoided the use of integer variables, was also considered. The advantage of this approach is a decreased problem complexity due to the elimination of integer variables. However, accurate model representation is lessened due to the inability to account for costs associated with shutdowns (besides production loss).

Current extensions to this work include the consideration of discrete process unit flow rates, common in manufacturing processes, as well as production and maintenance scheduling.

References

Balthazaar, A. K. S. (2005). Dynamic Optimization of Multi-Unit Systems Under Failure Conditions. M. A.Sc. thesis, McMaster University, Hamilton, Ontario, Canada.

Lee, E.S. and Reklaitis, G.V. (1989a). Intermediate storage and operation of batch processes under batch failure. Comput. Chem. Eng., 13, 491–498.

Lee, E.S. and Reklaitis, G.V. (1989b). Intermediate storage and the operation of periodic processes under equipment failure. Comput. Chem. Eng., 13, 1235–1243.

Tolio, T., Matta, A., and Gershwin, S.B. (2002). Analysis of two-machine lines with multiple failure modes. IIE Transactions, 34, 51–62.

641a Dynamics and Control of Vapor Recompression Distillation

Sujit S. Jogwar[1], Michael Baldea[2], and Prodromos Daoutidis[1]. (1) Chemical Engineering and Materials Science, University of Minnesota, 421 Washington Ave SE, 151 Amundson Hall, Minneapolis, MN 55455, (2) Praxair, Inc., 175 East Park Drive, Tonawanda, NY 14150

Distillation is one of the most energy consuming units in a chemical plant, motivating the need for energy integration. Vapor recompression distillation is one such energy integrated distillation configuration, wherein the vapor coming from the top of the distillation column is used to reboil the bottoms stream. Vapor recompression distillation is favored for the separations involving close-boiling liquids. Such separations result into large reflux ratios and small compressor duty is needed to facilitate heat transfer in the reboiler-condenser.

Most research on vapor recompression distillation has emphasized steady state economics focusing on capital costs, operating costs and optimal steady state operating conditions. Control objectives in these systems are the control of exit concentrations, pressure and holdups of various units. These integrated systems show strong control-loop interactions, because of the thermal coupling between units. However, most control studies have been within a multi-loop linear control framework. In this talk, we propose a comprehensive modeling analysis and control framework for such columns. Specifically, we illustrate that the discrepancies in material and energy flows in the column lead to a multi-time scale behavior. Large material recycle owing to large reflux ratio leads to time scale separation in the material balance dynamics. Energy recycle through the reboiler-condenser is dominated by latent heat effects, whose contribution to enthalpy is larger compared to sensible heat effects. This additional stiffness leads to further time scale separation in the energy dynamics, leading to overall three-time scale dynamic behavior.

Using singular perturbation arguments, reduced order non-stiff models valid in each time scale are obtained, which can be used for controller design. This analysis also leads to a classification of manipulated variables available in each time scale. The theoretical results will be illustrated via a simulation case study on a propane-propylene system.

641b Model Predictive Control of Feed Flow Reversal In a Reverse Osmosis Desalination Process

Panagiotis D. Christofides, Department of Chemical and Biomolecular Engineering, Department of Electrical Engineering, University of California, Los Angeles, Los Angeles, CA 90095, Alex Bartman, Chemical and Biomolecular Engineering, University of California Los Angeles, 420 Westwood plz, Los Angles, CA 90095, Charles McFall, Department of Chemical and Biomolecular Engineering, University of California, Los Angeles, Los Angeles, CA 90095, and Yoram Cohen, Chemical and Biomolecular Engineering Department, University of California, Los Angeles, 5531 Boelter Hall, Los Angeles, CA 90095-1592.

Reverse osmosis (RO) has emerged as one of the leading methods for water desalination due to the low cost and energy efficiency of the process [1]. Lack of fresh water sources has necessitated further development of these desalination plants, especially in areas with dry climates. In many reverse osmosis processes, particularly with brackish water feeds or systems running at a high level of recovery, dissolved ions can precipitate out of solution and crystallize on the membrane surface in a process called scaling. Scale formation on the membrane surface will lead to decreased permeate productivity, as well as permanent membrane damage if scaling is allowed to progress past its initial stages [2]. Several methods are currently used to prevent scale formation; addition of anti-scalant chemicals to the feed or flushing the membrane units with low-TDS (total dissolved solids) permeate water are two common procedures to accomplish this.

These current methods of scale mitigation have several disadvantages, anti-scalants can be expensive, and the addition of excess amounts can actually promote membrane scaling [3]. In the case of the permeate flush, this process will require the reverse osmosis operation to stop for a substantial amount of time to allow for the flushing cycle, eliminating any permeate production (even using up some of the previously produced permeate water). A novel technique called feed flow reversal has been developed which can prevent scale without the addition of expensive chemicals or extensive periods of system down-time [4]. This technique uses a system of actuated valves around the membrane modules configured specifically so that the direction of the feed flow through the membrane units can be reversed. This reversal of the feed flow also reverses the axial salt concentration profile at the surface of the membrane, effectively "resetting the induction clock" [4]. The reversal, if activated after crystals have already formed, also allows a substantial portion of scale deposited on the membrane surface to re-dissolve into solution. Model-predictive control algorithms are applied to a high capacity reverse osmosis system that utilizes feed flow-reversal in order to prevent and/or reverse scale crystal formation on the membrane surface. A dynamic non-linear model which incorporates feed concentration and membrane properties is used for simulation and control purposes [5]. Before flow reversal can take place on a high capacity RO plant, the flow into the membrane unit must be reduced to eliminate the risk of water hammer. A cost-function is formulated for the transition between the normal high flow steady-state operating point to a low flow steady-state operating point where it is safe to reverse the flow direction. Open-loop and closed-loop simulations demonstrate non-linear model-predictive control strategies that transition from the high-flow to low-flow steady-states in an optimal way while subjected to plant-model mismatch on the feed concentration, actuator constraints, and sampled measurements. These model-predictive control (MPC) strategies are compared to proportional integral (PI) control as well as a manually controlled transition.

References:

[1] A. Rahardianto, J. Gao, C. J. Gabelich, M. D. Williams, and Y. Cohen. High recovery membrane desalting of low-salinity brackish water: Integration of accelerated precipitation softening with membrane RO. Journal of Membrane Science, 289:123-137, 2007.

[2] D. Hasson, A. Drak, and R. Semiat. Inception of CaSO4 scaling on RO membranes at various water recovery levels. Desalination, 139:73-81, 2001.

[3] A. Rahardianto, W. Shih, R. Lee, and Y. Cohen. Diagnostic characterization of gypsum scale formation and control in RO membrane desalination of brackish water. Journal of Membrane Science, 279:655-668, 2006.

[4] N. Pomerantz, Y. Ladizhansky, E. Korin, M.Waisman, N. Daltrophe, and J. Gilron. Prevention of scaling of reverse osmosis membranes by "zeroing" the elapsed nucleation time. part i. calcium sulfate. Ind. & Eng. Chem. Res., 45:2008-2016, 2006.

[5] C. W. McFall, P. D. Christofides, Y. Cohen, and J. F. Davis. Fault-tolerant control of a reverse osmosis desalination process. In Proceedings of 8th IFAC Symposium on Dynamics and Control of Process Systems, volume 3, pages 163-168, Cancun, Mexico, 2007.

641c Networked Control and Scheduling of Distributed Energy Resources: Application to Biomass Gas-Fueled Fuel Cell Networks

Nael H. El-Farra, Yin Xu, Yulei Sun, and Sathyendra Ghantasala. Department of Chemical Engineering & Materials Science, University of California, Davis, One Shields Avenue, Davis, CA 95616

Distributed Energy Resources (DERs) are a suite of on-site, grid-connected or stand-alone technology systems that can be integrated into residential, commercial, or institutional buildings and/or industrial facilities. These energy systems include distributed generation, renewable energy sources, and hybrid generation technologies; energy storage; thermally activated technologies that use recoverable heat for cooling, heating, or power; transmission and delivery mechanisms; control and communication technologies; and demand-side energy management tools. Such distributed resources offer advantages over conventional grid electricity by offering end users a diversified fuel supply; higher power reliability, quality, and efficiency; lower emissions and greater flexibility to respond to changing energy needs. To fully realize the potential of distributed generation in enhancing the availability and reliability of electrical networks, advanced communication and control technologies are needed to enable the integration and inter-operability functions of a broad range of DERs, and meet the challenges of providing real-time, interactive, customer-managed service networks for an evolving electric infrastructure.

While managing DERs over a communication network offers an appealing modern solution to the control of distributed generation, it poses a number of challenges that must be addressed before DERs can be fully integrated into power networks. These challenges stem from the inherent limitations on the information transmission and processing capabilities of communication networks, such as bandwidth limitations, network-induced delays, data losses, signal quantization and real-time scheduling constraints, which can interrupt the connection between the central control authority, the distributed generation units and the loads, and thus degrade the overall control quality if not properly accounted for in the control and communication designs. Despite the availability of fast and reliable industry-standard communication networks, the fact that the distributed power market is primarily driven by the need for super-reliable, high-quality power implies that the impact of even a brief communication disruption (e. g., due to local network congestion or server outage) can be substantial. In sites such as hospitals, police stations, data centers and high-tech plants which cannot afford blackouts, millisecond outages that merely cause lights to flicker will cause costly computer crashes [1]. Such high-stakes risks provide a strong incentive for the development of robust control and communication strategies that achieve the desired levels of power supply and quality from DERs while minimizing reliance on the communication medium, which in turn minimizes the impact of data losses and disruptions on the power supply (see, for example, [2]).

One of the central problems in distributed energy generation is how to optimally dispatch and schedule DERs in real-time. This problem is typically addressed on the basis of economic considerations, such as minimizing the overall energy costs in real-time (see, for example, [3]). When a large number of DERs are deployed and communicate with a central control authority over a shared communication network, however, the intrinsic limitations on the number of DERs that can access the network at any given time become a critical factor in deciding the optimal scheduling policy. In practice, communication networks typically have a limited number of channels so that at any one time only some of the sensors and actuators can exchange information with the supervisor, while others must wait. Under these constraints, the performance of the control loops is dependent not only on the controller design for the individual DERs but also on the selection of the medium access strategy that would assign a transmission schedule to each transmission entity on the network based on a scheduling algorithm (i.e., a set of rules that, at any time, determines the order in which messages are transmitted). Communication-based scheduling also represents an important way of conserving the network's energy resources by selecting and dispatching only a subset of the deployed DERs at any given time.

To address this problem, we consider as a model system for our study a network of fuel cell plants in a power distribution system. The fuel cells are connected to a communication network over which they communicate with the supervisor (typically a computer equipped with suitable communication and control software) responsible for coordinating the levels of power supplied to the grid by each plant. Each fuel cell plant is composed of a Solid Oxide Fuel Cell (SOFC) fueled by biomass gas and interfaced with the electric utility grid through a power conditioning unit, a measuring unit and a control unit. While the utilization of a renewable energy source, such as biomass energy, is desirable to reduce both the consumption of fossil resources and the emission of carbon dioxide, operating a fuel cell on biomass gas can lead to an unstable power output. This instability is due to the fluctuation of gas composition in the fuel caused by the decomposition process of biomass resources (e.g., anaerobic digestion) and necessitates the use of feedback control to allow the fuel cell plant to rapidly follow changes in power demand and to also maintain the power output supplied to the grid at the desired level.

To illustrate the main ideas, we consider a configuration where the the supervisor has un-interrupted access to measurements from each SOFC plant and only the local actuator suites in these plants vie for access to the supervisor. Communication is limited in the sense that only one SOFC plant can receive control commands from the supervisor at any one time to stabilize the power output. The objective is to find a sequence for establishing and terminating communication without performance degradation in the collection. Both static scheduling, where the communication sequence is pre-determined off-line, and dynamic (i.e., feedback-based) scheduling policies are considered. The basic idea is to formulate the closed-loop system for each SOFC plant as a switched system that transitions between two modes of operation depending on whether access to the supervisor is granted or denied. By analyzing the behavior of each subsystem, we obtain an estimate of the maximum allowable rate of communication disruption that each plant can tolerate while maintaining its power output at the desired level. The estimate obtained depends on the time constant for each plant, the controller design parameters as well as the biomass feed composition fluctuations. Based on the various tolerance levels of the different plants, a static transmission schedule is then derived using the rate monotonic algorithm whereby priority is given to plants with faster dynamics and lower communication disruption tolerances. In addition to developing static schedules, we use Multiple Lyapunov Functions (MLF) techniques to devise dynamic (i.e., state-dependent) scheduling policies that are suitable for on-line implementation, are more robust when the power distribution system is subject to unpredictable disturbances, and allow the supervisor to respond quickly to a plant that requires immediate attention. The design, implementation and efficacy of the various scheduling policies are demonstrated using computer simulations of prototype SOFC plant models.

References:

[1] Borbely, A. and J. F. Kreider. Distributed Generation: The Power Paradigm of the New Millennium, CRC Press, Boca Raton, FL, 2001.

[2] Sun, Y., S. Ghantasala and N. H. El-Farra, ``Control of Distributed Energy Generation Over Communication Networks," AIChE Annual Meeting, paper 484g, Salt Lake City, Utah, 2007.

[3] Bae, I., J-O Kim, J-C Kim, and C. Singh, ``Optimal operating strategy for distributed generation considering hourly reliability worth," IEEE Transactions on Power Systems, 19, 2004.

641d Incipient Fault Detection In Offshore Oil and Gas Production Platforms: Hybrid Electro-Mechanical Monitoring of Rotating Equipment

Sathish Natarajan, Department of Chemical and Biomolecular Engineering, National University of Singapore, Singapore, Singapore, **Rajagopalan Srinivasan**, Chemical and Biomolecular Engineering, National University of Singapore, 10 Kent Ridge Crescent, Singapore 119260, Singapore, Singapore, and Ashwin Khambadkone, Electrical & Computer Engineering, National University of Singapore, Singapore, Singapore.

Offshore oil and gas production facilities are uniquely hazardous in that platform personnel have to work in a perilous environment surrounded by extremely flammable hydrocarbons. The harsh marine environment as well as oilfield disturbances makes the failure of production equipment common. Given the highly integrated nature and congestion of the processing facility, a failure in equipment could quickly propagate to others resulting in leaks, fires and explosions, causing loss of life, capital invested and production downtime. The failure of rotating equipments, in particular, on offshore platforms contributes significantly to the risk of an accident and hence their monitoring becomes important. The history of fault diagnosis and protection of rotating equipments is as archaic as the equipments themselves. The manufacturers and producers initially relied on simple acoustic and vibration observations to detect the presence of a fault which were sufficient when these equipments were first introduced at the start of the industrial revolution. However, nowadays, the tasks performed by these equipments and the processes they are utilized in have become increasingly complex. Hence, there is a need to develop intelligent monitoring tools capable of monitoring faults at their inception; as unscheduled downtime can upset deadlines and cause financial losses. Faults in rotating equipment in offshore platforms could be caused due to mechanical, electrical or process failures. Monitoring the overall process for the detection of such equipment faults may not be fruitful as faults may be detected too late and platform safety compromised. Hence, we propose to monitor the equipments characteristics to detect incipient faults early in their evolution.

Traditional techniques for monitoring of rotating equipment have relied on observations of periodically measured dynamic sampled response measurements such as vibration. Plenty of literature is available on rotating equipment monitoring using vibration analysis techniques, the most popular among them being frequency (Fourier, Hilbert, Cepstrum and Bispectrum) analysis and time-frequency (STFT, wavelet) analysis. Rotating equipments have also been monitored based on the performance of the electrical motor using current measurements. Similar to vibration analysis, motor current signature analysis (MCSA) also employ signal processing techniques like those mentioned above. But, there is not much literature on a combined electrical and mechanical fault detection technique for rotating equipments. For instance, it is well known that bearing faults induce eccentricities in both current and vibration signals but attempts to monitor the fault have relied on only either one. This paper describes a novel time-frequency signal processing technique along with a hybrid electro-mechanical approach towards monitoring of rotating equipment.

This method is evaluated on two distinct testbeds; one is an experimental setup in which the characteristic vibration and current signals were observed for a lab scale induction motor under normal and fault conditions. The other testbed is an electromechanical model of a typical pump and compressor, similar to those found on offshore oil and gas platforms. Electrical (broken rotor bar, stator winding) and mechanical (bearing) fault are introduced into this model and the characteristics observed. Using measurement data from these testbeds, the signal processing technique is used to monitor the signals and ascertain the condition of the equipment. The results of the combined electro-mechanical monitoring scheme is compared with monitoring either vibration or current signals alone. The proposed scheme is also benchmarked against other signal processing technique currently available in literature.

References

Thorsen, O.V., Dalva, M., (1995), "A survey of faults on induction motors in offshore oil industry, petrochemical industry, gas terminals and oil refineries" IEEE Transactions on Industry Applications, Vol.31, No 5.

Jardine, Andrew K.S., Lin, Daming and Banjevic, Dragan (2006), "A review on machinery diagnostics and prognostics implementing condition based maintenance", Mechanical Systems and Signal Processing, 20, 1483-1510

Nandi, S., Toliyat, H.A. and Li, Xiaodong (2005), "Condition monitoring and fault diagnosis of electrical motors", IEEE Transactions on Energy Conversion, Vol.20, No 4.

641e Successful Industrial Application of Robust Inferential Sensors for NOx Emissions Monitoring

Zdravko Stefanov[1], Leo H. Chiang[2], Arthur Kordon[1], James Edwards[3], David Roberts[3], Rodney Roubique[3], and Charles Harper[3]. (1) The Dow Chemical Company, Corporate R&D, 2301 Brazosport Blvd., B1463, Freeport, TX 77584, (2) Core R&D, The Dow Chemical Company, 2301 Brazosport Blvd., B1225, Freeport, TX 77584, (3) Energy Systems, The Dow Chemical Company, Plaquemine, LA 70765

As a large chemical company, Dow Chemical uses and maintains multiple power plants. One of the major environmental requirements on power plants is compliance of their NOx emissions with Environmental Protection Agency (EPA) regulations. To satisfy governmental regulation, Dow is obligated to maintain Emission Monitoring Systems. These systems are either hardware based (Continuous CEMS) or software based (Predictive PEMS).

The application of the robust inferential sensors was considered in 2004 when a vendor installed system was to be replaced and in-house solution was determined to be much more cost effective. Two modeling technologies were concurrently evaluated – Partial Least Squares (PLS) which is a linear approach and Genetic Programming (GP) which is a nonlinear approach. As a consequence GP models were proven to be superior and were implemented.

The robust implementation is achieved by sensor validation. Sensor validation provides redundant backup for the inputs of the inferential sensor, so the sensor continues to operate even if some of the inputs fails. This robustness is also an EPA requirement. Due to this implementation and the quality of the models, four years of EPA compliance are recorded as of 2008. Certain details on the inferential sensors maintenance and standardization will be discussed.

ON APPLYING ADVANCED PROCESS CONTROL FOR FCCU IN A PETROLEUM REFINERY

Xiaoming Jin, Zhejiang University, Hangzhou, China
Maying Yang, Zhejiang University of Technology, Hangzhou, China

Abstract

Fluid catalytic cracking unit (FCCU) is difficult for modeling and control due to the large process scale, complicated hydro-dynamics and complex kinetics of both cracking and coke burning reactions, highly heat and material interactions in its reactor-regenerator, main fractionator and absorber-stripper-stabilizer sections, and frequently changes of feed rate and feed composition. Since FCCU is capable of converting large quantities of heavy feed into valuable lighter products, any improvement in design, operation and control can result in substantial economics benefits. A commercial advanced process control (APC) suite was developed and implemented in a 1.8 million tons per year FCCU at Zhenhai Refinery of SINOPEC to help improve process operation and achieve process economics. In this paper the key points of APC technique and its FCCU Applications are highlighted with emphasis on discussion of FCCU control strategy design. Commercial results from the industrial implementation are also shown.
Keywords: FCCU, Advanced process control, Predictive control

1 Introduction

FCCU is one of the most important units in the petroleum refining industry for the conversion of heavy gas oil to gasoline and light hydrocarbons. With the escalating price of crude oil and increasingly heavier FCC feedstock, more and more refineries in China try to increase the blending ratio of residue in FCCU feedstock to improve the secondary processing benefits. Since the performance of the FCC units plays a major role on the overall economics of a refinery, there is a strong demand for advanced process control and real-time optimization with higher control quality to meet the challenges imposed by the growing technological and market competition.

The FCCU at Zhenhai refinery of SINOPEC was originally installed in 1978, designed for processing 1.2 million tons of VGO (vacuum gas oil) per year. It was expanded and revamped in 1997 and 2004 respectively. Based upon the special features of the two-stage riser FCC process, it has a handling capability of 1.8 million tons per year of the mixtures of VGO, CGO (delayed coker gas oil) and residual oil of the VDU (vacuum distillation unit), aiming at producing more propylene and improving gasoline quality. The APC system was implemented on the FCCU in 2005, using APC-Adcon and APC-Sensor technologies from Zhejiang Supcon Software Co., Ltd. It consisted of four multivariable controllers, covering reactor-regenerator section, main fractionator section, and absorber-stripper-stabilizer section, with seven soft sensors for catalyst circulation rate, product qualities of LPG (liquefied petroleum gas), stabilized gasoline, diesel and fuel gas. The main benefits of this application came from steady, safe and reliable FCCU operation while satisfying operating constraints, improvement of quantities and qualities of FCCU products, and continually maximization of residual oil feed against the available temperature limit of the second regenerator.

The paper is organized as follows. In Section 2 the FCC process and control requirements are briefly described. In Section 3 the APC structure and some important issues of the controllers are discussed. In Section 4 industrial application results are presented. The contrastive application results corresponding to conventional control and APC are given. The paper is finally concluded in Section 5.

2 FCC Process Description and Control Requirements

The FCCU at Zhenhai refinery was designed to convert low value mixtures of heavy distillate oil such as wax distillate and residual oil into high value light products, e.g. gasoline, light diesel, etc. by cracking the oil at high temperature in the presence of a catalyst. It consists of reactor-regenerator, main fractionator, absorber-stripper-stabilizer, main air blower and wet gas compressor, etc.

The catalyst type was chosen to maximize propylene based upon the special features of the two-stage riser FCC process. Reaction products and spent catalyst discharge in the reactor whose main function is to disengage the catalyst particles from the vapor product through a battery of cyclones. Catalyst regeneration is achieved by burning off the coke deposit in fluidized bed inside the two cascade regenerators. Catalyst in catalyst circulation rises from riser to reactor and from there flows to regenerator by stripper and stand-pipe and then rises up again from riser to reactor. Steam turbine driven air blowers supplies the oxygen to burn the coke deposit. The main fractionator separates the reactor products into bottoms slurry oil, cycle oil, light diesel, raw gasoline, and wet gas stream is separated in the absorber-stripper-stabilizer section. A simplified flow diagram of the FCCU is shown in Fig.1.

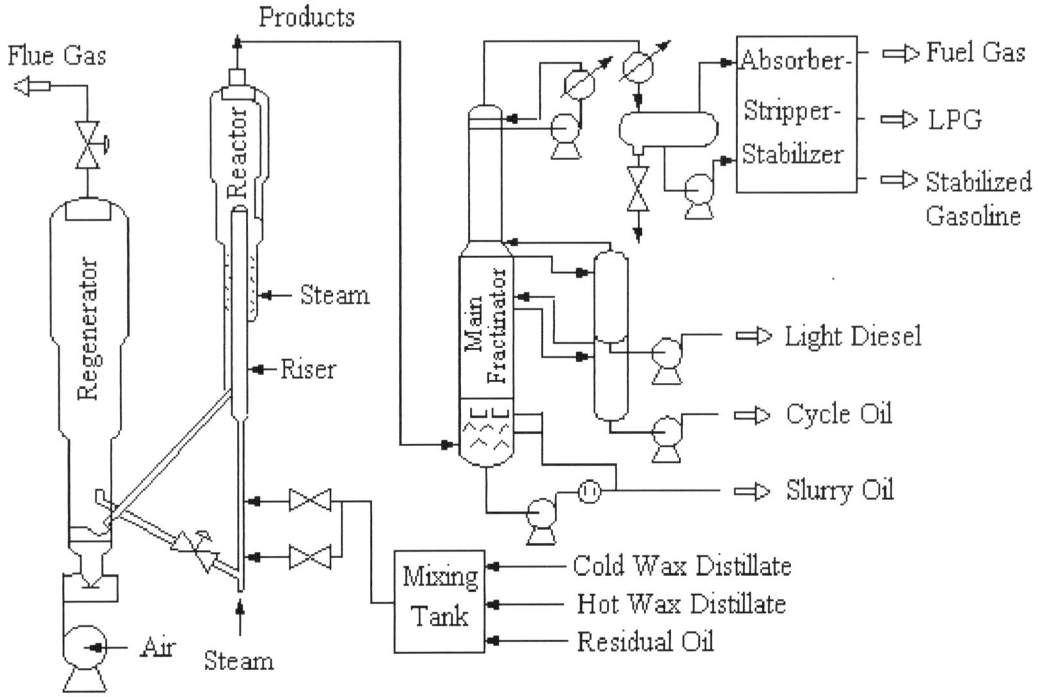

Fig.1 Simplified flow diagram of FCCU

The dynamics of the unit are complex, which consists of multiple sophisticated chemical and physical processes. The production goal is to maintain safe and reliable operation, to produce the cracked light products with cost minimization, i.e. improving the process operating stability, maximizing the process throughput and desirable products yield, improving the product quality, and minimizing energy consumption which in turn maximize the profit. From process control point of view, FCCU can be considered as a high dimension, highly nonlinear, interconnected, complex system. Furthermore, the unmeasurable, uncontrollable factors (such as feedstock and catalyst properties) and process constraints continuously disturb the unit making process control and optimization difficult.

The objective of the reactor-regenerator control is, to attain the best possible operation and gain most profits at large throughputs of residual oil and low energy consumption. It is sure that the process should be run at its optimized zones near constraints. The major independent reaction variables in the FCCU, operating in the maximum gasoline mode are: reactor temperature, combined feed rate, combined feed (feed preheat) temperature, reactor pressure, catalyst activity. The dependent variables, on the other hand, are responsive to changes in the independent variables. These include: the catalyst circulation rate, the catalyst/oil ratio, the regenerator temperature, and conversion.

The purpose of the main fractionator is to de-superheat and recover vapor. The hot-product vapors from the reactor flow into the main fractionator; these vapors enter the column near the base. The main function of the fractionator is to condense and separate the reaction products. The operation of the main column is similar to a crude tower but with two differences. First, the effluent vapors must be cooled before any fractionation begins. Second, large quantities of gases will go overhead with the un-stabilized gasoline for further separation.

Aside from slurry oil product and wet gas, the main fractionator has three side cuts: raw gasoline, light diesel and cycle oil. The recovered heat from the main fractionator is used to preheat the fresh feed, generate steam, and serve as a heating medium for re-boilers of the absorber-stripper-stabilizer section.

The control objectives in the absorber-stripper-stabilizer section are to maintain distillate composition at set point in the presence of disturbance. These disturbances may be characterized due to (i) loads, (ii) changes in cooling and heating medium supply condition and (iii) equipment fouling. Any distillation control system we design must be able to cope with the types of disturbances in the feed and in the supply conditions of the heat exchangers.

3 APC System Design

A hierarchical advanced process control and optimization system is developed as in shown in Fig 2, which integrates 4 sub-controllers based on model predictive control strategy and 1 intelligent controller for maximization of residual oil, 7 soft-sensors, by using APC-Adcon and APC-Sensor respectively. The system communicates with Honeywell TPS DCS by APP NODE.

There are more than ten operating variables to test the operation performance of the FCCU, mainly corresponding to product qualities, desirable products yield and energy consumption. The FCCU needs to be maintained close to optimal operating conditions, however, it's difficult to achieve favorable regulatory loop control performance all the time. Therefore, model predictive control (MPC) can be used to improve control performance with a reduction in the variability of the controlled variables through information gathering, process analysis, and constrained multivariable optimization.

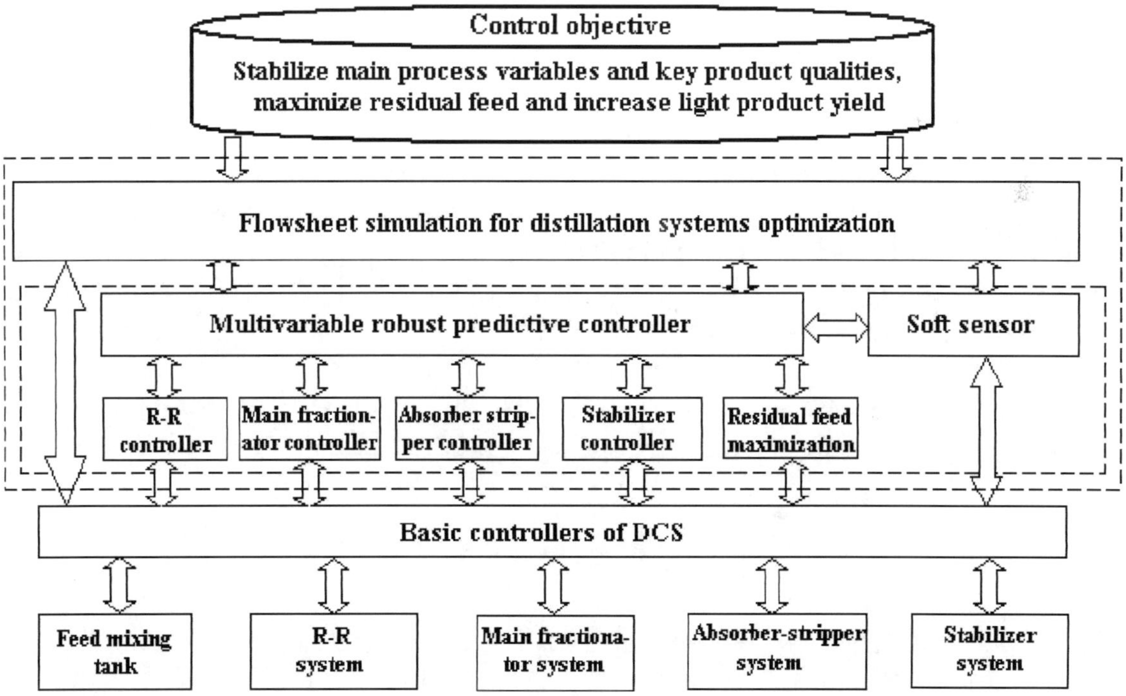

Fig 2 Advanced process control system hierarchy for RFCCU

3.1 Multivariable predictive controller

The advanced process control package of the FCCU at Zhenhai refinery comprises four sub-controllers and one intelligent controller. In the reactor-regenerator sub-controller, second regenerator temperature, feed tank level, second regenerator excess O2, and differential pressure of regenerated catalyst slide valve are the key controlled variables, combustion air flow residual feed flow, hot VGO flow, and combined flow of cold VGO, CGO and recycle oil are manipulated variables. In addition, there are eight disturbance variables. In main fractionator sub-controller, there are six controlled variables such as all product quality variables, ten manipulated variables included reactor riser temperature, and six disturbance variables. In the absorber-stripper-stabilizer section, we adopted two sub-controllers. The absorber-stripper sub-controller has five controlled variables, four manipulated variables and three disturbance variables, and the stabilizer has five controlled variables, three manipulated variables and three disturbance variables. The objective of the constraint controller is to maximize residual feed flow against the second regenerator metallurgical high temperature limit.

3.2 Soft sensor

In order to support the APC system of the unit, using APC-Sensor software, by incorporating principal models with fuzzy neural network, we developed 7 soft sensors to infer the key process variable values for process control and monitoring. which are: catalyst circulation, product properties for the main fractionator and the absorber-stripper-stabilizer such as the end point for raw gasoline, 95% point for LCO, the C2 and C5 concentrations in the LPG, the C3 concentration in fuel gas, and RVP for the stabilized gasoline, etc.

4 Industrial Application Results

After the implementation of the APC system, the deviation of the main process variables became one half of that before APC implementation, as shown in Fig 3 (TI521, LIC401, LI405, AND TI405 are the 2nd regenerator's temperature, level of separator of oil and gas, stripper bottom level, and stripper bottom temperature). The standard deviations of the main products qualities before and after APC are shown in Table 1.

As a result, The over economical merit from this implementation is approximately over four million RMB Yuan by increasing the production of light product, improving product quality, and minimizing operating cost. Industrial application results show that the APC software can maintain the best operation for a long time and realize ultimate operating potential of the RFCCU by reducing the consumption of energy, improving product quality, and minimizing operating cost.

5 Conclusions

This paper introduces the industrial application of APC for the FCCU in Zhenhai Refinery. The APC system is developed to deal with the constrained multivariable control of the reactor-regenerator, main fractionator, and absorber-stripper-stabilizer sections on-line. Several soft sensors are developed to implement direct control of product quality. It can be concluded from the control application example of the FCCU that along with continued implementation of model predictive control, well understanding of chemical engineering principle will help control the process subject to operating constraints, integrating flowsheet simulation and soft sensor technology improves operational effectiveness from the individual process level to the online optimization.

Acknowledgement

This work is supported by The National Creative Research Groups Science Foundation of China (NCRGSFC: 60421002). The authors would also like to express appreciations to Da Zhang, Junlin Shi and Zhenhai Refinery of SINOPEC for the cooperation.

References

1. R. E.Young(2006), "Petroleum Refining Process Control and Real-Time Optimization", *IEEE Control System Magazine*, Dec., pp. 73-83
2. A. C. Zanin, M. T. de Gouvea, D. Odloak(2002), "Integrating real-time optimization into the model predictive controller of the FCC system", *Control Engineering Practice*, 10, pp. 819-831
3. MY. Yang, et al(1998), "FCCU reactor-regenerator predictive control research", *ACTA Automatica Sinica*, 24(3), 359-362 (In Chinese)

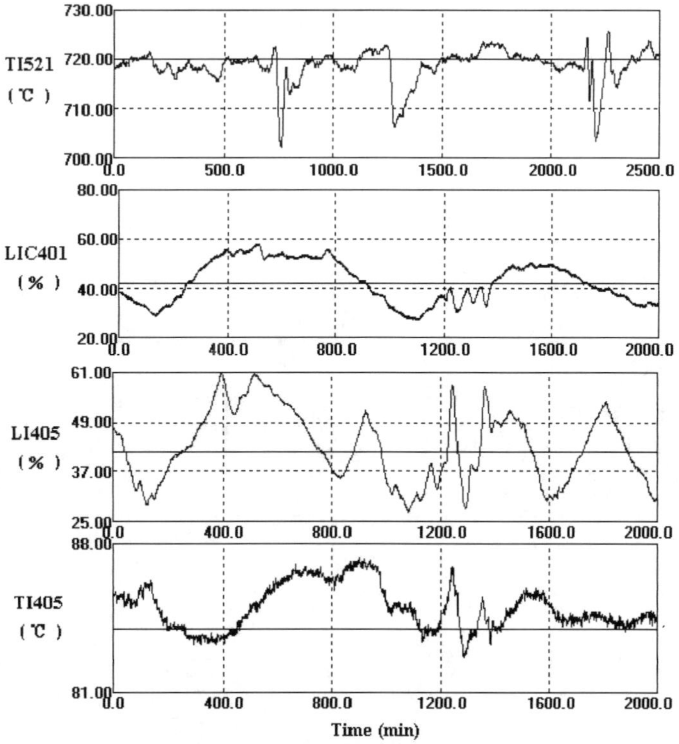

(a) before APC implementation

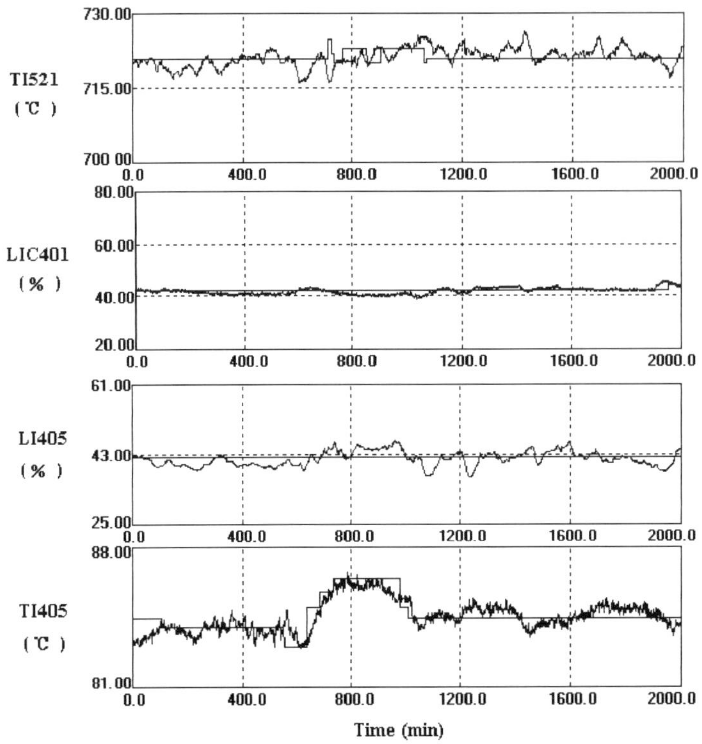

(b) after APC implementation

Fig 3 Comparison of process control performance under APC and classical control

Table 1 . Comparison of standard deviations of the main products qualities

standard deviations	Before APC	After APC	Reduction
end point for raw gasoline	2.57	4.18	38.52%
95% point for LCO	5.79	7.64	24.21%
C_3 concentration in fuel gas	0.368	0.514	28.40%
C_2 concentrations in the LPG	0.011	0.035	68.57%
C_5 concentrations in the LPG	0.167	0.21	20.48%
RVP for stabilized gasoline	1.3	1.58	17.72%

649a A History and View to the Future for Knowledge Management and Organizational Learning In the Process Industries

Vince Grassi, Air Products and Chemicals, Inc., 7201 Hamilton Blvd, Allentown, PA 18078

This presentation will discuss the history of how knowledge management has evolved from groupware information applications into an important contributor for talent management in the process industries. Evolutionary examples will be given of collaboration, best practice transfer, idea management, and replication of lessons learned through communities and networks. It will then be shown how KM has accelerated the use of informal learning reduce cost and provide greater effectiveness of learning across global enterprises. Finally we will discuss how further developments will occur as knowledge management integrates with talent management to deploy more effective global workforce networks.

649b Sustainability through Knowledge Management

James B. Porter Jr., DuPont, 1007 Market St, Wilmington, DE 19898-0001

Among the many tools and techniques that companies can implement to become more sustainable is effective management of their information assets. Sustainability requires a current, comprehensive awareness of the corporate "footprint" to properly assess alternatives for taking advantage of new opportunities and optimizing external impacts. In large organizations, where such information is both voluminous and complex, knowledge management can support more efficient and effective user access to and detailed analysis of information. In addition, effective dissemination of information about new and more sustainable business methods and offerings can lead to faster implementation across the enterprise. Knowledge management techniques are available to facilitate these communications and the extraction and synthesis of relevant knowledge from business performance data.

DEVELOPING A REPORTING SYSTEM FOR ENVIRONMENTAL COMPLIANCE

Author: Jannina Alvarez, International Specialty Products

The first step in developing a new environmental compliance reporting system is a detailed analysis of the company's goals and objectives. From this analysis, planners can define project requirements and establish an approach to making the crucial decision of whether to develop a custom application, purchase an off-the-shelf application, or work with a vendor to customize their off-the-shelf application to more closely address the company's needs.

The definition of project requirements may be further refined to identify user needs and system specifications. This paper will focus on a development process that ISP implemented to develop a new environmental reporting system. This paper will include "lessons learned" as well as insights that will be of interest to anyone facing a similar challenge in their organization.

1st Perform a Gap Analysis of the Environmental Data Management System

In order to develop a new Environmental Data Management System (EDMS) to meet stakeholders, corporate and site specific data needs, requirements were analyzed and evaluated in a top-down analysis. This analysis focused on stakeholder's needs, corporate needs, and site environmental management system (EMS) requirements.

- Step 1. Map current data storage and access requirements.

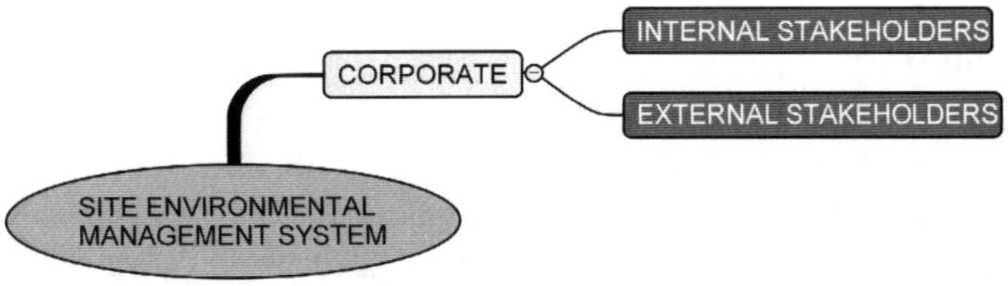

- Step 2. Identify corporate and site-specific environmental reporting requirements.

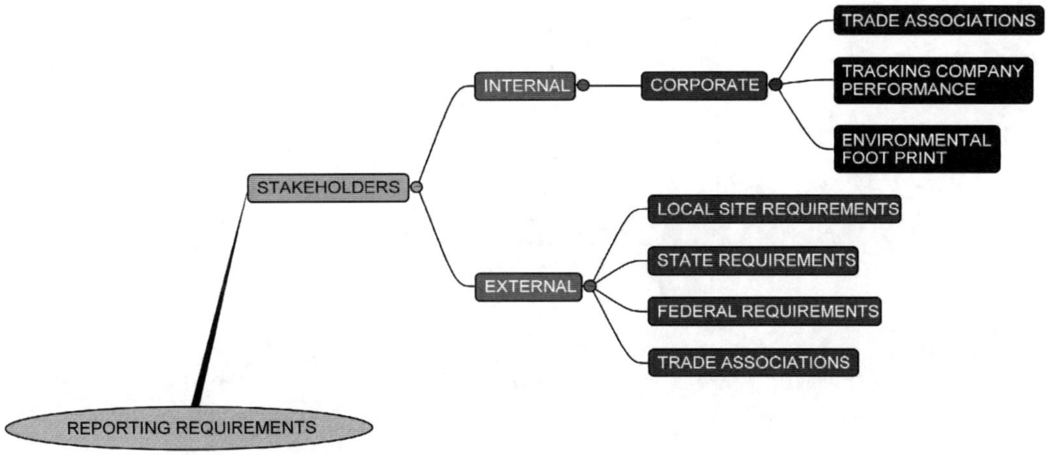

- Step 3. In order to evaluate the EDMS and identify core requirements, a gap analysis of the company's EDMS was performed. Based on this analysis, it was determined that the company's current EDMS was idle because plant personnel were not properly trained on utilizing this system and all users reported that it was difficult to implement. In addition, there were issues with data validation, data gathering, ability to create reports, lack of system support, and difficulty on connecting with other data libraries in the enterprise.

2nd Evaluate the Environmental Data Management Systems at Each Facility

ISP evaluated what each individual facility was doing in order to comply with their stakeholder's requirements and how their environmental data was being gathered and processed. The gap analyses determined that each facility was:

- Manually collecting data
- Excessively using excel sheets
- Manually entering data into excel and access databases
- Multiple handling of data
- Use of excel for data calculation
- Additional head count for data handling
- Numerous hours on data validation and reports generation

Based on the above information each method or form of data gathering and processing was map. These methods were essentially utilized by the facility in order to meet the stakeholder requirements and be in compliance with all local, state and federal regulatory requirements. By mapping each facility data handling, it allowed ISP to determine what kind of data management system the facility would benefit from.

The following data analysis was performed:

- What goes into MS Excel?
- What data is manually entered?
- What data is saved in compliance databases?
- What compliance reports are manually done?
- How much time goes into preparing reports and entering data?

Step 1.

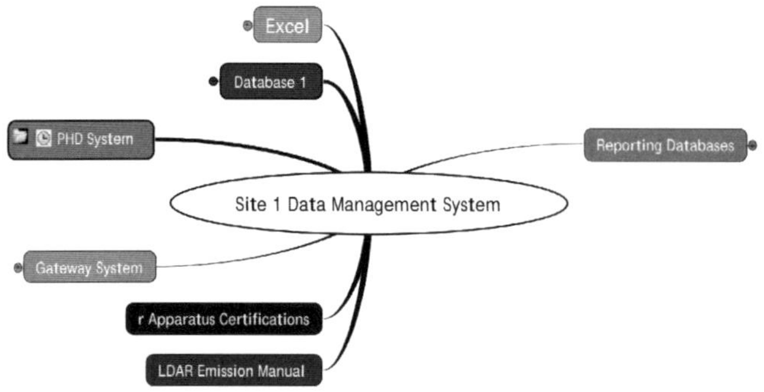

Step 2.

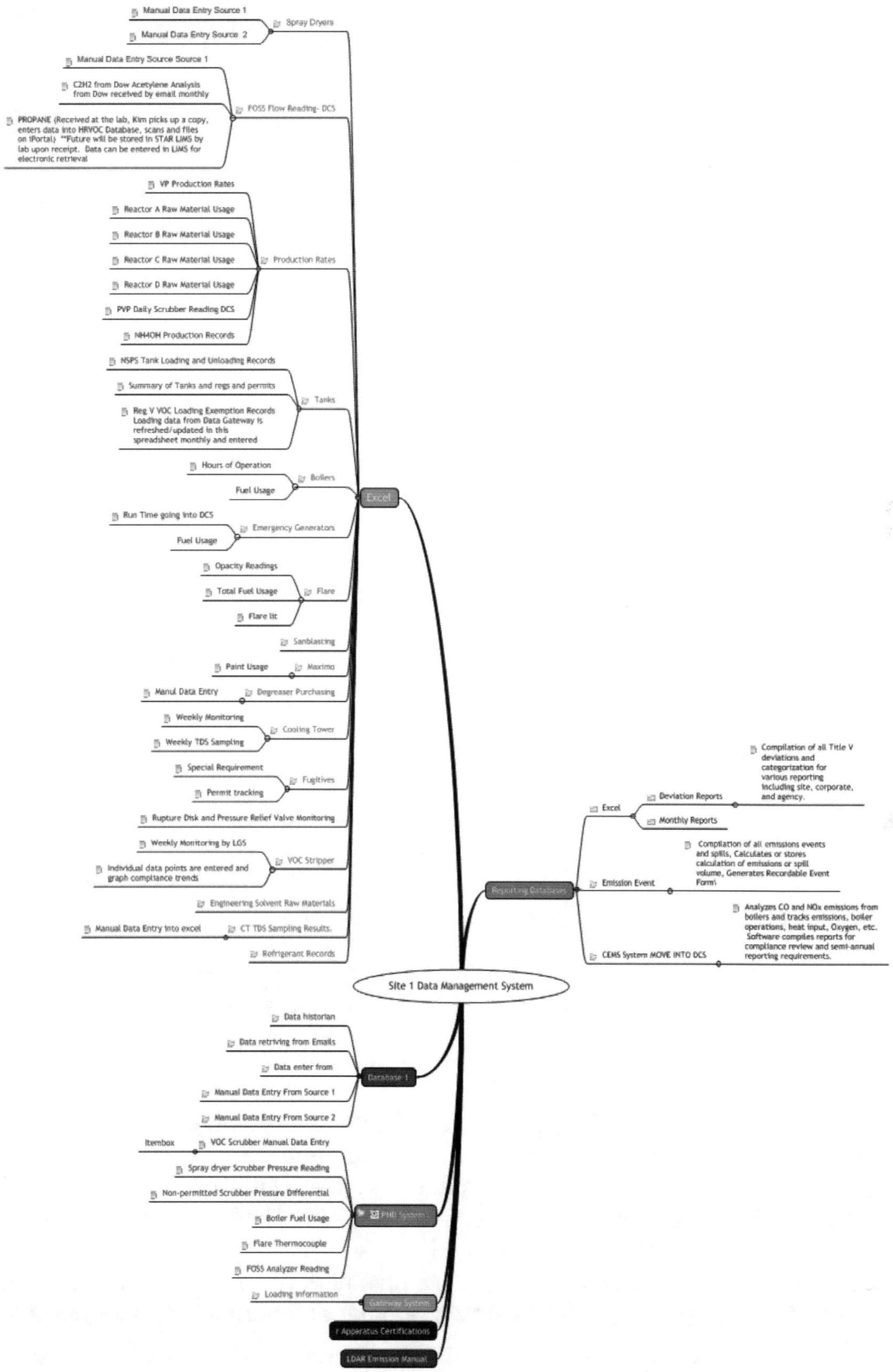
Site 1 Data Management System
Excel
Spray Dryers
Manual Data Entry Source 1
Manual Data Entry Source 2
FOSS Flow Reading- DCS
Manual Data Entry Source Source 1
C2H2 from Dow Acetylene Analysis from Dow received by email monthly
PROPANE (Received at the lab, Kim picks up a copy, enters data into HRVOC Database, scans and files on iPortal) **Future will be stored in STAR LIMS by lab upon receipt. Data can be entered in LIMS for electronic retrieval
Production Rates
VP Production Rates
Reactor A Raw Material Usage
Reactor B Raw Material Usage
Reactor C Raw Material Usage
Reactor D Raw Material Usage
PVP Daily Scrubber Reading DCS
NH4OH Production Records
Tanks
NSPS Tank Loading and Unloading Records
Summary of Tanks and regs and permits
Reg V VOC Loading Exemption Records Loading data from Data Gateway is refreshed/updated in this spreadsheet monthly and entered
Boilers
Hours of Operation
Fuel Usage
Emergency Generators
Run Time going into DCS
Fuel Usage
Flare
Opacity Readings
Total Fuel Usage
Flare lit
Sanblasting
Maximo
Paint Usage
Degreaser Purchasing
Manul Data Entry
Cooling Tower
Weekly Monitoring
Weekly TDS Sampling
Fugitives
Special Requirement
Permit tracking
Rupture Disk and Pressure Relief Valve Monitoring
VOC Stripper
Weekly Monitoring by LGS
Individual data points are entered and graph compliance trends
Engineering Solvent Raw Materials
CT TDS Sampling Results.
Manual Data Entry into excel
Refrigerant Records
Reporting Databases
Excel
Deviation Reports
Compilation of all Title V deviations and categorization for various reporting including site, corporate, and agency.
Monthly Reports
Emission Event
Compilation of all emissions events and spills, Calculates or stores calculation of emissions or spill volume, Generates Recordable Event Form\
CEMS System MOVE INTO DCS
Analyzes CO and NOx emissions from boilers and tracks emissions, boiler operations, heat input, Oxygen, etc. Software compiles reports for compliance review and semi-annual reporting requirements.
Database 1
Data historian
Data retriving from Emails
Data enter from
Manual Data Entry From Source 1
Manual Data Entry From Source 2
PHD System
VOC Scrubber Manual Data Entry
Itembox
Spray dryer Scrubber Pressure Reading
Non-permitted Scrubber Pressure Differential
Boiler Fuel Usage
Flare Thermocouple
FOSS Analyzer Reading
Gateway System
Loading information
r Apparatus Certifications
LDAR Emission Manual

3rd Identify a new Environmental Data Management System

After identifying the EDMS gaps at each facility, the following options were evaluated:

- Create a pilot study utilizing current EDMS by using an outside consultant firm
- Develop a EDMS system from scratch that could satisfy the Site needs
- Look for new data management systems
- What environmental data management systems are other companies using
- Cost Analysis of the new EDMS vs. Pilot study

Based on the above evaluation it was determined that the pilot study and developing a new data management system from scratch was going to be too expensive, so we decided to look for other options.

After contacting different industries we came across Process Data Control Corp. (PDC)'s and their software products (COMPASS Emission Inventory and Calculations, TaskTrakker, IntelliRegs, and Waste Modules), and decided to contact them. PDC's solutions were evaluated and found to be superior to EDMS tools and techniques that were currently in use, and they also offered a straightforward implementation strategy.

What PDC is currently doing for our facility is to streamline the data management process and minimize repetitive data handling and processing.

4th Define and Undertake Next Steps to Implement the New EDMS

After determining how to approach the need for a new EDMS at each facility, and for the corporate staff, the following strategy was developed:

- Roll-out the PDC Waste Module to other ISP facilities
- Identified the other PDC Modules that could be implemented in ISP facilities that would improve efficiency and effectiveness of environmental reporting
- Develop a platform that will allow corporate to track individual site compliance, and help to determine the Company's Environmental Performance

Conclusions

Developing an effective EDMS takes a lot of time and resources. It also requires focus on the issues related to compliance with applicable rules and permits. These issues cover a broad range of concerns, from high-level processes such as deviation reporting and review, to detailed issues such as personnel assignments to compliance tasks, recordkeeping, and audits. However, once the new EDMS implementation is completed, there is no better satisfaction than having a system that actively supports the attainment of key goals such as achieving compliance, minimizing the redundant handling of data, and reducing opportunities for error.

649d Understanding Root Causes of Abnormal Situations in Process Industries

Peggy A. Hewitt, Honeywell, 4-397 Brunswick Ave., Toronto, ON msr2z2, Canada

The Abnormal Situations Management (ASM) Consortium is a long-running and active research and development consortium of 13 companies and universities concerned about the negative effects of industrial plant incidents. The ASM Consortium has focused on improving abnormal situation management with a focus on the role of the operations personnel. From this perspective, the ASM Consortium has studied challenges to successful abnormal situation management and collaborated on the development of effective solutions spanning operations culture, organizational structure, management systems and use of technology to mitigate those challenges.

Understanding the root causes of abnormal situations in process industries is fundamental to eliminating and mitigating abnormal situations. This presentation will discuss research findings of the ASM Consortium on common problems and failure modes with an emphasis on the analysis of human factors in relation to incidents. We will share the top three causes identified in our research and discuss recommendations for companies to reduce abnormal situations.

649e The Importance of Human Factors in Control Systems and Knowledge Management

B. Wayne Bequette, Rensselaer Polytechnic Institute, Department of Chemical & Biological Engineering, 110 Eighth St., Troy, NY 12180-3590 and William Bosler, William Bosler, Moscow, Russia.

Process Control Systems enable operators of production units to operate their equipment and detect abnormal conditions. The knowledge acquired from these activities can enhance future operations if it is properly managed, and human factors have been found to play a significant role in this aspect of control system utilization. This paper will investigate how human factors should be considered in the design and engineering of control systems in a manufacturing environment.

649f Using Industry Consortiums for Knowledge Management and Technology Transfer

Ed Fiesinger, Zephyr Environmental Corp, 3000 Wilcrest Dr. Suite 145, Houston, TX 77042-3365

Voluntary Industry consortiums provide a unique opportunity to manage knowledge and the means by which knowledge may be shared among the members. Consortium can also develop new knowledge through the interaction of members within properly structured and resourced activities, such as roundtable discussions and collaborative design of data management systems. This paper will explore how consortiums can be developed and operated to maximize their benefits to knowledge management and organizational learning.

650a Quantifying Intrinsic Noise Effects In Biochemical Reaction Networks

Vassilios Sotiropoulos, Chemical Engineering and Material Science, University of Minnesota, 151 Amundson Hall, 421 Washington Ave SE, Minneapolis, MN 55455 and Yiannis N. Kaznessis, Chemical Engineering and Materials Science, University of Minnesota, 151 Amundson Hall, 421 Washington Ave SE, Minneapolis, MN 55454.

In many cases of biological phenomena the resulting cell phenotype is a direct result of the randomness of collisions between sparse interacting or reacting biological species. Consequently, the classical, well known and well studied continuous-deterministic modeling approach can be distinctive false and incapable of capturing the importance of intrinsic noise fluctuations. Instead a discrete-stochastic theoretical approach is more appropriate. The disadvantage is that the latter is less studied and more computationally demanding. In the classical approach the reaction networks used to describe cell functions are modeled using mass action kinetics resulting in systems of ordinary differential equations (ODEs). On the other hand, in the stochastic regime chemical kinetics models are described as jump (discrete) Markov processes governed by the chemical Master equation (CME). The question is when these two descriptions are equivalent, in which case the deterministic approach is preferable. In the simplest of cases, linear kinetics, it has been established that the mean of the probability distribution corresponds to the solution of the corresponding ODEs [1]. The two approaches become equivalent at the thermodynamic limit where the variance of the probability distribution vanishes. Conversely, for non-linear systems the correspondence of the mean of the probability distribution to the solution of the ODEs is just an approximation under the assumption that higher probability distribution moments (variance, third, fourth and so on moments) become negligible as the system's size increases, i.e. approaches the thermodynamic limit. Therefore the question reduces to when higher moments indeed become negligible. The challenge is that there are no analytical expressions for computing higher moments, except for the simplest of cases where the CME can be analytically solved. Thus a numerical method is a necessity for approximating their values. Our approach is based on the definition of jump moments or derivatives moments as originally called by Moyal [2]. We use a series of numerical experiments to compute (estimate) the values of jump moments for different biochemical networks and convey on the validity of the classical approach. Additionally, we try to relate our findings to the number of reacting species and probabilistic reaction rates and define thresholds above which higher moments of the probability distribution can be ignored for all practical purposes. Ultimately, we would like to explore the general applicability of such conditions, where one can determine for arbitrary networks a priori which regime, stochastic or deterministic, is appropriate.

[1]. Van Kampen NG: Stochastic Processes in Physics and Chemistry. Revised and enlarged edn. Amsterdam: Elsevier; 2004.

[2]. Moyal JE: Stochastic Processes and Statistical Physics. Journal of the Royal Statistical Society Series B (Methodological) 1949, 11:150-210.

650b Linking a Mitotic Oscillator to the Extracellular Environment

Ryan Vargo and Kyriacos Zygourakis. Chemical and Biomolecular Engineering, Rice University, Dept. of Chemical and Biomolecular Engineering, MS-362, P. O. Box 1892, Houston, TX 77251-1892

Cell proliferation and migration are two of the key processes regulating the growth rate and structure of engineered tissues. These important cellular processes are strongly influenced by the extracellular environment. Extracellular pH and the concentrations of dissolved oxygen, sugar sources, and growth factors are among the many factors that affect the dynamics of the cell cycle. They are all linked to the cell cycle by regulating the amount of energy and protein in the cell. ATP is an energy carrier that plays a crucial role in the cell cycle. It is necessary for phosphorylating the substrates of the key kinases that regulate the cell cycle. ATP is also necessary to synthesize proteins involved in the cell cycle.

In an earlier work, we utilized a bottom-up approach to develop a dynamic model that captured the ideal mitotic oscillator behavior while incorporating the current biological knowledge. Here, we will utilize a similar approach to link the mitotic oscillator to the extracellular environment by incorporating ATP into our model that now includes four coupled subsystems:

a)The subsystem responsible for activating the metaphase promoting factor (MPF) which triggers cellular division when it reaches some critical concentration.

b)The subsystem producing phosphorylated Cdc25A species that activate MPF and are stabilized by MPF, thereby forming a positive feedback loop.

c)The subsystem producing phosphorylated Wee1 species that inhibit MPF and are inhibited and destabilized by MPF, thereby forming a double negative or positive feedback loop.

d)The subsystem leading to the activation of the anaphase promoting complex (APC) which becomes activated by MPF and, in turn, contributes to the degradation of MPF and Cdc25A species, thus forming a negative feedback loop.

ATP is incorporated into the model in two different places. First, MPF and Wee1 must bind to ATP to phosphorylate their substrates. In addition, ATP is necessary for synthesizing MPF, Wee1, and Cdc25A.

A sensitivity analysis was performed to elucidate the effect of model parameters on the oscillatory characteristics of our system. Specifically, we investigated the effects of each parameter on the size of the oscillatory region, the amplitude and period of the oscillations, and whether or not the oscillations were frequency encoded. We also searched for the region of the parameter space that exhibited multiple steady states since hysteresis has been observed experimentally in the MPF subsystem.

While it is known that MPF inhibits Wee1 via two different pathways, there is still significant uncertainty about the structure of the underlying network. To elucidate the mechanism of Wee1 inhibition, we analyzed four different reaction networks that lead to Wee1 inhibition and studied the effect of these networks on the oscillatory characteristics of the mitotic oscillator.

Finally, we linked the intracellular ATP to extracellular glucose and incorporated substrate limitations into the protein synthesis rates. This model can now be tuned to multiple cell lines that grow under different extracellular glucose concentrations. Although glucose (or any single nutrient) is not the only environmental condition affecting cell proliferation, this work is an important first step in capturing the effects of the extracellular environment on the mitotic oscillator and advances our ability to accurately model tissue regeneration processes.

650c Designing Effective Trial Treatments against Cancerous Tumor Growth

Nemoy Rau and Ileana C. Carpen. Department of Chemical Engineering, Tennessee Tech University, Box 5013, Cookeville, TN 38505

Cellular adaptation to changing environments is intricately coordinated through many molecular and mechanical responses. As cells evolve, abnormal growth patterns occur (that cannot be controlled by normal mechanisms) leading to cancer. With mathematical modeling we can integrate different characteristics of tumor growth for a non-experimental study of cancer.[1] Through integration of some of the prior clinical, experimental, and mathematical studies, we utilize a cellular automata model to take into account multiple factors affecting tumor growth in healthy tissue.[2] This *in silico* simulation model of tumor growth is based upon molecular and life cycle features that effect the growth rates of cancer. The life cycle parameters used in this model include replication rate, nutrient concentrations for proliferation and survival, and possible drug effects.[3] This 3-D *in silico* model is used to study the effects of drugs upon the growth and development of the tumor and optimize a drug treatment protocol under various conditions.

650d Rapid Stochastic Simulation of Ultra-Coupled Chemical Reactions

Ian J. Laurenzi and Erdem Arslan. Chemical Engineering, Lehigh University, B330 Iacocca Hall, 111 Research Dr., Bethlehem, PA 18015

Nucleic acid-based technologies such as oligonucleotide microarrays typically feature thousands of chemical species, often designed to interact with only one other species such as cDNA, cRNA, or another nucleic acid polymer. Inasmuch as there are only four nucleotides employed in any given technological platform, and both probe and target species are serquences thereof, all DNA species may hybridize with each other, albeit weakly. Therefore, the design of oligonucleotide probes for microarrays that will preferentially hybridize to only one cDNA molecule must account for the possibility of thousands of confounding variables. Quantification of the population dynamics of such highly interconnected reaction networks, however, presents considerable computational obstacles. In this paper we present an exact method of characterizing the stochastic time evolution and equilibria of such highly-coupled reaction networks. Using this technique, we have simulated the hybridization of the entire S cerevisiae transcriptome to a genome of probes for the first time. Such types of simulations were heretofore impossible without the use of supercomputers, but may be performed on PCs within hours using our algorithm. Moreover, they may be used to characterize the robustness of a wide variety of genomic technologies including microarray designs, small-interfering RNA (siRNA) fidelity for RNA interference (RNAi), and massively parallel sequencing (MPSS).

650e An Adaptive Scheme for Continuously Inferring Blood Glucose Using Noninvasive Inputs

Nidhi Bhandari[1], Derrick Rollins[1], Nisarg Vyas[2], and Dave Andre[3]. (1) Chemical and Biological Engineering, Iowa State University, 2114 Sweeney Hall, Ames, IA 50011, (2) BodyMedia Inc., 4 Smithfield St., 11th floor, Pittsburgh, PA 15222, (3) BodyMedia, Inc.

The importance of tight glucose control in reducing the complications associated with diabetes is widely recognized. The primary ways that glucose has been managed include diet and exercise, stress management, insulin injections, and different types of drugs. In most cases, control has been open loop requiring diabetics to use test strip meters to monitor glucose behavior from just a few readings per day, a method that is not only painful and inconvenient, but also unreliable due to the guess work involved. While, there have been recent technological advancements such as glucose monitors that measure as often as every five minutes, twenty four hours a day, such monitors are invasive, have sensors that need to be replaced frequently, and costly. Such monitors can be useful in providing dynamic and frequent glucose information in the short term, but their long-term use can be highly inconvenient and expensive. Therefore, there is a need for an adaptive methodology for inferring blood glucose from non invasive input variables like food and activity, which is the goal of this work.

The modeling methodology chosen for this work is semi-empirical, block-oriented modeling based on Wiener structure, using a set of noninvasive variables that provides adequate information to explain a sufficient amount of the variation in glucose to be useful. A theoretical modeling approach is not practical due to inadequate first principle knowledge. It is common knowledge that food consumption alone is not enough to meet this requirement. While nutrient components, especially carbohydrates have a significant effect on blood glucose, they do not account for major behaviors particularly during times of low food consumption such as during sleep. Furthermore, stress and activity can have as great an impact on glucose levels as eating. Hence, an adequate data set must not only contain frequent and accurate sampling of food but also variables that measure activity and stress levels. Thus, we chose a device that showed much promise in meeting this requirement (SenseWear® Pro3 body monitoring system from BodyMedia Inc., Pittsburgh, PA). This device utilizes pattern detection algorithms that employ physiologic signals from a unique combination of sensors. To adequately cover the response space that a subject experiences, the input space must adequately cover the changes experienced by the subject. Thus, data for the study were collected under free-living conditions and for a prolonged period of time (several weeks). An empirical approach is not likely to succeed because, by necessity, data collection is under free-living conditions, leading to high input correlation, and thus, inhibiting cause and effect modeling.

The details of the study carried out in 2006 for four weeks on a real subject and used to develop the inferential model for predicting blood glucose are given briefly below:

Subject: The subject was 50 years old, in good health (except type two diabetes) with a body mass index

(BMI) of 27.9 kg/m2. He was not on diabetic medication or insulin.

Measures: To obtain the fast sampling rate necessary for dynamic modeling, the MiniMed Continuous Glucose Monitor, MMT-7102® (Medtronic, Minneapolis, Minnesota) was used to measure glucose at five minute intervals. The activity measurements were obtained using the SenseWear® Pro3 body monitoring system (BodyMedia Inc., Pittsburgh, PA). The glucose monitor requires the subcutaneous insertion of a sensor, typically in the torso, and assesses interstitial glucose at a reported rate of one sample every five minutes. The interstitial glucose measurements are used to infer blood glucose. The sensors were replaced weekly which resulted in one to two hours of no measurements for initialization. This monitor is self-calibrating but is referenced directly to measured blood glucose values obtained four times daily from a glucose meter. The subject in this study obtained these values from his personal One Touch Ultra® blood glucose meter (LifeScan, Inc., Milpitas, CA).

Protocol: Because this was a free-living study, no constraints were placed on diet or lifestyle. The subject recorded food ingested, the approximate serving sizes, and the eating times and durations in a food log. In addition, the subject also obtained at least four daily measurements of blood glucose using finger-stick measurement and entered that into the CGMS for calibration. Using nutrition tables the grams of carbohydrates, fats, and proteins ingested per meal or snack were determined and used in modeling

Model Development: The final Wiener model from this study consisted of the eleven variables and had 115 parameters. The selected inputs for this study have nonlinear, highly interactive, and dynamic affects on blood glucose. (By cause and effect modeling we mean model development that determines the independent and specific effect an input has on glucose response over the input space interest.) Therefore, given these limitations, we used a semi-empirical modeling approach. More specifically, the method was a unique and specific application of Wiener modeling that extends the method developed by Rollins and Bhandari [1] in a novel way to build accurate cause and effect models from free-living data.

The inferential model for glucose using non-invasive inputs only developed from this study has been made adaptive by using the glucose measurements obtained from the test-strip meter. On an average, the subject took four daily measurements from the test strip meter. (Note that the test strip meter is FDA-approved for monitoring blood glucose by diabetics at home). In this study 94 paired readings of monitor and meter were obtained in all (76 in training and 18 in test). These meter readings were used to improve the prediction accuracy of the inferential model by adaptively updating the model correction. The model correction term uses an exponentially-weighted value of the most current error (difference between the glucose meter and the monitor values). The optimum value of the weight was determined by minimizing the squared deviation between the monitor and the model values in training. This adaptive scheme was able to accurately infer blood glucose for several days of test data.

References:

1. D.K. Rollins, N. Bhandari Constrained MIMO Dynamic Discrete-Time Modeling Exploiting Optimal Experimental Design. Journal of Process Control: 14(6): 671-683, 2004.

Web Page: www.public.iastate.edu/~drollins/Publications

Stability and Structure of Amyloid-Forming Peptides from Computer Simulation

Jie Zheng
Chemical and Biomolecular Engineering,
University of Akron, Akron, OH 44325-3906

Proteins and peptides can fold into their unique 3-dimensional (3D) structures to perform their biological functions, or they can misfold to form insoluble amyloid fibrils, which are highly ordered protein aggregates currently known to be associated with more than 20 neurodegenerative diseases, including Alzheimer's, Parkinson's, Huntington's, diabetes type II, and various prion diseases (1-4). Regardless of the differences in size, function, sequence, and native structures of these amyloid-forming proteins/peptides, all amyloid fibrils adopt a common cross-β-sheet structure, in which the β-sheets are parallel to the fibril axis and the β-strands within a sheet are perpendicular to the fibril axis. Moreover, recent experiments found that even those disease-unrelated proteins/peptides also have an intrinsic tendency to form highly ordered amyloid fibrils under appropriate conditions (5). These findings imply that a general principle may govern amyloid fibril formation (6). Most importantly, although fibril formation is linked to neurotoxicity, accumulating evidences point to soluble oligomeric intermediates, rather than insoluble fibrils, as the primary toxic species (7-9), possibly due to their membrane-disruption abilities, while mature fibrils may represent an inert monomer reservoir. Amyloid oligomers are directly linked to infectivity as well arising from the instability of fibrils (10, 11) -- the higher tendency of the fibrils to break into oligomers, the more infection they are. Additionally, these prefibrillar intermediates display various discrete morphologies (micelle-like, annular-like, and linear-like structures) as observed by atomic force microscopy (AFM) and electron microscopy (EM) when exposed to different environmental conditions (12-14), suggesting that fibrillization/oligomerization may proceed through multiple assembly pathways. Despite substantial efforts and progresses have been made, the mechanism of amyloid formation and the origin of its toxicity are still not fully understood, primarily because little was known until recently about the molecular-level structures of fibrils, apart from the existence of the cross-β motif (15).

Obtaining atomic-level structures of various amyloid species is the first and important step towards understanding the principles of amyloid formation. Once the amyloid structures are available, the rational design of therapeutic agents and strategies becomes feasible to fight against neurodegenerative diseases. Many different experimental techniques are used to probe structural information and biological function of amyloidogenesis. Solid-state NMR and x-ray diffraction approaches are good at resolving atomic-level structural information (16-18), but the nature of protein aggregation (noncrystallization and insolubility of fibrils, small size and short-lived of oligomers, involvement of cell membrane) renders these experimental studies extremely challenging (19, 20). AFM and EM techniques can provide morphological data by capturing nanometer images over time (21-23), but detailed structural and kinetic information are not reliable. Surface plasmon resonance (SPR) is able to explore the kinetics of protein aggregation by measuring the mass amount of aggregates, but it has difficulty to provide structural information.

The difficulties and limitations of these experimental methods have inspired intensive computational studies as complement to experiments. Most computer simulations of amyloid-forming peptides fall into two categories at the atomic and coarse-grained levels with explicit and implicit solvent models. All-atom molecular dynamics (MD) simulations have been applied to study amyloid oligomer stability alone by testing different candidate β-sheet arrangements of preformed oligomers mimicking possible nucleus seeds at the very early stage of fibril formation (24-26). This approach can determine the most stable conformation for minimal nucleus seeds at the lowest free energy state, but can not provide the aggregation scenario of amyloid intermediates/fibril growth since aggregation is an extremely slow process on the timescale of minutes to days, which is typically beyond the timescale of nanoseconds for conventional MD simulations. To overcome computational limitations, alternative computer simulations using low-resolution models (e.g. coarse-grained protein models and implicit solvent models) have been used to directly study the formation of oligomers (small species) and even fibrils (large species) (19, 27). These simulations can qualitatively provide information on the kinetic pathways of protein aggregation, but can not adequately capture different detailed interactions, such as hydrophobic interactions, electrostatic interactions, and hydrogen bonding. Once the amyloid structure and its pathway are determined, various peptide inhibitors can be designed on the basis of complementarity of shape, hydrophobicity, charge, and hydrogen bonding against original oligomers. Then, molecular docking and binding simulations (28) can help to locate possible binding sites and evaluate binding strength for preventing amyloid fibril formation. Each experimental or theoretical method has its strengths and weakness, but a combination of experimental, theoretical, and computational methods can illustrate various aspects of fibril structure, formation, and toxicity at the molecular-level and thus provide an improved understanding of the complicated picture of protein aggregation.

In this work, a systemic analysis of preformed oligomeric structures is performed to examine their sequence and structural characteristics (10, 11, 29-35) using conventional all-atom molecular dynamics and replica exchange molecular dynamics with CHARMM force field. We identify several stable oligomeric structures with different structural morphology, size, and shape, delineate several common features in amyloid organizations and amyloid structures, and illustrate aggregation driving forces that stabilize these oligomeric structures using computational simulations. The structural comparison among different oligomers suggests that the aggregation mechanism leading to distinct morphologies and the aggregation pathways is sequence specific due to differences in side-chain packing arrangements, intermolecular driving forces, sequence composition, and residue positions. Furthermore, we systemically analyze two amyloid peptides which have available crystal structures and other amyloid sequences with proposed structures using computational simulations, we delineate three common features in amyloid organizations and amyloid structures: (i) Sheet-sheet recognition via a steric zipper characterized by complementarity: of shape, hydrophobicity, charge, and of hydrogen bonding; (ii) β-strand–loop-β-strand; (iii) twisted cross β-sheet. The common structural characteristics of fibril arrangement permit explanation of the observation that most peptide chains appear able to form amyloid fibrils under appropriate conditions, regardless of their sequences (33). A catalogue of these structural motifs is expected to be valuable targets in drug design for prevention and treatment of amyloid-related diseases.

1. Dobson, C. M. 2005. Structural biology: prying into prions. Nature 435:747-749.
2. Meredith, S. C. 2006. Protein denaturation and aggregation. cellular responses to denatured and aggregated proteins. Annals of the New York Academy of Sciences 1066:181-221.
3. Selkoe, D. J. 2001. Alzheimer's disease: genes, proteins, and therapy. Physiol. Rev. 81:741-766.
4. Dauer, W., and S. Przedborski. 2003. Parkinson's disease: mechanisms and models. Neuron 39:889-909.
5. Plakoutsi, G., F. Bemporad, M. Calamai, N. Taddei, C. M. Dobson, and F. Chiti. 2005. Evidence for a mechanism of amyloid formation involving molecular reorganisation within native-like precursor aggregates. Journal of Molecular Biology 351:910-922.
6. Thirumalai, D., D. K. Klimov, and R. I. Dima. 2003. Emerging ideas on the molecular basis of protein and peptide aggregation. Current Opinion in Structural Biology 13:146-159.
7. Klein, W. L., J. W. B. Stine, and D. B. Teplow. 2004. Small assemblies of unmodified amyloid β-protein are the proximate neurotoxin in Alzheimer's diease. Neurobiology of Aging 25:569-580.
8. Bucciantini, M., E. Giannoni, F. Chiti, F. Baroni, L. Formigli, J. Zurdo, N. Taddei, G. Ramponi, C. M. Dobson, and M. Stefani. 2002. Inherent toxicity of aggregates implies a common mechanism for protein misfolding diseases. Nature 416:507-511.
9. Kayed, R., E. Head, J. L. Thompson, T. M. McIntire, S. C. Milton, C. W. Cotman, and C. G. Glabe. 2003. Common structure of soluble amyloid oligomers implies common mechanism of pathogenesis. Science 300:486-489.
10. Jang, H., J. Zheng, R. Lal, and R. Nussinov. 2008. New structures help the modeling of toxic amyloid-[beta] ion channels. Trends in Biochemical Sciences 33:91-100.
11. Jang, H., J. Zheng, and R. Nussinov. 2007. Models of {beta}-amyloid ion-channels in the membrane suggest that channel formation in the bilayer is a dynamic process. Biophys. J. 93:1938-1949.
12. Gosal, W. S., I. J. Morten, E. W. Hewitt, D. A. Smith, N. H. Thomson, and S. E. Radford. 2005. Competing pathways determine fibril morphology in the self-assembly of [beta]2-microglobulin into amyloid. Journal of Molecular Biology 351:850-864.
13. Lashuel, H. A., D. Hartley, B. M. Petre, T. Walz, and P. T. Lansbury. 2002. Amyloid pores from pathogenic mutations. Nature 418:291-291.
14. Lashuel, H. A., B. M. Petre, J. Wall, M. Simon, R. J. Nowak, T. Walz, and J. P. T. Lansbury. 2002. Alpha-synuclein, especially the parkinson's disease-associated mutants, forms pore-like annular and tubular Protofibrils. Journal of Molecular Biology 322:1089-1102.

15. Tycko, R. 2004. Progress towards a molecular-level structural understanding of amyloid fibrils. Current Opinion in Structural Biology 14:96-103.
16. Chamberlain, A. K., V. Receveur, A. Spencer, C. Redfield, and C. M. Dobson. 2001. Characterization of the structure and dynamics of amyloidogenic variants of human lysozyme by NMR spectroscopy. Protein science 10:2525-2530.
17. Tycko, R. 2000. Solid-state NMR as a probe of amyloid fibril structure. Current Opinion in Chemical Biology 4:500-506.
18. Petkova, A. T., Y. Ishii, J. J. Balbach, O. N. Antzutkin, R. D. Leapman, F. Delaglio, and R. Tycko. 2002. A structural model for Alzheimer's beta -amyloid fibrils based on experimental constraints from solid state NMR. Proc. Natl. Acad. Sci. 99:16742-16747.
19. Mousseau, N., and P. Derreumaux. 2005. Exploring the early steps of amyloid peptide aggregation by computers. Acc. Chem. Res. 38:885-891.
20. Makabe, K., D. McElheny, V. Tereshko, A. Hilyard, G. Gawlak, S. Yan, A. Koide, and S. Koide. 2006. Atomic structures of peptide self-assembly mimics. Proc. Natl. Acad. Sci. 103:17753-17758.
21. Benseny-Cases, N., M. Cocera, and J. Cladera. 2007. Conversion of non-fibrillar [beta]-sheet oligomers into amyloid fibrils in Alzheimer's disease amyloid peptide aggregation. Biochemical and Biophysical Research Communications 361:916-921.
22. Legleiter, J., D. L. Czilli, B. Gitter, R. B. DeMattos, D. M. Holtzman, and T. Kowalewski. 2004. Effect of different anti-A[beta] antibodies on a[beta] fibrillogenesis as assessed by atomic force microscopy. Journal of Molecular Biology 335:997-1006.
23. Wang, Z., C. Zhou, C. Wang, L. Wan, X. Fang, and C. Bai. 2003. AFM and STM study of [beta]-amyloid aggregation on graphite. Ultramicroscopy 97:73-79.
24. Ma, B., and R. Nussinov. 2002. Stabilities and conformations of Alzheimer's beta -amyloid peptide oligomers (Abeta 16-22, Abeta 16-35, and Abeta 10-35): Sequence effects. Proc. Natl. Acad. Sci. 99:14126-14131.
25. Tsai, H.-H., M. Reches, C.-J. Tsai, K. Gunasekaran, E. Gazit, and R. Nussinov. 2005. Energy landscape of amyloidogenic peptide oligomerization by parallel-tempering molecular dynamics simulation: Significant role of Asn ladder. Proc. Natl. Acad. Sci. 102:8174-8179.
26. Zheng, J., D. Zanuy, N. Haspel, C. J. Tsai, C. Aleman, and R. Nussinov. 2007. Nanostructure design using protein building blocks enhanced by conformationally constrained synthetic residues. Biochemistry 46:1205-1218.
27. Nguyen, H. D., and C. K. Hall. 2006. Spontaneous fibril formation by polyalanines: discontinuous molecular dynamics simulations. J. Am. Chem. Soc. 128:1890-1901.
28. Soto, P., M. A. Griffin, and J.-E. Shea. 2007. New insights into the mechanism of alzheimer amyloid-{beta} fibrillogenesis inhibition by n-Methylated peptides. Biophys. J. 93:3015-3025.
29. Zheng, J., H. Jang, B. Ma, and R. Nussinov. 2008. Annular structures as intermediates in fibril formation of alzheimer A[beta]17-42. Journal of Physical Chemistry B 112:6856-6865.

30. Zheng, J., H. Jang, B. Ma, C.-J. Tsai, and R. Nussinov. 2007. Modeling the alzheimer a{beta}17-42 fibril architecture: tight intermolecular sheet-sheet association and intramolecular hydrated cavities. Biophys. J. 93:3046-3057.
31. Zheng, J., H. Jang, and R. Nussinov. 2008. [beta]2-microglobulin amyloid fragment organization and morphology and its comparison to A-[beta] suggests that amyloid aggregation pathways are sequence-specific. Biochemistry 47:2497 - 2509.
32. Zheng, J., B. Ma, C.-J. Tsai, and R. Nussinov. 2006. Structural stability and dynamics of an amyloid-forming peptide GNNQQNY from the yeast prion sup-35 Biophys. J. 91:824-833.
33. Zheng, J., B. Ma, and R. Nussinov. 2006. Consensus features in amyloid fibrils: sheet-sheet recognition via a (polar or nonpolar) zipper structure. Physical Biology 3:P1-P4.
34. Zheng, J., B. Ma, Y. Chang, and R. Nussinov. 2008. Molecular dynamics simulations of alzheimer's peptide A[beta]40 elongation and lateral association. Frontiers in Bioscience 13:3919-3930.
35. Li, L., and J. Zheng. 2008. Computational Modeling of Amyloid Oligomeric Structures. International Journal of Liquid State Sciences accepted.

650g Study of Tricyclic Cascade Networks Using Dynamic Optimization

Andrijana Radivojevic[1], **Benoît Chachuat**[1], Dominique Bonvin[1], and Vassily Hatzimanikatis[2]. (1) Automatic Control Laboratory, Ecole Polytechnique Federale de Lausanne (EPFL), Station 9, Lausanne, CH-1015, Switzerland, (2) Laboratory of Computational Systems Biotechnology, Ecole Polytechnique Federale de Lausanne (EPFL), CH-1015, Lausanne, Switzerland

The mitogen-activated protein kinase (MAPK) cascades are ubiquitous in eukaryotic signal transduction, and these pathways are conserved in cells from yeast to mammals [1]. They relay extracellular stimuli from the plasma membrane to targets in the cytoplasm and nucleus, initiating diverse responses involving cell growth, mitogenesis, differentiation and stress responses in mammalian cells.

Detailed kinetics models of MAPK cascades have been constructed in recent years that are comprised of mixed sets of differential and algebraic equations (DAEs) [2,3]. Such models typically involve many parameters, such as the kinetic rate constants and the concentration ratios between various kinases and phosphatases, the values of which are not directly accessible in vivo and are subject to large uncertainty. Dynamic optimization has proved to be a very useful tool to help relate the model parameters to functions in MAPK networks [4,5]. Large-scale, nonlinear DAE models can be handled within this framework, as well as a large variety of objective functions and constraints.

In a recent work [5], the response of an interconvertible monocyclic cascade (phosphorylation-dephosphorylation cycle) has been studied. It was shown, using dynamic optimization, that values of the kinetic parameters that confer, at the same time, (i) a short response time, (ii) a large amplification capability, and (iii) a steep response profile to a graded input (ultrasensitivity) can be found. However, it was also found that, in a monocyclic cascade, these properties are not robust towards variations in the ratio between signaling enzyme and substrate kinase concentrations as well as the ratio between phosphatase and substrate kinase concentrations.

In this presentation, we extend the analysis to the general case of multiple levels of cascades, with emphasis on a linear three-kinase model as described in [2]. The same response properties as in the monocyclic case are considered, and dynamic optimization is employed to identify parameter values that optimize these response properties. Special emphasis is placed on the robustness of the resulting tricyclic cascades in the face of variations in kinase and phosphatase concentration ratios. Comparisons with the monocyclic cascade case are also presented.

References

[1] E. Klipp, R. Herwig, A. Kowald, C. Wierling, and H. Lehrach. “Systems Biology in Practice. Concepts, Implementations and Applications,” Wiley-VCH, Weinheim (Germany), 2005.

[2] C. Y. Huang, and J. E. Ferrel Jr. “Ultrasensitivity in the mitogen-activated protein kinase cascade,”

Proc. Natl. Acad. Sci. U.S.A., 93:10078–10083, 1996.

[3] R. Heinrich, B. G. Neel, and T. A. Rapoport. “Mathematical models of protein kinase signal transduction,” Mol. Cell, 9:957–970, 2002.

[4] B. S. Adiwijaya, P. I. Barton, and B. Tidor. “Biological network design strategies: Discovery through dynamic optimization”, Mol. BioSyst., 2:650–659, 2006.

[5] B. Chachuat, D. Bonvin, and V. Hatzimanikatis. “Analysis of signal transduction pathways through dynamic optimization – Application to MAP kinase cascades,” AIChE Annual Meeting, Salt Lake City (UT), 2007.

652a A Dynamic Model for a Cupola Furnace

Sten Bay Jorgensen, of Chemical and Biochemical Engineering, CAPEC, Technical University of Denmark, Søltofts Plads, Kgs. Lyngby, DK 2800, Denmark and John Bagterp Jorgensen, Informatics and Mathematical Modelling, Technical University of Denmark, Richard Petersens Plads, Building 305 Office 109, Kgs. Lyngby, DK-2800, Denmark.

Cupola furnaces are widely used in the foundry industry and models have previously been developed. However, the focus of most previous cupola model developments have aimed at foundry related problems which are different from those of stone wool production.

A dynamic mathematical model of a mineral melting cupola furnace is used to illustrate how different process parameters affect the static and dynamic operation. A mathematical model of a mineral melting cupola furnace for stone wool production has been developed for improving cupola operation. The one space dimensional first engineering principle model includes mass and energy balances for the gas phase, solid phases for mineral and coal and a liquid mineral phase. The gas and solid/liquid phases flow counter currently. Several chemical

reactions account for conversion of coke, limestone and gaseous species. The heterogeneous reactions of coke conversion are limited by both kinetics and mass transport. Heat transfer between phases is modelled including both convection and radiation.

The model predicts gas concentration, mass flow rates and

temperature profiles of solid, melt and gas in the cupola and the heat loss to the water cooled walls. Input to the model is the coke, rock and blast air properties and blast air amount and coke percentage in the charge. The unknown model parameters have been estimated based on input/output measurements. Comparison of model predicted and measured concentration and temperature profiles inside

the cupola shows good agreement. The static model behaviour is compared to experimental data and rules of thumb for steady state cupola operation at stone wool factories. The comparison with the rules of thumb indicates that the model well captures the essential phenomena in the process including description of gas concentrations, temperatures and flow rates inside the cupola.

The dynamic model behaviour during start-up compares qualitatively reasonably well to industrial practice. The behaviours in response to step changes in air inlet temperature and coal loading also compare favourably with industrial experience. Thus the dynamic model is ready for further design and study of model predictive control.

652b Approximation of Actual Number Densities of Age-Structured Cell Division Models Using Stochastic Equations of Population Balances

Eric Sherer[1], Robert E. Hannemann[2], Ann E. Rundell[3], and Doraiswami Ramkrishna[2]. (1) e-Enterprise Center, Purdue University, West Lafayette, IN 47907, (2) Chemical Engineering, Purdue University, 1283 Forney Hall, West Lafayette, IN 47907, (3) Biomedical Engineering, Purdue University, 206 S. Martin Jischke Drive, West Lafayette, IN 47907

In previous work, age-structured population balances were used to capture leukemia cell dynamics in response to cell cycle specific chemotherapeutic agents (Sherer et al. 2006). The experimentally verified model was then used to optimize treatment scheduling. However, it was noted at the time that the objective criterion used in the optimization – minimizing the expected number of leukemia cells – was only an indicator of what truly quantifies chemotherapy outcomes: the probability of cure. That is, what is the likelihood that zero cancerous cells remain after a treatment regimen? In this presentation, we demonstrate a methodology – which utilizes stochastic equations of population balances – by which the likelihood of having an actual number of cells can be approximated for age-structured models. The technique is applied to a mother-daughter cell division model and a cell cycle model to predict cure rates for hypothetical chemotherapy treatments. The simulation results and computation times are also compared with Monte Carlo results to demonstrate the utility of the method.

Determining exactly when cure occurs is impossible because the timing of cell births, deaths, or other transitions are uncertain. So, the actual number density (the discrete list of each cell and its age) is only known probabilistically (Ramkrishna and Borwanker 1974). When large number of cells are considered, these fluctuations tend to average out and the expected number density (whose behavior is governed by the population balance equation) adequately describes the system. However, variations are inherently present and become more significant as the population size declines – the case in chemotherapy treatments. More generally, such information is necessary to quantify the state of any system in which significant variations from the expected behavior are anticipated.

In answering the question of the likelihood of cure, we show how the stochastic equations of the population balance are derived from the master (or Janossy) density where each higher order equation provides additional information – in the form of an additional moment - about the master density (Ramkrishna 2000). We then show how the behavior of the equations converges as the equation order increases. In fact, only the third or fourth order stochastic equations give a sufficient approximation of all higher order equations in the examples considered. The approximated solutions of the stochastic equations are used to approximate the moments of the actual number density which are then used to approximate the actual number density. Excellent agreement in the actual number density is seen between the approximate solution and multiple Monte Carlo simulation where the stochastic equation simulation have a considerable computation time advantage.

References:

Ramkrishna D and Borwanker JD. “A puristic analysis of population balance II”. Chemical Engineering Science 29: 1711-1721, 1974.

Ramkrishna D, Population balances: Theory and applications to particulate systems in engineering. Academic Press, San Diego, CA, 2000.

Sherer E, Hannemann RE, Rundell A, and Ramkrishna D, “Analysis of Resonance Chemotherapy in Leukemia Treatment via Multi-Staged Population Balance Models.” Journal of Theoretical Biology, 240: 648-661, 2006.

652c Model Predictive Control of Thin Film Porosity Using Stochastic Differential Equations

Gangshi Hu[1], Yiming Lou[2], Gerassimos Orkoulas[1], and Panagiotis D. Christofides[3]. (1) Department of Chemical and Biomolecular Engineering, University of California, Los Angeles, Los Angeles, CA 90095, (2) Advanced Projects Research, Inc, 1925 McKinley Ave., Suite B, La Verne, CA 91750, (3) Department of Chemical and Biomolecular Engineering, Department of Electrical Engineering, University of California, Los Angeles, Los Angeles, CA 90095

Thin film porosity strongly affects device performance and should be carefully controlled. For example, low-k dielectric films of high porosity are being used in current interconnect technologies to meet resistive capacitive (RC) delay goals and minimize cross-talk. However, increased porosity negatively affects the mechanical properties of dielectric films, increasing the risk of thermo-mechanical failures [1]. Furthermore, in the case of gate dielectrics, it is important to reduce thin film porosity as much as possible and eliminate the development of holes close to the interface.

Two different approaches, kinetic Monte Carlo (kMC) methods and stochastic differential equation (SDE) models, are employed to describe the evolution of film microscopic configurations and design feedback control laws. KMC simulation methods are widely used in microscopic modeling due to their efficiency in large temporal/spatial scale microscopic simulations. SDEs arise naturally in the modeling of surface morphology of thin film growth in a variety of material preparation processes [2, 3]. SDE models are effective as the basis for the design of control systems for regulating high-order moments of the microscopic distributions. Both linear and nonlinear controllers were developed based on SDE models to control surface roughness in various thin film growth processes, including deposition and sputtering process [4, 5]. However, there are no works that have focused on feedback control of film porosity.

Motivated by these considerations, a model predictive controller is designed using a linear SDE model of a porous film deposition process to regulate the film density to a desired value and minimize run-to-run fluctuations. The SDE model is constructed from the kMC model of a deposition process, which includes two types of microscopic processes, deposition and diffusion. Vacancies and overhangs are allowed in the kMC model to introduce porosity, which is mathematically represented by the thin film density. The predictive controller is formulated to minimize a cost function including the distance between the predicted density and the desired value. The manipulated input in the closed-loop system is the substrate temperature. Finally, the predictive controller is applied to the kMC simulation of the deposition process. Simulation results demonstrate the effectiveness of the proposed control method.

1. Kloster, G., Scherban, T., Xu, G., Blaine, J., Sun, B. and Zhou, Y. Porosity effects on low-k dielectric film strength and interfacial adhesion. Proceedings of the IEEE 2002 International; 242-244.

2. Edwards, S. F. and Wilkinson, D. R. The surface statistics of a granular aggregate. Proceedings of the

Royal Society of London. Series A, Mathematical and Physical Sciences 1982;381:17-31.

3. Lauritsen, K. B., Cuerno, R. and Makse, H. A. Noisy Kuramoto-Sivashinsky equation for an erosion model. Physical Review E 1996; 54:3577-3580.

4. Lou, Y. and Christofides, P. D. Feedback control of surface roughness using stochastic PDEs. AIChE Journal 2005;51:345-352.

5. Lou, Y. and Christofides, P. D. Nonlinear feedback control of surface roughness using a stochastic PDE: Design and application to a sputtering process. Industrial & Engineering Chemistry Research 2006;45:7177-7189.

652d Output Feedback Control of Transport-Reaction Processes Using Adaptive Model Reduction

Sivakumar Pitchaiah and **Antonios Armaou**. Department of Chemical Engineering, The Pennsylvania State University, University Park, PA 16802

Many industrially relevant diffusion-convection-reaction processes are characterized by the presence of strong spatial variations due to coupling between diffusive and convective mechanisms with chemical reactions. Typical examples of such processes include fluidized-bed and packed-bed reactors, rapid thermal processing, plasma reactors, and chemical vapor deposition. The feedback control problem for these processes is nontrivial owing to the spatially distributed nature of their dynamics. This leads to infinite dimensional dynamic descriptions in appropriate spaces which necessitate the use of model reduction in order to design practically implementable controllers. The key step in this approach is the computation of basis functions that are subsequently utilized to obtain finite dimensional approximations to the original infinite dimensional system. A common approach for this task, when the spatial dynamics are described by nonlinear operators, is the Karhunen-Loéve expansion (KLE) combined with method of snapshots. However, this approach requires the a priori availability of a sufficiently large ensemble of PDE solution data (snapshots) which excites all of the possible spatial modes in the solution of the PDE system, a requirement which may be difficult to satisfy. Recently, the adaptive model reduction methodology [1] was proposed as a solution to the above problem. In this methodology the eigenfunctions required for the model reduction were computed recursively as new snapshots from the process becomes available. Initially eigenfunctions were computed using a relatively small number of snapshots; we then kept track of the dominant eigenspace of the covariance matrix which was subsequently utilized to compute the empirical eigenfunctions required for model reduction.

In this work, we extend the applicability of the above methodology to processes where snapshots of the state evolution become available only periodically. We assume that point measurements from a limited number of sensors are continuously available. The reduced-order model of the PDE system is obtained using the empirical eigenfunctions. We then design a static output observer, based on the continuous point measurements, to estimate the modes of the reduced order model. The associated empirical eigenfunctions are recursively updated, when a new snapshot from the process becomes available, using the adaptive model reduction methodology. A nonlinear controller is designed using the estimated reduced-order model and the parameters of the controller are continuously revised as the process evolves based on changes in the estimated reduced-order model.

The proposed approach is applied to control temperature in a jacketed tubular reactor where first order chemical reaction is taking place. We design an output feedback controller, based on estimated modes from noisy point measurements, to successfully stabilize the process at an open-loop unstable steady state.

[1] A Varshney, S Pitchaiah and A Armaou, "Feedback control of dissipative PDEs using adaptive model reduction", submitted to AICHE journal, 2008.

652e Actuator Fault Detection and Reconfiguration In Distributed Processes with Measurement Sampling Constraints

Sathyendra Ghantasala and Nael H. El-Farra. Department of Chemical Engineering & Materials Science, University of California, Davis, One Shields Avenue, Davis, CA 95616

Many important engineering systems, such as transport-reaction processes and fluid flow systems, are characterized by spatial variations and are modeled by systems of Partial Differential Equations (PDEs). While the study of distributed parameter systems has been an area of active research within process control over the past few decades (see, for example, [1] and the references therein), the development of systematic methods for the diagnosis and handling of faults in distributed control systems has received only limited attention. This is an important problem given the vulnerability of automated industrial control systems to malfunctions in the control actuators, measurement sensors and process equipment, as well as the increasingly stringent requirements placed on safety and reliability in industrial process operation.

Recently, we developed in [2] and [3] a unified framework for the integration of model-based fault detection, isolation and control system reconfiguration for distributed processes modeled by nonlinear parabolic PDEs with control constraints and actuator faults. A key idea in these works is to tie the design of the fault detection and isolation (FDI) filters as well as the actuator reconfiguration logic, via singular perturbation techniques, to the intrinsic separation between the slow and fast eigenvalues of the differential operator of the infinite-dimensional system. This naturally leads to the derivation of explicit FDI thresholds and actuator reconfiguration rules that minimize false or missed alarms due to approximation errors when the low-order model-based architecture is implemented on the infinite-dimensional system. Practical implementation issues such as the presence of plant-model mismatch and the availability of measurements at a finite number of locations along the spatial domain were subsequently addressed in [4] and [5]. The central idea in these works is to shape the fault-free closed-loop behavior, via robust bounded feedback control, in a specific way that facilitates the derivation of FDI rules that are less sensitive to the uncertainty. Uniting robust control and FDI leads to an explicit characterization of the state space regions where FDI and actuator reconfiguration are feasible under uncertainty and constraints.

Beyond the problems of nonlinearities and model uncertainty, one of the key issues that needs to be accounted for in the design of monitoring and fault-tolerant control systems is the issue of measurement sampling. In practice, measurements of the process outputs are typically available from the sensors at discrete time instances and not continuously. The frequency at which the measurements are available is dictated by the sampling rate which is typically constrained by the inherent limitations on the data collection and processing capabilities of the measurement sensors. In some cases, constraints on the sampling rate may be imposed by the designer in order to limit the transfer of data over a bandwidth-limited communication channel that connects the sensor and the controller for the purpose of reducing network resource utilization. The limitations on the frequency of measurement availability imposes restrictions on the implementation of the feedback controller and can also erode the diagnostic and fault-tolerance capabilities of the fault-tolerant control architecture if not explicitly accounted for in the monitoring and control system design. Within the feedback control layer, for example, infrequent measurement sampling could result in substantial errors in the implemented control action leading to possible loss of stability or performance degradation. The lack of frequent measurements also limits our ability to accurately monitor the trajectory of the process variables rendering it difficult to evaluate the residuals or diagnose faults. At the control reconfiguration level, knowledge of the dependence of a given control configuration (i.e., the spatial placement of actuators and sensors) on the sampling rate is critical for identifying the appropriate backup configuration that should be activated following fault detection to preserve closed-loop. Unless the various components of the fault-tolerant control architecture are redesigned to account for the lack of continuous measurements, enforcing fault tolerance will be a difficult task.

Motivated by these considerations, we develop int his work a fault detection and fault-tolerant control structure for distributed processes modeled by parabolic PDEs with a limited number of measurements that are sampled at discrete time instances. The structure consists of a family of output feedback controllers, a fault detection filter that accounts for the discrete sampling of measurements and a switching law that orchestrates the transition from the faulty actuators to the healthy fall-backs following fault detection. The control, detection and reconfiguration components are designed on the basis of an approximate finite-dimensional system that captures the dominant dynamic modes of the PDE and is obtained using modal decomposition techniques. A key idea is to design a state observer that uses the available measurements to generate estimates of the states of the reduced-order system in the absence of faults, and to use these estimates both for controller implementation as well as fault detection. Fault detection is achieved by comparing the expected fault-free behavior of the reduced-order system with the actual process behavior, and using the discrepancy as a residual. Since the output measurements are available only at discrete time instances, a reduced-order model of the process is embedded with the state observer to provide it with estimates of the output measurements in between sampling instances when measurements are unavailable. The state of this model is then updated with the actual measurements whenever they become available from the sensors at discrete sampling times. By formulating the overall closed-loop system as a discrete jump system in which the model estimation error is re-set to zero at the sampling times, an explicit characterization of the minimum allowable sampling rate that guarantees both closed-loop stability and residual convergence in the absence of faults is obtained. The minimum sampling rate is characterized in terms of the model accuracy, the controller design parameters and the spatial placements of the actuators and sensors. This characterization leads to the derivation of (1) a time-varying threshold on the residual which can be used to detect faults under the given sampling constraint, and (2) an actuator reconfiguration law that determines the set of feasible fall-back actuators that preserve closed-loop stability under a given measurement sampling rate. Finally, the design and implementation of the proposed fault detection and fault-tolerant control architecture are demonstrated using a simulated model of a low-density polyethylene tubular reactor example.

References:

[1] Christofides, P. D. Nonlinear and Robust Control of PDE Systems: Methods and Applications to

Transport-Reaction Processes. Birkhauser, Boston, 2001.

[2] El-Farra, N. H., ``Integrated fault detection and fault-tolerant control architectures for distributed processes." Ind. & Eng. Chem. Res., 45:8338--8351, 2006.

[3] El-Farra, N. H. and S. Ghantasala, ``Actuator fault isolation and reconfiguration in transport-reaction processes," AIChE J., 53:1518--1537, 2007.

[4] Ghantasala, S. and N. H. El-Farra, ``Robust Detection and Handling of Actuator Faults in Control of Uncertain Parabolic PDEs," Dyn. Cont. Dis. & Impl. Syst. - Series A, 14 (S2), 203-208, 2007.

[5] Ghantasala, S. and N. H. El-Farra, ``Detection, isolation and management of actuator faults in parabolic PDEs under uncertainty and constraints," Proceedings of 46th IEEE Conference on Decision and Control, pages 878-884, New Orleans, LA, 2007.

652f Boundary Predictive Control of Kuramoto-Sivashinsky Equation

Stevan Dubljevic, Cardiology Department, Cardiovascular Research Laboratories David, Geffen School of Medicine, UCLA, 675 Charles E. Young Drive South Box 951760 Room#3645 MRL, Los Angeles, CA 90095-1760

This paper focuses on boundary predictive control of higher order linear dissipative PDEs, and in particular on predictive control of a dissipative fourth-order equation given by the Kuramoto-Sivashinsky equation, which describes a variety of physicochemical phenomena like long-wave motions of the liquid thin film over a vertical plate or the evolution of laminar fronts [1]. The problem of stabilization of Kuramoto-Sivashinsky equation has been addressed within the output feedback formulation by Christofides and Armaou [2] and within optimal control setting by Lee and Tran [3]. However, an important aspect of optimal control synthesis in which actuation is applied at the boundary has not be addressed.

In this work, modal model predictive control synthesis for control of Kuramoto-Sivashinsky equation has been developed. The evolution of a linear dissipative PDE is initially given by an abstract evolution equation in an appropriate Sobolev space. Modal decomposition technique is used to decompose the infinite dimensional system into an interconnection of a finite-dimensional (slow) subsystem with an infinite-dimensional (fast) subsystem. The predictive controller synthesis is then formulated in a way that the construction of the cost functional accounts only for the weighted evolution of slow (finite-dimensional) states, while in the state constraints a high-order (finite-dimensional) approximation of fast states is utilized. As an example of the proposed controller synthesis methodology, the optimal stabilization of spatially-uniform unstable steady state of Kuramoto-Sivashinsky equation subject to variety of boundary conditions is considered. Simulation results demonstrate successful application of the proposed predictive control technique within infinite-dimensional closed-loop system setting.

[1] L. H. Chen and H. C. Chang, "Nonlinear waves on liquid film surface-II. Bifurcation analysis of the long-wave equation," Chemical Engineering Science, vol. 41, pp. 2477-

2486, 1986.

2] A. Armaou and P. D. Christofides, "Feedback control of the Kuramoto-Sivashinsky equation," Physica D., vol. 137, pp. 49-61, 1999.

[3] P. D. Christofides and A. Armaou, "Global stabilization of the Kuramoto-Sivashinsky equation via distributed output feedback control," Syst. & Contr. Lett., vol. 39, pp. 283-294, 2000.

[4] C. H. Lee and H. T. Tran, "Reduced-order-based feedback control of the Kuramoto-Sivashinsky equation," Journal of Computational and Applied Mathematics, vol. 173, pp. 1-19, 2005.

652g Model Predictive Control of Granule Microstructure

Frantisek Stepanek[1], Vladislav Nevoral[1], and Mansoor A. Ansari[2]. (1) Department of Chemical Engineering, Institute of Chemical Technology, Prague, Technicka 5, Prague 6, 166 28, Czech Republic, (2) Department of Chemical Engineering, Imperial College London, South Kensington Campus, London, SW7 2AZ, United Kingdom

At present, particle size tends to be used as the main product attribute for the purpose of particulate processes control. However, in many particulate products there are other attributes – especially the internal distribution of formulation ingredients and phases within the particles – that strongly influence the product end-use properties. The ability to control the microstructure-related attributes is possibly more important than the ability to control particle size: while particle size distribution that deviates from requirements can be 'corrected' in the processing plant by the recycling of fines and the milling of over-size, internal attributes of particles are effectively impossible to revert once formed.

Predictive models of particle microstructure formation (Stepanek and Ansari, 2005) and structured particle dissolution (Stepanek, 2004) have been recently developed and used for the solution of the granule microstructure design problem (Ansari and Stepanek, 2006). In the present work we will address the problem of granule microstructure control. First, the control problem will be formulated and the issue of controllability addressed. By means of parametric sensitivity analysis, a reduced model will be derived, which will be used for model predictive control of granule microstructure in a fluid-bed wet granulation process, where the control objective is the dissolution time of the final granular product, the disturbances occur on the feed material properties space (primary particle size distribution etc.), and the manipulated variables are from the process parameters space (granulation process conditions).

References:

Stepanek, F., Ansari, M.A., 2005. Computer simulation of granule microstructure formation. Chem. Eng. Sci. 60, 4019-4029.
Stepanek, F., 2004. Computer-aided product design: granule dissolution. Chem. Eng. Res. Des. 82, 1458-1466.
Ansari, M.A., Stepanek, F., 2006. Design of granule structure: computational methods and experimental realisation. AIChE J. 52, 3762-3774.

653a Multiscale Modeling of Polymer Microsphere Drug Delivery

Ashlee N. Ford, Chemical & Biomolecular Engineering, University of Illinois, 600 South Mathews Avenue, 201 Roger Adams Laboratory, Box C-3, Urbana, IL 61801-3602, Daniel W. Pack, Chemical & Biomolecular Engineering, University of Illinois Urbana-Champaign, 600 S. Mathews, Box C-3 MC 712, Urbana, IL 61801, and Richard D. Braatz, Department of Chemical and Biomolecular Engineering, University of Illinois at Urbana-Champaign, 600 South Mathews Avenue, Box C-3, 293 Roger Adams Laboratory, Urbana, IL 61801-3602.

Proteins and pharmaceuticals can be encapsulated into biodegradable polymer microspheres of controlled size. Controlled-release drug delivery systems are being developed as an alternative to conventional medical drug therapy regimens which require frequent injections or oral dosages because some drugs become deactivated upon contact with other chemicals in the human body or are eliminated relatively quickly from the body. Also, some pharmaceuticals have poor oral bioavailability which means that the drug compounds are not readily absorbed by the body through the digestive system. Controlled-release systems have the potential to provide better control of drug concentrations and reduce the side effects of current drug therapies. These systems can be designed to provide specific drug concentrations over extended periods of time which are optimal for effectiveness of the drug and which avoid potentially harmful high and low levels of the drugs. They also can be administered locally to the part of the body where the drug is needed. Although there are clear advantages to using controlled-release systems for drug delivery, the design of controlled-release devices, such as biodegradable polymer microspheres, depends heavily on trial-and-error experiments due to incomplete understanding of the mechanisms that regulate the release processes. An accurate computational model of drug release from polymer microspheres would be useful for determining the optimal parameters for fabrication of the microspheres to yield a desired drug concentration profile. A successful mechanistic model of this type would be a useful tool for planning experiments, and once thoroughly validated, it could be used in the design of pharmaceutical manufacturing of microspheres.

This presentation describes a mechanistic model that includes the effects of autocatalytic degradation in polymer microspheres that models phenomena at length scales ranging from molecule to micropore to mesopore to hundreds of microns. Degradation effects of a large number of chemical species are considered along with bulk erosion which incorporates many highly coupled nonlinear partial differential equations for chemical reaction and diffusion. These equations are solved for small computational time steps for an extended period of time in order to capture an entire release profile. The high resolution simulation of the coupling between reaction and diffusion captures important dynamical phenomena observed in experiments that cannot be modeled with the models in the literature that have simpler numerical solution but do not take the coupling into account. By using mechanistic models for all time and length scales, the model has no empiricism or "fitting parameters."

Numerical techniques are described that reformulate discrete spatially-varying population balance equations that appear in the full-order mechanistic model into a continuous spatially-varying population balance equation. This extends a reformulation commonly used in modeling polymerization to spatially-varying depolymerization. Basis functions used to parameterize the spatially-varying continuous population distribution are selected to tune the resolution of the extrinsic size coordinate and an intrinsic size coordinate based on the locally appropriate scales, with the overall approach producing orders-of-magnitude speedup while reducing the memory requirements low enough for the computations to be carried out on a personal computer. A parallel implementation of the numerical algorithms is described which speeds the computation of the multiscale model enough for its use in iterative optimal design and control computations.

653b Principal Component Analysis for Estimating Population Density from Chord-Length Density

Stephanie Barthe, **Martha Grover Gallivan**, and Ronald W. Rousseau. School of Chemical & Biomolecular Engineering, Georgia Institute of Technology, 311 Ferst Drive, Atlanta, GA 30332-0100

Crystallization is a major separation technique used in numerous chemical processes, and its operation plays a major role in determining properties of crystalline products and the effectiveness of downstream processes. Real-time and in-line observation of the evolution of population density constitutes major asset in understanding crystallizer operation and the phenomena that influence product quality.

Focused-beam reflectance measurement (FBRM) is a technique based on backscattering of laser light that is useful in monitoring a representation of the crystalline population. Properly installed, the FBRM records the chord-length density (CLD), which is a function of crystal geometry and is related to the population density. While estimation of CLD from the population density is relatively straightforward, the inversion of this procedure is problematic because the problem may be ill-conditioned for non-spherical particles. The focus of the present work is on this inversion process.

The principal component analysis offers a straightforward, constraint-free and timely alternative to restore the population density accurately from chord-length density data. It also provides guidance on those aspects of the population density that can be restored accurately. Since the relationship between chord-length and population densities is a function of crystal geometry, this manuscript considers various non-spherical shapes, including highly challenging needles. We describe efficient methods to estimate the population density from the measured CLD, to analyze the inversion process, and to quantify its limits. Since noise is present in the actual measurements, the case of restoration of the population density from noisy CLDs also is demonstrated. It is thought that the presented methodology has great potential to be the basis for a control scheme that manipulates the population density produced from FBRM raw data.

653c Single Particle Modeling of Seed Polymerization Process

Weijie Lin, Lorenz T. Biegler, and Annette Jacobson. Department of Chemical Engineering, Carnegie Mellon University, 5000 Forbes Ave., Pittsburgh, PA 15213

Seed polymerization technology is recognized as a promising field for novel polymer material development. Its advantages in particle size control and versatile customized properties make it a competitive choice in many areas of application. However, due to the complexity of the seed particle growth phenomenon, many experimental observations remain unexplained, and the relationships between process modification and polymer characteristics are challenging to identify. Modeling intricate particle growth and morphology, which is significant for the design and operation of particulate processes, has not yet been developed.

In order to facilitate the seed polymerization progress, a comprehensive dynamic single particle model is discussed in this paper. To depict the special physical process of seed polymerization, the interactions of the outer mass transfer and the particle internal diffusion are considered with polymer kinetic models. A tie between macro-scale feeding strategy and single particle evolution is built for further purposes of optimal control. In contrast to traditional single particle models, the particle size growth is modeled in a more rigorous way, where the seed property and the dynamic density change are taken into account. More over, dynamic functions are adopted for polymer apparent kinetic rates, which reveal more insight about the morphology formation of these complex polymeric particles.

The model is consisted of a set of differential algebraic equations, and the numerical simulation is performed for a free radical seed polymerization process. The comparison of simulation results and experimental measurements shows consistent agreement both in time and space evolution of the particle growth. Further more, parameter estimation and process control problem are investigated based on the model formulation. Efficient algorithm could be applied with a direct nonlinear programming transcription method, showing its promising capacity for further industrial application interest.

653d Monte Carlo Simulation of Atomic Layer and Other Thin-Film Growth Processes

Raymond A. Adomaitis and Vivek Dwivedi. Chemical Engineering, University of Maryland, College Park, MD 20742

Atomic Layer Deposition (ALD) is a process crucial to the production of nanoscale electronic and photonic devices. In this process, thin films are deposited on a substrate through repeated exposure of the growth surface to a sequence of precursor species. This mode of operation allows for nearly perfect control of film thickness to atom-scale lengths, even over complex surface topographies.

Despite the performance that can be achieved in these processes and its current use in manufacturing (e. g., the HfO_2 gate oxide used in current Intel 45nm node technology [1]), fundamental gaps in our understanding of anomalous growth in even the highly studied Al_2O_3 ALD process [2] under non-ideal conditions persist. Therefore, we present in this paper our research on Monte Carlo (MC) based simulation of ALD and other thin film deposition processes to examine the fundamental growth mechanisms.

Our MC simulation methods are based on discretizing the growth surface, enumerating all possible states that can exist for each of the lattice regions and all possible transitions between these states during each of the ALD exposure half-cycles. Transition probabilities are quantified using a combination of published literature on Al_2O_3 ALD and from data obtained from our collaborators. In our work, the surface states and transitions are represented as nodes and edges of a graph, respectively. We will demonstrate the benefits of using this approach in terms of clarifying and classifying the relationships between the large number of chemical and physical processes at work in our deposition systems.

Furthermore, we consider the problem of directly computing the steady-state behavior of the ALD process, which consists of finding fixed points to limit cycle solutions. In our approach, we represent the growth surface both by the lattice of states and a lower-dimensional distribution function. At the start of a full ALD cycle, the distribution function is used to generate a microscopic realization of the growth surface, which is followed forward in time through each exposure half cycle using the Monte Carlo simulator. The result at the end of the cycle then is projected back down to the low-dimensional distribution (the *lift/simulate/restrict* approach of Kevrekidis and coworkers [3]), and a residual is computed from the difference between the distributions at the start and end of the full exposure cycle. This residual is used in a Newton-Raphson procedure to compute the steady-state.

This talk will focus mainly on Al_2O_3 ALD using TMA and water as precursors. The use of our fixed-point algorithm in conjunction with a reactor-scale transport model also will be presented as a demonstration of multiscale simulation of the complete ALD process.

[1] IEEE Spectrum, Oct. 2007.

[2] Rubloff, et al., 2007 AVS meeting.

[3] See, e.g., Cipria, SIAM News, June 2005, for an overview.

Web Page: www.isr.umd.edu/Labs/CACSE/A-Team

Multiscale stochastic simulation of emulsion polymerization

Hugo F. Hernandez and Klaus Tauer
Max Planck Institute of Colloids and Interfaces, Am Mühlenberg 1, 14476 Golm, Germany

Abstract

Heterophase polymerization is a highly complex dynamic process in which several simultaneous and usually competitive chemical (radical generation, propagation, termination, chain transfer) and physical events (diffusion, absorption, desorption, nucleation, coagulation) occur at very different time scales and dimensions. These events take place in a typical free-radical emulsion polymerization at rates ranging from about 10^0 to 10^9 s^{-1} and involving entities of very different length scales, such as ions and molecules (< 1 nm), macromolecules (1 – 10 nm), polymer particles (10 nm – 1 μm) and monomer droplets (>1 μm). A precise quantitative model of the kinetics of emulsion polymerization requires the detailed knowledge of the kinetics of all these physical and chemical processes.

In the present work, different events are investigated at different time and length scales using suitable simulation methods such as Molecular Dynamics (MD) simulation,[1,2] Brownian Dynamics (BD) simulation[3] and kinetic Monte Carlo (kMC) simulation.[4] The use of simulation techniques with spatial resolution such as MD and BD is advantageous when dealing with systems of increased geometrical complexity. This is the case of non-spherical polymer particles, structured polymer particles (e.g. core-shell, multilayered, etc.), non-uniform polymer particles (e.g. gradient in radial monomer composition), polydisperse particles and many others.

Molecular Dynamics simulation is used to estimate the diffusion coefficients of a given molecular species under very specific conditions. These diffusion coefficients are then used by the Brownian Dynamics simulation method to describe the molecular motion at a much larger time and length scale. The BD simulation of radicals in emulsion polymerization is used to determine the rate of radical capture by polymer particles,[5,6] as well as the rate of radical desorption from the particles to the continuous phase.[7] Similarly, the BD method can be used to simulate the absorption and desorption of monomer molecules by polymer particles, and therefore, to obtain a description of the monomer swelling dynamics.[8] The simulation of polymerization reactions under diffusion control, which usually takes place inside the polymer particles at low monomer volume fraction, is performed using a new modified version[9] of the stochastic simulation algorithm developed by Gillespie.[4] Finally, the integration of all these multi-scale processes into one single simulation is performed using a kinetic Monte Carlo (kMC) algorithm.

Introduction

The basic definition of heterogeneous polymerization is the presence of more than one phase at some moment during the polymerization process. In a typical emulsion polymerization the following phases may be found: an aqueous phase (continuous phase) which may contain dissolved ions and molecules, a monomer phase segregated as droplets in a wide range of sizes, micelles or surfactant aggregates, polymer particles and eventually, polymer clusters inside the particles. The size of each segregated phase in a typical emulsion polymerization may vary from a few angstroms (molecular aggregates) to some millimeters (macroscopic monomer drops). All these separate phases are clusters or aggregates of molecules, and all of them are correspondingly subjected to the processes of diffusion, aggregation (including absorption) and dissociation (including desorption), in addition to the corresponding polymerization reactions. This is graphically presented in Figure 1.

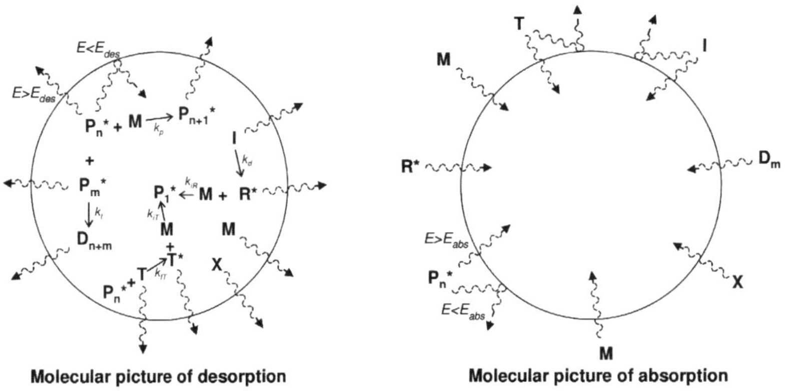

Figure **1** Molecular picture of the processes of reaction, desorption and absorption of molecules for a segregated phase. The transfer of any molecule through the interface is possible only if the energy (E) of the molecule is higher than the barrier for transfer (E_{des} for desorption, E_{abs} for absorption). M: Monomer, I: Initiator, R*: Primary radical, T: Chain transfer agent, T*: Transfer-derived radical, P_n*: Polymer radical, D_m: Dead polymer, X: Any other molecule.

The main factors determining the diffusion, dissociation and aggregation of clusters and individual molecules are the magnitude and direction of intermolecular forces acting between neighboring molecules. Because of this, it would be possible to get a comprehensive picture of heterophase polymerization using molecular dynamics simulation; however, the computational requirement for performing such a simulation for reasonable time and length-scales is extremely high and at present, it is practically impossible. Much larger time and length scales can be reached using Brownian Dynamics for many different processes taking place in emulsion polymerization. However, these methods are not enough to simulate processes at an industrial scale. For these purposes, stochastic kinetic Monte Carlo or deterministic numerical integration of differential equations can be used. Although deterministic methods offer fast estimations of average values, stochastic methods can handle the variability of the process.

In general, polymeric systems are difficult to simulate because of the wide spectrum of time and length scales characterizing their dynamics and structure.[10] If the scale of the simulation is increased in order to observe slower dynamics, the fast dynamics at lower scales must be either neglected or approximated. Depending on the type of approximation and its accuracy, the results obtained in the simulation can or cannot represent the real process. In this sense, a suitable multi-scale simulation algorithm capable of simultaneously considering all relevant dynamics would be desirable.

Multiscale simulation

In order to obtain a more complete picture of emulsion polymerization, the most relevant physical and chemical processes were investigated using different modeling strategies at their corresponding scales. These processes include: diffusion, chemical reaction under diffusion control, particle formation, radical capture, radical absorption, monomer swelling and particle aggregation.

Molecular diffusion was investigated using Molecular Dynamics simulation, considering Buckingham-type interatomic pair interaction potentials.[11] MD simulation is a numerical method used to solve Newton's classical equations of motion at the molecular level. Considering an initial configuration of a representative ensemble of molecules, and assuming a particular description of the interaction energy between different molecules it is possible to determine the positions and velocities of the molecules at any given future time. The results obtained from MD simulation were used to obtain an analytic expression for calculating the diffusion coefficient of the molecules. The diffusion coefficient was found to be proportional to the square root of the absolute temperature of the system, to be inversely proportional to the molecular mass, to depend exponentially on the density of system (in agreement with the free-volume theory)[12] and to decrease almost linearly with the steepness of the repulsive forces.

For the investigation of diffusion-controlled polymerization reactions, a modified stochastic simulation algorithm has been developed[9] which is able to incorporate imperfect mixing effects into the kinetic Monte Carlo scheme.[4] In the kMC method, all physical and chemical processes take place according to a certain probability distribution, usually a Poisson distribution, characterized by the corresponding rate of each process. In the traditional kMC algorithm, a small finite but representative volume is defined, which is assumed to be perfectly mixed and to remain constant during the whole simulation. In the modified algorithm, the perfectly mixed volume is not defined as an arbitrary and constant input parameter, but it is calculated at every simulation step as a function of the diffusivity of the reacting species. By means of this modified stochastic simulation algorithm, it was possible to describe mass transfer limitation effects (cage effect, gel effect and glass effect) in a typical bulk polymerization of methyl methacrylate, considering constant polymerization kinetics during the whole process.[9]

Another important mechanism in emulsion polymerization is the formation of polymer particles. Polymer particles formation is an aggregation process which involves not only the polymer chains but also monomer molecules and monomer droplets and any other molecule or molecular aggregate with a certain affinity with the polymer (e.g. surfactant molecules). It is for this reason that the most probable source of polymer particle formation comes from the interaction between growing polymer chains and small monomer droplets formed by the process of spontaneous emulsification.[13] However, given that the propagation of the radicals is a highly exothermic reaction, the chain length of the growing radical and the size of the aggregate are the limiting factors for the proper formation of a polymer particle. If the radical is too small, as well as the polymer-monomer aggregate, the energy released during a propagation step can be large enough to promote the desorption of the radical into the continuous phase.[14] For longer chains and larger aggregates, the thermal dissipation of the energy avoids the propagation-induced desorption of the radical. For these reasons, a new model of particle formation based on the propagation-induced desorption of the radicals has been developed. [11] According to this model, if the chain length of a growing radical is below a certain critical value, the energy released during propagation overcomes the energy barrier for desorption and the chain will return to the continuous phase. Only if the critical chain length is reached in the continuous phase and not inside an already existent particle, a new polymer particle is obtained.

The investigation of radical dynamics (absorption-desorption) in emulsion polymerization was performed using Brownian Dynamics simulation, achieving a precision which cannot be matched with the available experimental methods. BD simulation is a numerical technique used to solve Langevin's stochastic differential equation of Brownian motion. By means of this method, it is possible to determine the paths followed by Brownian entities (molecules, colloidal particles) as a result of the random collisions against the molecules in the medium. In the case of radical absorption or radical capture, it has been found that for very dilute polymer dispersions the rate of capture is in excellent agreement with the classical Smoluchowski's equation. However, for concentrated dispersions (>0.1%) the rate of radical capture deviates significantly this expression.[5] This deviation is represented by the dimensionless Smoluchowski number, and it has been found to increase almost linearly with respect to the volume fraction of the particles. According to these results, the dependence of the capture rate coefficient on the particle size ranges from a linear to a fourth order dependence depending on the volume fraction of particles in the system. This explains why different models of radical capture have been satisfactorily validated using experimental data.

In the case of radical desorption, BD simulation allowed a more precise determination of the desorption rate coefficient in excellent agreement with theoretical expressions. From the BD simulation of radicals in homogeneous spherical polymer particles, the rate coefficient of radical desorption was found to be: $k_0 \approx 60 D_r / d_p^2$, where D_r is the diffusion coefficient of the radical inside the particles and d_p is the particle diameter. Although this equation was found to be 5 to 30 times larger than the values previously reported, the correctness of the new expression was confirmed by a precise theoretical model.[7] BD simulation has been shown to be a very useful method for the determination of radical desorption kinetics especially when more complex but also more realistic situations are considered, like for example, in core-shell particles, in particles with radial monomer gradients, or for non-spherical polymer particles.[15]

It has been also evidenced that absorption/desorption dynamics and equilibrium are strongly affected by the presence of energy barriers for the mass transfer across the surface of the particles. Furthermore, the behavior described by BD simulation is not exclusive of the radicals but can be extended to all the molecules present in the system, including for example, the monomer molecules. Therefore, the swelling dynamics and equilibrium of polymer particles by monomer can also be described by this approach.[8]

Simulation Example

In this section, an approximation to the modeling of secondary particle nucleation in emulsion polymerization based on multi-scale simulation is presented. The system considered the semi-batch surfactant-free emulsion polymerization of vinyl acetate (VAc) in the presence of monodisperse polystyrene (PS) seed latex. This particular system is very interesting to investigate because:

- Vinyl acetate and styrene do not copolymerize. In fact, the presence of free styrene monomer inhibits the polymerization of vinyl acetate.
- These monomers have very different solubility in water. Vinyl acetate is hydrophilic (0.5 M) while styrene is hydrophobic (4.3×10^{-3} M).
- Their polymers are incompatible. Therefore, polymer phase separation is expected.

Table **1** Parameters used for the second example of multiscale kMC-BD simulation for the seeded emulsion polymerization of vinyl acetate

Parameter	Value	Parameter	Value
Temperature (°C)	80	Simulation volume (l)	$2x10^{-18}$ - $9x10^{-15}$
Seed volume fraction (%)	5 – 40	Seed particle diameter (nm)	100 – 1000
Particle number concentration (#/l)	$4x10^{14}$ – $4x10^{17}$	Aqueous monomer concentration (mol/l)	0.5
Monomer concentration inside the particles (mol/l)	7.5	Initial initiator concentration (mol/l)	$1x10^{-3}$
Initiator decomposition rate coefficient (s^{-1})	$8.6x10^{-5}$	Propagation rate coefficient (l/mol·s)	$1.29x10^{4}$
Chain transfer to monomer rate coefficient (l/mol·s)	2.32	Termination activation energy (J/mol)	2000
Primary radical diameter (nm)	0.526	Primary radical molar mass (g/mol)	96.16
Monomer unit diameter (nm)	0.664	Monomer unit molar mass (g/mol)	86.09
Energy barrier for desorption (J/mol)	12000	Energy barrier for absorption (J/mol)	8000
Enthalpy of propagation (J/mol)	88000	Minimum chain length for particle formation	18
Radical diffusion coefficient in water (m^2/s)	$1.4 x10^{-9}$	Radical diffusion coefficient inside the particles (m^2/s)	$1x10^{-11}$
Final simulation time (s)	0.42 – 390		

The dynamics of radicals around or inside a single structured polymer particle are simulated using a kMC-BD multiscale approach, where a BD simulation will be performed around one single particle achieving a spatial resolution of the motion of the radicals, while all other processes are periodically incorporated by means of the kMC approach. Not only radical absorption but also desorption are considered assuming energy barriers determined by non-conservative forces and by the chemical potential difference of the radicals between the particle and the continuous phase. In addition, radical desorption induced by the propagation of the growing chain and chain transfer to monomer are incorporated into the model, and the particles are assumed to be swollen by monomer before starting the polymerization. New particle formation is assumed to take place according to the propagation-induced desorption model, briefly described in the previous section. Since the time-scale that can be covered by BD simulation is relatively short compared to the total polymerization time, the simulation is performed only up to the first 400 seconds of the reaction. In order to investigate secondary particle formation this time is enough given that secondary nucleation, if ever takes place, is observed at the beginning of the process. In fact, since the existing polymer particles are continuously growing, the probability of radical capture increases with polymerization time and thus, the probability of particle formation is reduced. The parameters and conditions used in the simulation are summarized in Table 1. In a first case, the radicals are generated by thermal decomposition of initiator molecules which are assumed to be uniformly distributed in the continuous phase. A second set of simulation data was obtained using the same conditions of Table 1, but assuming that hydrophilic radicals are generated by a redox reaction between one hydrophilic reducing agent located in the continuous phase and one hydrophobic oxidizing agent present inside the polymer particles. In this way, the locus of radical formation can be restricted to the surface of the particles. A third set of simulation data was obtained using the same parameters and assuming that the radicals are generated by the thermal decomposition of inisurf molecules, that is, initiators with surface activity that are mainly located at the surface of the particles. The simulation results for all three situations are presented in Figure 2.

According to the simulation results, the probability of secondary particle formation decreases by increasing the seed volume fraction of the latex for a given seed particle size, and increases by increasing the seed size for a constant seed volume fraction. This is caused by the increased chain length of the polymers that can be achieved in the continuous phase. On the other hand, it is possible to clearly evidence the influence of the locus of radical generation on the probability of secondary nucleation. When a hydrophilic/hydrophobic redox initiation system is used, it is observed a complete suppression of secondary particle formation for seed volume fractions above 20% for a seed particle size of 500 nm, and for 20% volume fraction seed latexes, secondary nucleation is suppressed for seed particle sizes below 1 μm. In the case of the inisurf, the probability of particle formation is greatly reduced compared to normal initiators. Although both the redox initiator and the inisurf produce radicals close to the surface of the particles, the use of the redox system allowed a larger reduction in secondary particle formation compared to the inisurf. This result can be explained by the fact that the rate of radical generation in the redox system is much higher than for the thermal decomposition of the inisurf, leading to a higher concentration of radicals around the particles and to a larger number of termination events. Even though secondary nucleation is further suppressed with the redox system, the rate of polymerization can also be reduced and shorter chains may be produced. However, this multi-scale simulation approach showed that the control of the locus of radical generation is an excellent strategy for the control of secondary particle formation during the synthesis of structured particles by seeded emulsion polymerization.

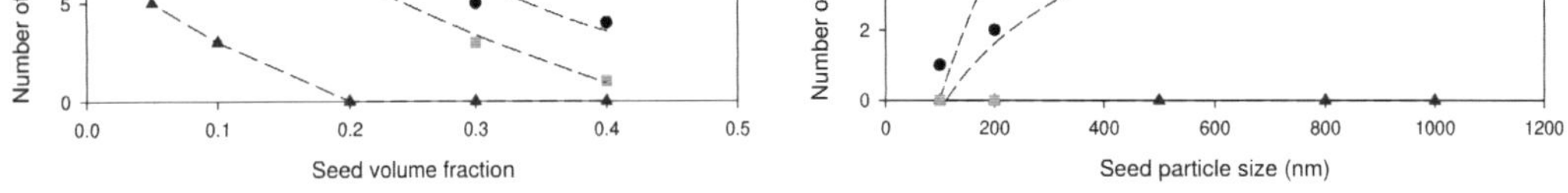

Figure **2** Number of new particles formed per 100 radicals as a function of the seed volume fraction for a seed latex of 500 nm in particle size (left) and as a function of the seed particle size for a seed latex of 20% volume fraction (right). Circles: Generation of radicals by thermal decomposition of water-soluble initiator. Triangles: Radical generation by redox interfacial reaction between a hydrophilic reducing agent and a hydrophobic oxidizer. Squares: Radicals generated by the thermal decomposition of an inisurf. Dashed lines: Fitted curves.

The experimental evidence of the success of this idea was presented by Soltan-Dehghan et al.[16] for the synthesis of structured polystyrene/poly(vinyl acetate) polymer particles. They used a two-stage polymerization process to prepare PS/PVAc nanocomposite polymer particles with a core-shell morphology in an emulsifier-free emulsion polymerization system, employing 2,2-azo bis(2-amidino propane) dihydrochloride as an inisurf initiator. The particle diameter of the polystyrene seed used was around 300-

400 nm, which is clearly beyond the limiting value of 150 nm obtained theoretically by Ferguson et al.[17] for this particular system.

Conclusion

The most important conclusion of the present work is that heterophase polymerization is a complex multi-scale stochastic process. For that reason, a very accurate description of this type of process requires the adequate integration of all these relevant scales, that is, requires the use of suitable multiscale dynamic simulation methods.

It has been shown that multiscale dynamic simulation methods can be successfully used for the investigation of the kinetics of emulsion polymerization. In particular, these methods were used to investigate the synthesis of structured polymer particles by semi-batch seeded emulsion polymerization as a representative case study. The results obtained in the simulation were used to find optimal conditions for the suppression of secondary nucleation in emulsion polymerization. It was demonstrated that by controlling the locus of radical generation at the surface of the particles it is possible to greatly reduce the probability of secondary particle formation in seeded emulsion polymerization systems. These conditions were successfully tested by combining Brownian Dynamics simulation with kinetic Monte Carlo simulation, and they could not be obtained using the traditional modeling techniques used in emulsion polymerization.

There are still enormous possibilities for the investigation and exploration of multi-scale integration in heterogeneous polymerization systems, involving different combination of scales ranging from the atomistic to the macroscopic scale. Hopefully, these ideas can be used to obtain an even more complete understanding of all the mechanisms taking place in heterophase polymerization at all relevant length scales, and also, that the fast increase in computational power can allow very soon to perform faster simulations and to span more easily a wider range of time scales.

References

[1] Frenkel, D. and B. Smit. "Understanding molecular simulation", Academic Press, San Diego 2002.

[2] Chen, J. C. and A. S. Kim. "Brownian Dynamics, Molecular Dynamics and Monte Carlo modeling of colloidal systems". Adv. Colloid Interf. Sci., 112, 159-173 (2004).

[3] Ermak, D. L. and J. A. McCammon. "Brownian dynamics with hydrodynamic interactions". J. Chem. Phys., 69, 1352-1360 (1978).

[4] Gillespie, D. T. "A general method for numerically simulating the stochastic time evolution of coupled chemical reactions". J. Comp. Phys., 22, 403-434 (1976).

[5] Hernández, H. F. and K. Tauer. "Brownian Dynamics Simulation of the capture of primary radicals in dispersions of colloidal polymer particles". Ind. Eng. Chem. Res., 46, 4480-4485 (2007).

[6] Hernández, H. F. and K. Tauer. "Brownian Dynamics Simulation Studies on Radical Capture in Emulsion Polymerization". Macromol. Symp., 259, 274-283 (2007).

[7] Hernández, H. F. and K. Tauer. "Radical desorption kinetics in emulsion polymerization 1. Theory and Simulation". Submitted to Ind. Eng. Chem. Res.

[8] Hernández, H. F. and K. Tauer. "Non-equilibrium molecular uptake by polymer particles in emulsion polymerization". In preparation.

[9] Hernández, H. F. and K. Tauer. "Stochastic simulation of imperfect mixing in free radical polymerization". Macromol. Symp. (2008). In press.

[10] Harmandaris, V. A., N. P. Adhikari, N. F. A. van der Vegt, K. Kremer, B. A. Mann, R. Voelkel, H. Weiss and C. C. Liew. "Ethylbenzene diffusion in polystyrene: United atom atomistic/coarse grained simulations and experiments". Macromolecules, 40, 7026-7035 (2007).

[11] Hernández, H. F. "Multiscale simulation of heterophase polymerization: Application to the synthesis of multicomponent colloidal polymer particles". Ph.D. Dissertation. Potsdam University. 2008.

[12] Fujita, H. "Diffusion in polymer-diluent systems". Adv. Polym. Sci., 3, 1-47 (1961).

[13] Tauer, K., H. F. Hernández, S. Kozempel, O. Lazareva and P. Nazaran. "Towards a consistent mechanism of emulsion polymerization – new experimental details". Colloid Polym. Sci., 286, 499-515 (2008).

[14] Harkins, W. D. "A general theory of the mechanism of emulsion polymerization". J. Am. Chem. Soc., 69, 1428-1444 (1947).

[15] Hernández, H. F. and K. Tauer. "Radical desorption kinetics in emulsion polymerization 2. Application of Brownian Dynamics Simulation". Submitted to Ind. Eng. Chem. Res.

[16] Soltan-Dehghan, M., N. Sharifi-Sanjani and N. Naderi. "Preparation of polystyrene/poly(vinyl acetate) nanocomposites with a core-shell structure via emulsifier-free emulsion polymerization". J. Appl. Polym. Sci., 100, 2409-2414 (2006).

[17] Ferguson, C. J., G. T. Russell and R. G. Gilbert. "Synthesis of latices with polystyrene cores and poly(vinyl acetate) shells. 1. Use of polystyrene seeds". Polymer, 43, 6371-6382 (2002).

653f A Novel Monte Carlo Algorithm for the Prediction of the Exact Topological Characteristics of Low Density Polyethylene

Dimitris Meimaroglou[1], Prokopis Pladis[2], Apostolos Baltsas[2], and **Costas Kiparissides**[1]. (1) Department of Chemical Engineering, Aristotle University of Thessaloniki & Chemical Process Engineering Research Institute, P.O. Box 472, Thessaloniki, 54124, Greece, (2) Chemical Process Engineering Research Institute, P.O. Box 60361, Thessaloniki, 57001, Greece

A well-known approach for the calculation of the distributed molecular properties of polymers (e.g., MWD, LCBD, CCD, etc.), is the use of bivariate population balance equations (PBEs), (Kiparissides, 2006). In principle, based on the polymerization kinetic mechanism, one can derive dynamic PBEs to describe the time evolution of the "live" and "dead" polymer chains in a polymerization reactor. The resulting PBEs can be solved numerically using well-established numerical methods (i.e., 2-D fixed pivot, orthogonal collocation on finite elements, etc.). A major limitation in the application of these methods is the fact that they are extremely complex mathematically and they cannot be extended to more than two dimensions (i.e., they cannot be applied to trivariate or multi-dimensional problems).

To deal with the above limitation, a novel Monte Carlo (MC) algorithm was developed. The proposed algorithm can be used for the accurate prediction of the complete topological characteristics of the polyethylene polymer chains produced in high pressure tubular and autoclave LDPE polymerization reactors. Via the MC algorithm, it is possible to predict the exact size and position of every single chain branch (either short- or long-) that is attached to a given polymer chain in the polymerization mixture, thus leading to the calculation of the complete topological structure of the polymer chains. Furthermore, no prior information on the distributional form of the "live" or "dead" polymer chains is required, neither any other kind of simplifying kinetic assumptions. The algorithm considers the evolution of both "live" and "dead" polymer chain populations and proceeds in complete conjunction with the polymerization kinetic mechanism.

The basic principles governing the proposed stochastic formulation are based on the original developments of Gillespie (1977). According to the above formulation, the desired molecular properties of the polymer chains are inferred by tracking the relevant changes occurring in a sample population, containing the different molecular species (Meimaroglou et al., 2007 and 2008). The net formation rates of all the chemical reactions in the reacting system are dictated by known reaction probabilities and are directly related to the polymerization kinetic mechanism.

The proposed MC algorithm was successfully applied to the free radical polymerization of LDPE, for the prediction of a series of distributed and topological properties of the polymer chains. The validity of the mathematical model was initially verified through a direct comparison of MC predictions with available experimental measurements on the molecular weight distribution (MWD). Moreover, the MC calculated average molecular properties (i.e., Mn, Mw, etc.) were directly compared with theoretical predictions from the application of the well-established method of moments (Kiparissides et al., 2005) in order to further verify the predictive capabilities of the method.

Via the MC calculations, a number of distributed topological properties can be obtained. These properties are directly related to the number (MW-LCBD, SCBs and LCBs per molecule, SCBs and LCBs per 1000C distribution), size (SCB and LCB length distribution) and order (branching order distribution) of the polymer chain branches. Additionally, the shape of the produced polymer chains can be inferred through the dynamic prediction of the seniority-priority distributions and the mean radius of gyration. The latter can lead to a straightforward calculation of the branching index of the polymer chains, g. Finally, all the information obtained by the MC algorithm can be further employed for the prediction of the viscoelastic properties of the polymer melts.

References

Gillespie D.T.J., Phys. Chem. 1977, 81, 2340.

Kiparissides C., J. Process Contr. 2006, 16, 205.

Kiparissides C., Baltsas A., Papadopoulos S., Congalidis J.P., Richards J.R., Kelly M.B. and Ye Y., Ind. Eng. Chem. Res. 2005, 44, 2592.

Meimaroglou D. and Kiparissides C., to be submitted to J. Polym. Sci. Polym. Phys. Ed. 2008

Meimaroglou D., Krallis A., Saliakas V. and Kiparissides C, Macromolecules 2007, 40, 2224.

653g Combined Run-to-Run and Lqg Control of a 12-Inch Rtp Equipment

Kwang Soon Lee[1], Wangyun Won[2], Woohyun Yoon[2], Sechan Lee[2], **Sanghyun Ji**[3], and Byungchul Na[3]. (1) Chemical & Biomolecular Engineering, Sogang Univ., 1-Shinsoodong, Mapogu, Seoul, 121-742, South Korea, (2) Chemical and Biomolecular Engineering, Sogang Univ., 1-Shinsoodong, Mapogu, Seoul, 121-742, South Korea, (3) Kornic Systems Co. Ltd., Hwasung, Kyunggido, South Korea

A combined run-to-run (R2R) and LQG control method has been proposed for rapid thermal processing (RTP) equipments for run-wise improvement together with real-time multivariable control of the temperature uniformity over the wafer. The standard LQG objective was modified to include a quadratic penalty term for input deviation from a bias value which is updated by the R2R control. The proposed method has been applied to a commercial 12-inch rotating RTP equipment which has ten circular groups of tungsten-halogen lamps and four pyrometers for manipulation and real-time measurements of wafer temperatures. Multivariable LQG control was evaluated under wafer rotation and found to show a quite accurate control performance. For testing the combined control technique, a wafer that has eight thermocouples (TC's) attached along the radical direction has been employed without wafer rotation considering the TC temperatures as the post-run uniformity measurements. Experiments obviously showed that the temperature uniformity is improved as the run number increases.

672a Control of the Maximum-Boiling Acetone/chloroform Azeotropic Distillation System

William L. Luyben, Chemical Engineering, Lehigh University, Bethlehem, PA 18015

The literature contains a number of papers that study the control of azeotropic distillation systems. Both homogeneous and heterogeneous azeotropes have been considered, but all of the systems considered in these control papers deal only with minimum-boiling azeotropes. These azeotropes are caused by molecular repulsion between different types of chemical components and are more common than systems in which molecular attractions occur that result in maximum-boiling azeotropes. There appears to be no papers that study the control of this type of azeotropic system.

This paper explores the design and control of the maximum-boiling azeotropic acetone/chloroform distillation system. The normal boiling points of the two pure components are 329.4 and 334.3 K, while the azeotropic boils at 337.6K at 1 atm with a composition of 34.09 mol% acetone.

A two-column extractive distillation process is used with dimethyl sulfoxide as the solvent. The two components are separated into 99.5 mol% pure products leaving in the distillate streams from two distillation columns. The solvent flowrate that minimizes total energy consumption is determined. A control structure that is capable of handling very large disturbances in throughput and feed composition is developed. The control of two tray temperatures in the extractive column is found to be necessary to handle feed composition disturbances.

672b Minimum Energy Distillation Column Sequences

Angelo Lucia, Bradley R. McCallum, and Amit Amale. University of Rhode Island, Dept of Chem. Engineering, Kingston, RI 02881

Finding distillation column sequences for sharp splits of multi-component mixtures that have overall minimum energy consumption is still an important problem, particularly with the rapidly rising cost of energy. While there is a vast literature on the subject for the case of (c – 1) columns, where c is the number of components in the mixture, virtually all papers disregard configurations with more than (c – 1) columns because of the capital costs associated with additional columns.

In this work, we study the problem of finding overall minimum energy sequences for non-interlinked columns using the shortest stripping line distance approach. A theoretical framework as well as applications involving mixtures modeled using a variety of phase models (i.e., constant relative volatility, ideal, non-ideal, and azeotropic behavior) are presented. Numerical results clearly show that the shortest stripping line distance approach can find the most energy efficient column sequence – even in cases where this sequence has more than (c – 1) columns – and thus show that stripping line distance correlates with overall energy consumption for non-interlinked column sequences. These results represent an important first step in defining a new and powerful methodology for energy efficient multi-column synthesis and design. Many geometric illustrations are used to highlight key points of our geometric design methodology. Finally, some additional remarks are made regarding interlinked column sequences.

672c Thermodynamics of Salt Lake System for Process Synthesis: Integration of Representation, Experiments and Visualization

Kui S. Kwok[1], Maria E. Taboada[2], Luis A. Cisternas[2], and Ka M. Ng[1]. (1) Department of Chemical Engineering, Hong Kong University of Science and Technology, Clear Water Bay, Kowloon, Hong Kong, China, (2) Departamento de Ingenieria Quimica, Universidad de Antofagasta, Casilla 170, Antofagasta, Chile

Inorganic salts have wide applications in diverse fields. One of their major sources is salt lakes, and crystallization-based techniques have been utilized to extract the desired salts from the salt lake brine. For a feed with a given composition, the determination of which particular salt (or a mixture of salts) crystallizes under a specific set of operating conditions, and the development of the corresponding crystallization-based extraction process, requires knowledge of the solid-liquid equilibrium (SLE) phase behavior of the salt lake system and process design methods. Despite of the well establishment of phase-diagram based process synthesis procedures and process optimization methods, however, dealing with a salt lake system poses special challenges. The existence of a huge number of different species in a salt lake greatly complicates the representation, experimental determination and visualization of the thermodynamic data of such system, which thus causes difficulty in the process synthesis.

To tackle this problem, a coherent, systematic approach for the determination of solid-liquid phase behavior of a multi-component salt lake system for use in the synthesis of crystallization process is presented. It centers on the thermodynamics of such a salt lake system and integrates three interrelated activities - representation of the system phase behavior as a phase diagram/thermodynamic model, experimental determination of the necessary data and visualization of the relevant crystallization regions. To illustrate this approach, the thermodynamics of a simplified salt lake system Li^+, Na^+, K^+, Mg^{2+}//Cl^-, SO_4^{2-} - H_2O at 25°C and 1 atm was determined. The identification of process alternatives using the resulting phase diagram for recovering $Li_2SO_4*H_2O$ was also demonstrated.

672d Parameter Estimation for Phase Equilibrium Problems Via Bilevel Programs

Alexander Mitsos, Aices, RWTH Aachen, Pauwelstrasse 12, Aachen, Germany, George M. Bollas, Department of Chemical Engineering - Process Systems Engineering Laboratory, Massachusetts Institute of Technology, 77 Massachusetts Ave., RM 66-363, Cambridge, MA 02139, and Paul I. Barton, Chemical Engineering, Massachusetts Institute of Technology, 77 Massachusetts Ave., RM 66-464, Cambridge, MA 02139.

In this talk a formulation for parameter estimation with activity coefficient models is proposed based on bilevel programs with nonconvex lower-level programs. The upper-level program is the minimization of prediction error. The lower-level programs represent the thermodynamic stability requirement. Case studies from liquid-liquid and vapor-liquid equilibria are given. Special emphasis is placed on formulation and solution methodology.

Good activity coefficient models are necessary for the reliable design of separation systems. There is a variety of activity coefficient models, such as the modified Wilson, NRTL and UNIQUAC models [1]. These models contain adjustable parameters and, provided that suitable parameter values are used, the models can accurately predict the phase behavior of virtually any relevant system. No direct method exists for the calculation of the parameters, and therefore these must be estimated based on measured phase-splits. In separations design temperature is not fixed, and therefore a temperature dependence for the parameters is needed, based on some postulated function, e.g., a linear or second order polynomial. The actual parameter estimation is to find the coefficients of this polynomial.

Current methods for this parameter estimation have several limitations. One limitation is that the isoactivity method can give unstable predictions with a ``perfect fit''. In other words, current methods can furnish parameter values for which the measured phase split is a local minimum of the Gibbs free energy, but not the global one, which is not physical. There are methods to check for this case in retrospect, but when stability is not satisfied, these methods don't offer a way of proceeding. A second limitation is that due to their flexibility the activity coefficient models can predict more phases than actually exist in the system [2,3,4]. Clearly, this case must also be excluded. There are methods to detect this situation a posteriori, but not a generic reliable method of excluding it a priori. Third, typically parameters are estimated at a fixed temperature and then the temperature coefficients are estimated in a second calculation; this is tedious and not guaranteed to give a good fit. Finally, since local optimization methods are used, it cannot be guaranteed that the best possible fit was obtained, which is particularly important in the case of model-experiment mismatch.

In this presentation parameter estimation for phase equilibrium problems is cast as a bilevel optimization problem. Bilevel programs are programs where one optimization problem (upper-level program) is constrained by the (global) solutions of another optimization problem (lower-level program). A major challenge in bilevel programs is nonconvexity in the lower-level program and only recently methods have been proposed to deal with this case rigorously [5]. In the proposed formulation, the upper-level program minimizes the error in predictions by adjusting the model parameters; typically a least squares error or absolute error objective function is used here. The lower-level programs represent the stability requirement, i.e., that the predictions must be the global minimum of the Gibbs free energy for the chosen parameter values; for computational reasons specialized stability criteria [6,7] are used instead. The lower-level programs are nonconvex and their global solution is necessary. The talk first motivates the formulations and then gives a formal statement. The solution method based on [5] is detailed and then case studies are presented. These examples chosen are known to be challenging. It is shown that existing methods can lead to spurious solutions that are avoided by the proposed approach.

[1] J. W. Tester and M. Modell. Thermodynamics and its Applications. Prentice Hall International Series in the Physical and Chemical Engineering Sciences. Prentice Hall, New Jersey, 3rd edition, 1997.

[2] R. A. Heidemann and J. M. Mandhane. Some properties of the NRTL equation in correlating liquid-liquid equilibrium data. Chemical Engineering Science, 28(5):1213--1221, 1973.

[3] A. C. Mattelin and L. A. Verhoeye. Correlation of binary miscibility data by means of NRTL equation. Chemical Engineering Science, 30(2):193--200, 1975.

[4] D. Tassios. Number of roots in NRTL and LEMF equations and effect on their performance. Industrial & Engineering Chemistry Process Design and Development, 18(1):182--186, 1979.

[5] A., P. Lemonidis, and P. I. Barton. Global solution of bilevel programs with a nonconvex inner program. In press: Journal of Global Optimization, October 12, 2007.

[6] L. E. Baker, A. C. Pierce, and K. D. Luks. Gibbs energy analysis of phase equilibria. Soc. Petrol. Engrs. J., 22:731--742, 1982.

[7] A. Mitsos and P. I. Barton. A dual extremum principle in thermodynamics. AIChE Journal, 53 (8):2131--2147, 2007.

672e The Effects of Kinetic and Hydrocarbon Distribution on the Multiplicities of Deep Hydrodesulfurization Via Catalytic Distillation

Arnulfo Rosales-Quintero, Matemáticas Aplicadas y Computación, Instituto Mexicano del Petroleo, Eje Central Lázaro Cárdenas 152, Col. San Bartolo Atepehuacan, Mexico City, 07330, Mexico and Felipe Vargas-Villamil, Matemáticas Aplicadas y Computación, Instituto Mexicano del Petróleo, Eje Central Lazaro Cárdenas 152, Col. San Bartolo Atepehuacan, Mexico City, 07730, Mexico.

Reactive distillation was first proposed by Backhaus (1921); since then, various processes have been proposed, and nowadays, it is becoming an important unit operation. A reactive distillation column integrates the reaction and separation in a single unit which makes its open loop behavior complex. Kienle and Marquardt (2002) classified its open-loop behavior as: i) processes with equilibrium-controlled/fractionation-controlled chemical reactions; ii) processes with kinetically controlled reactions, where all components have similar boiling points; and iii) kinetically controlled processes with large boiling point differences. Any of these may result in multiplicities and complex attracting sets. The physical mechanisms leading to multiplicities are different. Various efforts have been made to explain the multiplicities in these systems. For non-reactive distillation columns, multiplicities may arise as a result of: i) singularities in the mass-molar relationship, ii) the presence of an azeotrope, or iii) the dependence of the heat of vaporization on composition for certain input variables. However, a reactive distillation column is more complex, since chemical reactions take place and there are more operating and design variables that may cause multiplicities. In this system, the chemical reaction may produce a rapid change in the product composition without requiring a change in the feed-split or the energy balance (Chen et al. 2002). Thus, it is important to identify the region where a reactive column operates to detect possible operation problems from the design stage.

In this work, the effects of kinetic and hydrocarbon distribution on the multiplicities of a light gas oil (LGO) deep hydrodesulfurization (DHDS) two-bed catalytic distillation column (CDC) are studied. The physical mechanisms leading to multiplicities are identified and the multiplicity region is found for the reflux rate (***RR***) and bottom flow rate (***B***). These are the main operating variables and are generally used as control inputs. The Damköhler number (***Da***), which is the ratio of the characteristic residence time to the characteristic reaction time, is used to identify the mechanism leading to multiplicities. When ***Da*** > 1 the system is equilibrium-controlled/fractionation controlled, while when ***Da*** < 1 the system is kinetically controlled. The CDC studied in this work has 14 stages and two reactive zones. LGO is fed between the reactive sections while hydrogen is fed below the second reactive section.

To identify the multiplicity region and to evaluate the effect of the hydrocarbon distribution on the multiplicities, three hydrocarbon mixtures are used. The first is a real LGO fraction which is modeled as pseudocompounds; the second is a synthetic gas oil mixture (SGO), while the third is a lighter synthetic gas oil mixture (LSGO). The multiplicity regions for the CDC described above, for the three hydrocarbon mixtures, are identified and the bifurcation plots of the reflux rate (***RR***) and bottom flow rate (***B***) as a function of the ***Da*** number are presented.

The fractionating effects are clearly shown by the ***RR*(*Da*)** plots where double limit singularities are observed for the LGO. The multiplicity regions for the lighter mixtures (SGO and LSGO) are smaller than the multiplicities of the heavier mixture (LGO). For the LGO mixture, up to five multiplicities are found, while for the SGO and the LSGO mixtures only three steady states are observed. On the other hand, the ***B*(*Da*)** plots show a rich multiplicity behavior where up to seven multiplicity regions are found for the SGO mixture, while four multiplicity regions were found for the LGO fraction. Isola branches, 0/∞-disjoint, double limit singularities, and the classical S-shape multiplicities were found for both the LGO and SGO mixtures. Disjoint bifurcations result in non-feasible operating regions that separate non-closed disconnected steady-state solution branches. This behavior arises when there are bounds on the values of the parameters or state variables. Determining the feasible operating region is important during the design of feedback control.

Results show that this is an equilibrium-controlled/fractionation controlled chemical reactions system. This is, multiple solutions may be present for design and operating conditions that approach reaction equilibrium (***Da*** > *1*). Yet, a unique solution is expected at low Da numbers. In this system, the H_2 must be dissolved in the liquid phase so the reaction takes place. Yet, if the temperature decreases, the hydrogen diluted in the liquid phase increases. However, when the reaction section temperature decreases, the reaction rate also decreases. Here, the inverse correlation of these phenomena is clear. Therefore, we conclude that the multiplicities are the result of equilibrium-driven self-inhibition phenomena, which are the result of the presence of dissolved H_2 in the mixture. Results also show that the hydrocarbon distribution plays an important role in the occurrence of multiplicities. The hydrocarbon distribution in the mixture modifies the activity coefficients and thus, the boiling point of the mixture. This affects the internal heating/cooling flows and the reaction rates. In addition, it is shown that for ***RR* (*Da*)**, the multiplicity region is reduced for the lighter hydrocarbon mixtures, while for ***B*(*Da*)** it increases. It is also shown that the column operating point lies in a unique solution region that is far away from the multiplicity region. Yet, multiplicities may pose a problem during the start-up operations.

References

[1] A.A. Backhaus, (1921). Continuous processes for the manufacture of esters, *US patent 1400849*.

[2] F. Chen, R.S. Huss, M.F. Doherty, M.F. Malone, Multiple Steady States in Reactive Distillation: Kinetic Effects, Computer and Chemical Engineering, 26 (2002) 81-93.

[3] A. Kienle, W. Marquardt (2002), Nonlinear Dynamics and Control of Reactive Distillation Processes. In K. Sundmacher and A. Kienle (Eds), Reactive Distillation-Status and Future Directions, Wiley-VCH:Weinheim, 241-281.

Web Page: www.imp.mx

672f An Improved Group Method for Simulation and Optimization of Distillation Columns

Ravindra S. Kamath[1], Ignacio E. Grossmann[2], and Lorenz T. Biegler[1]. (1) Dept of Chemical Engineering, Carnegie Mellon University, 5000 Forbes Avenue, Pittsburgh, PA 15213, (2) Center for Advanced Process Decision-making, Dept of Chemical Engineering, Carnegie Mellon University, 5000 Forbes Avenue, Pittsburgh, PA 15213

Group methods are approximate calculation procedures for counter-current cascades that relate compositions of entering and exiting streams to the number of equilibrium stages. These methods were originally tailored for iterative or hand calculations and do not account for detailed changes in temperature and composition on individual stages. As a result, they have fewer variables and constraints as compared to rigorous models. Although rigorous models are capable of representing process behavior more accurately, they are more complex in terms of problem size and are often much more difficult to solve. In such cases, group methods may be used particularly if their performance closely matches that of rigorous models.

Group methods like Kremser and Edmister have been used to model absorbers, strippers, simple distillation columns etc. (Seader and Henley, 1981). The model equations for these methods often change depending upon whether the cascade acts as an absorber or stripper or is coupled with equipments like condenser (enricher) or reboiler (exhauster). Distillation columns with multiple feeds result in many counter-current cascades and it is not clear *a priori* whether a cascade of trays in a distillation column acts as an absorber or a stripper. Moreover, these models have not been systematically used in optimization studies of complex flowsheets where use of rigorous model either makes the problem intractable or computationally too difficult to solve.

In this paper, we propose a general cascade model that represents an aggregate model for a collection of trays in a section of distillation column. Our proposed model can be considered as an extension or improvement of the Edmister method. Some approximations in Edmister's method are replaced by more realistic constraints based on physical insight. Just like the rigorous tray-by-tray model, our model does not require knowledge about key components or about whether it behaves as an absorber or a stripper. Being an aggregate model, it can be connected to rigorous or aggregate models of other process models like feed stage of distillation column, condenser, reboiler or another such cascade. An interesting property of the proposed model is that if the number of stages in the cascade is set to zero, then the model forces the outlet vapor and liquid flow conditions to be exactly same as the corresponding inlet conditions i.e. it behaves as if the cascade does not exist. This is particularly helpful in topology optimization where sometimes minimization of capital cost may require removal of all the trays in a section (i.e. make the cascade disappear).

The capability of the proposed cascade model is demonstrated using both simulation and optimization case studies. Test problems involve distillation of binary or ternary systems where models for feed stages, condenser and reboiler are combined with cascade model to form the aggregate model and with model for all trays in a section to form the rigorous model. In the case of simulation case studies, the outlet conditions at the top and bottom are analyzed for a given set of input parameters (e.g. number of trays, column pressure, reflux ratio etc.). In the case of optimization case studies, the objective is to determine the optimal feed locations and/or optimal number of trays that minimize capital and/or energy costs while meeting the specified purity requirements. Such optimization problems have already been addressed using rigorous MINLP models by Viswanathan and Grossmann (1990, 1993a, 1993b). Their strategy involves use of binary variables which are associated with location of feed, reflux and boilup. On the other hand, our formulation using the cascade model requires that only the number of trays in each cascade (treated as integer variable) be optimized.

Results for the simulation case studies show that the outlet conditions predicted by the aggregate model are in close agreement with that of the rigorous model. For the optimization case studies, a rounding heuristic can be used i.e. the number of stages in the cascades can be relaxed as continuous variables and their optimal value can then be rounded to the nearest integer. In most cases, the integer solution obtained by this rounding heuristic was found to be the same as the integer solution of the MINLP model. Thus, use of integer variables can be eliminated for the aggregate model and its solution provides a good approximation to that of rigorous model (often missing by only one or two trays).

References

Henley, E. J. and Seader, J. D. (1981) *Equilibrium-Stage Separation Operations in Chemical Engineering*, Wiley, New York.

Viswanathan, J. and Grossmann, I. E. (1990) A combined penalty function and outer-approximation method for MINLP Optimization. *Computers and Chemical Engineering*, 14(7), 769-782.

Viswanathan, J. and Grossmann, I. E. (1993a) An alternate MINLP model for finding the number of trays required for a specified separation objective. *Computers and Chemical Engineering*, 17(9), 949-955.

Viswanathan, J. and Grossmann, I. E. (1993b) Optimal feed locations and number of trays for distillation columns with multiple feeds. *Industrial and Engineering Chemistry Research*, 32(11), 2942-2949.

672g Optimal Design of Multi-Cyclone Using Mixed-Integer Nonlinear Programming

Rajab Khalilpour, Chemical & Biomolecular Engineering, National University Of Singapore, E5-3-39, NUS, Singapore, 117576, Singapore and Ali Abbas, School of Chemical and Biomolecular Engineering, The University of Sydney, Sydney, NSW 2006, Australia, Sydney, Australia.

The cyclone is a simple device, which causes the centrifugal separation of materials in a fluid stream. These materials may be particles of solid, bubbles of gas or immiscible liquids. Cyclone separator, due to relative simplicity to fabricate, low cost to operate, and well adaptability to extremely harsh conditions has become one of the most important particle removal devices. Especially, nowadays because of stringent environmental regulations cyclone is becoming the best alternative for gas scrubbers and this has opened a wide application for this equipment. Many industries could then use this low investment and operating cost equipment with the required efficiency to avoid having to resort to equipments with much higher capital, operating, or maintenance costs, such as venturis, scrubbers and bag filters.

The cyclone utilizes the energy obtained from fluid pressure to create rotational fluid motion. This rotational motion causes the materials suspended in the fluid to separate from one another or from the fluid quickly due to the centrifugal force. The rotation is produced by the tangential or involuted introduction of fluid into the vessel.

There are many design and operation parameters that affect separation efficiency such as cyclone geometry, fluid velocity and stagnation velocity. The most important trade-off in cyclone design is that high efficiency separation results in high pressure drop and hence high operational costs. Based on fluid throughput when a single cyclone does not provide the required efficiency, one or multi cyclones are arranged to meet the separation requirements. In this study we did a mixed integer nonlinear programming to solve the cyclone problem in order to find the optimal multicyclone geometry and operational specifications in order to reach the desired separation requirement while having the minimum pressure drop.

672h Membrane Cascade Design Fundamentals for Energy Savings

Rugved Pathare, School of Chemical Engineering, Purdue University, 480 Stadium Mall Drive, West Lafayette, IN 47907 and Rakesh Agrawal, School of Chemical Engineering and Energy Center at Discovery Park, Purdue University, 480 Stadium Mall Drive, West Lafayette, IN 47907.

In a membrane separation unit with moderate permselectivity values, in order to recover products in high amount and high purity, it is important and sometimes necessary to cascade membrane units together in stages. The proper selection of such a cascade arrangement and operating parameters, to minimize capital cost and energy consumption is an important problem. The principal objective is to minimize the energy consumption in a cascade, but not use more than a few stages to keep capital costs low. Currently, all such design is done through innovative optimization technology, and there is no complete fundamental theoretical framework that guides the design. An important step in order to make membranes competitive is to develop a methodology to robustly design low energy consumption membrane separation schemes. Our approach focuses on developing a simple framework that may be easily automated by any user and guide them through the design of the economically optimal membrane cascades. In this work we present our analysis and limitations of several existing membrane cascade design strategies. Through our analysis we have also identified some novel, easy to implement and intuitive design ideas that could lead to high energy savings while using only a few stages in the membrane cascade.

687a A Two-Tier Architecture for Networked Process Control

Panagiotis D. Christofides[1], Jinfeng Liu[2], David Muñoz de la Peña[2], Ben Ohran[2], and James F. Davis[3]. (1) Department of Chemical and Biomolecular Engineering, Department of Electrical Engineering, University of California, Los Angeles, Los Angeles, CA 90095, (2) Department of Chemical and Biomolecular Engineering, University of California, Los Angeles, Los Angeles, CA 90095, (3) University of California - Los Angeles, Los Angeles, CA 90095

Increasingly faced with the requirements of safety, environmental sustainability, and profitability, chemical process operation is relying extensively on highly automated control systems. Traditionally, control systems utilize point-to-point wired communication links using a small number of sensors and actuators. The operation of chemical processes, therefore, could benefit from the deployment of control systems using hybrid communication networks that take advantage of an efficient integration of the existing, point-to-point communication networks (wire connections from each actuator/sensor to the control system using dedicated local area networks) and additional networked (wired or wireless) actuator/sensor devices. Such an augmentation in sensor information and network-based availability of wired and wireless data is now well underway in the process industries [1,2,3,4] and clearly has the potential to be transformative in the sense of dramatically improving the ability of the single-process and plantwide model-based control systems to optimize process and plant performance (in terms of achieving control objectives that go well beyond the ones that can be achieved with control systems using wired, point-to-point connections) and prevent or deal with adverse and emergency situations more quickly and effectively (fault-tolerance). Hybrid communication networks allow for easy modification of the control strategy by rerouting signals, having redundant systems that can be activated automatically when component failure occurs, and in general, they allow having a high-level supervisory control over the entire process [1,2,3,4]. However, augmenting existing control networks with real-time wired/wireless sensor and actuator networks challenges many of the assumptions in traditional process control methods dealing with dynamical systems linked through ideal channels with flawless, continuous communication. In the context of hybrid communication networks which utilize networked sensors and actuators, key issues that are important for process control include robustness, reliability and interference. These issues need to be carefully handled because integrated wired and wireless communication networks introduce more components in order to substantially improve closed-loop performance and fault-tolerance, and this increases the probability of missing data at any given point in time.

In this work, we introduce a two-tier control architecture for nonlinear process systems with both continuous and asynchronous sensing and/or actuation. This class of systems arises naturally in the context of process control systems based on hybrid communication networks (i.e, point-to-point wired links integrated with networked wired/wireless communication) and utilizing multiple heterogeneous measurements (e.g., temperature and concentration). Assuming that there exists a lower-tier control system which relies on point-to-point communication and continuous measurements to stabilize the closed-loop system, we propose to use Lyapunov-based model predictive control to design an upper-tier networked control system to profit from both the continuous and the asynchronous measurements as well as from additional networked control actuators. The proposed two-tier control system architecture preserves the stability properties of the lower-tier controller while improving the closed-loop performance. The theoretical results are demonstrated using two different chemical process examples.

1. Ydstie EB. New vistas for process control: Integrating physics and communication networks. AIChE Journal 2002; 48:422-426.

2. Davis JF. Report from NSF Workshop on Cyberinfrastructure in Chemical and Biological Systems: Impact and Directions, (see http://www.oit.ucla.edu/nsfci/NSFCIFullReport.pdf for the pdf file of this report). 2007.

3. Neumann P. Communication in industrial automation - what is going on? Control Engineering Practice 2007; 15:1332-1347.

4. Christofides PD, Davis JF, El-Farra NH, Clark D, Harris KRD, Gipson JN. Smart plant operations: Vision, progress and challenges. AIChE Journal, Perspective article 2007; 53:2734-2741.

687b A Subspace Identification Based Dynamic Soft Sensor Approach for Digester Control

Jin Wang[1], Qinghua (Peter) He[2], and Hector Galicia[1]. (1) Department of Chemical Engineering, Auburn University, Auburn, AL 36849, (2) Department of Chemical Engineering, Tuskegee University, Tuskegee, AL 36088

Kraft pulping is the commonly applied chemical pulping process that normally utilizes a (Kamyr) continuous digester. The most important quality variable in pulping is the Kappa number, which represents the residual lignin content of pulp, and it is desired to minimize the variations of the Kappa number of the pulp product. The continuous digester control problem is very challenging [1-6], and one of the reasons is that the main controlled variable, Kappa number, is measured infrequently and not in real time.

In this work, based on the formulation of the subspace identification [7-9], we derive a dynamic PLS model as the soft sensor to predict the Kappa number using process inputs and secondary measurements. By reducing the future horizon to one, the correlation between the future input and future disturbance can be eliminated. Therefore the developed method works well for closed-loop data. Because the developed PLS model is a reduced higher-order ARX model of the process, the soft sensor is not a static one but a dynamic one which captures the process dynamics; In addition, the model can be updated easily on line to capture any process changes.

Starting with a state-space representation of a linear system:

$$x(k+1) = Ax(k) + Bu(k) + Ke(k) \qquad (1)$$

$$y^s(k) = C^s x(k) + e(k) \qquad (2)$$

$$y^c(k) = C^c x(k) + e^c(k) \qquad (3)$$

where y^s and y^c are secondary and primary variables that are measured at different frequencies; process innovation $e(k)$ is assumed to be stationary white noise sequence. By representing the system described by Eqns (1) to (3) in its predictor form, we can obtain the extended state space model of the system, which is also the foundation for the subspace identification methods,

$$y_f^c(k) = \Gamma_f^{*c} L_p^* z_p(k) + G_f^{*c} z_{f-1}(k) + e_f^c(k) \qquad (4)$$

where subscript f denotes the future horizon, p denotes the past horizon, and $z(k) = \begin{bmatrix} u(k)^T & y^s(k)^T \end{bmatrix}^T$. The rest of the notations are the standard notation used in subspace identification (e.g. see [7]), and the superscript "*" indicates these matrices are constructed with predictor matrices instead of system matrices. Notice that many subspace identification methods do not apply to closed-loop data because the future input matrix contained in $z_f(k)$ are correlated to future disturbance matrix $e_f(k)$ due to feedback control because the future horizon is always larger than the system order. This limitation can be avoided by setting $f = 1$ and Eqn.(4) becomes the higher-order ARX model of the process,

$$y^c(k) = \left(C^c L_p^*\right) z_p(k) + e^c(k) \qquad (5)$$

Based on Eqn. (5), a linear regression between the current primary output and the past inputs and secondary outputs can capture the process dynamics because the future innovation $e^c(k)$ is independent of the past inputs and outputs. In this work, we apply partial least squares to capture the correlation between augmented input and output variables under closed-loop operation. PLS also naturally removes the noise and reduces the order of the ARX model.

In subspace identification formulation, the past horizon is usually much larger than the process order in order for the contribution from the initial state to be negligible. Consequently, a large number of regressor variables are included in the soft sensor while only a few of them contribute significantly to the prediction of the primary variable. Therefore, it is desirable to select most related variables to build the soft sensor. In this work, how to select process variables to be included in the soft sensor is also discussed.

Finally, the developed method is evaluated using both simulated digester process and industrial data. We used the fundamental model developed in [10] to simulate a single vessel digester, and the industrial data was provided by MeadWestvaco Corporation.

Reference

[1] J.H. Lee & A.K. Datta, Nonlinear inferential control of pulp digester, AIChE Journal, Vol. 40, 50-64, 1994.

[2] P.A. Wisnewski & F.J. Doyle, Control structure selection and model predictive control of the Weyerhaeuser digester problem, Journal of Process Control, Vol. 8, 487-495, 1998

[3] R. Amirthalingam & J.H. Lee, Subspace identification based inferential control applied to a continuous pulp disgester, Journal of Process Control, Vol. 9, 397-406, 1999.

[4] F.J. Doyle & F. Kayihan, Reaction profile control of the continuous pulp digester, Chemical Engineering Science, Vol. 54, 2679-2688, 1999.

[5] P.A. Wisnewski & F.J. Doyle, Model-based predictive control studies for a continuous pulp digester, IEEE Transactions on Control Systems Technology, Vol. 9, 435-444, 2001.

[6] J.J. Castro & F.J. Doyle, A pulp mill benchmark problem for control: application of plantwide control design, Journal of Process Control, Vol. 14, 329-347, 2004.

[7] J. Wang & S.J. Qin, Closed-loop subspace identification using the parity space, Automatica, vol. 42, 315-320, 2006.

[8] P. Van Overschee & B. De Moor, N4SID: subspace algorithms for the identification of combined deterministic-stochastic systems. Automatica, vol. 30, 75-93, 1994

[9] M. Verhaegen, Identification of the deterministic part of MIMo state space models given in innovations from input-output data. Aubomatica, Vol. 30: 61-74, 1994.

[10] P.A. Wisnewski, F.J. Doyle & F. Kayihan, Fundamental continuous pulp digester model for simulation and control, AIChE Journal, Vol. 43, 3175-3192.

687c Networked Control of Spatially Distributed Processes with Sensor-Controller Communication Constraints

Yulei Sun, Sathyendra Ghantasala, and **Nael H. El-Farra**. Department of Chemical Engineering & Materials Science, University of California, Davis, One Shields Avenue, Davis, CA 95616

With the significant growth in computing and networking abilities in recent times, as well as the rapid advances in actuator and sensor technologies, there has been an increased reliance in the process industry on sensor and control systems that are accessed over communication networks rather than dedicated links. The development of a control strategy that enforces the desired closed-loop objectives with minimal communication requirements between the control system components is an appealing goal since it reduces reliance on the communication medium and helps reduce network resource utilization. This is an important consideration particularly when the communication medium is a potentially unreliable wireless sensor network where conserving network resources (e.g., battery power) is key to prolonging the service life of the network. Beyond saving on communication costs, the study of this problem provides an assessment of the robustness of a given control system and allows the designers to identify the fundamental limits on the tolerance of a given networked control system to communication suspension. This can be a major consideration in deciding a priori whether the desired control objectives can be met with a certain kind of network.

Despite the substantial and growing body of research work on control over communication networks, the overwhelming majority of research studies in this area have focused on lumped parameter systems modeled by ordinary differential or difference equations. Many important engineering applications, however, are characterized by spatial variations owing to the underlying physical phenomena such as diffusion, convection, and phase-dispersion, and are naturally modeled by partial differential equations (PDEs). Typical examples of these systems include transport-reaction processes and fluid flow systems. Unlike spatially homogeneous processes, the control problem arising in the context of spatially-distributed processes often involves the regulation of spatially distributed variables (such as temperature and concentration spatial profiles) using spatially-distributed control actuators and measurement sensors. While the study of distributed parameter systems in process control has been an active area of research, the design and implementation of networked control systems for spatially distributed processes remain open problems that need to be investigated and addressed.

In this contribution, we present a methodology for the design of networked control systems for spatially distributed processes described by linear parabolic partial differential equations (PDEs) with measurement sensors that transmit their data to the controller/actuators over a bandwidth-limited communication network. The central design objective is to reduce the transfer of information from the sensors to the controller as much as possible without sacrificing closed-loop stability. This is achieved by (1) applying modal decomposition techniques to derive a finite-dimensional system that captures the PDE's dominant dynamics, (2) including a model of this system in the controller to provide it with an estimate of the dominant modes when sensor-controller communication is suspended, (3) updating the state of the model using the actual measurements when communication is re-established, and (4) explicitly characterizing the maximum allowable update period in terms of the model accuracy and the sensor and actuator spatial locations. We show that this characterization is exactly preserved when the networked control structure is implemented on the infinite-dimensional system under full state feedback. In the case of output feedback, an observer is included to generate estimates of the dominant modes from the available measurements which are then used to reset the model state. A singular perturbation formulation is then used to link the maximum stabilizing update period for the infinite-dimensional system with the separation between the slow and fast eigenvalues of the differential operator. We show that the stability criteria can be used to determine the optimal sensor and actuator configurations that maximize the networked closed-loop system's robustness to communication suspensions. Finally, the proposed methodology is illustrated using a simulation example.

687d Performance Evaluation of Two Different Decentralized Control Methods Applied to Reactor Networks

Arsun Artel[1], M. Derya Tetiker[1], Fouad Teymour[1], Ali Cinar[1], Mark R. Altaweel[2], and Michael North[2]. (1) Chemical and Biological Engineering, Illinois Institute of Technology, 10 W. 33rd Street, Chicago, IL 60616, (2) Argonne National Laboratory, Argonne, IL 60439

This paper proposes a management layer for performance evaluation of adaptive agent-based control systems for a Continuous Stirred Tank Reactor (CSTR) network, using different decentralized spatial reconfiguration techniques. In each reactor of this network, autocatalytic product species are competing against each other, providing a nice example for population dynamics and social networks. The approach used aims to select the best control method over other candidates, whenever a disturbance hits the system and/or a change is made in the desired grade (composition) of the product. The method is able to keep track of previous performances of different individual control methods by using a k-nearest neighbor (kNN) algorithm for regression, which then gives estimates for the performances of different algorithms depending on the similarity of the state of the network and the behavior exhibited by them in similar conditions.

The operation of highly nonlinear networks may benefit from evolutionary self-organizing control because the optimal operating regime and the required control strategies may not be known ahead of time. Agent-based control systems provide the capability for localized and global control strategies that are both reactive in controlling disturbances and proactive in searching for better operational solutions.

Large-scale spatially distributed systems have the properties of being nonlinear and mostly having a high order. Therefore, they present a unique and difficult control challenge. The control structure, which has to be implied to those systems, has a tendency to include continuous and discrete elements. Additionally, the structure itself is generally both discrete and distributed.

The control structures, the performances of which are evaluated in different operating conditions, are working in a decentralized manner and their objective is to maintain the desired grade. The decentralized control methods have some similarities in their operation. In both control techniques, whenever a control agent desires to change the dominant species within a reactor, it must transport some amount of the desired species from another neighboring reactor to the reactor, the concentration of which is being modified. If none of the immediate neighboring reactors contains the desired species, then the agent must negotiate with several other agents on the network in order to move the species from a more distant reactor. To minimize the disturbance to the operation of the network as a whole, the control agent attempts to find the shortest path between itself and a reactor containing the species that it needs to change the dominant species of its reactor. The similarities mostly end at the heuristics used to control the reactor network. One control approach is using an auction-based mechanism, where each control agent bids on the species that it want to produce on a particular reactor, whereas the other decentralized approach is using a perceptron-based learning for its decision mechanism, in which it observes its environment, compares the current grade with the objective one and calculates its species strength and comes up with a decision of transporting some amount from the reactor it is controlling to a neighboring one.

To acquire the ability to evaluate the performances and to calculate the estimates for picking up the control method suitable for the current state when a disturbance hits the reactor network, both methods should have completed a training phase. Training will continue, until they have enough data points to make the comparison meaningful. The training phase consists of introducing some pre-determined disturbances and picking the method for control, which has less data points in the space, until there are enough data points to make kNN work. We are using in our work 3-NN, hence the training phase continues until each method has been used three times on different scenarios.

The training examples are vectors in a multidimensional feature space. This vector is trying to capture the similarity of the reactor network to a previous one, which is measured by the current grade produced, a quantity called average neighborhood size, which is a measure of the amount of reactor clustering in the network producing same species and another quantity called average network strength, which is indicating how difficult for that reactor network it will be to make a transition from one grade to another. In order to calculate this quantity, individual concentrations of different species in each reactor are compared against each other and their average value is taken.

The training phase of the algorithm consists only of storing the feature vectors and the corresponding performance value, which is a weighted sum of the following measures:

- The algorithm meeting the objective and being able to maintain the desired grade.

- The amount of disturbance that the control method itself is giving to the system.

- The time that it needs to completion.

In the actual regression phase, the test sample is represented as a vector (showing the current reactor network condition) in the feature space. Distances from the new vector to all stored vectors are computed and three (k value in kNN algorithm) closest samples are selected. The performance values associated with those three vectors are weighed in an inversely proportional manner according to the Euclidean distance they have to the new vector location. That average will provide an estimate about how each method will behave on the current condition based on their individual histories of success and failure.

The results of the management layer for performance evaluation is illustrated in a case study where a controlled set of disturbances and set-point changes are introduced to the system for the training phase and others are used to test the performances of different decentralized control algorithms during actual regression phase. The approach enables online switching of different control algorithms based on their previous performances and provides a basis for updating the individual control mechanisms, in case they start behaving poorly because of some unforeseen changes in the network.

687e Safe-Parking of Nonlinear Process Systems: Handling Uncertainty and Lack of Measurements

Maaz Mahmood[1], Rahul Gandhi[2], and Prashant Mhaskar[2]. (1) Chemical Engineering, McMaster University, Hamilton, ON L8S 4L7, Canada, (2) Department of Chemical Engineering, McMaster University, Hamilton, ON L8S 4L7, Canada

The operation and control of chemical processes is experiencing increased reliance on automation with the enhanced availability of sensors and actuators and communication infrastructure. The increased availability of information has opened up several opportunities in making better use of the available resources and coordinating safety measures across the plant. The resulting interdependence of processing units, both materially and in terms of information flow, has also placed more emphasis on the need to preserve safe plant operation in the event of eventualities and contain the effects of faults to local processing units. Even under `normal' operating conditions, chemical processes exhibit significant complexity (manifested as nonlinearities) and have to account for operational issues such as constraints and disturbances and availability of limited measurements. Operation of chemical processes to meet profitability as well as safety requirements, therefore, relies on adequate design and maintenance, appropriate monitoring systems to detect and diagnose eventualities, and the presence of correcting mechanisms that, having been `informed' of an eventuality, prevent or minimize loss of performance, shutdowns, or hazardous situations while accounting for process nonlinearity, disturbances, constraints and availability of limited measurements.

The extensive economic, personnel and environmental damage caused by the faults (it is estimated that the U.S. petrochemical industry looses an estimated 20 billion per year due to faults [1]) as well as the practical inevitability of fault occurrence has motivated several researchers to consider the problem of handling of faults. The first step in handling of faults is the ability to detect and isolate the faults. Statistical and pattern recognition techniques for data analysis and interpretation (e.g., [2,3]) use historical plant-data to construct indicators that identify deviations from normal operation to detect faults. The problem of using fundamental process models for the purpose of detecting faults has been studied extensively in the context of both linear and nonlinear systems (e.g., [4,5]).

Having detected and isolated a fault, the existing results on fault handling have essentially focussed on continued operation at the nominal operating point, under the assumption of sufficiency of the depleted control action to preserve nominal operation. Under this assumption, one approach dictates fault-accommodation via robust/reliable control designs (e.g., see [6]) that allow continued operation at the nominal operating point. To handle the situation when the fault causes such significant depletion of the control action that prevents the handling of fault via controller re-tuning, other approaches assume the existence of redundant control configurations to preserve closed--loop stability. In the choice of redundant control configuration, however, the presence of nonlinearity, input constraints and uncertainty, as well as the hybrid nature of the closed--loop system must be accounted for.

The development of reconfiguration--based approaches has been facilitated by the development of extensive research on control of nonlinear and switched systems (see, e.g., [7, 8, 9]). These include Lyapunov-based nonlinear control designs (for a review, see [7]) that provide an explicit characterization of the stability region in the presence of constraints as well as model predictive control designs that allow incorporation of performance considerations in the control design and provide stability guarantees based on the assumption of initial feasibility of the optimization problem. Recently, model predictive controllers have been designed [8,9] that allow explicit characterization of the stability region, via mimicking the stability properties of Lyapunov-based bounded controllers, without assuming initial feasibility of the optimization problem. More recently, a model predictive controller has been designed [10] that by better utilizing the constraint handling capabilities of model predictive controllers enhances the set of initial conditions from where stability is achieved. The work in [10], however, does not explicitly consider uncertainty and assumes the availability of complete state information. One of the contributions of the present work is the generalization of the predictive controller of [10]to explicitly consider uncertainty and availability of limited measurements for subsequent use within a fault-handling framework.

In [11, 12, 13] reconfiguration-based fault-tolerant control structures have been developed that guarantee preserving of closed--loop stability, while accounting for process nonlinearity, constraints and performance. Specifically, closed--loop stability is preserved (having first detected and isolated the occurrence of a fault) via implementing a backup control configuration chosen such that 1) the state at the time of the failure resides in the stability region of the candidate backup control configuration and 2) the backup configuration does not use the failed control actuator. However, all the reconfiguration-based fault-tolerant control designs of [11, 12, 13] assume the existence of a backup, redundant control configuration. The scenario where a fault results in temporary loss of stability that cannot be handled by redundant control loops has not been explicitly addressed. In the absence of a framework for handling such faults, ad-hoc approaches could result in temporarily shutting down the process which can have substantially negative economic ramifications. Recently, in [14] a 'safe-parking' framework was developed to handle faults that necessitate fault rectification via identifying `safe-park' points where the process is stabilized accounting for process nonlinearity and constraints, and also ensures smooth resumption of nominal operation upon fault-recovery. Specifically, a candidate parking point is termed a safe-park point if 1) the process state at the time of failure resides in the stability region of the safe-park candidate (subject to depleted control action), and 2) the safe-park candidate resides within the stability region of the nominal control configuration. The safe-parking framework in [14], however assumes availability of the entire state information as well as precise process dynamics knowledge. Availability of limited measurements and the presence of disturbances and uncertainty, however, can destabilize even nominal operation and also invalidate the guarantees of safe-parking and resumption of smooth operation upon fault-recovery.

Motivated by the above considerations, this work considers the problem of handling faults in control of nonlinear process systems subject to input constraints, uncertainty and unavailability of measurements. A framework is developed to handle faults that preclude the possibility of continued operation at the nominal equilibrium point using robust or reconfiguration-based fault-tolerant control approaches and necessitate fault-rectification. The key consideration is to operate the plant using the depleted control at an appropriate `safe-park' point to prevent onset of hazardous situations as well as enable smooth resumption of nominal operation upon fault-recovery, while accounting for practical issues such as nonlinearity, constraints, uncertainty and availability of limited measurements. First, we consider the presence of constraints and uncertainty and develop a robust Lyapunov-based model predictive controller in designing the safe-parking algorithm that preserves closed--loop stability upon fault recovery. The framework utilizes the stability region characterization in selecting `safe-park' points from the safe-park candidates (equilibrium points subject to failed actuators). Then we consider the problem of availability of limited measurements. An output feedback Lyapunov-based model predictive controller, utilizing an appropriately designed state observer (to estimate the unmeasured states), is formulated and its stability region explicitly characterized. An algorithm is then presented that accounts for the unavailability of the state measurements in the safe-parking framework as well as controller design. The proposed framework is illustrated using a chemical reactor example and demonstrated on a styrene polymerization process.

References

[1] P. D. Christofides, J.F. Davis, N.H. El-Farra, D. Clark, K.R.D. Harris, and J. N. Gibson Jr. "Smart Plant Operations: Vision, Progress and Challenges," AIChE J., 2007, in press.

[2] J. V. Kresta, and J. F. Macgregor, and T. E. Marlin. "Multivariate statistical monitoring of process operating performance," Can. J. Chem. Eng., 69:35--47, 1991.

[3] P. Nomikos, and J. F. Macgregor. "Monitoring Batch Processes Using Multiway Principal Component Analysis," AIChE J., 40:1361--1375, 1994.

[4] P. M. Frank. "Fault Diagnosis in Dynamic Systems Using Analytical and Knowledge-based Redundancy -- A Survey and Some New Results," Automatica, 26:459--474, 1990.

[5] A. Saberi, and A. A. Stoorvogel, P. Sannuti, H. Niemann, P.M. Frank. "Fundamental problems in fault detection and identification," Int. J. Rob. & Non. Contr., 10:1209--1236, 2000.

[6] Z. D. Wang, B. Huang, and H. Unbehauen. "Robust reliable control for a class of uncertain nonlinear state-delayed systems," Automatica, 35:955--963, 1999.

[7] P. D. Christofides, N. H. El-Farra, "Control of Nonlinear and Hybrid Process Systems: Designs for Uncertainty, Constraints and Time-Delays," Springer-Verlag, 2005, Berlin, Germany.

[8] P. Mhaskar, N. H. El-Farra and P. D. Christofides. "Predictive Control of Switched Nonlinear Systems with Scheduled Mode Transitions," IEEE Trans. Automat. Contr., 50:1670--1680, 2005.

[9] P. Mhaskar, N. H. El-Farra and P. D. Christofides. "Stabilization of Nonlinear Systems with State and Control Constraints Using Lyapunov-Based Predictive Control," Syst. & Contr. Lett., 55:650--659,

2006.

[10] M. Mahmood, and P. Mhaskar. "Enhanced Stability Regions for Model Predictive Control of Nonlinear Process Systems," AIChE J., 2008, in press.

[11] P. Mhaskar. "Robust Model Predictive Control Design for Fault-Tolerant Control of Process Systems," Ind. & Eng. Chem. Res., 45:8565--8574, 2006.

[12] P. Mhaskar, A. Gani, N. H. El-Farra, C. McFall, P. D. Christofides, and J. F. Davis. "Integrated Fault-Detection and Fault-Tolerant Control for Process Systems," AIChE J., 52:2129--2148, 2006.

[13] P. Mhaskar, C. McFall, A. Gani, P. D. Christofides, and J. F. Davis. "Isolation and Handling of Actuator Faults in Nonlinear Systems," Automatica, in press.

[14] R. Gandhi, and P. Mhaskar. "Safe-Parking of Nonlinear Process Systems," Comp. & Chem. Eng., 2008, in press.

Detection and Control of Invading Species in Autocatalytic Networks

Sukanya Balasubramanian, Fouad Teymour, Ali Cinar

Department of Chemical and Biological Engineering,
Illinois institute of Technology, Chicago, IL-60616, USA
Email: [balasuk,teymour,cinar] @iit.edu

1 Introduction

The autocatalytic paradigm is gaining increasing popularity for its diverse range of application in population biology concepts, DNA replication and RNA polymerization studies. When autocatalytic species are populated in a continuous stirred tank reactor (CSTR) environment, they display rich static and dynamic phenomena. In the natural setting, most of the living species exhibit features of survival, coexistence and different degrees of interaction with other species. Birol and Teymour (1) showed that competition of multiple species populating in a CSTR leads to survival of one of the populations. Chaivorapoj (2; 3) extended the idea by considering a single CSTR hosting a *robust* species, perturbed by a disturbance of a *non-robust* or *invading species* in the inflow. When the invading species gets a foothold in the system, concentration of the host species decreases considerably. A feedback control design was implemented to eliminate the invading species and restore original concentration of host species. For simple reactor configurations, bifurcation analysis serves as a potential tool to provide insight into species behavior. As the number of reactors increases, controller tuning using bifurcation techniques fail owing to a geometric increase in steady states. The control problem is all the more complex because each reactor contains an inlet, exit stream and interconnections with its neighbors.

In this paper, a hierarchical agent based system is implemented for detection and elimination of an invading species from a network of CSTRs. An interconnected grid network of 25 reactors is designed to produce a host species. At some time instant, a pulse disturbance of invading species enters a specific location in the network. The "poison" spreads in the network, thereby causing deleterious effects to the system behavior. Agent based systems are a good option for use with large scale, distributed systems (4). Multi-agent systems exhibit intelligence through collective behavior of agents with simple functionalities (5; 6). In order to device detection and control strategies, the task of a subject matter expert is to perform a large number of simulations with pre-selected random values for parameters to collect information about process behavior. The process behavior is evaluated for conditions under normal operation and invasion. Depending on the magnitude of invasion, growth and death rates of the invading species, a variety of process behaviors evolve. A knowledge engineer characterizes the process behavior and stores the information in a knowledge base as rules. The knowledge structure is built using an expert system tool called "Knowledge Builder", where a Knowledge Base (KB) is configured and built using the tool's graphical user interface. Given a current state of the process, an agent invokes the expert system, which has the ability to make a decision on how to react, based on its configured knowledge. The output from the knowledge base in the form of actuator information is relayed to the system.

2 Mathematical Model

A network of *I* inter connected CSTRs (figure 1) is modeled by specifying material balances for each reactor *i* ($i = 1...N$). The reactors in the network are designed such that the volume ratio of the reactors on the outer edge to reactors in the core is 1:3. The design is robust to disturbances since the smaller reactors shield the inner reactors, where the bulk of the product is being produced. In order to ensure long term survival of the host species, small concentrations are always present in the feed. The cubic autocatalytic reaction for *N* host species and *M* invading species is

$$
\begin{aligned}
&R + 2P_n \rightarrow 3P_n \\
&P_n \rightarrow D_P \\
&R + 2Q_m \rightarrow 3Q_m \\
&Q_m \rightarrow D_Q
\end{aligned}
$$

where R is the resource, P_n is the nth host species, Q_m is m^{th} invading species and $D_{P(Q)}$ are the inert (dead) species. The growth rates and death rates define the strength of invading species. The rate of change of resource, host species and invading species in reactor *i* can be written as

$$
\begin{aligned}
V_i\frac{dR_i}{dt} &= F_0R_0 - F_{out,i}R_i + G(R' - R_i) - \sum_n k_{rp}R_iP_{n,i}^2 \\
V_i\frac{dP_{n,i}}{dt} &= F_0P_0 - F_{out,i}P_i + G(P' - P_{n,i}) - k_{rp}R_iP_{n,i}^2 - dAb_pAb_iP_{n,i} \\
V_i\frac{dP_{m,i}}{dt} &= F_0Q_0 - F_{out,i}Q_i + G(Q' - Q_{m,i}) - k_{rq}R_iQ_{m,i}^2 - dAb_qAb_iQ_{n,i} \\
V_i\frac{dAb_i}{dt} &= F_0Ab_0 - F_{out,i}Ab_i + G(Ab' - Ab_i) - \sum_n dAb_pP_{n,i} - \sum_m dAb_qQ_{m,i}
\end{aligned}
$$

where R_0 is the resource concentration in feed, P_0 is the host species concentration in feed, Q_0 is the temporary invasion disturbance in reactor *i*, R_i is resource concentration in reactor *i*, $P_{n,i}$ is n^{th} host species concentration in reactor *i*, $Q_{m,i}$ is the mth invading species concentration in reactor *i*, F_0 is the inlet flow rate, F_{out} is exit flow rate, G matrix defines the inter connection strength with neighboring reactors. Ab_i is the "antibiotic" concentration required to kill the invading species in reactor i. The presence of the antibiotic also leads to death of some of the desired species.

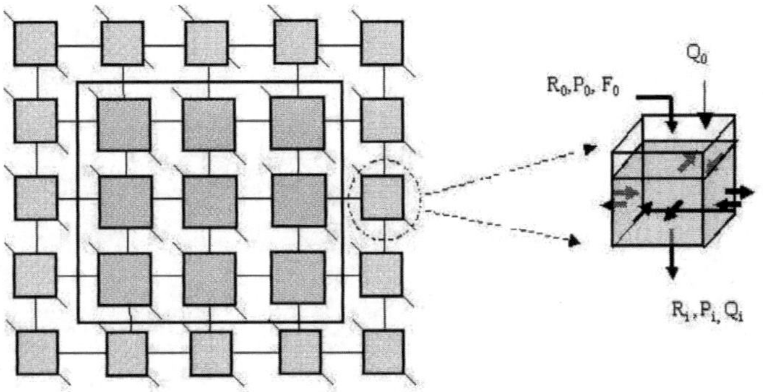

Fig. 1. (a) Network of interconnected reactors (b) Detail view of a reactor with inlet, outlet and interconnections.

3 Process Behavior

Detection and control strategies for a network poisoned by an external source cannot be designed from intuitive knowledge. Since the concentration of the invading species is an immeasurable quantity, a detection scheme should be designed based on diminishing resource and product species concentrations. The control challenge will be to choose among different control alternatives depending on the extent of invasion in the network. In order to device these strategies, it is necessary to collect information about the behavior of the system under invasion and non-invasion conditions offline. The information contribute to building heuristic knowledge. As a knowledge building exercise, a series of open loop dynamic case studies are carried out for various operating parameters to account for transient and steady state characteristics.

Normal Operation. 'Normal Operation' refers to the network behavior when only a host species populates in the system for different parameter values. In each run, the network parameters, feed flow rates, interaction flow rates and initial resource concentrations are chosen in the range $F_0(0.15-1)$, $g(0-3)$ and $R_0(0.8-3)$ respectively. A CVODE solver (7) is used to compute the time variation of resource and host species concentration states. A large number of such dynamic simulations are carried out to explore the boundaries of network behavior. The dynamic data organize themselves into a manifold, which relates the consumption of resource to production of host species (figure 2). The behavior for large values R, P correspond to large values of inlet resource concentration. In the ensuing sections we will focus on a fixed $R_0 = 1$ units. The steady states corresponding to lower values of R and P values concentrate on a continuous line with negative slope while the transient data appear as scatter points. A direct least squares algorithm (8) is used to fit an approximate ellipse around the cluster data. Since the dynamics are computed for random operating conditions, all the reactors produce the same control ellipse. For data points within the control ellipse, the gradient in concentration for host species and resource is often negative. A Matlab implementation of the ellipse-fitting algorithm proposed by Fitzgibon, results in an ellipse shown in figure 2.

Invasion. An invading species in the form of a concentration pulse disturbance in the feed is introduced at a specific location in the network. The invasion spreads to other reactors in the network because of inter connections with its neighbors. After the pulse disturbance is removed, the undesirable species can disappear from the environment, coexist with the host species, flush out the host species or propagate through the network as a disturbance. The extent to which the network is affected depends on the strength of invading species. A "strong invading species", characterized by large growth rates compared to the host species, gains a foothold in the system even after the disturbance is removed. Figure 3 illustrates the spectrum of network behavior under the effect of an strong invasion disturbance introduced in reactor 1. The initial conditions and final

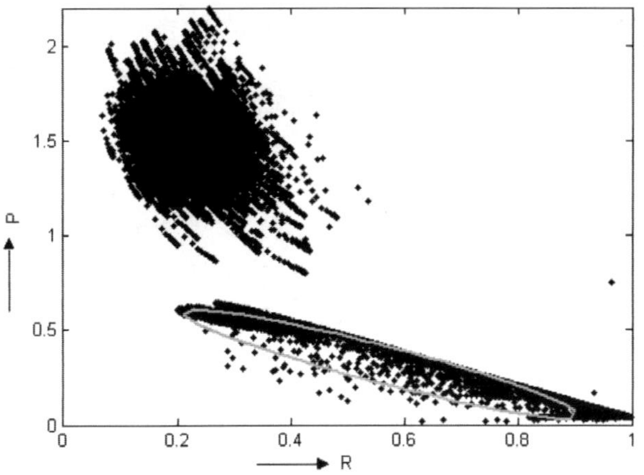

Fig. 2. Normal Operation behavior for operating parameters in the range $F_0(0.15-1), g(0-3)$ and $R_0(0.8-3)$

conditions for dynamics are represented by gray clusters and the transients are indicated by black dots. A pulse disturbance is introduced at $t = (150 - 200)$ time units with a magnitude of $Q_0 = 1$ units and strength $(k_{rq}, d_q) = (24.576, 0.3840)$ units . The initial conditions for dynamics are chosen to be normal operating steady states arising from random configurations of feed flow rates. The plots are constructed on a $R - P$ domain. Since Q is an immeasurable quantity, the growth and death of the invading species with measured in terms of consumption of resource and diminishing host species. In the invading reactor and its neighbors, Q gains a foothold in the system very quickly, whereas the distant reactors hit invasion after a delay. Figure 4 shows the response of the network to a "moderate invading species" with strength $(k_{rq}, d_q) = (10.0, 0.3840)$ introduced in reactor 1. The invasion responses (gray dots) are projected over the normal operating data (black dots). At the end of the simulation, the possible invasion responses are: the invading reactor may flush out the host species permanently or temporarily; the invading reactor neighbors hit invasion or contain traces of Q; in the distant reactors, the invasion propagates as a disturbance. A weak invading species with strength $(k_{rq}, d_q) = (2.0, 0.3840)$ stays within the normal operation thresholds.

Characterizing transients One striking observation from figure 4 is the invasion data deviate from normal operating regions. The presence of an invading species causes a drastic decrease in resource concentration and indirectly affects the production of the host species. The phenomenon is a strong indication of invasion and can be interpreted quantitatively as the occurrence of a positive concentration gradient. Starting from the normal operating steady states (continuous line), the large amount of network behavior data can be distilled into follow categories - the system immediately goes to an invasion

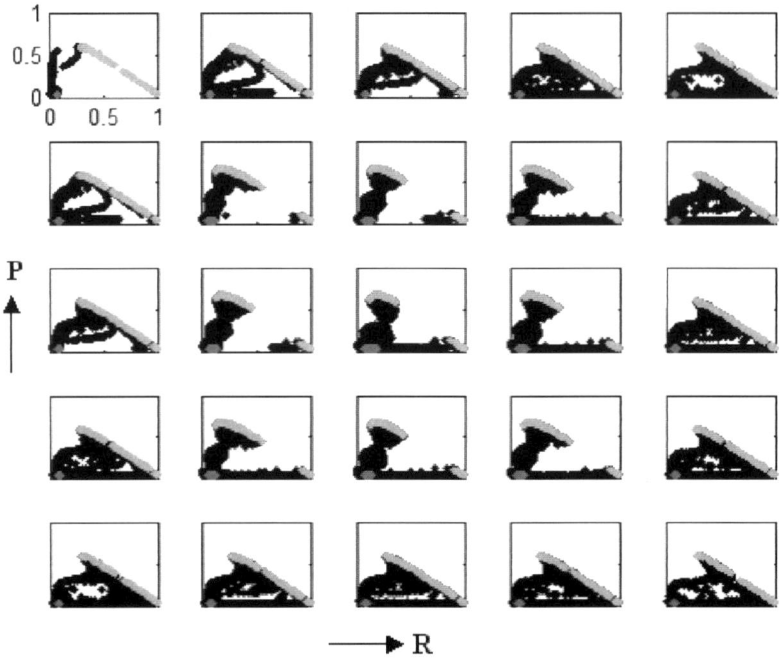

Fig. 3. Open loop responses when the network is attacked by a strong invading species.

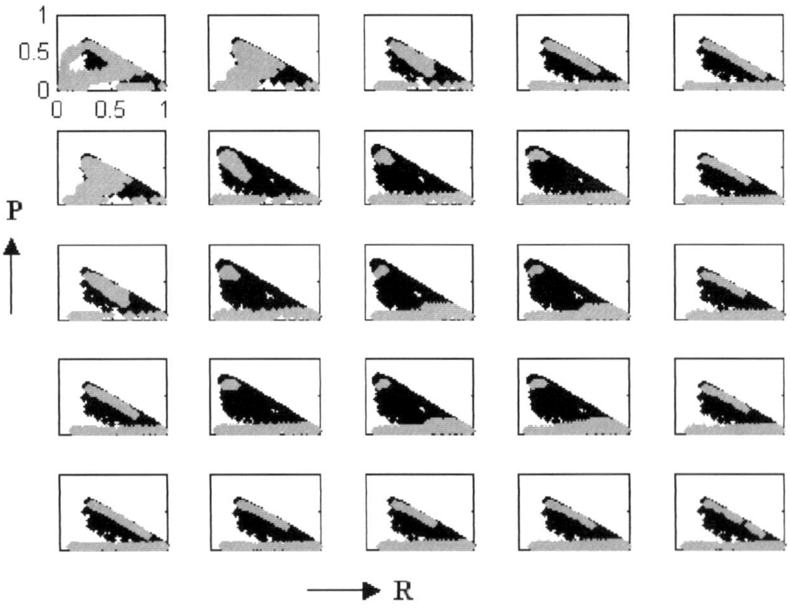

Fig. 4. Open loop responses under moderate invasion

steady state (figure 5a), heads to invasion state after a delay (figure 5b), instantly heads in invasion direction but loops back to its original steady state (figure 5c), loops back to its original state with a delay term (figure 5d) or settles down at coexistence steady state (figure 5e). The gradient sign changes are reflected in the figure.

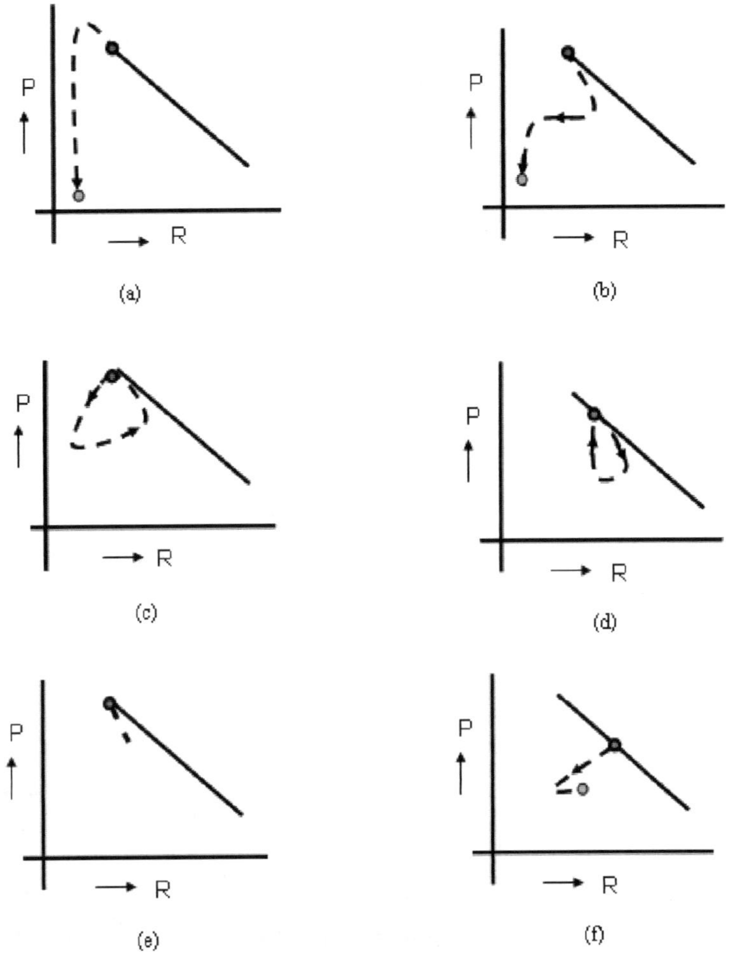

Fig. 5. Transients during invasion (a) Immediate invasion, $dpdr(+ve)$,(b) Invasion after delay $dpdr(-ve, +ve)$, (c) Immediate Looping $dpdr(+ve, -ve)$ (d) Looping after delay $dpdr(-ve, +ve, -ve)$ (e) Disturbance propagation, (f) Coexistence steady state

4 Knowledge Representation

Knowledge from the process must be translated into a more formalized representation to be implemented into a knowledge base. One type of knowledge representation scheme is "production systems" where rules are used as a mode of reasoning (9). In a rule-based system environment, an inference engine determines which rules are applicable and which of these candidate rules should executed in a given situation. The rules are expressed in the form

$$IF < conditions > THEN < actions >$$

The control mechanism is defined by a recognize-act cycle. The "recognize" step determines which rule to fire and the "act" step executes the conclusion of the rule.

To illustrate knowledge representations, we will build a complete program using "MADCABS Knowledge Builder", which is a novel tool for analyzing production systems. The knowledge builder overlooks process operation conditions and provides necessary detection and control.In the ensuing section, we will describe the detection and control strategies for our sample system, give an overview of knowledge builder, and discuss a very simple implementation and operation of the program.

System. The network of 25 reactors (figure 1) is desired be operating under "normal" conditions. At each time instant, a detection scheme checks whether reactors in the network give rise to any of the following flags,

- Total invasion flag - Raised when the system reaches an invasion steady state
- Potential invasion flag - Raised when the system proceeds in an invasion direction
- Indirect invasion alert- One of neighboring reactors is affected by invasion
- Invasion delay flag - Raised when the system proceeds in an invasion direction after some time delay.

The detection algorithm computes the invasion flags based on the relative location of the (R,P) projections wrt the control ellipse and gradient changes with time. The detection scheme accounts for minor fluctuations in gradient values and outliers in data. For example, a potential invasion flag is raised when an (R, P) projection returns "*outside the ellipse*" and the *dp/dr* gradient is positive for about 20 time units. The indirect invasion alerts facilitate faster detection. Figure 6 shows the flags raised in the R-P domain and concentration time domain in reactor 2.Table 1 shows the results from the detection algorithm designed using simple IF-THEN rules in MATLAB. We observe that the indirect invasion flags and potential flags provide faster detection capabilities.

The control strategies are designed based on the detection flags raised. In case of a potential invasion flag, the control strategy adopted is to isolate the reactor, add an "antibiotic" to flush out the undesirable species and reconnect the interactions. Other

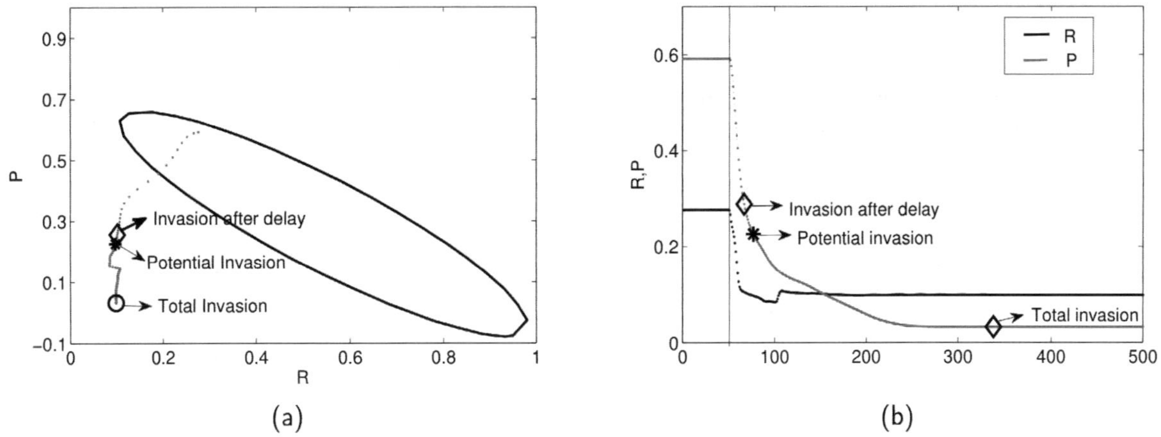

Fig. 6. Flags raised by the detection algorithm in reactor 2 in (a)R-P domain (b)Concentration vs time domain

#	Total	Potential	Indirect	#	Total	Potential	Indirect
1	329	71	-	**10**	338	177	170
2	338	77	71	**13**	419	321	200
3	364	110	95	**14**	398	293	205
4	338	154	141	**15**	366	198	189
5	334	170	157	**17**	332	194	169
7	333	134	84	**19**	363	290	205
8	398	200	112	**20**	339	220	211
9	332	194	169	**25**	334	232	228

Table 1
Invasion flags raised by the detection algorithm as time progresses

control strategies such as a slight tweaking of flow rates may be necessary to remove trace quantities of undesirable species. Figure 7 shows the implementation of the control strategy on the invaded reactor. We observe that the invading species introduced between $t = (150 - 200)$, kills the host species. At $t = 400$, an antibiotic addition, eliminates the invading species, while a small amount of host species is still present in the system. The original concentration of host species can be recovered by adding additional species in the feed.

Knowledge Builder and its implementation. In order to understand the working and implementation of Knowledge Builder, we shall apply it to a simple application, an interconnected network of 2 reactors. Each reactor contains an inlet and outlet feed stream and inter connections with its neighbors. Each reactor contains a stable configuration of

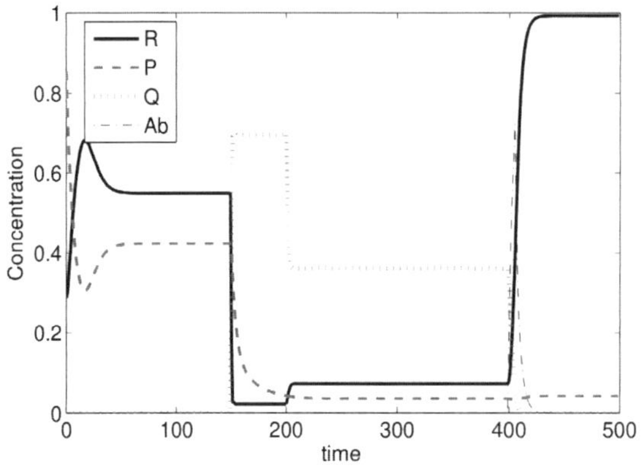

Fig. 7. Antibiobic action

different autocatalytic species. The control objective calls for replacing species in reactor 2. The reactor providing the desired species is designated as "source reactor" and the reactor to be changed is the "target reactor". The desired transformation can be achieved by (a) Increasing interconnection flow rate to target reactor (b) Incase the increase is not sufficient, seek assistance from target reactor.

The knowledge builder components for the source reactor is shown in figure 8. The knowledge builder's GUI consists of panels which explain different parts of the rule base. The "subject panel" shows the working memory or the current state of the process. State variables, input variables and local variables populate the knowledge base. The process simulator operates in a continuous mode. At each time instant, the simulator provides the state and input information. The knowledge builder program updates the state and input variables based on changes in a "watch file". Variables such as *controller status, arbitrator status* are local to the knowledge base and are set to default values. "Status Panel" shows the matching routines for the components. For example, in figure 8, the controller checks if the subject "controller status" is assigned to value "attack". The program will test conditions based on changes to the status of the system. It will then fire all the rules that depend on this status being true. For each of the rules to be fired, it will execute all of the rule actions that are assigned weight one. The "rule panel" shows what actions to execute based on the predicates specified in the "action panel". For example, the rule named "Rule controller" initiates the action "Proportional Controller" which increases interaction flow rate. The output from the agents is relayed to the system in the form of an "out file". The process reads the actuator settings from an out file. Each reactor has an instance of the knowledge base, which fire different rules based on the initial conditions. The knowledge base for each reactor is maintained by a "Knowledge Modeler" GUI figure

(9). The knowledge modeler allows negotiation protocols between the knowledge engines of each reactor.

The performance of the control system is demonstrated in the following example. Two species with identical growth and death rates $k_r = 25, d = 0.1$ populate in the reactor network. Initially, feed flow rates $f = 0.008$ and interaction flow rates $g = 0.001$ are identical in both reactors. When the interaction flow rates are manipulated, the outflows are adjusted to maintain a constant volume. The initial species concentration profile for reactor 2 is shown in region(a) of figure 10. In times $t = (0 - 1000)$, the knowledge builder only updates the process variables. At $time = 1000$ units, the *controller status* of the source reactor is set to *attack*. The knowledge base fires rules to increase the interaction flow rate from the source to the target reactor. We observe that this change is not sufficient to overtake reactor 2 (region (b)). In region (c), the target reactor reactor is exploited to increase its interaction flow rate to the source reactor. The target reactor requests the source reactor reactor to double its output. The interaction flow rate bounds are continuously updated and maintained in an arbitrator agent. Requests are accepted if the requested changes are within the bounds.

5 Research Ahead

The appropriate tools required for the scalability of knowledge builder are under development. A general template will be used to describe the basic knowledge of each reactor in the network. Once the network topology and inter connections are determined, an instance of the knowledge base is copied to every reactor. Additional information (subjects, statuses, rules) specific to each knowledge base can be incorporated into the builder. The knowledge builder should be able to support access settings such as private (local to a KB), protected (visible to its environment) and public (visible to the network) for the subject variables. The environment for a reactor is defined by its neighbors. A reactor engine can have access to protected and public subjects for its neighbors. Generalized request handling procedures for subjects need to be implemented. Another direction of research will be to explore the panorama of conflict resolution strategies (10). Some of the strategies available in literature are: rule selection based on a priority order, newly activated rules are placed above all activations, random activation of rules and many others.

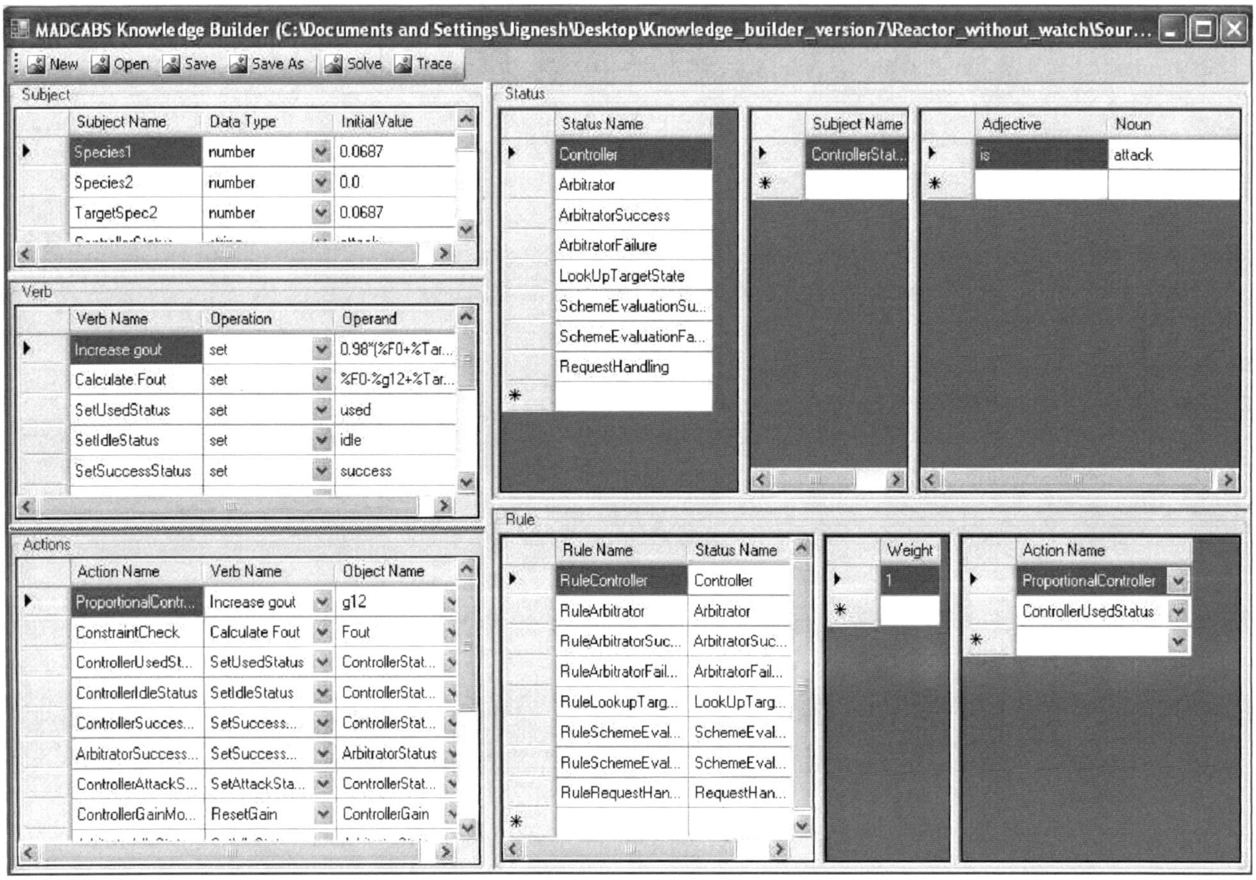

Fig. 8. Knowledge Builder representation for a 2 reactor system

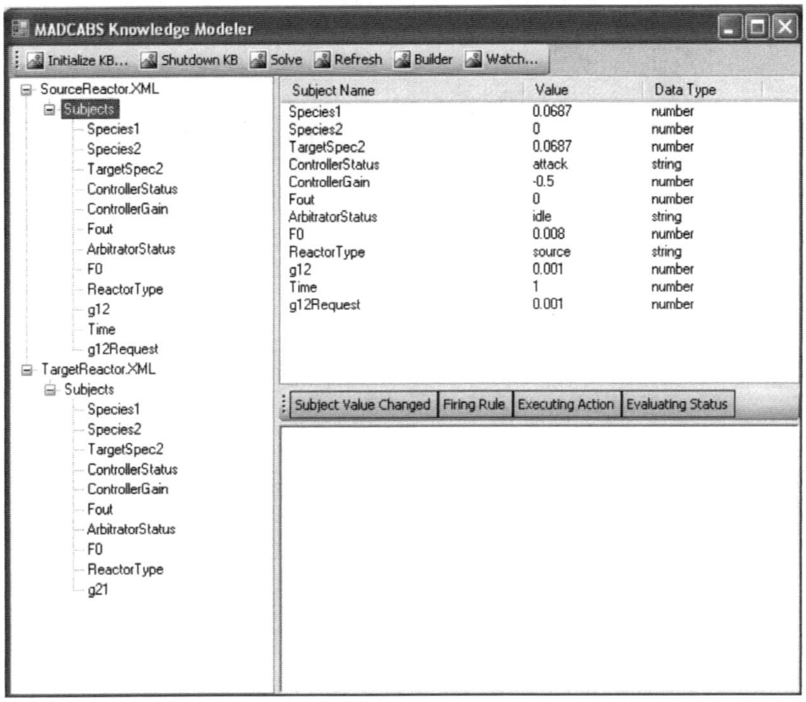

Fig. 9. Knowledge Modeler illustration for a SourceReactor and Target reactor

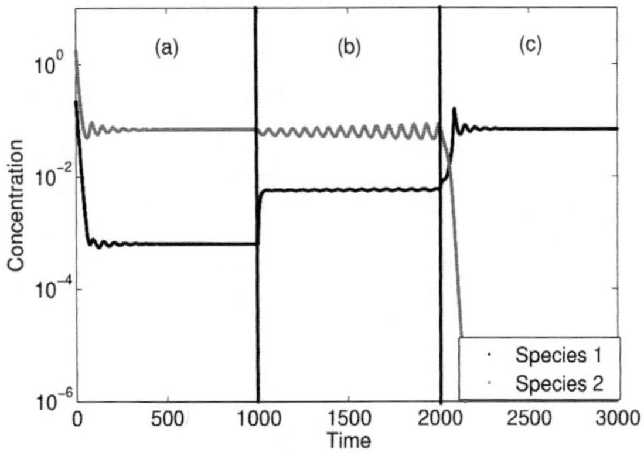

Fig. 10. Strategy for replacing autocatalytic species in reactor 2 (a) $(g_{12}, g_{21}) = (0.001, 0.001)$ (b)$(g_{12}, g_{21}) = (0.0088, 0.001))$,$(g_{12}, g_{21}) = (0.0165, 0.0088)$

References

[1] Birol, I., Teymour, F., "Statics and Dynamics of Multiple Cubic Autocatalytic Reactions," *Physica D*, Vol. 144, pp.

[2] Chaivorapoj, W., Birol, I., Cinar, A., and Teymour, F.,, "Feedback Control of a Continous-Flow Stirred Tank Reactor with Competing Autocatalators," *Ind. Eng. Chem. Res*, Vol. 42, pp. 3765-3785, 2003.

[3] Chaivorapoj, W., Birol, I., and Teymour, F.,, "Competition between Robust and Non-Robust Autocatalytic Replicators," *Ind. Eng. Chem. Res*, Vol. 42, pp. 3765-3785, 2003.

[4] Tatara,E., Birol,I.,Teymour,F.,Cinar,A.(2005),"Agent Based Control of Autocatalytic Replicators in Networks of Reactors," *Computers and Chemical Engg*,Vol 29, pp.807-815.

[5] Tatara,E.,Cinar,A.,Teymour,F.(2007),"Control of Complex distributed Systems with Distributed Agents," *Journal of Process Control*,Vol 17, pp.415-427.

[6] Tetikar,D.,Artel,A.,Tatara,E.,Teymour,F.,North,M.,Cinar.A.,(2006),"Agent Based System for Reconfiguration of Distributed Chemical Reactor Network Operation," *Proceedings of American Control Conference*,Minneapolis.

[7] Hindmarsh, A., Taylor (1998), A. PVODE and KINSOL: Parallel Software for Differential and Nonlinear Systems; Lawrence Livermore National Laboratory Technical Report UCRL-ID-129739; Lawrence Livermore National Laboratory: Livermore, CA.

[8] Fitzgibbon,A.,Pilu,M.,Fisher,R.(1999),"Direct Least Square Fitting of Ellipses," *IEEE transactions on Pattern Analysis and Machine Intelligence*,Vol 21, pp.476-480.

[9] Cronk,R.,Callahan,P.,Bernstein,L.(1988),"Rule based Expert Systems for Network Management and Operations: An Introduction," *IEEE Network*,September, pp.7-18.

[10] Hicks,R.,(2007),"The No-Inference Engine Theory - Performing Conflict Resolution during Development," *Decision Support Systems*,pp.434-444.

687g Safe Parking of Batch Processes

Siam Aumi, McMaster University, Hamilton, ON L8S 4L7, Canada and Prashant Mhaskar, Department of Chemical Engineering, McMaster University, Hamilton, ON L8S 4L7, Canada.

The operation of chemical processes often encounters faults. Owing to the frequency of fault operation and the significant economic, environmental and safety impact, several research efforts have focussed on developing methodologies for handling faults [1,2,3]. Most existing results on fault-tolerant control, however, have focussed on continuous processes, where the objective is to preserve operation around the nominal equilibrium point. More recently, a safe-parking framework [4] has been developed that addresses the problem of faults that preclude the possibility of continued operation at the nominal equilibrium point and prevents the onset of hazardous situations while enabling smooth resumption of nominal operation.

Existing fault handling techniques, and the safe-parking framework of [1], however, are not directly applicable to batch processes due to the fact that the desired process objective, that of producing the required product, is inherently not that of stabilization around an equilibrium point. Motivated by this consideration, this work considers the problem of developing safe-parking framework for batch processes. The key idea is to utilize available control action following a fault to maintain system conditions such that the desired end point can finally be reached upon fault rectification. The ideas are demonstrated through a CSTR example operated in batch mode with 2 inputs and a fed-batch bioreactor system with 1 input.

References

[1] Mhaskar, P., A. Gani, N. H. El-Farra, C. McFall, P. D. Christofides and J. F. Davis, "Integrated Fault Detection and Fault-Tolerant Control of Nonlinear Process Systems," AIChE J., 52, 2129-2148, 2006.

[2] P. Mhaskar, ``Robust Model Predictive Control Design for Fault-Tolerant Control of Process Systems," Ind. Eng. & Che. Res., 45, 8565-8574, 2006.

[3] Mhaskar, P., C. McFall, A. Gani, P. D. Christofides and J. F. Davis, "Isolation and Handling of Actuator Faults in Nonlinear Systems," Automatica, 44, 53-62, 2008.

[4] Mhaskar, P. AND Gandhi, R. (2008). Safe-parking of Nonlinear Process Systems. Computers and Chemical Engineering, 32, in press.

688a Synergistic Improvement of Process Safety and Product Quality

Ankur Pariyani[1], Warren D. Seider[1], Ulku Oktem[2], and Masoud Soroush[3]. (1) Chemical and Biomolecular Engineering, University of Pennsylvania, Philadelphia, PA 19104, (2) Risk Management and Decision Processes Center, Wharton School, University of Pennsylvania, Philadelphia, PA 19104, (3) Department of Chemical and Biological Engineering, Drexel University, Philadelphia, PA 19104

Major objectives to reduce the economic and human losses in the chemical process industries are to improve the mechanisms for process safety as well as product quality control. *Quality defects* are flagged by "special causes" (arising due to failures of hardware components, errors by operators in performing operating procedures, inefficient management [1], etc.) in the statistical process control mechanisms. In many respects, these are closely related to the *safety problems*, which are flagged by "near-misses" (indicators of potential accidents). Prior research has emphasized the importance of identifying these *near-misses*, high-probability, low-consequence events, in predicting and preventing the occurrence of catastrophic, low–probability, high-consequence events [2, 3]. This paper quantifies the relationship between *special-causes* and *near-misses* and proposes a novel approach to integrate and enhance techniques for safety and quality, thereby helping to improve the predictability of potential accidents.

An approach to identify potential special-causes detectable from product-defect information and determine the relative significance and probabilities of the occurrence of the near misses, including failure probabilities of safety systems, is presented. This approach involves utilizing dynamically available product-quality and safety data, and exploiting the pattern information from near-misses, often overlooked in product defects. Abnormal events that, otherwise, may lead to off-specification products and accidents, are often prevented using predictions of the occurrences and severity of potential abnormal events and product-quality defects. A multivariate Bayesian analysis framework is formulated for data analysis and predictive modeling. These predictions provide direction and emphasis to improve safety and product-quality performance. As a case study, an acrylic fed-batch polymerization reactor is considered to show the application and reliability of this approach.

References

1. Meel, A., W. D. Seider, and U. Oktem, "Analysis of Management Actions, Human Behavior, and Process Reliability in Chemical Plants. I. Impact of Management Actions," *Proc. Safety Prog.*, **27**, 1, 7-14 (2008).

2. Phimister J. R., U. Oktem, P. R. Kleindorfer, and H. Kunreuther, "Near-miss Incident Management in the Chemical Process Industry," *Risk Analysis*, **23**, 445-459 (2003).

3. Meel, A., and W. D. Seider, "Plant-Specific Dynamic Failure Assessment using Bayesian Theory," *Chem. Eng. Sci.*, **61**, 7036-7056 (2006).

688b Development of a New Pattern Recognition Method and Its Application to Just-In-Time Modeling

Koichi Fujiwara, Yosuke Mukai, Manabu Kano, and Shinji Hasebe. Chemical Engineering, Kyoto University, Nishikyo-ku, Kyoto, 615-8510, Japan

Linear regression is a simple and useful method to build process models for process monitoring and state estimation. However, linear models do not always function well in practice due to not only process nonlinearity but changes in process characteristics. Such changes are caused by catalyst deactivation and scale adhesion in the chemical industry or by equipment maintenance in the semiconductor industry, for example. To build good process models that can cope with the changes in process characteristics, unsupervised pattern recognition is a key methodology. A conventional unsupervised pattern recognition method is the k-nearest neighbor (k-NN) method, which classifies samples into several classes on the basis of the distance between samples. However, it is crucial for classifiers to take into account the correlation among variables to improve the model performance especially when process characteristics change, because such changes can be captured efficiently by the correlation instead of the distance. Self-organizing map (SOM) is an unsupervised learning algorithm that can construct clusters based on the sample similarity and visualize high dimensional data, but it is difficult to determine an appropriate map size and it requires a heavy computational load.

In the present work, a new pattern recognition method based on geometry of samples in a linear space is proposed. The proposed method can select samples whose correlations are similar to the query point without supervised information or other prior information. The proposed procedures are as follows:

1) Subtract the query point from all the other samples.

2) Calculate the correlation coefficient between all pairs of arbitrary two subtracted samples, and the pairs whose correlation coefficients are close to -1 are selected.

3) New vectors are created by subtracting one vector of the selected pairs from another.

4) Derive the subspace containing the query point from the created vectors by using principal component analysis (PCA).

5) The Q statistics between all samples and the derived subspace are calculated, and the samples having small Q statistics are selected as the similar samples to the query point.

In step 4), robust PCA (RPCA) can be used instead of ordinary PCA to cope with outliers, since the principal components derived by PCA are known to be very sensitive to outliers.

In addition, a new soft-sensor design method integrating the proposed pattern recognition method and Just-In-Time (JIT) modeling is proposed. The proposed method is referred to as Correlation-based JIT (C-JIT) modeling. The estimation performance of JIT modeling deteriorates when process characteristics change, because it selects the samples for local modeling based only on the distance between samples and the query point. By integrating JIT modeling and the proposed pattern recognition method, C-JIT modeling can select samples whose correlation are similar to each other and cope with the changes in process characteristics. The usefulness of the proposed pattern recognition method and C-JIT modeling is demonstrated through a case study of soft-sensor design for CSTR process. The proposed method can improve the estimation accuracy by 48% in comparison with conventional JIT modeling.

688c A New Algorithm for Spectrum Alignment by Maximizing Correlations among Signature Peaks

Qinghua (Peter) He, Department of Chemical Engineering, Tuskegee University, Tuskegee, AL 36088 and Jin Wang, Department of Chemical Engineering, Auburn University, Auburn, AL 36849.

Recent advances in mass spectrometry proteomic analysis have generated considerable research interest in bioinformatics community. Large data sets generated by high-throughput mass spectrometry techniques such as SELDI-TOF (surface enhanced laser desorption/ionization time-of-flight) and MALDI-TOF (matrix assisted laser desorption and ionization time-of-flight) have been analyzed for cancer diagnostic purposes. A mass spectrometry data set contains the information about the relative abundance of many protein/peptide sequences in the form of intensity (relative abundance) measurement and molecular mass-to-charge ratio (m/z) pairs. Ideally, the same protein/peptide sequence detected in the same mass spectrometer should have the same m/z value. However, in practice this is not the case due to experimental variations. Consequently, a series of data pre-processing steps are necessary before the mass spectrometry patterns can be analyzed comparatively. Such pre-processing steps include baseline subtraction (Kohli, et al. 2005), normalization (Callister, et al. 2006, Wang, et al. 2006), peak detection, peak alignment (Nielsen et al. 1998, Bylund, et al. 2002, Tibshirani, et al. 2004, Wang, et al. 2006) and peak selection. Among all pre-processing steps, peak alignment is one of the most challenging tasks. Successful peak alignment ensures that the same protein intensities are correctly identified in each sample spectrum to avoid errors in identifying the signals from peptides with similar molecular weight. Recent work (Semmes *et al.*, 2005) utilizing spectrometers located in different medical centers cites difficulties with alignment/calibration as major barriers that must be overcome to ensure that data from different centers are compatible. Baggerly *et al.* (2004) also identify alignment problems as a significant hindrance in achieving reproducibility from samples collected within the same lab. In the past decade, several algorithms have been developed in the literature of aligning data sets, such as dynamic time warping (DTW) (Kassidas et al. 1998), derivative dynamic time warping (DDTW) and Correlation Optimised Warping (COW) (Nielsen et al. 1998). In this work, a new approach for peak detection and alignment has been developed and applied to a prostate cancer and liver cancer proteomic data sets. One difference between the proposed method and existing alignment methods is that while in existing methods all data points are aligned with equal weight, the proposed method align major peaks first, and then align other points including between-peak points and insignificant peaks based on the major peak alignment. The alignment consists of three steps. In the first step, a reference spectrum is identified and major peaks in the reference spectrum are identified automatically. In the second step, major peaks in each spectrum in the data set are aligned against the reference spectrum. Potential mismatches among major peaks are eliminated in the second step as well by automatic detection of outliers in alignment shifts. In the last step, all other points than major peaks are aligned based on the alignment results obtained in step two. The proposed method is applied to two mass spectrometry data sets. One example of alignment is shown in Fig. 1 where (a) shows before alignment; (b) shows after alignment. The performance of the proposed alignment method is compared with the classical methods DTW, DDTW and COW. The results show that the proposed method is more robust and provides better alignment.

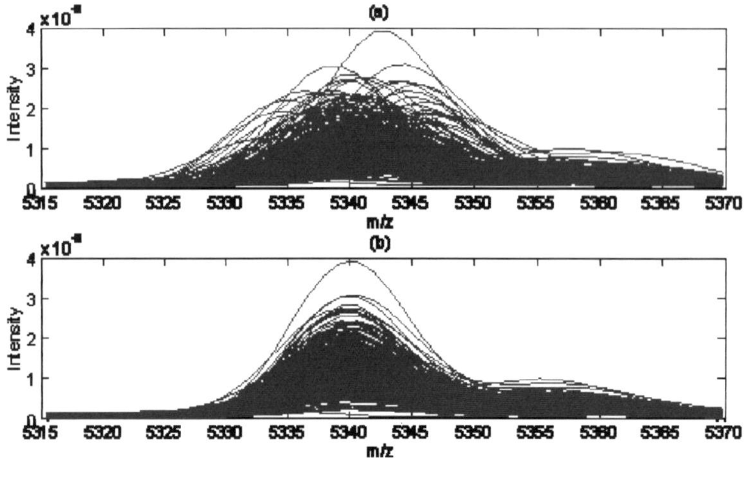

Fig.1. A segment of sample spectra before (a) and after (b) alignment

References:

Baggerly,K. *et al.* (2004) Reproducibility of seldi-tof protein patterns in serum comparing data sets from different experiments. *Bioinformatics*, **20**, 777–785.

Bylund, D et al. (2002) Chromatographic alignment by warping and dynamic programming as a pre-processing tool for parafac modelling of liquid chromatography-mass spectrometry data. *J. Chromatography*, 961:237-244.

Callister, SJ et al.(2006) Normalization approaches for removing systematic biases associated with mass spectrometry and label-free proteomics. *J. Proteome Res.*, 5:277-86.

Kassidas, A, MacGregor, JF, Taylor, PA.(1998) Synchronization of batch trajectories using dynamic time warping. *AIChE J.*; 44: 864–875.

Kohli, BM et al. (2005) An alternative sampling algorithm for use in liquid chromatography/ tandem mass spectrometry experiments. Rapid Commun. *Mass Spectrometry*, 19(5):589-596.

Nielsen, NPV, Carstensen, JM, Smedsgaard, J. (1998) Aligning of single and multiple wavelength chromatographic profiles for chemometric data analysis using correlation optimised warping. *J.*

Chromatogr. A; 805: 17–35.

Tibshirani, R et al. (2004) Sample classi_cation from protein mass spectrometry, by 'peak probability contrasts'. *Bioinformatics*, 20(17):3034-44.

Semmes,O. *et al.* (2005) Evaluation of serum protein profiling by surface-enhanced laser desorption/ ionization time-of-flight mass spectrometry for the detection of prostate cancer: I. assessment of platform reproducibility. *Clin. Chem.*, **51**, 102–112.

Wang, P et al. (2006) A statistical method for chromatographic alignment of LC-MS data. *Biostatistics*, doi:10.1093/biostatistics/kxl015.

688d Data Driven Neural Netowrk Based Models of Ro Desalination Plant Operation and Ro Membrane Performance

Francesc Giralt[1], Robert Rallo[2], Jaume Giralt[3], Dan Libotean[3], and Yoram Cohen[4]. (1) Fenòmens de Transport, Departament d'Enginyeria Química, Universitat Rovira i Virgili, Av. dels Països Catalans 26, Tarragona, 43007, Spain, (2) Universitat Rovira Virgili, Av. dels Paisos Catalans 26, Tarragona, 43007, Spain, (3) Departament d'Enginyeria Química, Universitat Rovira i Virgili, Av. dels Països Catalans 26, Tarragona, 43007, Spain, (4) Chemical & Biomolecular Engineering, University of California, Los Angeles, 5531 Boelter Hall, 420 Westwood Plaza, Los Angeles, CA 90095-1592

A machine learning modeling approach was investigated as a means of developing data-driven models for describing RO desalination plant performance (e.g., flux and salt rejection) and for evaluating the suitability of RO membranes for potable water applications requiring the rejection of a broad range of organic compounds. The concept of plant memory (incorporated to enable description of process fluctuations), along with basic input process operating parameters, formed the basis input variables for describing plant behavior. Cognitive neural network model training was carried out with the normalized permeate flux and salt passage for various model architectures and memory time intervals. Model results demonstrated that plant performance could be described to a reasonable level of accuracy with respect to both permeate flux and salt passage with a plant memory time interval. Forecasting of plant performance was shown to be feasible and with good accuracy with a reasonable memory time interval to enable practical process control.

The current approach is providing the basis for developing and incorporating neural network data-driven models in a control strategy and early-warning system of the deterioration of RO plant performance. In this regard, the passage of organics through RO membranes is particularly critical for applications that involve RO membranes in water treatment plants. Neural network models can be effective in generating Quantitative Structure-Property Relations (QSPR) for the organic passage (P), sorption (S) and rejection (R) using the most relevant set of molecular descriptors. In the present work, the approach was demonstrated based on an experimental data set of fifty organics with four different RO membranes. A number of feature selection methods were employed. Pre-screening was carried out, with Principal Components Analysis and SOM of the chemical domain for the study chemicals, as defined by chemical descriptors, to identify the applicability domain and chemical similarities. The QSPR models predicted organic passage, rejection, and sorption within the range of the standard deviations of measurements for the experimental data set of fifty compounds. The application for the approach for compounds of interest, for which experimental data were not available, demonstrated reasonable mass balance closures. The implications of the above approaches will be discussed with respect to the development of a comprehensive tool for assessing RO plant performance in RO water treatment processes.

688e Rectification of Multiscale Data with Reliability Assessment to Guide External Data Procurement In Life Cycle Assessment

Shweta Singh[1], Hang J. Kim[2], Bhavik R. Bakshi[1], and Prem K. Goel[2]. (1) Department of Chemical and Biomolecular Engineering, The Ohio State University, 125 Koffolt Laboratories 140 West, 19th Avenue, Columbus, OH 43210, (2) Department of Statistics, The Ohio State University, 404 Cockins Hall 1958 Neil Avenue, Columbus, OH 43210

Many engineering systems rely on interpretation of data collected at different spatial scales. For example, process engineering data may be available at scales such as specific equipment, plants and supply chains. Similarly, data for life cycle assessment (LCA) is often available at equipment, value chain and economy scales. Data rectification is a popular approach for reducing errors in process data based on using domain specific information at each scale (intra-scale models) and across scales (inter-scale models) [1]. Our work is motivated by the availability of multiscale data in life cycle assessment, but the proposed approach is general and can be applied to rectification of data from other sources. This presentation will focus on new methods for guiding the procurement of additional data at single or multiple scales to improve the quality of rectified results. This approach is relevant for making decisions about buying new data or installing additional sensors.

The reliability of results obtained from LCA depends on the quality of available life cycle inventory (LCI) data, which mandates the need for comprehensive and accurate databases. Among the other challenging steps in LCA, procurement of reliable dataset seems to be the most intriguing task. The present status of various LCI databases shows a need for rigorous techniques to enhance quality due to the presence of errors such as missing streams, random and gross errors in measurement, errors due to the combination of data from various sources and errors at multiple spatial and temporal scale. There has been much research in the field of data rectification for application in chemical process industries and in the past few years these methods have been implemented for rectification of LCI to improve their quality [1,2]. Despite the long history of data rectification methods and the recognition of their relevance to improving LCI data, applying these methods to LCI data introduces many new challenges. These include the need for information that is usually not included in LCI databases such as the underlying life cycle process network or process chemistry and inconsistencies arising due to aggregation of LCI databases at different scales for use in hybrid LCA. With the availability of several LCI databases another challenge is to find the most reliable dataset for a particular LCA study. Although recently there has been some work to address the need of multiscale rectification [2], it still has limitations in terms of excluding the specific process chemistry and extent of reactions as proper constraints. Including such information in data rectification is expected to have significant effect on the quality and reliability of results.

This work particularly focuses on developing a multiscale data rectification technique by using the process chemistry as constraints to provide a more reliable dataset for hybrid LCA studies. Reliable external data usually improve the LCI, but inclusion of the external data might not always be worth the cost and effort in terms of improvement of quality of LCI. The approach taken here uses "value of information" metrics to assess the marginal increase in quality of LCI by procurement of external data to guide decisions about purchasing the external dataset. Using the developed technique this work also illustrates the need of using multiple databases as complimentary to each other and then apply data rectification techniques to come up with a more reliable dataset. The results obtained by using the dataset developed by combination of data from two different datasets can give more reliable results from LCA studies. With the current scenario of environmental concerns it would prove worthy to put these extra efforts.

The need of multiscale data rectification and not relying on a single database to make decisions is illustrated by a case study based on data from commercial and public domain databases of chemical processes in a life cycle. which demonstrates the strength of approach taken here to improve the quality of LCI data by external data procurement. The case study also shows the augmentation of quality of LCI data available in a public database by use of fine scale data available from other database. These approaches are expected to add to the value of LCA studies in present day world which needs sustainability.

References

1.Hau, J. L., Yi, H., & Bakshi, B. R. (2007). Enhancing life-cycle inventories via reconciliation with the laws of thermodynamics. {JOURNAL OF INDUSTRIAL ECOLOGY}, {11}({4}), {5-25}.

2.Yi, H., & Bakshi, B. R. (2007). Rectification of multiscale data with application to life cycle inventories. {AIChE JOURNAL}, {53}({4}), {876-890}.

688f Latent Variable Batch Predictive Control

Masoud Golshan, Chemical Engineering, McMaster University, Hamilton, ON L8S 4L7, Canada, Mark-John Bruwer, ProSensus Inc., Hamilton, ON L8S 4L7, Canada, Prashant Mhaskar, Department of Chemical Engineering, McMaster University, Hamilton, ON L8S 4L7, Canada, and John MacGregor, Dept. of Chemical Engineering, McMaster University, Hamilton, ON L8S 4L7, Canada.

Batch processes play an important role in many industries such as pharmaceuticals and specialty chemicals. One of the reasons to use batch processes is their flexibility to produce a wide range of products due to the fact that they have a large range of process conditions within which a batch process can be operated.

The operation of a process in batch mode brings forth several distinguishing characteristics as opposed to continuous operation. Specifically, batch processes are characterized by a specific beginning and end time over the course of which the process variables often change by significant amounts. Additionally, the process variables at the end of the batch need not be at equilibrium, but rather correspond to desirable product properties. Both of these features have significant impact on the way batch processes are modeled, monitored and controlled. Another difficulty with the control of batch processes is that the nonlinearity and time varying characteristics cannot be ignored as the operating point is changing continuously. On the other hand the acceptable product quality is a tight range that makes the application of a high performance control methodology necessary. Predictive controllers are capable of doing this important job as long as a reliable model of the process is available. For nonlinear batch processes this has usually meant the use of fundamental nonlinear models embedded within an optimization. The problem with this approach is that the modeling effort is large, the computation time is large and the solvers are complex. In this paper we present an approach using Latent Variable models that capture the benefits of these nonlinear MPC's without implementation difficulties mentioned above.

Owing to the significant variation of the process variables over the batch duration and the fact that the process trajectory need not evolve around equilibrium points, traditional step test models identified around the nominal equilibrium point are generally inadequate to model batch processes. Several research efforts have focused on utilization of past batch data in building low dimensional models that adequately capture the process dynamics and have good predictive capability. The most promising among them are the Latent Variables Models based on PCA and PLS proposed by Nomikos and MacGregor (1994, 1995) and subsequent variations of these. These models are linear and hence lead to simple and fast on-line algorithms, but yet the capture the nonlinear behavior of batch processes by modeling the time varying covariance structure among all variables throughout the duration of the batch.

There are two classes of model predictive algorithms for batch processes. The first is the high level problem of controlling the final batch product properties, and the second is the lower level problem of trajectory tracking. Latent variable models have been used to effectively treat the former (high level) problem through model predictive mid-course correction approaches (e.g. Flores-Cerrillo and

MacGregor (2002), Yabuki and MacGregor (1997)). Flores-Cerrillo and MacGregor (2005) recently proposed a MPC algorithm for the lower level trajectory tracking problem based on PCA models.

In the Latent Variable MPC developed in this study several different latent variable modeling approaches are used to obtain models that better describes the nonlinearity and time varying characteristics of the process. The control methodology is also elaborated to yield better performance and to be capable of rejecting nonstationary disturbances.

The approach selected in this study is to develop a Principal Component Analysis (PCA) on the historical data gathered from past batches with an existing feedback control system plus a few batches with closed-loop designed experiments, and then to implement the control based on the predictions made by this model. Closed-loop identification is almost essential with batch processes to ensure that the product quality is not degraded. Since the data of a batch process is distributed in three dimensions (batches variables batch time), a suitable rearrangement of this cube of data set is required prior to developing a PCA model. Two data unfolding approaches are used: 1) Batch-wise unfolding and 2) Variable-wise unfolding with time lagging. These two methods have different characteristics that make each one appropriate for special conditions. Batch-wise unfolding shows better ability to explain the time varying properties of the process, but the dimensionality problem becomes evident in the algorithm as batch proceeds. On the other hand, column-wise time-lag unfolding is suitable for the conditions when fewer runs are allowed for data generation, but time varying characteristics are not well explained. In order to compensate the aforementioned problems, different phases along the batch are distinguished and multiphase modeling approach is performed. The MPC control formulation is developed to provide simultaneously for trajectory tracking and nonstationary disturbance rejection. Two control formulations are developed. In the first one the control optimization performed in the latent variable space and the manipulated variables then computed from the optimal scores using the PCA model. In the second one the control optimization is performed directly in the space of the manipulated variables.

The resulting algorithms are tested on a case study. The aim is to control a batch reactor temperature by manipulating the set point of the cascade controller built on the jacket temperature. The objective function is minimizing the deviation of the process output from the set point trajectory as well as penalizing the control movements. The two control formulations yield similar performances, but each of them has its own benefits. As an instance, the first algorithm has better performance and is more consistent with the concept of multivariate statistical modeling and it is faster than the second method. However, the second one is more familiar to the control community and it is easy to incorporate hard constraints on the manipulated variable trajectories.

References

[1] Flores-Cerrillo, J., MacGregor, J. F., (2005), Latent Variable MPC for trajectory tracking in batch processes, J. Process Cont. 15, 651-663.

[2] Flores-Cerrillo, J., MacGregor, J. F., (2002), Control of particle size distributions in emulsion

semibatch polymerization using midcourse correction policies, Ind. Eng. Res. 41, 1805–1814.

[3] Nomikos P., MacGregor J. F., (1994), "Monitoring batch processes using multiway principal component anakysis", AIChE Journal 40, 1361-1375.

[4] Nomikos P., MacGregor J. F., (1995), "Multiway partial least squares in monitoring batch processes", Chemometrics and intelligent laboratory systems 30, 97-108.

[5] Yabuki, Y., MacGregor, J. F., (1997), Product quality control in semibatch reactors using midcourse correction policies, Ind. Eng. Chem. Res. 36, 1268–1275.

688g Fractal Scaling In Crude Oil Price Evolution Via Time Series Analysis of Historical Data

Dimitrios I. Gerogiorgis, Department of Chemical Engineering, Massachusetts Institute of Technology (M.I.T.), 77 Massachusetts Avenue, Cambridge, MA 02139

The recent and ongoing surge in crude oil and fuel prices which has been continuously affecting national economies, multinational industries and international chemical process operations poses a fundamental question: do these extreme and unprecedented price increases still follow the established price statistics (falling within the existing variation pattern), or is the international oil economy entering a new and drastically different historical period?

The inherently stochastic nature of economic phenomena that evolve continuously at multiple frequencies – such as stock exchange, but also energy feedstock prices – poses difficulties and challenges to their systematic study, but also renders traditional deterministic analysis tools inadequate, especially inasmuch as quantitative understanding of intrinsic structure is an issue. Benoit Mandelbrot (1963), in a seminar paper, discovered the surprising fact that the seemingly random signals of cotton prices, when appropriately examined by means of advanced statistical techniques, possess inherent structure – the latter can be quantitatively characterized by means of both conventional statistical metrics (e.g. kurtosis) as well as novel indices (e.g. drift exponent). Essentially, Mandelbrot's fractal theory is impressively powerful and accurate for the purpose of price series analysis, because it allows for the study of statistical distributions that have a well-defined range of mean variation, but also infinite variance variation: this corresponds very well to the aforementioned stochastic signals, which are essentially bounded for a given period of study (thus finite mean variation), yet evolve across orders of frequency (thus infinite variance).

Time series data often arise when monitoring industrial processes or tracking corporate business metrics; in this case, we are interested in the evolution of crude oil prices over several decades. Its definitive characteristic is that time series analysis accounts for the fact that data points taken over time may have an internal structure (such as autocorrelation, trend or seasonal variation) that should be accounted for, and can actually be analyzed quantitatively. Methods and tools include concepts such as stationarity and seasonality, and tools such as autocorrelation function, single and multiple exponential smoothing, fractal dimension and intrinsic time.

In many statistical autocorrelation studies of intra-day financial time series data, it has been identified (Muller et al., 1995) that the absolute prices of price changes behave like the fractional noise of Mandelbrot and Van Ness (Mandelbrot, 1963). The mean absolute price change yields a linear correlation to the analysis time interval size, in logarithmic coordinates, effectively indicating that price changes obey a scaling law – and, while the frequencies involved in the phenomenon span several orders of magnitude, they can be conclusively and comprehensively identified and studied.

This paper presents historical price data for various crude oil types (Brent, Saudi, Iranian, Texas), applying different degrees of time resolution; results are interrelated in order to identify patterns and analyze variation timescales. A specific target of this study is to investigate the presence of fractal properties. In particular, the hypothesis that the mean size of the absolute values of price changes follows a “fractal” scaling law (a power law) as a function of the analysis time interval (the latter has been considered in this study as an independently varying parameter, ranging from a few minutes up to a year). The analysis reveals some interesting trends, which are useful for understanding the seasonality but also the intrinsic structure of crude oil markets.

697a A Framework for Generating An Optimal Set of Robust Product Design Alternatives

Beverly V. Smith, Department of Chemical and Biochemical Engineering, Rutgers - The State University of New Jersey, 98 Brett Road, Piscataway, NJ 08854 and M.G. Ierapetritou, Chemical & Biochemical Engineering Department, Rutgers University, the State University of New Jersey, 98 Brett Road, Piscataway, NJ 08854.

The chemical manufacturing industry is experiencing a shift from the development and manufacture of bulk commodity chemicals to the design and manufacture of specialty high value-added chemical products.[1] This emerging industry reality implies greater challenge to achieving the desired product quality while keeping cost at a minimum. The new market situation also creates new demands on existing product design approaches,since much of the work in chemical product design is multi-disciplinary in nature and involves time consuming experimentation[2]. Consequently, opportunities for cost reduction now extend beyond process design and optimization efforts to include product design and product development. In contrast to bulk chemicals, specialty chemicals are more susceptible to performance variability resulting from a number of factors such as lack of standard performance indices or quantitative methods for measuring end-use quality factors. Moreover, the emergence of global markets has brought about the need for rapid change and has significantly increased the range of uncertainty and variable application conditions for most products. This new environment has created a demand for an efficient product design framework that integrates robust enhancement techniques and ensures that optimal product performance is achieved at minimal cost.

The objective of this work is to develop a systematic framework that supports the generation of an optimal set of robust product design alternatives. The approach focuses on end-use application-based product design, and employs a mechanism that relates the functional requirements of the product to its properties. This capability, which relies on expert knowledge, is particularly useful when designing consumer-oriented products that involve satisfying a wide variety of functional requirements. An end-use application based product design emphasizes the interactions between technology and marketing domains throughout the product design and development process. This approach also ensures that the critical technical attributes of the product are strongly influenced by consumer preferences. The proposed framework also provides a procedure for deriving the utility function which incorporates consideration of the product attributes in determining the worth of the particular product alternative. A multi-objective optimization formulation is used to balance the tradeoff between product cost and product performance, while simultaneously ensuring minimization of performance variation.

In contrast to other approaches to chemical product design found in the literature[2] , the proposed framework incorporates robust design techniques, to address variation in the design domain, and considers the economic feasibility of the product alternative. Furthermore, the approach acknowledges the multidisciplinary nature of chemical product design[2] by integrating technology and marketing functions. This is done by accounting for both engineering performance and consumer preferences in the problem formulation that leads to the generation of a feasible set of product alternatives.

We illustrate this framework by reviewing the design of a consumer-oriented product. The result indicates that the proposed systematic framework attains optimal solutions that are robust to variation in design variables and noise parameters.

Reference:

(1)Cussler, E.L. and Moggride, G.D., *Chemical Product Design*, Cambridge University Press , Cambridge, 2001

(2)Gani R., *Chemical Product Design: Challenges and Opportunities*, Computer and Chemical Engineering 28 (2004) pp 2441-2457

697b The Virtual Product-Process Design Laboratory

Elisa Conte, Ricardo Morales-Rodriguez, and Rafiqul Gani. CAPEC, Department of Chemical and Biochemical Engineering, Technical University of Denmark, Søltoft Plads, Building 227, 2800 Kgs. Lyngby, Denmark

In design of chemical products and the processes that can make them, one first tries to find a candidate product that exhibits certain desirable or targeted behavior and then tries to find a process that can manufacture it with the specified qualities. The candidate may be a single chemical, a mixture, or a formulation of active ingredients and additives. For the later product type, additives are usually added to an identified active ingredient (molecule or mixture) to significantly enhance its desirable (target) properties. The common practice is experiment-based trial and error approach, supplemented sometimes with model-based computer-aided tools to speed-up some of the steps.

The objective of this paper is to present a "virtual" laboratory for chemical product-process design where the developer can test their ideas on model-based computer-aided tools rather than perform experiments. As the name suggests, the virtual lab contains methods and tools to allow the modeling and simulation of the needed experimental scenarios. For this to work the models need to be reliable and efficient and the architecture of the software needs to include the work-flows related to many different product-process design. Also, interfaces for efficient data-flow between different tools used in the work-flow needs to be defined. The architecture needs to be flexible to allow changes in work-flow, additions of new models and the addition of new data for future extension of the application range.

A prototype virtual PPD-lab has been developed using EXCEL as the simulation environment with proven and tested computer aided methods and tools developed by the authors. It includes databases of properties of chemicals, chemical product characteristics, of property models, and many more. It includes tools for molecular and mixture design to generate product alternatives. It includes tools for modeling to generate models for evaluation of process-product performance. It includes recipes for design of formulated products and mixture design. In principle, it can perform many of the virtual experiments that the product-process developer would need to perform, on a trial and error basis. Alternatively, the developer can ask for a generation of the short list of the best (or feasible) alternatives. The objective here is to quickly and efficiently generate a list of promising candidates for final testing (and selection) by experiments. In this way, rather than use the experiment-based trial and error approach from the start, the experimental resources are reserved for the final selection and verification step while the virtual PPD-lab is used for identifying the promising feasible candidates, thereby achieving faster to market at reduced cost for any potential chemical product-process.

The presentation will highlight the main features of the virtual PPD-lab, providing details of the available chemical product design recipes and the resident models, methods and tools that have been included. The applicability of the virtual PPD-lab will be highlighted through several chemical product-process design case studies from the pharmaceutical, agrochemical and petrochemical industries.

697c A Systematic Method for Integrating Product Attributes and Molecular Synthesis

Charles C. Solvason, Nishanth Chemmangattuvalappil, and Mario R. Eden. Department of Chemical Engineering, Auburn University, 230 Ross Hall, Auburn University, AL 36849-5127

The complete and efficient solution to the enumeration of candidate compounds and mixtures that meet specified consumer attributes is often a difficult mathematical programming problem. Most approaches to this problem involve the solution of a mixed integer nonlinear program (MINLP) which may achieve only local optima solutions. In this paper a proof-of-concept study is presented to show that empirical models can be used in a reverse problem formulation to ensure a complete set of candidate compounds and mixtures are found subject to the predictive power of the models. The method utilizes a transformation of consumer attributes to properties described by the group contribution method and solves the reverse problem formulation using the property clustering technique. The design of an environmentally benign refrigerant for the replacement of 1,1,1,2-tetrafluoroethane (R-134a) in a typical refrigeration cycle is used to highlight the method.

697d Effects of Solvent Selection In Optimal Separation Process Design and Controllability Properties Using a Nonlinear Sensitivity Analysis Approach

Athanasios I. Papadopoulos, Chemical Process Engineering Research Institute, Center for Research and Technology Hellas, 6th klm Harillaou-Thermi Road, P.O.Box 60361, Thermi-Thessaloniki, 57001, Greece and Panos Seferlis, Mechanical Engineering Department, Aristotle University of Thessaloniki, P. O. Box 424, Thessaloniki, 54124, Greece.

Numerous research efforts have been reported in published literature that propose the design of solvents as part of the broader flowsheet process design. Capturing the synergies and interactions between the designed solvents and the processes in which they are utilized leads to process-solvent schemes of optimum economic performance. Existing design methods focus on optimizing process design characteristics such as process structure and size, topology of recycle streams and solvent feed flowrates, to name a few. Subsequently, the performed optimization is combined with a method for the generation of solvent design alternatives hence the effects of various solvent options in the economic design of the particular process addressed are fully explored. However, processes are essentially dynamic environments susceptible to variations in operating parameters and influence from exogenous disturbances. For example, a solvent-process scheme designed for optimality under the assumption of certain purity levels in the process feed streams or constant process pressure, will not necessarily be optimal if any of those parameters vary during the process operation, as is often the case in industrial practice. Available approaches aiming to design solvent-process schemes of optimum performance often overlook the effects of such variations as is generally assumed that a properly designed control system will eventually compensate for such effects. While a number of methods have been developed to address process design optimality under operating parameter variations, the effects of using alternative solvent options in process design under operating variability that explicitly considers the controllability properties of the overall design have yet to be addressed systematically.

In this work, we address the economic and static controllability behavior of different solvent candidates introduced into a framework that enables optimal process design while considering variations in process parameters and disturbances in process operation. The employed process design framework consists of a design and a nonlinear sensitivity analysis stage. The separation process design stage is facilitated by use of orthogonal collocation on finite elements (OCFE) techniques for stagewise separation units that allows the reduction of the model size while maintaining a rigorous mathematical representation of the occurring phenomena. The nonlinear sensitivity analysis stage investigates the steady-state effects of process parameters and conditions in the optimal operating point that is imposed by an underlined control scheme. The imposed control scheme categorizes control objectives and preferences in the usage of manipulated variables in a multivariable centralized fashion (Seferlis and Grievink, 2001). As a result, a sensitivity matrix around optimal operating points is developed that reflects the variability of process variables in response to changes in process parameters. It is within this framework that a number of alternative solvents with varying chemical structure and physical properties are introduced in order to study their effects in optimal design under operating parameter variations. The solvents are selected and introduced into process design as discrete options drawn from a solvent set designed for optimality using a multi-objective optimization computer aided molecular design method, based on a previously developed approach by Papadopoulos and Linke (2005). In this respect, each solvent constitutes an additional design parameter in the formulated optimization problem, demonstrating alternating process design directions under the imposed operating parameter variations. The performed sensitivity analysis generates useful insights regarding the imposed control framework and the range of parameter variations within which the solvents demonstrate optimum performance. Furthermore, it allows the adaptive modification of the process design characteristics, inclusive of the available solvent options, in order to enhance the process behavior under the influence of operating variations. The proposed approach is demonstrated through case studies in separation through extractive distillation of a number of mixtures. The focus is maintained in assessing the effects of different solvents in process design and steady-state controllability properties under operating variability, while the impact of this variability at the solvent design stage is beyond the scope of this work.

References cited

Papadopoulos, A.I., Linke, P. (2005). Multi-objective molecular design for integrated process-solvent systems synthesis. AICHE J., 52(3), 1057

Seferlis, P., Grievink, J., (2001). Process Design and Control Structure Screening Based on Economic and Static Controllability Criteria, Comput. Chem. Eng., 25, 177.

697e Combined Property Clustering and GC+ Techniques for Process and Product Design

Nishanth Chemmangattuvalappil, Charles C. Solvason, and Mario R. Eden. Department of Chemical Engineering, Auburn University, 230 Ross Hall, Auburn University, AL 36849-5127

Traditionally process design and molecular design have been treated as two separate problems, with little or no feedback between the two approaches. Introduction of the property integration framework has allowed for simultaneous representation of processes and products and established a link between molecular and process design from a properties perspective. Utilizing this methodology enables identification of the desired properties by targeting the optimum process performance without committing to any components. The identified property targets can then be used as inputs for solving a molecular design problem. Earlier works have extended the property integration framework to include group contribution methods (GCM) for solving the molecular design problem. Yet, there are situations when the property contributions of some of the molecular groups of interest are not available in literature. Recently, techniques have been developed to create such missing groups and their contributions through a set of zero order and first order connectivity indices (GC+ approach). In this contribution, a methodology has been introduced to combine property clustering techniques and GC+ approach to enable simultaneous consideration of process performance requirements and molecular property constraints. The advantage of this method is we are not limited to the molecular groups whose GC properties are known. For the design of simple monofunctional molecules, a modified visual approach has been used where as for the design of more complicated structures and /or for treating more than three properties at a time, an algebraic method has been used. The algebraic method has been made more powerful by including the property contributions from third order groups and by developing guidelines for its use in complicated designs like fused ring compounds. This contribution will illustrate the developed methods and highlight their use through a case study.

697f Integrated Design of Self Heating Products and Processes

Aydin Sunol[1], Sermin G. Sunol, Brandon Smeltzer[2], and Raquel Carvallo[3]. (1) Department of Chemical Engineering, University of South Florida, 4202 East Fowler Avenue, ENB 118, Tampa, FL 33620, (2) Chemical Engineering, University of South Florida, 4202 E. Fowler Ave, ENB 118, Tampa, FL 33620, (3) Chemical Engineering Department, University of South Florida, 4202 East Fowler Avenue ENB 118, Tampa, FL 33620

This case study illustrates integrated use of Finite Element analysis and stochastic simulation in design of novel self heating products. The synthesis of reaction and transport processses within the product, design of the key particle encapsulation process, and integration of several manufacturing/packaging steps were accomplished with aid of advanced computation techniques to enable timely introduction of the initial set of products.

728a Bioprocess Synthesis, Design and Analysis through a Group-Contribution Approach

Merlín Alvarado-Morales[1], João Terra[1], Krist V. Gernaey[1], John Woodley[1], and Rafiqul Gani[2]. (1) Department of Chemical & Biochemical Engineering, Technical University of Denmark, Sølftofts Plads, Building 229, Kgs. Lyngby, Denmark, (2) CAPEC, Department of Chemical and Biochemical Engineering, Technical University of Denmark, Søltofts Plads, Building 229, 2800 Kgs. Lyngby, Denmark

Modelling and simulation of a process flowsheet usually involves identifying the structure of the flowsheet (defined by operations involved and their sequence), deriving model equations to represent each operation, and solving the resulting total model equations according to various simulation strategies. The flowsheet synthesis problem determines the type of operations and their sequence while the flowsheet design problem determines the optimal values for the conditions of operation and other operation/equipment related variables. The main idea here is to apply the group-contribution approach (from property estimation) to the synthesis, design and analysis of chemical and biochemical processes with groups representing process unit operations (reactor, distillation, flash, crystallization, etc.), bonds representing streams and/or recycles, rules for molecule feasibility representing process flowsheet feasibility and sum of group contributions representing the performance of the flowsheet.

The proposed paper will present the process groups based approach to model, design and analyze the bioethanol production process. The production route for the bioethanol process has been selected, due to the current interest in the use of renewable sources to produce important chemicals (the biorefiney concept). Also biofuels have become a priority to reduce the dependence on fossil fuels. The production of bioethanol has increased all over the world in the last few years through expansion of existing plants and construction of new facilities, but the economic competitiveness of bioethanol as a liquid fuel strongly depends on the energy resources used during its production. This implies the need for generation of sustainable alternatives to determine the optimal values for the conditions of operation and other operation/equipment related variables for the synthesized process flowsheet, which will be highlighted through the process group approach.

The sustainability of bioethanol production from alternative renewable sources will be analyzed together with a breakdown of the operational and equipment costs in different parts of the bioethanol process. Each part of the process (pretreatment, fermentation, ethanol recovery/purification) has been analyzed in detail. From this analysis, it has been identified that the economic feasibility of the bioethanol process can be improved through higher product yields and/or more efficient product recovery. Since the fermentor effluents are dilute aqueous solutions of the product, recovery is not straightforward. Furthermore, separation by distillation is both difficult and expensive because of the ethanol-water azeotrope. The option of increased product yield in the reaction is beyond the scope of this presentation as it involves the design/selection/testing of new catalysts or enzymes. In this presentation different schemes for ethanol recovery/purification will be presented using the process-groups approach. The optimal sustainable process for bioethanol production will be presented together with a lifecycle assessment. Finally, extension of the proposed approach to other chemicals within the biorefinery concept will be discussed.

728b Simultaneous Optimization of Cyclic Hoist Scheduling and Water-Reuse Network Design for Environmentally Benign Manufacturing

Chao-Wei Liu and Qiang Xu. Department of Chemical Engineering, Lamar University, Beaumont, TX 77710

Many chemical processing and manufacturing activities involve material handling in multi-stage process systems. Coating processes, such as electroplating and polymeric coating processes, are typical examples, where operations are performed by hoists. Each hoist is programmed to carry jobs from one unit to the other, based on a preset operational schedule. The cyclic hoist scheduling (CHS) has been well developed for productivity improvement. Water-reuse network design (WRND), on the other hand, targets the optimal design of water allocation network, such that fresh water consumption and wastewater generation could be the minimized. It is well known for its environmental significance in waste water minimization.

A recent work has proposed to integrate these two technologies by a two-layer optimization method, where the upper layer employs a MINLP CHS model to minimize the cycle time that responds to maximizing the production rate, and the lower layer utilizes a dynamic WRND model to minimize the total freshwater usage. Processing time information of each process unit operation is passed between the two layers, and the iterative computation between these two layers will eventually lead to the most desirable solution for both CHS and WRND.

In this paper, a new methodology for simultaneous consideration of CHS and WRND is developed. The CHS and WRND problems are integrated into one Mixed-Integer Dynamic Optimization (MIDO) model. The simultaneous approach, based on orthogonal collocation on finite elements, is used to transform MIDO model into a Mixed-Integer Nonlinear Programming (MINLP) model for solution identification. This developed MIDO model presents unique features as the multiple scheduling alternatives may exist for the same CHS problem, meanwhile the jump discontinuity may exist in the WRND problem.

The proposed methodology can successful tackle these difficulties. It can identify the optimal CHS schedule and water-reuse network design simultaneously. The proposed simultaneous method and its advancement have been demonstrated with an electroplating process. It demonstrates that economic and environmental thrusts can be simultaneously considered to create a win-win situation for manufacturing.

728c Global Optimization of Batch Direct-Recycle Networks

Viet Pham, Arwa Rabie, Carl D. Laird, and Mahmoud M. El-Halwagi. Chemical Engineering, Texas A&M University, College Station, TX 77843

Direct recycle networks are among the most attractive strategies for promoting sustainable processing. They lead to conserving material resources and mitigating negative environmental impact. While significant progress has been made in developing systematic techniques for the design of direct-recycle network in steady-state systems, much less work has been undertaken for batch systems. The objective of this work is to develop an optimization-based approach to the synthesis and scheduling of direct recycle networks for batch systems. A source-tank-sink representation is developed to embed potential configurations of interest. The representation allows for scheduling within the same cycle or between consecutive cycles. A two-stage approach is adopted: targeting followed by detailed solution. The targeting of fresh material usage and waste discharge is handled through a linear programming formulation. Network synthesis and scheduling involve the development of nonlinear program or a mixed-integer nonlinear program, depending on the choice of cost functions. In either case, the presence of bilinear terms induces non-convexity and necessitates the development of a global optimization approach. A global optimization technique is developed. The approach is based on three concepts: linearization by discretizing nonlinear variables, pre-processing using implicit enumeration of the discretization to form a convex-hull which limits the size of the search space, and application of integer cuts to insure compatibility between the original problem and the discretized formulation. A case study is solved to illustrate the computational aspects of the developed formulation and solution technique.

728d A Multiscale, Multiobjective and Thermodynamic Framework for Environmentally Conscious Process Engineering

Vikas Khanna, Department of Chemical and Biomolecular Engineering, The Ohio State University, Koffolt Laboratories, 140W, 19th Avenue, Columbus, OH 43210 and Bhavik R. Bakshi, Department of Chemical and Biomolecular Engineering, The Ohio State Unversity, Columbus, OH 43210.

Concerns over the depleting fossil energy sources, mounting regulatory compliance and the resulting push towards environmentally conscious process design in the last decade are forcing designers to consider reduced environmental impact as one of the product design objectives. With greater appreciation of such challenges, systems engineering has gradually broadened its analysis boundary from the equipment and manufacturing scale to the life cycle or value chain scale. Selecting the system boundary by expansion from the manufacturing to the value chain scale is appealing for environmentally conscious decision making, however the approach still relies on selecting the most relevant processes of the value chain. Such selection, as is the case with traditional life cycle assessment (LCA) is often quite arbitrary, subjective, and vulnerable to convenient drawings of the boundaries. On the other hand, expanding the system boundaries to include all interactions is computationally intractable and time consuming. In reality, data and models are available at multiple spatial scales ranging from individual equipment and processes, to the supply and demand chains, to the economy and ecosystems. The outstanding challenge here is the development and adoption of a broader multiscale and multiobjective approach that systematically accounts for all the conflicting design objectives as opposed to the narrowly focused cost-benefit analysis.

This work presents a novel multiscale and multiobjective optimization oriented approach to utilize the available information at multiple scales for gaining insights into the trade-off between ecological and economic aspects of manufacturing processes thus yielding a comprehensive scenario on which process alternatives can be evaluated. Economic aspects are accounted via traditional cost analysis. Ecological aspects are considered via exergy analysis of the inputs at each scale, and depend on the selected processes. Exergy or available energy accounts for the first and second laws and is the ultimate limiting resource since it is lost in all processes. This multiscale approach represents inputs and outputs in terms of cumulative exergy consumption (CEC). The scales considered in this work are the individual equipment or process manufacturing, value chain, economy, and ecosystem scales. The finest equipment or manufacturing scale corresponds to a traditional exergy analysis [1]. The trade-off between ecological and economic aspects is usually quite large at this scale because the narrow manufacturing boundary ignores the environmental impact of other processes in the life cycle that incur an economic cost such as capital equipment. The next coarser scale is at the value chain scale and corresponds to an exergetic life cycle analysis. The boundary is further expanded to include activity in the entire economy by combining exergy analysis with economic input-output analysis [2]. Finally, contributions from ecosystem goods and services are included at the coarsest ecosystem scale [3].

The utility of the proposed multiscale and multiobjective approach and its benefits over existing methods will be illustrated using the case studies of a heat exchanger and chemical product design. Trade-off between the economic and ecological objectives will be represented via a series of pareto optimal surfaces at various scales, thus avoiding arbitrary combinations until the final stages of decision making. The implications of the proposed approach in guiding ecologically and economically conscious methods and heuristics for broader engineering design will be described. Current work is in progress to explore the use of stochastic modeling for incorporating uncertainty information in the proposed methodology.

References

1. J. Szargut, D. R. Morris, and F. R. Steward. Exergy analysis of thermal, chemical and metallurgical processes. Hemisphere Pubs., New York, 1988.

2. N. U. Ukidwe and B. R. Bakshi. Industrial and ecological cumulative exergy consumption of the united states via the 1997 input-output benchmark model. Energy, 32(9):1560–1592, 2007.

3. J. L. Hau and B. R. Bakshi. 2004. Expanding exergy analysis to account for ecosystem products and services. Environmental Science and Technology. 38(13): 3768-3777, 2004.

Property-Based Optimization of Direct-Recycle Networks and Wastewater Treatment Processes

José M. Ponce-Ortega[a,b], Mahmoud M. El-Halwagi[c], Arturo Jiménez-Gutiérrez[a]

[a] *Instituto Tecnológico de Celaya, Celaya, Gto. México*
[b] *Universidad Michoacana de San Nicolás de Hidalgo, Morelia, Mich. México*
[c] *Texas A&M University, College Station, TX, USA*

Abstract

This paper presents a mathematical programming model to simultaneously optimize a recycle-reuse mass exchange network together with a wastewater treatment process. The model is based on disjunctive programming, and considers the technologies that are commonly used to treat wastewater streams so that they meet environmental regulations. In addition to standard mass balance and composition constraints, the model incorporates environmental constraint properties such as toxicity, theoretical oxygen demand, pH, color, and odor. The problem results in an MINLP model, and is used to minimize the total annual cost of the system, which includes the cost for the fresh sources, the piping cost for the mass integration and the wastewater treatment cost. The model is formulated with linear relationships in the disjunctions to avoid numerical complications. The solution of the proposed model shows that the simultaneous consideration of the mass integration and wastewater treatment sections yields important economic savings with respect to the solution that considers the mass integration first and then takes into account the wastewater treatment facilities. Also, the simultaneous optimization can yield solutions that require more consumption of the cleaner fresh sources than the one provided with the optimization of the mass exchange network alone, and the overall integration results in a lower total annual cost. The model did not show any numerical problems, and the CPU time required for the solution of the numerical examples was relatively small.

Introduction

Mass integration strategies have been the subject of study of much research efforts because of the economical and environmental benefits they provide. In the area of sustainable design, mass integration can be used to lower the consumption of fresh sources and to reduce the waste materials discharged to the environment. Recent surveys on mass integration can be found in literature.[1,2] A particular application of the mass integration is in the area of minimizing fresh water usage and wastewater discharge through recycle/reuse strategies. Bagajewicz[3] has published a paper review on the procedures reported to design water networks. Previous works have provided major contributions to the synthesis of water recycle/reuse networks. However, they have a couple of limitations. First, their focus has been given to in-plant recycle strategies with constraints limited to process units and without considering the effects of the resulting terminal wastewater streams and the environmental regulations. To satisfy the environmental constraints, the wastewater stream needs to be treated prior to be discharged to the environment; this situation may increase the total cost associated to the recycle-reuse mass integration process. For example, when the solution that minimizes fresh sources consumption yields a high wastewater treatment cost, then such a solution may not correspond to the overall optimal solution. In this case, it may be preferable to use more fresh sources and lower the wastewater treatment cost. Therefore, it is important to consider simultaneously the optimization of the process integration together with the wastewater treatment process to take into account the trade offs between both factors. Second, process characterization and recycle constraints have been based on compositions. It is worth noting that there are many industrial cases when such constraints should be based on properties. Examples include applications when the performance of the process units is affected by the properties of its feed and the cases when environmental regulations impose limits for specific properties of waste streams such as toxicity, theoretical oxygen demand (*ThOD*), *pH*, temperature, color, odor, viscosity and density. These properties are difficult to quantify as a function of composition because of the many components of the process streams. It is also difficult to track properties throughout the process because they are not conserved. To overcome this limitation, Shelley and El-Halwagi[4] introduced the concept of property-based componentless design by defining surrogate properties (called clusters) that enable the conserved tracking of properties.

This paper presents a mathematical programming model to simultaneously optimize the direct recycle networks together with the wastewater treatment process in order to satisfy a set of process and environmental constraints. An optimization formulation is developed based on disjunctive programming. The model is formulated to consider environmental-constrained properties such as toxicity, *ThOD*, *pH*, color, odor, and any other property that may cause pollution to the environment, in addition to the mass and composition constraints for hazardous compounds in the waste stream. The problem is formulated with proper constraints as an MINLP problem that minimizes the total annual cost of the system, which includes the cost for the fresh sources, the piping cost for the process integration and the wastewater treatment cost.

Model formulation

The problem addressed is defined as follows. Given is a set of process and fresh sources with known flowrates, compositions and properties. Also given is a set of sinks (process units) with constraints for the inlet flowrates and allowed compositions and properties. In addition, there is a set of constraints given by the environmental regulations for the waste streams discharged to the environment. The problem then consists of finding the optimal mass and property integration network that includes a direct recycle strategy that meets the environmental regulations for the waste streams and minimizes the total annual cost of the overall process (see Figure 1).

Splitting of process source streams.

$$W_i = \sum_{j \in NSINKS} w_{i,j} + waste_i, \qquad i \in NSOURCES \tag{1}$$

Splitting of fresh source streams.

$$F_r = \sum_{j \in NSINKS} f_{r,j}, \qquad r \in NFRESH \tag{2}$$

Total waste.

$$WASTE = \sum_{i \in NSOURCES} waste_i \tag{3}$$

Mass balances at the mixing point before each sink.

$$G_j = \sum_{i \in NSOURCES} w_{i,j} + \sum_{r \in NFRESH} f_{r,j}, \qquad j \in NSINKS \tag{4}$$

Component balances at the mixing point before each sink.

$$z_{j,c}^{in} G_j = \sum_{i \in NSOURCES} \left[z_{i,c} \, w_{i,j} \right] + \sum_{r \in NFRESH} \left[z_{r,c} \, f_{r,j} \right], \qquad j \in NSINKS, c \in NCOMP \tag{5}$$

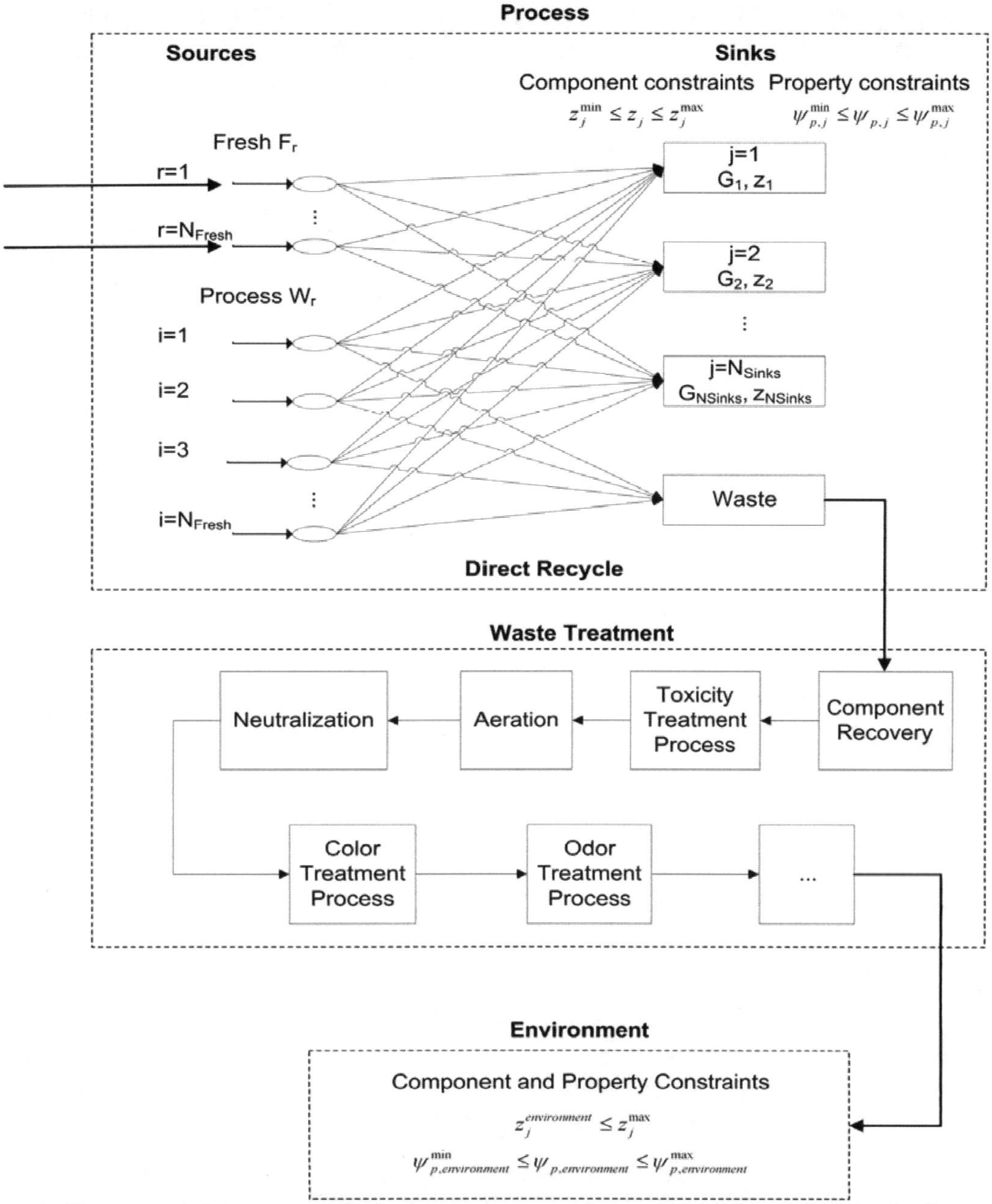

Figure 1. Source-sink representation for mass and property integration including waste treatment

Property balance at the mixing point before each sink.

$$\psi_p \left(p_{j,p}^{in} \right) G_j = \sum_{i \in NSOURCES} \left[\psi_p \left(p_{i,p} \right) w_{i,j} \right] + \sum_{r \in NFRESH} \left[\psi_p \left(p_{r,p} \right) f_{r,j} \right], \qquad j \in NSINKS, p \in NPROP \tag{6}$$

Sinks properties constraints.

$$p_{p,j}^{\min} \le p_{p.j}^{in} \le p_{p.j}^{\max}, \quad j \in NSINKS, \, p \in NPROP \tag{7}$$

Waste stream treatment constraints

A waste stream needs to satisfy the environmental regulations before it is discharged to the environment. Constraints are formulated from the environmental regulations for hazardous materials. Whenever possible, they are based on component constraints, but for properties that are difficult to characterize because of the large number of components in some waste streams, they are formulated as property constraints.

Ratio waste stream.

$$Ratio_i = \frac{waste_i}{\sum_{i \in SOURCES} waste_i} \tag{8}$$

Density for the waste stream.

$$\frac{1}{\rho_{waste}} = \sum_{i \in SOURCES} \left[\frac{Ratio_i}{\rho_i} \right] \tag{9}$$

General property treatment. For any property, the property integration principle yields,

$$f_{Property}\left(Property_{mean}\right) = \sum_{i \in NSOURCES} f_{Property}\left(Ratio_i \, Property_i\right) \tag{10}$$

$$Wn^{Property} = Wn\,Property_{mean} \tag{11}$$

$$Wn^{PropertyReg} = Wn\,Property_{Reg} \tag{12}$$

Accordingly, the following disjunction can be formulated,

$$\begin{bmatrix} Y^{Property} \\ Wn^{Property} \geq Wn^{PropertyReg} \\ Cost^{Property} = \left(Wn^{Property} - Wn^{PropertyReg}\right) Costu^{Property} H_Y \end{bmatrix} \vee \begin{bmatrix} \neg Y^{Property} \\ Wn^{Property} \leq Wn^{PropertyReg} \\ Cost^{Property} = 0 \end{bmatrix}$$

which can be modeled as follows,

$$Wn^{Property} = Wn^{Property^1} + Wn^{Property^2} \tag{13}$$

$$Wn^{PropertyReg} = Wn^{PropertyReg^1} + Wn^{PropertyReg^2} \tag{14}$$

$$Wn^{Property^1} \geq Wn^{PropertyReg^1} \tag{15}$$

$$Cost^{Property} = \left(Wn^{Property^1} - Wn^{PropertyReg^1}\right) Costu^{Property} H_Y \tag{16}$$

$$Wn^{Property^2} \leq Wn^{PropertyReg^2} \tag{17}$$

$$Wn^{Property^1} \leq M^{Wn^{Property}} y^{Property} \tag{18}$$

$$Wn^{PropertyReg^1} \leq M^{Wn^{Property}} y^{Property} \tag{19}$$

$$Wn^{Property^2} \leq M^{Wn^{Property}} \left(1 - y^{Property}\right) \tag{20}$$

$$Wn^{PropertyReg^2} \leq M^{Wn^{Property}} \left(1 - y^{Property}\right) \tag{21}$$

Objective function. The objective function consists of the minimization of the total annual cost associated with both the mass integration and the waste treatment processes. The costs of the mass integration process include the cost due to the fresh sources and the piping costs to build the mass integration network. Piping costs are important to be accounted for, and they are a function of the layout (which is known prior to the mass integration) and the flowrate of the streams. To meet the environmental regulations, it may be necessary to treat the waste streams. The costs associated with the waste treatment process include the cost for the recovery of specific compounds in the waste streams, the cost to remove the toxic compounds, the cost of aeration to satisfy the theoretical oxygen demand regulation, the cost for the neutralization process, the cost to decrease color, odor, and any other property that may cause contamination. The objective function can be formulated as follows,

$$
\begin{aligned}
TAC = & \sum_{r \in FRESH} Cost_r^{Fresh} F_r \, H_Y \\
& + \sum_{\substack{i \in SOURCES \\ j \in SINKS}} pip_{i,j} \, w_{i,j} + \sum_{\substack{r \in FRESH \\ j \in SINKS}} pip_{r,j} \, f_{r,j} \\
& + \sum_{u \in RECOV} Cost_u^{Recoveryu} + \sum_{t \in TOX} Cost_t^{Toxicity} \\
& + Cost^{ThoD} + Cost^{Neu} + Cost^{Color} + Cost^{Odor} + \cdots + Cost^{Property}
\end{aligned} \tag{22}
$$

Results

This section presents the application of the proposed algorithm. For the mass integration process, two fresh sources are available with a phenol impurity of 0 and 0.012 mass fraction, and with unit costs of \$0.0006/lb and \$0.0004/lb, respectively. In addition there are three process sources; the stream data for these sources are given in Table 1. The data for the sinks are given in Table 2.

Table 1. Sources data for the Example

Source, i	W_i [lb/hr]	$z_{i,\,phenol}$	$z_{i,\,acetone}$	$ThOD$ [gO_2/l]	pH	ρ [lb/l]
1	8,083	0.016	0.000	0.187	5.4	2.205
2	3,900	0.024	0.010	48.850	5.1	2.205
3	3,279	0.22	0.028	92.100	4.8	2.205

Table 2. Sinks data for the Example

Sink, j	G_j [lb/hr]	$z_j^{\max}$
1	6,000	0.013
2	4,400	0.013
3	2,490	0.1

The environmental regulations applied to this specific process are as follows. The maximum concentration allowed to be discharged to the environment for acetone and phenol is 0.005. The toxicity must be zero, and the maximum concentration and discharge load of phenol, the toxic material, are 0.0000011 (mass fraction) and 0.00541 lb/hr respectively. The theoretical oxygen demand of the waste stream must be lower than 75 mg O_2/l, and finally the *pH* must be between 5.5 and 9.0.

The unit costs for the recovery processes are 0.065 \$/lb and \$0.033 \$/lb for phenol and acetone, respectively. The unit cost for aeration is 0.006\$/lb air diffused, and the unit costs for the H_2SO_4 and NaOH used for the neutralization process are \$46/l and \$31/l, respectively.

Table 3 gives the unit piping cost for the mass integration process.

Table 3. Piping costs for the Example*

Sink, j	Sources				
	Process, i			Fresh, r	
	1	2	3	1	2
1	5	2	3	4.5	2.5
2	3.5	1	5	3	1
3	2	4	2	3.5	1.5

* Units in \$/year

Figure 2 shows the results for the case when only process constraints are considered, without including the waste treatment process. We refer to this structure as Solution A. Notice that the waste stream does not satisfy the environmental constraints; therefore, it will be necessary to include a waste treatment process to the solution obtained here.

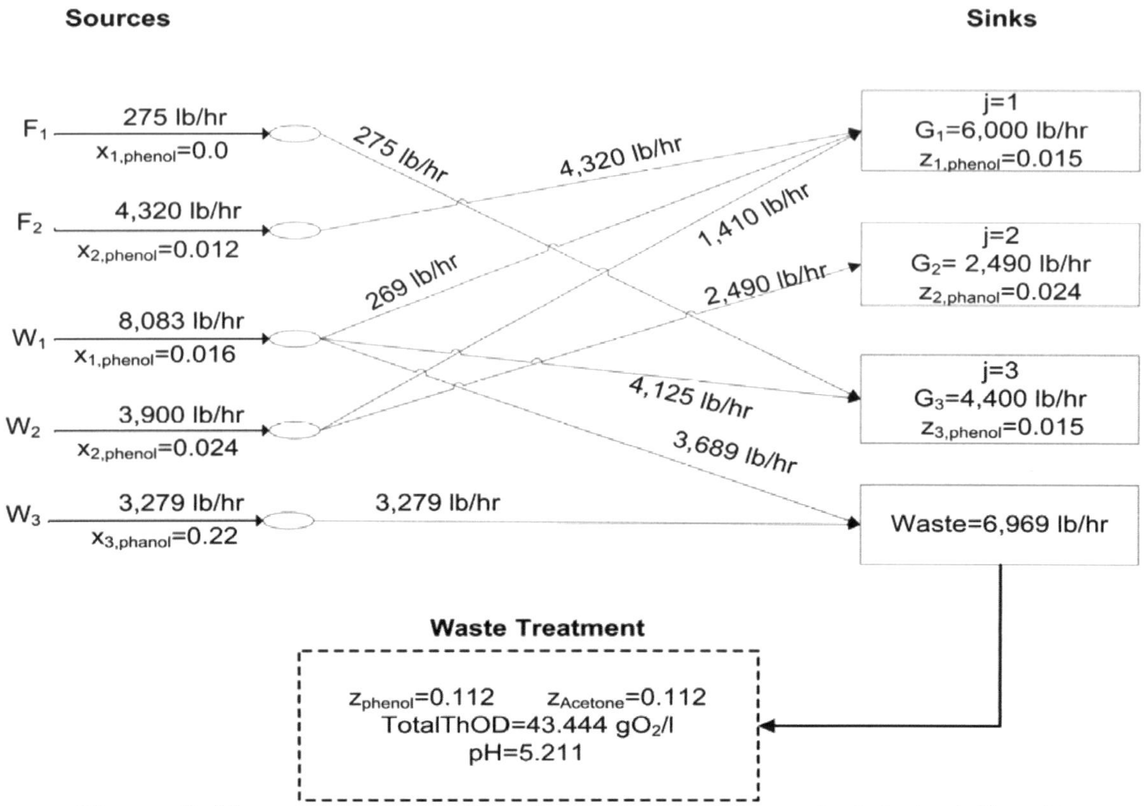

Figure 2. Mass integration without environmental constraints (solution A)

Figure 3 shows the solution of the example problem including all the environmental constraints for this example (Solution B). Notice that the total waste is reduced in the solution that incorporates the waste treatment process by 11.1% with respect to the solution that considered only the process constraints. In addition, in the optimal solution shown in Figure 3, the total consumption of fresh source 1 (with an impurity concentration of phenol of 0) is increased by 76.1% with respect to Solution A of Figure 2; this yields a decrease in the concentration of phenol and acetone in the waste stream by 22.2% and 30%, respectively. The theoretical oxygen demand in the waste stream for the optimal solution that considers the waste treatment process is reduced by 27.4% with respect to the simplified solution of Figure 2. The *pH* in the optimal solution B is increased by 0.93% with respect to the simplified Solution A.

Table 4 shows a summary of the costs associated with both solutions. For comparison purposes, the costs of the waste treatment processes that would be needed for Solution A were calculated after the optimization procedure. Notice in Table 4 that the sum of the costs associated with the process (fresh sources and piping costs) in solution B is 8% higher than those required for solution A. In addition, the total costs associated with the waste treatment process for the optimal solution B is 41.3% lower than those required by the simplified solution

A. Therefore, the solution that simultaneously optimize the process and the waste treatment process yields a total annual cost with savings of 25.4% with respect to the solution that only optimizes the process. This result confirms the advantage to simultaneously consider mass and property integration and to trade off direct recycle, usage of fresh resource, and treatment of waste streams while satisfying process constraints and environmental regulations.

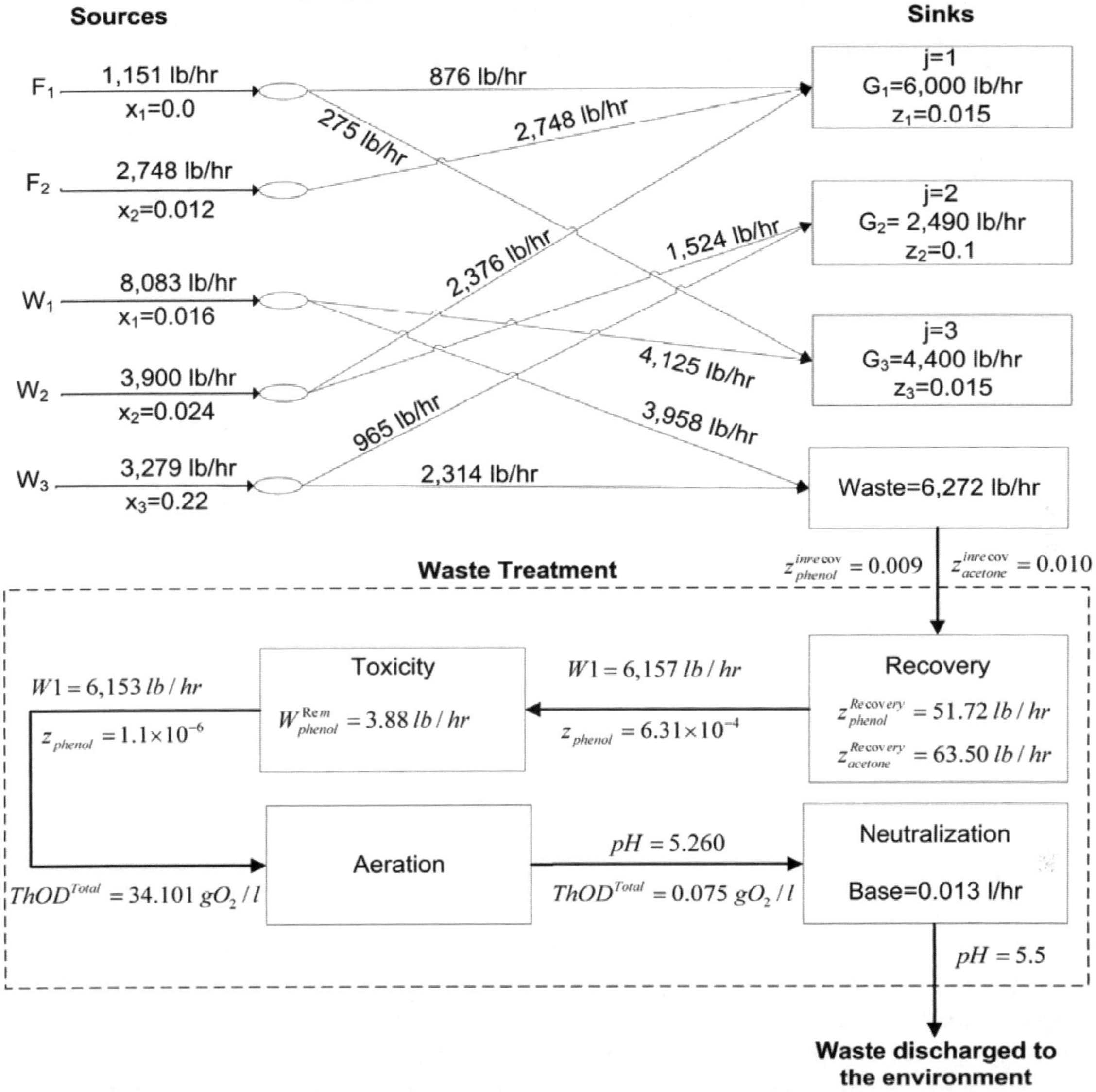

Figure 3. Mass integration including waste treatment process (solution B)

This example has shown that if one does not consider the waste treatment cost to satisfy the environmental regulations as part of the mass integration problem, the model can lead to sub-optimum solutions. We have shown in this paper how a simultaneous model can be formulated, which provides a better basis for the optimization of these types of systems.

Conclusions

A new framework has been introduced to (1) integrate the design of in-plant recycle/reuse networks with the end-of-pipe waste treatment facilities, (2) consider composition- and property-based constraints and process models. An optimization formulation has been developed to implement the devised framework. The model is based on disjunctive programming, and considers the technologies that are commonly used to treat wastewater streams so that they meet environmental regulations. The model is formulated with linear relationships within each one of the disjunctions, which aids its numerical solution. In addition to standard mass balance and composition constraints, the model incorporates property-based models, property mixing rules, and environmental constraints based on properties such as toxicity, theoretical oxygen demand, *pH*, color, and odor, and it is generalized to any other property that may cause pollution. The methodology has also shown how property integration can be used for cases in which a given property is difficult to estimate when there are many compounds in a waste stream. The formulation results in an MINLP model, which has been used to minimize the total annual cost of the system. The solution of the proposed model shows that the simultaneous consideration of mass and property integration and the tradeoff between in-plant recycle/reuse with wastewater treatment sections can yield important economic savings. The model did not show any numerical problems, and the CPU times required for the solution of the solutions of the example problem were relatively small.

Table 4. Comparison of results

Concept	With process constraints (Solution A)	With process and environmental constraints (Solution B)
Waste [lb/hr]	6,969	6,273
Fresh sources cost [$/yr]	15,146	14,318
Piping cost [$/yr]	26,669	31,128
Recovery cost [$/yr]	*61,859*	43,661
Toxicity cost [$/yr]	*7,226*	5,098
Aeration cost [$/yr]	*161,083*	43,363
Neutralization cost [$/yr]	*4,567*	3,237
Total process costs [$/yr]	41,816	45,446
Total waste treatment costs [$/yr]	*134,734*	95,360
TAC [$/yr]	*176,550*	140,807

Bibliography

1. El-Halwagi MM. (2006). *Process integration*. New York: Academic Press.
2. El-Halwagi MM. (1997). *Pollution prevention through process integration: Systematic design tools*. San Diego: Academic Press.
3. Bagajewicz M. (2000). A review of recent design procedure for water networks in refineries and process plants. *Computers and Chemical Engineering*. 24 (9-10):2093-2113.
4. Shelley MD and El-Halwagi MM. (2000). Componentless design of recovery and allocation systems: A functionality-based clustering approach. *Computer and Chemical Engineering*. 24 (9-10): 2081-2091.

728f Designing the Optimal Supply Chain for Biodiesel Production in Spain

Elisabeth Capón, Department of Chemical Engineering, Universitat Politècnica de Catalunya - ETSEIB, Diagonal, 647, Barcelona, E-08028, Spain, Gonzalo Guillén-Gosálbez, Chemical Engineering, University Rovira i Virgili, Av. Països Catalans, 26, Tarragona, 43002, Spain, **Laureano Jiménez-Esteller**, Department of Chemical Engineering, University Rovira i Virgili, Av. Països Catalans, 26, Tarragona, 43002, Spain, and Antonio Espuña, Chemical Engineering Department, Universitat Politècnica de Catalunya - ETSEIB, Diagonal, 647, Barcelona, E-08028, Spain.

The use of new sources of energy like fuel cells, solar energy or biofuels has recently become a priority. Among them, biofuels share some of the storage and distribution advantages of traditional fuels, as opposed to other energy carriers like hydrogen. Furthermore, they also avoid the economy dependence on oil price fluctuations and mitigate global warming. For these reasons, their production is expected to increase in the next years all over the world. Particularly, in Europe, the European Union has published the directive 2003/30/EC. This new legislation pursues the promotion of biofuels and other renewable fuels for transport, and stipulates that by 2010 a total of 5.75% of the total amount of fossil fuels must be replaced by biofuels.

In the transition process towards this new energy system, it is of crucial importance to design efficient supply chains capable of producing and delivering the necessary amount of biofuels to the exact locations and at the right time. This task can be accomplished by developing mathematical models that describe in an integrated manner all the elements of the supply chain within a single framework. Unfortunately, despite the effort made so far, the tools that currently exist in this area are still rather limited.

With the aim to contribute to this emerging research field, this work addresses the optimal design and planning of supply chains for biodiesel production. A mathematical superstructure is postulated, in which all the available manufacturing technologies are embedded, along with the possible locations for the production facilities, and the potential transportation links between the components of the network. The design task is mathematically formulated as a multi-objective mixed integer lineal program (MILP) that simultaneously accounts for the minimization of cost, environmental and social impacts. The Pareto solutions of the problem, which represent the optimal trade-off between the objectives considered in the analysis, are obtained via the epsilon constraint method. To expedite the application of such a method, we introduce a novel bi-level decomposition strategy, which decomposes the model into two hierarchical levels between which an iterative procedure takes place until a finalization criteria is satisfied.

The capabilities of the proposed modelling framework and solution strategy are illustrated through the design of the future supply chain for biodiesel production in Spain. The model is used to investigate a number of strategic decisions that configure the future supply chain that will satisfy the biodiesel demand in the Spanish markets. The data considered in the analysis have been taken from previous state-of-the-art contributions to the biodiesel production literature and other available sources, and has been validated with engineering-oriented methods to reflect the specific features of the country. The solutions obtained show that an inherent trade-off exists between the criteria considered in the study, and that significant environmental and social benefits can be achieved if the economic performance of the network is compromised. These solutions are intended to guide the national policy makers of the government towards the adoption of more sustainable strategic decisions from the environmental and social point of view.

References:

[1] DIRECTIVE 2003/30/EC OF THE EUROPEAN PARLIAMENT AND OF THE COUNCIL of 8 May 2003 on the promotion of the use of biofuels or other renewable fuels for transport

733a Optimal Model-Based Production Planning for Refinery Operations

Abdulrahman M. Alattas, Chemical Engineering, Carnegie Mellon University, 5000 Forbes Ave, Pittsburgh, PA 15213, **Ignacio E. Grossmann**, Center for Advanced Process Decision-making, Dept of Chemical Engineering, Carnegie Mellon University, 5000 Forbes Avenue, Pittsburgh, PA 15213, and Ignasi Palou-Rivera, Refining Technology, BP Products North America, Inc., 150 West Warrenwille Road, Mail code 603-3E, Naperville, IL 60563-8460.

Historically, the petroleum industry has used linear programming (LP) to address its planning and optimization needs (Favennec, 2001; Li et al 2005). The simplicity, robustness and convenience of this approach are tradeoffs for the true optimal and accurate solution to the planning model. In fact, planning technology is considered well developed and progress is expected in model refinement through the use of nonlinear programming (NLP) (Pelham & Pharris, 1996), which accommodates the use of nonlinear process models. These improvements are more pressing now with the changing market, increasing demands and limited refining capacities. This paper is an attempt to address this need with the goal to develop more accurate refinery planning models, using the latest NLP algorithms and implementing more accurate process modeling. The objectives are to establish the current status of planning models and propose nonlinear process model equations for implementation into a refinery planning model.

Starting with the configuration of a complex refinery, production planning model is designed to determine the types, quantities and mixing strategies for the different crude oils available for purchase, so that the refinery will meet the objective of maximizing profits while satisfying specific demands over a specified time period. Based on the available information for feedstock, products slate, unit capacities and conditions, the refinery planning model elements are the process units, separators, mixers, product blending and feedstock. The process units are modeled as a linear function to calculate the yields based on the feed streams. This approach satisfies the LP modeling requirements. The focus of this paper is the front end of the refinery, namely the crude distillation unit (CDU).

The CDU yield prediction is modeled using linear functions of the crude feed. However, it does not reflect true refinery operations where there are different operating modes (such as maximizing naphtha or light distillate, or to meet temporary limitations on utilities or maintenance). Each mode has its own set of coefficients for yield prediction. Therefore, the fixed-yield based model does not optimize the yield calculations or allow blending the discrete operating modes for optimum operation. The swing cut approach addresses this problem by allowing the exact cut or fraction to be optimized or “fine tuned”. After determining the desired product cuts of the crude, about 5% to 7% of the yield around adjacent fractions of the crude is allowed to change or “swing”, changing the yields so as to improve the objective function (Zhang, 2001, Trierwiler, & Tan, 2001). The minimum modifications required for the swing cuts approach allow more optimization opportunity and possible blending of different operating modes. Despite the improvement of the swing cut model, the model does not reflect the nonlinearity of the process, but it provides an incentive to further improve the planning model and calculate more accurate yields.

The CDU model can be upgraded from the linear swing cut equation using an aggregate model approach based on the work of Caballero & Grossmann (1999) for the synthesis of distillation columns. The principle of their approach is to treat the column sections above and below the feed tray as two integrated heat and mass exchangers. This aggregate representation, which includes a modest number of nonlinear equations, reflects the nonlinear nature of the process without the increasing computation or complexity of conventional short cut models or rigorous tray-by-tray model (Suphanit, 1999). Furthermore, this approach can serve as a precursor to more complex nonlinear models.

The aggregate model used applies to a typical distillation column, with a condenser, a reboiler and only top and bottom products. However, the CDU is a complex distillation column with multiple side streams, side strippers and condensers. It also depends on steam stripping and uses no bottom reboiler. Therefore, the aggregate model cannot be applied in its simple format directly to the CDU. This limitation applies also to the typical Fenske-Underwood-Gilliland (FUG) shortcut model (Suphanit, 1999). To overcome this limitation, the CDU is represented as a set of simple and thermodynamically equivalent cascaded distillation columns. The cascaded approach will be used for the aggregate model, as well as future nonlinear modeling approaches.

A key to a successful nonlinear programming model is good initialization. It helps the model to find a solution and converge. The initialization schemes are based on the understanding of the problem and foundation principles. For the CDU aggregate model, the adapted initialization scheme involves two stages, solving an LP model followed by a simple NLP model. The results of the second initialization stage serve as the initial data for the aggregate model. The two stage initialization scheme allows the aggregate model to converge and give a solution. Generating this initialization scheme allowed for analysis of the cascaded columns model and identifying additional constraints. These constraints are added to the aggregate model to further define the feasible region of the problem.

The developed NLP-based aggregate model for the CDU is integrated into a refinery production planning model. The results of the new NLP production planning model are compared with the current LP models, namely fixed yield and swing cuts based models. The benefits of introducing the NLP model are assessed in terms of accuracy, robustness and complexity.

[1] Caballero, J.A.; Grossmann, I.E. (1999). Aggregate models for integrated distillation systems. *Industrial & Engineering Chemistry Research, 38(6), 2330-2344.*

[2] Favennec, JP. (2001). Refinery Operation & Management. *Editions Technip.* Paris.

[3] Li, W.; Hui, C.W.; Li, A.X. (2005). Integrated CDU, FCC and product blending models into refinery planning. *Computers and Chemical* Engineering, *29, 2010-2028.*

[4] Pelham, R.; Pharris, C. (1996). Refinery operation and control: a future vision. *Hydrocarbon Processing, 75(7), 89-94.*

[5] Suphanit, B. (1999). The design of complex distillation systems, PhD Thesis. *UMIST.* Manchester.

[6] Trierwiler, D.; Tan, R.L. (2001) Advances in crude oil LP modeling. *Hydrocarbon Asia, 8, 52-58*

[7] Zhang, J.; Zhu, X.X.; Towler, G.P. (2001). A simultaneous optimization strategy for overall integration in refinery planning. *Industrial & Engineering Chemistry Research, 40, 2640-2653.*

733b Planning Parent Seed Production

Matt H. Bassett, Dow AgroSciences, 9330 Zionsville Rd., Indianapolis, IN 46268

Parent seed production involves the process of selecting and bulking up inbred (parent) seed in sufficient quantities to produce hybrid seed for sale to customers. Given the large number of inbreds being screened yearly to support the ever-increasing number of hybrids being sold, it was determined that a model was needed to track the whole production process. This talk will present a mixed-integer linear program that was developed to model the parent seed production process within Dow AgroSciences. Specific details of our actual inbreds and hybrids will not be discussed, but issues involving data collection/generation, model formulation, model solution, and interpretation of the results will be discussed.

733c Integrated Economic Optimization and Model Predictive Control of Bio-Fuel Production Plant

Carl A. Schweiger and Bijan Sayyar-Rodsari. Pavilion Technologies, 10415 Morado Circle, Building III, Suite 100, Austin, TX 78759-5638

The goal of the economic optimization of a bio-fuel manufacturing plant is to determine the most profitable way to operate the plant based on the current market and plant conditions. This optimization must take into account the complex interactions among the economic factors, the process interconnections, and the operational constraints. In order to achieve the economically optimal operation, the economic optimization must be integrated with the the underlying control system. Despite attracting significant interest from both academia and industry, the implementation of a truly integrated economic optimization and automated control solution for the production plant has proven a major challenge. The fundamental challenge lies in the development of consistent, accurate models for both the economic optimization and the automatic control layers in order to ensure that the optimal operational settings determined by the economic optimization can be realized by the controller.

This work describes the development of an integrated plant-wide economic optimization and Model Predictive Control (MPC) strategy for the economic optimization of an ethanol manufacturing plant. The plant-wide optimization takes into account the various sections of the ethanol plant: milling and slurry, fermentation, distillation, dehydration, separation, and drying. The goals of the economic optimization are to determine operational conditions that balance production (faster rates) with yield (slower rates) to maximize profit, utilize resources to minimize costs, and maintain feasible operation in all sections of the plant. The control layer uses nonlinear MPC to optimize the dynamic operation of the individual units given the operational directives provided by the economic optimization.

The approach uses a flexible, parametric hybrid modeling framework that enables the transparent inclusion of individual models developed for the various components of the process into a complex, modular plant-wide model that is suitable for the economic optimization and can be easily managed. This approach maintains the consistency between the plant-wide model used for economic optimization and the models developed for the MPC of the individual units while allowing additional process information to be included in the economic optimization model.

The solution is used for both online operation as well as for off-line scenario analysis and is implemented in the Pavilion8 software package. This tool allows direct control and provides tools for analyzing the behavior of the economic optimization and MPC systems.

733d Peninsular Malaysia Operation Gas Fields Flow Network Optimization Model

Shahrul Azman Zainal abidin and Wai Chong Low. Process Automation and Optimization, Group Technology Solution, PETRONAS, Level 4, Menara Dayabumi, Jalan Sultan Hishamuddin, 50050, Kuala Lumpur, Malaysia

A rigorous compositional analysis model was developed to optimize the gas fields flow network to cater for gas processing plant feed composition constraints. The Linear Programming (LP) technique model was developed and linked to a thermodynamic engine, iCON, a PETRONAS own steady state process simulation software to give accurate phase equilibrium and flashing calculations. The thermodynamic engine also can detect any condensation of traveling gasses in the networks and provide the physical properties of the fluid commingled from different wells.

The LP Optimization model was developed based on current gas network configuration and maximum gas network flexibilities to evaluate the value creations that the gas network facilities can generate. Process modifications to cater for different modes of operation together with the economic analysis are also discussed in this paper.

aspentech

Enabling model based decision making by sharing consistent equation oriented dynamic models between multiple simulation and optimization environments

Ajay Lakshmanan
Manager, R&D

aspentech

Consistent Model-Based Decision Making

Overview

- Why do we need to use consistent models?
- How do we enable use of consistent models?
  - Common computational engine
- How do we enable consistent dynamic models?
  - Enhance common computational engine
- Sample results of the implementation
  - Dynamic batch distillation model in Aspen Plus
  - Custom dynamic tank model in Aspen HYSYS and Aspen HYSYS Dynamics

aspenONE

aspentech

Model-Based Decision Making

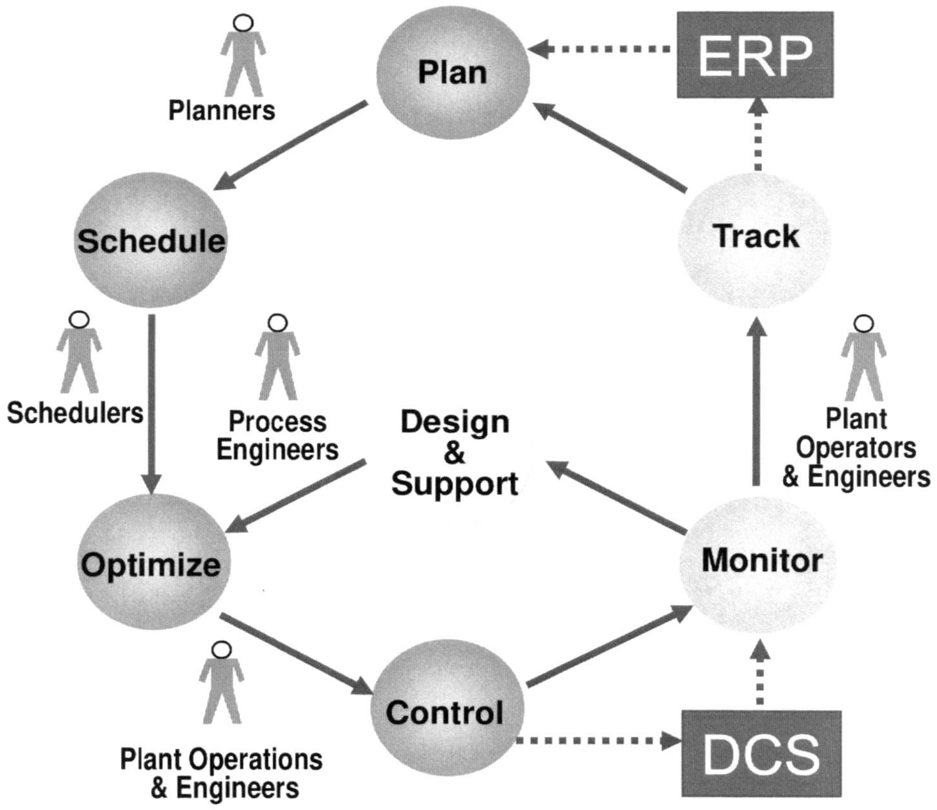

aspenONE

aspentech

Why Do We Need To Share Models?

- Consistent basis for making decisions
  - Reuse rigorous models.
- Reliability of the predictions
- Speed of deployment
- Better knowledge management
  - Standardized work process: Models created and maintained by a group; Distributed to all users
  - Lower cost. Reduces need to have experts at every site

aspenONE

aspentech

Consistent Model-Based Decision Making

Overview

- Why do we need to use consistent models?
- How do we enable use of consistent models?
  - Common computational engine
- How do we enable consistent dynamic models?
  - Enhance common computational engine
- Sample results of the implementation
  - Dynamic batch distillation model in Aspen Plus
  - Custom dynamic tank model in Aspen HYSYS and Aspen HYSYS Dynamics

aspenONE

aspentech

Consistent Model-Based Decision Making Common Computational Engine

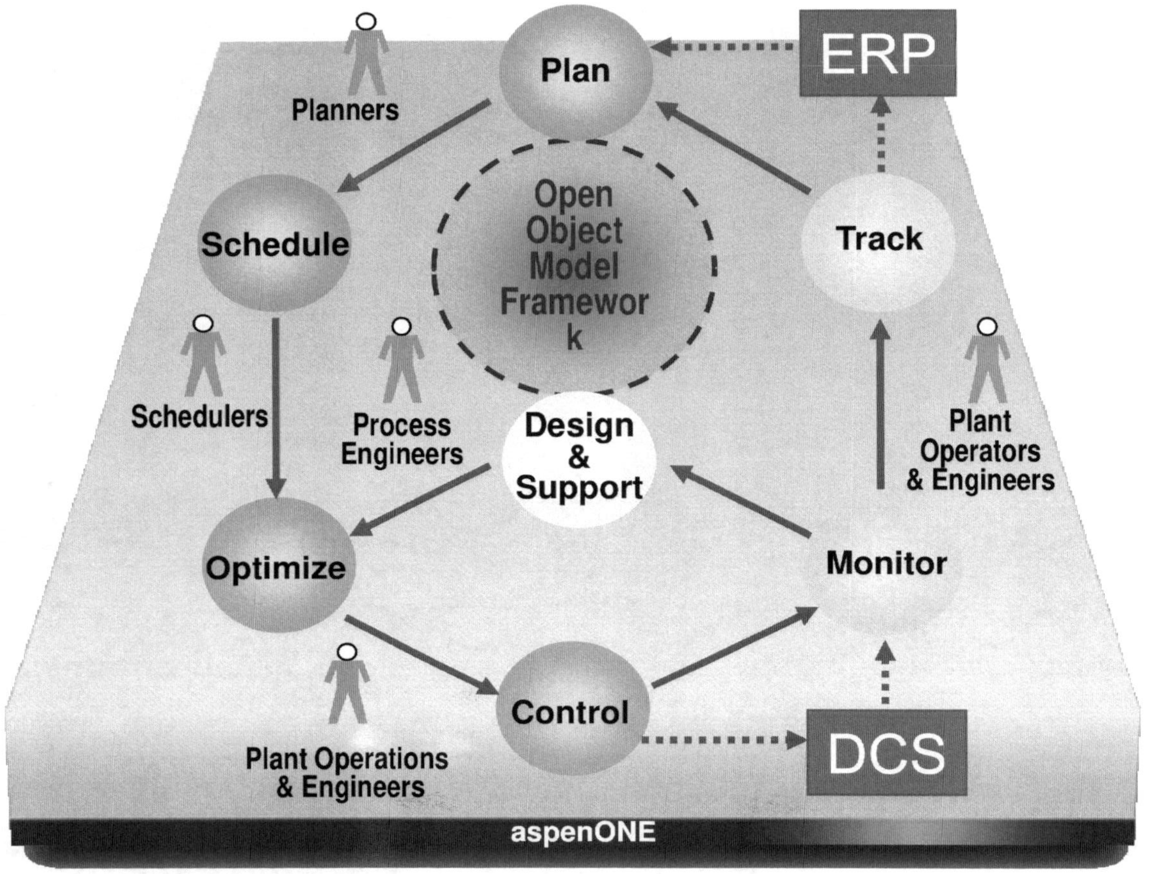

aspenONE

Consistent Model-Based Decision Making
Common Computational Engine

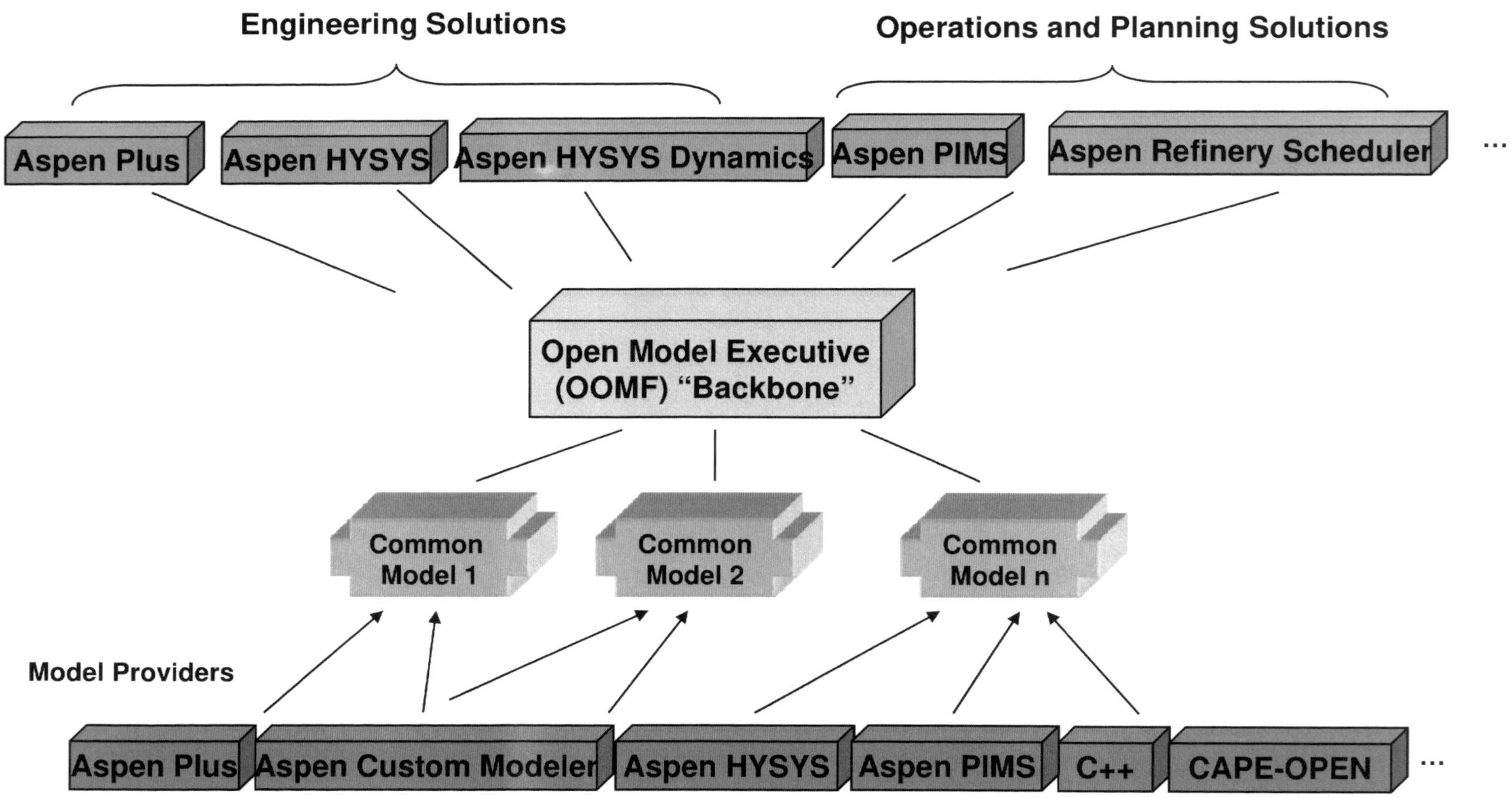

Consistent Model-Based Decision Making

Custom Steady State models can already be shared

- Since 2002 steady state models developed using Aspen Custom Modeler can be seamlessly integrated into process plant models in Aspen Plus, Aspen HYSYS etc.
  - Solve in Sequential Modular and Equation Oriented solution modes
  - Icons, tables, custom forms, Visual Basic scripts are exported and available in Aspen Plus, Aspen HYSYS
  - Model variables are accessible in design specifications, sensitivity calculations, calculator blocks

Consistent Model-Based Decision Making

Overview

- Why do we need to use consistent models?
- How do we enable use of consistent models?
  - Common computational engine
- ▶ How do we enable consistent dynamic models?
  - Enhance common computational engine
- Results of the implementation
  - Dynamic batch distillation model in Aspen Plus
  - Custom dynamic tank model in Aspen HYSYS and Aspen HYSYS Dynamics

Consistent Model-Based Decision Making

Rigorous Dynamic models can now be shared

- Author dynamic model in high level modeling environment
  - Aspen Custom Modeler
- Integrate the dynamic model with overall process model in
  - Aspen Plus and Aspen Plus Dynamics
  - Aspen HYSYS and Aspen HYSYS Dynamics
- Workflow enabled by common computational engine
  - Open Object Model Framework (OOMF)
  - Aspen Open Solvers (AOS)

Consistent Model-Based Decision Making

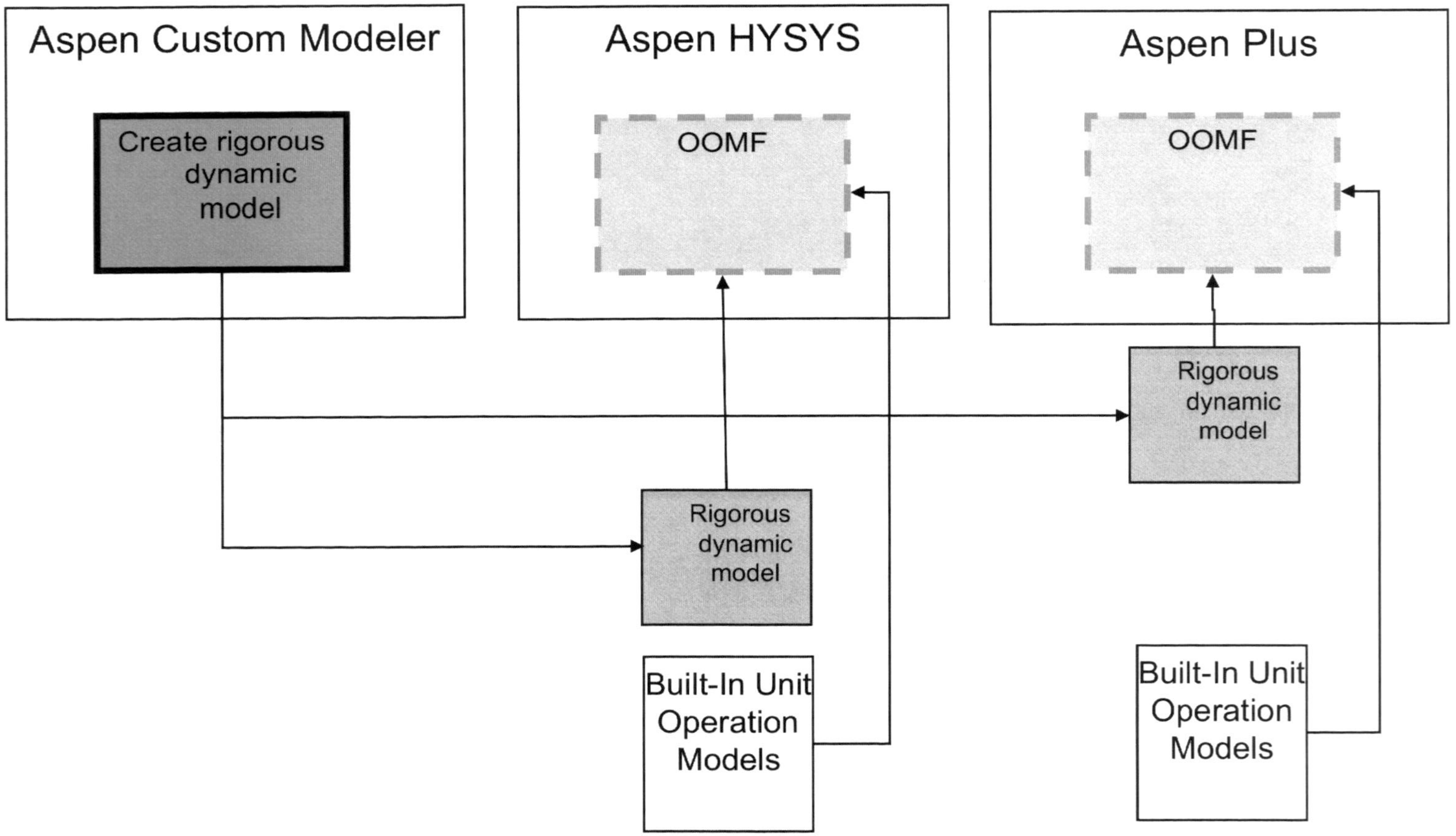

aspentech

Consistent Model-Based Decision Making
OOMF – Common Computational Engine

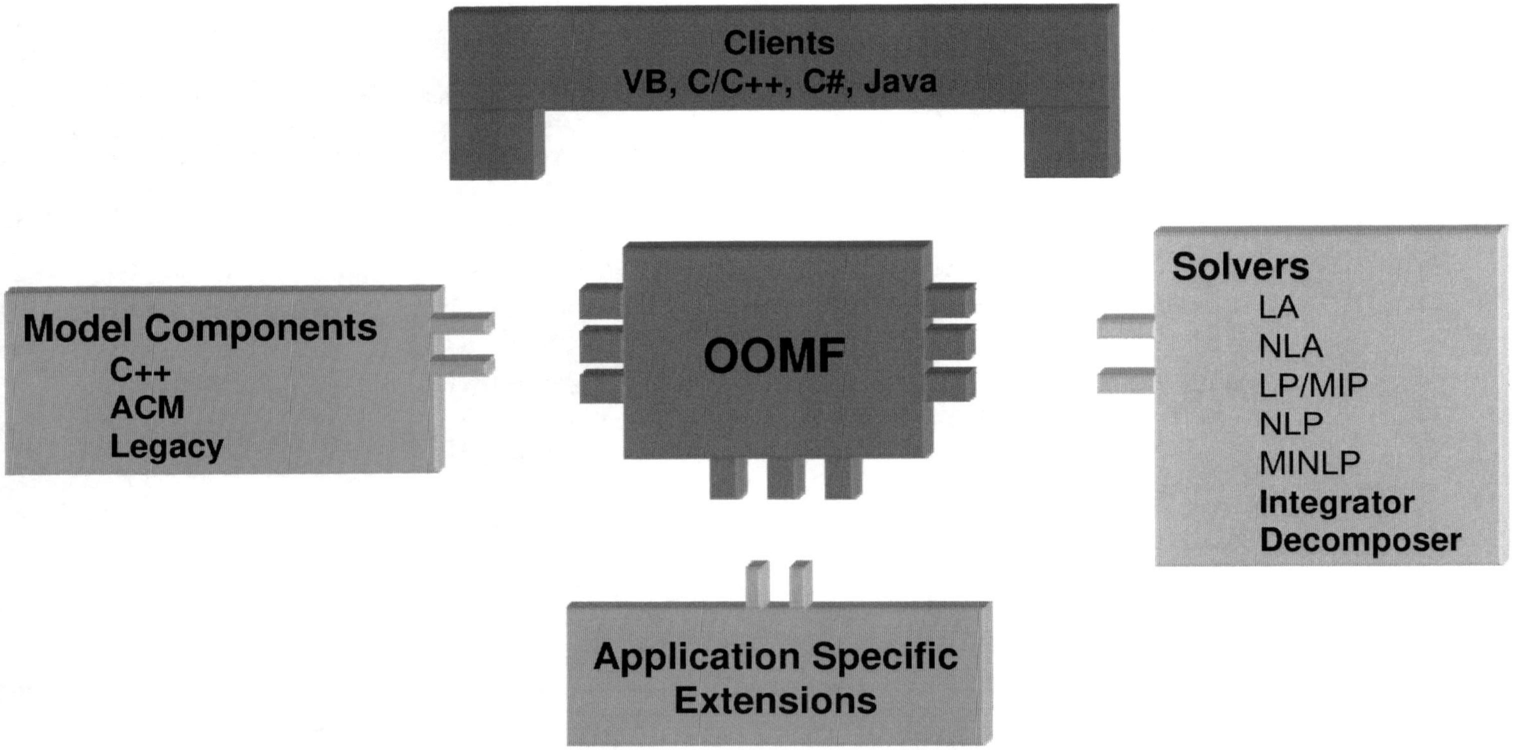

aspenONE

aspentech

Consistent Model-Based Decision Making OOMF – Dynamic Modeling Support

Drive the dynamic model through time

- Start, pause, re-start, and reset
- Finite State Machine
  - Manage and control the multiple states from start to end
- Task manager
  - Load, activate, parse and interpret configured tasks
- Event Manager
  - Explicit events - step the simulation through time
  - Implicit events - conditions, actions, monitors
- Data historian
  - Record, View variable time profiles; Save snapshots
- Open Solver driver
  - Differential Algebraic Equation System object

aspenONE

Consistent Model-Based Decision Making
Aspen Open Solvers

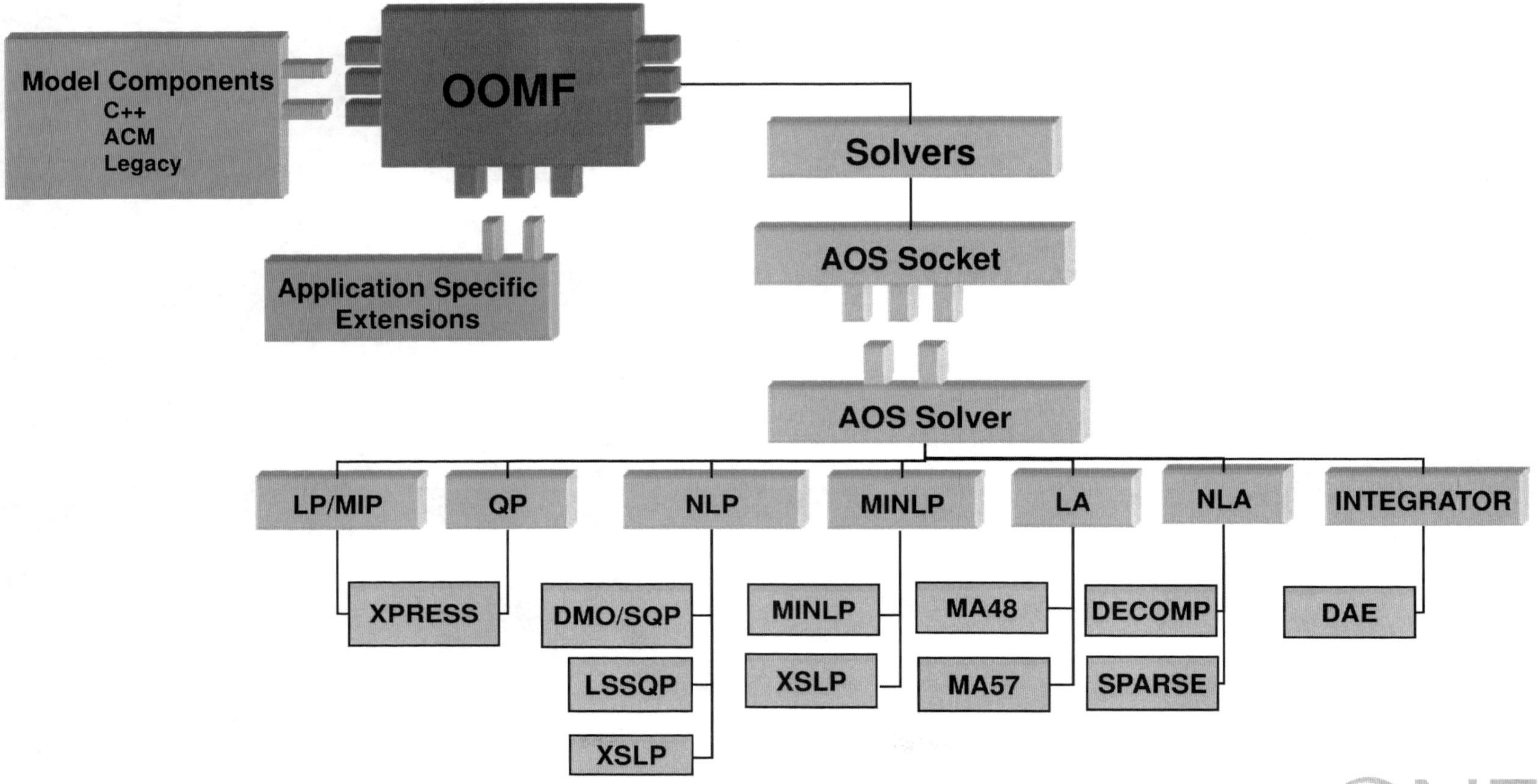

Consistent Model-Based Decision Making
Open Solver - Dynamic Modeling Support

Solve the dynamic model

- Aspen Open Solver Integrator
  - Explicit Euler, Implicit Euler, Runge-Kutta, Gear
- Interfaces and implementation for
  - Differential Algebraic Equation System Object
  - Group Decomposition
  - Tearing
  - Homotopy
- Event location
  - Discontinuities, Tasks
- Diagnostic Reports

Consistent Model-Based Decision Making

Overview

- Why do we need to use consistent models?
- How do we enable use of consistent models?
  - Common computational engine
- How do we enable consistent dynamic models?
  - Enhance common computational engine
- Results of the implementation
  - Dynamic batch distillation model in Aspen Plus
  - Custom dynamic tank model in Aspen HYSYS and Aspen HYSYS dynamics

Consistent Model-Based Decision Making
Results of the Implementation

Author the dynamic model in Aspen Custom Modeler.

- Create a dynamic model based on proprietary knowledge or chemical engineering literature
- Integrate dynamic model with overall process model in Aspen Plus and Aspen HYSYS.
- Two examples follow
  - Batch Distillation Model authored in Aspen Custom Modeler; integrated within an Aspen Plus process model
  - Custom tank model authored in Aspen Custom modeler; integrated within an Aspen HYSYS and Aspen HYSYS Dynamics model

Consistent Model-Based Decision Making Aspen Batch Distillation

- Aspen Batch Distillation is a dynamic model developed using Aspen Custom Modeler
- Used for design, analysis and optimization of batch distillation processes

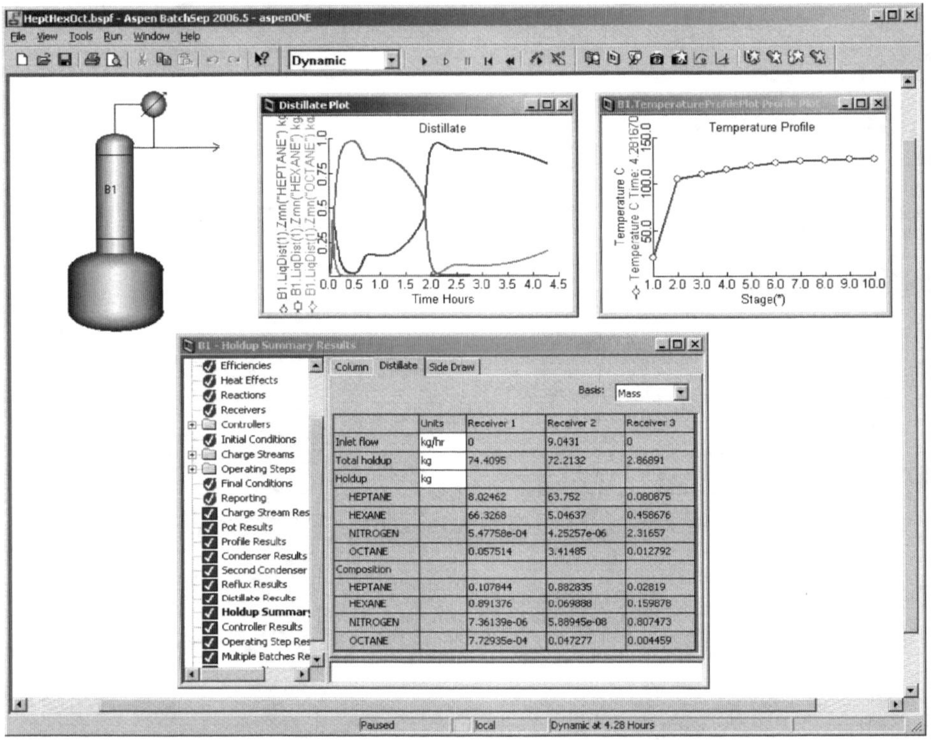

Consistent Model-Based Decision Making Aspen Batch Distillation

- Developed using Aspen Custom Modeler
- Uses Aspen Properties
  - Rigorously model two-phase and three-phase columns
- Integrates a large DAE system
  - 1000 to100000+ equations
  - Implicit Euler or Gear solution methods
- Uses tasks to model a sequence of batch operations
- Rich user interface for configuring column and operating steps

Consistent Model-Based Decision Making
Aspen Batch Distillation Sample Model

- Water-methanol separation

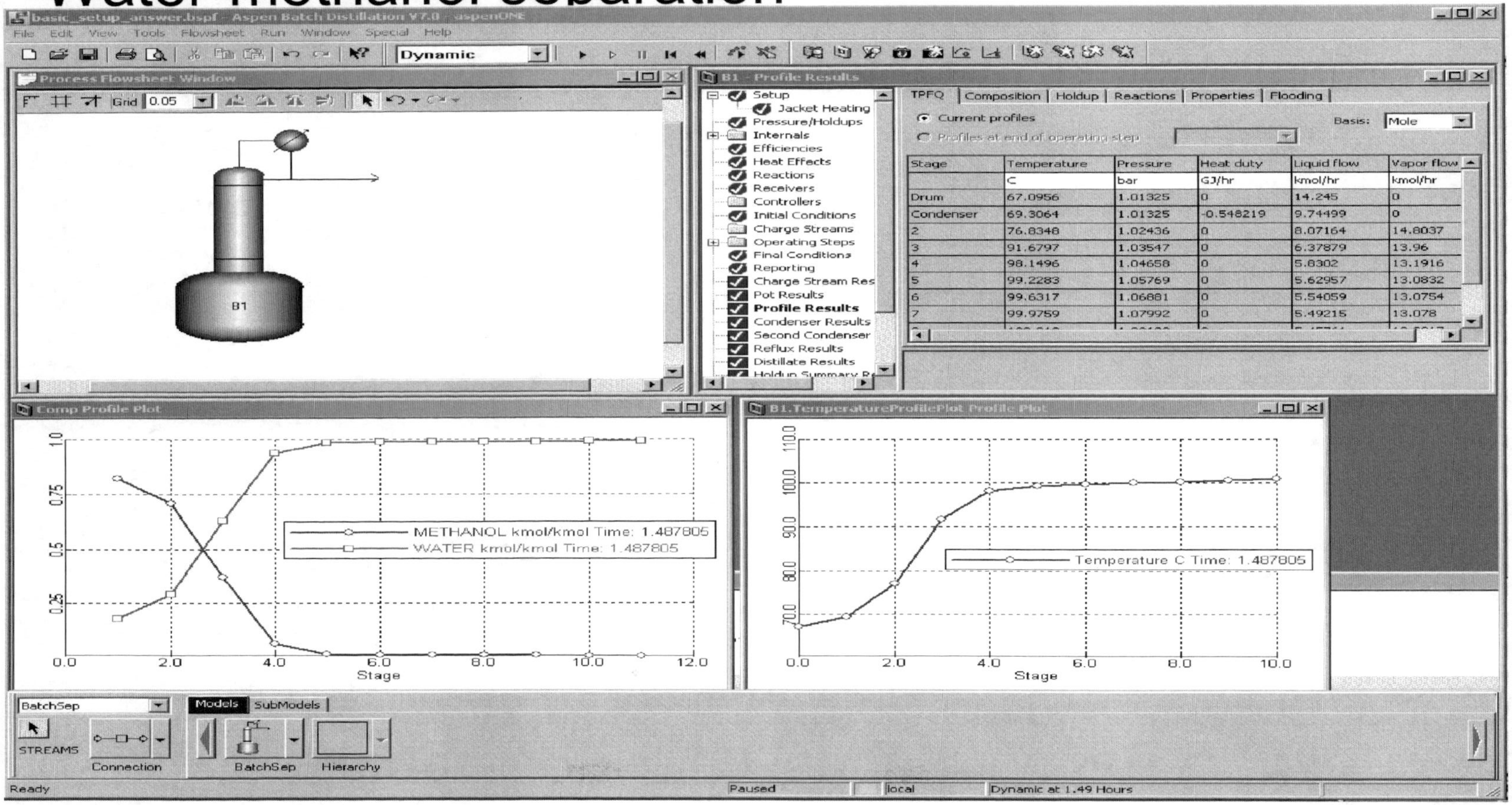

aspenONE

Consistent Model-Based Decision Making Integration into Aspen Plus Via OOMF

- Aspen Batch Distillation Model now also available as a unit operation in Aspen Plus

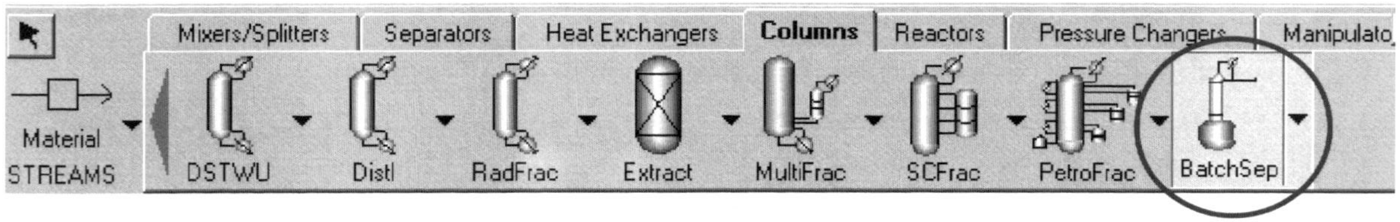

- Integration with Aspen Plus enables simulation and optimization of:
  - Batch distillation sequences
  - Entire process including other batch or continuous unit operations

Consistent Model-Based Decision Making Integration into Aspen Plus Via OOMF

- Connect to other models and optimize the process
  - Notional buffer tanks at inlets and outlets

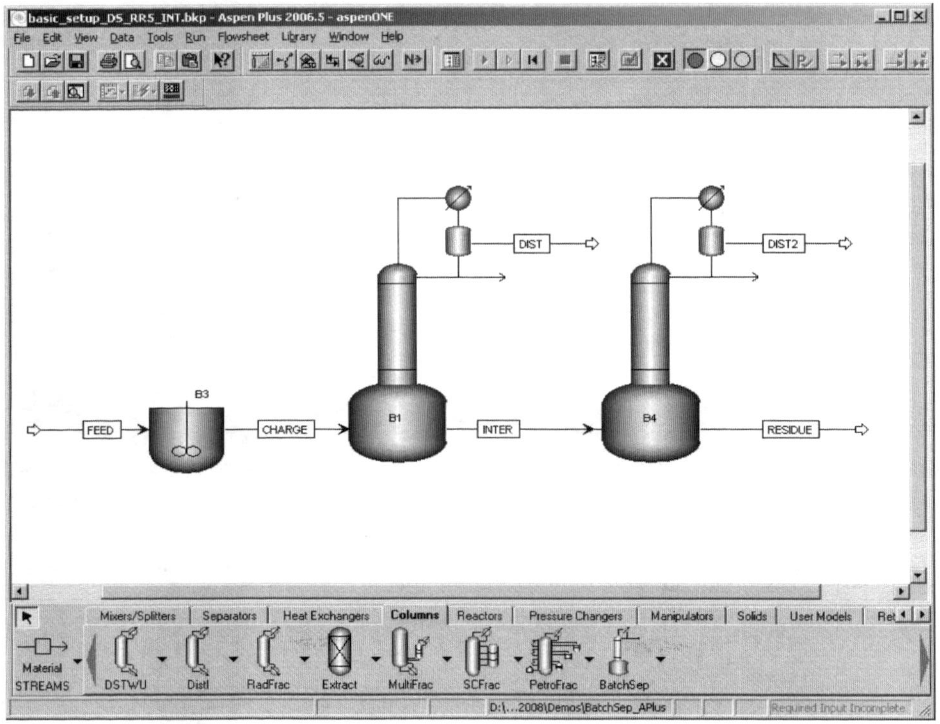

Consistent Model-Based Decision Making Integration into Aspen Plus Via OOMF

- Water-methanol separation using Aspen Plus

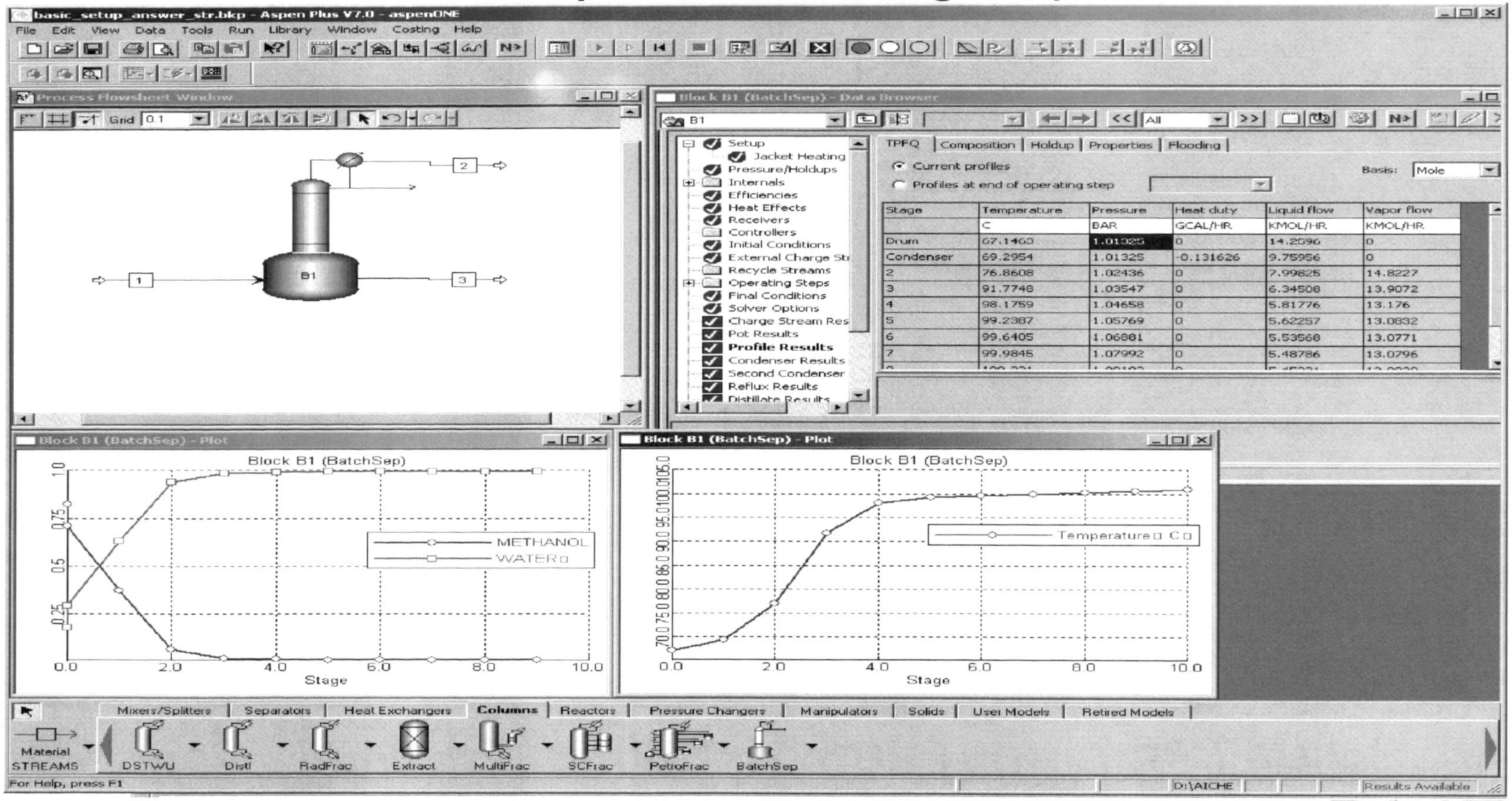

Consistent Model-Based Decision Making

Overview

- Why do we need to use consistent models?
- How do we enable use of consistent models?
  - Common computational engine
- How do we enable consistent dynamic models?
  - Enhance common computational engine
- Results of the implementation
  - Dynamic batch distillation model in Aspen Plus
  - Custom dynamic tank model in Aspen HYSYS and Aspen HYSYS dynamics

Consistent Model-Based Decision Making Custom Dynamic Tank Model

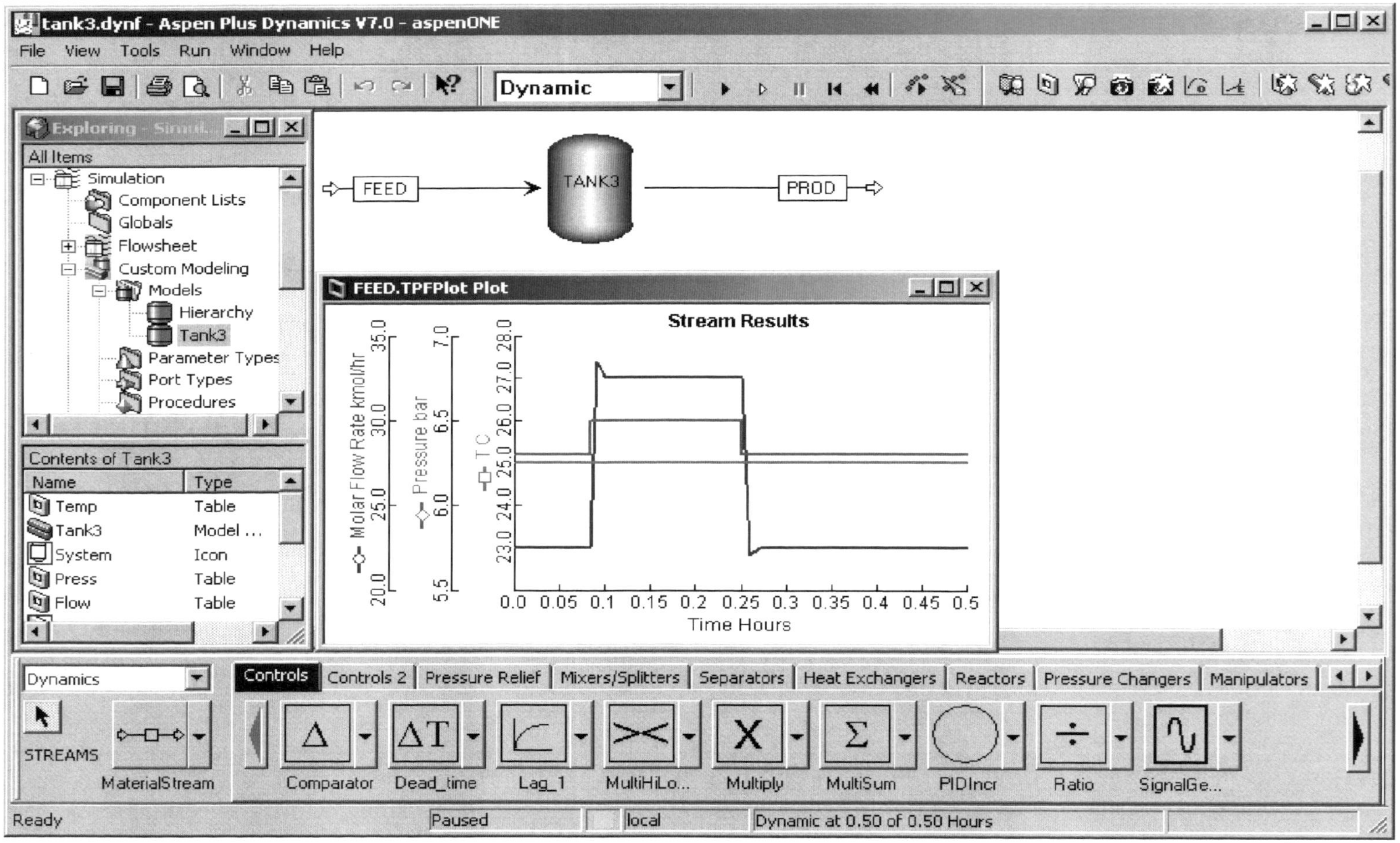

Consistent Model-Based Decision Making Integration into Aspen HYSYS Via OOMF

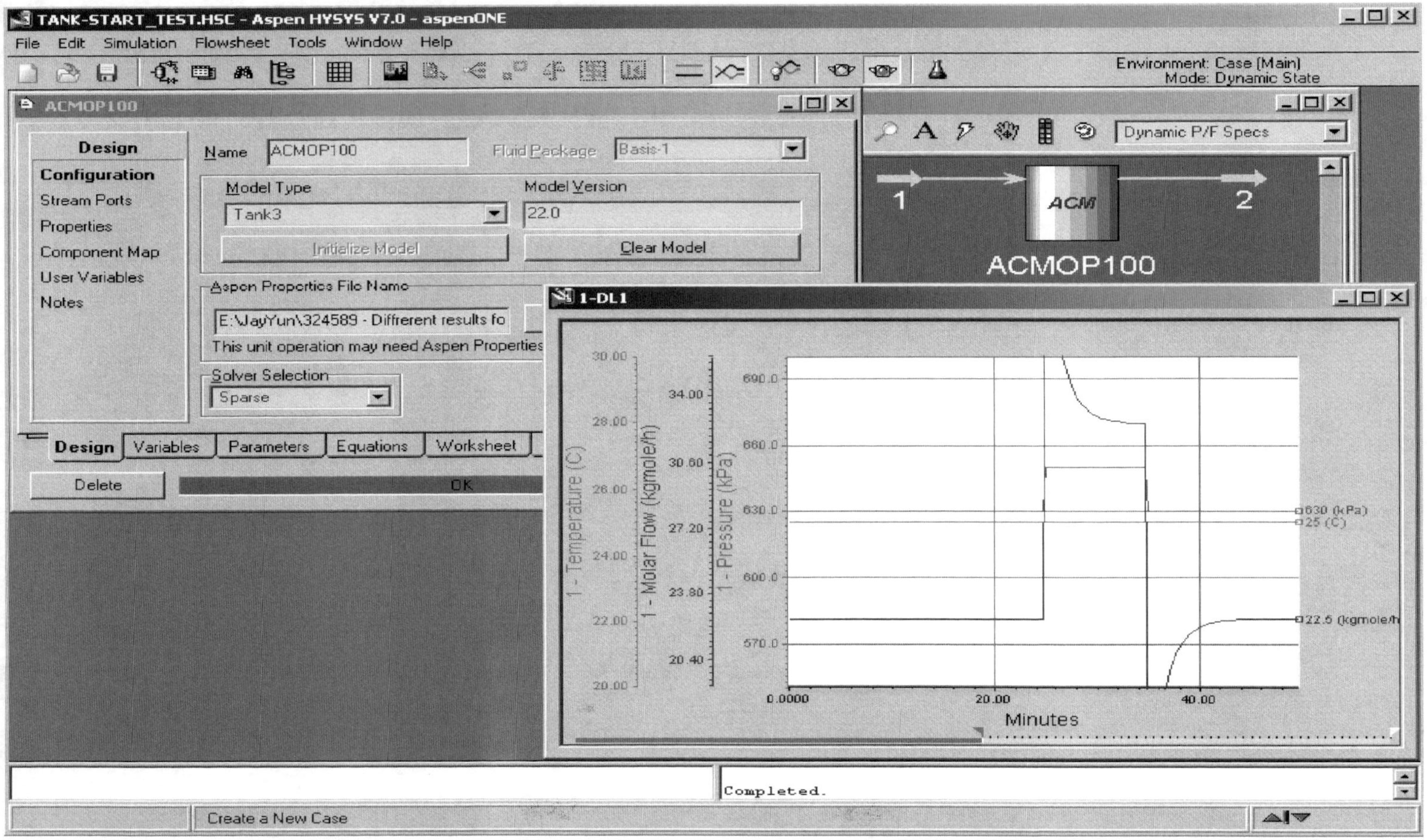

aspentech

Consistent Model-Based Decision Making

Dynamic models can be shared for better decision making.

- Author dynamic model in high level modeling environment
  - Aspen Custom Modeler
- Integrate the dynamic model with overall process model
  - Aspen Plus and Aspen Plus Dynamics
  - Aspen HYSYS and Aspen HYSYS Dynamics
- Improve process model accuracy, reliability and deployment.
  - Better knowledge management
- Workflow enabled by common computational engine
  - Open Object Model Framework (OOMF)
  - Aspen Open Solvers (AOS)

aspenONE

Consistent Model-Based Decision Making

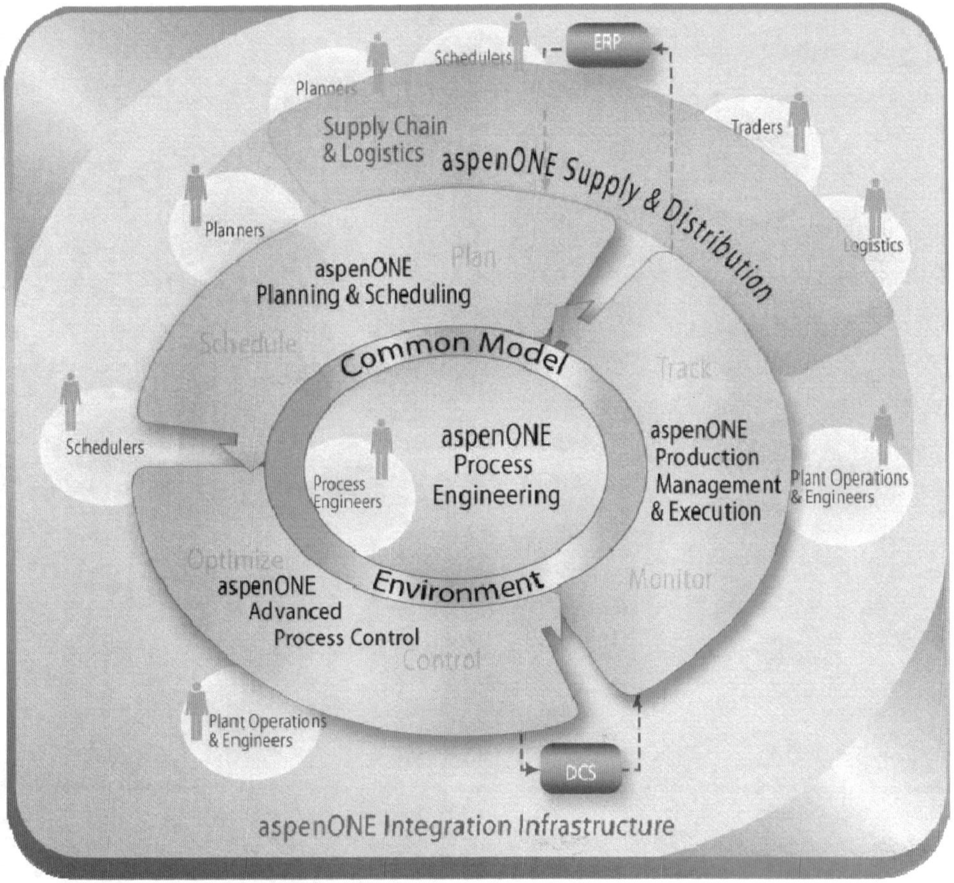

- Common Model Environment
  - Enables use of consistent models in Engineering, Planning and Scheduling, Advanced Control, Optimization, and Operations
- Sample implementations of sharing of dynamic model in engineering were described in this paper.
  - The next paper will describe how similar models are used across Engineering, Planning and Scheduling, and Operations in the refining industry

Acknowledgements

- Ashok Bhakta
- Gabriel Lopez-Calva
- James Fielding
- Ajay Modi
- James Goom

733f Refinery Lpg Transfer and Storage Scheduling. A Genetic Algorithm Approach

Roger Rocha, Logistic Division - Research and Development Center, PETROBRAS, Av. Horácio de Macedo 950, Ilha do Fundão, Rio de Janeiro, Brazil and Marcus V. S. Poggi de Aragão, Dept. of Informatics, Pontific Catholic University – PUC-Rio, R. Marques de São Vicente 225, Rio de Janeiro, Brazil.

A common refinery's objective is to ally operation safety and product quality to the process profitability. In this sense, computational resources in distribution logistics, planning and scheduling activities of petroleum refineries have been presented in the context of the Operational Research since the 1950's through the application of linear programming (LP) techniques [1], which remain as the most currently used mathematical programming approach for modeling this type of chemical plant. We tried to solve this problem using a Mixed-integer programming approach [2], but it turned out after incorporating all the specification of the problem this alternative became impracticable. In this work, we address the LPG transfer and storage scheduling in petroleum refinery by means of a genetic algorithm approach. Moreover a solution representation based on real numbers allows a continuous representation of the time leading to a more realistic solution.

The Liquefied Petroleum Gas (LPG) transfer and storage operations at the Capuava Refinery (RECAP) in São Paulo are a challenged problem due to the large amount of product transferred and its lack of storage capacity. Furthermore, the time windows constraint imposes an additional difficult to find a good solution without recourse to a computational tool.

The usual approach to solve scheduling problem by genetic algorithm is the discretization of the time and associate chromosomes to activities where each gene represents the state of an equipment in a given time [3]. Another important issue that arises when applying genetic algorithm is the feasibility of the solution generated. There are a lot of techniques to deal with infeasible solution, but as a rule of thumb this is something that should be avoided. Our algorithm overcomes these two problems pointed out by using a solution codification using real number and an algorithm for constructing solutions guaranteed feasible.

Our model was applied in some test instances of the problem and the results show a great improvement compared to a random search. Moreover, as the construction algorithm is based on the knowledge of the specialists, the solutions have more chance to be accepted in the refinery. Another advantage of our approach is the consideration of production specification in the constraint set of the problem can be made nearly in a transparent way.

References

[1] Symonds, G.H. (1955). Linear Programming: The Solution of Refinery Problems. New York, Esso

Standard Oil Co.

[2] Joly, M. ; Aires, M. ; Rocha, R. ; Smania Filho, P. (2005). Short-term Scheduling of the RECAP Refinery LPG area: A Modeling and Solution Algorithm. ENPROMER – 2nd Mercosur Congress on Chemical Engineering and 4th Mercosur Congress on Process Systems Engineering, Rio de Janeiro.

[3] Sikora, R. (1996). Integrating the Lot-sizing and Sequencing Decisions for Scheduling a Capacitated Flow Line. Computers Industrial Engineering, 30, 4, 969-981.

Implementation of Predictive Functional Control on a high pressure distillation column

B. Satriadarma, H. Arellano-Garcia, G. Wozny
Chair of Process Dynamics and Operation, Sekr. KWT-9, Berlin University of Technology,
Str. des 17. Juni 135, 10623 Berlin, Germany.
Tel. +49 (0)30 314 29640, Fax: +49 (0)30 314 25996, e-mail: bsa@zmms.tu-berlin.de

Introduction

In the last decade the meaning of **M**odel **P**redictive **C**ontrol (MPC) for the industries has been significantly increased, particularly in petrochemicals industries, as well as within the chemical engineering field. These based on the fact that MPC is suitable for the regulation of complex systems.
One of the advantages of the MPC is the robust control behaviour in term of controlling processes with a distinct ratio between its time constant and its dead time. Another advantage of MPC is its adaptable control algorithm. This means MPC can handle processes with changing steady state point due to the fluctuation of the resource composition.
Furthermore constraints and other additional operational conditions can be applicated in the control algorithm with small effort.
However, the implementation of the advance control algorithm fails in practice because of lack of acceptance from the plant operator. The complexity of the advance control algorithm and the low understanding of the control behaviour are presumed to be the root of the lack of acceptance.
In this work, we introduce a first approach of the linear PFC implementation on a complex chemical process. A binary azeotropic thermal separation process on a distillation column is addressed. Moreover, an alternative method of concentration control to preserve a given product specification is used, based on the thermodynamic character of the mixture.
The main objective of this work is the development of the PFC algorithm for complex control tasks in consideration of the influencing variable (disturbance handling), constraint handling and using a linear / nonlinear model. In addition, visualization of optimization-based process data with the consideration of multicriterion objectives will be discussed based on the process considered.

Problem formulation

As we all know rektification / distillation process is a high non linear process. A small changes on one variable (e.g. feed-flow rate or pressure drop) will affected the dynamic of the whole column. Maintaing the product quaity and quantity in a distillation process is a common problem that can be solved with controllers. The question is how good does the controller work ? A good controller can maintain the given set point during operation, and a bad (poorly adjust) controller can also achieve almost the same result. So what makes a good controller difference from a bad controller ? A good controler can preserve the process output or **C**ontroled **V**ariable (CV) close to the reference value (set point) the entire time, with minimum cost of the actuator / **M**anipulated **V**ariable (MV).
For example we look at a case in which the controller (PI) should maintain the temperatur at the given value 110°C. As we can see, the controller can hold the CV at the set point for most

time, although there is some changes in the dynamic (pressure lost changes). The problem appreas at ca. t = 120 *min,* when the column has reached a new steady state point. Regarding that the controller is not designed for this new steady state point, it can compensate the model missmatch for a certain time, with a high cost of the MV(heat duty)

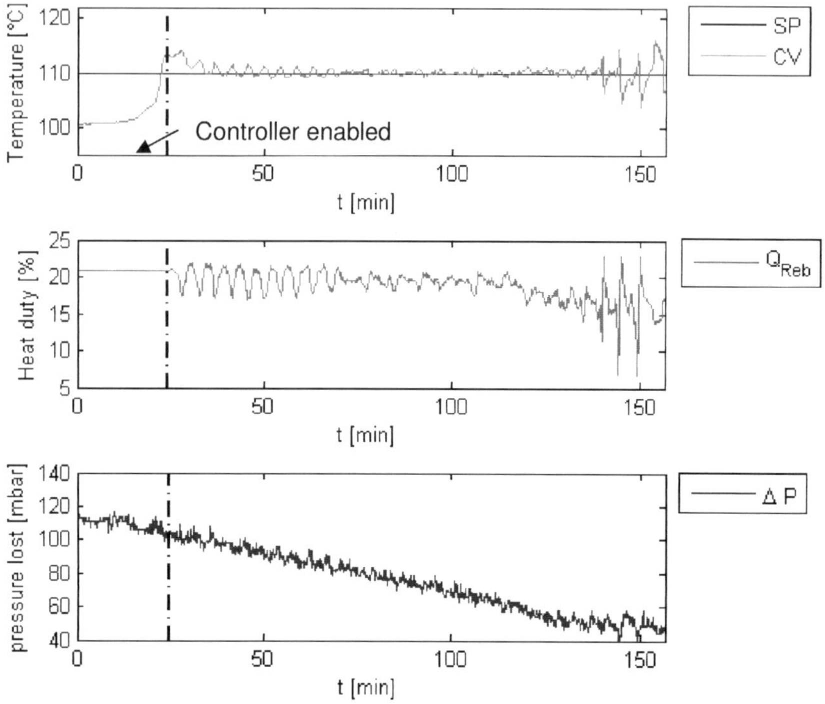

Figure 1. PI(D) control behaviour on a pressure drop change

MPC introduce an optimal control algorithm due its ability to predict the future behaviour of the systems and calculated a set of optimale MV.One disadvantage of MPC is its optimzation algorithm, which demands a high computational effort to predict the future and to calculate the MV. This lead to the second problem: how to modify the MPC algorithm, which is suitable for fast processes.
Finding a proper internal model for MPC is another problem that appears automatically with the implementation of MPC algorithm. In other words, this study focused on three main problems:

- Controlling product composition in a distillation process (concentration controlling)
- Investigation and analysis of the dynamic behaviour of the column (modelling)

However, the implementation of **A**dvanced **P**rocess **C**ontrol in the industry fails in the practice because of lack of acceptance from the plant operator.
The complexity of the advance control algorithm and not understanding the control behaviour presumed to be the root of the lack of acceptance. A further cause analysis has been done in [11].
While 80% of the conventional automation systems consist of standard blocks, however for the graphic integration of APC in operating specific control interface there are still no standard form established. As a consequence, the Human Machine Interface in the automation system has a limited usability, which is especially caused by the absence of process transparency and self explaining function.
The consequences of these problems lead up into two typical behaviours [10]:

- Distrust leads to neglect and/or non-usage of the systems.
- Excessive confidence can decrease the plant operateur's situation awareness, which lead up to losing his ability to take a proper action in case of a failure.

Solution

Predictive Functional Control (PFC)

In the late 80s, Richalet presented the Predictive Functional Control (PFC) method, which based on the model predictive control (MPC) principle. PFC has an undermanding algorithm, is faster, and offers higher control precision than normal model-based predictive control principles. Ability to use a linear or non linear model and the constraint handling are two from many essential advantages in comparison to the conventional controller. In praxis, PFC has been successfully applied to many different control areas [1],[2].
The main idea of PFC algorithm or the big difference between PFC and MPC algorithm is that, PFC calculate only one future MV, not like the common MPC , which calculate a set of future MV's behviour. So that the high cost of computation of the optimization algorithm and/or iterative methods, which is needed for the solution of QP-problems, is omitted [3]

Figure 2 shows, the structure of PFC in a control loop. It shows also that PFC algorithm consists of two main parts:

1. *internal model,*
   the internal model simulates the (future) behaviour of the process with the same given input MV_{lim}. A short introduction about the internal model will be discussed later in this chapter
2. *controller/control algorithm,*
   in this block, the controller calculates the optimal MV based on the current model output $Y_m(n)$, current system output $CV(n)$, past system output $CV(n\text{-}1)$ and set point $SP(n)$

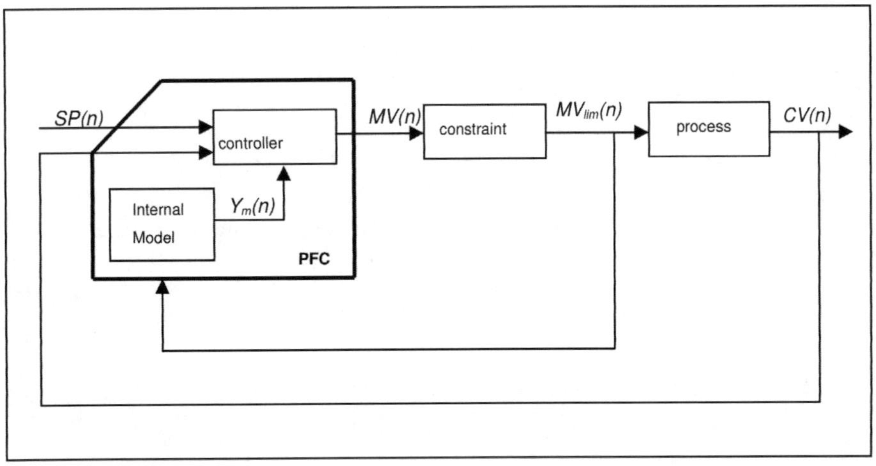

figure 2. PFC control loop

With:

$SP(n)$ - setpoint
$MV(n)$ - manipulated variable

$MV_{lim}(n)$	- limited / constrained MV
$CV(n)$	- controlled variable
$Y_m(n)$	- model output

internal model

PFC describes the internal model of the system as a discrete differentation equation. The main advantage of having the model in this state is, that the future behaviour of the system can be calculated/predicted using models's current and past output. In the example below a first order system will be described as ODE (**O**rdinary **D**ifferential **E**quation)

Example:

First order system in continuous state

$$\tau \frac{dy_m(t)}{dt} + y_m(t) = K \cdot MV(t) \tag{1}$$

Discretization through the Z-Transformation as a transfer function (eq.(2)) and as a discrete differential equation (eq.(3))

$$G(z) = \frac{Y_m(z)}{MV(z)} = \frac{K \cdot (1-\alpha) \cdot z^{-1}}{1-\alpha \cdot z^{-1}} \tag{2}$$

$$y_m(n) = y_m(n-1) \cdot \alpha + K \cdot (1-\alpha) \cdot MV(n-1) \tag{3}$$

with

$$\alpha = e^{-\frac{\tau_A}{\tau}} \tag{4}$$

Whereat τ_A the sampling time and τ the system's time constant

control algorithm

The assumption that the model output y_m by the time $(n+h)$ identical to the process output CV (eq. XX), is the initial point for calculating the future $MV(n+h)$, whereat the prediction horizon $h = 1$.

The current MV can be described as a function from the current set point, current process output CV, current model output ym and the decrement factor of the discrete reference trajectory β

$$MV(n) = \frac{[SP(n) - CV(n)] \cdot (1-\beta^h) - y_m(n) \cdot (1-\alpha^h)}{(1-\alpha^h) \cdot K} \tag{5}$$

For the specify calculation of the MV can be found in [3] and [4]

Prospective design of the Graphical User Interface (GUI)

As written above, the implementaion of APC fails in the practice due to low/lack of acceptance of the plant operateur, whose works should be actually supported by the APC. The initial point of this work is the hypothese

A transparent display of the system effects both the quality of process control and the plant operateur's acceptance positively

The goal of this work is to developt an innovative User Interface for a prospektive design of Human-Technology-Interaktion.

One important aspect in designing a Graphical User Interface is the trade of effect between the Workload and the Situation Awareness. Too high workload will lead the operatuers to neglect or abandon their second Task or even in a worst case, unaware of the incoming alarm. On the other hand, too low Workload indicating an excessive confidence in the control system, which may decrease the ability to react properly in case of system failure or emergency.

The first step in order to achieve this goal has been made. Figure 3 shows a prototype of an user interface for an advanced control system.

The development of the prototypes was limited to the Trend display, which give possibility for the plant operateur to see the future behaviour of the system at the given system input and its model uncertainty.

This kind of display has greatest usability for the interpretation of the dynamic behaviour of process variables [12].

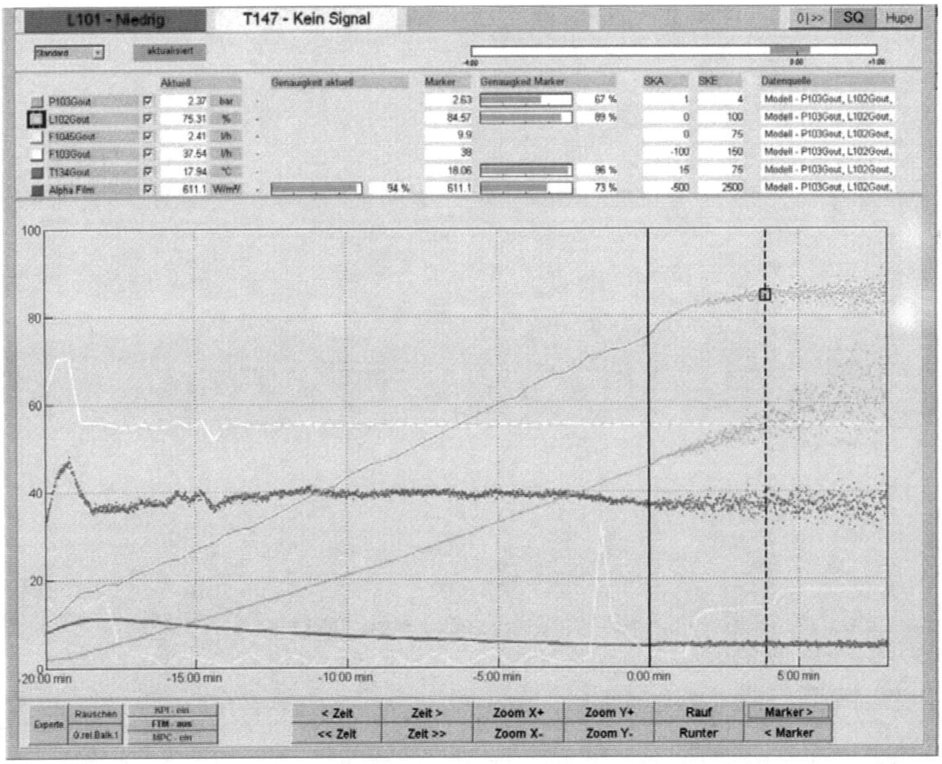

Figure 3. Graphical integration for Advanced Process Control [11]

Result and conclussion

Result

Some results of the PFC implementation on a distillation column to regulate the concentration of the bottom product can be seen in the figure 4, figure 5 und figure 6. Figure 4 shown the PFC control process with no disturbances. Meanwhile in the figure 5 and 6 is the result of the PI(D) and PFC to compensate a lasting disturbance, in this case feed flow rate changes.

Figure 4 below shows the PFCs control sequence during the pressure drop changes. The PFC can preserve the temperatur to the given set point for the entire time with a minimum cost of the heat duty. In direct comparison with the PI(D) control sequence, see figure 1, it is obvious the the PFC perform a better control performance during the operation without disturbances.

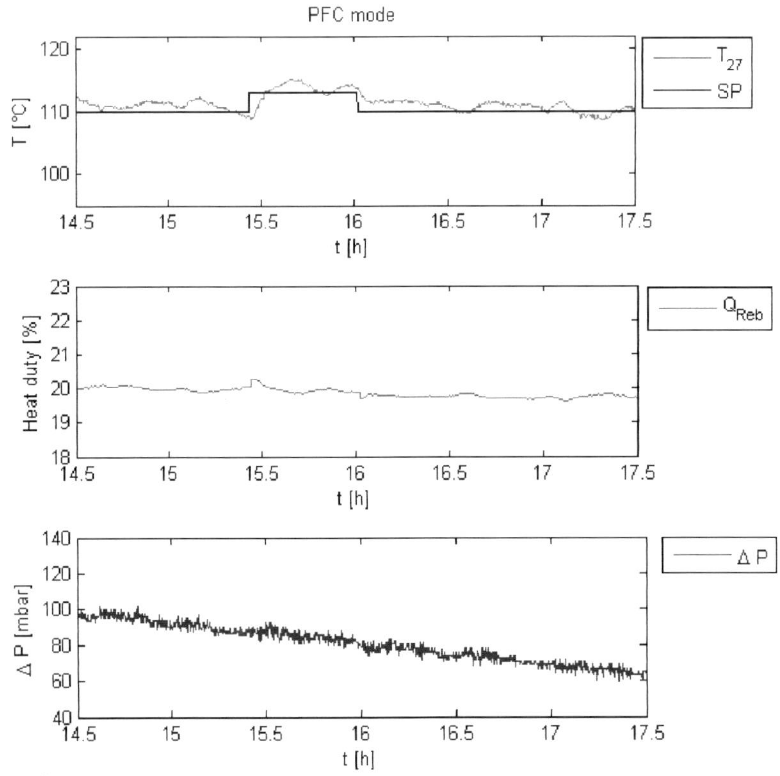

Figure 4. PFC: Temperature control with $CV = T_{27}$

Another interesting comparison shows the performance of both controllers in term of handling steady state changes during operation through increasing the feed load.

Figure 6 shows the control sequence from PI(D) controller. At approx $t = 68$ *min* the feed load increased from 16 *l/h* to 18 *l/h*. With higher feed load, the temperatur in lower parts of the feed tray will decreasing, beacause of the lower inlet feed temperatur, thus the heat duty should be increased, to maintain the temperatur at the given value.
Basically PFC can compensate the disturbance by increasing the heat duty to the new steady state point(from 14.5% to 17.5%), but it failed to fulfil the main task. It took about 60 minutes for the PFC to reach the new steady state point, and during this "action" the temperature couldn't be preserve to the given value. This cause concentration changes which lead to benefits lost.

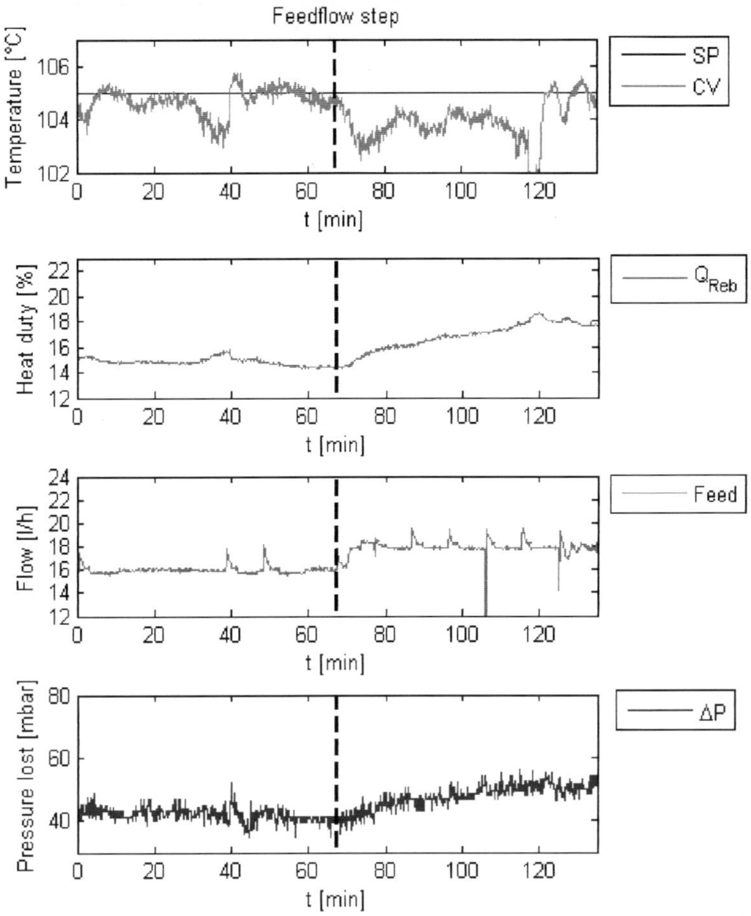

Figure 6. PI: steady state point changes

The same working condition as the PFC test-scheme is also applied to investigate theperformance of PFC.

As we can see in figure 6, PFC increase the heat duty a lot faster then PFC, it only takes about 3 minutes for the heat duty to reach the new steady state point. This action can be explained, because the PFC predict how the feed load change will influence the temperatur in the future and how to compensate this disturbance. In other words PFC acctually doesn't react based control error, but rather *prevent* the controlled variable leaving the setpoint

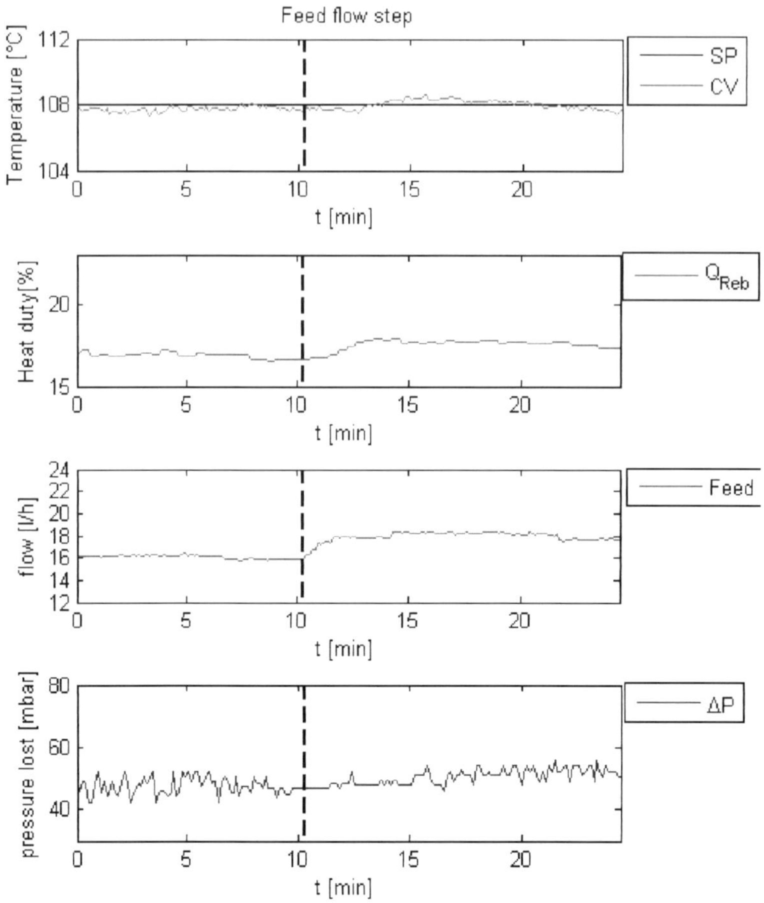

Figure 7. PFC: steady state point changes

This work shows that the implementation of **A**dvance **P**rocess **C**ontrol in order to optimize the porcess control strategy of complex system is successful. These results show also that with APC not only the bottom concentration can be maintained, which increase the benefits, but also the prodcution cost can be reduces due to manipulate variable'performance.

conclusion

The works shows, that the implementation of Predictive Functional Control has satisfied the expectation in optimizing the process control strategy. PFC not only maintain the bottom concentration at the given setpoint, but also reduces the operation cost due to manipulate variable'performance

On the other hand, model based control algorithm, particularly its model uncertainty has special contribution in developing a graphical integration of an Advanced Process Control. With the model uncertainty as an indicator, the plan operateur is demand to examine the process continuously, which affect the situation awareness of the plant operateur positively. With the display's transparency, which decrease the workload, and "in the loop" effect, which increases the Situation awareness the Graphical User Interface has completed its task to support and increase the performance of the plant operateur.

Literatur

1 RICHALET, J., ESTIVAL, J.L., FIANI, Ph.: *Industrial of Predictive Functional Control to metallurgical Industries.*

2 COMPAS, J.M., et.al: *Industrial applications of Predictive Functional Control to rolling mill, fast robot, river dam.*

3 SCHRAMM, B.: *Modellierung und prädiktive Regelung von Wärmetauscherprozessen.* Diplomarbeit (TU Berlin, Feb 2006).

4 SATRIADARMA, B.: *Predictiv Funtional Control an einer Destillationskolonne.* Diplomarbeit (TU Berlin, Sept 2007).

5 LÖWE, K., Li. Pu, WOZNY, G.: *Experimentelle Validierung einer optimalen Anfahrstrategie für hochrein trennende Bodenkolonnen, Chemie.* Ingenieur Technik (72, 3 Januar 2000).

6 DOHERTY, M. F., MALONE, M. F.: *Conceptual Design of Distillation Systems.* McGraw-Hill Chemical Engineering Series. (2001)

7 DÖRGES, F.: *Standardisierung des Einsatzes eines modellprädiktiven Regelungsverfahrens.* Diplomarbeit (FH Fulda, 2005)

8 GMEHLIN, J.: *Lehrbuch der technischen Chemie, Grundoperationen.*

9 GRASSMAN, P., WIDMER, F., SINN, H.: *Einführung in die thermische Verfahrenstechnik.* 3.Auflage , WdeG (1997).

10 WICKENS, C. D. & HOLLANDS, J. G. (2000): Engineering Psychology and Human Performance, Prentice Hall, 3 Upper Saddle River, NJ.

11 Barz, T., Satriadarma, B., Wozny, G. & Urbas, L. (2008). Visualisierung der Zuverlässigkeit modellgestützter Automatisierungslösungen. In: *Automation 2008*, 357-360

12 Kindsmüller, Martin Christof: Trend-Literacy - zur Interpretation von Kurvendarstellungen in der Prozessführung. Aachen: Shaker Verlag 2006

AUTHOR INDEX

AUTHOR INDEX

AUTHOR INDEX

AUTHOR INDEX

AUTHOR INDEX

AUTHOR INDEX

AUTHOR INDEX

AUTHOR INDEX

AUTHOR INDEX

AUTHOR INDEX

AUTHOR INDEX

AUTHOR INDEX